International Conference on New Energy and Future Energy System (NEFES 2017)

IOP Conference Series: Earth and Environmental Science
Volume 93

Kunming, China
22 - 25 September 2017

Editor:

Grigorios L. Kyriakopoulas

ISBN: 978-1-5108-5196-2
ISSN: 1755-1307

Printed from e-media with permission by:

Curran Associates, Inc.
57 Morehouse Lane
Red Hook, NY 12571

Some format issues inherent in the e-media version may also appear in this print version.

Printed by Curran Associates, Inc. (2017)

For permission requests, please contact the Institute of Physics
at the address below.

Institute of Physics
Dirac House, Temple Back
Bristol BS1 6BE UK

Phone: 44 1 17 929 7481
Fax: 44 1 17 920 0979

techtracking@iop.org

Additional copies of this publication are available from:

Curran Associates, Inc.
57 Morehouse Lane
Red Hook, NY 12571 USA
Phone: 845-758-0400
Fax: 845-758-2633
Email: curran@proceedings.com
Web: www.proceedings.com

Table of contents

Volume 93

2017 International Conference on New Energy and Future Energy System (NEFES 2017)

22–25 September 2017, Kunming, China

Accepted papers received: 24 October 2017
Published online: 9 November 2017

Preface

Papers

IOP Conf. Series: Earth and Environmental Science **93** (2017) 011001 doi:10.1088/1755-1315/93/1/011001

Preface

The 2nd International Conference on New Energy and Future Energy System (NEFES2017) was held from September 22nd to 25th, 2017 in Kunming, China. The technical program comprised three keynote speakers, fifty invited speakers, five oral presentation sessions and an interesting poster session.

NEFES 2017 is sponsored by Yunnan Normal University. It aims to provide a platform for researchers and practitioners in both industry and academia to have a brainstorm, share their latest achievements and discuss the possible challenges in terms of new energy and future energy system. This volume records the proceedings of NEFES 2017 and contains 81 rigorously selected manuscripts submitted to the NEFES 2017 conference. All the manuscripts were selected depending on their quality and relevance to the conference.

The Organizing Committee would like to thank all the authors who contributed to NEFES 2017 and also the reviewers who provided their valuable comments and suggestions.

The 3rd International Conference on New Energy and Future Energy System (NEFES 2018) will be held in Shanghai, China. Everybody is welcomed to submit high quality papers to NEFES 2018! More information will be available at the conference website: http://www.intergridconf.org/2018/

Editor

Dr. Grigorios L. Kyriakopoulos

National Technical University of Athens, Greece.

IOP Conf. Series: Earth and Environmental Science **93** (2017) 011002 doi:10.1088/1755-1315/93/1/011002

Peer review statement

All papers published in this volume of *IOP Conference Series: Earth and Environmental Science* have been peer reviewed through processes administered by the proceedings Editors. Reviews were conducted by expert referees to the professional and scientific standards expected of a proceedings journal published by IOP Publishing.

Experimental study on improving the drying uniformity in hot air cross-flow dryer

Y Bie[1, 2], M Li[1, 3], X Y Guo[2], J G Sun[2] and Y Qiu[1]

[1]Solar Energy Research Institute, Yunnan Normal University, Kunming 650500, China
[2]Faculty of Chemical Engineering, Kunming University of Science and Technology, Kunming 650500, China

E-mail: lmllldy@126.com

Abstract. Drying is one of the important steps in the deep processing of agricultural and sideline products. It is a main tendency to research and develop the high-efficient and energy-saving drying pattern and technology. However, it is found that drying uneven phenomenon often exists in the material-drying oven in the production practice of hot air flow drying process. In order to explore the causes of the formation of uneven phenomenon, and to propose the ways to eliminate the inhomogeneity, experimental research was conducted concerning a multifunctional drying equipment with the hot air circulation. The temperature distribution uniformity and drying efficiency in the oven were tested under no-loaded and loaded conditions, in aspect of oven tray structure improving, wind direction switching and moisture controlling of exhaust air. Experimental results show that the drying inhomogeneity and thermal efficiency could be improved by the ways of changing the wind direction from cross-flowing to cross-swept-flowing, and as well as proper moisture controlling of exhaust air.

1. Introduction

Drying is one of the important steps of agricultural and sideline products processing. Researching and developing the high-efficient and energy-saving drying pattern and technology is the main development tendency of drying technology. Different materials are suitable for different drying methods, among which the hot air drying is applied most in the actual producing. However, several problems might exist such as low drying efficiency, low product quality and high energy consumption due to uneven distribution of air flow in the hot air cross-flow drying oven. These problems attract the attention of the expert in the academic and industrial fields [1].

Zhu W X *et al* [2] conducted the experiment about regularities of air speed distribution in the hot air cross-flow drying oven. It is found that the drying evenness phenomenon were related to air speed distribution. The uniform distribution of air speed is decisive factor of temperature. What's more, there are many common points between drying distributions without load. It's been found that the measurements of temperature and air speed in the hot air drying oven were hard to get, which form Yin Y *et al* [3]. Some researchers [4,5] discussed the way to access the temperature in the vegetable drying oven and the level of material's uniformity. At the same time, they tested the effectiveness of this way, which can improve efficiency in the development of drying equipment. What's more, they also offered steps to improve the air distribution uniformity in the drying oven by adding baffle boards. These research contents are based on experiments. The measurement of speed was limited, which will have a bad influence on the whole oven's uniformity. Facing complicated drying processing, some scholars established the models about heat transfer theory and mass transfer theory. They predicted the case of changing in the temperature of the material and moisture content on the drying processing. As for the phenomenon of drying unevenness, Huan Z G and Rumsey *et al* [6] did a research on numerical simulation by CFD. Zhao J H [7] and other researchers [8] who are from China Agricultural University established the CFD theoretic model of air speed field in the drying oven with composite boards. Compared to the test result, they found that simulation result was viable. Also, they got the main reason why air speed distribution unevenness lead to drying low efficiency and uneven drying processing. Msthioulakis [9] and other researchers did a research on flow filed condition about intermittent tray in the hot air drying oven. By using phonics program. Meng W W [10], Gabites [11] and other scholars [12] did a simulation about drying model and flow field in the atomizing drying oven. Lu R [13] and other researchers simulated the flow field air speed distribution in the condition of whether carry boards with uniform air in the vertical type dryer. They thought that the effect of carrying was obvious. Also, the height and smooth level of materials were related to the air speed filed.

It's obvious that everyone in academic field has reached an agreement that the main factor of drying unevenness was caused by the flow of hot air's uneven distribution. However, few people thought out the way to improve the uniformity of drying besides adding baffle boards with uniform distribution wind. This article is about some ameliorative way from the structure of drying oven. Researchers studied the way to improve uneven drying and evaluated the improved effect by experiments.

2. Experiment setup

The research object is multi-functional drying equipment in which wet material can be dehydrated by circulated hot air as shown in figures 1 and 2. The drying device is mainly composed of hot air stoves, fans for smoke, commutators, pipelines for wind, exhaust drying boxes and other parts. The hot air oven provided hot air to drying boxes. Drying boxes were cubes, of which length is 1.01 m, width is 0.85 m, and the height is 1.18 m. There are no more than 14 drying plates from bottom to top, the entrance of the hot air is hot air distributor in the bottom of oven.

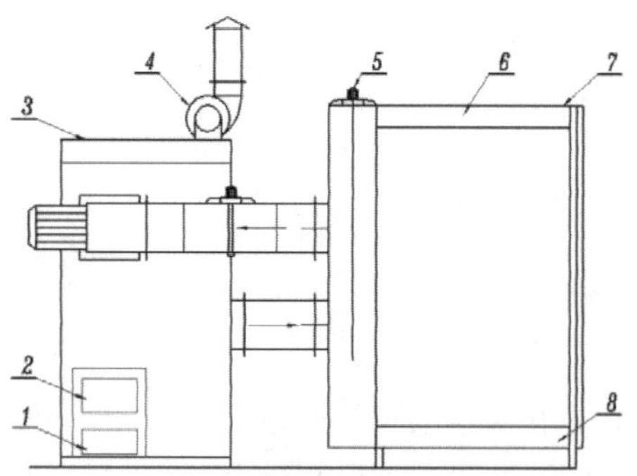

Figure 1. The structure diagram of hot air circulation dryer.

Figure 2. The photo of hot air circulation dryer.

The fresh broad bean, sliced fresh cabbage, cut new red peppers are used as drying material in the drying box's experiment. The measurement of temperature in the experiment came from Pt 100 hot resistance. And the measurement of humidity in the experiment came from PTS-3 transducer of air humidity. Both the hot resistance and the transducer are the parts of solar energy tester which is called TRM-2A produced by Jingzhou solar corporation. After 15 roads have been collected, the signal will be read by computers. And measurement of wind pressure came from K0603 aptitude pressure gauge.

3. Experiment methods

3.1. The experiment about unevenness in the primary drying oven

According to the research results, the fundamental cause of uneven drying is uneven distribution of internal flow field. And the uneven distribution is in the state of no load and cold condition. Next, difference is increasing after adding drying material, which may change the degree of this uneven distribution. Drying efficiency during given time can be obtained by measuring the dehydration rate of material on each plate, to study the unload condition and load condition by experiments, and get the way to improve the uniformity in the time and space. The way is about hot air baffle-flow.

3.2. The way to improve the efficiency of mixed hot air baffle-flow

To improve the drying plates, and change the effects of the hot air and material mass transfer, researchers changed the air direction from one-though cross-flow to the pattern of cross-flow mixed with baffle-flow. To explore whether different drying plate structures can bring different baffle-flow effects, the researchers did three ways to do experiments. The first was about single baffle-flow, the second was about double baffle-flow and the last was about double baffle-flow with micropore. Three hot air flow patterns can be shown as figure 3. Leaving 8 cm to lay cut-through flaps in the first level of drying plate's left space. Leave 8 cm to lay flaps in the second level of drying plate's right space. And do like this repeatedly, lay material in the remaining space to form double baffle-flow, trepan in the double baffle-flow boards and forming double baffle-flow with micropore, and lay drying plates in the 2th level, 5th level, 9th level and 13th level in the drying oven. And there were not yet drying cabbage which weighed 6 kg and its height was 5 cm probably in every plate. To represent the degree of the unevenness, researchers placed 3 temperature probes in every drying plate, which is shown in figure 4.

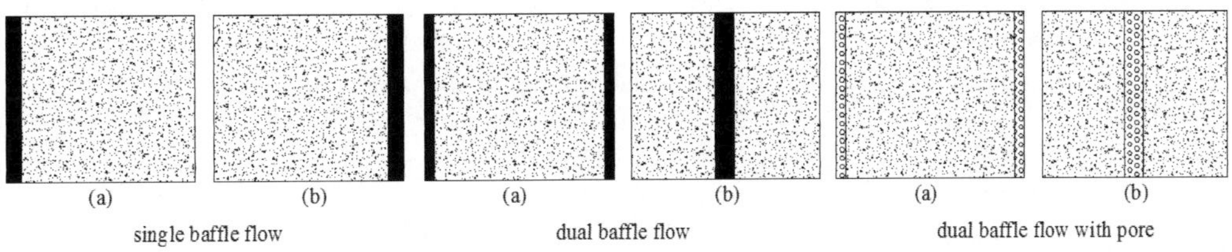

Figure 3. Three tray improvement schemes of baffle-flow in the dryer.

In the place of temperature probes in the plate, there are 3 temperature probes laid in upper left, central place and low right in every level as shown in figure 4.

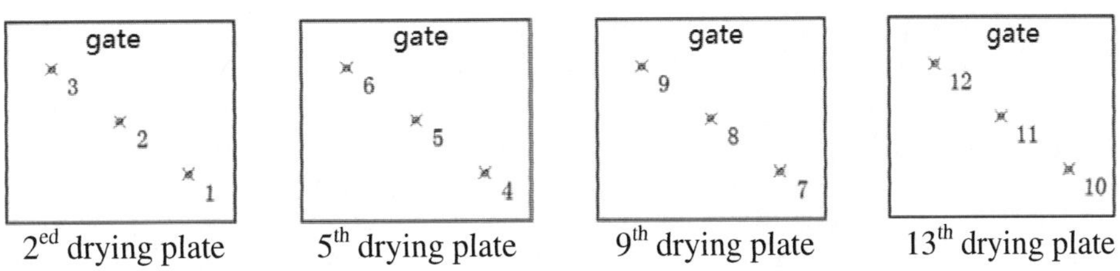

Figure 4. The location of temperature probes on the dryer tray.

4. Results and discussion

4.1. The unevenness about experiment in the primary drying ovens

To represent the degree of the unevenness in the no-load drying plates, the researchers placed temperature probes in five places of the last bottom in the drying plates, measuring the temperature change in a constant time and getting the collected wind data. The subscripts 1-5 in the temperature mark represent the top left corner, left bottom, central place, lower right corner and top right corner in turns as shown in figure 5. The recurrent air got into the entrance at the speed of 12 m/s, and it was heated continuously. The change about temperature in the 10 minutes is shown in figure 6. Obviously, when the temperature of hot air was from 40℃ to 65℃, the rising rate of temperature in central place was the lowest. And the result was in a condition of unload. The rate in central place was father lower than others. The temperatures of point 1, point 2 and point 4 were near, which indicated that the blast capacity in central place was lower than other places. Also, it provides an explanation why the place of wet island is in central place.

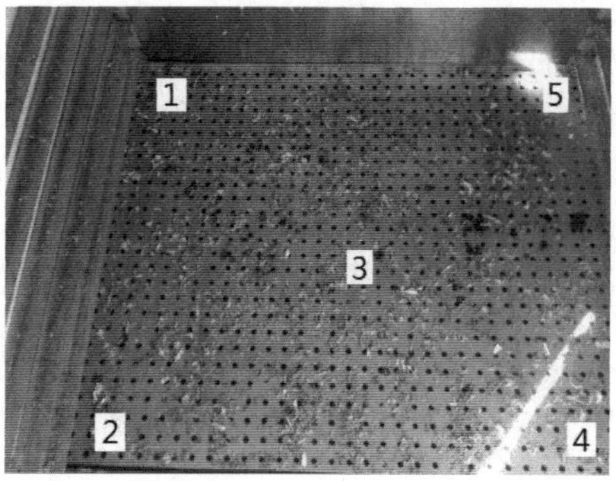

Figure 5. The location of temperature sensors under no-load.

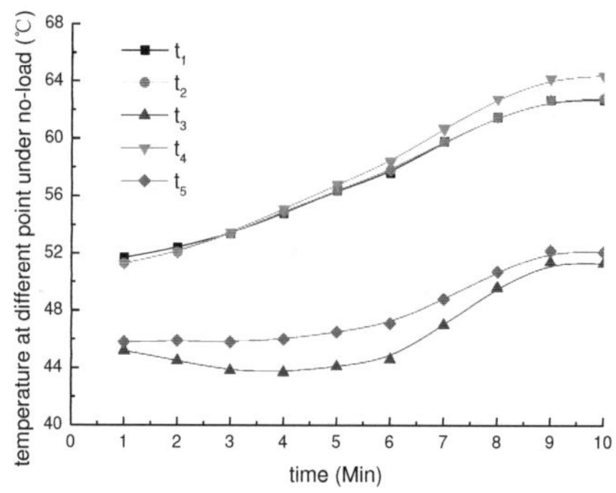

Figure 6. The history of temperature change on the dryer tray under no-load.

To improve the unevenness in every level's drying plate, researchers took the sliced fresh cabbage as not-yet-drying material. There are 100 kg wet materials totally in 14 levels. Besides the 6th level and the 7th level without plates, there are 8.33 kg material in every plate, and the material was laid averagely in the plates which are 9.6 m^2. The total time about drying was 425 minutes, the temperature of hot air was set as different levels, as shown in figure 7. What's more, in the former 3 hours, the patterns about wind access in the bottom and top was 6:4. In the aspect of wet ejectment, there are 150 minus carrying half wet ejectment later. The weights of material in the 2^{ed} level, 5th level, 9th level and 13th level drying plates were weighted in the drying processing, which was shown in figure 8. The change about dehydration rate with time is shown in figure 9. The loaded 12 levels material weights and dehydration rates in the end of drying processing are shown in figure 10. The wet ejectment temperature and wet gas in the drying processing are shown in figure 9. By analysing these figures, we know that the dehydration speeds in the 13th level and 2th level are near, and the speeds are father higher than the drying speeds of the 5th level and the 9th level. What's more, the drying speed of the 9th level is the lowest in four levels measured material. Taking measurement at the beginning of drying, researchers found that the dehydration rate of 9th level was less than 0, which indicated that the parts didn't lose water in former two hours. And it absorbed the water from other parts. At the end of drying, the weights of middle six levels' material were obviously higher than the weights of material which was near to the entrance in the bottom and top. And the dehydration rates of 9th level and 10th level were the lowest. The reason of this is that when hot wind crossed drying plates from one level to another, the accumulation of wind was increasing, and the speed of wind was lower and lower, the heat transfer and mass transfer speeds of hot air and material was slower and slower. However, switch the direction of wind entrance remitted the drying speed's reduction from one direction. And the drying speeds of two ends were higher than central level's. The 9th level and 10th level were near to the place of central top. Because the wind from top was lower, and the drying speeds of them were the slowest.

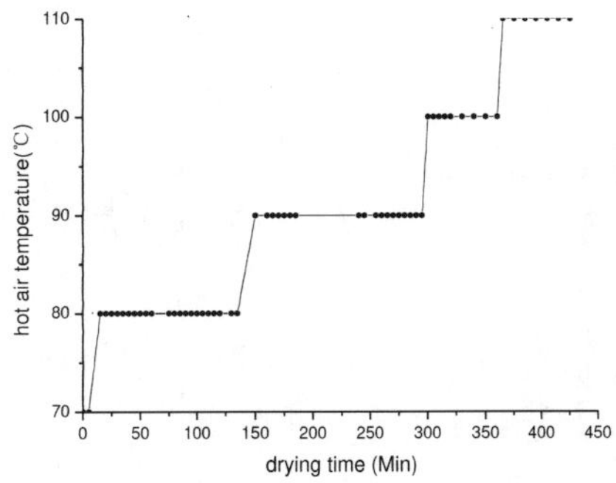

Figure 7. The temperature history of dryer inlet air.

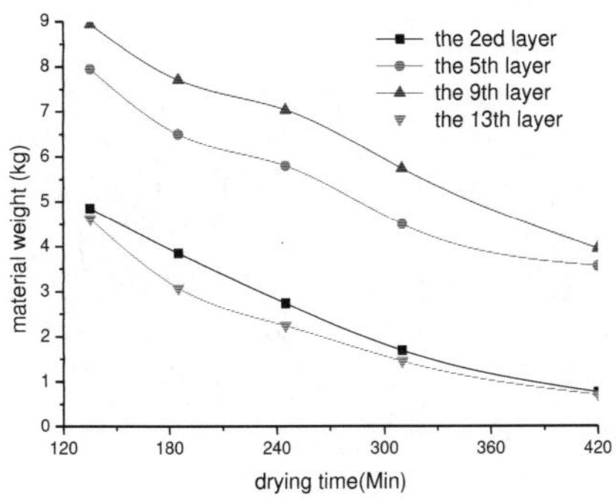

Figure 8. The material weight of the typical four layers.

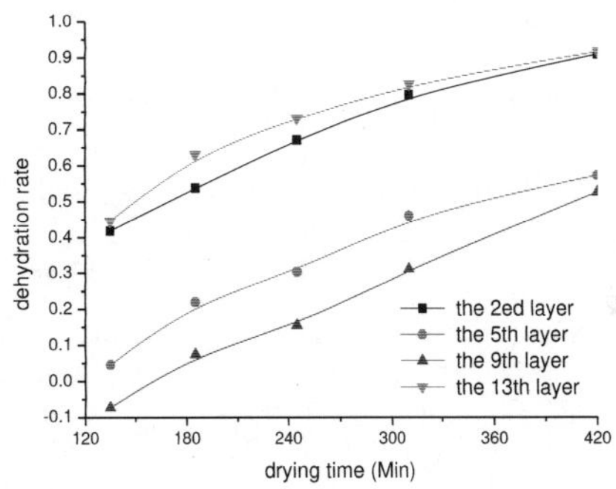

Figure 9. The material dehydration rate history of the typical four layers

In the aspect of material's dried condition, the average dehydration rate of 12th level material in drying plate was 71.1%. The average dehydration rates of the 2^{ed} level, the 5th level, the 9th level and the 13th level were 73.0% in end. And the result of this was near to the whole average dehydration rate. Therefore, the rate was representative. Researching the dehydration rate condition in drying processing deeply, the dehydration rate of drying material on 13th level plate reached 73.1% at the time of 345 minutes, the dehydration rate of the 2^{ed} level reached 73.0%. Therefore, the material in the top and bottom drying plates finished the drying processing nearly in the farmer 4 hours. And the later 3 hours were useless for the drying processing, which caused the low efficiency and high energy consumption doubtlessly.

It can be found from wet ejectment curve that the temperature of the dried waste gas was higher than 40℃, and its humidity was less than 40%, as shown the figure 11. In this case, there is a lot of recyclable wasted. Therefore, ejecting heat completely can't improve the unevenness and lower the utilizing efficiency. Also, the condition of ejecting heat in half was the same with the condition of ejecting heat completely.

What's more, researchers observed the form of the material and found that the cabbage in the edge of plates has been dried up. They found this by getting the weight of the material on the 2^{ed} level, the weight of the 5th level, the weight of the 9th level and the weight of 13th level. The division line about drying and wet qualitatively are drawn. And the moist island area of every level are shown in figure 12. The percentage of first gate area is probably 22%, the percentage of second gate area is probably 32%, the percentage of third gate area is probably 46%, the percentage of fourth gate area is probably 62%.

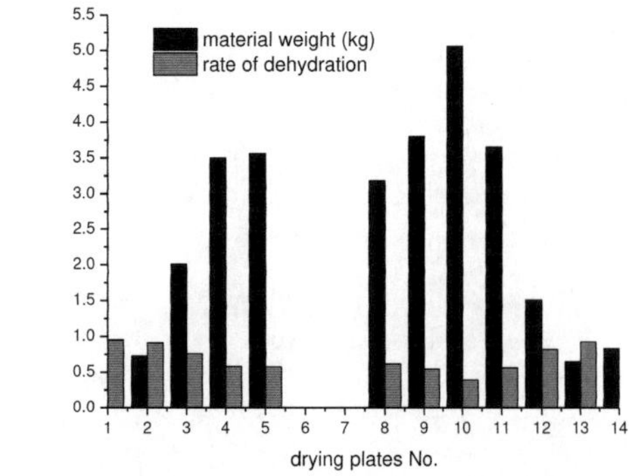

Figure 10. The distribution of dehydration rate at the drying end moment.

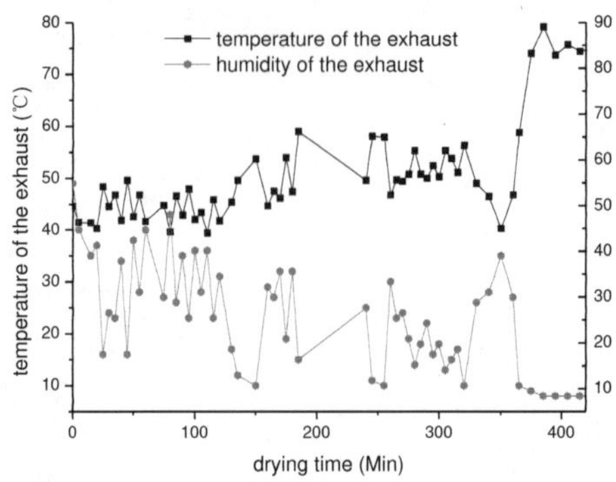

Figure 11. The temperature and humidity of exhaust air.

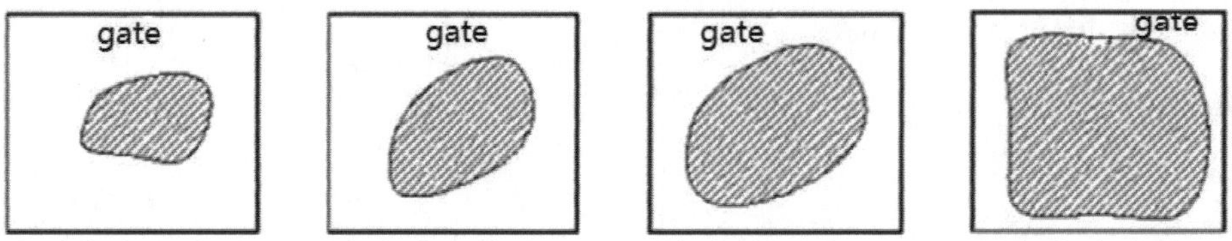

2ed drying plate 5th drying plate 9th drying plate 13th drying plate

Figure 12. The wet island area in the four typical layers.

4.2. The experimental results of baffle-flow evenness

Researchers did the experiment by 3 ways that were changing hot air flow to cross-flow and baffle-flow. They had been drying 13 minutes at the temperature of 80℃, and they got 12 temperature conditions of measuring points, which was shown in figure 13. Then, the drying plates were taken out, and next, the weight of the material in the plates was measured and dehydration rate was calculated. The temperature distribution and dehydration rate of every plate are shown in figures 14.

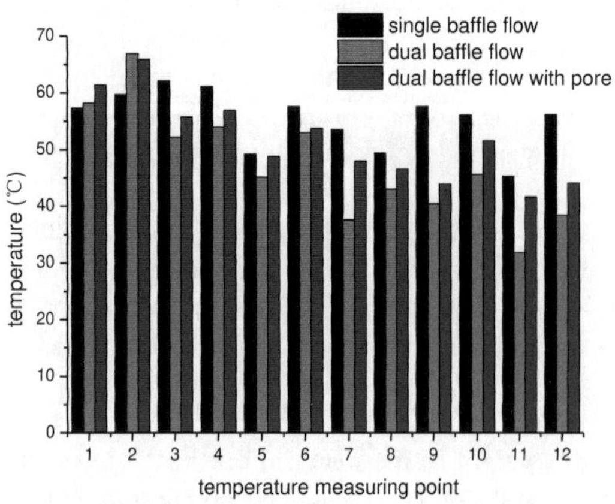

Figure 13. The temperature distribution under three baffle-flow scheme.

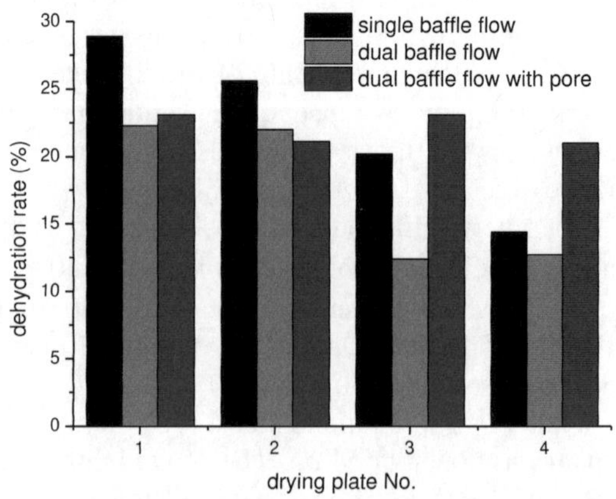

Figure 14. The dehydration rate distribution under three baffle-flow scheme.

The definition of unevenness as follows:

To remove the influence of different average data, RSD was defined as no-dimension parameter.

$$RSD = S/\bar{x} \times 100\% \quad (1)$$

The researchers calculated the uneven temperature and RSD in every plate by different ways of baffle-flow, and the results are shown in figures 15 and 16. In the condition of single baffle-flow, the uniformity of the first level plate was the best, the drying uniformity of other three level plates were the worst oppositely. In the condition of double baffle-flow, the uniformity of the 3th level was even, while the other plates' uniformity was worse. In the condition of double baffle-flow with micropore, except that the first level's uniformity was worse, the others were the best. In the aspect of the temperature distribution in the plates, the temperature of point 3, the temperature of point 4, the temperature of point 10 were the highest in the case of single baffle-flow. We can find that the highest temperature point is bended from figure 15, which indicated that the guide effect of convection current was obvious. In the condition of double baffle-flow the temperature of the point 2 and the point 8 in central place were highest. And the temperature of the point 5 and the point 11 were the lowest. However, the point 5 and point 11 were also in the place of central. From what discussed above, we know that the form of the case is double baffle-flow. In the condition of double baffle-flow with micropore, the case of the highest temperature and the lowest temperature were same to the double baffle-flow. However, the whole uniformity was obviously higher than the double baffle-flow. Because of the small different value of temperature mean value, the RSD change regular is same to the temperature unevenness. Only in the case of double baffle-flow, the unevenness was more obvious. From what has been discussed above, the unevenness of average temperature distribution was 3.4℃ in the case of single baffle-flow. The unevenness of average temperature distribution was 3.9℃ in the case of double baffle-flow. And the unevenness of average temperature distribution was 2.9℃ in the case of double baffle-flow with micropore. On the other hand, RSD were 6.26%, 8.18%, 5.73%, respectively. Obviously, in the aspect of temperature distribution, the best ways to improving drying plates' temperature unevenness and the temperature unevenness between plates were double baffle-flow with micropore.

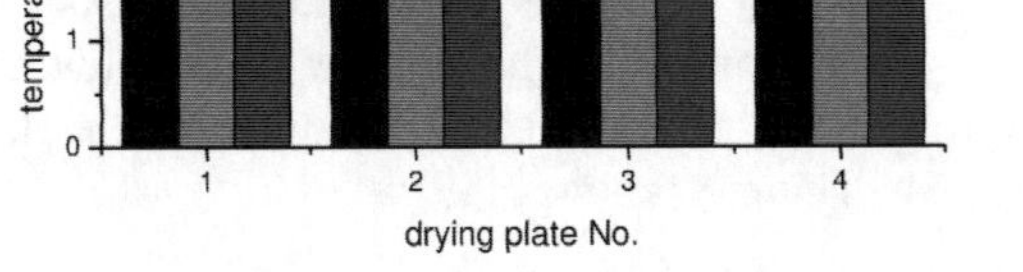

Figure 15. The temperature unevenness under the three baffle-flow scheme.

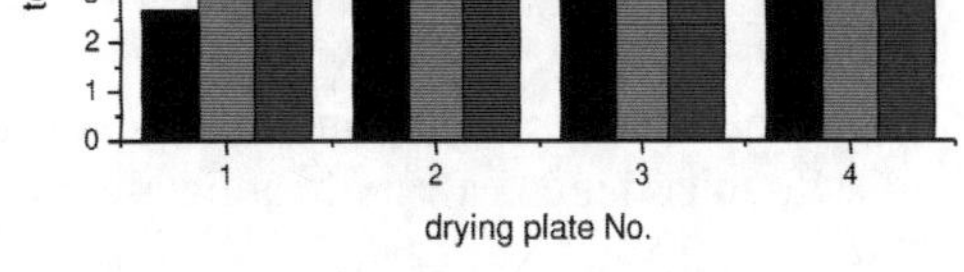

Figure 16. Temperature RSD under the three baffle-flow scheme.

By comparing the own dehydration rates 135 minutes later, the average dehydration ratesof single baffle-flow was 22.2%, the dehydration rate of double baffle-flow was 17.4%, and the average dehydration rate of double baffle-flow with micropore was 22.1%. RSD of single baffle-flow was 4.98%, RSD of double baffle-flow was 4.80%, and RSD of double baffle-flow with micropore was1.03%. However, before adding baffle-flow and in the case of hot air cross-flow, for the sane count of material, the average dehydration rate of the 2th level, 5th level, 9th level and 13th level was all 20.89% after 135 mins. And RSD was all 22.3%. Obviously, the absolute dehydration rate of double baffle-flow was the lowest, which even lower than the unimproved patterns. And the dehydration rates of other two ways were close, which was 5% higher than double baffle-flow probably, and 1.3% higher than unimproved way. However, in the aspect of uniformity of dehydration rate, the uniformity index after baffle-flow was improved largely. The index from 22.3% was up to 5%, and the index in the condition of double baffle-flow was up to 1%.

Judging from temperature distribution cases comprehensively, the best ways to improve the uniformity of drying plates and the uniformity between plates are double baffle-flow with micropore, although there are some bad effects from baffle-flow transformation, such as the reduction of drying area and the increasing of wind resistance. And by calculating, researchers found that the reduction of three ways was 0.8%, 1%, 1%, dividedly. The cases above can be accepted in engineering. However, the double baffle-flow with micropore can control the areas with micropore, which can deduce the wind resistance rather than increasing. Therefore, the bad effects of three ways can be ignored.

5. Conclusions

- For the original drying ovens, the temperature in central place of every level's drying plate was lower than bordered areas' temperature in the condition of load and unload. When hot air crossed different level's drying plates, the wind resistance increased gradually and the drying speed of material got more quickly, and the material was close to the entrance. Therefore, changing the entrance of wind and bringing wind commutatively can improve drying efficiency. Also, it can remit the efficiency deduction with levels. Drying unevenness not only caused the drying speed of several levels' material which was close to the entrance 40% lower than other levels' material, but also added the energy consumption. And it reduced the efficiency of drying.
- Researchers did a transformation on the drying plates in the drying ovens. They found the temperature extreme values appeared commutatively from the highest point and the lowest point. They also found the forms of single baffle-flow and double baffle-flow. And they found that these ways can guide the hot air flow direction well.
- The three ways of baffle-flow transformation improved the unevenness of drying. And the reduced areas no more than 1%. Comparing three ways of efficiency improvement transversely, and in the aspect of temperature distribution, researchers found that the double baffle-flow with micropore could improve the temperature uniformity in the drying plates. The double baffle-flow with micopore also can improve the unevenness between drying plates well. In the aspect of uniformity of dehydration rates, the index after baffle-flow was improved largely. And the improvement rate was from 22.3% up to 5%. What's more, under the pattern of double baffle-flow with micropore, the improvement rate was 3%.
- From comprehensive analysis, researchers found that the best way to improve the uniformity in drying plates and evenness between plates is double baffle-flow with micropore. And by controlling the area with micropore, the count of micropore in the double baffle-flow can reduce wind resistance and energy consumption of draught fan. Therefore, double baffle-flow with micropore is a very beneficial way to improve drying efficiency.

Acknowledgments

This work was supported by the Collaborative Innovation Center of Research, Development of Renewable Energy in the southwest area in China (No. 05300205020516009), the Project on Co-establishing China-Laos Joint Lab for Renewable Energy (No. 2015DFA60120) and Zhou Guofu Expert Work Station.

References

[1] Satyanarayan R S D and Vijaya G S R 2012 Advancements in drying techniques for food, fiber, and fuel *Dry Technol* **30** 1147-1159

[2] Zhu W X and Bai C R 1995 The uniformity of drying in the cross-flow dryer *J. China Agric. Univ.* **15** 35-40

[3] Yin Y, Xie X Y and Bai C R 1993 The study of the improvement way on the uniformity of drying in the cross-flow dryer. *Trans. Chin. Soc. Agric. Mach.* **24** 40-45

[4] Yang L and Tao B B 2005 The study of heat and mass transfer about the porous medium in the drying process *Trans. Chin. Soc. Agric. Eng.* **21** 27-31

[5] Guerman I E 2007 Drying kinetics derived from diffusion equation with flux-type boundary conditions *Dry Technol* **20** 55-66

[6] Rumsey T R and Rovedo C O 2001 Two-dimensional simulation model for dynamic cross-flow rice drying *Chem. Eng. Process* **40** 355-62

[7] Zhao J H, Yue X Y and Shi E J 2007 The CDF simulation and testing of the wind speed field with several layers in the shape of strap *Dry. Technol. Equip.* **2007** 15-21

[8] Defraeye T 2014 Advanced computational modelling for drying processes-A review *Appl. Energy* **131** 323-44

[9] Mathioulakis E, Karathanos V T and Belessiotis V G 1998 Simulation of air movement in a dryer by computational fluid dynamics: Application for the drying of fruits *J. Food Eng.* **36** 183-200

[10] Meng W W, Wan R W D Arun S M, *et al* 2008 CFD evaluation of droplet drying models in a spray dryer fitted with a rotary atomizer *Dry Technol* **26** 1180-1198

[11] Gabites J R, Abrahamson J and Winchester J A 2010 Air flow patterns in an industrial milk powder spray dryer Chem. Eng. Res. Des. 88 899-910

[12] Sha D O, Yang J, Lv H and Yi S L 2016 The numerical simulation and optimizing of the inner wind speed filed in the wood drying chamber *J. Nanjing U. Fore.: (Nat. Sci Ed.)* **2016** http://www.cnki.net/kcms/detail/32.1161.S.20160113.1528.042.html

[13] Lu R, Tan H Q and Huan Z M 2013 The study of the uniformity of the end speed field in the several units of the standing drying machine *Hubei Agr. Sci.* **52** 435-439

Study on convection improvement of standard vacuum tube

J H He[1, 4], W P Du[2], R R Qi[3] and J X He[2]

[1]School of Physics and Electronic Sciences, Chuxiong Normal University, Chuxiong, Yunnan, China
[2]Solar Energy Research Institute, Yunnan Normal University, Kunming, Yunnan, China
[3]ZheJiang University of Technology, Hangzhou, Zhejiang, China

E-mail: hejh@cxtc.edu.cn

Abstract. For the standard all-glass vacuum tube collector, enhancing the vacuum tube axial natural convection can improve its thermal efficiency. According to the study of the standard all-glass vacuum tube, three kinds of guide plates which can inhibit the radial convection and increase axial natural convection are designed, and theory model is established. Experiments were carried out on vacuum tubes with three types of baffles and standard vacuum tubes without the improvement. The results show that T-type guide plate is better than that of Y-type guide plate on restraining convection and increasing axial radial convection effect, Y type is better than that of flat plate type, all guide plates are better than no change; the thermal efficiency of the tube was 2.6% higher than that of the unmodified standard vacuum tube. The efficiency of the system in the experiment can be increased by 3.1%.

1. Introduction

Solar energy is the important content of the development and utilization of renewable energy and new energy [1]. Further study of solar photovoltaic and photothermal applications is of great significance for the effective collection, utilization, and storage of solar energy. In the solar energy application, standard all-glass vacuum tube collector has the advantages of lower heat loss coefficient and higher thermal efficiency [2]. Therefore, the research on the performance of all glass vacuum tubes has attracted much attention in solar thermal utilization. However, due to the complexity of the flow field in the vacuum tube, there are few studies on the flow field of the vacuum tube and the convection heat transfer of the internal fluid.

At present, the research on convection improvement of vacuum heat collecting tube is mainly based on the simulation with FLUENT software to study the flow field and convection heat transfer process in the tube. Then the simulation results are summarized and the factors affecting the convective heat transfer in the pipe are summarized.

The density difference in widely used vacuum tube is caused by the absorption of heat by the fluid in the pipe, which makes the fluid in the tube flow upper and lower reversely, and the hot water flow into the water tank along the axial direction of the pipe. But some of the hot water flows back before it reaches the top of the tube, the hot water flows back before it reaches the top of the tube, which can have disturbances or even eddy currents. In the process of hot and cold water circulation, there is no circulating water at the bottom of the pipe, forming a flowing stagnation zone, which has an adverse effect on convective heat transfer and reduces the efficiency of the collector.

According to the standard vacuum tube collector experimental study shows: In order to suppress the radial convection of the fluid in the vacuum tube, and to enhance axial convection to improve thermal efficiency of vacuum tubes, different shapes of baffles can be added into the vacuum tube to form a double-channel or even a multi-channel flow in the pipe. It can make the fluid ordered flow and stratification obvious, thus ameliorate the heat transfer of mixed cold and hot fluid, and improve its thermal efficiency.

2. Theoretical basis of convection heat transfer in vacuum heat collecting tube

2.1. The influence factors of vacuum tube collector efficiency

Wind direction, wind speed, optical power density, temperature, effective lighting area, the energy absorbed by the collector, convective heat transfer loss, flux, collector installation angle and its own performance and other parameters affect the efficiency of the collector. The paper studies the impact of vacuum tube heat transfer on heat efficiency.

The collector efficiency is defined as: Under steady-state conditions, the ratio of the energy output by the collector heat transfer material during the specified period to the area of the collector and the product of the solar irradiance incident on the collector at the same time [4].

$$\eta = \frac{\int Q_u dT}{A_c t \int G dt} \tag{1}$$

Take a small element dv in the vacuum tube, the temperature at the in element is:

$$T = T_1 + \frac{T_2 - T_1}{2} I \tag{2}$$

The energy absorbed by the element:

$$dQ = C\rho 2\pi r^2 (T_1 + \frac{T_2 - T_1}{2} I - T_0) dI \tag{3}$$

The heat absorbed by the whole tube is:

$$Q = \int_0^L C\rho 2\pi r^2 (T_1 + \frac{T_2 - T_1}{2} I - T_0) dI \tag{4}$$

In the above formulas, C—The specific heat capacity of the fluid J/(kg▪k); ρ—The density of fluid kg/m^3; G—Solar irradiance J/m^2; A_C—Effective lighting area m^2; T_2— Nozzle temperature ℃; T_1—Tube bottom temperature ℃; T_0—Initial water temperature ℃; t—Time S. The lighting area of vacuum tube selected in this paper is 0.093 m^2.

2.2. Convection heat transfer in vacuum tube

2.2.1. Convection heat transfer principle. Heat convection refers to the exchange of heat caused by relative motion of fluids of different temperatures. When the solar radiation intensity reaches a certain value, the natural convection occurs due to the different temperature of the cold and hot water in the vacuum tube. The natural convection in the pipe includes axial convection and radial convection. The basic formula is as follows:

$$\phi = hA(t_w - t_f) \tag{5}$$

$$q = \phi A = h(t_w - t_f) \tag{6}$$

h—surface heat transfer coefficient W/(m^2▪K)

2.2.2. Factors affected convection heat transfer of vacuum tube collector. The convective heat transfer of hot and cold fluid in vacuum tube has great influence on its thermal efficiency. The convective heat transfer of the internal fluid in the all-glass standard vacuum tube type collector is very complicated. The performance parameters of the vacuum tube, the installation angle, the stratification of the fluid temperature in the tank, the inlet and outlet temperature, the solar irradiance and other factors can have an impact on it. The analysis of the flow field in the tube has an important contribution to suppressing the radial convection of the fluid in the vacuum tube. The same also contribute to enhancing the axial convection, improving the convective heat transfer efficiency and improving the collector performance [4].

2.3. Convection heat transfer model and governing equations vacuum tube

2.3.1. Governing equation. Continuity equation (Derivation basis - conservation of mass): During the same time interval, the net mass of the inflow element is equal to the increment of the mass in the unit.

$$\frac{\partial \rho}{\partial t}+\frac{\partial(\rho u)}{\partial x}+\frac{\partial(\rho v)}{\partial y}+\frac{\partial(\rho \omega)}{\partial z}=0 \tag{7}$$

Vector representation: $\partial \rho / \partial t + div(\rho v) = 0$ (8)

ρ—fluid density; t—time; u, v, ω—the components of the velocity vector in the x, y, z directions, respectively.

Momentum conservation equation: According to Newton's second law, combined with the actual situation of this paper, the fluid is Newton's fluid, get the momentum conservation equation:

X-direction:

$$\rho\ (\frac{\partial \rho}{\partial t}+\frac{\partial(\rho u)}{\partial x}+\frac{\partial(\rho v)}{\partial y}+\frac{\partial(\rho \omega)}{\partial z})=-\frac{\partial p}{\partial x}+\mu\ (\frac{\partial^2 u}{\partial x^2}+\frac{\partial^2 u}{\partial y^2}+\frac{\partial^2 u}{\partial z^2})+F_X \tag{9}$$

Y-direction:

$$\rho\ (\frac{\partial v}{\partial t}+\frac{\partial v}{\partial x}+\frac{\partial v}{\partial y}+\frac{\partial v}{\partial z})=-\frac{\partial p}{\partial x}+\mu\ (\frac{\partial^2 v}{\partial x^2}+\frac{\partial^2 v}{\partial y^2}+\frac{\partial^2 v}{\partial z^2})+Fy \tag{10}$$

Z-direction:

$$\rho\ (\frac{\partial \omega}{\partial t}+u\frac{\partial \omega}{\partial x}+v\frac{\partial \omega}{\partial y}+\omega\frac{\partial \omega}{\partial z})=-\frac{\partial p}{\partial x}+\mu\ (\frac{\partial^2 \omega}{\partial x^2}+\frac{\partial^2 \omega}{\partial y^2}+\frac{\partial^2 \omega}{\partial z^2})+Fz \tag{11}$$

Vector form:

$$\rho\frac{dv}{dt}=F-\nabla \mathrm{p}+\mu\nabla 2\mathrm{V} \tag{12}$$

In the above formulas, μ—dynamic viscosity; p—pressure on a fluid element; $\frac{\partial p}{\partial x},\frac{\partial p}{\partial y},\frac{\partial p}{\partial z}$ —pressure gradient term; F_x, F_y, F_z —volume force terms; $\rho\left(\frac{\partial u}{\partial t}+u\frac{\partial u}{\partial x}+v\frac{\partial u}{\partial y}+\omega\frac{\partial u}{\partial z}\right),\rho\left(\frac{\partial v}{\partial t}+u\frac{\partial v}{\partial x}+v\frac{\partial v}{\partial y}+\omega\frac{\partial v}{\partial z}\right),\rho\left(\frac{\partial \omega}{\partial t}+u\frac{\partial \omega}{\partial x}+v\frac{\partial \omega}{\partial y}+\omega\frac{\partial \omega}{\partial z}\right)$ —inertial force term; $\mu\left(\frac{\partial^2 u}{\partial x^2}+\frac{\partial^2 u}{\partial y^2}+\frac{\partial^2 u}{\partial z^2}\right),\mu\left(\frac{\partial^2 v}{\partial x^2}+\frac{\partial^2 v}{\partial y^2}+\frac{\partial^2 v}{\partial z^2}\right),\mu\left(\frac{\partial^2 \omega}{\partial x^2}+\frac{\partial^2 \omega}{\partial y^2}+\frac{\partial^2 \omega}{\partial z^2}\right)$ —viscous force term.

Energy conservation equation: According to energy conservation equation, the increase of energy in the fluid microelement is the same as the net heat flux absorbed by the microelement plus the work done by physical force and surface force to infinitesimal bodies.

$$\frac{\partial t}{\partial \tau}+u\frac{\partial t}{\partial x}+v\frac{\partial t}{\partial y}+\omega\frac{\partial t}{\partial z})=a\ (\frac{\partial^2 t}{\partial x^2}+\frac{\partial^2 t}{\partial y^2}+\frac{\partial^2 t}{\partial z^2}) \tag{13}$$

$$\frac{dt}{d\tau}=a\cdot\nabla 2t \tag{14}$$

a—coefficient of temperature conductivity， a =λ/Cpφ.

2.3.2. Establishment of flow equation for single tube vacuum tube. In the vacuum tube to take an infinitesimal section dl, force analysis on dl. When DL flows in the pipe, it is affected by buoyancy, gravity, viscous force and the resistance along the tube wall [5]. Because the relative velocity of fluid flow in the pipe is not large, the viscous force can be neglected. Since the fluid is flowing at the microscopic conceptual, according to Newton's second law, it is known that the component of the buoyancy of dl along the tube axis is equal to the loss due to frictional resistance, and the flow of fluid can be solved mc [5]. Then:

$$F_\theta \sin\beta=\Delta\ P_f \tag{15}$$

$\Delta\ P_f$—The resistance along a fluid flow.

$$\Delta\ P_f=\lambda L\rho U_c^2/(4r) \tag{16}$$

λ—Resistance coefficient.

The flow state of the fluid in the vacuum tube is treated as a laminar flow; the coefficient of resistance λ is expressed as:

$$\lambda=64/R_e \tag{17}$$

$$\Delta\ P_f=2\mu L \tag{18}$$

μ— Dynamic viscosity.

The buoyancy force along the pipe axis of the infinitesimal section is [5]:

$$dF_\theta=(\rho_c-p_w)g(dv)\sin\beta \tag{19}$$

p_w—thermal fluid density of the infinitesimal sections,kg/m^3; ρ_c—Cold fluid density of the infinitesimal sections, kg/m^3; g—acceleration of gravity, m/s^2; β—Collector installation angle.

The relationship between density difference (ρ_c-p_w) and temperature difference (t_w-t_c) is [5]:

$$(\rho_c-p_w)=\rho\gamma(t_w-t_c) \tag{20}$$

ρ— the average density of the fluid infinitesimal sections, kg/m^3;

γ— coefficient of thermal expansion, 1/k.

From the analysis of the energy conservation equation for a single vacuum tube, the heat absorbed by the fluid in the vacuum tube is $m_c c(t_w-t_c)$, which is equal to the radiant energy absorbed by the heat fluid section [5]: $I_{eff}(\tau\alpha)_\theta\pi rL\cos\theta$. The flow of the fluid $m_c c$ can be expressed as $\varepsilon\rho U_c\pi r^2$.

When the infinitesimal section is small enough, the half of the heat absorption of a vacuum tube can be regarded as a rectangle. The absorption of solar radiation energy is: ($2I_{eff}\cdot r\cdot dr$)

$(\tau\alpha)_\theta$—The effective product of the transmittance ratio and absorption ratio. This paper selected the vacuum tube for τ is 0.93, α is 0.96.

$$\cos\theta=\sin\delta\sin\phi\cos\beta-\sin\delta\cos\phi\sin\beta\cos\gamma+\cos\delta\cos\phi\cos\beta\cos\omega \tag{21}$$

$$I_{eff}(\tau\alpha)_{\theta}\pi r L\cos\theta = m_c C_{pc}(tw-tc) = \varepsilon\rho U_c \pi r^2 C_{pc}(tw-tc) \tag{22}$$

U_c—Natural convection velocity of hot fluid, m/s.

ε: The ratio of the hot fluid flow cross-section to the vacuum tube cross-sectional area in the vacuum tube

ε=0.5 and according to formula (22) have:

$$t_w - t_c = 2 I_{eff}(\tau\tau)_{\theta} L\cos\theta\cos_{c} r C_{pc} \tag{23}$$

$$dF_{\theta} = 2\gamma g\, I_{eff}(\tau\alpha)_{\theta} L(dv)\cos\theta(in\beta/(U_c r C_{pc}) \tag{24}$$

Formula (9) integral from 0 to L, the buoyancy of the internal heat in the vacuum tube is:

$$F_{\theta} = \int_0^L dF_{\theta} = \gamma g\, I_{eff}(\tau\alpha)_{\theta} \pi r L^2\cos\theta os\theta\beta/(U_c C_{pc}) \tag{25}$$

The natural convection velocity of the hot fluid can be obtained according to equation (15):

$$U_c = \sqrt{\gamma g I_{eff}(\tau\alpha)_{\theta} r L^2\cos\theta sin\beta/(2\mu C_{pc})} \tag{26}$$

The mass flow rate of the hot fluid in the vacuum tube is:

$$m_c = 0.5\pi.5\rho\sqrt{\gamma g I_{eff}(\tau\alpha)_{\theta} r L^2\cos\theta sin\beta/(2\mu C_{pc})} \tag{27}$$

3. Experimental operation

3.1. Laboratory apparatus

Four vacuum tubes numbered A, B, C, and D; Four 4.5L water tanks; 12 temperature probes, 12 temperature transmitters; An information acquisition card; Irradiation instrument; Some wires; Support; Computer; Each one of the flat plate, T type, and the Y type guide plate.

3.2. The experimental process

This paper mainly studies the convective improvement of standard vacuum tubes. Three kinds of deflector are installed in the tube; they are a flat plate, a T-plate, and a Y-plate. The width of the flat plate is equal to the diameter of the inner tube of the vacuum tube; T-plate width 38 mm and high 34 mm; Y-plate three sides and the vacuum tube radius equal. Three types of deflector thickness are 2 mm, length is 1600 mm; Vacuum tube length is 1800 mm, outside diameter of the glass tube is 58 mm, and the internal diameter is 47 mm. Single tube glass tube placed south, tilt angle 35°. In the experiment, it is assumed that the vacuum tube is uniformly heated without considering the influence of air parameters (wind speed, ambient temperature, etc.).

In a vacuum tube, because the heat loss caused by heat conduction is smaller than the radiant heat loss, and the vacuum tube collector with a deflector has little effect on the loss due to heat conduction. Therefore, heat conduction losses can be neglected.

The deflector provided in the vacuum tube is shown as figure 1:

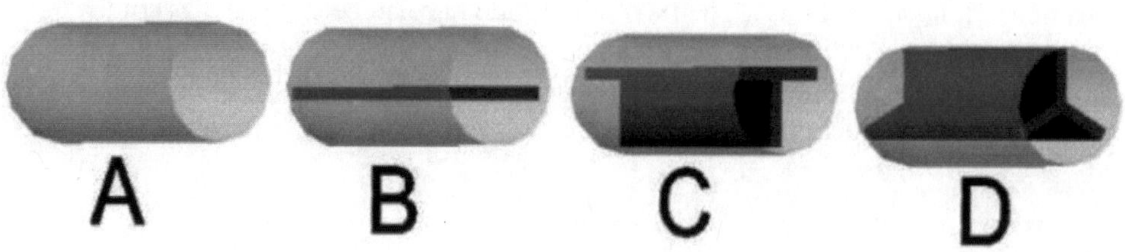

Figure 1. Schematic diagram of the design of the baffle.

3.2.1. Major experimental content. As shown in figure 2, to build a solar energy test platform, the first system test, A, B, C, D four vacuum tube on the bracket, A tube without the baffle, B tube plus plate baffle, C tube plus T-type baffle, D tube plus Y-type baffle. And at the top of each vacuum tube and the bottom of the water tank put a temperature probe, numbered respectively A1, A2, A3; B1, B2, B3; C1, C2, C3; D1, D2, D3. Then connect the temperature probe to the temperature sensor, the temperature sensor and then connected to the information acquisition card, the information acquisition card connected to the computer, and finally began to test.

And then the single tube boring test is done, the other steps are the same as the above steps apart from the temperature probe in the tank and the tank. The system test is on May 4, and the single tube boring test is on May 2.

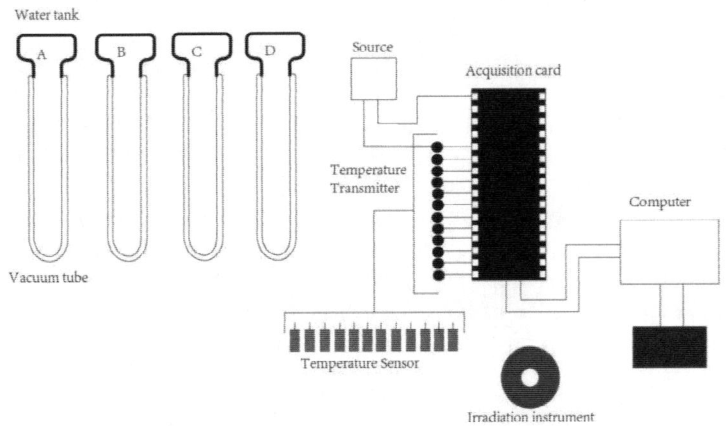

Figure 2. System test diagram.

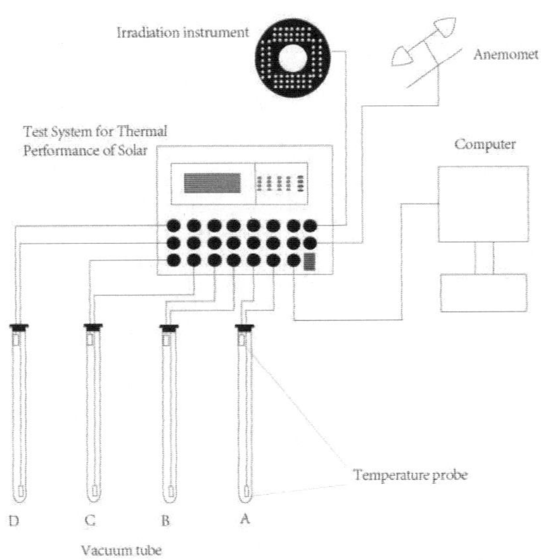

Figure 3. Schematic diagram of single tube boring test.

3.2.2. Problems encountered in the experiment and debugging

- Existing problems

IOP Conf. Series: Earth and Environmental Science **93** (2017) 012002 doi:10.1088/1755-1315/93/1/012002

1. The temperature difference between the 12 temperature probes is not significant.
2. The temperature measured by B1 is higher than that of B2.

- Debugging

1. As can be seen in figure 3, considering the influence of the position of the temperature probe, the temperature of the 12 temperature probes is approximately the same. Because the temperature probe is too light, the temperature probe at the bottom of the vacuum tube may float in the vacuum tube hot water layer. This problem is solved by adding a weight to the A1, B1, C1, D1 four temperature probes, and each temperature probe may have some error, so it is corrected by the thermometer.

2. First, check the connection; if the connection is right, there is something wrong with the temperature transmitter, and then the transmitter need to be replaced until make it coincide with the measured data.

3.3. Vacuum tube heat exchanger set characteristic analysis

3.3.1. Characteristic analysis of vacuum tube for system test. It can be seen from figures 4 and 5 that the temperature of the bottom of the four tubes and the temperature of the nozzle are increasing with the increase of time. At about 4:15, the temperature tends to stabilize and then gradually decreases. The reason why the temperature change is due to the change of the solar radiation intensity first enhanced and then weakened with the change of time. When the solar radiation intensity is strong, the energy absorbed by the fluid in the vacuum tube is higher and the temperature of the fluid in the tube is higher. As shown in figure 6, the solar radiation intensity is strong at 11 o'clock to 3 o'clock, and the temperature of the water in the vacuum tube is the highest at about 4 o'clock. This is because the temperature of the fluid in the tube needs a constant heat absorption process. And the collector itself has a certain heat capacity, which is consistent with the theory.

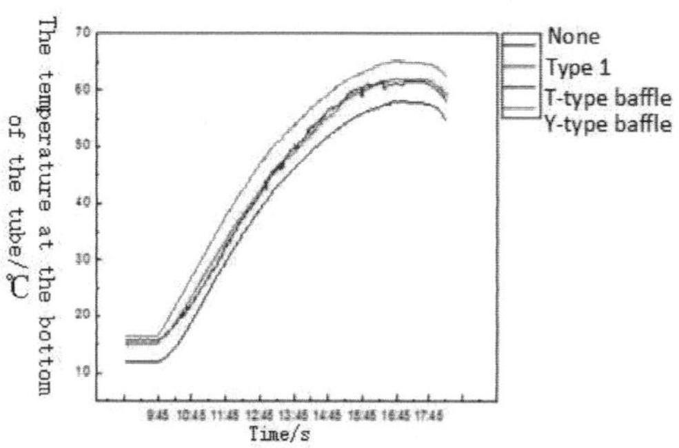

Figure 4. The temperature at the bottom of the system change with time.

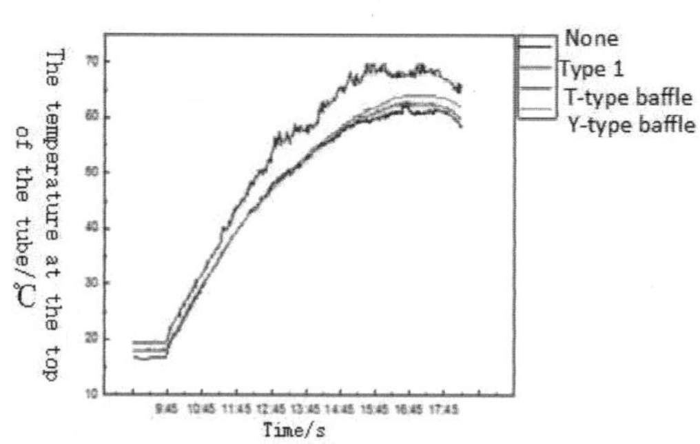

Figure 5. The temperature at the top of the system change with time.

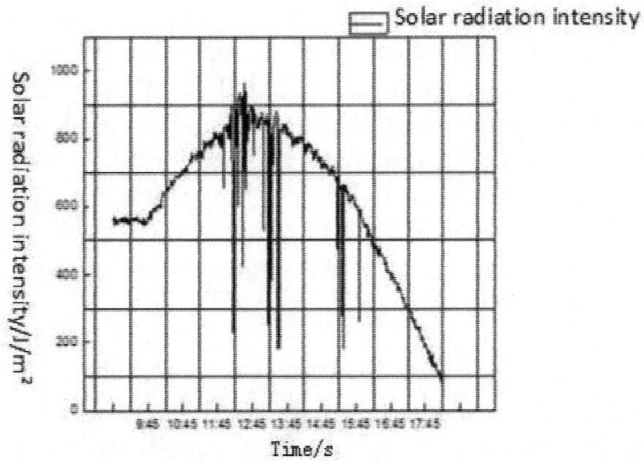

Figure 6. Solar radiation intensity and time diagram of the system.

As can be seen from figure 4, there is no deflector of the bottom temperature is the lowest, followed by the flat plate and T-plate, Y-plate has the highest temperature. The highest temperature of Y-plate reaches is 63℃, while the temperature without plate increases to 57℃, and the temperature of T-plate are similar to that of a flat plate, which is about 61℃. The plate type deflector is similar to the temperature of the T-type guide plate because the plate baffle plate divides the fluid in the vacuum tube into two channels to suppress lateral convection; However, the addition of a T plate divides the fluid in the vacuum tube into 3 channels, but the channel above the T deflector is the same as the flat plate. And the bottom of the two channels are divided by the cold fluid, so the lower part of the channel even if the separation of the cold fluid temperature is not large, it makes the two kinds of baffle temperature is similar. The Y-plate has the highest temperature because the Y-type guide plate divides the vacuum tube into 3 channels, and the channel occupied by the cold and hot fluid is more accurately separated, so the temperature is highest.

As can be seen from figure 5, the T-plate has the highest temperature, followed by the Y-plate and the flat plate, with the lowest without a board. T-plate highest temperature up to 68℃; the maximum temperature is 62℃ without plate; the temperature of the plate and the Y-plate was 62℃ and 65℃, respectively. According to figure 4 analysis we can see that adding guide plates in a vacuum tube, to suppress transverse convection to improve the collector efficiency have a certain effect, the effect of T-plate on the suppression of lateral convection is best, Y-plate second, plate type weakest.

The detailed data are shown in tables 1 and 2:

Table 1. Temperature rise at the bottom of the vacuum tube (℃).

Parameter	A_1	B_1	C_1	D_1
Bottom initial temperature	12.0115	14.995	15.785	16.604
Bottom equilibrium temperature	57.8265	61.635	61.014	63.882
T_C	45.815	46.64	45.229	47.278

Table 2. Temperature rise at the top of the vacuum tube (℃).

Parameter	A_1	B_1	C_1	D_1
Top initial temperature	16.573	18.005	19.435	17.948
Top equilibrium temperature	62.53	62.566	68.538	65.018
T_C	45.957	44. 561	49.103	47.07

The velocity of the fluid in the vacuum tube is inversely proportional to the temperature difference between the cold and hot fluid, and is proportional to the temperature at which the fluid rises at the inlet and outlet (Tc), the temperature at the inlet and outlet is approximately the temperature at the nozzle and the temperature at the bottom of the tube. It can be seen from tables 1 and 2 that the rising temperature of the fluid at the inlet and outlet of B, C, and D is higher than that of A, and the temperature difference between hot and cold fluid, B and D tubes, is lower than that of a tube. Although the C tube temperature is higher than that of the A tube, the temperature at which the fluid rises is much higher than that of the A tube at the inlet and outlet of the C tube, so the flow velocity of the B, C and D tubes is greater than that of the A tube in general. By type (2-21) can see greater velocity, more heat fluid absorption, which indicates that adding guide plates have the effect on inhibiting radial convection of vacuum tube, the water temperature in the water tank was stratified, improving heat transfer efficiency.

3.3.2. Characteristic analysis of single tube boring test. It can be seen from figure 7 that the temperature of the bottom of the four vacuum tubes gradually increases with time, and the maximum is about 4 o'clock, and it will gradually decrease after a certain period of time. Four reasons for temperature changes in the tube: First, because the vacuum tube collector continues to absorb solar radiation energy, and with the increase in light intensity, the more heat absorbed by the tube fluid, the temperature gradually increased, and later the intensity of solar radiation weakened, the temperature will be reduced, 11 o'clock to 2:30 between the strongest, and the temperature is about 4 o'clock when the highest, this is because the collector has a certain heat capacity.

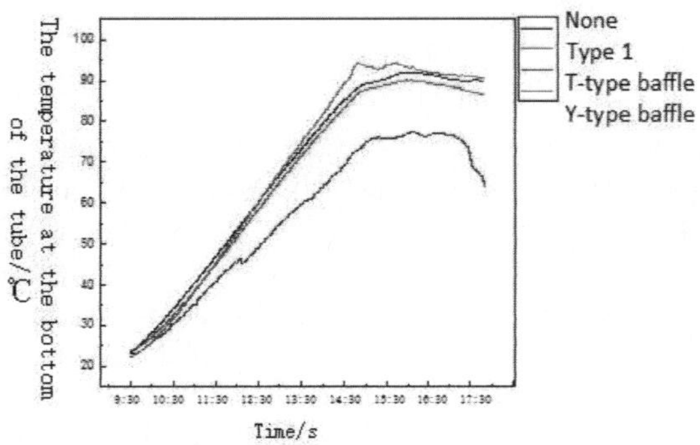

Figure 7. Bottom temperature change with time of single tube boring test.

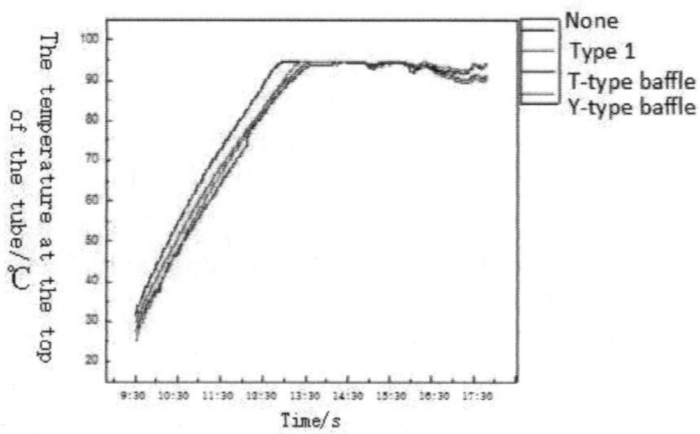

Figure 8. Top temperature change with time of single tube boring test.

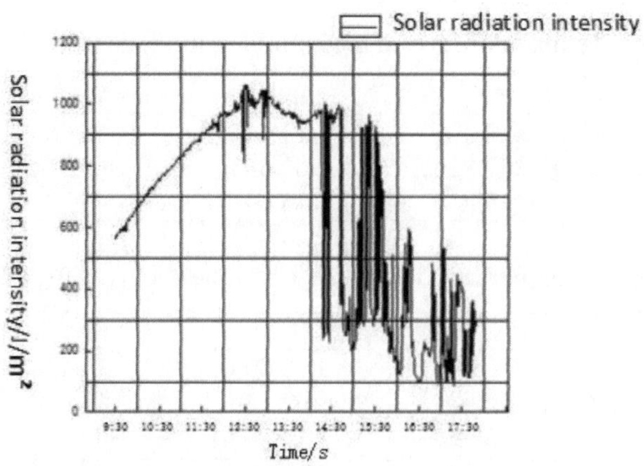

Figure 9. Solar radiation changes over time.

Figure 7 shows clearly that the Y-plate has the highest temperature; without the board and the plate is second; the lowest is T-plate. The maximum temperature of the Y plate reaches 93℃; the temperature without a plate and the temperature of the flat plate were 91.9℃and 90℃, respectively; plus T-plate reaches a maximum of 77℃.

As can be seen in figure 8, the temperature at the top of the four pipes gradually increases with time, rises to a certain extent, remains stable, and then decreases gradually. The tube without deflector reaches the highest at 1 o'clock, and the highest is 94.5℃. Three kinds of a vacuum tube with guide plate reached the highest in 1:30, the highest is 94.5℃. After 1:30, the temperature of the four tubes was almost equal, and there was a slight decrease after 4:30 o'clock. The reason for this phenomenon is that: As shown in figure 9, the vacuum tube collector absorbs the solar radiation energy continuously during the heat collection process so that the temperature of the fluid in the pipe increases continuously. When the fluid temperature is much higher than the ambient temperature, the solar collector's insulation performance deficiencies, bound outward by thermal radiation and convection heat transfer through the surface of the solar collector mode give off lots of heat, and after 3:00 solar radiation intensity also decreased. Tube fluid temperature in the 1:00 to 4:00 change is not more than 2℃. It is because, at 1 o'clock, the temperature of the fluid in the tube has reached the boring temperature, the temperature inside the tube will not be significantly increased.

According to the efficiency formula of the collector, the thermal efficiency of the system test and the single tube boring test are as follows:

Table 3. Maximum thermal efficiency of system and single tube.

Project	Standard vacuum tube	vacuum tube with a flat plate	vacuum tube with T-plate	vacuum tube with Y-plat
Single tube boring test	55%	55.4%	49%	57.6%
System test	29.1%	31.1%	32.2%	32%

From table 3 it can be concluded that single tube boring test plus Y-plate is the most efficient, plus flat plate is the next, and without adding plate is lower, plus T-plate is the minimum. The thermal efficiency of the Y-plate is 2.6% higher than that without the baffle, and the thermal efficiency of the T-plate is 6% lower than that without the baffle. Theoretically, the heat efficiency of a vacuum tube collector with a draft plate should be enhanced, which has a certain effect on suppressing radial convection and enhancing axial convection. But in the single tube stuffy test, the data added to the T plate decreased. The reason that may lead to this conclusion is that: Without the water tank, the water in the vacuum tube has no circulation; It is also possible that the heating time of the collector has not reached the maximum since the irradiation time of the T plate vacuum tube collector is not high enough, and the T plate has not reached the highest efficiency at the time of calculation.

In the system test, the efficiency of adding T-type plate is the highest, and the efficiency of adding y-type plate is next, plus the flat plate is lower, without the deflector is minimum.

4. Conclusion

In order to improve the convective heat transfer of the fluid in the standard all-glass vacuum tube collector, the guide plate that can inhibit the radial convection and enhance the axial natural convection were designed. The theoretical model was established. The thermal properties of various conditions were tested. The results show that:

The heat efficiency of the vacuum collector with the guide plate is improved due to the improvement of the internal flow field. For the system test, T-type guide plate, the effect of restraining the radial convection and enhancing the axial convection is better than Y-type guide plate, and the Y-type is better than the plate type. Vacuum tubes with guide plates are better than vacuum tubes without guide plates.

The efficiency of the vacuum tube collector with T, Y and plate type plate is increased by 3.1%, 2.9%, and 2% respectively compared with that without flow guide plate. The heat transfer efficiency of the vacuum tube with guide plate is 2.6% higher than that of the standard vacuum tube without modification.

The installation of the guide plate can restrain radial convection and enhance axial convection for both single tube and the system, improve the efficiency of the collector to improve the performance of the entire water heater, and it is more effective for the system.

Acknowledgments
This work was supported by the Scientific Research Project on the Cooperation of University and Enterprise in Chuxiong Prefecture of Yunnan Province (Grant No. XDXQ1503).

References
[1] Wang H, Hu X H and Cheng H Z 2013 *Solar Thermal Utilization* (Beijing: Tsinghua University Press) (in Chinese)
[2] Zhang T, Han J T, *et al* 2015 Solar vacuum tube heat transfer simulation and experimental study *J. Eng.Thermophys.* **36** 2027-31 (in Chinese)
[3] Ai N, Fan J H, *et al* 2007 CFD - PIV flow analysis technique is applied to the solar hot water system Research Progress, *Chem. Ind. Eng. Prog.* **26** 513-8 （in Chinese)
[4] Chen H X 1996 *Solar Thermal Utilization* (Beijing: Tsinghua University Press)
[5] Li Y M Model establishment and simulation analysis of heat transfer solar water heater (Ningxia University) （in Chinese)
[6] Yan S Y, Tian R, Yu W Y, *et al* 2010 factors on the fluid full glass solar vacuum solar water heater *J. Eng.Thermophys.* **2010** 641-3 (in Chinese)
[7] Wang Z F and Sun H W 2001 All glass vacuum tube solar air heating tubes in the numerical simulation of flow and heat transfer *Acta Energi. Sin.* **22** 35-9 (in Chinese)
[8] Zhang T, Tian R, Han J T and Yu Z T 2016 All glass vacuum tube solar collector convection heat test and simulation *J. Agr. Eng. Res.* **32** 206-12 (in Chinese)
[9] Zhong J L, Fu L X, *et al* 2005 All glass evacuated solar collector tube flow field and temperature field visualization *J.Zhejiang.Univ-SC.* A.**31** 351-4 (in Chinese)
[10] Redpath D A G 2012 Thermosyphon heat-pipe evacuated tube solar water heaters for northern maritime climates *Sol. Energy* **86** 705-15
[11] Budihardjo I and Morrison G L 2009 Performance of water-in-glass evacuated tube solar water heaters *Sol. Energy* **83** 49-56
[12] Li H, Dai Y J, Li Y, *et al* 2012 A case study of a two-stagerotary desiccant cooling/heating system driven by evacuated glass tube solar air collectors *Energ. Buildings* **47** 107-12

One-step preparation of biological aviation kerosene by catalytic hydrogenation of waste lard over Pt/SAPO-11

X Zhang[1], Y B Chen[1,3], X Y Li[1], D Souliyathai[1], S P Zhang[1], Q Wang[1], Q Liu[2], J C Du[2] and A M Zhang[2]

[1]School of Energy and Environment Science, Yunnan Normal University, Kunming, Yunnan, China
[2]Kunming Institute of Precious Metals, Kunming, Yunnan, China

E-mail: c20072007@163.com

Abstract. Biological aviation kerosene was produced by one-step catalytic hydrotreatment of *waste lard oil* over Pt/SAPO-11 in a high-pressure fixed bed micro reactor. The influence of reaction conditions such as temperature, pressure, hydrogen oil ratio, and space velocity on the deoxygenation rate, the selectivity of C_8-C_{16} hydrocarbons and the isomerization rate of C_8-C_{16} hydrocarbons have been investigated. The experimental results showed that the temperature of 400℃, pressure of 5 MPa, hydrogen oil ratio of 1000 and space velocity of 1.2 h^{-1} were the best experimental reaction conditions. Under these conditions, the conversion rate is 96.62%, the selectivity of C_8-C_{16} hydrocarbons is 50.25%, and the isomerization rate of C_8-C_{16} hydrocarbons is 35.68%.

1. Introduction

With the rapid development of economy and the improvement of people's living standard, the air transportation industry is playing an important role in the process of economic globalization [1]. As far as civil aviation concerned, the number of people and goods transported by air increased by 4.9% and 5.3%, respectively [2]. In 2005-2010, the total diesel fuel and jet fuel consumed 500-600 million barrels per day [3], and the global air transport industry consumes 15-17×10^8 barrels of aviation kerosene a year [4,5]. Conventional aviation kerosene burns on the aircraft's turbine engines, generating CO_2 directly into the atmosphere of the stratosphere, causing severe greenhouse effects; At the same time, it also affects the global climate change. As a result, the airline industry faces severe challenges in reducing greenhouse gas emissions. In the face of pressure reduction, IATA made three promises: from 2009-2020, the combustion efficiency of aviation kerosene average annual increase by 1.5%; by 2020, reach zero emissions targets; compared with the year 2050, emissions fell to 50% [6,7]. In addition to environmental problems, aviation kerosene prices soared, from jet fuel $320/t average cost in 2004 increased to $1005/t in 2011, it also forced the airline industry to pursue a clean, low price alternative fuel. In addition to environmental and cost considerations should also be considered such as aerospace safety work [8]. In addition, the air transport sector requires high energy density liquid hydrocarbon fuels [9]. Alternative fuels for aviation kerosene must have some special properties, for example, good cold flow, thermal stability and low freezing point [10], and fuel must be suitable for the currently designed aircraft engines. Biomass as the only carbon resource other than petrochemical energy [11] has a bright future in the production of clean energy. The biomass is converted into C_8-C_{16} isomeric alkanes, naphthenic hydrocarbons, olefins and aromatic hydrocarbons. It is the best way to substitute biochemical aviation kerosene for the replacement of petrochemical aviation kerosene by converting biomass into C_8-C_{16} isomeric alkanes, naphthenic hydrocarbons, olefins and aromatic hydrocarbons to prepare bio aviation kerosene [12]. Studies have shown that in the full life cycle, bio-aviation kerosene greenhouse gas emissions than Petrochemical aviation kerosene reduced by 50%-90% [13]. At present, many kinds of biological aviation kerosene have been developed, including Fischer Tropsch synthesis (F-T synthesis) method, biotransformation, thermochemical conversion, hydrogenation method, etc. In this paper, bio-fuel aviation kerosene was prepared by using one-step hydrogenation of the waste lard as raw material over Pt/SAPO-11 as catalyst. The effects of temperature, pressure, hydrogen oil ratio and space velocity on the deoxygenation rate, selectivity of C_8-C_{16} hydrocarbons and the isomerization rate of C_8-C_{16} hydrocarbons were investigated by one-step preparation of biological aviation kerosene.

2. Experimental

2.1. Materials and reagents

Waste lard: the main fatty acids were made up their fatty glyceride, as shown in table 1 and figure 1.

Table 1. The main fatty acid composition of lard.

Retention time	Fatty acid name	Molecular formula	Relative percentage content/%
7.04	myristic acid	$C_{14}H_{28}O_2$	1.2
8.84	hexadecenoic acid	$C_{16}H_{30}O_2$	1.88
9.05	cetylic acid	$C^{16}H_{32}O_2$	17.89
10.63	linoleic acid	$C_{18}H_{32}O_2$	15.59
10.69	oleic acid	$C_{18}H_{34}O_2$	50.38

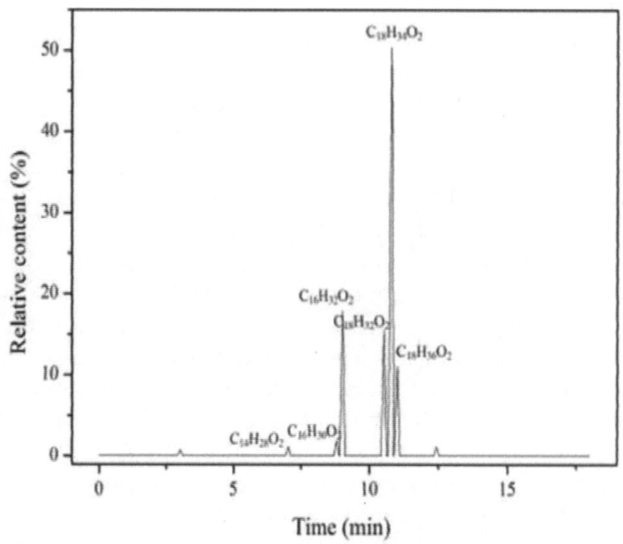

Figure 1. GC diagram of the main fatty acid composition of lard.

The chemical reagents are analytical pure (AR), Main: anhydrous sodium sulfate, methanol, concentrated sulfuric acid (Xilong Chemical Co., Ltd.). Petroleum ether, dichloromethane, anhydrous ethanol (Tianjin wind boat chemical reagent Technology Co., Ltd.)

Pt/SAPO-11 catalyst: the research group made the preparation by dipping method [14]. The functionalized SAPO-11 carrier was impregnated in H_2PtCl_6 solution with equal volume of 6 h, dried at room temperature for 15 h, dried at 110℃ for 4 h, calcined at 500℃ for 6 h, then cooled and ground to make 2%(WT) Pt/SAPO-11 catalyst [15,16]. The catalyst under 320℃/1MPa of hydrogen pressure was activated 6 h before using.

2.2. Main instruments

MRT-H00521JB type high-pressure micro fixed bed catalytic reaction evaluation equipment (research group design, commissioned by Beijing Aerospace century star laboratory equipment Co., Ltd.), Clarus 680-Clarus SQ8T GC-MS (American PerkinElmer Corporation), DHG-9203A electric blower drying box (Shanghai constant Scientific Instrument Co., Ltd.), HZQ-C double layer air bath constant temperature oscillator (Jintan Dadi automatic instrument factory), Analytical balance (Shanghai Hengping Scientific Instrument Co., Ltd.), Rectifying column (Tianjin friend Technology Co., Ltd.).

2.3. Method

2.3.1. Characterization of catalyst structure

- **X ray diffraction (XRD)**

The XRD test was carried out on the D/MAX-2000 type fully automatic X ray diffractometer.

The test conditions: source of radiation was CuKα, the wavelength was 0.15406 nm, the tube current was 40 mA, the tube voltage was 40 kV [17]. The X ray diffraction of the result are given in figure 2.

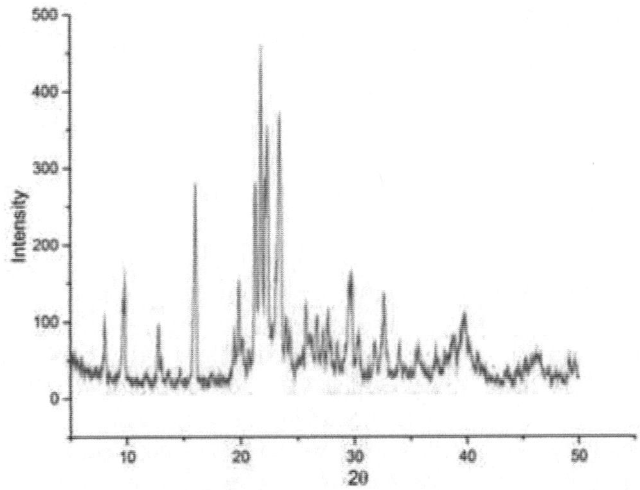

Figure 2. Pt/SAPO-11 catalyst.

Pt/SAPO-11 catalysts exhibit characteristic peaks of SAPO-11 molecular sieves at 9.82, 16.03, 21.84, 23.46, etc, and the peaks are sharp and high in intensity. Modification did not change the skeletal structure of the molecular sieve.

- **N_2-Physisorption**

Table 2. The catalyst BET analysis of the Pt/SPAO-11 molecular sieve.

sample	specific area (m^2/g)	pore volume (cm^3/g)	mean pore size (nm)
Pt/SPAO-11	223.17	0.2121	3.6

The surface area of the catalyst, pore capacity, the average aperture was detected in the United States Contador Company NOVA2000e surface area and porosity analyzer. The adsorption temperature of N_2 was 77 K [17]. The specific surface area, pore capacity and mean pore size were calculated by BET method. The catalyst BET analysis of the Pt/SPAO-11 molecular sieve is shown in table 2.

3. Results and discussion

3.1. Influence of temperature

At the space velocity of $1h^{-1}$, the hydrogen-oil ratio was 1200, and the pressure was 4 MPa, the temperature was heated to 360, 380, 400, 420 and 440 respectively, prepare aviation kerosene, in high pressure fixed bed micro evaluation device on one-step catalytic hydrogenation of waste lard. Figure 3 shows the influence of the temperature on conversion rate of the waste lard, the selectivity of C_8-C_{16} hydrocarbons, and the isomerization rate of C_8-C_{16} hydrocarbons under different reaction temperatures.

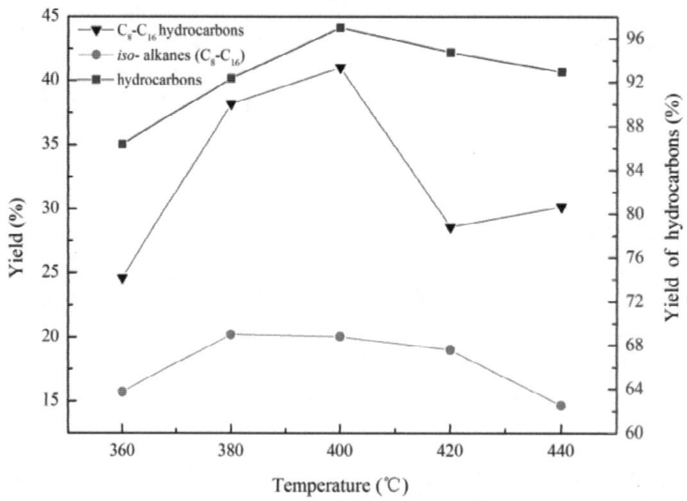

Figure 3. Effect of temperature on one step hydrogenation of waste lard.

Figure 3 shows that with the increase in temperature from 360 to 440℃, lard conversion rate obviously increased firstly and then decreased, and the 400℃ of temperature conversion rate reached 97.01%, for the selectivity of C_8-C_{16} hydrocarbons increases gradually. When the temperature was greater than 400, the selectivity of hydrocarbons was 41.02%, and then decreased by about 28%. It may be that the higher temperature causes the cracking reaction to increase and long chain heavy hydrocarbon are converted into short chain hydrocarbons, resulting in the selectivity of C_8-C_{16} hydrocarbons decreased. The isomerization rate of C_8-C_{16} hydrocarbons increases gradually with the increase of temperature, and the isomerization rate of C_8-C_{16} hydrocarbons decreases gradually when the temperature reaches 380℃. This is because the alkane isomerization reaction was exothermic reaction, the temperature was too high, which was not conducive to isomerization reaction, at 400℃ the selectivity of C8-C16 hydrocarbons was 20.02%. Above all, 400℃ was the best reaction temperature. The GC spectra of the catalytic products of waste lard under the optimum reaction temperature are shown in figure 4.

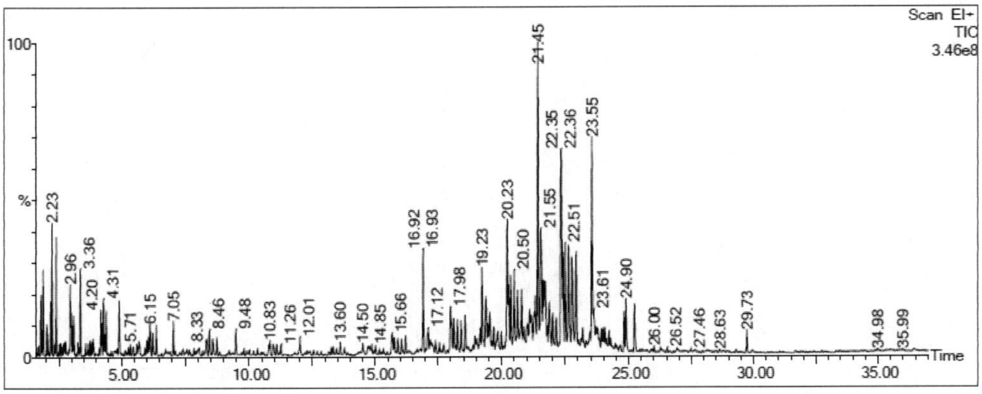

Figure 4. GC spectra of catalytic products of waste lard hydrogenation at the optimum reaction temperature.

3.2. Effect of space velocity

Figure 5 is the influence of space velocity on conversion, the selectivity of C_8-C_{16} hydrocarbons and the isomerization rate of C_8-C_{16} hydrocarbons.

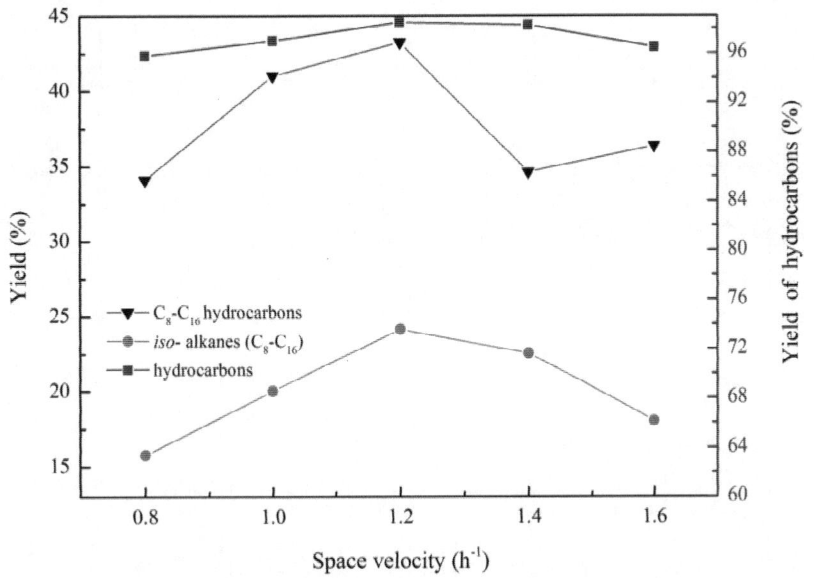

Figure 5. effect of space velocity on one step hydrogenation of waste lard.

Figure 5 shows that the conversion rate is small, and all of them remain above 95%. When the space velocity was 1.2h^{-1}, the selectivity of C_8-C_{16} hydrocarbons and the isomerization rate of C_8-C_{16} hydrocarbons increase gradually with the increase of space velocity, and reached the maximum of 43.24% and 24.14% respectively. In a certain range of space velocity, increasing the space velocity can make the raw oil more fully contact with the catalyst and increase the reaction speed, which is beneficial to the reaction. When the speed reaches a certain level, then increase the space velocity of raw oil attached to the catalyst surface hindered hydrogen and catalyst contact, which is not conducive to C_8-C_{16} hydrocarbons and C_8-C_{16} isoparaffin conversion, resulting in the selectivity of C_8-C_{16} hydrocarbons and the isomerization rate of C_8-C_{16} hydrocarbons are decreased. Therefore, the optimum space velocity was 1.2h^{-1}. The GC spectra of the catalytic products of waste lard under the optimum reaction space velocity are shown in figure 6.

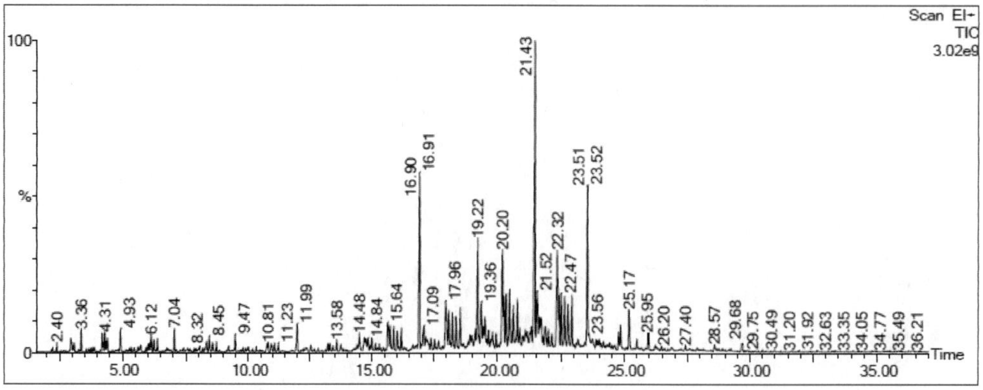

Figure 6. GC spectra of catalytic products of waste lard hydrogenation at the optimum reaction space velocity.

3.3. *Effect of hydrogen oil ratio*

Figure 7 is the influence of hydrogen oil ratio on conversion, the selectivity of C_8-C_{16} hydrocarbons and the isomerization rate of C_8-C_{16} hydrocarbons.

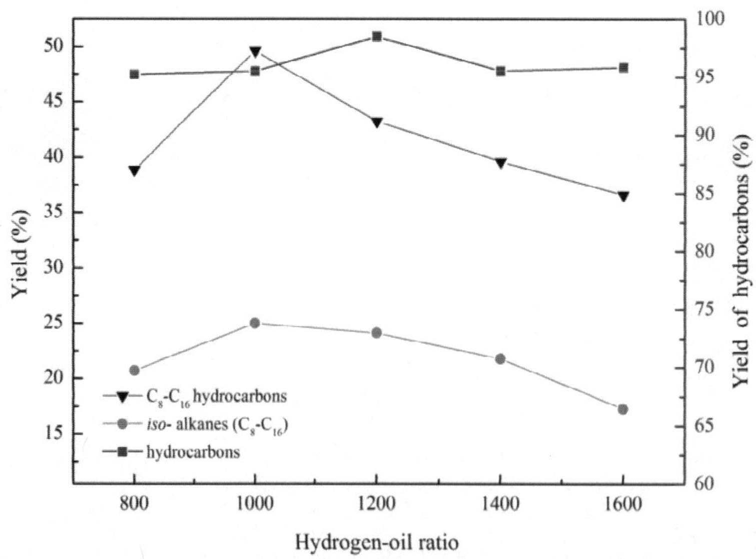

Figure 7. Effect of hydrogen oil ratio on one step hydrogenation of waste lard.

Figure 7 shows that the hydrocarbon content in liquid products, the selectivity of C_8-C_{16} hydrocarbons and the isomerization rate of C_8-C_{16} hydrocarbons tend to increase firstly and then decrease with the increase of hydrogen oil ratio. When the hydrogen oil ratio was 1000, the conversion rate, the selectivity of C_8-C_{16} hydrocarbon and the isomerization rate of C_8-C_{16} hydrocarbons reached the maximum values of 95.52%, 49.63%, 25.02% respectively. It may be because hydrogen oil ratio is relatively low in the entire reaction system, hydrogen, raw materials and catalysts are not fully exposed, resulting in slower reaction rate. Appropriately increasing the hydrogen oil ratio is beneficial to the conversion of hydrocarbons and isomerization of alkanes. But when the hydrogen oil ratio is too large, hydrogen surrounds the catalyst, so that the raw oil and catalyst can not be effectively contacted, leading to the rate of deoxygenation, the selectivity of C_8-C_{16} hydrocarbons and the isomerization rate of C_8-C_{16} hydrocarbons are decreased. So, the optimum hydrogen oil ratio was 1000. The GC spectra of the catalytic products of waste lard under the optimum reaction hydrogen oil ratio are shown in figure 8.

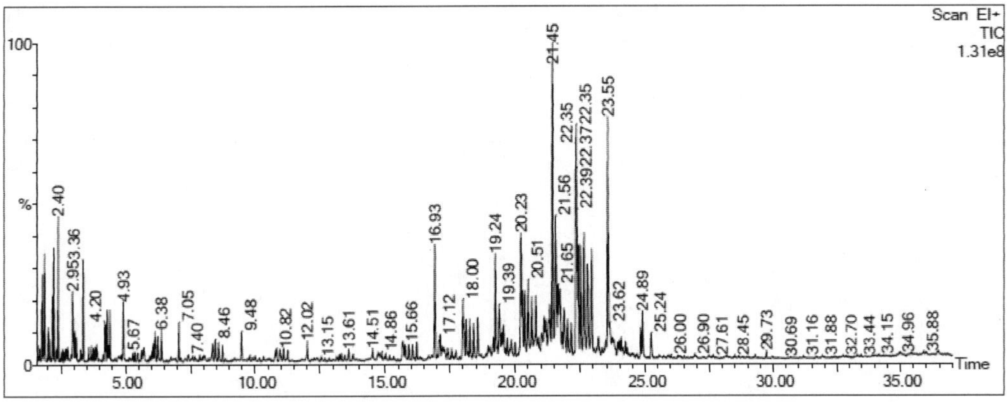

Figure 8. GC spectra of catalytic products of waste lard hydrogenation at the optimum reaction hydrogen oil ratio.

3.4. Effect of pressure

Figure 9 is the influence of pressure on conversion, the selectivity of C_8-C_{16} hydrocarbons and the isomerization rate of C_8-C_{16} hydrocarbons.

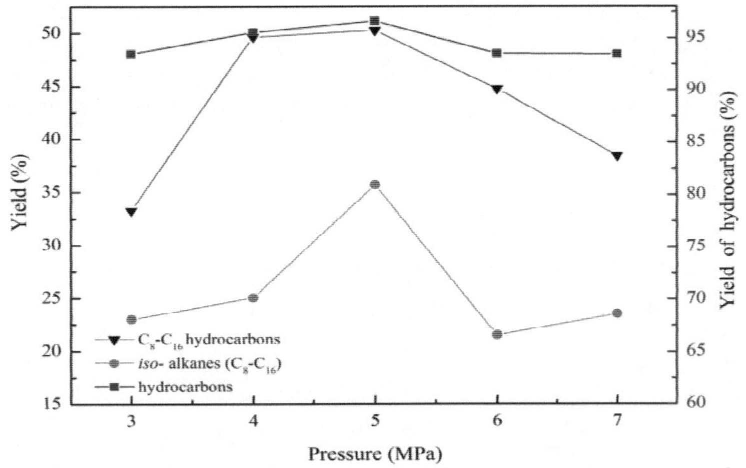

Figure 9. Effect of pressure on one step hydrogenation of waste lard.

Figure 9 shows that the conversion rate, the selectivity of C_8-C_{16} hydrocarbons and the isomerization rate of C_8-C_{16} hydrocarbons tend to increase firstly and then decrease with the increase of pressure. When the pressure was 5 MPa, the conversion rate, the selectivity of C_8-C_{16} hydrocarbon and the isomerization rate of C_8-C_{16} hydrocarbons reached the maximum values of 96.62%, 50.25%, 35.02% respectively. It may be due to the beginning of the reaction, as the pressure continues to rise, the reactant partial pressure increases, the reaction speed increases, resulting in deoxygenation rate, C_8-C_{16} hydrocarbon selectivity, and the isomerization rate of C_8-C_{16} hydrocarbons increases. When the pressure reaches a certain level, increasing the pressure is beneficial to the coke production, blocking the catalyst channel, resulting in deoxygenation rate, the selectivity of C_8-C_{16} hydrocarbons, and the isomerization rate of C_8-C_{16} hydrocarbons decreased. Therefore, the optimum reaction pressure was 5

MPa. The GC spectra of the catalytic products of waste lard under the optimum reaction pressure are shown in figure 10.

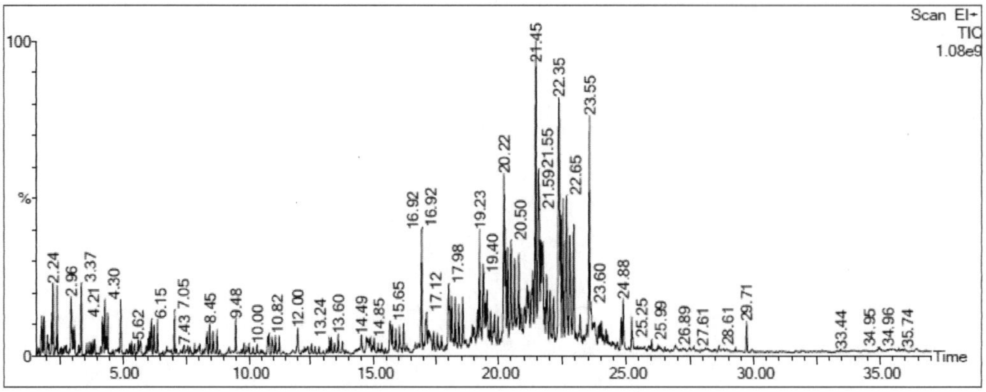

Figure 10. GC spectra of catalytic products of waste lard hydrogenation at the optimum reaction pressure.

4. Conclusions

- Temperature is the key factor affecting the conversion rate of waste lard. When the temperature is certain, other factors have little influence on the conversion rate of waste lard.
- The temperature and space velocity have great influence on the selectivity of C_8-C_{16} hydrocarbon. In the suitable range, increasing the temperature and increasing the space velocity are beneficial to the selectivity of C_8-C_{16} hydrocarbon; Space velocity and pressure have a great influence on the isomerization rate of C_8-C_{16} hydrocarbons. In a suitable range, increasing the space velocity and pressure is beneficial to the isomerization rate of C_8-C_{16} hydrocarbons.
- The optimum reaction conditions for the preparation of aviation kerosene by one-step hydrogenation of waste lard were: temperature of 400℃, space velocity of $1.2h^{-1}$, hydrogen oil ratio of 1000, and pressure of 5 MPa. Under these conditions, the conversion rate of waste lard, the selectivity of C_8-C_{16} hydrocarbon and the isomerization rate of C_8-C_{16} hydrocarbons were 96.62%, 50.25% and 35.68%, respectively.

Acknowledgements

The present study was supported by National Natural Science Foundation of China (No. 21266032) and International S&T Cooperation Program of China (No. 2015DFA60120).

References

[1] He Y F 2005 Study on the development strategy of China's air transport industry *Theory Discussion* **2** 71-3
[2] Sun X Y, Liu X and Zhao X B 2013 Research progress in preparation technology and application of aviation biofuels *Chin. J. Biotechnol* **29(3)** 285-98
[3] Faaij A and Dijk M V 2012 White paper on sustainable jet fuel *Copernicus. Ins. Utrecht. Univ*
[4] Zhou D 2014 Development of aviation bio fuel technology in CNPC *Sino-Global Energy* **19** 48
[5] Dong P and Tong H F 2005 Present situation and development suggestion of aviation bio fuel prepared by hydrogenation of oils and fats *Theory Discussion* **2** 71-3
[6] Li Y 2008 IATA Declaration on environmental protection *Chin. Civil. Aviation* **10** 23-4
[7] Hu X T, Qi B L and Fu X G 2012 Development background and application status of aviation bio fuel technology *Chem. Ind. Eng. Prog* **31(8)** 1625-30
[8] Hemighaus G, Boval T, Bosley C, *et al* 2006 Alternative jet fuels a Supplement to Chevron's

Aviation Fuels Technical Review *Chevy. Co. ltd*

[9] Yang Y, Liao B and Pang H 2015 The biomass platform compounds are converted into liquid alkane fuels by the construction of C-C bonds *Petrkchem. Techno* **44(2)** 137-43

[10] Mohammad M, Hari T K, Yaakob Z, *et al* 2013 Overview on the production of paraffin based-biofuels via catalytic hydrodeoxygenation *Renew. & Sust. Energy Rev.* **22(8)** 121-32

[11] Demirbas A 2007 Progress and recent trends in biofuels *Prog. Energy & Combustion Sci.* **33(1)** 1-18

[12] Nygren E, Aleklett K and Höök M 2009 Aviation fuel and future oil production scenarios *Energy Policy* **37(10)** 4003-10

[13] Fargione J, Hill J, Tilman D, *et al* 2008 Land clearing and the biofuel carbon debt *Sci.* **319(5867)** 1235-8

[14] Kabe T, Qian W, Hirai Y, *et al* 2000 Hydrodesulfurization and hydrogenation reactions on noble metal catalysts: I. Elucidation of the behavior of sulfur on alumina-supported platinum and palladium using the 35S radioisotope tracer method *J. Catal.* **190(1)** 191-8

[15] Gong S F, Chen N, Shirai H, *et al* 2013 Isomerization of n-alkanes derived from jatropha oil over bifunctional catalysts *J. Molecular Catal. A: Chemical* **370** 14-21

[16] Rabaev M, Landau M V, Herskowitz M, *et al* 2015 Conversion of vegetable oils on Pt/Al_2O_3/SAPO-11 to diesel and jet fuels containing aromatics *Fuel* **161** 287-94

[17] Du J C, Zhao Y Y, Xia W Z, *et al* 2015 Effects of precious metals and supports on catalytic hydrodeoxygenation and hydrocracking performance over catalysts *Rare Metal Materials Eng.* **44(9)** 2210-5

Study for the degumming pretreatment of rubber seed oil

X Y Li[1], Y B Chen[1, 2], X Zhang[1], D Souliyathai[1], S P Yang[1] and Q Wang[1]

[1]School of Energy and Environment Science, Yunnan Normal University, Kunming, Yunnan, PR China

E-mail: c20072007@163.com

Abstract. With the rapid development of the aviation industry, appearing of the aviation carbon tax and the increasingly serious environmental problems have forced the world to research the development of renewable bio-aviation fuel. Renewable biological aviation fuel contains phosphorus that could reduce the synthesis of noble metal catalysts such as Pd, Pt activity. In order to get low content of phosphorus in degummed oil of non-edible vegetable oil, in this paper, with rubber seed oil as raw material, making the experiment of single factor at the influence of temperature, stirring speed, adding amount of monoethanolamine (MEA) and water amount. The experimental results show that the added amount of MEA is 2.5% in the weight of oil, and temperature is 60℃, while the amount of added water is 2% in the weight of oil, reaction time is 40 min, and stirring speed is 200 r/min. Under these conditions, the phosphorus content of rubber seed oil can be reduced to below 3 mg/kg, degumming rate is 91.37%, and the degumming effects are obvious, which also provides some foundation for follow-up studies.

1. Introduction

With the development of the global economy, air transport occupies an important position and the growth rates of air and cargo air freight have arrived at 4.9% and 5.3% [1]. At the same time, the aviation industry is under double pressure to reduce greenhouse gas emissions and resource shortages [2]. At present, countries have carried out research on biomass energy, and searching for renewable bioaviation kerosene has become a hot topic in the research of scientists. With the oil of animals and vegetables as raw materials, the bio-aviation fuel technology with hydrogen treatment owns a simple process and a high maturity [3], the raw materials that China uses to prepare for bio-aviation fuel is mainly non-edible oil.

The oil that is not degummed with vegetable oil usually contains colloid, and most of it is phospholipid. The presence of phospholipids makes the oil bubbling while heating the oil, and the phospholipid becomes bitter, moreover, during the storage process, phospholipid can be easily rancid [4]. At high temperature, phospholipids produce confocal phenomena, so before entering the deodorization, it is required that the phosphorus content be less than 10 mg/kg, and it is ideal that the content of primary oil phosphorus is less than 5 mg/kg after refining [5].

In the preparation of bio-aviation fuel, the cost of raw materials accounts for more than 60% of the total [6]. Phosphorus in phospholipids seriously affects the performance and repeated use of precious metal catalysts, therefore, non-edible vegetable oil degumming pretreatment is an important part in the preparation of biofuel. At present, the degumming method mainly includes water degumming, acid degumming, enzymatic hydrolysis, membrane filtration, etc. Jiang X F and others [7] studied the technological conditions for the degumming of rapeseed oil, drawing a conclusion through response surface optimization: temperature 60℃, 50% citric acid content accounted for 0.22% of oil weight, water adding amount 3.34% of oil weight, the degummed rapeseed oil phosphorus content is 15.67 mg/kg when acid treatment time was 24.78 min, degumming rate was 93.76%. Wu Y J and others [8] studied for the cold pressed sesame oil with degumming acid method, taking the orthogonal experiment to optimize the best experimental conditions: the dosage of acid is 0.5%, temperature is 60℃, water content is 2.5%, and the reaction time is 45 min. In this condition, the phospholipid content of the degummed oil was 54.05 mg/kg and the stripping rate reached 72.9%.

In this paper, rubber seed oil is used as raw material, and on the basis of the existing acid method degumming, the defect of the acid method is not completely, MEA reagent was introduced in the process of degumming, the influence of various factors and the better experimental conditions are determined to provide references for further amplification experiments.

2. Materials and methods

2.1. *Experimental raw materials*

The rubber seed oil is produced in Pu'er, Yunnan province and its main properties are shown in table 1.

Table 1. The raw material oil properties.

material	Phosphorus content /mg/kg	Acid value /mg_{KOH}/g	Saponification value /$mg_{(KOH)}$/g	Iodine value /g/100g
rubber seed oil	34.76	38.57	186.64	159.58

2.2. *Experimental methods*

2.2.1. Preparation of rubber seed degumming oil. The preparation of rubber seed degumming oil mainly includes the following steps: I. The untreated rubber seed oil is heated in the setting temperature; II. Adjust the stirring speed, add MEA and deionized water to stir the reaction; III. After processing, cool to room temperature, then centrifuge separation; IV. After dehydration, get the oil of degummed. The experimental device is shown in figure 1.

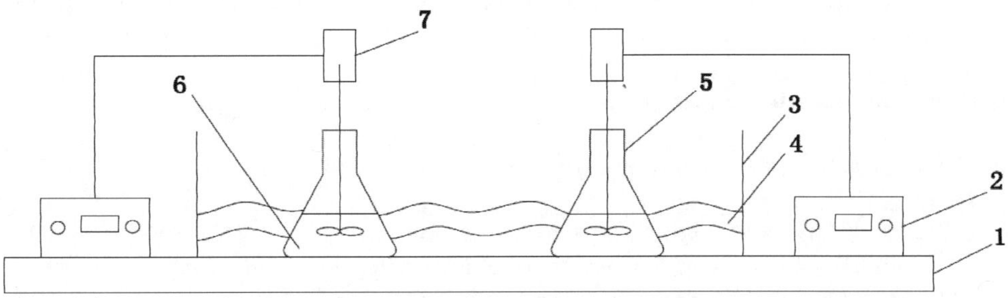

Figure 1. Experimental device diagram.
1. Experimental station; 2. JJ-1A digital display electric mixer controller; 3. Thermostatic water bath pan; 4. Water; 5. Triangular conical flask; 6. The oil; 7. Electric mixer motor

2.2.2. Single factors experiment design. Taking the temperature, stirring speed, adding amount of monoethanolamine (MEA), water amount and MEA processing time as variables, set 5 factors 5 levels and set 3 parallel experimental schemes for each level, to research the effects of each factor on phosphorus content in rubber seed oil degumming. The specific experimental parameters [7]: the temperature (40-80℃), stirring speed (100-300 r/min), added amount of monoethanolamine (MEA)

IOP Conf. Series: Earth and Environmental Science **93** (2017) 012004 doi:10.1088/1755-1315/93/1/012004

(1%-3%) of oil weight, water amount (1%-5%) of oil weight and MEA processing time (10-50 min).

2.2.3. Determination method. The phospholipid fore and after content the rubber seed oil degumming can be determined by GB/T 5537-2008 [9]. The calculation formula as follows:

$$X = \frac{P}{m} \times \frac{V_1}{V_2} \times 26.31$$

In the formula: X, the content of phospholipids measured, mg/g; P, the corresponding amount of phosphorus in the standard curve, mg; m, oil quality, g; V_1, the volume of diluted after sample ash, mL; V_2, the volume of the night being measured, mL; 26.31, the conversion coefficient between phosphorus content and phospholipid content.

The acid value was determined according to GB/T 5530-2005 [10]. The calculation formula is as follows:

$$S = \frac{56.1 \cdot V \cdot c}{m}$$

In the formula: *S*, Untested oil acid value, mg(KOH)/g; V, standard solution KOH volume, mL; c, standard solution KOH concentration, mol/L; m, oil quality, g; 56.1, KOH mole mass, g/mol.

The determination of the saponification value was determined according to the method of GB/T 5534-2008 [11]. The calculation formula is as follows:

$$I = \frac{(V_0 - V_1) \times c \times 56.1}{m}$$

In the formula: I, to test the saponification value of oil, mg(KOH)/g; V_0, blank standard solution HCl volume, mL; V_1, titration of the standard solution HCl volume, mL; *c*, standard solution HCl concentration, mol/L; m, oil quality, g.

The determination of iodine value was determined according to GB/T 5532-2008 [12]. The calculation formula is as follows:

$$\mathrm{W} = \frac{12.69 \times \mathrm{c} \times (\mathrm{V}_1 - \mathrm{V}_2)}{\mathrm{m}}$$

In the formula: I, to test the Iodine value of oil, g/100g; *c*, standard solution $Na_2S_2O_3$ concentration, mol/L; V_1, blank standard solution $Na_2S_2O_3$ volume, mL; V_2, titration of the standard solution $Na_2S_2O_3$ volume, mL; m, oil quality, g.

3. Experimental results and discussion

3.1. Results and discussion of single factor experiment

3.1.1. The influence of MEA addition on the content of phosphorus of rubber seed oil. At temperature of 60℃, added different amounts of MEA when the stirring speed is 200 r/min, at the same time, added the addition amount of 2% of the oil quality deionized water in the same temperature, continue to mixing process of 40 min then make centrifugal separation, take upper sample to determine phosphorus content, the result was shown in figure 2.

It can be seen from figure 2 that with the increasing of MEA addition, the phosphorus content of rubber seed oil decreased obviously. It is due to the rapid reaction of the phospholipid acid in the MEA reagent and oil to the phosphatidyl ethanolamine, which is easy to hydrate, leading to the rapid decrease of phosphorus content. When the addition of MEA was 2.5% of the oil weight, the addition of MEA was added, and the removal effect of phosphorus was not obvious, indicating that the phosphorus in rubber seed oil was mostly removed. Therefore, the best MEA addition to determine the experiment was 2.5%.

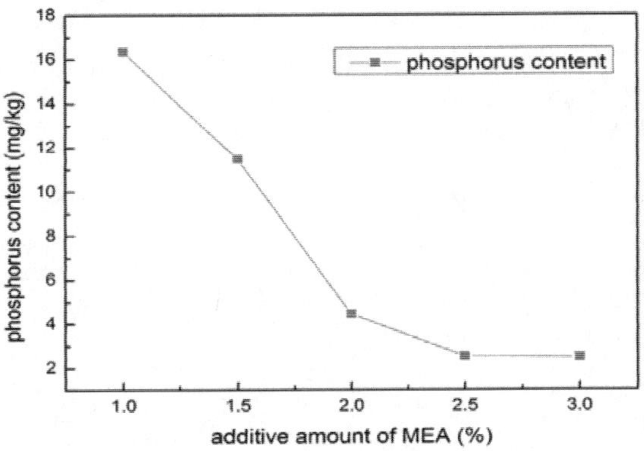

Figure 2. Effect of MEA addition on phosphorus content of rubber seed oil.

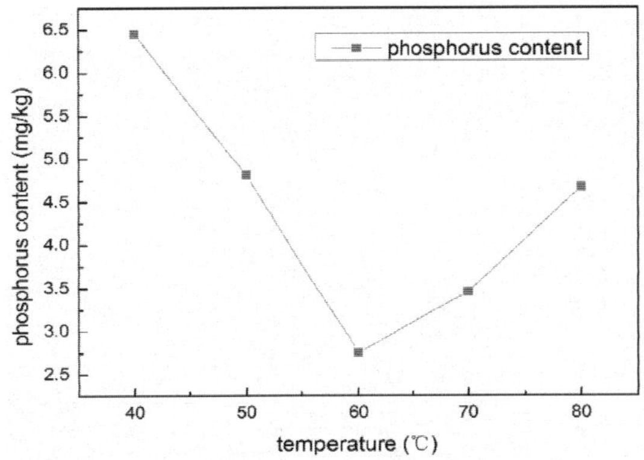

Figure 3. Effect of temperature on the phosphorus content of rubber seed oil.

3.1.2. Experimental effect of temperature on phosphorus content of rubber seed oil. Under stirring speed of 200 r/min, added the MEA that is 2.5% of the weight of oil, at the same time, added the deionized water that was 2% of oil by weight with the same temperature, respectively dealt with under the temperature of 40-80℃ for 40 min then made centrifugal separation, take upper sample determination of phosphorus content, the result was shown in figure 3.

It can be seen from figure 3 that in a certain temperature range, the increase of temperature was beneficial to the removal of phosphorus, because the raised temperature was conducive to the formation and separation of colloidal particles. When the temperature reached 60℃, the phosphorus content dropped to the lowest, continue to raise the temperature, phosphorus content had a rising trend. It was because the temperature was higher than the gelatinous mass of the condensate which results in the dispersion of the colloid. Comprehensively consideration, it can determine the optimum degumming temperature was 60℃.

3.1.3. The influence of water added content on the rubber seed oil phosphorus analyzed. At the temperature of 60℃, stirring speed of 200 r/min, added the MEA that was 2.5% of the weight of oil, added the deionized water of different amount but same temperature, after 40 minutes of continuous mixing, the centrifugal separation was done. And the phosphorus content was determined by the upper sample, and the result was shown in figure 4.

It can be seen from figure 4 that in a certain range, the increase of water content is beneficial to the removal of phosphorus. When the amount of added water with deionized water temperature was 2% of oil weight, the phosphorus content was minimized. The increase of phosphorus content was not decreased but increased a little while adding water, which may be due to the emulsification of excess water, which results in incomplete phospholipid separation. The suitable amount of water can make the phospholipids fully hydrated and absorb other gelatin. The granules were enlarged and easily removed. The absorption of water by phospholipid can result in emulsification and degumming effects. Therefore, it was determined that the optimal water addition of degumming of the experiment was 2% of the oil weight.

3.1.4. The experimental effect of MEA treatment time on phosphorus content of rubber seed oil. At the temperature of 60℃, the stirring speed of 200 r/min, add the MEA that was 2.5% of the weight of oil, at the same time, add the deionized water that was 2% of oil weight with same temperature. Continuous to mixing processing, the centrifugation was carried out according to different processing time, and the phosphorus content was determined by the upper sample, and the result was shown in figure 5.

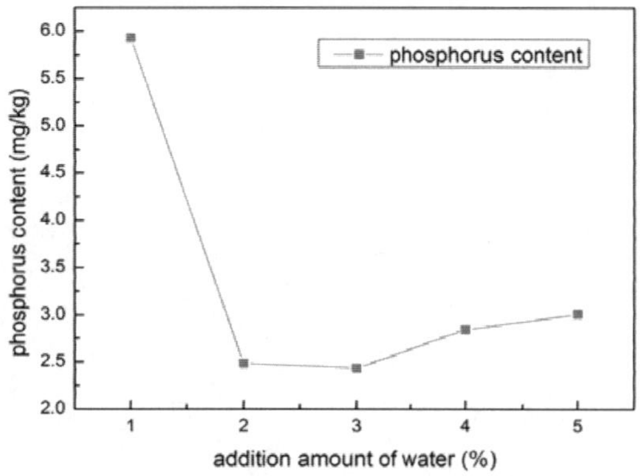

Figure 4. Effect of water addition amount on phosphorus content of rubber seed oil.

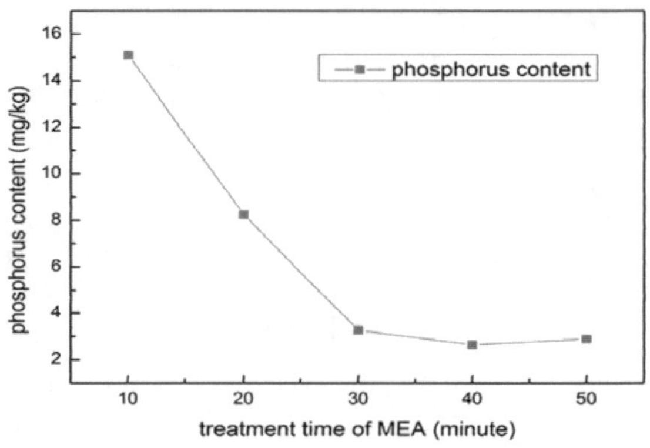

Figure 5. Effect of MEA treatment time on phosphorus content of rubber seed oil.

It can be seen from figure 5 that within 10-40 min, the phosphorus content gradually decreased with the increasing of MEA treatment time. When the treatment time of MEA was 40 min, the phosphorus content was minimized. Adding the MEA treatment time, the phosphorus content increased slightly. It was due to a certain amount of time to react with the phospholipid acid in the MEA reagent and oil. After a period of time, most of the phospholipids were converted to phosphatidyl ethanolamine, the MEA reagent no longer played a decisive role. The time was too long, the gelatinization and oil phase will occur local emulsification, resulting in a slight increase in phosphorus content. Comprehensively consideration, the best time to determine the experiment was 40 min.

3.1.5. The experimental effect of stirring speed on the phosphorus content of rubber seed oil. At the temperature of 60℃, added the MEA that was 2.5% of the weight of oil, at the same time, added the deionized water that was 2% weight of oil with the same temperature. Continuously, the centrifuge was separated after 40 min at different stirring speed, the phosphorus content was determined by the upper sample, and the result was shown in figure 6.

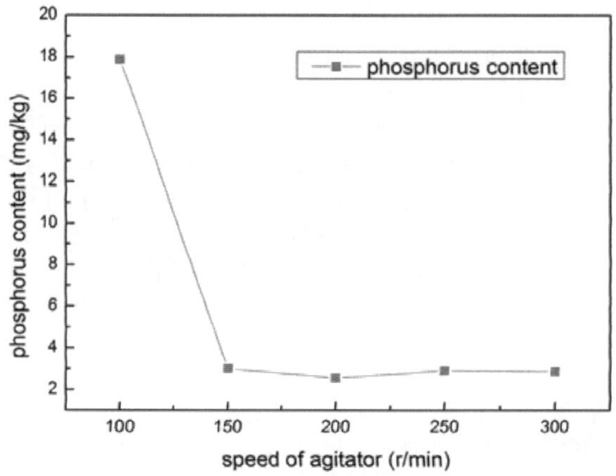

Figure 6. Effect of stirring speed on the phosphorus content of rubber seed oil.

As can be seen from figure 6, as the mixing speed increases, the phosphorus content starts to decrease obviously. It was due to the rapid reaction of the MEA reagent in the beginning that the rapid precipitation of phosphatidyl ethanolamine, after which the phospholipid content was less, the change of phosphorus content tended to be gentle. When the speed reached 200 r/min, the phosphorus content was minimized. Therefore, it was determined that the optimal stirring speed of the experiment was 200 r/min.

4. Conclusions

The experimental conditions of single factor were obtained by using rubber seed oil as raw material: the results show that addition of MEA is 2.5% of the weight of oil, temperature is 60℃, the amount of added water is 2% of oil by weight, reaction time is 40 min, and stirring speed is 200 r/min. Under these conditions, the phosphorus content of rubber seed oil can be reduced to below 3 mg/kg, degumming rate is 91.37%, the degumming effect is obvious. It provides some basis for the subsequent non-edible vegetable oil degumming experiment.

Acknowledgments

This work is partial fulfillment of funded research program 21266032, financially supported by Natural Science Foundation of China.

References

[1] Sun X Y, Liu X, Zhao X B, *et al* 2013 Progress in synthesis technologies and application of aviation biofuels *Chinese J. of Biotechnol.* **29(3)** 285-98
[2] Tao Z P 2011 Progress in alternative fuels for aviation and suggestions of ITS development in China *Petroleum Processing &Petrochmicals* **42(7)** 91-6
[3] Zhao G H, Jiang W, Niu X Y, *et al* 2014 Jey fuel preparation technology and application prospects *Sino-global Energy* **19(8)** 30-4
[4] He D P 2013 *Oil Chemistry* (Beijing: Chemical Industry Press) pp 191-2
[5] Narayana T, Kaimal B, Vail S R, *et al* 2002 Origin of problems encountered in rice bran oil processing *European J. of Lipid Sci. &Technol.* **104** 203-11
[6] Zhao G H, Niu X Y, Jiang W, *et al* 2014 Refining and degumming mechanism analysis of jatropha crude oil *Cereals &Oils Processing* **2** 47-9,52
[7] Jiang X F, Xia Y, Jin Q Z, *et al* 2013 Optimization of acid-assisted degumming of crude rapeseed oil *China Oils &Fats* **38(6)** 4-8
[8] Wu Y J, Zhang X, Song G H, *et al* 2014 Optimization of acid degumming process of cold pressed Sesame oil *Sci. &Technol. of Food Ind.* **17** 205-8
[9] GB/T 5537-2008 Inspection of grain and oils-Determination of phosphatide content
[10] GB/T 5530-2005 Animal and vegetable fats and oils-Determination of acid value and acidity
[11] GB/T 5534-2008 Animal and vegetable fats and oils-Determination of saponification value
[12] GB/T 5532-2008 Animal and vegetable fats and oils-Determination of Iodine Value

Design of solar adsorption refrigeration system with CPC and study on the heat and mass transfer performance

W P Du[1], M Li[1], Y F Wang[1,3], J H He[2] and J X He[1]

[1]Solar Energy Research Institute, Yunnan Normal University, Kunming, Yunnan, China

[2]School of Physics and Electronic Sciences, Chuxiong Normal University, Chuxiong, Yunnan, China

E-mail: wangyf@ynnu.edu.cn

Abstract. To overcome the problem that the heat source temperature is limited and the lower part of the adsorption tube cannot effectively absorb the solar radiation when solar radiation as the heat source of the adsorption refrigeration system. From the perspective of enhancing the adsorption refrigeration unit tube to absorb solar radiation, thereby strengthening the heat transfer characteristic of adsorption bed, which can improve the efficiency of the refrigeration unit refrigerating capacity and system refrigeration efficiency. Solar adsorption refrigeration system based on CPC was designed and constructed in this paper. The heat and mass transfer performance of the adsorption refrigeration system were studied. The experimental results show that the temperature of the adsorption bed with parabolic concentrating structure can rise to 100℃ under low irradiation condition. When the irradiation intensity is 600 w/m^2 and 400 w/m^2, the average temperature rising to desorption temperature reaches 0.67℃/min and 0.50℃/min, respectively. It can effectively solve the problem that the conventional adsorption bed is difficult to reach the required desorption temperature due to the low power density of the sunlight. In the experiment, the system COP were 0.166 and 0.143 when the system in the irradiance of 600 w/m^2 and 400 w/m^2.

1. Introduction

The use of renewable energy for refrigeration is one of the important ways to solve the energy and environmental problems [1]. The universal existence of solar energy, the greenhouse effect coefficient of solar refrigeration and the ozone depletion potential are zero. What's more, due to the change of solar radiation with time and the geographical distribution of solar radiation are highly consistent with air conditioning demand, solar refrigeration has been widely studied. Adsorption refrigeration can effectively use solar energy, industrial waste heat and other low-grade heat source and has no moving parts, low noise, long life and other advantages. So, solar adsorption refrigeration has been widely concerned [2]. However, the low refrigeration efficiency has become the bottleneck of commercial development of adsorption refrigeration technology currently [3].

For a long time, researchers have carried out extensive studies to improve the performance of the adsorption refrigeration system. In the refrigerant performance, in order to improve the heat transfer performance of the adsorbent and overcome the problem of expansion and caking in the process of adsorption cycle, scholars developed all kinds of high thermal conductivity of adsorbent [4-7]. In the improvement of the structure of the adsorbent bed, there are ways to increase the heat transfer fins, to promote heat exchange with heat pipes and to prepare heat exchange coatings [8,9]. Solar Energy Research Institute of Yunnan Normal University used activated carbon-methanol as working pair in the adsorption refrigeration system and designed a kind of fin tube solar adsorption collector bed for solar adsorption refrigeration system by increasing the diameter and adding heat transfer fins, have been obtained some research results [10,11,12]. But the fin tube adsorbent bed is not uniformly heated, only the upper part can receive sunlight during the desorption process, and the lower part of the fin tube inside of the insulating layer cannot effective absorption of solar radiation, so the refrigerant desorption is not sufficient, which result in affect the efficiency of system.

In order to overcome the problem that the heat source temperature is limited and the lower half of the fin tube cannot absorb solar radiation effectively and to increase the cooling capacity of the unit and the cooling efficiency of the system, the absorption of the solar by the adsorption refrigeration unit tube and the heat transfer characteristics of the adsorbent bed need to be enhanced. Combining no tracking focused composite parabolic (CPC) reflector panel with solar adsorption refrigeration, a solar adsorption refrigeration system based on CPC was constructed and its heat and mass transfer performance were studied.

2. Design of solar adsorption refrigeration system based on CPC

The solar adsorption refrigeration system based on CPC is shown in figure 1. The system is mainly composed of adsorption bed, condenser, evaporator and vacuum valve, vacuum pressure gauge and mass transfer channel. The main components of the system are introduced below.

Figure 1. The solar adsorption refrigeration system based on CPC.

2.1. Adsorption refrigeration tube with fin

The adsorption refrigeration tube adopts the fin adsorption tube as shown in figure 2, and the parameters of each part are shown in table 1. The length of adsorption refrigeration tube in the system is 1 meter, the average of each refrigeration tube filled 2.05 kg activated carbon. The fin adsorption refrigeration tube is shown in figure 1, and its structural parameters are shown in table 1.

Table 1. Structure parameters of finned tube.

Structure parameters	Size (mm)
Outer diameter /Inner diameter	90/40
Wall thickness (fin thickness)	1.5
Long fin / Short fin	25/10
Gap width of mass transfer	6

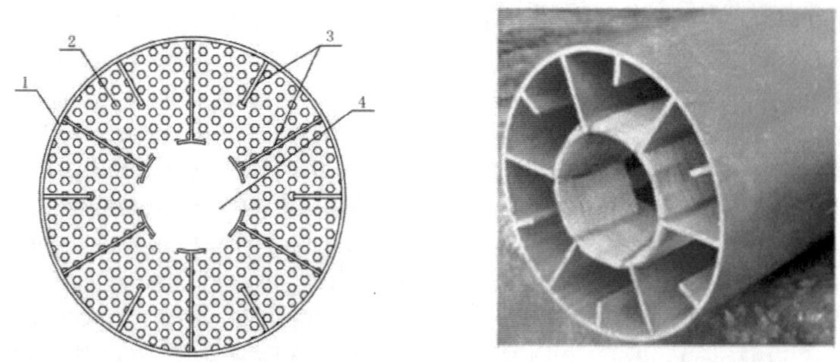

Figure 2. Structure diagram and physical map of finned adsorption tube.
1. Outer tube; 2. Adsorbent; 3. Heat transfer fins; 4. Mass transfer channel.

2.2. The adsorption bed design

In order to overcome the solar energy as heat source of the adsorption refrigeration system, heat source temperature is limited and the lower part of the fin tube cannot effectively absorb the solar radiation problems, by increasing the desorption temperature of the adsorption tube, the adsorption bed designed with a composite parabolic trough condenser structure (as shown in figure 3). In order to ensure that the light direct into the adsorbent bed can be full utilized, two parabolic focal points are located at F1 and F2 as shown in figure 4, and the two paraboloids are connected to the surface of the heat absorbing tube at point A. The urethane foam filled between the parabolic trough surface and the back of the adsorbent bed, which can effectively reduce the heat loss at the bottom of the adsorbent bed. According to this design, adsorption bed parabolic trough concentrating ratio ζ=1.24. The parameters of the condensing surface of the adsorption bed are shown in table 2.

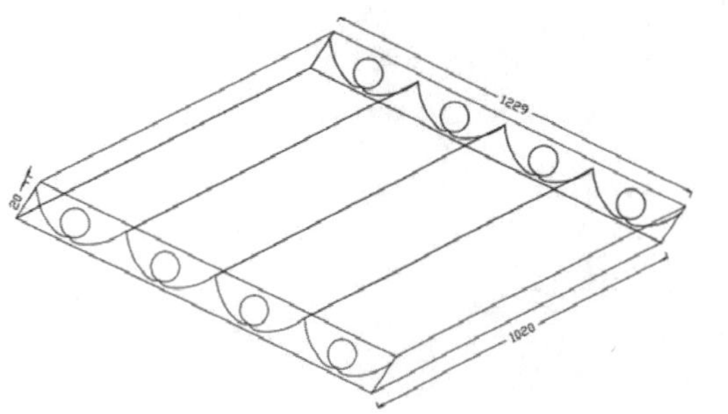

Figure 3. Adsorption bed of CPC solar adsorption refrigeration.

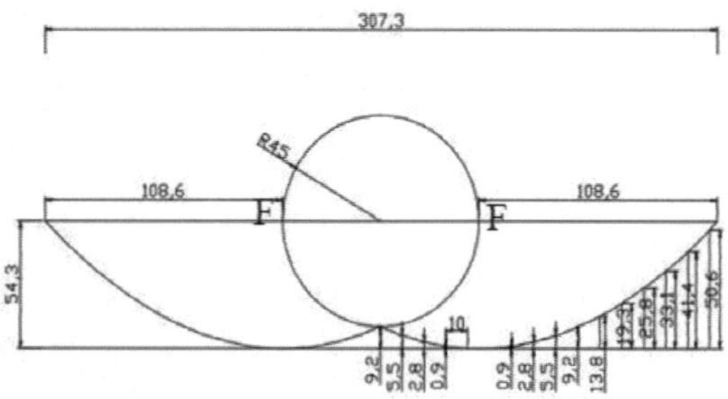

Figure 4. The unit tube of the adsorbed bed.

Table 2. Structural parameters of adsorption refrigeration unit tube and adsorbent bed.

Structural parameters	Size (mm)
Parabolic focal radius p	108.64
Parabolic trough depth H	99.32
Radiation width of a focusing unit D	307.28
The adsorption bed size (length * width * height)	1229*1020*80

2.3. Condenser design

The desorption of the refrigerant vapor is condensed into liquid through heat exchange with the condenser. Therefore, the condensing load of the system should be considered when designing the condenser. The calculation formula of condensing load is:

$$Q_{cond} = M(L_e + c_p \Delta t) \tag{1}$$

Q_{cond} is condensing load; M is the amount of desorbed refrigerant; L_e is the latent heat of vaporization of refrigerant; c_p is specific heat capacity of refrigerant steam at constant pressure; Δt is the temperature difference before and after condensation.

The required heat transfer area for condenser is:

$$A = \frac{Q_{cond}}{U_c \Delta t'} \tag{2}$$

U_c is the condenser heat transfer coefficient; $\Delta t'$ is a temperature difference between refrigerant and water; A is the required heat transfer area for the condenser.

The system uses the copper-aluminum composite finned tube water-cooled condenser, composed of four copper-aluminum composites finned tube and two copper tubes placed up and down, they all placed in the condensate tank. During the desorption process, the refrigerant vapor passes through the inner tube and exchanges heat with the cold water outside of the finned tube and then condense into liquid. The condenser structure is shown in figure 5. The structural parameters of the condenser are shown in table 3.

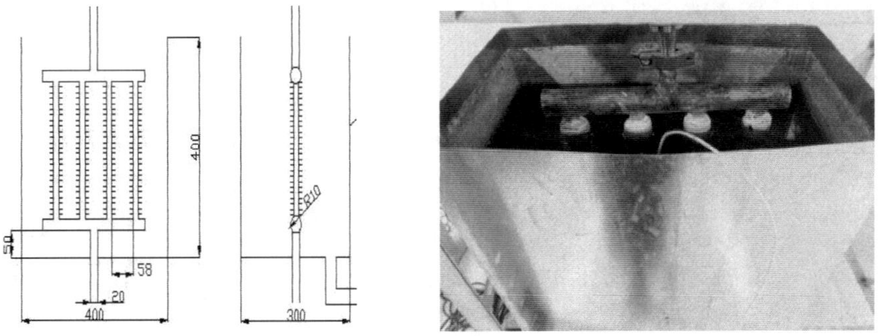

Figure 5. Condenser structure diagram and physical map.

Table 3. Structural parameters of condenser.

Name	Specifications
Fin tube inner / outer diameter	15.5 mm/32.5 mm
Fin net height / gap / thickness	8 mm/2.5 mm/0.3 mm
Condenser tank (length ×width× height)	400 mm×300 mm×400 mm

2.4. Evaporator design

While ensuring that the volume of the evaporator meets the requirements of the storage of the liquid refrigerant, the surface heat exchange area of the evaporator needs to be increased. It also needs to be considered that the convenience of ice taking and the maneuverability of manufacturing. The evaporator is designed as an aluminum alloy container as shown in figure 6. Evaporator parameters are as follows:

Table 4. Structural parameters of evaporator.

Project	Dimensions
Evaporator size	410mm*180mm*105mm
Evaporator volume	4.95L
Evaporator surface area	0.255 m^2

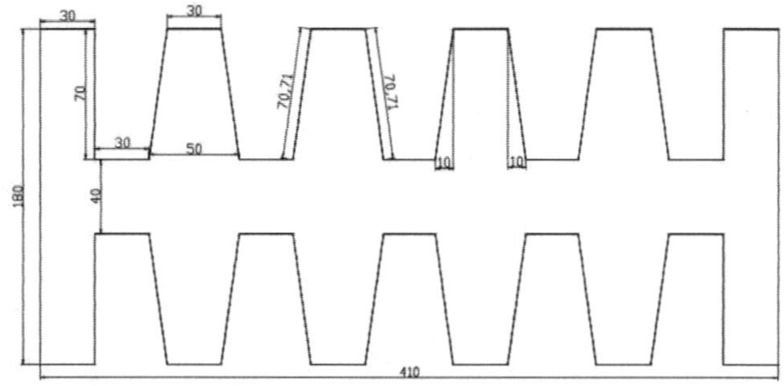

(a)

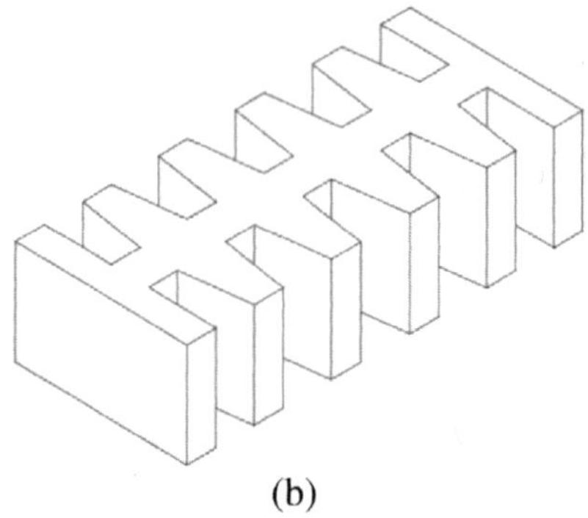

(b)

Figure 6. Evaporator structure diagram, (a) Axonometric drawing of evaporator and (b) Top view of evaporator.

3. Performance analysis

The process of the system is accompanied by the heat and mass transfer process, and the performance of heat and mass transfer determines the performance of the system. The performance of the system will be analyzed for two aspects of mass transfer performance and refrigeration performance.

3.1. Mass transfer model

The D-A equation applies to activated carbon-methanol working pair for adsorption refrigeration in equilibrium adsorption (Dubinin Astakhov equation) [13]

$$x = x_0 \exp\left[-K\left(\frac{T}{T_s}-1\right)^n\right] \tag{3}$$

X-Adsorption rate, $kg \cdot kg^{-1}$, the ratio of adsorbent refrigerant quality and adsorption quality; x_0 -Ultimate adsorption rate, $kg \cdot kg^{-1}$; K—A constant determined only by the structure of the adsorbent; T_s -The temperature of the liquid refrigerant, K; T —Adsorbent temperature, K; n —Characteristic parameters of working pairs.

In order to analyze desorption performance of the system conveniently, corresponding to adsorption rate x. The ratio of the mass of the desorbed refrigerant to the mass of the adsorbent charged in the system is defined as desorption rate λ, and it is expressed as:

$$\lambda = x_0 - x \tag{4}$$

Take equation (3) into equation (4):

$$\lambda = x_0\left\{1-\exp\left[-K\left(\frac{T}{T_s}-1\right)^n\right]\right\} \tag{5}$$

In the low-pressure adsorption refrigeration system with methanol as a refrigerant, when the methanol is in the liquid-gas phase equilibrium and the temperature range is not large, the change of pressure with temperature can be expressed by Clausius-Clapeyron equation:

$$\ln P = -\frac{A}{T_s} + B \tag{6}$$

A, B —The parameters of Clausius-Clapeyron equation of refrigerant.

By the deformation of the equation (6), T_s can be obtained:

$$T_s = \frac{A}{B - \ln P} \tag{7}$$

By taking formula (7) in (5), the equation of desorption rate is obtained:

$$\lambda = x_0\left\{1 - \exp\left\{-K\left[\frac{T(B - \ln P)}{A} - 1\right]^n\right\}\right\} \tag{8}$$

3.2. Refrigeration performance

Solar refrigeration system refrigerating efficiency COP is used to measure the performance of refrigeration system, which is expressed by the ratio of the effective refrigerating capacity of the system Q_{ref} to the solar radiation energy received by the adsorbent bed Q_s:

$$COP = \frac{Q_{ref}}{Q_s} \tag{9}$$

3.2.1. The effective refrigerating capacity of the system. After the desorption process, the adsorbent bed carries on radiation heat transfer and convection heat transfer with surroundings continuously. At the same time, the system pressure is also reduced, when the system pressure drops to the evaporation pressure, adsorption-evaporation stage start, the refrigerant is adsorbed to produce cooling effect Q_{ref}. The amount of cooling from refrigerant evaporation can be calculated by the following formula:

$$Q_{ref} = \Delta X Ma Le \tag{10}$$

$$\Delta X = X_{conl} - X_{dil} \tag{11}$$

Le—The latent heat of vaporization of refrigerant; Xconl—Adsorption amount of adsorbent to refrigerant before desorption; Xdil—Adsorption amount of adsorbent to refrigerant before adsorption.

The refrigerant is cooled from the condensing temperature TC to the evaporating temperature Te, The sensible heat of emission is:

$$Qcc = \int_{Te}^{Tc} M_a \Delta X C_{pl} dT \tag{12}$$

In this system with activated carbon-methanol as working pairs, the latent heat of vaporization of methanol can be calculated using the following equation:

$$L = 1252.43 - 1.596t - 0.0088t^2 \tag{13}$$

t: Temperature of methanol, ℃.

3.2.2. Solar radiation energy absorbed by adsorption bed. The energy received by the adsorbent bed during the heating process is used to increase the sensible heat of the metal adsorbent of the adsorbent bed itself, the sensible heat of the adsorbent, providing desorption heat and heat dissipation. The transient heat balance equation of the adsorption bed is:

$$A_e I(t)\tau\alpha = (M_m C_{pm} + M_a C_{pa})\frac{dT}{dt} + xM_a C_{pl}\frac{dT}{dt} + h_d M_a \frac{dx}{dt} + Q_{Hloss} \tag{14}$$

$I(t)$—The total solar radiation energy projected onto the unit surface area of the collector, which can be measured by a solar radiometer; A_e—heat collector area; τ —Sunlight transmittance of glass cover panels; α —Sunshine absorption rate of adsorption bed; M_m: The quality of a metal shell of adsorption collectors; C_{pm}: the specific heat of the metal shell adsorber; M_a: The quality of the adsorbent, C_{pa}: Specific heat of adsorbent; C_{pl}: The specific heat of refrigerant; $x(kg.kg^{-1})$: The adsorption quality of the refrigerant for the unit adsorbent; h_d: Heat desorption of refrigerant; Q_{Hloss}: The heat dissipation of the collector to the external environment.

4. Experimental results and analysis

In order to test the performance of solar adsorption refrigeration system based on CPC, the adsorption refrigeration system is placed under the TYD-PD4 matrix steady state solar simulator (The solar simulator consists of 15 xenon lamps, mainly used for solar water heaters, heat collectors and other products to test the light performance). The performance of the system is experimentally studied when the irradiance is 600 W/m^2 and 400 W/m^2. The experimental tests are shown in figures 7 and 8.

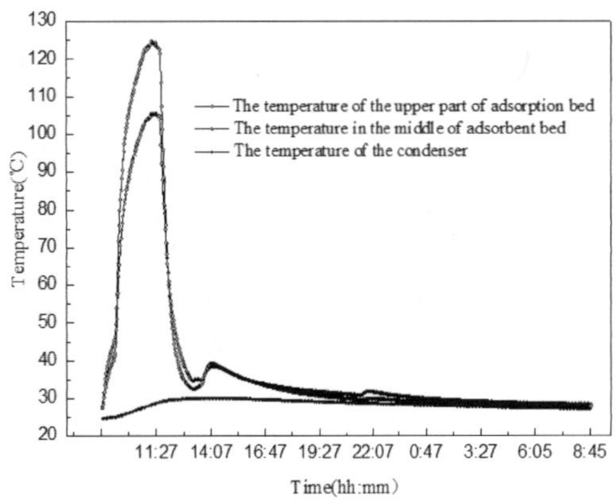

Figure 7. Temperature variation of adsorption bed and condenser at 600 W/m^2.

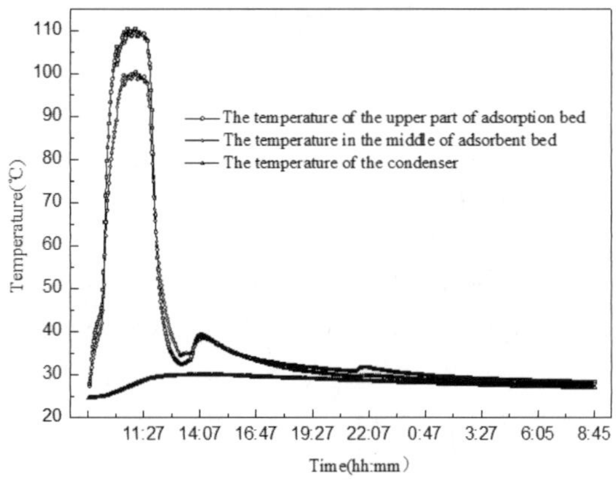

Figure 8. Temperature variation of adsorption bed and condenser at 400 W/m^2.

The experimental results show that the design of parabolic structure for adsorption bed condensing can make the adsorption bed temperature exceed 100℃ and the irradiation intensity is lower, while the non-concentrating adsorption bed generally needs more than 900 W/m^2 radiation intensity to achieve the same effect. In view of the low solar power density, conventional adsorption bed is difficult to achieve the required desorption temperature problem, condenser adsorption bed can be an effective solution. In 600 W/m^2 and 400 W/m^2 of the radiation intensity, the average temperature rise rate reached 0.67℃/min, 0.50℃/min to the desorption temperature. With the increase of the irradiation intensity, the difference between the top temperature and the bottom temperature of the adsorbent bed increase, which make the desorption rate in the adsorption tube different with the position. In the desorption process, due to the radiation power density fall too fast, so that in the desorption process there are two temperature fluctuations, moreover, the first one is larger than the second one. The temperature range of the condenser is very small, indicating that the condenser meets the cooling load.

In the desorption process, the desorption rate and the system pressure are changed in a similar trend, but the change of the desorption rate is lagging behind the change of the system pressure, in different irradiation experiments, the rate of change showed first slow, then quick, and then gradually stable trend. As can be seen from figures 9 and 10, the higher the irradiance is, the higher the final desorption rate reached. When the irradiance upgrade from 400 W/m^2 to 600 W/m^2, the desorption rate can be increased by 4.1%, and in the 600 w/ m^2 the experiments show the desorption process is long and stable, and finally the desorption rate reached 12.5%. Compared with the non-condensing adsorbent bed, the desorption rate of the adsorption bed is improved, and the cooling capacity of the unit tube is effectively improved. When the irradiation intensity is 600 W/m^2 and 400 W/m^2, it was calculated by the experimental data that the COP were 0.166 and 0.143, respectively.

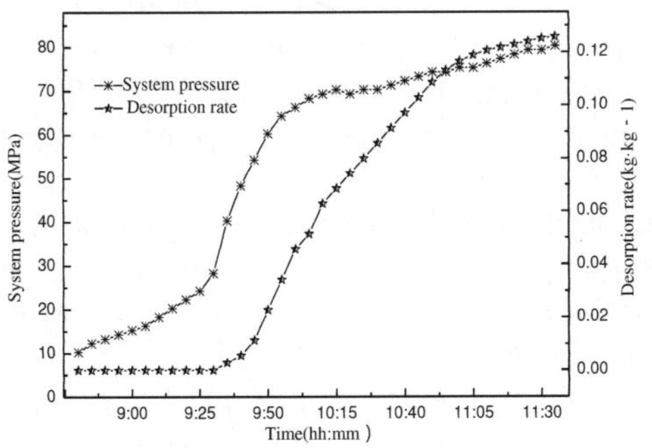

Figure 9. The process of desorption system pressure and desorption rate changes at 600W/m^2.

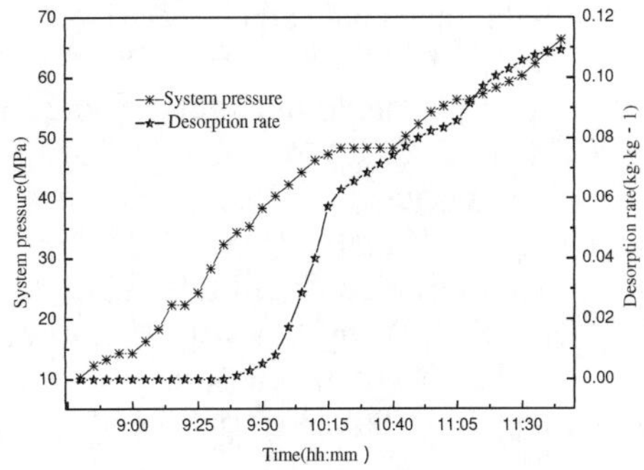

Figure 10. The process of desorption system pressure and desorption rate changes at 400 W/m^2.

5. Conclusion

Improving the heat and mass transfer performance in the adsorption refrigeration system is the key to improve the coefficient of performance of the system, in order to overcome the problem that the heat source temperature is limited and the lower part of the adsorption tube cannot effectively absorb the solar radiation when solar radiation as the heat source of the adsorption refrigeration system. From the perspective of enhancing the adsorption refrigeration unit tube to absorb solar radiation, thereby strengthening the heat transfer characteristic of adsorption bed, which can improve the efficiency of the refrigeration unit refrigerating capacity and system refrigeration efficiency. Solar adsorption refrigeration system based on CPC was designed and constructed in this paper. The heat and mass transfer performance of the adsorption refrigeration system were studied. The experimental results show that the temperature of the adsorption bed with parabolic concentrating structure can rise to 100℃ under low irradiation condition. When the irradiation intensity is 600 W/m^2 and 400 W/m^2, the average temperature rising to desorption temperature reaches 0.67℃/min and 0.50℃/min, respectively. It can effectively solve the problem that the conventional adsorption bed is difficult to reach the required desorption temperature due to the low power density of the sunlight. In the experiment, the system COP were 0.166 and 0.143 when the system in the irradiance of 600 W/m^2 and 400 W/m^2.

Acknowledgments

This work was founded by the National Natural Science Foundation of China (Grant No. 51466017) and Zhou Guofu Expert Work Station.

References

[1] Sarbu I and Sebarchievici C 2013 Review of solar refrigeration and cooling systems *Energ. Buildings* **67** 286-97
[2] Goyal P, Baredar P, Mittal A, *et al* 2016 Adsorption refrigeration technology-An overview of theory and its solar energy applications *Renew. Sust. Energ. Rev.* **53** 1389-410
[3] Wang D C, Li Y H, Li D, *et al* 2010 A review on adsorption refrigeration technology and adsorption deterioration in physical adsorption systems *Renew. Sust. Energ. Rev.* **14** 344-53
[4] Demir H, Mobedi M and Ülkü S 2010 The use of metal piece additives to enhance heat transfer rate through an unconsolidated adsorbent bed *Int. J. Refrig* **33** 714-20
[5] Wu W D, Wang C, Meng X W, *et al* 2016 Physical properties and refrigeration performance of

compound adsorbent composed of additive and zeolite molecular *Chem. Ind. Eng. Prog.* **35** 692-9

[6] Anirban S and Randip K D 2016 Review of technology used to improve heat and mass transfer characteristics of adsorption refrigeration system *Int. J .AC. Refrig.* **24** 1630003

[7] Fujioka K, Hatanaka K and Hirata Y 2008 Composite reactants of calcium chloride combined with functional carbon materials for chemical heat pump *Appl. Therm. Eng* **28** 304-10

[8] Ji X, Li M, Fan J Q, *et al* 2014 Structure optimization and performance experiments of a solar-powered finned-tube adsorption refrigeration system *Appl. Energ.* **113** 1293-300

[9] Huanxin C, Li W, Wei Z, *et al* 2013 Structure analysis and heat transfer performance study on spiral plate of the adsorption bed *J. Refrig.***34** 45-9

[10] Ji X, Li M, Fan J Q, *et al* 2014 Structure optimization and performance experiments of a solar-powered finned-tube adsorption refrigeration system *Appl. Energ.* **113** 1293-300

[11] Zhang S B, Wang Y F, Li M, *et al* 2016 Experimental study of solar adsorption refrigeration system based on intensified mass transfer function *Acta Energi. Sin.* **37** 2612-8

[12] Critoph R E 1998 Performance limitations of adsorption cycles for solar cooling *Sol. Energy* **14** 21-31

[13] Lu Z S, Wang R Z, Xia Z Z, *et al* 2013 Study of a novel solar adsorption cooling system and a solar absorption cooling system with new CPC collectors *Renew. Energ.* **50** 299-306

Comparison and analysis of organic components of biogas slurry from eichhornia crassipes solms and corn straw biogas slurry

Q Li[1], Y B Li[1], Z H Liu[1], J Min[1], Y[1] Cui and X H Gao[1, 2]

[1]Yunnan Normal University, School of energy and Environmental Sciences, Kunming 650500, China

E-mail: 771920989@qq.com

Abstract. Biogas slurry is one of anaerobic fermentations, and biomass fermentation biogas slurries with different compositions are different. This paper mainly presents through the anaerobic fermentation of Eichhornia crassipes solms biogas slurry and biogas slurry of corn straw, the organic components of two kinds of biogas slurry after extraction were compared by TLC, HPLC and spectrophotometric determination of nucleic acid and protein of two kinds of biogas slurry organic components, and analyzes the result of comparison.

1. Introduction

Biogas slurry is a product of anaerobic fermentation, the available nutrient ability, nutrients utilization rate is high, the crop can be quickly absorbed, not only can improve the yield and quality of crops, but also has the functions of preventing and anti-adversity effect, is a kind of high-quality organic liquid fertilizer [1]. Biogas slurry not only contains rich soluble inorganic salts, but also contains biochemical products of anaerobic fermentation, which has the functions of nutrition, bacteriostasis, stimulation and anti-adversity [2]. At the same time, biogas slurry can kill and inhibit crop diseases and insect pests, and has repellent effect on some insects. At present, the biogas slurry has the dual effect of biological fertilizer and biological pesticide and will not bring resistance and environmental pollution, but also reduces nitrate accumulation in crops [3], broad-spectrum known as anaerobic microorganism processing agent in the world [4].

Water hyacinth (Eichhornia crassipes Solms) is pontederiaceae water hyacinth is Chinese, name of Eichhornia crassipes. It has an erect oval gourd stem because of its dark green leaves, commonly known as water hyacinth. Water hyacinth is an aquatic herb, which can grow at a suitable temperature and absorb large amounts of N, P and K nutrients. It has been recognized as a water plant with strong N and P capacity in the world. At present, the research on water hyacinth mainly focuses on the research of water hyacinth nutrition, water hyacinth control and comprehensive utilization of research, now studies focus on water hyacinth removal in eutrophic water capacity of nitrogen and phosphorus and its application in water treatment, but its use in fertilizer has not been reported in details [5]. In 1901, it was introduced into China as a flower. In 30s, it was introduced into China's provinces as livestock and poultry feed, and promoted as a plant for ornamental and purifying water quality. Because of its asexual propagation speed, it has been widely distributed in 19 provinces of North China, East China, central China, Southern China and the southwest, especially in Yunnan (Kunming), Zhejiang, Fujian, the invasion of Jiangsu, Sichuan, Hunan, Hubei, Henan province is serious, and has spread to temperate regions. It caused great environmental problems and caused serious harm to the ecosystem [6]. Corn straw with C, H, O elements as the main content of the three part, accounting for more than 90% of the total elements, straw and carbohydrate rich material, the main ingredients of corn straw is wood fiber, wood fiber consists mainly of cellulose, hemicellulose and lignin [7]. At present, China's straw comprehensive utilization technologies include: organic fertilizer utilization technology, energy utilization technology, cultivation edible mushroom technology, industrial raw materials utilization technology and feed utilization technology. Although organic fertilizer utilization technology has made great progress, but the research is still not deep enough, and the research results cannot be effectively industrialized [8].

Study on water hyacinth liquid focused on the removal of N, P, and research on the application of water hyacinth also weakened, moreover, studies especially on the nutrients in the study of farmland utilization is still not seen [9]. This paper mainly presented through the anaerobic fermentation of water hyacinth biogas slurry and biogas slurry of corn straw, how two kinds of slurry the organic components extracted, using the analysis method of the analysis of the differences of organic components in biogas slurry.

2. Experiment

Experimental flow chart:

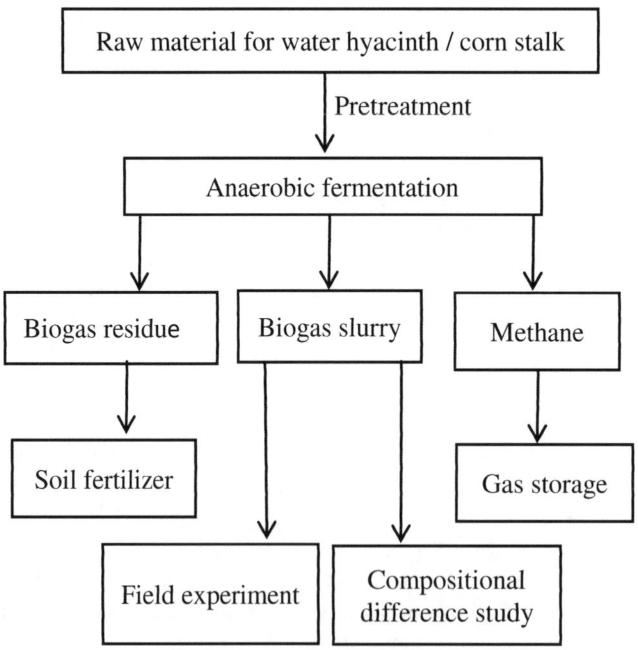

Figure 1. Chart of experiment.

2.1. Fermentation experiment of water hyacinth marsh

After the water hyacinth plant was removed, the roots were treated and cleaned to reach the required cleanliness at the experimental stage, and the shredder was used for processing. Take 100 g water hyacinth crushed in the three cervical fermentation flask, add 300 ml distilled water, adding 30 ml after experimental treatment of liquid manure, at room temperature (25 DEG C) mixed anaerobic fermentation, and the use of drainage gas gathering method of gas production record.

2.2. Fermentation experiment of corn stalk

After the corn stalks were removed, the roots and leaves were removed and cleaned, and the required cleanliness in the experimental stage was achieved, and the mills were treated. Remove the 100 g of corn straw in the three cervical fermentation flask, add 300 ml distilled water, adding 30 ml after experimental treatment of liquid manure, at room temperature (25 DEG C) mixed anaerobic fermentation, and the use of drainage gas gathering method of gas production record.

2.3. Extraction of organic components in biogas slurry

Take 100 ml water hyacinth biogas slurry in a separatory funnel, adding 100 ml petroleum ether first-order extraction, then separated with a separatory funnel, then add 50 ml of petroleum ether for two order extraction, the organic layer was poured into the flask preservation, water hyacinth biogas slurry of petroleum ether extract.

100 ml corn straw biogas slurry in a separatory funnel, adding 100 ml petroleum ether first-order extraction, then separated with a separatory funnel, and add 50 ml of petroleum ether for second order extraction, the organic layer is poured into the flask was saved, corn straw biogas slurry of petroleum ether extract.

Take 100 ml water hyacinth biogas slurry in a separatory funnel, with ethyl acetate 100 ml one-step extraction, and then use the separatory funnel water leakage, ethyl acetate were added to the two order 50 ml extraction, the organic layer is poured into the flask was saved, the ethyl acetate extract of water hyacinth biogas slurry.

Take 100 ml corn straw biogas slurry in a separatory funnel, with ethyl acetate 100 ml one-step extraction, and then use the separatory funnel water leakage, ethyl acetate were added to the two order 50 ml extraction, the organic layer is poured into the flask was preserved, ethyl acetate extract of corn straw biogas slurry.

Take 100 ml water hyacinth biogas slurry, pour it into a rotary evaporator for evaporation, with 5 ml methanol organic matter and dissolved, the liquid poured into the flask by water hyacinth biogas raw preservation, dry methanol extract.

Corn straw biogas slurry 100 ml, pour it into a rotary evaporator for evaporation, with 5 ml methanol organic matter and dissolved, the liquid poured into the flask was preserved, corn straw raw dry methanol extract of biogas slurry.

2.4. Comparative analysis of organic ingredients

The comparative analysis of the organic extracts of two kinds of biogas slurry extract mainly adopts the following analysis methods:

Thin layer chromatography (TLC) is to apply suitable fixed phases to glass plates, plastic or aluminium substrates to form a uniform layer. After sampling and unfolding, the method compares the specific shift value (Rf) of the chromatogram obtained by the same method according to the specific value of the shift value (Rf) and the appropriate contrast, and is used for the identification of the drug, the impurity inspection or the content determination method. Thin layer chromatography (TLC) is an important experimental technique for rapid separation and qualitative analysis of small amounts of substances. It is also used to track the reaction process. In the course of this experiment, two kinds of samples were analyzed by fluorescence detection, iodine vapor color detection and sulfuric acid color detection.

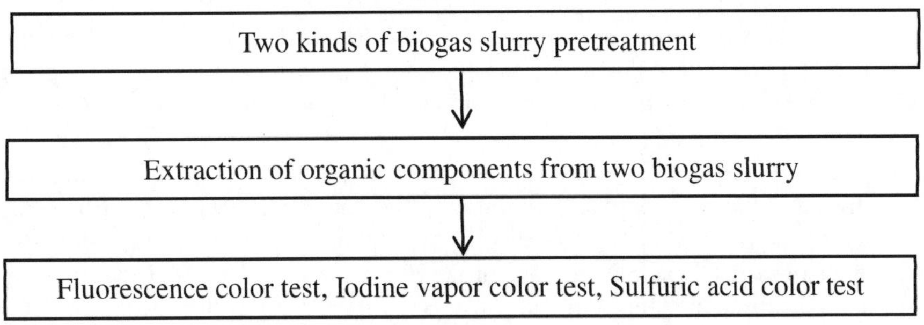

Figure 2. flow chart of experiment.

High Performance Liquid Chromatography (HPLC) is an important branch of chromatography, the mobile phase is liquid, using high pressure infusion system, mixed solvent, buffer flow will have a single solvent with different polarity or different proportion of the pump into the fixed phase column, the column the components are separated into. To detect, so as to realize the analysis of the sample.

Spectrophotometry is used to determine the absorbance of a light at a specific wavelength or a certain wavelength range. In this paper, the content of nucleic acid and protein in biogas slurry was analyzed by spectrophotometer.

2.5. Results and analysis

2.5.1. Experimental results and analysis of fermentation. Analysis of the fermentation of water hyacinth biogas slurry: in the process of experiment, it was found that the color of liquid in the fermentation device changed from light color to dark color, and the amount of drainage in this process also increased gradually. We further studied the fermentation broth of 120 h.The record of test is shown in Table 1.

Table 1. Fermentation of eichhornia crassipes solms.

Time	Gas production	Fermentation state
24 h	30 ml	Light green liquid
48 h	80 ml	Green liquid
72 h	140 ml	Dark green liquid
96 h	210 ml	Light yellow liquid
120 h	260 ml	Dark green liquid
144 h	250 ml	Dark green liquid

Analysis of the fermentation of corn straw biogas slurry: during the experiment, it was found that the color of the liquid in the fermentation plant changed from light color to dark color, and the amount of drainage in this process also increased gradually. We further studied the fermentation broth of 120 h.The record of test is shown in Table 2.

Table 2. Fermentation of corn straw slurry.

Time	Gas production	Fermentation state
24 h	15 ml	Light yellow liquid
48 h	40 ml	Yellow liquid
72 h	75 ml	Dark yellow liquid
96 h	145 ml	Dark yellow liquid
120 h	200 ml	Dark yellow liquid
144 h	205 ml	Dark yellow liquid

2.5.2. Results and analysis of thin layer chromatography. After TLC analysis of chromogenic results can be obtained in fluorescence color experiment, water hyacinth biogas slurry of petroleum ether extract, ethyl acetate extract of water hyacinth biogas slurry, water hyacinth Biogas Dry methanol extraction system of the original color liquid three subjects. The result is obvious in the color experiment in iodine vapor. Water hyacinth biogas slurry of the petroleum ether extract, ethyl acetate extract of water hyacinth biogas slurry, biogas slurry of the petroleum ether extract of corn straw color liquid three subjects. The result is obvious; the color experiment in concentrated sulfuric acid, alcohol, ethyl acetate extract of water hyacinth biogas slurry, biogas slurry of the petroleum ether extract of corn straw color liquid three subjects. The result is obvious. Through the results of color rendering experiments, we can preliminarily judge that there may be many organic compounds with high polarity. In view of this series of phenomena, the experimental subjects were analyzed by high performance liquid chromatography (HPLC).

2.5.3. Results and analysis of high performance liquid chromatography. Sample HPLC test results (methanol: water, 6:4, retention time, 40 min).

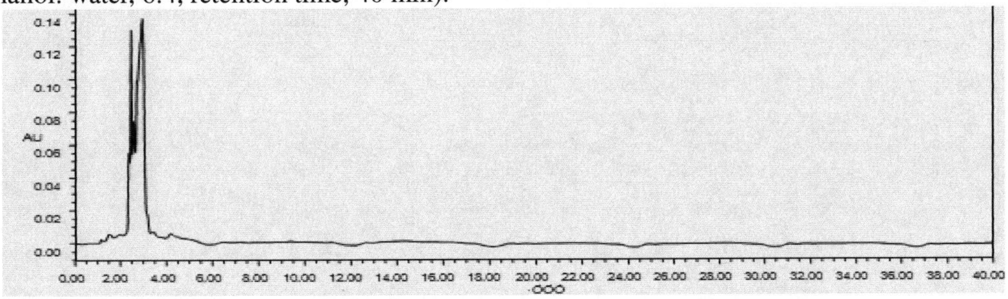

Figure 3. Eichhornia crassipes solms water extract solution.

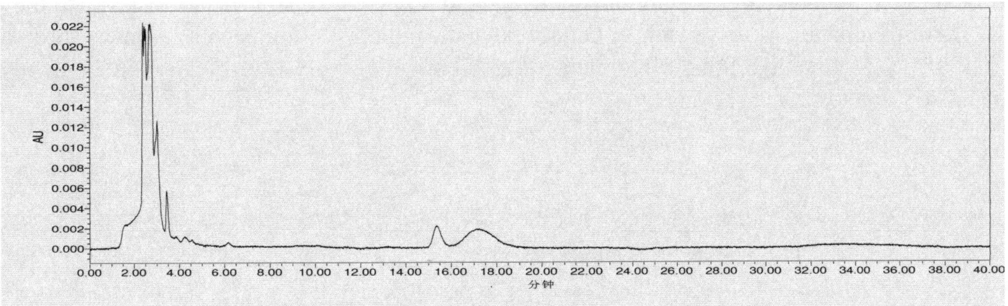

Figure 4. Corn straw water extract solution.

According to figure 3 and figure 4.Through the comparative analysis of water hyacinth and biogas slurry of corn straw and found that the two test objects in second minutes to peak area obviously fourth minutes in water hyacinth biogas slurry, first to second minutes is obvious in the peak district; corn straw biogas slurry, fifteenth to sixteenth minutes and seventeenth to nineteenth two minutes has obvious peak area. It can be explained that water hyacinth contains organic compounds with higher polarity, which can peak before corn stalk. From the map, it also reflects the obvious difference between the two kinds of stock solution. This series of differences makes the water hyacinth and corn stalk biogas slurry have different characteristics.

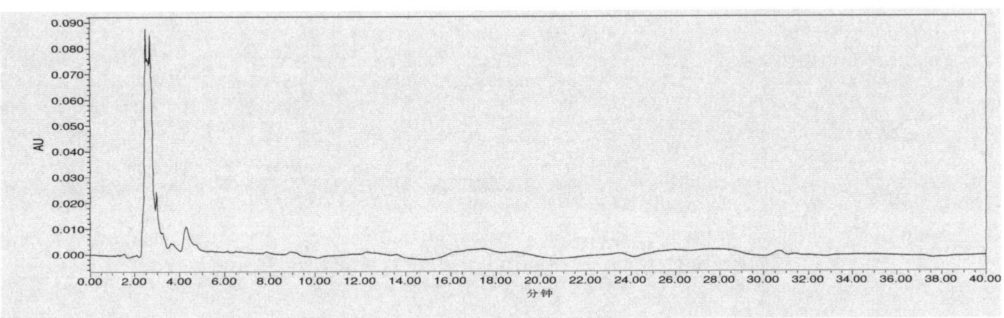

Figure 5. methanol extraction of eichhornia crassipes solms petroleum ether.

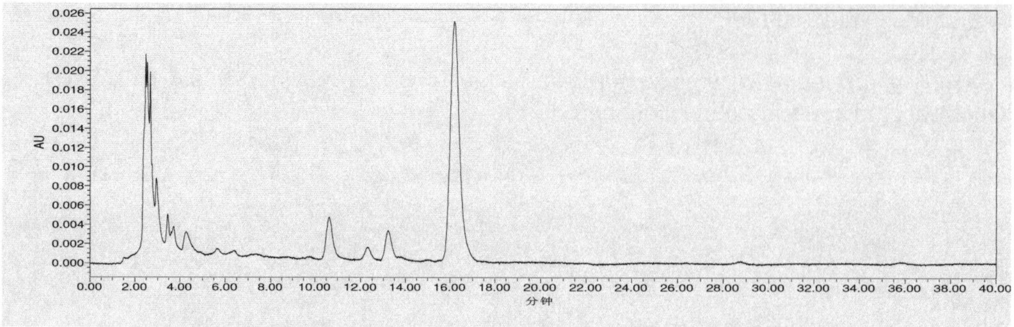

Figure 6. Methanol extraction of corn straw petroleum ether.

According to figure 5 and figure 6.Through the comparative analysis of the water hyacinth and corn straw biogas slurry of petroleum ether and methanol extract showed that two test objects in second minutes to the peak area, was fourth minutes and tenth minutes in the corn straw, to eleventh minutes, Twelfth minutes to fourteenth minutes, sixteenth minutes to eighteenth minutes, three hours there will be clear that the peak oil organic ingredients of corn straw biogas slurry in soluble components of larger organic species rich.

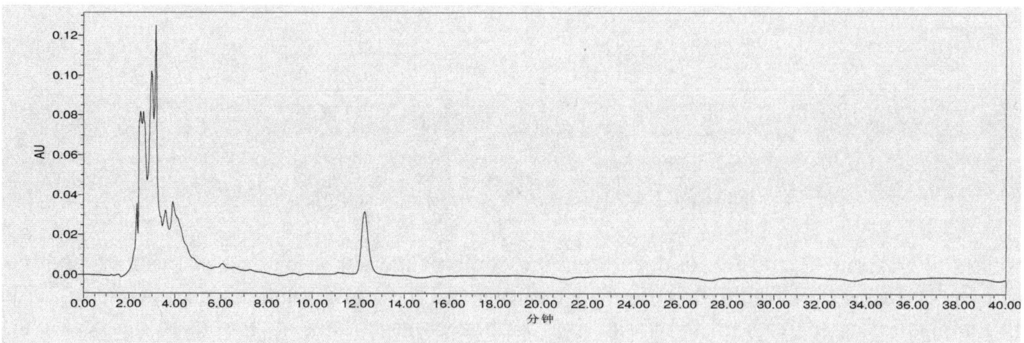

Figure 7. Extraction of ethyl acetate from eichhornia crassipes solms.

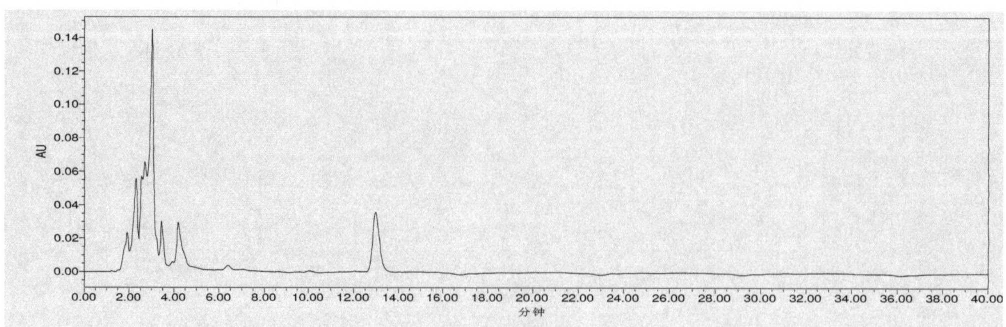

Figure 8. Extraction of ethyl acetate from corn straw.

According to figure 7 and figure 8.Through the comparative analysis of the water hyacinth and corn straw biogas ethyl acetate methanol extract showed that the two test objects in second minutes to peak area obviously in twelfth minutes to fourth minutes, fourteenth minutes have obvious peak area. In second minutes to fourth minutes, the difference between the peak value of the two subjects was large, and the peak time was also different. Among them, the extraction time of water hyacinth biogas slurry is earlier, the peak value is higher, and it also shows that the organic matter in water hyacinth slurry is good in ester solubility, organic matter composition is more, and polarity is higher.

2.5.4. Results and analysis of nucleic acids and proteins detected by spectrophotometer. By spectrophotometer in nucleic acid and protein and pH test analysis of water hyacinth and corn straw biogas slurry.

Table 3. Analysis of maize biogas slurry.

Test Name	Parameters	Wavelength(s)	Result	Displayed Units
DNA concentration	f 1=62.9	$A1=A_{260nm}$	A_1= 0.009	DNA mg/ml
	f 2=36.0	$A2=A_{280nm}$	A_2= 0.015	Protein mg/ml
	f 3=1552	$A_{ref}=A_{320nm}$	A_{ref}= 0.001	
	f 4=757.3		C-DNA: 0.097	
			C-Pro: 17.07	
			Ratio: 0.667	
Protein concentration	f 1=49.1	$A1=A_{260nm}$	A_1=0.002	DNA mg/ml
	f 2=3.48	$A2=A_{230nm}$	A_2= 0.006	Protein mg/ml
	f 3=183	$A_{ref}=A_{320nm}$	A_{ref}= 0.004	
	f 4=75.8		C-DNA: 0.081	
			C-Pro: 0.4224	
			Ratio: 0.912	

Table 4. Analysis of Eichhornia crassipes solms biogas slurry.

Test Name	Parameters	Wavelength(s)	Result	Displayed Units
DNA concentration	f 1=62.9	$A1=A_{260nm}$	A_1=0.007	DNA mg/ml
	f 2=36.0	$A2=A_{280nm}$	A_2= 0.014	Protein mg/ml
	f 3=1552	$A_{ref}=A_{320nm}$	A_{ref}= 0.013	
	f 4=757.3		C-DNA: 0.408	
			C-Pro: 5.643	
			Ratio: 9.361	
Protein concentration	f 1=49.1	$A1=A_{260nm}$	A_1=0.010	DNA mg/ml
	f 2=3.48	$A2=A_{230nm}$	A_2= 0.014	Protein mg/ml
	f 3=183	$A_{ref}=A_{320nm}$	A_{ref}=0.007	
	f 4=75.8		C-DNA: 0.770	
			C-Pro: 2.428	
			Ratio: 0.841	

In table, F1, F2, F3 and F4 are related calculation factors A1, A2, Aref, and the wavelength used in calculation.

From the two kinds of biogas slurries, composition contrast can be seen in the table that two kinds of slurries have obvious differences in the content of DNA and protein. Moreover, differences in the content of protein and nucleic acid are also relatively obvious. It also shows that the two kinds of functional differences in biogas slurry as fertilizer. After repeated measurements of pH meter, water hyacinth biogas slurry pH was 5.68. The pH value of corn straw biogas slurry was 5.08. Different acidity and alkalinity have different effects on fertilizer.

3. Conclusion

Eichhornia crassipes solms biogas slurry and biogas slurry of corn straw by anaerobic fermentation experiment, the two kinds of slurries after TLC analysis method can find that there are obvious differences in two kinds of organic components of biogas slurry, the analysis results in high performance liquid chromatography in corn straw biogas slurry which is rich in organic ingredients,

IOP Conf. Series: Earth and Environmental Science **93** (2017) 012006 doi:10.1088/1755-1315/93/1/012006

after the spectrophotometric determination of nucleic acid and protein, the two kinds of slurry organic components are obviously different, which was caused by the characteristics of two kinds of slurries with certain differences. According to the difference of two kinds of biogas slurries, the influence of application should be further studied.

References

[1] The Ministry of Agriculture 1990 *Biogas Technology Handbook* (Chengdu: Sichuan Science and Technology Press)

[2] Zhou M J, Zhang R L and Lin G S 2005 *Biogas Practical Technology* (Beijing: Chemical Industry Press)

[3] Zhang W D, Song H C and Ding Q 2001 Analysis of effect of biogas residue on crop diseases and insect pests *Agricultural Modernization Research* **22** 167-170

[4] Zhu B H 2014 Biogas slurry soaking technology *Anhui Agricultural Sciences* **2004**(9) 40-40.

[5] He C L, Wei Y Z and Ja Y B Research progress in the control and comprehensive utilization of water hyacinth *Bulletin of Sci. and Technol.* **21** 491-6

[6] Du J, Chang Z Z, Huang H Y, Ye X M, Ma Y, Xu Y D and Zhang J Y Study on optimization of dehydration technology parameters of water hyacinth *Jiangsu Agricultural Sci.* 2010(2):267-269.

[7] Sheng J, Chen L G, Zhu P P, Zhou W and Xue X H 2010 High nutrient enrichment plant Eichhornia crassipes farmland were studied by using *Chinese J. of Eco Agriculture* **18**(1) 46-49.

[8] Yang Z P 2010 General situation of corn stalk utilization and Prospect of straw fermentation M*odern Agricultural Machinery* 70-1

[9] Zha G J, Liu B and Zhang W D 2009 Study on water hyacinth juice after continuous fermentation solid-liquid separation *Renew. Energy* **27**(6) 54-57.

The membrane separation mechanism in protein concentration from the extract of waste press cake in biofuel manufacturing process of *Jatropha* seeds

T W Chung[1, 2], C K Chen[1] and S H Hsu[1]

[1]Department of Chemical Engineering, Chung Yuan Christian University, Taoyuan 32023, Taiwan

E-mail: twchung@cycu.edu.tw

Abstract. Protein concentration process using filter membrane has a significant advantage on energy saving compared to the traditional drying processes. However, fouling on large membrane area and frequent membrane cleaning will increase the energy consumption and operation cost for the protein concentration process with filter membrane. In this study, the membrane filtration for protein concentration will be conducted and compared with the recent protein concentration technology. The analysis of operating factors for protein concentration process using filter membrane was discussed. The separation mechanism of membrane filtration was developed according to the size difference between the pore of membrane and the particle of filter material. The Darcy's Law was applied to discuss the interaction on flux, TMP (transmembrane pressure) and resistance in this study. The effect of membrane pore size, pH value and TMP on the steady-state flux (J_{st}) and protein rejection (R) were studied. It is observed that the J_{st} increases with decreasing membrane pore size, the J_{st} increases with increasing TMP, and R increased with decreasing solution pH value. Compare to other variables, the pH value is the most significant variable for separation between protein and water.

1. Introduction

Jatropha curcas oil is gradually considered important to be the biofuel feedstock [1]. A large amount residue of pressed cake is generated after oil extraction. Both organic solvent and enzyme were used in various extractions [2-4]. The refined plant oil can be produced as raw material for biodiesel to replace traditional diesel fuel. However, the pressed cake has toxic substance of phorbol esters, resulting in the pressed cake is not suitable to be the animal feed. The pressed cake of *Jatropha curcas* contains high protein. The high protein meal can be used as fertilizer, biogas and adhesives to reduce the overall cost of biomass [5,6]. The pressed cake of *Jatropha curcas* is a mixture which containing fiber, protein and oil, therefore the separation process which include centrifugal [7-9], membrane and chromatography technologies were necessary to elevate usability of protein products [10,11]. The purpose of protein concentration is for the protein product preserving and transportation conveniently. The protein concentration was mostly carried out by using freeze-drying, which was a high energy-consumption process [12]. Therefore, membrane technology was applied as an improving method that was a simple operation and energy-saving separation process [13,14].

The separation mechanism of membrane filtration was the sieving, which is in accordance with the size difference between the membrane pore and the particle of filter material. The micro- and ultrafilter membranes were chosen for protein concentration usually [15, 16]. The pH value is a major factor for the protein concentration by membrane [17, 18]. The aggregation phenomenon of protein molecules was by electric interaction at specific range of pH value, resulting in the flux sharply fluctuation. In addition, it was found that the flux was influenced by transmembrane pressure (TMP) according to the Darcy's Law, which was usually applied to discuss the interaction on flux, TMP and resistance [19,20]. In this study, the membrane filtration for protein concentration will be conducted and the analysis of operating factors for protein concentration process using filter membrane was discussed. The separation mechanism of membrane filtration was developed according to the size difference between the pore of membrane and the particle of filter material.

2. Experimental section

2.1. Chemicals and raw materials

The materials and reagents used in this study include *Jatropha curcas* pressed cake, hydrochloric acid (HCl, 35 vol%, Showa), ammonium hydroxide (NaOH, ACS, Macron) and cellulose acetate filter membrane (CA, 0.2, 0.5 and 0.8 μm, Advance).

2.2. Experimental set up

An experimental schematic illustration, cross-flow microfiltration system is shown in figure 1.

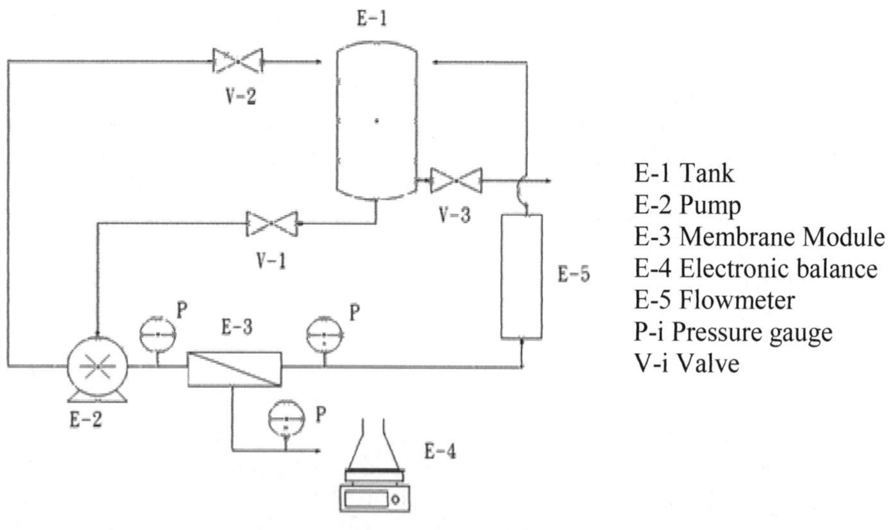

Figure 1. Experimental setup in this work.

The initial concentration of *Jatropha curcas* protein solution was 8.92 mg/mL. The pH values of protein solution were adjusted by HCl to 2, 5 and 8. Different pore sizes of flat-sheet cellulose acetate (CA) filter membrane were selected with 0.2, 0.5 and 0.8 μm. The effective filter area is 9.62×10^{-4} m^2. The transmembrane pressure (TMP) was controlled by V-2 in figure 1. The steady-state flux (J_{st}) and protein rejection (R) were defined as usual and expressed as following equation:

$$J_{st} = \frac{V_p}{A \times t} \quad (1)$$

where J_{st} means the steady-state flux, which is a constant value at final stage of filtration using L/m^2h as unit, A means an effective filter area, which is 9.62×10^{-4} m^2 in this study, t means an operation time, which control at 3 hours at a steady-state in this study, V_p means the accumulative permeation volume.

$$R = 1 - \frac{C_p}{C_F} \quad (2)$$

where R means the protein rejection, which is used to discuss the degree of separation using % as unit. ELISA technology [21, 22] was used to measure the concentration of protein in the solution. C_F and C_p express the protein concentration in the feed and permeation solution, using mg/mL as unit.

2.3. Theory

The filtration resistance was calculated by Darcy's law as following equation [23]:

$$J = \frac{\Delta P}{\mu Rt} \quad (3)$$

where J (m/s) means permeation flux, ΔP (Pa) means the transmembrane pressure (TMP), μ (Pa•S) means the dynamic viscosity of the solution in filtration process, and R_t (1/m) means the overall filtration resistance. The overall filtration resistance (R_t) can be expressed as the combination of various resistances as follow.

$$R_t = R_m + R_c + R_f \quad (4)$$

where R_m (1/m) means the resistance of membrane itself by measuring the steady-state DI water flux, R_c (1/m) means the resistance of cake formation onto the membrane surface at filtration process, R_f (1/m) means the resistance of fouling inside membrane pore.

On the other hand, the membrane filtration behavior can be expressed by the law of filtration blocking mode, which the cake layer heap on membrane surface at the last stage of filtration [24]. In the figure 2(a), the schematic diagram of cake blocking mode can be expressed as a function of time as following:

$$\frac{t}{V} = \frac{1}{Q_0} + \frac{K_c}{2} \quad (5)$$

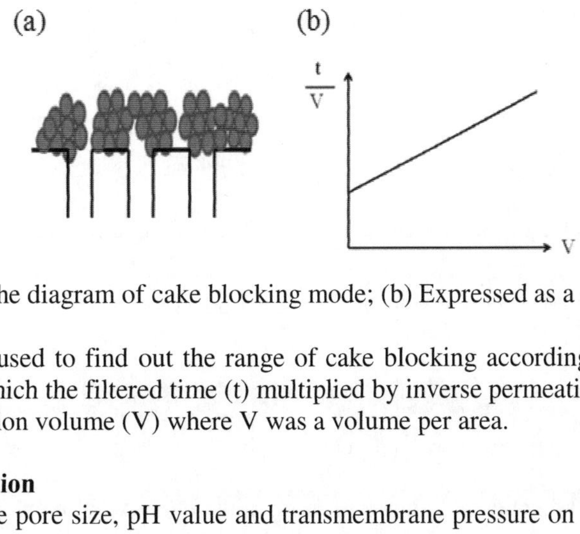

Figure 2. (a) The diagram of cake blocking mode; (b) Expressed as a function of time.

Figure 2(b) can be used to find out the range of cake blocking according to equation (5). It was found that the value, which the filtered time (t) multiplied by inverse permeation volume (1/V) linearly increased with permeation volume (V) where V was a volume per area.

3. Results and discussion

The effect of membrane pore size, pH value and transmembrane pressure on the steady-state flux (J_{st}) and protein rejection (R) were studied. It is observed that the steady-state flux (J_{st}) increases with decreasing membrane pore size. The membrane resistance (R_m) was smaller for the membrane with larger pore size, the protein molecules could pass easily through larger pore size membrane. However, the larger J_{st} was appeared on the membrane with smaller pore size. The filtration blocking law and Darcy's law were used to explain the relationship between J_{st} and membrane pore size.

In order to figure out this phenomenon, the cake formation time of filtration blocking stage (t) varied with membrane pore size of 0.2, 0.5 and 0.8 μm under 0.5 kg/m^2 TMP was observed first. It was found that a linear relationship (figure 3) between the t multiplied by inverse permeation volume

(1/V) and permeation volume (V) by graphing method of equation (5). It is integrated that the t increases with decreasing membrane pore size (table 1). It means the cake was formed slowly on the larger pore size of membrane surface, resulting in the smaller protein molecules easily adsorb onto the membrane to generate a larger cake resistance layer (figure 4). As a consequence, it was found that the cake resistance (R_c+R_f) increased with the larger membrane pore size operation, which given a poor result of J_{st}.

Table 1. The effect of membrane resistance (R_m), cake resistance (R_c+R_f) and the time of filtration blocking stage (t) on J_{st} varying membrane pore size.

pore size (um)	0.2	0.5	0.8
J_{st} (L/m^2-h)	43.36	41.76	38.75
R_mx10^{-9} (1/m)	14.22	7.44	3.79
(R_c+R_f)x10^{-12} (1/m)	6.84	7.13	7.68
t (min)	40-180	50-180	70-180

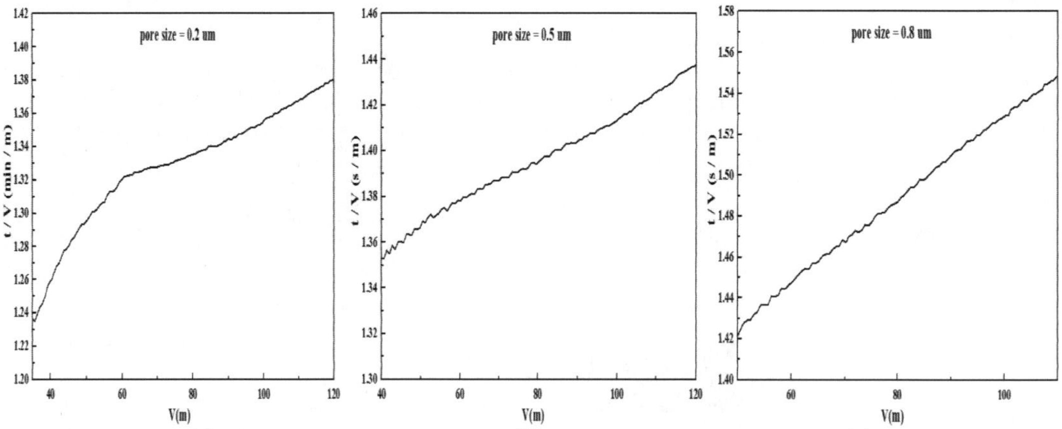

(a) (b) (c)

Figure 3. The cake blocking mode on varying membrane pore size with (a) 0.2 (b) 0.5 (c) 0.8 μm.

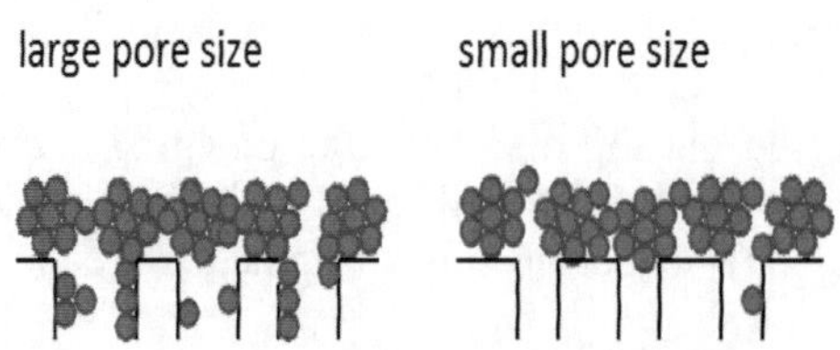

Figure 4. The schematic diagram of blocking mode on varying membrane pore size.

The protein aggregation phenomenon means the protein molecules aggregated to become the group of protein by electrostatic attraction (figure 5). For the smaller particle, non-aggregation protein would easily adsorb onto the membrane, resulting in the cake resistance (R_c+R_f) was larger. Therefore, it was observed that the larger value of J_{st} at lower pH of solution.

Similarly, it was found that R increased with decreasing solution pH value. Compared to other variables, the pH value is the most significant variable for separation between protein and water. The condition of non-aggregation protein has a lot of smaller protein molecules, resulting in these molecules can pass easily through the membrane to increase the protein concentration of permeation (figure 5). According to equation (2), C_p is smaller at the condition of protein aggregation, resulting in the R is higher than the condition of non-aggregation protein. Therefore, R increased with decreasing solution pH value.

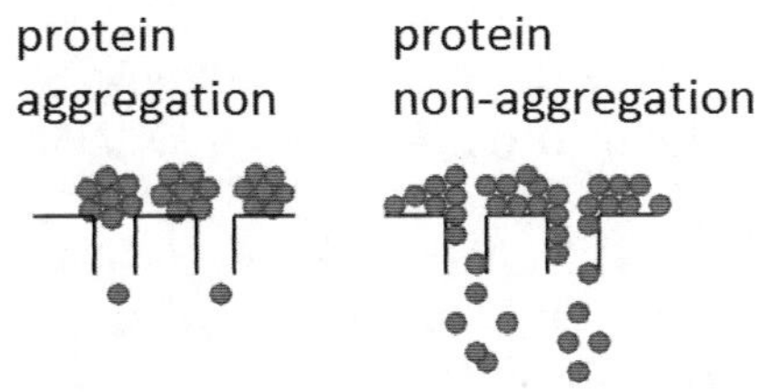

Figure 5. The schematic diagram of blocking mode on varying solution pH value.

It was found that J_{st} increases with increasing TMP which follows the Darcy's law. In the figure 6, it is observed that J_{st} increases linearly with increasing TMP due to the overall filtration resistance is quite small when the TMP below 0.5 kg/cm^2, and then the overall resistance sharply rise when the TMP above 0.5 kg/cm^2, resulting in the increasing of J_{st} is limited. On the other hand, it was observed that the protein rejection (R) was strongly dependent on the solution pH value. But, it was important that the R was independent on the interaction of membrane pore size and transmembrane pressure.

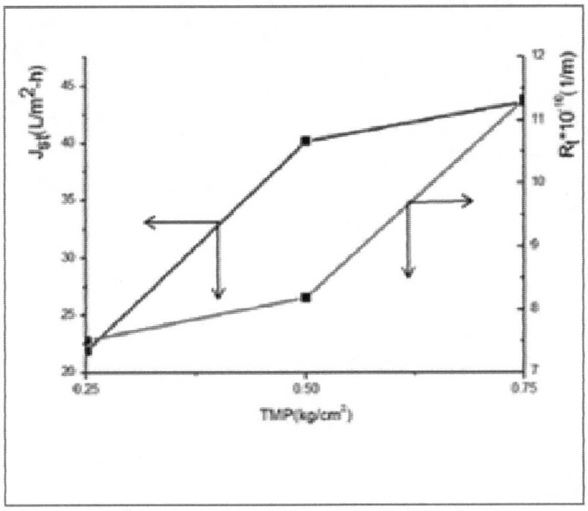

Figure 6. The effect of TMP on J_{st} and overall resistance.

4. Conclusion

On the basis of the energy-saving consideration, the membrane filtration process was chosen to replace the traditional heating of protein solution. In this study, the protein solution form extraction of *Jatropha curcus* pressed cake was conducted by the cellulose acetate filter membrane to concentrate. The effect of selected variables on the steady-state flux and protein rejection on the membrane were investigated. The higher steady-state flux was found at a condition of lower solution pH and higher transmembrane pressure. The higher protein rejection was found at a condition of lower solution pH and lager membrane pore size. The membrane filtration for protein concentration was conducted and compared with the recent protein concentration technology. The separation mechanism of membrane filtration was developed according to the size difference between the pore of membrane and the particle of filter material. The membrane filtration behavior can be expressed by the law of filtration blocking mode, which the cake layer heap on membrane surface at the last stage of filtration. The cake blocking mode was discussed with the varying membrane pore size, pH value, and TMP successfully.

References

[1] Hamarneh A, Heeres H, Broekhuis A and Picchioni F 2010 *Intl J. Adhesion and Adhesives* **30** 615-25
[2] Gofferje G, Klingele S, Stabler A, Weisz U and Floter E 2013 *Waste Biomass* **29** 301-8
[3] Lestari D, Mulder W and Sanders J 2010 *Biochemical Eng. J.* **50** 16-23
[4] Russin T, Arcand Y and Boye J 2007 *Biomass* **13** 308-19
[5] Giibitz G, Mittelbach M and Trabi M 1999 *Bioresource Tech.* **67** 73-82
[6] Deeba F, Kumar V, Gautam K, Saxena R and Sharma D 2012 *Biomass and Bioenergy* **40** 13-8
[7] Naderi N, House J and Pouliot Y 2014 *J. Food Eng.* **141** 85-92
[8] Lindner J, Menzel K and Nirschl H 2013 *Chemical Eng. Sci.* **97** 385-93
[9] Strixner T and Kulozik U 2013 *J. Food Eng.* **117** 89-98
[10] Zobel S, Helling C, Ditz R and Strube J 2014 *Ind. and Eng. Chem. Res.* **53** 9169-85
[11] Liu Y, Feng Z, Shao Z and Chen X 2012 *Materials Sci. Eng.* C **32** 1669-973
[12] Fan L, Ding S, Liu Y and Ai L 2012 *Intl J. Biological Macromolecules* **50** 1008-10
[13] Pouliot Y 2008 *Intl Dairy J.* **18** 735-40
[14] Daufin G, Escudier J, Carrere H, Berot S, Fillaudeau L and Decloux M 2001 *Trans IChemE* **79** 89-102
[15] Onics Inc. 2004 *Filtration & Separation* **41** 32-33
[16] Narsaiah K and Agarwal G P 2007 *J. Membrane Sci.* **287** 9-18
[17] Huisman H, Prádanos P and Hernández A 2000 *J.* Membrane *Sci.* **179** 79-90
[18] Ibanez R, Almecija M, Guadix A and Guadix E 2007 *Sepa. Purif. Tech.* **57** 314-320
[19] Ouammou M, Tijani N, Calvo J, Velasco C, Martin A, Martinez F, Tejerina F and Hernandez A 2007 *Colloids and Surfaces A: Physicochem. Eng. Aspects* **298** 267-73
[20] Ripperger S and Altmann J 2002 *Sepa. Purif. Tech.* **26** 19-31
[21] Chen V, Fane A, Madaeni S and Wenten I 1997 *J. Membrane Sci.* **125** 109-22
[22] Jiang W, Liu X, Wu D, Wang H, Wang Y, Chen H and Yuan L 2014 *Biotechnology and Applied Biochemistry* **20** 126-31
[23] Ecker C, Ertl A, Pulverer W, Nemes A, Szekely P, Petrasch A, Martin G and Markl M 2013 *Food Chem.* **141** 407-18
[24] Dizge N, Soydemir G, Karagunduz A and Keskinler B 2011 *J. Membrane Sci.* **366** 278-285

Peak reduction for commercial buildings using energy storage

K H Chua[1,2], Y S Lim[1] and S Morris[1]

[1]Universiti Tunku Abdul Rahman, Jalan Sungai Long, Bandar Sungai Long, Cheras, Kajang 43000, Selangor, Malaysia

E-mail: chuakh@utar.edu.my

Abstract. Battery-based energy storage has emerged as a cost-effective solution for peak reduction due to the decrement of battery's price. In this study, a battery-based energy storage system is developed and implemented to achieve an optimal peak reduction for commercial customers with the limited energy capacity of the energy storage. The energy storage system is formed by three bi-directional power converter rated at 5 kVA and a battery bank with capacity of 64 kWh. Three control algorithms, namely fixed-threshold, adaptive-threshold, and fuzzy-based control algorithms have been developed and implemented into the energy storage system in a campus building. The control algorithms are evaluated and compared under different load conditions. The overall experimental results show that the fuzzy-based controller is the most effective algorithm among the three controllers in peak reduction. The fuzzy-based control algorithm is capable of incorporating a priori qualitative knowledge and expertise about the load characteristic of the buildings as well as the useable energy without over-discharging the batteries.

1. Introduction

The growth of electrical energy demand has urged the utility companies to build more power plants when the peak demands approach the system's load carrying capacity. However, power plants are massive industrial complexes that can affect the environment by its construction as well as its operation. Hence, alternatives are sought to overcome the high peak demands growth.

Demand response is a tariff or program established to motivate customers to change their electric use pattern such a way to reduce the peak demands and reduce the need for expensive peaking plants [1-3]. Various price-based demand response such as time-of-use (TOU), real-time pricing (RTP), and critical-peak pricing (CPP) have been introduced by the utility companies to induce lower electricity use during peak demand hours. Besides, incentive-based demand response programs offer incentives to the customers who shift their peak period activities to off-peak periods [4-6]. However, there are still many technical and legal issues to be tackled for the implementation of demand response programs such as huge investment on advanced metering infrastructure and vague legislation [2,3].

Battery-based energy storage has emerged as a promising technology in reducing peak demands for both the utility and end users. Many large-scale battery-based storage systems have been installed for the purposes of peak reduction worldwide as illustrated in table 1.

A small-scale battery-based storage system is a good choice for the end users to shave their peak demands. High portability, short setup time, simple installation and commission, low capital cost investment, minimal space occupation and low maintenance and operating cost are the superiority of the battery-based storage system as compared to other storage technologies [10-12]. An assessment has been carried out to investigate the economic benefits of the ESS in providing peak reduction for

Table 1. Worldwide large-scale battery-based storage systems for peak reduction.

Project	Operational dates	Location	Battery type	System size	
				MW	MWh
Crescent Electric Membership Cooperative BESS [a]	1987 to 2002	Carolina, USA	Flooded cell, lead acid	0.5	0.5
Sumitomo Densetsu Office Battery System [a]	2000 to present	Japan	Vanadium redox flow	3	0.8
New York Bus Terminal Energy Storage Systems [a]	2008 to present	New York, USA	Sodium-sulphur	1.2	6.5
ZBB Energy Corporation Battery Storage Systems [b]	2005 to present	California, USA	Zinc-bromine	2	2
Zurich Battery Energy Storage System [c]	2012 to present	Dietikon, Switzerland	Li-ion	1	0.5

[a] Data collected from [7]
[b] Data collected from [8]
[c] Data collected from [9]

both the utility and end users [13]. A simple method has been proposed by [14] for peak-load reduction using battery ESS. The study compares the aggregated load profile with its average value during a specific period and operates the energy storage devices based on the proposed weighting factors. The average load reduction is 0.35 - 1%. A dynamic programming method was proposed by [15] to optimize the operation of the ESS and 8% of peak reduction has been achieved. In [16], the peak reductions of 27 – 45 % has been achieved using a demand tracking management model. Lu *et al* have proposed an optimal sizing and control of battery-based storage system for peak reduction at the utility side [17]. The study has diminished the difference between the peak load and the valley load via the optimal control of the energy storage system (ESS) and the peak reduction is 4.5 - 6.6%.

Although there are many studies carried out for peak reduction using battery-based storage systems, most of the research work are simulation-based [17-26]. However, there is a dearth of public research in real-time peak reduction for commercial buildings. The real-time peak reduction is more challenging than that of the simulation because the actual peak demands can be quite different from the forecasted load profiles. When the actual peak demands are unexpectedly high, the ESS may fail to reduce the peak due to the limited capacity of the power converters. Also, the actual peak can happen much earlier or later than the forecasted one. The ESS may fail to cut down the peak if it supplies the power at the wrong time. Another possibility is that the actual peak can happen for an extended period of time. The ESS may not have enough energy to sustain the supply of power. Considering the possible variation in the actual load profiles, any control strategies used for the ESS has to be evaluated empirically through experiments. Since the cost of the batteries is still high at present, the capacity of the ESS has to be limited or reduced in order to justify the financial viability of the system.

This paper presents three control algorithms, namely the fixed-threshold, adaptive-threshold, and fuzzy-based control algorithms for the ESS to shave the peak demands under various load conditions. The fixed-threshold controller instructs the ESS to supply power whenever the actual load demand exceeds a specified threshold. The adaptive-threshold controller adjusts the threshold when the peak demands are unexpectedly high or happen over an extended period. However, these two controllers do not take into account the available energy from batteries. This may result in insufficient energy for peak reduction. The fuzzy-based controller is capable of estimating the useable energy from the batteries and incorporating it into the fuzzy inference rules to determine the appropriate output power to be delivered to the load in order to prevent insufficient energy supply from the batteries during peak reduction. The performance of the three controllers was evaluated at two different buildings at Universiti Tunku Abdul Rahman (UTAR), Malaysia, to evaluate the effectiveness of the control algorithms in reducing peak demands under various load profiles.

2. Configuration of the ESS

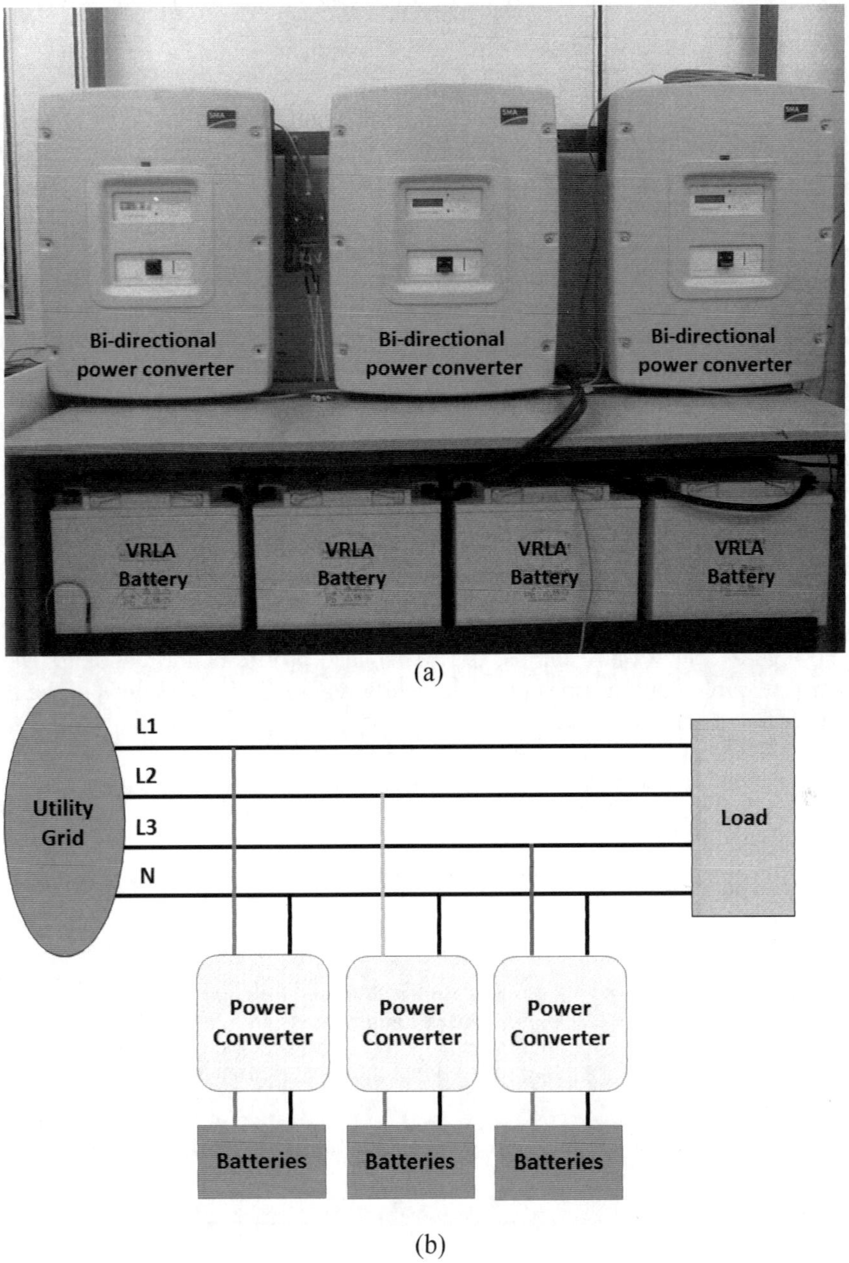

Figure 1. (a) Experimental ESS setup; (b) Electrical diagram of the ESS.

An experimental three-phase ESS is set up for peak reduction at two different buildings at two different time frames. The ESS consists of three single-phase bi-directional power converters rated at 5 kVA and 48 pieces of 12 V valve-regulated lead-acid (VRLA) batteries. Each bi-directional power converter is connected to 4 strings of batteries and each string of batteries is formed by 4 series-connected batteries to provide a 48 V DC supply to the converter. The total energy capacity of the

IOP Conf. Series: Earth and Environmental Science **93** (2017) 012008 doi:10.1088/1755-1315/93/1/012008

whole ESS is 64 kWh. A data acquisition system is developed using the C-series voltage and current measuring modules manufactured by National Instruments (NI). LabVIEW graphical programming software is used to develop the control system for the ESS. Figure 1(a) shows the experimental ESS setup and figure 1(b) shows the electrical diagram of the ESS connected to the 415 V, 50 Hz, 3-phase utility grid.

3. Formulation of control strategies for peak reduction

Usually, the commercial and industrial customers require a significant amount of power over a short time interval of about 2 to 4 hours during peak periods. The high peak demand during these short time intervals results into high maximum demand charges to the customers. In this study, the ESS is controlled to deliver power during peak periods and restore energy during off-peak periods. The ESS is installed in two different buildings, namely building A and building B to evaluate the effectiveness of the control algorithms in reduction peak demands under various load profiles. The load profiles of these two buildings are forecasted based on the historical load profiles. The predicted daily load profile provides information on the amount of energy required to shave the peak demands. The depth-of-discharge (*DOD*) of the batteries is limited to 50% based on the recommendation from the battery manufacturer. The grid power (P_{grid}) as measured by the utility power meter is equal to the sum of load demand (P_{load}) and the power of the energy storage (P_{ES}) as follows:

$$P_{grid} = P_{load} - P_{ES} \tag{1}$$

The sign of P_{ES} is positive when it delivers power and it is negative when it absorbs power.

3.1. Fixed-threshold control algorithm

Fixed-threshold control algorithm is the most basic control strategy developed to control the ESS for peak reduction. Figure 2 shows an example of a daily load profile of the building. The daily load profile can be assorted into three sections to reflect 3 different possible operations of the ESS, namely the charging, discharging and idle modes.

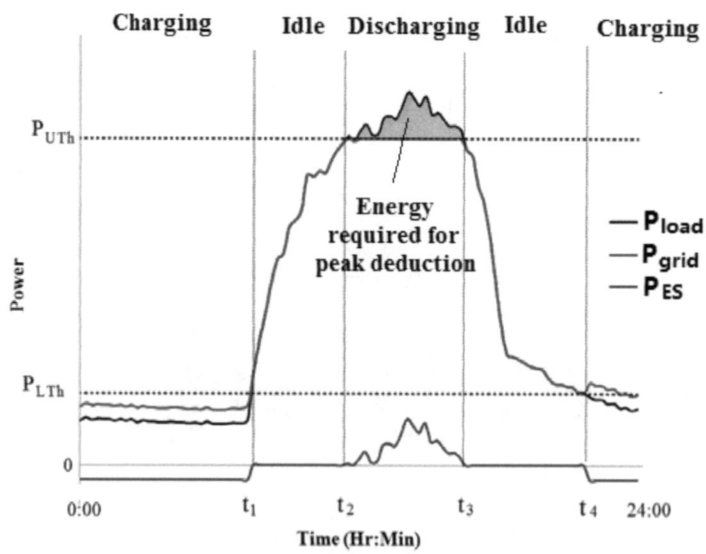

Figure 2. Daily load profile and the desired response of the ESS.

The lower power threshold (P_{LTh}) and an upper power threshold (P_{UTh}) are set such that the ESS operates when the load demand falls within a certain range of the thresholds. During the period from 0:00 to t_1 and t_4 to 24:00, P_{load} is lower than P_{LTh}. The ESS charges the batteries if it is not fully charged or stays in the idle mode if the batteries are fully charged. During the period from t_1 to t_2 and from t_3 to t_4, the ESS stays idle because P_{load} is in between P_{LTh} and P_{UTh}. During the period from t_2 to t_3, the ESS delivers power to the load because P_{load} is greater than P_{UTh}. The energy required to reduce the peak demands for each day can be determined by calculating the area under the graph between the load demand and P_{UTh} as illustrated in figure 2.

During the discharging mode, the load demand is greater than P_{UTh}. The ESS begins to supply power to the load if the state-of-charge (SOC) is greater than 50%. If the SOC is below 50%, the discharging of the batteries is forbidden to prevent the over-discharging of the batteries and prolong the service life of the batteries. If the difference between P_{load} and P_{UTh} is less than the rated power of the ESS (P_{ESmax}), then the ESS will inject power with the magnitude being the same as P_{load} minus P_{UTh} as shown in the following equation:

$$P_{ES} = P_{load} - P_{UTh} \tag{2}$$

If the difference between P_{load} and P_{UTh} is greater than the rating P_{ESmax}, then the ESS will inject its nominal power to the load as follows:

$$P_{ES} = P_{ESmax} \tag{3}$$

If the SOC is equal to or less than 50%, then the power converter will stop injecting power to the load.

During the charging mode, the ESS starts to charge the batteries when the actual load demand is lower than P_{LTh} in order to restore the batteries' energy for the next day of peak reduction. If the SOC of the batteries is less than 90%, then the batteries will be charged at a moderate rate of 6 kW. If the SOC is greater than or equal to 90%, then the battery is charged at a low rate of 3*[0.7 ln(t) - 2] kW. These charging rate expressions are chosen from the battery converter handbook with a view to prevent the overcharging of the batteries and to prolong the service life of the batteries.

During the idle mode, the ESS should output zero power as follows:

$$P_{ES} = 0 \tag{4}$$

3.2. Adaptive-threshold control algorithm

Adaptive-threshold controller is an improved version of the fixed-threshold controller because the output power of the ESS can be reduced to preserve the energy of the batteries when the actual peak demands occur for an extended period of time. The operation of the charging and idle modes of the adaptive-threshold controller is the same as that of the fixed-threshold control algorithm except for the operation of the discharging mode. During the discharging mode, if the difference between P_{load} and P_{UTh} is greater than P_{ESmax} within a specific duration of time (N) in minutes, then the high load demand is considered as temporary. The ESS will inject its nominal power of P_{ESmax}. When the difference between P_{load} and P_{UTh} is greater than P_{ESmax} for more than N minutes, the high load demands is no longer temporary. Hence, the new upper threshold ($P_{UTh'}$) will be moved to a new value based on the following equation:

$$P_{UTh'} = P_{load} - P_{ESmax} \tag{5}$$

The adaptive feature enables the ESS to deliver power to the load over the longest possible period.

3.3. Fuzzy-based control algorithm

The fuzzy-based control algorithm takes into account the useable energy (E_U) during the time of operation (t_{op}) to determine the amount of power to be delivered to the grid during the peak reduction. The t_{op} is defined as the time where the energy storage begins to supply its power. When the E_U has declined at the certain time, the power injection from the ESS is also decreased to ensure that the E_U can be sustained until the end of the peak reduction process. The input variables for the fuzzy-based controller are the SOC of the batteries and the t_{op} while the output variable is the power injection of the

ESS, P_{ES}. Table 2 shows the definition of the fuzzy sets of the SOC, t_{op}, and P_{ES} in linguistic terms. There are five membership functions for both the SOC and the P_{ES} while there are seven membership functions for the t_{op}.

Table 2. Definition of the fuzzy sets of the SOC, t_{op} and P_{ES} in linguistic variables.

SOC	Time of Operation (t_{op})	P_{ES}
VL – Very Low	EE – Extremely Early	VL – Very Low
L – Low	VE – Very Early	L – Low
M – Middle	E – Early	M – Middle
H – High	M – Middle	H – High
VH – Very High	L – Late	VH – Very High
	VL – Very Late	
	EL – Extremely Late	

The rules in the fuzzy inference system are derived from the desired responses of the ESS based on the predicted load demands. Table 3 shows the values of the fuzzificated *SOC*, t_{op}, and the corresponding P_{ES} of the fuzzy controller.

Table 3. The fuzzificated SOC, t_{op} and the corresponding P_{ES} of the fuzzy controller.

t_{op} \ SOC	VL	L	M	H	VH
EE	VL	VL	VL	VL	VL
VE	VL	VL	VL	VL	VL
E	VL	L	L	M	M
M	VL	L	M	M	H
L	L	M	H	VH	VH
VL	M	H	VH	VH	VH
EL	M	H	VH	VH	VH

4. Experimental results

4.1. Performance evaluation of the fixed-threshold control algorithm

Case study 1 evaluates the performance of the fixed-threshold control algorithm conducted at building A for the scenario where the load demand is higher than the predicted value and the actual peak occurs for a much longer duration than the predicted duration. The actual peak demands are unexpectedly higher than the predicted value where its maximum demand is 97.21 kW while the predicted maximum demand is 95 kW. Initially, P_{UTh} and the P_{LTh} are set at 80 kW and 30 kW, respectively according to the predicted value. Figure 3 shows the results of the peak reduction using the fixed-threshold control algorithm for the case study 1. It can be seen that when the load has exceeded 80 kW at 10:00, the ESS begins to deliver power to the load based on the difference between the P_{load} and P_{UTh}. The SOC of the batteries has declined to 50% at 13:00, triggering the ESS to stop delivering power to the load. As a result, the ESS fails to shave the peak after 13:00. In this case study, the ESS has failed to cut down the peak demand due to the limited rating of the power converters. It is also found that the actual peak occurs for a much longer duration than the prediction. Hence, the ESS does not have enough energy to sustain the delivery of its power over such a long duration.

Case study 2 evaluates the performance of the fixed-threshold control algorithm conducted at building B. The P_{UTh} is set at 70 kW according to the predicted value while the P_{LTh} is set at 50 kW. Figure 4 shows the results of the peak reduction. It can be seen that when the load has exceeded 70 kW at 7:45, the ESS begins to deliver power to the load based on the difference between the P_{load} and

P_{UTh}. The peak demand has been reduced from 78.4 kW to 70.9 kW. It is found that the peak reduction in this case study is 7.5 kW.

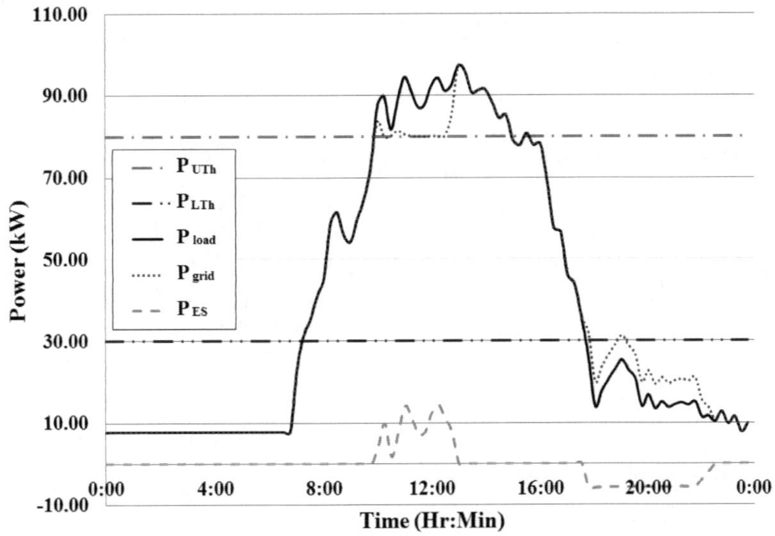

Figure 3. The peak reduction using the fixed-threshold control algorithm for case study 1.

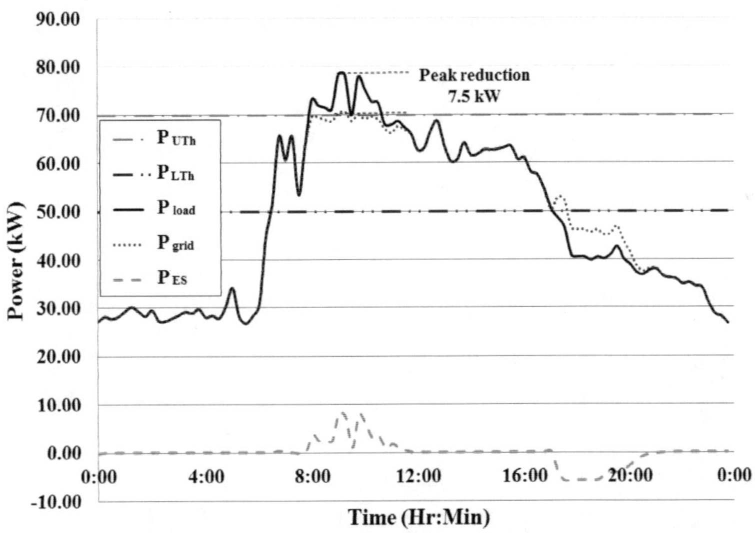

Figure 4. The peak reduction using the fixed-threshold control algorithm for the case study 2.

4.2. Performance evaluation of the adaptive-threshold control algorithm

Case study 3 evaluates the performance of the adaptive-threshold control algorithm conducted at building A. This case study illustrates the scenario where the actual peak is unexpectedly higher than the predicted value. Figure 5 shows the results of the peak reduction using the adaptive-threshold control algorithm for the case study 3. It is found that the maximum demand is 90.2 kW while the predicted maximum demand is 86 kW. Initially, the P_{UTh} is set at 71 kW according to the predicted value while the P_{LTh} is set at 30 kW. It can be seen that when the load demand has exceeded 71 kW at 12:00, the ESS begins to deliver power to the load. At 13:30, P_{load} has increased to 89.2 kW and prolonged for more than 10 minutes. The corrective mechanism in the adaptive-threshold algorithm adjusts the P_{UTh} to 74.5 kW at 13:45. At 14:00, P_{load} has increased to 90.2 kW and prolonged for more than 10 minutes and the P_{UTh} is adjusted to 77.3 kW at 14:15. The ESS begins to restore its energy at 17:00. The peak demand has been reduced from 90.2 kW to 78.5 kW In this case, the peak shaved is 11.7 kW.

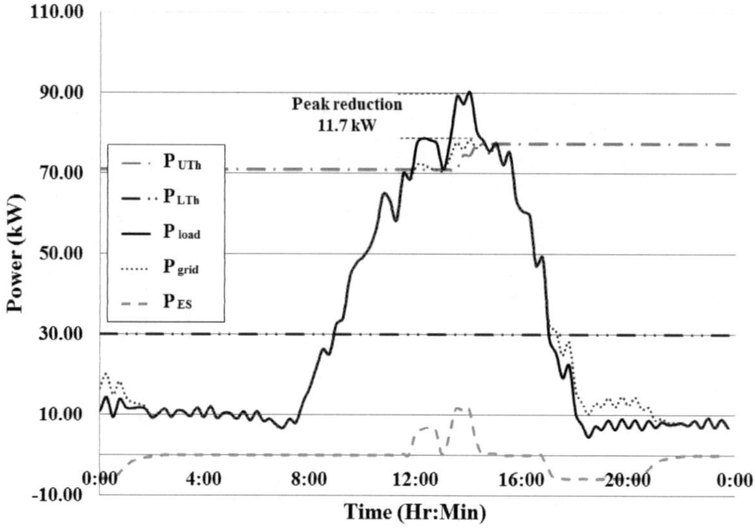

Figure 5. The peak reduction using the adaptive-threshold control algorithm for the case study 3.

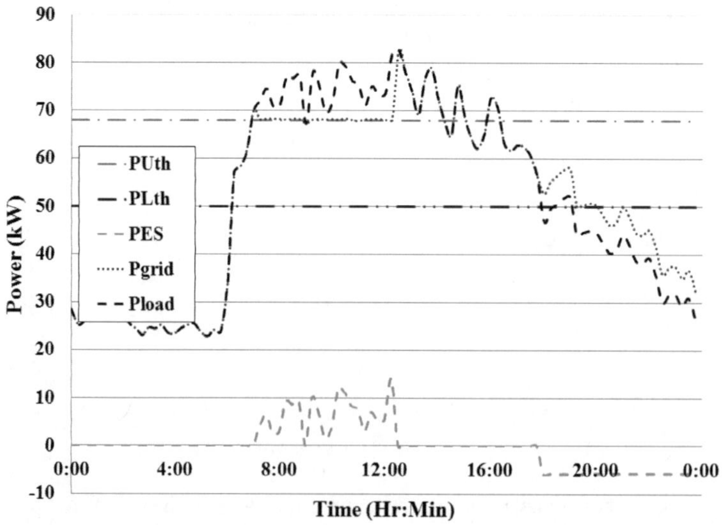

Figure 6. The peak reduction using the adaptive-threshold control algorithm for the case study 4.

Case study 4 evaluates the performance of the adaptive-threshold control algorithm conducted at building B. The P_{UTh} is set at 68 kW according to the predicted value while the P_{LTh} is set at 50 kW. Figure 6 shows the results of the peak reduction. The peak demand is 82.78 kW while the predicted P_{UTh} is 83 kW. It can be seen that the ESS starts to deliver power to the load at 7:15 when the P_{load} is greater than the P_{UTh}. At 12:15, the SOC of the ESS has dropped to 50%. Consequently, the ESS stops to deliver power to the load. As a result, the ESS failed to reduce the peak demand when the load has increased to 82.78 kW at 12:30. This case study shows that when the actual peak occurs for a much

IOP Conf. Series: Earth and Environmental Science **93** (2017) 012008 doi:10.1088/1755-1315/93/1/012008

longer duration than the predicted one, the ESS fails to deliver power to the load because it does not have enough energy to sustain the delivery of its power over such a long duration.

4.3. Performance evaluation of the fuzzy-based control algorithm

Case study 5 evaluates the performance of the fuzzy-based control algorithm conducted at building A. In this case, the P_{UTh} and P_{LTh} are set at 80 kW and 30 kW, respectively. The estimated peak demand is 95 kW. At 10:30, P_{load} is greater than P_{UTh} and the ESS begins to deliver power to the grid such that P_{grid} is not higher than P_{UTh}. It can be seen that at 13:45, P_{load} has reached 100.5 kW and it is higher than the estimated peak demand. The fuzzy-based controller delivers up to 12.1 kW instead of its full capacity of 15 kW to preserve energy so that the remaining energy is sufficient to shave the peak for the rest of time. Figure 7 shows the experimental results for the case study 5. The peak demand has been reduced from 100.5 kW to 88.4 kW. The peak demand reduction for this case is 12.1 kW.

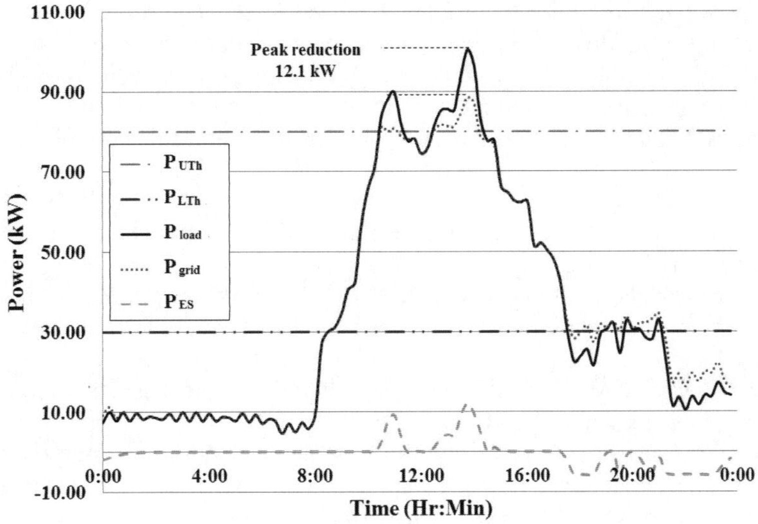

Figure 7. The peak reduction using the fuzzy-based control algorithm for the case study 5.

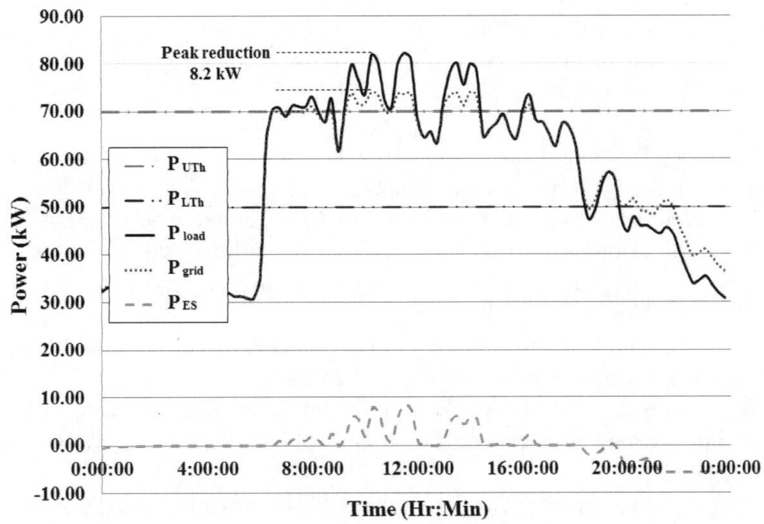

Figure 8. The peak reduction using the fuzzy-based control algorithm for the case study 6.

Case study 6 evaluates the performance of the fuzzy-based control algorithm conducted at building B. P_{UTh} and P_{LTh} are set at 70 kW and 50 kW, respectively. The peak demand occurs earlier than the predicted time of occurrence at 6:45. It can be seen that at 9:30 when P_{load} is 80 kW, the ESS restricts the power injection to only 5.9 kW in order to preserve energy for the rest of the time. Figure 8 shows the experimental results for the case study 6. The peak demand has been reduced from 82.2 kW to 74 kW. The peak demand shaved for this case is 8.2 kW.

5. Conclusion

The fixed-threshold and adaptive-threshold control algorithms may fail to shave the peak demand when the high peak demand is higher than that of the predicted value and when the actual peak demand occurs for a much longer duration than the predicted duration. By estimating the useable energy from the batteries and incorporating it into the fuzzy inference rules, the fuzzy-based control algorithm can determine the appropriate output power to be delivered to the load in order to prevent insufficient energy supply from the batteries. The fuzzy-based control algorithm is capable of incorporating a priori qualitative knowledge and expertise about the load characteristic of the buildings as well as the useable energy of the energy storage in peak reduction. The experimental results show that the fuzzy-based control algorithm has achieved the highest performance in peak reduction as compared to that of the fixed-threshold and adaptive-threshold control algorithms.

References

[1] Bradley P, Leach M and Torriti J 2013 A review of the costs and benefits of demand response for electricity in the UK *Energy Policy* **52** 312-27

[2] Cappers P, MacDonald J, Goldman C and Ma O 2013 An assessment of market and policy barriers for demand response providing ancillary services in U.S. electricity markets *Energy Policy* **62** 1031-9

[3] Nolan S and O'Malley M 2015 Challenges and barriers to demand response deployment and evaluation *Appl. Energy* **152** 1-10

[4] de Sa Ferreira R, Barroso L A, Rochinha Lino P, Carvalho M M and Valenzuela P 2013 Time-of-use tariff design under uncertainty in price-elasticities of electricity demand: A stochastic optimization approach *IEEE Trans. Smart Grid* **4** 2285-95

[5] He Y and Zhang J 2015 Real-time electricity pricing mechanism in China based on system dynamics *Energy Convers. Manag.* **94** 394-405

[6] Wang Z. and Li F 2011 Critical peak pricing tariff design for mass consumers in Great Britain *2011 IEEE Power and Energy Society General Meeting* 1-6

[7] Poullikkas A 2013 A comparative overview of large-scale battery systems for electricity storage *Renew. Sustain. Energy Rev.* **27** 778-88

[8] Peter L 2012 Demonstration of ZBB Energy Storage Systems

[9] Koller M, Borsche T, Ulbig A and Andersson G 2015 Review of grid applications with the Zurich 1 MW battery energy storage system *Electr. Power Syst. Res.* **120** 128-35

[10] Masoum A S, Deilami S, Moses P S, Masoum M A S and Abu-Siada A 2011 Smart load management of plug-in electric vehicles in distribution and residential networks with charging stations for peak shaving and loss minimisation considering voltage regulation *IET Gener. Transm. Distrib.* **5** 877-88

[11] Nair N-K C and Garimella N 2010 Battery energy storage systems: Assessment for small-scale renewable energy integration *Energy Build.* **4** 2124-30

[12] Arman N A, Logenthiran T and Woo W L 2015 Intelligent energy management of distributed energy storage systems in microgrid *Smart Grid Technologies-Asia (ISGT ASIA), 2015 IEEE Innovative* 1-6

[13] Chua K H, Lim Y S and Morris S 2015 Cost-benefit assessment of energy storage for utility and customers: A case study in Malaysia *Energy Convers. Manag.* **106** 1071-81

[14] Rahimi A, Zarghami M, Vaziri M and Vadhva S 2013 A simple and effective approach for peak

load shaving using Battery Storage Systems *North American Power Symposium (NAPS)* 1-5
[15] Oudalov A, Cherkaoui R and Beguin A 2007 Sizing and optimal operation of battery energy storage system for peak shaving application *Power Tech, 2007 IEEE Lausanne* 621-5
[16] Purvins A, Papaioannou I T and Debarberis L 2013 Application of battery-based storage systems in household-demand smoothening in electricity-distribution grids *Energy Convers. Manag.* **65** 272-84
[17] Lu C, Xu H, Pan X and Song J 2014 Optimal sizing and control of battery energy storage system for peak load shaving *Energies* **7** 8396-410
[18] Johnson M P, Bar-Noy A, Liu O and Feng Y 2011 Energy peak shaving with local storage *Sustain. Comput. Inform. Syst.* **1** 177-88
[19] Leadbetter J and Swan L 2012 Battery storage system for residential electricity peak demand shaving *Cool Roofs Cool Pavements Cool Cities Cool World* **55** 685-92
[20] Levron Y and Shmilovitz D 2012 Power systems' optimal peak-shaving applying secondary storage *Electr. Power Syst. Res.* **89** 80-4
[21] Oudalov A, Cherkaoui R and Beguin A 2007 Sizing and optimal operation of battery energy storage system for peak shaving application *Power Tech, 2007 IEEE Lausanne* 621-5
[22] Yang Y, Li H, Aichhorn A, Zheng J and Greenleaf M 2014 Sizing strategy of distributed battery Storage system with high penetration of photovoltaic for voltage regulation and peak load shaving *IEEE Trans. Smart Grid* **5** 982-91
[23] Garcia H E *et al* 2016 Dynamic performance analysis of two regional nuclear hybrid energy systems *Energy* **107** 234-58
[24] Wang X, Teichgraeber H, Palazoglu A and El-Farra N H 2014 An economic receding horizon optimization approach for energy management in the chlor-alkali process with hybrid renewable energy generation *J. Process Control* **24** 1318-27
[25] Chen J and Garcia H E 2016 Economic optimization of operations for hybrid energy systems under variable markets *Appl. Energy* **177** 11-24
[26] Upshaw C R, Rhodes J D and Webber M E 2015 Modeling peak load reduction and energy consumption enabled by an integrated thermal energy and water storage system for residential air conditioning systems in Austin, Texas *Energy Build.* **97** 21-32

IOP Conf. Series: Earth and Environmental Science **93** (2017) 012009 doi:10.1088/1755-1315/93/1/012009

Overview of offshore wind farm configurations

Q Wei[1,3], B Wu[1], D Xu[1] and N R Zargari[2]

[1]Ryerson University, 350 Victoria St, Toronto, ON, Canada
[2]Rockwell Automation, 135 Dundas St, Cambridge, ON, Canada

E-mail: qiang.wei@ryerson.ca

Abstract. Offshore wind energy has been attracting great attention. Compared with onshore wind power systems, offshore wind power applications present significantly greater economic challenges mainly due to the required bulky and costly offshore substation. To lower the cost of offshore wind power systems, various configurations are proposed in both industry and academia. The present work investigates existing offshore wind farm configurations.

1. Introduction

Offshore wind energy has been attracting increased attention [1]. There are generally two types of configuration for wind energy conversion systems (WECS), namely, fixed-speed and variable-speed configurations [1]. In the case of fixed-speed system, generator terminals are connected to the grid with no power converter being required. On the contrary, variable-speed system employs power converters for adjusting the generator speed to capture the maximum power from the wind. The variable-speed system features higher energy efficiency and lower mechanical stress, thus becoming the dominant technology in WECS [1].

In general, the whole offshore wind farm consists of two parts. One is the power generation system. The other is the power transmission system. Offshore wind farms are usually located far from the onshore grid connection point. To transmit the power of the offshore wind farm from the offshore collection point to the onshore collection point, both high voltage alternating current (HVAC) and direct current (HVDC) transmission links are used in practice, though with different features. HVAC is the simplest and most economic connection method when the distance between offshore and onshore connection points is less than 50 km. HVDC, on the other hand, is dominating the market when the distance is above 50 km [2].

Figure 1 shows the basic structure of a typical offshore wind farm where only main electrical components are considered [2]. As shown in figure 1, the output of turbine-generator-converter system, that is normally 690 V, is stepped up to 33 kV by a transformer. The 33-kV collection system is then transformed to an HVAC level by another transformer. The HVAC level is then transformed to an HVDC level by an centralized AC/DC power converter (modular multilevel converters (MMC) is dominating the market) [3]. The total captured wind power is then delivered to the onshore collection point by the transmission system. All the transformers, HVDC AC/DC converters, compensators, and energy storage components, if any, are housed in an offshore substation. The offshore substation is very bulky and costly.

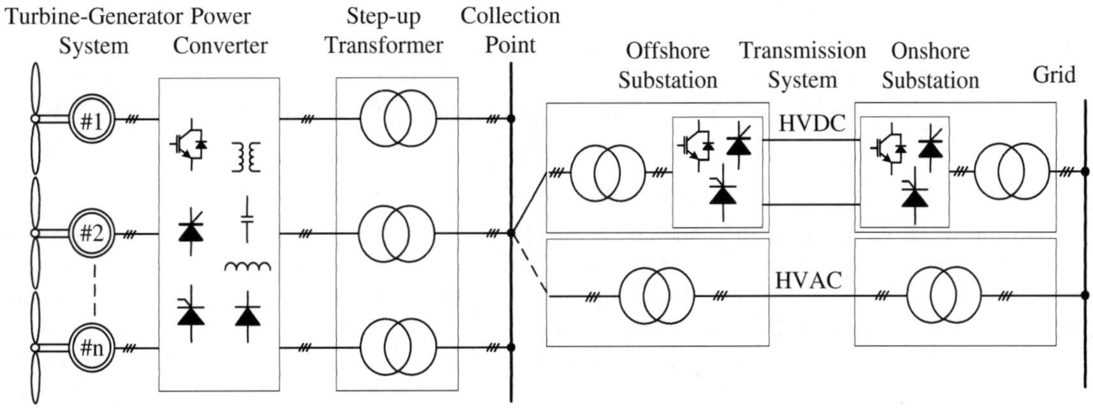

Figure 1. Basic structure of offshore wind farm.

2. Offshore wind farm configurations

The configurations of wind turbines directly affect the cost, efficiency, reliability and performance of an offshore wind farm. A couple of configurations have been proposed and reported and some of them have already been implemented in practice. In general, the offshore wind farm configurations proposed in literature and implemented practically can be classified into four types based on connection of wind turbines and the characteristics of the power to be delivered [4-7]: parallel AC connection with HVAC transmission system, parallel AC connection with HVDC transmission system, parallel DC connection with HVDC transmission system, and series DC connection with HVDC transmission system. In this section, these different types of configurations will be presented and analyzed thoroughly.

2.1. Configuration with parallel AC connection and HVAC transmission

Figure 2 illustrates the configuration with parallel ac connection and HVAC transmission system. The typical output voltage of wind turbines is 690 or 3000 (or 4000) V, which is converted to around 33 kV by the transformer. The outputs of the transformers are connected in parallel to form a medium voltage system, and then stepped up to an HVAC level by transformers housed on the offshore substation [8]. Between offshore and onshore converters is the HVAC transmission system.

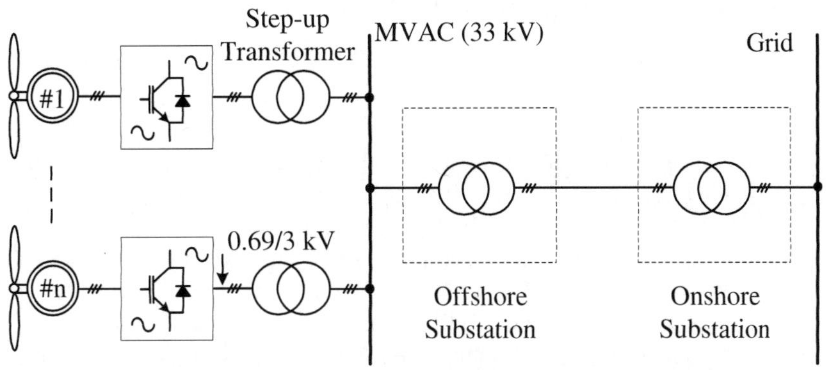

Figure 2. Parallel AC connection and HVAC transmission system.

This configuration is the simplest and most economic connection method when the distance between offshore and onshore connection points is less than 50 km. However, as the distance goes up, it has the following disadvantages [9]. First, the submarine ac cable produces large amount of reactive current due to its high capacitance. As a result, the active current-carrying capacity of the cable is significantly reduced with increasing transmission distances and voltage levels. Thus, reactive power compensations are normally required at both ends. Second, HVAC is unable to directly connect two ac power networks of different frequencies. In addition, the faults on the HVAC cables negatively affect the system and vice versa.

2.2. Configuration with parallel AC connection and HVDC transmission

Figure 3 illustrates the configuration with parallel AC connection and HVDC transmission system. It is the most popular configuration for wind farms which are located far away from onshore and have large power capacity [6] in practice. A step-up transformer is used to convert the MVAC system to HVAC level and then an AC/DC converter is used to convert HVAC to HVDC. All the transformers, AC/DC converters, and other related components are housed in the offshore substation. A centralized DC/AC converter and a step-down transformer are used at the grid side and located at onshore station.

Both voltage source converter (VSC) and current source converter (CSC) can be used in such a configuration. The classical line-commutated converter (LCC)-based configuration where line-commutated CSCs with naturally commutated thyristor valves are used features low cost. However, a couple of disadvantages are associated with this configuration [10]: (1) a relatively strong synchronous voltage source is needed to assist the communication of thyristor valves; (2) the conversion process demands reactive power which needs to be supplied from the large ac filter, shunt banks or series capacitor; and (3) the LCC-HVDC system cannot provide independent control of active and reactive powers. Although the reliability of the LCC-based HVDC is demonstrated by more than 50 years of service experience on-land, its large footprint due to the huge space requirements for the converter station and auxiliary makes it impractical for offshore application.

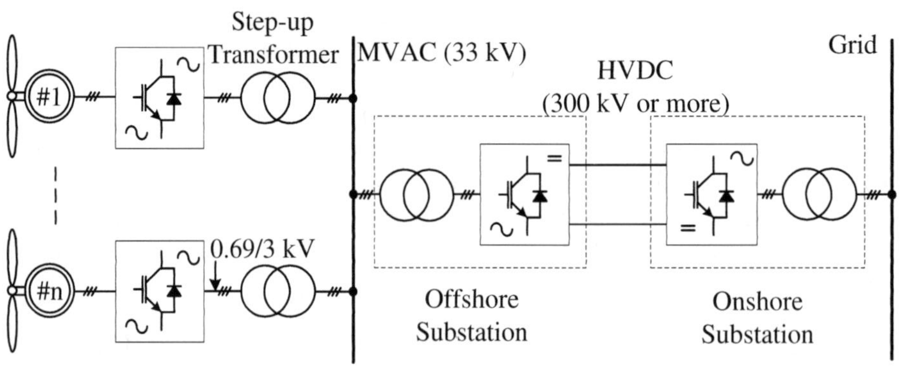

Figure 3. Parallel AC connection and HVDC transmission system.

VSC-based configuration featuring compact size, independent control of real and reactive power, and fast dynamic response [11], on the other hand, is now dominating the market of far located offshore wind farm applications. Two-level VSC with connected in series are mainly used in the traditional VSC-HVDC systems. The critical issue of such a configuration is the voltage balance control among series-connected switching devices. Modular multilevel converter (MMC)-based configuration has become a mainstream in HVDC-based offshore wind farm applications. MMC features modular structure, less derating of semiconductor switches, near sinusoidal output waveforms with low switching frequency, and high efficiency [12]. The main drawbacks for this configuration are significant initial and maintenance costs, especially for the very bulky and costly offshore substation required to house all the step-up transformers, power converters, batteries, and other related components.

2.3. Parallel DC connection and HVDC transmission

Figure 4 shows the configuration with parallel DC and HVDC system where an intermediate DC/AC/DC converter is used to step the low output voltage (1200/5000 V) of the AC/DC converter up to a MVDC level, that is ranging between 30 and 50 kV [13]. A MVDC collection system is formed by connecting the outputs of DC/DC converters in parallel. Then, the MVDC system is stepped up to an HVDC level with the help of a centralized DC/DC converter housed on an offshore substation. The DC/DC converters normally consist of three conversion stages, that are DC/AC, medium/high frequency transformer, and AC/DC. Compared with the parallel AC connection and HVDC system shown in figure 3, the size and weight of this configuration is smaller mainly thanks to the medium/high frequency transformer-based DC/DC converter. This configuration has not been implemented yet.

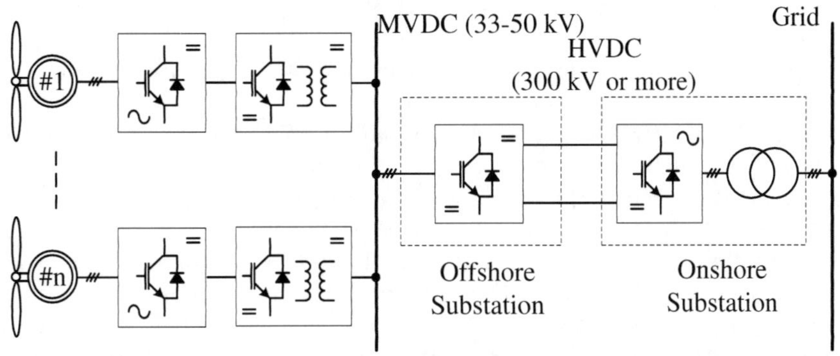

Figure 4. Parallel DC connection and HVDC transmission system.

2.4. Configuration with series DC connection and HVDC transmission

The configuration with series DC connection and HVDC transmission can eliminate the bulky and costly offshore substation, thus greatly lowering the cost of the whole offshore wind farm system. Both CSC- and VSC-based series-connected configurations have been developed for offshore wind farms in the literature. Figure 5 shows the LCC-based series-connected configuration [14]. Such a configuration is less suitable for the offshore wind farm as aforementioned, thus not repeated here.

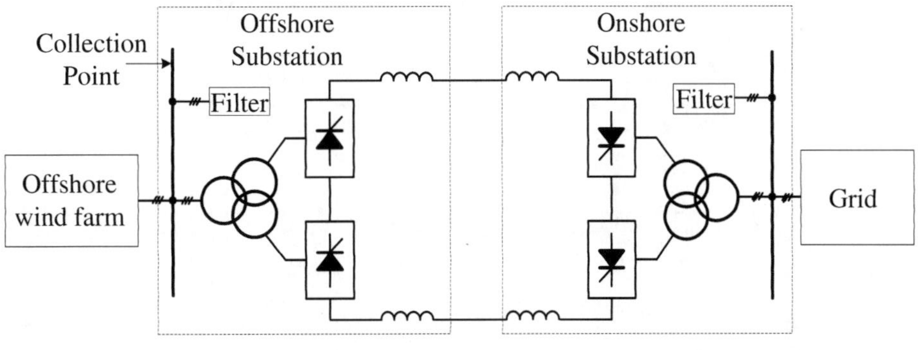

Figure 5. LCC-based series DC connection and HVDC transmission system.

Figure 6 shows the VSC-based series-connected configuration [15]. The DC outputs of turbine-generator AC/DC converters are connected directly in series to reach an HVDC level without any step-up transformers. Compared with the LCC-based configuration shown in figure 5, the VSC-based one features compact size, independent control of real and reactive power, and fast dynamic response.

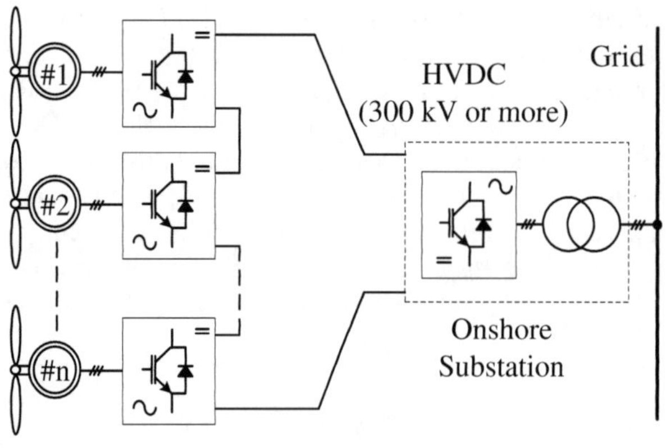

Figure 6. VSC-based series DC connection and HVDC transmission system.

Pulse width modulation (PWM) CSCs are well proven converters in applications of medium-voltage (MV) drives [11]. It features advantages like simple topology, grid-side friendly waveform, and reliable short-circuit protection. It is considered as highly promising converters for wind energy conversion systems. Figure 7 shows such a PWM CSC-based series-connected configuration developed for offshore wind farm [16]. A number of MV turbine-generator units are used offshore. The output DC voltage of each offshore CSC is connected directly in series to reach an HVDC level, then transferred to onshore by transmission cables. A same number of CSCs is connected directly in series to construct a DC/AC converter connected to the grid system through transformers.

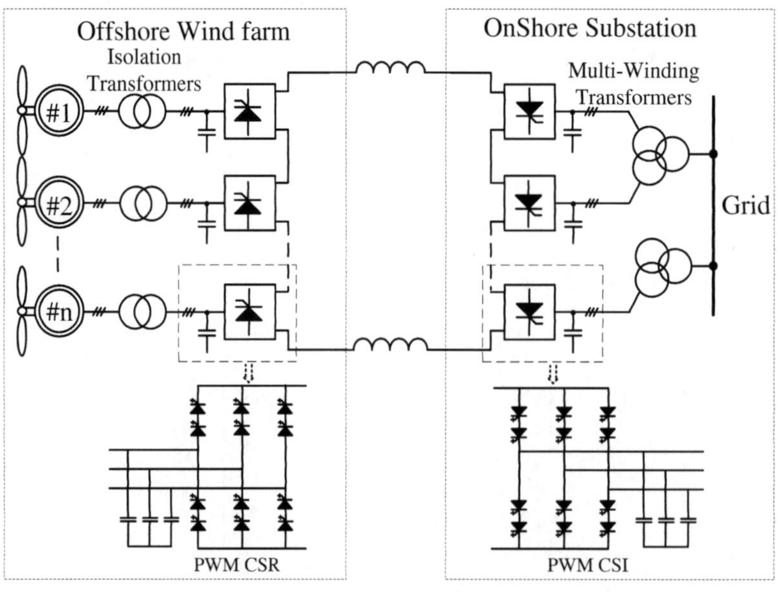

Figure 7. PWM CSC-based series DC connection and HVDC transmission system.

The most attracting feature for series-connected configuration is the elimination of the bulky and costly offshore substation used in existing offshore wind farms. However, the wind generators in this configuration need careful consideration when it comes to the system insulation. The wind generator furthermost from the ground needs to withstand the transmission level. To tackle this issue, a couple of methods are developed [17].

- insulate the generator winding, and offshore converter for high potential (the full transmission voltage) to the ground, or

- insulate the wind turbine tower for high potential (the full transmission voltage) to the ground and keep the nacelle on high potential, or
- using transformers

where the transformer-based solution seems more practical. As shown in figure 7, a three-phase isolation transformer is employed between generator and the front-end CSC. Such a three-phase transformer, however, is a high-power low-frequency transformer which is bulky, especially for the offshore application where the space in the wind turbine is limited.

To lower the size and weight along with enhancing the reliability of the system, a medium-frequency transformer (MFT)-based configuration is proposed for CSC-based offshore wind farm as shown in figure 8 [18]. For each offshore turbine-generator unit, it consists of an MV permanent magnet synchronous generator, a diode rectifier, a MFT-based cascaded DC-DC converter. MV CSC is used at onshore as same as that in figure 7. The MFT-based modular DC-DC converter is composed of a number of H-bridge converters with series input and series output. The three-phase diode rectifier interfacing the generator shown in figure 8 displays the advantages of high reliability, low cost, and small size and weight. The side effect of this passive converter is that it generates a relatively high torque ripple in the generator. However, various methods have been proposed in literature to solve this problem [19]. Furthermore, the synchronous inductance of a permanent magnet synchronous generator is usually above 0.4 per unit (pu) for high-power, low-speed wind applications, which further helps mitigate the torque ripple.

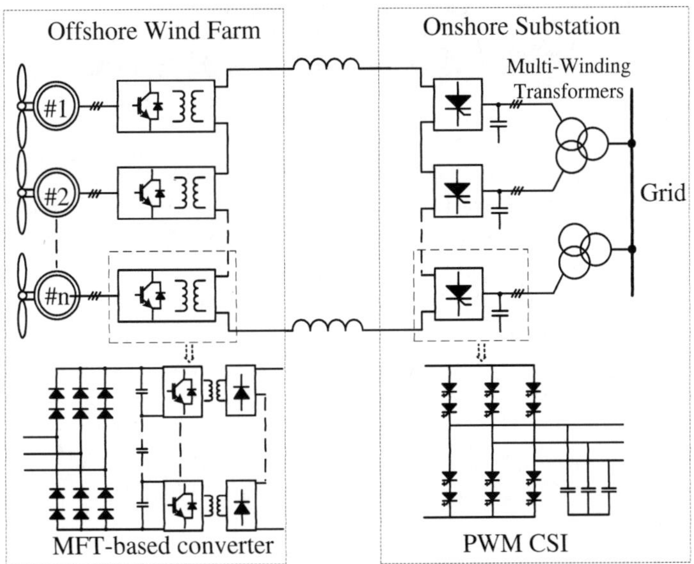

Figure 8. MFT-based series DC connection and HVDC transmission system.

The MFT-based modular DC-DC converter shown in figure 8 plays two roles [18]. First, it is for generator-side control. The primary objective for the generator-side control is to obtain the maximum power input from varying wind speeds. This can be achieved by regulating the modular MFT-based converter. Second, MFT is employed to solve the issue of generator insulation. For a series-connected wind farm, the farthest generator from the ground must withstand the full transmission level. In contrast to the low-frequency transformers shown in figure 7, MFT gives advantages of smaller size and weight. Furthermore, a modular design is implemented based on a number of cells that are connected in series at input and output. This helps reduce the burden of transformers implementation as one transformer only accounts for one part of a megawatt-level power. The modular design of the DC-DC converter also contributes to the choice of low-cost, low-voltage switching devices instead of high-voltage ones, and increasing the reliability and flexibility of the system.

3. Summary of different configurations

Apart from considering reliability and efficiency as the main requisites for all onshore conversion systems, the footprints and weights of the components are particularly important in offshore infrastructure [20]. In offshore wind farms, the offshore substation usually needs to accommodate step-up transformers, HVDC converters and compensator, depending on the type of wind farm configuration. These offshore substations are very bulky and very costly. Figure 9 shows one of the world's large offshore substations used for offshore wind farm "Dolwin2" commissioned by ABB [21]. The complete platform including substructure weighs around 23,000 tons and is around 100 meters long, 70 meters wide and 100 meters tall.

Figure 9. Offshore substation used in "Dolwin2" by ABB [21].

Compared with parallel-connected configurations shown in figures 2-4 where the offshore substation is required, the series-connected configuration shown in figures 5-8 is more attracting as the offshore substation can be eliminated which is more important for offshore applications.

Among the series-connected configurations, the thyristor-based one requires a relatively strong synchronous voltage source to assist the communication of thyristor valves, needs large ac filter, shunt banks or series capacitor for successful conversion process, and cannot provide independent control of active and reactive power. The PWM CSC- and VSC-based ones, on the other hand, are more promising due to their high dynamic performance and independent active and reactive power control.

However, both PWM CSC- and VSC-based series-connected configurations require low-frequency transformers to solve the issue of generator insulation. Such low-frequency transformers are high-power rated level, and are heavy and bulky, that adding burden to offshore infrastructure as the space in the wind turbine is limited.

Table 1. Summary of investigated offshore wind farm configurations.

Parallel AC connection with HVAC transmission	Parallel AC connection with HVDC transmission	Parallel DC connection with HVDC transmission	Parallel DC connection with HVDC transmission	
			Conventional configuration	MFT-based configuration
Offshore substation	Offshore substation	Offshore substation	No offshore substation	No offshore substation
Transmission distance <50 km	Transmission distance >50 km	Transmission distance >50 km	Transmission distance >50 km	Transmission distance >50 km
			Bulky and heavy low-frequency transformer	MFTs with smaller size and weight

The recently proposed MFT-based configuration helps solve the above issues. By employing an MFT-based converter interfacing the generator, the issue of the generator insulation is solved and the size and weight of the MFT are greatly smaller compared with the low-frequency transformer. In addition, the modular structure contributes to the implementation of MFTs in practice as each MFT only accounts for one part of megawatt-level power. In summary, the investigated configurations are summarized in table 1.

4. Conclusions

In this work, existing offshore wind farm configurations are investigated. Apart from considering reliability and efficiency as the main requisites for all onshore conversion systems, the footprints and weights of the components are particularly important in offshore infrastructure. On this basis, the comparison among investigated configurations is carried out that the series-connected one is an attractive choice for future offshore wind farms as the offshore substation used in existing offshore wind farms is not needed.

References

[1] Wu B, Lang Y, Zargari N and Kouro S 2011 *Power Conversion and Control of Wind Energy Systems* (New York: Wiley-IEEE Press)
[2] Friedrich K 2010 Modern HVDC PLUS application of VSC in modular multilevel converter topology *IEEE Int Symp on Ind Electron* 3807-10
[3] Das D, Pan J and Bala S 2012 HVDC light for large offshore wind farm integration *IEEE Symp (PEMWA)* 1-7
[4] Yaramasu V, Wu B and Sen P C 2015 High-power wind energy conversion systems state-of-the-art and emerging technologies *Proc. of the IEEE* **103** 740-88
[5] Lundberg S 2006 Evaluation of wind farm layouts *European Power Electronics and Drives Association* **16** 14
[6] Blaabjerg F, Chen Z and Kjaer S 2004 Power electronics as efficient interface in dispersed power generation systems *IEEE Trans Power Electron* **19** 1184-94
[7] Liserre M, Cardenas R, Molinas M and Rodriguez J 2011 Overview of multi-MW wind turbines and wind parks *IEEE Trans Ind Electron* **58** 1081-95
[8] Bresesti P, Kling W, Hendriks R and Vailati R 2007 HVDC connection of offshore wind farms to the transmission system *IEEE Trans Energy Con* **22** 37-43
[9] Chen Z, Guerrero J and Blaabjerg F 2009 A review of the state of the art of power electronics for wind turbines *IEEE Trans Power Electron* **24** 1859-75
[10] Carrasco J *et al* 2006 Power-electronic systems for the grid integration of renewable energy sources: A survey *IEEE Trans Ind Electron* **53** 1002-16
[11] Wu B 2006 *High-Power Converters and AC Drives* (NJ: Wiley/IEEE Press)
[12] Debnath S, Qin J, Bahrani B, Saeedifard M and Barbosa P 2015 Operation, control, and applications of the modular multilevel converter: A review *IEEE Trans Power Electron* **30** 37-53
[13] Meyer C, Hoing M, Peterson A and De Doncker R 2007 Control and design of DC grids for offshore wind farms *IEEE Trans Ind Appl* **43** 1475-82
[14] Bozhko S V, Blasco-Gimenez R, Li R, Clare J C and Asher G M 2007 Control of offshore DFIG-based wind farm grid with line-commutated HVDC connection *IEEE Trans Energy Con* **22** 71-8
[15] Xu L, Yao L Z and Sasse C 2006 Power electronics options for large wind farm integration: VSC-based HVDC transmission *IEEE Power Sys Conf and Expo (PSCE)* 760-7
[16] Popat M, Wu B, Liu F and Zargari N 2012 Coordinated control of cascaded current source converter based offshore wind farms *IEEE Trans Sus Energy* **3** 557-65
[17] Lundberg S 2003 Configuration study of large wind parks (Goteborg, Sweden: Chalmers Univ. Technol., Dept. of Elect, Power Engg)

[18] Wei Q, Wu B, Xu D and Zargari N 2017 A medium frequency transformer-based wind energy conversion system used for current source converter based offshore wind farm *IEEE Trans on Power Electronic* **32** 248-59
[19] Xia Y, Fletcher J, Finney S, Ahmed K and Williams B 2011 Torque ripple analysis and reduction for wind energy conversion systems using uncontrolled rectifier and boost converter *IET Renewable Power Generation* **5** 377-86
[20] Blaabjerg F and Ma K 2013 Future on power electronics for wind turbine systems *IEEE J Emerging and Selected Topics in Power Electronics* **1** 139-52
[21] Online: http://new.abb.com/systems/offshore-wind-connections

IOP Conf. Series: Earth and Environmental Science **93** (2017) 012010 doi:10.1088/1755-1315/93/1/012010

Strategy of investment in electricity sources--Market value of a power plant and the electricity market

R Bartnik[1], A Hnydiuk-Stefan[1,2] and Z Buryn[1]

[1]Opole University of Technology, Faculty of Production Engineering and Logistics, Department of Power Engineering Management

E-mail: a.hnydiuk-stefan@po.opole.pl

Abstract. This paper reports the results of the investment strategy analysis in different electricity sources. New methodology and theory of calculating the market value of the power plant and value of the electricity market supplied by it are presented. The financial gain forms the most important criteria in the assessment of an investment by an investor. An investment strategy has to involve a careful analysis of each considered project in order that the right decision and selection will be made while various components of the projects will be considered. The latter primarily includes the aspects of risk and uncertainty. Profitability of an investment in the electricity sources (as well as others) is offered by the measures applicable for the assessment of the economic effectiveness of an investment based on calculations e.g. power plant market value and the value of the electricity that is supplied by a power plant. The values of such measures decide on an investment strategy in the energy sources. This paper contains analysis of exemplary calculations results of power plant market value and the electricity market value supplied by it.

1. Introduction

The principal objective of running a business is to gain a profit. This goal of making a profit involves a need of following an investment strategy.

Investment strategies are decisions which play a fundamental role on the financial condition of the investors. Such strategies mean that an expense needs to be made, funds designed for an investment are frozen for a long period, an effect is obtained with a certain delay and they carry some degree of risk. An investment strategy has to involve a careful analysis of each considered project in order that the right decision and selection will be made while various components of the projects will be considered. The latter primarily includes the aspects of risk and insecurity, particularly in variable economic conditions. Financial gain forms the most important criterion in the assessment of an investment by an investor. Before making a decision about the investment of financial resources, an investor needs to ensure that the internal return rate on an investment will be high enough. If this is not the case, the investor will not make any decision. The answer to the question about the profitability of an investment in the electricity sources (as well as others) is offered by the measures applicable for the assessment of the economic effectiveness of an investment based on calculations. They include e.g. the market value of a power plant and the value of the electricity that is supplied by a power plant [1,2]. The values of such measures decide on an investment strategy in the energy sources.

The market value of a power plant J_{rw} is the price at which it can be sold/purchased by the owner/buyer, for which both the owner and the buyer can gain a profit that is satisfactory for them.

The value of the electricity market (and not only this) is expressed by the value of *NPV* in terms of financial resources, which can be gained from investing funds in it. The basic criterion applied for the assessment of the value of all enterprises (as well as energy sources) is associated with the Internal Rate of Return (*IRR*) obtained on the invested capital [1,2]. A satisfactory profit for the investor who purchases or builds a power plant (further called the *IPP – Independent Power Producer*) is expressed by the adequate value of the interest rate $\boldsymbol{IRR}_{rw}^{IPP}$ on the capital J_{rw} that is intended to be invested in it. This interest rate has to be higher than the profit that could be gained as a result of investing financial resources J_{rw} on the capital market. A satisfactory profit for the one who sells a power plant is represented by the adequately high share v_m (i.e. relative value of the market) in the joint ownership of a power plant which is preserved by them. The values of $\boldsymbol{IRR}_{rw}^{IPP}$ and v_m are very closely related. By purchasing a power plant, *IPP* also takes on the financial obligations arising from the loan that was obtained to finance the construction of the power according to the ratio of their stakes in the ownership $(1 - v_m)$.

2. Methodology of calculating the market value of the power plant and value of the electricity market supplied by it

The methodology applied to calculate the market value of the power plant and the value of the electricity market supplied by electricity production in it are presented in [1]. The formula applied to express the production cost of electricity $k_{el,av}$, which determines these values, as well as the formulae to calculate the market value of a power plant J_{rw} and the value of the market *NPV* supplied by it [1], are found below:

- mean specific cost $k_{el,av}$ of electricity production in a power plant

$$
\begin{aligned}
k_{el,av} = \Bigg\{ & (1+x_{sw,m,was})\frac{e_{fuel}^{t=0}}{a_{fuel}-r}[\mathrm{e}^{(a_{fuel}-r)T}-1]+\frac{\rho_{CO_2} p_{CO_2}^{t=0}}{a_{CO_2}-r}[\mathrm{e}^{(a_{CO_2}-r)T}-1]+ \\
& +\frac{\rho_{CO}\, p_{CO}^{t=0}}{a_{CO}-r}[\mathrm{e}^{(a_{CO}-r)T}-1]+\frac{\rho_{NO_X} p_{NO_X}^{t=0}}{a_{NO_X}-r}[\mathrm{e}^{(a_{NO_X}-r)T}-1]+\frac{\rho_{SO_2} p_{SO_2}^{t=0}}{a_{SO_2}-r}[\mathrm{e}^{(a_{SO_2}-r)T}-1]+ \\
& +\frac{\rho_{dust} p_{dust}^{t=0}}{a_{dust}-r}[\mathrm{e}^{(a_{dust}-r)T}-1]+(1-u)\frac{\rho_{CO_2} e_{CO_2}^{t=0}}{b_{CO_2}-r}[\mathrm{e}^{(b_{CO_2}-r)T}-1]+ \\
& +\frac{\eta_{el}}{rt_A}\left[(1+x_{sal,t,ins})\frac{J}{N_{el}}(1-\mathrm{e}^{-rT})\delta_{serv}+\frac{rzJ}{N_{el}}(\frac{1-\mathrm{e}^{-rT}}{T}+1)\right]\Bigg\}\frac{r}{\eta_{el}(1-\varepsilon_{el})(1-\mathrm{e}^{-rT})}
\end{aligned}
\quad (1)
$$

- market value of a power plant J_{rw}

$$J_{rw} = \Big\{ N_{el}(1-\varepsilon_{el})t_A e_{el}^{t=0}\frac{1}{a_{el}-IRR_{p,rw}^{IPP}}[e^{(a_{el}-IRR_{p,rw}^{IPP})T}-1] - \frac{N_{el}t_A}{\eta_{el}}(1+x_{sw,m,was})e_{fuel}^{t=0}\frac{1}{a_{fuel}-IRR_{p,rw}^{IPP}}[e^{(a_{fuel}-IRR_{p,rw}^{IPP})T}-1] +$$
$$-\frac{N_{el}t_A}{\eta_{el}}\rho_{CO_2}p_{CO_2}^{t=0}\frac{1}{a_{CO_2}-IRR_{p,rw}^{IPP}}[e^{(a_{CO_2}-IRR_{p,rw}^{IPP})T}-1] - \frac{N_{el}t_A}{\eta_{el}}\rho_{CO}p_{CO}^{t=0}\frac{1}{a_{CO}-IRR_{p,rw}^{IPP}}[e^{(a_{CO}-IRR_{p,rw}^{IPP})T}-1]+$$
$$-\frac{N_{el}t_A}{\eta_{el}}\rho_{NO_X}p_{NO_X}^{t=0}\frac{1}{a_{NO_X}-IRR_{p,rw}^{IPP}}[e^{(a_{NO_X}-IRR_{p,rw}^{IPP})T}-1] - \frac{N_{el}t_A}{\eta_{el}}\rho_{SO_2}p_{SO_2}^{t=0}\frac{1}{a_{SO_2}-IRR_{p,rw}^{IPP}}[e^{(a_{SO_2}-IRR_{p,rw}^{IPP})T}-1]+ \quad (2)$$
$$-\frac{N_{el}t_A}{\eta_{el}}\rho_{dust}p_{dust}^{t=0}\frac{1}{a_{dust}-IRR_{p,rw}^{IPP}}[e^{(a_{dust}-IRR_{p,rw}^{IPP})T}-1] - \frac{N_{el}t_A}{\eta_{el}}(1-u)\rho_{CO_2}e_{CO_2}^{t=0}\frac{1}{b_{CO_2}-IRR_{p,rw}^{IPP}}[e^{(b_{CO_2}-IRR_{p,rw}^{IPP})T}-1]+$$
$$-J(1+x_{sal,t,ins})\frac{\delta_{serv}}{IRR_{p,rw}^{IPP}}(1-e^{-IRR_{p,rw}^{IPP}T})\Big\}\frac{(1-p)(1-v_m)}{A}$$

where:

$$A = \frac{(1+IRR_{p,rw}^{IPP})^{b+1}-1}{(b+1)IRR_{p,rw}^{IPP}}\left(1+\frac{1-e^{-IRR_{p,rw}^{IPP}T}}{T}\right)+ \quad (3)$$
$$+\frac{(1+r)^{b+1}-1}{(b+1)r}\left[\frac{r}{IRR_{p,rw}^{IPP}}+\frac{1-e^{-IRR_{p,rw}^{IPP}T}}{T}\left(\frac{r}{IRR_{p,rw}^{IPP}}+\frac{1}{IRR_{p,rw}^{IPP}}-\frac{r}{(IRR_{p,rw}^{IPP})^2}\right)\right][(1-p)(1-v_m)-1]$$

- value of the market for IPP

$$NPV^{IPP} = \Big\{ N_{el}(1-\varepsilon_{el})t_A e_{el}^{t=0}\frac{1}{a_{el}-r}[e^{(a_{el}-r)T}-1] - \frac{N_{el}t_A}{\eta_{el}}(1+x_{sw,m,was})e_{fuel}^{t=0}\frac{1}{a_{fuel}-r}[e^{(a_{fuel}-r)T}-1]+$$
$$-\frac{N_{el}t_A}{\eta_{el}}\rho_{CO_2}p_{CO_2}^{t=0}\frac{1}{a_{CO_2}-r}[e^{(a_{CO_2}-r)T}-1] - \frac{N_{el}t_A}{\eta_{el}}\rho_{CO}p_{CO}^{t=0}\frac{1}{a_{CO}-r}[e^{(a_{CO}-r)T}-1]+$$
$$-\frac{N_{el}t_A}{\eta_{el}}\rho_{NO_X}p_{NO_X}^{t=0}\frac{1}{a_{NO_X}-r}[e^{(a_{NO_X}-r)T}-1] - \frac{N_{el}t_A}{\eta_{el}}\rho_{SO_2}p_{SO_2}^{t=0}\frac{1}{a_{SO_2}-r}[e^{(a_{SO_2}-r)T}-1]+ \quad (4)$$
$$-\frac{N_{el}t_A}{\eta_{el}}\rho_{dust}p_{dust}^{t=0}\frac{1}{a_{dust}-r}[e^{(a_{dust}-r)T}-1] - \frac{N_{el}t_A}{\eta_{el}}(1-u)\rho_{CO_2}e_{CO_2}^{t=0}\frac{1}{b_{CO_2}-r}[e^{(b_{CO_2}-r)T}-1]+$$
$$-J(1-e^{-rT})(1+x_{sal,t,ins})\frac{\delta_{serv}}{r} - Jz\frac{1-e^{-rT}}{T}+1)\Big\}(1-p)(1-v_m)$$

where: a_{el}, a_{fuel}, a_{CO_2}, a_{CO}, a_{SO_2}, a_{NO_X}, a_{dust}, b_{CO_2} – values of the exponents representing the variability in time of the prices of electricity, fuels, tariffs on the use of the environment and purchasing CO_2 emission allowances, b – duration of the construction period of a power plant in years, $e_{el}^{t=0}$, $e_{fuel}^{t=0}$, $e_{CO_2}^{t=0}$, $p_{CO_2}^{t=0}$, etc. – initial values of the of electricity, coal, CO_2 emission prices and tariffs on the use of the environment, J – actual investment made in the construction of a power plant, N_{el} – gross electrical capacity of the power plant, NPV – net present value of the market (total profit over T years of the operation of a plant), p – rate of income tax, r – interest rate on investment capital, t_A – annual exploitation time, T – calculated period for exploitation of a power plant (in years), u – ratio of chemical energy of coal in its total annual use, for which the purchase of CO_2 emission allowances is not required, v_m – relative value of the electricity market, $x_{sw,m,was}$ – coefficient used to account for the cost of supplementing water, auxiliary materials, disposal of wastewater, slag and waste storage, $x_{sal,t,ins}$ – coefficient used to account for the cost of remuneration, taxes, insurance, etc., z – discounting coefficient (coefficient of capital freezing) on the investment J at the instant of the completion of power plant construction, $_{el}$ – internal electrical load of a power plant, δ_{serv} – annual rate of fixed cost relative to the value of the investment (cost of maintenance and overhaul of equipment), η_{el} – energy efficiency of a power plant, ρ_{CO_2}, ρ_{CO}, ρ_{NO_x}, ρ_{SO_2}, ρ_{dust} – emission of CO_2, CO, NO_x, SO_2, particulate matter per specific unit of chemical energy of the coal.

The market value of a power plant J_{rw} is represented by the function of the relative value of the market v_m, rates: r and IRR_{rw}^{IPP}, prices of energy sources and environmental costs.

If we substitute the actual investment J associated in the construction of a power plant in the place of J_{rw} in formula (2) (according to the definition, i.e. under the assumption that no profit is obtained from the operation of a power plant and the rate of the income tax is equal to zero, $p = 0$), for $v_m = 0$ we obtain the total value of the interest rate *IRR*, which is obtained on the capital *J*.

The formula expressing the total value of the *NPV* market is derived from (4) for $v_m = 0$. For the seller of a power plant, the value of the market is equal to *NPV* v_m.

3. Discussion and analysis of results of exemplary calculations

The analysis reported here involves a newly built supercritical steam power plant unit as shown in figure 1.

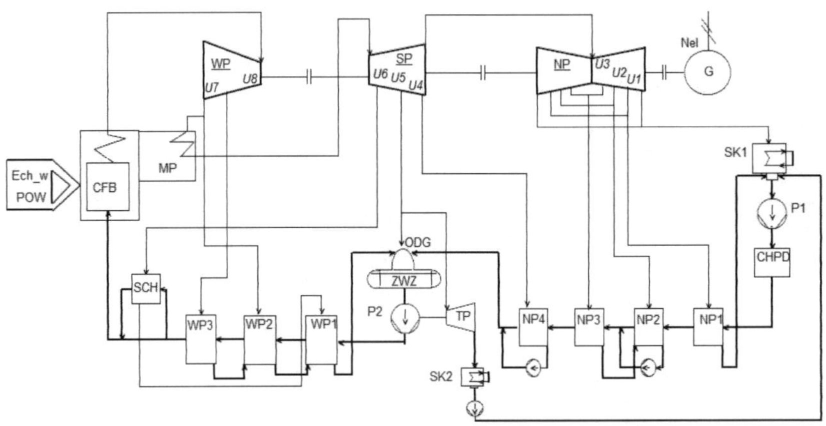

Figure 1. Schematic diagram of a supercritical pressure unit.
Ech_w – chemical energy of coal, POW – flow rate of air, CFB – fluidized boiler, MP – inter-stage steam reheater, WP – HP stage in the main turbine, SP – IP stage in the main turbine, NP – LP stage in the main turbine, U1-U8 – respective steam extractions from the main turbine, G – electricity generator, N_{el} – generator's active power, SK1 – main condenser, P1 – condensate pump, CHPD – cooler of steam from labyrinth seals, NP1 – NP4 – LP regeneration reheaters, TP – auxiliary turbine, SK2 – condenser in auxiliary turbine, ODG – deaerator, ZWZ – feedwater tank, P2 – feedwater pump, WP1 – WP3 – HP regeneration reheaters, SCH – extraction steam cooler.

Table 1 contains a summary of input data used for calculations.

Table 1. Summary of input data for the analyzed supercritical pressure power unit.

1.	Gross capacity of steam power unit N_{el} = 460 MW
2.	Investment per power unit J = 747.5 mln USD
3.	Annual exploitation time of power unit t_A = 7500 h
4.	Internal electrical load of power unit: ε_{el} = 7.6%
5.	Discount rate r = 7%, rate of income tax p = 19%
6.	Construction period of power unit b = 5 years

7.	Exploitation time $T = 20$ years
8.	Annual rate of maintenance and overhaul $\delta_{serv} = 3\%$.
9.	Total cost $x_{sal,t,ins} = 0.25$, total cost $x_{sw,m,was} = 0.02$.
10.	Specific coal price $e_{fuel} = 2.85$ USD/GJ
11.	Price per CO_2 emissions: $e_{CO_2} = 7.35$ USD/Mg$_{CO_2}$
12.	Tariff rates per emissions of: $p_{CO_2} = 0.0725$ USD/Mg$_{CO_2}$, $p_{CO} = 27.5$ USD/Mg$_{CO}$, $p_{NO_x} = 132.5$ USD/Mg$_{NO_X}$, $p_{SO_2} = 132.5$ USD/Mg$_{SO_2}$, $p_{dust} = 87.5$ USD/Mg$_{dust}$.
13.	Emissions from carbon coal combustion: $\rho_{CO_2} = 95$ kg/GJ, $\rho_{CO} = 0.01$ kg/GJ, $\rho_{NO_x} = 0.164$ kg/GJ, $\rho_{SO_2} = 0.056$ kg/GJ, $\rho_{dust} = 0.007$ kg/GJ.
14.	Ratio of chemical energy of coal in its total use for which the purchase of CO_2 allowances is not required $u = 0$.

3.1. Market value of critical parameter power plant and value of the electricity market supplied by it

In the calculations, we adopted zero values of the exponents: $a_{el} = 0$, $a_{fuel} = 0$, $a_{CO_2} = 0$, $a_{CO} = 0$, $a_{SO_2} = 0$, $a_{NO_X} = 0$, $a_{dust} = 0$, $b_{CO_2} = 0$. Therefore, it was assumed that all prices remain at a constant level throughout the T number of years over which a power plant is exploited. The results of exemplary calculations are presented in figures 2-7. The specific cost of electricity production in a power unit derived on the basis of formula (1) is equal to $k_{el,av}$ = 66.88 USD/MWh.

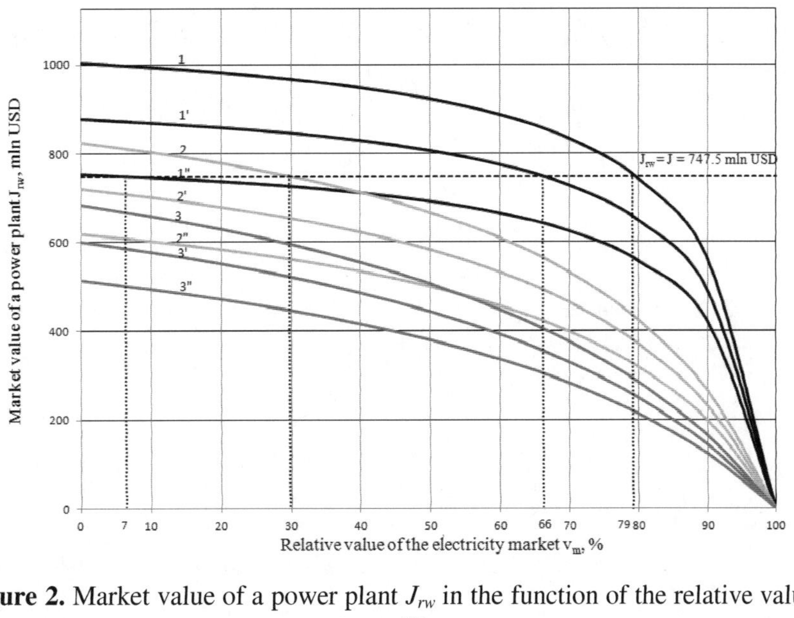

Figure 2. Market value of a power plant J_{rw} in the function of the relative value of the market v_m and the interest rate IRR_{rw}^{IPP} on capital J_{rw} demanded by IPP as a parameter.

1 – e_{el} =80 USD/MWh, $IRR_{rw}^{IPP} = 0.08$; 1' – e_{el} =75 USD/MWh, $IRR_{rw}^{IPP} = 0.08$; 1" – e_{el} =70 USD/MWh, $IRR_{rw}^{IPP} = 0.08$; 2 – e_{el} =80 USD/MWh, $IRR_{rw}^{IPP} = 0.1$; 2' – e_{el} =75 USD/MWh, $IRR_{rw}^{IPP} = 0.1$; 2" – e_{el} =70 USD/MWh, $IRR_{rw}^{IPP} = 0.1$; 3 – e_{el} =80 USD/MWh, $IRR_{rw}^{IPP} = 0.12$; 3' – e_{el} =75 USD/MWh, $IRR_{rw}^{IPP} = 0.12$; 3" – e_{el} =70 USD/MWh, $IRR_{rw}^{IPP} = 0.12$.

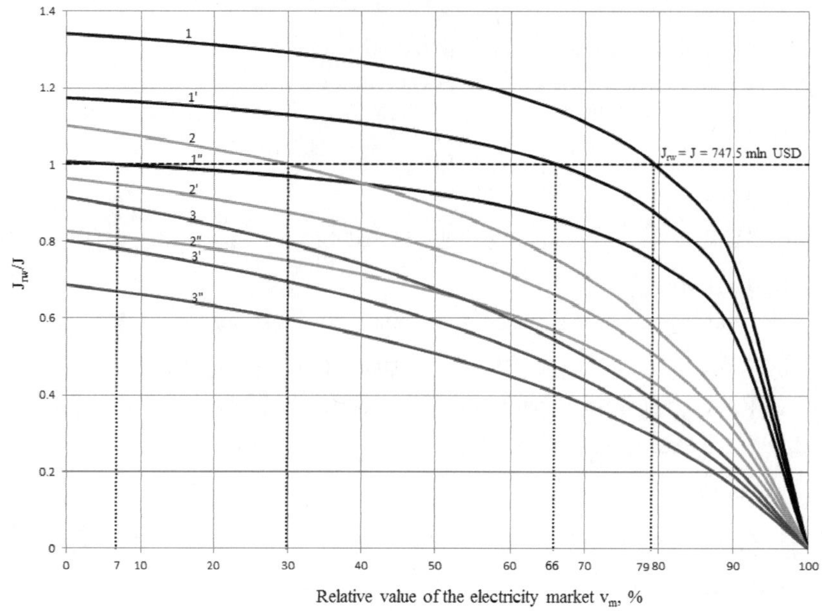

Figure 3. Ratio of the market value of a power plant J_{rw} in relation to the actual investment J made in its construction in the function of the relative value of the market v_m and the interest rate IRR_{rw}^{IPP} on capital J_{rw} demanded by IPP as a parameter.
1 – e_{el}=80 USD/MWh, $IRR_{rw}^{IPP} = 0.08$; 1' – e_{el}=75 USD/MWh, $IRR_{rw}^{IPP} = 0.08$; 1" – e_{el}=70 USD/MWh, $IRR_{rw}^{IPP} = 0.08$; 2 – e_{el}=80 USD/MWh, $IRR_{rw}^{IPP} = 0.1$; 2' – e_{el}=75 USD/MWh, $IRR_{rw}^{IPP} = 0.1$; 2" – e_{el}=70 USD/MWh, $IRR_{rw}^{IPP} = 0.1$; 3 – e_{el}=80 USD/MWh, $IRR_{rw}^{IPP} = 0.12$; 3' – e_{el}=75 USD/MWh, $IRR_{rw}^{IPP} = 0.12$; 3" – e_{el}=70 USD/MWh, $IRR_{rw}^{IPP} = 0.12$.

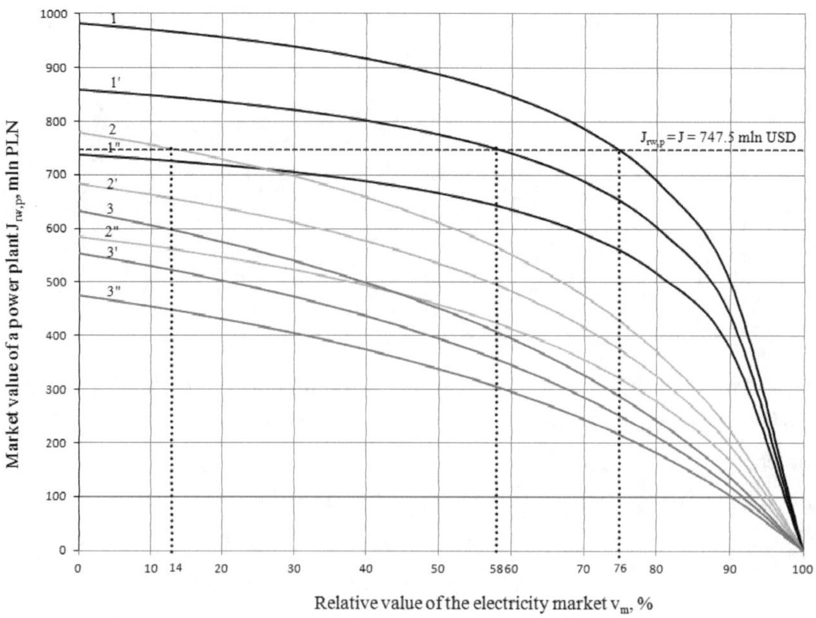

Figure 4. Market value of a power plant $J_{rw,p}$ in the function of the relative value of the market v_m and the interest rate $IRR_{p,rw}^{IPP}$ on capital $J_{rw,p}$ demanded by the investor (IPP) as a parameter.
1 – e_{el} = 80 USD/MWh, $IRR_{p,rw}^{IPP}$ = 0.08; 1' – e_{el} =75 USD/MWh, $IRR_{p,rw}^{IPP}$ = 0.08; 1" – e_{el} =70 USD/MWh, $IRR_{p,rw}^{IPP}$ = 0.08; 2 – e_{el} =80 USD/MWh, $IRR_{p,rw}^{IPP}$ = 0.1; 2' – e_{el} =75 USD/MWh, $IRR_{p,rw}^{IPP}$ = 0.1; 2" – e_{el} =70 USD/MWh, $IRR_{p,rw}^{IPP}$ = 0.1; 3 – e_{el} =80 USD/MWh, $IRR_{p,rw}^{IPP}$ = 0.12; 3' – e_{el} =75 USD/MWh, $IRR_{p,rw}^{IPP}$ = 0.12; 3" – e_{el} =70 USD/MWh, $IRR_{p,rw}^{IPP}$ = 0.12

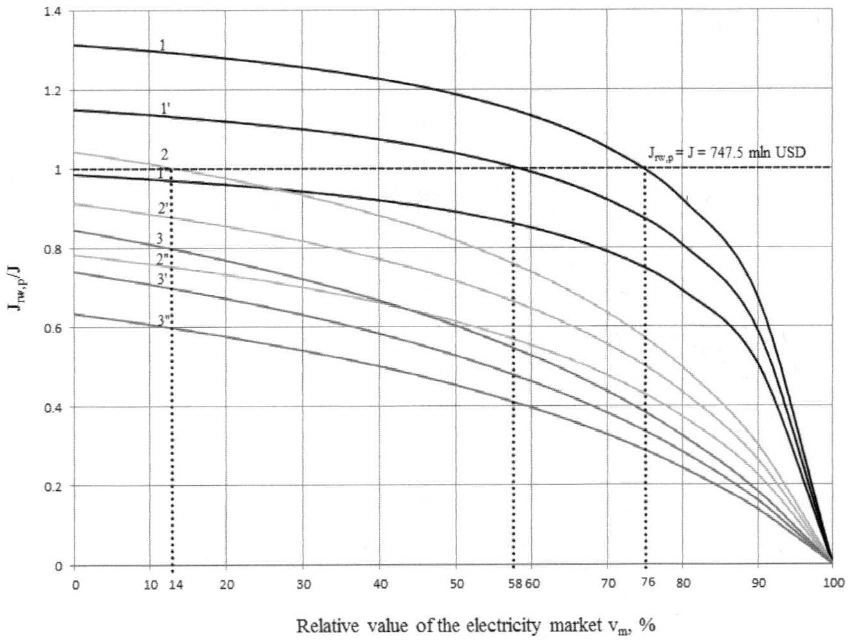

Figure 5. Ratio of the market value of a power plant $J_{rw,p}$ in relation to the actual investment J made in its construction in the function of the value of the market v_m and the interest rate $IRR_{p,rw}^{IPP}$ on capital J_{rw} demanded by the investor (IPP) as a parameter.
1 – e_{el} = 80 USD/MWh, $IRR_{p,rw}^{IPP}$ = 0.08; 1' – e_{el} = 75 USD/MWh, $IRR_{p,rw}^{IPP}$ = 0.08; 1" – e_{el} = 70 USD/MWh, $IRR_{p,rw}^{IPP}$ = 0.08; 2 – e_{el} = 80 USD/MWh, $IRR_{p,rw}^{IPP}$ = 0.1; 2' – e_{el} = 75 USD/MWh, $IRR_{p,rw}^{IPP}$ = 0.1; 2" – e_{el} = 70 USD/MWh, $IRR_{p,rw}^{IPP}$ = 0.1; 3 – e_{el} = 80 USD/MWh, $IRR_{p,rw}^{IPP}$ = 0.12; 3' – e_{el} = 75 USD/MWh, $IRR_{p,rw}^{IPP}$ = 0.12; 3" – e_{el} = 70 USD/MWh, $IRR_{p,rw}^{IPP}$ = 0.12

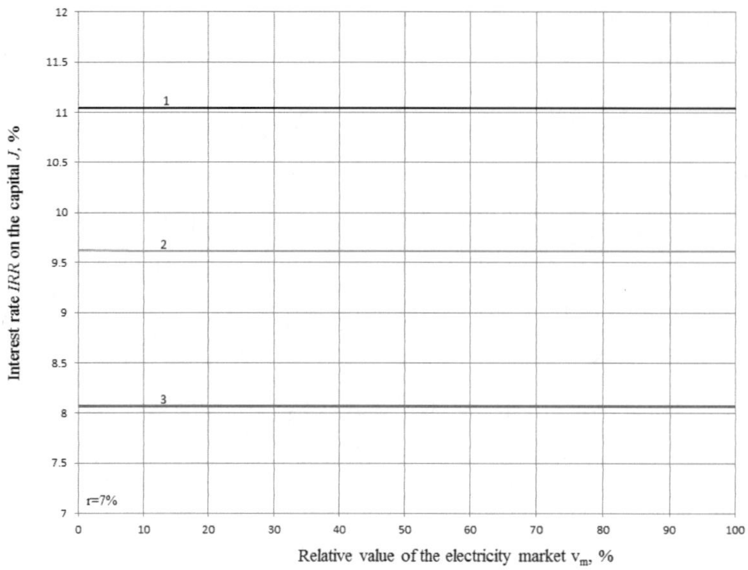

Figure 6. Interest rate *IRR* on capital *J*.
1 – price of electricity e_{el} = 80 USD/MWh; 2 – e_{el} = 75 USD/MWh; 3 – e_{el} = 70 USD/MWh; ($IRR \leq r$ for $e_{el} \leq k_{el}$).

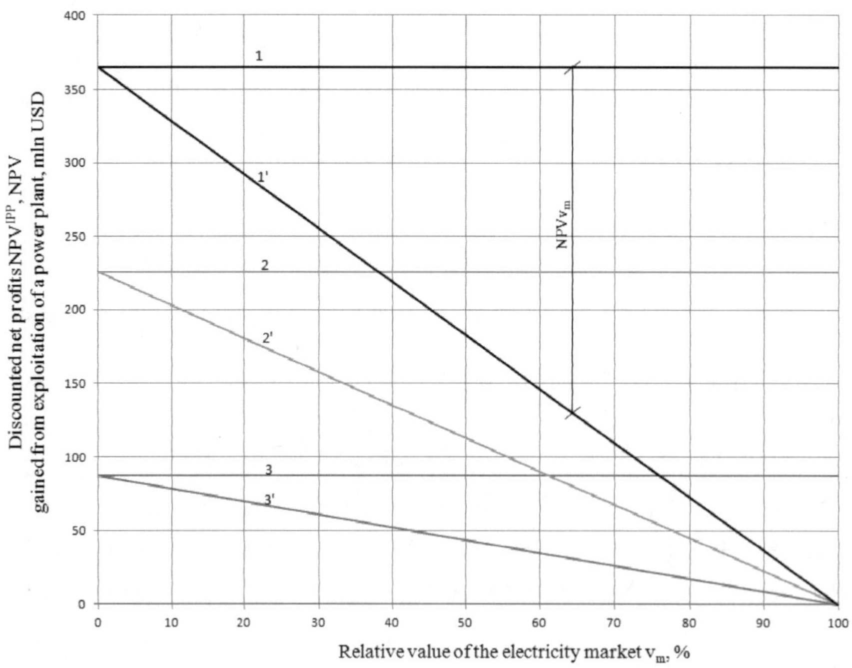

Figure 7. Discounted net profits NPV, NPV^{IPP}, NPVv$_m$ gained from exploitation of a power plant in the function of the relative value of the market v_m.
1 – data regarding NPV for e_{el} = 80 USD/MWh; 1'– data regarding NPV^{IPP} for e_{el} = 80 USD/MWh; 2 – data regarding NPV for e_{el} = 75 USD/MWh; 2'– data regarding NPV^{IPP} for e_{el} = 75 USD/MWh; 3 – data regarding NPV for e_{el} = 70 USD/MWh; 3'– data regarding NPV^{IPP} for e_{el} = 70 USD/MWh

The value J_{rw} is relative to the value IRR_{rw}^{IPP} demanded by *IPP* and the value v_m demanded by the seller (rate IRR_{rw}^{IPP} is applied in the case when we substitute zero, $p = 0$ in the place of the income tax in the formula in (2)). Figure 2 presents the market value of the power plant J_{rw} for various values of IRR_{rw}^{IPP} given the value of the rate $p = 0$ (as already mentioned before). Figure 3 illustrates the ratio J_{rw} / J, which demonstrates to what extent the market value J_{rw} is smaller or higher from the actual investment J incurred in the construction of the power plant.

The profit NPV^{IPP} (formula (4)) that is obtained by the investor (*IPP*) represents the profit without tax is deduced, *IPP* can demand that the price J_{rw} should be calculated in relation to this net profit as shown in figures 4 and 5. By accounting for the value of the income tax on the gross profit, the value $J_{rw,p}$ for *IPP* should be derived based of the relation in (2), which already accounts for this tax. Under such circumstances, the price of the power plant decreases from J_{rw} to the value $J_{rw,p}$ for the value $IRR_{p,rw}^{IPP}$ demanded by the investor, equal to IRR_{rw}^{IPP} and the same relative value of the market v_m. Whereas, for $J_{rw} = J_{rw,p}$ and $IRR_{rw}^{IPP} = IRR_{p,rw}^{IPP}$, the relative value of the market v_m decreases while the profit expressed as NPV^{IPP} for *IPP* increases as shown in figures 2-5 and 7.

The values J_{rw}, $J_{rw,p}$ are very sensitive to the variability of the value of the rates IRR_{rw}^{IPP}, $IRR_{p,rw}^{IPP}$. The variability in these values by one or two percentage points results in the change of the values J_{rw}, $J_{rw,p}$ by a dozen and even a few dozen percentage points.

The fundamental value, which decides on the market value of a power plant J_{rw}, $J_{rw,p}$ and values of the profit *NPV* and NPV^{IPP} gained on the electricity market supplied by a power plant is given by the difference between the sales price that can be obtained on this market and the specific cost associated with its generation. The greater the difference, the bigger the values of J_{rw}, $J_{rw,p}$, *NPV*, NPV^{IPP}. In addition, the lower the values IRR_{rw}^{IPP}, $IRR_{p,rw}^{IPP}$, the greater the market values J_{rw}, $J_{rw,p}$. For example, for the price of e_{el} = 80 USD/MWh, and the interest rate on the capital demanded by the investor (*IPP*) equal to $IRR_{p,rw}^{IPP}$= 8%, the market value of a power plant $J_{rw,p}$ that has to be paid by IPP to the seller should be equal to $J_{rw,p}$= 747.5 million USD under the assumption of the demanded stake in the ownership of a power plant equal to v_m = 76%. Hence, it is equal to the actual investment J made in its construction. For the values of v_m lower than 76%, the price $J_{rw,p}$ is higher from the actual investment. This price is also higher for the electricity price of e_{el} = 75 USD/MWh and $IRR_{p,rw}^{IPP}$= 8% for v_m < 58% and for e_{el} = 80 USD/MWh and $IRR_{p,rw}^{IPP}$= 10% for v_m < 14% as shown in figures 4 and 5. In these conditions, it would be more beneficial for *IPP* to build a new power plant instead of buying stakes in an existing one (1 – v_m). In this case, an answer that would have to be sought regards the question whether there would be a sufficient demand for the production of electricity in it in the already developed market. For the lower electricity prices, e.g. for e_{el} = 70 USD/MWh and $IRR_{p,rw}^{IPP}$= 12% and the exemplary value v_m = 30%, the market value of a power plant is only equal to $J_{rw,p}$= 405 m USD and the ratio is equal to $J_{rw,p}/J$ = 0.54. In this case, the market value of the power plant is equal to only a half of the initial investment. If the sales price of electricity were to fall below the specific cost $k_{el,av}$= 66.88 USD/MWh, the purchase of a power plant by the *IPP* is out of the question. In this case, the investor would make a loss.

Figure 6 illustrates the interest rate *IRR* on the capital *J* invested in the construction of a power plant.

As mentioned above, the value of interest *IRR* that is gained on the capital *J* is derived from the formula (2) for the income tax rate equal to zero, $p = 0$, and for $v_m = 0$ (since *IRR* is not relative to v_m). The revenues gained from the sales of electricity play a decisive role on the value of *IRR* (i.e. the first term of the formula in (2)), as it depends on its price and volume of production and the specific cost of its generation (formula (1)). If the difference in the price and, consequently, the specific cost is greater, the value of *IRR* increases as well. In this case, for the already set values of IRR_{rw}^{IPP}, $IRR_{p,rw}^{IPP}$ the market value J_{rw}, $J_{rw,p}$ increases as well. When the values of the demanded rates: IRR_{rw}^{IPP}, $IRR_{p,rw}^{IPP}$ exceed the value of *IRR*, the prices J_{rw}, $J_{rw,p}$ drop considerably below the investment *J* made in the construction of a power plant.

Figure 7 Illustrates the total value of *NPV* market, value of the market for *IPP* equal to $NPV^{IPP} = NPV\,(1 - v_m)$ and the value of the market for the seller of a power plant $NPV\,v_m$.

A comprehensive analysis of the market value of a power plant and value of electricity market (based on formulas (1)–(4)) needs to be conducted with the use of differential calculus. Such calculations can provide a number of additional, important information that could not be gathered without it or would be difficult to capture.

4. Summary and conclusions

An investment in energy sources requires the analysis of economic effectiveness of exploiting them. Prior to a strategic decision regarding the involvement of capital in an energy source, *IPP* needs to make sure that the interest rates: IRR_{rw}^{IPP}, $IRR_{p,rw}^{IPP}$ associated with the investment and the profit NPV^{IPP} from its exploitation will be sufficiently high to satisfy them.

The investment in a power plant is justified only on condition that the sales price that can be potentially gained from the generated electricity is higher from the specific cost of its production. For instance, in case of a supercritical power unit, this cost is equal to around $k_{el,av} = 67.5$ USD/MWh. At present, the sales price of electricity is considerably lower. When this price is exceeded, the production would bring a sufficient profit as shown in figure 7. Hence, the investment in a supercritical coal-fired power unit and coal fired units retrofitted by *CCS* installations, for which the value of the specific cost of electricity production is even greater [3], is not justified on the basis of economic calculation and would bring a loss to *IPP* unless the electricity from such sources was subsidized. The construction of wind turbines and photovoltaic installations renders the process particularly unviable from the economic perspective. The volume of electricity that can be derived from such sources is particularly small due to their short operating times. In particular, this applies to photovoltaics, which is coupled with the great investment in them and makes the electricity generated in them an expensive and luxury commodity [3].

The most rational investment strategy for electricity sector, i.e. one that could ensure energy security as well as stability, certainty and continuity of power supply, should be based on building nuclear power units and modernizing the existing coal-fired power plants. The necessary investment, even including the installation of new boilers and steam turbines with better thermal parameters of the fresh steam, is small compared with the investment needed to build supercritical steam units. In the latter case it would be necessary to build the infrastructure for the power plants, which in the retrofitted power units is already in place. At most, it would be necessary to modernize it. The cost of electricity production in such retrofitted units would be considerably smaller compared to new power units; hence, the price increases will be marginal. If they are coupled with the electricity cost in new nuclear power plants, the increases will be small. Another possibility associated with the modernization of the existing coal-fired units is associated with repowering them by a gas turbo generator. The energy efficiency of such retrofitted units will increase to around 50 %, whereas the emission of carbon dioxide per MWh of electricity will drop by a half; consequently, the cost of electricity production will also be small and equal to around 47.5 USD/MWh [4].

Acknowledgments
This research was financially supported by the scientific activity number 2017/01/X/ST8/00111 from the National Science Centre Poland (Polskie Narodowe Centrum Nauki - NCN).

References
[1] Bartnik R, Bartnik B and Hnydiuk-Stefan A 2016 Optimum investment strategy in the power industry *Mathematical Models* (New York: Wydawnictwo Springer)
[2] Bartnik R and Bartnik B 2014 *Economic Calculus in Power Engineering* (Warszawa: Wydawnictwo WNT)
[3] Bartnik R and Hnydiuk-Stefan A 2016 *Economic analysis of LCOE calculated for various generation technologies* (orginal title in Polish: Analiza ekonomiczna jednostkowych kosztów produkcji elektryczności w różnych technologiach jej wytwarzania], Energetyka [ISSN: 0013-7294] nr 5/69 pp 257-63 http://elektroenergetyka.pl/upload/file/2016/5/Bartnik.pdf
[4] Bartnik R 2013 The modernization potential of gas turbines in the coal-fired power industry *Thermal and Economic Effectiveness* (London: Wydawnictwo Springer)

Design and evaluation of a conical solar cooker in glazed and non-glazed configurations: Cooking tests

S Touré[1,2] and M Sidibé[1]

[1]Laboratoire d'Energie Solaire, Université Félix Houphouët Boigny, Abidjan Cocody 22 BP 582 Abidjan 22 Côte d'Ivoire, France

E-mail: siakaahtoure@yahoo.fr

Abstract. This paper reports the evaluation of a conical solar cooker. This prototype was designed in galvanized sheet with an opening diameter of 120 cm. The experiment was carried out on the roof of the building of the solar energy laboratory of FHB University. Tests of boiling eggs were carried out. During operation, two configurations of the conical cooker were adopted. In the first configuration, the cooker was used without modification. On the other hand, in the second configuration, a removable glass plate is placed in the cone and at a distance of 25 cm from the base. This contribution has the role of creating a greenhouse effect between the absorber and the glass. During the tests, the temperature of cooking, the ambient temperature as well as the solar illumination were measured. The tests were conducted in accordance with the maximum cooking time of some foods given by "Solar cookers international". Despite the observed cloudy periods, maximum temperatures obtained by the eggs were 82°C in 2 hours cooking for the first configuration and 100°C in cooking 1h10 for the second configuration. These results show that the addition of the glass plate in the conical solar cooker gives good satisfaction. In addition, in both cases, a perfect cooking of the eggs was observed.

1. Introduction

There is a need for cooking, everywhere in the world, in urban as well as in rural areas. So far, in developing countries, the energy requirements for cooking depend mainly upon conventional energy sources, such as firewood and coal. In urban areas, LPG is also used for cooking, as well as electricity. The collection of wood from forests is responsible for deforestation and other environmental damages. Solar energy is an alternative power source, which is clean, renewable and sustainable.

Therefore, several types of solar cookers have been developed. Several concentration systems of solar energy are used to reach high temperatures which are required for cooking. The paraboloid concentrator solar cooker is one of them. This type of solar cooker has been studied by several researchers [1-7]. Another type is the Pipe cooker [8]. This cooker is fitted with and oil filled pipe, which is connected to a cooking box. The hot box cooker is another type, which has been studied by several researchers [8,9]. Our paper is about the conical solar cooker. Some studies have been made about this type [10-12]. It has a conical concentrator. Sharaf E [10] has designed a conical solar cooker that concentrated solar radiation at the bottom end of the cone. Some characteristic parameters of one of his cookers are the following: apex angle $\theta = 30^{o}$; radius of the focus R_2 =14 cm; radius of opening R_1 =38.24 cm; length of the reflective part l= 93.68 cm; height of the reflective part h= 90.49 cm. He carried out some tests of boiling water to 100°C. He also performed tests of cooking chickens and red meat.

Kerdchang P and Puangsombur W [11] built a conical solar cooker. Its type is the same as the one designed by Sharaf E [10] with the following characteristic parameters: apex angle $\theta = 45^{o}$; radius of the focus R_2 =15 cm; radius of opening R_1 =36.2 cm; length of the reflective part l= 55.4 cm; the height of the cone is 87.4 cm. They performed cooking test of rice, eggs and pork without tracking of the sun. Mahdi K and Bellel N [12] made a study and a theoretical simulation of a conical solar collector for steam generation. The conical concentrator has a vertex angle equal to 90° and a cylindrical axial receiver. The theoretical study showed that the absorber can reach a temperature higher than 200°C.

In the present, a conical solar cooker is designed, built and experimented. It is the same type as the one designed by Sharaf E [10], but in the prototype built, the radius of opening R_1 is increased, in order to increase the solar radiation collected by the conical concentrator. As a consequence, the length of the reflecting part was increased. The reflecting conical surface is in galvanized steel. The conical solar cooking system is fitted with a manual tracking system, for the follow-up of the sun. Every 20 minutes, the conical concentrator is oriented in front of sun. The absorber is set at the bottom end of the cone and insulated. The experiments were carried out with and without a circular glazing installed in the cone above the absorber.

2. Design and description of the conical concentrator

2.1. Design

The design parameters of the conical concentrator are shown in figure 1, which is a cross-section of the reflective part of the cone. The cone is truncated along CC' which is the bottom and focus of the concentrator. Hence, the reflective frustum is a conical flask which is in galvanized steel. The absorber is located at the bottom of the conical concentrator.

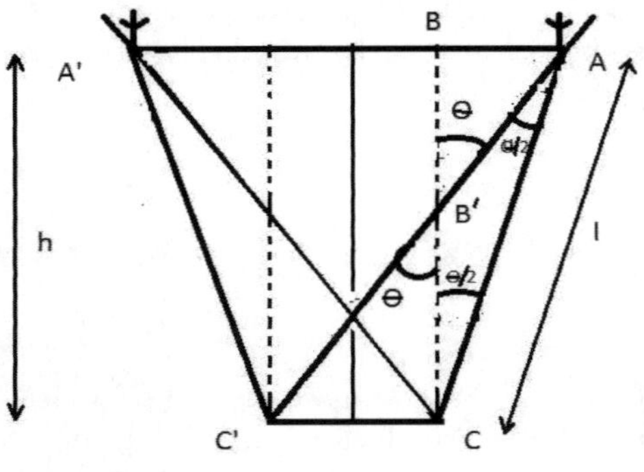

Figure 1. Cross-section of the conical concentrator.

As shown in figure 1, θ is the apex angle of the concentrator; CC'= $2R_2$ is the diameter of the focus; AA'= $2R_1$ is the diameter of the opening. Let us call A the opening area and S the focus area of the concentrator. Then the geometric concentration C is expressed as

$$C = \frac{A}{S} = \frac{\pi R_1^2}{\pi R_2^2} = \frac{R_1^2}{R_2^2} \tag{1}$$

From triangle C'B'C, one gets

$$\frac{C'C}{B'C} = \tan\theta \tag{2}$$

Triangle CB'A shows that $CB' = B'A$
Then from equation (2), one gets

$$\frac{C'C}{B'A} = \frac{2R_2}{B'A} = \tan\theta \quad (3)$$

Considering triangle BB'A, one gets

$$\frac{BA}{B'A} = \frac{R_1 - R_2}{B'A} = \sin\theta \quad (4)$$

From equations (3) and (4), the following equation is obtained

$$R_1 = R_2(2\cos\theta + 1) \quad (5)$$

Combining equations (5) and (1), one gets

$$C = (2\cos\theta + 1)^2 \quad (6)$$

The length l and the height h of the reflective part are obtained by considering triangle ABC. Since l= AC and h= BC, one gets

$$\frac{AB}{AC} = \frac{R_1 - R_2}{l} = \sin\frac{\theta}{2} \quad (7)$$

$$\frac{AB}{BC} = \frac{R_1 - R_2}{h} = \tan\frac{\theta}{2} \quad (8)$$

2.2. Description

The conical concentrator was built with the following dimensions: apex angle $\theta = 30^o$; radius of opening $R_1 = 40.99$ cm; radius of the focus $R_2 = 15$ cm; length of the reflective part l= 100.46 cm; height of the reflective part h = 97.03 cm. The geometrical concentration calculated by using equation (6) is C = 7.46.

Figure 2 shows a section of the conical solar cooker studied. The frame of the cone is a sheet of galvanized steel. The reflective inner surface of the cone is the surface of the galvanized sheet. The bottom of the cone is embedded in a circular vase in stainless steel painted black, called absorber. Thermal insulation of the absorber is ensured by wood, plywood, polystyrene and a thin layer of plaster. Some steel rods hold the cone.

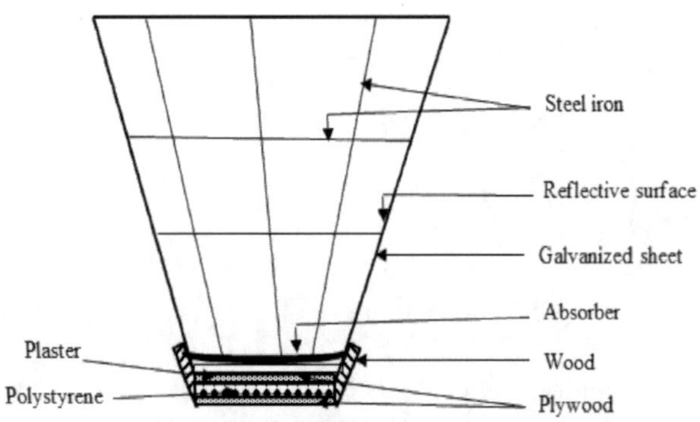

Figure 2. Schematic representation of the conical cooker.

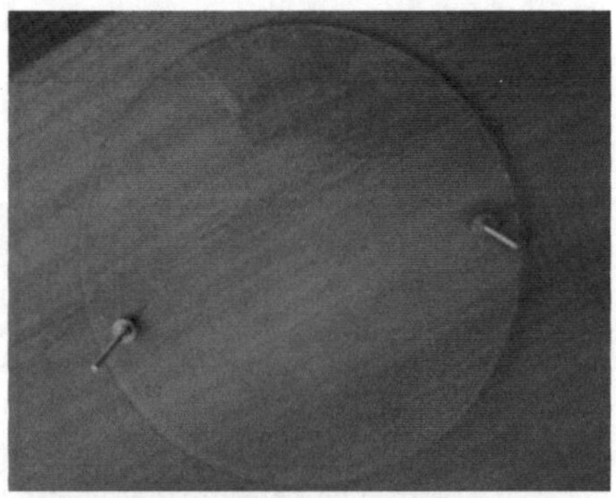

Figure 3. Circular glass cover.

The device thus represented is the non-glazed configuration. When the glass circular plate illustrated in figure 3 is placed in the cone 25 cm above the absorber, we have the glazed configuration. In this case, a greenhouse effect is created in the cone and an energy gain is observed.

The prototype is mounted on a support as shown in figure 4. The frame of the cone is fitted with a manual slider that guides the opening of the cone parallel to the solar radiation and stabilizes the cone. Figure 4 shows an overall view of the cooker with the accessories on the test site.

Figure 4. Overall view of the conical solar cooker with accessories on the experimentation site.

In operation, the solar rays falling parallel to the opening are reflected towards the bottom of the cone, as shown in figure 5. In the non-glazed configuration, the operating principle relies on two conjugate effects:

- The concentration of the incident solar rays reflected by the inner surface of the cone towards the absorber,
- The effect of the black body due to the absorber.

In its glazed configuration, the greenhouse effect is added by the insertion of the glass plate.

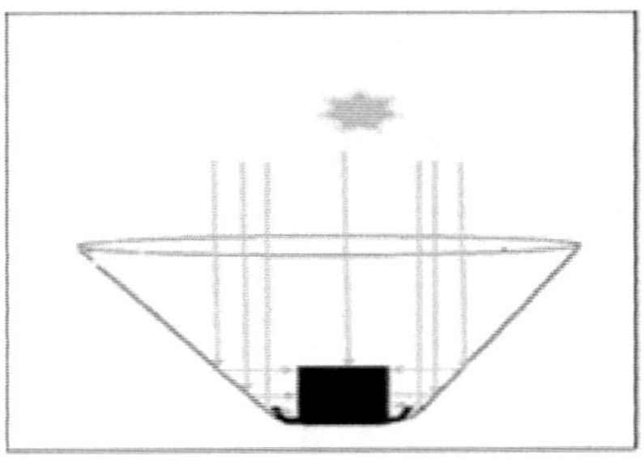

Figure 5. Solar radiation on the absorber in the cone.

3. Experimental study

The experimental tests were carried out on the roof of the Solar Energy Laboratory of the University Felix Houphouët Boigny.

These tests were carried out on November 8th, 2016 and on December 3th, 2016. The measured quantities were the temperatures (ambient, cooking) and the total solar radiation incident on a horizontal plane. The temperatures were measured by using platinum resistance thermometers. Using a manual switch, data was recorded at a time interval of 10 min. An Eppley-type pyranometer (precision $\pm$ 10 W / m^2) was used to record global solar radiation values.

Authors such as Adel M A *et al* [13], Sambo *et al* [14] cited by Adegoke [15] used the cooking duration and the efficiency, as parameters to evaluate the performance of the solar cooker. For cooking time, "Solar Cookers International" [16] has established approximate cooking times in sunny weather.

According to [16], perfectly cooking of rice, eggs, etc ..., requires a cooking duration of 2 hours. In addition, simple solar cookers, used under normal conditions, reach temperatures ranging from 82°C to 121°C or higher. Food starts to cook for temperatures between 82°C and 91°C, which are enough but not enough to burn or lose nutrients.

Regarding the efficiency of the solar cooker, the following expression is used [15]:

$$\eta_c = \frac{(m_p C_p + m_w C_w)(\bar{T}_w - \bar{T}_a)}{A_c I_G t} \quad (9)$$

where m_p and m_w represent mass of the pan and water respectively; C_p and C_w represent specific heat capacity of the pan (920 J / kg.°C) and the water (4200 J / kg.°C) respectively; T_w and T_a stand for average water and ambient temperatures respectively (°C); A_c refers to the area of the solar collector (m^2); I_G represents the average value of the intensity of the overall irradiation (W/m^2); t = Cooking time.

4. Results and discussion

Among the experiments carried out, two days of tests were chosen, namely November 8th and December 3th 2016. The average irradiance for those two days was 568 W / m^2 for November 8th 2016 and 644 W / m^2 for December 3th 2016. The tests consisted in boiling eggs in a pan painted black. During the first day, the conical stove operated in its unglazed configuration with 0.5 L of water and 5 eggs in the pan. On the second day, the conical stove operated in its glazed configuration with 0.5 L of water and 10 eggs in the pan. During these tests, maximum duration set was 2 hours.

Figures 6 and 7 show the temporal variations of solar irradiance, temperature of cooking and temperature of the ambient air versus time for November 8th and December 3th 2016 respectively.

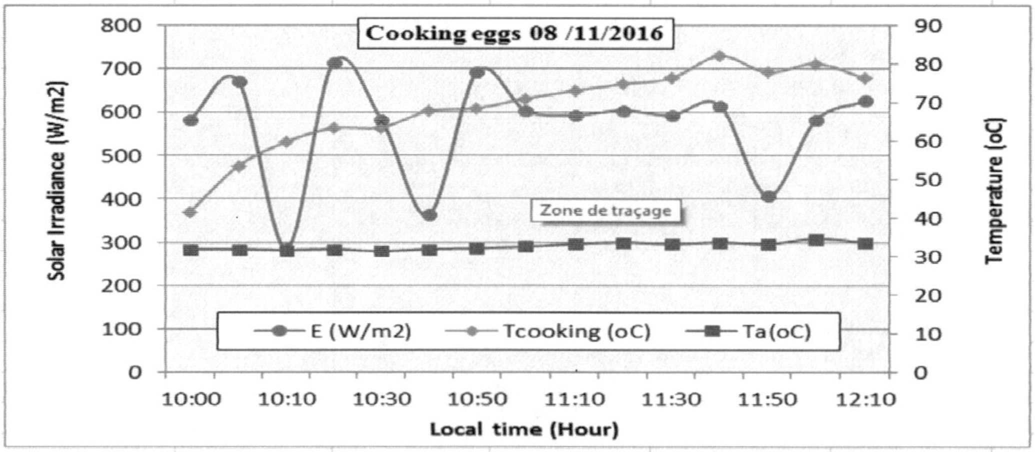

Figure 6. Temporal evolution curves of irradiance, ambient air and cooking temperatures, 08/11/2016.

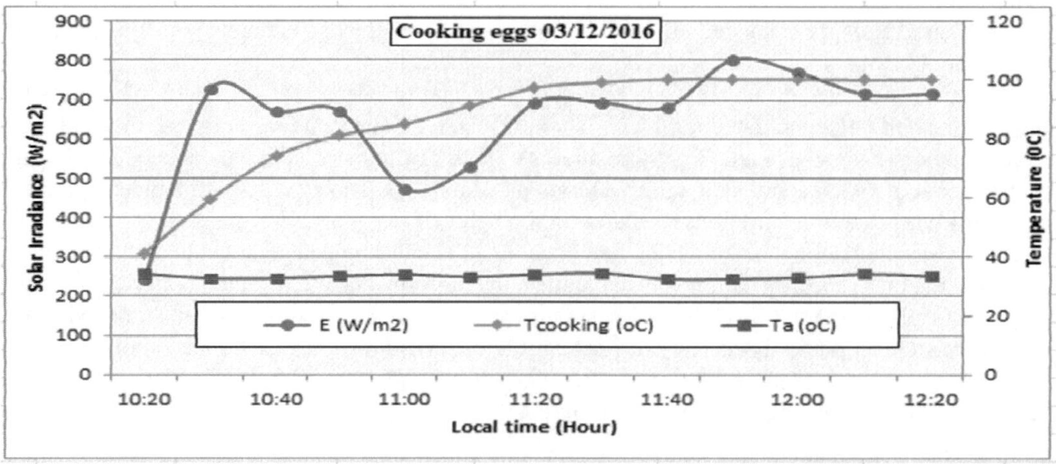

Figure 7. Temporal evolution curves of irradiance, ambient air and cooking temperatures, 03/12/2016.

In figures 6 and 7, E (W/m^2) is the solar irradiance; Tcooking (°C) is the temperature of cooking; Ta (°C) is the ambient temperature.

It should be noted that, at both configurations (glazed and non-glazed), a cloudy period lies approximately from 10:0 to 11.0. For the non-glazed configuration, another cloud passage is observed at 11:50.

So, the tests were carried out under intermittent cloud passages that are observed on the curves of solar irradiance by fluctuations. Despite this bad weather, the results show that cooking is achievable with both configurations of our conical solar cooker.

In the non-glazed configuration, the temperature of 82°C is reached in 2 hours of cooking the 5 eggs.

In the glazed configuration, this temperature of 82°C is reached within 40 minutes of cooking and the temperature of 100°C is reached within 70 min of cooking the 10 eggs. During these tests, the eggs were perfectly cooked despite intermittent cloud cover. So the insertion of the circular glass plate brings an energy gain in our conical solar cooker and accelerates the cooking of food.

Moreover, the influence of the cloud passages is more felt in the non-glazed configuration than in the glazed configuration. That seems logical because the presence of the glass reduces convection and radiation losses and promotes a gain of energy [17].

The determination of cooking efficiency gave 58 % for the non-glazed configuration and 70.89 % for the glazed configuration. These results show that the performance under glazed configuration is better in comparison to the non-glazed configuration.

- If compared with other solar cookers, the conical type as some advantages:
- It has a small opening area (0.53 m^2 in our case)
- The maximum time for cooking is shorter. In our case, 10 eggs were cooked in 70 minutes.

This cooking time will be made shorter when the reflective surface will be coated with mirror to make it more reflective.

5. Conclusion

In this work, a solar cooker was designed and built with the conical reflector in galvanized steel. The conical solar cooker was experienced in local climatic conditions. Two configurations of the solar cooker were tested and compared during cooking tests of eggs. The results showed that inserting a circular plate into the cone improved the performance of the prototype. Furthermore the cooking tests were carried out in conventional cooking durations given by "Solar Cookers International" [16].

Appendix

Nomenclature	
A_C	Area of the cooker (m^2)
C	Geometrical concentration
C_p	Specific heat of the pan (J/kg. °C)
C_W	Specific heat of the water (J/kg. °C)
h	Height of the reflective part (cm)
I_G	Overall irradiation (W/m^2)
l	Length of the reflective part (cm)
m_P	Mass of the pan (kg)
m_W	Mass of the water (kg)
R_1	Radius of opening (cm)
R_2	Radius of the focus (cm)
T_a	Ambient temperature (°C)
T_w	Temperature of the pan (°C)
t	Cooking time (s)
Greek letters	
θ	Apex angle

References

[1] Kassaby M 1991 New solar cooker of parabolic square dish design and simulation *Renew Energy* **1** 59-65
[2] Subbodh K T, Kandpal C and Mullick S C 1993 Heat losses from a paraboloid concentrator solar cooker experimental investigations on effect reflector orientation *Renew Energy* **3** 871-6
[3] Srinivasan M, Kulkarni L V and Pasupathy C S 1978 A simple technique of fabrication of parabolic concentrators *Solar Energy* **22** 463-5
[4] Hosn Y and Abou-Ziyan 1998 Experimental investigation of tracking parabolic and box solar cookers under Egyptian environment *Appl Therm Engin* **18** 1375-94
[5] Pohekar S D 2006 Utility assessment of Parabolic solar cooker as a domestic cooking device in India *Renew Energy* **31** 1827-38
[6] Gavisiddesha S P, Revenkar P P and Gorawar M B 2013 Evaluation of thermal performance of Paraboloid concentrator solar cooker *Int J Innov Res Technol & Sci* **1** 58-65
[7] El Ouederni A R, Dahmani A W, Askri F, Ben Salah M and Ben Nasrallah 2008 Experimental study of a parabolic solar concentrator *Revue des energies Renouvelables CICME 08 Sousse* 193-9
[8] Khalifa A M A, Taha M A and Akyurt M 1985 Solar cookers for outdoor and indoors *Energy* **10** 819-29
[9] Toure S, Sidibe M, Fasinou W, Gbane A and Salami H 2004 A performance study of two solar cookers experimented in Abidjan (Ivory Coast) *World Renw Energy CongressVIIi (WREC2014)* (Denver, Colorado USA)
[10] Sharaf E 2002 A new design for an economical, highly efficient, conical solar cooker *Renew Energy* **27** 599-619
[11] Kerdchang Pongsakom and Puangsombut Wichaya Conical Solar Cooker Hefat 2008 *6th Int Conf on Heat Transfer, Fluid Mechanics and Thermodynamics* (Pretoria, South Africa Paper number: KP1)
[12] Mahdi Khaled and Bellel Nadir 2016 Simulation of a conical collector for steam generation *Int Proceedings of Chemical, Biol and Environ Engin* **91** doi: 10.7763/ IPCBEE. V91.1
[13] Abel M A, Khalifa Taha M M and Ayurt M 1986 On prediction of solar cooker performance and cooking in pyrex *Pot. Solar and Wind Technol* **3** 13-9
[14] Sambo A S, Musa M and Danshehu B G 1993 Development and Performance of a flexible reflector box- type solar cooker *Niger J of Solar Energy* **12** 40-50
[15] Adegoke C O and Fasheun T A 1998 Performance evaluation of a solar cooker under a tropical humid climate *NigerJ of Revenue Energy* **6** 71-4
[16] Solar Cookers International www.solarcooking.org/francais/Plans-fr.pdf (consulté le 25 Décembre 2016)
[17] Chassériaux J M 1984 Conversion thermique du rayonnement solaire (Bordas Paris)

Design optimization of ultra-high concentrator photovoltaic system using two-stage non-imaging solar concentrator

C-W Wong[1,3], T-K Yew[1], K-K Chong[1], W-C Tan[1], M-H Tan[2] and B-H Lim[1]

[1]Lee Kong Chian Faculty of Engineering and Science, Universiti Tunku Abdul Rahman, Bandar Sungai Long, 43000 Kajang, Selangor, Malaysia
[2]Faculty of Engineering and Green Technology, Universiti Tunku Abdul Rahman, Jalan Universiti, Bandar Barat, 31900 Kampar, Perak, Malaysia

E-mail: wcwoon@utar.edu.my

Abstract. This paper presents a systematic approach for optimizing the design of ultra-high concentrator photovoltaic (UHCPV) system comprised of non-imaging dish concentrator (primary optical element) and crossed compound parabolic concentrator (secondary optical element). The optimization process includes the design of primary and secondary optics by considering the focal distance, spillage losses and rim angle of the dish concentrator. The imperfection factors, i.e. mirror reflectivity of 93%, lens' optical efficiency of 85%, circumsolar ratio of 0.2 and mirror surface slope error of 2 mrad, were considered in the simulation to avoid the overestimation of output power. The proposed UHCPV system is capable of attaining effective ultra-high solar concentration ratio of 1475 suns and DC system efficiency of 31.8%.

1. Introduction

Solar concentrator is important in the development of ultra-high concentrator photovoltaic (UHCPV) power generation system that transforms highly focused solar energy into electrical energy. In the UHCPV system, a solar concentrator that is made of relatively inexpensive optics (mirrors or lenses), has been deployed to concentrate the sunlight onto a receiver that is fitted out with multi-junction solar cells (MJSCs) at ultra-high solar concentration ratio (SCR) (> 1000 suns). This approach is capable of cutting down the use of expensive semiconductor materials. The higher the concentration ratio, the lower the levelized cost of electricity (LCOE) [1-3]. Therefore, the motivation behind developing solar concentrator for the UHCPV system is to further reduce the cost of solar electricity, whereby this system is more competitive than other renewable technologies.

To further reduce the cost of UHCPV system, the optimal optical design of solar concentrator is necessary so that maximum solar energy can be harnessed from the MJSCs. Several researchers have proposed several types of solar concentrators to attain ultra-high SCR and reasonably good optical performance [4-6]. For example, Perez-Higueras *et al* have proposed a two-optical-unit system comprised of two Fresnel lenses, two planar mirrors and a MJSC [7]. The geometrical concentration ratio of the proposed two-optical-unit system is 952 suns. However, the system efficiency is only 17.2% owing to poor optical performance and non-uniform illumination distributed on the MJSC. Besides, Ferrer-Rodriguez *et al* have introduced a 4-off-axis-unit Cassegrain-based concentrator to achieve effective concentration ratio of 1682 suns [8]. The presented concentrator consists of four paraboloid mirrors, four hyperboloid mirrors, a glass cover, a homogenizer and a MJSC. The optical efficiency of the three-stage optical system is limited to 73% due to shadowing, reflection and transmission losses in the system. Based on the above-mentioned systems, the performance of a UHCPV system is restricted by the optical design of solar concentrator and the non-uniform solar illumination distributed on the MJSCs.

In this study, a two-stage non-imaging solar concentrator that is capable of producing ultra-high SCR and reasonably uniform solar illumination focused on the MJSCs has been proposed. Instead of using the typical solar concentrators (Fresnel lens or parabolic dish), which are facing great challenge to produce uniform illumination, the primary optical element (POE), i.e. non-imaging dish concentrator (NIDC), comprises multi-faceted mirrors. Despite the employment of flat facet mirrors in the NIDC, the resultant flux distribution from simple super-positioning of all the reflected sunlight is inevitably non-uniform near the peripheral area. Therefore, a secondary optical element (SOE) is required for homogenizing the reflected sunlight from NIDC. Besides acts as a homogenizer, an array of crossed compound parabolic concentrator (CCPC) lenses, which are the SOEs, have been introduced to further focus the concentrated sunlight by the POE onto an array of MJSCs. This approach can reduce the use of expensive semiconductor materials. This new design maximizes the absorption of concentrated sunlight and further reduces the cost of UHCPV power generation system.

2. Overview of the UHCPV system

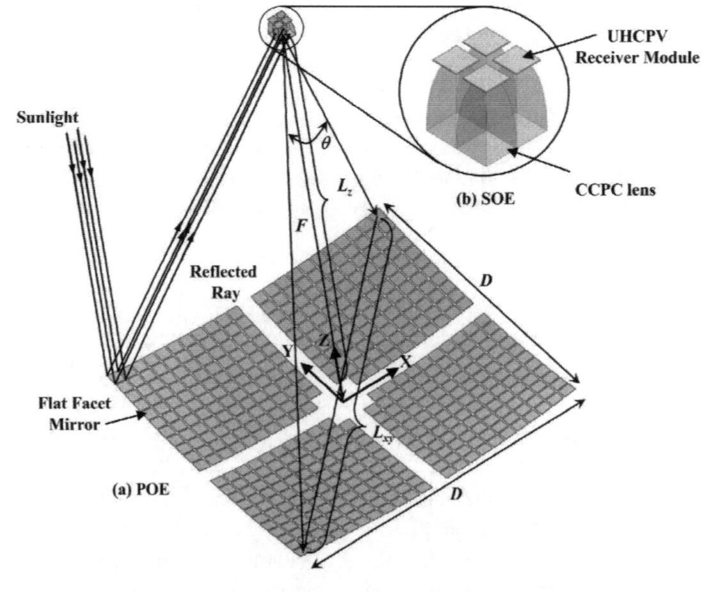

Figure 1. 3-D view of the proposed UHCPV system. (a) POE, non-imaging dish concentrator (NIDC); (b) SOE, crossed compound parabolic concentrator (CCPC) lens.

Figure 1 depicts the schematic diagram of the proposed UHCPV system comprised of two optical elements: a NIDC (POE) and an array of CCPC lenses (SOEs). By considering the compactness of the system, a NIDC with a width (D) of about 100 cm has been designed to focus the sunlight onto the CCPC lenses at any focal distance (F) along the Z-axis. In order to achieve high solar concentration ratio, the NIDC consists of 480 facet mirrors to be arranged into twenty-two rows and twenty-two columns on the mechanical structure. Four mirrors in the central region of concentrator have been removed because of the shading of receiver. Flat facet mirrors with a dimension of 4 cm × 4 cm are selected to form a total reflective area of 7680 cm^2 and a total projection area of 7590 cm^2. In order to avoid shadowing and blocking among the adjacent mirrors, the facet mirrors are gradually lifted from central to peripheral regions of the dish concentrator. Besides, a small gap of 0.5 cm between the facet mirrors is required to absorb the manufacturing defect by providing tolerance for the installation of facets. The detailed design of NIDC has been patented and described in our previous works [9-11].

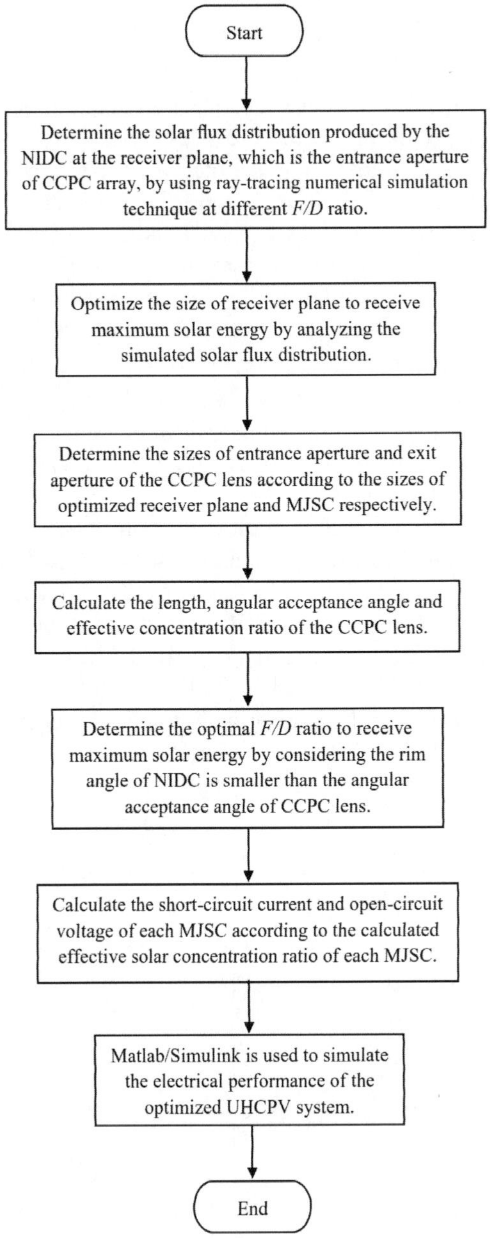

Figure 2. Flow chart shows the process of optimizing the design of UHCPV system.

3-D CCPC is designed by the intersection of two symmetric 2-D compound parabolic concentrators (CPCs) in orthogonal, and hence, it has a square-shaped entrance and exit apertures. Instead of using reflective CCPC, the CCPC lens that is made of B270 crown glass (refractive index, n = 1.52), has been utilized to further concentrate the sunlight reflected by NIDC from various angles onto the UHCPV receiver module that are equipped with a MJSC and a bypass diode. The advantages of dielectric-filled CCPC lens against reflective CCPC are twofold: (a) refraction at the air-dielectric interface improves the acceptance angle; (b) total internal reflection in the lens diminishes the reflection loss. In this paper, a systematic method has been proposed to optimize the optical design of NIDC and CCPC lens that are incorporated into the UHCPV system. Figure 2 presents the process of optimizing the design of UHCPV system.

3. Optical design and analysis

3.1. Design of POE: non-imaging dish concentrator (NIDC)

In order to optimize the design of NIDC, an optical characterization of the NIDC has been carried out with the use of ray-tracing numerical simulation technique as presented in our previous works [12,13]. By assuming each facet mirror has a reflectivity of 93%, solar illumination on the receiver plane that is the entrance aperture of CCPC array, has been simulated by varying the F/D ratio, where F is the longest focal distance from the centre of receiver plane to the centre of NIDC and D is the width of NIDC. In this study, the imperfection factors, i.e. circumsolar radiation and mirror surface slope error, are included in the simulation to avoid overestimation of output power. The seriousness of circumsolar radiation effect is rated in term of the circumsolar ratio (χ) and the mirror surface slope error is reported in standard deviation (δ) unit. The solar illumination on the receiver plane has been modelled for two different cases, namely ideal case (χ = 0 and δ = 0 mrad) and non-ideal case. The typical values of circumsolar ratio, χ = 0.2 and slope error, δ = 2 mrad are considered in the non-ideal case of this study. With the simulated solar illumination, the size of receiver can be optimized by minimizing the spillage loss of the NIDC.

Figures 3 and 4 present the simulation results of solar illumination in 2-D plots for both ideal and non-ideal cases respectively. The solar illumination is simulated at a F/D ratio ranging from 0.5 to 1.5 with the increment of 0.1. By omitting the imperfection factors (χ = 0 and δ = 0 mrad), the NIDC is capable of producing uniform illumination area. The SCR at the central region of solar illumination is nearly constant (see figure 3). The higher the F/D ratio the higher the SCR, ranging from 320 to 432 suns. Besides, the uniform illumination area reduces as the F/D ratio increases. According to the simulated results of the non-ideal case (χ = 0.2 and δ = 2 mrad), the uniform illumination region narrows while the base region of flux distribution broadens as the F/D ratio increases (see figure 4). For the non-ideal case, the SCR at the uniform illumination area does not significantly increases with the F/D ratio higher than 1.0. This phenomenon happened because the energy at the central region of flux distribution spread towards outside the flux distribution. Figure 5 shows the comparison between

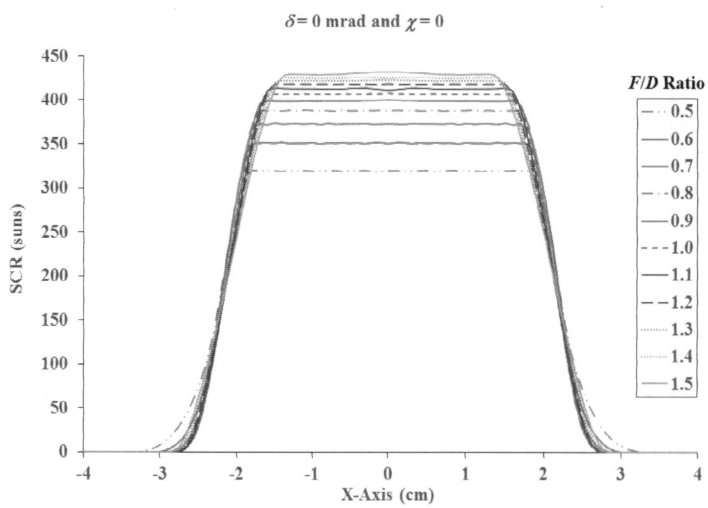

Figure 3. The solar illumination in 2-D plots for ideal case (χ = 0 and δ = 0 mrad) at F/D ratio ranging from 0.5 to 1.5.

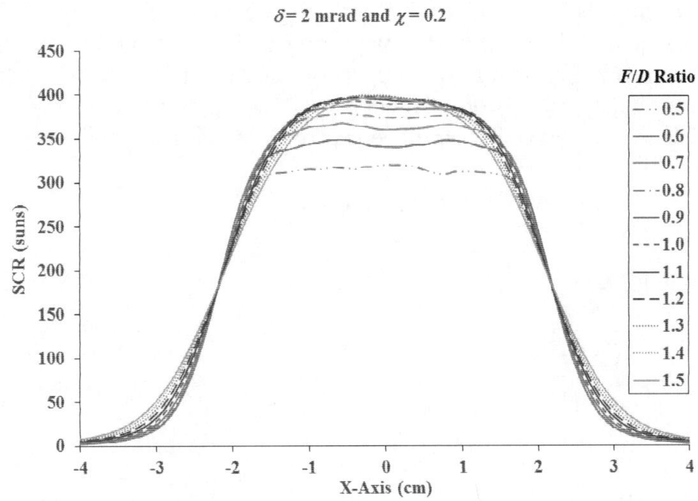

Figure 4. The solar illumination in 2-D plots for non-ideal case (χ = 0.2 and δ = 2 mrad) at *F*/*D* ratio ranging from 0.5 to 1.5.

the ideal case and the non-ideal case at *F*/*D* ratio of 1.5. For the non-ideal case at *F*/*D* ratio of 1.5, the SCR at the central region of flux distribution is much lower as compared to the ideal case. Therefore, overestimation of output power will be incurred if the imperfection factors are not considered in the concentrator design.

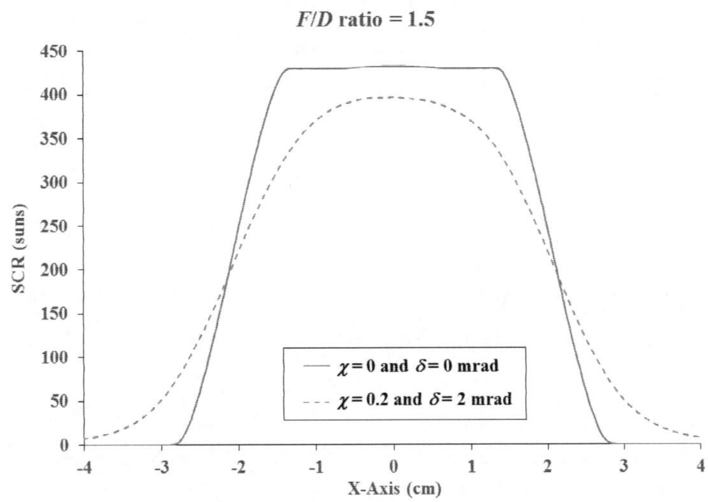

Figure 5. Comparison between the ideal case, i.e. χ = 0 and δ = 0 mrad (solid line) and the non-ideal case, i.e. χ = 0.2 and δ = 2 mrad (dotted line) at *F*/*D* ratio of 1.5.

Spillage loss (percentage of sunray falling beyond the receiver) was determined by adopting different dimensions of square receiver plane. Figures 6 and 7 illustrate the spillage loss versus the receiver plane dimension (square in shape) for both ideal and non-ideal cases respectively at *F*/*D* ratio ranging from 0.5 to 1.5. Practically, it is not a wise idea to design a receiver plane that just only covers the uniform illumination area, especially for the non-ideal case. According to the results, the spillage loss of non-ideal case for the receiver plane that just only covers the uniform illumination area becomes very severe, as high as 78.7%, if the *F*/*D* ratio increased to 1.5 (see figure 7). For the sake of minimizing the spillage loss and avoiding the oversizing of receiver, proper selection and detailed analysis of the receiver plane dimension are very important tasks in the concentrator design. In this study, the receiver plane with a dimension of 6 cm × 6 cm was chosen. As shown in figures 6 and 7, the selected receiver plane has a maximum spillage loss of 8.0% for both ideal and non-ideal cases.

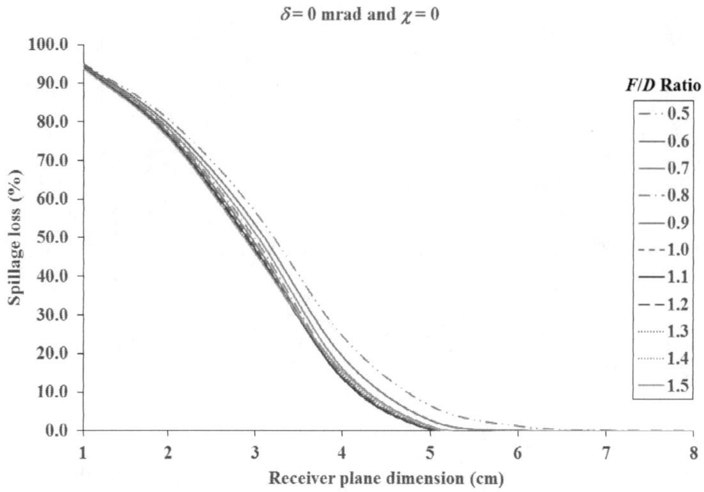

Figure 6. Spillage loss versus receiver plane dimension (square in shape) for ideal case ($\chi = 0$ and $\delta = 0$ mrad) at *F*/*D* ratio ranging from 0.5 to 1.5.

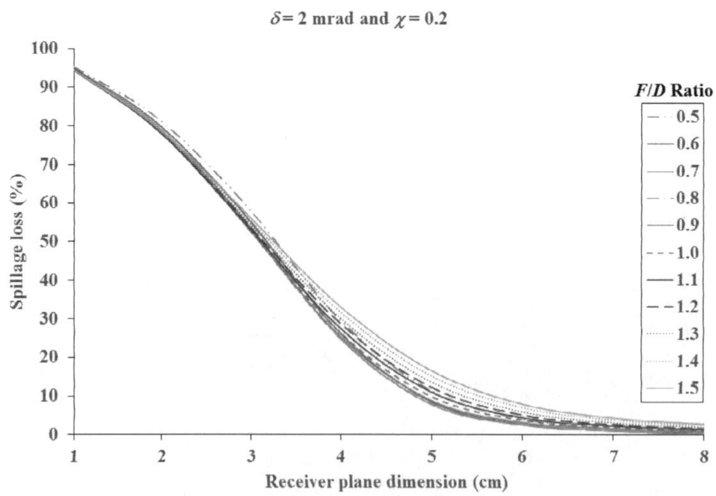

Figure 7. Spillage loss versus receiver plane dimension (square in shape) for non-ideal case ($\chi = 0.2$ and $\delta = 2$ mrad) at *F*/*D* ratio ranging from 0.5 to 1.5.

3.2. Design of SOE: dielectric-filled CCPC lens

In this study, the dielectric-filled CCPC has been integrated with the NIDC to homogenize the non-uniform illumination distributed on the receiver plane and to further focus the sunlight concentrated by the NIDC onto MJSC. According to the simulated results as shown in figures 3 and 4, the NIDC is capable of producing symmetrical flux distribution that is much larger than the active area of a MJSC. Therefore, 2 × 2 arrays of UHCPV receiver modules were deployed to convert the highly concentrated solar energy into electrical energy. This approach can diminish the current mismatch between the modules if they are interconnected in a series configuration. Figure 8 shows the cross-sectional view of the dielectric-filled CCPC lens. In order to match the dimension of optimized receiver plane, each UHCPV receiver module is bonded to a CCPC lens that has a square entrance aperture, $2a = 3$ cm. In addition, the exit aperture of CCPC lens needs to be matched with the dimension of MJSC, whereby the exit aperture of CCPC can be bonded to the MJSC with an optical adhesive. The MJSC, with a dimension of 1.0 cm × 1.0 cm, from Azur Space was deployed in this study [14]. By considering the practical limitations in assembly, the CCPC lens has an exit aperture, $2a' = 0.98$ cm, which is slightly smaller than the dimension of the MJSC. With a refractive index (n) of 1.52, the half acceptance angle of CCPC geometry, $\theta'_i = 19.06°$, the length of CCPC lens, $L = 5.76$ cm, the half acceptance angle of CCPC lens $\theta_i = 29.77°$, and the geometrical concentration ratio of CCPC lens, $C_{geo} = 9.37$ can be calculated by using the following equations as presented in our previous works [15,16].

$$\theta'_i = \sin^{-1}\left(\frac{a'}{a}\right) \tag{1}$$

$$L = \frac{a'(1+\sin\theta'_i)\cos\theta'_i}{\sin^2\theta'} \tag{2}$$

$$\theta_i = \sin^{-1}(n\sin\theta'_i) \tag{3}$$

$$C_{geo} = \left(\frac{a}{a'}\right)^2 = \left(\frac{1}{\sin\theta'_i}\right)^2 = \left(\frac{n}{\sin\theta_i}\right)^2 \tag{4}$$

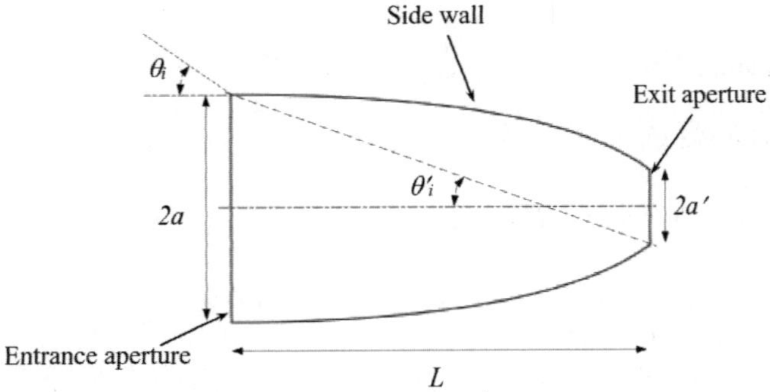

Figure 8. Schematic diagram to show the cross sectional view of the dielectric-filled CCPC lens.

By assuming constant optical losses of 15%, the effective concentration ratio of CCPC lens (C_{SOE}) is 7.97. The constant optical losses include the Fresnel reflection losses, absorption losses and some light rays escape from the side walls that are near to the exit aperture of CCPC lens.

As discussed in our previous works, the CCPC lens is designed in such a way that its angular acceptance angle ($2\theta_i$) is greater than the rim angle of NIDC. This way can make sure that the CCPC lens can guide all the concentrated sunlight from NIDC to MJSC [15,16]. The rim angle (θ) can be calculated by using the following formula

$$\theta = 2\tan^{-1}\left(\frac{L_{xy}}{2L_z}\right) \tag{5}$$

where L_{xy} is the distance of central points between the two most apart mirrors (the mirrors usually are positioned in opposite corners of the facet array), and L_z is the shortest distance in Z-direction from the central point of CCPC array to the distance line of L_{xy} (see figure 1).

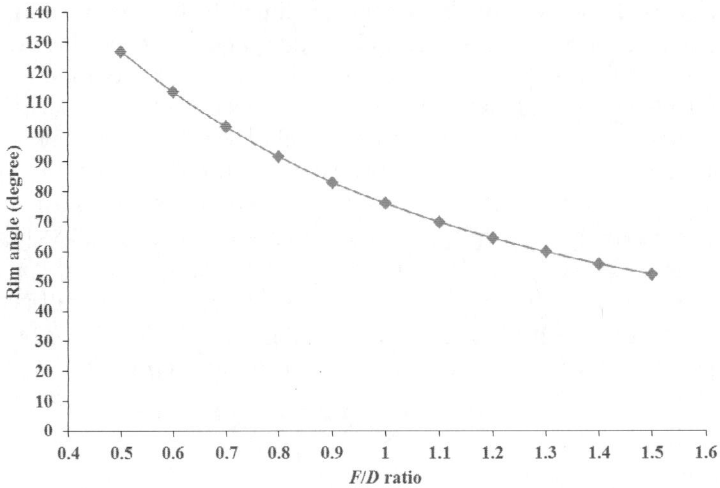

Figure 9. The calculated rim angles of NIDC at *F/D* ratio ranging from 0.5 to 1.5.

Table 1. The specification of the optimized UHCPV system.

Primary optical element (POE): Non-imaging dish concentrator (NIDC)	
Number of facet mirror	480 units (arranged in 2-D array)
Dimension of facet mirror	4 cm × 4 cm
Mirror Reflectivity	93%
F/D ratio	1.4
Total reflective area	7680 cm^2
Total projection area	7590 cm^2
Dimension of receiver plane	6 cm × 6 cm
Rim angle, θ	55.85°
Secondary optical elements (SOEs): dielectric-filled CCPC lens	
Dielectric material	B270 crown glass
Refractive index, n	1.52
Dimension of entrance aperture	3 cm × 3 cm
Dimension of exit aperture	0.98 cm × 0.98 cm
Length, L	5.76 cm
Half acceptance angle CCPC geometry, θ'_i	19.06°
Angular half acceptance angle CCPC lens, θ_i	29.77°
Optical losses	15%
Geometrical concentration ratio of CCPC lens, C_{geo}	9.37
Effective concentration ratio of CCPC lens, C_{SOE}	7.97
Number of CCPC lens	4 (arranged in 2 × 2 array)
Multi-junction solar cell (Azur Space product: 3C44 – 10 × 10 mm²)	
Typical efficiency at 1000 suns	40.5%
Dimension of cell aperture	10 mm × 10 mm
V_{oc} at 1000 suns (1000 W/m^2 irradiance @ T = 25°C)	3.14 V
I_{sc} at 1000 suns (1000 W/m^2 irradiance @ T = 25°C)	14.97 A
Maximum operating temperature	110°C

As shown in figure 9, the rim angle of NIDC at *F/D* ratio ranging from 0.5 to 1.5 has been calculated using equation (5). These calculated rim angles are essential for determine the optimal *F/D* ratio that is smaller than the angular acceptance angle of CCPC lens. According to the results, there are only two rim angles of NIDC that are less than the angular acceptance angle of CCPC lens, $2\theta_i$ = 59.54°, namely 55.85° and 52.34° at *F*/*D* ratio of 1.4 and 1.5 respectively. The *F*/*D* ratio of 1.4 was chosen to design the NIDC since its spillage loss is lesser than the *F*/*D* ratio of 1.5 for both ideal and non-ideal cases. Table 1 shows the specification of the optimized UHCPV system.

4. Electrical simulation and analysis

Before simulating the electrical performance of the UHCPV system, the effective SCR, C_{eff} of each MJSC for both ideal ($\chi = 0$ and $\delta = 0$ mrad) and non-ideal ($\chi = 0.2$ and $\delta = 2$ mrad) cases at *F*/*D* ratio of 1.4 was calculated by using equation (6). The MJSCs for both ideal and non-ideal cases have the effective SCR, C_{eff} of 1563 suns and 1457 suns respectively.

$$C_{eff} = C_{POE} \times C_{SOE} \tag{6}$$

where C_{POE} is the average SCR of NIDC at the entrance aperture of CCPC lens, and C_{SOE} is the effective concentration ratio of CCPC lens.

In this study, the electrical output power of the UHCPV system was simulated by using the Matlab/Simulink modeling. Both short-circuit current (I_{SC}) and open-circuit voltage (V_{OC}) of each MJSC can be modelled by the following equations:

$$I_{SC} = I_{SC}^{1} C_{eff} \tag{7}$$

$$V_{OC} = V_{OC}^{1} + N\frac{kT}{q}\ln C_{eff} \tag{8}$$

where I_{SC}^{1} and V_{OC}^{1} are the short-circuit current and open-circuit voltage of the MJSC under one sun respectively, C_{eff} is the effective SCR of the MJSC, N is the effective diode ideality factor, T is the operating temperature (Kelvin), k is the Boltzmann constant and q is the electronic charge.

For this study, the MJSCs are assumed to have zero value in series resistance, $R_s = 0\ \Omega$ and effective diode ideality factor, $N = 3$. According to the specification of Azur Space MJSC shown in table 1, the one sun short-circuit current (I_{SC}^{1}) and open-circuit voltage (V_{OC}^{1}) under standard test condition were calculated as 14.97 mA and 2.61 V respectively by using equations (7) and (8). The MJSCs are assumed to operate at the standard temperature of 298 Kelvin and the direct normal irradiance (DNI) level of 1000 W/m^2. After that, the maximum output power of the UHCPV system was simulated in MATLAB/Simulink platform system for both ideal and non-ideal cases. In this simulation, the UHCPV receiver modules are modelled to be connected in a series configuration.

Figure 10 shows the *I-V* and *P-V* curves of UHCPV system, which are simulated by using MATLAB/Simulink software, for the solar flux distribution of both ideal and non-ideal cases. As presented in figure 10, the maximum electrical output powers of both ideal and non-ideal cases are 259.3 W and 241.4 W respectively. There will be 7.4 % in overestimation of output power if the imperfection factors are not considered in the design of this UHCPV system. The DC system efficiency (η_{DC}) of UHCPV system can be calculated by the following equation:

$$\eta_{DC} = \frac{P_{elec\text{-}max}}{P_{in}} \times 100\% \tag{9}$$

where $P_{ele\text{-}max}$ is the maximum electrical output power of the UHCPV system and P_{in} is the input power of the UHCPV system. The input power of the system can be determined by multiplying both the total projection area of NIDC and DNI level. With a total projection area of 7590 cm^2 and DNI level of 1000 W/m^2, the DC system efficiency of both ideal and non-ideal cases are calculated as 34.2% and 31.8% respectively by using equation (9).

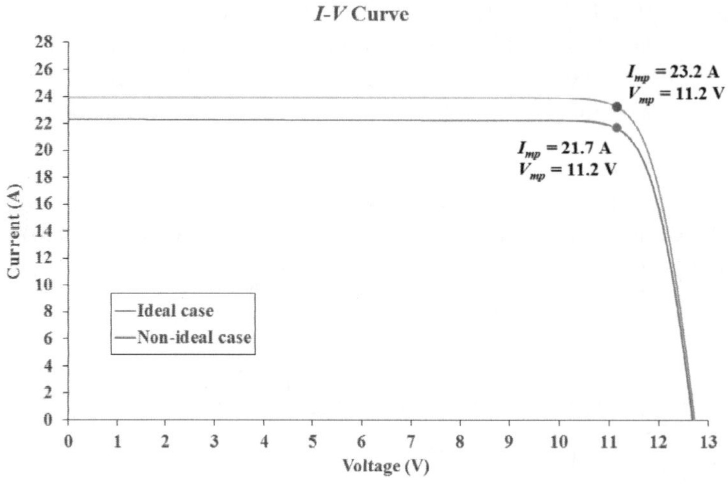

(a)

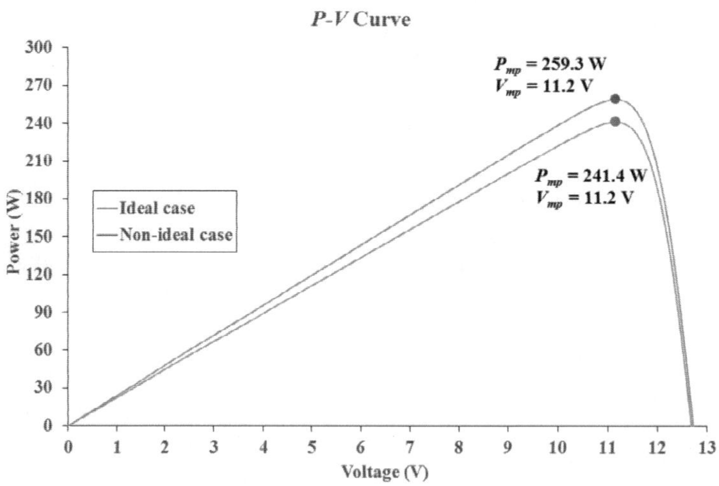

(b)

Figure 10. (a) *I-V* curve; (b) *P-V* curve for both ideal ($\chi = 0$ and $\delta = 0$ mrad) and non-ideal cases ($\chi = 0.2$ and $\delta = 2$ mrad) at F/D ratio of 1.4.

5. Conclusions

In this study, a systematic approach for optimizing the design of UHCPV system that integrated with two optical elements, i.e. NIDC and 3-D dielectric-filled CCPC lens. The NIDC can produce symmetrical and reasonably uniform solar illumination. The CCPC lens not only homogenizes the non-uniform flux distribution but also further focuses the reflected sunlight from NIDC onto MJSCs. The focal distance, spillage losses and rim angle of dish concentrator are the key parameters to get the optimal design of UHCPV system. In order to avoid the overestimation of output power, the simulation needs to consider the imperfection factors, i.e. mirror reflectivity of 93%, lens' optical efficiency of 85%, circumsolar ratio of 0.2 and mirror surface slope error of 2 mrad. According to the results, the proposed UHCPV system can obtain effective ultra-high solar concentration ratio of 1475 suns and DC system efficiency of 31.8%. The DC system efficiency can be further improved by applying anti-reflective coating at the entrance aperture of CCPC lens to diminish the Fresnel reflection losses.

IOP Conf. Series: Earth and Environmental Science **93** (2017) 012012 doi:10.1088/1755-1315/93/1/012012

Acknowledgments
The authors would like to express their gratitude to the Ministry of Energy, Green Technology and Water (AAIBE Trust Fund) with vote account 4356/001. Furthermore, authors would like to thank UTAR Research Fund with project number IPSR/RMC/UTARRF/2017-C1/W03 (vote account 6200/W61).

References
[1] Algora C and Rey-Stolle I 2012 *Next Generation of Photovoltaics*, ed A B Cristobal Lopez *et al* (Berlin: Springer) p 23
[2] Yamaguchi M, Takamoto T, Araki K, Imaizumi M, Kojima N and Ohshita Y 2011 *CLEO: 2011 - Laser Applications to Photonic Applications (Washington)* (Washington: OSA Publishing) pp 1-3
[3] Ortiz E and Algora C 2003 *Prog. Photovolt: Res. Appl.* **11** 155-63
[4] Katz1 E A, Gordon J M and Feuermann D 2006 *Prog. Photovolt: Res. Appl.* **14** 297-303
[5] Khonkar H, Wacaser B, Martin Y, Kirchner P, Alyahya A, Aljouad M, Halawani M and Kessel T V 2013 *9th Int. Conf. on Concentrator Photovoltaic Systems (Miyazaki)* (New York: AIP Publishing) pp 172-5
[6] Khonkar H, Alowais A, Sheikho A, Alyahya A, Alghamdi A, Alsaedan A, Eugenio N N, Alalweet F, Halawani M and Alsaferan A 2014 *10th Int. Conf. on Concentrator Photovoltaic Systems (New Mexico)* (New York: AIP Publishing) pp 16-20
[7] Pérez-Higueras P, Ferrer-Rodríguez J P, Shanks K, Almonacid F and Férnández E F 2015 *11th Int. Conf. on Concentrator Photovoltaic Systems (Aix-les-Bains)* (New York: AIP Publishing) p 130004-1
[8] Ferrer-Rodríguez J P, Férnández E F, Almonacid F and Pérez-Higueras 2016 *Opt. Lett.* **41** 1985-8
[9] Tan M-H, Chong K-K and Wong C-W 2014 *Appl. Opt.* **53** 475-86
[10] Chong K-K, Wong C-W, Yew T-K and Tan M-H 2013 *U.S. Patent Application 13/901* (pending)
[11] Chong K-K, Wong C-W, Yew T-K and Tan M-H 2012 *Malaysian Patent No. PI 2012002439* (pending)
[12] Wong C-W, Chong K-K and Tan M-H 2015 *Opt. Exp.* **23** A841
[13] Wong C-W and Chong K-K 2016 *Sol. Energy* **125** 86-98
[14] http://www.azurspace.com/images/products/0004355-00-01_3C44_AzurDesign_10x10.pdf
[15] Yew T-K, Chong K-K and Lim B-H 2015 *Sol. Energy* **120** 296-309
[16] Chong K-K, Yew T-K, Wong C-W, Tan M-H, Tan W-C, Lai A-C, Lim B-H, Lau S-L and Rahman F A 2014 *National Physics Conf. 2014 (Kuala Lumpur)* (New York: AIP Publishing) p 030009-1

Optimization of structure and control strategy for EV wireless charging systems

Y J Hou[1,3], T Hei[2], H M Tian[1], M Bai[1] and H Zeng[2]

[1]Energy Research Institute, Shandong Academy of Sciences, Jinan 250014, Shandong Province, China
[2]School of Electrical Engineering, Shandong University, Jinan 250061, Shandong Province, China

E-mail: 200412136@163.com

Abstract. Low efficiency is the key factor restricting the development of EV wireless charging systems in recent years. In order to improve the efficiency, the main circuit of the system is simplified by removing the DC/DC circuit and replacing full bridge inverter with half bridge inverter. And the output power control strategy is optimized by changing the duty cycle of PWM that controls the inverter. Besides, the peaks are reduced by changing the frequency of the system, and the output power can also be controlled by changing frequency. The performance of two most common magnetic couplers is compared and a direction of how to select couplers is given. The methods are all verified by simulations and experiments. By optimizing of the circuit and control strategy, the efficiency and the safety of the system are all improved.

1. Introduction

As for EV charging, wired charging is the most common at present. But compared to wireless charging, wired charging has a lot of weaknesses and deficiencies. For example, aged charging cable will endanger personal safety, and wired charging cannot realize charging automation and intelligence. But wireless charging works through a magnetic field between two coupling coils, and without direct electrical connection, which can be controlled intelligently and automatically, and electric shock can also be effectively avoided [1-3].

Although wireless charging has many advantages, the emerging technology is immature and there are also many problems to be solved. One of the biggest problems is the low efficiency. To increase the efficiency, many researchers have done some relevant works. Series resonant capacitive circuit and ZCS (Zero current switching, ZCS) have been designed to minimize converter switching losses [4]. Zhao *et al* [5] developed a method called phase-shift control to alleviate change of efficiency when input power changes. Zeng *et al* [6] and Zhang *et al* [7] optimized the magnetic core structures by FEM simulation and experiments, reducing the weight of the magnetic couplers as well as improving the efficiency, and Zeng *et al* [6] proposed a method of optimizing the core structure. The copper losses are analysed in [8] in order to reduce the losses in the coupler. A converter based on SiC-topology is introduced in [9], but its efficiency is a little bit low (85%). Deng *et al* [10] took a novel topology of a phase-controlled inverter and put it in a wireless charging system.

The low efficiency of the wireless charging system is mainly caused by the complexity of the circuit and low coupling coefficient. Thus, this paper will improve the efficiency by simplifying the circuit and optimizing the control strategy, which can reduce the number of devices and the impact of switching tube. Besides, choosing the best magnetic coupler by FEM simulation to achieve bigger coupling coefficient is also important.

2. Circuit simplification

2.1. The classic circuit design

Figure 1 shows a most common control circuit of transmitting coil, which mainly includes four parts: Rectifier bridge, PFC (Power Factor Corrector, PFC), buck-circuit and full bridge inverter.

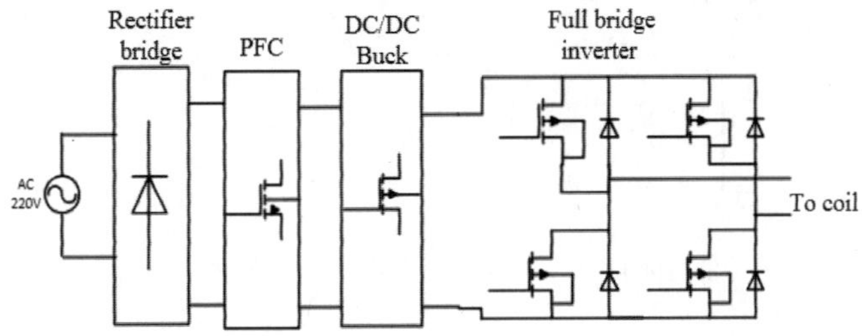

Figure 1. A classic circuit design of transmitting terminal of a wireless charging system.

The function of Buck-circuit is to control the output power of the wireless charging device by changing the duty cycle of the PWM waves. As shown in figure 1, the inverter is also controlled by the PWM, So if it is feasible to control the output power by controlling the duty cycle of the PWM that controls the inverter circuit, then the Buck-circuit can be discarded and the circuit is simplified.

Besides, as for inverter circuit, both full bridge inverter and half-bridge inverter can be used in charging system. It is necessary to explore their distinctions and try to use the half-bridge circuit mainly to reduce power consumption and devices used.

2.2. The simplified circuit and simulation

At first, the feasibility of removing DC/DC is researched by simulation. The toolbox called Simulink in MATLAB is used. Figure 2 shows the circuit without DC/DC, and figure 3 shows the connection type of capacitance and inductance. Figure 4 shows one of the simulation models in Simulink. With the same input and load, the efficiency curves and output power of the two circuits are shown in figure 5.

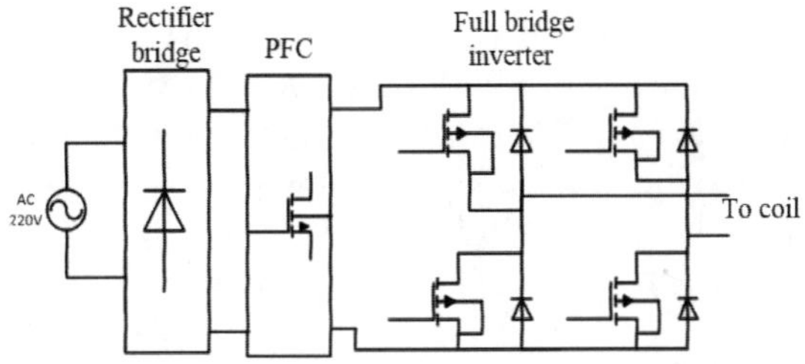

Figure 2. Circuit without DC/DC.

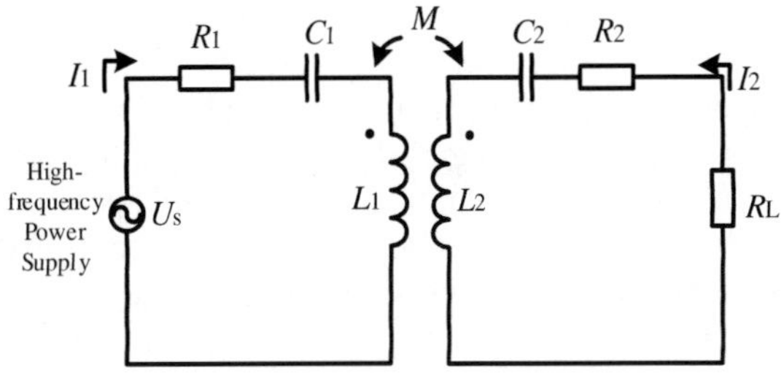

Figure 3. SS topology used in simulation.

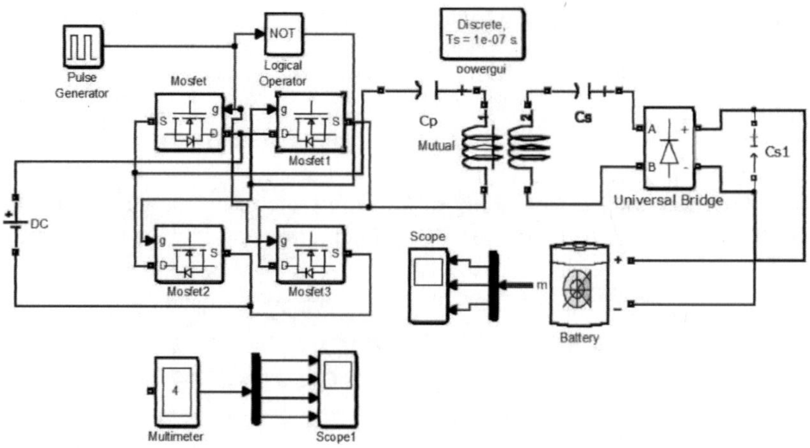

Figure 4. The simulation model without DC/DC in Simulink.

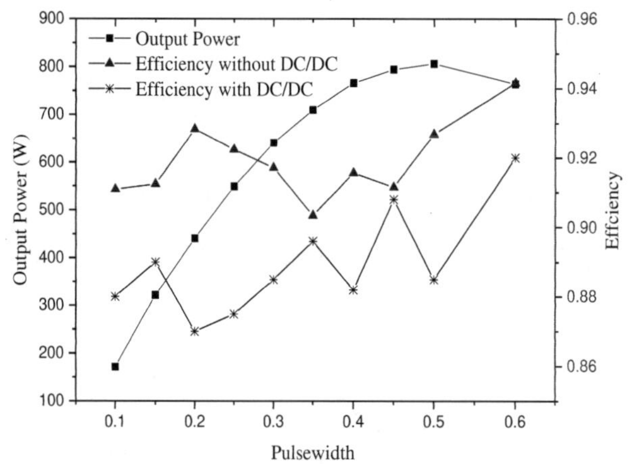

Figure 5. Changes in output power and efficiency when duty cycle of the invert is changed.

The load in simulation is 48 V, 24 Ah lithium battery. The DC voltage is 311 V, which is equal to the output voltage of single-phase (220 V) rectifier. The frequency of PWM that controls inverter is 85 kHz. As shown in figure 5, when the duty cycle varies between 0.1 and 0.5, the output power varies from 150 W to 800 W, which means, it is practicable removing the DC/DC. Besides, the efficiency is measured at t = 0.04 s, and the comparison of efficiency in figure 5 illustrates that removing DC/DC can improving the transmission efficiency.

From figure 4 and figure 5, the max output power is achieved when duty cycle is 0.5. Thus, if the max output power is not enough for the load, a Boost circuit is still needed to increase the output power.

Figure 6 shows the voltage and current waves of L1 when duty cycle is 0.2 and 0.5, respectively. With DC/DC, the duty cycle of the inverter circuit is always equal to 0.5 and the voltage and current waves are shown in figure 6(b). But if the duty cycle is changed to 0.2, the waves will be distorted (shown in figure 6(a)). So it is necessary to test it in experiment and make sure that the peak does no harm to the system.

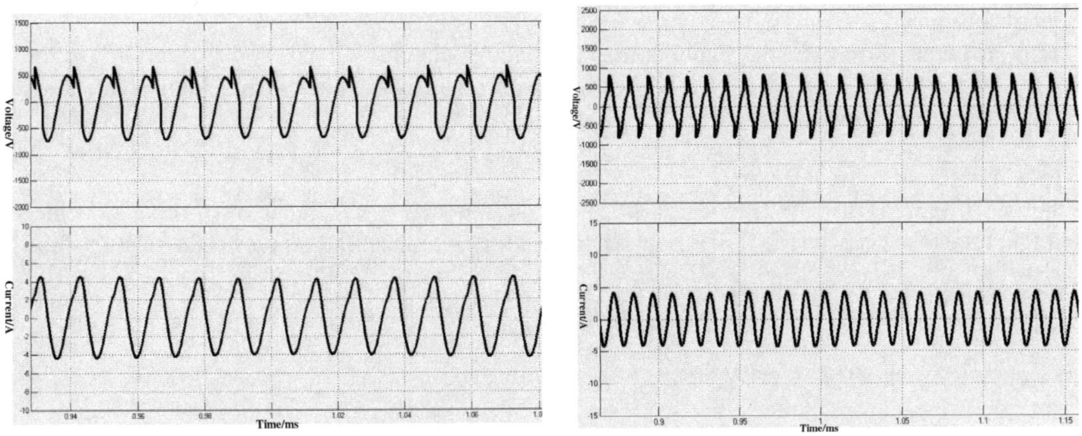

(a) Duty cycle=0.2 (b) Duty cycle=0.5

Figure 6. Voltage and current of L1 when duty cycle is different.

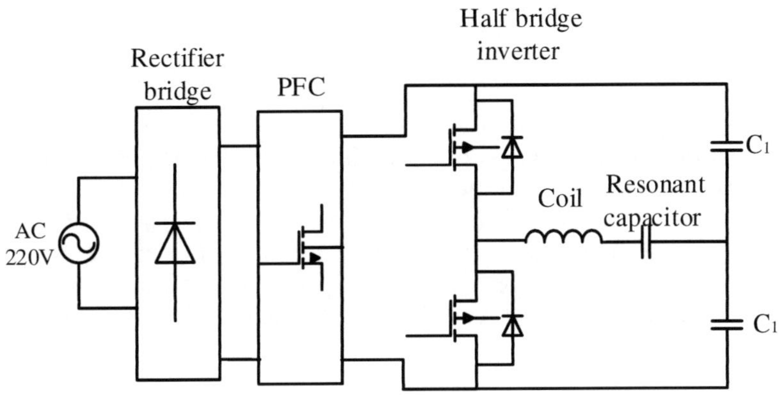

Figure 7. Circuit with half bridge inverter.

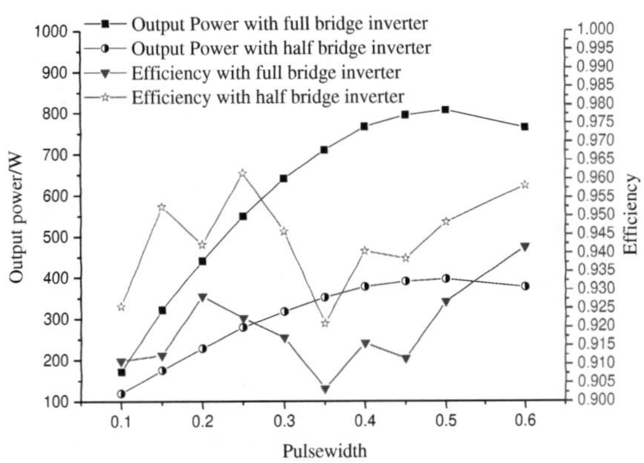

Figure 8. Performance comparison of circuit with full bridge inverter and half bridge inverter.

Then the full bridge inverter is replaced by a half bridge inverter, as shown in figure 7. Then, with the same input and load the simulation results are obtained and the curves about output power and efficiency are shown in figure 8.

From the comparison shown in figure 8, the efficiency of circuit with half bridge inverter is higher. However, its output power is just about half of the full bridge circuit. Thus, half bridge circuit can be used in low-power situations.

3. Optimization of control strategy

As shown in figure 3, usually compensation capacitor is resonant with inductor in order to achieve maximum transmission power [11]. Thus, in SS (series-series connection) topology, the relationship

$$\omega L = \frac{1}{\omega C} \tag{1}$$

is workable. Then, the frequency f is

$$f = \frac{1}{2\pi}\frac{1}{\sqrt{LC}} \tag{2}$$

And the output active power

$$P_{out} = I_2^2 R_L = \left(\frac{\omega M I_1}{R_2 + R_L}\right)^2 R_L \tag{3}$$

In EV wireless charging system, the output power is generally controlled by adjusting the duty cycle of the PWM. But the relative position of transmitting and receiving coils is varied as the position of EV is not stationary. Thus, when the position of car is changed, and the receiving coil is off-center, mutual inductance M will be decreased and equation (1) will be no longer valid. At this time, if improving the output power by adjusting the duty cycle of PWM, peaks will be enlarged on the output voltage waveform of inverter (shown in figure 9), and efficiency will be reduced. As shown in figure 9, the "Over" means overshoot. Then a new way to adjust the output power is come up with, which is frequency adjustment.

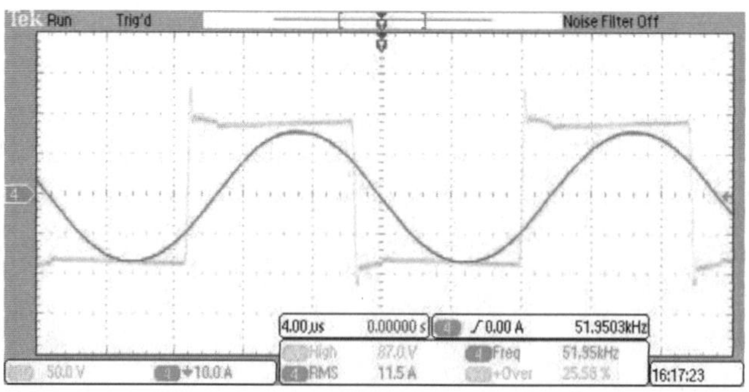

Figure 9. The enlarged peaks on inverter's output voltage.

3.1. Frequency adjustment to alleviate the peak

In EV wireless charging systems, when the receiving coil with no misalignment, the transmission power and efficiency will be maximum, and the impact will be minimal. This is because that inductance and capacitance resonate exactly at this time. But as the receiving coil with misalignment, mutual inductance M will be reduced and equivalent inductance will be reduced. So, the circuit is capacitive and there is no stream current when MOSFETs switch, which leads to larger peaks.

Solutions to the problem are to re-calculate the resonant frequency using equation (2) and change it by MPU. But we always don't know the change value of inductance, so higher frequency is set to alleviate the peaks as the circuit is inductive. And then change the frequency closer to the resonant frequency step by step.

The method described above is realized by MPU called STM32F103ZET6 in experiment, which realizes the peaks reduction and efficiency improvement.

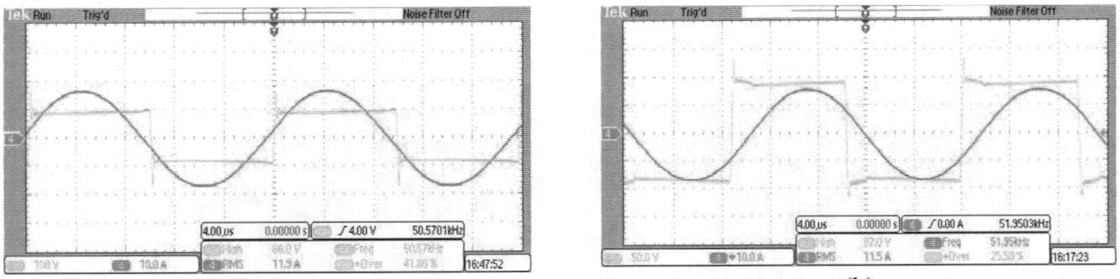

(a) (b)

Figure 10. The peaks comparison when the circuit is capacitive and inductive. (a) The output voltage and current of inverter when the circuit is capacitive; (b) The output voltage and current of inverter when the circuit is inductive.

As shown in figure 10, peaks in inductive circuit are smaller than in capacitive circuit. Thus, frequency adjustment is necessary to insure the safety of the circuit.

3.2. Frequency adjustment to adjust the output power

From equation (3) the output power is influenced by frequency. So, if the output power can be adjusted by changing frequency, there will be another reason to remove DC/DC. In order to let the circuit work in inductive condition, the frequency must be higher than resonant frequency. The experiment results are shown in figure 11.

From figure 11, the method changing frequency to adjust the output power is workable. As the output power changes, the efficiency decreases slightly, which conforms with reality. So, frequency adjustment has little influence on efficiency, which means the output power can be adjusted by changing the frequency, and the method can relieve the peaks.

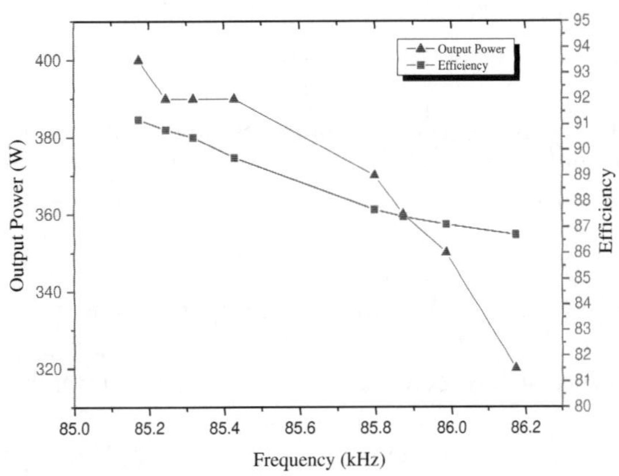

Figure 11. Changes in output power and efficiency as the frequency changes.

4. Optimization of magnetic coupler

There are so many different kinds of magnetic couplers in EV wireless charging systems. An ideal coupler is hoped to have high coupling coefficient and light weight, as well as good performance when misalignment occurs. At present the most commonly used are the circular magnetic coupler and DD coils. In this section, the performance of circular coils and DD coils is compared.

Supposing k represents the coupling coefficient, the coupling coefficients of circular coils and DD coils with different misalignments are shown in figure 13, which is analysed by FEM simulations.

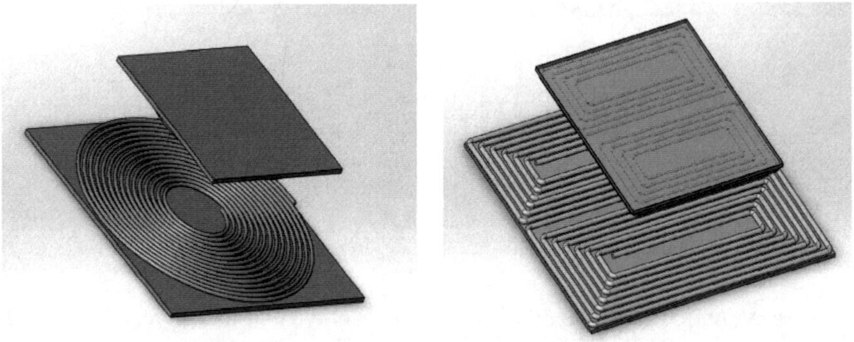

(a) (b)

Figure 12. The most commonly used couplers. (a) circular magnetic coupler; (b) DD coils.

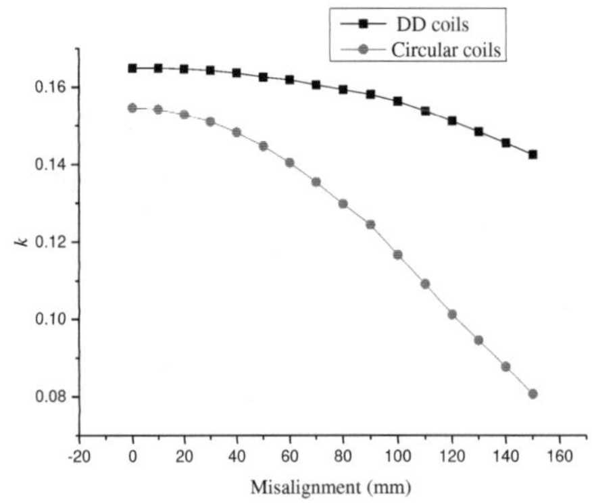

Figure 13. The performance of circular coil and DD coil.

From figure 13, when the receiving coil has no misalignment, the coupling coefficient is almost the same. But when misalignment is increased, the coupling coefficient of circular coils is falling faster than DD coils. Thus, the DD coils are much better applying in EV wireless charging systems.

5. Conclusion

In order to improve the efficiency of EV wireless charging system, DC/DC converter of the main circuit is removed and the output power control is realized by adjusting duty cycle of PWM that controls the inverter. Feasibility of this method is verified by simulation by a toolbox called Simulink. Besides, the full bridge inverter is replaced by half bridge inverter, but the limitation of the change is that circuit with half bridge inverter has a lower output power. With the two methods above, the frequency has been improved a lot.

To alleviate the peaks and insure safety as well as improve the efficiency, frequency adjustment is comp up with and by changing the working frequency of the system, the peaks are smaller and output power control can also be realized, which is verified by experiments.

At last, the two most common couplers are compared by FEM simulation. The conclusion is that the performance of DD coupler is much better than circular coupler, and DD coils are more suitable for EV wireless charging systems.

Acknowledgments
This research has received funding from the Major Science & Technology Program of Shandong Province (2015GSF121048).

References

[1] Musavi F, Edington M and Eberle W 2012 Wireless power transfer: A survey of EV battery charging technologies *2012 IEEE Energy Convers. Congr. Expo. (ECCE)* **11** 1804-10
[2] Wang Z H, Hu C, Sun Y and Dai X 2015 Design of magnetic coupler for inductive power transfer system based on output power and efficiency *Trans of CES* **30** 29-30
[3] Li S Q and Mi C C 2015 Wireless power transfer for electric vehicle applications *IEEE J. Emerg Sel Topic Circuits Syst* **3** 10-2
[4] Moradewicz A J and Kazmierkowski M P 2010 Contactless energy transfer system with FPGA-controlled resonant converter *IEEE Trans. Ind. Electron* **57** 3181-90
[5] Zhao C, Wang Z, Du J, Wu J, Zong S and He X 2014 Active resonance wireless power transfer system using phase shift control strategy *IEEE APEC* Fort Worth 1336-41
[6] Zeng H, Liu Z Z, Hou Y J, Hei T and Zhou B 2017 Optimization of magnetic core structure for wireless charging coupler *IEEE Trans Magn* **53** 1-4
[7] Zhang W, White J C, Abraham A M and Mi C C 2015 Loosely coupled transformer structure and interoperability study for EV wireless charging systems *IEEE Trans. Power Electron* **30** 6356-8
[8] Wang S and Dorrell D G 2015 Copper loss analysis of EV charging coupler *IEEE Trans Magn* **51** 1-3
[9] Bac N X, Vilathgamuwa D M and Madawala U K 2014 A SiC-based matrix converter topology for inductive power transfer system *IEEE Trans. Power. Electron* **29** 4029-38
[10] Deng Q J, Liu J T, Czarkowski D, Bojarski M, Asa E and Leon F 2016 Design of a wireless charging system with a phase-controlled inverter under varying parameters *IET Power Electron* **9** 2461-70
[11] Sun Y, Xia C Y, Dai X and Su Y G 2010 Analysis and optimization of mutual inductance for inductively coupled power transfer system *Proc CSEE* **33** 45-6

Deployment strategy for battery energy storage system in distribution network based on voltage violation regulation

H Wu[1], L Zhou[1], T Xu[1], W L Fang[2], W G He[3] and H M Liu[2, 4]

[1]Nanjing electric power company, Nanjing, 210019, China
[2]College of Energy and Electrical Engineering, Hohai University, Nanjing 211100, China
[3]China Electric Power Research Institute, Nanjing 210003, China

E-mail: liuhaom@hhu.edu.cn

Abstract. In order to improve the situation of voltage violation caused by the grid-connection of photovoltaic (PV) system in a distribution network, a bi-level programming model is proposed for battery energy storage system (BESS) deployment. The objective function of inner level programming is to minimize voltage violation, with the power of PV and BESS as the variables. The objective function of outer level programming is to minimize the comprehensive function originated from inner layer programming and all the BESS operating parameters, with the capacity and rated power of BESS as the variables. The differential evolution (DE) algorithm is applied to solve the model. Based on distribution network operation scenarios with photovoltaic generation under multiple alternative output modes, the simulation results of IEEE 33-bus system prove that the deployment strategy of BESS proposed in this paper is well adapted to voltage violation regulation invariable distribution network operation scenarios. It contributes to regulating voltage violation in distribution network, as well as to improve the utilization of PV systems.

1. Introduction

With growing trends in distributed photovoltaic (PV) system installations in the distribution network, the structural characteristics of traditional radiation distribution network have been changed. And it makes the voltage of the node change with the direction of the power flow, which usually leads to the situation of voltage violation. Therefore, it is necessary to regulate grid voltage more flexibly.

In order to address the voltage rise issue in distribution network caused by high penetration of distributed PV systems, various methods have been proposed. Using larger transformers and conductors to upgrade the existing distribution network is a straightforward but expensive approach [1,2]. Meanwhile, adjusting the load tap changer (LTC) positions of transformers is another way to regulate voltage in distribution network [3,4].

In addition to using transformers and conductors, real-time power control is also an effective way to manage voltage profiles in distribution networks. In [5], a new real-time power capping method for overvoltage regulation is proposed. In order to maintain voltage below a pre-set upper limit, the method is effective by maximizing the PV power output and allocating the real power curtailments among all the PV systems equally in the network.

PV and battery energy storage system (BESS) are capable of voltage regulation due to their active and reactive power regulation ability. In [6], a new strategy to achieve voltage regulation in distributed network with PV and BESS is proposed. The proposed method employs the network impedance matrix to obtain the solution of optimal sizing and allocation of BESS rather than use recursive algorithms.

When dealing with BESS deployment problem based on voltage violation regulation, the coordinated power control of PV and BESS are needed to be solved. Therefore, it is difficult to obtain the satisfactory solution when the model includes the coordinated power control of PV and BESS and the BESS deployment.

In this paper, a bi-level programming model is proposed for BESS deployment in the distribution network with PVs, so as to realize flexible voltage regulation. The questions how distribution network with PV under multiple output modes influence configuration of BESS and the coordination of PV and BESS are researched on. The rest of this paper is arranged as follows: in Section 2, a BESS deployment model based on bi-level programming is established. In Section 3, the solution to solve the mathematics model is presented. In Section 3.1, power control coordination is presented. In Section 3.2, augmented objective function and flow chart for solving this bi-level programming model are presented. Case study and results are demonstrated in Section 4, and finally, conclusions are drawn in Section 5.

2. Bi-level BESS optimal deployment model

To research on the BESS deployment in the distribution network with PV, bi-level programming model is established including inner optimization model and outer optimization model. Through inner optimization we get voltage violation bounds, then the voltage violation bounds are taken into consideration while we are seeking for optimal size and location of BESS. Therefore, via bi-level programming model which is effective in solving multi-goals programming problems, we can realize optimal size and allocation of BESS as well as voltage violation regulation simultaneously.

2.1. Inner optimization model

The active power and reactive power of PV and BESS are considered as variables which can be controlled to relieve voltage violation in the distribution network. The objective function of voltage violation regulation is

$$\min \quad f_1 = \sum_{t=1}^{T}\sum_{i=1}^{I} g(V_{i,t}) \tag{1}$$

where, $V_{i,t}$ denotes the voltage of ith node at time t; I is the number of nodes except the slack node; T is the total number of the periods in a whole day; $g(V_{i,t})$ denotes the function of voltage violation, which is demonstrated as follows

$$y(V_i) = \frac{1}{\sigma\sqrt{2\pi}}\left(1 - e^{-\frac{(V_i-\mu)^2}{2\sigma^2}}\right) \tag{2}$$

$$g(V_i)=\begin{cases} \omega_v \cdot \left(V_i - V_0\right)^2 & V_i \le V_{hl} \text{ or } V_i \ge V_{hu} \\ y(V_i) - y(V_{sl}) & V_{hl} < V_i < V_{sl} \\ y(V_i) - y(V_{su}) & V_{su} < V_i < V_{hu} \\ 0 & V_{sl} \le V_i \le V_{su} \end{cases} \tag{3}$$

where, μ and σ are the mean and standard deviation of Gaussian function respectively; $y(V_i)$ is the function obeying Gaussian function to describe the voltage violation; V_{sl} and V_{su} are the lower and upper soft bounds of node voltage respectively; V_{hl} and V_{hu} are the lower and upper hard bounds of node voltage respectively; V_0 is the rated voltage of node voltage; ω_v is the penalty factor of the hard voltage constraint, and it is usually set as a large positive number, e.g. 20000.

In order to minimize the active power fluctuation, two objective functions are established, which are shown in (4) and (5).

$$\min \quad f_2 = \tau \sum_{t=1}^{T} \sum_{s=1}^{S} \left| P_{s,t} \right| \tag{4}$$

$$\min \quad f_3 = \sum_{t=1}^{T} \tau \left(\sum_{l=1}^{L} P_{l,t} + \sum_{j=1}^{J} (P_{MPP,j,t} - P_{j,t}) \right) \tag{5}$$

where, τ denotes the length of time period, $P_{s,t}$ is the charging and discharging power of sth BESS at time t, S is the total number of BESS, $P_{l,t}$ is the active power loss of the lth line at time t, L is the number of the lines, $P_{MPP,j,t}$ and $P_{j,t}$ is the maximum active power and the actual active power of the jth PV at time t respectively, and J is the number of PV.

The objective function (4) is to minimize the operation of BESS so as to reduce the power loss, function (5) is to minimize the PV energy curtailment. Constraint conditions of these two objective functions are demonstrated as follows

$$\begin{cases} V_{hl} \le V_i \le V_{hu} \\ 0 \le \left| P_s \right| \le P_{r,s} \\ -Q_{lb,s} \le Q_s \le Q_{ub,s} \\ -\sqrt{P_{r,s}^2 - P_s^2} \le Q_s \le \sqrt{P_{r,s}^2 - P_s^2} \\ S_{OC,lb} \le S_{OC,t} - \dfrac{\eta_s \cdot P_s \cdot \tau}{C_{r,s}} \le S_{OC,ub} \\ S_{OC,0} - S_{OC,T} = 0 \\ -Q_{lb,j} \le Q_j \le Q_{ub,j} \\ \sqrt{Q_j^2 + P_j^2} \le S_j \end{cases} \tag{6}$$

where, P_s and Q_s are the active power and reactive power of the sth BESS respectively; $P_{r,s}$ denotes the rated power of the sth BESS; $Q_{lb,s}$ and $Q_{ub,s}$ are the maximum capacitive reactive power and maximum inductive reactive power of the sth BESS respectively; $S_{OC,t}$ is the state of charge of the sth BESS at time t; η_s is the charging/discharging efficiency of the sth BESS; $C_{r,s}$ is the rated capacity of the sth BESS; $S_{OC,lb}$ and $S_{OC,ub}$ are the lower and upper limits of SOC of the sth BESS; P_j and Q_j are the active power and reactive power of the jth PV respectively; S_j indicates the rated capacity of the jth PV; while PV is under the restricted reactive power output model, $Q_{lb,j}$ and $Q_{ub,j}$ are the maximum capacitive reactive power and maximum inductive reactive power of the jth PV respectively.

2.2. Outer optimization model

In order to achieve economic deployment of BESS, both minimal total rated capacity and power of BESS should be considered in objective functions. The outer optimization model is formulated as follows

$$\min \quad h_1 = \sum_{k=1}^{K} n_k F_k (f_1, f_2, f_3) \tag{7}$$

$$\min \quad h_2 = \sum_{s=1}^{S} C_{r,s} \tag{8}$$

$$\min \quad h_3 = \sum_{s=1}^{S} P_{r,s} \tag{9}$$

$$\text{s.t.} \quad \begin{aligned} &\sum_{s=1}^{S} C_{r,s} \le C_{r,ub} \\ &\sum_{s=1}^{S} P_{r,s} \le P_{r,ub} \\ &P_{r,s} \le c_{PC} \cdot C_{r,s} \end{aligned} \tag{10}$$

where, n_k is the number of days which is similar to the typical day k in a year; K is the number of kinds of typical days; F_k is the function concerning equation (1), (4) and (5), the specific expression of which is demonstrated in (12); $C_{r,ub}$ and $P_{r,ub}$ are the upper limit of the total capacity and power of BESS respectively; c_{PC} is the typical proportional coefficient between $C_{r,s}$ and $P_{r,s}$.

3. Application and solution for the BESS deployment model

3.1. Power control coordination

To realize coordinated power control of PV and BESS, several factors should be taken into consideration as the control variables

- The active power of the PV;
- The reactive power of the PV;
- The active power of the BESS;
- The reactive power of the BESS.

3.2. Augmented objective function

In order to turn the inner and outer optimization model into a single objective optimization model respectively, augmented objective functions of inner and outer optimization model are established in section 3.2.1 and section 3.2.2. The flow chart for solving the bi-level programming model is presented in section 3.2.3.

3.2.1. Augmented objective function of the inner optimization model. In order to observe the power control principles, the augmented objective function of inner optimization model is as follow

$$\min \quad H_{in} = g_{1,k} + p_{1,k} \tag{11}$$

$$p_{1,k} = f_{1.k} \tag{12}$$

$$g_{1,k} = \frac{c_1}{1+e^{-\log(\varepsilon+f_{1,k})}} + \frac{c_2}{1+e^{-\log(\varepsilon+f_{2,k})}} + \frac{c_3}{1+e^{-\log(\varepsilon+f_{3,k})}} \tag{13}$$

where, ε is a small positive variables; $f_{1,k}$, $f_{2,k}$ and $f_{3.k}$ are functions presented in equations (1), (4) and (5) on typical day k; c_1, c_2 and c_3 are the coefficients of f_1, f_2 and f_3 respectively; $g_{1,k}$ is the specific expression of F_k concerning typical day k; $p_{1,k}$ is the voltage penalty function on typical day k.

3.2.2. Augmented objective function of the outer optimization model. In order to turn the outer optimization model into a single objective optimization model, outer augmented objective function is as follows

$$\min \quad H_{ex} = g_2 + p_2 \tag{14}$$

$$g_2 = h_1 \cdot \left(\frac{a_1}{1 + e^{-\log(\varepsilon + h_2)}} + \frac{a_2}{1 + e^{-\log(\varepsilon + h_3)}} \right) \tag{15}$$

$$p_2 = \sum_{k=1}^{K} n_k \cdot p_{1,k} \tag{16}$$

$$h_1 = \sum_{k=1}^{K} n_k \cdot g_{1,k} \tag{17}$$

where, $p_{1,k}$ denotes the comprehensive penalty function of the typical day k ; a_1 and a_2 are the coefficients of h_2 and h_3 .

3.2.3. Solving procedure. The differential evolution (DE) algorithm [7,8] is applied to solve the bi-level programming model. A general flow chart is illustrated in figure 1, *G* and *N* are the numbers of the generation and individuals, *NG* is the pre-set maximum number of the generation; *NP* is the total number of individuals in a generation.

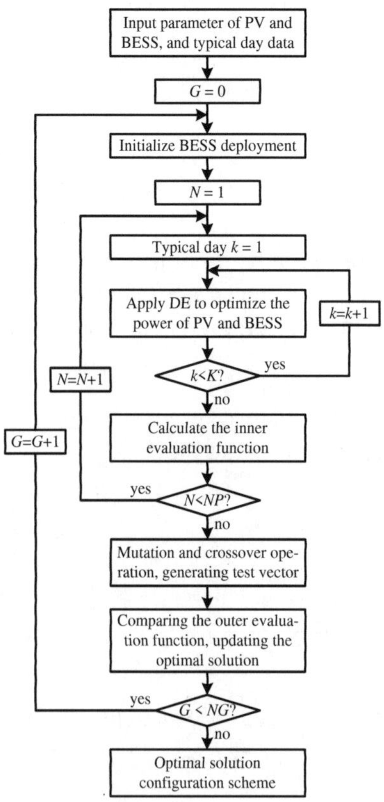

Figure 1. Flow chart for solving the BESS deployment model.

- Step 1: Input parameters of PV and BESS, and typical day data.
- Step 2: Generate the initial BESS deployment scheme.

- Step 3: Based on BESS deployment scheme, apply DE to optimize the power of PV and BESS in all typical days and then calculate the inner evaluation function H_{in}, which is presented in equation (11).
- Step 4: Apply DE to solve the outer optimization model. Based on the H_{in} in step 3, the outer evaluation function H_{ex} is calculated.
- Step 5: Determine whether the result satisfies convergence condition. If $G < NG$, output the configuration scheme. Otherwise, generate new configuration scheme of BESS through mutation crossover. Then go back to Step3 and repeat the process.

4. Case study

4.1. Distribution network operation scenarios

In order to test the validity of the proposed optimization model in this paper, the IEEE 33-buses distribution network is used. The single-line diagram is depicted in figure 2.

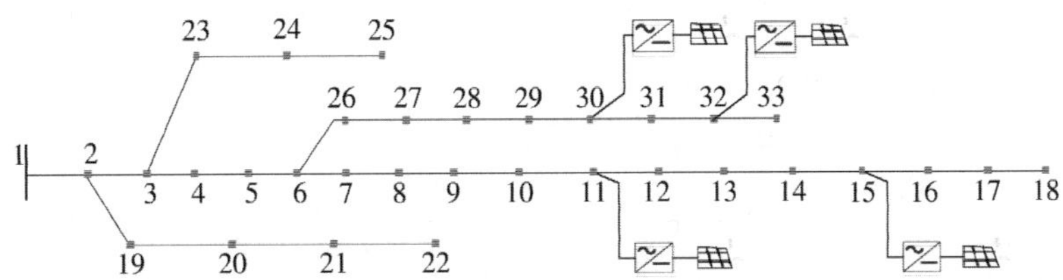

Figure 2. IEEE33-buses distribution network including distributed PV.

The rated voltage of the distribution network is 12.66 kV, and its reference capacity is 10 MVA. The nodes that PVs are connected to are 11, 15, 30 and 32, whose peak power are 0.1, 0.1, 0.1 and 0.12 p.u. respectively. The active power of PV connected to node 11 in 2 typical days is depicted in figure 3.

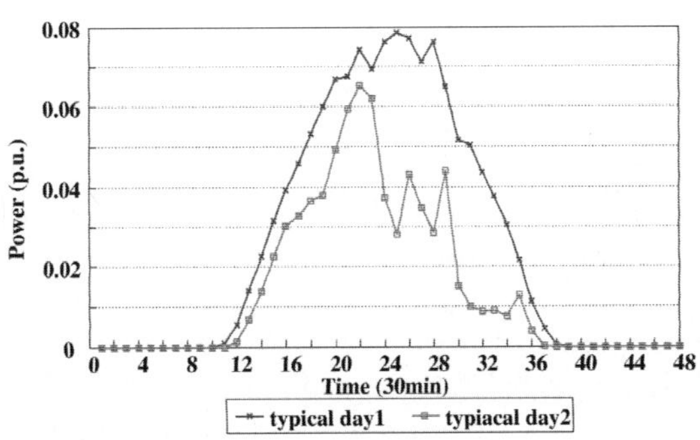

Figure 3. Curves of PV power output in two typical days.

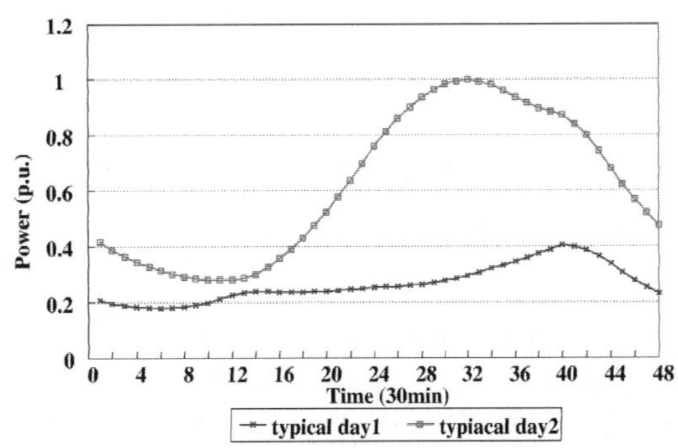

Figure 4. Normalization of load demand under operation scenarios in two typical days.

Loads of each node are depicted in figure 4. The selection of typical days depends on the time that voltage breaks the upper or the lower bound. In this case, η_s is 0.95, $SOC_0 = SOC_T = 0.4$, SOC_{lb} is 0.3 and SOC_{ub} is 0.9.

4.2. Multiple output modes of PV

Due to the different capability of reactive power output of different PV systems, there are several power output modes for PV. Details of each mode are further explained as follows:

- **Mode 1:** Unity Power Factor.

$$Q_j = 0 \tag{18}$$

In this mode, no reactive power is generated by PV, which means the value of Q_j is always zero.

- **Mode 2:** Constant power factor.

$$Q_j = \begin{cases} P_j/\tan\varphi & P_j \le S_j \sin\varphi \\ Q_j = 0 & P_j > S_j \sin\varphi \end{cases} \tag{19}$$

In this mode, the power factor of PV is constant, and its reactive power output is influenced by active power output. S_j is the rated power of jth PV system, φ is the power factor angle. The power factor is set as 0.98 here.

- **Mode 3:** Constrained reactive power.

$$Q_{min,j} \le Q_j \le Q_{max,j} \tag{20}$$

where, $Q_{\min,j}$ and $Q_{\max,j}$ are the maximum inductive and capacitive reactive power respectively of jth PV system, and they are set as $Q_{min,j}$=-0.3S_j,$Q_{max,j}$=0.3S_j.

4.3. Results and analysis

4.3.1. Analysis of energy storage configuration scheme. The configurations of BESS in three modes of PV are listed in table 1. The computational time of processing this algorithm is around 2 hours.

Table 1. Deployment plans of BESS under basic scenarios with different PV output modes.

		Node 11	Node 15	Node 30	Node 32	sum
Mode 1	C_r/p.u.	0.172	0.239	0.165	0.224	0.8
	P_r/p.u.	0.043	0.060	0.041	0.056	0.2
Mode 2	C_r/p.u.	0.168	0.244	0.160	0.228	0.8
	P_r/p.u.	0.042	0.061	0.040	0.057	0.2
Mode 3	C_r/p.u.	0.147	0.265	0.149	0.240	0.801
	P_r/p.u.	0.036	0.066	0.037	0.060	0.199

According to table 1, the total rated capacity and rated power of BESS are about 0.8 p.u. and 0.2 p.u. respectively. The mode of the PV operation has little influence on the configuration of BESS. Despite the large load demand at night in the typical day 2, the PVs aren't able to keep voltage within reasonable range whichever mode PV operates in because photovoltaic inverter stops working when the light intensity is lower than the threshold of PV inverter at night. Thus battery energy storage systems are dispatched to adjust the voltage. Besides, according to the table 1, the rated capacity and rated power of BESS at the end of the line are larger than those of BESS in other nodes.

4.3.2. Effect of voltage regulation. In this case, $V_{hl} = 0.9$ p.u., $V_{hu} = 1.1$ p.u., $V_{sl} = 0.952$ p.u., $V_{su} = 1.048$ p.u. Time sequence distributions of maximum node voltage under typical day 1 and minimum node voltage under typical day 2 after regulation are depicted in figures 5 and 6. The dotted line in figures 5 and 6 represent the lower and upper soft bounds of node voltage respectively.

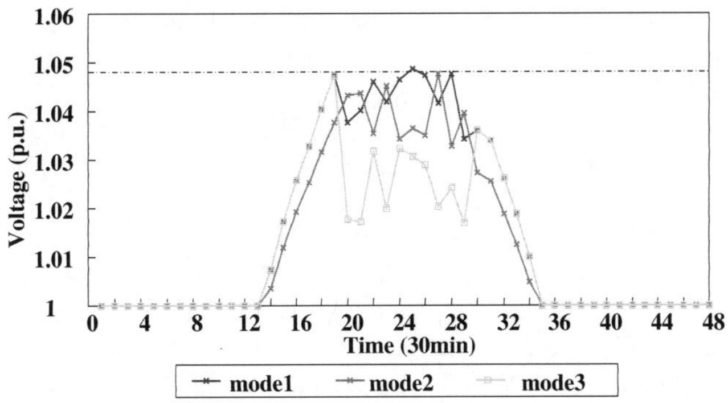

Figure 5. maximum node voltage after regulation in typical day 1.

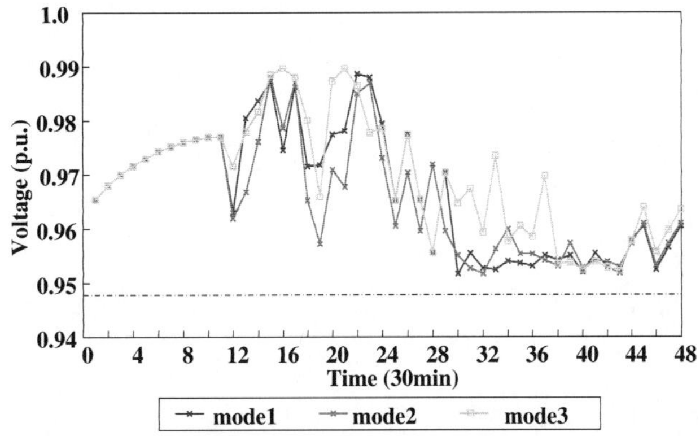

Figure 6. Minimum node voltage after regulation in typical day 2.

From figures 5 and 6 we can see that, the voltage basically meets the soft restriction under all PV output modes during the whole day after deploying BESS. Besides, the voltage always meets the hard restriction under all PV output modes during the whole day. Thus, it can be seen that the coordinated power control between PV and grid-connected BESS has a positive effect on voltage regulation.

5. Conclusions

In this paper, a bi-level programming based BESS deployment model is proposed, aiming at regulating voltage violation in the distribution network considering the coordinated power control based on both BESS and distributed PV with multi-modes. From the case study, main conclusions can be drawn as follows:

- In a typical day scenario based on different PV output modes in the distribution network, BESS deployment scheme with large capacity and large rated power at or near the end of the line is preferred, in order to diminish the voltage violation.
- The coordinated power control between grid-connected PV and BESS contributes to better effect dealing with voltage violation.

References

[1] Wang Y, Zhang P, Li W, Xiao W and Abdollahi A 2012 Online overvoltage prevention control of photovoltaic generators in microgrids *IEEE Trans. Smart Grid* **3** 2071-8

[2] Tonkoski R, Lopes L A and El-Fouly T H 2011 Coordinated active power curtailment of grid connected PV inverters for overvoltage prevention *IEEE Trans. Sustain. Energy* **2** 139-47

[3] Hiscock N, Hazel T G and Hiscock J 2008 Voltage regulation at sites with distributed generation *IEEE Trans. Ind. Appl.* **44** 445-54

[4] Sarimuthu C R, Ramachandaramurthy V K, Agileswari K R, *et al* 2016 A review on voltage control methods using on-load tap changer transformers for networks with renewable energy sources *Renew Sustain Energy Rev* **62** 1154-61
[5] Alyami S, Wang Y, Wang C S, Zhao J H and Zhao B 2015 Adaptive real power capping method for fair overvoltage regulation of distribution networks with high penetration of PV systems *2015 IEEE Power & Energy Society General Meeting* (Denver, USA) p 1
[6] Nazaripouya H, Wang Y, Chu P, Pota H R and Gadh R 2015 Optimal sizing and placement of battery energy storage in distribution system based on solar size for voltage regulation *IEEE Power & Energy Society General Meeting* (Denver, USA) pp 1-5
[7] Fan Q and Yan X 2016 Self-adaptive differential evolution algorithm with zoning evolution of control parameters and adaptive mutation strategies *IEEE Trans Cybern* **46** 219-32
[8] Li H, Zhang L and Jiao Y 2016 Discrete differential evolution algorithm for integer linear bilevel programming problems *J Syst Engrg Electron* **27** 912-9

Energy-saving management modelling and optimization for lead-acid battery formation process

T Wang[1,2], Z Chen[1,2], J Y Xu[1,2], F Y Wang[3] and H M Liu[3,4]

[1]State Grid Energy Conservation Service Co. Ltd., Beijing 100052, China
[2]State Grid (Beijing) Energy Conservation Design and Research Institute Co. Ltd., Beijing 100083, China
[3]College of Energy and Electrical Engineering, Hohai University, Nanjing, 211100, China

E-mail: liuhaom@hhu.edu.cn

Abstract. In this context, a typical lead-acid battery producing process is introduced. Based on the formation process, an efficiency management method is proposed. An optimization model with the objective to minimize the formation electricity cost in a single period is established. This optimization model considers several related constraints, together with two influencing factors including the transformation efficiency of IGBT charge-and-discharge machine and the time-of-use price. An example simulation is shown using PSO algorithm to solve this mathematic model, and the proposed optimization strategy is proved to be effective and learnable for energy-saving and efficiency optimization in battery producing industries.

1. Introduction

With the increasing challenges of energy and climate change, it is realistic and urgent to figure out how to manage power consumption loads in more optimal and better ways to make electrical power consumption more efficient.

As those industries developed vigorously, which have a great demand of batteries such as wind power, energy storage and electric vehicle industries, as a result, there has been a sharp rise in battery production amounts in recent years. The lead-acid battery has an extensive application in the fields of communication, railway transportation, power energy, aircraft industry and shipbuilding industries because of its prominent advantages: simple production process, cheap raw material, better power characteristic, low self-discharge rate and better high-and-low temperature performance. The lead-acid battery industry will still enjoy a great development and market potential in the next several decades [1].

As a high energy-consumption industry, the battery production gains more and more attention from the public about energy loss in battery production process [2,3]. The battery formation process, which is led by power to enable the internal chemical reaction of new produced batteries to proceed repeatedly and circularly in order to motivate the inside active substance, is one of the important steps in lead-acid battery manufacturing. The formation process takes up about 60-65% of the whole power consumption in battery production. For many researchers, it becomes an intensive problem to increase the utilization factor of power in formation process. An energy-saving manner to release the extra electric energy of the formation process to the power grid and utilize it again is put forward to solve this issue and has been applied abroad in actual production. However, releasing power directly from lower voltage grid to higher voltage grid is specifically forbidden in China. A strategy is presented in [4], where energy discharged during the formation process is transmitted conversely to power grid by inverters in order to avoid being wasted on high-power resistors. Nevertheless, manufacturers may need to pay a lot for constructing appropriate power sub-networks to prevent the shock wave towards high voltage grid. As a result, the production cost cannot be reduced and the implementation will be difficult. Another method based on DC-bus topological structure has been come up with in [5] to decrease sub-networks construction cost by using the parallel high-capacity energy storage device as an energy buffer. This method will bring about extra production cost, as well as maintenance expense. A formation step control strategy is explained in [6] to optimize the massive batteries formation procedure and accomplish energy feedback without any pollution to the power grid. To be specific, the energy released in the charge process is directly utilized in the parallel discharge process.

In this paper, a comprehensive energy consumption optimization model considering charge and discharge machine (CDM) efficiency and time-of-use (TOU) price is established, aiming at reducing electricity cost and increasing energy utilization rate.

The paper is organized as following: a general introduction of lead-acid battery formation process is presented in Section 2, then an optimal model of battery formation process to realize minimum total electricity expense is established in Section 3. The optimal algorithm and a case study are shown in Section 4 and Section 5 respectively, and finally main conclusions are drawn in Section 6.

2. Analysis on lead-acid battery formation process

Nowadays, lead-acid batteries are among the most popular batteries, and their manufacturing is a typical high energy-consumption industry. The raw material of lead-acid battery includes lead ingot, lead powder, vitriol, battery lead plate, battery holder and other active substance and so on [7]. The electricity expense occupies about 90% of the overall energy cost of the enterprise, therefore there is an enormous energy-saving potential in the electricity consumption process.

2.1. The lead-acid battery formation process

The formation process is indispensable in the lead-acid battery production before finished products leaving the factory. The difference among batteries in series connection can be reduced by optimizing their formation process and using advanced electric facilities. Through these ways the quality of batteries can be improved including service life and initial capacity.

Basically, the formation process is to enable the internal reaction of new batteries by charging and discharging circularly in order to motivate the active substance such as PbO_2 and spongy lead so that the finished battery is capable of storing and generating electric power [8]. The battery formation process varies among different types of batteries and different manufacturers. Anyway, batteries always charge with constant current or constant voltage, and discharge in a constant-current way. The only difference in formation process of batteries with various types and capacities is the charge-and-discharge time.

2.2. The CDM for formation process

An energy saving formation process is dependent on an ideal CDM. The CDMs for lead-acid battery can be regarded as a power source that supplies batteries with constant-current electricity and a power consumption load as well that transfers energy back to power grid to achieve energy regeneration. In consideration of the current situation that the equipment is not advanced enough and energy consumption is huge in most battery manufacture factories, it is imperative to upgrade electrical facilities and update the manufacture techniques to save energy and reduce emission. It is more efficient in formation process and satisfies the demand for future development of battery producing industries to adapt Insulated Gate Bipolar Transistor (IGBT) charge and discharge device than Silicon Controlled Rectifier (SCR).

The electricity transformation efficiency is the most important factor for IGBT CDM in battery producing process. In the formation process, transformation efficiency has a direct relationship with electric power consumption. If the transformation efficiency is not perfect, the economy cost will be increased accordingly. As a result, a high level transformation efficiency is a key-point for energy saving.

$$\eta = \frac{P_k}{P_s} \times 100\% \tag{1}$$

where, η means the transformation efficiency of CDM, P_k and P_s mean the output power and input power of CDM respectively. This transformation efficiency of IGBT is relevant to its load rate and temperature. The relationship is pictured in figure 1.

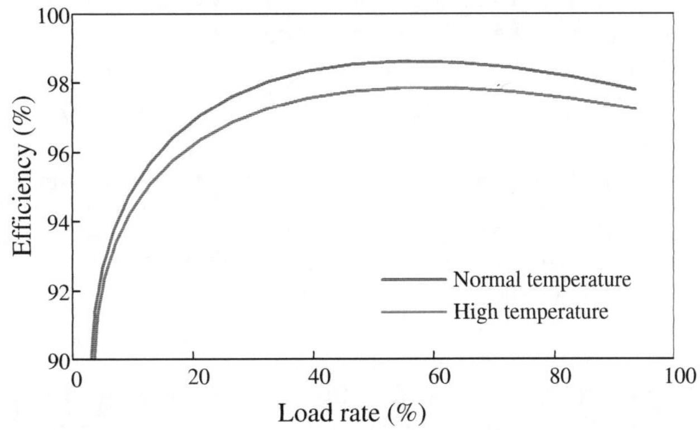

Figure 1. Power efficiency curve of IGBT CDM.

The load rate β of IGBT is

$$\beta = \frac{P}{P_0} \times 100\% \tag{2}$$

where, P and P_0 mean the load power and rated power of CDM.

Figure 1 indicates that the transformation efficiency η trend is not constant with the load rate β. There is a sharp increase in transformation efficiency within the section where the load rate ranges from 0 to 30%. While the load rate grows from 30% to 100%, transformation efficiency keeps at a high level and even goes a little lower near the end. Moreover, ignoring other factors, the increase of operating temperature will make transformation efficiency drop.

3. Construction of optimal model on formation process

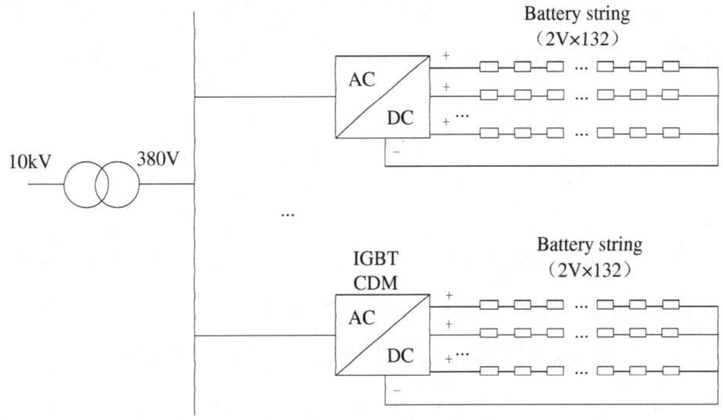

Figure 2. Formation progress topology.

The main equipment of formation process is CDM. Each machine carries more than one string of batteries. Supposing that the charging voltage is 400 V, meaning that every string has 132 batteries and the finished product is 2V-battery. Obviously every battery in the string can get a voltage about 3 V. The concrete structure of formation is shown in figure 2.

3.1. Assumptions

To simplify the model, we make following assumptions:

- Every CDM can start or stop at any time, with no time limit and no times limit. Overlooking the temperature increase and the service life reduce caused by the start-stop, we just take the start-stop loss into consideration when every start-stop occurs.
- Regard every string of batteries as one target, and ignore their internal difference [9]. Never would a battery break down when formation proceed.
- Do not consider electricity quality.
- Every string of batteries has the same formation period, while their formation power curve may be different.
- Overlook the time of replacing new batteries. The target time section is any complete period of formation.
- The efficiency of IGBT CDM is influenced by temperature and load rate, but in this paper only the load rate is considered.

3.2. The objective function

Because the formation curves of every string are different, batteries in different strings can hardly charge at the same time. Either all strings charging or standing is low efficient. To avoid this, we take several actions:

- According to TOU electricity price, we change the formation start moment of every string in order that more batteries can charge during bottom load periods and keep standing or discharge during peak load periods.
- Optimize the formation start moment of every string to avoid all strings batteries discharging at the same time.
- Since high load rate makes high transformation efficiency, we artificially change the string numbers connected to every IGBT machine to keep the load rate at a high level.

In the paper, the formation period T is the targeted time section. An optimal model is established to minimize the electricity cost of formation process

$$C = \min_{u_j(t)}\{\rho(t)\sum_{t=1}^{T}\sum_{j=1}^{J}(\frac{u_j(t)P_j(t)}{\eta(P_j(t))} + C_j(t))\} \tag{3}$$

where, every interval is numbered by t, t=1,…,T, and $\rho(t)$ means TOU electricity price. For each CDM indexed by j, $P_j(t)$ means the load power, $C_j(t)$ means the start-stop loss and $u_j(t)$ means a state variable which equals 1 when machine works and equals 0 when machine stops. Here the load power of CDM j is

$$P_j(t) = \sum_{k_i(t)=j} P_i(t), \quad k_i(t) = \{1,\cdots,J\} \tag{4}$$

where, i means the index of string of batteries $(i = 1, 2, \cdots\cdots, I)$. $k_i(t)$ means the number of CDM that the ith string connected to in t^{th} interval. $P_i(t)$ means power of the ith string of batteries.

The charging power of each string is related to the formation power and formation start moment.

$$P_i(t) = f_i(t - t_{0i}) \tag{5}$$

where, for the ith string, $f_i(t)$ means the formation power curve function, and t_{0i} means the formation start moment.

We overlook the time for replacing new battery string, hence the power curve of every string is a periodic function with a period T.

$$f_i(t+T) = f_i(t) \tag{6}$$

where, the period T is determined by a given battery power curve. In this context, the formation period of each string of batteries is the same, although their power curves may be different.

Every CDM will bring about some energy loss when a start or a stop occurs. The start-stop loss is defined as:

$$C_j(t) = [1-u_j(t-1)] \cdot u_j(t) \cdot C_s + u_j(t-1) \cdot [1-u(t)] \cdot C_t \tag{7}$$

where, C_s means the start loss and C_t means stop loss. Assume that C_s=C_t.

3.3. The constraint conditions

- Load rate constraint

$$\alpha P_{j0} \le P_j(t) \le P_{j0}, \quad \forall t \ge 0 \tag{8}$$

where, $0 \le \alpha \le 1$. To keep the CDM within a high efficiency range, we set $\alpha \ge 0.3$.

- Overall power constraint

$$\sum_{i=1}^{I} f(t-t_{0i}) = \sum_{j=1}^{J} u_j(t) P_j(t) > 0, \quad \forall t \ge 0 \tag{9}$$

which indicates that the overall power is positive at any moment in the dispatch period T.

- Formation start moment constraint

$$0 \le t_{0i} \le T \tag{10}$$

which shows that the formation start moment of each string must be within the dispatch period T.

- Switch intervals constraint

$$|t_{i1} - t_{i2}| \ge t_{is}, \quad \forall t_{i1} \ne t_{i2} \in A, \quad A = \{t_i | k_i(t) - k_i(t-1) \ne 0\}, \quad t = 2, \cdots, T \tag{11}$$

where, A means the set of switch intervals and t_{is} means the minimum switch interval time.

- Total switch times constraint

$$\sum_{t=1}^{T} m_{i,t} \le n_i \tag{12}$$

where, $m_{i,t}$ represents whether the ith string of batteries switches within time section t. $m_{i,t} = 1$ if $k_i(t) - k_i(t-1) \ne 0$. n_i means the maximum switch times for each string within the dispatch period T.

4. Algorithm for model solving

The model is solved by Particle Swarm Optimization (PSO) algorithm. In the optimization model, $k_i(t)$ and t_{0i} mean optimized variables, and the fitness function is the objective function C. To simplify the computing process, we use special battery strings allocation method to optimize $k_i(t)$:

- Firstly, sort all strings of batteries at every time interval according to their power value;
- Then, combine the biggest two strings and the smallest two strings as a pair, and the rest pairs can be combined in the same way;

- At last, connect the string pairs to every CDM.

The advantage of this method is that every CDM has similar load power and accordingly power releasing will hardly occur.

The PSO is applied like that: firstly, we pre-set a random number for t_{0i}, and figure out $k_i(t)$. Then let computer judge whether the combination $(t_{0i}, k_i(t))$ satisfies the given constraints. If not, pre-set a number for t_{0i} again until the combination meet all constraints. These combinations are particle swarm. The most optimal particle will be picked out from the swarm to be applied into the optimization algorithm through a number of iterations. Finally, we can get the optimum solution that makes the overall formation operation cost minimum.

The value of key parameters are given as follows: the particle swarm size M=100, the inertia factor (the velocity of every particle) w=0.6, learning factor $c_1 = c_2 = 2$, convergence index ε=0.0005.

The formation period of batteries is divided into T intervals as the dispatch period. If let the minimum switch interval equal dispatch period, then two constraints can be removed: (1) the switch intervals constraint; (2) the switch time constraint.

5. Case study

5.1. The parameters of case

The management model and strategy for formation process in a certain lead-acid battery enterprise is taken as an example. Main products in this enterprise are 2 V batteries, which are charged by IGBT CDM in a constant-current formation mode. The charging voltage is about 400 V and there are 132 batteries in every string. The whole formation process lasts more than 100 hours while the charging, discharging, standing processes are conducted alternately. Also, the formation current and voltage are precisely set in order to ensure the quality of batteries.

Two kinds of 2 V batteries are taken as examples which have different capacity (400 Ah and 500 Ah) and the same formation period. Based on the practical operation data, the information of formation process (voltage, current, duration, etc.) are neatened in tables 1 and 2.

Table 1. Formation information of 2 V battery with capacity of 400 Ah.

Steps	Formation process	Current (A)	Time interval (h)	Voltage (V)
1	Charging	16	4	210~215
2	Charging	48	40	215~265
3	Charging	20	8	265~268
4	Standing	0	1	-
5	Discharging	40	8	204~197
6	Charging	48	8	206~272
7	Charging	24	36	272~275
8	Standing	0	1	-
9	Discharging	40	10	207~191
10	Discharging	48	11	216~260
11	Charging	20	4	260~271
12	Charging	4	4	271

The company is equipped with two kinds IGBT CDMs. This study takes one of the two IGBT CDMs with a capacity of 60 kW into consideration. Based on the operation manual, the data pertaining to the relationship between the transmission efficiency and load rate of IGBT CDM is presented in table 3, according to which the transmission efficiency function curve can be fitted. We can also get the transmission efficiency value of every load rate from the fitting function curve in figure 1.

Table 2. Formation information of 2 V battery with capacity 500 Ah.

Steps	Formation process	Current (A)	Time interval (h)	Voltage (V)
1	Charging	20	4	211
2	Charging	50	45	221~268
3	Charging	25	8	253~264
4	Standing	0	1	-
5	Discharging	50	8	202~205
6	Charging	50	10	206~273
7	Charging	30	32	273~274
8	Standing	0	1	-
9	Discharging	50	8	207~191
10	Charging	50	12	216~260
11	Charging	25	4	271
12	Charging	5	2	271

Table 3. Transformation parameter of 60 kW rated capacity IGBT CDM in 23 centigrade.

Load rate (%)	Efficiency (%)
5	94.80
10	97.04
15	97.66
20	98.01
25	98.18
30	98.32
50	98.61
75	98.63
100	98.58

The parameters in the model are as follows:

- The start-stop loss is set to be 0.5 kW, that is to say $C_j(t)=0.5$.
- 5 CDMs work in parallel and 18 strings of batteries are connected to these CDMs.
- The capacity of every CDM is 60 kW.
- Assuming that the first 8 strings of batteries follow the first kind battery formation model while the rest 10 strings follow the second kind.
- 4 strings of batteries can be connected to every CDM at most.
- TOU price: peak load time lasts for 16 hours with a price of 1.2 RMB/kWh while bottom load time lasts for 8 hours with a price of 0.56 RMB/kWh.
- Punishing price is set to be 3 RMB/kWh.
- The formation period T=135 hours is divided into 135 dispatch intervals.

5.2. Interpretation for the result of optimization model

According to the steps presented above, we can solve the formation management optimization model. For comparison, two optimizations are carried out: in optimization I, only t_{0i} is optimized while $k_i(t)$ remains unchanged; in optimization II, two optimized variables both t_{0i} and $k_i(t)$ are all taken into consideration. That is to say, we optimize not only the start moment of every battery string but also the battery strings allocation method. The power curve before optimization is shown in figure 3, and the optimal formation power curves are shown in figures 4 and 5.

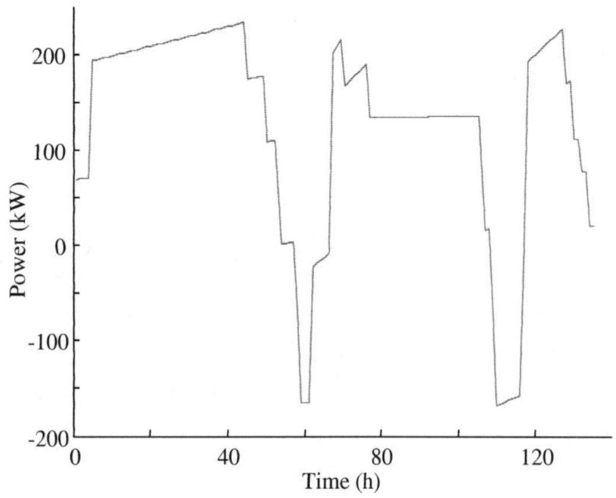

Figure 3. Power curve before optimization.

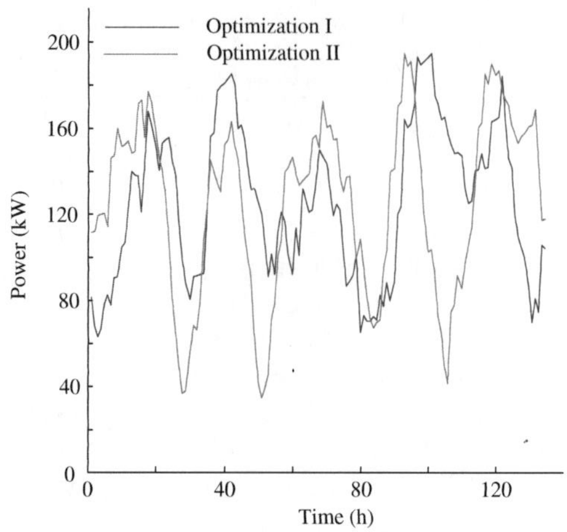

Figure 4. Power curve of optimization I and II.

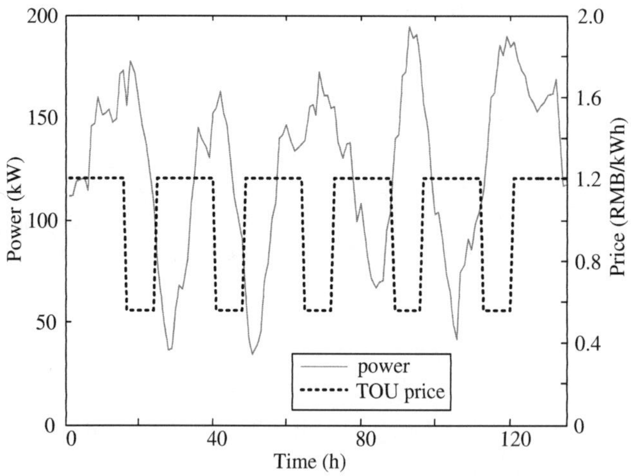

Figure 5. Power curve of optimization II and TOU price.

Table 4. Results comparison before and after optimization.

Results comparison	Before optimization	Optimization I	Optimization II
Formation operation cost (RMB)	27994.03	18887.31	18617.10
Maximum power (kW)	233.09	194.71	194.22
Minimum power (kW)	-168.03	62.99	34.35
The peak valley power difference (kW)	401.12	131.72	159.87
Average efficiency of CDMs	87.17%	87.96%	87.96%
Maximum time of power releasing (h)	18	0	0

As table 4 indicates, the electricity costs after optimization I and II are reduced, meaning that the batteries producing cost is decreased. Similarly, we find that power releasing caused by formation discharging does not occur after optimization any more. That is because we changed the start moment of formation process and used a special batteries strings allocation optimization method. The peak valley power difference before and after the optimization also decreases. From figure 4, we infer that the overall formation power distribution is skewing to the bottom load time section. The characteristic of overall power consumption presents to be higher when electricity price is low while lower when within peak electricity price, which is in accordance with the purpose of the objective function.

However, comparing optimization I with II, the difference between peak and bottom power value is widen from 131.72 kW to 159.87 kW when the battery allocation method is adapted. That is because the more optimal battery allocation is, the more influential the TOU electricity price is by inducing power consumers move load to lower-price time sections in order to save their electricity costs. Therefore, a new load peak will appear and is likely to bring about power crash. This is the feature of this proposed optimization strategy: for the sake of lead-acid batteries manufacturers, it does not make any constraints upon the difference between peak and valley value of power curve.

6. Conclusions

In this paper, a comprehensive formation process optimization model considering transformation efficiency of CDMs and TOU price was built to reduce the overall formation electricity costs. Based on the TOU price, the formation start moment was optimized to make heavy current in bottom price time while slight current or standing in peak price time sections. Meanwhile, based on the CDM transformation efficiency curve, a reasonable allocation method for battery strings connected to CDMs was applied in order to avoid lower efficiency for CDMs caused by slight load rate. In a more general situation, the proposed optimization model will be of great benefit to the lead-acid battery manufacturer to decrease the total cost.

Acknowledgments

This work is supported by SGCC Project SGRI-DL-71-15-006.

References

[1] http://wenku.baidu.com/view/264183a069dc5022aaea00ce.html
[2] Maharjan L, Yamagishi T and Akagi H 2012 Active-Power control of individual converter cells for a battery energy storage system based on a multilevel cascade PWM converter *IEEE Trans. Power Electron* **27** 1099-107
[3] Zheng Z, Zhou W and He H 2013 Research on the novel charge-discharge system of storage battery *J Comput* **2** 485-92
[4] Liu Z 2010 Research on monitoring system of high capacity li-ion battery formation equipment (Beijing: Beijing Jiaotong University)
[5] Wang S, Xi X and You L R 2011 Research on formation energy feedback control system of lithium power battery *Chin J Power Sources* **35** 393-5
[6] Zhang C J 2014 Large-scale and flexible control technology research for battery formation process equipment (Wuhan: Huazhong University of Science and Technology)
[7] Li J J and Chen F A 2015 Simulation and experimental study on charge and discharge characteristics of battery *Auto Applied Technol* **8** 89-91
[8] Sun H J 2011 Research on charging and discharging control system for power battery formation (Harbin : Harbin Institute of Technology)
[9] Chiang Y H, Sean W Y and Ke J C 2011 Online estimation of internal resistance and open-circuit voltage of lithium-ion batteries in electric vehicles *Chin J Power Sources* **8** 3921-32

IOP Conf. Series: Earth and Environmental Science **93** (2017) 012016 doi:10.1088/1755-1315/93/1/012016

The effects of sun-tracking on rooftop photovoltaic systems

D Wang[1,2]

[1]Business School, China University of Political Science and Law, 25 Xitucheng Road, Beijing, 100088, China

E-mail: dwang@cupl.edu.cn

Abstract. Sun-tracking is an important mechanism to boost the electricity generation from photovoltaic systems by adjusting the orientation of the systems to receive more solar radiation. However, compared to the fixed-array systems, the use of the single-axis and dual-axis sun-tracking systems is usually impeded by increased cost, decreased reliability, and cumbersome maintenance. In this paper, we compare the performance of fixed-array, single-axis, and dual-axis rooftop photovoltaic systems in California using the data envelopment analysis method. For each system, an efficiency score is computed based on capacity, electricity generation, system cost, module use, solar irradiance and ambient temperature. We find the single-axis systems perform significantly better than the fixed-array and dual-axis systems. The fixed-array systems are more efficient than the dual-axis systems but the difference is not significant. The results imply the policymakers and solar installers should promote the single-axis sun-tracking and be cautious of using the dual-axis systems.

1. Introduction

The electricity generation from a solar photovoltaic (PV) system depends on the amount of solar radiation received by the solar panels. The solar radiation imparted on the panels is maximized when the sun lights hit the panels at a perpendicular angle. Since the position of the sun in the sky changes over time, the orientation of PV panels should be dynamically adjusted to produce the maximum power output. Sun-tracking is an important mechanism to boost the electricity generation by adjusting PV orientation [1,2]. The gain in electricity output from sun-tracking compared to the fixed-array system depends on many factors. Literature has reported that annual output gain of sun-tracking systems over the fixed-array systems lies in the range between 20% and 40% [1].

In terms of degree of freedom, there are two types of sun-tracking systems, i.e., single-axis and dual-axis. The single-axis system has one degree of freedom and usually tracks the sun's east-to-west movement [3]. The dual-axis system follows the sun's east-to-west movement and changing altitude angle [4]. While the sun-tracking can increase the electricity output compared to the traditional fixed-array systems, it suffers from cost, reliability, and maintenance problems as a result of more complex design [5].

In this paper, we would like to evaluate the performance of sun-tracking using data from the California Solar Initiative (CSI), the most prominent solar incentive program in the state. Specifically, we apply the data envelopment analysis (DEA) method to evaluate a dataset of 1,030 rooftop PV systems. DEA is a non-parametric mathematical programming approach that has been demonstrated as a powerful tool to evaluate the performance of electricity generation facilities, including hydro plants, thermal plants, and wind farms [6-9]. However, its application to solar PV systems is rather scarce except for a few recent papers [10-12].

2. Model

DEA constructs an efficiency frontier by fitting piecewise linear segments to enclose all the units to be evaluated. For each unit, DEA computes an efficiency score to capture the unit's relative distance to the frontier. Units on the frontier are deemed efficient and assigned efficiency of one. Units within the frontier are deemed inefficient and assigned efficiency less than one. To apply DEA, it is important to discern discretionary inputs and non-discretionary inputs. The environmental factors are exogenous to the PV systems and hence are non-discretionary. To incorporate the non-discretionary inputs, we use the DEA model developed in Banker and Morey (1986). The model has been demonstrated to be useful for PV installation assessment [10,11].

We let j denote the index of the PV system for $j = 1, \dots, n$; k denotes the system under evaluation; ϕ_k denotes the efficiency score for system k; $\lambda = (\lambda_1, \lambda_2, \dots \lambda_n)$ is the vector of weights for the units; $x_j = (x_{1j}, x_{2j}, \dots, x_{mj})$ denotes the vector of m discretionary inputs for system j ; $z_j = (z_{1j}, z_{2j}, \dots, z_{rj})$ denotes the vector of r non-discretionary inputs for system j; $g_j = (g_{1j}, g_{2j}, \dots, g_{sj})$ denotes the vector of s outputs for system j. The model is as follows.

$$
\begin{aligned}
&\text{Efficiency}_k = \text{Min} \quad \phi_k \\
&s.t. \quad \phi_k x_{ik} - \sum_{j=1}^{n} x_{ij}\,\lambda_j \geq 0, \quad i = 1, \dots, m \\
&z_{lk} - \sum_{j=1}^{n} z_{lj}\,\lambda_j \geq 0, \quad l = 1, \dots, r \\
&-g_{hk} + \sum_{j=1}^{n} g_{hj}\,\lambda_j \geq 0, \quad h = 1, \dots, s \\
&\sum_{j=1}^{n} \lambda_j = 1 \\
&\lambda_j \geq 0, \quad \forall j = 1, \dots, n; \ \phi_k \text{ unconstrained}
\end{aligned} \tag{1}
$$

To capture the performance of PV systems, we employ the following inputs and outputs to capture the most salient features of solar installation performance. There are two variables for the non-discretionary inputs $\{z_{lj}\}$ in model (1):

- Solar irradiance: Solar irradiance is the most important factor in PV electricity generation. We use the global horizontal irradiance (GHI), measured in kW/m^2/day. GHI captures the solar resource available at the installation site.
- Ambient air temperature: This is the ambient air temperature measured in Celsius (°C) at the site of the system. Because the PV cell energy conversion efficiency decreases in temperature [13], ambient temperature can be regarded an undesirable input and we employ the inverse of it in computing model (1). Using the inverse of an undesirable input is a common technique in DEA [14].

We construct the following two variables for discretionary inputs $\{x_{ij}\}$ in model (1):

- Modules: This is the total number of PV modules used in the system. The module quantity can affect capacity and electricity generation. We note that number of modules has been used as an input in PV assessment in literature [10,11].
- Cost of the system: It is measured in thousand dollars and usually includes the module, inverter, labor and overhead costs.

We construct the following two variables for the outputs $\{g_{hj}\}$ in model (1):

- Capacity: We use the CSI rating as capacity of the system, measured in kW. The CSI rating is measured upon 1,000 W/m^2 solar irradiance, 20°C ambient temperature and 1 m/s wind speed. It also factors in inefficiency of inverter and design factors (e.g., shading, mounting, orientation). The CSI rating is on average 85% of the nameplate capacity.
- Electricity generation: This is the total electricity generated by the PV system measured in MWh for a designated time period.

In computation, we adjust for inflation for system costs by converting the costs of installations at different times to January 2013 value, using the monthly consumer price index for all urban consumers from the US Bureau of Labor Statistics (http://www.bls.gov/cpi/data.htm).

3. Data and variables

From the PV system data, we obtain the CSI working dataset and the production dataset (both datasets are available at https://www.californiasolarstatistics.ca.gov/data_downloads). From the data, we can identify four tracking types, i.e., fixed-array (no tracking), single-axis, dual-axis, and mixed-array. Because the CSI does not specify what tracking methods the mixed type is exactly using, we delete all mixed systems in data. For each system, we can observe the location (city, county and ZIP code), the capacity, the cost, and the quantity of the modules. For a subset of systems, CSI reports the monthly production data. By January 2016, a total of 157,802 system-month observations have been entered into the production dataset. Following previous literature that evaluates the sun-tracking performance in one annual period [1], we assess the impact of sun-tracking from January 1st 2013 to December 31st 2013. PV systems without production data have to be removed. Finally, we obtain a sample of 1,030 PV systems.

We obtain the solar irradiation data from NREL (http://www.nrel.gov/gis/data_solar). The data provides monthly average daily solar irradiation at 10-kilometer resolution for the entire United States. The ZIP code for each 10×10 kilometer grid cell is also given in the data. Therefore, we assign the irradiance (converted from irradiation) to a PV system by matching the ZIP code. Note that higher irradiation leads to higher electricity generation and among the three types of PV systems, on average the dual-axis systems are installed at sites with the highest irradiation of 5.35kWh/m^2/day. We obtain the ambient air temperature data from the National Climatic Data by the National Oceanic and Atmospheric Administration (NOAA, https://www.ncdc.noaa.gov/cdo-web/datasets). The NOAA records the temperature at a group of weather stations in California. For each PV system, we use the inverse distance weighting interpolation to derive the ambient temperature from weather station data [15]. Table 1 shows the descriptive statistics of all variables. There are 959 fixed-array systems, 56 single-axis systems, and 15 dual-axis systems. Based on table 1, the average cost-to-capacity ratios are 6.59$/W for fixed-array, 5.22$/W for single-axis, and 5.89$/W for dual-axis. Hence, by simple economic analysis, single-axis is the most cost-effective tracking method in terms of capacity expansion. This is further validated by the DEA results.

Table 1. Descriptive statistics for the variables.

	Unit	Fixed-array	Single-axis	Dual-axis
No. of systems	No.	959	56	15
Irradiation	kWh/m^2/day	5.30	5.34	5.35
Temperature	°C	17.66	16.06	18.76
Modules	No.	986.97	2038.29	362.20
Cost	thousand$	1294.80	2792.53	573.49
Capacity	kW	196.40	534.49	97.36
Generation	MWh	367.33	1015.55	159.90

4. Results

The DEA results obtained from model (1) are summarized in table 2 with the mean, standard deviation, minimum, maximum, 25th quartile and 75th quartile of the efficiencies reported for each sun-tracking type. Overall, the fixed-array systems attain a mean efficiency score of 0.58. The dual-axis systems have a mean efficiency of 0.54, the lowest among the three tracking types. The single-axis systems have a substantially better mean efficiency of 0.75, which is 29.24% higher than the fixed-array performance and 38.77% higher than the dual-axis performance. The maximum efficiency is 1.00 for both fixed-array and single-axis systems, while the highest efficiency for dual-axis is only 0.89. The efficiency scores of the single-axis systems strictly exceed the fixed-array and dual-axis at the minimum, 25th quartile and 75th quartile. The results seem to indicate that single-axis sun-tracking is the best tracking mechanism, while dual-axis is the worst one.

Table 2. Sun-tracking methods and efficiency measures.

	Fixed-array	Single-axis	Dual-axis
No. of systems	959	56	15
Mean	0.58	0.75	0.54
Std. Dev.	0.17	0.20	0.14
Min	0.20	0.42	0.28
Max	1.00	1.00	0.89
25th quartile	0.47	0.57	0.47
75th quartile	0.65	0.98	0.60

Since the DEA results do not provide statistical inference, we resort to statistical methods to test the differences. To compare the performance of the sun-tracking methods, we run the nonparametric Wilcoxon rank sum test and the t-test. The Wilcoxon rank sum test does not need the normality assumption and serves to complement the t-test. The test results are reported in table 3. The efficiency gain of the single-axis systems over the fixed-array and dual-axis systems is strongly significant in both t-test ($p<0.001$) and Wilcoxon rank sum test ($p<0.0001$). The difference between the fixed-array and the dual-axis systems turns out to be insignificant. The reason may be that the increased generation from a dual-axis system is compromised by the complexity and cost of the system compared to fixed-array. The results imply that adding one degree of freedom to the orientation of fixed-array PV system through the single-axis mechanism can substantially improve the PV performance. However, compared to the single-axis design, the PV performance deteriorates after a second degree of freedom is introduced through the dual-axis system. This is because compared to the single-axis system, the increment in solar radiation by adding a second degree of freedom is not sufficient to compensate for the increased cost from complex design and installation procedure [16].

Table 3. Test of differences.

	Single-axis vs. Fixed-array		Dual-axis vs. Fixed-array		Single-axis vs. Dual-axis	
t-test	7.1855	(0.0000)	-0.9022	(0.3672)	3.8250	(0.0003)
Wilcoxon rank sum test	5.9539	(0.0000)	-0.5402	(0.5891)	3.3798	(0.0007)

[a] The table reports the t-statistics and z-statistics followed by the p-value in parentheses.

An intriguing problem to explore is how the efficiencies of different sun-tracking methods have evolved over time. The CSI dataset does not specify the installation date but we can proxy it using the interconnection application date. It turns out all the systems in sample were installed in 2008-2012. We plot the annual mean efficiencies of the three tracking types in figure 1 over time. The single-axis tracking is substantially better than the fixed-array and the dual-axis for every single year in 2008-2012, and the gap has been widening from 2010 to 2012. Therefore, as time goes by, single-axis has been an increasingly favorable tracking method. The fixed-array is slightly better than the dual-axis in 2008 and 2009. But from 2010 onwards, the fixed-array and dual-axis have almost identical mean efficiencies.

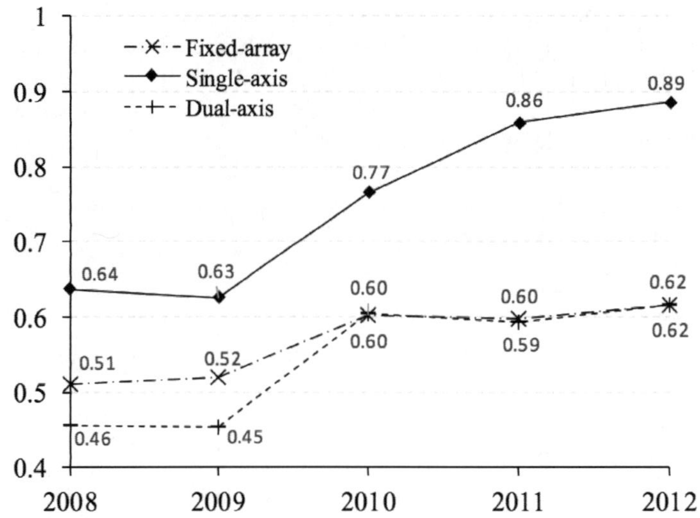

Figure 1. Change of performance of sun-tracking methods over time.

5. Conclusions

In this paper we study the performance of different sun-tracking mechanisms by applying the DEA method to rooftop PV systems in California. After factoring in capacity, electricity generation, cost, module use, solar irradiance and ambient air temperature, we find the single-axis systems perform significantly better than the fixed-array systems with a roughly 30% gain in efficiency. The dual-axis system does not show significant improvement over the fixed-array system. Moreover, the efficiency gap between single-axis and other tracking methods has been enlarging from 2010 to 2012. Our results imply that policymakers and solar installers should promote the use of the single-axis sun-tracking and be cautious of using the dual-axis systems. Finally, we would like to point out that the aforementioned results are derived from California data. In other regions with different climate and installation cost, the validity of the results should be examined through further tests [17]. However, the methodology developed in this study is general and can be adapted for different scenarios.

References

[1] Mousazadeh H, Keyhani A, Javadi A, *et al* 2009 A review of principle and sun-tracking methods for maximizing solar systems output *Renew Sustain Energy Rev* **13** 1800-18
[2] Singh G K 2013 Solar power generation by PV (photovoltaic) technology: A review *Energy* **53** 1-13
[3] Chang T P 2009 Output energy of a photovoltaic module mounted on a single-axis tracking system *Appl Energy* **86** 2071-8
[4] Al-Mohamad A 2004 Efficiency improvements of photo-voltaic panels using a sun-tracking system *Appl Energy* **79** 345-54
[5] Tania M H and Alam M S 2014 Sun tracking schemes for photovoltaic panels *2014 3rd Int Conf Dev Renew Energy Technol* 1-5
[6] Jha D K and Shrestha R 2006 Measuring efficiency of hydropower plants in Nepal using data envelopment analysis *IEEE Trans Power Syst* **21** 1502-11

[7] Sarica K and Or I 2007 Efficiency assessment of Turkish power plants using data envelopment analysis *Energy* **32** 1484-99
[8] Sueyoshi T, Goto M and Ueno T 2010 Performance analysis of US coal-fired power plants by measuring three DEA efficiencies *Energy Policy* **38** 1675-88
[9] Iribarren D, Vázquez-Rowe I, Rugani B, *et al* 2014 On the feasibility of using emergy analysis as a source of benchmarking criteria through data envelopment analysis: A case study for wind energy *Energy* **67** 527-37
[10] Wang D D and Sueyoshi T 2017 Assessment of large commercial rooftop photovoltaic system installations: Evidence from California *Appl Energy* **188** 45-55
[11] Sueyoshi T and Goto M 2014 Photovoltaic power stations in Germany and the United States: A comparative study by data envelopment analysis *Energy Econ* **42** 271-88
[12] Banker R D and Morey R C 1986 Efficiency analysis for exogenously fixed inputs and outputs *Oper Res* **34** 513-21
[13] Tyagi V V, Rahim N A A, Rahim N A, *et al* 2013 Progress in solar PV technology: Research and achievement *Renew Sustain Energy Rev* **20** 443-61
[14] Knox Lovell C A, Pastor J T and Turner J A 1995 Measuring macroeconomic performance in the OECD: A comparison of European and non-European countries *Eur J Oper Res* **87** 507-18
[15] Shepard D 1968 A two-dimensional interpolation function for irregularly-spaced data *23rd ACM Natl Conf* 517-24
[16] Abu-Khader M M, Badran O O and Abdallah S 2008 Evaluating multi-axes sun-tracking system at different modes of operation in Jordan *Renew Sustain Energy Rev* **12** 864-73
[17] Semprini S, Sánchez D and De Pascale A 2016 Performance analysis of a micro gas turbine and solar dish integrated system under different solar-only and hybrid operating conditions *Sol Energy* **132** 279-93

IOP Conf. Series: Earth and Environmental Science **93** (2017) 012017 doi:10.1088/1755-1315/93/1/012017

New design of a PEFC cathode separator of for water management

K Sugiura[1,2], N Takahashi[1] and T Kamimura[1]

[1]Department of Technological Systems, Mechanical Engineering Course, Osaka Prefecture University College of Technology, Neyagawa, Osaka 5728572, Japan

E-mail: sugiura@osaka-pct.ac.jp

Abstract. Generally, polymer electrolyte fuel cells (PEFCs) need humidifiers to prevent the drying of the membrane, but this use of humidifiers creates water management issues, such as the flooding/plugging phenomena and decreased system efficiency because of an increase in the electric energy needed for auxiliary equipment. Although most researchers have developed high-temperature membranes that do not need humidifiers, a lot of time is necessary for the development of these membranes, and these membranes drive up costs. Therefore, we propose a new cathode separator design that can recycle water generated by power generation in the same cell and a stack structure that can redistribute water collected in the cathode outlet manifold to drying cells. Because the new cathode separator has a bypass channel from the gas outlet to the gas inlet to transport excess water, a dry part in the gas inlet is supplied with excess water in the gas outlet through the bypass channel even if the PEFC is operated under dry conditions. Excess water in the PEFC stack can be transported from the cell with excess water to the drying cell through the cathode outlet manifold with a porous wall. Therefore, we confirm the influence of the plugging phenomenon in the cathode gas outlet manifold on the cell performance of each cell in the stack. As a result, the cell performance of the new cathode separator design is better than that of the standard separator under the low humidity conditions. We confirm that the plugging phenomenon in the cathode outlet manifold affects the cell performance of each cell in the stack.

1. Introduction

Recently, polymer electrolyte fuel cells (PEFCs) are beginning to spread to a wide range of fields, such as "ENE-FARM", as a domestic use co-generation systems, and "Mirai" and "Clarity Fuel Cell" as fuel cell vehicles. However, PEFCs are not spreading rapidly because of the cost. Although PEFCs need a humidification system to avoid membrane drying during operation, the use of humidifiers causes a decrease in the system's efficiency and the flooding/plugging phenomena. Hakenjos A, Litster S, Dunbar Z W, Jao T C *et al* have studied these issues on the cathode side using various visualization, such as optical images, MRI, and x-ray [1-4]. K. Takaya *et al* have elucidated the flooding phenomenon in GDL by a computational simulation [5]. Tanuma T, Eller J, Nishida K, Kitahara T, Kinumoto T, Date T *et al* proposed water repellency processing to GDL [6-11] as a solution to these issues. However, this method lacks durability and is not effective. Although Rastedt M and Wippermann K *et al* have also studied membranes for a high-temperature PEFC (HT-PEFC) that operates at a temperature where water is not condensed [12,13], and a lot of time is necessary for the practical use of HT-PEFC. Because the gas inlets of both electrodes are located at the top of the cell, the flooding/plugging phenomena occur downstream of the cell as shown in figure 1 (100% RH condition). We understand that the flooding/plugging phenomena in the cathode do not occur alone, and these flooding/plugging phenomena affect the anode channel through the membrane. The plugging water transport phenomenon through the membrane is affected by the location of the cause of the flooding/plugging phenomena in each gas channel [14]. Therefore, we have developed a self-water management separator with a water absorption layer (WAL), which absorbs the plugging water and drains to the water drain to eliminate the plugging phenomenon as shown in figure 2. As a result, although several cathode channels of the cell with a WAL of 14 μm of mean pore diameter (MPD) were plugged by excess water, the plugging phenomenon hardly occurred on the cell with a WAL of 0.9 μm MPD under 100% RH condition as shown in figure 3. Although the cell voltage and the plugging ratio were unstable on the cell with a WAL of 14 μm MPD, that of the 0.9 μm MPD was extremely stable as shown in figure 4 [15]. Although the self-water management separator with a WAL was effective under a high humidity operation, the WAL promoted the drying of a membrane under a low humidity operation because the WAL absorbed too much water in the membrane. A recent trend is a decrease in the amount of humidity to reduce the power required for accessory units. Because there is a large difference in the humidity between the gas inlet and the outlet mentioned above, if the amount of humidity of both electrodes decreases, cell performance deteriorates due to drying of the membrane near the gas inlet. Therefore, we propose a new self-water management separator with a bypass channel as shown in figure 5. Scilicet, this self-water management separator (SWMS), has a bypass channel connecting the supersaturation area and the low humidity area to move excess water by the capillary phenomenon and water concentration diffusion for effective use of the produced water. Because this improvement only adds to the present PEFC composition, the development time is negligible compared to HT-PEFC. The SWMS can be introduced to the present PEFC at once if the effectiveness of the SWMS on cell performance can be confirmed.

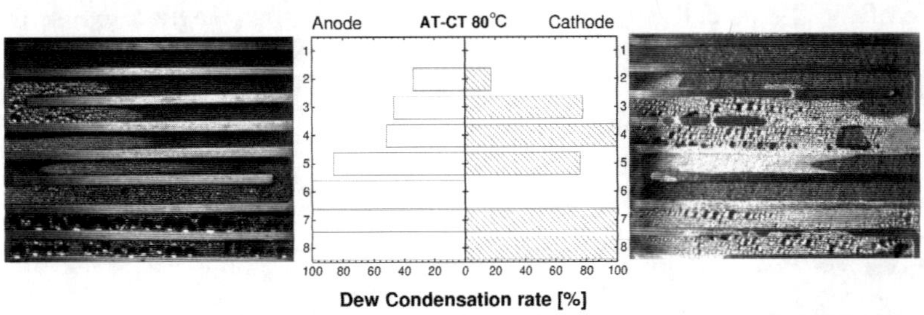

Figure 1. Plugging phenomenon in both gas channel under Tc/Tda/Tdc=80°C/80°C/80°C (100% RH)

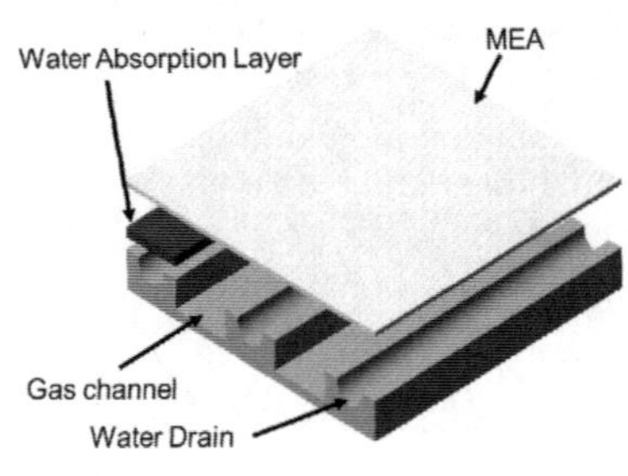

Figure 2. Schematic diagram of the self-water management separator with WAL

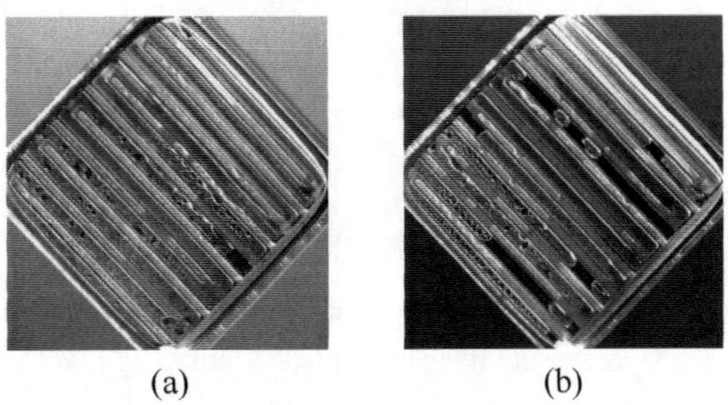

Figure 3. Comparison of plugging phenomena: (a) 0.9 μm MPD and (b) 14 μm MPD.

However, PEFC stacks have specific issues, such as the flow rate distribution to each cell and the temperature distribution in the stack. The supply flow rate to each cell of the stack depends on the fluid resistance of each cell; the temperature at the middle of the stack increases due to an exothermic reaction (cell reaction). Even if only one cell of the stack is deteriorated, the performance of each cell influences the other cells, and finally the performance of the entire stack decreases. In the PEFC stack, the humidity difference also occurs because of the peculiar issues of a stack and originates in the difference of the stacked cell positions [16]. If it is possible to move excess water from the supersaturation cells to the low humidity cells through the cathode outlet manifold, the system efficiency can be improved by reducing energy consumption. Therefore, we also propose a self-water management PEFC with an absorption layer made of porous materials in the cathode outlet manifold inner wall as also shown in figure 5. After applying the SWMS to the PEFC stack and verifying the effect of the SWMS on cell performance, this paper also confirms the influence of the plugging phenomenon of the cathode outlet manifold on the cell performance of each cell in the stack.

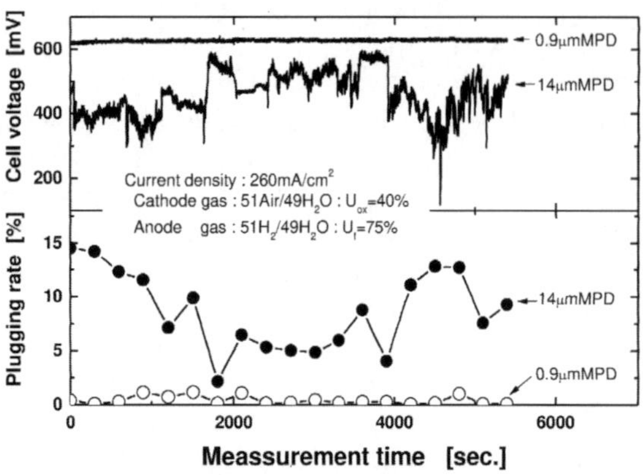

Figure 4. Influence of MPD of WAL on cell voltage variation and plugging ratio.

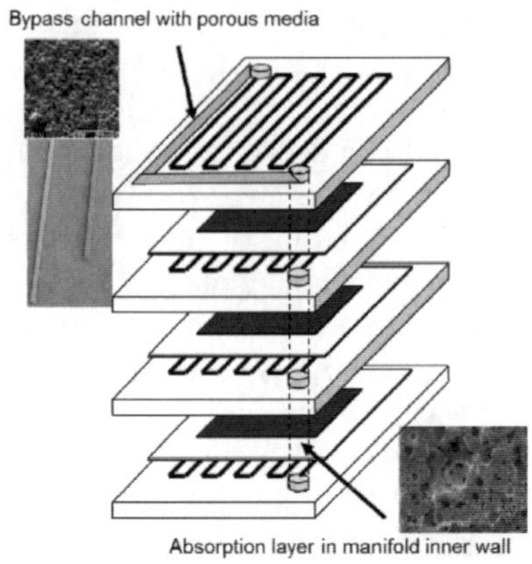

Figure 5. Schematic diagram of self-water management PEFC stack.

2. Effect of self-water management separator on cell performances

2.1. Experimental apparatus and conditions

Figure 6. Photograph of SWMS-V made of acrylic fiber for visualization.

Figure 7. Photograph of SWMS-P made of carbon for verifying cell performances.

The SWMS had a bypass channel that connected the cathode gas outlet to the gas inlet and was installed in a place different from the gas channel as shown in figures 6 and 7. The SWMS-V made of the acrylic fiber for visualization had a 3-serpentine gas channel; the gas channel was 2.2 mm in width, 1 mm in depth, and 50 mm in length. Another SWMS-P made of carbon for verifying cell performance had a single serpentine gas channel; its geometry was 2.2 mm in width, and 1 mm in depth, and the geometry of the bypass drainage channel was 1.8 mm in width, and 1.2 mm in depth. A standard separator made of carbon without the bypass channel was prepared to compare with the effect of the SWMS. This standard separator was the same as the SWMS excluding the bypass channel. The gas channel area of all separators was 50 mm × 50 mm, which was the same as the effective electrode area of 25 cm^2. In the visualization experiment, MEA was placed between the SWMS-A and the acrylic fiber plate, and water transportation in the bypass channel was measured from the front of SWMS-A by the 12-bit CCD camera. Cotton containing red ink was packed into the bypass channel, and the water transport phenomenon in the cell was confirmed by the change in the color of the cotton in the bypass channel when the cathode standard gas (oxidant gas utilization: 50%, cathode humidifier temperature: 80°C) was supplied to the SWMS-A.

In the power generation experiment, SWMS-P was installed in as the cathode separator, and the standard separator was installed in as the anode separator. The standard cell in which the standard separator was installed in both poles was evaluated as the reference data of the effect of the SWMS-P. MEA was the Primer®5580/5580CARBEL by Gore-Tex®. Figure 8 shows the schematic diagram of the experimental apparatus for evaluating power generation characteristics. Hydrogen and air were supplied to PEFC single cell through each humidifier. The cell resistance was measured by a milliohm meter with AC 4 probe, and I-V performance was measured by a programmable electronic load device. The cell voltage and temperature were recorded by the data logger. The Cole-Cole plot was also measured by an impedance meter to evaluate the effect of the SWMS-P on the cell performance.

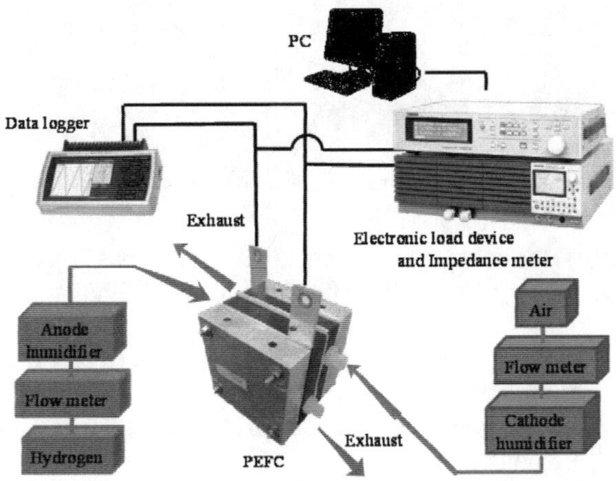

Figure 8. Schematic diagram of experimental apparatus.

There were two experimental conditions: the high humidity condition and the low humidity condition. The standard experimental condition was the high humidity condition called the full humidity condition (100% RH) because the cell temperature (Tc) was the same as 80°C of the anode/cathode humidifier temperatures (T_{da}/T_{dc}). The low humidity condition was 42% RH in the cathode channel because the cathode gas was supplied to the cell through the humidifier at 60°C, and this condition was used to verify the effect of SWMS-P on cell performance under low humidity conditions. The fuel gas utilization was 60%, the oxidant gas utilization was 35%, the standard current density was 0.24 Acm^{-2}, and the experimental pressure was atmospheric pressure.

2.2. *Results and discussions*

2.2.1. Confirmation of the water transportation using the capillary phenomenon. The effectiveness of the water transport phenomenon was confirmed using the SWMS-A with a visualization technique. Before the experiment, the cotton colored by a red ink was installed in the bypass drainage channel of the SWMS-A. Air humidified by passing the humidifier at 80°C was supplied to the mock cell in which the SWMS-A was installed; the flow rate was the standard flow rate of 334 mL/min that corresponds to an oxidant gas utilization of 40%. Because the concentration difference of water between the gas inlet and the gas outlet does not occur only by supplying the humidified gas to the gas channel, there is no water transport through the bypass-channel, as shown in figure 9(a). Therefore, to imitate the plugging phenomenon, water was injected near the gas outlet with a syringe. The injected water is absorbed by the cotton packed into the bypass channel by the capillary phenomenon at the same time as water is being injected near the gas outlet, and this absorbed water is transported from the gas outlet to the gas inlet through the bypass channel by the concentration diffusion of water. This is proven because the red ink in the cotton moves from the outlet to the inlet, as shown in figure 9(b) [17]. Because the SWMS-A was able to transport the imitated plugging water from the gas outlet to the inlet through the bypass channel by the capillary phenomenon and the concentration difference of water, the SWMS-P can be expected to eliminate the plugging phenomenon in the actual cell.

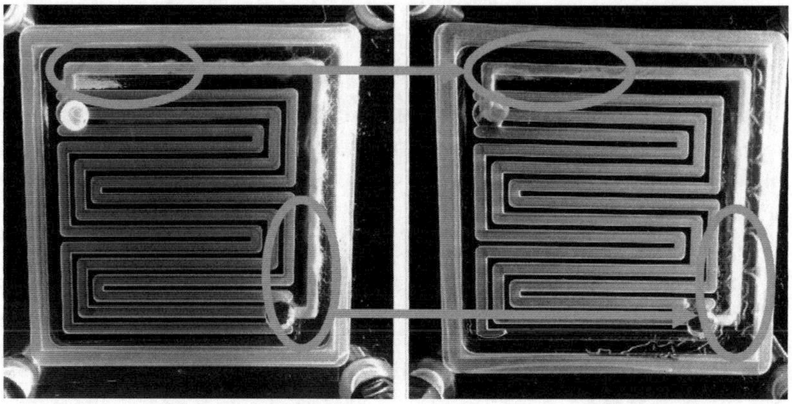

(a) (b)

Figure 9. Image of water transport phenomenon in the bypass of the SWMS-A: (a) Before imitated plugging and (b) After imitated plugging.

2.2.2. Effect of the SWMS-P on cell performance. Because the effect of the SWMS could be confirmed by the experiment without the power generation, this effect was confirmed by installing the SWMS-P in the actual cell. Figure 10 shows the influence of the SWMS-P and humidity conditions on each polarization as the Cole-Cole plots.

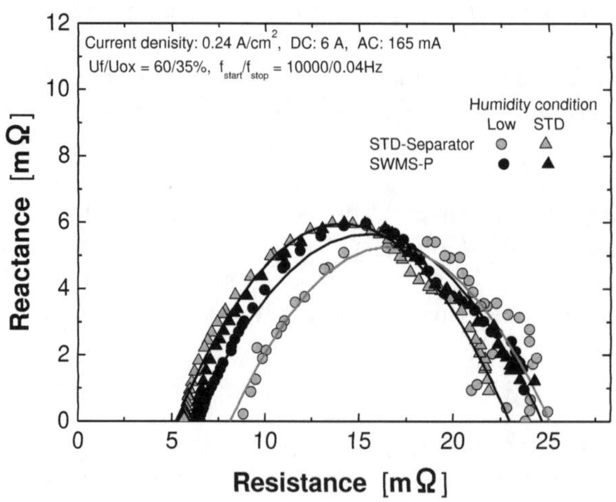

Figure 10. Influence of SWMS-P and humidity condition on each polarization (Cole-Cole plot).

The load current was 6 A, the alternative current was 165 mA, and the measurement frequency varied from 10,000 Hz to 0.04 Hz. From this figure, the activation polarizations of the cells under the low humidity condition are slightly larger than that under the standard conditions despite using the same MEA. Generally, a three-phase-interface forms as the reaction field is formed by catalysts, reactant gas and water, which is a proton conductor. Under the low humidity condition of the cathode side, although the three-phase-interface of the anode side is formed by supplying water in the anode gas, the interface on the cathode side is not easily formed because the water content in the cathode gas is small, as shown in figure 11(a). Therefore, the activation polarization under low humidity condition becomes a little larger. The resistance polarization of the cell with a standard separator becomes larger under the low humidity condition than under the standard condition. Because the humidity difference between the membrane (100%) and the cathode gas (19.7%) is 80.3 %, if the membrane is wet enough by feed water from the anode side, water in the membrane of the cathode side evaporates to the cathode gas, and consequently the resistance polarization becomes large. The resistance polarization of the cell with the SWMS-P is almost equal to that of the standard separator under the standard condition despite the low humidity, and that hardly changes even if the humidity varies. Therefore, because the membrane is wet by transporting excess water at the gas outlet to the gas inlet through the bypass channel, the resistance polarization of the cell with the SWMS-P hardly changes despite of low humidity. Although the diffusion polarization of the cell with the standard separator begins growing under the standard condition, that of the SWMS-P hardly occurs. As the humidities of MEA, the anode gas and the cathode gas are the same under the standard condition, the cathode system will be supersaturated by adding water generated by a cell reaction. Therefore, oxygen cannot reach the platinum of the cathode catalyst, which is submerged by the condensate, as shown in figure 11(b); thereafter, the diffusion polarization of the cell with the standard separator increases as the power generation reaction progresses. However, the diffusion polarization of the cell with the SWMS-P hardly grows for the reason mentioned above. From these results, the effectiveness of the SWMS-P on the cell performance is confirmed under various humidity conditions.

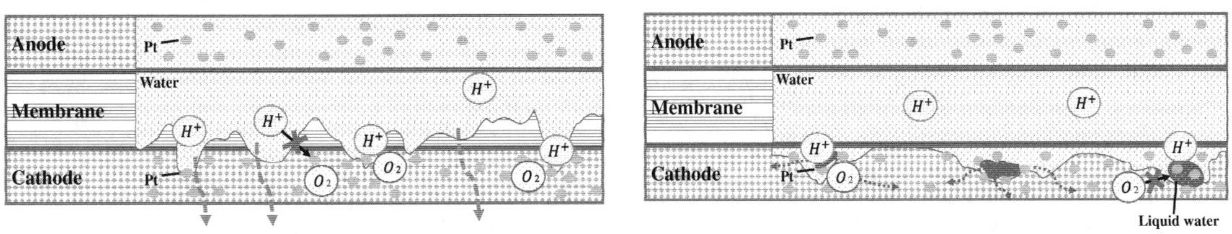

(a) (b)

Figure 11. Schematic diagram of the comparison of the three-phase-interface state with a difference in humidity: (a) Under the low humidity condition and (b) Under the standard condition.

3. Confirmation of the influence of the plugging phenomenon in the cathode outlet manifold on cell performance of each cell in the stack

Because the effectiveness of the SWMS for the self-water management was clarified, next the influence of the plugging phenomenon in the cathode outlet manifold on cell performance of each cell in a stack was confirmed using the visualization technique for applying this SWMS to a stack. To imitate the plugging phenomenon, plugging was caused by packing cotton into the cathode outlet manifold near each cell exit of a three-cell stack.

3.1. Experimental apparatus and conditions

The visualization PEFC stack was composed of two standard separators and one visualization separator as shown in table 1. The thickness of both separators was thicker than that of a conventional separator because these separators have a cartridge heater to change the temperature of each separator. The visualization separator has seven observation windows of 0.8 cm^2; three windows were installed in the cell to observe the cathode gas channel, and four windows were installed in the side of the visualization separator to observe the inside of the gas manifold of the cell-1, as shown in figure 12. Because this paper focused on only the cathode outlet manifold, other windows were not observed. The plugging phenomenon inside of the manifold was measured by the 12-bit CCD camera. Compulsory plugging was caused by packing cotton into the cathode outlet manifold near each cell exit of a three-cell stack, as shown in figure 13; pseudo-plugging-1 was between cell-1 and cell-2, pseudo-plugging-2 was between cell-2 and cell-3, and pseudo-plugging-3 was the outlet of the stack.

Table 1. Specification of each separator for the visualization of the PEFC stack.

Description	(a) Non-visualized separator	(b) Visualized separator
Separator	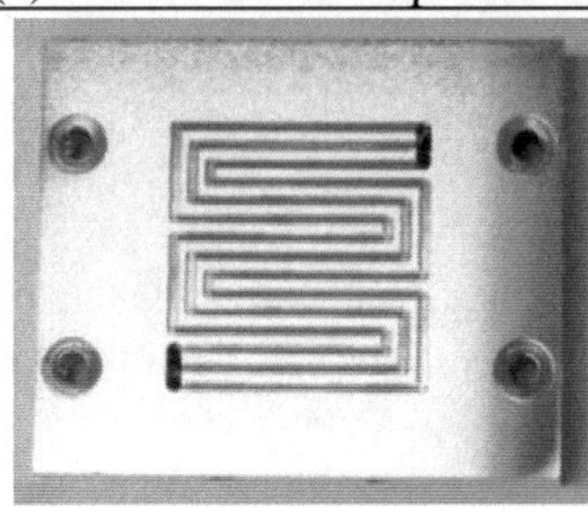	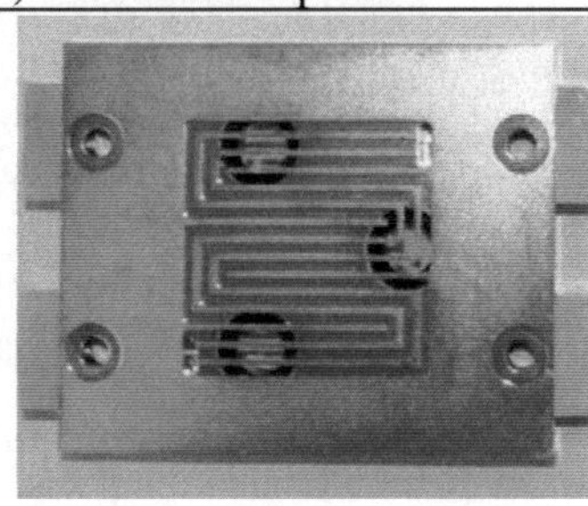
Material of the separator	Gold-plated stainless steel	Gold-plated stainless steel
Visualized area	-	0.8 cm^2 × 3parts (2.4 cm^2) inside 0.8 cm^2 × 4parts(2.4 cm^2) outside
Thickness	20 mm	30 mm
Resistance	1.815 mΩ	4.22 mΩ

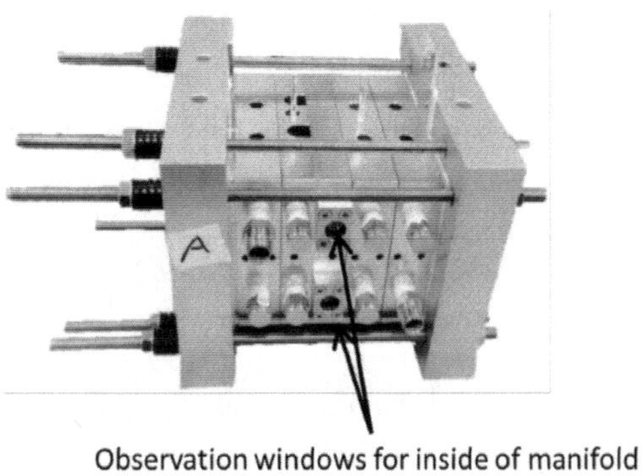

Figure 12. Photograph of the assembly of the visualization PEFC stack.

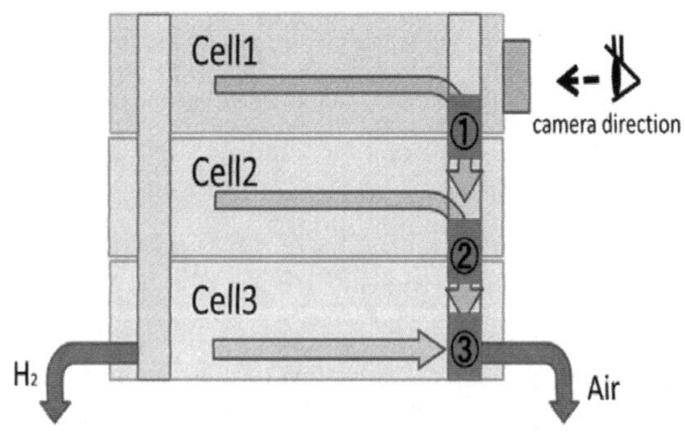

Figure 13. Schematic diagram of the plugging position in the cathode outlet manifold and the position of the observation window.

The flow rate of each gas was controlled by a mass flow controller, and each gas was humidified by passing each humidifier. The temperature of the pipe between the humidifier and the stack was maintained at 80°C by a ribbon heater. The cell voltage was measured and recorded by a data logger. A milliohm meter that utilized four-terminal probes measured the cell resistance.

As the experimental conditions, the temperatures of the cell, the anode humidifier and the cathode humidifier were 80°C, the fuel/oxidant gas utilization was 70%/40%, and the standard current density was 0.3 Acm^{-2}.

3.2. Results and discussions

Figure 14 shows the initial I-V performance in pseudo-plugging-1. Although all cell voltages are quite stable until 0.06 Acm^{-2}, the cell voltage of cell-1 decreases drastically, and all cells thereafter are not able to generate because of the deterioration of cell-1. The reason is that the stack shorted by the short of cell-1 because the backpressure of cell-1 rises and cell-1 is not supplied air by the plugging. Therefore, it becomes clear that the deterioration of only one cell influences the entire stack.

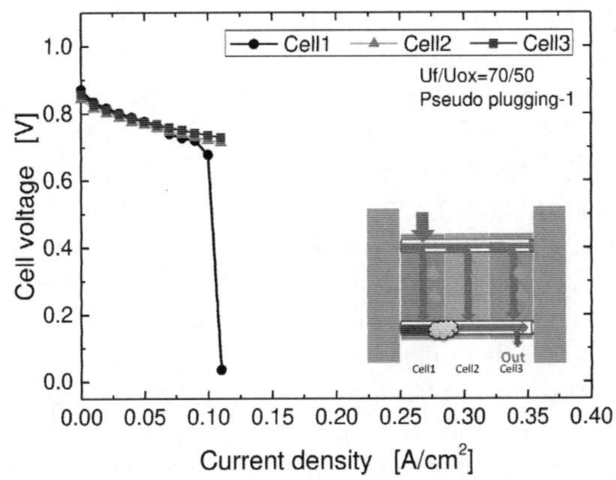

Figure 14. Influence of pseudo-plugging-1 on each initial cell performance of three-cell stack.

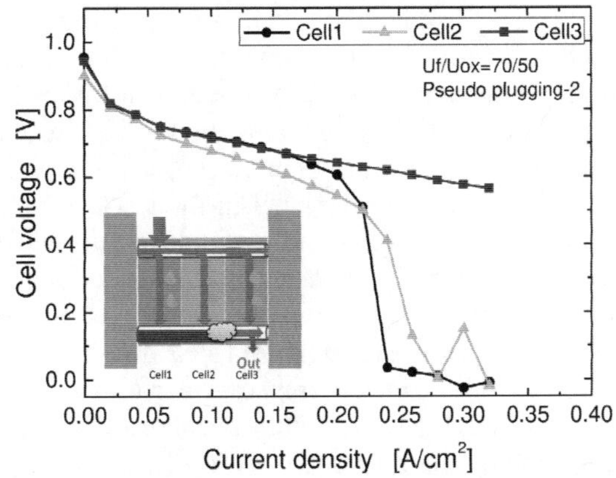

Figure 15. Influence of pseudo-plugging-2 on each initial cell performance of three-cell stack.

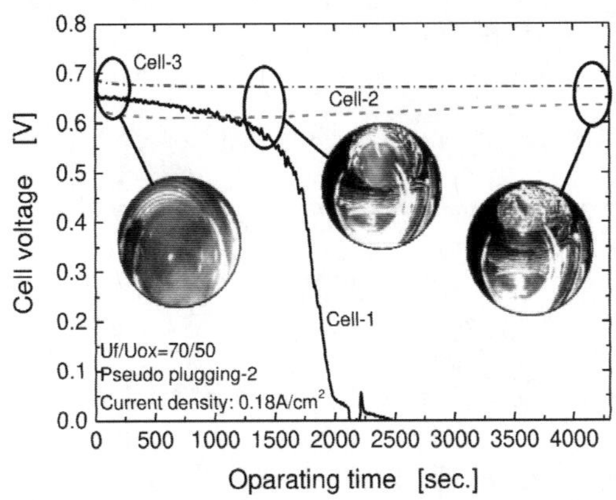

Figure 16. Lapse of time of each cell voltage and images of the inside of the cathode outlet manifold near the cell-1 outlet at the current density of 0.18 mAcm^{-2}.

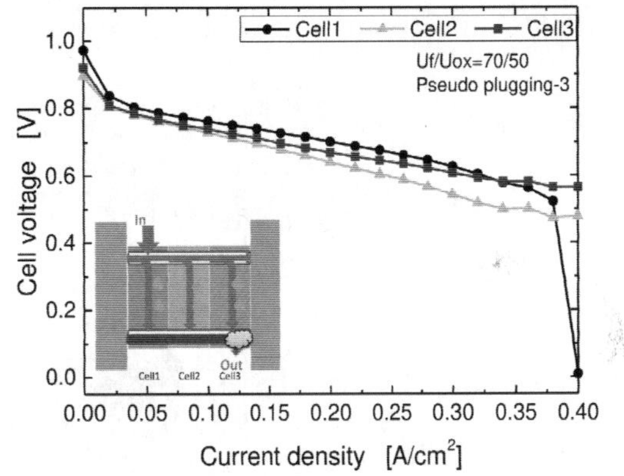

Figure 17. Influence of the pseudo plugging-3 on each initial cell performance of three cells stack.

In pseudo-plugging-2, the initial I-V performance of each cell is shown in figure 15. The cell voltages of the cells-1 and -2 located in the upstream part of the plugging position decrease drastically when the current density exceeds 0.18 Acm^{-2} because cotton was packed into the cathode outlet manifold between cell-1 and cell-2 in pseudo-plugging. However, cell-3 located in the downstream part of the plugging position is hardly influenced by the plugging, and the performance of cell-3 is improved because most of the cathode gas is supplied to cell-3 by the plugging. Figure 16 shows the changes with the lapse of time of each cell voltage under 0.18 Acm^{-2} of current density and the images of the inside of the cathode outlet manifold near the cell-1 outlet. The voltage of cell-1 immediately after beginning the measurement is steady; water does not exist in the cathode outlet manifold. However, the voltage of cell-1 decreases drastically after 25 minutes, and water reaches half of the observation window. Because the cathode outlet manifold between cell-1 and cell-2 was completely occluded by water, cell-1 was not supplied air. However, air was supplied to cell-2 because cotton was moved to the cell-1 side at an early stage when cotton began to absorb water. Although the voltage of cell-2 begins to decrease a little immediately after beginning measuring, it improves gradually after moving the cotton. Cell-3, which is located in the downstream part of the plugging position, is hardly influenced by the plugging.

Figure 17 shows the initial I-V performance of each cell in pseudo-plugging-3. A big voltage drop such as in pseudo-plugging-1 and -2 is not caused until the high current density region because the buffer volume in the cathode outlet manifold is larger than that of pseudo-plugging-1 and -2. However, cell-1 located in a place that is the furthest from the plugging position decreases drastically compared with other cells in the high current density region. If the plugging was caused by cotton absorbing water, the entire gas channel in the stack becomes a dead water region without the gas in the stack flowing. Therefore, we understand that the plugging phenomenon was caused first in cell-1, which was the most active to react because water generated by a cell reaction remained in each cell. Figure 18 shows the lapse of time of each cell voltage under a current density of 0.18 $mAcm^{-2}$ and the image of the cathode outlet manifold near the cell-1 exit. Each cell voltage is stable until 3000 seconds, although there is a difference in performance. The voltage of cell-3 drastically decreases when the operating time exceeds 3000 seconds, and it recovers 1000 seconds later. The reason is that the water absorbed by the cotton drains outside the stack and is easily compared with pseudo-plugging-1 and -2 because pseudo-plugging is located near the outlet of the stack. However, the cathode outlet manifold becomes an obstruction at once because the condensate flows from each cell into the cotton even if water in the cotton drains temporarily [18]. Therefore, because the plugging phenomenon was caused anywhere in the stack, if the plugging phenomenon was caused near the exit of the cathode outlet manifold, it is necessary to avoid the plugging phenomenon in this area.

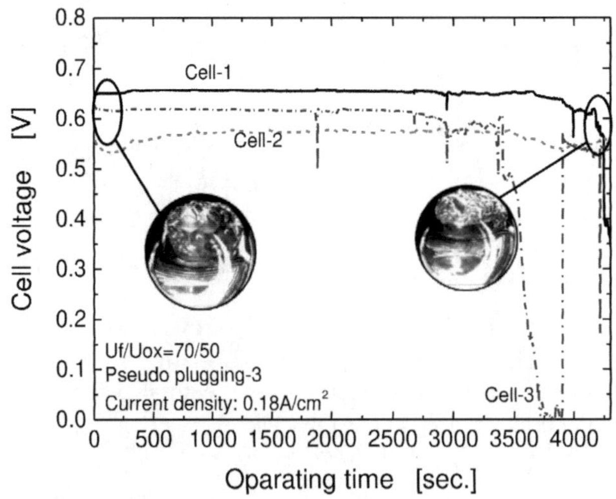

Figure 18. Lapse of time of each cell voltage and images of the inside of the cathode outlet manifold near the cell-1 outlet at the current density of 0.18 $mAcm^{-2}$.

These results show that plugging in the manifold greatly influences the performance of each cell in the stack. We could suggest that it is also possible to move water between cells in the stack through this manifold if water management in this manifold was done appropriately.

4. Conclusions

The objective of this study was to propose a PEFC with water management functions. The results of this study are summarized as follows:

- The bypass drainage channel of the SWMS absorbed water by the capillary phenomenon and transported the absorbed water from the gas outlet to the inlet using water concentration diffusion.
- The SWMS contributes to the increase of the three-phase interface by managing water inside the cell and improves cell performance despite low humidity.
- The plugging phenomenon in the cathode outlet manifold affects the cell performance of each cell in the stack.
- We suggest that it could also be possible to move water between cells in the stack through this manifold if the water management in this manifold is done appropriately.

Acknowledgments
We wish to express our gratitude to the late Dr. MakotoYamauchi who promoted this research. This work was supported by JKA and its promotion funds from KEIRIN RACE.

References
[1] Hakenjos A, Muenter H, Wittstadt U and Hebling C 2004 *J Power Sources* **131** 213-6
[2] Litster S, Sinton D and Djilali N 2006 *J Power Sources* **154** 95-105
[3] Dunbar Z W and Masel R I 2008 *ECS Transactions* **16(2)** 1001-8
[4] Jao T C, Sasabe T, Uemura S, Yoshida T and Hirai S 2016 *ECS Transactions* **75(14)** 179-88
[5] Takaya K and Araki T 2016 *ECS Transactions* **75(14)** 563-72
[6] Tanuma T, Kawamoto M and Kinoshita S 2017 *J Electrochemical Society* **164(6)** F499-503
[7] Eller J, Roth J, Marone F, Stampanoni M and Buchi F N 2017 *J Electrochemical Society* **164(2)** F115-26
[8] Nishida K, Kono Y, Sato M and Mizuguchi D 2016 *ECS Transactions* **75(14)** 227-36
[9] Kitahara T, Nakajima H and Ishikawa K 2016 *ECS Transactions* **75(14)** 209-17
[10] Kinumoto T, Matsumura T, Miyoshi H, Matsuoka M, Arai Y, Tsumura T and Toyoda M 2017 *ECS Transactions* **75(52)** 11-6
[11] Date T, Yamauchi M and Sugiura K 2013 *ECS Transactions* **51(1)** 207-14
[12] Rastedt M, Pinar F J, Wagner P, García H R, Steenberg T, Hjuler H A, Paidar M and Bouzek K 2016 *ECS Transactions* **75(14)** 301-15
[13] Wippermann K, Wackerl J, Lehnert W, Huber B and Korte C 2016 *J Electrochemical Society* **163(2)** F25-37
[14] Yamauchi M, Ueda T, Shimatani S and Sugiura K 2010 *ECS Transactions* **26(1)** 123-30
[15] Sugiura K, Shiramizu T, Yamauchi T, Yamauchi M, Matsuzaki S, Kada N and Itoh Y 2008 *ECS Transactions* **12(1)** 131-8
[16] Takeda A, Yamauchi M and Sugiura K 2013 *ECS Transactions* **51** 201-6
[17] Takahashi N, Sugiura K and Kamimura T 2016 *Proc. of the 27th International Symposium on Transport Phenomena, 092* (Honolulu, USA))
[18] Fujiwara T and Sugiura K 2015 *Materials and Technologies for Energy Efficiency* (Madrid, Spain: Brown Walker Press) pp 69-72

IOP Conf. Series: Earth and Environmental Science **93** (2017) 012018 doi:10.1088/1755-1315/93/1/012018

The onset of nucleate boiling of self-rewetting fluids in microchannels

A Sitar[1,3], M Zupančič[1], M Crivellari[2] and I Golobič[1]

[1]University of Ljubljana, Faculty of Mechanical Engineering, Aškerčeva 6, 1000 Ljubljana, Slovenia
[2]Fondazione Bruno Kessler, Center for Materials and Microsystems, Via Sommarive 18, I - 38123 Povo, Trento, Italy

E-mail: anze.sitar@fs.uni-lj.si

Abstract. The comparative experimental analysis of flow boiling of pure water and four different self-rewetting fluids (6% butanol, 2% pentanol, 0.6% hexanol and 0.15% heptanol aqueous solution) was performed in an array of interconnected 50×50 μm microchannels. The results show that the onset of boiling was postponed to higher surface superheats when SRFs were used instead of water. In addition, the visualization of the flow boiling of self-rewetting fluids showed a substantially more abrupt and unstable flow, which was observed as a misty two-phase flow in the microchannel array. Both occurrences are attributable to (i) the absence of properly sized nucleation cavities in the microchannels and (ii) higher achieved temperatures during boiling of self-rewetting fluids, which lead to a higher momentum force of the emerging vapor phase. The measured static and dynamic contact angles and therefore also the surface tension of the used self-rewetting fluids are lower in comparison with water.

1. Introduction

Flow boiling in larger channels as well as in microchannels introduces some advantages compared to the single-phase flow: improved heat transfer coefficient, more uniform temperature distribution, smaller mass flux required, etc. However, there are also some drawbacks of employing boiling in microchannels, which are extensively described in the literature: flow instabilities [1], high superheats required for the ONB [2], vapor backflows [3,4]. One of the possible methods of the boiling heat transfer enhancement is the improvement of the channels' construction (nucleation sites [5], fins and other surface improvements [6], constructional theory designs [7]), whereas the use of an appropriate working fluid presents a simpler and an inexpensive way of increasing the heat transfer performance of an existent heat transfer device. It is difficult to substitute water in an attempt to improve the heat transfer coefficient during boiling, due to its' large latent heat. However, the use of the so-called self-rewetting fluids (SRF) already confirmed some benefits over water. The most distinguishing thermodynamic property of the SRFs is the increasing surface tension at increasing temperature, which is a unique and highly beneficial quality. Namely, SRF used as the working fluid promotes faster wetting of the hot and dried spots, which inevitably occur during boiling [8].

The self-rewetting property relies on the Marangoni effect and the positive surface tension gradient at increasing temperatures, which was presented in Savino *et al* [9] for butanol and heptanol aqueous solution along with some ternary mixtures. The performance of a heat pipe was improved in Senthilkumar *et al* [10] by adding n-butanol and especially n-pentanol to water. Authors claim that the

SRF showed a more stable and higher thermal efficiency and also lower thermal resistance compared to the experiments conducted with water as the working fluid. The comparison of the heat transfer performance of a loop heat pipe using various concentrations of three different alcohols (butanol, pentanol and hexanol) in water was presented by Wu [11]. The experimental results indicated that the optimal SRF for the heat transfer in a heat pipe at the selected working conditions was a 6 % aqueous butanol solution. Another experimental investigation of the SRF heat transfer in heat pipes was presented in Su *et al* [12], where a combination of n-butanol aqueous solution and graphene oxide nanofluid was used. Authors demonstrated a 16 % improvement in the heat transfer performance with the addition of an optimal amount of the nanofluid to the SRF, compared to the heat transfer performance of the SRF alone.

Other researchers have focused on enhancing the heat transfer in pool boiling, which is more similar to flow boiling in microchannels compared to the heat transfer in heat pipes. Recently, Hu et al. [13,14] improved the heat transfer during pool boiling of heptanol-water mixture and attributed the enhancement to: (i) the accelerated wetting of the dried spots, due to the reversed Marangoni effect (positive surface tension gradient), (ii) the more difficult coalescence of the emerging bubbles, (iii) thicker liquid layer on the heated area and (iv) smaller bubbles formation on the heater (micro-bubble emission). Zhou *et al* [15] performed pool boiling experiments with aqueous solutions of butanol, pentanol and hexanol with the mass concentrations of 7.3%, 2.2%, and 0.6%, respectively. The authors showed that with increasing the alcohols' molecular weight, the heat transfer during boiling also enhances.

The transition of SRF to a microcavity was presented in a numerical study by Zhou *et al* [16] with using a n-butanol aqueous solution. The results show that the heat transfer of the employed SRF is superior to water. Experimental investigation of a n-butanol solution in a microchannel flow boiling was presented in Sitar and Golobič [17]. The SRF provided a 10 K decrease of the maximum temperature compared to the pure water boiling at the same heat flux. Additional publications of SRF in microchannels were not found in the open literature. Therefore, our current study aims to disclose the effects of various self-rewetting fluids on boiling in microchannels. We are focused on four alcohol-water binary mixtures with a known positive surface gradient, which promotes the Marangoni flow of the liquid to the dried-out areas.

2. Materials and methods

2.1. Experimental setup and measurement procedure

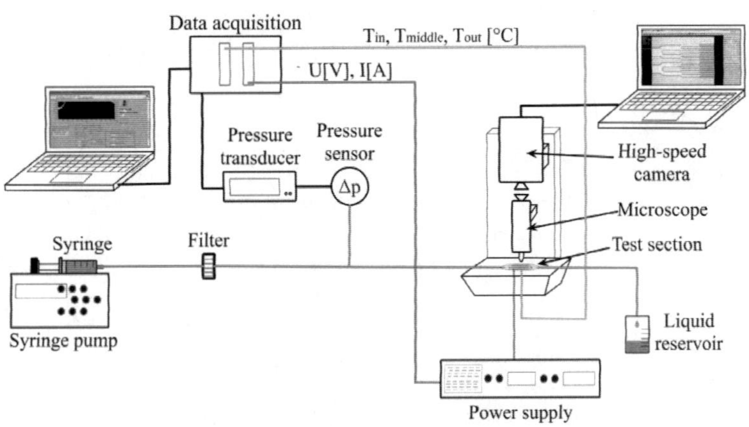

Figure 1. Experimental setup.

The experimental setup was comprised of a visualization system and a measuring system as presented

in figure 1. Temperatures were measured at three locations with a Pt100 class A resistive temperature detectors with an accuracy of 0.35°C at 100°C. Pressure was determined with a Validyne DP15 sensor, which was connected to the pressure transducer. Voltage drop across the heater elements was measured directly with the data acquisition system, whereas the electrical current was determined indirectly with a voltage drop measurement on a reference shunt with a 0.05 Ω resistance. A detailed description of the used experimental setup is available in Sitar *et al* [18].

The comparative analysis of the different working fluids is substantially affected by the operating conditions of the heat transfer experiments. The heat losses and the temperatures of the surrounding air and surfaces have an important impact on the measured temperatures during the boiling experiments. Considering these influential parameters, we utilized the same heating procedure for all of the conducted experiments (e.g. see table 1). The only controlled quantity was the heater's voltage, whereas all the following parameter values are approximate and vary a little from one to another experimental run. The subordinate quantities depend on the applied voltage and also on the heaters' temperature, which affects its electrical resistance and consequentially also the electrical current as well as the heating power.

Table 1. Power supply heating cycle and heater characteristics.

Time (min)	Electrical voltage (V)	Heating power (W)	Electrical current (A)	Heater's electrical resistance (Ω)
0	0	0	0.0	-
1.5	1	0.3	0.3	3.3
3	2	1.0	0.5	4.0
4.5	3	2.2	0.7	4.1
6	4	3.7	0.9	4.3
7.5	5	5.6	1.1	4.5
9	6	7.6	1.3	4.7
10.5	6.5	8.7	1.3	4.9
12	7	9.8	1.4	5.0
13.5	7.5	10.9	1.5	5.2
15	8	12.0	1.5	5.3
16.5	8.5	13.0	1.5	5.6
18	9	14.3	1.6	5.7
19.5	9.5	15.6	1.6	5.8
21	10	16.8	1.7	6.0
22.5	10.5	18.2	1.7	6.1
24	11	19.6	1.8	6.2
25.5	10.5	18.2	1.7	6.1
27	10	16.9	1.7	5.9
28.5	9.5	15.5	1.6	5.8
30	9	14.2	1.6	5.7
31.5	8.5	12.9	1.5	5.6
33	8	11.7	1.5	5.5
34.5	7.5	10.5	1.4	5.4
36	7	9.4	1.3	5.2
37.5	6.5	8.3	1.3	5.1
39	6	7.3	1.2	4.9
40.5	0	0	0	-

2.2. Microchannel test section and the self-rewetting fluids

The microchannel test section was comprised of five microchannel arrays with different features and designs. We used an array of 32 interconnected microchannels with a cross-section of 50×50 μm with a borosilicate glass covering the upper side of the silicon. The temperature sensors and the heater were positioned at the bottom side of the microchannel array, which is schematically presented in figure 2. All of the walls and bottom of the microchannels were smooth after the etching process with the roughness well below 1 μm, and more importantly no artificial nucleation cavities were added.

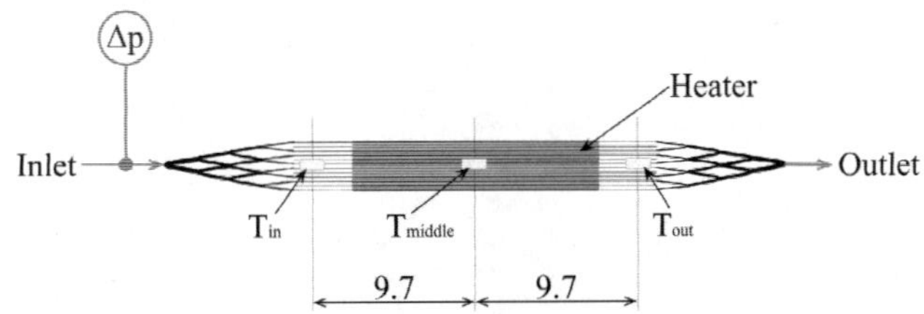

Figure 2. Array of microchannels.

The maximal solubility of long chain alcohols was found in the open literature, however the solubilities were not unambiguously defined. The ranges of maximal solubilities of alcohols in water at 25°C are presented in table 2. The weight concentrations of alcohols used in our experimental work were fixed at approximately 80 % of the average maximal solubility at 25°C found in Yalkowsky *et al* [19]. Therefore, the mass concentrations of aqueous solutions of 1-butanol, 1-pentanol, 1-hexanol and 1-heptanol were 6.0%, 2.0%, 0.60% and 0.15%, respectively.

Table 2. The solubility of alcohols in water.

Alcohol	**Lowest solubility (g/L)**	**Lowest solubility (wt%)**	**Largest solubility (g/L)**	**Largest solubility (wt%)**	**Average maximal solubility (wt%)**	**Used concentration (wt%)**
1-butanol	65	6.52	83	8.32	7.42	6.0
1-pentanol	17	1.71	33	3.31	2.51	2.0
1-hexanol	5.8	0.58	9.1	0.91	0.75	0.60
1-heptanol	1.7	0.17	1.8	0.18	0.18	0.15

3. Results and discussion

3.1. Acquired experimental data and visualization

The volumetric flow of the working fluid was fixed at 0.5 mL/min and 1 mL/min, which corresponds to the mass flux of 104 kg/m^{2}s and 208 kg/m^{2}s, respectively. The effectiveness of heat transfer is commonly evaluated with three parameters: the heat transfer coefficient (HTC), the critical heat flux (CHF), and the superheat needed for the onset of boiling (ONB). The comparative analysis of various working fluids presented in this paper was focused primarily on the ONB. The gauge pressure measurements allow a distinct determination of boiling incipience, as it is seen in the figure 3. The summary of the pressure and temperature measurements at the ONB is given in table 3, as the actual values are difficult to obtain from the figures included. The maximum temperature was measured in the middle of microchannels at all experimental runs. Lower temperature of the ONB corresponds well with a lower gauge pressure after the boiling incipience at both volumetric flows for all working fluids.

The lowest temperatures of the ONB were measured with pure water as the working fluid at both analyzed volumetric flows. The pressure drop is lowering gradually with the increase of the fluid temperatures, whereas a sharp increase of the pressure occurs at the ONB due to the vapor backflows. At the volumetric flow of 0.5 mL/min the ONB was seen at first in water (102°C), and followed by hexanol (103°C), heptanol (108°C), pentanol (133°C) and butanol (133°C) aqueous solution.

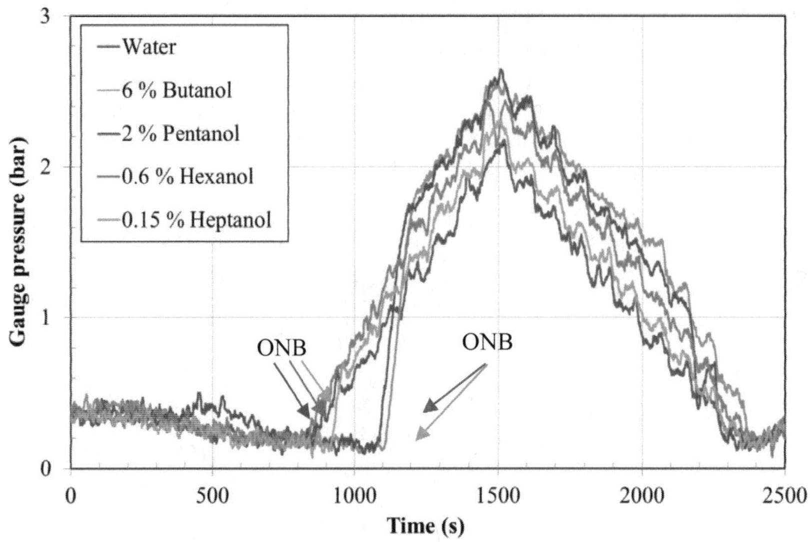

Figure 3. Gauge pressure at the volumetric flow of 0.5 mL/min.

Table 3. The summary of pressures and temperatures at the onset of boiling in microchannels.

	Maximum temperature at ONB (°C)		**Gauge pressure prior to ONB (bar)**		**Gauge pressure after ONB (bar)**	
Volumetric flow	0.5 mL/min	1 mL/min	0.5 mL/min	1 mL/min	0.5 mL/min	1 mL/min
Water	102	105	0.17	0.50	0.37	0.69
6% butanol	133	163	0.13	0.30	1.83	2.92
2% pentanol	131	129	0.16	0.33	1.73	2.02
0.6% hexanol	103	136	0.14	0.28	0.43	2.09
0.15% heptanol	108	108	0.13	0.38	0.66	0.79

Elevated temperatures of the ONB resulted in larger pressure jump. It can be derived from the table 3 that the pressure has risen for a factor of 2.2 at water and for a factor of 14.1 at the butanol solution, which is proportional to the superheat needed for commencement of boiling. The measured pressures presented in figure 3 show that the pressure drop of water was lower at almost all times and temperatures. Therefore, the surface of the microchannels was not appropriate for boiling of SRF. Firstly, the superheat needed for the ONB was higher for all of the SRFs used and secondly, the pressure drop during boiling was the lowest when water was used, which implicates that the boiling surface encouraged abrupt bubble nucleation of the SRFs.

The temperatures at the inlet, middle and outlet of the microchannels are presented in figures 4(a)-4(c), respectively. The temperatures at the middle of the microchannels were the highest, followed by the outlet and the inlet temperatures. Distinctive differences in temperatures are seen at the ONB, as the single-phase liquid flow transfers heat substantially less effective as the two-phase flow. Consequentially, the temperature plunges at the boiling incipience. Some differences are also seen during boiling of pentanol aqueous solution at 2200 s. This is the result of a more vigorous boiling of the SRF, which momentarily reduces the upstream temperatures.

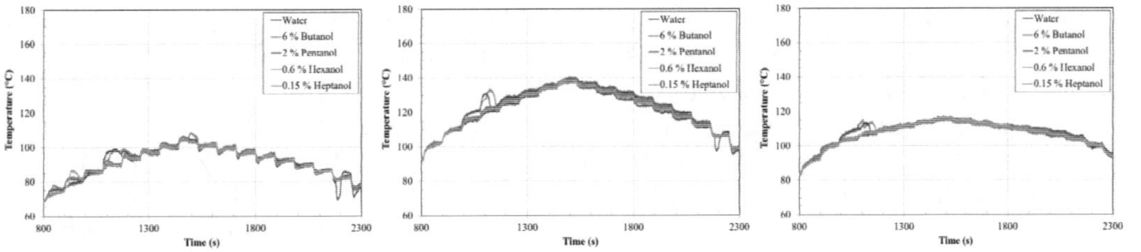

(a) (b) (c)

Figure 4. Temperatures at the (a) inlet; (b) middle and (c) outlet of the microchannels at 0.5 mL/min.

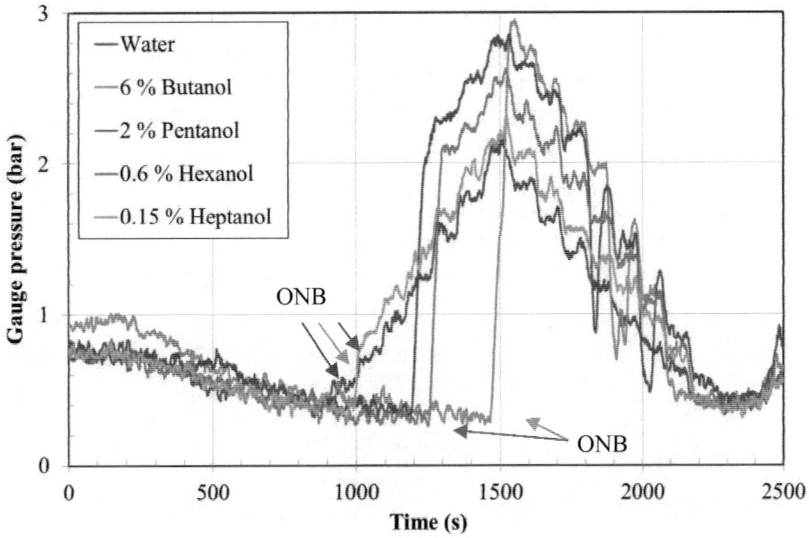

Figure 5. Gauge pressure at the volumetric flow of 1 mL/min.

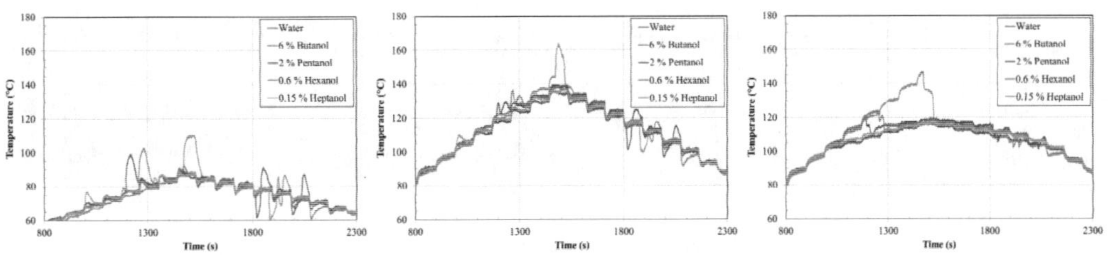

(a) (b) (c)

Figure 6. Temperatures at the (a) inlet; (b) middle and (c) outlet of the microchannels at 1 mL/min.

Similar results were measured at the larger volumetric flow of 1 mL/min, as it is seen from figures 5 and 6. The ONB was achieved in water at 105°C, heptanol solution at 108°C, pentanol solution at 129°C, hexanol solution at 136 °C and butanol solution at 163°C. The pressure drop at ONB has risen from 0.50 bar to 0.69 bar for water and from 0.30 bar to 2.92 bar for 6% butanol solution. The pressures were again the lowest when water was used as the working fluid, whereas the butanol and pentanol solutions show severe pressure fluctuations during the decrease of the heating power. This phenomenon is visually seen as an abrupt boiling of a mist like two-phase flow, which occurs due to the highly superheated smooth surface.

The temperatures at the inlet, middle and outlet of the channels are presented in figures 6(a)-6(c).

The distinctive temperature differences are seen at the ONB, when pentanol, hexanol and especially butanol solution were used as the working fluids. The temperature of the boiling incipience was 163°C for the butanol solution, whereas pure water started boiling at merely 105°C at the same volumetric flow and the same microchannel array. The absence of artificial nucleation cavities was the origin of the high superheat needed for the ONB of the butanol solution, as its surface tension is much lower compared to water. The addition of long-chain alcohols lowered the contact angle on the silicon, which postponed the boiling commencement at all the SRFs used. The temperature measurements further indicate the presence of an abrupt nucleation and vigorous boiling, which is more pronounced during butanol and pentanol solution boiling at a decreasing heat flux. The temperatures were oscillating up to 30°C during a constant heating power. The temperature fluctuations are less distinctive at the outlet, as there was mostly vapor present for all of the working fluids. The upstream portion of the microchannels were filled mostly with liquid water, hexanol or heptanol solutions, whereas the same location contained a vapor-liquid mixture of pentanol or butanol solutions during the temperature oscillations in the range of 1800 s to 2100 s.

Sample images during flow boiling of water and 6% butanol solution boiling are presented in figure 7. The sets of images that were selected from 4000 fps high-speed videos are depicting the most distinctive differences between the boiling phenomenon of water and SRFs in the microchannel array without any artificially etched nucleation cavities. The boiling of water was stable without any abrupt occurrences, whereas the boiling of the butanol solution displays an eruptive boiling behavior, which results in a mist two-phase flow with high pressure and temperature oscillations.

t = 0.00 ms t = 2.25 ms t = 5.00 ms

(a)

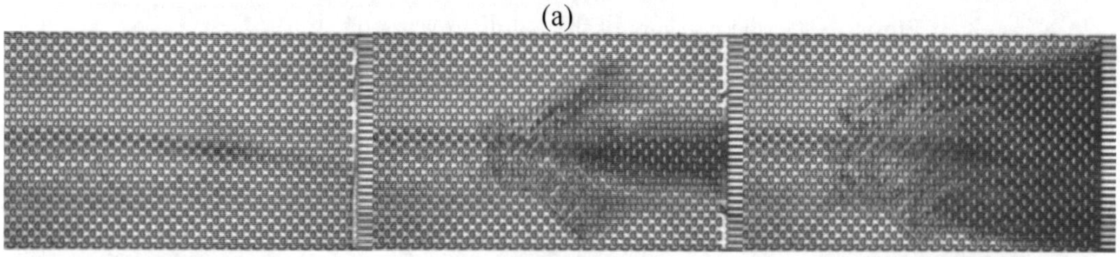

t = 0.00 ms t = 3.00 ms t = 5.75 ms

(b)

Figure 7. Flow boiling of (a) pure water and (b) 6 % aqueous butanol solution at 1 mL/min and 15.6 W.

3.2. Wettability of the working fluids

The wettability of a fluid is highly dependent on the exact type of material and the actual surface topography. Wettability itself was already proven to have significant effect on pool boiling and flow boiling performance [20,21]. For that reason, we have performed the measurements of static and dynamic contact angles for all the working fluids on a silicon wafer. Dynamic contact angles are especially important, since the three-phase contact line around which high heat transfer occurs, is constantly advancing and receding during the boiling process. The contact angle measurements were performed at 25°C on goniometer equipped with a high-speed video camera, a microscopic objective and micrometer syringe with a steel size 22 capillary tube. The droplet volumes for static measurements were constantly 5 μL. Advancing and receding contact angles were measured as the maximum and the minimum contact angles while adding or removing the fluid from the surface, respectively. The measured values of the contact angles presented in table 4 and figure 8 are the averaged values of ten measurements, whereas one sample image for each contact angle and fluid is additionally given in figure 8.

Table 4. The average values of static, advancing and receding contact angle.

	Static contact angle (°)	**Advancing contact angle (°)**	**Receding contact angle (°)**
Water	44	54	30
6% butanol	31	44	9
2% pentanol	38	50	11
0.6% hexanol	40	53	14
0.15% heptanol	43	54	12

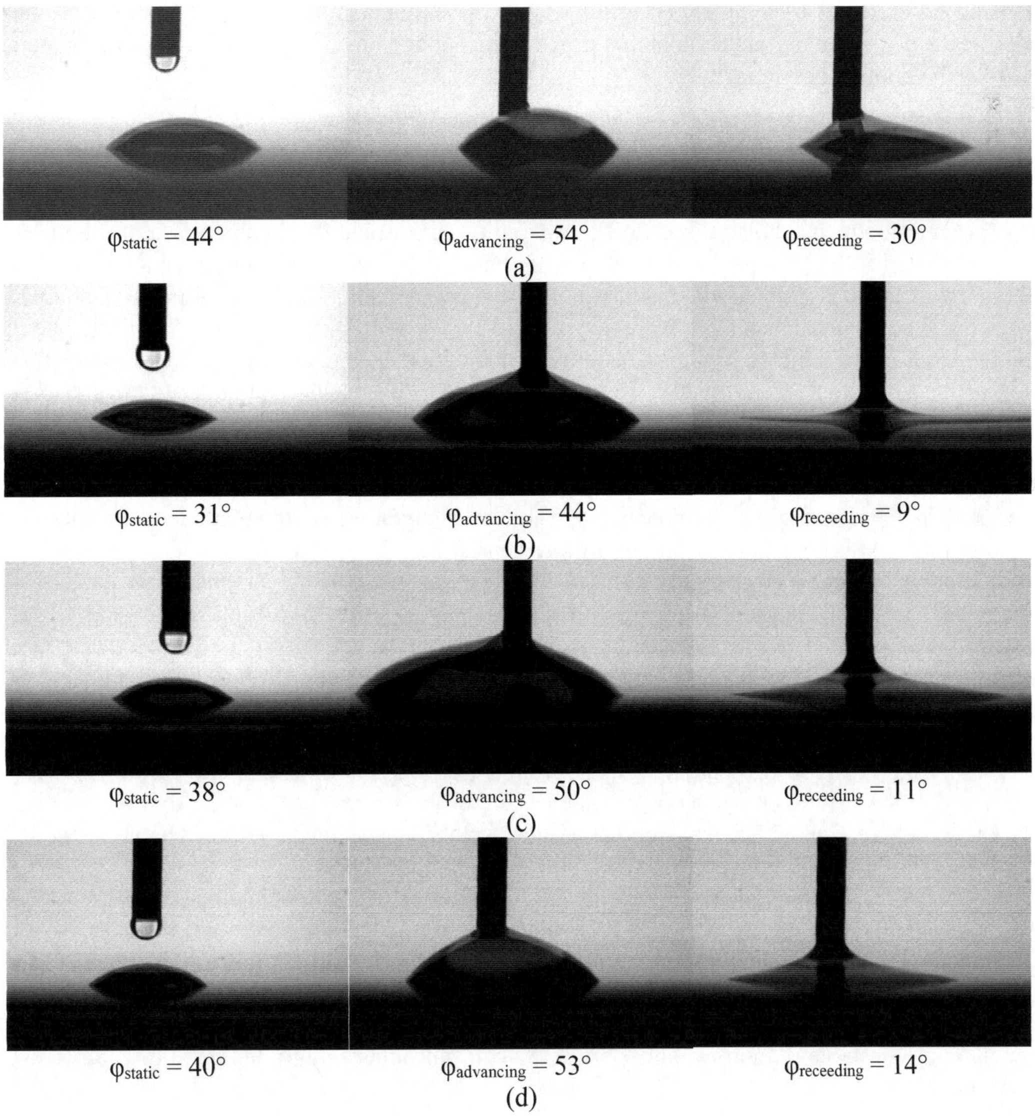

$\varphi_{static} = 43°$ $\varphi_{advancing} = 54°$ $\varphi_{receeding} = 12°$

(e)

Figure 8. Static and dynamic contact angles of (a) water, (b) 6% butanol, (c) 2% pentanol, (d) 0.6% hexanol and (e) 0.15% heptanol solution on a silicon wafer.

The largest contact angles were measured at a water drop placed on the silicon wafer, whereas the lowest angles were determined for the 6% butanol solution. The contact angles of all the other analyzed SRFs were measured between these two outermost angles. The wettability results were anticipated on the basis of temperature and pressure measurements at the ONB. Namely, the larger contact angle of the fluids on the silicon surface coincide extremely well with the lower superheat needed for the boiling incipience.

4. Conclusions

The presented experimental results show that the onset of nucleate boiling was postponed to higher temperatures for the self-rewetting fluids. The superheats required for the ONB were up to 57 K higher for the 6% aqueous butanol solution compared to pure water. The temperature of the ONB for other investigated SRFs (2% pentanol, 0.6% hexanol and 0.15% heptanol solution) increased as well, however it was less pronounced in comparison with the butanol solution. The comparative analysis of the working fluids is consistent at both employed mass fluxes of 104 kg/m^{2}s and 208 kg/m^{2}s. The most probable origin of the boiling suppression is the combination of smooth surfaces and lower surface tension of the SRFs compared to water. Results showed that 6% butanol solution provides the lowest static and dynamic contact angles among all the tested fluids. The wall temperatures decreased substantially after the incipience of boiling, as the heat transfer coefficient increased at the single-phase to two-phase flow transition. The temperature drop and the rise of the pressure were well pronounced in the experiments performed, due to the interconnected microchannels in the arrays, which promoted the propagation of boiling to a larger surface of the microchannels. Consequentially, the temperatures dropped up to 25 K and the pressure has risen up to 14 times at the ONB. The comparative analysis of various working fluids was enabled with the utilization of same heating procedure for all of the experimental runs. Therefore, the temperature time dependence was measured at approximately the same heating power and heat losses for all of the fluids. The analyzed temperatures after the incipience of boiling indicate that pure water was transferring heat superiorly compared to the SRFs. Although the positive surface gradient of the SRFs improves the wettability of the dried areas, it was not the deciding factor for an improved heat transfer during flow boiling. The visualization of the boiling phenomenon revealed that the SRFs were boiling abruptly in an oscillating mist like two-phase flow. The most probable cause of the deteriorated boiling heat transfer is the absence of properly sized nucleation sites, which would stabilize the two-phase flow.

The presented study demonstrated that the introduction of SRFs to a microchannel heat transfer device has some pitfalls, which should be recognized and mitigated already at the design phase of the microchannel fabrication. The combination of the working fluid and microchannel surface is crucial not only for an earlier transition to boiling, but also for a more stable and efficient nucleation of bubbles. In the opposite case, the temperatures of the ONB are higher and the following flow boiling becomes more unstable, vigorous and abrupt, which introduces high temperature and pressure fluctuations.

References

[1] Kandlikar S G, Colin S, Peles Y, Garimella S, Pease R F, Brandner J J and Tuckerman D B 2013 Heat transfer in microchannels—2012 status and research needs *J Heat Transfer* **135** 091001-01-18
[2] Sitar A and Golobic I 2016 Effect of nucleation cavities on enhanced boiling heat transfer in microchannels *Nanosc Microsc Therm* **20** 33-50
[3] Kuo C J and Peles Y 2007 Local measurement of flow boiling in structured surface microchannels *Int J Heat Mass Transfer* **50** 4513-26
[4] Wang G, Cheng P and Bergles A E 2008 Effects of inlet/outlet configurations on flow boiling instability in parallel microchannels *Int J Heat Mass Transfer* **51** 2267-81
[5] Lee J Y, Kim M-H, Kaviany M and Son S Y 2011 Bubble nucleation in microchannel flow boiling using single artificial cavity *Int J Heat Mass Transfer* **54** 5139-48
[6] Law M, Lee P-S and Balasubramanian K 2014 Experimental investigation of flow boiling heat transfer in novel oblique-finned microchannels *Int J Heat Mass Transfer* **76** 419-31
[7] Ghaedamini H, Salimpour M R and Campo A 2011 Constructal design of reverting microchannels for convective cooling of a circular disc *Int J Therm Sci* **50** 1051-61
[8] Petkovsek J, Heng Y, Zupancic M, Gjerkes H, Cimerman F and Golobic I 2016 Ir thermographic investigation of nucleate pool boiling at high heat flux *Int J Refrig* **61** 127-39
[9] Savino R, Cecere A and Di Paola R 2009 Surface tension-driven flow in wickless heat pipes with self-rewetting fluids *Int J Heat Fluid Fl* **30** 380-8
[10] Senthilkumar R, Vaidyanathan S and Balasubramanian S 2011 Thermal analysis of heat pipe using self rewetting fluids *Therm Sci* **15** 879-88
[11] Wu S-C 2015 Study of self-rewetting fluid applied to loop heat pipe *Int J Therm Sci* **98** 374-80
[12] Su X, Zhang M, Han W and Guo X 2016 Experimental study on the heat transfer performance of an oscillating heat pipe with self-rewetting nanofluid *Int J Heat Mass Transfer* **100** 378-85
[13] Hu Y, Zhang S, Li X and Wang S 2014 Heat transfer enhancement mechanism of pool boiling with self-rewetting fluid *Int J Heat Mass Transfer* **79** 309-13
[14] Hu Y, Zhang S, Li X and Wang S 2015 Heat transfer enhancement of subcooled pool boiling with self-rewetting fluid *Int J Heat Mass Transfer* **83** 64-8
[15] Zhou L, Wang Z, Du X and Yang Y 2015 Boiling characteristics of water and self-rewetting fluids in packed bed of spherical glass beads *Exp Therm Fluid Sci* **68** 537-44
[16] Zhou S, Zhou L, Du X and Yang Y 2017 Heat transfer characteristics of evaporating thin liquid film in closed microcavity for self-rewetting binary fluid *Int J Heat Mass Transfer* **108A** 136-45
[17] Sitar A and Golobic I 2015 Heat transfer enhancement of self-rewetting aqueous n-butanol solutions boiling in microchannels *Int J Heat Mass Transfer* **81** 198-206
[18] Sitar A, Sedmak I and Golobic I 2012 Boiling of water and fc-72 in microchannels enhanced with novel features *Int J Heat Mass Transfer* **55** 6446-57
[19] Yalkowsky S H, He J and Jain P 2010 *Handbook of Aqueous Solubility Data* 2nd Ed. (Boca Raton, FL, USA: CRC Press)
[20] Shojaeian M and Koşar A 2015 Pool boiling and flow boiling on micro- and nanostructured surfaces *Exp Therm Fluid Sci* **63** 45-73
[21] Bourdon B, Di Marco P, Rioboo R, Marengo M and De Coninck J 2013 Enhancing the onset of pool boiling by wettability modification on nanometrically smooth surfaces *Int Commun Heat Mass Transfer* **45** 11-5

Development turbine blade for ultramicro hydro power generation by 3D printer system

T Kamimura[1,2], H Itoh[1] and K Sugiura[1]

[1]Osaka Prefecture University College of Technology, 26-12 Saiwai-cho, Neyagawa, Osaka 572-8572, Japan

E-mail: kamimura@osaka-pct.ac.jp

Abstract. We have developed micro generation system for effective use of unutilized energy and the spread of a self-controlled dispersion energy supply system. The turbine blade was designed for achieving high performance by special shape. The turbine type was called quasi-Peace turbine type. Turbine with a diameter of 30cm is made of metal, it was created by the 5-axis milling machine. The experimental apparatus was fabricated by the 3D printer. An experiment was carried out in the scale down model. The specific speed of this turbine was much lower than that of existing turbines.

1. Introduction

For global environmental problems are serious, it is desired that a low-carbon society is realized soon. Thus, governments over the world support developing the renewable energy. Since the Great East Japan Earthquake, for the operation of nuclear power generation facilities are stopping, the demand for power generation by the other system has been increasing. Micro-generator apparatus of practical use is not widespread. In Japan, which has many rivers with abundant water because most of the land is occupied by mountains and forests, micro-hydropower system is an effective means. Since it has potential to generate green energy, the development cost is very low and without dam or water storage is needed, the micro hydro turbine technology gives good effectiveness even for unutilized agricultural water way and sand sliding defending dams. The realization of constructing the energy-providing local network, what is called local production for local consumption, will be very robust system against extensive disasters [1,2].

However, to design a high efficient hydraulic turbine is a very complex task, which requires a lot of engineering efforts. In a large-scale development for 1000 KW over class, as profitability can be taken, development specialized for a site is possible. Scale downed model experiment technologies are being adopted as effective way for micro turbine design allowing shortening the development time and saving money, instead of the custom fabricate.

In this research, the availability of method by using scale-downed micro turbine model experimental system, whose components are made by 3D printer, is considered, compared with real scale model experimental results. Design time of the turbine by the effective use of 3D printer can be expected to be significantly reduced. This method will make it possible to predict the actual performance without a large-scale experiment.

2. Quasi-peace turbine

2.1. Turbine type

Turbine is the main parts in the micro hydro system, where the task is to convert hydro power to rotational force. Hydro turbine can be classified into two types; namely impulse turbine and reaction turbine. Most of the impulse turbines are suitable for high head and medium head with low flow site. In contrast, a reaction turbine is used for low head site with high flow water, without taking into consideration whether it is horizontal or vertical arrangement. In this study, it is investigated about the quasi-Peace turbine, which is developing for micro hydro system in our laboratory [3]. The circumference of the turbine is filled with water and it is a reaction type.

The turbine is designed that the outer diameter of the turbine of the inlet part so that it can receive the drag of the running water to the maximum. In addition, the exhaust part of the turbine is designed so that the diameter of the turbine becomes small, and achieve high lift from high flow velocity. The turbine should be as small as possible, because the location for using our system is supposed to be narrow space such as a mountain area. From above, by using the newly developed turbine, it is expected to achieve high energy conversion efficiency even in places with low head and small amount of water.

The turbine shape is very complicated because it has conflicting features. We associate turbine shape with tornado phenomenon which encompasses enormous energy. We make full use of the CAD/CAM system to produce the turbine from designed shape. We also use a rapid prototype system that makes rapidly many type models to find out optimal shapes.

2.2. Turbine shape

Turbine modeling is achievement by utilizing 3D-CAD system (Pro/ENGINEER, PTC, Japan). An outline of the turbine assembled in casing by using Pro/ENGINEER is shown in figure 1. Real scale turbine is made of aluminum metal. The shape forming is used with 5-axis milling machine. Milling time is about 24 hours. In the focus of processing time, case of using the milling machine is more than 3D-printer (Mojo, Stratasys, USA) [4].

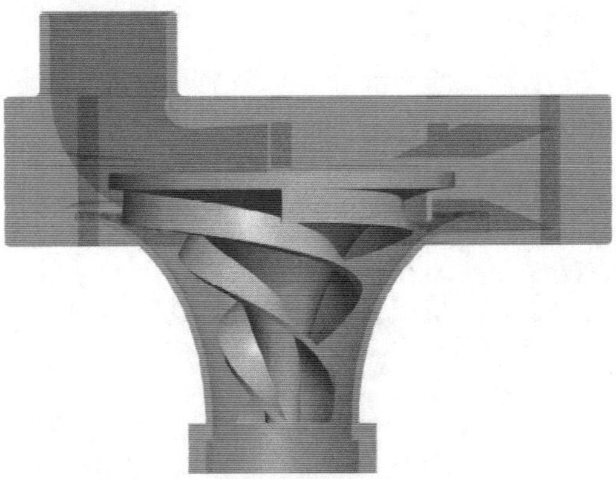

Figure 1. Outline of turbine part by 3D-CAD.

The procedure of modeling by the 3D-CAD is shown, and corresponding figures are shown in figure 2. Figure 3 is a drawing of the developed turbine, D indicates the diameter, and L indicates the height of the turbine.

Themodeling procedure is as follows:

- Cylindrical shape whose height is accordance with the turbine's one, is formed by extrude command.
- Ridge line of base part is elliptical arc, which is drawn by sweep command. The arc is quarter of ellipse.
- Twelve virtual planes, is located, which equally divide the cylinder circumferentially by 30

degrees. The arc is also divided into 12. The sketch line segment of blade height is drawn. The angle of the line segment is normal direction of the arc.

- End points of line segments are drawn with smooth curves.
- A number of blades are drawn by copy command.
- Surplus portion of the blade is removed by the cut command.

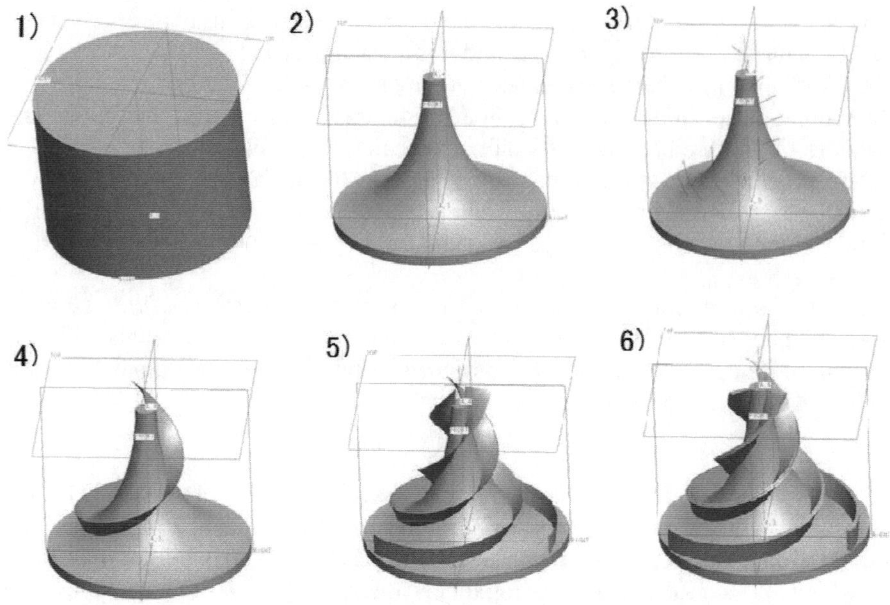

Figure 2. Outline of 3D shape of each feature.

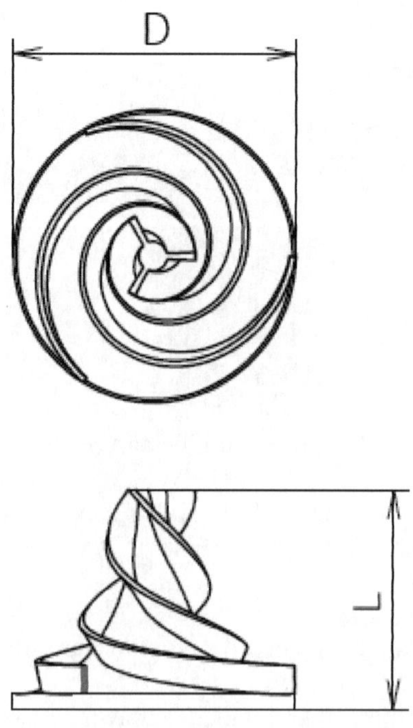

Figure 3. Front and top view of quasi-Peace turbine.

2.3. Scale downed model

The development goal is our micro power generation device that achieves about 500 W output at 10 meters gross head. Because of being able to keep running on all day and night, it is assumed that is equivalent to 3 kW class solar power generation unit, which is used for 4 hours a day. The flow rate is desired to be more than 0.005 m^3/s. In the head range of 1 m to 10 m, the flow rate range is expected to 0.001 m^3/s to 0.01 m^3/s. Our study is carried out under the low head and low flow condition [5].

In this study, the developing device is tested in not real field but our laboratory test bench. For restriction of the space, we must experiment under scale downed environmental condition. The head of experimental apparatus is adjusted by using the head tank. When we fill the tank, the maximum head is 2.5 m. At the same time, we adjust the manufacturing time and materials cost is less and use scale down model. As a result, the volume of the turbine is very small. The theoretical water power can be obtained by the following equation (1).

$$P_h = \rho g Q H \quad (1)$$

where P_h means Water power [W], ρ means Water density [kg/m^3], g means Gravitational acceleration [m/s^2], Q means Flow rate [m^3/s] and H means Effective head [m].

Since the target of our device is set to 500 W with an actual device level, it is expected that about 7 W calculated from the scale-down model experiment formula will be obtained. The specification of our turbine is shown in table 1.

Table 1. The specification of our turbine.

Parameter	Real Model	Experimental Model
Size D×L	φ300 mm × 225 mm	φ100 mm × 75 mm
Material	Aluminum	Acrylonitrile Butadiene Styrene Resin
Output P (target value)	500 W	7 W
Head H	10m	2.5 m

3. Experimental method

Figure 4 shows an overview of experimental setup in this study. The experiment is conducted with using the tap water from tank in our laboratory. The water in the lower tank is pumped up to the upper head tank and falls from bottom of the head tank. A water level of head tank is constant by spilling over. The level is variable parameter. The quasi-Peace turbine installed at the casing part receives the falling water, which rotates the turbine. The flow rate is measured with mounted ultrasound flow meter (FD-Q50C, Keyence, Japan) downstream of the valve. The rotational speed is controlled by loading the rotational axis, which is connected with DC motor and measured by optical tachometer. It is short-circuited by resistance between the terminals of the motor. We use the variable resistance to change a resistance level easily. We can measure the torque by a rotational torque detection device, which is coupled with rotational axis. The shaft power output characteristics are evaluated by measurement of the rotational number and torque of the rotor axis (equation (2)). The measurement is started from stationary condition and torque and rotation number are measured to calculate output as the load are decreased by the variable resistance.

$$P_S = T\omega \quad (2)$$

$$\eta = \frac{P_S}{P_h} \quad (3)$$

where P_S means Shaft power [W], T means Torque by axis [Nm], ω means Angular velocity of turbine [rad/s] and η means Turbine efficiency.

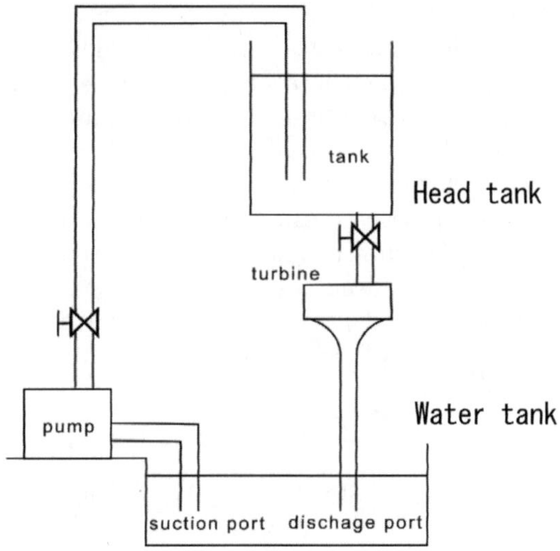

Figure 4. Experimental setup.

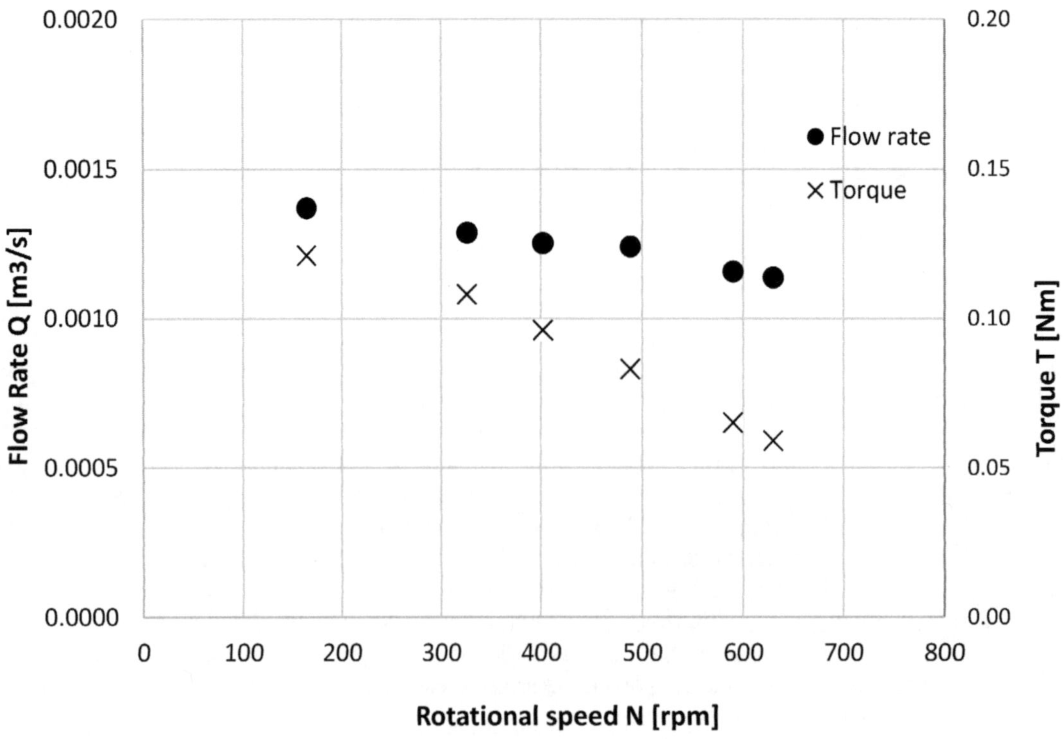

Figure 5. Measured values of flow rate and torque.

4. Results and discussion

The flow rate range measured at the inlet part was from 0.0011 m^3/s to 0.0014 m^3/s. The data of rotational speed N and rotation torque T were measured by the experiment. These measured values were shown in figure 5. The shaft powers, which were produced by turbine, were calculated by measured values of each point. The calculated power output characteristics are shown in figure 6. Maximum outputs were recognized in the case of around about 500 rpm. Highest output of our experiment were 4.24 W. This result was greatly less than the targeted value. An efficiency of this system could be calculated as the ratio of shaft power to water power (equation (3)). The calculated efficiency is shown in figure 6, too. If improvement of the efficiency is planned, the output may greatly exceed it from the targeted value.

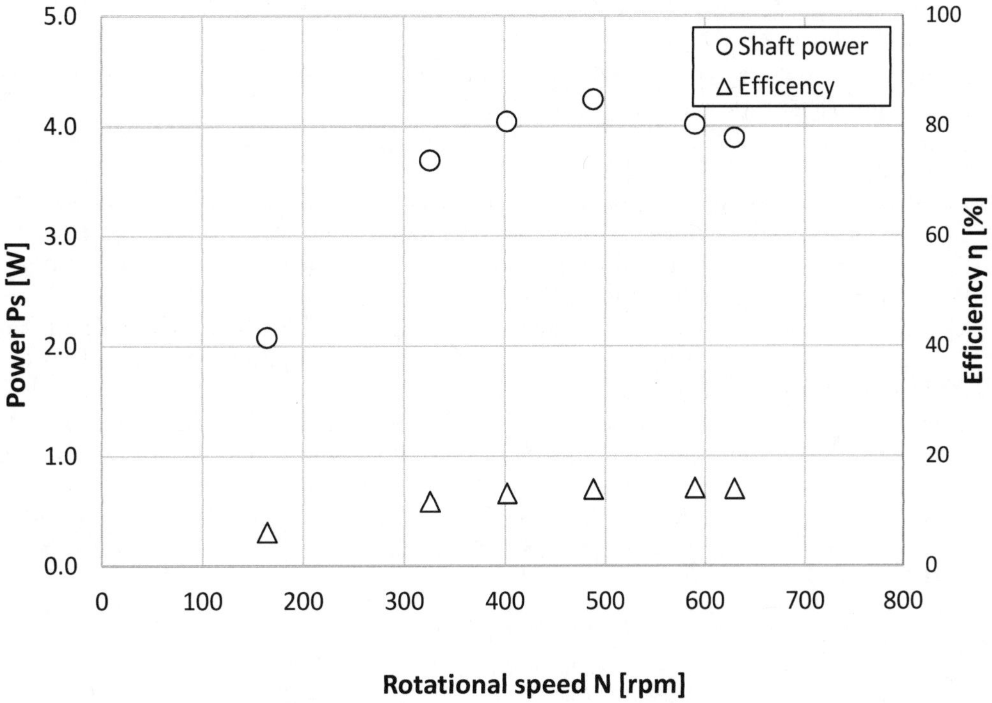

Figure 6. Distribution of shaft power and efficiency.

We calculated the specific speeds from my experimental data by following equation (4). P_S in the equation indicates the shaft power. The power was given by the product of measured torque and rotational speed.

The clearance between the casing wall surface and the turbine tip is very narrow. Therefore, it can be expected that the fluid flows between the turbine blades, but it was difficult to visually check the fluid. We were able to observe the swirling flow in the discharge section.

$$N_s = N\frac{\sqrt{P_S}}{H^{\frac{5}{4}}} \quad (4)$$

where $\boldsymbol{N_s}$ means Specific speed (m-kW) and $\boldsymbol{N}$ means Rotational speed [rpm].

In table 2, the specific speed of Quasi-Peace model and the other type is recorded.

Table 2. Specific speed.

Water Turbine		Specific Speed
Impulse Turbine	Pelton	8 ~ 25
Reaction Turbine	Francis	50 ~ 300
	Kaplan	200 ~ 900
	Tubular	500 >
-	Quasi-Peace (our model)	2~12

5. Conclusions

We designed a new type turbine for micro hydro power generation. We could get the basic performance of the turbine, which was fabricated by the 3D printer, in scale downed model experiment.

The following conclusions can be drawn:

- In the measurement of shaft power characteristics, the maximum power is Ps=4.24 W.
- Calculated specific speed of our model turbine is 2 – 12.

We continue our study on the new type turbine for optimum performance with taking into consideration the blade shape and casing shape.

References

[1] Ministry of Economy 2014 Trade and Industry ed. *Strategic Energy Plan* http://www.enecho.meti.go.jp/en/category/others/basic_plan/pdf/4th_strategic_ energy_plan.pdf

[2] Campbell R J 2010 Small hydro and low-head hydro power technologies and prospects Congressional Research Service R41089 1-15

[3] Okada S, Satonaka N, Fujiwara T, Kamimura T, Suga K and Shibata M 2012 Development of high efficiency turbine of high– performance micro water-power generation Proc of JSME Design and Systems Division **22** 3406-1-8

[4] Kamimura T 2015 Micro Turbine Flow in 3D Printer System (Abstracts book of International Conference on Advanced Technology in Experimental Mechanics 2015 (ATEM'15) and The 14th Asian Conference on Experimental Mechanics (ACEM14)) p 295

[5] Farriza M B, Boejang H, Masjuri M, Arasc M S M, Razik N H A, Mate S and Sopian K 2015 Evolution of simple reaction type turbines for pico-hydro applications J Teknologi (Sci & Eng) **77** 1-9

IOP Conf. Series: Earth and Environmental Science **93** (2017) 012020 doi:10.1088/1755-1315/93/1/012020

Wind speed time series reconstruction using a hybrid neural genetic approach

H Rodriguez[1,5], J J Flores[2], V Puig[3], L Morales[4], A Guerra[1] and F Calderon[2]

[1]Instituto Tecnológico de Culiacán, División de Estudios de Posgrado e Investigación, Juan de Dios Bátiz 310 pte. Col Guadalupe C.P. 80220, Culiacán, México
[2]División de Estudios de Postgrado, Facultad de Ingeniería Eléctrica, Universidad Michoacana de San Nicolás de Hidalgo, Morelia, Michoacán, México
[3]Institut de Robótica i Informática Industrial (CSIC-UPC), Carrer LLorens Artigas, 4-6, Barcelona (Spain), 08028
[4]Facultad de Ingeniería Civil, CONACYT-Universidad Michoacana de San Nicolás de Hidalgo, Morelia, Michoacán, México

E-mail: hrodriguez@itculiacan.edu.mx

Abstract. Currently, electric energy is used in practically all modern human activities. Most of the energy produced came from fossil fuels, making irreversible damage to the environment. Lately, there has been an effort by nations to produce energy using clean methods, such as solar and wind energy, among others. Wind energy is one of the cleanest alternatives. However, the wind speed is not constant, making the planning and operation at electric power systems a difficult activity. Knowing in advance the amount of raw material (wind speed) used for energy production allows us to estimate the energy to be generated by the power plant, helping the maintenance planning, the operational management, optimal operational cost. For these reasons, the forecast of wind speed becomes a necessary task. The forecast process involves the use of past observations from the variable to forecast (wind speed). To measure wind speed, weather stations use devices called anemometers, but due to poor maintenance, connection error, or natural wear, they may present false or missing data. In this work, a hybrid methodology is proposed, and it uses a compact genetic algorithm with an artificial neural network to reconstruct wind speed time series. The proposed methodology reconstructs the time series using a ANN defined by a Compact Genetic Algorithm.

1. Introduction

Energy plays a very important role in human activities. Thus, energy production has become one of the main economic and environmental issues worldwide [1]. In addition, the energy demand has increased steadily over time, causing environment damage when producing it. According to the census conducted by Observ'ER & Foundation Énergies pour le Monde in 2012, the dominant production of energy is the one produced through fossil fuels, covering more than two-thirds of total energy production (68.1%) [1]. This type of energy, besides being non-renewable, generates irreversible environmental damage to our planet. Due to the large amounts of carbon dioxide (CO_2) emitted into the atmosphere, thereby increasing greenhouse gases.

A mechanism that seeks to counteract environmental damage and reduce energy production costs is the use of renewable energy, also called clean energy. A wide range of options within the renewable energies such as: solar, wind, biomass, geothermal, non-renewable waste and hydro, among others were found. Of the above, solar and wind energy have presented an annual growth between 2002 and 2012 of 50.6% and 26.1%, respectively [1].

Taking into account the great and continuous growth of these renewable energies, many challenges have been encountered (meeting demand, lowering production costs, improve energy plant planning, etc.). So to reach a full adoption of this type of energy is necessary to find effective solutions to these challenges. One of the main ones is the necessity of making predictions of the variables involved in the production of energy. That is, to have a certainty of the amount of energy to be produced in the next few minutes, hours, days, or months [2]. Some of the variables involved in the production of clean energy could be the following: water flows, tides and internal heat of the earth, etc. This work deals with wind speed forecast.

The prediction of these variables is usually performed in the short, medium and long term. There is no definite time frame for each of the periods [3]. But regularly medium and long-term forecasting is used for tactical and strategic planning, respectively. The short-term prediction is used at the operational level and its periodicity could be in time scale of minutes, hours, days [3,4].

The ability to forecast wind speed is essential for the correct integration of wind energy into electrical systems. According to Barber *et al* [5], the importance of wind forecasts for the wind energy industry stems from three facts:

- The aggregate power produced and consumed throughout a power system must be nearly in balance at all times to guarantee supply reliability and safety.
- The power output of a wind farm is highly variable since it depends strongly on wind speed and direction.
- An efficient and cost effective energy storage mechanism does not exist.

Recently, an estimation for the year 2030 mentions that a perfect forecast will be valued at 3 Billion US dollars annually [6], for the United States power system; the estimation was performed by the Department of Energy [5]. Therefore, wind speed forecasting still plays an important role in the task of energy supply [4].

In literature different approaches employed for this task were found [7-10]. Chang presents [4] a survey that categorizes different approaches that deal with the forecasting problem establishing Persistence, Physical, Statistical, Spatial Correlations, Artificial Intelligence, and Hybrid Methods. The Persistence method (or Naïve) is a basic forecasting method, where $Y_{t+\Delta} = Y_t$ [11]. This particular method is often used as a baseline method. The Physical methods are numerical weather predictions developed by meteorologists for a large scale weather prediction [12]. The Statistical methods (SM) uses historical data to find relationships inside the wind speed time series. Within the Statistical methods are the Auto Regressive (AR), Auto Regressive Moving Average (ARMA), Auto Regressive Integrate Moving Average (ARIMA) [13-15]. Artificial Intelligence methods (AIM) uses Artificial Neural Networks (ANN) [14,16], Support Vector Machine (SVM) [17], k-Nearest Neighbor (kNN) [18], among others as a forecast method. AIM improve in some cases the results obtained using statistical methods. Hybrid Methods are a combination of forecasting methods [14] (e.g. combination of statistical and artificial intelligence approaches).

SM (e.g. ARIMA, ARMA, Holt-Winters, etc.) or AIM (e.g. ANN, SVM, KNN) usually perform the forecast using Historical data. But, if the dataset presents missing values inside, the SM may have problems to perform an adequate forecast. In reality, wind time series present missing data due to problems with the sensors (e.g. communication problems, natural wear, lack of maintenance, etc.). The wind speed is irregular and intermittent [19] such that added to the fact that the time series are incomplete, the task of short term forecast is even harder. This work deals with the problem of the wind speed time series reconstruction. This work starts by characterizing the temporal behavior of the times series, creating a pattern database. Then, it uses a hybrid method to model the time series. This method uses an ANN as a forecasting approach and Genetic Algorithms (GA) to define the best topology of the ANN. Once, the ANN was trained, it is possible to reconstruct the wind speed time series.

The rest of the paper is organized as follows: Section 2 describes the methodology proposed to reconstruct a wind speed time series, Section 3 presents the experiments performed with our proposal, and finally the conclusions of this work are presented in Section 4.

2. Time series reconstruction

Dealing with the problem of forecasting wind speed time series is not something new. It is a complicated task since there is a wide range of different behavior patterns in the data to model. Adding the problem that the time series could be incomplete, the problem becomes even more complicated, although, nowadays there are a number of algorithms and techniques that offers an effective forecasting [4,11]. Nevertheless, in most of the literature, the problem of wind speed forecasting is addressed assuming that the time series is complete (without any missing data). However, this is not the case in reality. Regularly, there are problems with sensors giving as a result missing data within the wind speed time series. Thus, to deal with this problem using Statistical Methods as ARIMA or Holt-Winters becomes a hard work, or even impossible.

To deal with the missing values that are within the time series, a temporality removal of the time series was proposed. Instead of having a sliding window of size *k*, the Temporality Removal Process creates a database where k features of the time series are recorded, associated with it respective forecasting value. With the database created, the Construct the Time Series Model (CTSM) process uses an ANN for the regression problem and instead of using the trial error method for selecting the optimal topology, a Compact Genetic Algorithm was used to define the optimal topology. Once the optimal ANN structure is defined, the Data Reconstruction Process ends the process using the ANN trained to reconstruct the missing data. Figure 1 shows the flow chart of the wind speed time series reconstruction process.

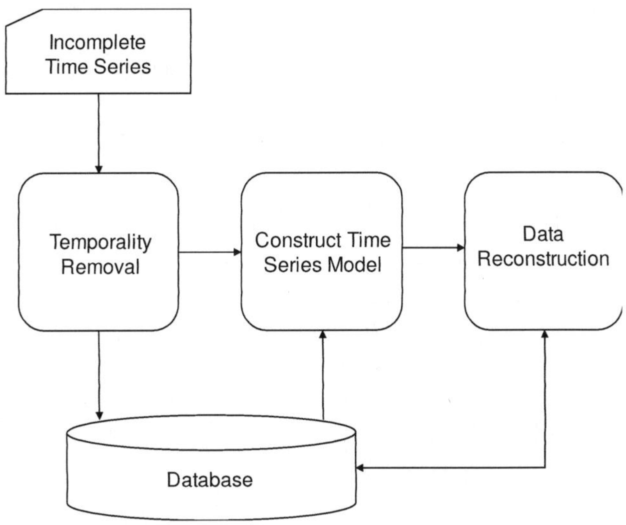

Figure 1. Flow chart of the wind speed time series reconstruction process.

2.1. Temporality removal process

Regularly, for the forecasting problem, there is a window of size *k* that slides through the time series. But because of the missing values inside the time series, this window can no longer slide in the same way. Therefore, a database with the time series temporal patterns is created. The database records the k (window size) characteristics and associates them with their respective values forecasting value (y_{t+1}). After creating the database, all records where missing data is present are deleted, leaving aside the time dimension.

Table 1 shows the result of creating the database and extracting the k features associated with an expected output. Each record inside the database represents the last k observations at a defined time (t), associated to the expected output y_{t+1} (for training purposes). Table 1 shows records with the presence of missing values (φ), this records (φ) are deleted. With the database defined (without missing data), it is possible to start searching for an optimal ANN that models the behavior of the wind speed time series.

Table 1. Transformation the time series into a database structure.

Inputs (K_t)							$\hat{y}_{t+1}$
k	y_1	y_2	y_3	…	y_k		y_{k+1}
$k+1$	y_2	y_3	y_4	…	y_{k+1}		y_{k+2}
$k+2$	y_3	y_4	y_5	…	y_{k+2}		φ
$k+3$	y_4	y_5	y_6	…	φ		y_{k+4}
$k+4$	y_5	y_6	y_7	…	y_{k+4}		y_{k+5}
$k+5$	y_6	y_7	φ	…	y_{k+5}		y_{k+6}
$k+6$	y_7	φ	y_9	…	y_{k+6}		y_{k+7}
$k+7$	φ	y_9	y_{10}	…	y_{k+7}		y_{k+8}
$k+8$	y_9	y_{10}	y_{11}	…	y_{k+8}		y_{k+9}
$k+9$	y_{10}	y_{11}	y_{12}	…	y_{k+9}		y_{k+10}
		…				…	
n-1	y_{n-k}	y_{n-k+1}	y_{n-k+2}	…	y_{n-1}		y_n

2.2. Construction of a time series model

In recent works, ANN has been proved to present great classification an regressions qualities. ANN have their basis in the study of the human brain, trying to simulate the process of neuron interconnections. ANNs are thus able to learn from previous knowledge provided at the training phase, regardless of the type of task assigned to them. That is why they are useful in different fields, such as industry and science [4].

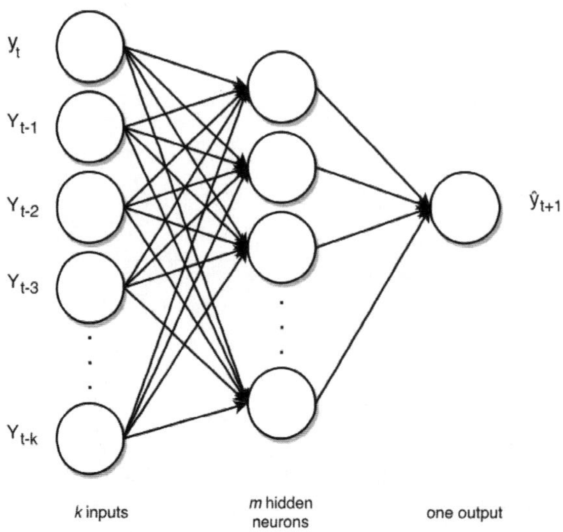

Figure 2. ANN architecture with k+1 past observations, m hidden neurons and a single output ($\hat{y}_{t+1}$).

Here, the ANN architecture proposed is a Feedforward Multilayer Perceptron trained, by gradient based methods. Figure 2 shows the architecture of the MLP ANN proposed in this work. An ANN, as a universal approximator can learn any given function. A set of k past observations are considered as the input data, the hidden layer consists of a m hidden neurons, the output layer corresponds to the forecast value $\hat{y}_{t+1}$; A sigmoid is used as the activation function. The ANN that models the wind speed time series $\hat{y}_{t+1}$ can be defined as:

$$\hat{y}_{t+i} = f_1\left(\sum_{l}^{k+1} w_l x_l\right)$$
$$x_l = f_2\left(\sum_{j}^{m} w_{lj} y_t - l_j\right) \quad (1)$$

where f_1 and f_2 are the activation functions, and w are the coefficients (also known weights connection).

In the wind speed time series modelling problem, it is necessary to provide an accurate model that defines the behavior of the wind speed. To define the accuracy of the optimal model, the Mean Square Error statistical measure is used, which is defined as

$$MSE = \frac{1}{n}\sum_{i=1}^{n}(\hat{y}_i - y_i)^2 \quad (2)$$

This measure will be minimized during the training process. GA is an optimization technique inspired on Darwin's principle of evolution. That is, it mimics a simplistic version of the process of biological evolution, which consists of creating a population of individuals, where each individual represents a prospective solution of the problem being solved. GA modifies this population using genetic operators: selection, mutation, recombination, etc. [20]. cGA is a variant of the simple GA in which a probability distribution is used to represent the population. Therefore, it requires less memory than the simple GA. The cGA can be used to give a quick answer to define an optimal architecture of an ANN [21].

Table 2. Training algorithms.

1	Gradient descent BP
2	Gradient descent with momentum BP
3	Gradient descent with adaptive learning rate
4	Gradient descent with momentum BP and adaptive
5	Resilient BP
6	Broyden-Fletcher-Goldfarb-Shanno
7	Newton conjugate gradient
8	Conjugate gradient

Determining the best ANN architecture and training algorithm is an optimization problem. This is, defining the architecture of the ANN by determining the number of inputs, the number of neurons in the hidden layer, and and the training algorithm. This is a critical step in the training process that is usually performed by trial and error. In previous works, the use of GA has been tested yielding excellent results [13]. The difference here is the use of a cGA, which is an improvement of a simple GA, also several training algorithms are considered. Each individual (the chromosome) in the GA is defined as a binary vector that is coding the number of inputs, number of hidden neurons and the training algorithm of the ANN. Algorithm uses GA to finds the best ANN topology and describes step by step the used process. The cGA starts by defining the probability vector, from this vector two individuals are generated and evaluated. The evaluation process is used to test the topology proposed by the cGA. The training algorithms used in this approach are defined in table 2. The result of the training algorithm corresponds to the individual's evaluation. Both individuals compete, and according to the winner individual, the probability vector is updated. This procedure repeats until a convergence criterion is achieved.

The genetic algorithm basically provides as the result of how many inputs will be optimal for the model, how many neurons must be considered for the hidden layer of the network and which training algorithm should be used. At the end of the process, the best individual (i.e., the one presenting best fitting) found during the evolution is returned as the solution of the problem (topology of the ANN). After defining the topology of the ANN, a refinement process is started. Refinement is the same process as the evaluation process, but the number of epochs is increased significantly.

2.3. Data reconstruction

For the reconstruction process, the ANN that models the behavior of the wind speed time series is used. Using this model, it is possible to start with the reconstruction process.

Algorithm 1 Finding the best topology algorithm using cGA

```
1:  procedure GA(n as population size, l as chromosome length)
2:      Initialize probability vector
3:      for i:= 1 to do p[i] :=0.5;
4:      Generate two individuals from the vector
5:      a:= generate(p);
6:      b:= generate(p);
7:      evaluate(a);
8:      evaluate(b);
9:      winner,loser := compete(a,b);
10:     //Update the probability vector towards the better one
11:     for i := 1 to l do
12:         if winner[i] ≠ loser[i] then
13:             if winner[i] = 1 then p[i] := p[i] + 1/n
14:             else p[i] := p[i] − 1/n;
15:     //Check if the vector has converged
16:     for i := 1 to l do
17:         if p[i] > 0andp[i] < 1 then return to step 4;
18:     //p return the final solution
```

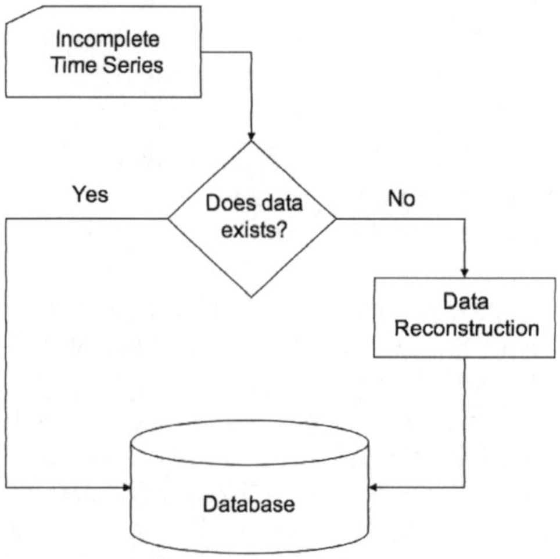

Figure 3. Reconstruction process for wind time series.

This process starts using a vector ordered observations that are evaluated sequentially. When a missing value in the time series is found, the reconstruction data process starts. This process uses the k last observations of the vector used as input. The output of the ANN already trained in previous steps is used to fill the missing value. The process continues until the end of the observation vector is reached. Figure 3 shows the flow chart of the reconstruction process.

3. Results

In order to test the suitability of our proposal, experiments were carried out with time series of the wind speed coming from anemometers located in different locations inside the the state of Michoacán, México recorded hourly. Since the sensors are susceptible to miss-functions and reading errors, the problem of obtaining time series with missing data was encountered. In some cases, this could be a quite important problem since the time series contains more that 5,000 adjacent missing data.

We perform several experiments applying the approach described in figure 1. For the lack of space, in this section, the results of one of the experiments are described. Each experiment starts by removing the temporality of the time series by creating a database following the procedure described in the previous section. Once the database has been obtained, an optimal ANN that models the behavior of the wind speed time series is defined. The optimal topology obtained at this stage is described in table 3

Table 3. The winner topology obtained by the cGA.

Input Neurons	25
Hidden Neurons	28
Training Algorithm	Conjugate

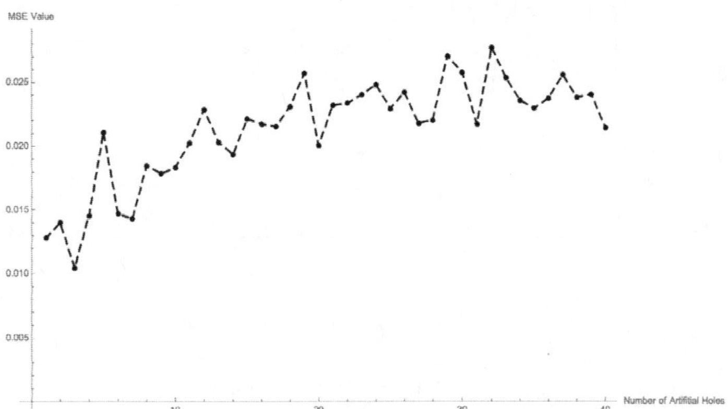

Figure 4. Experiment No. 1 MSE results for 1 to 40 artificial missing values.

The optimal topology obtained in this stage is used in a refinement process, which is only increasing the number of epochs of the evaluation function of the cGA. In this case, 3,000 epochs were used.

In order to test the accuracy of the reconstruction process proposed in this work, a new training set in which we artificially place missing values of size h was created. Once the missing values were generated, the reconstruction process is started. With this action were able to compare the results of the reconstruction with real information (reconstruction simulation). In the experiments, the proposed approach was tested by generating artificial training sets with missing data up to size 40. Figure 4 shows the MSE values for the performed experiments. Table 4 shows numerically the results. From figure 4, we observe that as the size of the missing data grows, the accuracy is gradually lost, but this accuracy loss is not very significant. The implementations have been developed under Python platform using the library NeuroLab 0.3.5 [22].

Table 4. Experiment No. 2 time series MSE for 1 to 40 missing values.

Number of holes	MSE	Number holes	MSE	Number of holes	MSE	Number of holes	MSE
1	0.012802972	11	0.020267953	21	0.02319457	31	0.021700934
2	0.013979674	12	0.022868016	22	0.023364267	32	0.027749779
3	0.010446051	13	0.020316086	23	0.024026512	33	0.025341943
4	0.01453436	14	0.019358147	24	0.024815368	34	0.023582028
5	0.021092693	15	0.022149374	25	0.022944667	35	0.022948563
6	0.014692938	16	0.021738649	26	0.024217165	36	0.023763214
7	0.014295101	17	0.021536263	27	0.021791693	37	0.025565695
8	0.018491887	18	0.023112223	28	0.022043973	38	0.023803695
9	0.017864363	19	0.02571133	29	0.027073025	39	0.024064536
10	0.018317789	20	0.020054946	30	0.025773363	40	0.02142727

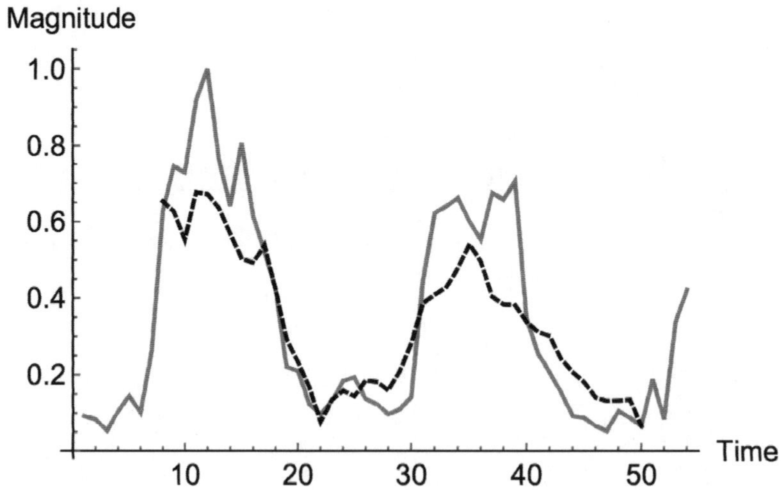

Figure 5. Reconstructed time series.

A graphical example of the reconstruction process is shown in figure 5. In the figure 5, the continuous line represents the real data, and the dotted ones represent the reconstruction.

4. Conclusions

This paper has proposed a time-series reconstruction approach based on a hybrid neural- evolutionary methodology. This methodology allows to determine the optimal ANN architectures to model incomplete time series and later to be able to reconstruct them. Experiments were performed considering multiple time series with artificially added missing data in the different phases: model generation and data reconstruction. A correlation was determined between the size of the missing data and the error obtained from the data validation, showing a positive trend between the size of the missing values present and the error obtained. The experiments were performed using the under Python platform using the Neurolab library [22] for the ANN implementations.

References

[1] Observ E 2013 *Worldwide Electricity Production from Renewable Energy Sources* (Paris,

France: Ninth Inventory)
[2] Rodriguez Rangel H, Puig V, Farias R L and Flores J J 2016 Short-term demand forecast using a bank of neural network models trained using genetic algorithms for the optimal management of drinking water networks *J Hydroinformatics* **19** 1-16
[3] Giacomoni M, Kanta L and Zechman E 2013 Complex adaptive systems approach to simulate the sustainability of water resources and urbanization *J Water Resour Plan Manage* **139** 554-64
[4] Chang W-Y A 2014 literature review of wind forecasting methods *J Power Energy Engin* **2** 161
[5] Barber C, Bockhorst J and Roebber P 2010 Auto-regressive hmm inference with incomplete data for short-horizon wind forecasting *Advances in Neural Information Processing Systems* (California, USA: Curran Associates, Inc.) pp 136-44
[6] Piwko D and Jordan G 2010 The economic value of day-ahead wind forecasts for power grid operations *2010 UWIG Workshop on Wind Forecasting*
[7] Cassola F and Burlando M 2012 Wind speed and wind energy forecast through kalman filtering of numerical weather prediction model output *Appl Energy* **99** 154-66
[8] Liu H, Tian H-Q, Pan D-F and Li Y-F 2013 Forecasting models for wind speed using wavelet, wavelet packet, time series and artificial neural networks *Appl Energy* **107** 191-208
[9] Foley A M, Leahy P G, Marvuglia A and McKeogh E J 2012 Current methods and advances in forecasting of wind power generation *Renew Energy* **37** 1-8
[10] Alexiadis M, Dokopoulos P, Sahsamanoglou H and Manousaridis I 1998 Short-term forecasting of wind speed and related electrical power *Solar Energy* **63** 61-8
[11] Zhao X, Wang S and Li T 2011 Review of evaluation criteria and main methods of wind power forecasting *Energy Procedia* **12** 761-9
[12] Ma L, Luan S Y, Jiang C W, Liu H L and Zhang Y 2009 A review on the forecasting of wind speed and generated power *Renewable and Sustainable Energy Reviews* **13** 915-20
[13] Flores J J, Loaeza R, Rodriguez H and Cadenas E 2009 Wind speed forecasting using a hybrid neural-evolutive approach *MICAI 2009: Advances in Artificial Intelligence* (Guanajuato, México: Springer) pp 600-9
[14] Flores J J, Graff M and Rodriguez H 2012 Evolutive design of arma and ann models for time series forecasting *Renew Energy* **44** 225-30
[15] Erdem E and Shi J 2011 Arma based approaches for forecasting the tuple of wind speed and direction *Appl Energy* **88** 1405-14
[16] Barbounis T G, Theocharis J B, Alexiadis M C and Dokopoulos P S 2006 Long-term wind speed and power forecasting using local recurrent neural network models *IEEE Trans Energy Conversion* **21** 273-84
[17] Zhou J, Shi J and Li G 2011 Fine tuning support vector machines for short-term wind speed forecasting *Energy Conver Manage* **52** 1990-8
[18] Yesilbudak M, Sagiroglu S and Colak I 2013 A new approach to very short term wind speed prediction using k-nearest neighbor classification *Energy Conver Manage* **69** 77-86
[19] Hayashi M and Kermanshahi B 2001 Application of artificial neural network for wind speed prediction and determination of wind power generation output *Proceedings of ICEE* pp 12-5
[20] Rangel H R, Puig V, Farias R L and Flores J J 2016 Short-term demand forecast using a bank of neural network models trained using genetic algorithms for the optimal management of drinking water networks *J Hydroinformatics* doi:10.2166/hydro.2016.199.
[21] Harik G R, Lobo F G and Goldberg D E 1997 The compact genetic algorithm, Urbana 5161801
[22] https://pypi.python.org/pypi/neurolab

Experiment and ANSYS simulation analysis for metal aluminum solid and fluid conversion

Y-Y Wang[1,2], P Guo[1], Y Wu[1], Z-L Zhang[1] and S-M Jiang[1]

[1]College of Electromechanical Engineering, Qingdao University of Science and Technology, Qingdao 266061, China

E-mail: 787388073@qq.com

Abstract. In this paper, study on metal aluminum solid and fluid conversion was carried out by using crucible resistance furnace, and observing the phenomenon of metal aluminum solid and fluid conversion. In the experiment, the same shape aluminum block was kept under the same heating rate and heated by the resistance furnace. The experimental results show that the melting point of metal aluminum is between 650°C and 660°C, and after the melting point, the metal aluminum began to melt when it maintained for a long period of time, however, when the temperature is higher than the melting point, the aluminum will melt very quickly. In addition, in ANSYS simulation, the solid aluminum melted completely at 670°C in 5430 seconds, much longer than the actual experiment, it due to the heating rate was faster, not in an ideal experimental environment and there is heat exchange with the outside world and convection, at the same time, the aluminum block may contain impurities, so the actual melting time could be shorter than the simulation. In this paper, it was explored for the liquid and solid conversion in depth, and had a certain actual value.

1. Introduction

With development and progress of the society, the demand that people need energy was growing gradually, and people faced the current situation that shallow resources are exhausted [1]. Therefore, it is very urgent to explore deep energy, but researches of domestic and foreign scholar suggested that the deep mining environment problem was very complex, there exist factors such as high temperature, complex stress field and it makes the difficulty of deep mining increase [2-4]. To solve these problems, we carry out simulation experiments with various kinds of existing materials to study the problem of solid flow transformation of deep soft rock, thus reducing the difficulty of mining.

The solid and fluid material conversion refers to material change process between solid and fluid under effect of the external environment factors, such as the storage of foods like sugar and chocolate [5,6], the melting of ice, snow [7] and metal [8]. In this paper, solid and flow conversion experiment for metal aluminum was carried out using the crucible resistance furnace, the phenomenon of solid and fluid conversion was particularly described through the molten experiment of metal aluminum. The finite element simulation experiment of the solid flow transformation of metal aluminum has also been carried out in this paper, and some valuable conclusions have been obtained [9-11].

2. Transformation experiment between solid and flow of aluminum

It is common to metal melting and crystallization in the daily life and industrial production, its essence is that the internal metal mutual conversion between solid and liquid crystal. However, because the metal melting temperature is higher, there are almost no special experiments about metal conversion of solid and fluid, and in this paper, the metal aluminum solid and fluid conversion experiment was carried out by using the crucible resistance furnace.

2.1. Experimental equipment

6061 aluminum, DRZ-4 resistance furnace temperature controller, G2-5-10 crucible resistance furnace, crucible, 50 mm (long) X 50mm (wide) X 80 mm (high) aluminum block, hand-held infrared thermometer, ruler, sandpaper, experimental units, thermometer, screwdriver, wire, socket, etc.

2.2. Experimental scheme

Experimental scheme 1: Put the aluminum blocks into the graphite crucible, carry out melting experiments in crucible resistance furnace, find the aluminum melt temperature through slowly heating up, rerun the experiment, make the aluminum blocks melt under the temperature when the aluminum started to melt, observe the phenomenon of the aluminum melt, record experimental data, and record the time required that the aluminum melted completely under this temperature and the time required for complete solidification of the liquid aluminum at room temperature is then recorded.

Experimental scheme 2: put the same shape aluminum blocks into a crucible, carry out melting experiments in resistance furnace, make aluminum blocks separately heat to the temperature that 10 degrees higher than the temperature of aluminum blocks started to melt, observe the phenomenon of the aluminum melting, record experimental data, and the time needed for complete melting of the aluminum block is recorded separately.

2.3. Experimental steps

Equipment connection: connect the DRZ-4 resistance furnace temperature controller to the SG2-5-10 crucible resistance furnace with wires; the temperature sensor is placed in the crucible. Place the installation in ventilation safety position, as shown in figure 1.

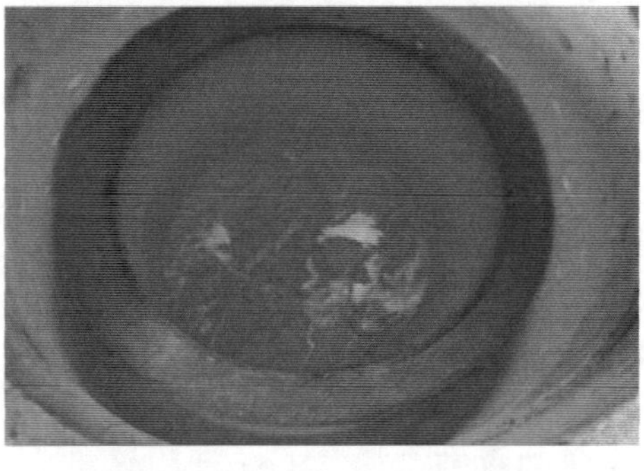

Figure 1. Device connection.

Figure 2. Melt finished.

Material placement: First, putting aluminum and crucible into the electrical resistance furnace, covering the furnace cover, and then switch on the power supply.

Heating: Making the crucible resistance furnace heat through temperature controller, and keep a constant temperature, observe whether aluminum has be soft and started to melt or not, and recording the experimental data. Finally finding the time points of aluminum started to melt, and recording the temperature at this time [12-14].

When aluminum completely melt into liquid, as shown in figure 2, close the device and observing the process which aluminum changes from liquid to solid, make the liquid aluminum with extremely slow speed to cool in the crucible, in the process of cooling, every once in a while measuring a temperature, and record it .When the aluminum becomes solid and the crucible furnace drop to room temperature, take out the crucible and the solidified aluminum block.

3. Experimental data and experimental conclusions

3.1. Experimental results

The experimental results are shown in tables 1-3.

Table 1. The relationship between the melting time, temperature and the state of molten aluminum.

	Time (min)	Temperature (°C)	Status
1	0	0	solid
2	10	100	solid
3	15	200	solid
4	19	300	solid
5	23	400	solid
6	28	500	solid
7	35	600	solid
8	40	610	solid
9	45	620	solid
10	50	630	solid
11	55	640	solid
12	60	650	softening
13	65	660	solid-liquid mixed
14	84	660	liquid
15 Close the resistance	85	660	liquid
furnace	90	652	solid-liquid mixed
16	100	652	solid-liquid mixed
17	114	594	solid
18			

Table 2. The relationship between the melting time, temperature and the state of molten aluminum.

	Time (min)	Temperature (°C)	Status
1	0	0	solid
2	10	100	solid
3	14	200	solid
4	20	300	solid
5	25	400	solid
6	32	500	solid
7	40	600	solid
8	43	650	softening
9	45	660	solid-liquid mixed
10	55	660	solid-liquid mixed
11	65	660	liquid
12 Close the resistance	65	660	liquid
furnace	75	651	solid-liquid mixed
13	85	651	solid-liquid mixed
14	90	615	solid
15			

Table 3. The relationship between the melting time, temperature and state of molten aluminum.

	Time (min)	Temperature (°C)	Status

1	0	0	solid
2	9	100	solid
3	13	200	solid
4	18	300	solid
5	25	400	solid
6	32	500	solid
7	38	600	solid
8	42	650	solid
9	47	670	solid-liquid mixed
10	55	670	solid-liquid mixed
11	59	670	liquid
12 Close the resistance	60	670	liquid
furnace 13	65	648	solid-liquid mixed
14	80	648	solid-liquid mixed
15	89	625	solid

3.2. Experimental conclusions

- From table 1, it can be seen that the rise of temperature is linear. The metal aluminum softening phenomenon began to appear at 650°C, and it is in a solid-liquid mixed state at 660 °C. It indicates that the melting point of metal aluminum is in the middle of 650-660°C, which was consistent with the theory.
- From table 2, it can be seen that the metal aluminum began to melt at 660°C. It took 19 minutes to complete the melting. From the beginning of the heating to the end of the melting, it lasted 65 minutes. The liquid aluminum solidified into solid aluminum in about 25 minutes, and the temperature of the aluminum in solid-liquid mixed state is lower than the melting point of aluminum, this phenomenon is what we call supercooling.
- Comparing tables 2 with 3, it can be seen that the metal aluminum cost a total of 12 minutes from the beginning to the end of the melting at 670°C, less than the melting time at 660°C, which indicates that the higher the heating temperature is, the faster the aluminum melts.

4. Numerical simulation analysis

The phase change analysis module in ANSYS software is adopted in this paper, it mainly regarded the enthalpy as material attribute definition, the phase could be distinguished through temperature. By phase change analysis, the temperature distribution of the material at each moment can be obtained. Through the temperature nephogram, we could get the time required for complete phase change (melting or solidification time). The analysis process consists of the following steps: (1) Pre-processing, namely creating finite element model of metal aluminum, dividing grid units, carrying out temperature load and constraint condition; (2) Solving, set the solution option and time steps, and then solve; (3) Post-processing, checking the temperature distribution nephogram, inspecting test results.

4.1. Description of problems in metal aluminum melting test

The melting process of aluminum is analyzed, putting the aluminum block with size of 50 mm (long) X 50 mm (wide) X 80 mm (high) into the crucible and then melt in the crucible resistance furnace, the initial temperature of the aluminum and environment are room temperature, from time is equal to zero start to load 670°C for metal aluminum, ignoring the influence of air and crucible. The material parameters of metallic aluminum are shown in table 4, when analyzing, the temperature unit used °C, other units used pharmaceutical units. Through Phase transformation analysis of ANSYS, the time how long does the metal aluminum blocks melt complete need under this temperature was obtained.

The phase transition analysis of metal aluminum block is carried out using SOLID70 thermal analysis unit of three-dimensional eight node, an imperial unit is used during the analysis.

Table 4. The material parameters of aluminum.

Temperature /°C	Enthalpy /(J/m^3)	Solid specific heat capacity /(J/kg•°C)	Liquid specific heat capacity /(J/kg•°C)	Density /(kg/m^3)	Melt point /°C
0	0				
659	1.5984e9	896	1050	2707	660
661	2.6747e9				
1000	3.6382e9				

4.2. Analysis of the aluminum melting time course

From figure 3, it could be obtained when the metal aluminum melting time were 2000 s, the highest temperature was 433°C and the minimum temperature was 358°C. And it still in solid state.

From figure 4, it could be obtained when the metal aluminum melting time were 3000 s, the maximum temperature was 631°C and the lowest temperature was 576°C. At this time, it still in solid state, but it already very close to the melting point of aluminum blocks.

From figure 5, it could obtained when the metal aluminum melting time were 3600 s, the maximum temperature of 660°C and the lowest temperature was 610°C, at this time, the solid state aluminum has already started to melt.

From figure 6, it could obtained when the metal aluminum melting time were 5430 s, the highest temperature was 694°C and the minimum temperature was 652°C, the melting of solid state aluminum had accomplished basically.

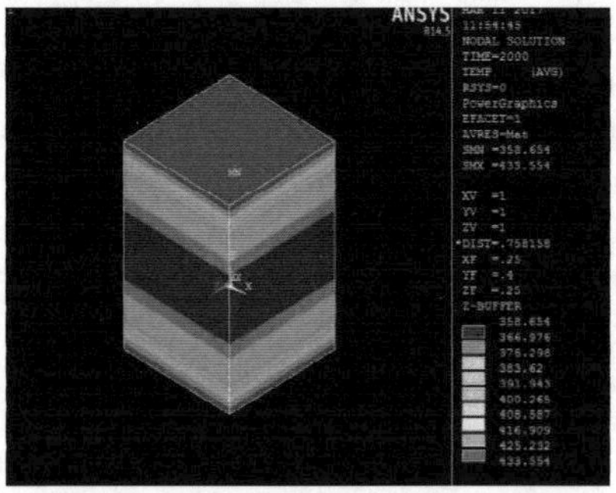

Figure 3. Temperature distribution of the aluminum block in 2000 seconds.

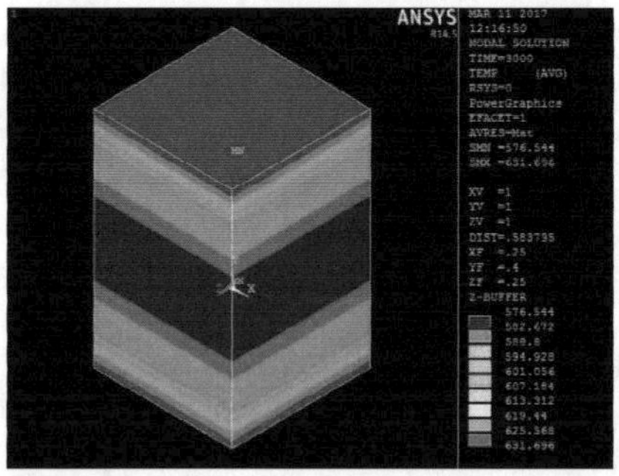

Figure 4. Temperature distribution of the aluminum block in 3000 seconds.

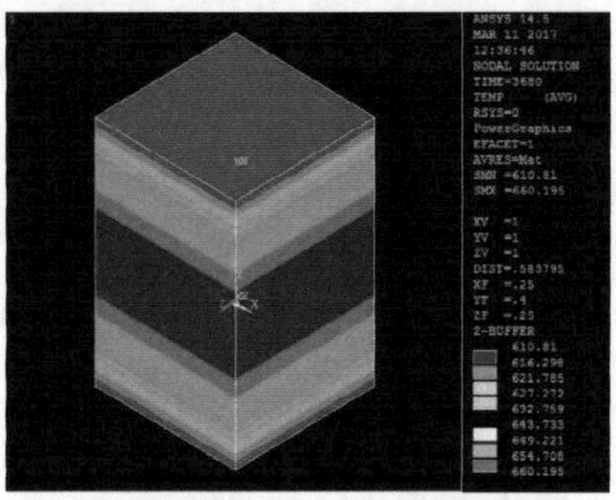

Figure 5. Temperature distribution of the aluminum block in 3600 seconds.

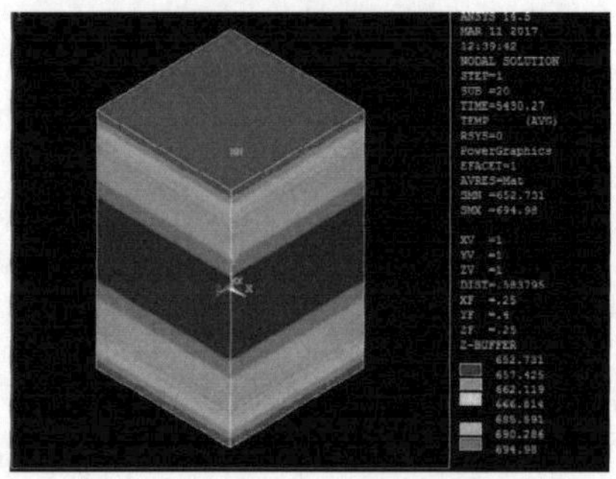

Figure 6. Temperature distribution of the aluminum block in 5430 seconds.

In summary, it could be obtained through the whole temperature distribution nephogram of metal aluminum under different melt time, when the melting time was 1000 s, due to the temperature is lower, the aluminum blocks have not yet started to melt; When the melting time was 3600 s, the aluminum blocks began to melt, and the face loaded of the aluminum blocks started to melt; When the melting time were 5430 s, the minimum temperature closed to 660°C, it shown that the aluminum blocks have melted completely.

5. Conclusions

In this paper, we got the time and temperature parameters of aluminum block melting and crystallization through the comparative analysis by the solid-flow transformation experiment of aluminum and ANSYS numerical simulation analysis. It was founded that the important factor influenced the aluminum solid and fluid conversion was temperature, the metal aluminum melting temperature is between 650°C and 660°C. The time were 20 minutes or so, which the solid metal aluminum need to transformed to liquid completely under 660 temperature, it also suggested that the melting rate will be faster when the aluminum melting temperature higher than the melting point. It is useful for the research in this field and has important application value.

References

[1] Hu S R. Peng J C and Huang C 2011 The mining research status and progress of Km above the mine *China Min Mag* **20** 105-10
[2] Wang Y Y, Lv Y M and Xiao Z J 2010 The creep modulus of sandstone in the chemical corrosion *J China Coal Soc* **7** 1095-8
[3] Wang Y Y and Wang Y C 2012 The creep law of deep soft rock tunnel excavation simulation under the action of temperature-stress-three chemical coupling *J China Coal Soc* **37(S2)** 275-9
[4] Zhang X D. Fu Q and Zheng X F 2011 The determination of coal and mudstone rheological model identification and support method *Mech Pract* **33** 30-5
[5] Xie S C, Cao D and Shi S J 2009 Cocoa butter crystallization and melting characteristics of research under the condition of non-isothermal *F Indus Tech* **12** 152-4
[6] Dong J and Liu H P 2013 The influence of melting process on quality of remanufacturing chese, when it is kept on the normal temperature *China D Indus* **8** 33-5
[7] Wang S Y and Li D W 2014 The heave characteristic test of the deep surface soil under the condition of different load of the cold end *Cryo Eng* **6** 63-7
[8] Zhou C D 2004 The experimental study of melting about vanadium and titanium iron ore *Angang Tech* **2** 1-3
[9] Poduval V S and Mistry V V 1994 Manufacture of reduced-fat mozzarella Cheese using ultraltered sweet buttermilk and homogenized cream *J Dairy Sci* **82** 1-9
[10] Glenn T A, Daubert C R, Farkas B E, *et al* 2003 A statistical analysis of creaming variables impacting process cheese melt quality *J. Food Qual* **26** 299-321
[11] Qin N 2011 Experimental study and numerical simulation of ice water solid flow transformation (Qingdao: Qingdao University of Science and Technology)
[12] Aluminum Association 1984 *Aluminum: Properties and Physical Metallurgy* ed J E Hatch (ASM International)
[13] Mondolfo L F 2013 *Aluminum Alloys: Structure and Properties* (Geauga County, Ohio: Elsevier)
[14] Cannon J F 1974 Behavior of the elements at high pressures *J Phys Chem Ref Data* **3** 781-824
[15] Chase M W 1986 JANAF thermochemical tables by Chase (MW Washington, DC: American Chemical Society; New York: American Institute of Physics for the National Bureau of Standards, c1986; United States: National Bureau of Standards)

Study of the temperature dependence of the uniaxial creep property of similar material of new soft rock

Y Y Wang[1], Y Wu[1,2], X Y Fan[1], J L Zhang[1], P Guo[1] and J G Li[1]

[1]College of Electromechanical Engineering, Qingdao University of Science and Technology, Qingdao 266061, China

E-mail: 403751612@qq.com

Abstract. Using the experimental method, the experimental research of creep properties were conducted under different temperature ranging from 10℃ to 60℃. The similar material of new soft rock consists of paraffin, which can obtain that the deformation contains the instantaneous elastic deformation and creep deformation through the uniaxial creep experimental results. And thus the increase of temperature has great influence on the creep characteristics of similar soft rock according to the creep curve of similar soft rock at 10℃ and 60℃. With the increase of temperature, the slope of the stress-strain curve of similar soft rock is increasing, while the average of the creep modulus is decreasing, which means that the capacity of resist deformation is reduced. Therefore, the creeps law of high-temperature and short-time can be shown the creep phenomenon of low-temperature and long-time, and further shorten the creep experimental cycle.

1. Introduction

Soft rock engineering stability control problem is always one of the major problems of coal mine production and construction in China. With the increase of mining depth and the complexity of geological conditions, the problem of soft rock tunnel support becomes increasingly serious. Soft rock rheology mainly led to the instability of surrounding rock of soft rock supporting engineering. Rock creep law is the main law reflecting its timeliness deformation. With the further development of deep engineering, the underground environment becomes more complex, temperature is closely related to the time effect of soft rock deformation [1-5]. The effect of temperature on the rock creep properties has great significance for engineering safety assessment and prediction. Researches of the temperature effect on rock creep at present are as follows: Wawersik W R [6] pointed that high temperature increased creep strain and creep rate of rock salt. Li J G [7] took sandrock as an example and analyzed that temperature effect should be considered in the design of deep soft rock engineering by experiment. Wang Y C [8] analyzed the study on creep law of deep soft rock under thermal-mechanical-chemical coupling effect and explored its mechanism, and thus established visco-elastic-plastic nonlinear creep model of considering temperature and chemical PH value influencing factor deep rock. Xi B P [9] pointed that steady-state creep rate and temperature of layered salt rocks comply with index correlation. Liu Q S [10] proposed the temperature has large influence degree for different rock rheological according to uniaxial compression test under three gorges granite specimen.

The experimental research of temperature on creep property mainly concentrated on the hard rock or great-intensity soft rock. While the creep has long experimental cycle and requires high temperature, which can't ensure the experiment safety. The paper chooses a paraffin wax as cementing agent,

40-mesh river sand as aggregate to configure a kind of similar soft rock material instead of soft rock prototype material, which can reflect the creep law under lower temperature and shorter time and greatly save the experiment period. And thus study the experiment research for creep property of similar soft rock under different temperature, which can provide an important basis to predict long-term stability and security in prototype engineering of soft rock.

2. The experimental process

2.1. The preparation of similar soft rock specimen and determination of physical parameters

Due to better thermal sensitive, the paper chooses weak cementation named paraffin as similar material cementing agent, which can simulate the lower intensity and bigger deformation the plastic of plastic destructive rock mass. The aggregate [11] chooses 40 mesh river sand; the auxiliary material chooses gypsum and industrial iron power, the proportion of paraffin, river sand, gypsum, iron powder is respectively 1:6:1:1. The experiment quickly put the material into the prepared mold and compacted after stirring evenly the ingredients under the constant 100℃, then put the specimen in the constant temperature and humidity environment, and cool and cure them under room temperature to avoid temperature effect on the forming specimen. The specimens are the cylinder of 50 mm diameter and 100 mm height.

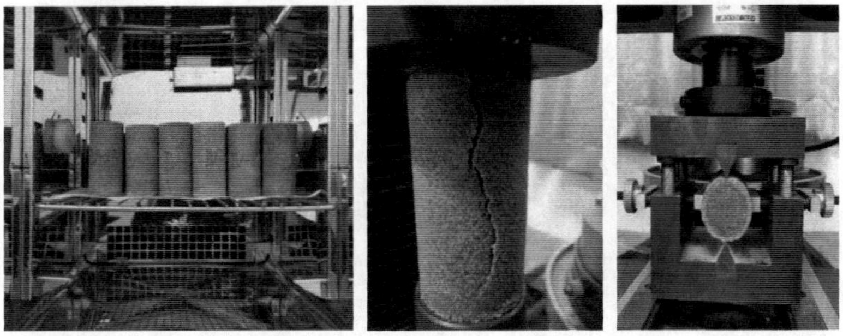

Figure 1. Part of the preparation of similar specimens in the incubator curing.

As shown in figure 1, similar specimens are done uniaxial and triaxial compression experiments under room temperature by the TAW - 200 electronic multi-function material mechanics testing machine and obtained the experimental test value of compressive strength, elastic modulus, Poisson's ratio. The experiment uses KD2 Pro Thermal characteristics analyzer (America), which can measure the thermal conductivity coefficient of similar soft rock. Take the visual density of the prototype material $\rho = 2.4g/cm^3$, the visual density of the model material $\rho = 2g/cm^3$, the volume-weight similar scale $C_\rho = \rho_P / \rho_M = 1.2$ and the geometry similar scale $C_L = 16$ [12] and can be obtained the calculated value of similar material physical parameters based on the similar theory [13,14]. As shown in table 1, in this paper, the preparation of the similar material can better satisfy the similarity criterion, which means that the prototype materials have high similarity.

Table 1. Calculation and experimental values of physical parameters of similar materials.

	compressive strength /MPa	tensile strength /MPa	elasticity modulus /Gpa	Poisson's ratio	Cohesion /MPa	internal friction angle /°	Density /g/cm^3	heat conductivity coefficient W/m.k
Calculated value	0.23~4.1	0.05~2.06	0.46~1.18	0.09~0.35	0.1~1.56	20~40	2	0.479
experimen	3.12	0.76	0.48	0.31	1.17	38	1.987	0.472

tal value

2.2. Creep experiment

- Laboratory equipment

The experiment uses TAW - 200 electronic multi-function material mechanics testing machine and the high-low temperature environment test chamber developed by mechanics research center of Qingdao University of Science and Technology (figure 2) can provide high and low temperature environment, stress and strain material data, which can realize automatic temperature control, the highest temperature can reach 200℃. Set the experiment temperature is respectively 10℃, 20℃, 30℃, 40℃, 50℃, 60℃, put every three specimens in temperature box. Heat up rapidly and do creep experiments when reaching the set temperature, the experimental results take the average of the three specimens and ignore the influence on error specimen.

Figure 2. TAW-200 Electronic multifunctional materials mechanical testing machine, temperature box.

Table 2. Rated loading at different temperatures.

loading method	temperature/℃	load/MPa
step loading	10	1, 1.5, 2, 2.5
	20	1, 1.5, 2, 2.5
	30	1, 1.5, 2
	40	0.25 0.5, 0.75, 1
	50	0.15, 0.2, 0.25, 0.3, 0.35, 0.5
	60	0.1, 0.15, 0.2

- The experiment method and parameter setting

Load and time on the same sample keeps constinuous step by step and obtains creep curves of specimen under different stress level by hierarchical load creep experiment scheme. This experiment scheme can avoid discrete factors of the specimens and maintenance. The hierarchical load conditions under different temperatures are shown in table 2 and each load can keep 2 h.

3. The experimental results and analysis

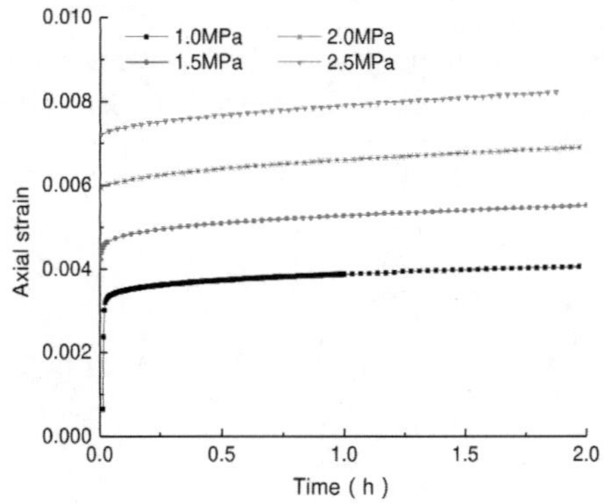

(a) 10℃

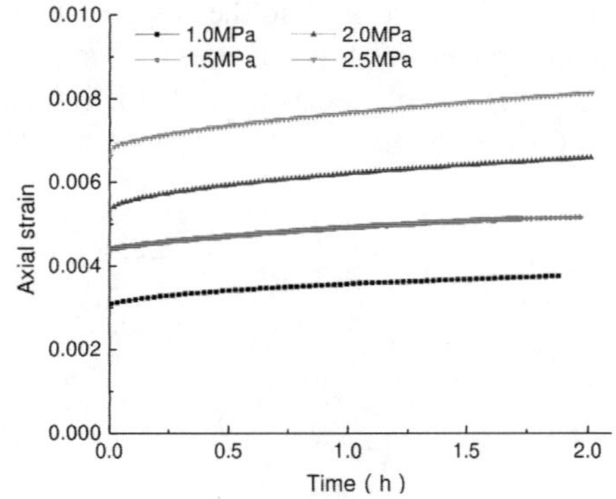

(b) 20℃

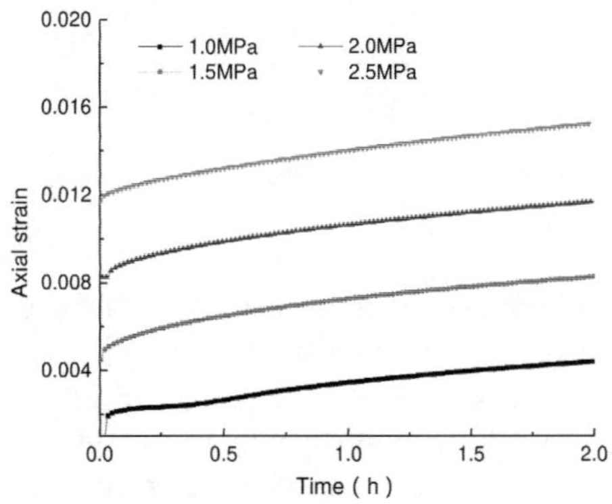

(c) 30℃

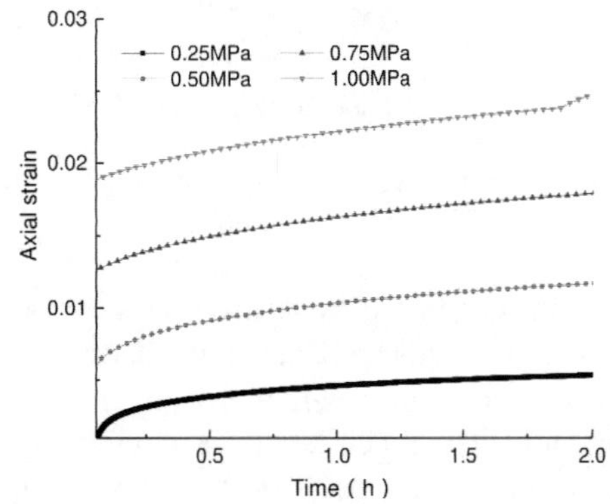

(d) 40℃

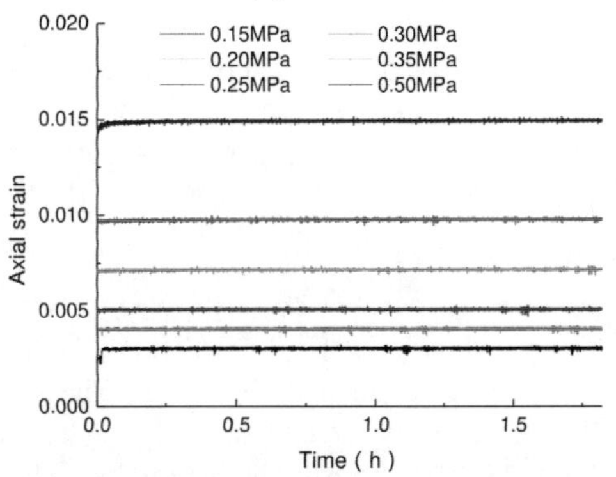

(e) 50℃

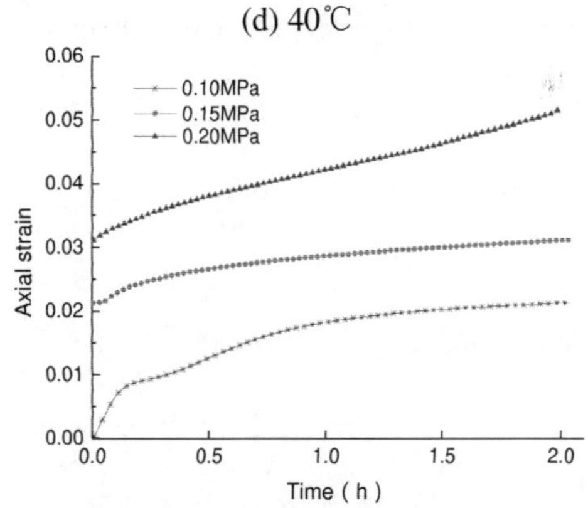

(f) 60℃

Figure 3. Creep curves of similar soft rock at different temperatures and loads.

From figure 3 can be seen the creep curve of similar material specimen at different temperature. Take 10℃ and 60℃ as example to better describe the creep properties of similar soft rock and can be seen from figure 3(a) that the axial strain of similar soft rock showed the law of decelerating creep and stable creep stage in the four stress level in 2 h, and with the increase of the loading stress level, the creep strain increases, the creep value for stress level of 2.5 MPa (0.0083) is more than 2 times to 1.0 MPa (0.004). Similar soft rock showed the deceleration creep and stable creep stage and deceleration creep accounted for the main part in the low-stress level of 0.1 MPa and 0.15 MPa from figure 3(f) at 60℃. In the level of 0.25 MPa, similar soft rock started showing accelerated creep stage and exhibited plastic failure even if the stress was still low. Compared the creep curves at two temperature, the temperature has a huge impact on the creep properties of similar soft rock and the creep value under 0.1 MPa in 2 h is 2.5 times to the creep value under 2.5 MPa in 2 h.

The creep data such as the stress value, strain value, time and temperature after 1 h were extracted from the similar soft rock creep characteristic data, and then the stress-strain curves under different temperatures were plotted. As can be seen in the figures 4 and 5:

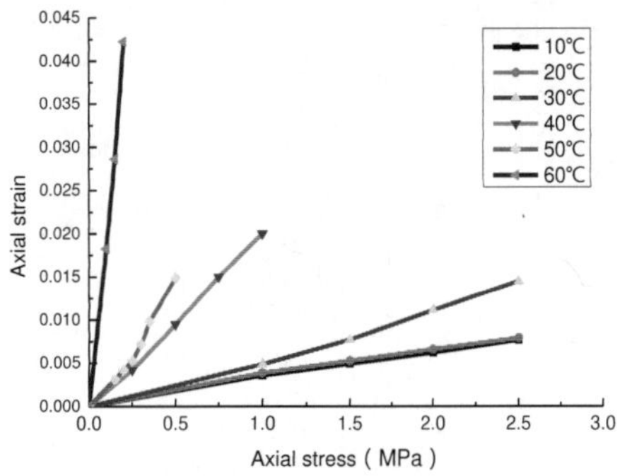

Figure 4. Creep stress-strain curves of similar soft rock at different temperatures (1 h).

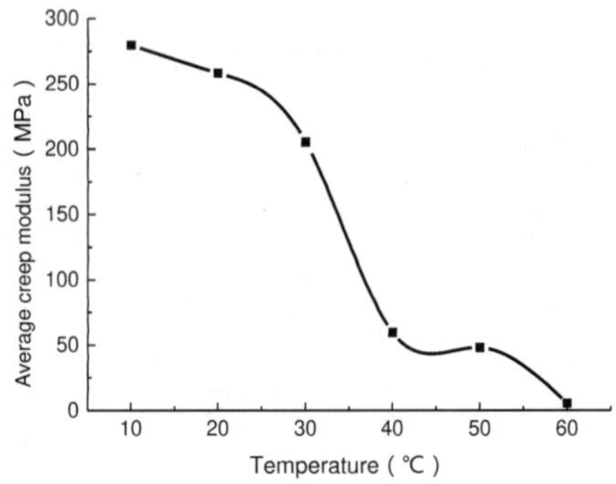

Figure 5. The average creep modulus varies with temperature (1 h).

The stress-strain curve of similar soft rock is linear, and the temperature has little effect on the stress-strain relationship of similar soft rock at 20℃and below; the stress-strain curve is upward migration with the temperature rising, and the higher the temperature is, the greater the magnitude of the offset is. To reflect the effect of temperature on the creep deformation capacity of the similar soft rock, define the reciprocal of the straight slope of stress-strain curves as the average creep modulus. Obviously, as the average creep modulus are constantly reduced, the decrease in 30-40℃is particularly evident with the temperature increasing.

During the experiment, the similar soft rock showed large deformation and crept damage under high-temperature environment and low-stress level within a short time. Due to this feature, we can characteristic the creep condition in high temperature and short time instead of low temperature and long time, which greatly shortens the test cycle of indoor model test and can better simulate the rheological properties of soft rock.

4. Conclusion

- The uniaxial creep experimental results of similar rock show that the deformation of similar soft rock contains the instantaneous elastic deformation and creep deformation. The creep curve of similar soft rock only shows decay creep and constant creep law within 2 h under 10℃and 2.5 MPa stress, while under 60℃and 0.25 MPa stress, the creep law shows acceleration within 2 h, and it can be seen that the increase of temperature has great influence on the creep characteristics of similar soft rock.
- Under normal temperature (20℃), the strain curve of similar soft rock is basically linear, with the increase of temperature, the slope of stress-strain curve of soft rock is increasing, and the soft rock creep is more sensitive to the load. With the increase of temperature, the average of the creep modulus is decreasing. It means that the capacity of resist deformation is reduced. According to this feature, the creeps law of high-temperature and short-time can be shown the creep phenomenon of low-temperature and long-time, and further shorten the creep experimental cycle.

Acknowledgments

This research was financially supported by the National Natural Science Foundation of "Research of theory and technology on rheological, creep and solid-fluid transformation of deep soft rock" (51674149).

References

[1] He M C 2004 *Present Situation and Prospect of Rock Mechanics in Deep Mining Engineering* (Beijing: Science Press) pp 88-94
[2] Wang Y Y and Su R H 2001 Predicting research of using gray model for the deformation of the soft rock around tunnels in the initial period of tunnel excavation *Chin J Appl Mech* **18(2)** 41-7
[3] He M C, Jing H H and Sun X M 2003 E*ngineering Mechanics of Soft Rock* (Beijing: Science and Technology Press)
[4] He M C, Zou Z S and Zou Y F 1993 *Introduction to Engineering of Soft Rock Tunnel* (Xuzhou: China Mining University Press)
[5] Zhang Y 2013 Transient temperature field of surrounding rock of the high geothermal roadway and its heat control mechanism by heat insulation (Xu zhou: China Mining University)
[6] Wawersik W R 1984 Alternatives to a power-law creep model for rock salt at temperature below 160℃ The Mechanical Behavior of Salt, Proceedings of the Second Conference (Clausthal: Trans. Tech. Publications) pp 103-31
[7] Li J G and Wang Y Y 2012 Experimental analysis of temperature effect in creep of soft rock *J China Coal Soc* **37(S1)** 81-5
[8] Wang Y C 2013 Study on creep law of deep soft rock under thermal-mechanical-chemical couping effect (Qingdao: Qing Dao University of Science & Technology)
[9] Xi B P, Zhao Y S *et al* 2009 Study of constitutive equation of granite rheological model with thermo-mechanical coupling effects *Chin J Rock Mech Eng* **28** 957-66
[10] Liu Q S 2001 Testing study on mechanical properties of the three gorges granite concerning temperature and time *Chin J Rock Mech Eng* **20** 715-9
[11] Zhang J and Hou Z J 2004 Experimental study of simulation materials for solid-liquid coupling *Chin J Rock Mech Eng* **23** 3157-61
[12] Li J G 2015 Study on strength and creep characteristics of composite rock mass with inclined soft-weak interlayer (Qingdao: Qing Dao University of Science & Technology)
[13] Wang Y and Li Y 2012 Deduction and application of similarity criteria in soft rock rheological model experiment *J Liaoning Techn Univ* **31** 354-7
[14] Yuan W Z 1998 *The Similarity Theory and the Statics Model Test* (Chengdu: Southwest Jiaotong University Press)

Properties and heat transfer coefficients of four molten-salt high temperature heat transfer fluid candidates for concentrating solar power plants

T L Liu[1], W R Liu[1] and X H Xu[1,2]

[1]School of Mechanical Engineering and Automation, Harbin Institute of Technology, (Shenzhen), Shenzhen 518055, Guangdong Province, China

E-mail: xuxinhai@hit.edu.cn

Abstract. Heat transfer fluid is one critical component for transferring and storing heat energy in concentrating solar power systems. Molten-salt mixtures can be used as high temperature heat transfer fluids because of their thermophysical properties. This paper studied the thermophysical properties of Li_2CO_3-Na_2CO_3-K_2CO_3 eutectic salt and three eutectic chloride salts NaCl-KCl-$ZnCl_2$ with different compositions in the range of 450-600°C and 250-800°C, respectively. Properties including specific heat capacity, thermal conductivity, density and viscosity were determined based on imperial correlations and compared at different operating temperatures. The heat transfer coefficients of using different eutectic salts as heat transfer fluids were also calculated and compared in their operating temperature range. It is concluded that all the four eutectic salts can satisfy the requirements of a high-temperature heat transfer fluid.

1. Introduction

Our society is facing with two major challenges: power shortage and environmental pollution. It is an urgent task for human beings to study renewable and sustainable energy. Concentrating solar power (CSP) has aroused many researchers' interests due to its potential to meet basic load applications, especially for applications with a large power output higher than 100 MW [1,2]. There are four most commonly used technologies in a CSP system: linear Fresnel reflector (LFR), solar power tower (SPT), parabolic trough collector (PTC), and parabolic dish system (PDS) [3]. Among these, less than 18% of the CSP plants utilize SPT, LFR and PDS, others utilize PTC to collect the solar energy. However, the most advanced CSP plants are the SPT-based plants, because they are more suitable for achieving very high temperature compared to other CSP plants and thereby can enhance the efficiency of converting heat into electricity [1,4]. A typical CSP plant consists of four major components: the solar collection system, the receiver, the heat transfer fluid (HTF) system and the traditional power block. The sunrays are concentrated onto the receiver by the reflector configuration in the solar collection system. The receiver absorbs the solar energy and transmits it to a HTF. The HTF acts as the thermal energy carrier transferring the concentrated heat to the power block which converts the heat into electricity [5].

HTF is one of the most important components in a CSP plant [5]. Due to the characteristics of low vapor pressure, high heat capacity, and excellent thermal stability, molten-salt mixtures have been widely used as HTF candidates for high temperature (higher than 500°C) applications [6-10]. Operating a CSP plant requires a large quantity of HTF, so that it is important to select the most appropriate HTF to minimize the cost and at the same time maximizing its performance [3]. The U.S. Department of Energy has funded for developing a new high-temperature HTF with a working temperature of at least 800°C by using eutectic molten salts [11]. Other important requirements for the thermophysical properties of the HTF include: vapor pressure below 1.0 atm at 800°C, specific heat capacity ≥1.5 kJ/(kg·K), thermal conductivity ≥0.51 W/(m·K), density ≤5400 kg/m^3, viscosity ≤0.012 Pa·s at 300°C and ≤0.004 Pa·s at 600°C.

The ability to exchange heat is one important characteristic of the HTF, which can be represented by the heat transfer coefficient *h*. In a theoretical study, *h* can only be determined based on its correlation with Nusselt number. The main objective of this paper is to compare the thermophysical properties and heat transfer coefficients of four eutectic molten-salt mixtures which could be used in CSP plants as high temperature HTFs. All the four kinds of molten-salt mixtures which include Li_2CO_3-Na_2CO_3-K_2CO_3 salts and three NaCl-KCl-$ZnCl_2$ salts with different composition ratios have thermally stable points higher than 650°C. The innovation points of this paper are the calculation and comparison of heat transfer coefficient *h* for these eutectic molten salts. In section 2, equations to calculate the specific heat capacity, thermal conductivity, density and viscosity of the four molten salts were presented. In section 3, the importance of the heat transfer coefficient *h* and the methodology employed to calculate the heat transfer coefficient *h* based on Nusselt number was discussed. In section 4, the thermophysical properties and heat transfer coefficients of the four molten-salt mixtures at different operating temperatures were compared.

2. Thermophysical properties

In this study, four different eutectic molten-salt mixtures are considered, including one carbonate salts [12] and three chloride salts [13]. Compositions and thermophysical properties of them are listed in tables 1 and 2. They are all potential HTF candidates for a CSP tower power operating in the temperature above 550°C because their thermal stability temperatures are all higher than 600°C. Vignarooban *et al* [3] summarized the currently available molten-salt HTF candidates in a recent review, and the selected four eutectic molten salts are the only candidates which satisfy the thermal stability requirement.

Table 1. Compositions, melting points and stability points of heat transfer fluids.

	Carbonate Salts[a]	Chloride salts[b]		
Compositions	Li_2CO_3-Na_2CO_3-K_2CO_3	NaCl-KCl-$ZnCl_2$		
wt. %	32.1-33.4-34.5	7.5-23.9-68.6	8.1-31.3-60.6	10.0-15.1-74.9
Melting point (°C)	400	204	229	213
Stability point (°C)	658	> 800	> 800	> 800

[a] [12,14,15]
[b] [13]

Table 2. Heat transfer fluids' thermophysical properties.

Properties	Units	Temperature range (°C)
Li_2CO_3-Na_2CO_3-K_2CO_3[a]		
$C_p = 1.61 \pm 0.08$	J/(g·K)	450 - 600
$\lambda = C_p \cdot \rho \cdot \alpha$	W/(m·K)	450 - 600
$\rho = 2.27 - 4.34 \times 10^{-4} T$	g/cm^3	450 - 600
$\mu = 0.0852 \exp(\frac{3.51 \times 10^4}{RT})$	mPa·s	450 - 650

$NaCl$-KCl-$ZnCl_2$ [b]		
$C_p = 0.9 - 0.92$	J/(g·K)	230 - 350
$\lambda = 0.5145 - 2.3309\times10^{-4}T$	W/(m·K)	300 - 800
$\rho = 2.8783 - 0.9263\times10^{-3}T$	g/cm³	300 - 800
$\mu = [0.12055\exp(\frac{-T}{204.70939}) + 497613.0848\exp(\frac{-T}{29.9196})]\times10^3 + 3.41$	mPa·s	250 - 800
$NaCl$-KCl-$ZnCl_2$ [b]		
$C_p = 0.9 - 0.92$	J/(g·K)	230 - 350
$\lambda = 0.4372 - 1.2301\times10^{-4}T$	W/(m·K)	300 - 800
$\rho = 2.5417 - 0.5302\times10^{-3}T$	g/cm³	300 - 800
$\mu = [152.3679\exp(\frac{-T}{56.03143}) + 0.05994\exp(\frac{-T}{235.78682})]\times10^3 + 2.97$	mPa·s	250 - 800
$NaCl$-KCl-$ZnCl_2$ [b]		
$C_p = 0.9 - 0.92$	J/(g·K)	230 - 350
$\lambda = 0.3895 - 8.1686\times10^{-5}T$	W/(m·K)	300 - 800
$\rho = 2.5811 - 0.4321\times10^{-3}T$	g/cm³	300 - 800
$\mu = 131.0731\exp(\frac{-T}{62.36328})\times10^3 + 4.46$	mPa·s	250 - 800

[a] [12,14,15]
[b] [13]

The ternary LiNaK carbonate salts (Li_2CO_3-Na_2CO_3-K_2CO_3, 32.1-33.4-34.5 wt.%) has a melting point of 400°C and a thermal stability temperature of 658°C. An *et al* [**错误!未定义书签。**] studied the thermal conductivity, viscosity and density of the ternary LiNaK carbonate salts. Ejima *et al* also examined the viscosity of this eutectic salt mixture.

Li *et al* [16] proposed to use the $NaCl$-KCl-$ZnCl_2$ ternary salt mixture to replace the commercial nitrate based salts as high temperature HTFs. Three composition ratios of the $NaCl$-KCl-$ZnCl_2$ were founded to form eutectic mixtures and the thermophysical properties were all studied [13]. For the sake of convenient reference, this study used salt #1, salt #2 and salt #3 to denote the $NaCl$-KCl-$ZnCl_2$ salt mixtures with mass fraction ratios of 7.5%-23.9%-68.6%, 8.1%-31.3%-60.6%, and 10.0%-15.1%-74.9%, respectively. The working temperature ranges of the three chloride salt mixtures are from 250°C to at least 800°C.

3. Heat transfer coefficient calculation

The convection heat transfer coefficient *h* is expressed as,

$$h = -\lambda \frac{\left(dT_f / dy\right)_w}{T_w - T_b} \tag{1}$$

Where T_b is the fluid temperature, T_w is the wall temperature, λ is the thermal conductivity and $(dT_f/dy)_w$ is the fluid temperature gradient [5]. To have a high CSP plant's efficiency, the heat flux exchanged between the tube wall and the HTF must be high. The heat flux can be increased by increasing the heat transfer coefficient or increasing the temperature difference between the fluid and the wall. However, molten salts could decompose or vaporize at too high temperatures, so the temperature difference between the fluid and the wall is limited if using molten salts as HTFs. Therefore, the heat transfer coefficient h is critical. Equation (1) shows that the heat transfer coefficient h can be improved by two methods: increasing the thermal conductivity λ or the temperature gradient $(dT_f/dy)_w$. For a given HTF, the thermal conductivity λ is directly related to the HTF, and the temperature gradient $(dT_f/dy)_w$ can only be increased by increasing the velocity of the flowing HTF inside the tube. However, high velocity results in a high pumping power loss. To reach a maximized heat transfer performance for a specific HTF, a trade-off is necessary between these two effects [5].

In practice, h is calculated by the Nusselt number Nu,

$$h = \frac{Nu\lambda}{d} \tag{2}$$

Where d is the characteristic length, and λ is the fluid thermal conductivity.

Generally the imperial equation proposed by Wu *et al* [17] can be used to determine the Nusselt number for the molten-salt HTFs, which is shown in equation (3).

$$Nu = 0.02948\,\mathrm{Re}^{0.787}\,\mathrm{Pr}^{1/3} \tag{3}$$

Which is valid for $10^4 \leq \mathrm{Re} \leq 4.6\times10^4$, $1.6 \leq \mathrm{Pr} \leq 23.9$. Re and Pr are calculated as,

$$\mathrm{Re} = \frac{\rho V D}{\mu} \tag{4}$$

$$\mathrm{Pr} = \frac{C_p \mu}{\lambda} \tag{5}$$

Where ρ is the density, V is the fluid velocity, D is the characteristic length, C_p is the specific heat capacity, μ is the dynamic viscosity and λ is the thermal conductivity.

The usual tube size of 25 cm internal diameter was selected, and the selected fluid velocity is 2 m/s for the HTFs [5].

4. Results and discussion

4.1. The comparison of specific heat capacity C_p

The specific heat capacity of the Li_2CO_3-Na_2CO_3-K_2CO_3 molten-salt mixture can be taken as a constant of 1.61 J/(g·K) at the temperature range of 450°C to 600°C [12]. The heat capacities of the three eutectic chloride salts were measured in the temperature range from 230°C to 350°C. In this temperature range, the heat capacities varying between 0.9 and 0.92 J/(g·K) did not show appreciable change with temperature, and thus constant heat capacities can be assumed for engineering applications [13]. It is noticed that the heat capacity of Li_2CO_3-Na_2CO_3-K_2CO_3 is 43.5% larger than those of the NaCl-KCl-$ZnCl_2$ systems. The larger heat capacity of the Li_2CO_3-Na_2CO_3-K_2CO_3 compared to that of the chloride salt system demonstrates that the carbonate salts have better energy storage capacity.

4.2. The comparison of thermal conductivity λ

Thermal conductivity of a heat transfer fluid determines the thermal energy transfer efficiency and thus is considered as a significant parameter for practical applications [18]. The thermal conductivities of the Li_2CO_3-Na_2CO_3-K_2CO_3 eutectic salt mixture are 0.454, 0.458, 0.470 and 0.492 W/(m·K) at 450, 500, 550, and 600°C, respectively. It slightly increases with increasing temperature. The thermal conductivities of the NaCl-KCl-$ZnCl_2$ systems were measured at temperatures between 300°C and 800°C. Figure 1 shows that the thermal conductivities of all the three eutectic salts are in the range of 0.26-0.38 W/(m·K) and they all slightly decrease with increasing temperature. The thermal conductivity of Li_2CO_3-Na_2CO_3-K_2CO_3 is higher than those of NaCl-KCl-$ZnCl_2$, indicating that the LiNaK carbonate salt can transfer and store energy in a more efficient way.

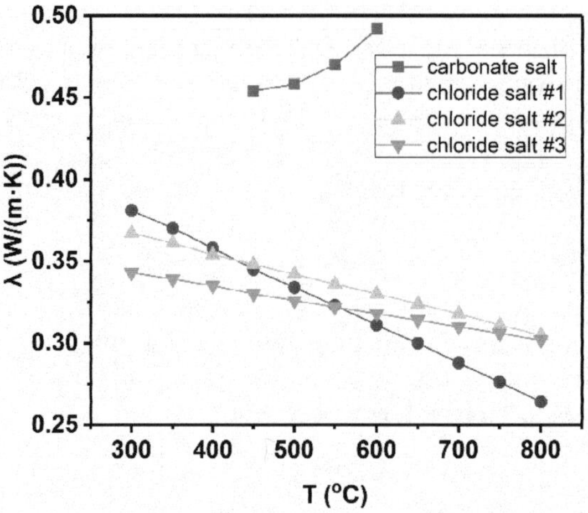

Figure 1. Thermal conductivity of the four molten salts as function of temperature.

4.3. The comparison of density ρ
Figure 2 shows comparison of densities at the temperature between 300 and 800°C. All the densities linearly decrease with increasing temperature. The density of LiNaK carbonate salt is in the range of 1.9-2.0 g/cm^3 and the three eutectic chloride salts is in the range of 1.9-2.3 g/cm^3. Because of the relatively higher density of $ZnCl_2$, the salt with a higher fraction of $ZnCl_2$ has higher density.

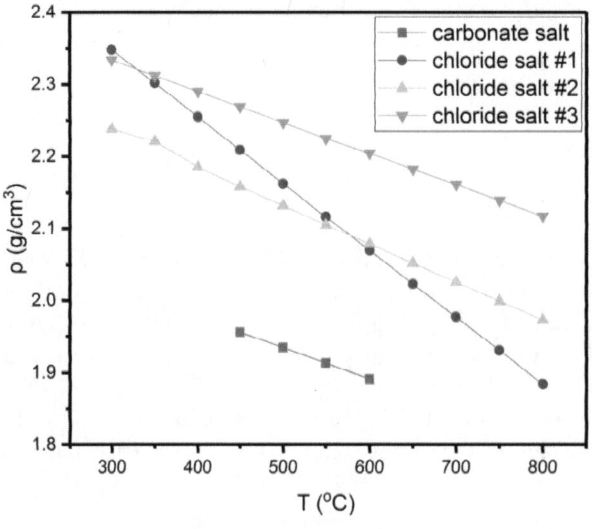

Figure 2. Density of the four molten salts as function of temperature.

4.4. The comparison of viscosity μ
The viscosity of the Li_2CO_3-Na_2CO_3-K_2CO_3 exponentially decreases as the temperature increases. The viscosities of the NaCl-KCl-$ZnCl_2$ mixtures were measured in the temperature range from 250°C to 800°C. Figure 3 shows comparison of the viscosities. The results show that the viscosity of the LiNaK carbonate salt is significantly higher than those of the chloride salt mixtures in its operating temperature range. As for the chloride salt mixtures, their viscosities are quite low at relatively high

temperature above 400°C. The viscosities are around 15, 7 and 4 mPa·s for the chloride salt mixtures at 300, 400 and 700°C, respectively. It can be seen that the low mass fraction of $ZnCl_2$ in salt #2 contributes to its low viscosity. However, the differences in the viscosities of the three eutectic molten salts are trivial.

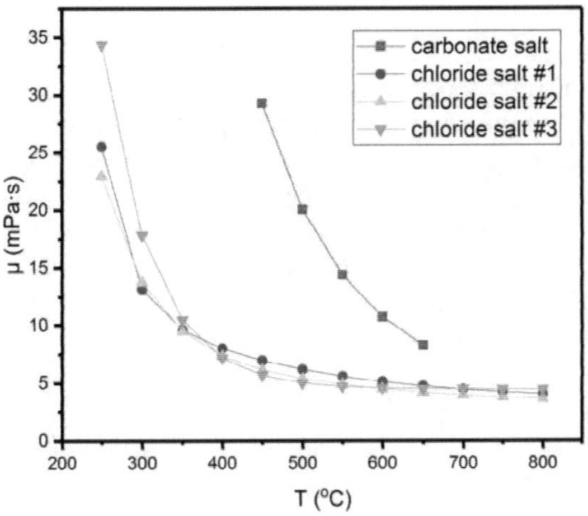

Figure 3. Viscosity of the four molten salts as function of temperature.

4.5. The comparison of heat transfer coefficient h

Figure 4 plots the heat transfer coefficients of the four molten salts as functions of temperature. For LiNaK carbonate salt, *h* increases from 1640 to 2520 W/(m^2·K) as its working temperature increases from 450°C to 600°C. For chloride salt #1, *h* ranges between 1800 and 2210 W/(m^2·K). It increases with *T* up to a maximum value at 550°C and then decreases. For salt #2 and salt #3, *h* ranges between 1680 and 2500 W/(m^2·K) and between 1470 and 2490 W/(m^2·K), respectively. The maximum heat transfer coefficient value is reached at 700°C and 550°C, respectively. The trend of the heat transfer coefficients of the chloride salts are mainly due to that the thermal conductivity decreases with increasing temperature.

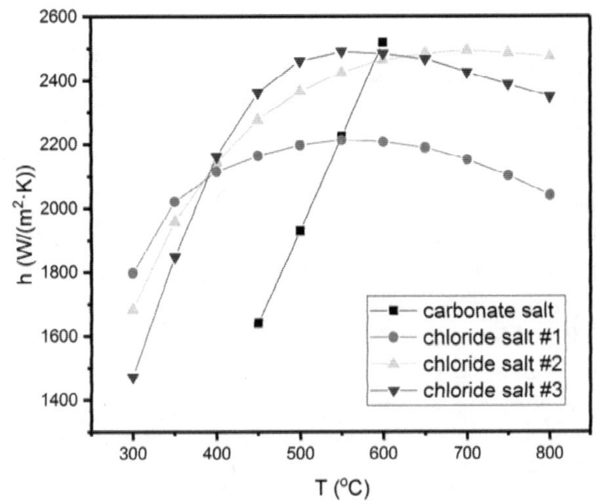

Figure 4. Heat transfer coefficient of the four molten salts as function of temperature.

5. Conclusion

In this paper, the thermophysical properties of the LiNaK carbonate salt and three chloride salts were compared. The specific heat capacity C_p can be regarded as constants for all the salts at their separate operating temperature ranges. However, C_p of the LiNaK carbonate salt is larger than those of the three chloride salts. The thermal conductivity λ of the LiNaK carbonate salt slightly increases with increasing temperature. In the contrary, λ slightly decreases with increasing temperature of the three chloride salts. Besides, it is lower than that of the LiNaK carbonate salt. All the densities linearly decrease with increasing temperature. Densities of the three chloride salts are slightly higher than that of the LiNaK carbonate salt. In addition, density of the chloride salt #3 is slightly higher than those of the other chloride salts due to a higher fraction of $ZnCl_2$ in its compositions. All the viscosities of the four eutectic salts exponentially decrease as the temperature increases. However, the viscosity μ of the LiNaK carbonate salt is significantly higher than that of the three chloride salts in its operating temperature range. For the convection heat transfer coefficient h, the chloride salts generally have higher h than the carbonate salts. In the temperature range of 400 – 600°C, chloride salts #3 has the highest heat transfer coefficient. While the temperature is over 600°C, the heat transfer coefficient of chloride salts #2 has the maximum value.

Acknowledgments
Support from Shenzhen Science and Technology Research and development Funds (JCYJ20160318094917454) is gratefully acknowledged.

References

[1] Xu X, Vignarooban K, Xu B, Hsu K and Kannan A M 2016 *Renew. Sust. Energy. Rev.* **53** 1106-31
[2] Cooper C and Sovacool B K 2013 *Renew. Energy.* **50** 628-36
[3] Vignarooban K, Xu X, Arvay A, Hsu K and Kannan A M 2015 *Appl. Energy.* **146** 383-96
[4] Moser M, Trieb F and Fichter T 2013 *J. Sust. Dev. Energy Water. Environ. Syst.* **1** 122-40
[5] Benoit H, Spreafico L, Gauthier D and Flamant G 2016 *Renew. Sust. Energy. Rev.* **55** 298-315
[6] Peng Q, Ding J, Wei X, Yang J and Yang X 2010 *Appl. Energy* **87** 2812-7
[7] Wang T, Mantha D and Reddy R G 2012 *Sol. Energy. Mater. Sol. Cells.* **100** 162-8
[8] Reddy R G, Wang T and Mantha D 2012 *Therm. Acta.* **531** 6-11
[9] Pacio J and Wetzel T 2013 *Sol. Energy.* **93** 11-22
[10] Cordaro J G, Rubin N C and Bradshaw R W 2011 *J. Sol. Energy. Eng-Trans ASME.* **133** 011014
[11] Stekli J, Irwin L and Pitchumani R 2013 *J. Therm. Sci. Eng. Appl.* **5** 021011
[12] An X, Cheng J, Zhang P, Tang Z and Wang J 2016 *Faraday. Discuss.* **190** 327-38
[13] Li P W, Molina E, Wang K, Xu X, Dehghani G, Kohli A, Hao Q, Kassaee M H, Jeter S M and Teja A S 2016 *J. Sol. Energy. Eng.* **138** 054501
[14] Olivares R I, Chen C and Wright S 2012 *J. Sol. Energy. Eng.* **134** 041002
[15] Ejima T, Sato Y, Yamamura T, Tamai K, Hasebe M, Bohn M S and Janz G J 1987 *J. Chem. Eng. Data.* **32** 180-2
[16] Li P W, Chan C L, Hao Q, Deymier P A, Muralidharan K, Gervasio D F, Lucas P, Momayez M, Jeter S, Teja A and Kannan A M 2013 *SunShot. CSP. Prog. Rev.* 85-6
[17] Wu Y T, Chen C, Liu B and Ma C F 2012 *Int. Commun. Heat. Mass. Transfer.* **39** 1550-5
[18] Wang T, Mantha D and Reddy R G 2015 *Sol. Energy. Mater. Sol. Cells.* **140** 366-75

Application of clustering analysis in the prediction of photovoltaic power generation based on neural network

K Cheng[1,2], L M Guo[1], Y K Wang[1] and M T Zafar[1]

[1]School of Power and Energy, North-western Polytechnical University, Xi'an 710072, Shaanxi province, China

E-mail: cksolar@163.com

Abstract. In order to select effective samples in the large number of data of PV power generation years and improve the accuracy of PV power generation forecasting model, this paper studies the application of clustering analysis in this field and establishes forecasting model based on neural network. Based on three different types of weather on sunny, cloudy and rainy days, this research screens samples of historical data by the clustering analysis method. After screening, it establishes BP neural network prediction models using screened data as training data. Then, compare the six types of photovoltaic power generation prediction models before and after the data screening. Results show that the prediction model combining with clustering analysis and BP neural networks is an effective method to improve the precision of photovoltaic power generation.

1. Introduction

With the expansion of installed capacity of photovoltaic power generation, its proportion in the power grid is also increasing these days. The random fluctuation characteristics of its output will cause the considerable impact on the power grid system, which indirectly affects the operation of the security and stability of the power system. Therefore, it is extremely important to predict the photovoltaic power generation effectively to ensure the security and stability of the power system. At present, some forecasting models for photovoltaic power generation are SVM model [1,2], ARMA model [3], gray theoretical model [4] and neural network model, etc. [5,6]. However, the neural network model generally has better performance among them [7]. Therefore, this study uses this method to predict the photovoltaic power generation.

In the process of BP neural network prediction, the quality of the input samples will directly affect the accuracy of the prediction model. Practically, when the photovoltaic power plant collects and stores the running data, the errors occur which form the abnormal samples. These anomaly samples deviate from the factual basis and the corresponding weather patterns are not correctly described according to its established model. Therefore, it is necessary to screen out these anomalous samples.

Manually, it is very difficult to screen out anomalous samples because of a large amount of data recorded in the past years. Therefore, most of the research papers use raw data collected from the untreated power plants [4-8], and only a few papers deal with filtered data [9]. However, these papers do not focus on the specific screening method. There are some papers, which use the wavelet analysis method, the mean fill method and other methods to deal with the original data [10,11]. However; there is no specified method to verify the accuracy and effectiveness of the power generation forecasting model.

Therefore, based on the statistical data of different kinds of weather, this paper cleans the historical data using the clustering analysis. Then, based on the BP neural network, 6 different kinds of photovoltaic power generation forecasting models are established and verified in the sunny, cloudy and rainy days in MATLAB software [12,13]. The results show the validity and accuracy of the clustering analysis of historical data.

2. Prediction model of photovoltaic power generation

2.1. Prediction model of photovoltaic power generation

In this paper, the clustering analysis and BP neural network are combined to process the raw data collected by the power station. The processed data is used as the input of the neural network to establish a reliable photovoltaic power generation forecasting model. The data in the research process comes from the actual record of the data collection device of Xi'an Jiayang New Energy Co., Ltd. PV power plant. The model is designed based on the photovoltaic system installed on the roof of the building. The total installed capacity of the photovoltaic power station is 1.2 MW. The measured data of the power station in the time period of 2012.01-2015.01 is selected as the original sample. The selected samples consist of total 868 data samples.

The main process of the model is shown in figure 1.

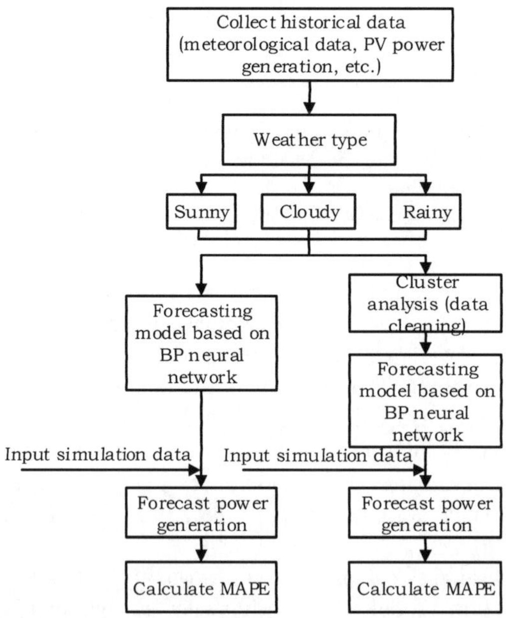

Figure 1. The framework of photovoltaic power generation forecasting model.

- Firstly, 3 typical weather types of sunny, cloudy and rainy days are taken as examples to analyze the theoretical radiation, actual radiation and power generation. The outliers are filtered out according to the cluster pedigree.
- Secondly, the historical data of 2012.01~2014.01 is used as the training sample. The theoretical radiation and the highest and the lowest temperatures are used as the input of BP neural network. The historical data collected by the data collector of PV power plant are used as the output of BP neural network. The neural network model is trained to establish the prediction models of 6 kinds of photovoltaic power generation
- Thirdly, the simulation data of 2014.02 ~ 2015.01 is put into the prediction model of photovoltaic power generation. The purpose is to make use of the BP neural network's

IOP Conf. Series: Earth and Environmental Science **93** (2017) 012024 doi:10.1088/1755-1315/93/1/012024

nonlinear learning ability to predict the daily power generation under different weather types

- Finally, in order to verify the need of data screening, the MAPE (the mean absolute percentage error) value is calculated for the predicted power generation and the actual generation and the performance is evaluated.

2.2. Clustering analysis model

Clustering analysis is a process of dividing an object of observation into several groups or classes based on certain quantitative features. The division is done in such a way that the same class within the data object has a high degree of similarity and low degree of similarity between various types of data object. The main purpose of clustering is to bring similar things together.

2.2.1. Clustering method. According to the different clustering methods, clustering analysis can be divided into system clustering method, adding method, decomposition method and dynamic classification method, etc. [14]. In this paper, the purpose of clustering analysis is to deal with the outlier data (abnormal samples) from the original data to improve the quality of the original data. Therefore, this paper uses the system clustering method. The main algorithm steps are as follow:

- To convert the objects in the sample set;
- The processed objects are divided into n classifications and every class contains an object;
- Calculating the distance between each of the two classes;
- Merging the nearest two classes as a new class;
- Calculating the distance between the new class and the current class. If the distance is close enough, then merge into a class. This process continues until all classes (or samples) are combined. If this step works well then we will move further to step 6, otherwise, to step 4.
- Drawing the clustering pedigree of the sample set, as shown in figure 2;
- According to the number of classes to be classified, the corresponding classification results are obtained.

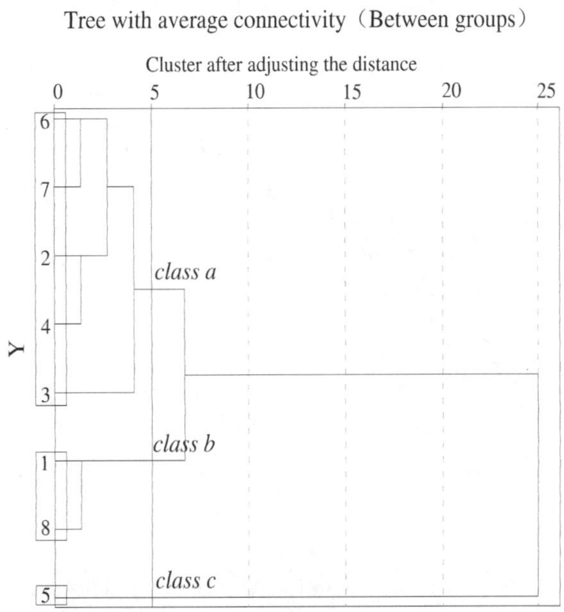

Figure 2. Clustering pedigree chart.

Figure 2 shows the data clustering results for randomly selected eight groups on rainy day. The sample can be clustered into two, three, four or five categories. As shown in the figure, when the class distance is 5, the samples can be clustered into three classes. The samples 6, 7, 2, 3 and 4 are class a, sample 1 and sample 8 are b, and sample 5 is c.

2.2.2. Select the metric. The metric directly affects the quality of the cluster, so the statistical analysis of the historical data of the database is carried out to determine the metric. In the figures 3-5 of cloudy weather, some samples are found away from the main data sets which are outliers. While, the data distribution exhibits a distinct oblique strip.

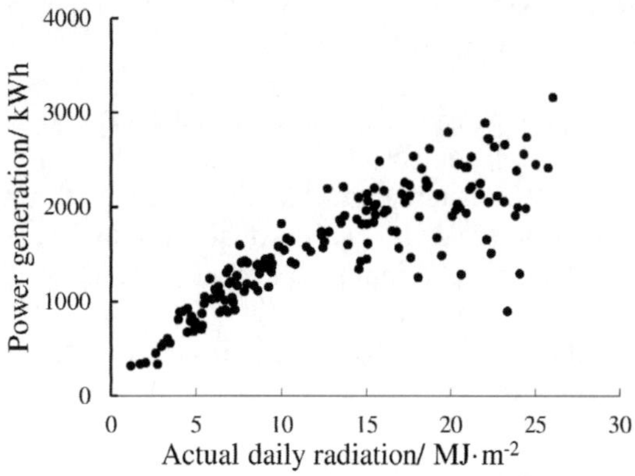

Figure 3. Scatter plots of actual radiation and power generation.

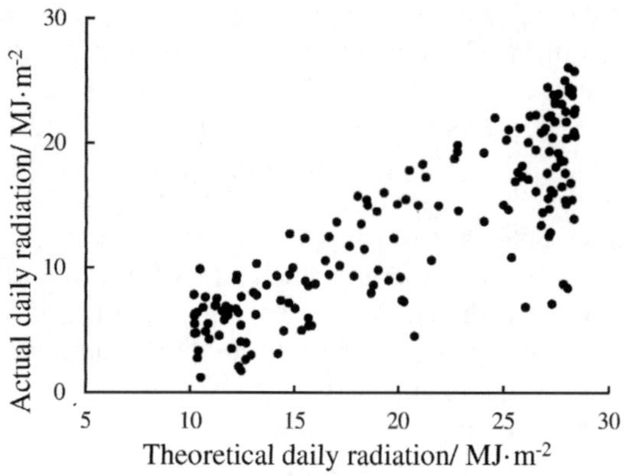

Figure 4. Scatter plots of theoretical and actual radiation.

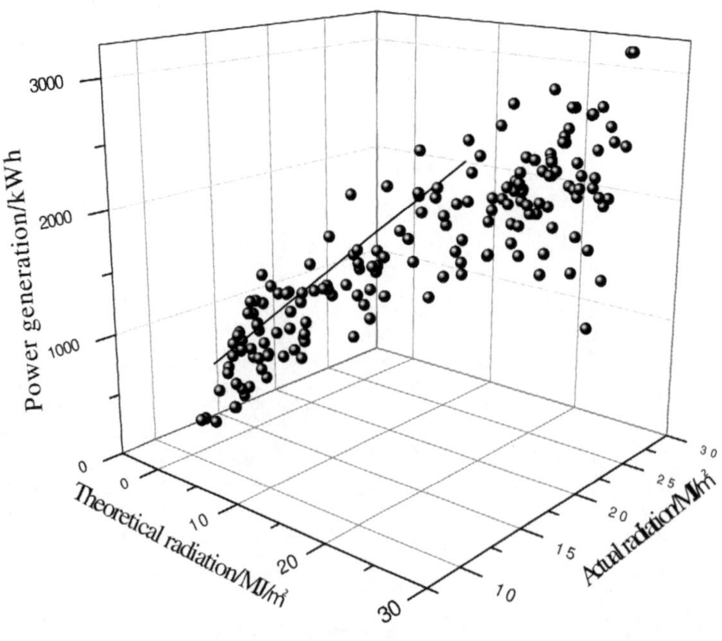

Figure 5. Spatial scatter plots of actual radiation, theoretical radiation and power generation.

According to the previous study results [15], the data covered by the oblique range belong to a class. Selecting the cosine coefficient as the criteria, the data on the oblique bar is classified as valid data and other data is classified as the outlier which is far from the oblique bar. After the outlier deletion, the discrete degree of the sample is optimized. The cosine distance is measured according to formula (1):

$$d_{ij} = \cos(\theta_{ij}) = \frac{\sum_{k=1}^{0} X_{ik} X_{jk}}{\sqrt{\sum_{k=1}^{n} X_{ik}^2 \sum_{k=1}^{n} X_{jk}^2}} \tag{1}$$

In the above formula, d_{ij} -the distance between simples X_{ik} and X_{jk}; X_{ik} - the k-the variable of the *i* sample; X_{jk} - the k-the variable of the *j* sample; *n* - number of samples.

2.3. Neural network model

Fundamentals. The neural network is a network system that performs the parallel processing and non-linear transformation of information same as human brain. The BP neural network chosen in this paper is a forward process using error back propagation algorithm of the network. Its structure is shown in figure 6.

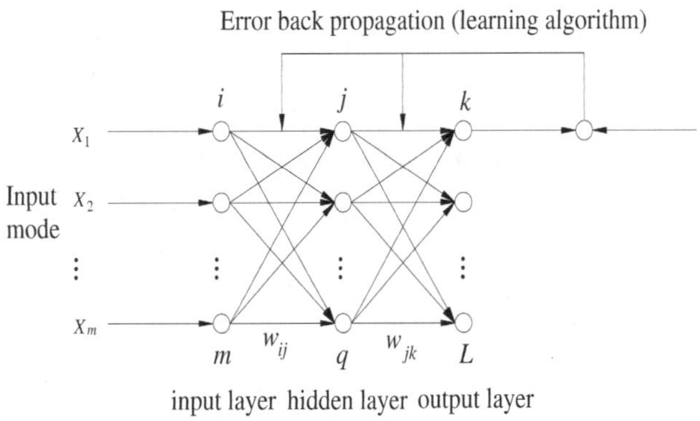

Figure 6. Structure of BP neural network.

Where w_{ij} is the connection weight between the input layer node and the hidden layer node, and w_{jk} is the connection weight between the hidden layer node and the output layer node. The input of the hidden layer and the output layer node is the weighted sum of the output of the previous node [16]. In this paper, a three-layer neural network is used which contains only one hidden layer.

Evaluation of photovoltaic power generation forecasting model. In this paper, a two-layer forward neural network based on Levenberg-Marquardt (LM) algorithm is adopted. The neural network toolbox of Matlab 2010b is used to build and train the network and obtain the simulation results. The training data and test data are randomly selected by neural network as 90% and 10% respectively of the total data. In order to ensure the credibility of the results, test data is not used in the training part.

The simulation results use the average absolute percentage error (MAPE) to evaluate the entire system, which is calculated as (2):

$$MAPE = \left(\frac{100}{N}\right) \sum_{i=1}^{N} \left| \frac{P_f^i - P_a^i}{P_a^i} \right| \% \tag{2}$$

In the above formula, *N* - the total number of data; P_f - the actual power generation; P_a - predicted power generation; *i* - the data serial number. The lower MAPE value ensures the better predictive effects.

3. Clustering process and results

The original data of three kinds of weather types such as sunny, cloudy and rainy days are extracted by SPSS (statistical analysis software). The cosine coefficient is used as the criteria to cluster and analyze the different weather types respectively. When the number of clusters is 4, for sunny days, according to the clustering pedigree, a_s class has 119, b_s class has 23, c_s class has 5, and d_s class has 3; for rainy days, according to the cluster spectrum, a_r class has 4, b_r class has 2, c_r class has 19, and d_r class has 36; for cloudy days, a_c class has 10, b_c class has 30, c_c class has 61, and d_c class has 67, comparing the characteristics of the number of samples contained in each cluster and screening a few samples with the least number of clusters. Figure 7 shows the filtered cloudy sample scatter plots.

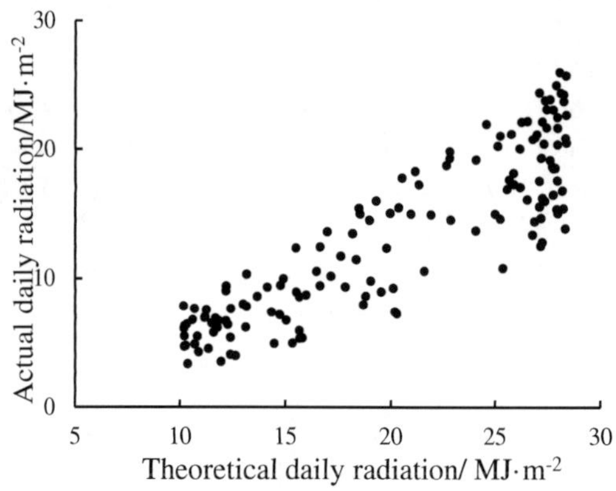

Figure 7. The theoretical and actual daily radiation scatter plots after screening in cloudy.

The clustering results of the various data samples are shown in table 1. In the sunny days, the atmospheric state is relatively stable and the reverse radiation of the atmosphere is weak. Therefore, the theoretical daily radiation is close to the actual amount of radiation in the whole day, and the degree of data discrimination is small and the screening ratio is 10.7%. In the cloudy days will have large random variation in terms of clouds and the similarity of the data is low, the screening ratio is 13.1%. In the rainy days, the atmospheric state fluctuates at a higher degree and the screening ratio is also large which is 21.3%.

Table 1. The clustering result statistics for all kinds of sample data.

Weather type	Sunny	Cloudy	Rainy	Total
Before screening	150	168	61	379
After screening	134	146	48	328
Screening ratio	10.7%	13.1%	21.3%	13.5%

4. Predictive results and analysis

4.1. Forecast results

In order to verify the effectiveness of clustering analysis for the prediction of photovoltaic power generation, the prediction model of PV generation is established with raw data as a comparison. Firstly, cluster analysis is used in data cleaning for different kinds of weather. Then, corresponding type of model is selected from forecasting models and simulation data is entered. Finally, the power generation is predicted.

Figures 8-10 shows the predicted power generation, actual power generation, and the data for the three kinds of weather before and after the clean-up process (the day number in the x axis is the data of the forecast in a year). From the results, it can be seen that the difference between the predicted value and the actual value of the model established by the weather type is relatively large before the data is cleaned up. After data clearing, the difference between the predicted value and the actual value become small. The predicted value of the model after the cluster analysis is closer to the actual value.

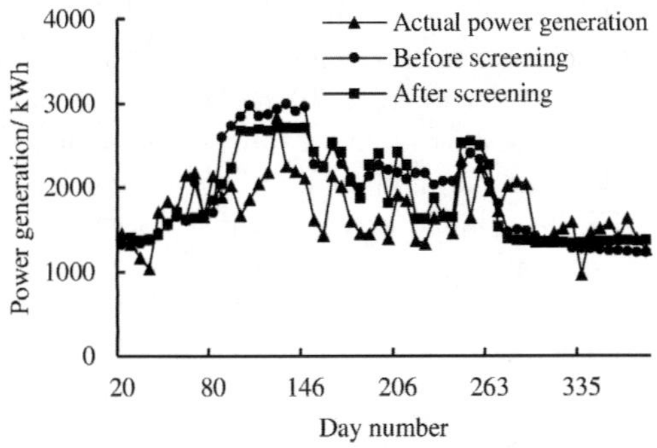

Figure 8. The relationship between the actual power generation and the prediction of power generation in sunny weather.

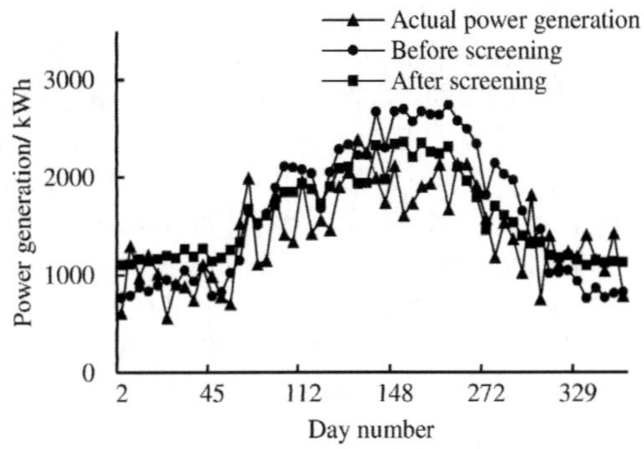

Figure 9. The relationship between the actual power generation and the prediction of power generation in cloudy weather.

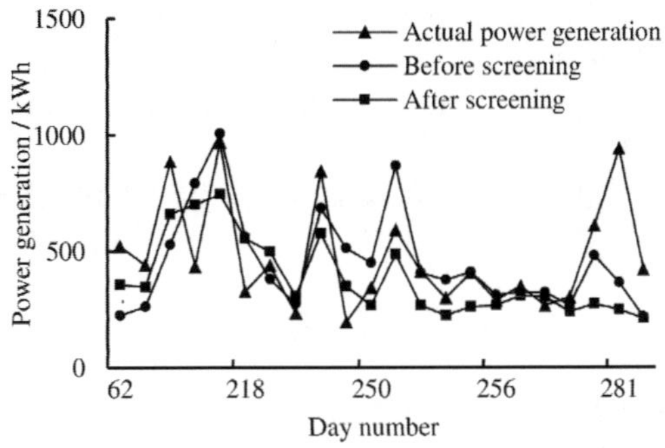

Figure 10. The relationship between the actual power generation and the prediction of power generation in rainy weather.

4.2. Evaluation of the results

Table 2. MAPE values of various weather neural network models.

	Weather type	sunny	cloudy	rainy	average value
Prediction error	Before screening	32.4%	45.6%	85.8%	46.8%
	After screening	20.7%	22.6%	34.5%	23.6%

Table 2 shows the difference between the predicted results of the six neural network models. Compared with the MAPE mean value, it can be seen that the predicted value of the model after the cluster analysis is closer to the actual value in each case. The average MAPE value of the model after clustering analysis is 23.2% which is much smaller than the conventional models. Indicating that the clustering analysis is effective, and the prediction model of photovoltaic generation is more accurate base on the data after clustering analysis. Therefore, it is necessary to screen data.

In the case of sunny days, the prediction results of two models are closer, because the sunny weather atmospheric state is relatively stable and the degree of data discrete is small, the MAPE value of the prediction model that after screening is 11.7% lower than that before screening, the prediction effect is better. In the case of cloudy days, the thickness and location of the clouds is difficult to predict, the predicted results of the two models are relatively different, the MAPE value of the prediction model that after screening is 23% lower than that before screening. In the case of rainy days, the MAPE value of the prediction model that after screening is 51.3% lower than that before screening, which difference is the biggest. It can be seen from the table 2, the prediction of rainy days is far less accurate than sunny and cloudy days. The reason is the rain weather atmospheric changes are more complex, in different rainy days, clouds, rain and vapor will be very different, and the absorption of solar radiation will be very different, this will produce a large relative error, so it is difficult to achieve the desired prediction effect in rainy days.

Based on overall assessment, the MAPE value of the predicted model is 23.6% which is less than 46.8% of the pre-screening prediction model. The comparison results show that photovoltaic power generation models are more accurate when using the clustering analysis of the original data. Therefore, it is necessary to carry out data cleanup.

5. Conclusion

In this paper, the prediction model is established by combining the clustering analysis and the BP neural network. The abnormal samples in the historical monitoring data of the PV power plant are screened out, and the prediction results are evaluated and analyzed. Research indicates:

- Using the cosine coefficient as the evaluation criteria, the abnormal samples are removed using clustering analysis and screening of historical data, which can effectively filter the discrete operating data.
- After the data is cleaned up as the training data, the BP neural network prediction models can be established which can improve the accuracy of the prediction results.
- The data cleaning is necessary for establishing the prediction model using the historical data. The cluster analysis can be effectively applied to the photovoltaic power generation forecasting system.

The above research results can provide an effective method for the forecasting technology of PV power plants, which is beneficial to the popularization and application of photovoltaic power generation technology.

References

[1] Li R and Li G M 2008 Photovoltaic power generation output forecasting based on support vector machine regression technique *Electr Pow* **41(2)** 74-8

[2] Shi J, Lee W J, Liu Y Q, *et al* 2012 Forecasting power output of photovoltaic systems based on weather classification and support vector machines *IEEE Trans Ind Appl* **48(3)** 1064-9

[3] Lan H, Liao Z M and Zhao Y 2011 ARMA model of the solar power station based on output prediction *Electr Meas Instrum* **48(54)** 31-4

[4] Wang S X and Zhang N 2012 Short-term output power forecast of photovoltaic based on a grey and neral network hybrid model *Autom Electr Pow Syst* **36(19)** 1-5

[5] Christophe P, Cyril V, Marc M *et al* 2010 Forecasting of preprocessed daily solar radiation time series using neural networks *Sol Energy* **84(12)** 2146-60

[6] Zhang Y X and Zhao J 2011 Application of recurrent neural networks to generated power forecasting for photovoltaic system *Pow Syst Prot Control* **39(15)** 96-101

[7] Şenkal O and Kuleli T 2009 Estimation of solar radiation over Turkey using artificial neural network and satellite data *Appl Energ* **86(7-8)** 1222-8

[8] Wang F, Mi Z Q, Zhen Z *et al* 2013 A classified forecasting approach of power generation for

photovoltaic plants based on weather condition pattern recognition *Proc CSEE* **33(34)** 75-82
[9] Lu J, Zhai H Q and Liu C 2010 Study on statistical method for predicting photovoltaic generation power *East China Electr Pow* **38(4)** 563-7
[10] Lang Y and Zhang W T 2016 Research on the application of MATLAB in the data processing of photovoltaic power generation *Electr Technol Softw Eng* **2016** 87
[11] Zhang H N, Zhang J T, Yang L B *et al* 2016 Date processing method for PV power station data acquisition system *Electr Energ ManageTechnol* **2016** 9-13
[12] Chen C S, Duan S X, Cai T *et al* 2011 Online 24-h solar power forecasting based on weather type classification using artificial neural network *Sol Energy* **85(11)** 2856-70
[13] Dai Q, Duan S X, Cai T *et al* 2011 Short-term PV generation system forecasting model without irradiation based on weather type clusting *Proc CSEE* **31(34)** 28-35
[14] Wu S, Pan F M *et al* 2014 *SPSS Statistical Analysis* (Tsinghua University Press) pp 315-30
[15] Mohamed A and Badrul C 2015 Solar power forecasting using artificial neural networks *Proc of The 47th Annual North American Power Symp* (USA: Charlotte, North Carolina)
[16] Chen C S, Duan S X and Yin J J 2009 Design of photovoltaic array power forecasting model based on neutral network *Trans China Electrotech Soc* **24(9)** 153-8

Prediction of energy balance and utilization for solar electric cars

K Cheng[1,2], L M Guo[1], Y K Wang[1] and M T Zafar[1]

[1]School of Power and Energy, Northwestern Polytechnical University, Xi'an 710072, Shaanxi province, China

E-mail: cksolar@163.com

Abstract. Solar irradiation and ambient temperature are characterized by region, season and time-domain, which directly affects the performance of solar energy based car system. In this paper, the model of solar electric cars used was based in Xi'an. Firstly, the meteorological data are modelled to simulate the change of solar irradiation and ambient temperature, and then the temperature change of solar cell is calculated using the thermal equilibrium relation. The above work is based on the driving resistance and solar cell power generation model, which is simulated under the varying radiation conditions in a day. The daily power generation and solar electric car cruise mileage can be predicted by calculating solar cell efficiency and power. The above theoretical approach and research results can be used in the future for solar electric car program design and optimization for the future developments.

1. Introduction

Solar electric cars rely on the photovoltaic effect to produce electricity. In these types of cars, the controller and battery are used to drive the motor by overcoming the driving resistance. The solar energy characteristics are clean, pollution-free and inexhaustible. The application of solar energy in the renewable based cars has broad prospects and has emerged to be a hot topic [1,2]. In 1984, the solar electric car of Japan's Tokai University won the competition named as World Solar Challenge by using an average speed of 100 km/h [3]. Domestically, the research institutions of the universities including Tsinghua University, Shanghai Jiao Tong University and Jilin University have also carried out relevant research, and successfully produced a number of solar electric cars [4].

The core components of the solar electric car energy system are solar cells. The performance of the battery is affected by solar radiation and temperature, and the power generation capacity changes with time. Domestic scholars have done some research in proposing the corresponding solution to improve the matching of the power system and the energy efficiency [5-7]. The prediction of energy balance and utilization is still very complex because of many contributing factors such as change in solar irradiation, temperature and environmental factors. These above-mentioned factors change in the form of season, time domain, regional and domestic variations. Therefore, the theoretical based performance prediction of the solar electric car energy system is carried out by using driving resistance, solar cell power generation model, metrological data and thermal variations.

2. Running resistance and drive power

Solar electric cars rely on driving force to overcome the driving resistance to be able to drive. Driving resistance is the most important energy load of solar cars. Driving resistance mainly includes rolling resistance R_r (N), air resistance R_a (N), acceleration resistance R_{ac} (N) and slope resistance R_c (N), the formulae are given in preceding sections [8-10]:

2.1. Rolling resistance
The wheel in the car driving process continues to rotate and the resistance generated in the wheel is rolling resistance which is expressed as:

$$R_r = \mu_r \cdot W \tag{1}$$

In the above formula, μ_r - rolling resistance coefficient and W - total car weight (kg).

2.2. Air resistance
The car bears the resistance against the air while moving and it is named as the air resistance written as:

$$R_a = \frac{1}{2} C_d \rho A v^2 \tag{2}$$

In the above formula, C_d - air resistance coefficient; ρ - air density (kg/m^3); A - windward area (m^2); v - travel speed (m/s).

2.3. Slope resistance
When the car moves uphill, the car's gravity component along the slope is expressed as the slope of the car resistance, expressed as:

$$R_c = m \cdot g \cdot \sin\alpha \tag{3}$$

In the above formula, m - car mass (kg), g - acceleration due to gravity (m/s^2), α - slope angle (°).

2.4. Accelerate resistance
When the car accelerates, it is necessary to overcome the inertia force which is the same as to accelerate the resistance. The mass of the car is divided into two parts: translation mass and accelerated mass. Acceleration is not only the translation of the mass of inertial force, rotation mass also has inertial force. The acceleration resistance of the car can be expressed as:

$$R_{ac} = \delta \cdot m \cdot a \tag{4}$$

In the above formula, δ - car rotation mass conversion factor, m- car mass (kg), a -car acceleration (m/s^2).

2.5. Drive power
The air resistance is mainly determined by the design of the car. However, the acceleration resistance is determined by the way of running and slope resistance is determined by the driving the car on the road. Therefore, in the overall design, the car is considered runningon a flat road with the uniform speed but the impact of acceleration and slope resistance is not considered.

Knowing the rolling resistance and air resistance, the calculation of the drive power P_D (W) of the solar electric car power system can be done as follows:

$$P_D = (R_r + R_a) \cdot v \tag{5}$$

3. Solar cell power generation model

The principle of solar cell power generation is based on the semiconductor photovoltaic effect of solar radiation directly converted to electricity. The maximum power generation P_{mpp} of the solar cell is obtained under standard test conditions where the atmospheric thickness is AM 1.5, the temperature is 25°C, and the solar irradiance is 1000W/m^2. Since, the actual performance of the solar cell is affected by solar radiation and temperature, it is necessary to use the maximum power temperature coefficient to correct the power at the actual temperature and obtain the final actual power according to the approximate relationship with the solar irradiance [11].

3.1. Temperature correction

When the battery temperature is T (°C), the maximum power P_T (W) of the solar cell is:

$$P_T = [1 + (T - 25) \cdot \alpha'] \cdot P_{mpp} \tag{6}$$

In the above formula, α' - the maximum power temperature coefficient is generally around - 0.5% /°C.

3.2. Irradiation correction

In order to simplify the calculation, the solar cell power generation P is estimated to be proportional to the incident solar irradiance,

$$P = G_{on} \cdot P_T \cdot 1000^{-1} \tag{7}$$

In the above formula, G_{on} - the solar irradiance received by the solar cell (kWh/m^2), its value can be obtained according to the hourly meteorological data.

4. Hourly weather data

The location of the operation is located in Xi'an at 34.5° north latitude, 108.5° east longitude and 693 m above sea level. Since so far, China has not yet established a unified standard and national hourly weather data, especially solar radiation data. Therefore, the use of already known method of time based meteorological data generation is more appropriate at this stage. The weather data for solar and ambient temperatures is derived from NASA's historical data [12], as detailed in table 1:

Table 1. Solar energy and temperature.

month	daily total solar irradiation kWh/m^2/d	sunny total solar irradiation kWh/m^2/d	Daily average temperature °C	daily range of temperature °C
1	2.84	3.83	-5.49	7.44
2	3.34	4.86	-2.15	7.26
3	3.85	6.21	3.20	8.36
4	4.80	7.17	11.0	9.41
5	5.29	7.80	16.6	9.61
6	5.27	7.76	20.4	9.06
7	5.11	7.53	22.1	8.47
8	4.57	5.98	20.7	7.90
9	3.69	5.98	16.2	7.33
10	3.06	4.87	9.80	6.63
11	2.71	3.71	2.98	6.43
12	2.61	3.44	-3.18	5.80

The average daily sun irradiation is the daily average of the total radiation for the month, and the sunny day total solar irradiation is the amount of radiation in the sunny day, and the following formula is used [13]:

$$r_t = \pi/24 \cdot (a + b\cos\omega)(\cos\omega - \cos\omega_s)(\sin\omega_s - \omega\cdot\cos\omega_s)^{-1} \quad (8)$$

In the above formula, r_t - the ratio of total radiation in hour to total radiation, ω_s - sunset angle, ω – hour angle; the coefficient a, b are calculated as follows:

$$a = 0.409 + 0.5016\cdot\sin(\omega_s - \pi/3) \quad (9)$$

$$b = 0.6609 - 0.4767\sin(\omega_s - \pi/3) \quad (10)$$

The solar irradiance at different time can be obtained by calculating r_t of different time and the daily total radiation.

In June and December, for example, the daily total solar irradiation and sunny total solar irradiation distribution with time is shown in figure 1:

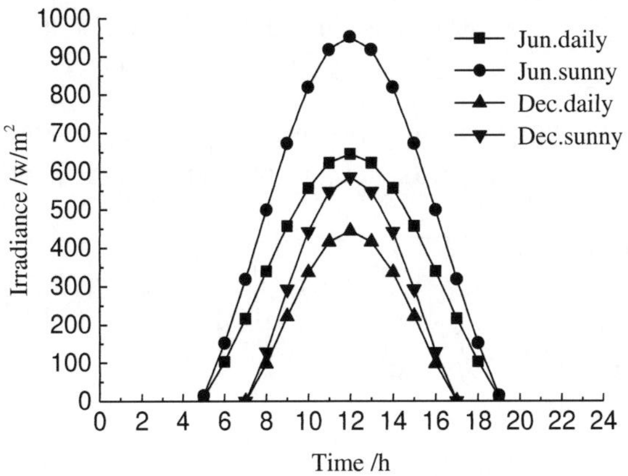

Figure 1. Hourly solar irradiance.

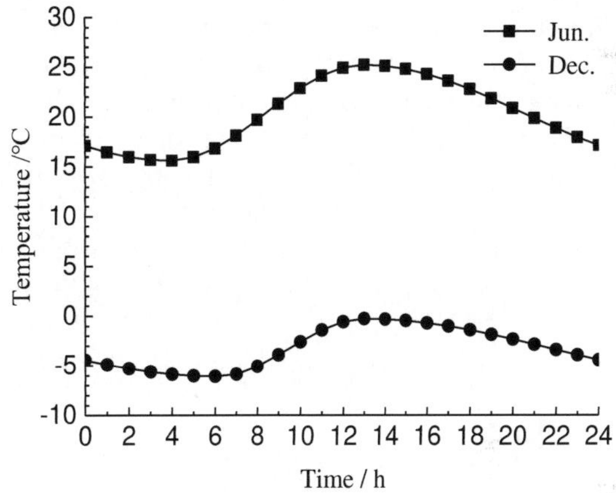

Figure 2. Hourly ambient temperature.

The data show that [14,15], the ambient temperature in the day can be approximated with a cosine function. Moreover, the maximum temperature generally occurs at 14:00, and the lowest temperature appears in the 1 hour before the sunrise. According to the daily average temperature, temperature differences and the change in sunrise time in the months of June and December are shown in figure 2.

5. Prediction of energy balance and utilization

5.1. Solar electric car conceptual model

The solar electric car model is shown in figure 3, which has length of 4 m, width of 1.7 m, height of 1 m, and weight of 350 kg. Solar electric cars rely on the photovoltaic effect to produce electricity. In these types of cars, the controller and battery are used to drive the motor by overcoming the driving resistance. For maximum power output, the system uses MPPT solar controller. When the intensity of light is large, the electric energy can drive the motor directly through the control system and drive the wheel. When the vehicle is at low speed or stagnant state, excess electricity can be stored in the battery. Otherwise, the battery through the control system discharge drives motor and then drives the car. The speed of the car is dependent on the control system which adjusts the current of the motor.

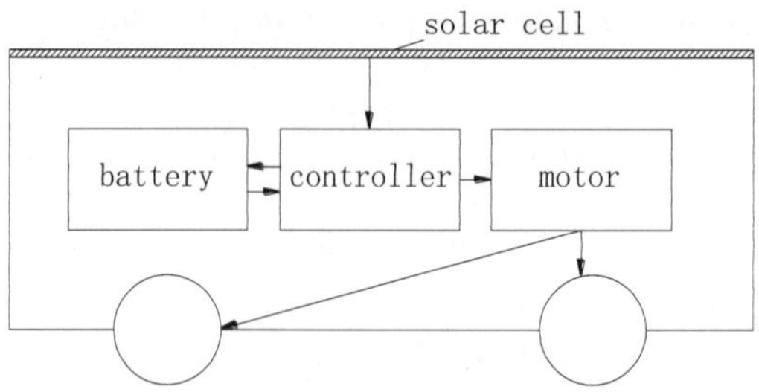

Figure 3. Solar electric car model.

The surface of solar electric car is equipped with 432 pieces of HH125M165-S2 monocrystalline silicon solar cells. In addition, 144 pieces of these cells constitute a PV module and a total of three PV modules are used. Under standard conditions, the power comes out to be 1223 Wp. Under the standard condition, the efficiency φ of HD125M165-S2 monocrystalline silicon solar cell is 18.2%, peak power P_{mpp} is 2.833Wp, and the maximum power temperature coefficient α is -0.46%/°C.

The four NCR18650A lithium ion battery stacks are used in series to form a power storage unit. In addition, sub string is made by joining 10 groups of these power storage units and 45 groups of sub-string are then connected in parallel. The total capacity of the battery pack is 2.09 kWh, the maximum discharge current is 1395 A and the total weight is 81.9 kg.

The information relevant to the solar electric model car include temporary cruise speed of 30 km/h (8.33 m/s), rolling resistance coefficient of 0.015, the wind area of 0.5 m^2, air resistance coefficient 0.35, and air density is 1.205 kg/m^3.

According to the formulas (1)-(6), the rolling resistance is 51.5 N and the air resistance is 7.3 N. Therefore, the total resistance is 58.8N and the required power is 489.8 W. The different efficiencies come out to be 95%, 90%, 95% and 75% for motor, mechanical transmission, MPPT solar controller and battery respectively. In addition, the required solar power supply is 804 W.

5.2. Solar cell temperature

Solar cell temperature is generally higher than the ambient temperature. However, sunshine makes the temperature rise but the wind blows the temperature down. The solar cell temperature T is simplified according to the thermal equilibrium relation [16].

$$\varepsilon \cdot G_{on} \cdot (1-\phi) = \varepsilon \cdot \sigma \cdot (T^4 - T_{sky}^4) + h(T - T_a) \tag{11}$$

The formula on the left side is the heat energy absorbed by the solar cell and the right side shows the release of heat which includes the radiant heat to the sky and the convective heat transferring with the environment. In the above formula, ε - solar cell heat radiation emissivity is set to 0.95; G_{on} - solar irradiance obtained by solar cells. According to meteorological data obtained by time; φ - solar cell photoelectric conversion efficiency and σ - Stephen-Boltzmann Constant, 5.76×10-8W/m^2·K^4. T_{sky} - Sky temperature, calculated as follows:

$$T_{sky} = 0.0552 T_a^{1.5} \tag{12}$$

T_a is the ambient temperature according to the meteorological data obtained by time. While, h - air convection heat transfer coefficient is calculated by the following formula:

$$h = 2.8 + 3.0 v_w \tag{13}$$

Where v_w is the surface wind speed which is approximately equal to the cruising speed of 8.33 m/s.

In June and December, for example, by using the formula (11) the changes in the solar cell temperature T with the time are shown in the results in figure 4.

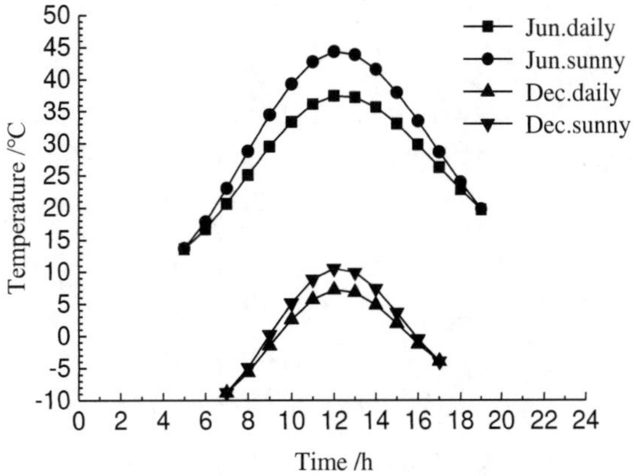

Figure 4. Hourly solar cell temperature.

5.3. Prediction results

The change in the solar cell temperature with the time will result in the change of efficiency which is shown in figure 5. Further, with the change of solar irradiance with time, the peak power changes are shown in figure 6.

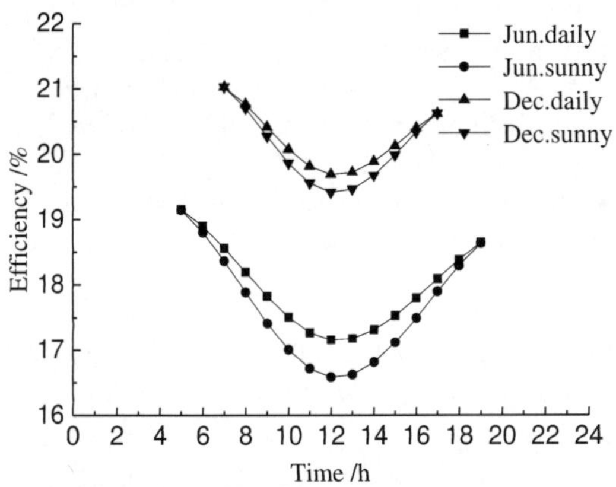

Figure 5. Hourly solar cell efficiency.

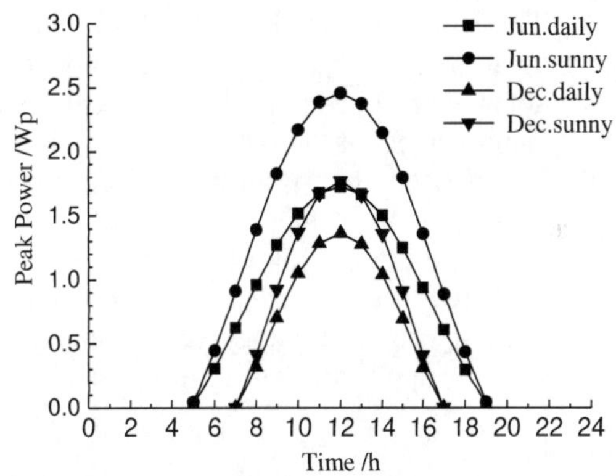

Figure 6. Hourly peak power of solar cell.

The time of 12 o'clock can be taken as symmetry for the solar power produced all around the day. By the impact of ambient temperature hysteresis, the power of the morning session is slightly higher than the afternoon. The average efficiency in June was 18.0% and 17.7% with average power of 0.960 W and 1.378 W respectively. On the other hand, the average efficiency in December was 20.1% and 19.9%, with the average power of 0.893 W and 1.164 W respectively.

The solar cell efficiency and power change with time from January to December. The total theoretical solar power generation from January to December is shown in figure 7. Further, according to the needs of the solar electric car photovoltaic power supply, from January to December, the solar electric car daily cruise stroke is shown in figure 8.

In the whole year, the maximum daily generation and cruise travel took place in May and the minimum occurred in December. In June, for example, the performance prediction results show that the average daily generating capacity of 432 solar modules is 6.33 KWh and 9.10 KWh under normal daily radiation and sunny day radiation respectively. According to the cruise speed of 30 km/h and the required solar power supply capacity of 804.0 W, the maximum theoretical travel is 232.0 km and

333.15 km with cruising time of 5.5 hours and 7.9 hours respectively. It should be noted that the actual cruise travel and time will be less than the above forecasted results due to acceleration and slope resistance.

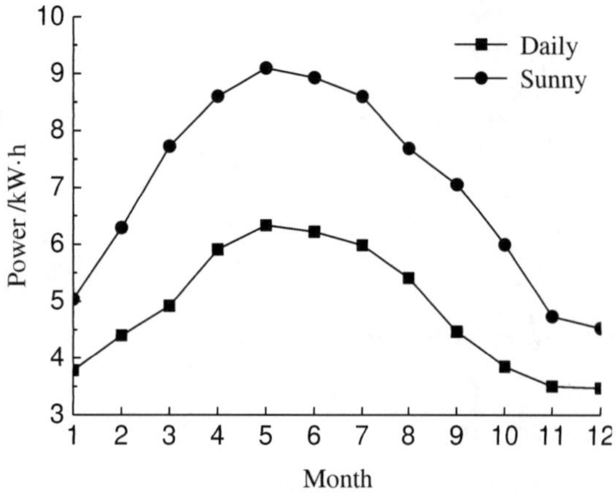

Figure 7. Theory daily power generation.

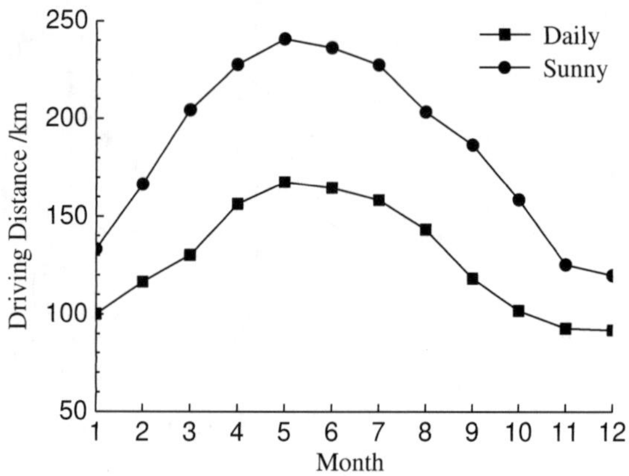

Figure 8. Daily driving distance.

6. Conclusion

Based on the concept model of solar electric car, this paper studies the performance prediction of energy system, and can get the following conclusions:

- The model and method established in this paper can be used to predict the performance of solar energy car energy system in different months. It can be used to verify and analyze the solar electric car in the developing stage. It can provide base for the selection and improvement of the scheme.
- The performance of the solar electric car energy system changes in a day with the time of 12 o'clock as the center. The 12 o'clock can be considered as symmetrical point. However, the lag of environmental temperature results into the morning value slightly greater than afternoon.
- Affected by changes in solar energy and ambient temperature in different months, the working capacity of solar energy vehicle energy system has nearly doubled the scope of the design. And the applications of the solar electric car system will have greater impact in the future.

References

[1] Kelly 2002 Raycing with the sun: Solar cars thrive in summer heat *JOM Journal of the Minerals, Metals and Materials Society* **54** 14-7
[2] Peter P and Phil H 2002 Critical speed control of a solar car *Optimization and Engineering* **3** 97-107
[3] Surhone L M, Tennoe M T, Henssonow S F, Racing S C and Highway S 2010 *World Solar Challenge* (Saarbrücken: Betascript Publishing)
[4] Zhao Y F, Wan J, Zhu Z P, Chen J, Le Z G, Ma F W, Ding Y and Zhao F Q 2011 The application of solar cells on vehicle *Agricultural Equipment and Vehicle Engineering* **237** 39-42
[5] Shi W K, Lv D H and Liang T Y 2008 A foundation investigation of parameters matching for power train smission of solar electric car *Acta Energlae Solaris Sinica* **29** 959-64
[6] Wang X C, Zhang P S, Zhang L H and Jia W 2009 Research on power control system of solar electric vehicle *Machinery and Electronics* **12** 36-9
[7] Wang X Y and Kang L Y 2010 The research and development of maximum power point tracker for solar vehicle *Power Electronics* **44** 12-3
[8] Guerra-zubiaga D A, Ramon-Raygoza E D, Rios-Soltero E F, Tomovic M and Molina A A

2009 PLM tools taxonomy to support product realization process: A solar racing car case study *Product Realization* **1** 1-26
[9] Japan Solar Energy Association 2010 *Mechanism and Production of Eco-electric Vehicle* (Xi'an: Xi'an Jiaotong University Press)
[10] Hiroyuki O, Sumio N and Dai H 1998 Development of aerodynamics for a solar race car *JSAE Review* **19** 343-9
[11] Luo Y J, He Z N and Wang C G 2005 *Solar Energy Technology* (Beijing: Chemical Industry Press)
[12] NASA: Surface meteorology and Solar Energy [EB/OL] http://eosweb.larc.nasa.gov/cgi-bin/sse/sse.cgi?na
[13] Zhang H F 2004 *Solar Energy Heat Utilization Principle and Computer Simulatio* (Xi'an: North-western Polytechnical University Press)
[14] Akasaka H *et al* 2000 Development of expanded AMeDAS weather data for building energy calculation in Japan *ASHRAE Transactions* **206** 455-6
[15] Tsinghua University, Tongji University, Xi'an Institute of metallurgy and architecture and Chongqing Institute of civil engineering and architecture 1986 *Air Conditioning* (Beijing: China Architecture and Building Press)
[16] Yang S M and Tao W Q 2006 *Heat Transfer* (Beijing: Higher Education Press)

IOP Conf. Series: Earth and Environmental Science **93** (2017) 012026 doi:10.1088/1755-1315/93/1/012026

Causes analysis on the failure of government environmental responsibility
——Based on the perspective of law and economics

L N Fa[1,2]

[1]Shanghai University of Political Science and Law (SHUPL), Post-doctor in Law, Fudan University, Shanghai, China

E-mail: falina12@163.com

Abstract. As the important Environmental Interests of Subjects, government behooves to undertake the corresponding responsibility of Pollution Control and Environmental Protection. The current situations in our country, however, appear as government environmental responsibility failure. Based on the analysis of law and economics, this article reaches the conclusion through game analysis, principle-agency relationship and utility theory that the prisoners dilemma of environmental interest game between government and enterprise, and the inherent defect of the principal-agency relationship between central government and local government are the inherent causes of government environmental responsibility failure. Many officials tends to graft and corrupt to maximum their own benefit, thus leading to the government failure among environmental pollution treatment and the environmental responsibility to undertake.

1. Introduction

The development and changes of modern society make the environmental interest subject diversified, and the environmental interest also shows the features of diversity and multi-level nature at the same time. According to the differences between the parties of the right configuration, the environmental interest subject can be divided into government, enterprises and the public. The parties have different demands for environmental interest. The government and enterprises try to obtain more creative environmental interest from natural resources and pay more attention to the economic benefits brought by the environment. However, the public always pursue better natural environment from their own welfare perspective. As the important Environmental Interests of Subjects, government, enterprise and public behoove to undertake the corresponding responsibility of Pollution Control and Environmental Protection. The current situations in our country, however, appear as government environmental responsibility failure, the serious lack of enterprise environmental responsibility and the weak awareness of the public environmental rights. The opening of the 8th part in the 18th CPC National Congress have pointed out the importance of building ecological civilization, that building ecological civilization is a long-term task of vital importance to the people's well-being and China's future. Faced with increasing resource constraints, severe environmental pollution and a deteriorating ecosystem, we must raise our ecological awareness of the need to respect, accommodate to and protect nature. We must give high priority to the ecological civilization. By this token, it is obvious that the achievement of ecological civilization construction is closely related with the environmental responsibility government to undertake. However, there is serious vacancy of government during the process of ecological civilization construction in our country. Hereby, in this article, we research the inherent causes mainly for the failure of government environmental responsibility based on the legal and economic methods.

2. Environmental benefit game easily falls into prisoners dilemma between government and enterprise

Suppose economic interest subjects of each party involved in the game – both government and enterprise are rational completely and the basis they make decision on is always maximum benefit. Also, there is no imbalanced information between each subject. They know their strategic space and income condition well when they make decision on the affairs related to their environmental interest, but they don't concern the detail about others' decisions. Based on the assumption above, we research the inherent causes of the failure of government environmental responsibility at present by the game analysis, starting from the environmental interest game between government and enterprise [1].

As an example concerning the environmental interest game between local government and enterprise, respectively, the decisions they possibly make are issuing a tacit permission of discharging pollution directly to enterprises. In order to pursue the rapid economic development, sometimes it is difficult to take care of environmental protection at the same time, in which case some governments may take the environmental pollution of enterprises as "acquiescence" to a certain extent. The matrix below can reflects their game benefit conditions in more detail:

Table 1. Profit matrix of environmental interest game between local government and enterprise.

		Enterprise	
		Direct pollution discharge	Processed emission
Local government	Tacit permission	$(t\times I, I)$	—
	oppose	—	$(t\times (I-C), I-C)$

As shown in the table above, imagine local officials performance assessment is directly related to regional economic index. In table 1, productive profit is I without waste treatment, the cost of waste treatment is C, the benefits are always t times higher than profits of enterprises whether government issues tacit permission to enterprises' behavior or not. In this condition, when enterprises choose to discharge pollution directly and governments choose tacit permission, both of them get benefits as tI and I; but if governments oppose discharging pollution directly and enterprises conduct waste treatment, then both local governments and enterprises will lose a combination benefits, and that is $(t\times (I-C), I-C)$.

Hereby, just refer to the matrix shown in table 1, when enterprises choose direct emission and governments choose tacit permission, both of them get maximum benefits, and they achieve Nash Equilibrium under this combination strategy. The officers focus more on personal interests when they make decision as the representative of local governments, and possibly, they don't perform governmental responsibilities in order to achieve personal achievements and promotion. In the process of environmental pollution treatment, if government was treated as a similar organization like enterprise, for members of organization – officers, especially local officers, in a large extent, the goal they pursue is conflicted with the goal set by government – provide excellent public service and push the construction of ecological civilization by policy making and legal methods, this results in the act that officers tend to short-term for their own goals, and this behavior substantively become a big barrier to achieve government macroscopic target.

3. The inherent defect for the principal-agency relationship of environmental responsibility among governments

Social products can be divided into private, public and quasi-public products in accordance with the difference of various products, service exclusiveness and competitiveness provided in market. In theory, as stronger public character and bigger external effect, environmental benefit is more similar with public and quasi-public products in nature. For the affairs related supply of environmental benefit, governmental supply or increasing governmental funding input are imperative. But in current governmental structure of country, the below follow the behavior of the above on environmental protection and management policies are very general, which is quite weak for local government to take environmental responsibilities. In the principal-agency theory, for the affairs related environmental benefit, just refer to figure 1, the principal-agency relationship exists between the public and government at all levels [2,4].

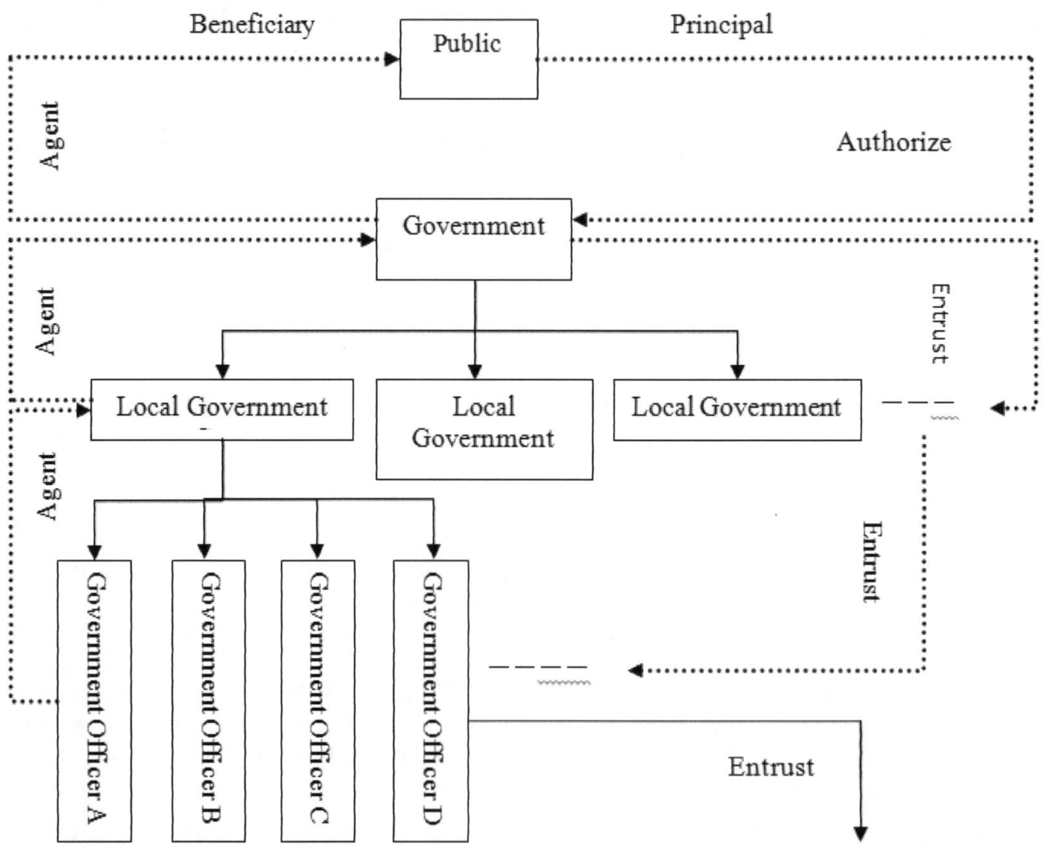

Figure 1. The principal-agent relationship of environmental responsibility between central government and local government.

As shown above, the principal-agency relationship of environmental responsibility exists essentially among citizen, government and officers. Public authorize the government and supervise its work, thus to benefit from it. The central government entrusts local governments at all levels to assist its work. With the division of labor above, can a complete principal-agent relationship of environmental responsibility between central and local government exists. Government is authorized by citizen to manage environmental pollution and protect natural environment instead of the public, and award the capital to government in tax. Meanwhile, government will have the rights to use the capital. In principle, the public should enjoy the environmental protection service provided by government as the beneficial owners.

In the meantime, government is a set of multiple identities containing many individuals, it forms a dual principal-agency relationship, that is the central government entrusts the local government and the local government entrust their representative - local officers. There is a complex principal-agency relationship between the central and local government, each interest subject in the chain has double identity. Theoretically, the central government plays the roles of agency and principal at the same time, on the one hand, according to the authorization of managing the public property, it should take the environmental responsibility in the process of building ecological civilization on behalf of people, and push the construction of ecological civilization from the perspective of policy and legal construction; on the other hand, as the principal, central government entrusts its subordinate body to perform its environmental responsibilities, transfer its environmental responsibilities to local governments at all levels. Therefore, local government should take relevant responsibility as an agent in the process of environmental protection, pollution control and building ecological civilization because they accept the environmental responsibility awarded by the central government. But at the same time, it is a same conception with government that local government is also a theoretic entity which consists of many individuals [3].

Because of the different status the principal and agent have, central government, local government and every government officers separately pursue different political goals, thus, facing different utility function. If regarding environmental interests and economic benefits as commodities, central government will undertake the responsibility of ecology civilization construction, and pursue the improvement of environmental interests in total, thus, preferring more to environmental interests. The improvement of environmental interests makes the better satisfaction, that is to say, as far as central government is concerned, the substitution effect of environmental interest utility is stronger than the economic one. For central government, concretely, the marginal substitution rate environmental interests to economic benefits are:

$$MRS_{ec,en}(C) = -\frac{\Delta en}{\Delta ec} \tag{1}$$

At the same time, as its superior government agent of environmental responsibility and rights, local government officers are the final assignee for all people and behoove to perform its own environmental responsibility. But in consideration of the officials' performance assessment mechanism with the main economic index our country currently in effect, enterprises will be limited as soon as the local government enhance the management of environment. Spontaneously, local government get fewer income, the development of Gross Domestic Product (GDP) and employment increase slowly, the improvement of environmental interests are obvious smaller than the central government. In other words, local government takes more utilities subjectively when consuming less. Thus, local government prefers more to economic benefits. As for local government, accordingly, the marginal substitution rate environmental interest to economic benefits is:

$$MRS_{ec,en}(G) = -\frac{\Delta en'}{\Delta ec'} \tag{2}$$

According to the analysis above, as for the preference of environmental interests, the central government is higher than the local government, that is to say, the relationship of marginal substitution rate between the two is:

$$MRS_{ec,en}(C) > MRS_{ec,en}(G) \tag{3}$$

Reflecting in the figure, the difference curve of central government utility is gentler than the local government one, as shown in figure 2.

IOP Conf. Series: Earth and Environmental Science **93** (2017) 012026 doi:10.1088/1755-1315/93/1/012026

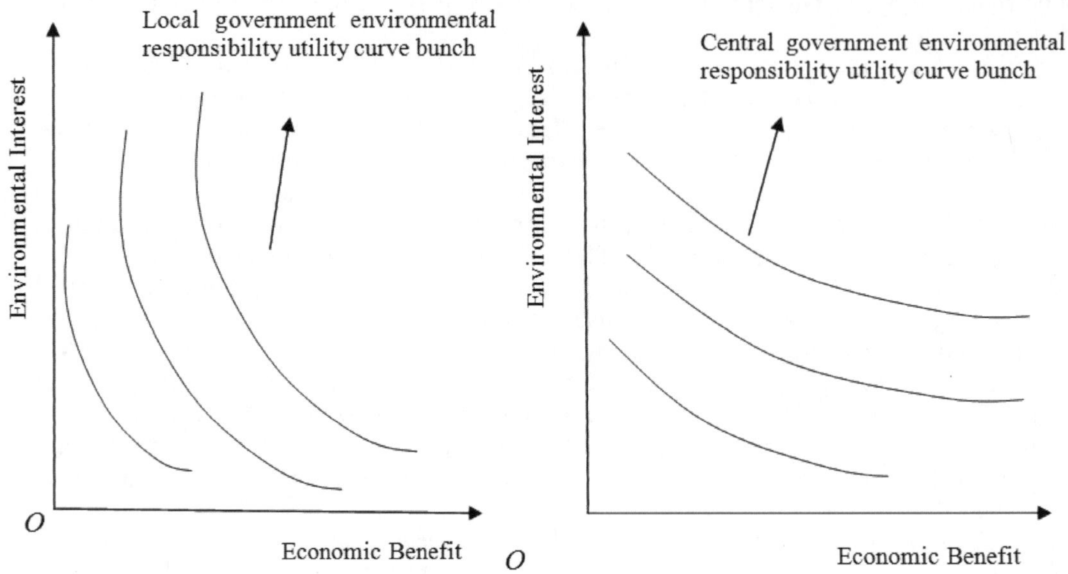

Figure 2. The utility curve of environmental interests of local government and central government.

Local officers often combined with enterprises to become an interest group, conniving or even supporting the emissions of enterprises without limitation because their performance appraisal is associated with local economic index, local GDP gets increasing and employment is promoted while government obtains revenue. This result in serious atmospheric pollution and haze, and citizen become the uppermost victims caused by this interest group, they suffer from the diseases without any economic benefits. Officers will try every means to do the rent-seeking activity, which means government intervenes and controls the economic activities by their administration power, hindering the role of competition in the market, and thus, creating an opportunity for few privileges to get excess income.by their privilege in order to earn personal benefits, finally resulting in the government failure among environmental pollution treatment and the environmental responsibility to undertake due to corruption [5].

As mentioned above, because of the inherent defect of the principal-agency relationship, in the principal-agency chain of the public and government, the central government and local government, local government and various officers, there is efficiency loss of information delivery and policy implementation. If the central and local government pursue the maximum utility respectively then the agent will pursue his maximum utility function in order to his own interests inevitably, the agent doesn't insure the interests of the principal, and sometimes will injure the interests of the principal possibly, it results in the failure of governmental environment responsibility in the end.

Therefore, in order to compensate for this shortcoming, we should improve the efficiency of information transmission and governance in the main chain of public and government, central and local governments, local governments and officials at all levels, which can be improved by the development of strict laws and regulations. Define the level of information delivery. At the same time, do reward and punishment. The appropriate reward for officials can improve the efficiency of the ruling; corresponding, combat the wrong ruling vigorously can also play a warning role.

4. Conclusion

As the important Environmental Interests of Subjects, government behooves to undertake the corresponding responsibility of Pollution Control and Environmental Protection. The current situations in our country, however, appear as government environmental responsibility failure. The issues of ecological damage and environmental pollution get worse and worse and the total environmental interests get injured seriously because of the government environmental responsibility failure. According to this, this article make the conclusions based on the perspective of law and economy that, as for government, because the information is incomplete and asymmetry, central government, as a principal, would never know what action its agent – the final agent of local governments and its environmental responsibility – would take, or even if the central government knows how its agent will assume his awarded environmental responsibility, there will also be abuses in supervising its agent's effort. Meanwhile, there are interest connections between central government and local governments as well as local governments and government officers, they share the environment corresponding interests and economic benefits, which will make the agent deviate from the principal's will with impulse, and finally lead to the issue of principal-agency. But the prisoner dilemma of environmental interests game between government and enterprises has further aggravate the failure of government's environmental responsibility, it is the internal cause for government not performing environmental responsibility well. And concluded a solution according to the analysis above, that is to develop strict laws and regulations, define the level of information delivery and do reward and punishment.

Acknowledgments
This article is funded by "China National Institute for SCO International Exchange and Judicial Cooperation Research Fund Project" and "2017 Shanghai University of Political Science and Law (SHUPL) Youth Science Fund Program" (2017XQN21).

References
[1] Liu H R and Yuan Y H 2013 Criticism on the distribution of environmental benefits *Social Sciences of Shandong Province* **2013(04)** 34-5
[2] Gong G 2008 Theoretical basis analysis on the government environmental responsibility *Journal of China University of Geosciences (Social Sciences Edition)* **2008(04)** 19-21
[3] Dai J 2012 Analysis on environmental crime (Shanghai: East China University of Politics and Law)
[4] Yan P Y 2013 A study on the accountability system of government environmental responsibility (Chongqing: Chongqing University)
[5] Zhang J W 2008 Some thoughts on perfecting government environmental responsibility *Hebei Law Science* **2008(03)** 50-3

Causes analysis on the serious defect of enterprise environmental responsibility
——Based on the perspective of law and economics

L N Fa[1,2]

[1]Shanghai University of Political Science and Law (SHUPL), Post-doctor in Law, Fudan University, Shanghai, China

E-mail: falina12@163.com

Abstract. As the important Environmental Interests of Subjects, enterprises behoove to undertake the corresponding responsibility of Pollution Control and Environmental Protection. The current situations in our country, however, appear as the serious lack of enterprise environmental responsibility. Based on the analysis of law and economics, this article reaches the conclusion through game analysis and cost-benefit analysis that the prisoners dilemma of environmental interest game between enterprises is the inherent causes for the serious defect of enterprise environmental responsibility. Meanwhile, at the point of cost-benefit, the externality of environment illegal act results in the imbalanced cost-benefit, lacking of the motivation to control pollution and protect environment in an active way.

1. Introduction

The development and changes of modern society makes the environmental interest subject grow diversified increasingly, and the environmental interest also shows the features of diversity and multi-level nature at the same time. According to the differences between the parties of the right configuration, the environmental interest subject can be divided into government, enterprises and the public. The responsibility of public mentioned in [1,2] are just for consultation so that the author can better understand the differences between the three parties and pick out the special points of enterprises, which are the main focus of this text. The parties have different demands for environmental interest. The government and enterprises try to obtain more creative environmental interest from natural resources and pay more attention to the economic benefits brought by the environment, while the public are always pursuing better natural environment from their own welfare perspective. As the important Environmental Interests of Subjects, government, enterprise and public behoove to undertake the corresponding responsibility of Pollution Control and Environmental Protection. The current situations in our country, however, appear as government environmental responsibility failure, the serious lack of enterprise environmental responsibility and the weak awareness of the public environmental rights. I've consulted references 3 to 5 and attached the importance to detailed clauses on national environmental laws and how they work. There are detailed clauses on national environmental laws for the performing methods of enterprises' environmental responsibility and the punishment to polluted-enterprises, currently, the national regulation in use are mainly the standard emission, environmental influence evaluation, three simultaneousness and emission license, all above are the fixed environmental legal obligations designed for enterprises directly. However, enterprises haven't really taken their environmental responsibilities they should do yet as a double subject containing environmental and economic interest. Hereby, in this article, we research the inherent causes mainly for the serious defect of enterprise environmental responsibility.

2. Environmental benefit game easily falls into prisoners dilemma among enterprises

Take the process of the environmental interests game between two enterprises as an example, suppose A and B manufacture the homogeneous products with the same marginal cost, the profit for an unit product is R, the additional payment costs which are used to equipment and labor input for waste treatment is D. Then the competitive relationship is formed between enterprises, concretely, the profit situation of the environmental interests game between enterprise A and B can be reflected in table 1 below.

Table 1. Profit matrix of environmental interest game among enterprises.

		Enterprise B	
		Direct pollution discharge	Processed emission
Enterprise A	Direct pollution discharge	(R, R)	$(R, R-D)$
	Processed emission	$(R-D, R)$	$(R-D, R-D)$

As shown in table 1, when enterprise A and B face the choices of direct emission and waste treatment, the possible results of the game may as follows:

- When both A and B don't implement the waste treatment actively, that is, both of them don't input the equipment, labors and do not pay the cost D, then both parts will earn total productive profits R;
- If one of them input the treatment cost, and implement the waste treatment actively, then his profits will be decreased while the profits for another enterprise still keeps unchanged, in this condition, the integrated benefits of A and B is $(R, R-D)$ or $(R-D, R)$, their benefits are unequal, the enterprise who implements the waste treatment actively always undertakes the payment cost;
- If both A and B take their environmental responsibilities actively and input the costs to implement the waste treatment at the same time, then the benefits they get are equal, all are $R-D$.

As mentioned above, in the case of no governmental regulation and supports, the enterprises, as the rational subject, prefer not to implement the waste treatment and keep polluting without any costs input on pollution treatment in order to pursue the maximum profit. Meanwhile, if one of enterprises (for example A) pay for the pollution treatment payment costs, then it will make up for the losses of profits by putting up the product's price, thus, the consumers will choose the product B which is homogeneous and lower price definitely, and then A will lose its competitiveness in the market due to the waste treatment. Besides, if both A and B input the equipment and labors to deal with the waste, then both of them will suffer from the loss of profits. In this condition, any of enterprises doesn't have the motivation to control pollution actively. Therefore, for enterprise A and B, no matter what choice the other makes, its best strategy is always discharging pollution directly, this time, both parties of the game go into Nash Equilibrium, which will be reached when participants are not willing to change their strategies in giving other participants an optimal strategy. From total benefits, this tacit understanding will be kept between enterprises if no external force destroy this balance, enterprises will expand their production and never take any action to waste treatment in order to get much more profits, it makes the local ecologic environment get worse and worse and causes the public environmental interests injured seriously, finally falling into the prisoners dilemma [3,4] That is a special game between two arrested prisoners, illustrating why it is still difficult to maintain cooperation even when this cooperation is beneficial for both parties. It is the most representative case in the game theory, reflecting that personal optimum choice is not group optimum choice.

3. Enterprises perform the imbalanced cost-benefit of environmental responsibility

The cost for enterprises performing environmental responsibility is from the expenditure which is used to waste treatment and environmental protection activities and so on in order to respond to the environmental protection goal of country, including the additional cost input which is used to improve productive technology, introduce advanced equipment for emission reduction and the relevant managing cost. In addition, view from the opportunity cost, the cost of increased production and the related competitiveness enhanced caused by purchasing equipment and developing environmental protection activities belong to the cost of performing environmental responsibility, too. The benefits of performing environmental responsibility are the promotion of total social interests including the quality improvement of air and water and so on in a certain place when enterprises take their environmental responsibility completely. Accordingly, when enterprises don't take environmental responsibility, they will have the emission against the legal standard and not pay the relevant sewage charge in the process of production and management, this time, the benefits they obtain are the cost saving that they never get punished and sanctioned and the price advantage that they win in the marketable competition.

Suppose the output of enterprises is constant in a period, then the emission load and required pollution charge will keep unchanged. As shown in figure 1, the point A in figure shows the budget for enterprises' pollution charge, the point F shows the fine to be equivalent to pollution charge A. Assume the cost budget for enterprises taking environmental responsibility (including the pollution abatement cost and the fine without pollution abatement) is M, then the constraint formula of polluter for budget cost is $A+F \leq M$.

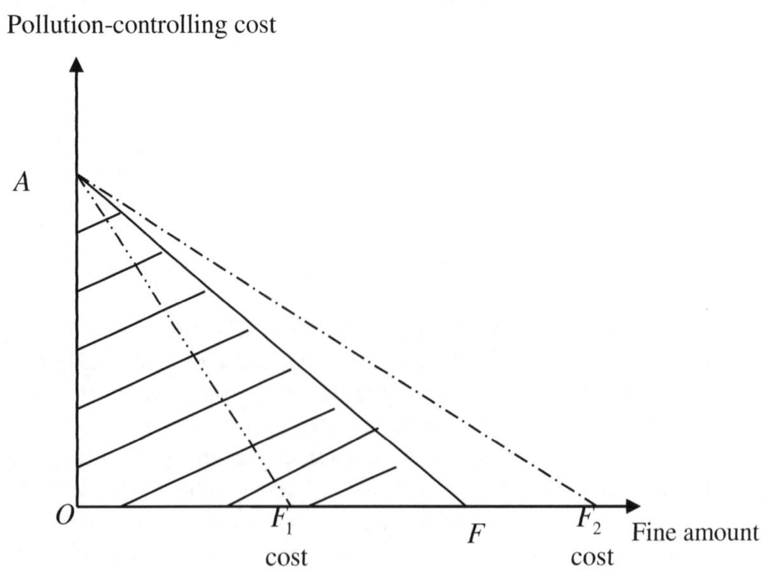

Figure 1. Pollution-controlling cost and fine budget for enterprises.

Under the premise of the above assumption, the budget range of enterprises for pollution abatement cost and fine is equivalent to the shaded area of △ AOF. When the fine is F1 (F1<F), the budget of enterprises are surplus, thus the enterprises get benefits and will not input the pollution abatement cost. When the fines are out of budget constraint AF, for example, it is on point F2 (F2>F), the enterprises not only input the pollution abatement cost A but also need to input the additional fines in order to perform their environmental responsibility, in this condition, the cost for performing environmental responsibility exceed the original budget, the enterprises will face the choices for increasing the pollution abatement cost to reduce the fines or not performing environmental responsibility.

From the significant environmental pollution events happened in recent years, it is a commonplaceness that the cost-benefit unbalance occurs for enterprises' lost performing environmental responsibility [5]. For example, in the significant water pollution events happened in Tuojiang River, Sichuan province, Sichuan chemical company Ltd. produced and discharged the pollutant excessively to cause TuoJiang River polluted seriously, this resulted in 200 million yuan economic losses, but environmental law enforcement only issued one million fine according to the relevant maximum fines. In nature, the punishment this enterprise took for its illegal emission and emission cost saving was far from making up the losses to ecological environment and public life for these events. This shows that the tiny fines encourage enterprises to take to the woods actually, the legal maximum fines results in the cost that the enterprises don't perform environmental responsibility is much less than the benefits they get by emission without limitation and not managing wastes, the serious unbalanced cost-benefit existed for the enterprises' performing their environmental responsibility, it results in serious responsibility vacancy of the enterprises' environmental responsibility in the end and become an important barrier in the process of the total environmental interests and ecological civilization construction.

4. Conclusion

As the important Environmental Interests of Subjects, enterprises behoove to undertake the corresponding responsibility of Pollution Control and Environmental Protection. The current situations in our country, however, appear as the defect of enterprise environmental responsibility. The issues of ecological damage and environmental pollution get worse and worse and the total environmental interests get injured seriously because of the defect of enterprise environmental responsibility. According to this, this article make the conclusions based on the perspective of law and economy that, as for enterprises, there are Nash Equilibrium existing in environmental interests game between enterprises, in the view of total benefits, enterprises will maintain tacit understanding if there are not any external force to break this equilibrium, they will expand their production without waste disposal in order to make as much profit as possible, thus, local ecological will get worse and worse and the collective environmental interests get injured seriously. Based on perspective of cost-benefit, there are severe imbalanced cost and benefit in environmental illegal action, thus makes all the enterprises lack of the motivation to treatment pollution and protect environment actively. The tiny fines encourage enterprises to take to the woods actually, the legal maximum fines results in the cost that the enterprises don't perform environmental responsibility is much less than the benefits they get by emission without limitation and not managing wastes, the serious unbalanced cost-benefit existed for the enterprises' performing their environmental responsibility, it results in serious responsibility vacancy of the enterprises' environmental responsibility in the end and become an important barrier in the process of the total environmental interests and ecological civilization construction.

Acknowledgments

This article is funded by "China National Institute for SCO International Exchange and Judicial Cooperation Research Fund Project" and "2017 Shanghai University of Political Science and Law (SHUPL) Youth Science Fund Program" (2017XQN21).

References

[1] Sun Y H 2012 Research on the legal thinking and the rule of law to promote ecological civilization construction *Chinese Environmental Law* **2012(03)** 61-3
[2] Cai S Q 2014 On reconstruction of the ways of our citizens' environmental interests law protection *Legal Expo* **2014(04)** 51-4
[3] Song J B and Li D N 2013 Theoretical research on environmental responsibility performance of

enterprises and its practical enlightenment *Journal of Renmin University of China* **2013(05)** 43-5

[4] Wang X D 2013 Research on system security of the environmental responsibility of enterprises (Jilin: Jilin University of Finance and Economics)

[5] Chen H X 2010 On environmental responsibility of enterprises (Suzhou: Suzhou University)

The realistic dilemma of environmental interest subject of legal responsibility ——Based on the perspective of law and economics

L N Fa[1,2]

[1]Shanghai University of Political Science and Law (SHUPL), Post-doctor in Law, Fudan University, Shanghai, China

E-mail: falina12@163.com

Abstract. As the important environmental interests subject, enterprises, public and government should assume the corresponding responsibility of pollution control and environmental protection. However, in the aspect of environment, there are failure existing in performing government responsibility, the financial expense and investment our government has participated in the aspects of pollution control and environmental protection are serious insufficient. In the meantime, in spite of the clear definition of the range and principles of enterprises' environmental responsibility according to some corresponding law documents, in view of our country's condition, enterprises always fail to assume their own environmental responsibility, and there are cases existing in pollution control and environmental protection that the investment is insufficient and the treatment effect is not obvious. In addition, it is especially outstanding in our country that the awareness of public environmental rights is pretty weak. The issues of ecological damage and environmental pollution get worse and worse and the total environmental interests get injured seriously because of the failure and vacancy of environmental responsibility of different subjects of right.

1. Introduction

The development and changes of modern society make the environmental interest subject grow diversified increasingly, and the environmental interest also shows the features of diversity and multi-level nature at the same time. According to the differences between the parties of the right configuration, the environmental interest subject can be divided into government, enterprises and the public. In general, these three environmental interest subjects - the government, enterprises and the public - still follow the rational person identity to make decisions when participating in the economic affairs of environment benefits, and many different subjects of rights are always pursuing their own long-term promotion and maximum of environment benefits, pursuing sustainable development. However, this non-differentiation of environmental benefit promotion goal is not equal to the non-differentiation of environmental right to assume between the government, enterprises and the public. From the situations of our country, the situations that failure or absence of environmental rights to assume and the weak awareness of environmental rights, which exist in different interest subjects, have become the dilemmas of Chinese environmental interest subject of legal responsibility of failure.

2. The failure of government environmental responsibility

State Environmental Protection Administration (SEPA) takes the most basic and important responsibility for environmental protection which is the specialized agent of environmental protection in accordance with Chinese environmental basic law, however, with the condition getting worse and worse on ecology environment and pollution, more and more requirements from environmental benefit are increasing for each interest subject, the SEPA's capability to improve environmental condition is limited by labors, resources and communication with other environmental agencies, in general, administrative capacity is insufficient for SEPA [1].

Nowadays, as GDP is the main index on official performance assessment mechanism, on the one hand, local governments, as a rational person, are always willing to combine with the enterprises and become interest group with whom can make a lot of job opportunities and productive profit but may cause very serious pollution and life-threatening condition to region. Perhaps so as to they can achieve the improvement of local economic indicators such as GDP growth, employment rate raising and people's income increasing and so on [2]. On the other hand, environmental benefit is a public good base on its external feature, in principle, government should provide it as fiscal expenditure, in detail, the governmental spending on energy saving and environmental protection projects during 2007 to 2012 is shown as table 1:

Table 1. Local fiscal expenditure in environmental protection (2007~2012).

year	Local energy-saving expenditure in environmental protection in RMB (billion)	Local energy-saving expenditure in environmental protection/ Local fiscal expenditure in environmental protection
2007	961.23	2.51%
2008	1385.15	2.81%
2009	1896.13	3.11%
2010	2372.5	3.21%
2011	2566.79	2.77%
2012	2899.81	2.71%
2013	3383	2.42%

As it is shown in the table 1, the local energy-saving expenditure in environmental protection shows an increasing trend in total, but the local energy-saving expenditure in environmental protection accounts for the proportion of local fiscal expenditure in environmental protection shows a downward trend in general. Among them, the energy-saving expenditure in environmental protection in 2013 was 554.9 billion dollars, accounting for 2.42% of the total expenditure. Compared with other items of fiscal expenditure, the proportion of energy-saving expenditure in environmental protection is relatively low. From the research reporting the SEPA published in 2006, severe shortage of investment in environmental protection in China has pointed out that China needs to spend 13.5 trillion dollars to clean up all of the industrial pollutants and waste household appliances, but the actual investment was only 23.8 billion dollars. From the reality in our country, the government hasn't been able to undertake the environmental responsibility yet, existing serious phenomenon in ecological civilization construction [3].

3. The serious vacancy of enterprises environmental responsibility

Zijin Mining Group Co., Ltd. is a large-scale mineral group based on gold and other non-ferrous metal mineral resources survey and development. Started from 1993, after more than 10 years of development, it became a well-known domestic gold mining and key high-tech enterprise. But with the high-speed development of Zijin Mining, environmental pollution problems occurred frequently. July 3rd, 2010, a sewage leakage accident occurred at Zijinshan copper mine wet plant, located in Fujian Shanghang County. 9100 cubic meters of sewage leaked into Tingjiang, resulting in a serious pollution and a large number of fish was killed. February 1st, 2011, Zijin Mining announced that the accident was guilty of major environmental pollution accident and sentenced to 4.75 million dollars fine, the other five defendants were sentenced to 3 years to 3 years and 6 months imprisonment.

At present, our country has made clear legal regulations for environmental responsibility of enterprises. For example, the special legislation of environmental protection has stipulated that the range of undertaking of enterprises environmental responsibility consists of location, design, construction, production, etc., but for these enterprises who have already caused environmental pollution are supposed to follow the principle that anyone who created pollution should be responsible to control it and undertake their own environmental responsibility. Moreover, the legal document of environmental pollution control of single item which is specialized in particular pollutants such as Water Pollution Control Law, Marine Environmental Protection Law, Air Pollution Prevention and Control Law, Water Law, Sand Prevention and Control Law and so on, also definite the demands of the environmental responsibility of enterprises to undertake and the enterprises' responsibility to decline environmental pollution [4].

Nevertheless, although the range and principle of enterprises environmental responsibility have already been definited in corresponding legal document, in view of the reality of our country, enterprises are always losing responsibility of undertaking their own environmental responsibility. And there is always lack of investment and not obvious effects in environmental protection and environmental pollution control. Specifically, we can see from the view of China's industrial pollution control investment during 2000 to 2012:

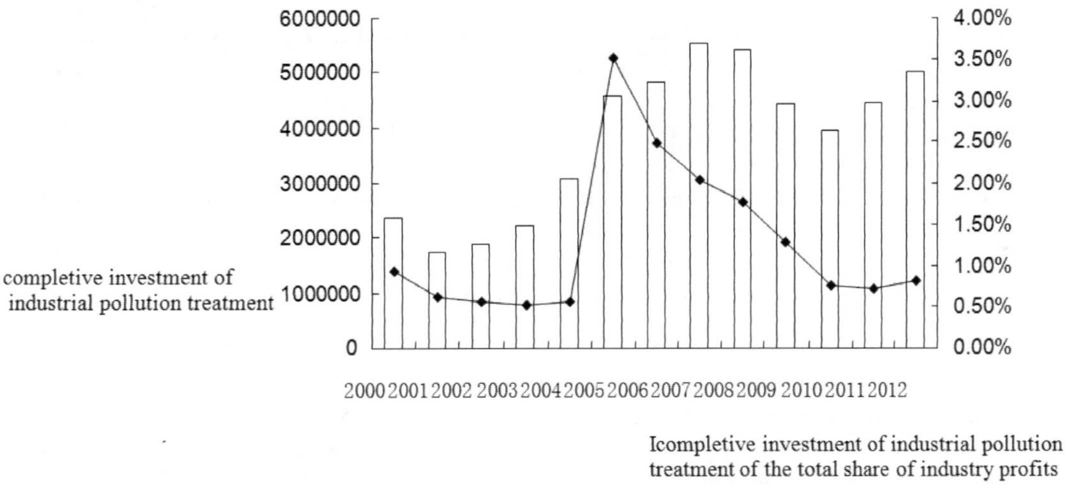

Figure 1. Industrial pollution of investment in China (2000~2012).

View from status of industrial enterprise on figure 1, total investment of industrial pollution treatment increased markedly in 2012 comparing with 2000, but the proportion of completive amount among industrial total profit went down in the mass. According to the statistic data from the people's Republic of China Ministry of environmental protection, total completive investment of industrial pollution treatment was 7928 million dollars, separately, the treatment cost were 2223.1 million dollars, 4082.6 million dollars, 392.1 million dollars and 18.4 million dollars on waste water, waste gas, waste solid and noise. Although total investment increased 10% comparing with 2011 and 110% addition comparing with 2000, but it only held 0.81% of total industrial output value, it is still a lower level investment on industrial pollution treatment [5].

As is mentioned above, current investment of industrial pollution treatment in our country still keeps in a lower level. Oppositely, the situation of industrial pollution is getting worse and worse. Based on the recent five years' data of waste gas emissions, the emissions of waste gas appear an increasing trend. Details as below figure 2:

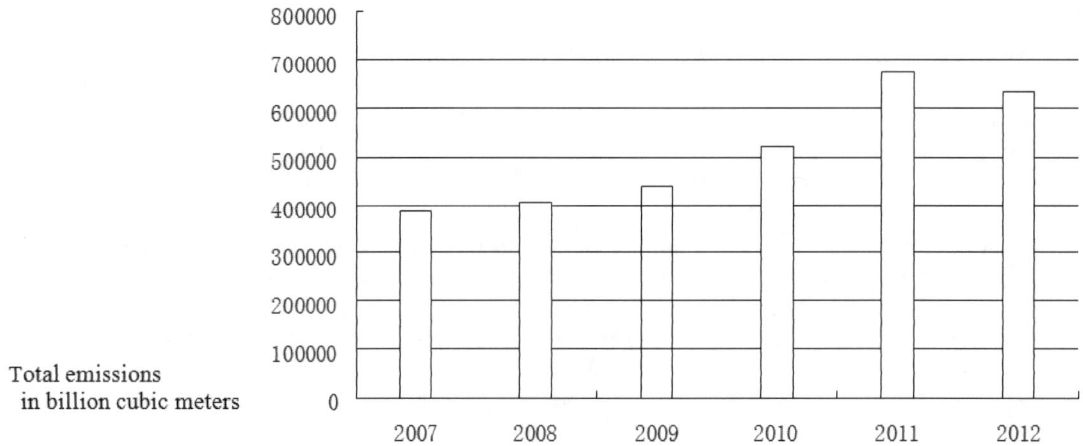

Figure 2. Total industrial emissions of China (2007~2012).

As one of important environmental interest subjects, enterprises benefit directly from environmental benefits. On the one hand, they gain profits by natural environment and resource, on the other hand, their productive activities will destroy ecology and cause environmental pollution, so they should take the related responsibilities for environment. For enterprises, environmental protection, waste treatment and emission reduction are their specific social responsibilities in environmental protection field, and are also detailed method of performance based on environmental law. Enterprises should take their environmental responsibilities in the aspect of law by improving productive technology, upgrading productive facilities and other methods while they are seeking for maximum profits so as to insure ecology environment protected, saving natural resource and maintaining public environmental benefits. Nevertheless, in our country, the actual situation is that enterprises never really took their bounden duty on environment, and environmental pollution caused by enterprises is still an urgent issue to be solved in the present stage.

4. The weak awareness of public environmental rights

At the point of its property, environmental benefit is a type of public benefit corresponding to private benefit, public - all unspecific - all benefit from it, the indivisible feature it has becomes the realistic block to achieve a balanced environmental benefit between different subjects of right [6].

Actually, the environmental benefit is always lacking of the protection of law and system. And because of this, on the one hand, the public have not built the awareness of environmental benefit yet; on the other hand, the public can not gain the benefit relief via effective methods when they are under average. In this case, whether the government blowdowns directly or processes of polluting substances before blowing down, the public would become the weak subjects finally and face the risks of losing environmental benefits under the situation that they can't have any economic benefits.

As is shown in the figure 3 above (where NPC refers to the National People's Congress and CPPCC refers to the Chinese People's Political Consultative Conference), from the conditions of national environmental letter visits from 1995 to 2012, people did not realize that environmental benefit is a legal right they should have in nature due to public feature of environmental benefits under the lack of effective legal system, more often, people only treat environmental benefits as an auxiliary effect of their economic and social activities, in another word, it is a positive external effect. In parallel, economic benefits almost pass through all human activities, it is first and the premise and foundation of other benefits in the real life. The public haven't built the awareness of environmental rights yet as the important subject of environmental benefits, and they are lacking of motivation to defend their environmental benefits actively, on the other hand, it is often difficult for the public to express their demands regarding environmental benefits because of limited open channels [7].

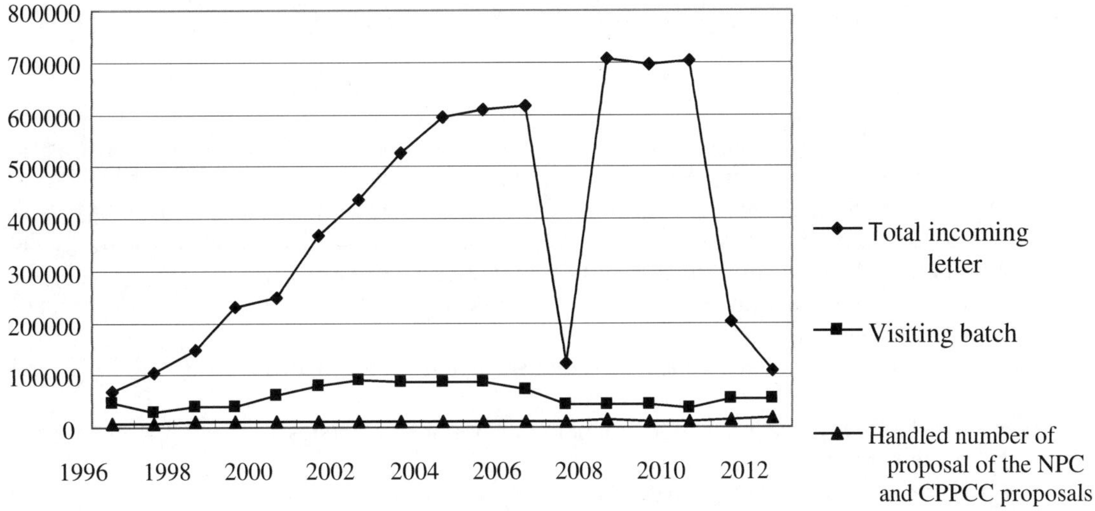

Figure 3. Conditions of national environmental letter visits (1995~2012).

5. Conclusion

As the important environmental interests subject, enterprises, public and government should assume the corresponding responsibility of pollution control and environmental protection. However, in the aspect of environment, there are failure existing in performing government responsibility, the financial expense and investment our government has participated in the aspects of pollution control and environmental protection are serious insufficient. The administrative capacity is insufficient for SEPA, and compared with other items of fiscal expenditure, the proportion of energy-saving expenditure in environmental protection is relatively low. From the reality in our country, the government hasn't been able to undertake the environmental responsibility yet, existing serious phenomenon in ecological civilization construction. In the meantime, in spite of the clear definition of the range and principles of enterprises' environmental responsibility according to some corresponding law documents, but in view of our country's condition, enterprises always fail to assume their own environmental responsibility, and there are cases existing in pollution control and environmental protection that the investment is insufficient and the treatment effect is not obvious. For enterprises, environmental protection, waste treatment and emission reduction are their specific social responsibilities in environmental protection field, and are also detail method of performance based on environmental law. Nevertheless, in our country, the actual situation is that enterprises never really took their bounden duty on environment, and environmental pollution caused by enterprises is still a urgent issue to be solved in the present stage. In addition, it is especially outstanding in our country that the awareness of public environmental rights is pretty weak. Actually, the environmental benefit is always lacking of the protection of law and system. And because of this, on the one hand, the public have not built the awareness of environmental benefit yet; on the other hand, the public can not gain the benefit relief via effective methods when they are under average. The public haven't built the awareness of environmental rights yet as the important subject of environmental benefits, and they are lacking of motivation to defend their environmental benefits actively, on the other hand, it is often difficult for the public to express their demands regarding environmental benefits because of limited open channels. The issues of ecological damage and environmental pollution get worse and worse and the total environmental interests get injured seriously because of the *failure* and vacancy of environmental responsibility of different subjects of right.

Acknowledgments
This article is funded by "China National Institute for SCO International Exchange and Judicial Cooperation Research Fund Project" and "2017 Shanghai University of Political Science and Law (SHUPL) Youth Science Fund Program" (2017XQN21).

References

[1] Meng Q L 2008 The theory of environmental responsibility – Reflection on some basic theoretical problems on environmental law (Shandong: Ocean University of China)
[2] Kahn M E, Zheng S Q, Sun W Z and Luo D L 2013 Incentivising China's urban mayors to mitigate pollution externalities: The role of the central government and public environmentalism (National Bureau of Economic Research) 18872
[3] Wang C L 2005 On the legal protection of environmental interests (Beijing: China University of Political Science and Law)
[4] Dai J 2012 Analysis on environmental crime (Shanghai: East China University of Politics and Law)
[5] Wang Y 2011 On the legal protection of farmers' environmental interests (Chongqing: Southwest University of Political Science and Law)
[6] SEPA: Research Report of Green National Economic Accounting 2004
[7] The Environment Protection Law of the People's Republic of China 2014-04-24

Measures to reinforce the legal liability of the environmental interest subject
——Based on the perspective of law and economics

L N Fa[1,2]

[1]Shanghai University of Political Science and Law (SHUPL), Fudan University, Shanghai, China

E-mail: falina12@163.com

Abstract. Local government should be regarded as the main subject to be stipulated by environmental law, thus to avoid local government's alignment with commercial interests. Such a shift would, furthermore, discourage collusion against environment law or speculative behaviors motivated by maximizing production at the expense of environment pollution. Moreover, whether companies make proactive decisions to prevent pollution or not depends on the severity of appropriate environment legal system's sanctions for their action. It would encourage enterprises to undertake their own environmental responsibility if environmental law could further enhance their environmental liability. In addition, public environmental rights should be embedded into environmental law. In this way, the public may become more aware of their environmental rights as well as the positivity of total environmental interests.

1. Introduction
The development and changes of modern society have increased and diversified environmental issues, and social interest reflects corresponding features of diversity and multi-level nature at the same time. Roles related to environmental protection can be divided into government, enterprises and the public. Heretofore, environmental issues could not be resolved effectively because of shortcomings in the substance of environmental liability, the active behavior of enterprises, and ineffective governmental regulation. However, as a quasi-public good, environmental factors have external effects with wider results including economic sustainable development, the long-term stability of the society, and so on; all the people can enjoy the environment as a matter of public welfare at the same time. Accordingly, the existing vacancy in legal liability with respect to environmental interest subjects, cannot be blamed simply on any one of government, enterprises and the public. In order to promote social overall environmental interests, we should revise legislation on the environmental liability of government, enterprises and publics.

2. Government: To bring the environmental liability of local government into legal regulation
The central government is generally unaware of what behaviors their proxies, the subordinate levels of government – or their end proxies of environmental liability, the officials – will take due to the incompleteness and asymmetry of information; nor do they even he know what kind of behavior the proxies will take to perform their assigned environmental responsibilities. Malpractice occurs at the working effort level of supervising proxies. In contrast, in some western countries, local governments

provide the source of authority to the federal government. Such local governments are accountable to the people who elect them as opposed to owing an allegiance to a centralized sponsor. Meanwhile, there are coordinating relationships between the central government and local government as well as local government and officials, as they share the environmental and economic interests. In such a case, local authorities may actually go against higher-level government's expectations. When local officials take regulatory action based on citizen welfare implications rather than strict code compliance, it will cause the principal-agency issues accordingly. The prisoner dilemma caused by the environmental interest game between government and enterprises can further aggravate the environmental liability failure of government [1].

In fact, if the government, especially local government, fails to enforce environmental responsibilities, they may become culpable for injuring the environmental interests of all the people through environmental pollution and ecological damage. This issue of governmental environmental liability should be incorporated into environment law. To view from the existing Environmental Protection Law, Atmospheric Pollution Prevention Law, Marine Environment Protection Law, and Solid Waste Pollution Environment Protection Law which are slip laws that involve in environmental pollution regulations and environmental protection, the main subjects of regulation are enterprises. The environmental liability of local government is not clear, so local government may align with enterprises as an interest group. Substantial economic interests may be exchanged, with correspondingly serious losses of regional environmental interests. Local officers sometimes trade off improvements of economic indicators for damage to regional environment and ecology - as well as personal promotion. This internalized response to personal interests worsens and is not effectively regulated by law directly [2].

Therefore, according to the legislative purpose of Environmental Protection Law, which is the first rule of Environmental Protection Laws of the People's Republic of China, in order to protect and improve life and ecologic environment, prevent pollution and other public hazard, ensure human health and promote the development of socialist modernization, this law is established, the rule of otherness should be issued as the law for the responsibilities taken by the central and local government in the affairs which are related to environmental interests. Rules should stipulate the local government within the regulation of environment law to avoid local government's alignment with enterprises to be an interest group, and further conniving activities against environment law, or some speculative behaviors that favor production and profits over environment protection. More specifically, with the increasingly comprehensive environment protection act and other environmental legislation related to environmental pollution regulation and environmental protection, the primary regulator should be local government.

Firstly, as mentioned above, government responsibilities for active prosecution of environmental liability, including environment protection, pollution abatement and emission supervision etc. should be detailed clearly in the Environmental Protection Law. Further, in order to solve the principal-agency issue between the central government and local government on taking environmental liability, the central government should assume a role of providing more recommendations and support rather than prescribing mandatory measures when delegating the environmental liability to local government or officials. We can refer to the example of existing central-local practices for environmental management in America to clarify the allocation of environmental rights and responsibilities for all levels governments. At the same time, central government focus would shift to supporting local governments through financial power, providing guidance to local governments on enforcing environmental liability, researching environment protection, and supervising the polluting behaviors and so on through preferential fiscal and tax policy, appropriate industrial policy tilt, and approval for important construction projects, etc [3].

3. Enterprises: To strengthen the environmental obligation of enterprises by law

Firstly, in theory, there is a natural profitability motive for the enterprises - personal profit maximization is always the goal that they pursue when producing. On the other hand, the enterprises must comply with the laws, administrative regulations, social morals, commercial morality and honesty, accept the supervisions of government and publics and undertake the social responsibility (The fifth rule in the first chapter of the Company law of the People's Republic of China) as an important object of environmental interests when they work on business operation. Enterprises must take their own environmental liability into account when they encounter environmental issues during operation. However, per previous analysis, we know that there is Nash equilibrium exists for the environmental interest game of enterprises. When enterprises reach tacit understanding, they naturally choose resource maximization and operation without waste treatment at the expense of environmental interests, environment pollution and serous ecological damage. Although the enterprises achieve the personal economic interest maximization, it results in huge losses to overall social interests due to the prisoner dilemma. Beyond this, it is clear that a cost-benefit imbalance exists in the process of the enterprises performing environmental liability; this results in a lack of motivation for the enterprises' environmental protection and waste treatment.

As a powerful tool of influence, the law's mechanism of action focuses on influence of the regulated subjects, accordingly, through economic incentives for environmental responsibility or failure by the enterprises. Enterprises make the decision of pollution or active abatement, motivated by the severity of sanctions on their behavior through the environmental legal system. Further strengthening the environmental obligation of enterprises in the environmental legal system will encourage enterprises to take their environmental responsibility more seriously [4].

In the process of the ecological civilization construction, the main goals of environmental legislation should focus on the adjustment of environmental interests among the different stakeholders. The legal regulation of rights and obligations create relationships which lead to compromise and harmonization among each other to reach the goal of reducing and eliminating the conflicts. For example, with respect to significant environment pollution affairs, the criminal responsibilities of the polluters and relevant principals should be called to account to reduce risk of environmental pollution, and mitigate the expected effects when the enterprises carry out the polluting behavior or inadequately deal with wastes, therefore reducing the cost of compliance while increasing costs of noncompliance - leading the polluters to adjust their emissions duly. In addition, it is imperative to clarify the environmental responsibility of the enterprises as obligation in environment law, especially the environmental obligation of the enterprises should be clarified in the slip laws related to environment pollution and environment protection. It is beneficial to break the Nash equilibrium of an enterprise and other enterprises, citizens and enterprises as well as government and enterprises and correct the thumb a lift mentality of enterprises, letting environmental law to play a necessary role of ensuring that enterprises' meet environmental responsibilities and control their polluting behavior.

4. The public: To clear the environmental rights of publics by law

As mentioned above, the public should collectively advocate for environmental interests and consider personal environmental rights protection as an important component of environmental interests. All the people are beneficiaries of environmental protection and victims of environmental losses, due to the public character of environmental interests and the positive externality of environmental tights protection. Inevitably, the level of environmental interests promotion obtained by the individual rights protection and litigation is lower than the force of overall social requirements; therefore, the positive externality of individuals protecting environmental rights is much less efficient than social rights protection – essentially, ineffective.

As mentioned above, nowadays, the public lacks awareness and effective channels for environmental rights protection. Alternatively, a democratic environmental protection process would inform and empower citizens to recognize and advocate for responsible environmental practices. That is to say, we can confirm the programmatic position of environmental rights in the environmental law in theory. In the meantime, because environmental issues are getting worse and worse, the environmental rights should be embedded into environmental law to empower public rights. In this way, the public could bring reasonable litigation with respect to environmental pollution behavior of enterprises, government and interest group when their own environmental rights are injured. It is therefore beneficial to

IOP Conf. Series: Earth and Environmental Science **93** (2017) 012029 doi:10.1088/1755-1315/93/1/012029

strengthen the environmental awareness of the public and excite them to protect their environmental rights in environmental law as well as the positivity of total environmental interests [5].

Using haze issue as an example, personal health and life are threatened by unregulated emission of vehicle exhaust and waste gases, representing a market failure. Overall social environmental interests are menaced by the market. Individual citizens could organize as a social group voluntarily, based on underlying interests, actively advocating for relevant legislation for haze control or ask enterprises to take real methods for emission reduction. But currently, the environment does not support such social groups playing a role; civil awareness is insufficient and corporate strength is weak. Therefore, for the haze control, we should make an effort to build a civil society in the future so as to ensure an environment in which such social groups and organizations can play a role.

Furthermore, personal and public interests should cooperate through a principle of public participation in environmental legislation. This means that environmental law should ensure the wide participation of the different interests and mechanism they advocate. When the environmental legislation involves divers interests and significant legal influence, all stakeholders can participate in sufficient communication and negotiation before making final decisions. Meanwhile, they should ensure the wide range and the relative equal status of the participators as possible as from the legal level, and make the interests demands of all stakeholders - especially those in weak positions – are heard. Thence the supply of environmental legislation is perfected accordingly basing on this benchmark - just as the NEPA process in the US, which provides a framework for citizens' involvement and stakeholder conversations.

To establish and perfect the principle of public participation, environmental legislation should be modified in several ways. Firstly, public participation should be established as a legislative principle of the environmental law: clarify the environmental rights of the public in the legal level, make sure the reasonable rights of the public and all interests subjects in the process of environmental legislation such as the rights of supervision and the rights to know, and excite the environmental rights and the environmental legal awareness of the public. Secondly, the relevant legal regulations should be equipped to ensure the rights of publics effectively, perfect the expression mechanism of the public to the interests demands, make the interests demands of the public being able to go up to the legal level effectively, push the supply of the environmental legislation to be perfected in the standard of demands, and provide additional force for the improvement of imbalance between supply and demand on the environmental legislation. In addition, initiating from the legal interests relief function, establish and perfect the system of environmental public benefit lawsuit, provide the platform to protect and relieve the vulnerable interests for the vulnerable groups, excite the public to protect the vulnerable interests, and guide the public to pay attention to and protect their own environmental interests and rights [6].

5. Conclusion

Local government should be regarded as the focal point to be stipulated by environmental law, as the direct authority with respect to enterprises' illegal emission. In the past, the government, especially local government, already was party to injuring the environmental interests of all the people and caused environmental pollution and ecological damage; therefore governmental environmental liability should be incorporated into the environment law. The law should also clarify the corresponding responsibilities of local government and central government. It should fit the local government into the regulation of environment law to avoid local government's alignment with enterprises as an interest group, and further conniving activities against environment law, or undue emphasis on maximizing production at the cost of environment pollution. Moreover, as an important subject of environmental interests when they work on business operation, the enterprises should manage their own liabilities when they encounter the environmental issues during operation. Whether companies make proactive decisions to treatment pollution or not depends on the severity of appropriate environment legal system's sanctions for their action; it will be better for enterprises to undertake their own environmental responsibility if environmental law can further enhance enterprises' environmental liability. In the process of the ecological civilization construction, the main goals of environmental legislation should focus on the adjustment of environmental interests. Conflict among the different interest subjects and legal regulations of rights and obligations require the environmental interests to compromise and harmonize with each other to reach the goal of reducing and eliminating the conflicts. In addition, as the important subject of environmental interests, the public are beneficiaries of environmental interest promotion and undertakers of environmental interest losses at the same time. Because environmental issues are getting worse and worse, the environmental rights should be fitted into environmental law to clear up the environmental rights of publics in law. In this way, the public can make reasonable litigation to the environmental pollution behavior of enterprises, government and interest group when their own environmental rights are injured. It is beneficial to strengthen the environmental awareness of the public and excite the public to protect their environmental rights in environmental law as well as the positivity of total environmental interests, reinforcing the interests of all interests subjects, especially who is in weak position. Environmental legislation may be clarified and therefore improved accordingly, based on this benchmark.

Acknowledgments

This article is funded by "China National Institute for SCO International Exchange and Judicial Cooperation Research Fund Project" and "2017 Shanghai University of Political Science and Law (SHUPL) Youth Science Fund Program" (2017XQN21).

References

[1] Gong L L and Wang G F 2012 Public and political dimension of environmental justice--Institutional arrangements for environmental goods distribution system *Study and Explore* **2012(5)** 10-2
[2] Wu Y X 2013 Research on the basic legal responsibility of environmental damage identification *Environment and Sustainable Development* **2013(03)** 22-5
[3] Zhang W 2011 Research on legal responsibility of environmental administrative subject (Harbin: Northeast Forestry University)
[4] Naohiko H 1999 *Environmental Law -- Law Books of the Waseda University in Japan* (Law Press)
[5] Morck R, Yeung B, Wu J, Deng Y H and Huang J 2013 Incentives and outcomes: China's environmental policy (National Bureau of Economic Research Working Paper) 18754
[6] Chen D D 2017 A study on the legal responsibility of environmental administrative subject in China *Legality Vision* **2017(02)** 65-7

A method of optimized neural network by L-M algorithm to transformer winding hot spot temperature forecasting

B G Wei[1,2,4], X Y Wu[3], Z F Yao[1] and H Huang[1]

[1]Shanghai Municipal Electric Power Company Electric Power Research Institute, Hongkou 200437, Shanghai, China
[2]Shanghai SEPRI Power Technology Co., LTD., Hongkou 200437, Shanghai, China
[3]State Grid Shanghai Municipal Electric Power Company, Pudong 200120, Shanghai, China

E-mail: wbgsj@126.com

Abstract. Transformers are essential devices of the power system. The accurate computation of the highest temperature (HST) of a transformer's windings is very significant, as for the HST is a fundamental parameter in controlling the load operation mode and influencing the life time of the insulation. Based on the analysis of the heat transfer processes and the thermal characteristics inside transformers, there is taken into consideration the influence of factors like the sunshine, external wind speed etc. on the oil-immersed transformers. Experimental data and the neural network are used for modeling and protesting of the HST, and furthermore, investigations are conducted on the optimization of the structure and algorithms of neutral network are conducted. Comparison is made between the measured values and calculated values by using the recommended algorithm of IEC60076 and by using the neural network algorithm proposed by the authors; comparison that shows that the value computed with the neural network algorithm approximates better the measured value than the value computed with the algorithm proposed by IEC60076.

1. Introduction

Electric transformers are widely used in all levels of the power grid and play an important role in all processes of the economical transmission, flexible distribution and safe consumption of power [1]. The capacity of transformer continues getting larger with the increasing electricity demand, however, it can't keep getting large correspondently, which will cause excessive heat generated by windings and more eddy loss, and eventually it will lead to the failures [2]. The oil-immersed transformers take a proportion of 75% in all transformers in distribution network. Exposed to high temperature environment, the oil-paper insulation degradation fastened and thus the life expansion of transformers lessened [3]. According to the law of 6℃, when the temperature of hot-spot reaches 98℃, the relative aging rates is 1 [4]. For every 6℃'s increase of the temperature, the life cycle halved, namely the relative aging rate double. Therefore, it has a great meaning to accurately compute the hottest temperature (HST) of windings under different load modes and surrounding temperatures The most accurate method to get the hot-spot temperature is to lay temperature sensor near the conductor and measure directly, it is commonly used in tests despite its high cost and difficult to maintain [5,6]. For transformers in operation, indirect methods are adopted which generally are the temperature rise calculation formulas of the hot-spot recommended in the standard IEEE C57.91 [7] and the IEC 60076

[8]. The recommended formulas are limited since they neglected the influences of top-oil temperature and hot-spot temperature of factors like sunlight, wind speed and etc. In domestic and abroad field, investigations are done to compute the top-oil temperature and hot-spot temperature with new methods. Lesieutre B C, *et al* [9] built the top-oil temperature calculation model which the surrounding temperature are added to it as a variable. The backward Euler discrete equation and the linear regression are used to calculate the top-oil temperature and the identifying parameters of the model; however, the model remains relatively high error. Susa D, *et al* [10] took the variation of oil viscosity and the loss with temperature into consideration, and equaled to the heat transferring process inside the transformer with two thermal lumped-parameter circuits in series. The time constant of top-oil temperature calculated by this model is less than the recommended value in the IEC 60076. However, the model neglected the temperature difference between the oil and the tank, and the complex heat transferring process on the external surface of the tank. Susa D and Nordman H, Zeng H and Zhou X X and Chen W G and Zhao T [11-13] also made much research work about it.

In this paper, firstly, the heat mechanism and the thermal behavior inside transformers were analyzed, the influence on the oil thermal characteristics parameters of surroundings' factors and change of temperature were taken into consideration. And then the parameters that have more influence on the hot-spot temperature were selected through using formulas recommended in guidance. The neural network optimized with L-M algorithm was trained to get proper oil thermal characteristics parameters using data from the cases of the guidance. Finally, the hot-spot temperature was calculated by using the improved equation and the proper oil thermal characteristics parameters obtained with the neural network. Comparison between the measured value and the calculated values that obtained by using the recommended algorithm of IEC60076 and the neural network algorithm showed that the value computed by neural network algorithm approximated more to the measured value.

2. The computational algorithm of winding hot-spot temperature

The computation equation of hot-spot temperature is given in IEC 60076 guideline [8].

$$\theta_h = \theta_a + \Delta\theta_o + \Delta\theta_h \tag{1}$$

When the load coefficient increases,

$$\theta_h(t) = \theta_a + \Delta\theta_{oi} + \{\Delta\theta_{or} \times [\frac{1+R\times K^2}{1+R}]^x - \Delta\theta_{oi}\} \times f_1(t) + \Delta\theta_{hi} + \{Hg_r K^y - \Delta\theta_{hi}\} \times f_2(t) \tag{2}$$

Where,

$$f_1(t) = (1 - e^{(-t)/(k_{11}\times\tau_o)}) \tag{3}$$

$$f_2(t) = k_{21} \times (1 - e^{(-t)/(k_{22}\times\tau_w)}) - (k_{21} - 1) \times (1 - e^{(-t)/(\tau_o/k_{22})}) \tag{4}$$

When the load coefficient decreases,

$$\theta_h(t) = \theta_a + \Delta\theta_{or} \times [\frac{1+R\times K^2}{1+R}]^x + \{\Delta\theta_{oi} - \Delta\theta_{or} \times [\frac{1+R\times K^2}{1+R}]^x\} \times f_3(t) + Hg_r K^y \tag{5}$$

Where,

$$f_3(t) = e^{(-t)/(k_{11}\times\tau_o)} \tag{6}$$

3. Analysis of transformer heating mechanism

Transformers transfer the electromagnetic energy into heat during operation. The core and windings all produce losses and the windings are the main heat source. The heat transfer to the surroundings raises the temperature of transformer.

The losses of transformers can be expressed as follows:

$$P_T = P_C + P_L \tag{7}$$

In the equation (7), P_T means the total loss, W; P_C means no-load loss, W; P_L means load loss, W.

No-load losses are the active power transformer absorption when the rated voltage is applied on one winding while the other windings are open. The no-load losses include eddy current loss and hysteresis losses. The on-load loss includes resistant loss caused by windings, eddy current loss and the stray loss caused by components like holdings and tank. There are sleeves outside the iron core in order to separate the oil-duct of core from the oil duct of windings inside the transformer, so on-load loss is the main cause of temperature rise.

No-load loss can be expressed as follows:

$$P_C = P_1 + P_2 = \delta_h f B_m^{\ 2} + \delta_e f^2 B_m^{\ 2} \tag{8}$$

In the equation (8), P_1 means the hysteresis loss of core; P_2 means the eddy current loss of core; δ_h means the hysteresis loss coefficient; δ_e means the eddy current loss coefficient; B_m means the amplitude of magnetic flux density, Wb/m^2. According to the equation (8), load loss is directly proportional to the square of B_m. The value B_m of transformers is designed to be nearing saturation zone of magnetization curve for the saturation characteristic of the magnetization curve. The B_m will not increase a lot with the increasing load current.

On-load loss can be expressed as follows:

$$P_L = I^2 R + P_E + P_S \tag{9}$$

In equation (9), I^2R means the direct-current loss of resistance; P_E means the eddy current loss of winding; P_S means the stray loss of metal components.

There are some parameters have tiny impact on hot-spot temperature that can be ignored, such as $\Delta\theta_{oi}$, R and $\Delta\theta_{hi}$. The parameters which have large impact on hot-spot temperature are related with the oil and windings, for example x, τ_o, y and τ_ω. Recommended values of the transformer thermal characteristic parameters are according to empirical constant [8]. However, the thermal characteristic parameters especially the exponent of oil and winding (x and y) will change a lot with the changing temperature. So it is important to get the actual dynamic parameters to predict the hot-spot temperature.

4. The determination of hot-spot temperature parameter

There is an example of 250MVA ONAF transformer in IEC 60076. According to the example and equations (1)-(6), the variation of the hot-spot temperature based on one parameter changing are calculated, the relationship between the single parameter and hot-spot temperature are shown in table 1:

Table 1. Relationship between change of parameters and hot-spot temperature.

Parameter name	Relationship expression
Ambient temperature θ_a	$\Delta T = \Delta\theta_a$
Top-oil temperature rise $\Delta\theta_{or}$	$\Delta T = 1.1(d\Delta\theta_{or})$
Top-oil temperature rise at start $\Delta\theta_{oi}$	$\Delta T \approx 0$
Average time constant of oil τ_0	$\Delta T = 2.79e^{(-\Delta\tau_0/44.62)} - 2.764$
Ratio (R) of on-load loss to no-load losses at rated current	$\Delta T \approx 0$
Time constant of winding τ_ω	$\Delta T = 1.43 + 19.4197 * \sin(\pi(\Delta\tau_\omega - 14.68)/14.28*180 \div \pi)$
Exponent of oil x	$\Delta T = 21.12e^{(\Delta x/0.48)} - 20.954$
Exponent winding y	$\Delta T = 87.02e^{(\Delta y/1.52)} - 87$

Temperature gradient from hot-spot to top-oil at start $\Delta\theta_{hi}$	$\Delta T \approx 0$

Parameters like ambient temperature, top-oil temperature rise at start, ratio of on-load loss to no-load loss at rated current and top-oil temperature can be directly acquired. These parameters are necessary to obtain accurate thermal characteristic parameters. In equation (1), $\theta_a + \Delta\theta_o$ is the sum of environmental temperature and top-oil temperature rise, namely the top-oil temperature θ_o. In this way, we use the more accurately measured value θ_o and the part related to hot-spot and top-oil temperature in equations (2) and (5) are ignored. It is helpful to get more accurate oil thermal characteristic parameters.

5. The structure design of the neural network

The neural network consists of three layers which are the input layer, the hidden layer and the output layer. These three layers connect to each other but the nodes of the same layer don't connect. As shown in figure 1, the input of network is $X = [x_1, x_2, ..., x_n]^T$, the output of network is $Y = [y_1, y_2, ..., y_m]^T$. Supposing the number of learning samples is P and the desired output are respectively $d^1, d^2, ..., d^P$, through adjusting the weights of error correction, the purpose of neural network is to minimize the errors between y^P and d^P. To simplify the process of computation, we added the thresholds of each nodes into the weight vectors. That is to assume that: $\theta'_l = \omega'_{n_2 l}$, $\theta'_k = \omega'_{n_1 k}$, $\theta_j = \omega_{nj}$, $x'_{n_1} = x'_n = -1$.

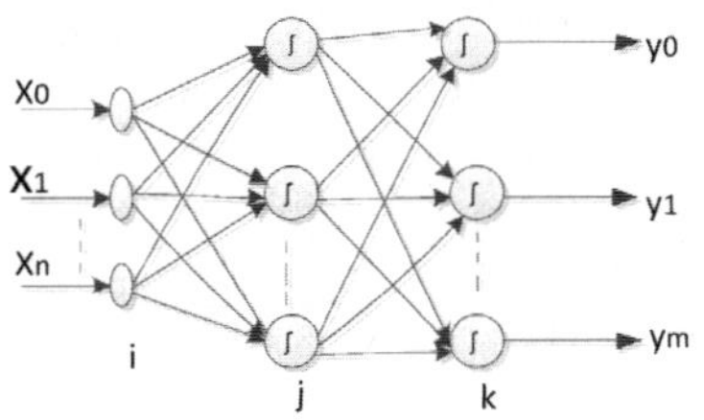

Figure 1. Topological graph of BP neural network.

The learning rule is derived from the principle of Least-Mean-Square (LMS). When a sample is input into the network and an output is generated, the mean-square-error is the sum of the square of each output unit error as shown in equation (10).

$$E^{(p)} = \frac{1}{2}\sum_{k=0}^{m-1}(d_k^{(p)} - y_k^{(p)})^2 \tag{10}$$

Then the total error is:

$$E_A = \sum_{p=1}^{P} E^{(p)} = \frac{1}{2}\sum_{p=1}^{P}\sum_{k=0}^{m-1}(d_k^{(p)} - y_k^{(p)})^2 \tag{11}$$

Assuming the ω_{sp} is a connecting weight, based on the gradient descent algorithm, the weight correction is:

$$\Delta\omega_{sp} = -\eta\frac{\partial E_A}{\partial \omega_{sp}} \tag{12}$$

Since the neural network adapt the steepest descent method which is based on the gradient, problems like that the low convergence speed and the existence of local minimum occurred. Improvements can be made by adapting Lenvenberg-Marquardt (L-M) method. This method is the combination of gradient descent method and gauss-newton method, the new weight adjustment equation is:

$$\omega(k+1)=\omega(k)-[J^T J+\mu I]^{-1} J^T e \tag{13}$$

Where, J means Jacobian matrix namely the network error' derivative of weight; e means error vector; I means unit matrix; μ means scalar quantity which is self-adaptive in the process of calculation. When μ is large, the equation is approximate to the gradient method; when μ is small, the equation gets close gauss-newton method [14].

According to Kolmogorov principle, any continuous function can be realized by a three layers neural network. So the three layers BP neural network is selected to predict the oil thermal characteristic parameters in this paper.

To use the examples given in the IEC guidance which only include six groups of data at six time points, we selected one hundred groups of data by changing the exponent of oil x and the average oil time constant τ_o in a reasonable range ($\pm 10\%$), and the six groups of data in guidance are not included in the one hundred groups of data. According to the equation $\theta_o=\theta_a+\Delta\theta_o$, and with the other parameters in guidance are adapted into the calculation, we obtained the hot-spot temperature, then the database includes 100 groups of data was obtained.

As the variables of input and output were determined by the real situation, we designated the ambient temperature θ_a, top-oil temperature rise at start $\Delta\theta_{oi}$, load factor K and top-oil temperature θ_o as variables of input; and the exponent of oil x and average oil time constant τ_o as variables of output. In order to turn the BP neural network, which is a static network into a memory system, the delay unit is used to save the state value of input at previous moment as the input of the next moment. The initial oil temperature of every time step is the calculated top-oil temperature of its previous moment, consequently the static network was turned into a dynamic network.

The selection of hidden layer directly matters the training of neural network. Thus many empirical equations are referenced to determine the nodes number of the hidden layer by using the heuristics [15]. The nodes values of hidden layer adapted in BP networks are shown in table 2.

$$l=\sqrt{n+m}+\alpha \tag{14}$$

Where, l means the number of hidden unit; n means the number of input unit, m means the number of output unit, α means a constant ranges between 1 to10.

Table 2. The node number of hidden layer of neural network.

Number	1	2	3	4	5	6	7	8	9	10
Node number of the hidden layer	4	5	6	7	8	9	10	11	12	13

The networks were trained for 10 times at each situation, at each time the node number of the hidden layer are different. The node number that corresponded to the minimum average training error was selected as the optimal node number, thus we determined the optimal node number was 10.

In order to make accurate predict, the sample data should be pre-processed. For the standard function, the range of input and output were both limited between [0,1], so the data were normalized as the following equation (15):

$$y=0.1+\frac{x_i-x_{\min}}{x_{\max}-x_{\min}}\times 0.9 \tag{15}$$

Where, y_i is the normalized input value, x_j is the actual input value, $x_{\max}$, $x_{\min}$ is the maximum and minimum of actual input. The output of neural network ranges between [0.1, 0.9], and the hot-spot temperature can be acquired after the inverse transformation of output.

These data is divided into two groups, one group was named by training set, which is used to train neural network and another group named test set is used to test the trained network. The training set includes 80 groups of data while the test set includes 20 groups of data. After the network being properly trained, the thermal characteristic parameters were calculated as shown equation (1), and based on the improved algorithm in reference [10], the comparison are done between the hot-spot temperature predicted and the temperature in guidance.

In order to make the prediction more accurate, the L-M algorithm was used to optimize the traditional BP neural network. The train target was 0.001, the train step was 50. The output layer adopted the logarithmic function, and the logarithmic function was also used in hidden layer to maintain the nonlinear characteristic of the hidden layer.

The results of neural network training are shown in figures 2 and 3. Figure 2 is the performance Figure which indicates the process of network training. The final error is only 0.00188, which shows that the good training result. Figure 3 is the regression figure which reflects the fitting ability of the network. Since the value of R is 0.99449, which is very close to one, it shows the network matches the sample well.

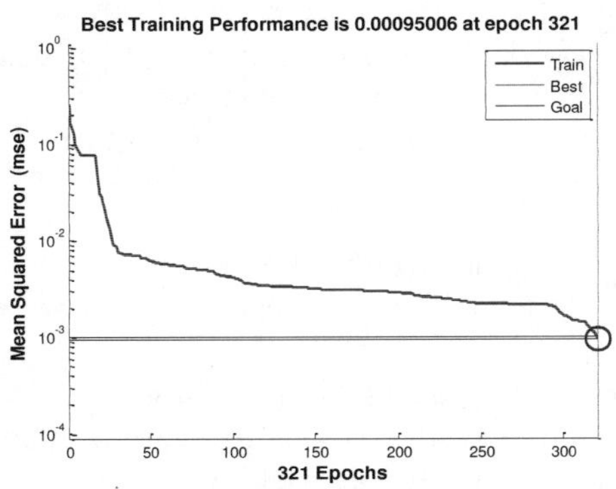

Figure 2. Performance of network.

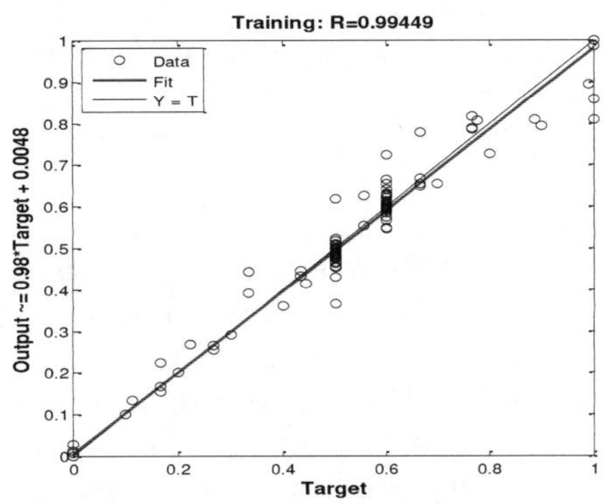

Figure 3. Regression of network.

Predicting the six groups of data given by guideline with the trained network, the results are shown in figure 4 and table 3:

Figure 4 shows the changes bring by the load and time's variation of the measured value and calculation values using the recommended algorithm of IEC60076 and the neural network algorithm. The value computed with network algorithm approximates better to the measured value than the value computed with algorithm of the guidance. Table 3 is the hot-spot temperature at the end of each load level; it also shows that the results of optimized neural network matched the measured results better.

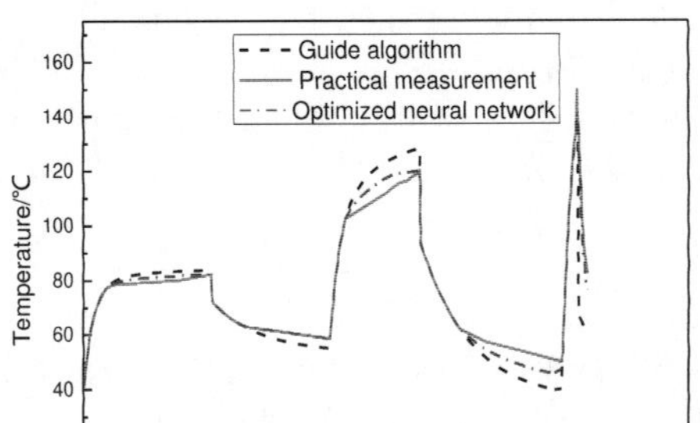

Figure 4. The temperature variation.

Table 3. The hot-spot temperature at the end of each load level.

Time (min)/Load factor	Hot-spot temperature/℃		
	Guide algorithm	Optimized neural network	Practical measurement
190/1.0	83.8	82.2187	82.2
365/0.6	54.9	58.5822	58.6
500/1.5	127.5	119.3922	119.2
710/0.3	39.5	46.2803	49.8
735/2.1	138.2	140.5979	140.7
750/0.0	59.5	75.5228	82.4

6. Conclusions

According to the heat mechanism of transformers, the influence of the oil thermal characteristics parameters of surroundings' factors and change of temperature are taken into consideration. Based on the improved temperature calculation algorithm recommended in guidance, and combined the optimized neural network with L-M algorithm， the temperature of hot-spot are calculated by adopting proper functions and parameters and training the neural network. Comparison between the calculation values using the recommended algorithm of IEC60076 and the neural network algorithm was made, it shows that the value computed by network algorithm approximates more to the measured value.

References

[1] Xie Y C 2003 *Power Transformer Handbook* (Beijing: China Machine Press)
[2] Peng H 2004 Computation of leakage magnetic field and tank eddy current loss of transformer (Beijing: Tsinghua University)
[3] Wang X Y 2004 *Transformer Fault and Monitoring* (Beijing: China Machine Press)
[4] Mao Y Z and Wang X C 2004 Analysis of fesibility and necessity to measure transformer winding temperature by winding thermometric device *Transformer* **41** 13-7
[5] Qian Z and Sun J D 2003 On-line monitoring of hot-spot temperature in transformer winding *High Voltage Engineering* **29** 26-8
[6] Chen S J and Wang S G 2000 Development and application of on-line monitoring device of transformer winding heat spot *Transformer* **37** 41-5
[7] IEEE 1995 IEEE Std C57.91-1995 IEEE guide for loading mineral-oil- immersed transformers
[8] IEC 2005 IEC60076-7 Loading guide for oil-immersed power transformers
[9] Lesieutre B C, *et al* 1997 An improved transformer top oil temperature model for use in an on-line monitoring and diagnostic system *IEEE Trans Power Delivery* **12** 249-56
[10] Susa D, *et al* 2005 Dynamic thermal modeling of power transformers *IEEE Trans Power Delivery* **20** 197-204
[11] Susa D and Nordman H 2009 A simple model for calculating transformer hot-spot temperature *IEEE Trans Power Delivery* **24** 1257-65
[12] Zeng H and Zhou X X 2000 Establish dynamic power transformer temperature model with intelligent learning methods *China Power International Conference* (Shanghai, China) pp 471-3
[13] Chen W G and Zhao T 2009 Improved method for calculation of hot-spot temperature in power transformer *High Voltage Application* **45** 53-6
[14] Li M L 2010 Studies on the prediction method of transformer winding hot-spot temperature based on BPNN (Chongqing: Chongqing University)
[15] Chen H Q 2003 Theory and application of genetic algorithm based on artificial neural network (Wuhan: Wuhan University of Technology)

Information entropy and its application in heat conduction

C Q Su[1], Y Wu[1,2], X Y Fan[1] and P Guo[1]

[1]College of Electromechanical Engineering, Qingdao University of Science and Technology, Qingdao 266061, China

E-mail: 403751612@qq.com

Abstract. In this paper, heat conduction problems are analyzed from the view of information entropy. The variation law of temperature with time in the lumped-heat-capacity system can be expressed in the form of exponential distribution. It is proved that exponential distribution is the solution of a conditional extreme value problem with information entropy as an objective function. The general analysis procedure of heat conduction problems from the view of information entropy is proposed. Fundamental solution of heat conduction equation can be written in the form of normal distribution, and normal distribution can be derived from the maximum entropy principle.

1. Introduction

The concept of information entropy was firstly put forward by the American electrical engineer Shannon C E [1] in 1948. It is the generalization of physics entropy (thermodynamic entropy) and they are both the description of randomness and uncertainty of systems [2,3]. Jaynes E T [4] proposed the maximum entropy principle to determine the probability distribution in 1957. In 1967, Burg J P put forward the maximum entropy spectral estimation in time series analysis based on the maximum entropy principle [5]. Li X S [6] adopted the maximum entropy principle in structural optimization design. In this paper, we will study the application of information entropy in heat conduction.

2. Lumped-heat-capacity system and information entropy

The information entropy can be calculated if the probability density is given. For one-dimensional random variable, suppose its probability density is $f(x)$, the information entropy is defined as [7]

$$S[f(x)] = -\int_{-\infty}^{+\infty} f(x) \ln f(x) dx \tag{1}$$

It can be seen from the above expression that information entropy is a functional of probability density.

According to [8], the analytical solution of temperature field based on lumped parameter method is

$$\Theta = \frac{\theta}{\theta_0} = \frac{t - t_\infty}{t_0 - t_\infty} = \exp(-\frac{hA}{\rho c V}\tau) = \exp(-\frac{\tau}{\tau_c}) \tag{2}$$

Where τ_c is called the time constant of the lumped-heat-capacity system.

Transient heat flow can be expressed as

$$\Phi = -\rho c V \frac{dt}{d\tau} = (t_0 - t_\infty) hA \exp(-\frac{\tau}{\tau_c}) \tag{3}$$

Let $(t_0 - t_\infty)hA = \Phi_0$, then dimensionless transient heat flow can be written

$$\frac{\Phi}{\Phi_0} = \exp(-\frac{hA}{\rho c V}\tau) = \exp(-\frac{\tau}{\tau_c}) \tag{4}$$

From formulas (2) and (4), we can see that dimensionless temperature difference and transient heat flow have the same variation law and they are both in exponential attenuation with time.

Divided by τ_c of both sides of formulas (2) and (4), we can get

$$f(\tau) = \frac{1}{\tau_c}\exp(-\frac{\tau}{\tau_c}), \tau > 0 \tag{5}$$

According to [9], $f(\tau)$ obeys exponential distribution with parameter τ_c, where time constant τ_c represents mathematical expectation.

The quantity of heat exchange can be obtained by integrating equation (3)

$$Q = \int_0^{\tau} \Phi d\tau = (t_0 - t_\infty)\rho c V[1 - \exp(-\frac{hA}{\rho c V}\tau)] \tag{6}$$

Let $(t_0 - t_\infty)\rho c V = Q_\infty$, then dimensionless quantity of heat exchange can be expressed as

$$\frac{Q}{Q_\infty} = 1 - \exp(-\frac{hA}{\rho c V}\tau) = 1 - \exp(-\frac{\tau}{\tau_c}) \tag{7}$$

Formula (7) is distribution function of exponential distribution with parameter τ_c.

From the above discussion, we can find the variation law of physical quantities in lumped-heat-capacity system is closely related with exponential distribution.

Now we prove that exponential distribution with parameter μ is the solution of the following conditional extreme value problem.

$$\begin{cases} \max\limits_{f(x)} S[f(x)] = -\int_0^{+\infty} f(x)\ln f(x)dx \\ s.t. \quad \int_0^{+\infty} f(x)dx = 1 \\ \qquad \int_0^{+\infty} xf(x)dx = \mu \end{cases} \tag{8}$$

Proof: Introduce the auxiliary function

$$H = -f(x)\ln f(x) + \lambda_1 f(x) + \lambda_2 xf(x) \tag{9}$$

Where λ_1 and λ_2 are the Lagrange multipliers.

Introduce the auxiliary functional

$$S^* = \int_0^{+\infty} \{-f(x)\ln f(x) + \lambda_1 f(x) + \lambda_2 xf(x)\}dx \tag{10}$$

By Euler-Lagrange Equation, we can get

$$-\ln f(x) - 1 + \lambda_1 + \lambda_2 x = 0 \tag{11}$$

So

$$f(x) = e^{\lambda_1 - 1 + \lambda_2 x} \tag{12}$$

Substituting formula (12) in limiting conditions, we can finally get

$$f(x) = \frac{1}{\mu} e^{x/\mu} \tag{13}$$

From the above discussion, we can get the general analysis procedure of heat conduction problems from the view of information entropy:

- Expressing physical quantities in the dimensionless form;
- Normalization of physical quantities, i.e., transforming physical quantities into a probability density function;
- Finding the conditional extreme value problem corresponding to probability density function with information entropy as an objective function.

It is easy to verify that exponential distribution is the solution of the following initial value problem of ODE.

$$\begin{cases} y' + \dfrac{1}{\mu} y = 0 \\ y(0) = \dfrac{1}{\mu} \end{cases} \tag{14}$$

That is to say, initial value problem (14) is equivalent to conditional extreme value problem (8).

According to [10,11], there is no classical variational principle corresponding to initial value problem (14) because of the appearance of the first-order derivative. The introduction of information entropy concept may bring some new ideas to solve this problem.

3. Fundamental solution of heat conduction equation and information entropy

The solution of the following definite solution problem is defined as fundamental solution of one-dimensional heat conduction equation

$$\begin{cases} u_t - a^2 u_{xx} = 0, & -\infty < x < +\infty, t > 0, \\ u(x,0) = \delta(x), & -\infty < x < +\infty \end{cases} \tag{15}$$

Where $\delta(x)$ is one-dimensional Dirac Delta function.

By Fourier transform method, fundamental solution of one-dimensional heat conduction equation can be expressed as

$$u(x,t) = \frac{1}{2a\sqrt{\pi t}} e^{-\frac{x^2}{4a^2 t}}, \quad -\infty < x < +\infty, t > 0 \tag{16}$$

Let $2a^2 t = \sigma^2$, then formula (16) can be rewritten as

$$f(x) = \frac{1}{\sqrt{2\pi}\sigma} e^{-\frac{x^2}{2\sigma^2}} \tag{17}$$

Formula (17) is one-dimensional normal distribution whose mathematical expectation is zero and variance is σ^2. Variance σ^2 is proportional to time t.

By variational method [12], we can prove that normal distribution (17) is the solution of the following conditional extreme value problem.

$$\begin{cases} \max\limits_{f(x)} S[f(x)] = -\int_{-\infty}^{+\infty} f(x) \ln f(x) dx \\ s.t. \quad \int_{-\infty}^{+\infty} f(x) dx = 1 \\ \qquad \int_{-\infty}^{+\infty} x^2 f(x) dx = \sigma^2 \end{cases} \tag{18}$$

Substituting formula (17) into equation (1), we have

$$S[f(x)] = \ln \sqrt{2\pi e \sigma^2} = \frac{1}{2} + \ln \sqrt{2\pi} + \ln \sqrt{\sigma^2} \tag{19}$$

Therefore, information entropy of normal distribution is proportional to its variance σ^2.

The above discussion can also be extended to multi-dimensional heat conduction problems.

4. Conclusion

- The variation law of temperature and transient heat flow in lumped-heat-capacity system can be expressed in the form of exponential distribution.
- The general analysis procedure of heat conduction problems from the view of information entropy is proposed.
- The fundamental solution of heat conduction equation can be written in the form of normal distribution, and normal distribution can be derived from the maximum entropy principle.

References

[1] Shannon C E 1978 *J Tech Bell Sys* **27** 379-433
[2] Xing X S 2009 *Sci China* **31** 77-84
[3] Zou B Y 2004 *Thermodynamics and Molecular Physics* (Beijing, China: Central China Normal University Press)
[4] Jaynes E T 1957 *The Physical Review* **106** 620-30
[5] Li P and Lu W X 2008 *Global Geology* **27** 63-7
[6] Li X S 1989 *Computational Structural Mechanics and Applications* **6** 36-46
[7] Jia S L 2001 *Fundamentals of Information Theory* (Heilongjiang, China: Harbin Institute of Technology Press)
[8] Yang S M and Tao W Q 2006 *Heat Transfer* (Beijing, China: Higher Education Press)
[9] Sheng Z and Xie S Q 2001 *Probability and Statistics* (Beijing, China: Higher Education Press)
[10] Cheng X G, Li Z X and Guo Z Y 2004 *J Eng Thermophys* **25** 457-9
[11] Liu G L 1999 *Acta Mech Sinica* **31** 165-71
[12] Lao D Z 2010 *Fundamentals of Variational Method* (Beijing, China: National Defence Industry Press)

Ore sintering flue gas desulfurization and its resoureilization by using pyrolusite

S E Li[1,2,5], Z G Long[3], F Z Wu[1,2], H Y Li[1,2], T M Cui[1,2] and X Z Zhou[4]

[1]College of Materials and Metallurgy, Guizhou University, Guiyang, 550003, China
[2]Guizhou Province Key Laboratory of Metallurgical Engineering and Process Energy Saving, Guizhou University, Guiyang, 550003, China
[3]School of Marxism, Guizhou University, Guiyang, 550003, China
[4]Library, Guizhou University of Finance and Economics, Guiyang, 550003, China

E-mail: 568144269@qq.com

Abstract. Flue gas desulfurization (FGD) has been implemented for sulfur dioxide gas emission reduction by pyrolusite in iron ore sintering. However, the mechanism of SO_2 reduction through FGD is still not fully clear. And in present work, the effects of operating conditions on desulfurization rate and Mn^{2+} leaching rate of pyrolusite were investigated. Six hours later, the desulfurization rate and Mn^{2+} leaching rate all can be higher than 70%. And a higher absorption temperature was good for desulfurization rate, while a middle temperature was good for Mn^{2+} leaching rate. A higher manganese ore granularity and SO_2 concentration were good for desulfurization rate and Mn^{2+} leaching rate. However, a higher liquid-solid rate was only good for desulfurization rate, but Mn^{2+} leaching rate. The results demonstrate that the pyrolusite is a kind of very promising adsorbent in industrial flue gas desulfurization application due to its low cost and good desulfurization capacity.

1. Introduction

China's western regions, such as Sichuan, Guangxi, Guizhou, Yunnan, Gansu, Ningxia, and Qinghai, have vast deposits of pyrolusite. A large number of pyrolusite is found in the area of Hunan and Guangdong. The characteristics of pyrolusite resources in our country are poor, miscellaneous, and fine. The proportion of poor Mn is large, which is about 93% of the country's total reserve. The direct industrial application of these low-grade Mn ore resources is difficult. In the production process, iron and steel enterprises consume large amounts of sulfur resources and energy. Meanwhile, SO_2 emissions from these industries have important effects on the environment. The use of pyrolusite leads to SO_2 absorption and by-product $MnSO_4$ and other products that are widely used and have a good market value. Therefore, the technique in China can not only effectively remove sulfur and control pollution but also can recover and result in resource optimization. Numerous studies have been conducted on flue gas desulfurization by using Mn [1-9], but only a few studies have been conducted on sintering flue gas desulfurization.

2. Materials and process

2.1. Materials

The composition and industrial analysis of the materials (pyrolusite, iron concentrate, and powdered coal) are given in tables 1 to 3.

2.1.1. Pyrolusite. Mn is obtained from manganese ore, which was ground to an average particle size of 0.075 mm after being broken. The ore was obtained from a mountain in Guizhou Province.

Table 1. Components of pyrolusite.

Ingredient	MnO_2	SiO_2	Al_2O_3	Fe	CaO	Others
Quality content of low-grade pyrolusite (%)	35.20	28.22	11.90	11.52	5.63	8.6

2.1.2. Iron ore. Iron ore is the concentrate used in sintering in a domestic steel company.

Table 2. Components of iron-ore fines.

Ingredient	TFe	SiO_2	H_2O	Al_2O_3	CaO	S	MgO
Quality content (%)	67.64	3.75	8.29	0.32	0.59	0.049	0.48

2.1.3. Powdered coal and coke.

Table 3. Proximate and ultimate analysis of powdered coke and coal.

Material	Quality content (wt.%)			Industrial analysis (wt.%)				
	Ash content	Moisture	Fixed carbon	S	C	H	N	O
Coke powder	9.59	11.5	77.1	0.63	77.1	4.31	0.8 0	4.86
Steam coal	20.86	7.66	59.35	0.54	59.38	3.88	0.56	7.12

2.2. Conditions and process
The experimental conditions are shown in table 4.

Table 4. Experimental conditions.

Order number	Project	parameter
1	Concentration of sulfur dioxide in flue gas	~3000 mg/m^3
2	Temperature (℃)	20
3	Stirring speed (r/mine)	300
4	Liquid-solid ratio (kg water/kg ore)	5:1
5	Average mineral powder particle size (mm)	0.075
6	Fume volume (m^3/h)	10
7	Ore grade (%)	Low grade, 35.20% MnO_2
8	Pulp total volume (L)	5

Pyrolusite slurry was used in flue gas desulfurization and to prepare the $MnSO_4$. Pyrolusite ore was crushed [10] into a specific particle size, and the pyrolusite slurry was based on a particular proportion with water. Flue gas flowed into the absorber after removing dust. The desulfurization and leaching of Mn^{2+} were performed in the absorber. The mother liquid was used to prepare the $MnSO_4$ by filtering, purifing, crystallizing, and drying.

3. Experimental

Flue gas desulfurization by pyrolusite is based on the reducibility of SO_2 and the oxidation of MnO_2.The effects of absorption temperature, liquid-solid ratio, particle size of pyrolusite, and SO_2 concentration on the desulfurization rate and Mn^{2+} leaching rate were determined.

4. Results and discussion

4.1. Absorption temperature

The temperature is 20℃, 30℃, 40℃, 50℃, and 60℃. The other conditions are the same as those in table 4. The effects of temperature on the desulfurization rate and Mn^{2+} leaching rate, as well as the results, are shown in figures 1 and 2.

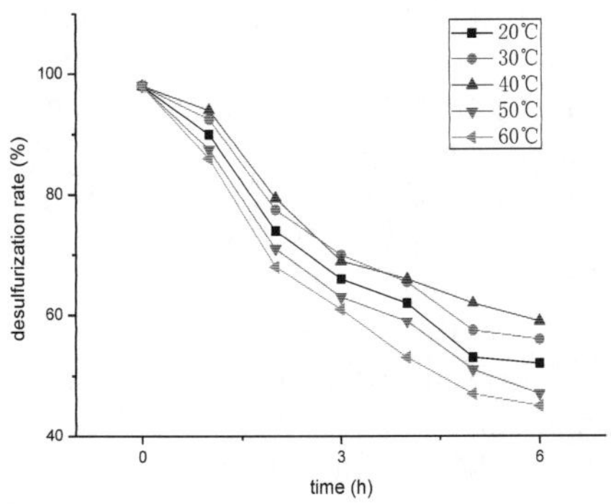

Figure 1. Temperature influence on desulfurization rate.

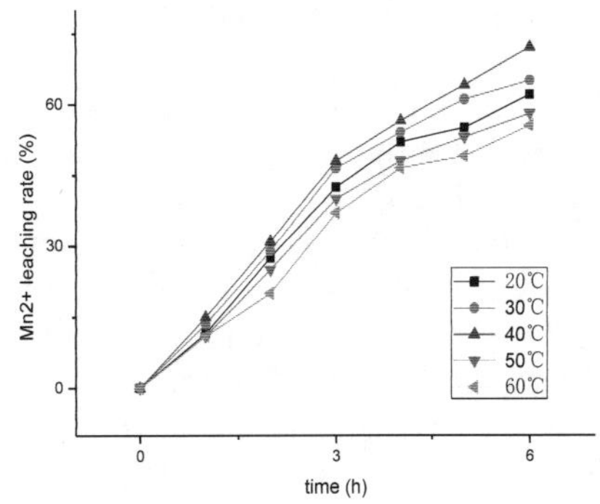

Figure 2. Temperature influence on the rate of leaching Mn^{2+}.

During the desulfurization by pyrolusite, the initial effect is almost unnoticeable in figure 1. When the temperature is 30℃~40℃, the desulfurization effect is improved, but the difference is not significant. With the decrease (e.g., at 20℃) or increase (e.g., at 50℃ and 60℃), the desulfurization rate decreases to a particular extent. This phenomenon is mainly due to the two aspects of temperature on flue gas desulfurization by pyrolusite. When the temperature increases, SO_2 solubility in the liquid phase makes it difficult to the reaction of MnO_2 and SO_2. On the other hand, the high temperature is favorable for desulfurization with reduced liquid viscosity, increasing diffusion coefficient, and increasing reaction rate constant. It shows that the temperature has a similar effect on both the desulfurization rate and Mn^{2+} leaching rate in figure 2.

4.2. Liquid-solid ratio

The liquid-solid ratio in pyrolusite pulp is the ratio of the quantity of water and mineral powder, which is expressed in kg water/kg mine. The liquid-solid ratio is an important parameter, which has an effect on the desulfurization rate, the leaching rate of Mn^{2+}, as well as the equipment investment costs and operating expenses ,and so on. The condition of the liquid-solid ratio was 3:1, 5:1, and 7:1. The effects of liquid-solid ratio at these three conditions on the desulfurization rate and Mn^{2+} leaching rate were investigated; other conditions were the same as those in table 4. The results are shown in figures 3 and 4.

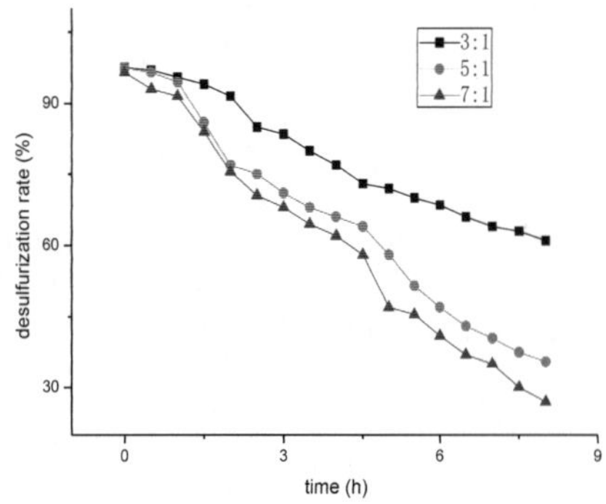

Figure 3 Rate of liquid quantity and solid quantity influence on desulfurization rate.

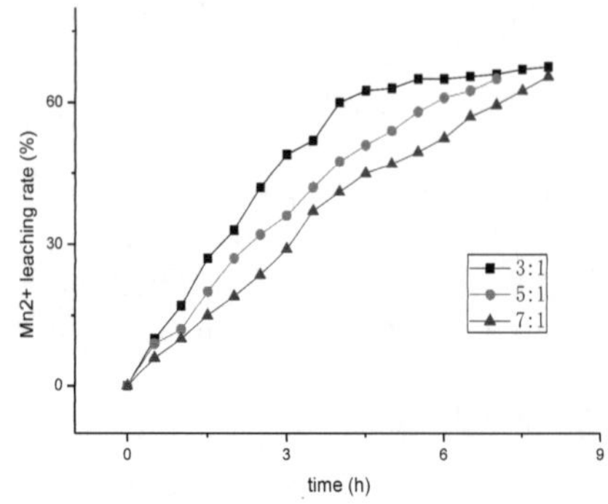

Figure 4 Rate of liquid quantity and solid quantity influence on the rate of leaching Mn^{2+}.

The desulfurization rate gap is significantly small in the initial stage in figure 3. The desulfurization rate decreases with increasing liquid-solid ratio of the pyrolusite system after 1 h. When the liquid-solid ratio is 3:1, the desulfurization rate is higher in a significantly longer period of time than those of 5:1 and 7:1. When the liquid-solid ratio is 5:1 and 7:1 and after 4.5 h, the desulfurization rate decreases observably. When the variation is mainly due to the low liquid-solid ratio, both the proportion of MnO_2/SO_2 in the liquid phase and the concentration of reactants increase.These are favorable to the absorption of SO_2. Meanwhile, the numerous particles are suspended in the liquid phase of packed column, which increase the contact area of gas-liquid. It is conducive for SO_2 to permeate through the liquid and solid phases. So, the absorption rate of SO_2 is improved.

The leaching rate of Mn^{2+} decreases when the liquid-solid ratio increases in the same reaction time in figure 4. This phenomenon is mainly due to the high liquid-solid ratio (7:1), in which the molar concentration of MnO_2 in the pulp is low, the proportion of MnO_2/SO_2 is small, and both the reaction rate of MnO_2 in the pulp and the depletion rate is slow. Thus, the leaching rate of Mn^{2+} decreases.

4.3. Manganese ore granularity

The average particle size of mineral powder was 0.30 mm, 0.15 mm, and 0.075 mm. The other conditions are the same as those in table 4. The effects of the particle size on the desulfurization rate and leaching rate of Mn^{2+} were studied. The results are shown in figures 5 and 6.

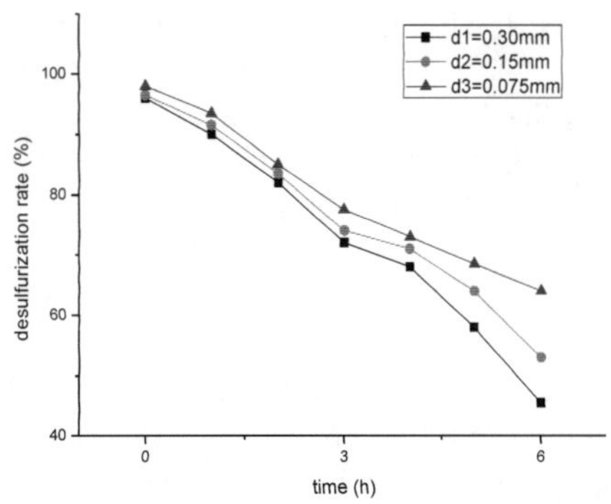

Figure 5. Granularity of manganese influence on desulfurization rate.

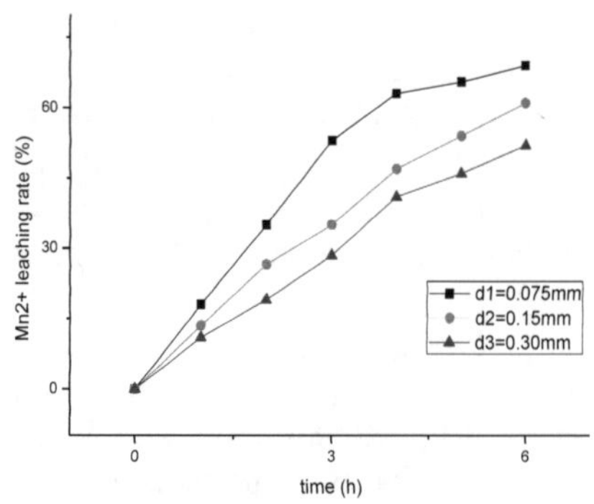

Figure 6. Granularity of manganese influence on the rate of leaching Mn^{2+}.

From kinetics, the particle size of pyrolusite has a relatively significant impact on the reaction rate.

Therefore, it is also an important factor that affects the desulfurization rate and leaching rate of Mn^{2+}. The desulfurization rate and leaching rate of Mn^{2+} increase with decreasing particle size in figures 5 and 6. When the concentration is constant, finer particles provide larger liquid-solid contact area. The large contact area is conducive to mass transfer, in which the internal diffusion resistance from the gas phase to the solid particles is reduced, thereby the reaction rate is accelerating. So, the finer the size of manganese ore is, the higher the rate of desulphurization and leaching of Mn^{2+} is.

4.4. SO_2 concentration

Experiments with the control samples of SO_2 concentration (3000, 4000, and 5000 mg/m^3) were conducted. The other conditions are the same as those in table 4. The effects of SO_2 concentration on the desulfurization rate and leaching rate of Mn^{2+} were investigated, and the results are shown in figures 7 and 8.

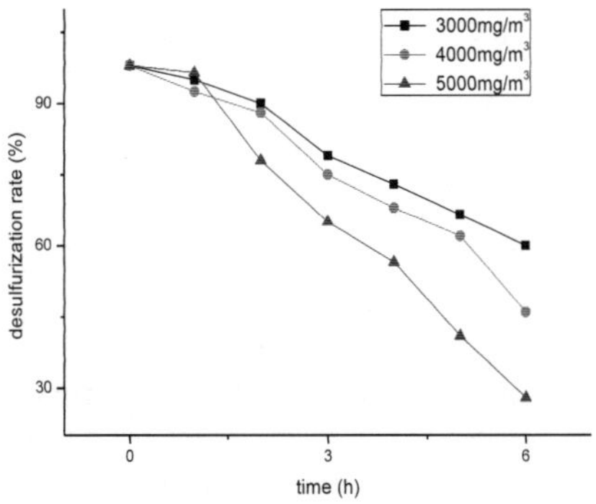

Figure 7. Inletconcentration influence on desulfurization rate.

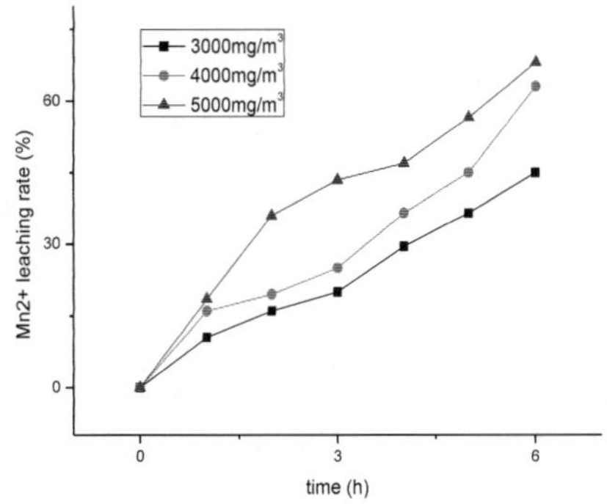

Figure 8. Inletconcentration influence on the rate of leaching Mn^{2+}.

In the initial reaction, the difference in desulfurization rates is low in figures 7 and 8. The flue gas desulfurization rate decreases with increasing SO_2 concentration for 1.5 h. While the SO_2 concentration increases, the gas-liquid mass transfer force also increases. Thus, the transmission rate of SO_2 to the liquid phase increases, Mn consumption is fast, the desulfurization rate rapidly decreases, and the leaching rate of Mn^{2+} rapidly increases. These phenomena are due to the continuous cycle of SO_2. The total amount of MnO_2 in the slurry is constant. At the same desulfurization rate, the higher SO_2 concentration needs more consumption of MnO_2. In the late reaction, while the quantity of MnO_2 decreases, both the pH value and the desulfurization rate significantly decreases. When the SO_2 concentration of the reaction tower is low, the Mn/S value in the liquid phase becomes relatively large and the consumption of MnO_2 becomes slower. Thus, the time for the same desulfurization rate is longer. Some MnO_2 particles exist in the later period, wherein the curve of desulfurization rate slightly declines. On the other hand, for the leaching rate of Mn^{2+}, a higher concentration of inlet SO_2 results in a faster consumption of MnO_2. Therefore, to achieve the same leaching rate of Mn^{2+} for SO_2 at a higher inlet concentration, the required time must be shortened.

This test does not limit the SO_2 content but deepens the understanding of the mechanism of desulfurization. The high concentration of SO_2 is advantageous for pyrolusite because the high SO_2 concentration increases mass transfer force, thereby increasing the amount of dissolved SO_2 in the liquid phase per unit time, both the reaction rate of MnO_2 and the leaching rate of Mn^{2+} increase. In addition, the purification and crystallization of the by-product of desulfurization has specific requirements on the concentration of $MnSO_4$. Therefore, the high SO_2 concentration may be appropriate in increasing the circulation amount and pH value of the slurry. Meanwhile, it is high for desulfurization rate and leaching rate of Mn^{2+} to reduce the liquid-solid ratio, increase the MnO_2 content in the pulp, and prolong the absorption time.

5. Conclusions

When the temperature decreases (e.g., 20℃) or increases (e.g., 50℃ and 60℃), the desulfurization rate decreases, and temperature has no significant effect on the desulfurization rate and leaching rate of Mn^{2+}. When the temperature is at 40℃, desulfurization and leaching rate of Mn^{2+} are slightly higher than those at 30℃. At the same reaction time, the desulfurization rate of pyrolusite slurry system decreases and the leaching rate of Mn^{2+} increases with increasing liquid–solid ratio. When the reaction proceeds within the same reaction time, the finer the size of manganese ore is, the higher the rate of desulphurization and leaching of Mn^{2+} is. The Desulfurization rate in the flue gas decreases with increasing inlet SO_2 concentration, whereas the leaching rate of Mn^{2+} rapidly increases.Six hours later, the desulfurization rate and Mn^{2+} leaching rate all can be higher than 70%.

Acknowledgments

This work was financially supported by the National Natural Science Foundation of China (No. 51574094) and the Guizhou Provincial Department of Education Items (Qian Jiao He KYword[2014]222).

References

[1] Wei S J, Wang S and Zhou A 2014 Research on present situation of desulfurization and denitrication technology for sintering flue gas Environ Eng **2** 95-7
[2] Chun T J, Long H M, Di Z X, Zhang X Y, Wu X J and Qian L X 2017 Novel technology of reducing SO_2 emission in the iron ore sintering Process Saf Environ Prot **1** 297-302
[3] Sun J, Ding S L and Su S J 2016 Reaction process of wet flue gas desulfurization with manganese carbonate ore slurry Journal of Chengdu University (Natural Science) **3** 296-9
[4] Dong J Q 2016 Study on operational parameters of wet flue gas desulfurization for sintering Science and Technology of Baotou Steel **2** 75-7
[5] Li S E, Wu F Z and Wang Q 2013 pH control of flue gas desulfurization in sintering process with pyrolusite Adv Mater Res **864-867** 1592-7
[6] Tian Y, Jing Y N and Sang Y 2014 Discussions on approach for comprehensive utilization of by-products for flue gas desulfurization of sintering Science and Technology of Baotou Steel **2** 72-5
[7] Zhang Y, Li S E, Wu F Z and Hu Y L 2013 The design of washing tower with liner for off－gas washing Guangzhou Chemical Industry **1** 1-2+7
[8] Zhao C L, Wu T and Bo X 2014 The status quo of sintering flue gas desulfurization of iron & steel industry and pollutants co-treatment suggestion Environ Eng **10** 76-78+103
[9] Wang S P 2013 Study on energy saving and environmental protection in iron ore sintering process (Wuhan: Wuhan University Of Technology)
[10] Wu F Z, Li S E and Jin H X 2014 Study on desulfurization of sintering flue gas with pyrolusite slurry Nonferrous Meta (Metallurgy) **4** 4-6

Study on optimum technological conditions of ore sintering flue gas desulfurization by using poor manganese

H Y Li[1,2], S E Li[1,2,5], Z G Long[3], F Z Wu[1,2], T M Cui[1,2] and X Z Zhou[4]

[1]College of Materials and Metallurgy, Guizhou University, Guiyang, 550003, China
[2]Guizhou Province Key Laboratory of Metallurgical Engineering and Process Energy Saving, Guizhou University, Guiyang, 550003, China
[3]School of Marxism, Guizhou University, Guiyang, 550003, China
[4]Library, Guizhou University of Finance and Economics, Guiyang, 550003, China

E-mail: 568144269@qq.com

Abstract. Orthogonal experiments were conducted to study the effect of each single factor on the desulfurization rate and leaching rate of Mn^{2+} to obtain improved process parameters. The results showed that the use of pyrolusite flue gas and the process method of by-product $MnSO_4$ can not only effectively remove the sulfur in the gas, thereby controlling environmental pollution, but can also recover sulfur.

1. Introduction

The amount of SO_2 produced by iron and steel industry in China is second only to that of the coal power industry. And SO_2 produced by sintering process accounts for more than 40% of iron and steel industry, while it is one of the main sources of air pollution [1-5]. Therefore, the research on how to develop a desulfurizing technique with less investment, less operating costs and resource regeneration, is very important [6-9]. It is certainly worth studying on optimum technological conditions of sintering flue gas desulfurization.

2. Materials

2.1. Materials

The composition and industrial analysis of the materials (pyrolusite, iron concentrate, and powdered coal).

- Mn is obtained from Mn ore, which was ground to an average particle size of 0.075 mm after breaking. The ore was obtained from a mountain in Guizhou province.
- Iron ore is the concentrate used in sintering in a domestic steel company.
- There is powdered coal and coke too.

2.2. Conditions and process

Pyrolusite slurry was used in flue gas desulfurization and to prepare the $MnSO_4$ products. Pyrolusite ore was crushed [10] into a specific particle size, and the pyrolusite slurry was prepared based on a particular proportion with water. Flue gas was applied to the absorber after dust removal. The desulfurization and leaching of Mn^{2+} were performed in the absorber. The mother liquor was obtained via filtration, purification, crystallization, and drying to prepare the $MnSO_4$ products.

3. Experiment

The absorption temperature, liquid-solid ratio, pyrolusite granularity, and SO_2 inlet concentration have specific effects on the desulfurization rate and leaching rate of Mn^{2+}. The first three factors have a direct impact on the desulfurization and leaching rate of Mn^{2+} in the pyrolusite pulp. The effect of SO_2 inlet concentration on the desulfurization and leaching rate of Mn^{2+} can be controlled by changing the pH value of the slurry, the slurry circulation volume, and the length of absorption time. Therefore, we focused on studying the effects of the interaction of various factors, namely, absorption temperature, liquid-solid ratio, and particle size of pyrolusite. To perform the analysis with fewer tests and to determine the significance of each factor, the orthogonal table test was performed. Each experiment was conducted for 6 h. The desulfurization rate in the table is the average desulfurization rate, and the leaching rate of Mn^{2+} is the total leaching rate at the end of the cycle.

4. Results and discussion

Three factors were included in the orthogonal experiment. Each factor takes three levels, and their interactions with one another were excluded. The orthogonal table L9 (34) was selected to arrange the orthogonal test for the study of the desulfurization and leaching rates of Mn^{2+}. The factors and levels selected in the test are shown in table 1, and the orthogonal experiment table is shown in table 2. Variance analysis is shown in table 3.

Table 1. Factors in the orthogonal experiment.

Level	Impact factor		
	A (Absorbent temperature,°C)	B (liquid-solid ratio; kg water: kg ore)	C (Average Mn ore granularity, mm)
1	40	3:1	0.30
2	60	5:1	0.15
3	80	7:1	0.075

Table 2. Results of the experiment on the desulfurization and leaching rates of Mn^{2+}.

Test number	Factor			Target	
	A	B	C	Average desulfurization rate η (%)	Total Mn leaching rate χ (%)
1	40	3:1	0.300	70.13	67.56
2	40	5:1	0.150	74.57	76.20
3	40	7:1	0.075	76.48	71.45
4	60	3:1	0.150	65.57	69.93
5	60	5:1	0.075	77.33	74.52
6	60	7:1	0.300	61.43	64.56
7	80	3:1	0.075	72.45	63.34
8	80	5:1	0.300	56.66	58.42

	9	80	7:1	0.150	61.38	73.35
η	I	221.18	208.15	188.22	T_η==616.00	
	II	204.33	208.56	201.52		
	III	190.49	199.29	226.26	$\bar{\eta}=\frac{1}{9}\sum_{i=1}^{9}\eta_i$ =68.44	
	$\overline{\text{I}}$	73.73	69.38	62.74		
	$\overline{\text{II}}$	68.11	69.52	67.17		
	$\overline{\text{III}}$	63.49	66.43	75.42		
	Range	10.23	3.09	12.68		
	I^2	48920.5943326.4235426.77				
	II^2	41750.7543497.2740610.31				
	III^2	36286.4439716.5051193.59				
χ	I	215.21	200.83	190.54	T_χ==619.33	
	II	209.01	209.14	219.48		
	III	195.07	209.36	209.31	$\bar{x}=\frac{1}{9}\sum_{i=1}^{9}x_i$ =68.81	
	$\overline{\text{I}}$	71.74	66.94	63.51		
	$\overline{\text{II}}$	69.67	69.71	67.17		
	$\overline{\text{III}}$	65.04	69.79	69.77		
	Range	6.70	2.85	6.26		
	I^2	46315.3440332.6936305.49				
	II^2	43685.1843739.5448171.47				
	III^2	38067.9143831.6143810.68				

Table 3. Validated results of experiment of the optimum conditions.

Test number	Reaction temperature(°C)	liquid-solid ratio(kg:/kg)	Mn ore granularity (mm)	Average desulfurization rate (%)	Total Mn leaching rate (%)
1	40	5□1	0.075	78.30	77.35
2	40	5□1	0.075	78.26	77.52
3	40	5□1	0.075	79.15	78.64
Average value	–			78.57	77.84

4.1. Effects of various factors on the indicators

For the average desulfurization rate, the range R of the data in the third column of the orthogonal experimental results is at maximum (table 2), followed by the first column and the second column. The orthogonal test describes that the effect on the desulfurization rate of pyrolusite system is at maximum when the process conditions of the particle size of Mn changes. By contrast, the effect on desulfurization rate is at minimum when the liquid-solid ratio changes. Thus, the primary and secondary orders of the factors that affect the desulfurization rate can be arranged as follows:

$$\text{primary} \xrightarrow{\text{C} \quad \text{A} \quad \text{B}} \text{secondary}$$

For the total leaching rate of Mn^{2+}, the range of the data of the third column in the orthogonal experimental results (table 2) is at maximum, followed by the first column and the second column. The orthogonal test describes that the effect on desulfurization rate of the pyrolusite system is at maximum when the process conditions of the particle size of Mn^{2+} changes. By contrast, the effect on desulfurization rate is at minimum when the reaction temperature changes. Thus, the primary and secondary orders of the factors that affect the total leaching rate of Mn^{2+} can be arranged as follows:

$$\text{primary} \xrightarrow{\text{A} \quad \text{C} \quad \text{B}} \text{secondary}$$

4.2. Analysis of the optimal reaction conditions

Comparisons among the values I, II, and III in table 2 show that for the average desulfurization rate, the reaction temperature level Al is better than A2 and A3, the liquid-solid ratio level B2 is slightly better than B1 and B3, and the granularity level of the Mn ore C3 is better than C1 and C2. Therefore, the optimal factor combination in this experiment is A1B2C3, that is, a reaction temperature of 40°C, a liquid-solid ratio of 5:1, and a particle size of Mn ore of 0.075 mm. For the total leaching rate of Mn^{2+}, the reaction temperature level A1 is slightly better than A2 and A3, the level of liquid-solid ratio B2 is better than B1 and B3, and the granularity level of Mn ore C3 is better than C1 and C2. Therefore, the optimal factor combination in this experiment is A1B2C3, which is the same with that for the desulfurization rate.

This group of tests does not appear in the orthogonal experiment. Additional experiments are needed for verification. The verification tests were repeated for three times, and the experimental results are shown in table 3.

The verification results of the optimum technological experiments show that the average desulfurization rate and the total leaching rate of Mn^{2+} are higher than the test results of each group in the orthogonal tests. Thus, the level of the derived $A_1B_2C_3$ combination guarantees higher desulfurization and leaching rates of Mn^{2+}.

Further analysis was conducted at the temperature of the slurry cycle (40°C) because the average outlet temperature of the flue gas is at ~140°C. Therefore, at 40°C, when pyrolusite was used to absorb SO_2, the slurry does not need additional heating. For wet desulfurization, the liquid-solid ratio is essentially the content indicator of the desulfurizing agent in a liquid phase. The desulfurizing agent in the pyrolusite desulfurization system is not only MnO_2 but also CaO, MgO, or other substances. Therefore, the selection of the liquid-solid ratio cannot guarantee a high desulfurization rate. And, process conditions and equipment choice should also be considered. The optimum liquid-solid ratio obtained from the orthogonal experimental results is 5:1, which was determined and based on the composition of low-grade pyrolusite. The economic and safe operation of the absorption tower and the recycling equipment were also considered. If the particle size of pyrolusite is too small, the ore reduction and grinding times both increase, thereby increasing power consumption. In addition, liquid-solid separation also becomes more difficult. Considering the grinding, solid-liquid separation, SO_2 absorption rate, and leaching rate of Mn^{2+}, the most appropriate average particle size of Mn is 0.075 mm.

5. Conclusion

The optimum process parameters, which were obtained from orthogonal and optimal experiments, for the flue gas desulfurization of pyrolusite slurry in the packed column are as follows: the temperature is at 40°C, liquid–solid ratio is 5:1, and the average particle size of Mn is 0.075 mm. In these conditions, higher desulfurization rate and leaching rate of Mn^{2+} can be achieved.

Acknowledgments
This work was financially supported by the National Natural Science Foundation of China (No. 51574094) and the Guizhou Provincial Department of Education Items (Qian Jiao He KYword[2014]222).

References

[1] Deng Z J 2015 Discussion on design parameters selection of sintering flue gas desulfurization *Resource Economizatio Environmental Protection* **3** 19
[2] Pan D, Hao W and Yang L 2017 Investigation on the relationship between the fine particle emission and crystallization characteristics of gypsum during wet flue gas desulfurization process *J Environ Sci* **5** 303-10
[3] Wu F Z, Li J Q, Cai J J, Jin H X and Chan X H 2009 Study on flue gas desulfurization of sintering in pilot-scale experiment *J Chin Inst Eng* **19** 65-71
[4] Wu F Z, Cai J J and Li J Q 2008 Flue gas desulphurization of iron and steel industry and circular economy *Industrial Furnace* **30** 8-12
[5] Dong J Q 2016 Study on operational parameters of wet flue gas desulfurization for sintering *Science & Technology of Baotou Steel* **2** 75-7
[6] Dang Y H, Qi Y H and Wang H F 2010 Technology of flue gas desulfurization *J Iron & Steel Res* **22** 1-6
[7] Yu Z Y, Fan X H, Gan M, *et al* 2016 Reaction behavior of SO_2 in the sintering process with flue gas recirculation *Journal of the Air & Waste Management Association* **66** 687
[8] Liu Z J, Zhang J L and Yang T J 2009 Research and development of sintering flue gas desulphurization technology *China Metallurgy* **19** 1-5
[9] Li S E, Li W, Pan FF, *et al* 2016 Enrichment and purification process of by-product manganese sulfate in the flue gas desulphurization *Guangzhou Chemical Industry* **1** 53-5
[10] Wu F Z, Li S E and Jin H X 2014 Study on desulfurization of sintering flue gas with pyrolusite slurry *Nonferrous Met (Metallurgy)* **4** 4-6

Volatility spillover between crude oil and exchange rate: A copula-CARR approach

Y J Pu[1] and M Y Guo[1,2]

[1]College of Management and Economics, Tianjin University, Tianjin 300072, China

E-mail: yingjian_pu@126.com

Abstract. Oil provides a powerful impetus for modern society's production and life. The influences of oil price fluctuations on socio-economic development are obvious, and it draws more attention from scholars. However, the distribution of oil is highly centralized, which leads to the vast majority of oil trading through foreign trade. As a result, exchange rate plays an important role in the oil business. Study on the relationship between exchange rate and crude oil gradually becomes a hot research topic in recent years. In this paper, we use copula and CARR model to study correlation structure and relationship between crude oil price and exchange rate. We establish CARR models as marginal models and use five copulas which are Gaussian Copula, Student-t Copula, Gumbel Copula, Clayton Copula and Frank Copula to study the correlation structure between NYMEX crude oil price range and U. S. Dollar Index range. Furthermore, we use Copula-CARR model with structural breaks to detect the change points in the correlation structure between NYMEX crude oil price range and U. S. Dollar Index range. Empirical results show that the change points are closely related to the actual economic events.

1. Introduction

20th century's several oil crises have crippled some countries' national economy, so government agencies and scholars focus on oil price's fluctuation. The world trade is one of the major channels of oil deals, while foreign exchange is also related to a country's national income. Thus, some scholars' researches focus on the oil prices and exchange rate's relationship and correlation structure.

In general, these empirical studies can be categorized into two types. One type discusses the relationship between the oil price and exchange rate. This type of analysis is typically based on VAR model, GARCH model or ECM model. For example, Ying Huang *et al* [1] analyze the impact of oil price fluctuations on the real exchange rate of RMB using four-dimensional VAR model. Adeniyi O, *et al* [2] construct GARCH model and EGARCH model to estimate the impact of oil price on the nominal exchange rate of daily data from January 2, 2009 to September 28, 2010. Oriavwote V E, *et al* [3] use the co-integration test and Granger causality test to study the relationship between the real oil price and the real exchange rate by GARCH model. Beckmann J, *et al* [4] construct Markov-switching vector error correction model (MS-VECM) model to analyze the relationship between oil price and the exchange rate. Chen H T, *et al* [5] investigate the impacts of oil price shocks on the bilateral exchange rate of the U.S. dollar against currencies in 16 OECD countries. Chou K W, *et al* [6] use the asymmetric autoregressive distributed lag model to evaluate the influence of oil price and exchange rate fluctuations on retail gasoline price in Taiwan. Yang L, *et al* [7] study the co-movement between the crude oil price and the exchange rate markets by studying their dynamics in the time and frequency

IOP Conf. Series: Earth and Environmental Science **93** (2017) 012034 doi:10.1088/1755-1315/93/1/012034

domain. These studies demonstrate the relationship between the oil price and the exchange rate in the empirical.

The other type studies the correlation structure between oil prices and exchange rate by copula which is proposed by Sklar A [8]. Copula can describe variables' relationship and correlation structures. Rémillard B, *et al* [9] analyze the relationship between Canadian / US dollar exchange rate and oil futures' price with Elliptic copulas and Archimedean copulas. Reboredo J C [10] uses copula-ARMA model to study correlation structure and relationship between oil price and exchange rate. Alouia R *et al* [11] apply a vine copula approach to investigate the dynamic relationship between energy, stock and currency markets.

It is important to note that the Copula model has change points in some cases. Some scholars have studied on copula models with structural breaks. For example, Dias A and Embrechts P [12,13] use the method put forward by Gombay E and Horváth L [14] to study the changes in the German mark and Japanese yen's relation structure using Copula models with structural breaks. This method can detect the change in parameters without assuming possible date of the change point, and it can be used in risk management, asset pricing and asset allocation. Cai X, *et al* [15] use change point detection method to describe the dependence of English Pound and Eurodollar by Mixed Gumbel Copula and generalized Pareto distribution. Boubaker H, *et al* [16] use three Archimedean copulas (Gumbel copula, Clayton copula and Frank copula) to investigate the dependence structure between daily oil price changes and stock market returns in six GCC countries (Bahrain, Kuwait, Oman, Qatar, Saudi Arabia and United Arab Emirates). Zhu X Q, *et al* [17] propose a multiple change point detection approach from the perspective of risk dependence by using copula function, and detect the start point, end point and peak period at the subprime crisis in American banking. Therefore, we should pay attention to the structural change point when we use copula function to study the relationship between oil price and exchange rate.

To choose a right marginal distribution model is very important in Copula method. In general, most literature used ARCH models, GARCH models and SV models as the marginal distribution model. Conditional Autoregressive Price range Model (CARR model henceforth) is proposed by Ray Yeutien Chou [18] which use daily price range to model and forecast variables' volatility. In terms of volatility forecasting, CARR model can be used as the preferred option because it covers more information than GARCH model. CARR models have been widely used in the research of financial volatility. For example, Cai Y J, *et al* [19] apply DSTCC- CARR model to study the fluctuation in the stock market and CPI. Chou R Y, *et al* [20] combine CARR with a DCC structure and conclude that the price range-based volatility models have more significant economic value than the return-based volatility models. Sin C Y [21] uses CARRX model to study the factors affecting the volatilities of Asian equity markets. These empirical results show that it is feasible to apply the CARR model to the study oil price and exchange rate's volatility.

Previous studies have shown that copula function has some advantages in depicting related relation and corresponding structure, and CARR model is a better choice in volatility modeling. Therefore, this paper uses the Copula-CARR model to analyze the correlation and dependency structure of NYMEX crude oil price range and U. S. Dollar Index range, and the optimal copula model is used to detect the optimal Copula model. In addition, this paper discusses the changes in the dependency structure of NYMEX crude oil price range and U. S. Dollar Index range.

This paper proceeds as follows: Section 2 and Section 3 present a briefly introduction of CARR model and Copula. Section 4 presents a change point detectection method. We use Copula-CARR model to analyze New York Mercantile Exchange (NYMEX henceforth) crude oil price range and U. S. Dollar Index range's correlation structure and detect change points for the optimal Copula-CARR model in Section 5. Finally, we conclude in Section 6.

2. CARR model

CARR model is proposed by Chou (2005), which is used to characterize the dynamic structure of asset extremum within a fixed time period. This model is similar to GARCH model.

The basic form of CARR (p, q) is:

$$\begin{aligned} &R_t = \lambda_t \varepsilon_t \\ &\lambda_t = \omega + \sum_{i=1}^{p} \alpha_i R_{t-i} + \sum_{j=1}^{q} \beta_j \lambda_{t-j} \\ &\varepsilon_t \sim iidf(\bullet) \\ &R_t = \ln P_t^{\max} - \ln P_t^{\min} \end{aligned} \tag{1}$$

Where, P_t represents the price of stock observed at time t, R_t is the difference of maximum and minimum logarithm P_t; λ_t is the conditional mean of price range based on all information up to time t, and $\lambda_t = E(R_t / I_{t-1})$, $\lambda_t \geq 0$; The distribution of ε_t, or the normalized price range $\varepsilon_t = R_t / \lambda_t$ is assumed to be distributed with a density function $f(\bullet)$ with a unit mean; ω represents the initial level of the price range, and also represents the uncertainty of price range; α_i is the coefficient of price range, represents the short-term effects of conditional mean; β_j is the coefficient of conditional mean, represents the long-term effects of conditional mean. For the process to be stationary, a condition is that the characteristic roots of the polynomial are outside the unit circle, or $\sum_{i=1}^{p} \alpha_i + \sum_{j=1}^{q} \beta_j < 1$.

When we assume ε_t follows exponential distribution with unit mean, such a model is called the ECARR model. The density functions are respectively:

$$f(R_t \mid \lambda_t) = \frac{1}{\lambda_t} \exp\left\{-\frac{R_t}{\lambda_t}\right\} \tag{2}$$

When we assume $\varepsilon_t \sim Gamma(\theta, 1/\theta)$, such a model is called the WCARR model. The density functions are respectively:

$$f(R_t \mid \lambda_t, \theta) = \theta \left[\Gamma\left(1 + \frac{1}{\theta}\right)\right]^{\theta} \frac{R_t^{\theta-1}}{\lambda_t^{\theta}} \exp\left\{-\left[\Gamma\left(1 + \frac{1}{\theta}\right) \frac{R_t}{\lambda_t}\right]^{\theta}\right\} \tag{3}$$

When we assume $\varepsilon_t \sim Gamma(\kappa)$, such a model is called the GCARR model. The density functions are respectively:

$$f(R_t \mid \lambda_t, \theta, \kappa) = \frac{1}{\theta_t^{\kappa} \Gamma(\kappa)} R_t^{\kappa-1} \exp\left\{-\frac{R_t}{\lambda_t}\right\} \tag{4}$$

3. Copula

Copula can be seen as a connection function of marginal distributions and joint distributions. If $F(x_1, \cdots, x_n, \cdots, x_N)$ is a joint distribution function with marginal distribution functions $F_1(\cdot), \cdots, F_N(\cdot)$, then there exists a copula C such that:

$$F(x_1, \cdots, x_n, \cdots, x_N) = C(F_1(x_1), \cdots, F_n(x_n), \cdots, F_N(x_N)) \tag{5}$$

The function $F(x_1,\cdots,x_n,\cdots,x_N)$ which is defined by equation (5) is joint distribution function of marginal distribution functions $F_1(\cdot),\cdots,F_N(\cdot)$.

3.1. Elliptic copulas

In general, Elliptic Copulas include Gaussian copula and Student-t copula.

3.1.1. Gaussian copula. Nelsen R B [22] defines Gaussian copula, which distribution function and density function are as follows:

$$C(u_t,...,u_n,..u_N;\rho)=\Phi_\rho\left(\Phi^{-1}(u_t),...,\Phi^{-1}(u_n),...,\Phi^{-1}(u_N)\right) \tag{6}$$

$$c(u_1,\cdots,u_n,\cdots,u_N;\rho)=|\rho|^{-\frac{1}{2}}\exp\left(-\frac{1}{2}\varsigma^T\left(\rho^{-1}-\mathrm{I}\right)\varsigma\right) \tag{7}$$

Where, ρ means symmetric positive definite matrix; Φ_ρ means a multivariate normal distribution with correlation matrix ρ; $\Phi^{-1}(\cdot)$ means the inverse function of normal distribution; $\varsigma_n=\Phi^{-1}(u_n)$, I means a unit matrix.

3.1.2. Student-t Copula. Nelsen defines Student-t copula, of which distribution function and density function are as follows:

$$C(u_1,\cdots,u_n,\cdots,u_N;\rho,v)=T_{\rho,v}\left(t_v^{-1}(u_1),\cdots,t_v^{-1}(u_n),\cdots,t_v^{-1}(u_N)\right) \tag{8}$$

$$c(u_1,\cdots,u_n,\cdots,u_N;\rho,v)=|\rho|^{-\frac{1}{2}}\frac{\Gamma\left(\frac{v+N}{2}\right)\left[\Gamma\left(\frac{v}{2}\right)\right]^N}{\left[\Gamma\left(\frac{v+1}{2}\right)\right]^N\Gamma\left(\frac{v}{2}\right)}\frac{\left(1+\frac{1}{v}\varsigma^{-1}\rho^{-1}\varsigma\right)^{-\frac{v+N}{2}}}{\prod_{n=1}^{N}\left(1+\frac{\varsigma_n^2}{v}\right)^{-\frac{v+1}{2}}} \tag{9}$$

Where, ρ means symmetric positive definite matrix; $T_{\rho,v}$ means a multivariate standard Student's t distribution with v degrees of freedom and correlation matrix ρ; $t_v^{-1}(\cdot)$ means the inverse function of univariate Student's t distribution; $\varsigma_n=t_v^{-1}(u_n)$.

3.2. Archimedean copulas

In general, Archimedean Copulas include Gumbel copula, Clayton copula and Frank copula.

3.2.1. Gumbel copula. Gumbel Copula's distribution function and density function are as follows:

$$C_G(u,v;\alpha)=\exp\left\{-\left[(-\log u)^{\frac{1}{\alpha}}+(-\log v)^{\frac{1}{\alpha}}\right]^\alpha\right\} \tag{10}$$

$$c_G=\frac{C_G(u,v;\alpha)(\log u\cdot\log v)^{\frac{1}{\alpha}-1}}{uv\left[(-\log u)^{\frac{1}{\alpha}}+(-\log v)^{\frac{1}{\alpha}}\right]^{2-\alpha}}\left\{\left[(-\log u)^{\frac{1}{\alpha}}+(-\log v)^{\frac{1}{\alpha}}\right]^\alpha+\frac{1}{\alpha}-1\right\} \tag{11}$$

Where, $\alpha\in(0,1]$, relevant parameter.

3.2.2. Clayton copula. Clayton Copula's distribution function and density function are as follows:

$$C_C(u,v;\theta)=\left(u^{-\theta}+v^{-\theta}-1\right)^{1/\theta} \tag{12}$$

$$c_C(u,v;\theta)=(1+\theta)(uv)^{-\theta-1}\left(u^{-\theta}+v^{-\theta}-1\right)^{-2-1/\theta} \tag{13}$$

Where, $\theta\in(0,\infty)$, relevant parameter.

3.2.3. Frank copula. Frank Copula's distribution function and density function are as follows:

$$C_F(u,v;\lambda)=-\frac{1}{\lambda}\log\left[1-\frac{\left(1-e^{-\lambda u}\right)\left(1-e^{-\lambda v}\right)}{1-e^{-\lambda}}\right] \tag{14}$$

$$c_F(u,v;\lambda)=\frac{\lambda\left(1-e^{-\lambda}\right)e^{-\lambda(u+v)}}{\left[\left(1-e^{-\lambda}\right)-\left(1-e^{-\lambda u}\right)\left(1-e^{-\lambda v}\right)\right]^2} \tag{15}$$

Where, $\lambda\neq 0$, relevant parameter; $\lambda>0$, random variables u and v has a positive correlation; $\lambda\to 0$, u and v tend to be independent; $\lambda<0$, u and v has a negative correlation.

All kinds of copulas have their own advantages to describe correlation between variables. Density function of bivariate Gaussian copula has symmetry which can capture the correlation change between variables in central, not in tail. If variables are asymptotically independent in the tail, Gaussian copula can't catch variable's "fat tail" feature. Compared to bivariate Gaussian Copula, t-Copula has thick tail, which means it can better describe tail dependence between variables. Density function of Gumbel copula has asymmetry, and its density distribution is "J" type. Gumbel copula is more sensitive to the change in upper tails, which can be used to describe upper tail dependence between variables. Clayton copula also has asymmetry, but in contrast to Gumbel copula, its density distribution is "J" type that more sensitive to the change in lower tail and can be used to describe lower tail dependence between variables. Frank copula's density distribution is "U" type, which means it is symmetrical. In contrast to Clayton copula and Gumbel copula, Frank copula can describe negative correlation between variables. Compared to Gaussian copula, Frank copula is more emphasis on tail dependence, and its upper tail dependence and lower tail dependence's growth is symmetry.

4. Diagnosis of structural breaks of Copula model: maximum likelihood function method [12]

Maximum likelihood function method can be used to detect structural breaks of copula. The maximum likelihood function method regards change point as a parameter, by constructing statistics to find the maximum value of likelihood function and estimate value of the change point.

Let (X_t,Y_t)'s joint distribution function is $C(x,y;\theta,\eta)$.

Where, θ and η is Copula's parameters and their marginal distribution functions are $F(x)$ and $G(y)$. We detect change points of θ, whereas η will be regarded as nuisance parameter. To detect whether the correlation structure between the variables in the time of $k=k^*$ has changed significantly, we can use the following test:

The null hypothesis:

$$H_0:\theta_1=\theta_2=\cdots=\theta_n \text{ and } \eta_1=\eta_2=\cdots=\eta_n \tag{16}$$

The alternative hypothesis:

$$H_1:\theta_1=\theta_2=\cdots=\theta_{k^*}\neq\theta_{k^*+1}=\cdots=\theta_n \text{ and } \eta_1=\eta_2=\cdots=\eta_n \tag{17}$$

k^* is the location or the time of the single change point. All the parameters are supposed to be unknown under both hypotheses. Assuming that $k = k^*$ is known, we need to test whether the two samples come from the same population. If the value of Λ_k is significant, the null hypothesis will be rejected.

$$\Lambda_k = \frac{\sup \prod_{1 \le i \le n} c(u_i, v_i; \theta, \eta)}{\sup \prod_{1 \le i \le k} c(u_i, v_i; \theta, \eta) \prod_{k \le i \le n} c(u_i, v_i; \theta', \eta)} \tag{18}$$

Where, c is the density function of C, and

$$c(u_i, v_i; \theta, \eta) = \frac{\partial^2 C(u_i, v_i; \theta, \eta)}{\partial u_i \partial v_i} \tag{19}$$

Λ_k can be gotten by maximum likelihood estimates.

If we denote

$$\begin{aligned} L_k(\theta, \eta) &= \sum_{1 \le j \le k} \log c(u_i, v_i; \theta, \eta) \\ L_k^*(\theta, \eta) &= \sum_{k \le j \le n} \log c(u_i, v_i; \theta, \eta) \end{aligned} \tag{20}$$

For each $k(1 \le k \le n)$, there is an unique corresponding - $\hat{\theta}_k$. The likelihood ratio can be written as:

$$-2\log(\Lambda_k) = 2[L_k\left(\hat{\theta}_k, \hat{\eta}_k\right) + L_k^*\left(\hat{\theta}_k^*, \hat{\eta}_k\right) - L_T\left(\hat{\theta}_T, \hat{\eta}_T\right)] \tag{21}$$

If Z_n's value is greater than the critical value of 9, the null hypothesis will be rejected. It means there exists a structural break.

$$Z_n = \max_{1 \le k \le n} (-2\log(\Lambda_k)) \tag{22}$$

5. Empirical research

5.1. Data and summary statistics

Light, sweet crude futures on the NYMEX is often used as a major international pricing benchmark because of its good trading liquidity and price transparency. U. S. Dollar Index (USDX henceforth) is a measure of the value of the U. S. dollar relative to majority of its most significant trading partners, which can comprehensively reflect the exchange rate of the dollar in the international foreign exchange market. Therefore, we choose NYMEX Light Crude (CL) and USDX to represent crude oil

Table 1. Descriptive statistics for price ranges of NYMEX crude oil and USDX.

	Mean	Standard deviation	Skewness	Kurtosis	Jarque-Bera	ADF	
						Test Value	P Value
NYMEX crude oil price	2.3682	1.4560	1.1329	5.4991	965.3706	-6.0276	0.0000
USDX	0.7607	0.4006	1.7821	9.3249	4471.2950	-4.9318	0.0000

prices and exchange rates. We use daily data from the period 2005/5/31 to 2014/5/30.

According to equation (1), we calculate price ranges of NYMEX crude oil and USDX. Summary statistics of the price range series and ADF test are presented in table 1.

All ADF test statistics reject the hypothesis of a unit root at the 1% confidence level, which means that they are stationary time series and we can model them directly.

5.2. Analysis based on copula-CARR models

5.2.1. Analysis based on CARR models. Previous empirical study indicates that it is not easy to determine the orders of the model. As a result, researchers mostly used low orders' model, such as GARCH (1,1) model, GARCH (2,1) model and GARCH (1,2) model, etc. Based on the order selection experience of GARCH model, in this paper, we construct CARR models with 2 orders for the price range of NYMEX crude oil and the price range of USDX. We assume the random error ratio term is respectively assumed to follow standard exponential distribution, standard Weibull distribution and standardized generalized Gamma distribution, respectively denoted by ECARR, WCARR, GCARR.

Assuming that the residual items respectively follow standard exponential distribution, standard Weibull distribution and standardized generalized Gamma distribution, we construct CARR model for the price range of NYMEX crude oil and the price range of USDX.

CARR (p, q) model for the price range of NYMEX crude oil is as follows:

$$\begin{aligned} R_{Nt} &= \lambda_{Nt}\varepsilon_{Nt} \\ \lambda_{Nt} &= \omega + \sum_{i=1}^{p}\alpha_i R_{Nt-i} + \sum_{j=1}^{q}\beta_j\lambda_{Nt-j} \\ \varepsilon_{Nt} &\sim iidf(\bullet) \\ R_{Nt} &= \ln P_{Nt}^{\max} - \ln P_{Nt}^{\min} \end{aligned} \tag{23}$$

Where, R_{Nt} represents price range of NYMEX crude oil; $P_{Nt}^{\max}$ and $P_{Nt}^{\min}$ represents NYMEX crude oil's daily maximum price and daily minimum price; λ_{Nt} is the conditional mean of NYMEX crude oil's price range based on all information up to time t; ε_{Nt} is residual item.

According to p-values (significance at the 5% level) in table 2, all the parameters in WCARR (1, 1), GCARR (1, 1) and GCARR (2, 2) are significant. As a result, we need to use Akaike information criterion (AIC henceforth) to select the optimal model. Akaike H [23] proposed the Akaike information criterion to measure the statistical model's goodness of fit, which is built on the basis of the concept of entropy. Statistical models often appear over-fitting, which means use the extra parameter to make the model has higher goodness of fit. AIC is an evaluation standard to choose the optimal model in alternative models which can best explain the data and with the minimum free parameter. In the multiple alternative models, the minimum AIC value model is the best choice.

AIC is defined as follows:

$$\begin{aligned} AIC = &-2\ln(\text{Maximum likelihood function value of the model}) \\ &+2(\text{The number of parameters of the model}) \end{aligned} \tag{24}$$

Table 2. CARR models using daily price range of NYMEX crude oil, T=2036.

Model	Parameters							
	ω	α_1	α_2	β_1	β_2	θ	κ	Log-likelihood
ECARR(1, 1)	0.0179	0.0832		0.9091				
	[0.7074]	[3.3378]		[30.6324]				-3666.2423

	-0.4793	-0.0008		0.0000				
ECARR(1, 2)	0.0207	0.0983		0.6483	0.2445			
	[0.6618]	[1.9335]		[0.9405]	[0.3800]			-3661.4211
	-0.5081	-0.0532		-0.3470	-0.7041			
ECARR(2, 1)	0.0157	0.1036	-0.0283	0.9179				
	[0.6849]	[1.7897]	[-0.4705]	[31.414]				-3661.3747
	-0.4934	-0.0735	-0.6380	0.0000				
ECARR(2, 2)	0.0368	0.0961	0.0684	-0.0653	0.8851			
	[0.7395]	[3.0251]	[2.2266]	[-1.2219]	[17.6205]			-3661.6390
	-0.4596	-0.0025	-0.0260	-0.2217	0.0000			
WCARR(1, 1)	1.6057	0.3050		0.0449		1.9444		
	[36.8627]	[15.0821]		[1.9379]		[76.2278]		-3114.0022
	0.0000	0.0000		-0.0526		0.0000		
WCARR(1, 2)	1.2064	0.2944		-0.0627	0.2844	1.9665		
	[30.3869]	[15.8392]		[-2.5195]	[10.2929]	[73.3458]		-3085.2257
	0.0000	0.0000		-0.0118	0.0000	0.0000		
WCARR(2, 1)	1.4221	0.2215	0.2526	-0.0509		2.0264		
	[31.7628]	[11.9713]	[14.8120]	[-1.9676]		[73.9352]		-3035.8639
	0.0000	0.0000	0.0000	-0.0491		0.0000		
WCARR(2, 2)	1.2599	0.2365	0.2088	-0.1207	0.1648	2.0265		
	[27.8028]	[12.7464]	[10.7632]	[-4.9549]	[5.2210]	[74.0252]		-3029.3285
	0.0000	0.0000	0.0000	0.0000	0.0000	0.0000		
GCARR(1, 1)	1.4884	0.3252		0.0765		1.4700	1.6806	
	[30.4592]	[13.2419]		[2.6685]		[23.5340]	[14.4002]	-3097.9737
	0.0000	0.0000		-0.0076		0.0000	0.0000	
GCARR(1, 2)	1.0345	0.3075		-0.0445	0.3271	1.4375	1.7919	
	[24.8558]	[13.8125]		[-1.4211]	[9.5506]	[21.9289]	[13.6415]	-3064.5605
	0.0000	0.0000		-0.1553	0.0000	0.0000	0.0000	
GCARR(2, 1)	1.3325	0.2350	0.2576	-0.0281		1.5638	1.6127	
	[26.8919]	[10.7259]	[12.5846]	[-0.8679]		[22.3464]	[14.1013]	-3022.9812
	0.0000	0.0000	0.0000	0.0000		0.0000	0.0000	
GCARR(2, 2)	1.0806	0.2553	0.2000	-0.1363	0.2486	1.5030	1.7403	
	[22.0897]	[11.8241]	[9.3941]	[-4.3541]	[6.4125]	[22.2672]	[13.9825]	-3011.0769
	0.0000	0.0000	0.0000	0.0000	0.0000	0.0000	0.0000	

Note: [] is t statistic; () is the p-values.

Table 3. The alternative models' AIC value of NYMEX crude oil' CARR model.

Model	AIC Value
WCARRX(1, 1)	6066.6569
GCARRX(1, 1)	6199.9475
GCARRX(2, 2)	6030.1538

From table 3, GCARR (2, 2) model with the smallest AIC value is an optimal CARR model for price range of NYMEX crude oil.

$$R_{Nt} = \lambda_{Nt}\varepsilon_{Nt}$$
$$\lambda_{Nt} = 1.0806 + 0.2553\, R_{Nt-1} + 0.2000\, R_{Nt-2} - 0.1363\, \lambda_{Nt-1} + 0.2486\, \lambda_{Nt-2} \qquad (25)$$
$$\varepsilon_{Nt} \sim Gamma(1.7403)$$

Where, R_{Nt} represents price range of NYMEX crude oil; λ_{Nt} is the conditional mean of NYMEX crude oil's price range based on all information up to time t; ε_{Nt} is residual item.

The price range of USDX's CARR (p, q) model is as follows:

$$\begin{aligned} R_{Dt} &= \lambda_{Dt}\varepsilon_{Dt} \\ \lambda_{Dt} &= \omega + \sum_{i=1}^{p}\alpha_i R_{Dt-i} + \sum_{j=1}^{q}\beta_j \lambda_{Dt-j} \\ \varepsilon_{Dt} &\sim iidf(\bullet) \\ R_{Dt} &= \ln P_{Dt}^{\max} - \ln P_{Dt}^{\min} \end{aligned} \tag{26}$$

Where, R_{Dt} represents price range of USDX; $P_{Dt}^{\max}$ and $P_{Dt}^{\min}$ represents USDX's maximum and minimum values of the day; λ_{Dt} is the conditional mean of USDX's price range based on all information up to time t; ε_{Dt} is residual item.

Table 4. CARR model using daily price range of USDX, T=2036.

Model	Parameters							
	ω	α_1	α_2	β_1	β_2	θ	κ	Log-likelihood
ECARR(1, 1)	0.0024	0.0782		0.9183				
	[0.2418]	[1.3456]		[14.5865]				-1375.4225
	-0.8090	-0.1784		0.0000				
ECARR(1, 2)	0.0028	0.0743		0.9558	-0.0344			
	[0.2569]	[0.5184]		[0.4757]	[-0.0184]			-1374.3413
	-0.7972	-0.6042		-0.6343	-0.9853			
ECARR(2, 1)	0.0024	0.0665	0.0143	0.9157				
	[0.2407]	[0.4652]	[0.0919]	[13.2726]				-1374.3093
	-0.8098	-0.6418	-0.9268	0.0000				
ECARR(2, 2)	0.0046	0.0767	0.0722	0.0596	0.7848			
	[0.1979]	[0.7269]	[0.1925]	[0.0110]	[0.1576]			-1373.5884
	-0.8432	-0.4673	-0.8473	-0.9912	-0.8748			
WCARR(1, 1)	0.0060	0.0859		0.9055		2.6178		
	[4.1530]	[11.7003]		[109.2228]		[67.2153]		-291.3359
	0.0000	0.0000		0.0000		0.0000		
WCARR(1, 2)	0.0056	0.0800		0.9962	-0.0842	2.6179		
	[3.0611]	[4.4504]		[4.2336]	[-0.3884]	[66.2398]		-291.0326
	-0.0022	0.0000		0.0000	-0.6977	0.0000		
WCARR(2, 1)	0.0062	0.0730	0.0170	0.9011		2.6182		
	[4.1081]	[3.9738]	[0.8082]	[93.7680]		[65.8385]		-290.9305
	0.0000	-0.0001	-0.4190	0.0000		0.0000		
WCARR(2, 2)	0.0119	0.0817	0.0881	-0.0898	0.9029	2.6245		
	[4.1319]	[9.7869]	[11.9773]	[-9.7552]	[106.7649]	[64.4726]		-288.5163
	0.0000	0.0000	0.0000	0.0000	0.0000	0.0000		
GCARR(1, 1)	0.0021	0.0779		0.9191		0.8020	10.4213	
	[1.4006]	[8.7979]		[96.2944]		[16.2917	[9.7143]	-171.4014

						]		
	-0.1613	0.0000		-0.0526		0.0000	0.0000	
GCARR(1, 2)	0.0019	0.0719		1.0142	-0.0890	0.8051	10.3347	
	[1.3212]	[3.3875]		[3.4880]	[-0.3292]	[16.2684]	[9.7030]	-171.1503
	-0.1864	-0.0007		-0.0005	-0.7420	0.0000	0.0000	
GCARR(2, 1)	0.0021	0.0653	0.0153	0.9162		0.8005	10.4564	
	[1.3940]	[3.0407]	[0.6580]	[87.5041]		[16.2053]	[9.6721]	-171.0746
	-0.1633	-0.0024	-0.5106	0.0000		0.0000	0.0000	
GCARR(2, 2)	0.0035	0.0690	0.0606	0.2945	0.5708	0.8239	9.8379	
	[1.0976]	[3.4294]	[0.9296]	[0.3123]	[0.6577]	[16.2767]	[9.7053]	-171.1206
	-0.2724	-0.0006	-0.3526	-0.7548	-0.5107	0.0000	0.0000	

Note: [] is t statistic; ()is the p-values.

Table 5. The alternative models' AIC value of USDX' CARR model.

Model	AIC Value
WCARR(1, 1)	586.6718
WCARR(2, 2)	585.0327

According to p-values (significance at the 5% level) in table 4, all the parameters in WCARR (1, 1) and WCARR (2, 2) are significant.

From table 5, WCARR (2, 2) with the smallest AIC value is an optimal CARR model for price range of USDX.

$$\begin{aligned} R_{Dt} &= \lambda_{Dt}\varepsilon_{Dt} \\ \lambda_{Dt} &= 0.0119 + 0.0817\, R_{Dt-1} + 0.0881\, R_{Dt-2} - 0.0898\, \lambda_{Dt-1} + 0.9029\, \lambda_{Dt-2} \\ \varepsilon_{Dt} &\sim Weibull(2.6245) \end{aligned} \quad (27)$$

Where, R_{Dt} represents price range of USDX; λ_{Dt} is the conditional mean of USDX's price range based on all information up to time t; ε_{Dt} is residual item.

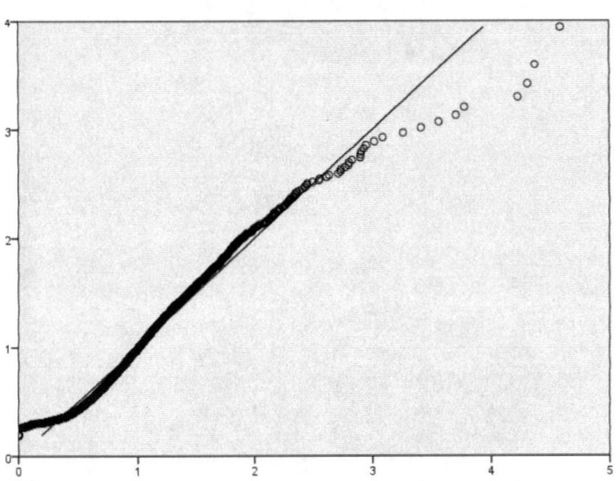

Figure 1. QQ-plots of NYMEX crude oil returns against Gamma distribution.

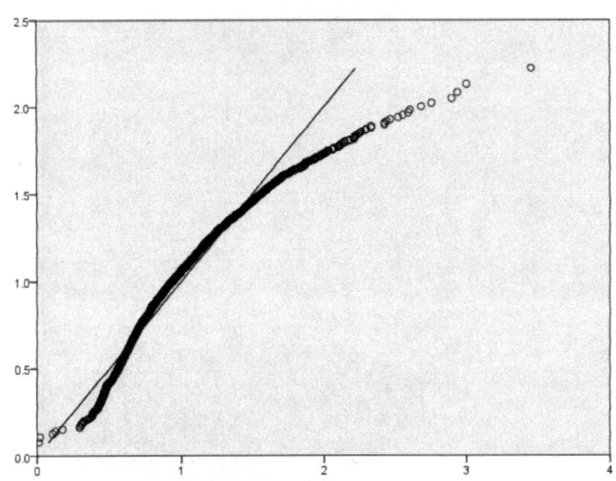

Figure 2. QQ-plots of USDX returns against Weibull distribution.

We employ Q-Q plot to test whether the models are applicable. The results in figures 1 and 2 shows Q-Q plot of NYMEX crude oil and USDX. From figures 1 and 2, we apply equation (25) and equation (27) as marginal distribution models are appropriate.

5.2.2. Analysis based on Copula-CARR models. Table 6 shows estimation results of Copula for NYMEX crude oil and USDX. Z value and its p value can be used to measure whether the parameters of various types of Copula are significant. Kendall rank correlation coefficient (τ) is used to measure the consistency degree of their change direction. Spearman rank correlation (ρ) is used to measure the degree of correlation.

Table 6. Estimates of the dependence parameters of different copula models for NYMEX crude oil and USDX.

	Parameters		Z value		Goodness	τ	ρ	Log-
	Name	Value	Test value	P value	of fit			likelihood
Gaussian Copula	ρ^N	0.1522	6.9930	0.0000	0.0599	0.0957	0.1431	23.0400
t-Copula	ρ^t	0.1569	6.6870	0.0000	0.0796	0.1058	0.1582	-7.6560
Gumbel Copula	α	1.0966	65.2200	0.0000	0.0792	0.0892	0.1331	22.1100
Clayton Copula	θ	0.1401	4.5870	0.0000	0.2027	0.0774	0.1159	12.1400
Frank Copula	λ	1.0095	7.4820	0.0000	0.0337	0.1143	0.1709	28.0700

From Z value and its p value in table 6, we can see that all the parameters of Copula are significant under 5% significance. In addition, goodness of fit of Clayton Copula is the largest one. So Clayton Copula is the best copula for NYMEX crude oil and USDX's correlation structure. Clayton copula's density distribution is "J" type, which means that it is more sensitive to lower tail change of variables and can be used to describe lower tail dependence. Figures 3 and 4 show the density function diagram and distribution function diagram of copulas for NYMEX crude oil and USDX.

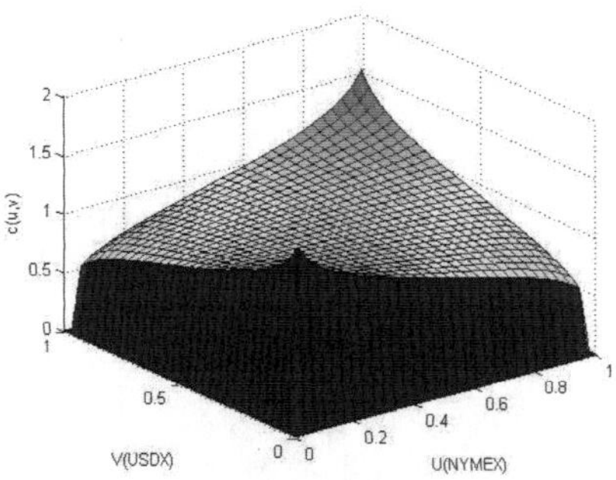

Gaussian Copula

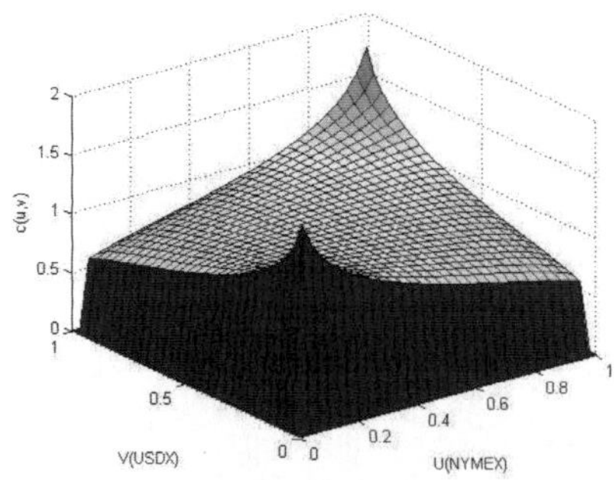

t-Copula

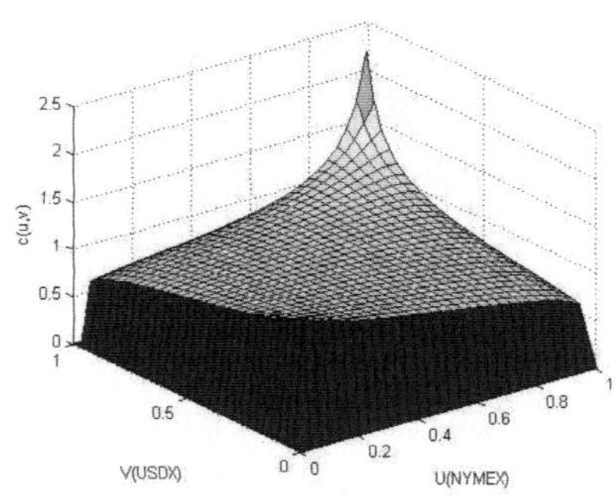

Gumble Copula

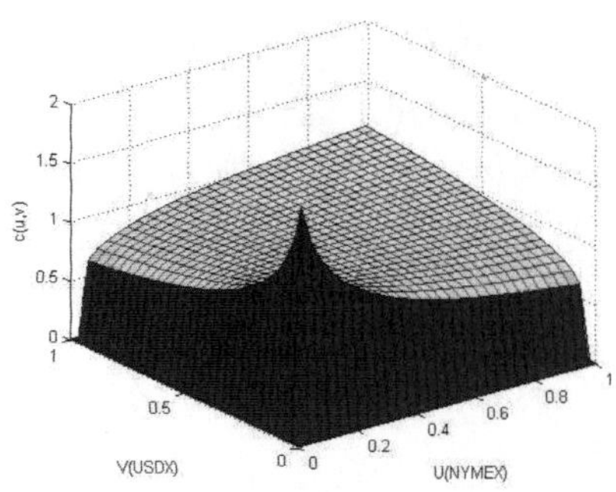

Clayton Copula

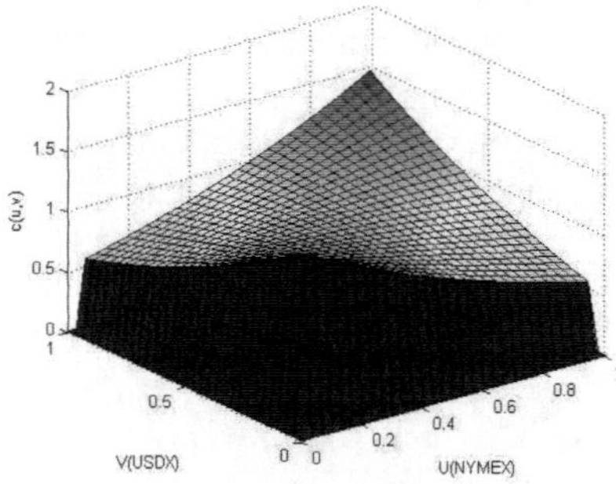

Frank Copula

Figure 3. The density function diagram of copulas for NYMEX crude oil and USDX.

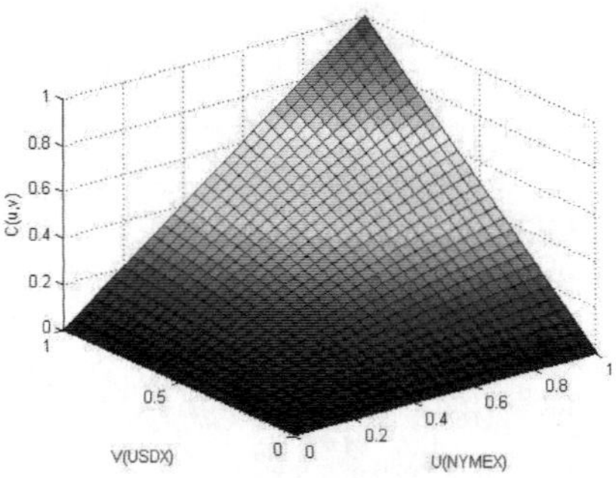

Gaussian Copula

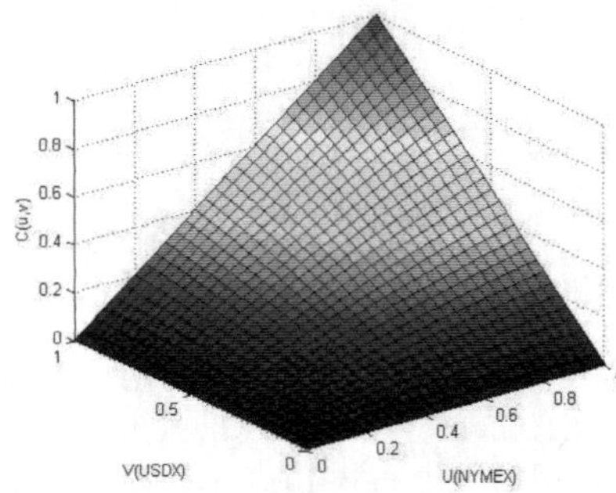

t-Copula

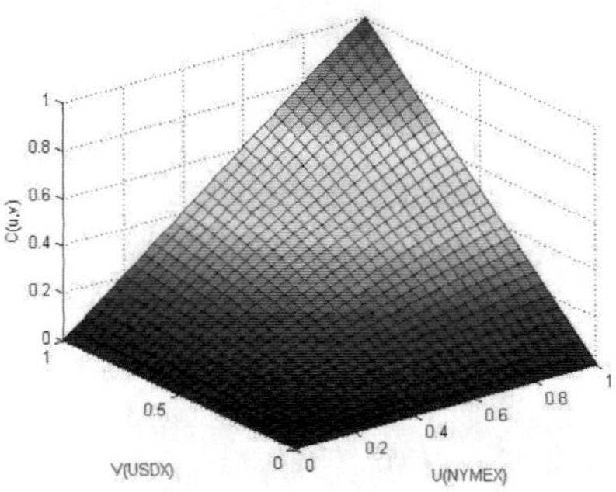

Gumble Copula

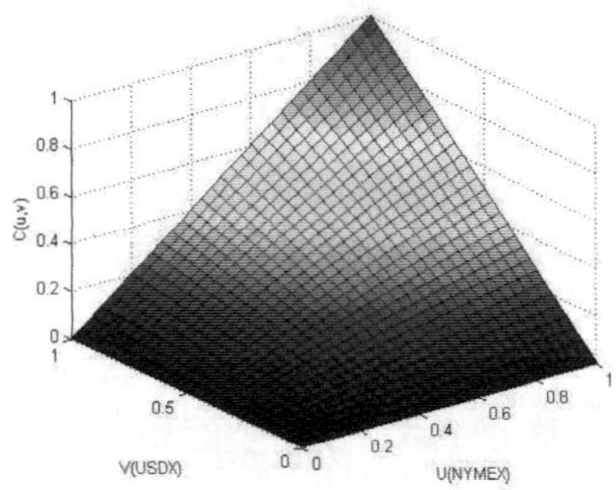

Clayton Copula

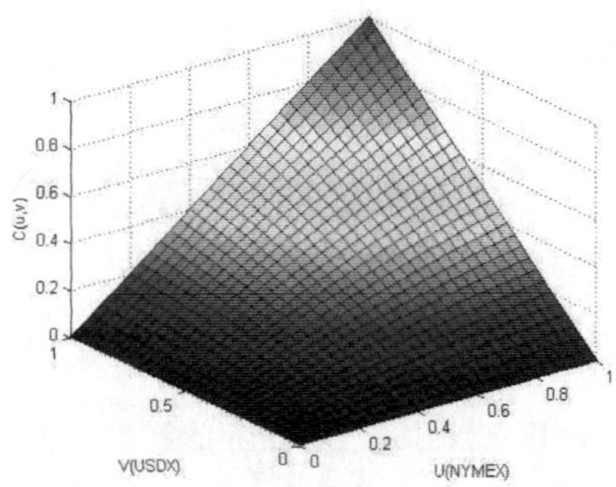

Frank Copula

Figure 4. The distribution function diagram of copulas for NYMEX crude oil and USDX.

IOP Conf. Series: Earth and Environmental Science **93** (2017) 012034 doi:10.1088/1755-1315/93/1/012034

5.3. Copula-CARR model with structural breaks

We analyze the correlation between NYMEX crude oil and USDX using Copula-CARR model with structural breaks.

We will detect change points of Clayton Copula-CARR model, which is the optimal Copula-CARR model for NYMEX crude oil's price range and USDX's price range.

Firstly, we get Z_n value of the overall sample shown in figure 5. From figure 5, we can see that Z_n obtains the maximum value in the 1005th data and it is over 9. It means the 1005th data is a change-point.

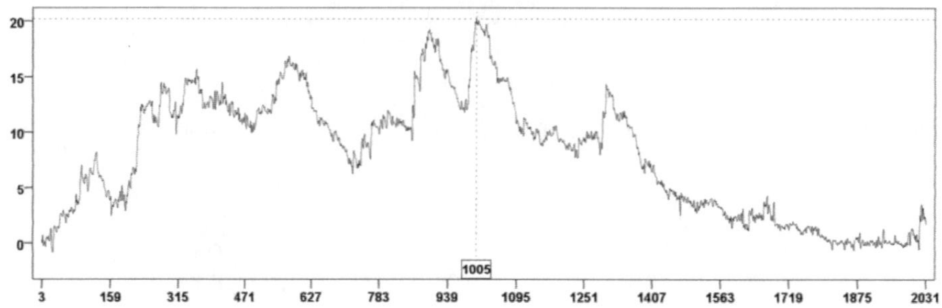

Figure 5. Z_n values of sample 1-2036.

Secondly, the overall sample is divided into two sub-samples by the 1005th data. Figure 6 depicts Z_n value of the sample from 1 to 1005. From figure 6, we can get the second change point - the 731th data (Z_n value is 10. 2416). Figure 7 depicts Z_n value of the sample from 1005-2036. From figure 7, we can get the third change point - the 731th data (Z_n value is 9. 1728).

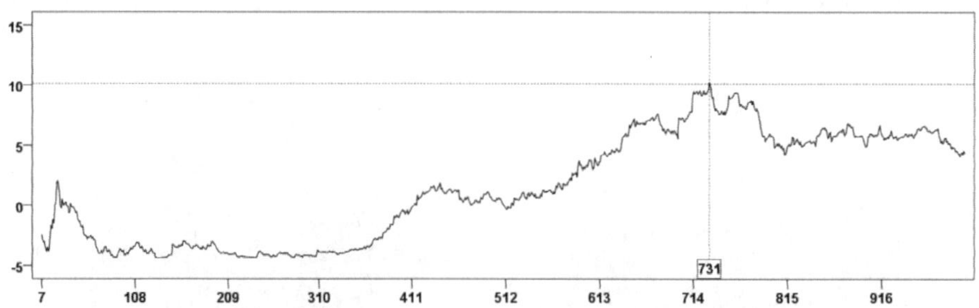

Figure 6. Z_n values of sample 1-1005.

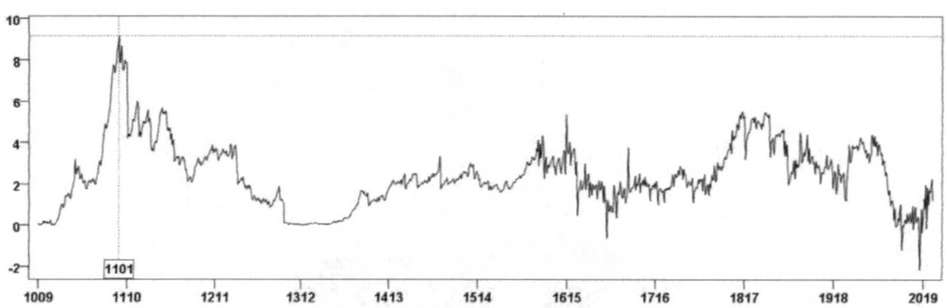

Figure 7. Z_n values of sample 1005-2036.

Thirdly, the sample from 1 to 1005 is divided into two sub-samples by the 731th data. Figure 8 depicts the Z_n value of the sample from 1 to 731. From figure 8, we can get the fourth change point -

the 605th data (Z_n value is 15. 1194). Figure 9 depicts the Z_n value of the sample from 731 to 1005. From figure 9, we can see that all Z_n values are less than 9, which means there doesn't exist change point of the sample from 731 to 1005.

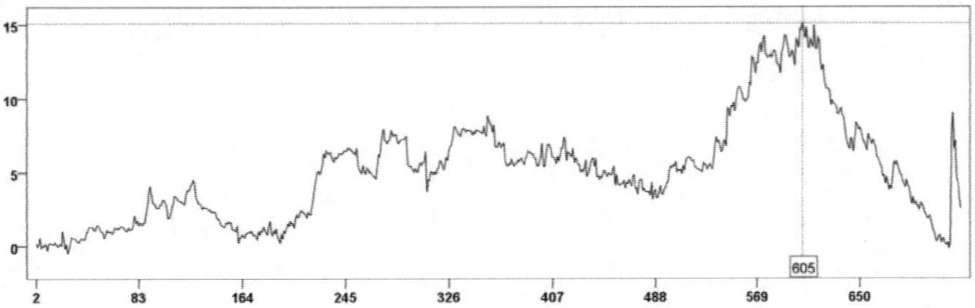

Figure 8. Z_n values of sample 1-731.

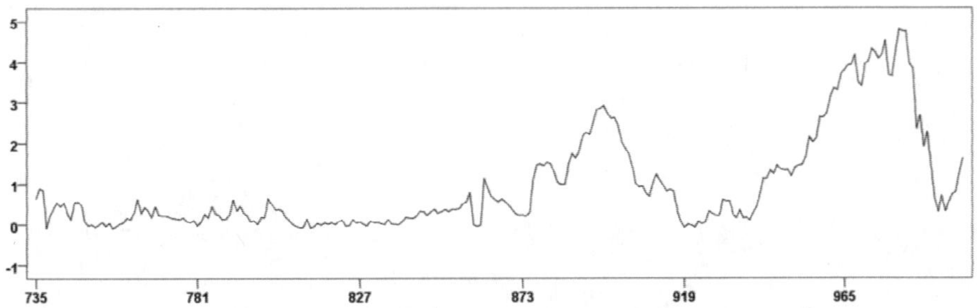

Figure 9. Z_n values of sample 731-1005.

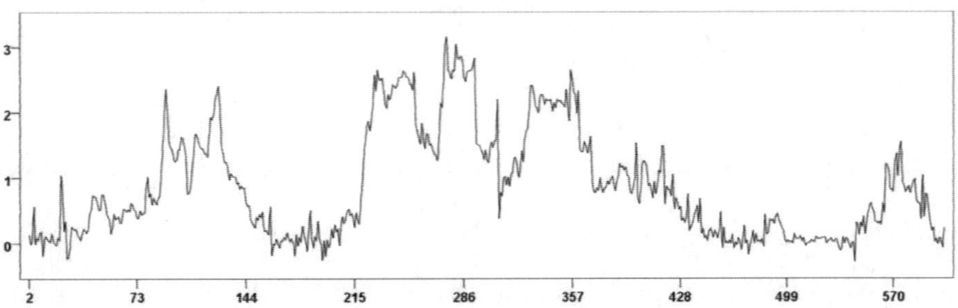

Figure 10. Z_n values of sample 1-605.

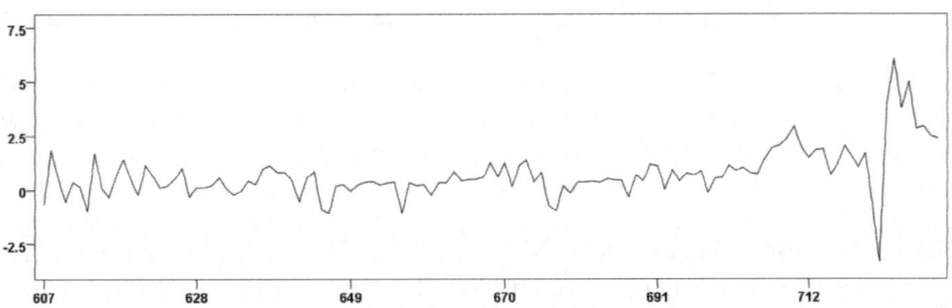

Figure 11. Z_n values of sample 605-731.

Figure 10 depicts the Z_n value of the sample from 1 to 605. Figure 11 depicts the Z_n value of the sample from 605 to 731. From figure 10 and figure 11, we can see that all Z_n values are less than 9, which means there doesn't exist change point.

Finally, the sample from 1005 to 2036 is divided into two sub-samples by the 1101th data. Figure 12 depicts the Z_n value of the sample from 1005 to 1101. Figure 13 depicts the Z_n value of the sample from 1101 to 2036. From figure 12 and figure 13, we can see that all Z_n values are less than 9, which means there doesn't exist a change point.

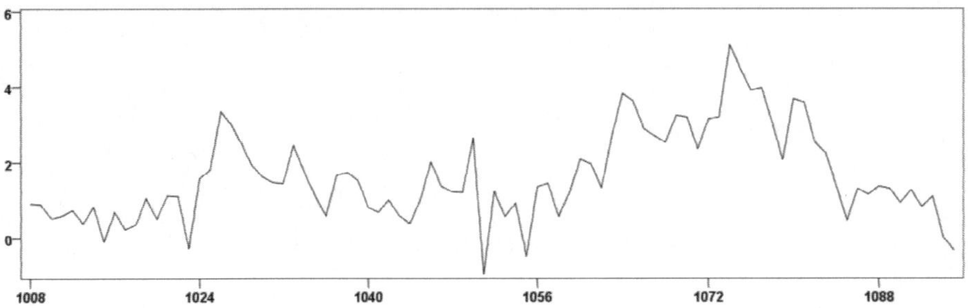

Figure 12. Z_n values of sample 1005-1101.

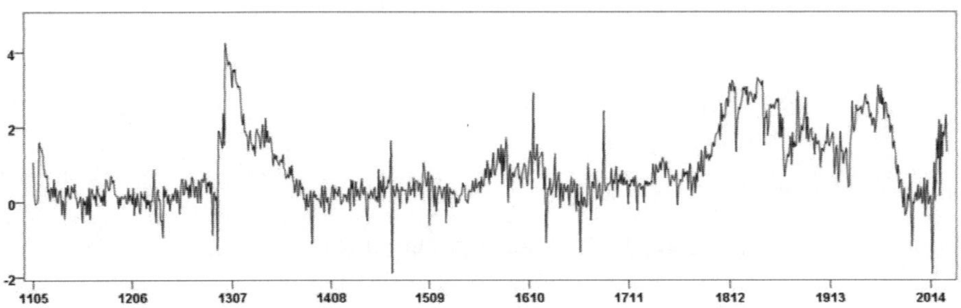

Figure 13. Z_n values of sample 1101-2036.

We get four change points in the optimal Copula-CARR model of NYMEX crude oil price's range and USDX's price range, which correspond to the dates of March 17, 2008, October 30, 2008, January 4, 2010 and June 1, 2010.

The dates of change points are closely related to the economic events.

- On March 12, 2008, the Federal Reserve announced that it will expand the securities lending program. On March 16, the Federal Reserve decided to lower the discount rate from 3. 5% to 3. 25%, and create a new discount window for primary dealers. International oil prices continued to rise in early 2008, impacted by the situation continued unrest in Nigeria, a major oil producer in Africa. On March 12, New York oil prices hit a record high, breaking the $ 110 / barrel.
- On September 15, 2008, the fourth largest US investment bank- Lehman Brothers declared bankruptcy. On the same day, the US investment bank - Merrill lynch & co. was acquired by BOA (Bank of America). These events had a direct impact on the subsequent global stock market's crash. Due to the rise in the US dollar exchange rate and the US financial crisis, Oil demand in the United States and other developed countries fell sharply. International oil prices tumbled. When the international financial crisis intensified, oil prices fell more than 30 US dollars / barrel. On October 16, WTI crude oil fell to $ 69. 85 / barrel, down 51. 9%.
- In December, 2009, world's three major credit rating agencies (Standard & Poor's, Fitch and Moody's) downgraded Greece's sovereign debt rating, which meant the beginning of the European debt crisis. Euro exchange rate fell sharply. In early 2010, due to the high northern latitudes' temperature dropped to a historic low in winter, U. S. oil inventories continued to

fall in December. The international oil market continued to push on. WTI crude oil price hit its peak since the middle of October 2008.

- Crude oil futures were gradually overshadowed by the high U. S. crude oil inventories in April, 2010. With the unfavorable economic data, the international oil's price went down. Due to the serious European debt crisis and the rising oil inventories, crude oil market's situation worsened in May, 2010. Until the end of May, 2010, as gasoline consumption season and the Atlantic hurricane season came, larger rebound appeared in international oil prices.

6. Conclusion

In recent years, the impact of crude oil price fluctuations and exchange rate fluctuations on the national economy have been closely concerned by governments around the world. Many scholars have found that there is a correlation between the two fluctuations in the empirical study. The study of the two fluctuations can provide some implications for the government departments to formulate relevant policies and measures to deal with the crisis. Based on these considerations, we use Copula-CARR model with structural breaks to study yhe correlation structure and relationship of NYMEX crude oil price range and U. S. Dollar Index range.

Different from previous studies, we propose a new approach--CARR models to discuss the fluctuations of NYMEX crude oil price range and U. S. Dollar Index range. That is to say, we study their ranges' fluctuations. Then, based on CARR model, we use Gaussian Copula, Student-t Copula, Gumbel Copula, Clayton Copula and Frank Copula to study the correlation structure for NYMEX crude oil price range and U. S. Dollar Index range. From the empirical results, we can see that Clayton Copula is the optimal copula function for NYMEX crude oil price range and U. S. Dollar Index range's correlation structure. There is an obvious lower tail dependence between them.

In addition, we detect change points for NYMEX crude oil price range and U. S. Dollar Index range using Copula-CARR model with structural breaks. We get four change points and prove that they are closely related to the economic events, especially the 2008 financial crisis and the 2010 European debt crisis.

The empirical results of this paper offer us several policy implications:

Firstly, significant fluctuations of oil price give us a sign of exchange rate's abnormal fluctuation. Thus, enterprises and governments should take timely measures to manage exchange rate risk when the signal appears.

Secondly, the fluctuations of U.S. dollar, which is the world's currency, will have great effect on other currency's exchange rate. For other countries, precautionary measures should be taken in advance to manage their currency's risk by paying close attention to oil price fluctuations.

Thirdly, financial crisis will lead to the fluctuations of oil prices and exchange rate. As a result, enterprises and governments should manage their financial risk and prevent financial crisis.

Acknowledgments

The work was supported by National Natural Science Foundation of China under Grant number 70901055; National Social Science Foundation of China, under Grant number 14CTJ012; and the Independent Innovation Fund project of Tianjin University, China.

References

[1] Huang Y and Guo F 2007 The role of oil price shocks on China's real exchange rate *Chin Eco Rev* **18** 403-16
[2] Adeniyi O, Omisakin O, Yaqub J and Oyinlola A 2012 Oil price-exchange rate nexus in Nigeria: Further evidence from an oil exporting economy *Int J Human Soc Sci* **4** 113-21
[3] Oriavwote V E and Eriemo N O 2012 Oil prices and the real exchange rate in Nigeria *Int J Eco Fine* **4** 198-205
[4] Chen H T, Liu L, Wang Y D and Zhu Y M 2016 Oil price shocks and U.S. dollar exchange rates *Energy* **112** 1036-48

[5] Chou K W and Tseng Y H 2016 Oil prices, exchange rate, and the price asymmetry in the Taiwanese retail gasoline market *Eco Model* **52** 733-41
[6] Yang L, Cai X J and Hamori S 2017 Does the crude oil price influence the exchange rates of oilimporting and oil-exporting countries differently? A wavelet coherence analysis *Int Rev Eco Fin* **49** 536-47
[7] Sklar A 1959 Fonctions de repartitionàn dimensions et leurs marges *Publication Institut Statistique Universitéde Paris* **8** 229-31
[8] Rémillard B, Papageorgiou N and Soustra F 2012 Copula-based semiparametric models for multivariate time series *J Multivariate Analysis* **110** 30-42
[9] Reboredo J C 2012 Modelling oil price and exchange rate co-movements *J Policy Modeling* **34** 419-40
[10] Alouia R and Aïssab M S B 2016 Relationship between oil, stock prices and exchange rates: A vine copula based GARCH method *North American J Eco Fin* **37** 458-71
[11] Dias A and Embrechts P 2004 Change-point analysis for dependence structures in finance and insurance *Risk Measures for the 21st Century* ed. Giorgio Szegö (Hoboken: John Wiley and Sons) Chapter 16 pp 321-35
[12] Dias A and Embrechts P 2009 Testing for structural change in exchange rates' dependence beyond linear correlation *The European J Finance* **15** 619-37
[13] Gombay E and Horváth L 1996 On the rate of approximations for maximum likelihood tests in change-point models *J Multivariate Analysis* **56** 120-52
[14] Cai X, He G T, Guan J and Li X M 2009 Multiple change-points analysis for the dependence structures of exchange rate *Systems Engineering-Theory & Practice* **29** 48-53
[15] Boubaker H and Sghaier N 2013 Instability and time-varying dependence structure between oil prices and stock markets in GCC countries *Working Paper Series* **23** 1-18
[16] Zhu X Q, Xie Y J, Li J P and Wu D S 2015 Change point detection for subprime crisis in American banking: From the perspective of risk dependence *Int Rev Eco Fin* **38** 18-28
[17] Chou R Y 2005 Forecasting financial volatilities with extreme values: The conditional autoregressive range(CARR) model *J Money Credit Banking* **37** 561-82
[18] Cai Y J, Chou R Y and Li D 2009 Explaining international stock correlations with CPI fluctuationsand market volatility *J Banking & Finance* **33** 2026-35
[19] Chou R Y and Liu N 2010 The economic value of volatility timing using a range-based volatility model *J Ecoc Dynamics & Control* **34** 2288-301
[20] Sin C Y 2013 Using CARRX models to study factors affecting the volatilities of Asian equity markets *North American J Economics and Finance* **26** 552-64
[21] Nelsen R B 1998 *An Introduction to Copulas* (New York: Springer) pp 456-77
[22] Akaike H 1969 Fittring autoregreseive models for prediction *Selected Papers of Hirotugu Akaike* ed Emanuel Parzen, Kunio Tanabe and Genshiro Kitagawa (New York: Springer) pp 131-5

The analysis of energy efficiency in water electrolysis under high temperature and high pressure

L W Hourng[1,3], T T Tsai[1] and M Y Lin[2]

[1]Department of Mechanical Engineering, National Central University, Taoyuan, Taiwan
[2]Department of Mechanical Engineering, Army Academy, Taoyuan, Taiwan

E-mail: lwhong@cc.ncu.edu.tw

Abstract. This paper aims to analyze the energy efficiency of water electrolysis under high pressure and high temperature conditions. The effects of temperature and pressure on four different kinds of reaction mechanisms, namely, reversible voltage, activation polarization, ohmic polarization, and concentration polarization, are investigated in details. Results show that the ohmic and concentration over-potentials are increased as temperature is increased, however, the reversible and activation over-potentials are decreased as temperature is increased. Therefore, the net efficiency is enhanced as temperature is increased. The efficiency of water electrolysis at 350°C/100 bars is increased about 17%, compared with that at 80°C/1bar.

1. Introduction

Hydrogen is considered to be the most promising energy carrier in the future. Most of the hydrogen nowadays is produced through the steam reforming of natural gas. However, the need for hydrogen produced by other processes, which include electrochemical, biomass, biological, and solar hydrogen productions, is increased. Water electrolysis belongs to the electrochemical hydrogen production. The purity of hydrogen produced by water electrolysis is higher than other hydrogen produced methods, and there is no carbon monoxide emitted during the production and consumption processes. Besides, the investment of facility on water electrolysis is rather reasonable. These advantages make water electrolysis being a potential method for hydrogen production.

From electrochemistry and thermodynamics, the semi-reactions on anode and cathode electrodes at 25°C and 1bar are

Cathode: $2H_2O + 2e^- \rightarrow H_2 + 2OH^-; \quad E_{rev} = -0.828\,V$

Anode: $2OH^- \rightarrow \frac{1}{2}O_2 + H_2O + 2e^-; E_{rev} = 0.401\,V$

Therefore, the net reaction in the electrolysis cell is

$$H_2O \rightarrow H_2 + \frac{1}{2}O_2; \quad E_{rev} = 1.23\,V$$

Under 25°C/1bar, the reversible voltage (E_{rev}) is 1.23 V, which is the minimum voltage needed for water electrolysis process as the reaction occurs ideally. However, when electric current passes through an electrochemical battery, there always exists a potential difference between practical and ideal situations. This phenomenon is called electrode polarization [1], and the degree of electrode polarization is defined as over-potential.

$$\eta = E - E_{rev} \tag{1}$$

where η is the over-potential. E is the applied potential, and E_{rev} is the reversible potential or reversible voltage. There are three kinds of polarization, namely, concentration polarization, activation polarization, and ohmic polarization. Basically, the reaction rate and polarizations of water electrolysis depend on the working temperature and pressure. However, there is few researches dealing with this issue, and the present article aims therefore at exploring the influences of temperature and pressure on the water electrolysis theoretically. Three kinds of polarization are described briefly as below.

1.1. Reversible voltage

The relation between the reversible voltage and the Gibbs free energy is

$$E_{rev} = -\Delta G/nF \tag{2}$$

where E_{rev} is the reversible voltage, ΔG the change of Gibbs free energy, n the equivalent number of material, and F the Faraday's constant. Gibbs free energy of the liquid water at different temperature and pressure can be found in the reference book JANAF [2]. The reversible voltage can thus be obtained by equation (2).

1.2. Activation polarization

All chemical reactions, which occurred on the interface between electrodes and electrolyte, needs extra electric potential in addition to the reversible potential to cross over an energy obstacle. This kind of extra electric potential difference to yield chemical reactions is called the activation over-potential. The relation between activation over-potential and current density is governed by Butler-Volmer equation as below.

$$i = i_0[exp(\frac{\alpha zF\eta_a}{RT}) - exp(\frac{-(1-\alpha)zF\eta_a}{RT}) \tag{3}$$

η_a is the activation over-potential, and i_0 is the exchange current density. α is the transfer coefficient, which is 0.5 for water electrolysis. R is the universal gas constant and z is electrochemical equivalent. From Thampan, *et al* [3], the relation between exchange current density i_0 and temperature is given as

$$i_0 = {}_M \exp[\ \frac{E}{R}(\frac{1}{T}\ \ \frac{1}{T_{ref}})]i_{0.ref}$$

For anode and cathode, it yields

$$i_{A.0} = {}_M \exp[\ \frac{E_A}{R}(\frac{1}{T}\ \ \frac{1}{T_{ref}})]i_{A.0.ref}$$

$$i_{C.0} = \gamma_M \exp[-\frac{E_C}{R}(\frac{1}{T} - \frac{1}{T_{ref}})]i_{C.0.ref}$$

where $i_{A,0}$ and $i_{c,0}$ are exchange current density of anode, and cathode, respectively. $i_{A,0,ref}$ and $i_{c,0,ref}$ are reference exchange current density of anode, and cathode respectively. For platinum electrode, the corresponding values are 10^{-3} A/cm^2 and 10^{-11}A/cm^2. γ_M is the roughness factor, and taken to be 1.8 for platinum electrode [4].

1.3. Ohmic polarization

Ohmic polarization may be caused by electrode line segment or generated bubbles. Line segment of electrodes is the wire joining the electrode slice to pass the electric current through. Electrode slice is very thin with large cross-sectional area compared with that of line segment. Therefore, the resistance of electrode, compared with resistance of line segment, is very small and can be neglected. In the present high temperature circumstances, metal ohmic resistance of line segment becomes large and can't be neglected anymore. The ohmic over-potential for line segment obeys the ohmic law as

$$\eta_{ohm} = i\Omega \tag{4}$$

Where η_{ohm} is ohmic over-potential of line segment, and Ω is the resistance of line segment. The resistance is determined by

$$\Omega = \rho l/A \tag{5}$$

Where ρ is the specific resistance, l the length parallel to current direction, and A is the cross-sectional area of the line segment. The relation between specific resistance ρand temperatures is as below.

$$\rho = \rho_o(1+aT) \tag{6}$$

Where ρ_o is the coefficient of specific resistance and a is the percentage change in resistivity per unit temperature. For platinum, $\rho_o = 1.1\times10^{-7}$ and a = 0.00392 [5]. The ohmic resistance due to bubbles will be discussed in more detail later.

1.4. Concentration polarization

For water electrolysis in alkaline electrolyte, the hydroxyl produced in cathode will be transported toward anode by electric force. There is a diffusion layer near the surface of electrode, that is, there exists a concentration gradient from the surface of electrode to the edge of diffusion layer. In a still electrolyte without fluid flow, the thickness of the layer (δ) is about 0.05 cm. The hydroxyl concentration is not high enough near anode, which is unfavourable to reaction on anode. To maintain certain electric current, more electric potential is needed. This phenomenon is called concentration polarization.

In the electrolysis reaction, as the ion thickness near the electrode surface approaches to zero, the over-potential of water electrolysis will be suddenly increased. The corresponding electric current is called the limiting current. As the current reaches its limiting current, no matter how much the applied voltage is added, the electric current will not increase anymore. The limiting current density is related to ion properties as follows.

$$i_L = \frac{D_j a_j F z}{(1-t_j)\delta}$$

where i_L is the limiting current density, a_j the activity of ion j, D_j the diffusion coefficient of ion j, δ the thickness of diffusion layer, and t_j the transition coefficient of ion j. The relation between concentration over-potential and electric current is

$$\eta_{con} = \frac{RT}{ZF} ln \frac{i_L}{i_L - i} \tag{7}$$

Where η_{con} is the concentration over-potential and i is the electric current.

2. Results and discussions

Four different kinds of reaction mechanisms, namely, reversible voltage, activation polarization, ohmic polarization, and concentration polarizations, are investigated here in details. Then the electric energy consumption at 1bar/80°C and 500bar/370°C are compared to show the effect of high temperature and pressure on the water electrolysis.

The reversible voltages at different temperatures and pressures are shown in figure 1. It indicates that the higher the temperature and the pressure are, the lower the reversible voltage will be. That is, considering only the reversible voltage, the efficiency of water electrolysis will be improved by raising the operating temperature and pressure. From the figure, it seems that increasing temperature enhances the efficiency of water electrolysis significantly, while the effect of pressure is actually very small. However, the high pressure is still required to maintain water in its liquid phase when the operating temperature is very high. Furthermore, the hydrogen gas produced with high pressure can be stored in the tank directly, and no compression procedure is required.

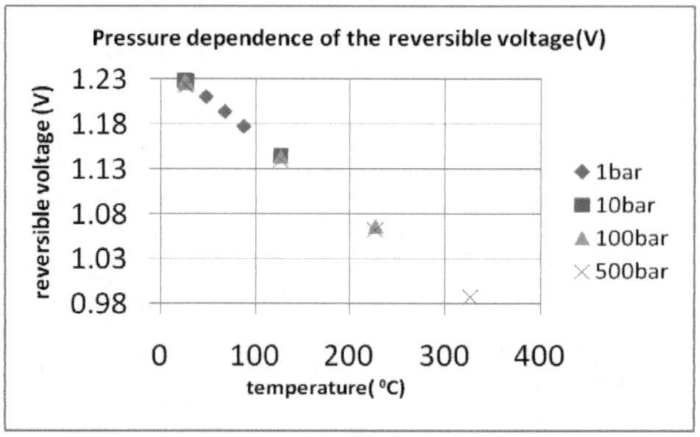

Figure 1. Diagram of reversible voltage via temperature and pressure.

The activation polarization reaction is due to desorption between hydrogen ion and electrode, which is a metal-non-metal bonding or a typical ionic bond. The rate of desorption is determined by ionic bond energy. Since ionic bond energy is almost not influenced by pressure, we can conclude that activation over-potential is also not influenced by pressure. The activation over-potential is thus assumed depending on temperature only in this article.

Figure 2 shows the variation of activation over-potential via temperature as the exchange current density $i_{A.0}$ and $i_{C.0}$ is fixed at certain value. It indicates that the activation over-potential will increase as the current density is increased. It also shows that the activation over-potential increases when temperature increases. This result is not compatible with experimental observation, in which the supplied voltage will decrease as the temperature increases. If the current density is set to be function of temperature, the variation of activation over-potential via temperature is plotted as shown in figure 3. The results show that activation voltage decreases as temperature is increased, which agrees qualitatively with experimental observations. Therefore, low current density and electrolyte temperature is preferred to reduce the activation over-potential during water electrolysis.

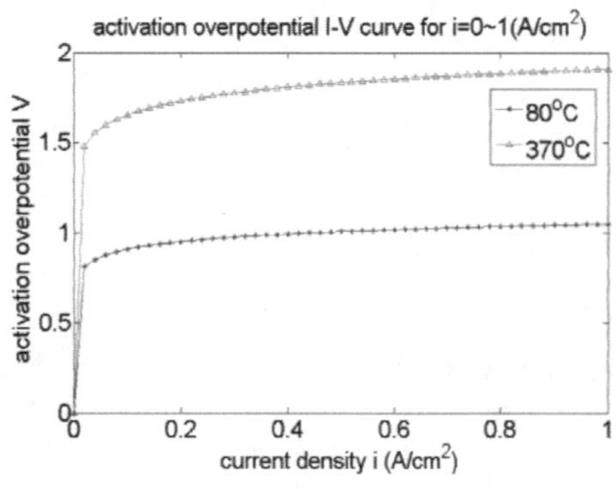

Figure 2. Relation between activation over-potential and current at 80°C and 370°C. The exchange current density is considered to be constants.

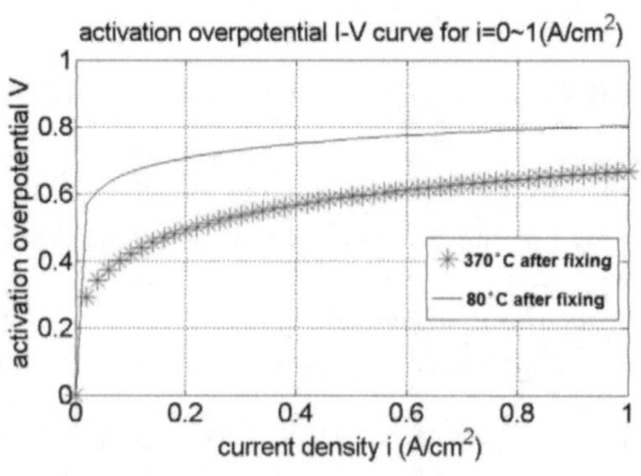

Figure 3. Relation between activation over-potential and current at 80°C and 370°C. The exchange current density is considered to be function of temperature.

The ohmic resistance due to line segment can be calculated based on the ohmic law as described before. The corresponding ohmic resistance is twice higher at 370℃ than that at 80℃, thus it should not be neglected in high temperature water electrolysis. An efficient way to reduce the ohmic resistance is to increase the cross-sectional area of the wire.

Ohmic over-potential due to bubbles in electrolyte made by Kuhn, *et al* [6] is shown in figure 4, and the empirical results of ohmic over-potential due to bubbles in electrolyte immersed with proton exchange membrane made by Roy, *et al* [7] is also shown in figure 5. The former seems to have better performance than the latter. The reason may due to that bubbles are easy to be trapped in the porous media such as proton exchange membrane, and therefore increase the resistance significantly. The water electrolysis in the present study is closer to the experiment of Kuhn, *et al* as shown in figure 4, which indicates that the ohmic over-potential due to bubbles can be neglected when the current density is smaller than 1 A/cm^2.

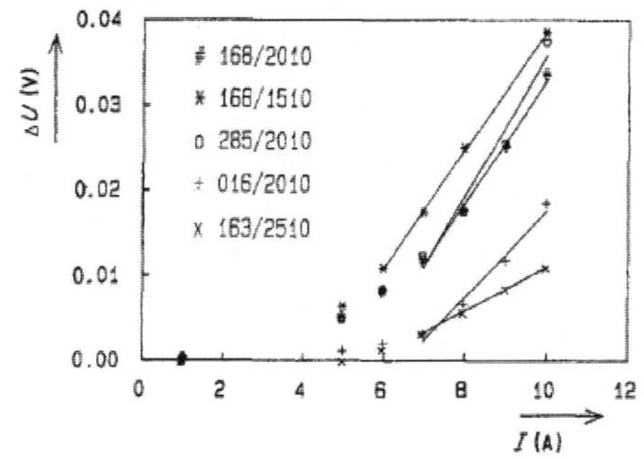

Figure 4. Polarization curve of ohmic over-potential due to bubbles [6].

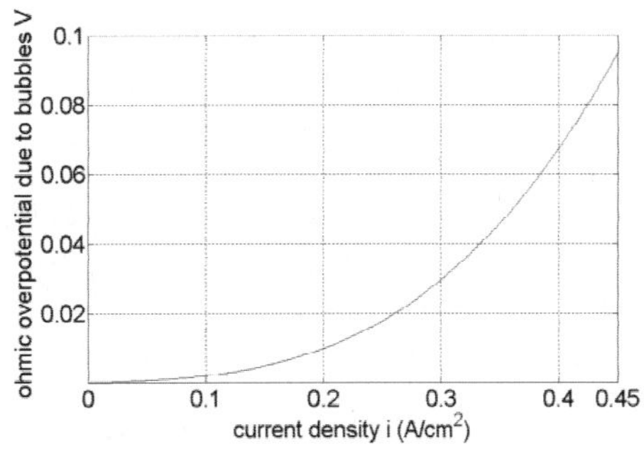

Figure 5. Polarization curve of ohmic over-potential due to bubbles, based on the empirical equation of Roy, *et al* [7].

There is no theory to relate the concentration over-potential with temperature and pressure. However, the diffusion of ions in water electrolysis at standard pressure and temperature can be found in table 1. Substituting the data in table 1 into equation (7), the relation between concentration over-potential and current can be plotted as shown in figure 6. It can be found the concentration over-potential can be neglected when the current density is below then 1 A/cm^2. At high temperature water electrolysis, as we discuss here, the concentration over-potential should be even less. Because the equivalent conductivity and ionic mobility rise with temperature, ions are more easily to transport and the corresponding resistance will decrease as shown in table 2. In addition, found from [8], the effect of pressure on ionic transportation is limited such that the effect of pressure on concentration over-potential can also be neglected.

Table 1. Properties of 0.5 M NaOH electrolytes at 80°C/1bar.

	Parameters	Dimensions	Values
Activation	α_j	mol/ml	0.693
Diffusion coefficient	D_j	none	2.632 x 10^{-9}
Transport number	t_j	0.109	0.600
Diffusion layer thickness	d	m	0.0005
Mobility	u_j	m^2/s V	20.5 x 10^{-8}

Table 2. The effect of temperature on ion properties.

OH^-	dimensions	18°C	25°C
Equivalent conductivity	cm^2/Ω	171	198.3
Mobility of ion	m^2/V s	17.7 x 10^{-8}	20.5 x 10^{-8}

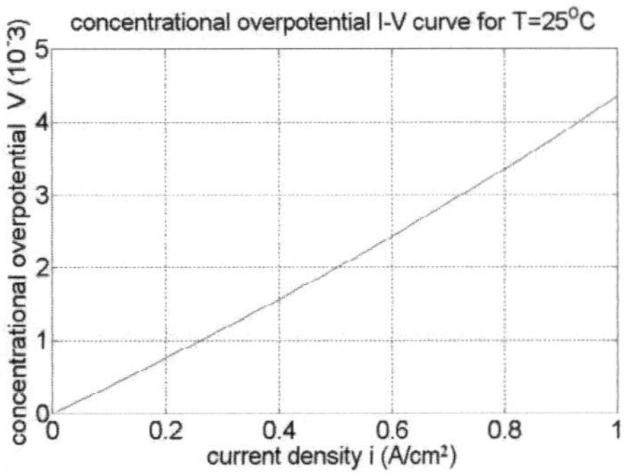

Figure 6. Polarization curve of concentration over-potential at 80°C/1bar.

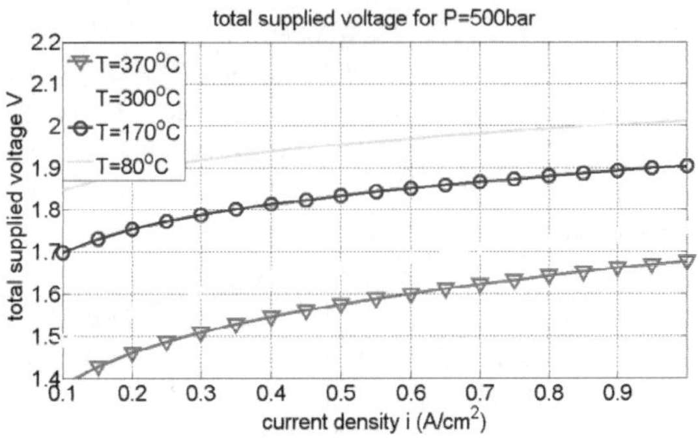

Figure 7. Total supplied voltage for different current density under 500 bar.

Because current density below 1 A/cm^2 is considered here, the ohmic over-potential due to bubble and concentration over-potential are neglected as mentioned above. The total applied potential thus includes reversible voltage, activation over-potential, and ohmic over-potential due to line segment of electrode. The influence of temperature on the supplied voltage is shown in figure 7. It indicates that the supplied voltage can be reduced as the temperature is increased.

The influence of line segment area on applied voltage is shown in figures 8 and 9. In figure 8, the ohmic over-potential due to line segment with cross-sectional area of 1mm^2 is about 23% of supplied voltage, and should not be neglected. The percentage of ohmic over-potential may be reduced if a large cross-sectional area is used as shown in figure 9. However, in order to save equipment costs, line segment with cross-sectional area of 1 mm^2 is generally used.

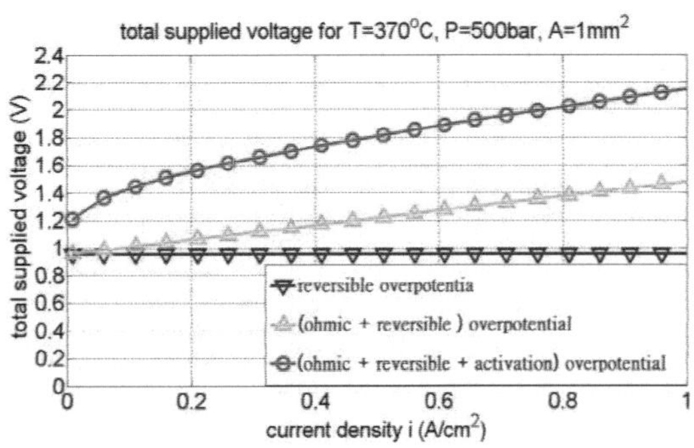

Figure 8. Comparison of over-potentials at 370°C/500 bar. The cross-sectional area of line segment is 1 mm^2.

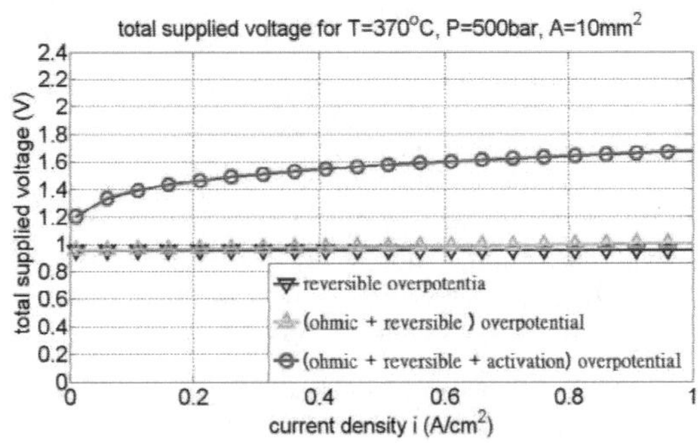

Figure 9. Comparison of over-potentials at 370°C/500 bar. The cross-sectional area of line segment is 10 mm^2.

The efficiency of electrolysis at 370°C/500 bar and 80°C/1bar are compared in figures 9 and 10. The supplied voltage is 1.678 V at 370°C/500 bar, 1 A/cm^2 and 2.016 V at 80°C/1 bar, 1 A/cm^2, thus the efficiency at 370°C/500 bar is increased about 17% for water electrolysis. Furthermore, if the consumption of energy for product hydrogen to elevate pressure from 1 bar to 500 bars is also considered, the efficiency of water electrolysis is increased about 22%. The calculated results are in the consistent trend with the experimental results of Marangio, *et al* [9].

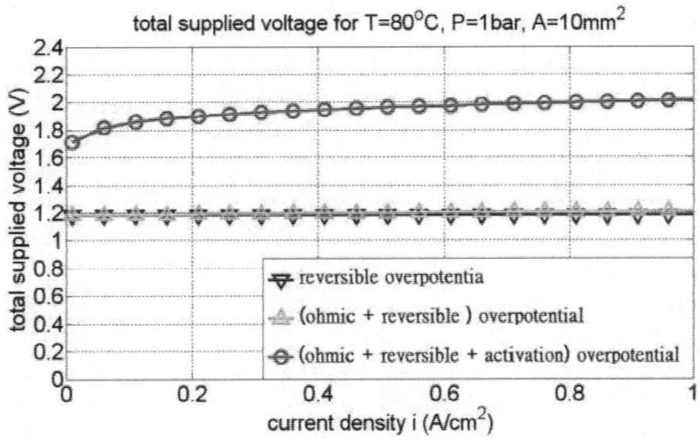

Figure 10. Comparison of over-potentials under 80°C/1bar. The cross-sectional area of line segment is 10 mm^2.

3. Conclusions

Based on the results and discussions, several conclusions can be drawn as below.

- As temperature or pressure rises, the reversible voltage of water electrolysis is reduced.
- The activation over-potential is reduced as temperature rises, and nearly not affected by pressure.
- Under high temperature of 370°C, the ohmic over-potential of the line segment should be considered. However, it may be reduced by the increase of the cross-sectional area or decrease of the length of line segment.
- The ohmic over-potential due to bubbles is increased with the increase of temperature, as well as the decrease of pressure. However, this effect can be neglected while current density is below 1A/cm^2.
- The concentration over-potential is increased with the increase of temperature rise. This effect can be neglected while current density is lower than 1 A/cm^2.
- The efficiency of water electrolysis at 350°C/100 bars is increased about 17%, compared with that at 80°C/1 bar. If the produced hydrogen is to be used at 1 bar and the expansion work from pressure 500 bars to 1 bar is also fully converted, the efficiency of water electrolysis is increased about 22%.

4. Nomenclatures

a	percentage change in resistivity per unit temperature
a_j	activity of ion j
A	cross-sectional area
D_j	diffusion coefficient of ion j
E	applied potential
E_{rev}	reversible potential
F	Faraday's constant
G	Gibbs free energy
i	electric current
$i_{A.0}$	exchange current density of anode
$i_{C.0}$	exchange current density of cathode
$i_{A.0.ref}$	reference exchange current density of anode and
$i_{C.0.ref}$	reference exchange current density of cathode
i_L	limiting current density
l	length of wire or rod
n	equivalent number of material

IOP Conf. Series: Earth and Environmental Science **93** (2017) 012035 doi:10.1088/1755-1315/93/1/012035

R	universal gas constant
T	temperature
t_j	transition coefficient of ion j.
Z	electrochemical equivalent
η	over-potential
η_{act}	activation over-potential
η_{con}	concentration over-potential
η_{ohm}	ohmic over-potential
ρ_o	coefficient of specific resistance
δ	thickness of diffusion layer
Ω	resistance of line segment
ρ	specific resistance
γ_M	roughness factor

References

[1] Koryta J, Dvořák J and Kavan L 1993 *Principles of Eelectrochemistry* (New York: John Wiley)
[2] Chase M W 1985 *JANAF Thermochemical Tables* (Maryland: National Institute of Standard and Technology)
[3] Thampan T S, Malhotra S, Zhang J X and Datta R 2001 *Catalysis Today* **67** 15-32
[4] Wang H, Wingender C, Baltruschat H, Lopez M and Reetz M T 2001 *J. Electroanalytical Chemistry* **509** 163-9
[5] Serway R A 1998 *Principles of Physics* (Philadelphia: Saunders College Pub)
[6] Kuhn M and Kreysa G 1989 *J. of Applied Electrochemistry* **19** 720-8
[7] Roy A, Watson S and Infield D 2006 *Int. J. of Hydrogen Energy* **31** 1964-79
[8] Hamann S D 1957 *Physico-chemical Effects of Pressure* (London: Butterworths Scientific Publications)
[9] Marangio F, Santarelli M and Cali M 2009 *Int. J. Hydrogen Energy* **34** 1143-58

Management decision of optimal recharge water in groundwater artificial recharge conditions- A case study in an artificial recharge test site

H Y He[1,2,3], X F Shi[3], W Zhu[3], C Q Wang[3], H W Ma[3] and W J Zhang[1,2,4]

[1]Key Laboratory of Groundwater Resources and Environment, Ministry of Education, Jilin University, Changchun 130021, Jilin Province, China
[2]College of Environment and Resources, Jilin University, Changchun 130021, Jilin Province, China
[3]Shenyang Center China Geological Survey Bureau, Shenyang 110034, Liaoning Province, China

E-mail: zhangwenjing80@126.com/ zhuwei_1114@163.com

Abstract. The city conducted groundwater artificial recharge test which was taken a typical site as an example, and the purpose is to prevent and control land subsidence, increase the amount of groundwater resources. To protect groundwater environmental quality and safety, the city chose tap water as recharge water, however, the high cost makes it not conducive to the optimal allocation of water resources and not suitable to popularize widely. To solve this, the city selects two major surface water of River A and B as the proposed recharge water, to explore its feasibility. According to a comprehensive analysis of the cost of recharge, the distance of the water transport, the quality of recharge water and others. Entropy weight Fuzzy Comprehensive Evaluation Method is used to prefer tap water and water of River A and B. Evaluation results show that water of River B is the optimal recharge water, if used; recharge cost will be from \$0.4724/m^3 to \$0.3696/m^3. Using Entropy weight Fuzzy Comprehensive Evaluation Method to confirm water of River B as optimal water is scientific and reasonable. The optimal water management decisions can provide technical support for the city to carry out overall groundwater artificial recharge engineering in deep aquifer.

1. Introduction

In many developing countries, groundwater plays a major life support to mankind, as it is the major source to support domestic needs and irrigation purposes. Groundwater occurs in a wide range of rock types and usually requires little or no treatment; therefore it is often the cheapest and simplest water supply option. However, the rising demand for water worldwide, mostly for production and irrigation, can lead to problems of over exploitation of these resources and conflicts with competing demands [1]. With the irrational development and utilization of groundwater resource, it makes a series of environmental and geological problems, such as groundwater resource depletion, land subsidence, etc [2].

Artificial recharge is one of the effective means to increase the amount of groundwater recharge, prevent and control of land subsidence, prevent seawater intrusion and storage energy [3]. In the nineteenth century, the United States and Europe had been carrying out research of artificial recharge [4], and after the 1930s, it had been widely used in more than 30 countries and regions in the world [5].

Now, the technique of artificial recharge is very mature, and more widely applied.

With the rapid economic growth, the over-exploitation of groundwater resources has existed for a long time, which leads to the serious land subsidence and groundwater resources depletion in the city. To solve the above environmental geology problems, the city conducted groundwater artificial recharge test which was taken a typical site as an example, because of water quality of the target aquifer is good, and the aquifer will be a reserve after recharge, therefore, tap water with good water quality and easily recharged is used to be the first choice of recharge water. However, the high cost makes it not conducive to the optimal allocation of water resources and not suitable to popularize widely. To solve this, the city selects two major surface water of River A and B as the proposed recharge water, to explore its feasibility.

In order to prefer tap water and water of River A and B, it should be a comprehensive analysis a variety of factors, such as the cost of recharge, the distance of the water transport, the quality of recharge water and others, the factors is independent that is difficult to determine which factors play a leading role. Therefore, it needs to carry out a multi-criteria decision-making to evaluate the optimal water, so that the result of evaluation will be more objective and reasonable. Domestic and foreign experts and scholars made numerous studies on multi-criteria decision-making evaluation; fuzzy theory is most widely applied, the weights of indicators are to measure their impact on the final decision, and they are of great importance [6]. Obviously the greater the weight of an indicator is, the greater its impact is on the decision [7]. However, they have many limitations of the method to calculate the definition of weights, such as tedious calculation, heavy workload, and no considering the link between multiple indicators, and so on [8]. Fuzzy Clustering Method, Analytic Hierarchy Process (AHP), and Fuzzy Comprehensive Evaluation Method are common and traditional analysis methods in the fuzzy theory; especially Fuzzy Comprehensive Evaluation Method is widely used in optimal allocation of water resources, groundwater resource management decisions [9-12], etc. Subjective methods of determining the weight, such as Fuzzy Comprehensive Evaluation Method, are also in use, but this may cause the bias of evaluation results because of subjective factors [13]. To solve the problems above, Zou, *et al* [14] proposed a new weight evaluation process using entropy method. It is based on actual data of indicators which reflects the changes of objective information [15]. Thus it is an objective method, and can reduce effects of subjective factors. It is a simple calculation, which simplifies fuzzy evaluation process greatly. The workload can be reduced evidently. Another remarkable character is to avoid equalization of weights distribution [16]. By using the fuzzy mathematics and entropy theory, the evaluation results of the fuzzy synthetic evaluation would be more scientific and reasonable than the normal methods.

In summary, Entropy weight Fuzzy Comprehensive Evaluation Method will be used to determine the optimal recharge water in the paper.

2. Materials and methods

2.1. Study area

The study area is located in the city; north-east China, as a result of the rapid development of the city's economy, the 1st to 4th aquifers were exploited with a large number while water quality of 1st to 3rd aquifers were contaminated by human activity. In order to control land subsidence and save water resources, the 4th aquifer is used as a backwater source for the city after recharge. The city chooses the artificial recharge test site as typical, if successful, it will be fully promoted in the city. The target aquifer is the fourth confined aquifer which will be the city's water reserves after recharge in figures 1 and 2 (the maps were obtained from Shanghai Institute of Geological Survey). The key aquifer is the 4th confined aquifer, a homogeneous isotropic aquifer, about 50 m thick. Lithology is mainly gray gravel with the fine sand and the coarse sand. Aquifer elevation and hydraulic conductivity are -170 m and 50 m/d respectively, and the hydraulic gradient along the flow direction is about 0.3‰. The city has two major rivers with plenty of water, taking River A and B as the proposed recharge water. Water points of River A and B were selected as the nearest water intake, the distance from the water intake of

River A and River B to water treatment plant are separately 14.4 km and 13.8 km, water treatment plant is 2.4 km away from recharge test site. Test results of water of River A and B show that components of concentration in excess are total iron, manganese and TOC. According to the most stringent specifications, if the two alternative water will be recharged, water of River A and B must be transferred to water purification plant, the quality of the water is up to the standard (GB/T14848-2007), the recharge water was selected with the lowest cost, according to the cost of water taken and treatment, and then be recharged to the target aquifer.

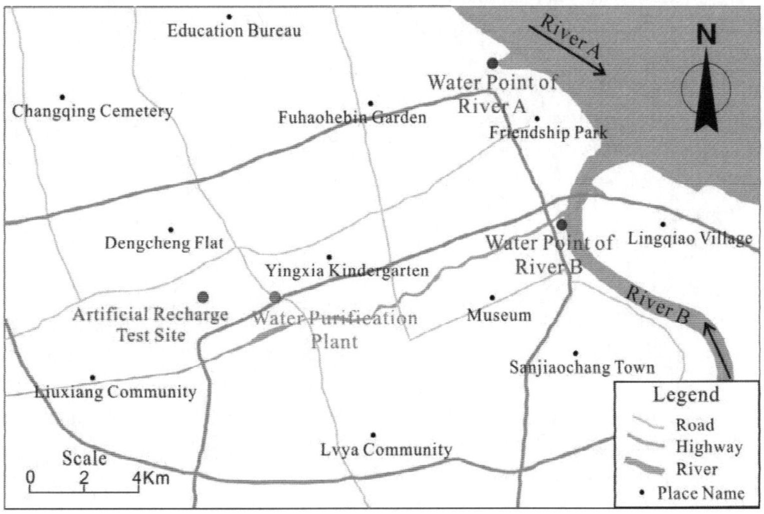

Figure 1. Location of the study area.

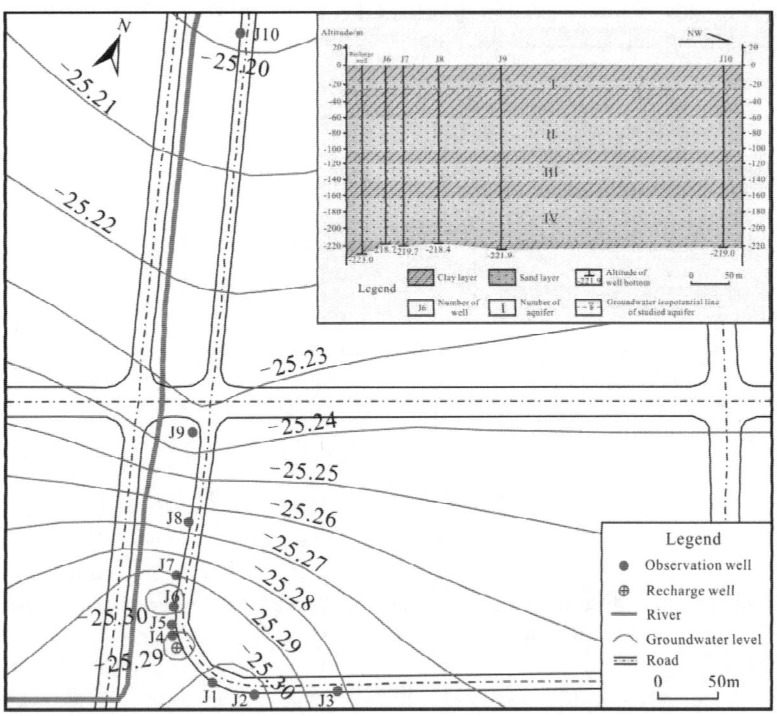

Figure 2. Schematic diagram of the groundwater isopotential line in the 4th confined aquifer in the test field.

2.2. Methods

How much information obtained from the evaluation and decision-making is one of the decisive factors about the accuracy and reliability of evaluation [17]. In information theory, entropy is the measure of the disorder degree to the system, it can also measure the data provided by the information effectively. Therefore, entropy can be used to determine the weight [18]. When value of the evaluation objects on an index is larger, the entropy value is smaller, shows that the larger amount of effective information provided by the index, the bigger weight, otherwise, if the difference between the values is smaller, the entropy value is larger, means that the index provides a smaller amount of information, the weight is smaller [19]. Therefore, the entropy theory is a kind of objective weighting method.

2.2.1. Normalization of the original evaluating matrix. Suppose there are evaluating indicators counted m, evaluating objects counted n, then forms an original indicators value matrix $U = (u_{ij})_{m \times n}$

$$U = \begin{bmatrix} u_{11} & u_{12} & \cdots & u_{1n} \\ u_{21} & u_{22} & \cdots & u_{2n} \\ & & & \\ u_{m1} & u_{m2} & \cdots & u_{mn} \end{bmatrix} \tag{1}$$

The selection of the optimal recharge water is a comprehensive evaluation result. According to the principle of the evaluation index of the testability, reliability and adequacy, therefore, U contains indicators of economic cost, technical feasibility and water quality. All calculations were done by computing platform which is spss18.0.

Normalization this matrix to get equation (2):

$$R = (r_{ij})_{m \times n} \tag{2}$$

where, r_{ij} means the data of the jth evaluating object on the indicator, and $r_{ij} \in [0,1]$. Among these indicators, to which the bigger the better, there are

$$r_{i,j} = \frac{x_{i,j} - \min\{x_{i,j}\}}{\max\{x_{i,j}\} - \min\{x_{i,j}\}} \tag{3}$$

while, the smaller the better, there are

$$r_{i,j} = \frac{\max\{x_{i,j}\} - x_{i,j}}{\max\{x_{i,j}\} - \min\{x_{i,j}\}} \tag{4}$$

Normalization matrix is

$$R = \begin{bmatrix} r_{11} & r_{12} & \cdots & r_{1n} \\ r_{21} & r_{22} & \cdots & r_{2n} \\ & & & \\ r_{m1} & r_{m2} & \cdots & r_{mn} \end{bmatrix} \tag{5}$$

Initial parameters is obtained by the factual research and test data. Since each element is composed of different dimensions, they cannot be directly compared, and it is necessary to normalize, the assumptions are that all the different parameters can be normalized. The key of the fuzzy evaluation model is that parameters must be normalized.

2.2.2. Definition of the entropy. In the m indicators, n evaluating objects evaluation problem, the entropy of jth indicator is defined as

$$H_j = -k\sum_{j=1}^{n} f_{ij} \ln f_{ij} \quad (i=1,2,\ldots,m) \tag{6}$$

In the equation: $f_{ij} = r_{i,j} \Big/ \sum_{j=1}^{n} r_{i,j}$, $k = 1/\ln n$, when $f_{ij} = 0$, $\ln f_{ij}$ has no mathematical sense, therefore need to be revised to the f_{ij}, $f_{ij} = (1+r_{i,j}) \Big/ \sum_{j=1}^{n}(1+r_{i,j})$.

The coefficient of H indicates that the larger the value, the greater influence. According to the aim, the optimal water is determined by the minimizing economic cost, maximizing technical feasibility, and optimizing water quality.

2.2.3. Definition of the weight of entropy. Weight of entropy is a method of determining the weight based on the degree of discretization, which can avoid subjective human disturbance. The effect of different evaluation indexes on optimal water source selection is different. Therefore, in order to make the evaluation results more objective and true, this paper uses the weight of entropy to determine the index weight.

The weight of entropy of jth indicator could be defined as:

$$w_j = \frac{1-H_j}{n-\sum_{j=1}^{n} H_j} \tag{7}$$

in which $0 \le w_j \le 1$, $\sum_{j=1}^{n} w_j = 1$. Weight of entropy matrix is

$$W = \begin{pmatrix} w_1 & w_2 & \ldots & w_n \end{pmatrix} \tag{8}$$

2.2.4. Entropy weight fuzzy comprehensive evaluation method. The entropy weight and fuzzy normalization matrix are combined to Entropy weight Fuzzy Comprehensive Evaluation Method. The influence factors of the optimal water source were determined and judged, the judgment matrix is equation (9):

$$Q = W \otimes R = \begin{pmatrix} w_1 & w_2 & \ldots & w_n \end{pmatrix} \begin{bmatrix} r_{11} & r_{12} & \ldots & r_{1n} \\ r_{21} & r_{22} & \ldots & r_{2n} \\ \\ r_{m1} & r_{m2} & \ldots & r_{mn} \end{bmatrix} \tag{9}$$

According to the calculation result of the judgment matrix, combining the principles of minimizing economic costs, maximizing technical feasibility, and optimizing water quality, and then the optimal recharge water source is selected to recharge to the 4th confined aquifer to control land subsidence and save water resources. After the successfully field test, the optimal recharge water will be fully promoted in the city, and a comprehensive artificial recharge program and requirements will be put forward.

3. Results and discussion

According to a comprehensive analysis feasibility of tap water and water of River A and B as recharge water, three indicators of recharge water management are identified, indicators of economic cost, technical feasibility, water quality, respectively. According to the actual situation and combining the local water policies, the indicator of economic cost can be disassembled to treatment cost, transportation cost and additional cost of recharge water. Water from river to aquifer need to be transported and purified, the indicator of technical feasibility consists of three parts, ease of water treatment; the distance of water transport and water storage condition. According to the test reports of water of River A and B, components of concentration in excess are total iron; manganese and TOC, so, the indicator of water quality consist of total iron, manganese and TOC.

3.1. Economic cost

The tap water price for administrative business in the city is \$0.3258/m^3, treatment and operating cost are \$0.1955/m^3 and \$0.1303/m^3, municipal water recharge needs to pay the recycled water fee which is \$0.1466/m^3, and additional cost is \$0.2769/m^3 which is the sum of operating and recycled water fee.

As recharge water, water of River A and B need to remove iron, manganese and TOC, water treatment process is determined in the water purification test field as figure 3:

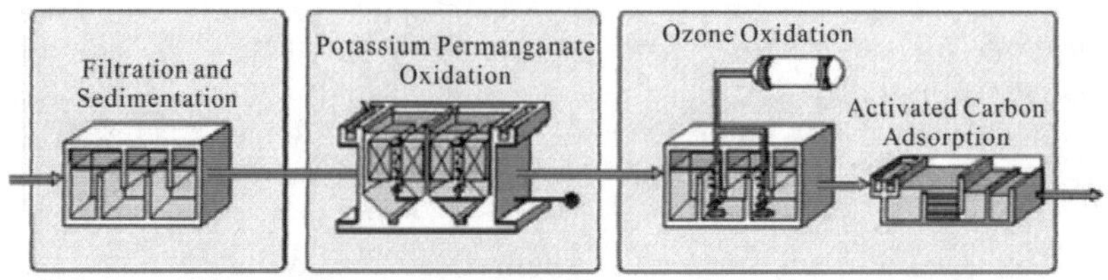

Primary Water Purification Removal of Total Iron and Manganese Removal of Total Organic Carbon

Figure 3. Water treatment process flow diagram.

The treatment cost of water of River A and B is consist of the primary treatment cost and removing iron, manganese and TOC cost, the sum are separately \$0.1565/m^3 and \$0.1583/m^3. Additional cost is operating cost which is the same to tap water that contains equipment cost, maintenance cost, management cost, etc. The transportation cost of water of River A and B are separately about \$0.08438/m^3 and \$0.08096/m^3 [20].

3.2. Technical feasibility

Technical feasibility index cannot be used to evaluate the superiority of index through quantitative assignment, thus, the subset assignment need to use theory and method of dealing with fuzzy system regularity, it takes only two values of ordinary set theory which are 0 and 1, the variable relationship is "one or the other" [21].

Tap water was recharged to aquifer without any treatment, the process is the simplest, assignment is as curtained as 0; the above-mentioned treatment technology which is used to purify water of River A and B is more complicated than tap water, thus the assignment of the two water source is 1. The distance from the water intake of River A and River B to the test site are separately 14.4 km and 13.8 km (figure 1). Tap water was recharged without storage, the assignment of the distance is 0. After removing iron, manganese and TOC, water of River A and B need to be stored separately rather than drinking water, the storage condition is more complex than tap water; thus, the assignment is 1.

3.3. Water quality

According to the test data of total iron, total manganese and TOC of three kinds of water which were tested by the Lab of Shanghai Institute of Geological Survey, the total iron concentration of tap water, water of River A and B are separately 0.14, 2.33 and 2.18 mg/L; the total manganese concentration are separately 0.005, 0.32 and 0.26 mg/L; and the TOC concentration are separately 1.885, 5.9 and 6.7 mg/L.

3.4. Discussion

Entropy weight Fuzzy Comprehensive Evaluation Method is built according to the factual research and test data, proper order of the matrix is tap water and water of River A and B.

$$U = \begin{bmatrix} 0.1955 & 0 & 0.2769 & 0 & 0 & 0 & 0.14 & 0.005 & 1.885 \\ 0.1565 & 0.08438 & 0.1303 & 1 & 14.4 & 1 & 2.33 & 0.32 & 5.9 \\ 0.1583 & 0.08096 & 0.1303 & 1 & 13.8 & 1 & 2.18 & 0.26 & 6.7 \end{bmatrix}$$

First of all, normalize the original matrix by equations (2)-(5). The entropy and its weight could be calculated according to equations (6) and (7), the calculated results of entropy and its weight as below:

$$H = (0.2408 \quad 0.3155 \quad 0.4530 \quad 0.4507 \quad 0.4531 \quad 0.4507 \quad 0.4624 \quad 0.4616 \quad 0.4619)$$
$$W = (0.1446 \quad 0.1302 \quad 0.1044 \quad 0.1046 \quad 0.1042 \quad 0.1046 \quad 0.1024 \quad 0.1026 \quad 0.1025)$$

It can be observed from the calculated value of entropy and its weight that the difference between each index is large, especially the first two indicators which are treatment cost and transportation cost. Among the results, the two smallest value of entropy are 0.2408 and 0.3155, and the largest value of weight are 0.1446 and 0.1302, according to the theory of entropy and its weight, it points out that the two indexes occupy the most important position and have the largest impact for selecting optimal water source. It means that economic cost factors also play the decisive role.

Through adding the results of entropy weight into fuzzy normalization matrix, judgment result is made to be more precise to show the impact of each indicator for selecting optimal water source. According to equation (8), the calculation as below:

$$Q = (0.249 \quad 0.734 \quad 0.8215)$$

The calculated results show the comprehensive score of tap water, water of River A and B are separately 0.249, 0.734 and 0.8215. Fuzzy comprehensive judgment model take the rule that the water with the larger management indicator is the better selection. Therefore, the optimal water is water of River B. The results of entropy and its weight indicate that, if water treatment technology is feasible and quality of water is up to standards, economic cost index is the main factor to decide whether recharge water is optimal. Taking the distance of getting water, water treatment cost, water transportation cost, additional cost and other factors into comprehensive consideration, the cost of water of River B is much less than tap water, the distance of River B is shorter than River A, so, selecting water of River B as the optimal water is truthful. If water of River B is used as recharge water, the cost will decline from \$0.4724/m^3 to \$0.3696/m^3, it will save \$0.1028 every ton. In the present quantity of recharge water (35 m^3/h), after recharging five years, recharge cost of \$157.57 thousand can be saved. When the city starts to carry out the artificial recharge engineering in deep aquifer, the saving of recharge cost will be more considerable. Thus, using Entropy weight Fuzzy Comprehensive Evaluation Method to confirm water of River B as optimal water is correct and reliable. The optimal water management decisions can provide technical support for the city to carry out overall groundwater artificial recharge engineering in deep aquifer.

4. Conclusions

Using Entropy weight Fuzzy Comprehensive Evaluation Method to confirm water of River B as optimal water, the optimal water is not only a technical feasible but also the lowest cost, the cost will decline from \$0.4724/m^3 to \$0.3696/m^3. By using the fuzzy mathematics and entropy theory, the evaluation results would be scientific and reasonable. The optimal water management decisions can provide technical support for the city to carry out overall groundwater artificial recharge engineering in deep aquifer.

Acknowledgments

The research was financially supported by National Natural Science Foundation of China (41103045). The authors would like to thank the journal editors and the reviewers, their comments help to improve the paper considerably. We are also thankful to the staff at Shanghai Institute of Geological Survey for their assistance in the field.

References

[1] Thangarajan M 2007 Groundwater models and their role in assessment and management of groundwater resources and pollution *Groundwater* **4** 189-90

[2] Li M and Roy F S 1997 Effects of artificial recharge on ground water quality and aquifer storage recovery *J Am Water Resour As* **33(3)** 561-72

[3] Juliet S J, Lawrence A B and Peter F 1999 Geochemical tranformations during artificial groundwater recharge: Soil-water interactions of inorganic constituents *Water Resour* **33(1)** 196-206

[4] Andrew L H, Karen J R, Peter P, Dillon J, Paul P and Karen E B 2004 Geochemical processes during five years of aquifer storage recovery *Groundwater* **42(3)** 438-45

[5] Salem B and Hamed B D 2010 A thirty-year artificial recharge experiment in a coastal aquifer in an arid zone: The Teboulba aquifer system (Tunisian Sahel) *C R Geosci* **34(2)** 60-74

[6] Liu L, Zhou J Z, An X L, Zhang Y C and Yang L 2010 Using fuzzy theory and information entropy for water quality assessment in Three Gorges region, China *Expert Syst Appl* **37** 2517-21

[7] Zhou J H, Zhang B, Hao C L and Zhu C J 2009 The fuzzy evaluation of surface water quality in Liujiazhuang of Qingzhang River *Proc Environ Sci Info Appl Tech Int Conf* (Wuhan, China) **1** 265-7

[8] Moslem S, Mohammad J S, Mostafa P and Foroogh P 2014 Environmental comprehensive assessment of agricultural systems at the farm level using fuzzy logic: A case study in cane farms in Iran *Environ Modell Softw* **58** 95-108

[9] Zhao R H and Chen S Y 2008 Fuzzy pricing for urban water resources: Model construction and application *J Environ Manage* **88** 458-66

[10] Zhang K J, Li H and Gopal A 2009 Fuzzy-stochastic characterization of site uncertainty and variability in groundwater flow and contaminant transport through a heterogeneous aquifer *J Contam Hydrol* **106** 73-82

[11] Chen H W and Chang N B 2010 Using fuzzy operators to address the complexity in decision making of water resources redistribution in two neighboring river basins *Adv Water Resour* **33** 652-66

[12] Ajay S 2012 An overview of the optimization modelling applications *J Hydrol* **466** 167-82

[13] Evangelos T, Andreas P and George A 2014 Development of an operational index of water quality (PoS) as a versatile tool to assist groundwater resources management and strategic planning *J Hydrol* **517** 339-50

[14] Zou Z H, Yun Y and Sun J N 2006 Entropy method for determination of weight of evaluating in fuzzy synthetic evaluation for water quality assessment indicators *J Environ Sci* **18(5)** 1020-3

[15] Xun J X and Yan Y J 2007 Application of gray correlative decision-making model for the optimization of water supply schemebased on the variation coeffcient method *J Water Resour Water Eng* **18(4)** 9-11

[16] Andre L, Lidia Y, Mihail L and Maria A S M 2009 River quality analysis using fuzzy water quality index: Ribeira do Iguape river watershed, Brazil *Ecol Indic* **9(6)** 1188-97

[17] Hamid R P, Majid M and Biswajeet P 2012 Landslide susceptibility mapping using index of entropy and conditional probabilitymodels in GIS: Safarood Basin, Iran *Catena* **97** 71-84

[18] Agung W, Mariana B, Eberhard K and Patrick B 2013 Maximum entropy estimation of a Benzene contaminated plume using ecotoxicological assays *Environ Pollut* **172** 170-9

[19] Bithin D, Dibakar C and Anirban D 2011 Identification of unknown groundwater pollution sources using classical optimization with linked simulation *J Hydro-Environ Res* **5(1)** 25-36

[20] Wu J Z, Wang H M and Yang T L 2009 Experimental research on artificial recharge to shallow aquifer to control land subsidence due to construction in Shanghai city *Geoscence* **23(6)** 1194-1200

[21] Jia Z Y, Wang C M, Huang Z W and Zhang G 2011 Evaluation research of regional power grid

IOP Conf. Series: Earth and Environmental Science **93** (2017) 012036 doi:10.1088/1755-1315/93/1/012036

companies' operation capacity based on entropy weight fuzzy comprehensive model *Proc Eng* **15** 4626-30

Construction technology of high-rise pile cap foundation of offshore wind power in Taiwan Strait

C Y Li[1,3], Y Chi[1], X Q Sun[1], Y P Han[1], X Chen[2], L C Zhao[2] and H Zhang[2]

[1]Institute of Offshore Engineering Technology, Zhongneng Power-Tech Development Company Limited, Beijing 100034, China
[2]Department of offshore wind power project, Fujian Longyuan Wind Power Company Limited, Fujian 350103, China

E-mail: licunyi@clypg.com.cn

Abstract. Offshore wind farms promise to become an important source of energy in the near future. The high-rise pile cap foundation is one of the typical foundation types for offshore wind turbine. This paper introduces the structural characteristics and construction technology of high-rise pile cap foundation, aiming at the characteristics of the sea area of Taiwan Strait and combining with engineering examples. The construction technology of high-rise pile cap foundation is expounded emphatically from the manufacture and transportation of steel pipe piles, pile foundation construction and bearing platform construction. Compared with the traditional construction technology, the construction technologies used in this project are safer and more reliable. The construction period of piles cap foundation is shortened by 10 ~ 48 days. The construction technology provides reference for offshore wind power foundation construction.

1. Introduction

In the globalized market, wind power has been considered as the most competitive price advantage of green energy technology, while leading the global energy transformation [1]. The total installed capacity of global wind power is expected to reach 2000GW by 2030. By 2050, wind energy accounted for 25% to 30% of global electricity generation [2]. Offshore wind farm has become a hot spot in the development of new energy sources, because of its abundant wind energy resources, good stability, low turbulence intensity, small wind cutting edge, less occupation of land resources, and small noise pollution [3].

Compared with the construction on the land, offshore wind farm construction has the characteristics of large investment, difficult construction and high risk. Offshore wind turbine foundation accounts for 20% ~ 30% of the total cost of wind farm, according to the proportion of investment in offshore wind farm which has been established at home and abroad [4]. The offshore wind turbine foundation mainly includes gravity based, monopile, pile group cap foundation, suction bucket, and floating type. The construction technology of different foundation forms has great differences [5-8]. The high-rise pile cap foundation is widely used in the construction of offshore wind power, accounting for 7 per cent of the 6496 offshore wind turbine foundations worldwide [9]. The design method of high-rise pile cap foundation mainly refers to the design method of cross-sea bridge and oceanographic engineering [10-12]. The construction technology of bridge pile foundation can be used for the construction of high-rise pile cap foundation [13]. Large-diameter monopile foundation is a common choice for most offshore wind farms in Europe, while conversely, Asian market is dominated by other fixed-grounded foundation concepts, especially of type high-rise pile cap [14]. According to the characteristics of different seas, different construction technologies of pile cap foundation are needed. The Donghai Bridge offshore wind farm is Asia's first large-scale offshore wind farm to use a high-rise cap pile group foundation successfully [15]. Its geological conditions are silt formation and sand strata. In the Taiwan Strait, the basement of the individual rock in the field is relatively shallow, new construction techniques are needed. The rock-socketed foundation is a new form of high-rise pile cap foundation, which is the first time using in the bedrock seabed in China and Asia. Based on an example of wind farm construction in Taiwan Strait, this paper analyzes and discusses the scheme, construction technology and construction efficiency of high-rise pile cap foundation.

2. Project overview

2.1. Wind energy resources and marine hydrological conditions

The project is located in the middle of the Taiwan Strait. Due to the great influence of "narrow-tube effect" in the Taiwan Strait, this region has a large annual average wind speed, stable wind direction and very rich wind resources. Annual north-north-east is prevailing wind direction, the annual average wind speed is 7.2 m/s, and the annual average number of days without wind only accounts for 5%. The tide of the sea area belongs to the regular semidiurnal tide and the tidal range is about 8 m. The region is a typhoon-prone area and affected by typhoons each year from July to October.

2.2. Engineering geology characteristics

The geological conditions of the area are complex: the upper layer is Quaternary marine deposit and the lower layer is Yanshan period granite. The Quaternary marine deposit from top to bottom respectively is shown in table 1.

Table 1. The Quaternary marine deposit.

Number	Stratigraphic subdivision	Layer thickness (m)	Standard Penetration Test (SPT)
1	silt	0.6 ~ 1.9	
2	fine sand, medium-dense to dense	7.8 ~ 8.25	17 ~ 32
3	medium coarse sand, slightly dense to medium-dense	3.15 ~ 7.2	10 ~ 23
4	clay	4.6 ~ 11.9	
5	medium coarse sand, slightly wet	1 ~ 3	20 ~ 23

The layers of Yanshan period granite from top to bottom respectively are shown in table 2.

Table 2. The layers of Yanshan period granite.

Number	Stratigraphic subdivision	Layer thickness (m)	Drillability level
1	completely weathered granite, rock structure is basically destroyed, feldspar and other minerals are seriously weathered	3.2 ~ 5.1	2
2	granular highly weathered granite, most of the rock structure is destroyed	3.4 ~ 9.2	5
3	cataclastic semi-weathered granite, jointed fracture development	0.35 ~ 4.8	6

4	weak-weathered granite	top elevation -23.64 ~ -44.93 m	7 ~ 9

2.3. Construction characteristics

Analyses of the wind energy, marine hydrological conditions and engineering geology conditions in the sea area, the main construction characteristics of the project are as follows:

- The sea is rough, the number of effective operation days is relatively small. The area is a typhoon prone area, and the construction equipment and construction technology requirements are high.
- There are a number of islands and reefs in the field, the seabed topography fluctuation is large, the stratigraphic distribution is uneven and the sediment thickness varies greatly. Different construction techniques should be chosen according to different sea conditions. The basement of the individual rock in the field is relatively shallow and the depth of the pile foundation cannot meet the design requirements, so the rock-socketed construction is necessary. The pile foundation diameter is large, all of them are inclined piles, the rock drillability level is high, and so the construction is very difficult.
- The moulding of the cap is a one-time large volume of concrete pouring. Concrete pouring and maintenance are more difficult at sea.
- The situation is complicated. There are obstructions such as water pipes, submarine cables and sunken ships in the field, which should be probed, avoided and cleaned before construction.

3. High pile cap foundation design

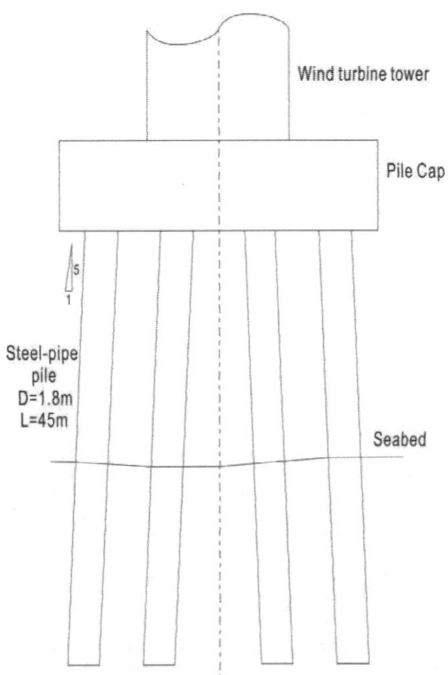

Figure 1. Schematic diagram of friction piles foundation.

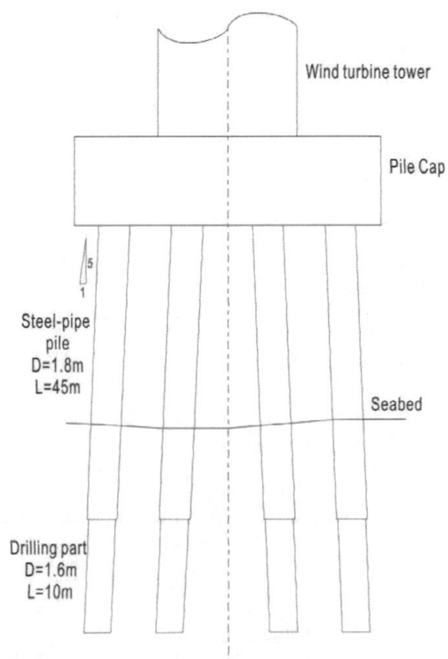

Figure 2. Schematic diagram of rock-socketed piles foundation.

The high pile cap foundation mainly consists of pile and cap. The pile foundation mainly carries axial load, horizontal load and bending moment. The platform combines the wind turbine tower with the pile foundation, which plays the role of load transfer and bearing. The pile foundation forms of high pile cap include friction piles and rock-socketed piles. The foundation of the project adopts the high pile cap and the pile foundation has both friction piles and rock-socketed piles. Eight steel pipe piles with a diameter of 1.8 m (wall thickness 28 ~ 30 mm) are used as foundation piles. Eight foundation piles are arranged evenly along the circumference of 10 meters in diameter on the bottom of the base and their slope is 5:1. Granular highly weathered or cataclastic weathered granite is designed to be the bearing layer of the friction piles, as shown in figure 1. Weak-weathered granite is designed to be the bearing layer of the rock-socketed piles. The rock-socketed part is 10 m in length, 1.6 m in diameter and adopts C40 reinforced concrete structure, as shown in figure 2.

4. Construction technologies

According to the design requirements and construction characteristics of the high pile cap, the construction technologies used in this project include the production and transportation of steel pipe piles, pile foundation construction, and pile cap construction.

4.1. Production and transportation of steel pipe piles

Steel pipe pile is delegated to the professional steel processing enterprises for production and transported to the construction site. The production processes of steel pipe pile include steel pipe pile section, welding and corrosion. The steel pipe piles are transported by 3000 t square barge, which is equipped with anchor system.

4.2. Pile foundation construction

4.2.1. The construction technologies of the friction pile foundation. The construction technologies of the friction pile foundation are shown in figure 3.

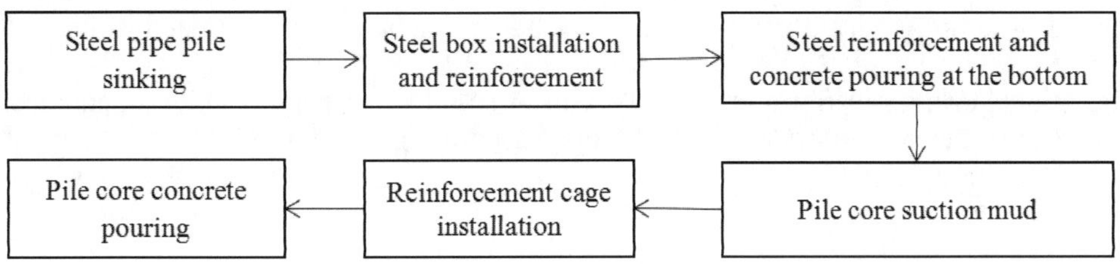

Figure 3. The construction technologies of the friction pile foundation flow diagram.

The S800 hydraulic hammer and piling guide rack are used to carry out pile-sinking operations. The hydraulic hammer is fixed to the piling rack, which can effectively avoid the influence of the eccentric force on the sinking pile and ensure the inclination of the pile body. This process can be used to sink the steel pipe piles directly to the design elevation. The construction process of sinking piles is shown in figure 4.

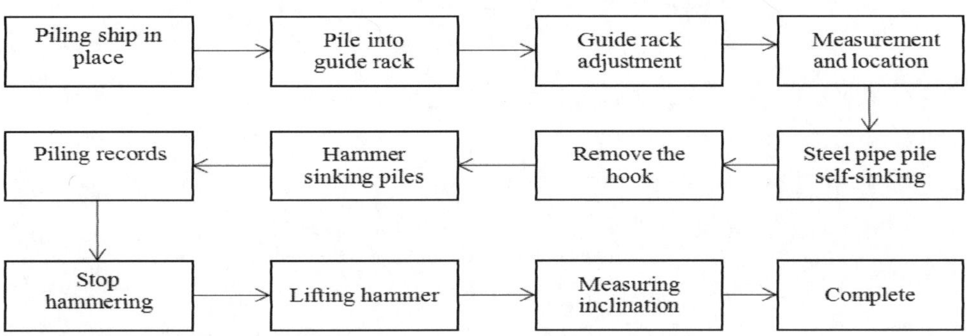

Figure 4. Sinking pile construction process flow diagram.

Friction pile core construction does not need to set up special construction platform. After the completion of the pile sinking, steel boxed cofferdam is directly installed into the top of the steel pipe pile, then reinforcement and concrete pouring are carried out. The bottom concrete structure is the platform for the construction of subsequent pile cores. In order to improve the overall rigidity of the foundation, increase uplift resistance of single pile foundation, mud suction operation and C40 micro-expansion concrete pouring are carried out inside piles.

4.2.2. The construction technologies of the rock-socketed pile foundation. The construction technologies of rock-socketed pile foundation are shown in figure 5.

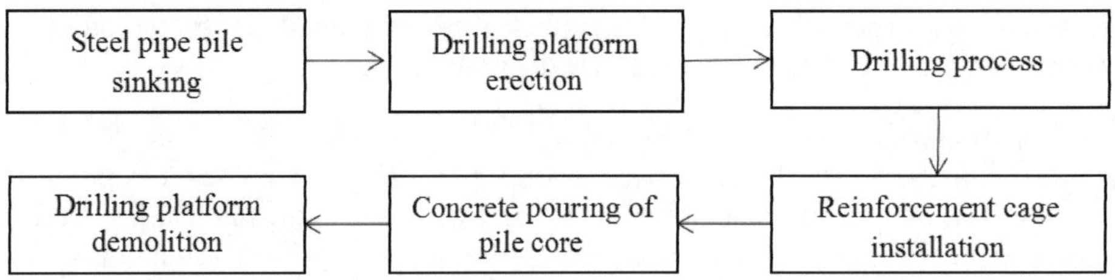

Figure 5. The construction technologies of rock-socketed pile foundation flow diagram.

Friction piles and rock-socketed piles have the same construction technologies of sinking piles. In the position of rock-socketed wind turbine, the installation of integral drilling platform is required when steel pipe piles are sunk. The drilling platform is anchored on the steel pile through the hydraulic system. Use ZJD-3000 drilling rig to carry out drilling construction by gas-lift reverse circulation drilling technology, drill to design depth and then carry out pile core concrete pouring and drilling platform demolition.

The traditional drilling platform erection is built at sea. Due to poor working conditions, small working area, the efficiency is very low. In this project the drilling platform is built on land and then the whole drilling platform is lifted to the top of steel pipe piles, so the construction efficiency improves a lot.

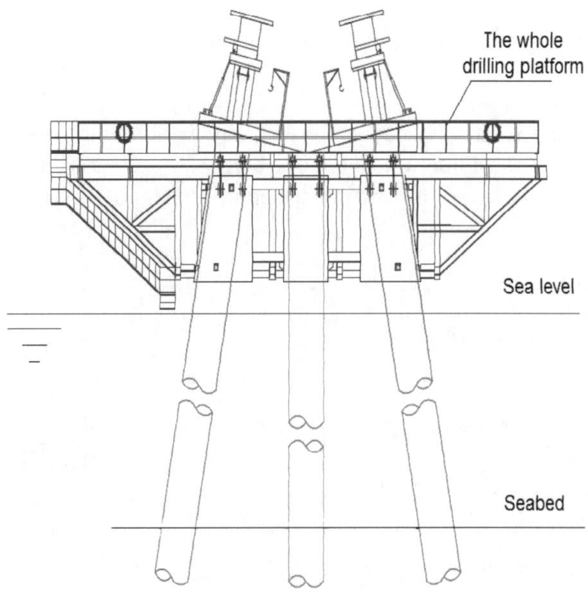

Figure 6. Schematic diagram of the whole drilling platform.

The main support structure of the assembled drilling platform is the steel frame. The drilling platform is 24 m in length and 19.2 m in width. The roof is the lattice type structure, which directly bears the drilling rig weight. The slope of the steel structure at the bottom of the drilling platform is the same as that of the steel pipe pile. The platform and steel pipe piles are anchored by hydraulic system, as shown in figure 6.

Drilling processes mainly include drilling rig in place and installation, drilling, hole cleaning, reinforcement cage hanging and concrete pouring.

- Drilling rig in place and installation. When the drilling platform is installed, the rig is hoisted to the position by crane ship. To ensure that the center of the drill pipe and the steel pipe pile are in a straight line, adjust the slope of the drilling rig and fix the drilling rig on the platform. Deviation is no more than 2 cm.
- Drilling. In the drilling process, the air-lift reverse circulation drilling technology is adopted with the cone bit and φ 320 mm drilling pipe. In order to ensure the inclination of the drilling hole, the centralizer is arranged on the drill pipe near the drilling bit. The compressed air is sent to the bottom of the hole through the air supply pipe, which is arranged on the drill pipe, and the high pressure air and water are mixed at the bottom of the drilling pipe, forming a low density inflatable drilling fluid. The negative pressure is formed in the drilling pipe. Under the pressure difference, the drilling fluid carries the rock debris at the bottom of the hole to the ground. High speed and low pressure drilling process are used when drilling in clay layer. When drilling in sand and pebble, the drilling pressure increases slightly and the air volume increases. When drilling in rock formation, the large drilling press and low speed are used. Drill into the design depth and drilling hole completion.
- Hole cleaning. The air-lift reverse circulation process is used to clear the hole. The compressor sends compressed air through a pipe to the bottom of the hole. Rock debris on the bottom of the hole is suspended by the low density mixed liquid. Circular clearance between drill pipe and steel pipe pile is filled with clear water to drain out the rock debris. The thickness of sediment at the bottom of the hole should be less than 3cm after cleaning. If the sediment is larger than the design requirement, secondary hole cleaning shall be carried out.
- Steel reinforcement cage hanging. The steel reinforcement cage is made and completed on the pier, and then it will be transported to the construction site by transport ship. When placed, it should sink slowly and reduce the friction with the steel pipe. In order to guarantee the vertical degree of the reinforcement cage and the thickness of the concrete protection layer of the outside of the steel cage, after placed in the pile hole, the cage should be fixed.
- Concrete pouring. Before pouring concrete, check hole depth and sediment thickness. To ensure the quality of the concrete reinforcement, the concrete pouring must be done at one time continuously. The internal diameter of the catheter is 250 mm and the spacing between the tubes is 3 meters. The height between the bottom of the tube and the bottom of the hole (the design specification is 300 ~ 400 mm) and the verticality must be ensured to meet the design requirements. The concrete is provided by the concrete mixing ship at the scene. Calculate the amount of concrete in the first pot and then place enough concrete in the funnel. At the beginning of pouring, quickly open the valve. Under concrete's own gravity, the water within the catheter and circular clearance is squeezed out so that the concrete quickly seals bottom and the buried pipes are more than 1m. During the process of continuous concrete pouring, the catheter is slowly lifted with a range of about 15 ~ 60 cm each time. Concrete pouring must be done within initial setting time of concrete. Pouring speed should be moderate, too slow will lead to the formation of hard shell on surface of concrete, too fast will lead to collapse of the hole [16].

4.3. Pile cap construction

The processes of the pile cap construction mainly include manufacture and installation of steel boxed cofferdam, bottom concrete pouring, installation of foundation ring, cap assembling reinforcement, cap concrete pouring, concrete curing and formwork removal of steel boxed cofferdam. The pouring of cap structure is carried out at the sea. The pile cap structure is a cylindrical structure with a diameter of 14 m and a height of 5.6 m. The bottom concrete (a height of 0.8 m) and the cap concrete (a height of 4.8 m) are both poured at once. Pouring concrete pile cap belongs to large volume concrete pouring, which must be constructed strictly according to the design requirements and the code for construction of mass concrete. The water cooling system is used for the curing of concrete in the inner and top of the cap, and the curing days are not less than 14 days. The internal temperature of concrete cap is monitored by the large temperature monitoring system.

Compared with the completed Donghai Bridge Wind power project, the main feature of the project is the need for construction of rock-socketed. At the beginning of the project, the traditional construction technology similar to the Donghai Bridge Wind power project was adopted, but the efficiency was low. The construction period of the friction and rock-socketed pile foundation were 39 days and 122 days respectively. Therefore, it is necessary to optimize the construction technology of high-pile cap foundation. Compared with the traditional construction technology, the construction technologies used in this project have the advantages of low cost, simple operation, high efficiency, and high security. In this project the technical optimizations are as follows: 1) the piling hammer is fixed on the guide rack to improve the efficiency of sinking piles; 2) the drilling platform adopts an integral platform which is assembled as a whole; 3) the steel boxed cofferdam is assembled at the wharf which is holistically lifted and dismantled. Most of the steel bars and embedded parts are arranged in the shore in advance and the rest is completed at sea. After the optimization of the construction technology the construction efficiency is greatly improved. The construction period of the friction pile foundation is shortened by 10 days, and the construction period of the rock-socketed pile foundation is shortened by 48 days.

5. Environmental impact of foundation construction

Wind power construction will have an impact on the environment. The specific impact is as follows:

- The sea is occupied and destroyed. Engineering construction will cause local sea water pollution.
- Marine life. Destroys the habitats of marine life, damages and disturbs marine life.
- Bird. Risks to birds from offshore wind construction arise from four main potential impacts: direct loss of habitat; displacement; collision risk; avoidance behaviour / barrier effect [17].
- Traditional fisheries.
- Impact of maritime traffic.

The new high-rise pile foundation construction time is short, the impact is a local change in species composition. During the construction period, corresponding measures should be taken to reduce the impact on the environment.

- The monitoring and management of environmental quality should be strengthened.
- Adopt ecological remediation measures. Increase the restocking of fish and shrimp.

6. Conclusion

Offshore wind power is the future trend of wind power industry. China's offshore wind power is in a rapid development stage. The foundation of high pile cap is widely used in offshore wind farm construction. In this study, the construction technology of friction and rock-socketed high-rise pile cap foundation are introduced in detail, and technological process of the pile foundation and pile cap construction are emphasized. Practice has proved that construction technique of pile hammer, integral drilling platform and integral steel boxed cofferdam can meet the design requirements in Taiwan Strait. After process optimization, the efficiency of friction piles foundation and rock-socketed piles foundation are shortened by 10 days and 48 days respectively. The high-rise pile cap foundation construction technology is safe, reliable and economical and it will provides some reference value for the construction of offshore wind farm in China.

References

[1] China continues to lead wind power growth 2020 global installed or up to 800 million http://www.cnelc.com/Article/1/160512/AD100415420_1.html
[2] Luke N 2014 Global wind power to hit 2000 GW by 2030 https://www.edie.net/news/6/Global-wind-power-to-hit-2000-GW-by-2030/
[3] Li X Y and Yu Z 2004 Development of offshore wind power *Acta Energiae Solaris Sinica* **25** 78-84
[4] Shang J H, Luo R and Zhang L 2009 Foundation structure selection and construction process for offshore wind turbines *Appl Sci Tech* **36** 6-10
[5] Sanjeev M 2011 Selection, design and construction of offshore wind turbine foundations *Wind Turbines* **10** 231-64
[6] IRENA 2016 Innovation outlook: Offshore wind *Int Renew Energ Agenc* **2** 145-78
[7] Al-Karim G, Rhodri J and Adriana C 2014 Appraisal of the offshore wind industry in Japan https://www.carbontrust.com/media/566323/ctc834-detailed-appraisal-of-the-offshore-wind-industry-in-japan.pdf
[8] Malhotra A S 2011 Selection, design and construction of offshore wind turbine Foundations *AL-BAHADLY I. Wind turbines. InTech* Doi: 10.5772/154610
[9] Wang W and Yang M 2013 Design theory and applications of offshore wind turbine foundation *China construction industry press* **B** 11-41
[10] American Petroleum Institute (API) 1997 Recommended Practice for Planning, Designing and Constructing Fixed Offshore Platforms *Load and Resistance Factor Design* RP 2A-LRFD
[11] ISO 13819-1 1995 Petroleum and natural gas industries – Offshore structures – Part 1: General Requirements, ISO TC 67 (ISO 19900)
[12] Norwegian Technology Center (NTC) 2000 NORSOK standard N-001 *Structural Design* **3**
[13] Li B Y, Guan H G and Yuan J W 2005 Construction of large diameter hundred meters ultra deep pile foundation of main pier of Zhanjiang Gulf Bridge *Highway* **5** 206-8
[14] Qi W G, Tian J K, Zheng H Y, *et al* 2014 Bearing capacity of the high-rise pile cap foundation for offshore wind turbines *Sustainable Development of Critical Infrastructure* **8** 413-20
[15] Lang R Q, Liu R, Lian J J and Ding H Y 2014 Study on load-bearing characteristics of a new pile group foundation for an offshore wind turbine *The Scientific World Journal* **1** 11
[16] Li C J 2011 Analysis of construction principle and quality control measures for pouring underwater concrete by conduit method *Urban Const Theor Res* **35** 119-22
[17] OSPAR Commission 2008 Assessment of the environmental impact of offshore wind-farms http://qsr2010.ospar.org/media/assessments/p00385_Wind-farms_assessment_final.pdf

Prediction model of energy consumption in Jiangsu Province based on constraint condition of carbon emission

Z G Chang[1,3], T T Xue[2,3], Y J Chen[1] and X H Chao[1]

[1]Department of Mining, Xinjiang Institute of Engineering, Urumqi, 830000, China
[2]Department of basis teaching, Xinjiang Institute of Engineering, Urumqi, 830000, China

E-mail: xuett@cumt.edu.cn

Abstract. In order to achieve the targets for energy conservation and economic development goals in Jiangsu Province under the constraint of carbon emission, this paper uses the gray GM (1,1) model to predict and optimize the consumption structure of major energy sources (coal, oil, natural gas, etc.) in Jiangsu province in the "13th Five-Year" period and the next seven years. The predictions meet the requirement of reducing carbon dioxide emissions per unit GDP of China by 50%. The results show that the proposed approach and model is effective. Finally, we put forward opinions and suggestions on the way of energy-saving and emission-reduction, the adjustment of energy structure and the policy of coal consumption in Jiangsu Province.

1. Carbon emission constraint

Since the industrial revolution, the international economy has changed enormously. However, the rapid growth of the global economy is based on the high input, high consumption and low efficiency of the extensive production. A large number of carbon emissions have become the chief culprit of global warming. In order to curb global warming, China has actively cooperated with other governments to promote low-carbon economy with the core of reducing carbon emissions [1].

Jiangsu is a province with large energy consumption. The rapid development of the economy also causes great pressure to the environmental quality, which causes that the major pollutant emissions in Jiangsu economic output have exceeded the national average. The energy bottleneck, environmental carrying capacity and other environmental resources constraints faced by Jiangsu province are increasing day by day [2]. Reasonable control of carbon dioxide and other greenhouse gas emissions, scientific and accurate prediction of coal consumption in the future, can play the key roles in the formulation of relevant development plans and policies by the state and regional government related functional departments [3]. Accurate grasp of the trend of coal consumption, accurate prediction of the future demand of coal can guide the production and supply of coal in a timely manner. It is of great significance to ensure the safety of coal supply in China [4].

The calculation of CO_2 emissions usually adopts the carbon emission coefficient method, which refers to the amount of carbon emissions per unit energy produced during the process of burning or using for each kind of energy [5]. The relationship of carbon emissions between energy consumption and carbon emission coefficient is showed as follows:

$$A_{n,i} = \frac{T_{n,i} \cdot \beta_i \cdot \varphi}{\alpha_i}$$

Where, i =1, 2, 3, respectively represent raw coal, oil and gas. $A_{n,i}$ is carbon emission and $T_{n,i}$ is energy consumption for raw coal, oil and gas. $\alpha_i = [0.7143, 1.4286, 1.3300]$ is the coefficient of equivalent coal for energy determined by the amount of carbon in the chemical formula. $\beta_i = [0.7476, 0.5825, 0.4435]$ is energy carbon emission coefficient set by the national development and reform commission. φ is the conversion factor of carbon emissions to CO_2 emissions, $\varphi=44/12=3.67$.

Therefore, the total carbon emissions over the years are:

$$A_n = \sum_{i=1}^{3} A_{n,i}$$

The carbon emission restriction is mainly embodied in the reduction rate of carbon dioxide emissions per unit area, that is to say, the decrease in carbon dioxide emissions per unit area of GDP compared with the previous year's emissions, abbreviated as carbon dioxide emission reduction rate per unit GDP. Relationships are as follows:

$$\eta_{n+t} = \frac{\frac{A_{n+t}}{GDP_{n+t}} - \frac{A_n}{GDP_n}}{\frac{A_n}{GDP_n}} \quad (1)$$

Use ρ_n to indicate the carbon dioxide emission rate per unit GDP in the upper model, the formula (1) can be expressed as:

$$\eta_{n+t} = \frac{\rho_{n+t} - \rho_n}{\rho_n} \quad (2)$$

According to the data from the National Bureau of statistics, the total energy carbon emissions of Jiangsu and the carbon dioxide emission per unit GDP since 1990 have been calculated, which are shown in table 1. As can be seen from table 1, China's total energy consumption, total carbon emissions and regional GDP values have shown a steady upward trend in recent years. However, due to the continuous change of the proportion of raw coal, oil and natural gas in total energy consumption, the ratio of rising between A_n and GDP_n is inconsistent, which eventually leads to a trend of decreasing annual carbon dioxide emission per unit GDP as shown in figure 1.

Table 1. Calculation for carbon dioxide emissions rate per unit of GDP.

year	Total energy consumption (tec)	$A_{n,1}$--Carbon consumption of raw coal (10^4 t)	$A_{n,2}$--Carbon consumption of oil (10^4 t)	$A_{n,3}$--Carbon consumption of gas (10^4 t)	A_n--Total carbon emissions (10^4 t)	Regional GDP gross (Billionyuan)	Carbon dioxide emissions rate per unit of GDP (t/yuan×10^{-4})
1990	4123.10	12278.57	1091.08	5.04	13374.68	651.82	20.52
1991	4382.20	13117.50	1179.30	5.36	14302.16	744.94	19.20
1992	4922.30	14696.41	1464.48	6.02	16166.90	922.33	17.53
1993	5508.10	16635.51	1663.46	6.74	18305.70	1208.85	15.14

1994	5586.50	17000.89	1687.12	6.83	18694.85	1321.85	14.14
1995	5509.00	16807.36	1762.55	6.74	18576.64	1416.50	13.11
1996	5780.80	17348.19	1858.16	7.07	19213.42	1601.38	12.00
1997	6296.50	19113.41	1958.04	7.70	21079.16	2136.02	9.87
1998	6625.80	20087.47	2000.99	8.11	22096.57	2998.16	7.37
1999	7357.70	22645.12	2024.02	3.25	24672.39	4057.39	6.08
2000	8047.20	24396.68	2153.55	3.09	26553.32	5155.25	5.15
2001	8111.20	24217.35	2231.32	2.27	26450.95	6004.21	4.41
2002	7991.10	23613.27	2365.52	1.96	25980.75	6680.34	3.89
2003	8118.00	23614.35	2403.09	2.76	26020.20	7199.95	3.61
2004	8163.50	24028.88	2526.42	3.58	26558.88	7697.82	3.45
2005	8612.40	24226.44	2935.73	3.90	27166.08	8553.69	3.18
2006	8881.40	24710.40	2814.98	3.41	27528.79	9456.84	2.91
2007	9608.60	26770.51	3002.37	11.75	29784.62	10606.85	2.81
2008	11060.70	29967.23	3654.52	13.52	33635.28	12442.87	2.70
2009	13651.70	36620.44	4000.36	50.08	40670.88	15003.60	2.71
2010	16895.40	46683.22	4824.57	227.24	51735.02	18598.69	2.78
2011	18742.20	51354.50	4931.63	504.15	56790.28	21742.05	2.61
2012	20604.40	55508.28	5236.80	730.59	61475.67	26018.48	2.36
2013	21775.50	56406.93	4948.45	1038.36	62393.74	30981.98	2.01
2014	23709.00	57575.79	5684.39	1031.48	64291.66	34457.30	1.87
2015	25774.00	63321.72	6404.41	1163.85	70889.97	41425.48	1.71

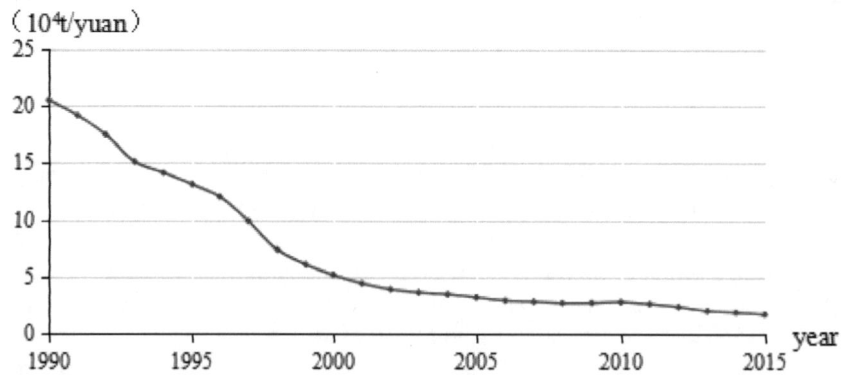

Figure 1. Change trend chart for carbon dioxide emissions rate per unit of GDP.

2. Establishment of grey GM (1,1) model of energy consumption structure

GM (1, 1) model is essentially a differential equation based on the original data sequence, which is the most representative in the gray modeling model [6]. It directly transforms the event sequence data into differential equations, and uses the system information to quantify the abstract model [7]. In the GM model, the original data sequence is accumulated at first, so that the accumulated data presents a certain rule, and then the typical curve is used to fit the graph. The specific steps are as follows:

- First test the data, and require the series ratio to meet the following conditions

$$\sigma(i) = \frac{x^{(0)}(i-1)}{x^{(0)}(i)} \in (e^{-\frac{2}{n+1}}, e^{\frac{2}{n+1}})(i = 2,3,\cdots,n)$$

After inspection, if all series ratios belong to $(e^{-\frac{2}{n+1}}, e^{\frac{2}{n+1}})$, then the GM (1, 1) model can be established.

- Assume there exists the time series data:

$$x^{(0)} = \left\{x^{(0)}(1), x^{(0)}(2), \cdots, x^{(0)}(n)\right\}$$

Make a cumulative for sequence $x^{(0)}$ to generate sequence $x^{(1)}$, so that the item t of the new data sequence $x^{(1)}$ is the sum of the first t items of the original data sequence $x^{(0)}$, i.e.

$$x^{(1)} = \left\{x^{(1)}(1), x^{(1)}(2), \cdots, x^{(1)}(n)\right\} = \left\{x^{(0)}(1), \sum_{t=1}^{2} x^{(0)}(t), \cdots, \sum_{t=1}^{n} x^{(0)}(t)\right\}$$

- the mean series $z^{(1)} = (z^{(1)}(2), z^{(1)}(3), \cdots, z^{(1)}(n))$ can be obtained by averaging the cumulative sequence $x^{(1)}$

$$z^{(1)}(k) = \frac{\sum_{t=1}^{k} x^{(0)}(\mathrm{t})}{k}$$

If there is no difference in mean series, that is

$$z^{(1)}(k) = \alpha z^{(1)}(k) + (1-\alpha) z^{(1)}(k-1) \quad (k = 2, 3, \cdots, n)$$

It indicates that the predicted sequence is less relevant to the original sequence. The further cumulative sequence is required

- The corresponding differential equations of GM (1, 1) model are established

$$\frac{dx^{(1)}}{dt} + \delta x^{(1)} = \mu$$

Where, δ represents development gray number and μ represents the endogenous control of gray number

- Parameter estimation for δ, μ (least square method)

$$\hat{\delta} = \left(B^T B\right)^{-1} B^T Y_n$$

- Establish the cumulative time series forecasting model.

$$\hat{x}^{(1)}(k+1) = \left[x^{(0)}(1) - \frac{\mu}{\delta}\right] e^{-\delta k} + \frac{\mu}{\delta}, \quad k = 0, 1, 2 \ldots, n$$

- Make the difference to get the original forecast

$$\hat{x}^{(0)}(k+1) = \hat{x}^{(1)}(k+1) - \hat{x}^{(1)}(k) = (x^{(0)}(1) - \frac{\mu}{\delta}) \cdot (e^{-\delta t} + e^{-\delta(k-1)})$$

- After the reduction value is derived from the model, the accuracy of the gray model must be tested by residuals:

$$\xi(k) = \frac{x^{(0)}(k) - \hat{x}^{(0)}(k)}{\hat{x}^{(0)}(k)} (\mathrm{k} = 1, 2, \cdots, n)$$

If $\xi(k)$ <0.2, the general requirements are met. If $\xi(k)$ <0.1, the effect is better.

3. The solution of GDP prediction model

In this section, we establish the GM (1, 1) prediction model based on the discussion of the above constraints on carbon emissions. First of all, we forecast the growth of energy consumption and GDP in Jiangsu Province for the "13th Five-Year" and the next five years. Then according to the indicators of constraints by provisions of the state that in 2020, the carbon dioxide emissions per unit GDP of Jiangsu were reduced by 50%, the consumption structure of coal, oil, gas is optimized.

Firstly, taking the data of GDP samples in Jiangsu Province from 1990 to 2015 to predict by using GM (1, 1) model. Then, the model is analyzed and tested by the relative error and residual value between the real value and the predicted value. Finally, the GDP of Jiangsu Province in 2016 -2022 is predicted by the model.

Establish a set of primitive sequences based on GDP statistics:

$$x^{(0)} = (651.82, 744.94, \cdots, 34457.30, 41425.48)$$

In this paper, by using MATLAB software, it is easy to obtain model parameters:

$$\delta = -0.1550, \ \mu = 796.0442$$

Thus, the GDP prediction model is:

$$y = -5137.23 + 5789.05 * \exp(0.154956 * t)$$

Through the model, the prediction of GDP in 1990 -2015 is calculated, and the relative error and residual error are analyzed which are shown in table 2.

Table 2. Relative error and residual index of real value and predicted value.

year	GDP	relative error	residual	year	GDP	relative error	residual	year	GDP	relative error	residual
1990	651	0	0	1999	4057	0.1739	4.29E-05	2008	12442	-0.0864	6.95E-06
1991	744	-0.3024	0.000406	2000	5155	0.2408	4.67E-05	2009	15003	-0.0520	3.47E-06
1992	922	-0.2283	0.000247	2001	6004	0.2389	3.98E-05	2010	18598	0.0090	4.87E-07
1993	1208	-0.0942	7.80E-05	2002	6680	0.2013	3.01E-05	2011	21742	0.0102	4.71E-07
1994	1321	-0.1684	0.000127	2003	7199	0.1347	1.87E-05	2012	26018	0.0342	1.32E-06
1995	1416	-0.2731	0.000193	2004	7697	0.0551	7.16E-06	2013	30981	0.0530	1.71E-06
1996	1601	-0.3148	0.000197	2005	8553	0.007	8.33E-07	2014	34457	0.005883	1.71E-07
1997	2136	-0.1509	7.07E-05	2006	9456	-0.0485	5.14E-06	2015	41425	0.0345	8.33E-07
1998	2998	0.0425	1.42E-05	2007	10606	-0.0915	8.63E-06				

The relative error diagrams are showed in figures 2 and 3 as follows:

As can be seen from the diagram, GM (1, 1) model prediction is very accurate, especially after fifteenth years of prediction, that is, the GDP data after 2005, the relative error between the predicted and the true values is basically between -0.1 and 0.1, and the overall mean error is 0.146. From years' relative error, the fluctuation of GDP is small, and it is basically in steady growth. It can be seen that GM (1, 1) model achieves good fitting effect and can predict GDP value very well.

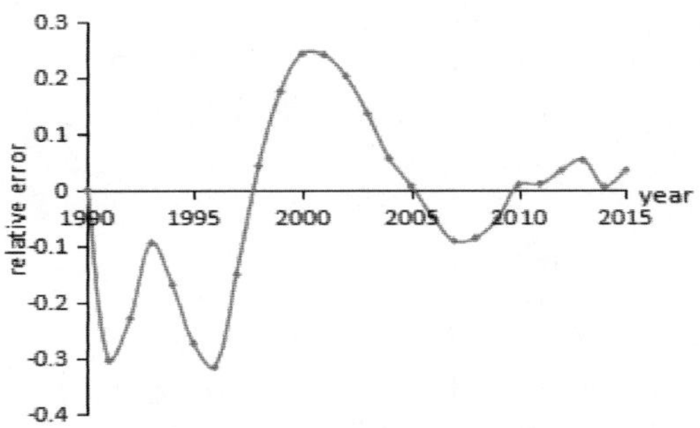

Figure 2. Relative error of prediction results.

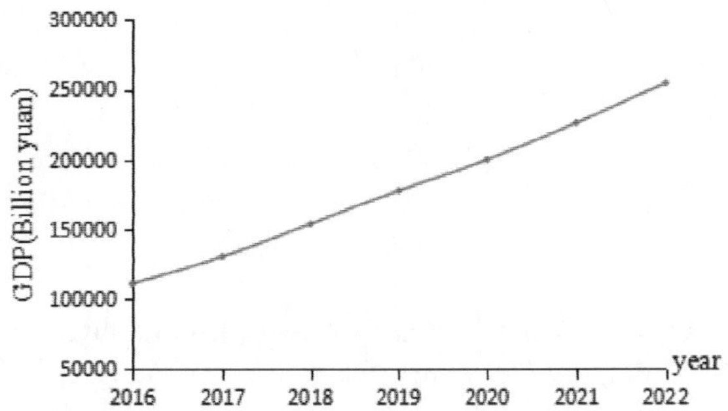

Figure 3. The GDP forecast value from 2016 to 2022.

Therefore, the GDP forecast value by GM (1, 1) model from 2016 to 2022 is in table 3 as follows:

Table 3. The GDP forecast value from 2016 to 2022.

year	2016	2017	2018	2019	2020	2021	2022
GDP (Billion yuan)	110876.98	130248.41	153654.78	177654.58	199857.56	225986.57	254471.38

4. Energy consumption prediction model

The method and process of the energy consumption forecasting model is basically similar to the third section in this paper, so it is omitted. Only prediction model and prediction results are given here:

$$\begin{cases} y_1 = -34418.3 + 37617.8 * \exp(0.713545e-1 * t) \\ y_2 = -10037.9 + 10767.7 * \exp(0.687776e-1 * t) \\ y_3 = -63.809700 + 67.92970 * \exp(0.326690 * t) \end{cases}$$

Using GM (1, 1) model, the prediction of energy consumption in Jiangsu province from 2016 to 2022 is shown in figure 4 and table 4 as follows:

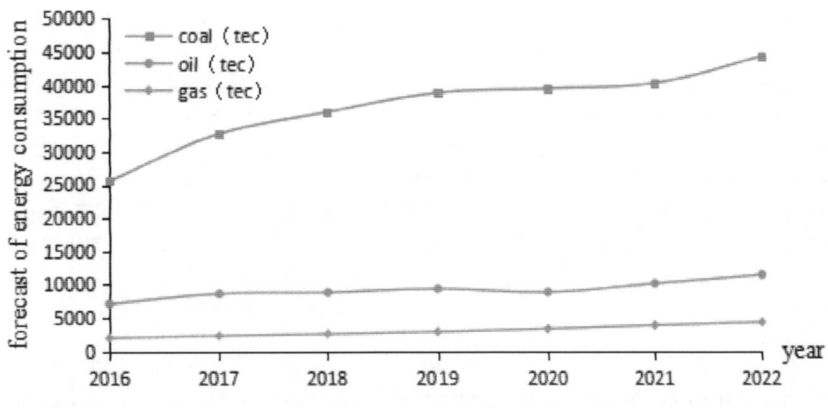

Figure 4. Forecast of energy consumption in Jiangsu from 2016 to 2022.

Table 4. Forecast of energy consumption in Jiangsu from 2016 to 2022.

year	Coal (tec)	Oil (tec)	Gas (tec)
2016	25588.23	7131.013	2028.254

2017	32619.51	8600.24	2347.357
2018	35883.53	8791.086	2618.979
2019	38785.96	9335.081	2917.801
2020	39413.88	8821.068	3333.539
2021	40230.61	10132.95	3836.462
2022	44245.53	11416.45	4293.969

5. Optimization of energy consumption structure

According to the above GDP forecast of Jiangsu during the period of 15th Five-Year and over the next seven years, it can be obtained that the GDP value in 2020 is 199857.56 Billion yuan. We can see from the formula (2) that the carbon emission restriction is the carbon dioxide emission rate per unit GDP not greater than 1.39 in 2020, which can predict the carbon dioxide emissions of that year:

$$\sum_{i=1}^{3} A_{n,i} = \rho_n \cdot GDP_n$$

Also, total carbon emissions are made up of carbon emissions from coal, oil and natural gas. Therefore, the upper form can be translated into:

$$A_{n,1} + A_{n,2} + A_{n,3} \leq \rho_n \cdot GDP_n$$

Combining formula 1, An inequality of consumption of coal, oil and gas can be obtained:

$$3.7349T_{n,1} + 1.4951T_{n,2} + 1.2227T_{n,3} \leq \rho_n \cdot GDP_n$$

Taking the $T_{n,1}, T_{n,2}, T_{n,3}, \rho_n, GDP_n$ of 2020 into the above formula, we can get

$$164472.3 < 199857.56$$

So the consumption predictions of major energy sources in Jiangsu in the next seven years through the GM (1, 1) model meet the requirement of reducing carbon dioxide emissions per unit GDP of China by 50%. The results show that the proposed approach and model are effective.

6. Conclusion

From the prediction results of energy consumption, in the "13th Five-Year" period, the short-term energy consumption structure in Jiangsu province is still dominated by coal, the energy consumption accounts for between 71.7%-76.4%, and the average is 74.2%. Analysis of the data shows that in the next few years, the proportion of oil in the consumption structure is decreasing at a steady rate, while the proportion of demand for natural gas is increasing year by year, which indicates that China's energy structure is shifting towards clean energy and new energy sources. In short, as China's industrialization process, the reserves of non-renewable energy sources are not able to meet the growing demand for energy. Therefore, how to improve energy efficiency and find alternative energy sources will become a major problem facing China's energy industry in the future.

Acknowledgments

This study was supported by the Foundation of Urumqi Science & Technology Bureau (Y141310017) and the Research Foundation of Xinjiang Institute of Engineering (2015xgy331712). The authors are grateful for the supports.

References

[1] Zha J P, Tang F F and Fu H 2011 An empirical study on technical efficiency of China's industry and its determinants--under carbon emission constraints *J Shanxi Financ Econ Univ* **33** 44-52

[2] Mo J L, Agnolucci P and Jiang M R 2016 The impact of Chinese carbon emission trading scheme (ETS) on low carbon energy (LCE) investment *Energ Policy* **89** 271-83
[3] Jiang F L, Huang W B and Ma G W 2016 Factor decomposition of carbon emission from energy consumption of Sichuan Province *China Popul Resour Environ* **26** 45-8
[4] Zhang Y J and Da Y B 2015 The decomposition of energy-related carbon emission and its decoupling with economic growth in China *Renew Sust Energ Rev* **41** 1255-66
[5] Gustavsson L, Joelsson A and Sathre R 2010 Life cycle primary energy use and carbon emission of an eight-storey wood-framed apartment building *Energ Buildings* **42** 230-42
[6] Tao P and Qi Z Y 2010 Gray GM(1,1) forecasting model of investment and financing for ecological environment infrastructure projects *J Eng Manage* **24** 294-8
[7] Li L, Sun Y R, Huang Q, Zhang S H and Xiao Z J 2005 Combined energy forecasting model of gray GM(1,1) and neural networks *Energ Res Util* **1** 10-3

Analysis of optimal design of low temperature economizer

J H Song[1,2,3] and S Wang[1,2,3]

[1]Guangdong Electric Power Grid Co. Ltd. Electric Power Research Institute, Guangdong, Guangzhou, China
[2]School of energy and power engineering, North China Electric Power University, Hebei, Baoding, China

E-mail: songjinghui2006@163.com/1322251803@qq.com

Abstract. This paper has studied the Off-design characteristic of low temperature economizer system based on thermodynamics analysis. Based on the data from one 1000 MW coal-fired unit, two modes of operation are contrasted and analyzed. One is to fix exhaust gas temperature and the other one is to take into account both of the average temperature difference and the exhaust gas temperature. Meanwhile, the cause of energy saving effect change is explored. Result shows that: in mode 1,the amount of decrease in coal consumption reduces from 1.11 g/kWh (under full load) to 0.54 g/kWh (under half load), and in mode 2, when the load decreases from 90% to 50%, the decrease in coal consumption reduces from 1.29 g/kWh to 0.84 g/kWh. From the result, under high load, the energy saving effect is superior, and under lower work load, energy saving effect declines rapidly when load is reduced. When load changes, the temperature difference of heat transfer, gas flow, the flue gas heat rejection and the waste heat recovery change. The energy saving effect corresponding changes result in that the energy saving effect under high load is superior and more stable. However, rational adjustment to the temperature of outlet gas can alleviate the decline of the energy saving effect under low load. The result provides theoretical analysis data for the optimal design and operation of low temperature economizer system of power plant.

1. Introduction

As the energy usage of coal-fired plants constitutes the largest proportion of the national energy consumption, conserving energy and reducing emissions in coal-fired plant are meaningful to the energy-saving strategy of China, which commonly resort to recovering waste heat of flow gas in utility boilers for being beneficial to energy conservation, economy and environment [1].

Implementing low temperature economizer systems is one of the most sophisticated and prevailing ways of recovering waste heat. For improving heat utilizing efficiency and reducing coal consumption, low temperature economizer systems are added behind air preheaters installed at the end of boilers to recover the waste heat of gas flow used to heat condensate water and then facilitate the stream-extracting action of turbines [2,3]. Currently, low temperature economizer systems have been successfully applied in the Secondary Power Plant of Changchun, the Third Power Plant of Shanghai Waigaoqiao, Longkou Power Plant and etc [4-6]. Meanwhile, many relative scholars have deeply researched energy saving effect arched the characteristics of low temperature economizer systems in various aspects; Preciously, Lin W C, *et al* proposed that analysing low temperature economizer systems and its thermally-economic efficiency can be achieved by introducing equivalent enthalpy drop method [7]. Han Z H, *et al* analyzed the thermal economic of adding low-temperature economizer in flue duct of a boiler's tail section to use flue gas to heat condensed water [8]. Huang X Y, *et al* thoroughly analysed operation optimization of low temperature economizer systems [9,10]. For low temperature economizer systems design, Ding L Q, *et al* carried out thermal calculations, systemic calculations and economic requirement calculations in a lot of projects and acquired a standard calculation model to calculate optimal heating squares of low temperature economizer systems [11,12]. Li B, *et al* adopted distributed parameter method to discover the dynamic characteristics of low temperature economizer systems and estimated safety and reliability of the systems [13]. Wang Z J, *et al* put forward a segmented low-temperature economizer flue gas waste heat recovery system to improve the quality of the use of extraction steam [14]. Zhang H D, *et al* introduced the design features and advantages of the combined system of low pressure economizer and air heater [15].

However, many existed researching energy saving effect arches only refer to the engineering application, economic efficiency and optimization of low temperature economizer systems [16-20], there is few of work that describes off-design characteristics of this system. Therefore, by selecting a domestic 1000 MW plant as a case plant, this paper investigates the variation of energy saving effect when low temperature economizer systems works in different operation conditions, and provides important data reference to designing low temperature economizer systems, recovering waste heat of boilers, and even optimizing or modifying the entire boilers.

2. The integrated way of low temperature economizer systems

2.1. Introduction of the case plant

This paper selected a typical 1000 MW ultra supercritical plant as researching energy saving effect arch objective, which includes 1013-28/600/620 single – regeneration condensing turbine with 9-stage regenerator and 1-stage externally-installed stream cooler. Figure 1 symbols the schematic thermal system of the case plant. The main parameters of coal quality analysis are provided in table 1, the main thermodynamic parameters of the boiler are in table 2 and those of Regenerative Heating System in tables 3-6.

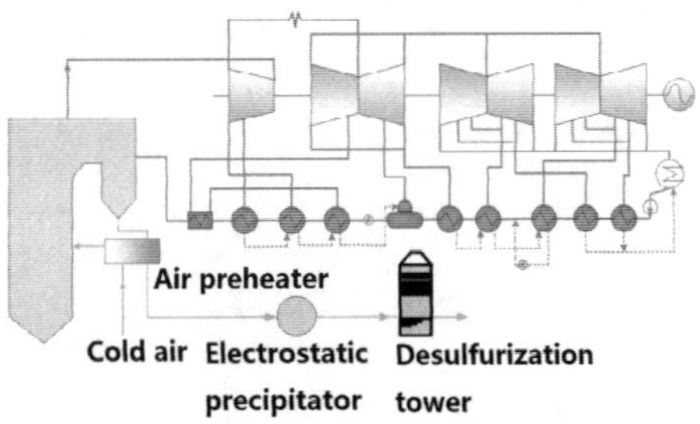

Figure 1. A schematic thermal system of the case plant.

Table 1. Main parameters of coal quality analysis.

Item	Car (%)	Har (%)	Oar (%)	Nar (%)	Sar (%)	Mar (%)	Qnet, ar MJ/kg
Value	57.33	3.26	9.43	0.61	0.63	10.66	21.23

Car-coal as received coal; Har- Hydrogen as received coal Oar-Oxygen as received coal; Nar-Nitrogen as received coal; Sar-Sulfur as received coal Mar-Moisture as received coal

Table 2. The main thermodynamic parameters of the boiler.

Load	100%	90%	75%	50%
Fuel consumption/t/h	366.35	331.44	278.81	193.15

Boiler thermal efficiency/%	94.59	94.48	94.45	94.25
Excess air coefficient	1.15	1.18	1.24	1.35
Air preheater outlet flue gas flow/104 Nm^3/h	393.53	365.112	319.39	239.51
Air preheater outlet flue gas temperature/℃	121	120	115	109

Table 3. The main thermodynamic parameters of the regenerative heating system at 100% load.

Feedwater heater number	5#	6#	7#	8#	9#
Extraction temperature /°C	297.1	247.5	192.2	88.1	63.1
Extraction pressure/MPa	0.60	0.38	0.23	0.07	0.02
Outlet condensed water temperature/°C	153.8	137.1	120.0	84.0	59.2
Outlet condensed water pressure/MPa	1.21	1.26	1.30	1.34	1.39
Condensed water flow/kg/s (lb/s)	569.0 (1251.8)	569.0 (1251.8)	569.0 (1251.8)	503.6 (1107.92)	503.6 (1107.92)
Dewatering temperature/°C	142.7	125.6	122.5	86.8	38.9

Table 4. The main thermodynamic parameters of the Regenerative Heating System at 90% load.

Feedwater heater number	5#	6#	7#	8#	9#
Extraction temperature /°C	298.4	249.0	193.8	84.4	60.4
Extraction pressure/MPa	0.51	0.33	0.20	0.06	0.02
Outlet condensed water temperature/°C	150.1	133.7	117.0	81.6	57.60
Outlet condensed water pressure/MPa	1.12	1.15	1.19	1.23	1.26
Condensed water flow/kg/s(lb/s)	510.7 (1123.54)	510.7 (1123.54)	510.7 (1123.54)	453.5 (0.9977)	453.5 (0.9977)
Dewatering temperature/°C	139.3	122.6	119.5	84.4	39.1

Table 5. The main thermodynamic parameters of the regenerative heating system at 75% load.

Feedwater heater number	5#	6#	7#	8#	9#
Extraction temperature /°C	301.2	252.1	197.1	83.7	57.4
Extraction pressure/MPa	0.43	0.28	0.17	0.05	0.02
Outlet condensed water temperature/°C	143.8	128.1	112.1	77.6	54.7
Outlet condensed water pressure/MPa	0.98	1.01	1.03	1.06	1.08
Condensed water flow/kg/s(lb/s)	425.1 (935.22)	425.1 (935.22)	425.1 (935.22)	379.6 (835.12)	379.6 (835.12)
Dewatering temperature/°C	133.7	117.7	114.6	80.4	39.3

Table 6. The main thermodynamic parameters of the regenerative heating system at 50% load.

Feedwater heater number	5#	6#	7#	8#	9#
Extraction temperature /°C	306.1	257.9	203.6	88.7	50.9
Extraction pressure/MPa	0.30	0.19	0.12	0.03	0.01
Outlet condensed water temperature/°C	130.9	116.5	101.7	69.0	48.1
Outlet condensed water pressure/MPa	0.74	0.75	0.77	0.78	0.79
Condensed water flow/kg/s(lb/s)	287.4 (632.8)	287.4 (632.8)	287.4 (632.8)	259.2 (570.24)	259.2 (570.24)
Dewatering temperature/°C	122.1	107.3	104.3	71.8	39.8

As shown in table 2 the outlet gas temperature of air preheater is relatively high, correspondingly, the heat loss of gas flow is significant, which means that heat energy cannot be efficiently used. According to tables 3-6, with the decreases in working load, the extraction pressure of each stage in regenerative heater gradually decreased, consequently, the amount of utilizable thermal energy in extracted stream reduced. Meanwhile, in the regenerative heater, the outlet temperatures and the amount of condensate water in each grades decreased as the working load reduced.

2.2. The integrating way and optimizing law of low temperature economizer systems

Given that in the boiler of the case plant the heat loss of outlet gas flow is significant, implementing low temperature economizer systems at the flue (behind air preheater) to heat condensate water would successfully decrease the gas temperature and heat loss of gas flow, and then would facilitate the stream-extracting action of turbine. Consequently, rising plant efficiency and reducing coal consumption would be achieved. In addition, adding low temperature economizer systems would decrease inlet gas temperature of low-electrostatic-precipitators and desulfurization tower, which satisfies the requirement of improving dust-removing efficiency and reducing water consumption during desulfurization process.

After comprehensively considering the energy saving effect and practical circumstances of case plant under various operating conditions, LETS was installed in series between Stage 7 and Stage 8 of low-pressure regenerative heater, as shown in figure 2.

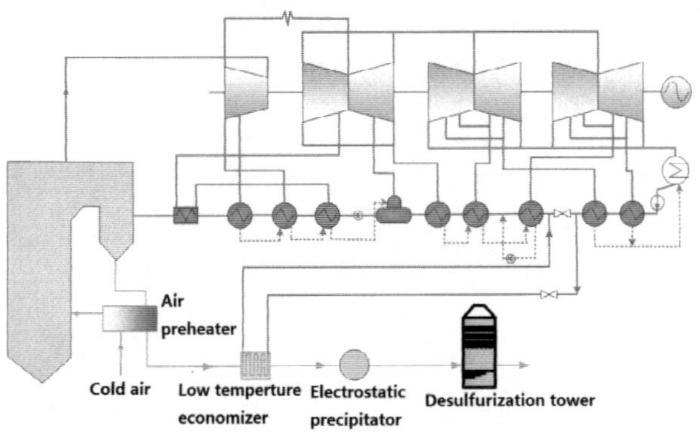

Figure 2. Integrated schematic of low temperature economizer of the case plant.

There are two basic laws which should be followed during system operation optimization:

- for avoiding low-temperature corrosion, the designed outlet temperature of low temperature economizer systems should not be lower than 95℃.
- for feasibility consideration, the Heat Transfer Temperature Difference of low temperature economizer systems in each operating condition should be above 20℃.

3. Thermodynamic analysis of low temperature economizer systems under off-design working conditions

3.1. Thermodynamic calculation of low temperature economizer systems

Based on the integration way and operation optimization laws mentioned in Chapter 1.2, this paper mainly analyses comprehensive performances of low temperature economizer systems under two working modes listed as follows.

Mode 1: low temperature economizer systems operate with fixed outlet gas temperature, which means that the temperatures of each working condition are fixed as 100℃.

Mode 2: low temperature economizer systems operates with consideration of both average temperature differences and outlet gas temperatures, which means that the average temperature difference of heat transfer in low temperature economizer systems is assumed 20 ℃ with the precondition that outlet gas temperature would not be lower than 95°C. Firstly, this paper carries out thermodynamic calculations under the assumption that the average temperature difference of heat transfer in low temperature economizer systems is 20°C. Then, if the outlet gas temperature is higher than 95°C, results are acceptable. If the outlet gas temperature is lower than 95°C, calculate under the assumption that outlet gas temperature is 95°C.

This paper carries out off-design thermodynamic calculations of low temperature economizer systems under the two modes and acquires the corresponding parameters respectively listed in tables 7 and 8.

Table 7. Main parameters under each load of the low temperature economizer system while operating mode 1.

Load	100%	90%	75%	50%
Inlet flue gas temperature/℃	121	120	115	109
Outlet flue gas temperature /℃	100	100	100	100
inlet condensed water temperature/°C	83.98	81.56	77.54	68.94
Outlet condensed water temperature/°C	94.17	91.56	85.47	74.22
Heat transfer temperature difference/℃	20.96	23.08	25.83	32.88
Heat transfer amount/kW	21.87	19.30	12.75	5.76

Table 8. Main parameters under each load of the low temperature economizer system while operating mode 2.

Load	100%	90%	75%	50%
Inlet flue gas temperature/℃	121	120	115	109
Outlet flue gas temperature /℃	98.8	96.2	95	95
inlet condensed water temperature/°C	84.0	81.6	77.5	68.9
Outlet condensed water temperature/°C	94.7	93.4	88.1	77.1
Heat transfer temperature difference/℃	20.0	20.0	21.9	28.9
Heat transfer amount/kW	23.12	22.95	16.99	8.95

Table 7 shows the results of Mode 1 that the outlet gas temperatures under each working condition are above 100℃ and the Heat Transfer Temperature Difference above 20℃, which guarantees the feasibility of the system. Meanwhile, the temperature difference gradually increases as a result of the decrease in working load, then the amount of transferred heat decreases; Transferred heat can reach 21.87 MW under full working load, while it is merely 5.76 MW under half load.

Table 8 suggests the results in Mode 2, the outlet gas temperatures under each working load are respectively 98.8℃/96.2℃/95℃/95℃. Regarding Heat Transfer Temperature Difference, when the working loads are 100% or 90%, temperature differences keep at 20℃. When the loads are decreased at 75% or 50%, temperature differences are separately 21.9℃ and 28.9℃. All of the parameters satisfy the operation optimization law and guarantee its feasibility. As the working load gradually becomes half from its full value, the temperature difference does not change at the first stage and then increases. Meanwhile, with the reduction of working load, the amount of transferred heat slowly decreases, while the range of decreasing is not as large as that in Mode 1.

3.2. Energy saving effect analysis

Carrying out heat balance analysis of power plant which has installed low temperature economizer systems, the energy saving effect of the two operating modes under different working conditions can be acquired and collected in tables 9 and 10.

Table 9. Calculation results of each load while operating mode 1.

Load	100%	90%	75%	50%
Initial generator power/MW	1012.0	910.8	759.0	506.0

Low Temperature Economizer System generator power/MW	1016.3	914.6	761.4	507.0
generator power increment/MW	4.3	3.8	2.4	1.0
Initial system thermal efficiency/%	46.88	46.70	46.32	44.86
Low Temperature Economizer System thermal efficiency/%	47.08	46.89	46.47	44.95
Thermal efficiency increment/%	0.20	0.19	0.15	0.09
Initial coal consumption g/kWh*(l b/kWh)	261.97 (0.576334)	262.96 (0.578512)	265.10 (0.58322)	273.75 (0.60225)
Low Temperature Economizer System coal consumption/g/kWh* (l b/kWh)	260.86 (0.573892)	261.87 (0.576114)	264.25 (0.58135)	273.21 (0.601062)
coal consumption reduce/ g/kWh*(l b/kWh)	1.11 (0.002442)	1.09 (0.002398)	0.85 (0.00187)	0.54 (0.001188)

Table 10. Calculation results of each load while operating mode 2.

Load	100%	90%	75%	50%
Initial generator power/MW	1012.0	910.8	759.0	506.0
Low Temperature Economizer System generator power/MW	1016.6	915.4	762.2	507.6
generator power increment/MW	4.6	4.6	3.2	1.6
Initial system thermal efficiency/%	46.88	46.70	46.32	44.86
Low Temperature Economizer System thermal efficiency/%	47.09	46.93	46.52	45
Thermal efficiency increment/%	0.21	0.23	0.20	0.14
Initial coal consumption g/kWh*(l b/kWh)	261.97 (0.576334)	262.96 (0.578512)	265.10 (0.58322)	273.75 (0.60225)
Low Temperature Economizer System coal consumption/ g/kWh*(l b/kWh)	260.79 (0.573738)	261.67 (0.575674)	263.97 (0.580734)	272.91 (0.600402)
coal consumption reduce/ g/kWh*(l b/kWh)	1.18 (0.002596)	1.29 (0.002838)	1.13 (0.002486)	0.84 (0.001848)

According to table 9, compared with the original system (without low temperature economizer systems), all of the systems which installed low temperature economizer systems operate with significant energy saving effect under Mode 1. This effect was alleviated by reducing working loads; energy saving effect is salient under full load condition, with an increase of 0.2% in heat efficiency of case plant and a decrease of 1.11 g/kWh (0.0024421 b/kWh) in coal consumption. This effect was slightly abated when the load becomes 90%, with the decrease of 1.09 g/kWh (0.0023981 b/kWh) in coal consumption. The influence of reducing load, which would abate energy saving effect, can clearly be seen when the load decreases to 75% where coal consumption is 0.75 g/kWh (0.001651 b/kWh). Finally, energy saving effect in half load is the worst, since the increase in heat efficiency becomes 0.09% and the decrease in coal consumption is only 0.54 g/kWh (0.0011881 b/kWh).

It can be seen in table 10 that under Mode 2, low temperature economizer systems operates with considerable energy saving effect in different working loads; the performance of saving energy is optimized under the 90% load. The effects are similar between the load of 100% and that of 75%, and the amount of decreases in coal consumption are 1.18 g/kWh (0.0025961 b/kWh)and1.13 g/kWh

(0.0024861 b/kWh) respectively. Under the condition of half loading, the value becomes 0.84 g//kWh (0.0018481 b/kWh), which is smaller than that under the condition of 90% loading by 0.45 g/Kwh (0.000991 b/kWh). Generally, the performance of Mode 2 under each working condition is superior to that of Mode 1. In addition, with the increasingly decreased working load, the priority of Mode 2 in saving energy became apparent; Under 75% and 50% loading, the value of decrease of coal consumption in Mode 2 are 1.13 g/kWh and 0.84 g/kWh (0.0018481 b/kWh) respectively, which are larger than those in Mode 1 by 33% and 56% separately.

3.3. Comparison of energy-saving effect in off-design circumstances

As of the results about energy saving effect shown in 2.2, the original case plant is compared with the system which adopts low temperature economizer systems, and then the curve of the reduction of coal consumption rate along with the changing of the load is acquired in figure 3.

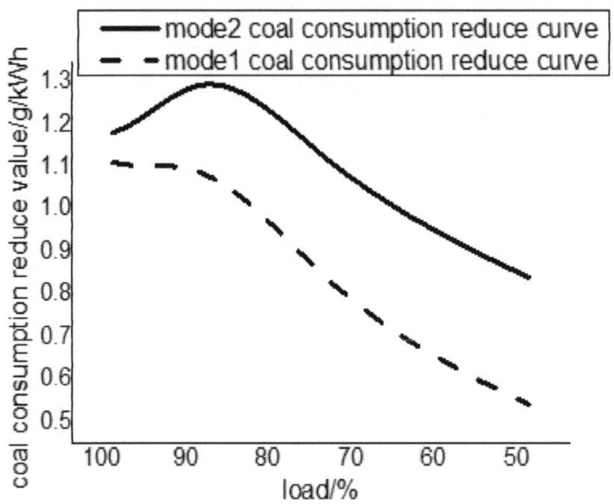

Figure 3. The curve of the reduction of coal consumption rate along with the changing of the load.

From figure 3, apparently, system operates with energy saving effect in both Mode 1 and Mode 2. Comparing the two curves, with the decrease in working load, the curve which symbols the decrease in coal consumption gradually declines in Mode 1; when the load is above 90%, the curve is relatively plain while it sharply decreases once the load is below 90%. In Mode 2, the curve which symbols the reduction of coal consumption rises firstly and then falls with the load decrease. The optimal energy-saving performance occurred when the load is 90%, which means that the reduction of coal consumption is highest. When the load is below 90%, there is a more significant decrease in the reduction of coal consumption. And when the load becomes 75%, the value of coal consumption reduction value is as same as that in full loading condition. Generally, the decreasing trend of the curve which symbols the reduction of coal consumption in Mode 2 is more gentle than that in Mode 1, and energy saving effect of Mode 2 is superior to that of Mode 1 under each working condition. Accordingly, energy saving effect of low temperature economizer systems is similar under high loading conditions, while it would be sharply abated during low loading circumstance. Therefore, energy saving effect of low temperature economizer systems would affect better if the plant keeps high loading operation. Meanwhile, after comparing the two curves, it has been discovered that energy saving effect under low loading circumstances can be improved by adjusting the outlet gas temperature of low temperature economizer systems.

4. Cause analysis of energy saving effect changes in off-design conditions

4.1. Heat transfer temperature differences and temperature ranges of flue gas heat rejection

The level of thermodynamic completion of low temperature economizer systems significantly influences the energy saving performance in off-design working conditions, which is often reflected by the values of Heat Transfer Temperature Difference. Besides, temperature range of flue gas heat rejection reflects the amount of remained thermal energy to some extent, which is closely relative to the energy-saving performance of system. Figures 4 and 5 suggest the off-design Heat Transfer Temperature Difference and the temperature ranges of Flue Gas Heat Rejection under both the two operation modes.

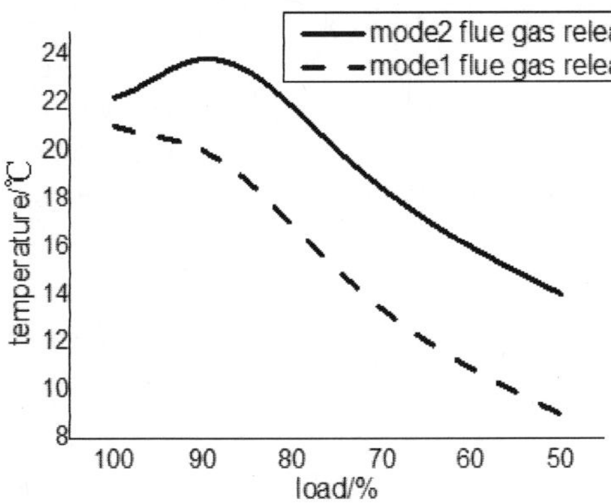

Figure 4. The curve of heat transfer temperature difference with the changing of the load.

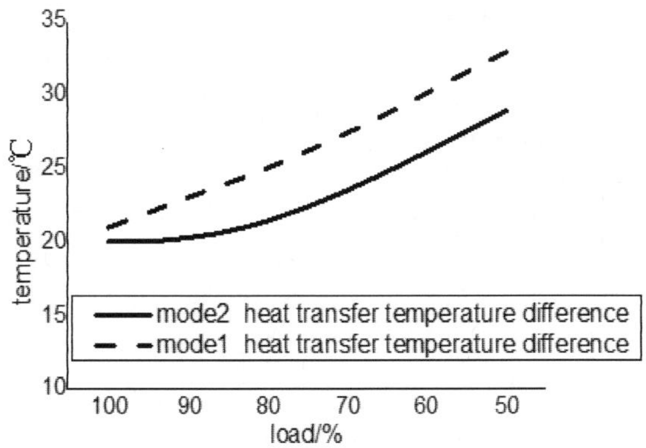

Figure 5. The curve of flue gas heat release temperature with the changing of the load.

As of figure 4, with the decrease in working load, heat transfer temperature difference gradually increases, energy loss correspondingly increases, and the degree of thermodynamic completion drops. In Mode 2, Heat Transfer Temperature Difference is in plateau firstly with the loading reduction and then increases. When the load is below 90%, as the load decreased continuously, Heat Transfer Temperature Difference slowly increases, as well as energy loss. Consequently, the degree of thermodynamic completion of low temperature economizer systems decreases, which influences the energy saving effect of the entire system.

Figure 5 suggests the relation between temperature ranges of Flue Gas Heat Rejection and working loads. In Mode 1, when the working load is reducing, the temperature difference is decreasing as well, then the recoverable waste heat is decreased. In Mode 2, the temperature range firstly rises and then declines with the increasingly reduced working load. The peak value, which means there is the greatest amount of recoverable waste heat, occurs when the load is 90%. Comparing the two curves of temperature range, it can be seen that the ranges in Mode 2 are always higher than those in Mode 1. Therefore, energy saving effect of Mode 2 is superior to that of Mode 1. Generally, when the load keeps above 75%, the variation of heat transfer temperature difference is slight, and there is not significant difference in the degree of thermodynamic completion of low temperature economizer systems. Meanwhile, given that temperature range of Flue Gas Heat Rejection is relatively high, which means that the amount of recoverable waste heat is relatively high, system operates with efficient energy-saving performance. While under low-loading conditions, the variation of Heat Transfer Temperature Difference becomes greater than that in high load condition, then the degree of thermodynamic completion is abated. Moreover, since the temperature range of Flue Gas Heat Rejection is relatively low as well, system cannot operate with highly-efficient energy-saving performance. Through analysing the two modes and operating the optimized system, it can be discovered that adjusting outlet gas temperature would improve the performance of low temperature economizer systems in low-loading conditions, which means that system can still keep well energy saving effect in low-loading operation. Importantly, safety should not be compromised in all circumstances.

4.2. The variation of gas flow rates

The rate of gas flow, which passes through low temperature economizer systems, significantly influences the amount of transferred heat. Figure 6 shows the outlet Gas Flow Rate of air preheater under each working condition of the case plant.

From figure 6, with the decrease in working load, the curve which symbols the outlet Gas Flow Rate of air preheater proportionally decreases; in both Mode 1 and 2, the rates of gas flow, which passes through the air preheater, gradually reduce as the load decreases, then the total amount of heat rejection in gas flow decreases as well. Therefore, the influence of Gas Flow Rate on the two modes would not be taken into account. However, the influence of this flow rate on the energy saving effect of low temperature economizer systems is unneglectable. With the constantly decreased load, the flow rate proportionally reduces, which causes a gradually declined amount of transferred heat in low temperature economizer systems. As a consequence, energy saving effect is abated.

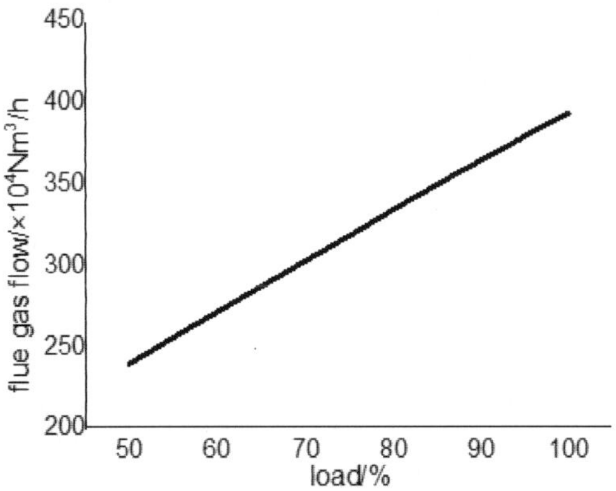

Figure 6. The curve of flue gas flow rate of air preheater outlet with the changing of the load.

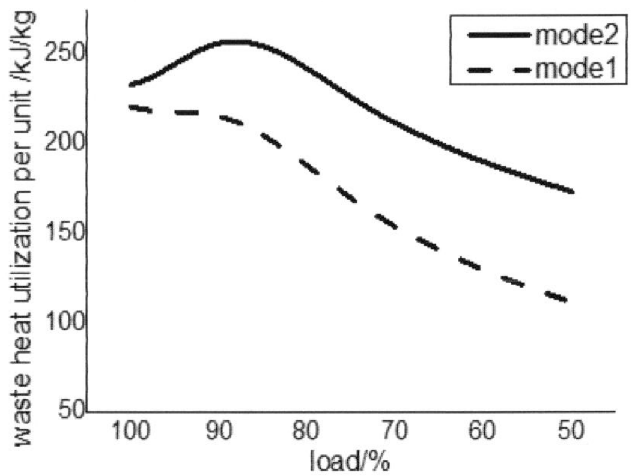

Figure 7. The curve of the amount of waste heat recovery of fuel unite with the changing of the load.

4.3. The amount of waste heat recovery (Kilojoules per Kilograms)

The amount of waste heat recovery should be considered during analysing energy saving effect of low temperature economizer systems, although it is incomprehensive that only comparing waste heat recovery amount under different working conditions. With the changes in working load, there are also variations of the required fuel amounts and the required outlet gas temperatures of the boiler. Therefore, it is unreasonable that only considering the parameter – the amount of waste heat recovery – to analyse the causes of the change in energy saving effect. Consequently, a standardised unit (kilojoules per kilograms) has been introduced for further r energy saving effect arches. Figure 7 shows the amount of waste heat recovery of low temperature economizer systems in off-design conditions under both the two modes.

In figure 7, both the two curves are plain under high working loads and are sharp under low loads. The heat recovery amount is actually the superposition of the Gas Flow Rate variation under each load and the temperature range variation of Flue Gas Heat Rejection. In Mode 1, the amount of waste heat recovery (kilojoules per kilograms) declines as the load decreases; in high loading condition there is only slight change in the heat recovery amount, while the change that being decreased in this amount gets increasingly fast as the load reduces. In Mode 2, as working load declines as well, the recovery amount rises firstly and then declines. Furthermore, this figure insignificantly changes under high load operation, and dramatically drops once the load is below 75%. As of the results, in both the two modes, the amounts of recoverable waste heat (per kilograms) are higher under high load conditions than those under low loads, which can substitute more extracted stream from turbines to achieve better energy-saving performance. Above all, more waste heat can be recovered by keeping low temperature economizer systems operates under high load conditions.

As of the comparison between figures 4 and 7, the variation trend of waste heat recovery amount is as same as that of temperature ranges, which verifies that being narrow of the heat rejection temperature range under low working loads is the main cause of being small of the waste heat recovery amount. Furthermore, the comparison between figures 3 and 7 suggests that the variation trend of heat recovery amount is as approximately same as that of the final energy saving effect (the decrease in coal consumption), which means that the decrease in heat recovery amount (per kilograms) caused by being narrow of Flue Gas Heat Rejection temperature range and being reduced of the gas flow amount is the main reason of the abated energy-saving performance under low working loads in low temperature economizer systems.

5. Conclusion

This paper discusses the off-design characteristics of low temperature economizer systems under two operation modes – one is to fix exhaust gas temperature, the other mode is taking the average temperature difference and the exhaust gas temperature into account – by thermodynamically analysing and comparing the energy-saving performances under off-design working conditions. Besides, this paper also investigates causes of the changes in energy saving effect of low temperature economizer systems under off-design working conditions. In conclusion, the results demonstrate that:

- Under the operating mode with fixed outlet gas temperature, in the case plant which installs low temperature economizer systems, when the working load is halved from full state energy saving effect is apparently abated. Meanwhile, the amount of decrease in coal consumption reduces from 1.11 g/kWh (0.0024421 b/kWh) (under full load) to 0.54 g/kWh (0.0011881 b/kWh) (under half load). The effect of energy saving in half load is only 45% of that in full load. Regarding the changing speed of energy saving effect, in the high loading circumstances, this effect slowly changes while it sharply drops once the load becomes low.
- Under the operating mode which takes both the average temperature difference and the exhaust gas temperature into account, as the load also declines, the energy-saving effect is firstly intensified and then abated. The optimal energy-saving performance occurs when the load is 90%, which also symbols an increase in heat efficiency of 0.23% and a decrease in coal consumption of 1.29 g/kWh (0.0028381 b/kWh). When the load decreases from 90% to 50%, the decrease in coal consumption reduces from 1.29 g/kWh (0.0028381 b/kWh) to 0.84 g/kWh (0.0018481 b/kWh). Under high-loading conditions, the energy-saving effects of low temperature economizer systems are insignificantly influenced by the load variations. As the load continuously decreases, the energy-saving performance sharply abates. Therefore, well energy-saving performances can be achieved by keeping low temperature economizer systems operating under high working loads.
- In conclusion, the energy-saving performance is relatively well under higher-loading operations in both the two modes, which means that it would be considerably abated under low-loading operations. There are two main causes of this phenomenon: Firstly, the inlet gas flow temperature of low temperature economizer systems significantly decreases when the working load becomes low, meanwhile, the outlet temperature of low temperature economizer systems should be high enough to prevent from low-temperature corrosion. Secondly, the amount of flue gas reduces under increasingly low working load. Both the two points would cause a considerable decrease in the amount of recoverable waste heat which is exhausted from air preheater, furthermore, would abate the energy-saving performance of the system.
- Both Heat Transfer Temperature Difference of low temperature economizer systems and the outlet gas temperatures are considered in Mode 2, with the precondition of guaranteeing safety and minimum heat transfer temperature difference, slightly reducing the outlet gas temperature of low temperature economizer systems (under low loading condition) would extend the

temperature range of flue gas heat rejection in low temperature economizer systems. Compared with the energy-saving performance in Mode 1, this adjustment improves that in Mode 2, particularly under low working load conditions.

In conclusion, when the system is needed to operate under low working load conditions, properly reducing the final outlet gas temperature would considerably improve the working performance of low temperature economizer systems, with the precondition of safety and minimum heat transfer temperature difference.

References

[1] Xu G, Xu C, Yang Y P, *et al* 2013 Deep utilization of exhaust heat and comprehensive optimization of tail heating surfaces for utility boilers *Proceedings of the CSEE* **33** 1-8
[2] Huang X Y, Hi Y T, Sun F Z, *et al* 2008 Practice of low pressure economizer installed in 670 t/h coal fired to reduce the exhaust gas temperature *Elect Power* **2008** 55-8
[3] Kang X N, Ma W J, Tao M A, *et al* 2012 Economic study on low-temperature economizer on a 320 MW unit *Therm Power Gener* **2012** 8-11
[4] Deng J L, Huang S W, Xu G, *et al* 2013 Efficient waste heat recovery system of power plant boiler East China *Elect Power* **41** 200-4
[5] Liu H Z and Lian Z Q 2010 Application of low temperature saving coal device in generate electricity power plant *Elect Power Surv & Des* **4** 32-8
[6] Huang X Y 1998 The optimum design of low pressure economizer system of boiler No. 1 in Longkou power plant *Boiler Technol* **1998** 22-5
[7] Lin W C and Sun S W 1991 Energy saving technology of coal-fired power plant thermal system and its application *Elect Power Technol* **9** 17-20
[8] Han Z H and Li P 2016 Thermal economic analysis for adding low-temperature economizer in boilers *Therm Power Gener* **2016** 20-5
[9] Huang X Y and Wang L P 1999 Optimal general mathematical model of LP economizer system in steam power *Plant Power Syst Eng* **15** 20-5
[10] Huang X Y, Ping Y M and Sun F Z 2003 Optimum water flow distribution for a low pressure economizer system in power stations *J Hydrodyn* **18** 526-31
[11] Ding L Q, Jie H L, Dai W, *et al* 1997 Optimization model for design of parameters of boiler waste heat recovery installations *Power Engin* **17** 16-9
[12] Ding L Q and Zhang L 2006 Economic analysis of installing low pressure economizer in thermal power plant *J Northeast Dianli Univ* **2006** 26-30
[13] Li B and Dang Z L 2014 Design and dynamic analysis of a low teperature economizer *Therm Power Gener* **42** 25-9
[14] Wang Z J, Xiong Y Q, Su Y H and Song J 2016 Application of segmented low-temperature economizer in the flue gas heat recovery system of a boiler *Power Eq* **2016** 34-7
[15] Zhang H D, Zhang Z X, Jia Z P, Liang S G, Xu D Q, Zhang G C and Hou X 2016 Application of combined system of low pressure economizer and air heater *Therm Power Gener* **2016** 26-30
[16] Xue N, Yin X Y and Wang C C 2015 Combined application of hot primary air heater and low pressure economizer *Therm Power Gener* **44** 65-8
[17] Zhang H F, Wang Y, Tian S F, *et al* 2011 Economic analysis of low pressure economizer system based on the equivalent enthalpy drop method *Energy Conserv Technol* **29** 457-61 (in Chinese)
[18] Niu S Y 2012 Adding low temperature economizer to a CFB boiler for energy saving *Therm Power Gener* **2012** 9-12
[19] Xie H L, Ding L Q, *et al* 1995 Analysis and comparison about methods of waste heat recovery in power plant boiler *Boiler Technol* **1995** 15-9 (in Chinese)
[20] Wen S, Yan W P, *et al* 2013 Two-stage low pressure economizer technology and it's economic analysis *Therm Power Gener* **42** 7-11

A review on prognostics approaches for remaining useful life of lithium-ion battery

C Su[1, 2] and H J Chen[1]

[1]School of Mechanical Engineering, Southeast University, Nanjing 211189, China

E-mail: suchun@seu.edu.cn

Abstract. Lithium-ion (Li-ion) battery is a core component for various industrial systems, including satellite, spacecraft and electric vehicle, etc. The mechanism of performance degradation and remaining useful life (RUL) estimation correlate closely to the operating state and reliability of the aforementioned systems. Furthermore, RUL prediction of Li-ion battery is crucial for the operation scheduling, spare parts management and maintenance decision for such kinds of systems. In recent years, performance degradation prognostics and RUL estimation approaches have become a focus of the research concerning with Li-ion battery. This paper summarizes the approaches used in Li-ion battery RUL estimation. Three categories are classified accordingly, i.e. model-based approach, data-based approach and hybrid approach. The key issues and future trends for battery RUL estimation are also discussed.

1. Introduction

With the significant advantages, such as high energy density, high galvanic potential, wide temperature range, no memory effect and long service life, lithium-ion (Li-ion) battery has been widely used in industrial areas, including communication, aviation and vehicle [1]. However, the performance of battery will deteriorate with time gradually, and failure will occur from time to time when the performance exceeds a given threshold. It will lead to a reduction in system's performance, and result in increased cost and even catastrophic failures [2]. By now, various studies have been carried out on performance degradation and remaining useful life (RUL) estimation of Li-ion battery.

RUL is usually defined as the time from the present health state to failure. RUL estimation tries to use previous and current performance to predict system's future state. In addition, reliable forecast information can provide alarm before failures occur, thus timely repair and maintenance schedule can be made in advance to prevent performance degradation, malfunction or even catastrophic failures [3]. As for Li-ion battery, RUL can be defined as the number of cycles remaining until the battery's capacity falls below a predetermined threshold [4].

Generally, the approaches for RUL estimation are divided into model-based approach and data-driven approach [3]. However, some novel estimation approaches belong to neither of them. Therefore, in this study we classify the prediction methods into three categories, i.e. model-based approach, data-driven approach and hybrid approach.

2. Model-based approach

Model-based approach is to establish physical model of equipment's life cycle, including load condition and failure mechanism. Therefore, it is necessary to study system's failure mechanism, build a right model, select appropriate parameters which can reflect system characteristics, revise and adjust the prediction model in time.

According to [5] and our findings, model-based prediction methods of Li-ion battery can be classified into physical-based model, equivalent circuit model and filtering approach. In some literatures, filtering approach is also grouped into data-driven method [6]. In this study, three types of model-based approaches are considered:

- Physical-based model: describe and model the battery physical performance evolution.
- Equivalent circuit model: use equivalent circuit model to simulate the operation process.
- Filtering approach: provide a sequential estimation to the system states.

2.1. Physical-based model

Physical-based model seeks to quantify the factors that influence battery's performance, and obtains the description of performance evolution. However, considering that several factors may interact with each other to impact the degradation of the performance, it is not easy to make a reliable and precise model to simulate battery system. Thus, this approach usually focuses on the specific physical and chemical phenomena occurring during the utilizations [5].

Phenomenological model is extensively used in performance prediction. It can provide dynamic description of battery cell, while it does not involve in the electrodes materials. Christensen, *et al* [7,8] linked the capacity decay with side reactions and resistance increase, they also developed a model to represent the solid electrolyte interphase (SEI) which causes a significant increase in impedance. Safari, *et al* [9] introduced a phenomenological model to estimate the health state of a Li-ion cell and assessed the impact on battery performance caused by capacity loss and the increase of SEI.

In some phenomenological models, the intrinsic properties related to electrode and electrolyte materials cannot be precisely involved and considered. In order to further understand the phenomena appearing during degenerative process at a nanoscale level and with a higher prediction accuracy, some models have adopted atomic and molecular methods. By using a multi-interface superlattice method and calculated through density functional theory (DFT), Dalverny, *et al* [10] investigated the interface electrochemistry in conversion reactions. Tasaki, *et al* [11] linked the capacity fade andlithium salts dissolution. In addition, molecular dynamic is also an effective way to understand the degradation mechanism.

The performance degradation of battery is dynamic and non-linear. Due to the complexity of electrochemical system, it is difficult to monitor a battery's internal state in real time, thus an exact physical model is not so easy to obtain. Furthermore, accurate estimation requires a variety of parameters and complex calculations due to the various uncertain factors. Therefore, although physical-based method has significant achievements in mechanism analysis, technically it is unavailable for industrial applications at present.

2.2. Equivalent circuit model

Equivalent circuit model is widely adopted to simulate battery's running process and estimate parameters in models [5]. This method requires large and diverse data set.

Yoshida, *et al* [12] investigated the mechanism of capacity fade with experimental data from Li-ion cells on satellite, and build a prediction model to fit the capacity decay data. Later, they improved the previous estimation model where the SEI growth blocking mechanism was taken into account [13]. The result shows that the improved model can fit long-term capacity decay data better. Andre, *et al* [14] proposed a structured neural network (SNN) algorithm to describe the mathematical function of battery. The SNN is achieved through the mathematical description of the electrical equivalent circuit model.

2.3. Filtering approach

Filtering approach is one of the most common-used methods in RUL estimation. Usually, the calculation of filtering approach includes two steps, i.e. prediction and correction.

Kalman filter (KF) is a recursive filter especially for time-varying linear systems. It estimates the state of a process by combining the past measurement estimation errors into new measurement errors so as to estimate the future errors. Wang, *et al* [15] presented a battery state estimation method based on KF. Considering that the extended Kalman filter (EKF) can deal with nonlinear Gauss problem better, Xu, *et al* [16] adopted expectation maximization (EM) and EKF to estimate the parameters o in the model and update the states. Given that the unscented Kalman filter (UKF) has higher precision than EKF, He, *et al* [17] proposed a joint coulomb counting method, where UKF is used to adjust parameters, estimate health states and predict RUL of the battery.

Particle filter (PF) is based on Monte Carlo method with sequential importance sampling technique. The core idea of PF is to express the distribution by extracting random state particles from a posteriori probability. Yu, *et al* [18] integrated PF and logistic regression to predict RUL of Li-ion battery with the degradation experiment data. Wang, *et al* [19] constructed a state-space model to assess the capacity degradation, and then presented a spherical cubature particle filter to solve the state-space. The analysis result shows that this algorithm is more effective than the other PF-based prediction approaches.

Model-based method can reflect physicochemical characteristics of the battery to some extent. However, due to the complex reaction mechanism as well as the calculation, model-based method is not usually used to estimate the RUL of Li-ion battery alone, some other methods are mixed together to improve the prediction performance.

3. Data-driven approach

Data-driven approach extracts effective feature information to form system's operational data obtained by advanced sensor technology, and then constructs a degradation model to predict RUL. Thus, accurate descriptions of the system mechanism are not essential for it. In other words, it describes the degradation-inherent relationship and the trend based on data. Generally, data-driven approach includes two categories, i.e. artificial intelligence (AI) technique and life expectancy model [6,20].

3.1. Artificial intelligence

AI method uses monitoring data to fit a degradation model, and estimate RUL through extrapolating the characteristics variables. AI approach mainly includes neural network (NN), support vector machine (SVM) and relevance vector machine (RVM). Table 1 gives a comparison for the three kinds of AI methods.

Table 1. Comparison of AI methods.

Item	NN	SVM	RVM
Principle	A non-linear and self-adaptive information processing system formed	A supervised learning model in high dimensional feature space.	Based on Bayesian framework
Advantage	1. Non-linear approximation ability 2. High robustness and strong fault-tolerant capacity 3. Strong capacity of synthetically processing information	1. Avoid the curse of dimensionality to a certain extent 2. Require less training data than NN 3. Effectively avoid a local minimum 4. High precision, and convergence	1. Effectively avoid over-fit or less-fit 2. Better generalization performance 3. Acquire point estimation and interval estimation 4. Support uncertainty representation 5. Do not need many parameters

Disadvantage	1. Require lots of data 2. Network structure is complex 3. Training may lead to over-fitting and high computation 4. Computational complexity is large 5. Lack of uncertainty representation	1. Require lots of data 2. Kernel function must satisfy Mercer conditions 3. Parameters selection lack of guidance and hard to determined, 4. Lack of uncertainty representation	1. Require lots of data 2. Parameters selection lack of guidance and hard to determined 3. Poor long-term prediction capability 4. Low stability
References	Wu, *et al* [21] used modified Elman network to predict the remaining capacity, and analysed the relationship among varying characteristics, internal resistance and open-circuit voltage.	Wang [22] devised an iterative and multi-step linear SVM model using energy efficiency and working temperature as input and estimated the RUL at room temperature.	Liu, *et al* [1] used an online training strategy in RVM to improve the ability of dynamic model updating and enhance prediction ability.
	Andre, *et al* [14] proposed a SNN algorithm with prior knowledge to describe the mathematical function of battery voltage, current and resistance with the equivalent circuit.	Dong, *et al* [23] modelled the relationship among the battery cycle times, capacity, and internal resistance. They combined SVM with PF to achieve parameters estimation and predict the RUL.	Zhang, *et al* [24] used wavelet decomposition to reduce noise, and RVM improved by dierential evolution algorithm is used to estimate battery's RUL based on the denoised data.

3.2. Life expectancy model
Life expectancy model determines RUL of individual component based on the expected deterioration risk under known operating conditions. It can also be divided into two types, i.e., stochastic model and statistical model [20].

3.2.1. Stochastic model. Stochastic model is used to characterize the performance degradation process, and infer the distribution of the RUL. Given that the degradation process of Li-ion battery is essentially an uncertain stochastic process, various stochastic process models have been proposed, including Gaussian process (GP), Wiener process (WP), hidden Markov model (HMM).

Liu, *et al* [25] adopted Gaussian process regression (GPR) to obtain the uncertain interval of RUL prediction results, and built a model to predict RUL of the battery online. However, by now this method has not been fully explored. Li, *et al* [2] constructed an improved Gaussian degradation model by monitoring data in different battery conditions, and PF was used for the parameters identification. Tang, *et al* [26] presented a WP-based method with measurement error, which has higher accuracy of battery RUL prediction than the simple WP-based method. Si, *et al* [27] presented a WP model with recursive filter algorithm, both the recursive filter and EM algorithm were used to update drift coefficient and re-estimate unknown parameters in time. Pattipati, *et al* [28] applied the HMM model into SVM to represent the uncertainty of prognostic.

According to these studies, stochastic model can describe the degradation process of Li-ion battery. However, some dynamic factors have not been taken into consideration, such as time-varying environment, random varying current, self-recharge characteristics [6].

3.2.2. Statistical model. Statistical model predicts the damage initiation and progression according to the previous inspection results on similar machines. This method does not need knowledge on the ageing mechanism, while it requires a large number of effective data set. Among them, autoregressive (AR), autoregressive and moving average (ARMA), autoregressive integrated moving average (ARIMA) models are widely used for the modeling and prediction of the time series data.

Long, *et al* [29] adopted AR model to track battery's capacity degradation process, particle swarm optimization (PSO) algorithm and a new error criterion were used to decide the order of AR model. Liu, *et al* [3] used an optimized nonlinear degradation AR model and regularized PF to estimate the RUL. Saha, *et al* [30] adopted an ARIMA algorithm to model battery internal parameters and its capacity. Besides, they compared the prediction ability of several data-driven methods on Li-ion battery, including ARIMA, RVM, EKF and PF.

Furthermore, some data-driven prognostics approaches have not been covered in this paper, for example grey model, Markov chains, similarity matching. In summary, data-driven approach considers only the features of data, and establishes prediction model based on historical and monitoring data. Thus, the adaptability and robustness are always the challenges for the application of this kind of method. In some situations, the sensitivity of parameters setting is another major concern. Among the various data-driven prognostic methods, the methods supporting uncertainty representation and management are hot issues for battery applications.

4. Hybrid approach

Hybrid approach is proposed via combining two or more model-based or data-driven methods to strengthen their advantages, overcome their limitations and thus improve the prediction performance. Generally, there are two categories of hybrid approach, one is the fusion method containing both model-based and data-driven methods, the other one integrates different data-driven methods [4]. Here we focus mainly on the first category.

According to [31] and our study, hybrid approach combining model-based and data-driven methods for battery RUL prediction can be divided into four types.

- Use data-driven method to infer/compensate the physical model. It is acceptable to displace or compensate the complicated physical model by data-driven model when the degradation model is tedious to obtain or the system state cannot be measured directly. In [32], a NN model was developed to estimate the battery's state of charge, and then UKF was used to reduce the errors and improve the prediction accuracy.
- Use data-driven method to estimate future measurements for model-based method. The predicted measurements from data-driven method can be regarded as new measurements in model-based method when lack of measurements during long-term prediction. Wang, *et al* [33] adopted a data-driven predictor to predict battery's future measurements, and these predicted measurements are incorporated into PF to update the prediction model parameters.
- Use data-driven method to estimate/adjust the parameters of physical-based method. Data-driven method describes the degradation-inherent relationship and trend based on data, and uses it to estimate the parameters in the model. In [34], RVM was used to obtain the most representative relevant vectors and to determine the parameters of the capacity degradation model.
- Use filtering approach to estimate/adjust the parameters of data-driven method. Filtering is often used to reduce noise in data and estimate the parameters of the model. In [35], PF approach was adopted to adjust the parameters of a fused regression model online in order to track the degradation trend of the battery.

In the future, the ensemble learning can be used for the fusion and integration of different data-driven prognostic methods. The fusion of online algorithm and uncertainty may be one key issue in hybrid approach for battery's RUL estimation.

5. Conclusions

Various RUL prediction methods for Li-ion battery are explored in this paper, and their characteristics are also discussed and compared. Figure 1 gives a summary.

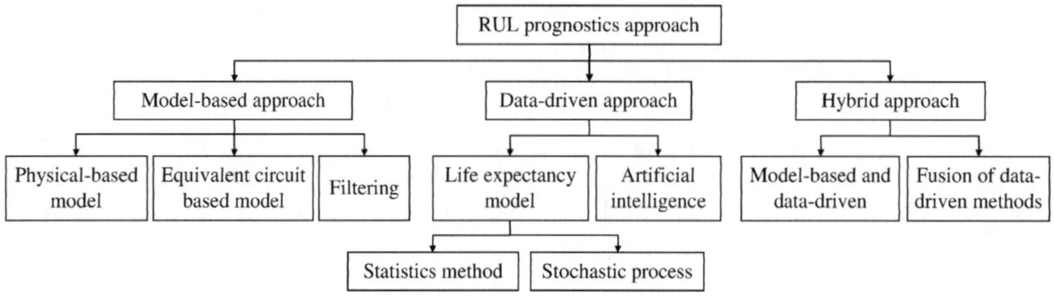

Figure 1. Prediction methods for RUL of Li-ion battery.

As we know that each kind of method has its own advantages and disadvantages. Table 2 lists the pros and cons of proposed methods.

Table 2. Comparison of the proposed methods.

Categories	Advantages	Disadvantages	References
Physical model	1. Give the details of degradation 2. Provide confidence limits 3. More accurate estimation	1. Require detailed knowledge of system 2. The prediction accuracy is affected by experimental conditions 3. Not suitable for practical application	[7-11]
Filtering approach	1. Suitable for any form of state-space model 2. Suitable for nonlinear, systems	1. Need data mode 2. Point estimation 3. (PF)Initialization process complicated	[15-19]
Artificial intelligence	1. No data model required 2. Simple and feasible calculation 3. Suitable for nonlinear systems	1. Point estimation 2. Lack the uncertainty of prediction results 3. Need large amount of data is needed	[1,14,21-24]
Life expectancy model	1. Consider the time-dependence of the degradation process 2. Support uncertainty representation	1. Complex calculation 2. Consider uncertain factors	[25-30]
Hybrid approach	1. Combine the merits of different methods 2. Effective, accurate and stable	1. Huge calculation	[31-35]

For Li-ion battery, the key issues during RUL prediction include time-varying environment, random-variable current, self-healing feature, etc. [6]. Various uncertain factors caused by these issues have a significant impact on the accuracy of the prediction. The uncertainty has become a hotspot in the Li-ion battery RUL prediction research.

In the future, RUL estimation approaches for Li-ion battery can combine monitoring with prognostics techniques. The following aspects can be considered:

- RUL estimation under dynamic operating conditions.

- Fusion and integrated methods for Li-ion battery health status assessment.
- Related issues on battery pack.
- Integrated prognostic framework involving state monitoring, anomaly detection, intelligent diagnostics and prognostics.

Acknowledgments
This research was supported by the National Natural Science Foundation of China under No. 71671035 and the Open Fund of Jiangsu Wind Power Engineering Technology Center, China under No. ZK15-03-01 and ZK16-03-07.

References

[1] Liu D, Zhou J, Pan D, Peng Y and Peng X 2015 Lithium-ion battery remaining useful life estimation with an optimized relevance vector machine algorithm with incremental learning *Measurement* **63** 143-51
[2] Li F and Xu J 2015 A new prognostics method for state of health estimation of lithium-ion batteriliues based on a mixture of gaussian process models and particle filter *Microelectron. Reliab.* **55** 1035-45
[3] Liu D, Luo Y, Liu J, Peng Y, Guo L and Pecht M 2014 Lithium-ion battery remaining useful life estimation based on fusion nonlinear degradation AR model and RPF algorithm *Neural Comput. Appl.* **25** 557-72
[4] Liu D, Zhou J and Peng Y 2014 Data-driven prognostics and remaining useful life estimation for lithium-ion battery: A review *Instrumentation* **1** 59-70
[5] Barré A, Deguilhem B, Grolleau S, Gérard M, Suard F and Riu D 2013 A review on lithium-ion battery ageing mechanisms and estimations for automotive applications *J. Power Sources* **241** 680-9
[6] Wu L, Fu X and Guan Y 2016 Review of the remaining useful life prognostics of vehicle lithium-ion batteries using data-driven methodologies *Appl. Sci.* **6** 166
[7] Christensen J and Newman J 2003 Effect of anode film resistance on the charge/discharge capacity of a lithium-ion battery *J. Electrochem. Soc.* **150** 1416-20
[8] Christensen J and Newman J 2004 A mathematical model for the lithium-ion negative electrode solid electrolyte interphase *J. Electrochem. Soc.* **151** 1977-88
[9] Safari M and Delacourt C 2011 Mathematical modeling of lithium iron phosphate electrode: galvanostatic charge/discharge and path dependence *J. Electrochem. Soc.* **158** 63-73
[10] Dalverny A L, Filhol J S and Doublet M L 2011 Interface electrochemistry in conversion materials for li-ion batteries *J. Mater. Chem.* **21** 10134-42
[11] Tasaki K and Harris S J 2010 Computational study on the solubility of lithium salts formed on lithium ion battery negative electrode in organic solvents *J. Phys. Chem. C* **114** 8076-83
[12] Yoshida H, Imamura N, Inoue T and Komada K 2003 Capacity loss mechanism of space lithium-ion cells and its life estimation method *Electrochemistry* **71** 1018-24
[13] Yoshida H, Imamura N, Inoue T, Takeda K and Naito H 2010 Verification of life estimation model for space lithium-ion cells *Electrochemistry* **78** 482-8.
[14] Andre D, Nuhic A, Soczka-Guth T and Sauer D U 2013 Comparative study of a structured neural network and an extended kalman filter for state of health determination of lithium-ion batteries in hybrid electricvehicles *Eng. Appl. Artif. Intell.* **26** 951-61
[15] Wang S, Shang L, Li Z, Deng H and Ma Y 2015 Lithium-ion battery security guaranteeing method study based on the state of charge estimation *Int. J. Electrochem. Sci.* **10** 5130-51
[16] Xu X and Chen N 2016 A state-space-based prognostics model for lithium-ion battery degradation *Reliab. Eng. Syst. Safe.* **159** 47-57
[17] He W, Williard N, Chen C and Pecht M 2013 State of charge estimation for electric vehicle batteries using unscented kalman filtering *Microelectron. Reliab.* **53** 840-7.
[18] Yu J 2015 State-of-health monitoring and prediction of lithium-ion battery using probabilistic

indication and state-space model *IEEE T. Instrum. Meas.* **64** 1
[19] Wang D, Yang F, Tsui K L, Zhou Q and Bae S J 2016 Remaining useful life prediction of lithium-ion batteries based on spherical cubature particle filter *IEEE T. Instrum. Meas.* **65** 1282-91
[20] Sikorska J Z, Hodkiewicz M and Ma L 2011 Prognostic modelling options for remaining useful life estimation by industry *Mech. Syst. Signal Process.* **25** 1803-36
[21] Wu Z Y, Cao L H, Tang C, Tian L, Zeng L B and Bin H U 2012 Research of modified elman neural network elman networks in the lithium-ion battery capacity prediction method *J. Seu Sci. Technol.* **27** 65-9
[22] Wang S, Zhao L, Su X and Ma P 2014 Prognostics of lithium-ion batteries based on battery performance analysis and flexible support vector regression *Energies* **7** 6492-508
[23] Dong H, Jin X, Lou Y and Wang C 2014 Lithium-ion battery state of health monitoring and remaining useful life prediction based on support vector regression-particle filter *J. Power Sources* **271** 114-23
[24] Zhang C, He Y, Yuan L, Xiang S and Wang J 2015 Prognostics of lithium-ion batteries based on wavelet denoising and DE-RVM *Comput. Intell. Neurosci.* **2015** 14
[25] Liu D, Pang J, Zhou J, Peng Y and Pecht M 2013 Prognostics for state of health estimation of lithium-ion batteries based on combination gaussian process functional regression *Microelectron. Reliab.* **53** 832-9
[26] Tang S, Yu C, Wang X, Guo X and Si X 2014 Remaining useful life prediction of lithium-ion batteries based on the wiener process with measurement error *Energies* **7** 520-47
[27] Si X S, Wang W, Hu C H, Chen M Y and Zhou D H 2013 A wiener-process-based degradation model with a recursive filter algorithm for remaining useful life estimation *Mechan. Syst. Signal Process.* **35** 219-37
[28] Pattipati B, Sankavaram C and Pattipati K 2011 System identification and estimation framework for pivotal automotive battery management system characteristics *IEEE T Syst. Man Cy. C* **41** 869-84
[29] Long B, Xian W, Jiang L and Liu Z 2013 An improved autoregressive model by particle swarm optimization for prognostics of lithium-ion batteries *Microelectron. Reliab.* **53** 821-31
[30] Saha B, Kai G and Christophersen J 2009 Comparison of prognostic algorithms for estimating remaining useful life of batteries *T. Inst. Meas. Control* **31** 293-308
[31] Zheng X and Fang H 2015 An integrated unscented kalman filter and relevance vector regression approach for lithium-ion battery remaining useful life and short-term capacity prediction *Reliab. Eng. Syst. Safe.* **144** 74-82
[32] He W, Williard N, Chen C and Pecht M 2014 State of charge estimation for li-ion batteries using neural network modeling and unscented kalman filter-based error cancellation *Int. J. Electr. Power Energy Syst.* **62** 783-91
[33] Wang J, Wang W, Ma F, Yang Y B and Yang C S 2012 A data-model-fusion prognostic framework for dynamic system state forecasting *Eng. Appl. Artif. Intell.* **25** 814-23
[34] Wang D, Miao Q and Pecht M 2013 Prognostics of lithium-ion batteries based on relevance vectors and a conditional three-parameter capacity degradation model *J. Power Sources* **239** 253-64
[35] Xing Y, Ma E W M, Tsui K L and Pecht M 2013 An ensemble model for predicting the remaining useful performance of lithium-ion batteries *Microelectron. Reliab.* **53** 811-20

Effects of setting angle and chord length on performance of four blades bionic wind turbine

Z X Yang[1,2], G S Li[1,2,3], L Song[1,2] and Y F Bai[1,2]

[1]Institute of Systems Science and Engineering, Henan Engineering Laboratory of Wind Power Systems, Luoyang, Henan, China
[2]Henan University of Science and Technology, No. 48 Xiyuan Road, Luoyang 471003, China

E-mail: lgs911@126.com

Abstract. With the energy crisis and the increasing environmental pollution, more and more efforts have been made about wind power development. In this paper, a four blades bionic wind turbine was proposed, and the outline of wind turbine was constructed by the fitted curve. This paper attempted to research the effects of setting angle and chord length on performance of four blades bionic wind turbine by computational fluid dynamics (CFD) simulation. The results showed that the setting angle and chord length of the bionic wind turbine has some significant effects on the efficiency of the wind turbine, and within the range of wind speed from 7 m/s to 15 m/s, the wind turbine achieved maximum efficiency when the setting angle is 31 degree and the chord length is 125 mm. The conclusion will work as a guideline for the improvement of wind turbine design.

1. Introduction

Being clean and renewable, wind energy has great reserves in nature, and it has great potential as a new energy with good prospects for development. In China, the demand for energy is increasing rapidly with the development of economy [1-3]. The traditional oil and coal has caused serious environmental pollution, while renewable wind energy can provide complement to modern society without pollution. Therefore more and more countries have recognized the importance of wind energy. Wind drive the wind turbine, and wind turbine drive the generator, then wind energy into electricity, this is the most important way to use the wind energy, and its installed capacity is increasing year by year.

The structure of wind turbines is very simple, but the aerodynamic characteristics of the work are extremely complex. At present, we have two main methods to research wind turbine flow field characteristics: numerical analysis [4-7,8] and wind tunnel testing [9-12].

According to the direction of wind and axis, the wind turbines can be divided into horizontal axis wind turbine [13-15] and vertical axis wind turbine [8,16-19] Compared with the traditional horizontal axis wind turbine, the vertical axis wind turbine has the advantage of advanced design methods, high wind power utilization, low starting wind speed, low noise, and broad market prospects. Vertical axis wind turbine includes Savonius type [20-22], Darrieus type [23], and H-type vertical axis wind turbine.

This paper introduces a new type of four blades bionic wind turbine, and researches into the relationship between the setting angle and chord length and the wind turbine characteristics by CFD.

2. Structure and principle of fish- bionic wind turbine

As is shown in figure 1, fish swimming in the water is similar to air flowing through something since fish tail swing generates driving force. According to the principle of bionics, the paper improves the utilization of wind energy by optimizing the end of the blade, changing the chord length and the setting angle.

Figure 1. Fish swimming in the water.

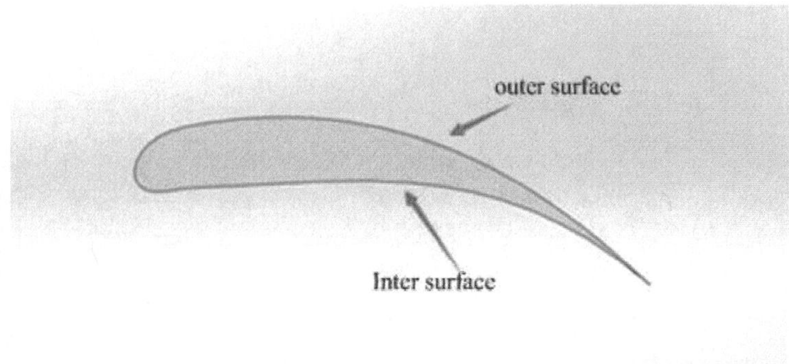

Figure 2. Sample points and approximate curve fitting of bionic wind turbine.

We get wind turbine rotor blades sample from swimming fish show in figure 2, and use approximate equation to describe sample by curve fitting, then the Fourier fitting of intersection line between upper cover and inner surface is:

$$f_1(x) = \left(1.935x - 2.69x^2\right)\sin\left(0.05232x - 0.2124\right) \quad (1)$$

The Fourier fitting of intersection line between upper cover and outer surface is:

$$f_2(x) = \left(0.6611x - 1.337x^2\right)\sin\left(0.5234x - 0.2651\right) - 0.02462 \quad (2)$$

Where the x is in the range [0-1].

According to the results of curve fitting, we have designed the rotor blades shown in figure 3.

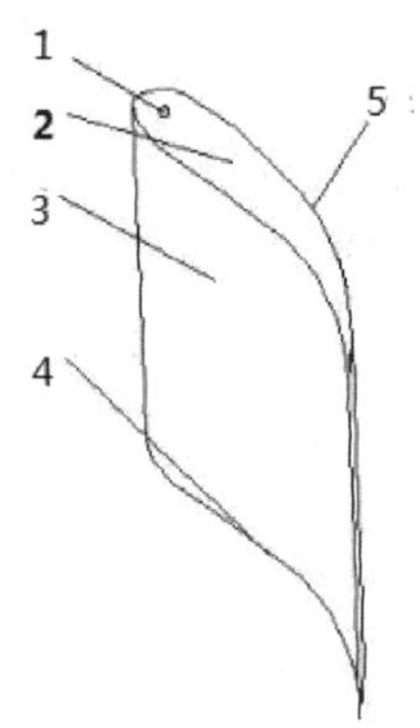

Figure 3. Wind turbine blade structure.

In figure 3, (1) is mounting hole; the two mounting holes play a role in fixing the upper and lower cover. Both are located on the chord length. (3) is inner surface, (5) is outer surface, (2) is upper cover, (4) is lower cover, the upper cover and lower cover are parallel to each other and the projection of the covers are coincident, and inside surface and outside surface are perpendicular to the upper cover and lower cover, the thickness of the blade reduces gradually and intersects at a point from the front edge point to the rear edge point.

In this paper, the outer diameter of the wind turbine is 500mm, the wind turbine height is 500 mm. the ratio between outer diameter and the wind turbine height is 1. The bionic wind turbine has two circular cover plates at up and down. As shown in figure 4, there are four blades evenly distributed along the circumference between the upper and lower cover. The blades can be rotated around point A to adjust the wind turbine setting angle.

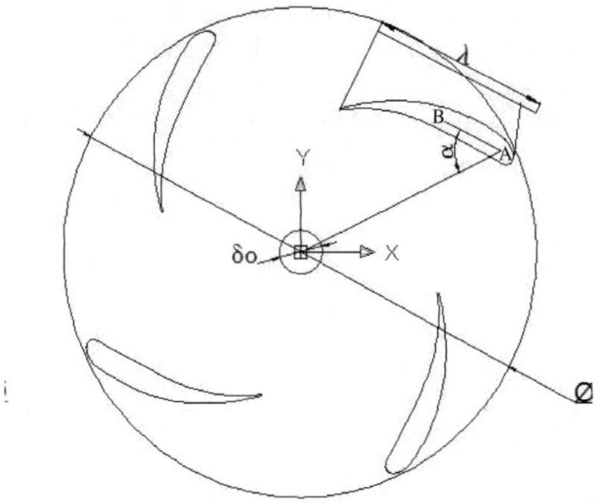

Figure 4. Wind turbine.

Figure 5. Wind turbine.

In figure 4, Φ is outside diameter of wind turbine, δo is wind turbine shaft diameter, A is the centre of curvature of the head part of blade; B is the centroid of the blade, λ is the chord length, α is the setting angle. Figure 5 is the three-dimensional structure of the wind turbine.

3. CFD model

Computational fluid dynamics use discrete mathematical methods and it can simulate the running state of model in theory, and uses powerful computing capability of computer to solve various issues of hydrodynamics calculations. CFD software simulations can greatly reduce the use of pre-investment projects, save time and improve work efficiency, thus it has become an important modern engineering software tool.

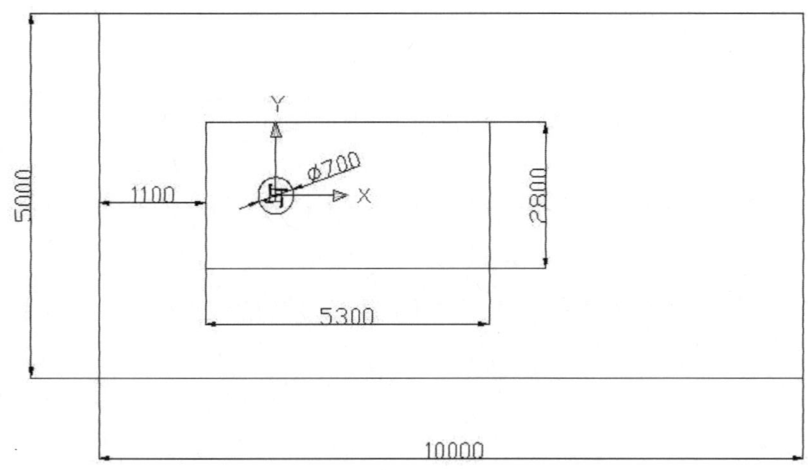

Figure 6. CAD drawing of wind turbine computational domain (unit: mm).

The bionic wind turbine was constructed and computational domain was chosen according to equation (1) and equation (2), then transient and dynamic performance of wind turbine was analysed by sliding mesh method of computational fluid dynamics.

As shown in figure 6, wind turbine diameter and height were 500 mm, the distance between the wind turbine and left boundary was five times of the diameter of the wind turbine, the distance between the wind turbine and upper and lower bounds was five times of the diameter of the wind turbine, the distance between the wind turbine and right boundary was fifteen times of the diameter of the wind turbine. Mesh quality is the main factor affecting the results of the simulation experiment.

According to the simulation results, the last cycle of the wind turbine torque was selected, and the average torque of last cycle was used to calculate the efficiency of wind turbines by following equation.

Wind power input

$$P_{\mathrm{i}}=\frac{\rho HD\upsilon^3}{2\times10^3\times10^3} \tag{3}$$

Wind turbine power output

$$P_{\mathrm{o}}=\frac{2\pi nT}{60} \tag{4}$$

Wind turbine efficiency

$$\eta=\frac{P_{\mathrm{o}}}{P_{\mathrm{i}}} \tag{5}$$

Where υ represents Wind velocity (m/s), ρ represents Air density (1.225 kg/m3), H represents Height of wind turbine (500 mm), D represents Diameter of wind turbine (500 mm), n represents Rotor speed (r/min), and T represents Wind turbine output torque (N.m).

4. CFD simulation

The simulation conditions are shown in table 1 and 2 as follows:

Table 1. Setting angles.

Setting angles	Chord length	Wind speed
25	125	11
27		
29		
31		
33		
35		
37		
39		

Table 2. Chord length.

Setting angles	Chord length	Wind speed
31	65	11
	80	
	95	
	110	
	125	

140
155
170

In the simulation, the setting angles were between 25° and 40°, simulate every 2°. The chord length was between 62.5 mm and 187.5 mm, simulate every 15 mm. The simulation results are as follows.

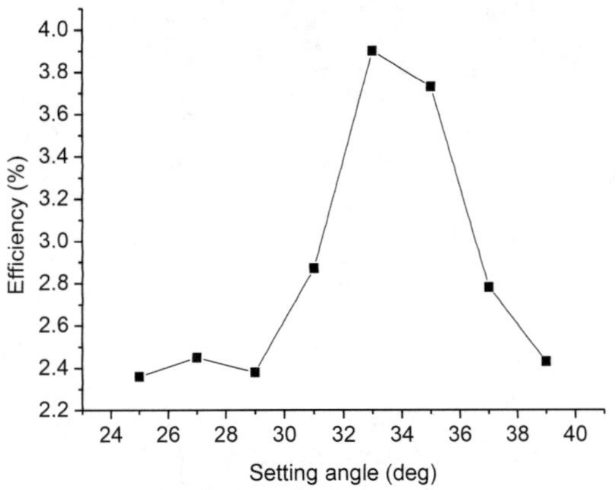

Figure 7. The efficiency at different setting angles.

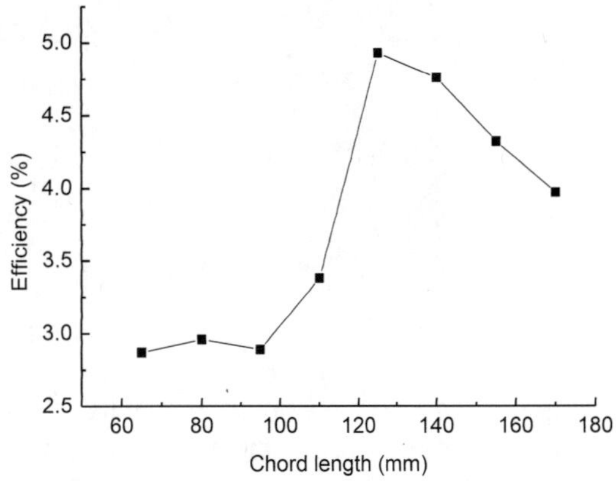

Figure 8. The efficiency at different chord length.

From the simulation results figures 7 and 8, we can see that the best installation angle is 31°, the best chord length is 125 mm.

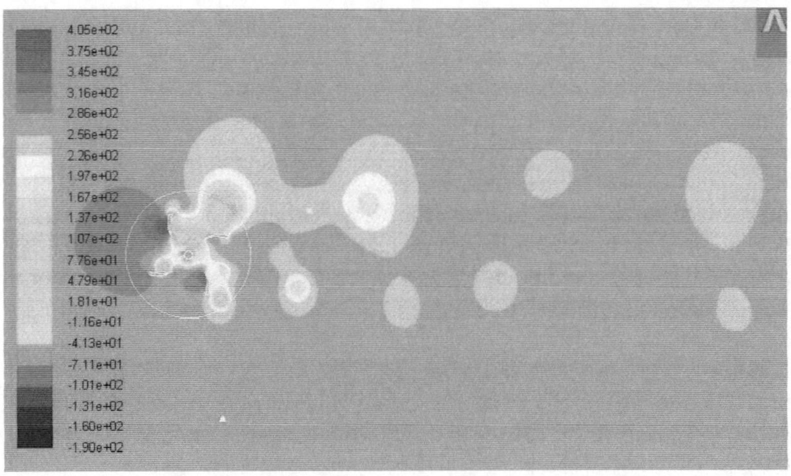

Figure 9. The pressure field.

As shown in figure 9, from the distribution of the pressure field, it can be seen that the lower right end of the wind turbine is beneficial to the secondary use of wind energy, a low pressure zone appears in the outlet of the airflow channel of the wind turbine, this is because the gas flowing out from the gas flow channels is faster, leading to pressure reduction, this makes the wind turbine show lift-typed wind turbine characteristics. In addition, when the wind through the wind turbine inlet, it will form a larger pressure between the up and down surface of wind turbine due to the special structure of the head of the circular surface. The pressure difference will promote the rotation of the wind turbine, and overcome the lift type wind turbine is difficult to start.

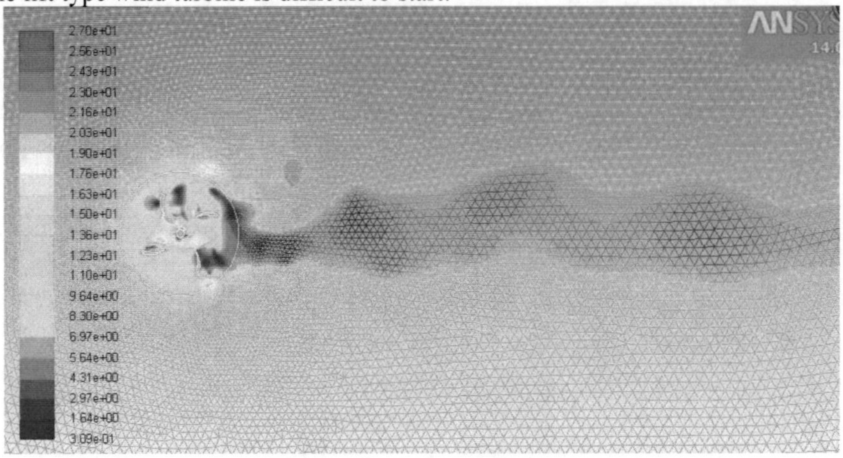

Figure 10. The velocity field.

Figure 10 shows the velocity field when the wind turbine mounting angle is 31° and chord length is 125 mm. From the figure, there are many low-speed areas in the back of the wind turbine, which is caused by the rotation of the wind turbine. The low-wind area is not only conducive to the rapid circulation of the wind between the blades, more conducive to the wind turbine rotation. In the process of the wind turbine movement, the blade tail shows the resistance type wind turbine characteristics. The blade tail is enclosed in a wind turbine with a secondary drive function, which increases the speed and output torque of the wind turbine.

5. Concluding remarks

The paper described a kind of higher efficiency four blades bionic wind turbine, and established the mathematical model of the proposed wind turbine. The influence of setting angle and chord length on the transient performance and dynamic performance of wind turbine is analysed by using computational fluid dynamics. According to the simulation results, the features of the proposed wind turbines are drawn as following:

- Four blades bionic wind turbine has a unique structure and the optimized blade, which can use wind energy twice, so the wind turbine efficiency is higher.
- The setting angle and chord length are key factors that affect the efficiency of the four blades wind turbine, because the setting angle and chord length can affect the secondary use of wind energy inside the wind turbine.

Acknowledgments

The authors were supported financially by the National Natural Science Foundation of China (NSFC) under grant No. 71071078 & No. 70671035, and the 2017 Henan science and technology plan project No. 172102210246.

References

[1] Feng-lin W and Chuang-lin F 2009 Wind Power Resource Appraisal and Development Stage Regional Division of China *Journal of Natural Resources* **24** 1412-22
[2] Heng X, Ruizhao Z, Zhenbin Y and Chunhong Y 2001 Assessment of wind energy RESERVES in China *Acta Energiae Solaris Sinica* **22** 167-71
[3] Ke L, Fanneng H and Jianchao X 2010 An Analysis of utilization potential distribution of wind power in Mainland China *Resources Science* **32** 1672-9
[4] Cheng U and Yan-rong W A 2009 Research of 3D finiteelement modelling on windwheel blade of wind power generator *Journal of Machine Design* **26** 50-4

[5] Sheng C R-, da Z L-, Xu W and nan Z Y- 2009 The ANSYS- based study of dynamic characteristics of a wind- power machine's wind wheel blade *Yunnan Water Power* **25** 28-32
[6] Zhong-yun X, Zhu Z, Zhuo-bin C and Gang L 2009 Numerical simulation of stall regulated wind turbine *Acta Aerodynamica Sinica* **27** 405-11
[7] Xiong L, Yan C and Quan Y Z 2006 Research on the HAWT rotor blades optimization *Journal of Shantou University(Natural Science)* **21** 44-50
[8] Li G S, Yang Z X, Song L, Chen Q, Li Y B and Chen W 2016 Effects of setting angle on performance of fish-bionic wind wheel. In: *2016 International Conference on New Energy and Future Energy System* (Beijing: IOP Publishing) pp 1-8
[9] Gensheng L, Zongxiao Y and Lei S 2016 Performance test bench of vertical axis wind turbines *Journal of Henan University of Science and Technology(Natural Science)* **37** 16-22
[10] Rui-Tao D, Zhi-Yuan C, Xiao-Guang Y, Lei S and Zong-Xiao Y 2012 Research and development of a simple vertical-axis wind turbine test equipment *Journal of Henan University of Science and Technology(Natural Science)* **6** 20-6
[11] Karki R, Po H and Billinton R 2006 A simplified wind power generation model for reliability evaluation *IEEE Transactions on Energy Conversion* **21** 533-40
[12] Lu M S, Chang C L, Lee W J and Wang L 2009 Combining the wind power generation system with energy storage equipment *IEEE Transactions on Industry Applications* **45** 2109-15
[13] Guo Y, Hosseini S H, Tang C Y, Jiang J N and Ramakumar R G 2013 An approximate wind turbine control system model for wind farm power control *IEEE Transactions on Sustainable Energy* **4** 262-74
[14] Kusiak A, Zhang Z and Li M 2010 Optimization of wind turbine performance with data-driven models *IEEE Transactions on Sustainable Energy* **1** 66-76
[15] Petru T and Thiringer T 2002 Modeling of wind turbines for power system studies *IEEE Transactions on Power Systems* **17** 1132-9
[16] Dong L, Liao M, Li Y, Song X and Xu K 2009 Study on Aerodynamic Design of Horizontal Axis Wind Turbine Generator System *2009 ICEET '09. International Conference on Energy and Environment Technology* (Guilin) pp 841-4
[17] Mesemanolis A and Mademlis C 2014 Combined maximum power point and yaw control strategy for a horizontal axis wind turbine *2014 International Conference on Electrical Machines (ICEM)* (Berlin) pp 1704-10
[18] Murshed M, Arafat M Y and Razzak M A 2016 Analysis of air foils and design of blades for a low-speed 250W horizontal axis wind turbine suitable for coastal areas of Bangladesh *2016 4th International Conference on the Development in the in Renewable Energy Technology (ICDRET)* (Kuala Lumpur) pp 1-6
[19] Schreck S J and Robinson M C 2007 Horizontal Axis Wind Turbine Blade Aerodynamics in Experiments and Modeling *IEEE Transactions on Energy Conversion* **22** 61-70
[20] Preen R J and Bull L 2015 Toward the coevolution of novel vertical-axis wind turbines *IEEE Transactions on Evolutionary Computation* **19** 284-94
[21] Shahariar G M H and Hasan M R 2014 Design and construction of a Vertical Axis Wind Turbine*2014 9th International Forum on Strategic Technology (IFOST)* (Cox's Bazar) pp 326-9
[22] Sharma M, Rieger B, Jou C and Butka B 2012 Development of helical vertical axis wind turbine *2012 Proceedings of IEEE Southeastcon* (Orlando) pp 1-5
[23] Lei S, Hong-Zhao L and Zong-Xiao Y 2015 Performance comparison for savonius type wind turbines by numerical analysis approaches *2015 International Conference on Advanced Mechatronic Systems (ICAMechS)* (Beijing) pp 402-7

Study on detection method of distorted voltage sag in distribution network under the background of incorporating new energy

H Lei[1,3], R C Fang[1], D J Yang[1], L Wu[2], J Y Xu[1] and Y J Zhou[1]

[1]SG HBEPC Economic & Technology Research Institute, Wuhan 430077, Hubei Province, China
[2]SG Huangshi Electric Power Supply Company, Huangshi 435000, Hubei Province, China

E-mail: regallyhust@hotmail.com

Abstract. An advanced voltage sag detection method and compensation technique can improve the power quality of the distribution network to a entirely new level. A high performance voltage sag detection method demands fast response and good accuracy, even detecting distorted grid voltage sag caused by new energy accessing to distribution network. The voltage sag detection method, based on sliding window iterative discrete fourier transfer (DFT) algorithm, has excellent filtering ability but long detection time. On the other hand, the voltage sag detection method, based on nonrecursive DFT algorithm, has short detection time but poor filtering ability. Aiming to meet the requirements of fast response and good accuracy, this paper proposed a voltage sag detection method based on improved nonrecursive DFT algorithm. By means of disturbance filter, which is combined with several notch filters and a low-pass filter, the filter characteristic of nonrecursive DFT algorithm has been improved, and the detection delay time is reduced. The proposed algorithm can guarantee shorter detection time and higher detection accuracy.

1. Introduction

Voltage sag has become the main interference to the safe and stable operation of distribution network under the background of incorporating new energy [1]. When the distributed photovoltaic (PV) suddenly loses power, voltage sag will appear in the distribution network, and the shunt capacitor compensation device should be connected as fast as possible, to keep the bus voltage stable [2]. In addition, voltage sag will lead to some sensitive equipment malfunction or failure, such as computer restart [3]. A large amount of grid connection converter will cause distorted grid voltage, under the situation of new energy accessing to distribution network. The distorted grid voltage will be a challenge to voltage sag detection method. Therefore, rapid and accurate detection of voltage sag is worthy of study.

The d-q transform sag detection methods, by means of transformation of the three-phase voltages to a rotating reference, are widely used [4,5]. However, single-phase voltage sag is more common, which means that the detection method needs to be suitable for single-phase voltage sag [6]. To solve this problem, several single-phase voltage sag detection methods based on rms calculation and discrete fourier transfer (DFT) have been proposed [7-11]. Rms calculation algorithm is a most common way for its simplicity, but the algorithm will be inaccurate when the voltage is distorted or the sag contains phase-angle jump [7,8]. The DFT algorithm has excellent performance of suppressing harmonic disturbance, but poor convergence speed [9]. The speed of voltage sag detection based on DFT algorithm depends on the length of window [10]. The rms voltage will be refreshed faster when the window is sufficiently short, but at the same time the harmonic components will result in inaccuracy of the rms value. In [11], the sliding-window rms calculation method based on Fourier transform simultaneously caters for the speed and accuracy. However, if the voltage sag contains phase-angle jump, the effect of proposed method will be limited.

Voltage sag detection method proposed in this paper is based on non-recursive DFT algorithm, and disturbance filter is combined to suppress harmonic interference. Through the rational design of filter parameters, the proposed method can achieve shorter detection time, but also to ensure higher detection accuracy.

2. Voltage sag detection method based on DFT algorithm

The DFT algorithm can accurately extract the fundamental component of distorted grid voltage, but the convergence of the algorithm is relatively slow, which cannot meet the requirement of fast detection. Therefore, there are some literatures on this basis to make improvements.

2.1. Voltage sag detection method based on non-recursive DFT algorithm

The fundamental component of the sampled signal f(t) is defined as

$$f^*(t) = X_a \sin(\omega t) + X_b \cos(\omega t) \tag{1}$$

Where, ω represents fundamental angular velocity. Sampled waveform f(t) may be approximated by f*(t) to minimize the mean square error. By means of discretizing f(t), the total square error E2 is written as

$$E^2 = \sum_{n=0}^{N-1} (f_n - X_a \sin(n\omega T_s) - X_b \cos(n\omega T_s))^2 \tag{2}$$

Where, fn represents discrete sampled signal, N represents length of window which is adjustable, and Ts is sampled time.

The partial derivative with respect to Xa and Xb should be equal to zero for the reason that E2 should be as small as possible [10].

$$\frac{\partial E^2}{\partial X_a} = -2\sum_{n=0}^{N-1} \sin(n\omega T_s)[f_n - X_a \sin(n\omega T_s) - X_b \cos(n\omega T_s)] = 0 \tag{3}$$

$$\frac{\partial E^2}{\partial X_b} = -2\sum_{n=0}^{N-1} \cos(n\omega T_s)[f_n - X_a \sin(n\omega T_s) - X_b \cos(n\omega T_s)] = 0 \tag{4}$$

Equations (3) and (4) can be transformed to (5) and (6),

$$X_b \sum_{n=0}^{N-1} \sin(n\omega T_s)\cos(n\omega T_s) + X_a \sum_{n=0}^{N-1} \sin^2(n\omega T_s) = \sum_{n=0}^{N-1} f_n \sin(n\omega T_s) \tag{5}$$

$$X_b \sum_{n=0}^{N-1} \cos^2(n\omega T_s) + X_a \sum_{n=0}^{N-1} \sin(n\omega T_s)\cos(n\omega T_s) = \sum_{n=0}^{N-1} f_n \cos(n\omega T_s) \tag{6}$$

Substituting

$$A = \sum_{n=0}^{N-1} \sin^2(n\omega T_s) \tag{7}$$

$$B = \sum_{n=0}^{N-1} \sin(n\omega T_s)\cos(n\omega T_s) \tag{8}$$

$$C = \sum_{n=0}^{N-1} \cos^2(n\omega T_s) \tag{9}$$

The following equations can be given as

$$X_a = \frac{1}{AC - B^2} \sum_{n=0}^{N-1} f_n [C \sin(n\omega T_s) - B \cos(n\omega T_s)] \tag{10}$$

$$X_b = \frac{1}{AC - B^2} \sum_{n=0}^{N-1} f_n [A \cos(n\omega T_s) - B \sin(n\omega T_s)] \tag{11}$$

Substituting

$$K_{an} = \frac{1}{AC - B^2} [C \sin(n\omega T_s) - B \cos(n\omega T_s)] \tag{12}$$

$$K_{bn} = \frac{1}{AC - B^2} [A \cos(n\omega T_s) - B \sin(n\omega T_s)] \tag{13}$$

X_a and X_b can be calculated as

$$X_a = \sum_{n=0}^{N-1} K_{an} f_n \tag{14}$$

$$X_b = \sum_{n=0}^{N-1} K_{bn} f_n \tag{15}$$

The fundamental rms X is derived as

$$X = \sqrt{X_a^2 + X_b^2} \tag{16}$$

As shown in figure 1, there is no harmonics, and voltage sag occurs in 0.047s. The depth is 0.2 p.u., and the phase jump angle is 10°. The width of window is set to 1/8 of the fundamental period, and the convergence of the fundamental rms is shown in figure 1. According to IEEE 1159 standard, when the fundamental rms is less than 90% of its rated value, which is considered to voltage sag. Therefore, the voltage sag detection time is 2.2 ms, meeting the requirements of fast detection.

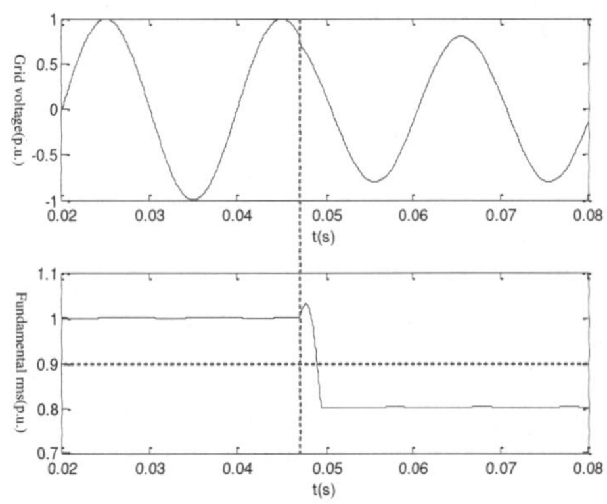

Figure 1. Nonrecursive DFT algorithm (Non-distorted voltage sag).

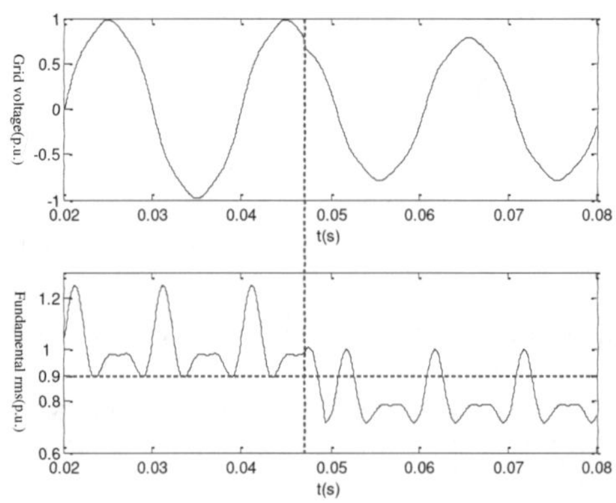

Figure 2. Nonrecursive DFT algorithm (Distorted voltage sag).

As shown in figure 2, there are 5% 3rd harmonic and 3.5% 5th harmonic components in the grid voltage, the other conditions are consistent with figure 1. It can be found that the fundamental rms calculated by non-recursive DFT algorithm has poor accuracy when the grid voltage is distorted.

In order to improve the detection accuracy of distorted grid, the common approach is to combine with a low-pass filter to suppress harmonic disturbances. Due to the lowest harmonic frequency of fundamental rms is 100 Hz, the cut-off frequency of low-pass filter is set as 50 Hz, the convergence of the fundamental rms is shown in figure 3. Although the steady error is reduced to less than 2%, but the detection time is longer, about 7.7 ms.

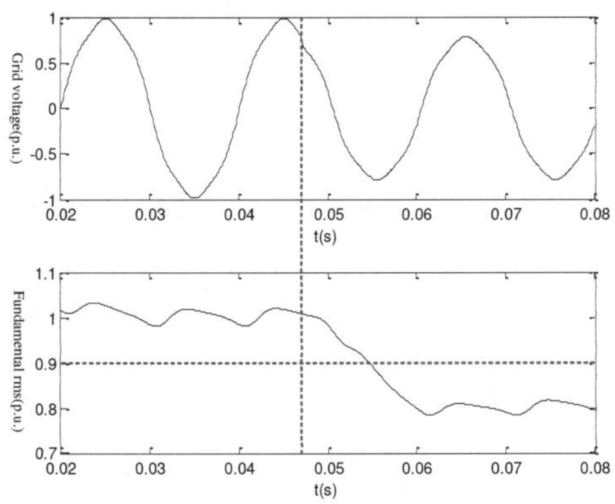

Figure 3. Nonrecursive DFT algorithm based on low-pass filter (Distorted voltage sag).

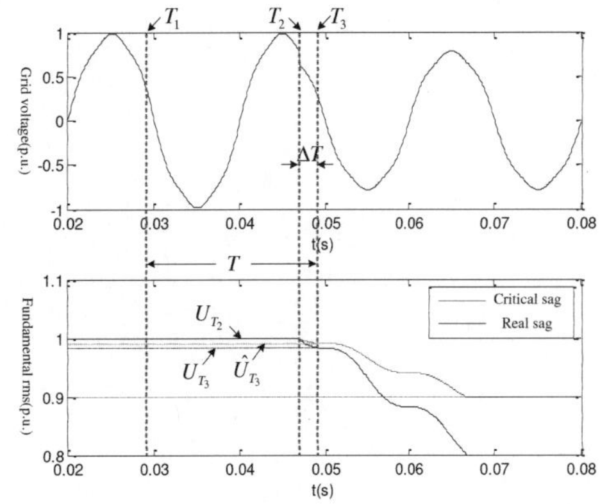

Figure 4. Improved sliding-window iterative DFT algorithm (Distorted voltage sag without phase-angle jump).

2.2. Improved sliding-window iterative DFT algorithm

The sliding-window iterative algorithm can ensure the detection accuracy even if the grid is distorted, but the detection rate is slow [9]. The test conditions in figure 2 are almost consistent with figure 4, except that there is no phase jump. As shown in figure 4, the steady error of fundamental rms is very small, but the detection time is 10 ms.

In order to ensure high detection accuracy and short detection time, some researcher proposed a improved sliding window iterative algorithm, which modified the sag detection criteria [11]. Assuming the critical sag occurs in T2, and the depth is 0.1 p.u. The convergence of the critical fundamental rms is shown in figure 4. U_{T_2} is the fundamental rms before sag occurs. Suppose the desired detection time is ΔT, then the fundamental rms in T3 (T3=T2+ΔT) is $\hat{U}_{T_3}$, which is defined as

$$\hat{U}_{T_3}=\sqrt{\frac{1}{T}\left\{\int_{T_1}^{T_2}\left[\sqrt{2}U_{T_2}\sin(\omega t)\right]^2 dt+\int_{T_2}^{T_3}\left[0.9*\sqrt{2}U_{T_2}\sin(\omega t)\right]^2 dt\right\}} \tag{17}$$

The new sag criterion is written as

$$\Delta\hat{U}=U_{T_2}-\hat{U}_{T_3} \tag{18}$$

When the voltage sag occurs, the fundamental rms in T3 can be obtained by sliding-window iterative DFT algorithm. The fundamental rms difference between T2 and T3 is expressed as

$$\Delta U=U_{T_2}-U_{T_3} \tag{19}$$

If $\Delta U>\Delta\hat{U}$, it is determined that the sag occurs; If $\Delta U\le\Delta\hat{U}$, it is determined that the sag not occurs. According to the new sag criterion, the detection time is 2 ms.

Although the improved sliding-window DFT algorithm greatly reduced the detection time, but it did not take into account the influence of phase jump. Besides the voltage drop during a sag, there is also a phase-angle jump associated with it. Thus, the formula (17) may be rendered inaccurate, except that phase jump angle can be obtained in advance to modify the formula, which is hard to achieve.

As shown in figure 5, the test conditions are almost consistent with figure 4, except that the phase jump angle is -10°. It can be found that the convergence of actual fundamental rms is not always lower than critical one. It means that ΔT should be longer than 5 ms to accurately determine the voltage sag. The phase jump angle will vary with different types of voltage sag, which means ΔT is difficult is set.

IOP Conf. Series: Earth and Environmental Science **93** (2017) 012042 doi:10.1088/1755-1315/93/1/012042

Therefore, the method proposed in literature [11] is not suitable for voltage sag with phase-angle jump.

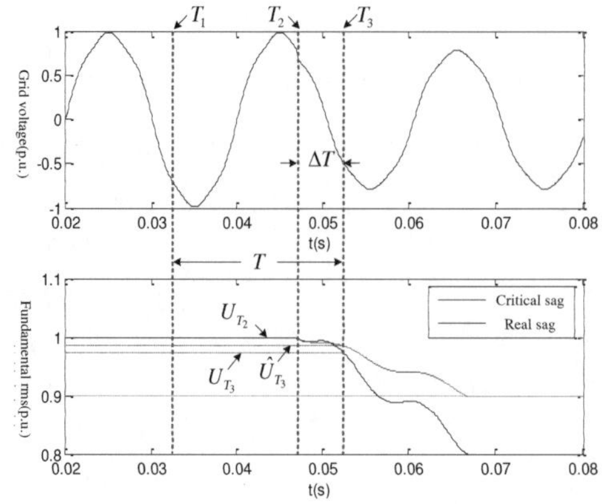

Figure 5. Improved sliding-window iterative DFT algorithm (Distorted voltage sag with phase-angle jump).

2.3. Comparison of different voltage sag detection methods

Table 1 shows the comparison of different detection methods for distorted grid voltage sag. It can be seen that, the current voltage sag detection methods based on DFT algorithm cannot meet the requirements of dynamic voltage restorer (DVR).

Table 1. Comparison of different detection method for distorted voltage sag.

t	Detection time	Detection accuracy	Phase-angle jump
Nonrecursive DFT	short	low	suitable
Nonrecursive DFT +low-pass filter	long	high	suitable
Sliding-window iterative DFT	long	high	suitable
Improved sliding-window iterative DFT	short	high	unsuitable

3. The application of disturbance filter in voltage sag detection method based on nonrecursive DFT algorithm

3.1. Analysis of disturbance filter

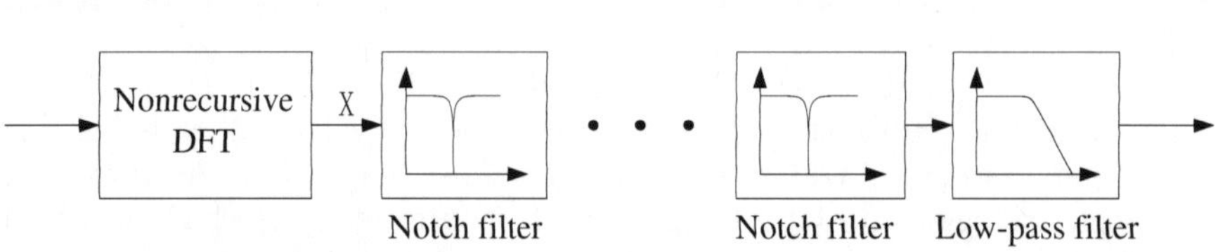

Figure 6. Schematic diagram of nonrecursive DFT algorithm based on disturbance filter.

The disturbance filter contains several second order notch filter and a second order low-pass filter. The notch filter is used to suppress specific low frequency harmonic disturbance, and the low-pass filter is used to suppress high frequency harmonic disturbance. Compared to traditional low-pass filter, the disturbance filter can improve the response speed, but also to guarantee small steady error. The control block diagram is shown in figure 6.

The transfer function of second order notch filter is

$$G_{nf}(s)=\frac{s^2+\omega_{nf}^2}{s^2+\frac{\omega_{nf}}{Q_{nf}}s+\omega_{nf}^2} \tag{20}$$

ωnf represents resonant frequency, Qnf represents quality factor. The transfer function of second order low-pass filter is

$$G_{lf}(s)=\frac{\omega_{lf}^2}{s^2+\frac{\omega_{lf}}{Q_{lf}}s+\omega_{lf}^2} \tag{21}$$

Ωlf represents resonant frequency, Qlf represents quality factor.

3.2. Design of disturbance filter

The fundamental rms calculated by nonrecursive DFT algorithm may contain 2nd, 4th and 6th harmonic components, and so on. The more notch filters cause to better low-frequency harmonic disturbance rejection, but slower response speed. Therefore, two notch filters are chosen and the respective resonant frequency is set as 2nd and 4th harmonic frequency. The respective quality factor of notch filter is both set as 1. The cut-off frequency of low-pass filter is set as 4th harmonic frequency, and the quality factor is set as 0.5.

In order to be applied in digital control, the transfer function of disturbance filter need to be discrete. Traditional discretization method, such as zero-order hold, bilinear transform, cannot be adapted to grid frequency fluctuation. The performance of disturbance filter can be influenced by frequency fluctuation, especially notch filter. To solve this problem, some scholars made a structure conversion of the second-order generalized integrator, and then discretized the pure integrator [12]. The resonant frequency can be adjusted with a phase-locked loop to ensure a high performance. According to this thought, the block diagram of disturbance filter is shown in figure 7.

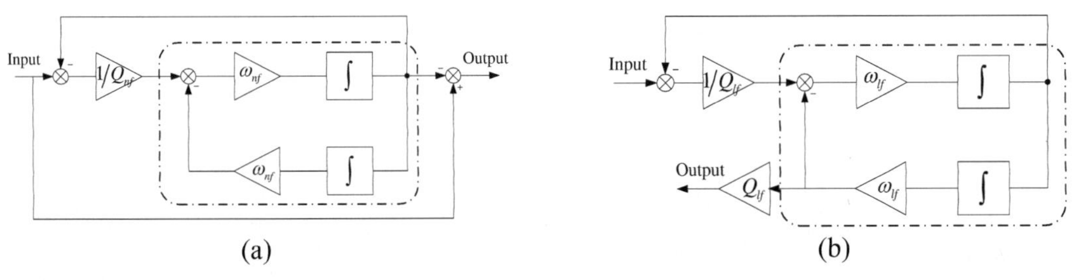

Figure 7. The block diagram of disturbance filter. (a) Second-order notch filter and (b) Second-order low-pass filter.

As shown in figure 8, the test conditions are consistent with figure 2, it can be found that the steady state error of fundamental rms is less than 2% and the detection time is 5.2 ms. The harmonic disturbance is well suppressed to improve the reliability of the detection, and the detection time just delays a little.

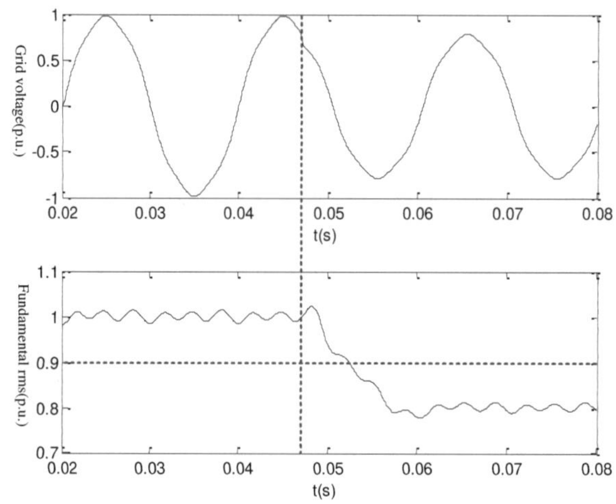

Figure 8. Non-recursive DFT algorithm based on disturbance filter (Distorted voltage sag).

4. Experimental results

4.1. Non-distorted voltage sag detection

In order to verify the effectiveness of the non-recursive DFT algorithm based on disturbance filter presented in this paper, four different depths of voltage sag are tested. The total harmonic distortion (THD) of grid voltage is 4.89%, as shown in figure 9.

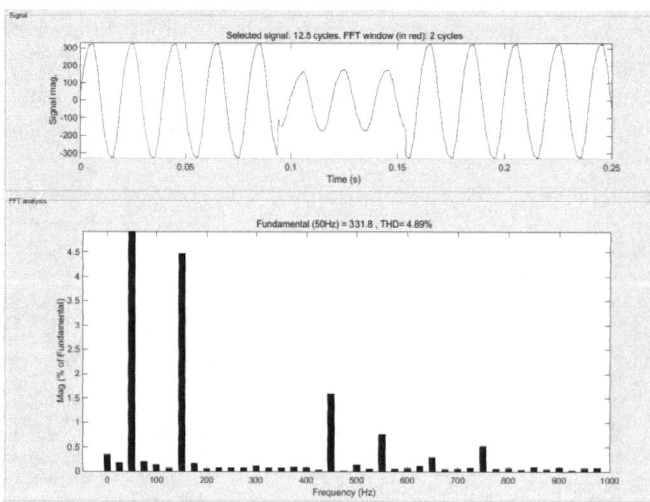

Figure 9. Harmonic analysis of undistorted grid voltage.

As shown in figure 10, the depth of voltage sag in figure 10(a) is 0.2 p.u., the depth of voltage in figure 10(b) is 0.5 p.u., the depth of voltage in figure 10(c) is 0.7 p.u., the depth of voltage in figure 10(d) is 1 p.u.. Channel 1 shows the voltage sag, and channel 2 shows the change of detection signal. When the detection signal is high, it means that the voltage is stable. When the detection signal turns low, it means that the voltage sag begins.

As shown in figure 10, the detection time are 4.8 ms, 3.9 ms, 3.2 ms and 2.1 ms, respectively. Deeper sag means faster detection because of higher rms declining slope. Therefore, the proposed algorithm is suitable for non-distorted voltage sag detection.

IOP Conf. Series: Earth and Environmental Science **93** (2017) 012042 doi:10.1088/1755-1315/93/1/012042

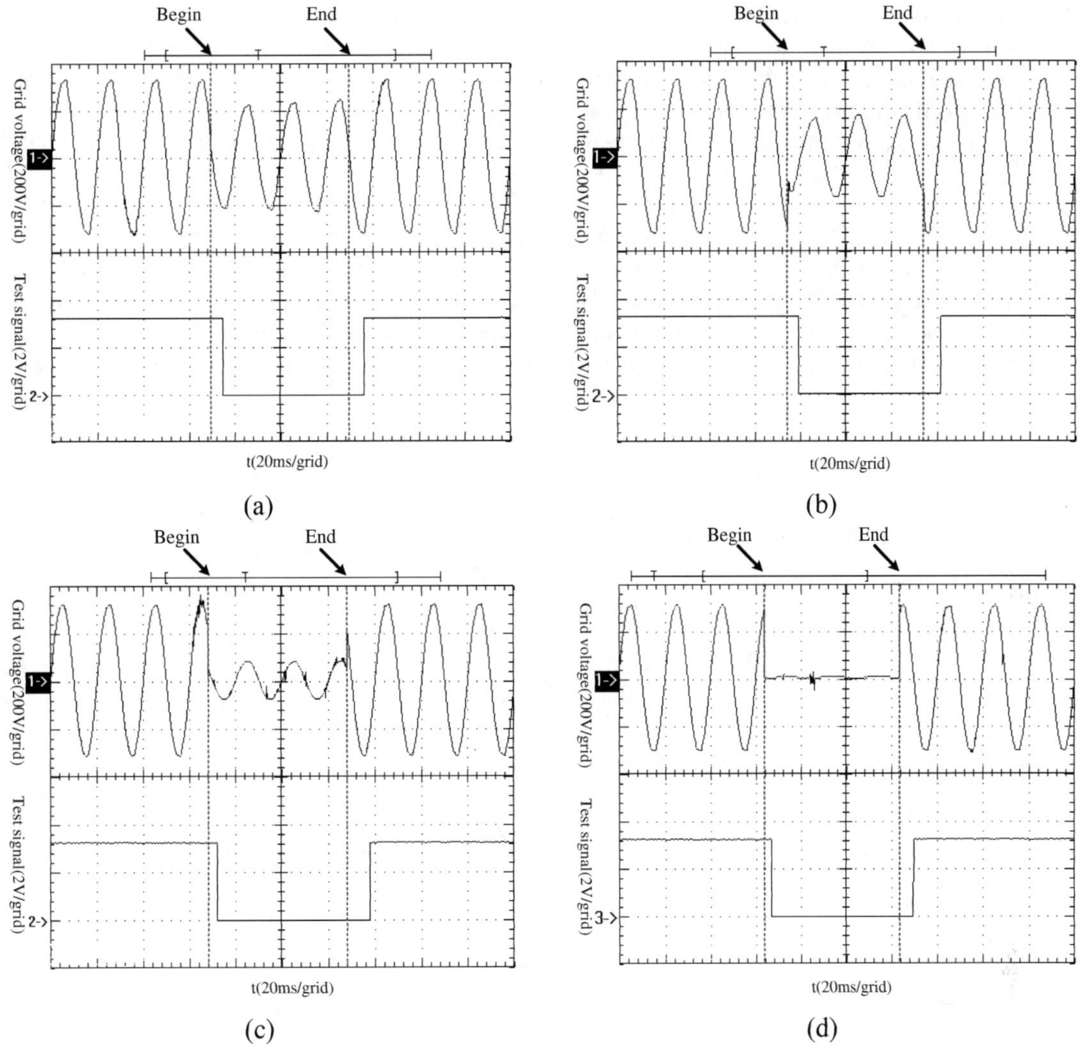

Figure 10. Waveforms of different voltage sags. (a) The depth of voltage sag: 0.2 p.u., (b) The depth of voltage sag: 0.5 p.u., (c) The depth of voltage sag: 0.7 p.u. and (d) The depth of voltage sag: 1 p.u..

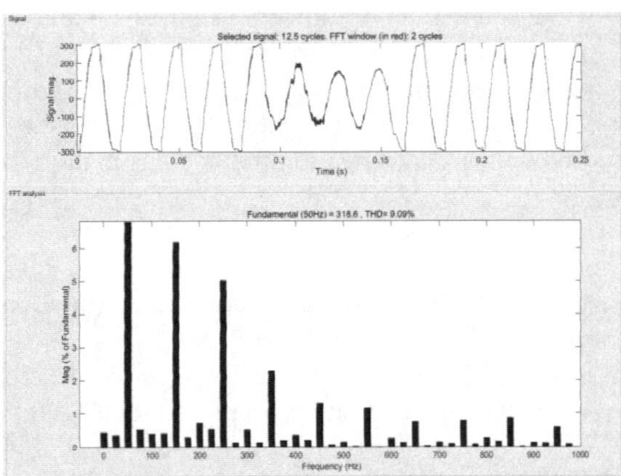

Figure 11. Harmonic analysis of distorted grid voltage.

4.2. Distorted voltage sag detection

To further verify the distorted voltage sag detection capability of the proposed algorithm, some non-linear loads are connected to the grid, which will increase the harmonic components in the grid voltage. The THD of grid voltage is 9.09%, as shown in figure 11.

Figure 12(a) shows the distorted voltage sag detection based on simple non-recursive DFT algorithm without disturbance filter. When the grid voltage is distorted, the detection signal is always low, even if the grid voltage is stable. Due to the poor harmonic suppression ability of non-recursive DFT algorithm, there will be false judgement when the grid voltage is distorted.

Figure 12(b) shows the distorted voltage sag detection based on non-recursive DFT algorithm with disturbance filter. The voltage begins in 0.06 s, and the depth of voltage sag is 0.3.p.u., and the detection time is 4.2 ms. The addition of disturbance filter compensates for the defect of non-recursive DFT algorithm, therefore, the proposed algorithm is also suitable for distorted voltage sag detection.

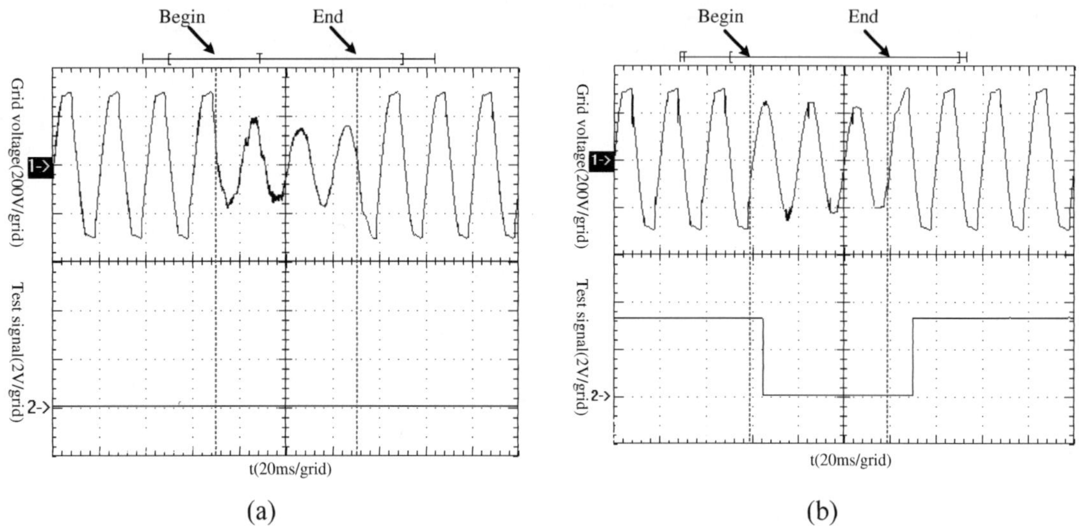

Figure 12. Waveforms of distorted voltage sag. (a) Distorted voltage sag detection (without disturbance filter) and (b) Distorted voltage sag detection (with disturbance filter).

5. Conclusion

Due to existing voltage sag detection methods based on DFT algorithm cannot simultaneously meet the requirements of detection time, detection accuracy and adaptability of phase-angle jump, this paper presents an improved short-window DFT algorithm based on disturbance filter. Two second-order notch filters are designed to suppress 2nd and 4th harmonic disturbances, and a second-order low-pass filter is designed to suppress high frequency harmonic disturbance. Although the disturbance filter slightly delays detection time, but in exchange for higher detection accuracy. Experiments show that, in general situation, the proposed voltage sag detection method has high performance whether the grid voltage is distorted or non-distorted.

References

[1] Martinez J A and Arnedo J M 2006 Voltage sag studies in distribution networks-part I: System modeling *IEEE Trans Power Delivery* **21(3)** 338-45

[2] Nguyen H V and Akihko Y 2013 Influence of penetration levels and fault ride-through characteristics of photovoltaic generation on voltage stability *J Int Council Electri Eng* **3(4)** 283-9

[3] Abdul Mannan Rauf and Vinod Khadkikar 2015 An enhanced voltage sag compensation scheme for dynamic voltage restorer *IEEE Trans Ind Electron* **62(5)** 2683-92

IOP Conf. Series: Earth and Environmental Science **93** (2017) 012042 doi:10.1088/1755-1315/93/1/012042

[4] Kumsuwan Y and Sillapawicharn Y 2013 An application of improved synchronous reference frame-based voltage sag detection in voltage sag compensation system *Proc IEEE IEEE EPE* (Lille, France) 1-8

[5] Cong M, *et al* 2014 Fast voltage detection method for the voltage ride through operation of grid-tied renewable energy generation systems *Proc IEEE APEC* (Fort Worth, USA) pp 386-91

[6] Bollen M H J 1997 Characterisation of voltage sags experienced by three-phase adjustable-speed drives *IEEE Trans Power Delivery* **12(4)** 1666-71

[7] Jose M, *et al* 2014 A fast RMS meter for detect ing sag events in household environments *IEEE International Conference on Consumer Electronics(ICCE)* (Las Vegas, USA) pp 321-2

[8] Kim G T and Kim R Y 2008 A cost-effective ac-ac voltage sag protector with a fast dynamic response using a new RMS averaging method *Proceedings of IEEE APEC* (Austin, USA) pp 443-8

[9] El-Habrouk M and Darwish M K 2001 Design and implementation of a modified Fourier analysis harmonic current computation technique for power active filters using DSPs *Proc IEEE EPA* **148(1)** 21-8

[10] Amaris H, *et al* 2009 Computation of voltage sag initiation with Fourier based algorithm, Kalman filter and Wavelets *Proc IEEE PowerTech* (Burcharest, Romania) pp 1-6

[11] Bae B, *et al* 2010 Line-interactive single-phase dynamic voltage restorer with novel sag detection algorithm *IEEE Trans Power Delivery* **25(4)** 2702-9

[12] Ciobotaru M, Teodorescu R and Blaabjerg F 2006 A new single-phase PLL structure based on second order generalized integrator *Proc IEEE PESC* (Jeju, Korea) pp 1-6

IOP Conf. Series: Earth and Environmental Science **93** (2017) 012043 doi:10.1088/1755-1315/93/1/012043

Rolling scheduling of electric power system with wind power based on improved NNIA algorithm

Q S Xu[1,2,3], C J Luo[1], D J Yang[1,2], Y H Fan[1,2], Z X Sang[1,2] and H Lei[1,2]

[1]State Grid Hubei Economic Research Institute, Wuhan 430077, Hubei Province, China

[2]State Grid Laboratory for Hydro-thermal Power Resources Optimal Allocation & Simulation Technology, Wuhan 430077, Hubei Province, China

E-mail: xu_qiushi@163.com

Abstract. This paper puts forth a rolling modification strategy for day-ahead scheduling of electric power system with wind power, which takes the operation cost increment of unit and curtailed wind power of power grid as double modification functions. Additionally, an improved Nondominated Neighbor Immune Algorithm (NNIA) is proposed for solution. The proposed rolling scheduling model has further improved the operation cost of system in the intra-day generation process, enhanced the system's accommodation capacity of wind power, and modified the key transmission section power flow in a rolling manner to satisfy the security constraint of power grid. The improved NNIA algorithm has defined an antibody preference relation model based on equal incremental rate, regulation deviation constraints and maximum & minimum technical outputs of units. The model can noticeably guide the direction of antibody evolution, and significantly speed up the process of algorithm convergence to final solution, and enhance the local search capability.

1. Introduction

Power generation with renewable energy has been developing dramatically owing to the global shortage of energy and the public attention to environmental issues [1]. As the best developed way of power generation in light of technology and economy, wind power has now gradually escalated to the significant role of substitute energy. However, as a primary energy, wind energy cannot be converted into a sustainable, stable and controllable form of energy like coal consumption of coal-fired power generation unit and flow rate of reservoir at hydropower plant. Natural wind is affected by multiple factors including temperature, topography, climate and weather, so the power output of wind turbine generator (WTG) set is highly stochastic, intermittent and fluctuating, making it much more difficult to carry out the wind power prediction. For this reason, the traditional economical and safe operation of power grid based on reliable power sources is severely challenged [2].

Advanced scheduling technology [3-6] is one of the significant ways to lower the influence of stochastic and intermittent wind power on power grid and increase the system's accommodation capacity of wind power. Nevertheless, along with the increasing amount of wind power connected to grid, the "unpredicted" fluctuation of wind power caused by the deviation of day-ahead power prediction will significantly affect the economy and practicality of scheduled output of conventional unit, and will also result in the heavy burden of regulation on AGC unit. The traditional scheduling model combining 2 temporal scales, i.e. day-ahead scheduling and AGC [7-9], has been unable to

IOP Conf. Series: Earth and Environmental Science **93** (2017) 012043 doi:10.1088/1755-1315/93/1/012043

satisfy the needs, so it is necessary to develop more accurate scheduling model. The rolling modification of generation schedule is one of the key tasks in the elaborate scheduling [10]. Zhang B M *et al* [11] put forward the scheduling idea of "multilevel coordination, gradual detailing", but the study was conducted mainly based on the extended short-term load forecast, but overlooked the modification for curtailed wind power of power grid. Wang K *et al* [12] modified the day-ahead scheduled output directly, but the coal-fired power generation unit was adjusted significantly due to the influence of day-ahead prediction deviation. This paper builds a rolling scheduling model with modification based on double objectives, e.g. the operation cost increment of unit and curtailed wind power of power grid curtailed. By continuously updating and modifying the output curve of conventional unit and the maximum output curve of wind farm, the economical and safe operation of power grid will be guaranteed while reducing curtailed wind power as much as possible.

Due to the complexity of model solution, this paper presents a solution strategy for the improved NNIA algorithm based on the information on antibody preference. The NNIA algorithm is an evolutionary multi-objective optimization (EMO) algorithm based on artificial immune system. Compared with three algorithms representing the development level of EMO, i.e. NGSA2, SPEA2 and PESA2, the NNIA algorithm can obtain the optimal solution of evenly distributed Pareto, and it is less complicated to calculate, making it a very effective EMO algorithm [13]. The proposed improvement solution strategy, which has defined an antibody preference relation model based on equal incremental rate, regulation deviation constraints and maximum & minimum technical outputs of units, will guide the evolutionary direction of antibodies, speed up the search for global optimal solution, and offer stronger local search capability. In a 10-unit system with wind farm, the feasibility and effectiveness of the model and algorithm proposed in this paper have been verified.

2. Rolling optimization scheduling model with wind farm

2.1. Power balance relationship between time periods in the rolling scheduling

Assuming that the initial time of rolling scheduling is $[t_0,\ldots,t_{\mathrm{H}-1},\ldots,t_{2\mathrm{H}-1},\ldots,t_{h-1},\ldots]$, the power balance at the time periods of rolling scheduling should satisfy the following constraints:

$$
\begin{aligned}
&Day-ahead && \sum_{i=1}^{N_G} p_{i,\,t}^{0} + \overline{W}_t - D_t^{*} = 0,\ t = 1,2,\quad ,96 \\
&t_0 && \sum_{i=1}^{N_G} p_{i,\,t}^{1} + W_t^{1} - D_t^{1*} = 0,\ t = 1,2,\quad ,96 \\
&t_{\mathrm{H}-1} && \sum_{i=1}^{N_G} p_{i,\,t}^{\mathrm{H}} + W_t^{\mathrm{H}} - D_t^{\mathrm{H}*} = 0,\ t = \mathrm{H},\mathrm{H}+1,\quad ,96 \\
&t_{2\mathrm{H}-1} && \sum_{i=1}^{N_G} p_{i,\,t}^{2\mathrm{H}} + W_t^{2\mathrm{H}} - D_t^{2\mathrm{H}*} = 0,\ t = 2\mathrm{H},2\mathrm{H}+1,\quad ,96 \\
&t_{h-1} && \sum_{i=1}^{N_G} p_{i,\,t}^{h} + W_t^{h} - D_t^{h*} = 0,\ t = h,h+1,\quad ,96
\end{aligned}
\tag{1}
$$

In which, N_G is the total number of units subject to rolling scheduling; i is unit No.; H stands for the cycle of rolling scheduling; $h = \mathrm{H}, 2\mathrm{H}, \ldots$; t_{h-1} is the initial time of the hth time period; $p_{i,\,t}^{h}$ is the rolling scheduled output of unit i in the tth time period after modification at the time t_{h-1}; $\overline{W}_t$ is the day-ahead scheduled output of wind power in the tth time period; W_t^h is the scheduled output of wind power in the tth time period after modification at the time t_{h-1}; and D_t^{h*} is the latest predicted value of system load at the tth time period before modification at the time t_{h-1}.

2.2. Objective of operation cost increment of unit

Normally, the operation cost of unit is fitted into the secondary function of generation power. The objective of operation cost increment of unit during modification at the time t_{h-1} can be represented as follows:

$$f_1(\Delta p_{i,\,t}^{h}) = \sum_{t=h}^{T}\sum_{i=1}^{N_G}[a_i \Delta p_{i,\,t}^{h\;2} + (2a_i p_{i,\,t}^{h-\mathrm{H}} + b_i)\Delta p_{i,\,t}^{h}] \tag{2}$$

In which: a_i, b_i, c_i are the secondary characteristic coefficients for operation cost of coal-fired power unit. $p_{i,\,t}^{h-\mathrm{H}}$ is the rolling scheduled output of unit i in the tth time period before modification; $\Delta p_{i,\,t}^{h} = p_{i,\,t}^{h} - p_{i,\,t}^{h-\mathrm{H}}$.

2.3. Objective of curtailed wind power of power grid

As the percentage of wind power in minimum load is increasing gradually, it is necessary to restrict the output of wind power, so as to maintain the system operation safety and stably after the derated output of conventional power sources reaches its limit. In this paper, it is considered that the output of conventional units may be adjusted economically and reasonably based on some logic principle reflected in the rolling scheduling model to coordinate the change of wind power and accommodate the wind power to the maximum.

The matrix E^* presents the latest predicted output of wind power in the remaining time periods, in which the element W_t^{h*} stands for the latest predicted output of wind power in the tth time period before modification at the time t_{h-1} .; the matrix E presents the scheduled output of wind power from the rolling scheduling. Taking the cycle of rolling scheduling H =1hour as an example, it is known that:

$$\mathrm{E}^* = \begin{bmatrix} W_1^{1*} & W_2^{1*} & W_3^{1*} & W_{\mathrm{H}}^{1*} & & W_{2\mathrm{H}}^{1*} & & W_h^{1*} & W_{96}^{1*} \\ 0 & 0 & 0 & W_{\mathrm{H}}^{\mathrm{H}*} & & W_{2\mathrm{H}}^{\mathrm{H}*} & & W_h^{\mathrm{H}*} & W_{96}^{\mathrm{H}*} \\ 0 & 0 & 0 & 0 & 0 & W_{2\mathrm{H}}^{2\mathrm{H}*} & & W_h^{2\mathrm{H}*} & W_{96}^{2\mathrm{H}*} \\ 0 & 0 & 0 & 0 & 0 & 0 & & & \\ 0 & 0 & 0 & 0 & 0 & 0 & 0 & W_h^{h*} & W_{96}^{h*} \\ 0 & 0 & 0 & 0 & 0 & 0 & 0 & 0 & \\ 0 & 0 & 0 & 0 & 0 & 0 & 0 & 0 & W_{96}^{24\mathrm{H}*} \end{bmatrix} \tag{3}$$

$$\mathrm{E} = \begin{bmatrix} W_1^{1} & W_2^{1} & W_3^{1} & W_{\mathrm{H}}^{1} & & W_{2\mathrm{H}}^{1} & & W_h^{1} & W_{96}^{1} \\ 0 & 0 & 0 & W_{\mathrm{H}}^{\mathrm{H}} & & W_{2\mathrm{H}}^{\mathrm{H}} & & W_h^{\mathrm{H}} & W_{96}^{\mathrm{H}} \\ 0 & 0 & 0 & 0 & 0 & W_{2\mathrm{H}}^{2\mathrm{H}} & & W_h^{2\mathrm{H}} & W_{96}^{2\mathrm{H}} \\ 0 & 0 & 0 & 0 & 0 & 0 & & & \\ 0 & 0 & 0 & 0 & 0 & 0 & 0 & W_h^{h} & W_{96}^{h} \\ 0 & 0 & 0 & 0 & 0 & 0 & 0 & 0 & \\ 0 & 0 & 0 & 0 & 0 & 0 & 0 & 0 & W_{96}^{24\mathrm{H}} \end{bmatrix} \tag{4}$$

Let $\mathrm{E}_h^*, \mathrm{E}_h$ stand for the vectors in the h/H line of the matrixes E^* and E respectively. Thus, the objective of curtailed wind power of power grid during modification at the time t_{h-1} can be represented as follows:

$$f_2(W_t^h) = \left\| \mathrm{E}_h^* - \mathrm{E}_h \right\|_1 \tag{5}$$

In which, $\left\| \mathrm{E}_h^* - \mathrm{E}_h \right\|_1$ stands for the 1-norm of the vector.

2.4. Other constraints

- Constraint of Output of wind power

$$0 \le W_t^h \le W_t^{h*}, \quad t = h, h+1, \quad 96 \tag{6}$$

- Constraint of Ramp rate of conventional unit

$$\begin{cases} p_{i,\ t+1}^h - p_{i,\ t}^h \le \Delta p_{i,up} \Delta T, & t = h, h+1, \quad 96 \\ p_{i,\ t}^h - p_{i,\ t+1}^h \le \Delta p_{i,down} \Delta T, & t = h, h+1, \quad 96 \end{cases} \tag{7}$$

In which: $\Delta p_{i,up}$ and $\Delta p_{i,down}$ stand for the maximum and minimum ramp rate per minute of unit i respectively; ΔT is 15 min.

- Constraint of Output of conventional unit

$$p_{i\min} \le p_{i,\ t}^h \le p_{i\max}, \quad t = h, h+1, \quad 96 \tag{8}$$

In which, $p_{i\max}$ and $p_{i\min}$ stand for the maximum and minimum technical outputs of unit i respectively.

- Constraint of Regulation deviation of conventional unit

To guarantee the relevancy between rolling scheduled and day-ahead scheduled output of conventional unit, the limit of regulation deviation for each unit is set, to be controlled within a certain scope. The regulation deviation constraint of conventional unit during modification at the time t_{h-1} is as follows:

$$\left| p_{i,\ t}^h - p_{i,\ t}^0 \right| \le \gamma_i, \quad t = h, h+1, \quad 96 \tag{9}$$

In which, γ_i is the permitted maximum deviation of unit i.

- Constraint of Transmission Section power flow of power grid

With regard to the key transmission section of power grid affecting the system's security and stability seriously, the restrictions over the output of wind power and conventional units in the transmission section should be taken into account. The output of conventional units in the transmission section can be modified by means of rolling scheduling, so as to effectively alleviate curtailed wind power of power grid due to its insufficient transmission capacity.

$$\delta_{W,l} W_t^h + \sum_{i=1}^{N_l} \rho_{l,i} p_{i,\ t}^h \le \overline{S_{l,t}}, \quad t = h, h+1, \quad 96 \tag{10}$$

In which, l represents the transmission section; N_l is the number of conventional units in l; $\rho_{l,i}$ is the sensitivity of the output of unit i to l; $\delta_{W,l}$ is the sensitivity of the output of wind power to l; and $\overline{S_{l,t}}$ is the upper power limit of l in the t th time period.

2.5. Model transformation

Penalty function method is employed to transform the above model into an unconstrained single-objective optimization problem. The rolling scheduling model of electric power system with wind power is presented as follows:

$$\min f = f_1 + \kappa f_2 + \sum_{j=1}^{K} \kappa_j g_j \tag{11}$$

In which, κ is the transformation coefficient to convert the wind power into the operation cost increment of equivalent coal-fired power unit, and its value can be set as: $\max(a_i \gamma_i + 2a_i p_{i\max} + b_i)$, $i = 1, 2, \quad , N_G$; g_j stands for the j th constraint; κ_j is the penalty coefficient of the j th constraint; and K is the number of constraints.

3. Solving strategy based on improved NNIA algorithm

NNIA algorithm is an EMO algorithm based on artificial immune system. When the NNIA algorithm is employed to solve the above rolling optimization scheduling model, antibody refers to the decision variable contained in the model, that is, the rolling scheduled output of each unit in every time period and the scheduled output of wind power in each time period. The NNIA algorithm employs the proportional cloning, recombination, super mutation, and other tactics to well guarantee the diversity and integrity of the antibody population [14]. Meanwhile, it also increases the size of antibody population and the complexity of algorithm. This part will build an antibody preference relation model that taking the starting and ending points to determine the preference area and guiding the evolution of antibodies.

3.1. Strategy for improvement of algorithm through rectification based on equal incremental rate

The approach of equal incremental rate has very strict mathematical optimality in the economical scheduling of electric power system. According to the principles of equal incremental rate, equations (12) and (13) are used to obtain the theoretical optimal value of output of unit i in the tth time period.

$$\sum_{i=1}^{N_G} p_{i,\,t}^{h\ *} + W_t^{h*} - D_t^h = 0,\ \ t = h, h+1,\ \ ,96 \tag{12}$$

$$\frac{dF_1}{dp_{1,\,t}^{h\ *}} = \frac{dF_2}{dp_{2,\,t}^{h\ *}} = \ \ = \frac{dF_{NG}}{dp_{NG,\,t}^{h\ *}} = \lambda_t^h,\ \ t = h, h+1,\ \ ,96 \tag{13}$$

In which, λ_t^h stands for the equal consumption incremental value in the tth time period; $p_{i,\,t}^{h\ *}$ stands for the theoretical optimal value of output of unit i in the tth time period; and F stands for the operation cost function of unit.

According to the scheduled output of unit before modification and the theoretical optimal value of output of unit, i.e. $p_{i,\,t}^{h-\mathrm{H}}$ and $p_{i,\,t}^{h\ *}$, the search direction in the algorithm for the output of unit can be determined for the algorithm, as presented in figure 1, which is divided into two cases a and b:

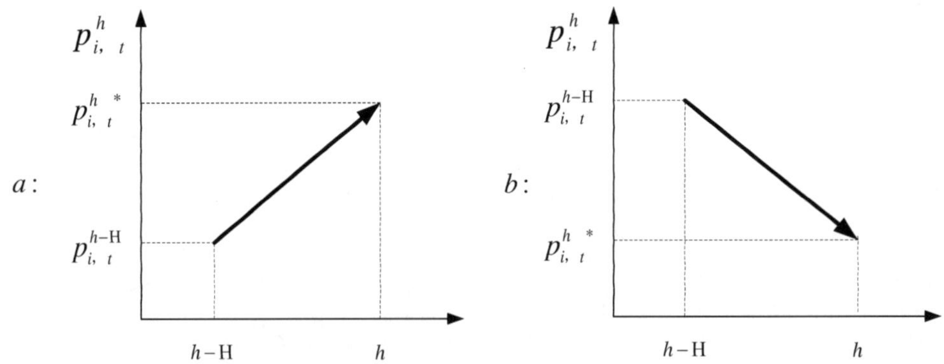

Figure 1. Search direction for the output of unit.

Due to the rolling scheduled output of unit i , $p_{i,\,t}^{h}$ must satisfies equations (8) and (9), there are six different cases as follows, i.e. c, d, e, f, g, h, based on the positional relations between $p_{i,\,t}^{h\ *}$ and the intervals determined by equations (8) and (9), and the rectification search direction for the output of unit is presented in figure 2.

As revealed in figure 2, the starting and ending points of each antibody can be determined, i.e. the starting and ending points indicated by "→" in figure 2.

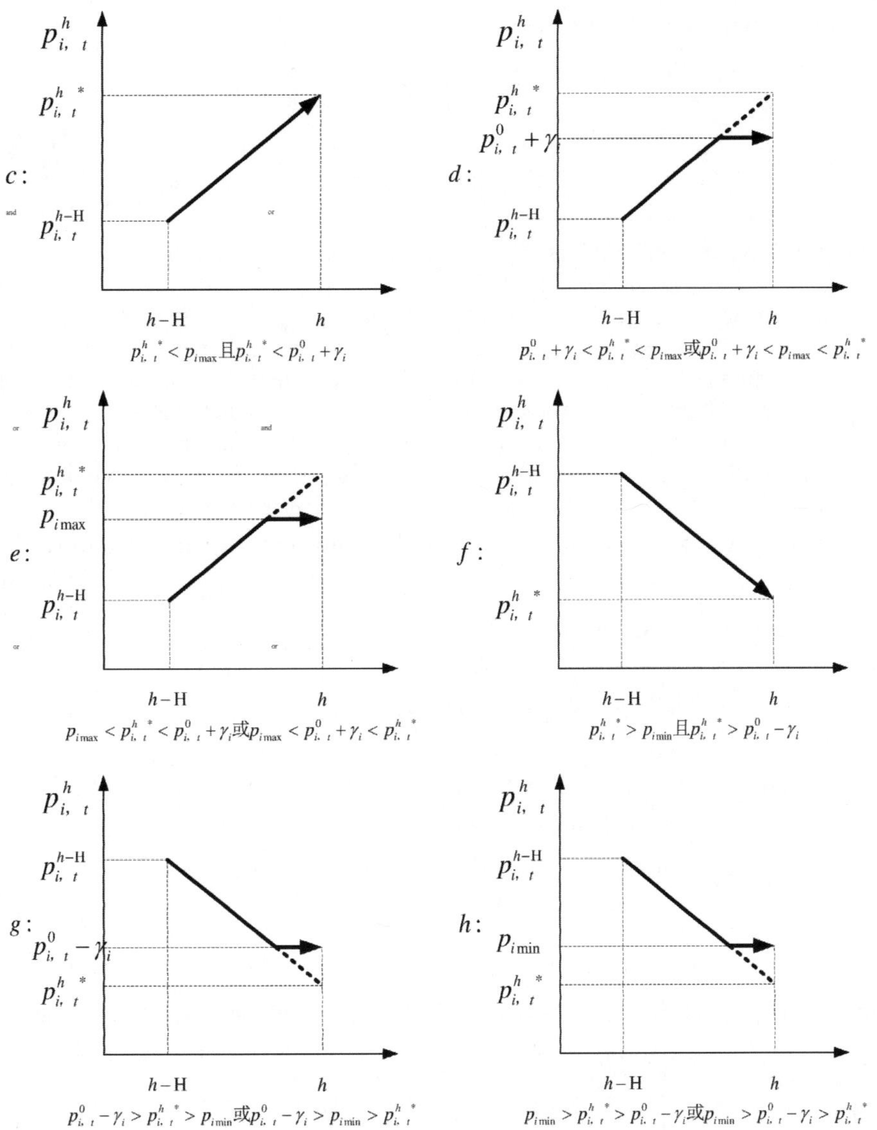

Figure 2. Rectification search direction for the output of unit.

It is assumed that each antibody contains m decision variables, the antibody preference relation model describes a preference area which is a m-dimension ultra-long square space, and the value on each dimension must fall into the scope demarcated by the values of the antibodies at the starting and ending points. It is assumed that $x_a \in X_f$ is any feasible antibody for the rolling scheduling in this paper, X_f stands for antibody population, $x_a = \left(x_a^1 \; x_a^2 \quad x_a^j \quad x_a^m\right)$, $j=1,2,\cdots,m$, x_v is the antibody at the given starting point, and x_r is the antibody at the given ending point. φ_a^j is calculated as follows:

$$\begin{cases} \varphi_a^j = 1, & \left|x_a^j - x_v^j\right| \le \left|x_v^j - x_r^j\right| \wedge \left|x_a^j - x_r^j\right| \le \left|x_v^j - x_r^j\right| \\ \varphi_a^j = 0, & else \end{cases} \tag{14}$$

The judgment formula is obtained as follows:

$$\varphi_a = \prod_{j=1}^{m} \varphi_a^j \tag{15}$$

As we know, $\varphi_a = 1$, so the antibody x_a falls into the preference area; $\varphi_a = 0$, so the antibody x_a does not fall into the preference area. At the time of solution, after the basic NNIA algorithm is employed to generate the antibody population, the antibodies outside the preference area can be removed.

With regard to every feasible antibody in the current generation, the antibody in the current generation falling into the preference area and closest to the ending point is called most preferred solution, or "Intermediate Antibody". According to the "Intermediate Antibody" and the "Optimal Antibody" in the current generation (i.e. the optimal solution in the current generation with the lowest objective function value), the preference area of the next-generation antibody population can be determined by taking the "Optimal Antibody" as the starting point and the "Intermediate Antibody" as the ending point, so as to constantly direct the evolutionary direction of antibodies.

The "Intermediate Antibody" can be obtained in the following way. It is assumed that x_c is any feasible antibody in the current generation and d_c stands for the distance from the antibody x_c to the ending point, so there is the following definition:

$$d_c = \sum_{j=1}^{m} \left\{ \left| x_c^j - x_v^j \right| + (\rho + 1) \left| x_c^j - x_r^j \right| \right\} \tag{16}$$

In which, ρ is a positive real number lower than $1/m$. The "Intermediate Antibody" in the current generation is the antibody with the minimum value of d_c.

3.2. *Procedure of the improved NNIA algorithm*

- Step 1 Initialize

Initialize the predicted wind power W_t^{h*}, predicted load D_t^{h*}, and the previous rolling scheduling interpolation of unit $p_{i,\ t}^{h-\mathrm{H}}$ at the current time (e.g. the time t_{h-1}). Initiate other fixed parameters.

The initial antibody population B_0 of the size N is generated. Let the preference antibody population E_0, active population A_0 and cloned population C_0 be empty, and $k = 0$.

- Step 2 Update preference antibody population

Determine all the antibodies of B_k in the antibody population. Based on the equations (14) and (15), judge whether the antibodies belong to the current antibody preference area (the k th generation). Keep the antibodies inside the antibody preference area, and place them in the temporary preference antibody population ET_{k+1}. If the size of ET_{k+1} is lower than N, let $E_{k+1} = ET_{k+1}$. Otherwise, calculate the crowding distance of all individuals in ET_{k+1} [14], sequence the individuals from large to small based on the crowding distance, and select the first N individuals to form E_{k+1}. Based on equation (16), calculate the "Intermediate Antibody" in E_{k+1}, and then the "Optimal Antibody" of E_{k+1}, and determine the antibody preference area of the $k+1$th generation.

- Step 3 Calculate the conditions for termination

If $k \geq g_{-\max}$, terminate the search; or $k = k + 1$.

- Step 4 Select the nondominated neighbor

If the size of E_k is lower than the maximum size of active population N_A, let $A_k = E_k$; or calculate the crowding distance of all individuals in E_k, sequence the individuals in descending order based on the crowding distance, and select the front N_A individuals to form the active population A_k.

- Step 5 Proportional cloning

Implement the proportional cloning of A_k to obtain the cloned population C_k.

- Step 6 Recombination and super mutation

Implement the recombination and super mutation of the cloned population C_k to obtain a new population C_k'.

- Step 7 Combine C_k' and E_k to gain the antibody population B_k, and then move to step 2.

4. Analysis with calculation example

Due to the high speed development of wind power and the relative lag of construction of power grid in China, a mass of curtailed wind power of many provinces have appeared in recent years. Based on this research background, the 10-unit test system in 96 time periods of a day is taken for the simulative calculation to verify the effectiveness of the proposed rolling scheduling model and improved NNIA algorithm in this paper. The parameters of conventional units are shown in table 1. The maximum and minimum ramp rate per minute of unit are set to 1% and -1% of the unit's maximum output respectively; γ_i is set to $0.1p_{i,t}^0$; the calculated value of κ is 202.8 in this calculation example; and κ_j is consistently set to 10^5. The system contains a wind farm connected to grid. The data of actual operation from a wind farm in a region of China are used, including fifty-six 2.5 MW WTG sets with the total installed capacity of 140 MW. Among them, unit 1 and wind farm are combined into a transmission section, and its upper power limit is 500 MW.

Table 1. Parameters of conventional units.

Unit No.	$p_{i\max}$ /MW	$p_{i\min}$ /MW	a_i /(\$/(MW2·h))	b_i /(\$/(MW·h))	c_i /(\$/h)
1	470	150	0.043	21.60	958
2	460	135	0.063	21.05	1313
3	340	73	0.039	20.80	604
4	300	60	0.070	23.90	471
5	260	57	0.056	17.87	601
6	243	73	0.079	21.62	480
7	130	20	0.211	16.51	502
8	120	15	0.480	23.23	639
9	80	10	1.091	19.58	455
10	55	55	0.951	22.54	692

4.1. Analysis on effect of rolling scheduling model

The calculated effect of rolling scheduling model is presented by taking the time t_{31} as an example. The latest predicted loads of the system are presented in table 2.

Table 2. Latest predicted loads.

Time (hour)	Load/MW	Time (hour)	Load/MW
9	1984	17	1540
10	2132	18	1688

11	2206	19	1836
12	2280	20	2132
13	2132	21	1984
14	1984	22	1688
15	1836	23	1392
16	1614	24	1244

Table 3. Latest predicted wind power.

Time Period (15 min)	Wind Power /MW	Time Period (15 min)	Wind Power /MW	Time Period (15 min)	Wind Power /MW	Time Period (15 min)	Wind Power /MW
33	108.4	49	75.0	65	91.7	81	75.0
34	108.3	50	83.3	66	75.0	82	75.0
35	108.3	51	83.3	67	75.0	83	91.7
36	100.0	52	83.3	68	75.0	84	83.3
37	91.7	53	83.3	69	66.7	85	83.3
38	91.7	54	83.3	70	66.7	86	100.0
39	91.7	55	75.0	71	58.3	87	100.0
40	91.7	56	75.0	72	58.33	88	100.0
41	91.7	57	75.0	73	58.3	89	100.0
42	83.3	58	75.0	74	66.7	90	108.3
43	83.3	59	75.0	75	66.7	91	108.3
44	83.3	60	83.3	76	75.0	92	108.3
45	83.3	61	83.3	77	75.0	93	108.3
46	75.0	62	91.7	78	83.3	94	108.3
47	75.0	63	91.7	79	75.0	95	108.3
48	75.0	64	75.0	80	75.0	96	108.3

The latest predicted wind power is presented in table 3. After modification by rolling scheduling, it is calculated that curtailed wind power is reduced by 15.30 MW in average every time period. Meanwhile, accompanied by more accommodation of wind power it also reduces the system's consumption of primary energy remarkably, and saves the operation cost by -52082USD. The objective function values before and after modification are presented in table 4.

Table 4. Objective function values before and after modification.

	Average Reduction of Curtailed Wind Power /MW	Operation Cost / USD	Incremental Operation Cost / USD	General Objective/ USD
Before	17.2031	5016455	/	59574
After	1.9029	4964373	-52082	-44267

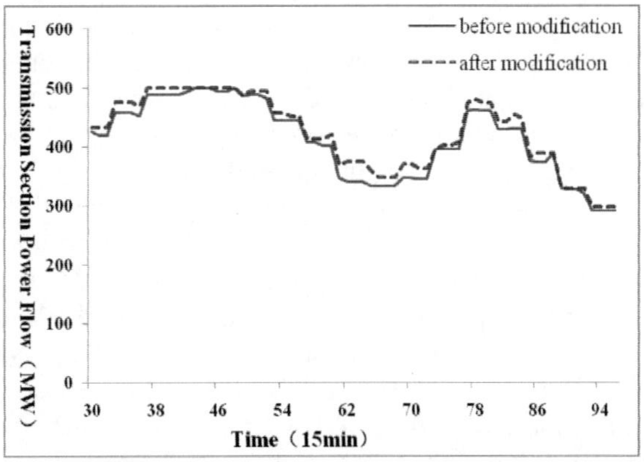

Figure 3. Curves of transmission section power flow before and after modification.

Figure 3 presents the curves of the transmission section power flow before and after modification. Clearly, the rolling scheduled outputs of wind farm and unit 1 after modification can still satisfy the transmission section power flow constraint although the output of wind power has increased.

Figure 4 presents the rolling scheduled output of unit 1 to 9 before and after modification in every time period. Unit 10 represents the fixed output unit, so its curve is no need to be drawn into the figure. The figure shows that the adjustment of output of conventional units is very small after modification due to the inhibitive effect of the constraint of regulation deviation of unit.

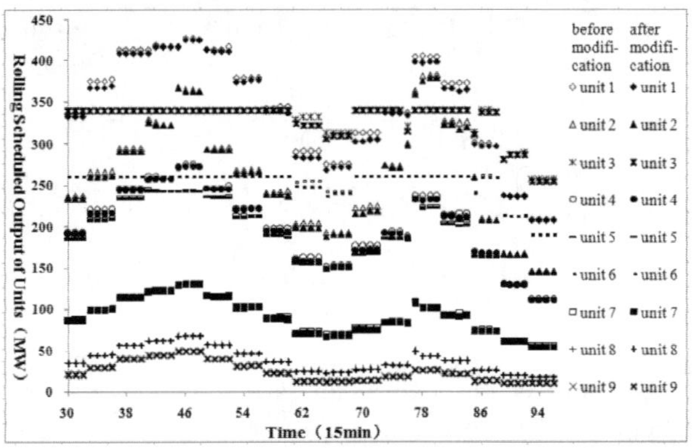

Figure 4. Rolling scheduled output of conventional units before and after modification.

4.2. Analysis on performance of algorithm

To verify the superiority of the solving strategy based on improved NNIA algorithm proposed, genetic algorithm, NNIA algorithm and improved NNIA algorithm have been comparatively analyzed for solving the rolling scheduling model.

Table 5 gives the optimization objective value and calculation time after three algorithms running for 1,000 times respectively.

Table 5. Comparison of optimization objective of three algorithms.

Algorithm	Optimization Objective/ USD	Calculation Time /s
Genetic algorithm	-41151	446.2
NNIA algorithm	-44209	549.6
Improved NNIA algorithm	-44267	177.5

Compared with genetic algorithm, NNIA algorithm obtains more ideal optimization objective value, while genetic algorithm leads to the local optimal solution. NNIA algorithm demonstrates its superiority in global search for solving the high-dimensional optimization problem. Furthermore, improved NNIA algorithm not only obtains better optimization objective value, but also reduces the calculation time significantly. Hence, it can be concluded that improved NNIA algorithm is extremely helpful to removing the non-optimal antibodies and guiding the evolution of antibody population towards the optimal antibody that significantly improve the search performance of the algorithm.

Table 6 presents the variation process of the optimization objective value with the increase of iteration count of NNIA algorithm and improved NNIA algorithm. It is clear that improved NNIA algorithm converges towards the optimal solution much faster than NNIA algorithm, and its optimization objective value is basically stable in the late iterative calculation. If the maximum iteration count is set to 800 times, improved NNIA algorithm can obtain the same optimization objective value. If the maximum iteration count is set to 600 times, improved NNIA algorithm's optimization objective value decrease by 1.4% only, while NNIA algorithm's optimization objective value decrease by 10.7%. Moreover, when it tends to be stable in the late iteration, improved NNIA algorithm's optimization objective value is still better than NNIA algorithm, which means that improved NNIA algorithm has better local search capability.

Table 6. Optimization objective value of NNIA algorithm and improved NNIA algorithm for different iteration counts.

NNIA Algorithm			Improved NNIA Algorithm		
Maximum Iteration Count/ times	Optimization Objective /USD	Calculation Time /s	Maximum Iteration Count/ times	Optimization Objective /USD	Calculation Time /s
400	-24151	227.5	400	-41151	88.9
600	-39484	339.1	600	-43656	117.3
800	-42027	442.7	800	-44267	142.5
1000	-44209	549.6	1000	-44267	177.5

5. Conclusion

The rolling scheduled model built in this paper can further optimize the operation cost of electric power system in the implementation process of intra-day rolling scheduling, and modified the key transmission section power flow the transmission power of key sections in a rolling manner to satisfy the security constraint of power grid. As the modification function for curtailed wind power of power grid is taken into account, it can also enhance the system's ability to accommodate wind power. The proposed strategy for solving the problem based on improved NNIA algorithm is characterized by quicker convergence and stronger local search capability.

References

[1] European Smart Grid Technology Platform 2006 "Vision and Strategy for Europe's Electricity Networks of the Future," EU Commission, Directorate-General for Research, Information and Communication Unit, Brussel, Belgium

[2] Lei Y Z 2003 Studies on wind farm integration into powersystem *Automat of Electr Power Syst* **27(8)** 84-9 (in Chinese)

[3] Jin M, Feng W, Liu P, Marnay C and Spanos C 2017 MOD-DR: Microgrid optimal dispatch with demand response *Appl Energy* **187** 758-76

[4] Jin M, Feng W, Liu P, Marnay C and Spanos C 2017 Microgrid to enable optimal distributed energy retail and end-user demand response *Appl Energy* http://dx.doi.org/10.1016/j.apenergy.2017.05.103.

[5] Mashayekh S, Stadler M, Cardoso G and Heleno M 2017 A mixed integer linear programming approach for optimal DER portfolio, sizing, and placement in multi-energy microgrids *Appl Energy* **187** 154-68
[6] Tulabing R, Rong X Y, DeForest N, *et al* 2016 Modeling study on flexible load's demand response potentials for providing ancillary services at the substation level *Electr Power Syst Res* **140** 240-52
[7] Lee T Y 2007 Optimal spinning reserve for a wind-thermal power system using EIPSO *IEEE Trans. Power Syst* **22** 1612-21
[8] Hetzer J and Yu D C 2008 An economic dispatch model incorporating wind power *IEEE Trans. Energy Convers* **23** 603-11
[9] Siahkali H and Vakilian M 2010 Stochastic unit commitment of wind farms integrated in power system *Electr Power Syst Res* **80** 1006-17
[10] Shen W, Wu W C, Zhang B M, *et al* 2011 An on-line rilling generation dispatch method and model for accommodating large-scale wind power *Automat of Electr Power Syst* **35(22)** 136-40 (in Chinese)
[11] Zhang B M, Wu W C, Zhen T Y, *et al* 2011 Design of a multi- time scale coordinated active power dispatching system for accommodating large scale wind power penetration *Automat of Electr Power Syst* **35(11)** 1-6 (in Chinese)
[12] Wang K, Zhang B H, *et al* 2011 Areal-time dispatch in wind power integrated system based on choas quantum-behaved particle swarm optimization *Automat of Electr Power Syst* **35(22)** 141-6 (in Chinese)
[13] Freschi F and Repetto M 2006 VIS: An altificial immune network for multi-objective optimization *Eng Optimiz* **38(8)** 975-96
[14] Coello C A and Cortes N C 2005 Solving multi-objective optimization problems using an artificial immune system *Genet Program and Evol Mach* **6(2)** 163-90

Review of the development of multi-terminal HVDC and DC power grid

Y X Chen[1,2]

[1]School of Electrical and Electronic Engineering, North China Electric Power University, Beijing 102206, China

E-mail: cyx@ncepu.edu.cn

Abstract. Traditional power equipment, power-grid structures, and operation technology are becoming increasingly powerless with the large-scale renewable energy access to the grid. Thus, we must adopt new technologies, new equipment, and new grid structure to satisfy future requirements in energy patterns. Accordingly, the multiterminal direct current (MTDC) transmission system is receiving increasing attention. This paper starts with a brief description of current developments in MTDC worldwide. The MTDC project, which has been placed into practical operation, is introduced by the Italian–Corsica–Sardinian three-terminal high-voltage DC (HVDC) project. We then describe the basic characteristics and regulations of multiterminal DC transmission. The current mainstream of several control methods are described. In the third chapter, the key to the development of MTDC system or hardware and software technology that restricts the development of multiterminal DC transmission is discussed. This chapter focuses on the comparison of double-ended HVDC and multiterminal HVDC in most aspects and subsequently elaborates the key and difficult point of MTDC development. Finally, this paper summarizes the prospect of a DC power grid. In a few decades, China can build a strong cross-strait AC-DC hybrid power grid.

1. Introduction

The cross-regional and long-distance DC transmission is an important method to identify the optimal allocation of regional resources. In 1954, Sweden built the first HVDC project, and thus far the world has more than 100 HVDC projects that have been introduced into commercial operation [1]. However, most of them are double - ended HVDC system, which cannot only achieve point-to-point DC power transmission, and when using DC interconnection between the multiple AC systems, we need multiple point-to-point DC transmission lines, which will significantly improve the investment costs and operating costs. Importantly, wind power, solar, and other new energy power generations are intermittent and strongly random. In a word, they are intermittent power supplies. With the large-scale renewable energy access to the grid, the traditional power equipment, power-grid structure, and operation technology are becoming increasingly powerless. As a result, we must adopt new technologies, new equipment, and new grid structure to meet the profound changes in the future energy pattern [2]. Therefore, the multiterminal direct current (MTDC) transmission system developed on double-ended HVDC transmission system has received increasing attention.

2. Research status at home and abroad

As the DC transmission technology continuously develops and becomes increasingly perfect, it becomes widely used. That is, many countries have become very active on the research of multiterminal DC transmission technology. At present, multiterminal HVDC projects, namely, Corsican – Sardinia (three-ended) and Quebec – New England (five-ended, actually three-terminal on operation) has been introduced into operation; Japanese Chikumagawa three-terminal DC transmission system is the world's first back-to-back system; in addition, Canada's Nelson River bipolar 1 and bipolar 2 DC transmission project and the United States Pacific interconnection DC transmission projects both have the operating mode and characteristics of four-terminal DC transmission [3].

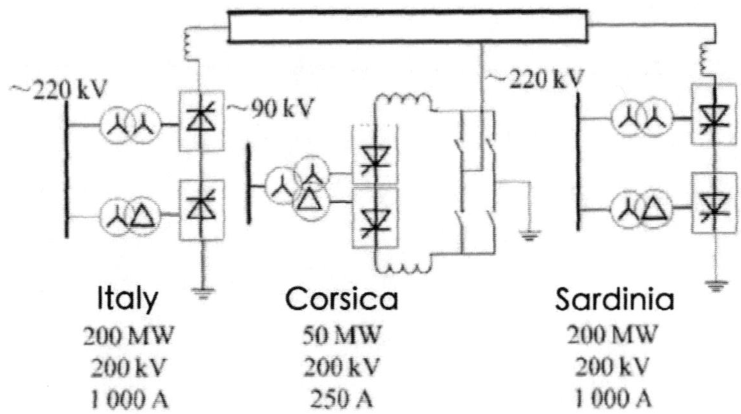

Figure 1. Italy - Corsica - Sardinian three - terminal DC transmission project connection diagram.

According to figure 1, Italy – Corsica – Sardinia three-terminal HVDC project is the world's first official operation of the multiterminal DC transmission project, of which the mode of operation is FM-based. Canada's Quebec–New England five-terminal DC system is the world's largest multiterminal HVDC project, which delivers cheap electricity from the northern Berkshire cascade to the New England power grid in the northeastern United States and the load center in southern Quebec. The project consists of five parts as follows: Radisson, Nikolai, DesCantons, Comerford, and Sundries. Whereas the converter in Radisson can only be operated as a rectifier station, the other converter stations can be operated as either a rectifier station or an inverter station; the project can vary in using two-, three-, four-, or five-terminal operation (usually three-ended operation) based on system requirements.

Japanese Shinano's three-terminal HVDC project is the world's first back-to-back voltage source converter (VSC) multiterminal DC project, which aims to achieve the interconnection of 50 Hz power grid in eastern Japan and the 60 Hz power grid in western Japan. Canada Nelson River four-terminal DC transmission is composed of two bipolar DC transmission systems (Nelson River bipolar 1 and bipolar 2). Under normal circumstances, the bipolar system is under independent operation. If necessary, bipolar line or bipolar converter can be changed to be under parallel operation, which has the characteristics of a multiterminal DC system. In 1970, the US Pacific line at both ends of the DC project was completed, and then new bipolar converter stations are added on either end in the expanded project in 1989. The new stations and old stations are operated in parallel to form a four-terminal DC transmission system [4].

3. Basic characteristics and adjustment methods

3.1. Main form and operating characteristics

In double-ended HVDC system, the two converter stations are parallel because they have a common DC voltage and the converter stations are also considered in series because they pass the same DC current. When adding new converter stations a problem on whether maintaining a common DC voltage or maintaining the same DC current arises. Therefore, according to the different operating conditions and design requirements, a variety of topologies of the wiring can be formed. In general, wiring can be divided into parallel connection, series connection, and mixed connection, as shown in figure 2, where the parallel can be divided into radial and ring network [3]. Each type of structure has different operating control characteristics, and power allocation can be made through the control of different electricity in the converter station [1].

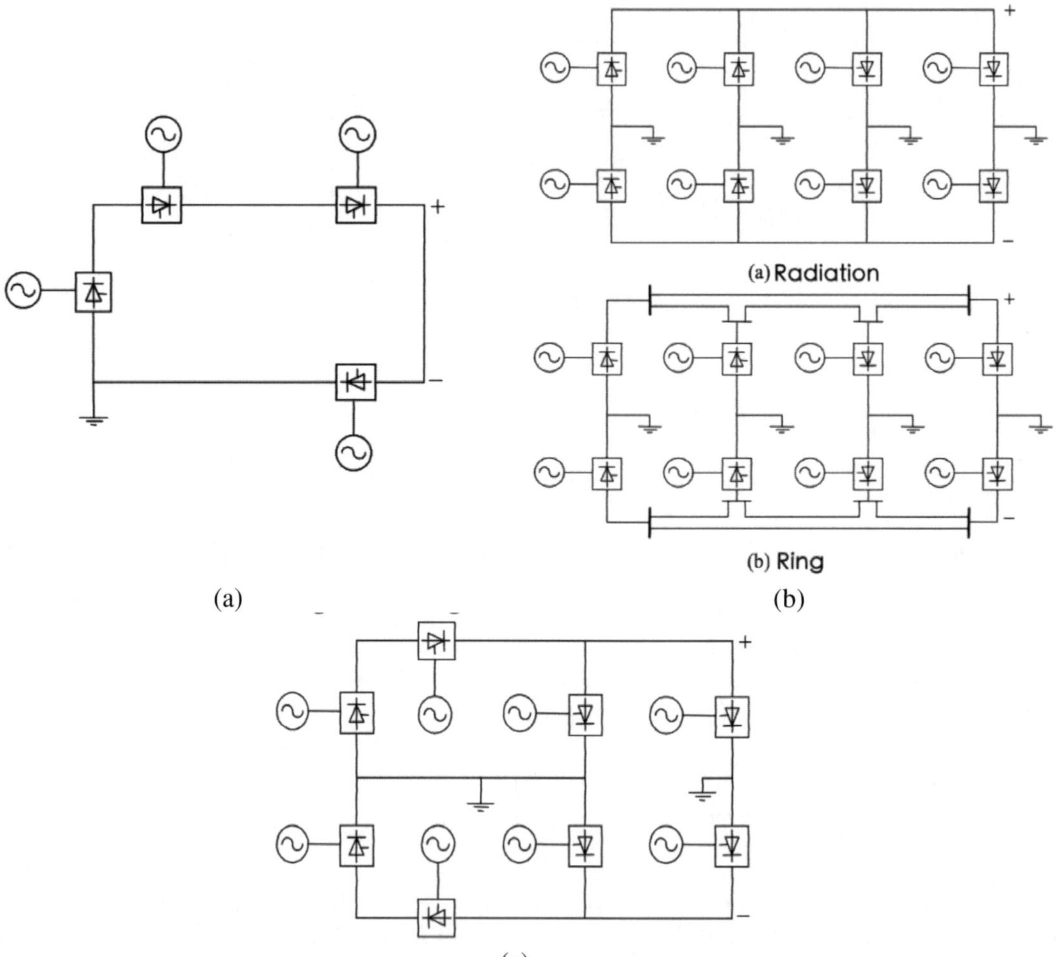

Figure 2. Topology of series, parallel and mixed connection multiterminal DC power transmission. (a) series connection, (b)parallel connection and (c) mixed connection.

The following are the different operation characteristics: parallel converter station run of the same level of DC voltage, power distribution achieving by changing the current of each converter station; series converter station run with the same level of DC current, power distribution achieving by changing the DC voltage of each converter; and mixed type increases the flexibility of the multiterminal DC connection. In the design period, wiring connection should be selected based on investment, loss, reliability, flexibility, and the specific requirements of specific projects [4].

In brief, compared with the series connection, the parallel connection has a smaller line loss, a larger adjustment range, easier to achieve the insulation with a more flexible manner of expansion and outstanding economy. Thus, the current operation of the multiterminal DC transmission project is used in parallel connection method [1]. The detailed comparison of MTDC system of different topology are listed in table 1.

Table 1. Comparison of parallel and series connected MTDC system.

compared aspect	parallel connection	series connection
active loss	small	large
reactive power demand	small	large
Economy	excellent	not good
Fault recovery speed	fast	slow
energy flow reverse	slow	fast
Extensibility	flexible	complex

3.2. Adjustment method

When the MTDC transmission system uses different wiring methods, the adjustment methods are also different [5]. We are discussing certain basic adjustment methods in this part.

3.2.1. Adjustment of current balance control. This method is a double-ended DC transmission system constant current regulation mode extension, that is, a converter station was used to control the entire DC network's DC operating voltage, all other converter stations are regulated by constant current so that the DC system has a stable operation point.

3.2.2. Constant current and constant voltage regulation mode. As shown in figure 3, when mode I is used, the converter station, which is responsible for voltage control, is running on the constant or constant feature, so that the DC voltage changes when the AC side voltage fluctuates. Although the automatic adjustment function of the converter transformer tap can reduce the range of this fluctuation to a certain extent, the adjustment speed is slow and discontinuous. If the system has strict requirements on the DC voltage, then a DC voltage regulator should be added at the control station of the control voltage and should be changed to constant voltage operating mode during normal conditions.

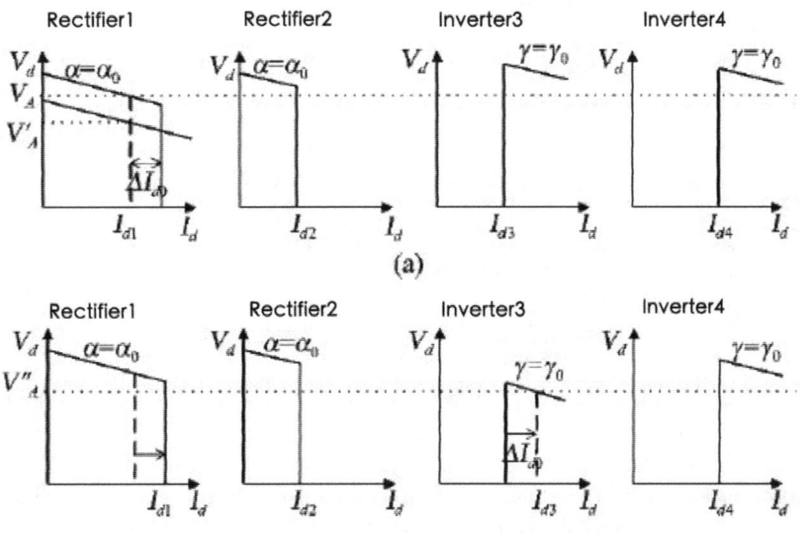

Figure 3. Current margin control [6].

3.2.3. Limited voltage regulation mode. Limited voltage adjustment mode is suitable for situations where a large number of converter stations exist, and the characteristics of the converter station are composed of both constant voltage and constant current (inverter station also set δ) and other multi-segment composition. The voltage of the DC system is controlled by a rectifier station and is maintained at the rated value. The remaining convertor stations are operated in the constant current mode. If the converter station that controls the DC voltage broke down, which leads to the decrease in the DC voltage at this end, the control of the DC voltage will be given to other station to complete, and this station is changed to the DC current control mode.

3.2.4. Basic adjustment method of series connection MTDC transmission system. Series connection multi-terminal DC transmission system is characterized by various converter stations and lines of the same DC current; the use of constant current operation mode is appropriate. To ensure that the DC system has a stable operating point, we frequently selected a converter station to be the constant current mode, with the remaining converter stations being set by constant or constant adjustment. This adjustment method is relatively simple, and the total amount of reactive power consumption in the converter station is also minimal.

4. Key technology overview and characteristic analysis

MTDC transmission system is a kind of DC transmission mode developed on the basis of double-ended DC transmission. Therefore, most of the technologies used in traditional DC transmission can be applied to MTDC transmission system. However, the particularity of MTDC transmission network topology and function is different from the double-ended DC transmission in the selection of circuit breaker equipment, basic control mode, coordinated control, power flow calculation, stability, and reliability analysis. This occurrence must be further explored and analyzed. The MTDC transmission technology research also focused on these aspects at home and abroad. The following will review several key issues including circuit breaker equipment, basic control and protection, coordination control and stability analysis of the domestic and foreign MTDC transmission technology.

4.1. Circuit breaker equipment

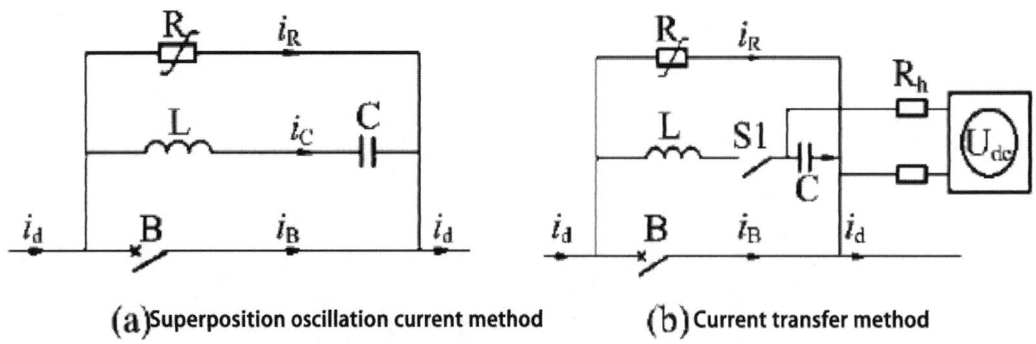

Figure 4. Principle of DC circuit breaker.

The thyristor rectifier has the ability to rapidly cut-off the current. Thus, in the double-ended DC transmission system, the DC outage can be completed through the rectifier rather than the DC circuit breaker. However, if traditional method of processing is used for multiterminal HVDC systems, we should make short-term outage of the entire multiterminal DC system to clear the fault, which will lead to its connected AC system influenced by a greater effect especially on the weak exchange system. Therefore, a high-voltage DC circuit breaker should be installed on a multiterminal DC system, similar to the AC system, to cut-off the fault current and clear the fault. This occurrence will significantly reduce the recovery time after the failure, and shutting down the entire multiterminal DC system is not necessary. However, the DC current does not have natural zero-crossing, which should be forced to zero, and at simultaneously we should consider both the arc time and system over-voltage, thus breaking the AC current is much more difficult than breaking the DC current. High - voltage DC circuit breakers are now the bottleneck of the development and application of multi - terminal HVDC technology [4].

Two main exist for high-voltage DC circuit breaker to break the DC current [7,8] as shown in figure 4.

4.1.1. Superposition oscillation current method. The method utilizes the negative resistance characteristic of the arc and superimposes an oscillating current with increasing amplitude in the DC current to make the "artificial current zero point" to complete the DC current breaking. However, when the arc current increased to a certain extent, its negative resistance characteristics become less obvious, which cannot guarantee that the oscillation current can reach the amplitude that can make zero point. Thus, the type of circuit breaker's breaking current capacity has certain restrictions. However, it has become the most practical application of the current type of DC circuit breaker because of its simple structure and easy-to-control characteristic.

4.1.2. Current transfer method. This method produces a "manual current zero" by a precharged capacitor discharging to produce a current that is opposite to the direction of the system current. Using this principle, the circuit breaker can break the larger DC current with shorter breaking time, but the control of this type of circuit breaker is more complex, which brings less reliability.

4.2. Basic control and protection

Compared with AC transmission, DC transmission has the ability to quickly adjust the power flow distribution, single pole failure in another pole's overload conditions can induce power transfer that can be achieved provided that single pole fails with another pole's overload ability to minimize the loss of transmission power. However, this rapid adjustment characteristic cannot be separated from the controller with superior performance. The traditional DC transmission controller usually provides the converter station with a constant current controller as the main controller and is equipped with a constant control (minimum) as a backup control to regulate the DC current under normal operating conditions. The inverter station is equipped with a constant set-off angle or a constant voltage controller as the main controller and is equipped with a constant current controller that considers the current margin. It is used to control the DC voltage under normal operating conditions. The basic control characteristics are shown in figure 5.

In terms of protection, the multiterminal DC transmission is more complicated than the double-terminal DC transmission. This complexity is reflected in the reasonable setting of the measuring point position, the accurate judgment of the fault type and fault location, and the choice of the system running mode after failure. A good protection program should ensure that the system after the failure can ensure non-fault system to continue to run to reduce the load loss as much as possible while ensuring the impact of the equipment within the limitation [1].

4.3. Coordination control

The biggest difference between MTDC system controller and the double-ended DC transmission system is the coordinated control. Different from the pairs of converter stations of DC transmission, the control methods of MTDC system multiple converter station in the aspects of start and stop, normal operation, and accident under the coordination become a new problem. Take the parallel-connected MTDC system for example, during the process of start and stop process in synchronization of the whole station, you should consider how to meet the trigger time of all the converter stations and how to adjust the lift speed of the current setting value; in the single station start and stop processes, the power distribution problem with station stopping and the other station start and stop process not yet started. Under normal operating conditions, only one converter station controls the DC voltage, whereas the remaining stations control the DC current flowing through the station. When the DC current setting value changes, the balance of the power distribution of the whole system is necessarily reconsidered. In the fault conditions, the reallocation of power after a station or a pole quitting should also needed be considered. The coordination of such systems is the new problem of MTDC system controller design, which should be based on a power coordination program, and a coordination controller should coordinate the station setting value and operation mode.

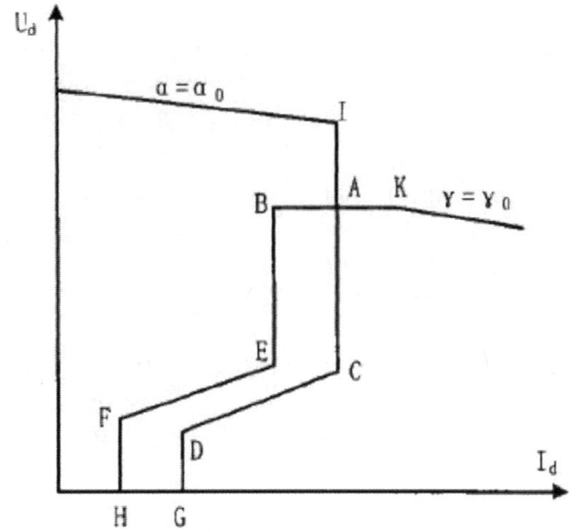

Figure 5. The basic control characteristic of double-ended DC transmission system.

Four-terminal HVDC system take two steps to transit from the outage state to the running state:

- synchronous start-up: all converter station normally unlocks simultaneously and transit to the running state;
- asynchronous start-up: a rectifier station together with an inverter station unlocks at first, when the double-ended system run into a stable state, another rectifier station or inverter station unlocks to form a three-terminal system, and finally the remaining converter station unlocks to constitute a four-terminal system.

The purpose of the system-starting control is to make the whole system start the power transmission under the minimum transient fluctuation and make the power flow of the AC system connected with the rectifier station and the inverter station increase steadily.

Valve-unlocking sequence control design: the inverter side unlocks at first, obeying the order to shift phase for 164 degrees to unlock corner; receiving the message that the inverter side has unlocked, the rectifier side should operate similar to the inverter side, which indicates that it should place the smallest filter group and unlock corner with 164 degrees.

Four-terminal HVDC system takes one step to transit from the running and transit into the outage state:

- multi-station asynchronous outage: a rectifier station or inverter station locks at first to form a three-terminal system, and then the remaining converter station locks to ensure that the remaining stations can work on double-ended operation. Finally, according to conventional DC system, the rectifier station and the inverter station lock simultaneously.

4.4. Stability analysis

The stability analysis method of multiterminal HVDC can be analogous to the steady-state analysis method of power system. The MTDC transmission system is linearized near the operating point, and then the nonlinear equation of state of the whole system is transformed into the state equation. The stability of the system is examined by eigenvalue analysis in algebraic equation analysis. However, this method is more of deviation; the practice use is infrequent [9].

5. Fault characteristics

5.1. DC fault characteristics

- Conducting fault location based only on local data. In the AC system, the fault location is measured by distance protection that measures the impedance value. However, the DC system fault position cannot be determined by measuring the impedance.
- For VSC-MTDC systems, the ability of the IGBT to withstand overcurrent and over- voltage is limited. At the same time, given that the DC network will also supply fault current in the AC system failure, possibly causing DC protection malfunction, which is contrary to the principle of selective characteristic.

5.2. DC fault solution

Option 1: IGBT circuit breaker with high-speed DC switch to form a DC network. In this topology, all VSCs are blocked during DC faults and are isolated from the faulty lines by IGBT circuit breakers. The DC fault is removed by high-speed DC switch, and then IGBT circuit breaker closes, VSC restarts, and the system resumes operation.

Option 2: AC circuit breaker with high-speed DC switch. The installation of costly IGBT DC circuit breakers is not required under this type of topology, but all VSCs must also be locked under DC faults. The following operation is similar to Option 1.

Option 3: Hybrid networks. A small number of IGBT circuit breakers are only installed on some of the selected DC line connection points, faulty subsystem will be isolated under DC faults to ensure the normal operation of other subsystems. This scheme is suitable for large-scale DC networks (such as large offshore wind farms). However, even if the faulty subsystem is isolated, whether the non-defective subsystem has sufficient voltage support should still be explored.

6. DC power grid

The possible topology of the future development of a multiterminal DC system is shown in figure 6, which is the simplest form of a multiterminal HVDC system. It introduces multiple converter stations out of the AC system, which connects different AC systems with multiple point-to-point DC interconnections. Multiterminal DC does not form a grid thus shows no redundancy; as a result, it is improper to be called a network. When any of the converter stations or lines in the topology malfunction, the entire line and the converter stations connected to both sides of the line will be completely out of operation, and thus the reliability will be low.

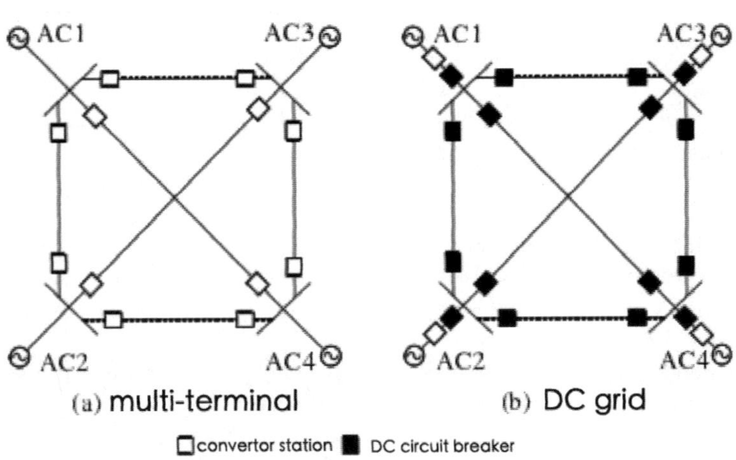

Figure 6. Topology of multi-terminal DC power transmission.

If the DC transmission line is connected in the DC side, the formation of "one-to-many" and "many-to-one" will be formed, and finally exact DC power grid is formed. As shown in figure 6(b), each AC system connects to the DC network through a converter station. Each converter station is connected to each other in multiple DC lines with high voltage DC circuit breakers. When a fault occurs, the circuit breaker can be used to selectively remove the line or converter station. The real DC power grid has the following characteristics:

- The number of converter stations can be significantly reduced. Only one converter station is required to set up at each connection point with the AC power grid, which cannot only significantly reduce the construction costs, but also reduce the overall transmission loss;
- Each converter station can separately transmit (transmit or receive) the power and can change their transmission status from transmission / reception to reception / transmission without affecting the transmission status of the other converter stations;
- More redundancy. Though in a line outage, other lines can still be used to ensure reliable power delivery.

The DC power grid will be an intelligent stable AC-DC hybrid wide-area transmission network with advanced energy management systems. In this network, different clients, existing transmission networks, microgrids, and different power supplies can obtain effective management, optimization, monitoring, control, and timely response to any power problems. It integrates multiple power supplies and transmits and distributes electrical energy over thousands of kilometers with minimal loss and maximum efficiency [2].

7. Conclusion

Compared with the double-ended DC transmission, multiterminal DC transmission system in the following occasions will be more economical and more flexible:

- From the energy base to a number of load centers in the distance;
- Connecting load or power supply to the middle of the line of the DC transmission;
- Several isolated AC systems that use DC transmission lines to achieve non-synchronous communication between the grids. In addition, with the power electronic full-controlling switching device technology and the new DC transmission technology becoming increasingly mature, multiterminal DC transmission in the distributed power generation, renewable energy power generation, and urban DC power distribution..

R & D, manufacture, and improvement of the practical DC circuit breaker in HVDC transmission project is the key problem that must be solved urgently in the development of multiterminal HVDC technology.

Multiterminal DC transmission is a phase of the development of DC grid, which helps achieve multi-power supply and multi-drop power. The DC transmission line is connected to each other in the DC side to form an exact DC power grid. The DC power grid has significantly reduced the number of converter stations that can transmit power separately, and the transmission status can be flexibly switched. In addition, the DC power grid has high reliability.

References

[1] Hu Y R 2013 Study on some problems of multi - terminal HVDC transmission system (Zhejiang: Zhejiang University)
[2] Tang G F, Luo X and Wei X G 2013 Proceedings of the Chinese Society for Electrical Engineering **33** 8-17
[3] Han X 2011 Study on steady - state characteristics of multi - terminal HVDC transmission system (Beijing: North China Electric Power University)
[4] Zhang W L, Tang Y and Zeng N C 2010 *Power System Technology* **1** 1-6
[5] Yuan X F and Cheng S J 2006 *Power System Protection and Control* **34** 61-7
[6] Tu Q R and Xu Z 2009 *East China Electric Power* **37** 267-71
[7] Xu G Z 2000 Principle and Application of High Voltage Circuit Breaker (Beijing: Tsinghua

University Press)
[8] Zheng Z F, Zou J Y, Dong E Y and Duan X Y 2007 *High Voltage Apparatus* **42** 445-9
[9] Padiyar K and Geetha M 1991 Study of torsional interactions in multiterminal DC systems through small signal stability analysis *AC and DC Power Transmission* **29** 411-3

Research on automatic network based on low voltage power line carrier

T J Zhu[1], H B Guo[1,2,3] and Y Xu[1]

[1]Cogeneration Power Supply Center of Shengli Petroleum Administration Bureau, Shandong, Dongying, China
[2]Beijing Institute of Residential Building Design & Research Co. Ltd, Beijing, China

E-mail: zhutiejun196.slyt@sinopec.com

Abstract. To improve the power line communication distance, shorten the networking time and the reliability of the system communication, a method based on the clustering algorithm to change the network time and based on the distance between nodes is proposed in this paper. Firstly, the physical structure and logical topology of the low-voltage power line network are described in details. Then, the clustering algorithm and the method of route reconstruction are described. Finally, the method is simulated and analyzed. The results show that the proposed method can not only reduce network time, but also improve the communication stability of nodes within the network.

1. Introduction

Compared with other wired communication, power line communication has the following advantages: wide distribution, no re-wiring and small construction difficulty. There is a protective layer on the outside of the power line, such as walls or leather tube, small damage and easy post-maintenance. Of course, power lines also have some problems, such as limited transmission band, small baud rate, access equipment random, channel of line noise, signal attenuation [1-3] and so on. At present, many people at home and abroad use the dynamic relay to achieve the expansion of the network, while improving the stability of the network, such as References [4-7]. Ran Q H, *et al* [8] combined with the carrier meter reading system centralized network structure, and proposed a cross-clustering path search algorithm, Zhao J W, *et al* [9] proposed an ant colony network algorithm based on channel quality and service requirements, and Peng Q W and Zhang H [10] proposed a routing algorithm based on neighbor relation table. But few people on the basis of the network time to optimize the study. To improve the stability of the network and the reliability of network communication, this paper proposes a method to randomly change the communication time slice based on the distance between nodes, which can reduce the time of networking and improve the number of networking times in per unit time.

2. Topological structure of low voltage power line carrier communication network

The low voltage power line carrier communication network consists of a concentrator (master node), terminal collectors (slave nodes) and terminal collectors (power line network). In the meter reading process, the concentrator and terminal collectors are of a master-slave relationship, and the communication mode is half-duplex mode. The system is placed on the low voltage side of the zone transformer and the data collection and processing for the entire power line network is shown in figure 1. As the three-phase transformer makes the secondary side of the three-phase circuit between the independent, one of the communication reliability and communication time need be studied in the research process [11].

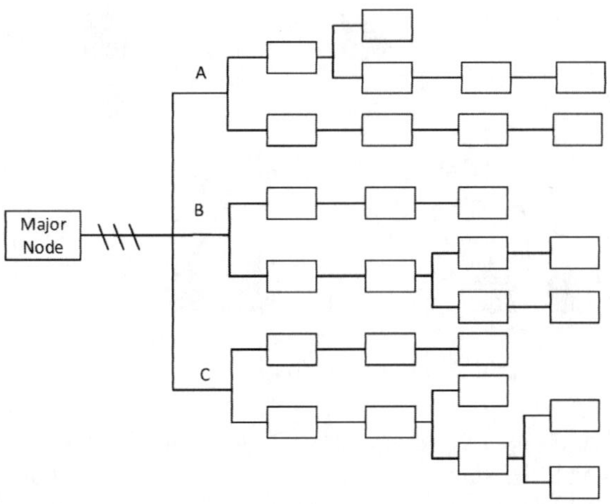

Figure 1. Three-phase low-voltage power line communication network physical topology.

3. Low voltage power line ad hoc network theory

3.1. Networking instructions

Supposing that a low-voltage power line network has (n≥1) nodes, for the introduction of the initialization algorithm, it needs to define the following some aspects: (1) the physical address of the master node is 0, slave of nodes physical address from 1 to n; (2) as much as possible to make the nodes within the network to achieve the connection, which reduces the network time and the number of clusters and layers; (3) after obtaining the logical ID, exit the current logical address allocation competition [11]; (4) node and node communication time is called time slice. Each node has its own time slice, and it has a one-to-one correspondence with the logical number of the node [12]; (5) the smaller the logical ID number, the higher the priority of its communication; (6) there is no solitary phenomenon in the network.

3.2. Networking steps

Step 1: The master node sends a routed broadcast (route broadcast), with (1 <m≤n) nodes receiving the broadcast. The node response determines the order of the response based on the distance traveled by the node to the central node (channel quality is the same). According to the above points, when the master node receives a response from a slave node, then it assigns a logical ID to the slave node, and the node exits the round address assignment. The remaining m-1 slave nodes also allocate logical addresses in this way. Specify this slave node as logical layer 1.

Step 2: Initialize Layer 2. Since the smaller the logical ID of the slave node, the higher the communication priority. When the network is configured for the second layer, the master node broadcasts the slave node from logical ID of 1, and then the slave node responds to the master node and the cluster head logical ID is 1. The logical ID allocation is performed for the nodes of the logical layer 2 according to the allocation method of step 1 [13].

Step 3: If the second layer of the implementation of the network, all nodes in the network are put to join, it means that the initialization is completed, the performance of the phenomenon when the re-stratification is empty response. At this point, the network initialization is complete and the logical layer of the network is 2. The cluster head in the network is the slave node of the first layer, and the number of the corresponding cluster members is $g_1, g_2, \cdots, g_m$.

If the slave node is not all added to the network at this time, it needs to continue the initialization of the network until all the slave nodes are found. In this way, all nodes in the network establish a routing relationship. The cluster structure is shown in figure 2.

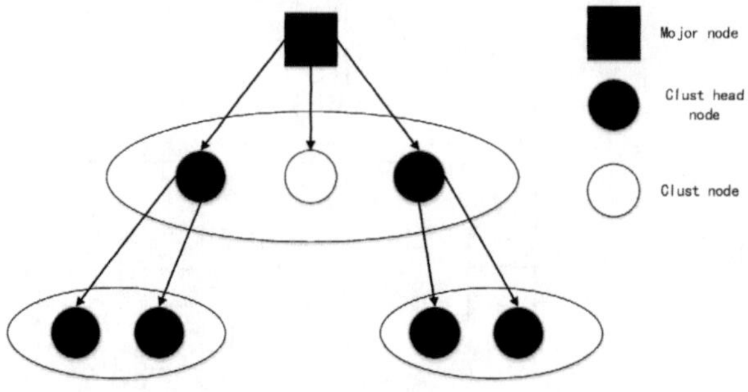

Figure 2. Schematic diagram of clustering.

According to the above analysis, two special cases will appear after the initialization: (1) the master node can communicate with all the slave nodes directly, without relay, and the logical structure of the network is star; (2) and the cluster head can only communicate with neighboring nodes, and the nodes of the whole network are in a series state.

3.3. Network maintenance principles

As the equipment can be arbitrarily inserted and removed, the physical and logical structure of the node changes constantly, resulting in the network routing relationship's change, and when the network channel is good or bad, the network maximum effective communication distance will accordingly change, which will also make the previous the routing relationship change. In order to ensure the network of robustness and improve the network's survivability, the need for network maintenance and maintenance principles are as follows:

(1) When the network is running normally, if the cluster does not has the data communication, the cluster head will initiate the local route detection to judge whether there is any abnormality in its own cluster; (2) when the network is in the idle state, the master node will initiate the entire network detection command; (3) when the node appears to exit or join the network, the central node makes the local network carry on routing reconstruction.

3.4. Principles of network reconfiguration

It is assumed that there are n nodes in the power line, and there are m clusters after initialization. If one of the cluster heads finds that the cluster member which belongs to the cluster is not responding, it indicates that the cluster member is not in the network. The cluster head will report this information to the master node, and the master node will delete all the routing tables associated with the node. The master node reorganizes the cluster [8].

3.5. Simulation results and analysis

It is assuming that, in a power line system, there are 59 sets of electrical equipment which can be simplified as a node by Matlab simulation. The network is simplified to (-10,10), (-10,10) area. (1) After the network is initialized, the master node logical ID is 0 and the logical number of the slave node is from 1 to 58; (2) as the power line channel quality follows the number of access to the power equipment changes, which will cause the maximum effective communication distance change (3) when the maximum communication distance is constant, the closer the distance between the nodes, the more the communication will form the link, so that the logical ID is smaller; (4) when the channel between two nodes is poor, it can be considered that the communication distance between the two nodes increases; (5) the nodes are randomly distributed in the region, and each node, at least, can communicate with one node.

The following is simulation of the different communication distance of the situation. When the maximum effective communication distances of the nodes are 5 m and 7 m, the clustering algorithm initialization structure is shown in figure 3. The two labels in the figure represent their physical ID and logical ID (in parentheses).

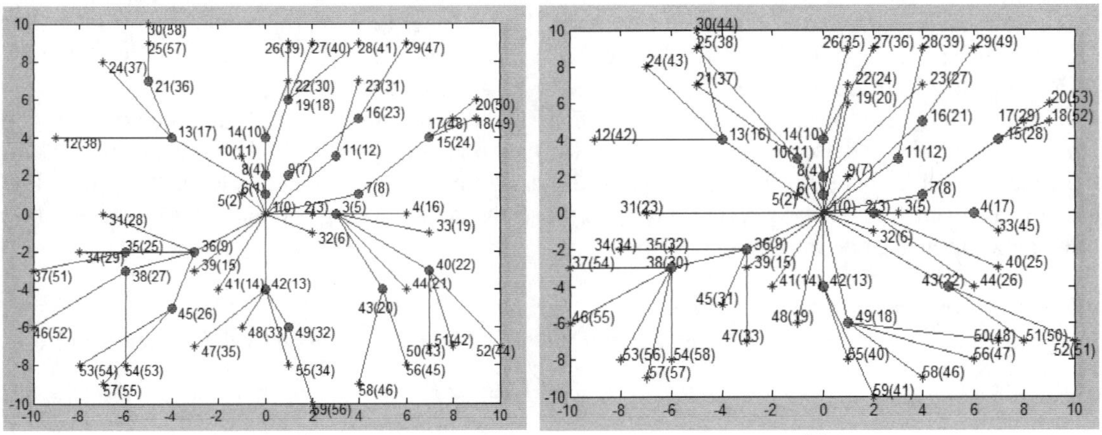

(a) (b)

Figure 3. Simulation results of different maximum effective communication distances: (a) Maximum communication distance 5 m and (b) Maximum communication distance 7 m.

The following conclusions can be drawn: (1) The use of non-overlapping clustering method to achieve low-voltage power line routing, including node relay, receive and send all the required functions; (2) when the effective communication distance of the nodes on the power line changes, the original routing relationship changes dynamically according to the distance; (3) when the channel of nodes are good, the maximum communication distance will become larger, which will make the cluster of the level and the number of network reduce. As shown in figure 3(a), the node path with the physical address of 30 is: 0 (0) → 6 (1) → 13 (17) → 21 (36) → 30 (58), a total of 5 layers of network, and in figure 3(b), it is in the range of 0 (0) → 13 (16) → 30 (44), a total of 3 layers of network. The cluster head is 20, and the number of network is 36 in the figure 3(a), while they are 16 and 30 in the figure 3(b); (4) as shown in figure 3(a), the physical ID of 49 and 48 of the two nodes in the network initialization phase, the two competing to report the physical ID of the probability of cluster should be the same, but if the 48 → 42 channel is poor, then 49 will be reported ahead of 48. Therefore, according to the simulation results, clustering algorithm can achieve automatic relay and dynamic routing.

4. Improved clustered routing timing assignment

4.1. Traditional timing assignments

In the initialization phase of the cluster algorithm, it is necessary to consider the allocation of the logical number according to the node. A non-overlapping clustering algorithm is adopted by this paper. In the case of single channel, which only one node can transmit data in the channel to avoid the collision of signals in the channel. The algorithm requires that the time interval at which the adjacent nodes of any logical ID send channel should meet:

$$T_i = T_1 + T_2 + T_3 \tag{1}$$

T_1 represents the time information sent; T_2 is the data processing time; $T_3$is the redundancy time. Where the time consumed by T_2 and the settling time of T_3 are in the order of microseconds. Therefore, in order to simplify the algorithm, the carrier signal transmission time is taken as the total time interval.

The problem of channel timing assignment in clustering algorithm is mainly to solve the fastness of networking, and to ensure that the channels are not collided, and the timing protocol is simple.

The traditional algorithm is meant that after the central node sends the broadcast, the node receiving the network broadcast will send the response frame to the master node according to certain time rules, and then the master node sends the logical ID number to the corresponding slave node. For example, nodes 1, 3 need 1 $\times T_i$, 3 $\times T_i$ time interval to send a reply, and obtain the master node assigned logical ID. Similarly, the rest of the logic layer is such a distribution process.

From the above allocation process, we found that the method based on the non-overlapping clustering algorithm is based on the expense of the network time to simplify the agreement to ensure the reliability of communication. However, in order to be able to more quickly in the power line environment and faster network, and have to ensure the reliability of communication, the existing timing allocation must be improved.

4.2. Change the timing of the allocation

Improved algorithm: It can be seen from the time slice T_i, which was decided by the carrier signal,

$$L = v \times T_T = v \times T_i \quad (2)$$

Where v is the communication speed of the carrier and L is the communication distance between the two nodes. For a carrier, the speed of the carrier is constant, so we can be between the two points of the communication distance L instead of time as a research object, of course, the time can also be representative of the distance as a research object.

The maximum communication distance of figure 3(a) is 5 m, and the corresponding communication time interval is T_i. And the communication distance between nodes is randomly distributed within 5 m, regardless of whether the communication distance is the largest 5 m or other, the time interval will be T_i, so in order to speed up the network time, the node can communicate with the node to do the distance: When communication distant between nodes is $0 \leq L < 2.5$ m, then the communication time interval is changed to $T_i/2$; when communication distance between nodes is $2.5 \leq L \leq 5$ m, the time interval is the original T_i. Thus, for nodes with small node spacing, the communication time of the communication channel extension system will not be occupied for a long time, and the communication protocol will not be complicated and the original network channel will not be changed.

4.3. Simulation results and analysis

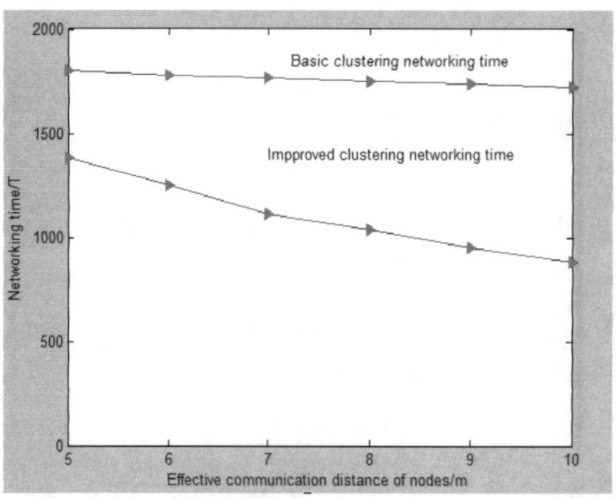

Figure 4. Time-consuming comparison of two timing assignments at different maximum communication distances.

Figure 4 shows the simulation results under the conditions with different maximum effective communication distances. The Y-axis represents the number of time slices T, and the X-axis indicates the maximum communication effective distance.

According to the simulation results, we can conclude that: (1) After the improved timing allocation, the networking time is much smaller than that before the improvement, but the complexity of the routing protocol increases, and it is necessary to add a definite time whether T_i or T_i / 2. Because each time the node and the communication between the nodes, you need to return to the time required to determine the time slice value. (2) Whether it is improved before or improved after the timing of the distribution method, the power line network communication channel interference by the smaller time, the network will be shorter. (3) Time-consuming time of basic clustering time as the communication distance between nodes increases, the clustering time is time-consuming is not significant. This is because when the node's effective communication distance is 14.1 m, the network structure is only the main node and slave nodes, no cluster head. All nodes will be in

$$(1+2+\cdots+58)\times T_i = 1711T_i \tag{3}$$

of the time to complete network. When the effective communication distance of the node is 5 m, according to figure 3(a), the distributions of nodes in the network are the first layer of 15 nodes, the second layer of 16 nodes, the third layer of 16 nodes, and the fourth layer has a node. If the second layer of the node to the main node to return frame data, such as 13 (17), need to go through 17 $\times T_i +$ T_i time to return to the main node. This shows that the second layer of the node from the main node to communicate with the time than the first layer of a T_i, the same reason, the third layer to more than $2T_i$, and so on. However, the number of nodes in the second and following clusters is small, resulting in a significant reduction in the time-consuming time when the communication distance increases.

When the network time is shortened, the number of initiating networking of the primary node increases in a certain period of time, which increases the update of the master node from the nodes in the network, such as deletion, addition and other updates. This can be used to update the nodes routing relationship, improve the master node and slave node communication success times, and improve the network anti-jamming.

5. Conclusions

By analyzing the characteristics of low voltage power line network, the idea of clustering stratification is used to improve the communication distance of low voltage distribution network. Based on the initialization of the network, the routing relationship of all the nodes in the network is obtained. When the network appears abnormal circumstances, the network maintenance strategy is started to achieve the stability of the relationship between network nodes. The method is mainly used in the data link layer, so it does not need to consider the original physical characteristics, only needs to consider the algorithm of programming, with a strong versatility. Although the traditional non-overlapping clustering algorithm's timing allocation method can make the communication some simple, to avoid the channel communication conflict, but the network time there is too much waste, in order to improve the network anti-jamming, proposed a node distance of the improved timing allocation method. According to the theory and simulation, it is proven that the improved timing allocation method can not only complicate the communication protocol, but also shorten the networking time obviously.

References

[1] Fang S G 2007 Characteristics of low voltage power line communication channels *Relay* **35** 53-6
[2] Li P, Zhao Z H and Zhang Z R 2007 Modeling and simulation of the impulse noise in low voltage power line communication channel *Relay* **35** 58-62
[3] Papaleonidopoulos I C, Capsalis C N, Karagiannopouls C G, *et al* 2003 Statistical analysis and simulation of indoor single-phase low voltage power-line communication channels on the basis of multi-path propagation *IEEE Trans on Consum Electron* **49** 89-99

[4] Hou S Z, Yang L and Chen L 2008 Research on automatic routing algorithm for remote meter reading in low voltage power line *Low Voltage Apparatus* **2008** 28-31
[5] Liu X S, Zhou Y and Qi J J 2006 Method study of automatic routing for power line communication *Proceedings of the CSEE* **26** 76-81
[6] Qi J J, Xu D G, Zhou Y, *et al* 2009 Characteristics model and routing algorithm of power-line communications over low-voltage distributions *Proceedings of the CSEE* **29** 56-62
[7] Liu Z, Wang X Y and Cai S L 2009 Method of low-voltage PLC networking *Telecommun for Electr Power Syst* **20** 17-20
[8] Ran Q H, Wu Y C and Qi M J 2011 Research on automatic routing method of low-voltage power line carrier *Netw* **39** 53-63
[9] Zhao J W, Lu W B and Li X L 2007 Research on automatic routing for remote meter reading in power line carrier *Telecommun for Electr Power Syst* **28** 1-5
[10] Peng Q W and Zhang H 2009 Medium - voltage distribution line carrier based on multiplic traversal adaptive relay algorithm *Power Syst Commun* **30** 81-4
[11] Li Q and Cui J R 2013 Method of carrier wavemaking based on low voltage power line *Res. and Explor. in Lab.* **2013**
[12] Qi J J, Liu X S, Xu D G, *et al* 2008 Simulation study on cluster-based routing algorithm and reconstruction method of power line communication over lower-voltage distribution *Proceedings of the CSEE* **28** 65-71
[13] Cao Y T, He C and Jiang L G 2007 Clustering algorithm based on adaptive timer strategy in wireless sensor networks *J of Electron* **35** 1719-23

Two-stage energy storage equalization system for lithium-ion battery pack

W Chen[1], Z X Yang[1,2], G Q Dong[1], Y B Li[1] and Q Y He[1]

[1]Institute of Systems Science and Engineering, Henan Engineering Laboratory of Wind Power Systems, Henan University of Science and Technology, No. 48 Xiyuan Road, Luoyang 471003, China

E-mail: zxyang@haust.edu.cn

Abstract. How to raise the efficiency of energy storage and maximize storage capacity is a core problem in current energy storage management. For that, two-stage energy storage equalization system which contains two-stage equalization topology and control strategy based on a symmetric multi-winding transformer and DC-DC (direct current-direct current) converter is proposed with bidirectional active equalization theory, in order to realize the objectives of consistent lithium-ion battery packs voltages and cells voltages inside packs by using a method of the Range. Modeling analysis demonstrates that the voltage dispersion of lithium-ion battery packs and cells inside packs can be kept within 2 percent during charging and discharging. Equalization time was 0.5 ms, which shortened equalization time of 33.3 percent compared with DC-DC converter. Therefore, the proposed two-stage lithium-ion battery equalization system can achieve maximum storage capacity between lithium-ion battery packs and cells inside packs, meanwhile efficiency of energy storage is significantly improved.

1. Introduction

Energy storage technology is a crucial part of six steps containing adoption- generate power-transmission- distribution- utilization of power- energy storage in power grid operation, implementing effectively management of supply side, peak shaving and load shifting, which can efficiently utilize power equipment and reduce the power cost. The development of lithium-ion battery for large-scale energy storage systems enable this technology to address high energy density, long cycle life, high standard voltage, and other factors, which are in line with the current green requirements of sustainable development. Energy storage technology is available for large-scale applications, such as electric vehicle, light railway vehicle, smart building, etc [1-3]. The success of these applications of energy storage will depend on how to maximize storage energy and raise the efficiency of stored energy [4], therefore, energy storage equalization technology is a key expectation at present.

Many scholars have made in-depth research and design for energy storage equalization topology of the Lithium-ion battery at home and abroad. Two main categories are as follows: dissipative equalization and non-dissipative equalization. The dissipative equalization refers to dissipate redundant energy in the form of heat through the parallel resistance on both ends of the battery, and it is with advantages of simplicity, speediness, high reliability, but low energy efficiency and thermal management issues [5,6]. Non-dissipative equalization is that energy is transferred between the cells through the inductor or capacitor, which has various circuit forms: Ye Y M [7], Bruen T [8] and their fellows have proposed the combination usage of switch and capacitor to achieve energy transfer between the cells. Guo X W [9] used energy storage inductors as energy storage devices, the redundant energy of cells are transferred to the necessarily balanced cells, in order to achieve the consistent battery energy. Zhu H [10], etc. took into account the characteristics of reliable passive equalization and active equalization without energy loss, and proposed the balanced solution of combination of passive equalization and active equalization under the low cost. Gui Y [11] and Park H S [12] designed a balanced circuit topology based on fly-back transformer, redundant energy among cells is transferred through the conversion between electric energy and magnetic energy. All these research results about the equalization topology are accomplished under the conditions of less cells, however, large-scale energy storage equalization systems often need to load thousands of series cells, which can bring quite complex control strategies in the course of energy transfer [13].

From what has been discussed above, there are obvious advantages of hierarchical equalization for simplifying the control process based on large scale cells. Therefore, this paper proposes two-stage energy storage equalization system which contains two-stage equalization topology and control strategy based on a symmetric multi-winding transformer and DC-DC converter, in order to realize the objectives of consistent lithium-ion battery packs voltages and cells voltages by using a method of the Range. Equalization effect of the two-stage energy storage equalization system will be verified compared with DC-DC converter finally.

2. Design of energy storage equalization system

2.1. *Structure of energy storage equalization system*

Application pattern of energy storage technology for the lithium-ion battery pack in the field of distributed generation of micro-grid is shown in figure 1, where energy storage system based lithium-ion battery pack is an imperative part.

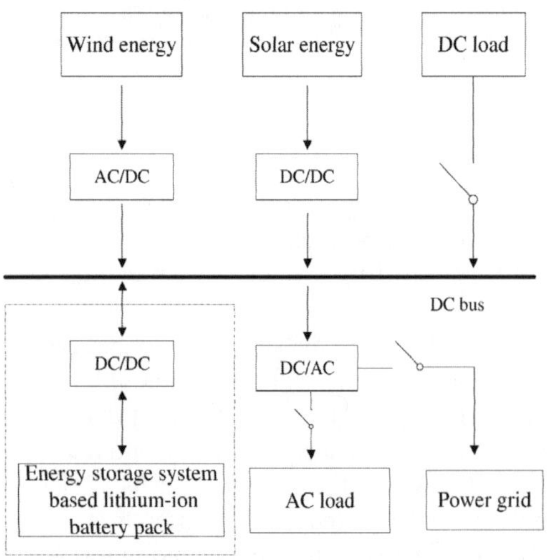

Figure 1. Application pattern of energy storage technology for the lithium-ion battery pack in the micro-grid.

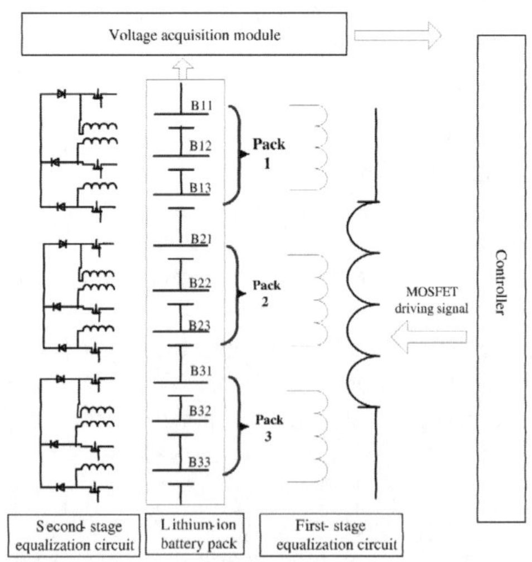

Figure 2. Structure of energy storage equalization system.

The energy storage equalization system proposed in this paper consists of lithium-ion battery pack, first-stage equalization circuit, second-stage equalization circuit, voltage acquisition module and controller, as shown in figure 2. The whole lithium-ion battery pack is composed of a number of single cells, of which four single cells form a lithium-ion battery pack. The first-stage equalization circuit between the lithium-ion battery packs is used for energy equalization of lithium-ion battery packs; The second-stage equalization circuit between cells in the pack acts as energy equalization of cells in the pack; The voltage acquisition module collects the voltages of the single cells in real time and transmits the voltage signals to the controller; The controller calculates and processes the voltage signals collected by the voltage acquisition module. If it is judged that the energy between the lithium-ion battery packs and the cells in the pack is not consistent, the controller outputs the control signals according to the two-stage energy storage equalization control strategy, the energy equalization is achieved between the lithium-ion battery packs and the cells in the pack.

2.2. *The first-stage equalization circuit*

The first-stage equalization circuit is characterized by a symmetrical multi-winding transformer possessing an iron core, a primary winding and three secondary windings. Each secondary winding connects to two ends of lithium battery pack, a MOSFET (metal-oxide semiconductor field-effect transistor) Sp and a filter inductor Lf correspondingly, and the number of turns of winding ratio of 1: 1: 1 symmetrical design. The circuit has the advantages of accurate equalization, high efficiency, and low magnetic loss. The equalization circuit schematic diagram is shown in figure 3.

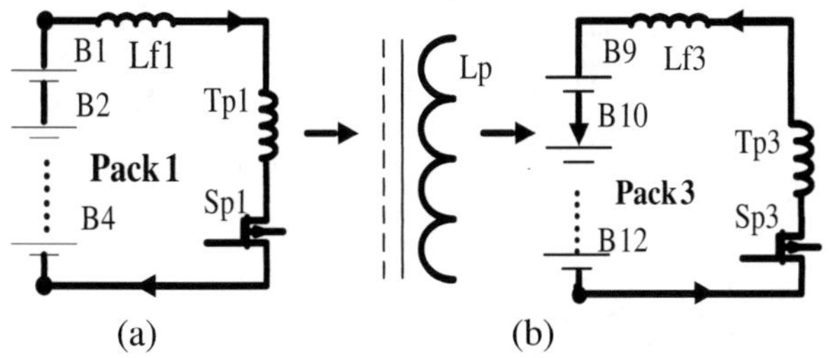

Figure 3. Schematic diagram of the first-stageequalization circuit, (a) is the course of pack1 release energy, (b) is the course of pack3 storage energy

Course of pack1 energy release: symmetrical multi-winding transformer uses for energy equalization of the lithium-ion battery packs, if the controller judges minimum voltage lithium-ion battery pack3 by processing voltage signals collected and meets the condition of V_{pack1}-V_{pack3}>Δ, where the difference of lithium-ion battery pack1 voltage and lithium-ion battery pack3 voltage is greater than the equalization threshold Δ, the lithium-ion battery pack1 externally releases energy to maintain the energy equalization between the lithium-ion battery packs, then the MOSFET Sp1 is on.

During the Sp1 switched on, the current flowing through Sp1[14], $I_{sp(t)}$, is gave by:

$$I_{sp\,(t)} = \frac{V_{pack1}}{T_{p1}} t, \quad \left(0 \le t \le t_{on}\right) \tag{1}$$

When t=t_{on}, the current of secondary winding T_{p1} reaches the maximum:

$$I_{sp\max} = \frac{V_{pack1}}{T_{p1}} t_{on} \tag{2}$$

Transformer secondary winding T_{p1} energy stored is marked as W_{p1}:

$$W_{p1} = \frac{V_{pack1} I_{sp\max}}{2} t_{on} \tag{3}$$

Where T_{p1} is the secondary winding inductance, and t is the time of secondary winding T_{p1} storage energy.

Course of pack3 energy storage: When the lithium-ion battery packs voltages meet the condition of V_{pack1}-V_{pack3}>Δ, the lithium-ion battery pack3 needs to receive energy from other lithium-ion battery pack to keep the system balanced, then commutating diode conducts in the MOSFET Sp3, and secondary winding T_{p3} generates current. T_{p3} winding current peak is:

$$I_{\text{tpmax}} = \frac{V_{\text{pack1}} + V_{\text{pack2}} + V_{\text{pack3}}}{T_{\text{p3}}} n^2 \left(t_s - t_{\text{on}}\right), \quad (t_{\text{on}} < t < t_s) \tag{4}$$

The energy absorbed by the lithium-ion battery pack3 is:

$$W_{\text{p3}} = \frac{V_{\text{pack1}} + V_{\text{pack2}} + V_{\text{pack3}}}{2} I_{\text{tpmax}} \left(t_s - t_{\text{on}}\right) = \frac{\left(V_{\text{pack1}} + V_{\text{pack2}} + V_{\text{pack3}}\right)^2}{2T_{\text{p3}}} n^2 \left(t_s - t_{\text{on}}\right)^2 \tag{5}$$

Where n is the winding turns, T_{p3} is the secondary winding inductance, t_s-t_{on} is the time of secondary winding T_{p3} energy released.

2.3. The second-stage equalization circuit

The second-stage equalization circuit is characterized by a DC-DC converter containing a storage energy inductor *L*, a MOSFET S and a commutating diode *D*. The equalization circuit has a complex energy storage equalization control strategy for large numbers of single cells, but with the advantage of modularity. Equalization schematic diagram is shown in figure 4.

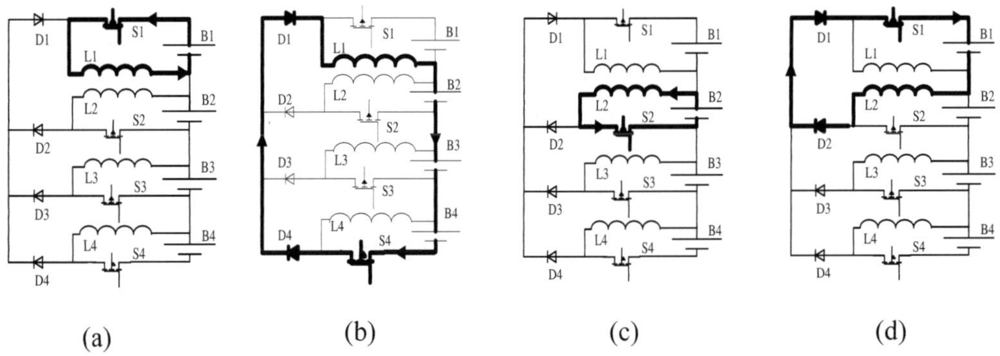

Figure 4. Schematic diagram of the second-stage equalization circuit. (a) B1 release energy, (b) B2, B3, B4 storage energy, (c) B2 release energy and (d) B1 storage energy.

DC-DC converter uses for energy equalization of cells in the pack, the principle of specific equalization shown in figure 4, Taking the lithium-ion battery pack1 for example, voltage signals collected by the data acquisition module from the cells are transmitted to the controller, which judges the state of cells according to voltage signals. If the minimum cell voltage meets the condition of V_{B1}-V_{Bmin}>Φ, where Φ is equalization threshold of cells in the pack, that is, the difference of B1 voltage and the minimum cell voltage is greater than the equalization threshold, the MOSFET S1 is on, and then B1 releases energy that is transferred to inductor L_1, as shown in figure 4(a).

Voltage transferred by B1 can be expressed as [15,16]:

$$\Delta V_{\text{B1}} = L_1 \frac{\text{d}i}{\text{d}t}, \quad \begin{cases} \text{d}i = I_1 \\ \text{d}t = DT \end{cases} \tag{6}$$

The peak current of inductor L_1, I_1, is gave by:

$$I_1 = \frac{\Delta V_{\text{B1}} DT}{L_1} \tag{7}$$

The energy released by B1 is deduced from the formula $W=U\,I\,t$:

$$W_1 = \frac{1}{2}\Delta V_{B1} I_1 DT = \frac{\Delta V_{B1}^2 D^2 T^2}{2L_1} \tag{8}$$

Where L_1 is the energy storage inductance, T is the circuit work cycle, D is MOSFET duty ratio.

As shown in figure 4(b), if the any voltage of B12, B13, B14 is less than the equalization threshold, the energy stored in inductor L_1 is transferred to B2, B3 and B4 during the S1 switch-off period. In order to release completely the energy stored in the inductor and avoid saturation hysteresis, DC-DC converter operates under the condition of the DCM (discontinuous current mode). As shown in figure 4(c), if the controller judges that the voltage difference between the cell B2 and the minimum voltage cell is larger than the equalization threshold, the MOSFET S2 is turned on, then B2 releases energy in the period of storing energy of L_2. As shown in figure 4(d), if B1 voltage is less than the equalization threshold, B1 stores energy in the period of releasing energy of L_2. If the controller determines that the cell B3 voltage meets the equalization condition, the MOSFET S3 is turned on, then B3 releases energy in the period of storing energy of L_3; if any voltage of B1, B2 is less than the equalization threshold, L_3 releases energy towards B1, B2. If the controller determines that the cell B4 voltage meets the equalization condition, the MOSFET S4 is turned on, then B4 releases energy in the period of storing energy of L_4; if any voltage of B1, B2, B3 is smaller than the equalization threshold, L_4 releases energy to B1, B2, B3. When all the cells in the lithium-ion battery pack1 are equalized, the energy released by the cells during each switching cycle is equal to the energy absorbed from the other cells.

3. Equalization control strategy

Two-stage energy storage equalization control strategy for lithium-ion battery pack is divided into two stages, the specific process shown in figure 5.

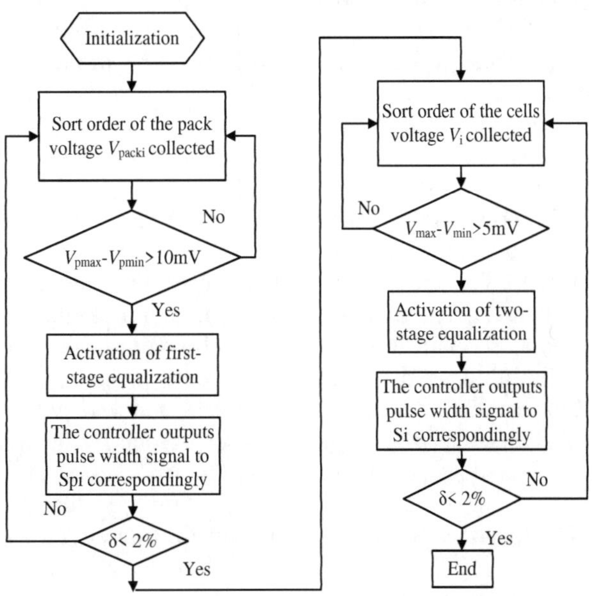

Figure 5. Flow diagram of two-stage energy storage equalization control strategy.

Phase 1: Firstly, initializing the equalization system, and then sorting the lithium-ion battery packs voltages collected by the voltage acquisition module. If the Range between the maximum lithium-ion battery pack voltage and the minimum lithium-ion battery pack voltage is more than 0.01 V, the first-stage equalization is activated. The controller outputs pulse width signal to drive the

IOP Conf. Series: Earth and Environmental Science **93** (2017) 012046 doi:10.1088/1755-1315/93/1/012046

corresponding MOSFET Spi, performing energy transfer for lithium-ion battery packs based on symmetrical multi-winding transformer. If the Dispersion is greater than 2%, the system will continue the energy equalization between the lithium-ion battery packs; if the Dispersion is less than 2%, the system will complete the energy equalization for lithium-ion battery packs.

Define the Dispersion δ:

$$\delta=\sqrt{\frac{\sum_{i=1}^{n}\left(V_i-\overline{V}\right)^2}{n-1}},\quad \overline{V}=\frac{1}{n}\sum_{i=1}^{n}V_i \tag{9}$$

where $\overline{V}$ is the average value, V_i is voltage of lithium-ion battery pack or cell, and n is the number of lithium-ion battery pack or cell.

Phase 2: After finishing the first-stage energy storage equalization for lithium-ion battery packs, the voltage acquisition module collects the single cells voltage, which are sorted in the meantime. If the Range of maximum cell voltage and the minimum cell voltage is greater than 0.005 V, the second-stage equalization is activated, the controller outputs pulse width signal to drive the corresponding MOSFET Si, equalizing energy for cells based on DC-DC converter. If the Dispersion is greater than 2%, the system will continue the energy equalization between the cells; if the Dispersion is less than 2%, the system will complete the energy equalization for the cells in the pack.

4. System simulation and results analysis

4.1. Establish of simulation model

In order to verify the feasibility and effectiveness of the two-stage energy storage equalization system designed in this paper, lithium-ion battery model and equalized topology circuit model are set up in the simulation environment of MATLAB / SIMULINK [17,18], where Bi, packi are cells and battery pack modules respectively, Vi/ Ci collects battery voltages or current in real time on behalf of voltage/current acquisition modules, Si represents MOSFET switch that are used for controlling the state of circuit model, Strategy is the same as controller which processes the voltage signals collected by Vi. If it is judged that the energy between the lithium-ion battery packs and the cells in the pack is not consistent, the controller outputs the control signals according to control strategy. In order to simulate energy release and energy storage characteristics and reduce the non-linear characteristics of the lithium-ion battery, RC equivalent circuit are selected, where R is the internal resistance of the lithium-ion battery, the capacitor C is the open circuit voltage of the lithium-ion battery. The different capacitance values in the circuit simulate the energy differences of the cells in practice.

The first-stage energy storage equalization circuit model is built in the simulation environment as shown in figure 6. The three lithium-ion battery pack models are composed of four cells in series respectively, the initial voltages of which are 13.6 V, 13.4 V, and 13.2 V, respectively. Symmetrical multi-winding transformer is characterized by rated power, frequency, winding leakage inductance, transformer inductance, filter inductance with 24 W, 20 kHz, 20.8 μH, 30 H, and 20.7 μH, correspondingly, regardless of core saturation. The simulation step length is set to ode23tb, and powergui is continuous. The first-stage energy storage equalization circuit model is simulated in the process of energy storage and energy release by using a controlled constant current source and a 100 Ω resistive load for the lithium-ion battery packs respectively.

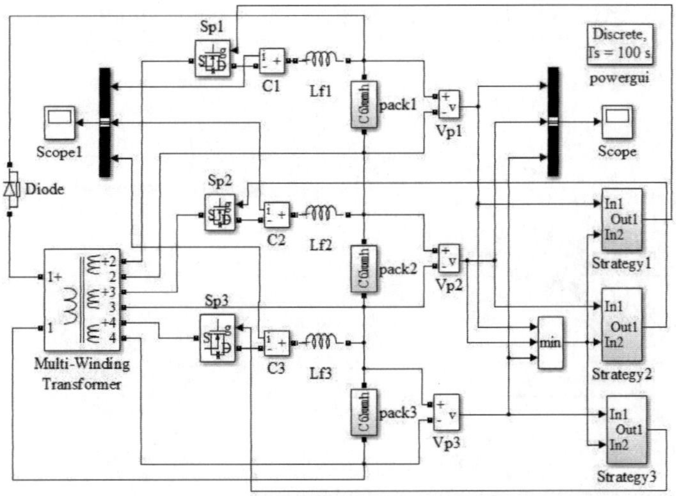

Figure 6. The first-stage equalization circuit model.

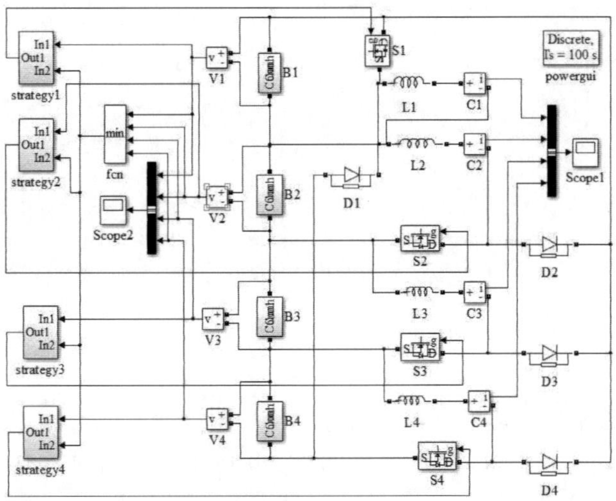

Figure 7. The second-stage equalization circuit model.

The second-stage energy storage equalization circuit model is built in the simulation environment as shown in figure 7. The lithium-ion battery models are composed of four cells in series in the pack1, the initial voltages of which are 3.4 V, 3.3 V, 3.0 V and 2.9 V. The energy storage inductance are 11.1 μH, 16.7 μH, 12.5 μH, and 11.1 μH correspondingly, meanwhile, the parasitic resistances of MOSFET and diode are set as the default value. The simulation step length is set to ode23tb and powergui is continuous. The second-stage energy storage equalization circuit model is simulated in the process of energy storage and energy release by using a controlled constant current source and a 100 Ω resistive load for the cells in the pack1 respectively.

4.2. Analysis of simulation results

The voltage variation curves for the lithium-ion battery packs and the cells in the pack1 are obtained about the simulation results, as shown from figure 8 to figure 12.

As shown in figure 8, voltage curves of the 12 cells which contain B1, B2, B3...B12 are acquired based on DC-DC converter in the process of energy storage, where voltage fluctuations are fairly obvious. Voltage values of 12 cells have achieved a consistent state simultaneously after equalization time of 0.75 ms.

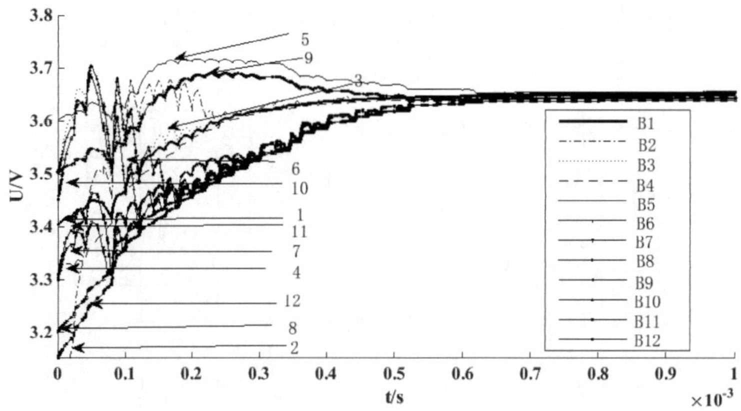

Figure 8. Voltage curves of the 12 cells based DC-DC converter.

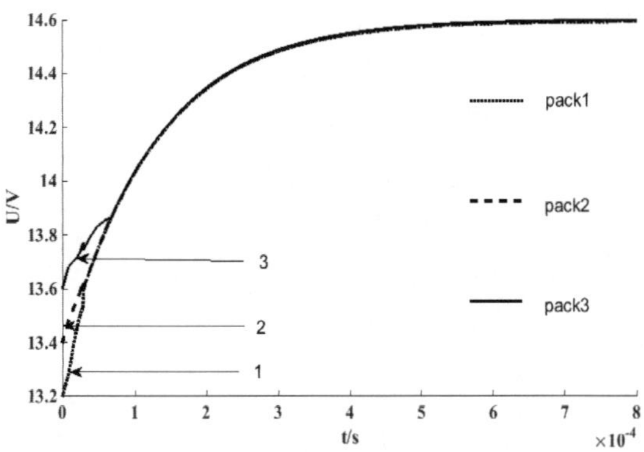

Figure 9. Voltage curves of three battery packs based the first-stage equalization in the period of energy storage.

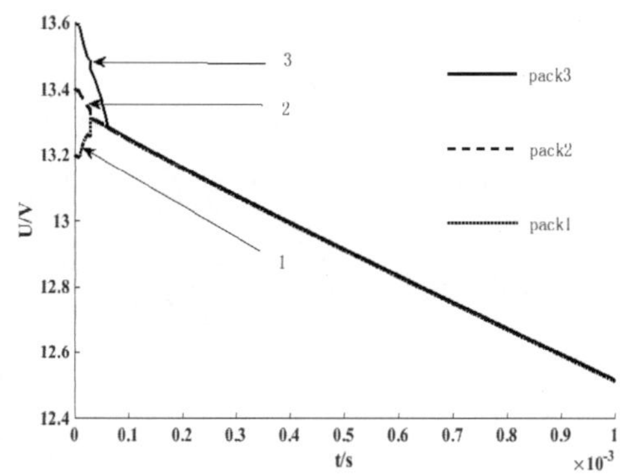

Figure 10. Voltage curves of three battery packs based the first-stage equalization in the period of energy release.

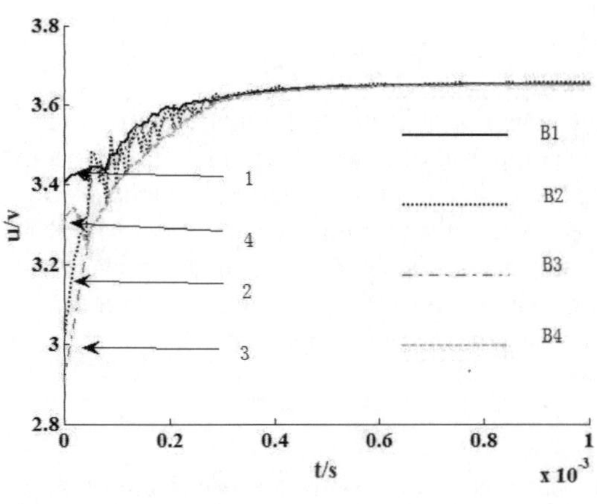

Figure 11. Voltage curves of 4 cells in the pack1 based the second-stage equalization in the period of energy storage.

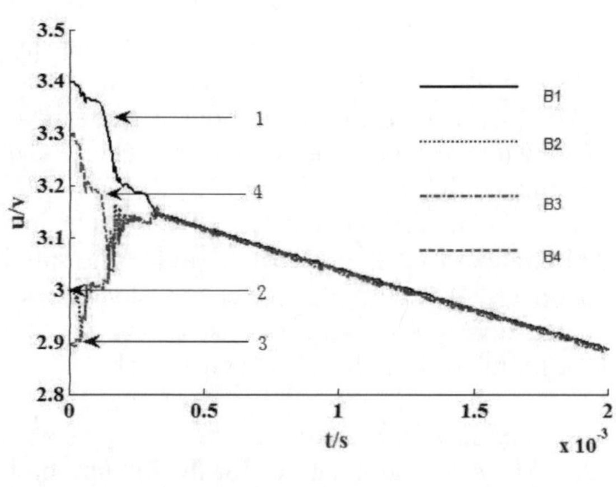

Figure 12. Voltage curves of 4 cells in the pack1 based the second-stage equalization in the period of energy release.

As shown in figures 9 and 10, the voltage differences between the lithium-ion battery pack1, pack2, and pack3 have been eliminated within equalization time of 0.1 ms, and voltage curves changes smoothly. As shown in table 1, the Dispersion of the lithium-ion battery packs is reduced from 20% to 0.07% and 0.47% respectively in the process of energy storage and energy release, and the Range of voltages is also kept within 10mV, therefore, the two parameters of dispersion and range reaches the evaluation index.

Table 1. The state of lithium-ion battery packs and cells in the pack1.

State	The lithium-ion battery packs		Cells in the pack1	
	Range/V	Dispersion /%	Range/V	Dispersion /%
Initial state	0.4000	20.00	0.5000	23.00
After energy storage	0.0049	0.07	0.0046	0.22
After energy release	0.0094	0.47	0.0051	0.27

As shown in figures 11 and 12, volumes of four cells in the pack1 are stored energy fully when equalization time is 0.5 ms. Voltage fluctuations are slighter than that 12 cells based on DC-DC converter shown in figure 8, meanwhile, equalization time is shorter significantly. As shown in table 1, the Dispersion of the cells in the pack 1 is reduced from 23% to 0.22% and 0.27% respectively in the process of energy storage and energy release, and the Range of voltages is also maintained at 5 mV, therefore, the evaluation indexes of dispersion and range are met totally.

From what has been analyzed above, the two-stage energy storage equalization system can achieve the equalization control objective accurately and quickly. The energy transfer for the lithium-ion battery packs and the cells in the pack is accomplished within equalization time of 0.5 ms, which eliminates the problem of voltage inconsistency and achieves maximum storage capacity.

5. Conclusion

Two-stage energy storage equalization system proposed which contains two-stage equalization topology and control strategy based on a symmetric multi-winding transformer and DC-DC converter is designed to reach consistent lithium-ion battery packs voltage and cells voltage during energy storage and energy release, and that raises the efficiency of energy storage and maximize storage capacity. The simulation results show that the method achieves an efficient and quick equalization with the Dispersion of 2% and equalization time of 0.5ms for the lithium-ion battery packs and the cells in the pack. The idea of hierarchical energy storage equalization simplifies the control process and shortens the equalization time of 33.3%compared with the DC-DC converter, therefore, the efficiency of energy storage is improved further.

Acknowledgments

The financial support from the National Natural Science Foundation of China (NSFC) under the grant No. 70671035 and No. 71071078 is acknowledged gratefully.

References

[1] Farhadi M and Mohammed O 2016 Energy storage technologies for high-power applications *IEEE Trans Ind Appl* **52** 1953-61

[2] Ciccarelli F, Pizzo A D and Iannuzzi D 2014 Improvement of energy efficiency in light railway vehicles based on power management control of wayside lithium-ion capacitor storage *IEEE Trans Power Electr* **29** 275-86

[3] Uno M and Kukita A 2015 Bidirectional PWM converter integrating cell voltage equalizer using series-resonant voltage multiplier for series-connected energy storage cells *IEEE Trans Power Electr* **30** 3077-90

[4] Zhong L and Zhang C B 2014 A method for the estimation of the battery pack state of charge based on in-pack cells uniformity analysis *Appl Energ* **113** 558-64

[5] Javier G L and Enrique R C 2014 Battery equalization active methods *J Power Sources* **246** 934-49
[6] Federico B, Roberto R 2014 Performance comparison of active balancing techniques for lithium-ion batteries *J Power Sources* **267** 603-09
[7] Ye Y M and Eric K W 2016 An automatic switched-capacitor cell balancing circuit for series-connected battery strings *Energies* **9** 138
[8] Bruen T and Marco J 2015 Model based design of balancing systems of electric vehicle battery packs *Science Direct* **48** 395-402
[9] Guo X W and Kang L Y 2015 Research on a novel power inductor-based bidirectional lossless equalization circuit for series-connected battery packs *Energies* **8** 5555-76
[10] Zhu H, Zhao C, Liu Y F and Huang K L 2015 A new type of power battery energy balance system research *J Power Technol* **35** 2387-90
[11] Gui Y and Liu H W 2014 Two-direction balanced control strategy research and system development base on pure electric car lithium-ion battery pack (Changchun: Jilin University)
[12] Park H S and Park K B 2009 Design of a charge equalier based on battery modularization *IEEE Trans Veh Technol* **58** 3216-23
[13] Ju F, Deng W W and Li J F 2016 Performance evaluation of modularized global equalization system for lithium-ion battery packs *IEEE Trans Autom Sci Eng* **13** 986-96
[14] Xu J, Li S Q and Chris M 2013 SOC based battery cell balancing with a novel topology and reduced component count *Energies* **6** 2726-40
[15] Wang S L and Shang L P 2016 Online dynamic equalization adjustment of high-power lithium-ion battery packs based on the state of balance estimation *Appl Energ* **166** 44-58
[16] Zhang D H, Zhu G R and He S J 2015 Balancing control Strategy for a Li-Ion batteries string based on the dynamic balanced point *Energies* **8** 1830-47
[17] Phung T H, Collet A and Crebier J C 2015 An optimized topology for next-to-next balancing of series-connected lithium-ion cells *Energies* **8** 1830-47
[18] Liang G S, Fu G Q and Lai C P 2016 Simulation research of battery balance control technology based on the SOC *J Power Technol* **40** 332-5

IOP Conf. Series: Earth and Environmental Science **93** (2017) 012047 doi:10.1088/1755-1315/93/1/012047

Solarising tropical Africa's rural homes to sustainably overcome energy poverty

K E Kanyarusoke

Mechanical Engineering Department, Cape Peninsula University of Technology, Cape Town, South Africa

E-mail: kanyarusokek@cput.ac.za

Abstract. At less than 30% electrification, Tropical Africa is the most energy-poor electrified region of the world. At home level, the annual per-capita electric energy consumption ranges between 0 and 150 kWh in rural areas, where 83% of the population reside. This is well below the 250 kWh recommended by the International Energy Agency (IEA) as the threshold for exiting rural 'Energy Poverty'. Some governments have tried to extend the grid to such areas but these efforts have not yielded much. The approaches of rural electrification – as is being done now have therefore failed – and they may not be able to electrify every home in the countries concerned. An alternative approach promoting stand-alone photovoltaic (PV) and other solar powered heat and mass transfer systems at home level is proposed. An example of the approach in a village home in rural Uganda, East Africa is given. It is estimated that the combined unit energy cost over the systems' lifespan would be just about US 3 cents. Health, Education, and Sustainability in all its forms would be greatly improved. The main recommendation is for policy makers to adopt this approach for rural homes while sparing grid supply only for commercial and industrial activities.

1. Introduction

The energy poverty paradox in Africa is well documented. On one hand, here is a continent with 7.7, 7.6, 3.6, and 13 percentage points of the world's known oil, gas, coal and hydro electric energy reserves [1] on the other, is a 15.4% world population consuming 3.1% of the world's electric energy output [2]. If relatively 'developed' South Africa and semi tropical to Mediterranean North Africa are removed from the energy statistics, we get a 2017 Tropical Africa with 681 out of 951 Million people having no access to electricity at all [3]. The concerns in this paper are not that levels of industrialization and electrified transport in the region are low - which they are, anyway [4]. Rather, the absence of convenient, safe, clean, reliable and sustainable forms of energy in the homes of the 681 Million people above amidst a plentiful unharnessed solar energy resource [5]. This, at a time when literacy in the region is rising and when technology is diffusing faster, incentivizes a contribution in addressing the paradox in a way that could impact significantly on the lives of these people.

As of now, in non-electrified tropical African homes, non-muscle energy is mainly required for cooking food, boiling water, lighting and limited space heating at night. Excluding the latter, Williams estimated these activities to consume approximately 1 GJ or 277.8 kWh per month per household in rural South Africa [6]. In most of tropical Africa, this energy comes from incomplete combustion of bio fuels - mostly wood fuel and sometimes charcoal, in poorly ventilated kitchens and living spaces

[7]. Daily chores like water sourcing, all forms of washing, animal tending, cultivation, crop harvesting and processing, etc. are handled manually. Many of these activities are left to women and children [8]. This is unlike modern homes where running water, washing machines, electric stoves, water geysers, televisions, electric/electronic security systems etc. may be the norm [9]. Doesn't it matter that in spite of limitations facing women and children of the first group, they are largely responsible for primary manual production of food even for the second group? It does, principally because of reasons to do with health and sustainability. These are now briefly reviewed as a basis for the reported work in subsequent sections.

1.1. Some health and environmental concerns

The incomplete combustion of biofuels mentioned above produces large quantities of particulates and poisonous gases which affect both the environment and the health of fire attendants. Nussbaumer *et al* for example, report presence of wood tar in soot and of total particulate matter concentrations of 5000 mg per m^3 in flue products of poorly combusted wood fuel [10]. This is two orders of magnitude as high as emissions from a well-run combustion unit. The World Health Organization (WHO) gives emissions guidelines on two particle aerodynamic size limits: the 10 μm (PM10) threshold particles, which can be filtered by the upper respiratory tract, and the 2.5 μm (PM2.5) limit defining the observed smallest particle to exhibit a 95% chance of damaging lung and heart functions upon prolonged exposure [11]. The annual mean limits are given as 20 and 10 μg/m^3 respectively while the daily values are 50 and 25. Poor wood fuel combustion leads to exceeding these limits as reported in [7] and [12]. Figure 1 is a modified illustration on how the concentration of particles, and hence of mass, between the two limits increases with different combustion modes.

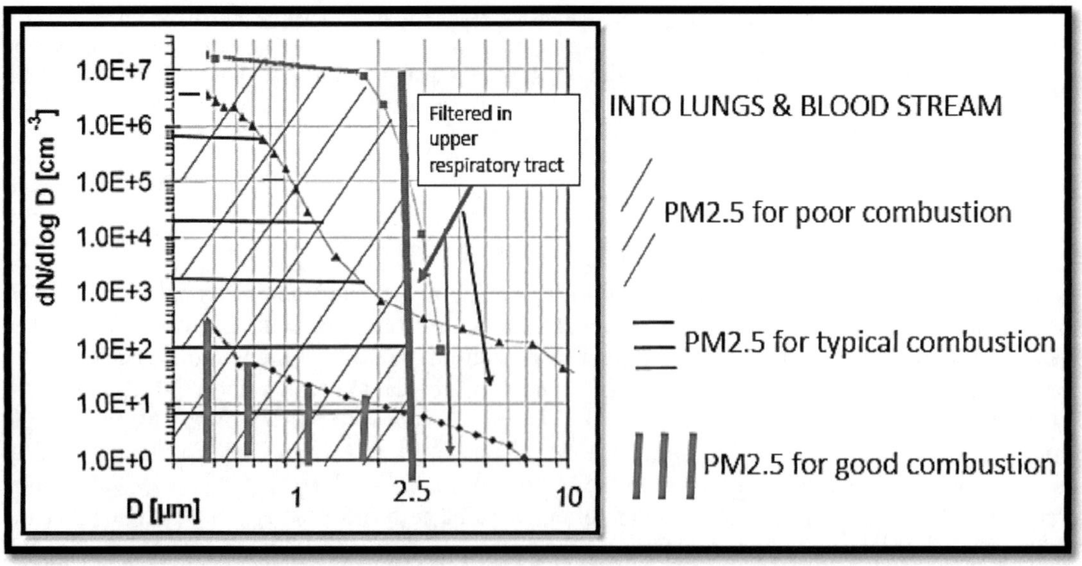

Figure 1. Illustrating effects of incomplete combustion on PM2.5 and PM10 values (modified from [10]).

The fact that most primary production of food and export agricultural products in tropical African countries is from efforts of people who are subjected to the above off limit exposures means food security and economic stability could be compromised if timely action is not made on addressing this situation. Moreover, the inefficient combustion additionally implies that more of the increasingly scarce fuel has to be burnt for a given task. This means more man hours have to be spent in the wild to fetch the wood or to produce charcoal. The latter is believed to lead to environmental degradation and adverse climate change. On the former, the United Nations Department of Economic and Social Affairs (UNDESA) reports that between 1990 and 2003, rural women were spending 2 to 4 hours a day, looking for firewood [13]. It adds that worldwide, 3.1 Million people die annually of indoor and outdoor pollution.

Following the above background information, the rest of the paper is organised as follows: Section 2 gives a commentary on current approaches to addressing the energy problem in the region. Section 3 reviews systems that were developed by the author to augment the above efforts. In section 4, previously unpublished results of the work are given and they are then discussed in the penultimate section. The paper concludes with an overall summary and makes one key recommendation.

2. Some approaches to alleviating the problems

In this section, a short commentary on efforts by different people, organizations and governments to address the energy problem for the 681 Million energy-poor tropical Africans. The efforts include those on: combustion improvements, rural electrification, and Concentrated solar power (CSP) systems developments. Some research on Solar Thermal Systems (STS) has been done especially in Nigeria and Kenya as exemplified by Arekete [14], Mwithiga and Kigo [15] and Ozuombo *et al* [16].

2.1. Actions on combustion

On combustion, methods used to tackle the problem include use of less polluting fuels and of improved stoves. Actions on fuels are however more expensive as some require capital expenditure in addition to running fuel costs. Thus, ethanol gels, plant oils, kerosene, natural and bio gas, - all require upfront investment in a suitable burner and/or a safe fuel storage system. More effort has been put in improved stoves partly due to a low investment cost (e.g. equivalent of US$ 2.00 in Ethiopia [17]) but mainly because of continuity in use of the familiar wood/charcoal fuel. Human behaviour is such that it is easier to accept one change at a time rather than several, simultaneously. Thus, different stove designs capable of reducing fuel consumption and consequent emissions by up to 50% on 'typical' combustion mode of figure 1 have been introduced in Ethiopia, Kenya, Rwanda, Uganda, Nigeria and many other tropical African countries [17-20]. Both the change of fuel and change of stove, however do not remove the other less talked about risk: that of fires in grass thatched houses that many of Africa's rural populations reside in. The author for example recalls that between 1966 and 1977, his homestead in rural Uganda lost most household items on three occasions, two of which were due to in-house made fires for lighting, and a third was due to day cooking needs. Today, even in rural South African homes - where paraffin has been used to augment solid fuels – ref [21] reports of homes which have been burnt on annual basis.

2.2. Rural electrification

Out of social political considerations, many African governments are trying to address these problems by extending national electric grid systems to rural areas. For example, as early as 1989, Ghana started a series of 5 year rural electrification programs, intended to avail electricity to everyone by 2020 [22]. Starting at 28% access rate then, they had nearly doubled that percentage by 2005. As figure 2 below shows, they are now at 75%, well on course to achieving their 2020 target. Nigeria set up a rural electrification agency in 2006 but it had continuity problems and 1946 electrification projects were temporarily put on hold midway. They resumed in 2012 [20]. The first decade of this century saw other tropical countries setting up similar agencies or authorities. Thus, Zambia set up an authority in 2006, when rural electrification was at 3.1%. Because of challenges on tariffs, population distribution and incomes, staffing and funding they have since struggled to maintain the rural access rate at 3% [18]. Zimbabwe's agency targeted rural income generating activities like Small to Medium scale Manufacturing Enterprises (SMMEs) for sustainability [24]. Kenya set up an authority in 2006 which started work in 2008. Unlike Zimbabwe, they targeted public facilities and rural trading centres. As a result, by May 2013, they had added 25873 schools, health and trading centres to the grid. This brought the national household electrification rate to 26% [17]. Between 2009 and 2012, Rwanda perhaps outperformed expectations in relative terms. Starting with a generating capacity of 69.5 MW and 110000 connections, they had 165 MW connected to 350000 users by 2012. All administration centres, health facilities and 50% of all primary and secondary schools were connected to the grid during the period [19]. This raised overall national access from 6 to 16%.

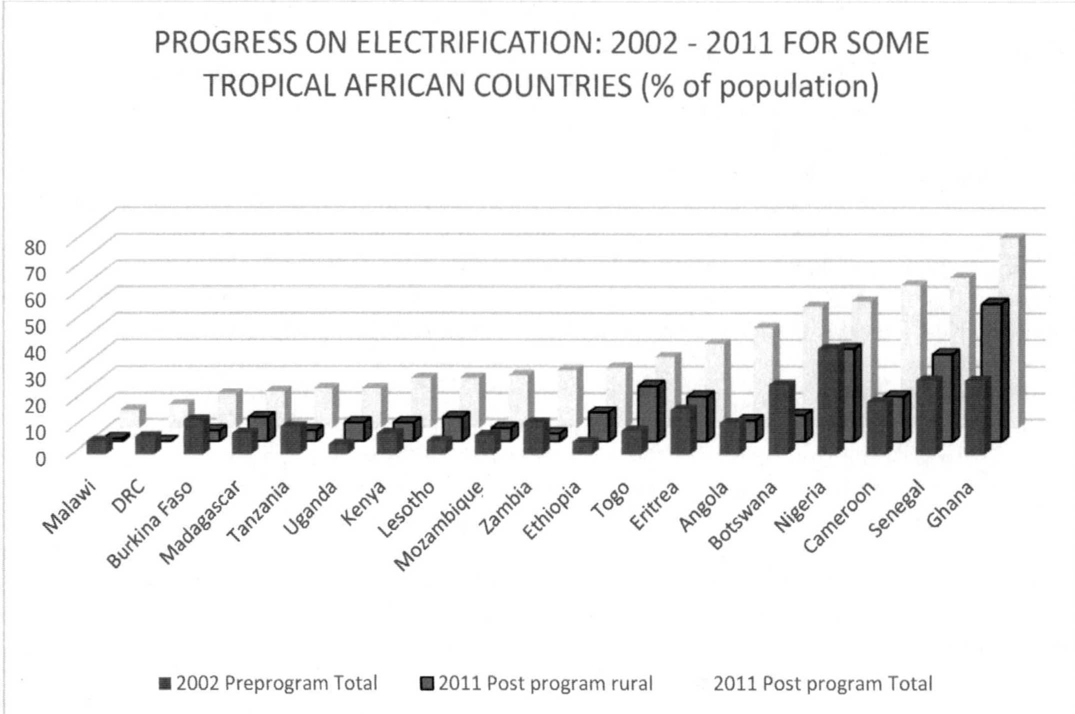

Figure 2. Changes in electrification for some countries between 2002 and 2011: Data compiled from national governments publications and the IEA [2].

The above relative achievements notwithstanding, there are problems with centralized efforts to rural electrify. In addition to those faced by Zambia mentioned above, Abeeku and Kemausuor mention lack of energy marketing strategies for small systems as a hindrance to full exploitation to narrow the supply-demand gap [24]. Mulugetta *et al* observe that more effort was being directed to a "Technology push" than to a Market pull [25]. Yet, if there were to be a shift in focus, the demand-supply gap could only get bigger as the population is growing rapidly. For example, it is estimated in [26] that tropical countries in the Nile basin alone will need to generate an additional 375 GW to attain present day South Africa's per capita energy consumption by 2050. It is shown that this is well beyond present day hydroelectric potential of the basin. It is perhaps because of such limitations that attention is being drawn to other renewables. In fact, some of the successes mentioned above would not have been possible were it not for inclusion of solar and geothermal forms of energy in the programs. Kenya in particular has had 200 MW added to the grid from geothermal sources [27]. Ghana, Zambia, Zimbabwe, Tanzania agencies/authorities have included photovoltaic (PV) power in their achievements. Kenya has gone a step further. The Naivasha based assembly plant is the only one from tropical Africa - of eleven reported by Enfsolar to be making PV panels in Africa [28]. To the extent that PV connections in Kenya are reported to be outpacing new normal grid connections now [29]. Rwanda's success is partly attributed to inclusion of 8.5 MW of PV electricity and the exploitation of Methane gas under Lake Kivu waters.

The above experiences illustrate the importance of using a mixed strategy on rural electrification. Central planning may be necessary for connecting public facilities and rural trading centres to the grid.

Dispersed individual homes, however, can be difficult and expensive to access. Kenya's example on PV panels shows that these are better left to private individuals and organizations. In Uganda, the rural electrification program has connected many trading and district administration centres [30]. But recent efforts to overstep this and freely connect homes in some areas involve a bureaucracy and create potential grounds for corruption as stated by Kanyarusoke *et al* [31]. In addition, they might be politically misunderstood or economically unsustainable – as the recipients will have to pay the high recurrent bills. That said, we now turn to efforts in non-electrical applications of solar energy.

2.3. Actions on CSP systems

Solar radiation reaches the earth as a composite mixture of beam and diffuse radiation. The beam part, coming directly from the sun, about 150 Million km away, arrives as parallel rays. CSP systems use refraction and/or reflection to concentrate these rays on a smaller area, thus increasing their intensity and effect either in terms of electricity generation or of heating ability. Here, we comment on efforts that use the latter.

Although concentration could be achieved using convex refractors, flat and concave reflectors, documented work shows that researchers and users in Tropical Africa have so far focused on reflectors. At domestic level, perhaps a most significant flat reflector application has been the CooKit cardboard cooker, designed in 1994 by Roger Bernard [32]. Figure 3 shows an illustration of the cooker. It is reported to be: effective in heating (70-90°C); affordable by the poor (US$10); light (0.5 kg); easy to construct (about 2 hours with support of an instruction manual); and easy to store or transport (foldable). With these attributes, the cooker has been popularised in West African countries and in refugee camps of East and Central Africa.

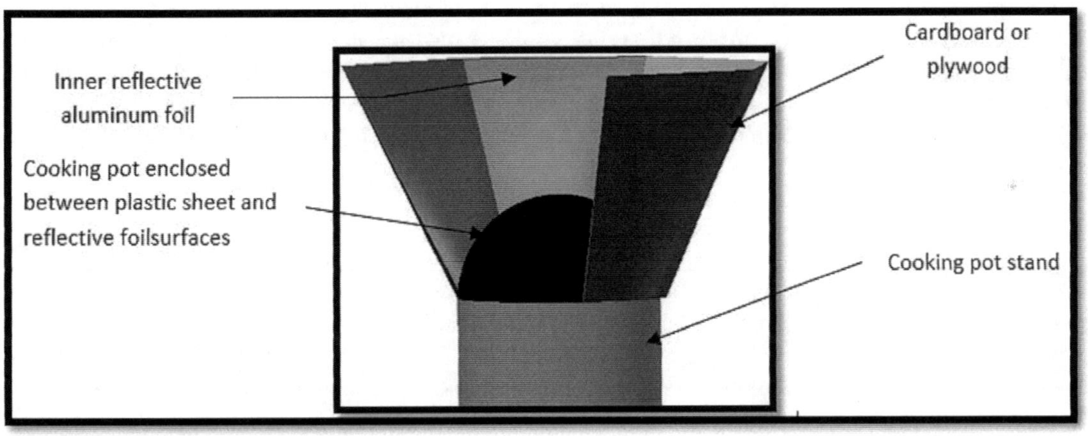

Figure 3. A 3-D illustration of the CooKit panel cooker.

Kimambo gives a summary of studies on solar cookers being used in Tanzania [33]. He classifies them as: Panel (e.g. CooKit and plane glass reflectors), Box (e.g. Sunstove) and Parabolic (polished and unpolished aluminium surfaces). Research results on six cookers in these categories are given and a recommendation that glass reflectors and polished aluminium cookers should be used in clear sky, high beam radiation areas is made. Unpolished aluminium types are discouraged. Panel types are recommended where there is medium sky cover. Box types are reported to be most advantageous in cloudy areas because of their ability to use diffuse radiation.

In Nigeria, much effort has gone into parabolic units. Notable is Dasin's solar tracking parabolic cooker [34] and Abdulrahim's bifocal solar tracking parabolic collector [35,36]. While most other solar cookers in the region require manual adjustment every 15 to 20 minutes to face the sun, these two have a mechanical clock – weight system to automatically keep the paraboloids' foci at the cooking pot position. Figure 4 shows Abdulrahim's cooker.

Figure 4. Abdulrahim's bifocal parabolic cooker [36].

Widespread acceptance of solar cookers faces problems mainly because of variability of solar radiation and other weather components, longer cooking time when compared to combustion assisted cooking and inability to perform from late afternoon onwards. The Tanzania research above [33] appears to suggest that most kitchens would require more than one solar cooker type if reliance on this resource is to be increased. Variability and cooking time problems could be mitigated by augmentation of the incident radiation using boosters. Elsewhere in the world, cooking temperatures are reported to have been raised by values of up to 17°C relative to those of ordinary cookers by augmentation with extra reflecting surfaces and/or finned pot surfaces [37]. Thus, cooking times can be reduced. Use of heat storage materials, including sand jackets and phase change materials like Magnesium nitrate hexahydrate ($Mg(NO_3)_2.6H_2O$), Stearic acid, Acetalmide among others have enabled cooking in the evenings. However, all these improvement methods do not seem to have been adopted in tropical Africa to any reported measure yet.

3. Developing sustainable energy systems for rural homes

In this section, examples of affordable and sustainable engineering actions that could be taken to overcome rural homes' energy poverty are given in areas of electricity, heat and mass transfer. These three areas form the bulk of energy requirements in a household and they have profound effect on the social-economic wellbeing of rural communities. Other needs such as transportation and land tilling in agriculture are also important in rural homes but no work has as yet been done by the author on these. They are, therefore, not discussed.

3.1. Photo electricity

The most basic requirement for electricity in a rural home is lighting at night – so that the useful work day can be extended. Also, communication and entertainment through respective use of radios and fully charged cell phones are enabled. Work done in this area is reported in refs. [31], [38] and [39]. In [31], a MATLAB® program is used along with ASHRAE weather data [40] to predict manufacturer photovoltaic (PV) panels' performances in Tropical Africa. Ref [38] uses validated TRNSYS modelling [41] to give optimal fixed panel tilts and optimised PV equipment selections for sub-Sahara Africa's rural homes in colour-coded maps. In [39], performance testing of a patented solar tracking device suitable for rural Africa's conditions of low technical skills and disposable incomes is described. Outcomes of most of this work have been put to use in a rural home in Kyegegwa, Uganda since June 2016. Figure 5 shows the installation. Costs and performance indications of the installation after about one year are given in sections 4 and 5 below.

PUTTING UP THE 200 Wp PANEL | PANEL AT 5 DEG SLOPE, 0 DEG AZIMUTH | WORKING INSIDE NEWLY CONSTRUCTED ROOF

Figure 5. Typical rural village house trying to emerge from energy poverty.

3.2. Heating and cooling

Heating in a rural home is a major consumer of energy because of cooking. Prevalent usage of biomass [42] has also affected the environment through cutting of trees for firewood and/or charcoal [43]. Cooling in rural homes has not been a big energy consumer though the need for food preservation and air-conditioning is evident in hot and humid climatic regions - such as the coastal areas, and in warmer hinterlands [44]. Work done on heating has so far been on water heating using solar thermal units. That on cooling has been on use of PV- powered refrigeration systems.

3.2.1. Water heating. For rural home use, a series of low cost fixed slope solar syphon systems ranging from 1 to 2 m^2 and heating 50 to 150 litre of water have been developed. Figure 6 shows the construction and installation of one such unit in a household compound – as opposed to being on a roof since there is no pressurised water to fill the tanks in most rural areas. Suitable optimal slopes for 2 m^2 units have also been determined for the entire Tropical Africa region using TRNSYS modelling. They are different from those of PV panels because of added thermal necessity to generate a sufficient hydraulic head to circulate the water. Other developments on solar syphon systems have included azimuthal and inclined axes solar tracking. The full theoretical development of the latter tracking is given in [45]. Figure 7 shows the two tracking units.

Figure 6. Installing a fixed slope solar collector in a household compound.

Figure 7. Solar tracking collectors.

3.2.2. Refrigeration. An inverter-less PV run refrigeration system was developed for rural small scale farm daily fruit and vegetable harvests. The system was able to cool 20 kg of straw berry fruits from 30°C to 5°C in 5 hours on successive days as described and tested in [46].

3.3. Mass transfer: Water purification and crop drying

Rural homes have to use some of their heating energy resource for boiling water to make it relatively safe to drink. The alternative is to have high susceptibility to water borne diseases because most water sources are microbiologically contaminated [47]. In some cases, water as collected from the sources has dissolved salts which affect healthy development of the overall human body [48]. A low cost solar water purifier was therefore developed to address these problems. It is slightly different from ordinary solar stills in that the evaporated steam is condensed outside the still, thereby improving the effectiveness of the still. Figure 8 shows a unit that can be installed in a homestead compound for purifying at least 5 litre of water per day in most tropical African areas.

Another low-cost mass transfer device developed for rural homes is the crop dryer. Homesteads normally do not expend heating energy on drying of crops. However, they use both manual and mental energy to dry crops and protect them from not only weather elements but from birds, rodents and even human thieves as well. Mental effort, in this case, is deployed to plan when and where to place the crop for ease of covering/removing it in case of rain. Also, judgment as to the attainment of a desired final moisture content depends on a mental recollection and observation of crop physical features that mark the desired end point. The crop dryer such as in figure 9 was developed to address these problems.

Figure 8. A rural solar water purifier.

Figure 9. A rural solar crop dryer.

4. Results

In this section, key results of the above work which have not been discussed in other publications are given. These include costing of PV systems, energy yields of solar syphon systems and typical water yields from the water purifying system. Costing of heat and mass transfer systems is based on current developmental costs since there are at present no known players on the market.

4.1. Costing of PV electricity

Costs of the installation of figure 5 in Kyegegwa, Uganda (nearest TRNSYS weather station – Mbarara) are given in table 1. For comparison purposes, they have been worked out on a present net value basis after taking into consideration of different inflation rates on various products and services over a system life time of 25 years. A full development of these costs is available in the appendix.

Table 1. Approximate energy cost comparisons between fixed and tracking panel using different versions of the solar tracker.

	FIXED SLOPE	MANUAL TRACK	SEMI AUTO	FULLY AUTO
SOLAR PANEL REQUIRED (Wp)	200	150	150	150
TOTAL 'POSSIBLE' ELECTRICITY IN 25 YEARS (kWh)	6568	6600	6600	6600
CAPITAL COST (excluding non-material tracker costs) (US$)	**685.02**	**856.64**	**903.15**	**931.47**
PRESENT VALUE 25 YEAR OPERATING COST (US$)	527.58	527.58	527.58	527.58
PRESENT VALUE 25 YEAR MAINTENANCE COST (US$)	406.12	507.87	535.45	552.24
PRESENT VALUE RUNNING UNIT ENERGY COST (US$/kWh)	**0.14**	**0.16**	**0.16**	**0.16**
PRESENT VALUE TOTAL UNIT ENERGY COST (US$/kWh)	**0.25**	**0.29**	**0.30**	**0.30**

4.2. Energy yield from a 2 m² solar syphon system

A TRNSYS simulation for a 2 m² collector with slope fixed at 5° gives daily energy yields as shown in figure 10 when heating 50 litres of water for the Kyegegwa site. This gives an annual energy harvest of 2434 kWh. Peak water temperatures are at 57°C as shown in figure 11.

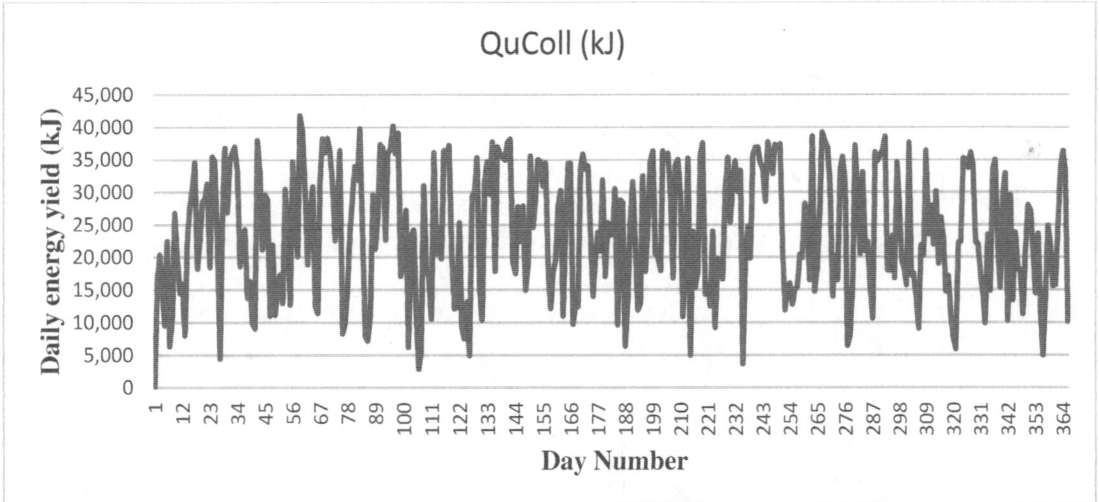

Figure 10. TRNSYS predicted Energy yield from a 2 m² solar syphon system heating 50 Litre water in Kyegegwa, Uganda.

4.3. Water yield from a solar water purifier

Typical performance on a sunny day of the newly developed water purifier of figure 8 is shown in figure 12. Along the distillation data are shown environmental conditions as well. On this particular day (22 Nov 2014), 2.8 Litre was produced from the prototype purifier. A larger unit producing about 5 Litre from an average 'dirty' water temperature of 20°C would save 3.6 kWh of direct heating energy per day.

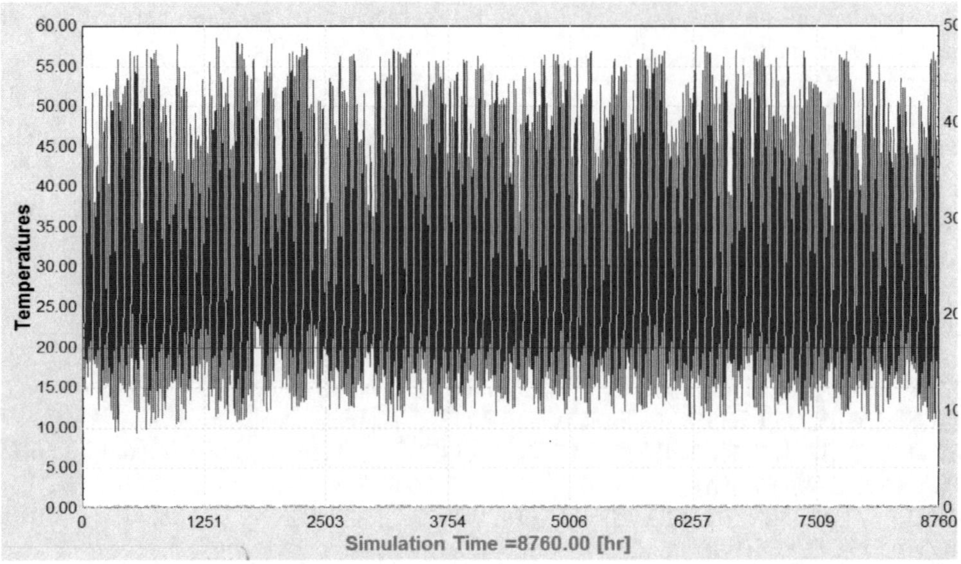

Figure 11. TRNSYS predicted water temperature variation from a 2 m^2 solar syphon system heating 50 Litre water in Kyegegwa, Uganda.

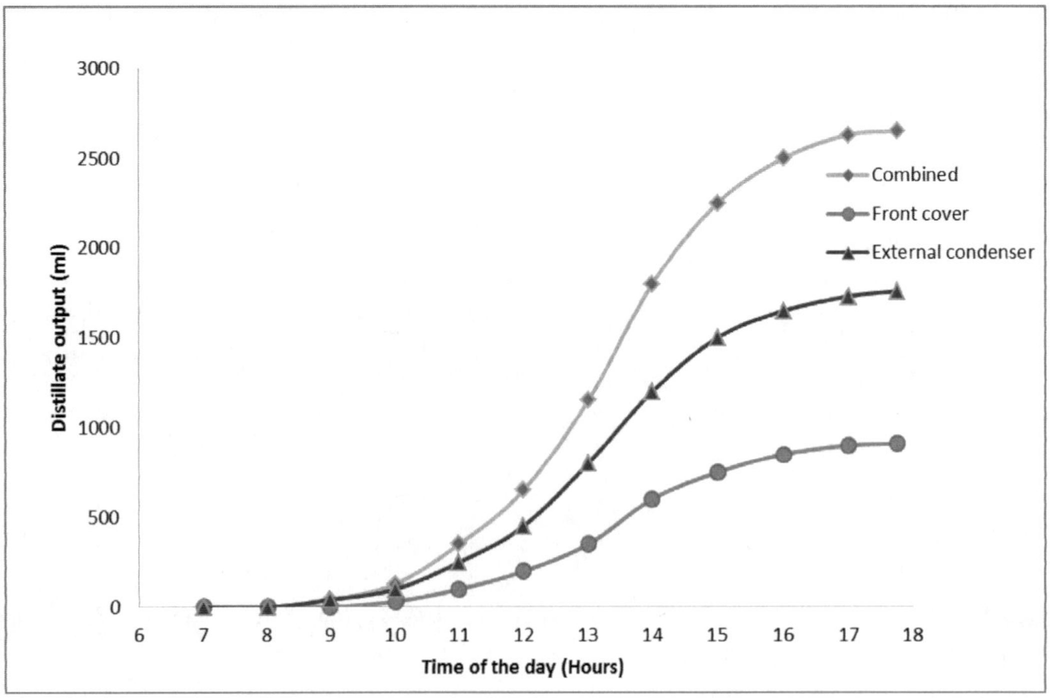

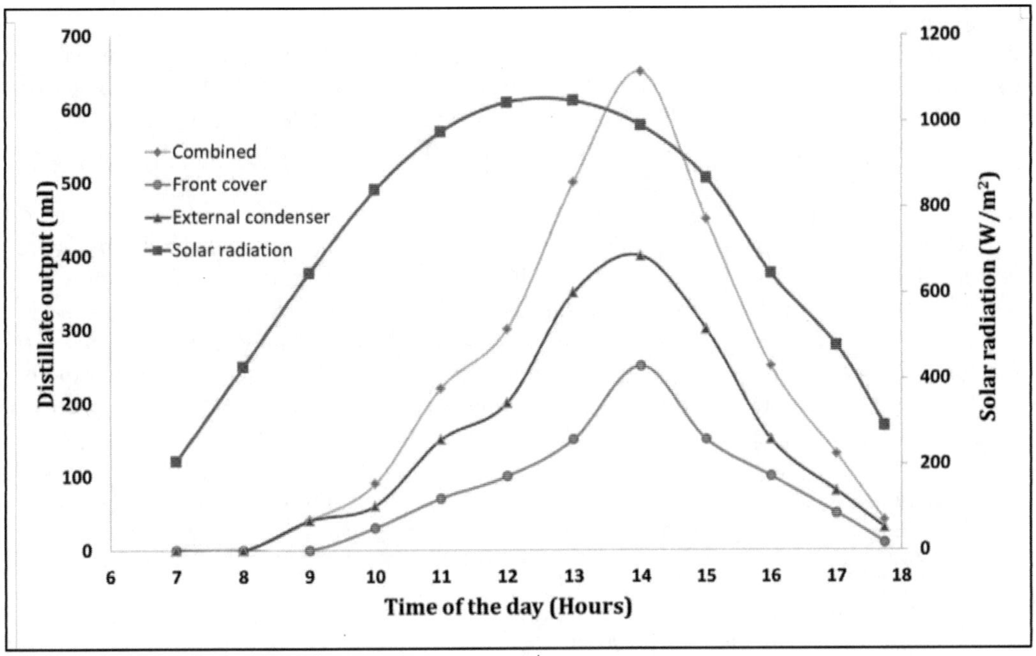

Figure 12. Measured typical performance and solar radiation data in the newly developed solar water purifier [49].

5. Discussion

The above results are discussed from perspectives of: energy utility, energy costs, social utility and sustainability. The discussion is concluded with a look at whether or not, centralising the provision of these energy services would be a worthy effort at accelerating reduction of energy poverty in Tropical Africa.

5.1. Energy utility

The most important benefit of the systems is the availability of useful energy in all places to people who would otherwise be constrained to live without it, and therefore miss opportunities availed by it. For a system life time of say 25 years in Kyegegwa, estimates of quantity of energy so availed by the systems are given in table 2.

Table 2. Useful energy availed by the solar resource in Kyegegwa, Uganda.

	Annual quantity (kWh)	Life time quantity (MWh)
Electricity (for 200 Wp panel)	263	6.6
Solar water heating (for 2 m^2 collector)	2434	60.9
Solar water distillation (at 5 L/day)	1314	32.9
Total energy	4011	100.3

The IEA defines energy poverty for rural areas at a per capita annual consumption of below 250 kWh electricity and below 500 kWh for urban dwellers [2]. Since electricity is used mostly for lighting and heating in homes, it could be said that the 4011 kWh in table 2 is more than adequate to change the status of homesteads of 8 to 16 people from energy poverty. Given that the household size in Uganda averages 4.7 with rural districts averaging between 2.5 and 7.6 persons [50], it seems clear that the systems would fundamentally alter the status. Quality of this energy may be judged from what it helps the homesteads achieve and how. These are discussed in section 5.3 below.

5.2. Energy costs

In table 1, the unit energy cost of US $0.25 from a fixed axis PV panel may seem high when compared with the April 2017 quoted value of US$ 0.19 for Uganda's grid electricity [51]. But the comparison is erroneous. The former cost includes both capital and running costs while the latter refers only to the bills paid to the grid supply company. This is part of the running costs which should include maintenance (e.g. bulbs and heaters replacements). In addition, the cost of acquisition should be amortised over the installation period. In 2012, the indicative acquisition cost for a home about 200 m from the nearest low voltage pole in Uganda's capital city, Kampala was US$ 800 [31]. This is greater than the capital cost of the PV system of figure 5. For a Kyegegwa rural home 5 km from the nearest pole, the cost can for practical purposes be taken as infinite. Hence, there cannot be a basis for comparison with grid supply cost. In like manner, there is no basis for comparing costs of the other two systems with grid supply. To estimate their energy cost however, actual costs of prototype construction materials were simply doubled to get an estimate of selling prices of commercial units. On treatment of these estimates like for PV, unit energy costs of table 3 were obtained.

Table 3. Estimated Solar energy cost of different systems in Kyegegwa, Uganda.

	Energy cost (US$/kWh)
Electricity (for 200 Wp panel)	0.25
Solar water heating (for 2 m^2 collector)	0.017
Solar water distillation (at 5 L/day)	0.013
Total solar energy	0.031

It is seen that the heating and mass transfer energy costs are very small, under 2 US cents. This arises from higher thermal efficiencies (40 to 60%) of conversion of solar energy to heat in the systems compared to the 15-17% of converting the same to electricity in crystalline silicon PV panels. Because of the large quantities of energy used in heating relative to lighting, the overall energy cost is very low, at about 3 US cents. The importance of this fact will become more apparent in section 5.5 below.

5.3. Social utility

From a health point of view, reduction in cardio pulmonary dysfunction in women is the most prominent possibility because of reduced exposure to smoke. Studies by Mills *et al* [52] and Hunter and Mills [53] among many others, explain how combustion derived pollutants lead to cardiovascular disease. But also among children, it is noted that poor lighting from paraffin lanterns and candles at night has two immediate implications. One – that for reading, the children would have to be closer to the lanterns to increase the intensity. This increases chances of airborne disease transmission between them, aggravating the effects of incomplete combustion of paraffin and wick. Secondly, the low intensity can lead to eyestrain and headaches. These problems are eliminated by the PV system of figure 5. Heating water either for hygiene purposes such as baths, laundry, dish washing helps reduce possibilities of water borne infections, since it is seen in figure 11 that temperatures achievable on many days are high enough to deactivate microbes as reported in [54]. Education can be enhanced by the systems in at least two ways. First, the children are able to read and do their homework at night in good light (and possibly health). Secondly, they would spend less time collecting firewood on weekends, perhaps devoting more time to private study. A real possibility is that all rural children would grow up knowing what 'electricity' and 'solar energy utility' are. This can stimulate study of Science among a bigger fraction of them than at present.

5.4. Sustainability

Replacing paraffin lanterns with PV lighting eliminates acid rains and millions of Tonnes of CO_2 from these sources. This addresses global warming and soil degradation issues. Solar water heaters lower the CO_2 load if adopted in urban centres. Action on either of the systems in rural areas could lead to growth in the latter. That way, the CO_2, acid rain gases and deforestation increases would be checked.

The other aspect of sustainability is the ability of users to continue affording to use the systems after initial trials. This problem does not arise for the water heating and mass transfer systems because the running costs are almost non-existent. For PV panels, occasional replacements of LED bulbs and replacements of batteries every after 5 or more years is the major expense. But because the periods are so long, there is enough time to prepare for them. For the Kyegegwa case of figure 5, approximately one year since installation, there has been no expense on the unit.

5.5. Why not centralise solar energy sourcing?

Centralising PV electricity for distribution as has been done by the Uganda government in some areas [55], and by Rwanda [56] is beset with problems. First, the dc power has to be inverted and then, the voltage stepped up before feeding into a long transmission line. Each of these steps has some of the generated electricity being converted to thermal energy in line with the second law of Thermodynamics. For initial efficiencies of 15-17%, the overall efficiency even before transmission drops towards 10 -12%. Transmission line and distribution network losses in many African countries are high (e.g. Nigeria, 12% in 2014 [57], Uganda 33% in 2011 [58], etc.). This means less than 10% of the original energy received by the panel eventually becomes 'useful' electricity. The second problem has to do with multiphase ac supply – which requires either multiple inverters or a single three phase inverter. The problem of line balancing in such cases become more critical [59]. The third problem on electricity is simply that most households in rural Africa are dispersed. This makes it difficult - if not impossible - to make electricity generated centrally to reach every homestead.
For heat and mass transfer systems, centralised systems need a lot of space, management and possibly conflict resolution practitioners to effect. These professionals are not available in most rural areas.

6. Conclusion

To conclude, this paper tried to show how a social economic problem of energy poverty in rural Tropical Africa could sustainably be addressed using solar energy technologies. There are some health and environment degradation issues attendant to this poverty which could be addressed by the technologies. Current alternative efforts by different countries to address the problem were explored and it was found that success was limited – mainly to electrification of rural public sectors. Reaching individual rural homesteads from centralised supply was fraught with capacity, cost and distribution problems because of the homesteads' dispersal. Efforts to address health issues through improved combustion and stove designs were also not yet universal. Besides, they did not remove the danger of fires in rural homesteads.

The paper then proceeded to suggest and illustrate simpler ways to address these and other problems at the household level. On lighting, previous work on photo-lighting rural homes was cited and an example, implementing results of that work at one place in rural Uganda was given. A US$ 700 home installation was reported to have been running in line with performance expectation but with nil cash outflow for about one year. It was also shown that water heating for hygiene purposes could be achieved using simply constructed, low cost solar syphon systems. Most importantly, a system for converting both micro-bio and chemically contaminated water into safe potable water was illustrated. It was shown to save significant amounts of energy otherwise required to boil the water in attempt to destroy biological contaminants. A solar crop dryer employing similar principles of mass transfer as the water purifier was illustrated. So were references to work on inverter-less solar-assisted refrigeration.

Cost estimates based on actual PV commercial prices, doubled construction material costs of other systems and life time running costs were done and shown to give a very low unit energy utility cost of the order of US$ 0.03. Because of universal availability of solar energy, unlike other sources - especially grid electricity, there was no basis for comparing unit costs.

To summarise therefore, in presence of the illustrated devices, it could be said that solar energy usage by rural households in lighting, hygiene water heating and family drinking water should be strongly advocated to reduce energy poverty. When the three are taken together, possibilities of health improvement, focus on educational work by children, and positive national productivity changes are availed. Both household capital and recurrent expenses are much lower than with a grid supply. There is less demand for wood or other carbon-based fuel, and hence, less consequent environmental harm. Using solar energy for much larger projects such as centralised services may however be problematic. There is less certainty about its potential effectiveness, and therefore this paper presents no strong support for it. It is suggested that such services as would be required at a rural community or commercial centre might be better served with grid supply like is being done in some countries. Therefore, the main recommendation of this work in respect of rural areas is that governments, development partners and private sectors should be looking at a dual system where homes are served with solar energy systems while grid electricity is reserved for commercial and bigger institutions at commercial and industrial centres.

Acknowledgments
Most of this research and development work was funded by the Cape Peninsula University of Technology over a period of about 5 years through its URF mechanism. The author is most grateful especially to the University Research staff.

Appendix

A1: Computation of photo electric energy costs: an approximation

A1.1. Development of costing

The cost of electricity from the systems in this paper – namely the optimised fixed slope 200 Wp panel and its equivalent inclined axis solar tracking 150 Wp panel – is estimated for a rural home near the equator in Uganda. Actual June 2016 prices of components are used in the estimates. The other data on the economy used is as follows:

General inflation rate i_r = 6% (based on long term value over the past 20 years)
Bank interest rate over inflation, i_b = 10% (based on a savings interest rate of 16% from Bank of Uganda)

Special item inflation rates over and above general rate were assumed as follows (i_{pdct}):

Maintenance labour:	=	5%	Maint. cost rate =		4% of initial capital
Batteries:	=	2.5%	Life	=	5 years
Charge controllers:	=	5%		=	15 years
LED bulbs:	=	-2.5%		=	15 years
System life expectancy	=	25 years;	Maintenance	=	1 year

For an item costing an amount ***C*** today, the present value cost after ***n*** years is given by the equation [60].

$$C_n = C\left[\frac{1+i_r+i_{pdct}}{1+i_r+i_b}\right]^n \tag{A1}$$

Also, the present value $\boldsymbol{E_n}$ of total annual expense ***E*** growing at a rate $\boldsymbol{r_p}$ over and above general inflation after ***n*** years is given as:

$$E_n = E\left[\frac{1+i_r+r_p}{r_p-i_b}\right]\left[\left[\frac{1+i_r+r_p}{1+i_r+i_b}\right]^n - 1\right] \tag{A2}$$

For the actual costs of items as purchased, equations (A1) and (A2) were laid out in an Excel spreadsheet and present day costs evaluated as given in A1.2.

Different data can be in-put into the grey-shaded cells to reflect given economic conditions and/or prices but all unshaded cells are automatically computed.

A1.2. Energy costing for fixed slope 200 Wp and 150 Wp solar tracking panels

Capital costs:	SIZE	LOCAL PURCH. PRICE (US$)	QTY.	LOCAL TRANSPORT	**Fixed slope** DELIVERED COST	**Manual Track** 150 Wp panel	**Semi auto** 150 Wp panel	**Auto track** 150 Wp panel
Solar Panel	200 Wp	141.6	1	0	141.60	106.20	106.20	106.20
Batteries	105 Ah	121.21	2	0	242.42	242.42	242.42	242.42
Charge Controller	15 A	120	1	0	120.00	120.00	120.00	120.00
Bulbs (LED)	11 W	7	12	0	84.00	84.00	84.00	84.00
Installation Materials	Lot	67	1	0	67.00	67.00	67.00	67.00
Sub Total basics					655.02	619.62	619.62	619.62
Installation labour					30.00	35.00	40.00	40.00
Solar Tracker (materials only)					0	202.02	243.53	271.85
Total capital costs					685.02	856.64	903.15	931.47

Running costs:		Running costs: Batteries	Running costs: LED Bulbs	Running costs: Charger	System Maintenance costs
System expected lifespan: years	25				
Expected charger lifespan: years	15				
Expected LED Bulbs Lifespan: years	15				
Expected Battery Lifespan: years	5				
Anticipated general annual inflation rate(%)	6				
Excess alternative investment return rate relative to inflation on general items: e.g. Bank saving interest rates over and above inflation (%)	10				
Annual Maintenance charge as % of initial capital	4				
Anticipated excess price inflation over and above that of general items (%)		2.5	**-2.5**	**5**	**5**
	Repl. No.				
Number of replacements expected over entire life time		4	1	1	25
Present worth of Operational expenses for 1st replacement (US$)	1	173.55	15.2	61.97	
Present worth of Operational expenses for 2nd replacement (US$)	2	124.25	0.00	0.00	
Present worth of Operational expenses for 3rd replacement (US$)	3	88.95	0.00	0.00	
Present worth of Operational expenses for 4th replacement (US$)	4	63.68	0.00	0.00	
Present worth of Operational Expenses: US$		**450.43**	**15.2**	**61.97**	

	Fixed slope	Manual Track	Semi auto	Auto track
Present worth of Operational Expenses: US$	**406.12**	**507.87**	**535.45**	**552.24**
Total Present Worth Capital + Running costs (US$)	**1618.73**	**1892.1**	**1966.2**	**2011.3**
Total Present Worth Running costs (US$)	**933.71**	**1035..5**	**1063.0**	**1079.8**
Total useful energy produced over lifecycle: kWh	**6568.00**	**6600**	**6600**	**6600**
Present worth Running unit energy cost: US$/kWh	**0.14**	**0.16**	**0.16**	**0.16**
Present worth Capital unit energy cost: US$/kWh	**0.10**	**0.13**	**0.14**	**0.14**
Present worth Total unit energy cost: US$/kWh	**0.25**	**0.29**	**0.30**	**0.30**

References

[1] BP 2014 *BP Statistical Review of World Energy* https://www.bp.com/content/dam/bp/en/corporate/pdf/energy-economics/statistical-review-2017/bp-statistical-review-of-world-energy-2017-full-report.pdf
[2] IEA 2014 *Key World Energy Statistics* http://www.iea.org/publications/freepublications/publication/KeyWorld2014.pdf
[3] IEA 2016 *World Energy Outlook 2016* http://www.iea.org/publications/freepublications/publication/WorldEnergyOutlookSpecialReport2016EnergyandAirPollution.pdf
[4] African Development Bank, Organization for Economic Co-operation and Development, United Nations Development Programme 2014 *African Economic Outlook 2014*: Global value chains and Africa's industrialization
[5] European Commission and Heliochem-1 2008 Africa - Photovoltaic solar electricity potential http://www.soda-is.com/eng/map/maps_for_free.html
[6] Williams 1994 Energy supply options for low income urban households (University of Cape Town)
[7] Ayazika W 2002 Health related effects of traditional biomass fuels: the Ugandan case study (NEMA DfID KaR Project R8019)
[8] UN 2010 *The World's women 2010* http://unstats.un.org/unsd/demographic/products/Worldswomen/WW2010%20Report_by%20chapter(pdf)/Environment.pdf
[9] Statistics South Africa 2013 *GHS Series, volume V, Energy*, 2002 -2012. Report 03-18-04 18-9
[10] Nussbaumer T, Czasch C, Klippel N, Johansson L and Tullin C 2008 Particulate combustion in IEA countries Survey on measurements and emission factors *IEA Bioenergy Task 32* (Zurich, Swiss Federal Office of Energy (SFOE))
[11] WHO 2006 *Air quality Guidelines for particulate matter, ozone, nitrogen, dioxide and sulphur dioxide, Global Update* 2005
[12] Klippel N and Nussbaumer T 2007 Health relevance of particles from wood combustion in comparison to Diesel soot *15th Eur. Biomass Conf.* (International Conference Centre, Berlin) pp 7-11
[13] UNDESA 2014 A survey of international activities in rural energy access and electrification May 2014 "Global tracking framework," *Sustainable Energy for All* (2013), 262
[14] Arekete A S 2013 *Journal of Energy Technologies and Policy* **3** 1-8
[15] Mwithiga G and Kigo S N 2006 *J. Food Eng.* **74** 247-52
[16] Ozuomba J O, Edebeatu C C, Opara M F, Udoye M C and Okonkwo N A 2012 *Research Journal of Chemical Sciences* **2** 64-7
[17] Ayieko O Z 2011 Rural electrification programme in Kenya *AEI Practitioner workshop* Dakar Senegal, Nov 2011 http://siteresources.worldbank.org/EXTAFRREGTOPENERGY/Resources/717305-1327690230600/8397692-1327691237767/Rural_Electrification_in_Kenya_presentation_Final_11thNov2011.pdf
[18] Lyalabi S 2011 REA's experience in implementing rural electrification projects *AEI Practitioner workshop* Dakar Senegal Nov 2011 http://siteresources.worldbank.org/EXTAFRREGTOPENERGY/Resources/717305-1327690230600/8397692-1327691245128/Implementing_REAs_ZAMBIA.pdf
[19] Uwamahoro Y 2013 Rural electrification in Rwanda *PACEAA*
[20] Elusakin J E, Ajide O O and Diji J C 2014 *African Journal of Engineering Research* **2** 51-7
[21] Lloyd P 2014 *JESA* **25** 2-8
[22] Abavana C G 2011 *Electricity access progress in Ghana* (Noxie Consult, Ghana)
[23] Mapako M and Prasad G 2007 *Rural electrification in Zimbabwe reduces poverty by targeting income generating activities* CSIR ECR University of Cape Town http://hdl.handle.net/10204/871
[24] Abeeku B-H and Kemausuor F 2009 *Current Opinion in Environmental Sustainability* **1** 83-

8
[25] Mulugetta Y, Nhete T and Jackson T 2000 *Energy Policy* **28** 1069-80
[26] Kanyarusoke K 2013 Water and energy needs in a developing Nile Basin *The New Vision* (Kampala Uganda)
[27] Energy Information Administration 2014 Kenya country analysis note http://www.eia.gov/countries/country-data.cfm?fips=KE
[28] Enfsolar 2014 Solar panel Manufacturers in Africa www.enfsolar.com/directory/panel/africa
[29] Eveleens I 2011 EAfrica's first solar-panel plant supports Kenya's clean energy push *AlertNet*, 23rd Nov 2011
[30] Uganda Rural Electrification Agency 2012 *Annual Rural Electrification Report* 2010/2011 http://www.rea.or.ug/
[31] Kanyarusoke K E, Gryzagoridis J and Oliver G 2012 Predicting photovoltaic panel yields in sub-Sahara Africa *Proc. ICEAS* (Beijing China July 2012) pp 223-55
[32] Toonen H M 2009 *Phys Chem Earth* **34** 65-71
[33] Kimambo C Z M 2007 JESA **18** 41-51
[34] Dasin D Y, Asere A A and Habou D 2011 *Nigeria Journal of Solar Energy* **22** 7-13
[35] Abdulrahim A T, Diso I S and Abdulraheem A S 2012 *JMER* **4** 136-41
[36] Abdulrahim A T, Diso I S and Ngala G M 2011 *Nigeria Journal of Solar Energy* **22** 29-33
[37] Cuce E and Cuce P M 2013 *J. Ap Energy* **102** 1399-421
[38] Kanyarusoke K E, Gryzagoridis J and Oliver G 2016 *Turk J Elec Eng & Comp Sci* **24** 4763-72
[39] Kanyarusoke K E and Gryzagoridis J 2016 The new hydro-mechanical solar tracker: Performance testing with a PV panel *Proc. 24th DUE Conf.* (Cape Town Mar 29-31, 2016) pp 178-85
[40] ASHRAE 2009 *ASHRAE Handbook-Fundamentals* Atlanta: American Society of Heating Refrigeration and Air Conditioning Engineers, Inc. Ch. 14
[41] Klein S A, *et al* 2012 TRNSYS 17: *A transient system simulation program* Madison, WI, USA: Solar Energy Laboratory, University of Wisconsin
[42] VENRO 2009 Rethinking Biomass Energy in Sub-Sahara Africa http://www.venro.org/fileadmin/redaktion_afrikas_perspektive/publikationen/Projekt-Publikationen/091124_Arfikas-Perspektive_Bioenergiestudie_Final.pdf
[43] Malhi Y, Adu-Bredu S, Asare R A, Lewis S L and Mayaux P 2013 African rainforests: past, present and future. *Phil Trans R Soc* B 368 20120312
[44] Kula N, Haines A and Fryatt R 2013 Reducing vulnerability to climate change in Sub-Saharan Africa: The need for better evidence *PLOS Medicine* 10 e1001374 https://doi.org/10.1371/journal.pmed.1001374
[45] Kanyarusoke K E, Gryzagoridis J and Oliver G 2013 Modelling Annual Yields of a solar-tracking Solar syphon Using ASHRAE's Weather Data for Tropical Africa. *Trans. 2013 ASHRAE* July 2013 Denver CO http://www.thefreelibrary.com/Modeling+annual+yields+of+a+solar-tracking+solar+syphon+using...-a0371285197
[46] Alkilani F & Kanyarusoke K E 2016 Inverter-less solar assisted farm level refrigeration system for fruits and vegetables *2016 Proc. of 24th ICUE Conf.* (Cape Town 15-17 August 2016. Cape Town) 178-84
[47] Pan Africa Chemistry Network 2010 *Africa's water quality* http://www.rsc.org/globalassets/04-campaigning-outreach/tackling-the-worlds-challenges/water/rsc-pacn-water-report.pdf
[48] World Health Organisation 1996 Health criteria and other supporting information *Guidelines for drinking-water quality* 2nd ed. Vol. 2. http://www.who.int/water_sanitation_health/dwq/chemicals/tds.pdf
[49] Mbadinga P J K 2015 A solar water purification system for rural areas *MTech dissertation* (Cape Peninsula University of Technology)

[50] Uganda Bureau of Statistics 2016 *National Population and Housing Census 2014 report* Kampala, p 12
[51] UMEME 2017 Electricity end user tariffs for the period April to June 2017 http://www.umeme.co.ug/assets/resources/Tariff%20Adjustments/Tariff_Adjustment.pdf
[52] Mills N L, Donaldson K, Hadoke P W, Boon N A, MacNee W, Cassee F R, Sandström T, Blomberg A and Newby D E 2009 *Nature Reviews Cardiology* **6** 36-44
[53] Hunter A, Mills N and Newby D 2012 *Br J Hosp Med (Lond).* **73** 492-7
[54] Kanyarusoke K E 2016 Seeding energy sustainability through transformative teaching: any way forward for sub-Sahara Africa? Delfino E & Vezzoli C Eds. *Proceeding of 2016 LeNSes Conference on Sustainable Energy for All* (28 – 30 September 2016, Cape Town) pp 167-76 Online e-book https://www.scribd.com/document/324775328/LeNSes-Conference-Proceedings
[55] ESI Africa 2016 Uganda: Soroti solar power plant comes online https://www.esi-africa.com/news/uganda-soroti-solar-power-plant-comes-online/
[56] GigaWatt Global 2016 *Rwanda gets a solar field* https://gigawattglobal.com/projects3/rwanda/
[57] Amadi, Hachimenum N, Ephraim N C, Okafor and Izuegbunam F I 2016 *AJEEE* **4** 123-30
[58] Parsons Brinckerhoff Africa 2011 *Study on distribution system losses and collection rates by UMEME Ltd* Kampala Uganda Electricity Regulatory Authority 32
[59] Peacock Media Group 2017 Phase Solar: What you need to know about connecting solar to your 3 phase supply https://www.solarquotes.com.au/blog/3-phase-solar-what-you-need-to-know-about-connecting-solar-to-your-3-phase-supply/
[60] Blank L P E and Tarquin A P E 2012 *Engineering Economy* 7th ed. (Singapore Mc Graw Hill) pp 367-74

Low-cost data acquisition systems for photovoltaic system monitoring and usage statistics

S Fanourakis[1], K Wang[2], P McCarthy[1] and L Jiao[1,3]

[1]School of Engineering, Grand Valley State University, Grand Rapids, Mi, 49504 USA
[2]Forest Hills Central High School, Grand Rapids, MI 49546 USA

E-mail: jiaoh@gvsu.edu

Abstract. This paper presents the design of a low-cost data acquisition system for monitoring a photovoltaic system's electrical quantities, battery temperatures, and state of charge of the battery. The electrical quantities are the voltages and currents of the solar panels, the battery, and the system loads. The system uses an Atmega328p microcontroller to acquire data from the photovoltaic system's charge controller. It also records individual load information using current sensing resistors along with a voltage amplification circuit and an analog to digital converter. The system is used in conjunction with a wall power data acquisition system for the recording of regional power outages. Both data acquisition systems record data in micro SD cards. The data has been successfully acquired from both systems and has been used to monitor the status of the PV system and the local power grid. As more data is gathered it can be used for the maintenance and improvement of the photovoltaic system through analysis of the photovoltaic system's parameters and usage statistics.

1. Introduction

As a result of increasing energy demand and gradually shrinking fossil fuel supplies, efforts have been made to explore alternative energy resources. Among these resources, solar energy is the most abundant, clean, and renewable alternative energy source available. Solar photovoltaic (PV) is deemed as a scalable and sustainable solution to the energy shortage problem. PV system performance is largely dependent on the location and the condition of the PV panels. Panel temperature, shading, and dust can affect the system performance; in some regions up to 70% of power losses could be from panel soiling [1]. Degradation of the panels can also reduce the system performance [2]. Thus, continuous monitoring of PV systems is important in maintaining and improving their performance.

In recent years, a number of data-logging devices have been developed to monitor PV systems. Commercially available PV monitoring systems capable of monitoring PV system health and usage are costly. For example, systems such as the TED Pro Home Energy Monitor or the Outback Power Flexware FN-DC Flexnet Advanced DC System Monitor can cost more than $300. While these components are optional, they are a valuable tool in ensuring the longevity of the PV system as a whole. PV monitoring systems allow for timely replacement of faulty components as well as the replacement of underperforming parts. Thus far, low cost data acquisition systems have been developed in the research setting [2-9]. As many PV systems are located in areas that are difficult to access, some devices have incorporated concepts of low-maintenance and automated data collection into their designs. Ammar Mahjoubi *et al* presented a low cost data acquisition system (DAS) for a remote PV water pumping system. A 10-bit 16F877 microcontroller was used to collect information which was stored in an EEPROM memory and then transferred to a personal computer through the Short text Message Service (SMS) [3]. Mukaro *et al* designed a DAS for solar radiation and environmental monitoring [5]. The system was based on the ST62E20 8-bit microcontroller and the SolData silicon-cell pyranometer. The solar radiation information was stored in a serial EEPROM and later transferred to the computer through the RS232 serial port. In the paper presented by Akposionu *et al*, a low-cost data logger for solar energy parameters was designed and fabricated [6]. The data logger used the Microchip PIC18F4550 microcontroller to monitor the solar panel voltages and the ambient temperature. Purwadi *et al* prototyped a low cost data logger for a PV based LED street lighting system [9]. Their data logger was developed based on Atmel Atmega 8 microcontroller, Dallas DS 1307 Real Time Clock and a serial EEPROM from Microchip. Information (the status of PV cell, battery, charge controller, and LED lamp) was saved in the memory and sent to the computer through RS232 serial port.

Many other features were unique to specific data loggers. Chiandone M *et al* used an Ethernet/TCP-IP communication network to collect and transfer data from PV fields [11]. An IP-enabled data acquisition device was used to successfully transfer data [12]. The data logger developed by Akposionu *et al* focused on the accuracy of each measurement, and thus created a data logger capable of collecting and storing data every three seconds [6]. While real-time access to data may be an important feature in some systems, in others it is not as critical; in such cases, local data storage can be used. To this end, SD cards provide an effective storage method and using multiple cards can ensure the data is recorded without corruption [13].

As each device is developed for a different, specific scenario, each contains deficiencies within their designs when employed for a different scenario. The system to be monitored in this paper (the Solar Suction Surgery System, or S4) is located in Malawi. It is used to provide power for lighting, phones, tablets, and a suction machine at the Embangweni Mission hospital. The battery is charged using a solar panel, and power is provided to five auxiliary ports. Due to its remote nature, local resources, and cost considerations, the data acquisition system described in this paper employs microSD cards for its data transfer; this allows for large data sets to be stored as well as the intercontinental transfer of the data. A microcontroller is utilized to acquire PV system parameters such as voltages and currents of the solar panel, the battery, the loads, as well as the ambient and battery temperatures. Furthermore, auxiliary sensors were used for the acquisition of specific usage information.

The data acquisition module detailed within this paper merges many past ideas while also improving on ease of integration. The data logger was designed with low cost as its core concept. Along with this concept, the system takes into account the difficulty of constant physical collection of data. To remedy this issue as well as simplify intercontinental transfer of data, microSD cards were used for data storage. Additionally, the designed data logging system directly accesses the charge controller, so no additional sensors are necessary to acquire parameters from solar panel and battery. Because of this simplification, the data logger can be easily integrated to PV systems and is capable of commercialization and operating in a real world environment. The monitored system is used in a healthcare environment. As such, it is essential to acquire information on the system parameters and usage in order to provide maintenance and future improvements. To this end, the data logger of the S4 is used in conjunction with a data logger monitoring the power outages of the region; the data is compared in order to examine the correlation between system usage (of the S4) and regional power outages.

2. Design

2.1. Hardware design

2.1.1. S4 Data logging system. The main hardware components of the S4 data logging system include an Atmaga328 microcontroller (KOOKYE Mini USB Arduino Nano V3.0 ATMEGA328P), a MAX485 RS485 transceiver, a micro SD card reader breakout board, two LEDs, and two buttons. Figure 1 shows a high level block diagram of the system.

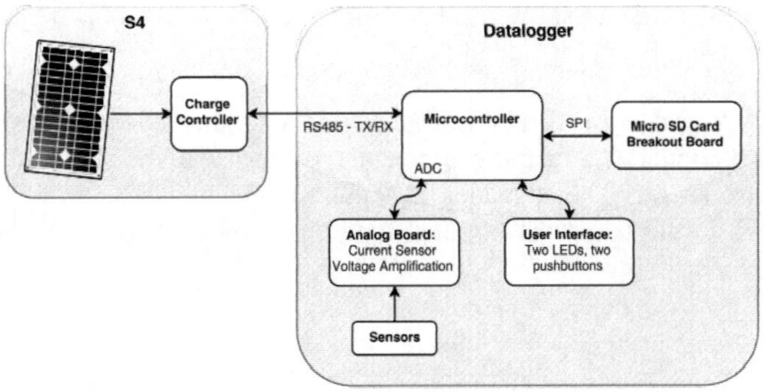

Figure 1. S4 data logging system block diagram.

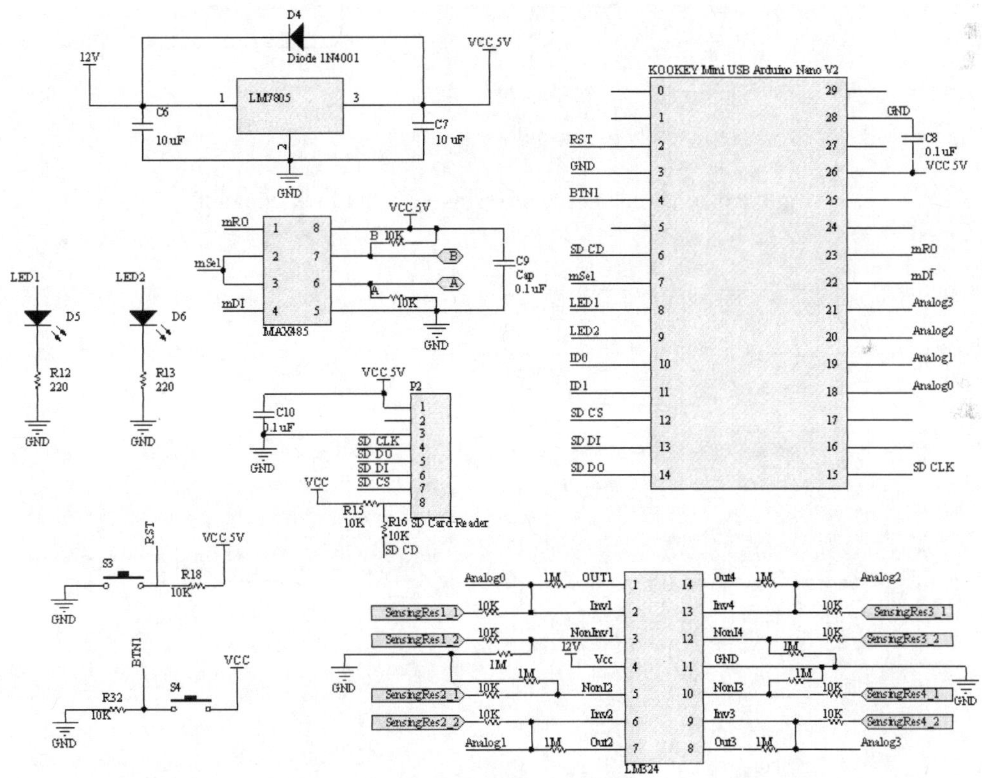

Figure 2. Schematic of the S4 data logging system.

The charge controller records the total load current; thus, additional circuitry was designed to enable the acquisition of the individual load currents. Four current sensing circuits were added at each load. The fifth current is calculated from the difference of the total load current and the sum of the four recorded currents. A differential transimpedance amplifier circuit with a gain of approximately 100 was included for each sensing circuit to convert current to voltage to be recorded by the microcontroller's analog read ports. To store the charge controller data as well as the four load information, a microSD card breakout board was added. Due to the remote nature of the solar unit it was determined that utilizing SD cards for the transfer of data would be the most cost effective method since it requires little maintenance. The file size is small enough to email with a PC or send via a messaging app on a mobile phone. Two LEDs and two pushbuttons were incorporated to allow the user to safely eject the SD card and allow for the communication of errors and system status. Figure 2 shows the schematic of the S4 data logging system.

2.1.2. Wall power data logging system. The wall power data logging system was similar in structure to the S4 data logging system. The main components of the wall power data logger include an Atmega328 microcontroller, an SD card reader breakout board, a real time clock (RTC DS3231), a 4.7 mF capacitor, two LEDs, and two push buttons. Figure 3 shows a high level block diagram of the system.

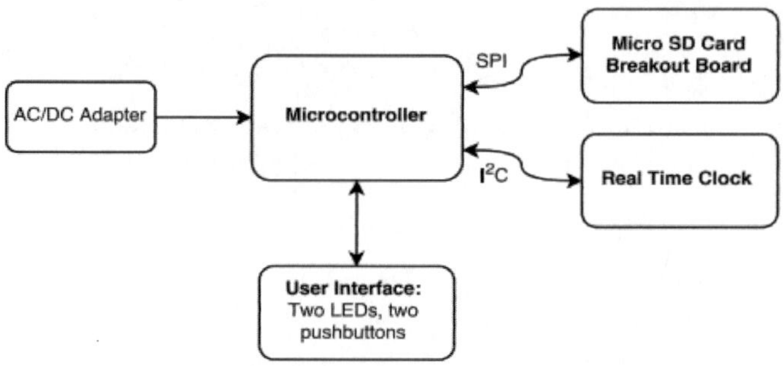

Figure 3. Wall power data logging system block diagram.

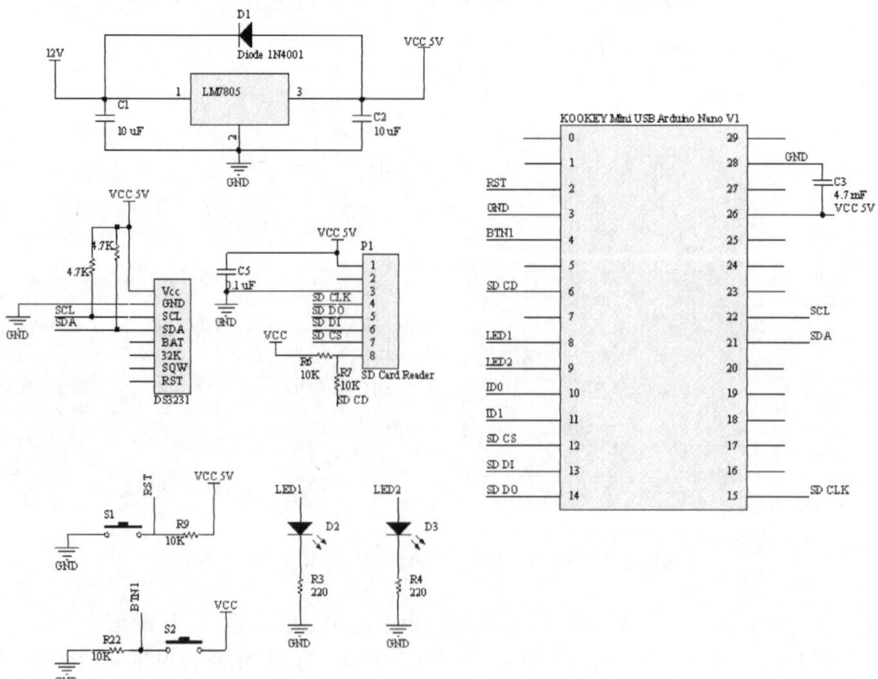

Figure 4. Schematic of the wall power data logging system.

The wall power data logger was designed to record a timestamp to the SD card on every boot, and at 5 minute intervals once connected to a power outlet. Power outage data may be gleaned from the timestamps. To protect the system and allow for uncorrupted data to be written to the microSD card, a 4.7 mF capacitor was added. This enabled the stalling of the loss of power, allowing for one last data line to be written to the microSD card. The date and time was determined using an RTC. The RTC was equipped with a rechargeable battery to retain the date and time data regardless of the availability of power. Figure 4 shows the schematic of the wall power data logging system.

2.2. Software design

2.2.1. S4 Data logging system. An Atmega328 microcontroller was utilized due to its versatility and readily available libraries enabling serial and SPI communication. The MAX485 transceiver chip was necessary for the microcontroller to communicate with the solar charge controller. Charge controllers such as the Renogy VS1024BN charge controller are equipped with RJ45 ports enabling communication with a computer or an external display. The Modbus RTU protocol is utilized via an RS-485 communication bus interface for communication. The MAX485 chip was selected to enable the RS-485 serial communication. Functions written with the aid of the Serial library in accordance with the Modbus protocol and the charge controller's register information. The microcontroller was programmed to acquire the following information from the charge controller every five minutes: date, time, solar panel voltage, solar panel current, battery voltage, battery current, load voltage, load current, battery state of charge, remote temperature sensor temperature, ambient temperature, and internal temperature. Figure 5 displays the code flowchart for the S4 data logging unit.

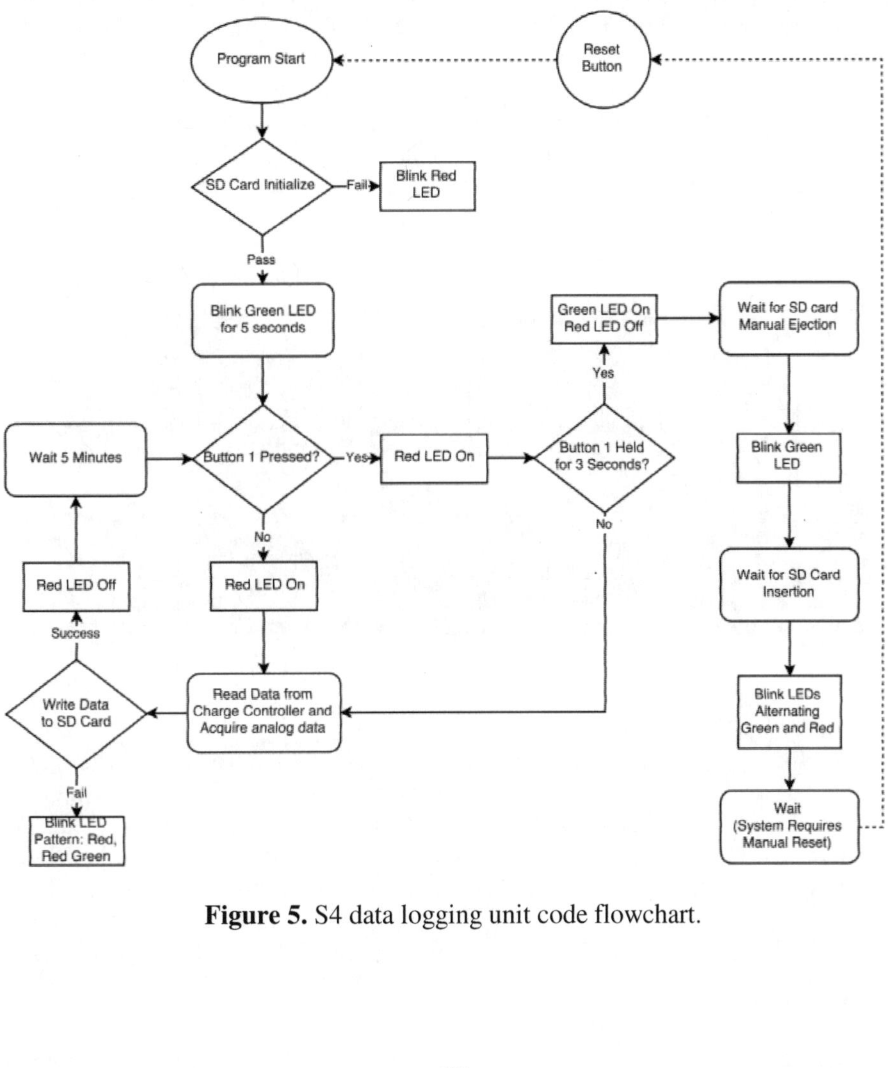

Figure 5. S4 data logging unit code flowchart.

2.2.2. Wall power data logging system. The wall power data logging system had a nearly identical software design to the S4 data logging system. However, instead of the microcontroller requesting data from the charge controller, the microcontroller requested data from the real time clock. For this functionality, the I^2C communication was enabled and functions were written to request and receive date and time data from the real time clock.

3. Implementation

Two data logging systems were built: the S4 data logger and the wall power data logger. The S4 data logger consisted of a digital board containing the microcontroller, Modbus communication, microSD card reader, the user interface and the communication systems (total cost was $30.15), and an analog board containing the voltage amplification circuitry for the current sensors (total cost of the board was ~$5). The cost of this PV data monitoring system is close to 10% of the cost of the commercially available systems. The wall power data logger consisted of one circuit board and an AC/DC adapter (total cost was $46.82). It included a microcontroller, SD card reader, real time clock (RTC) breakout board, a 4.7 mF capacitor across the power input and ground of the microcontroller, the user interface, and power was supplied via an AC/DC converter.

4. Integration and testing

4.1. S4 data logging system

To enable data transfer, an RJ45 cable was used to connect the data logger to the S4. The data written on the microSD card was able to be viewed on a computer to ensure consistency between the recorded data and the data displayed directly on the charge controller. To integrate the analog board, the small sensing resistors were soldered onto copper plates, which were inserted at the ground wire of each of four loads. Heat shrink was applied to insulate each current sensing resistor. Due to the variability in resistance in each differential amplifier circuit, calibration plots were acquired. Voltage was measured at each amplifier circuit output while varying the load current from 0 to 2 A in 0.1 A steps. Each had a slightly different gain but the relationship between the current and output voltage was linear as expected. Figure 6 shows the integrated data logging system.

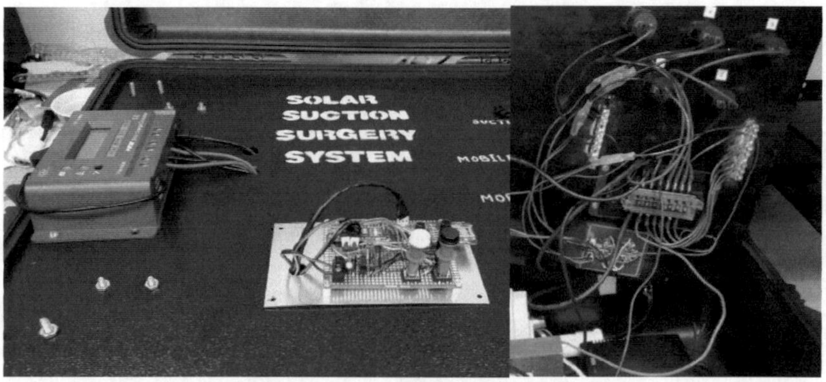

Figure 6. S4 data logging circuits.

4.2. Wall power data logging system

Correct date and time recordings as USB power was removed and reintroduced was ensured. The 4.7 mF capacitor was able to extend the power sufficiently. Once power was removed the microcontroller, microSD card board, and RTC had power for approximately one second. The time extension of one second was determined to be sufficient for one last recording to the microSD card. The data logger was installed in a location that was not supplied with the same power as the hospital. If diesel is available at the hospital, the hospital is able to be powered via a diesel generator. Thus, if the wall power data logger was to be plugged at the hospital, the data may not have been accurate.

5. Results

Since installation, data has been acquired from both data loggers. The errors in the data were minimal at 2.78%. The data was analysed using the MATLAB software. The monitored voltages and currents as a function of time for a typical day in October can be seen in figures 7(a) and 7(b), respectively. Figure 7(a) illustrates the voltages of the PV panel and the battery. PV panel generates voltage between 5:30 am and 5:30 pm when there is sunlight. The battery voltage increases in the morning when the battery is charging and stays constant throughout the rest of the day. The battery voltage decreases after sunset when it powers the loads, which have the same voltage as the battery. Figure 7(b) shows the currents of the PV panel and the loads. Consistent with the voltage case, the current of the PV panel increases in the morning when the battery is charged and starts to decrease when the battery voltage reaches the maximum voltage. The load current is sourced by the PV panel during the daytime and by the battery after sunset.

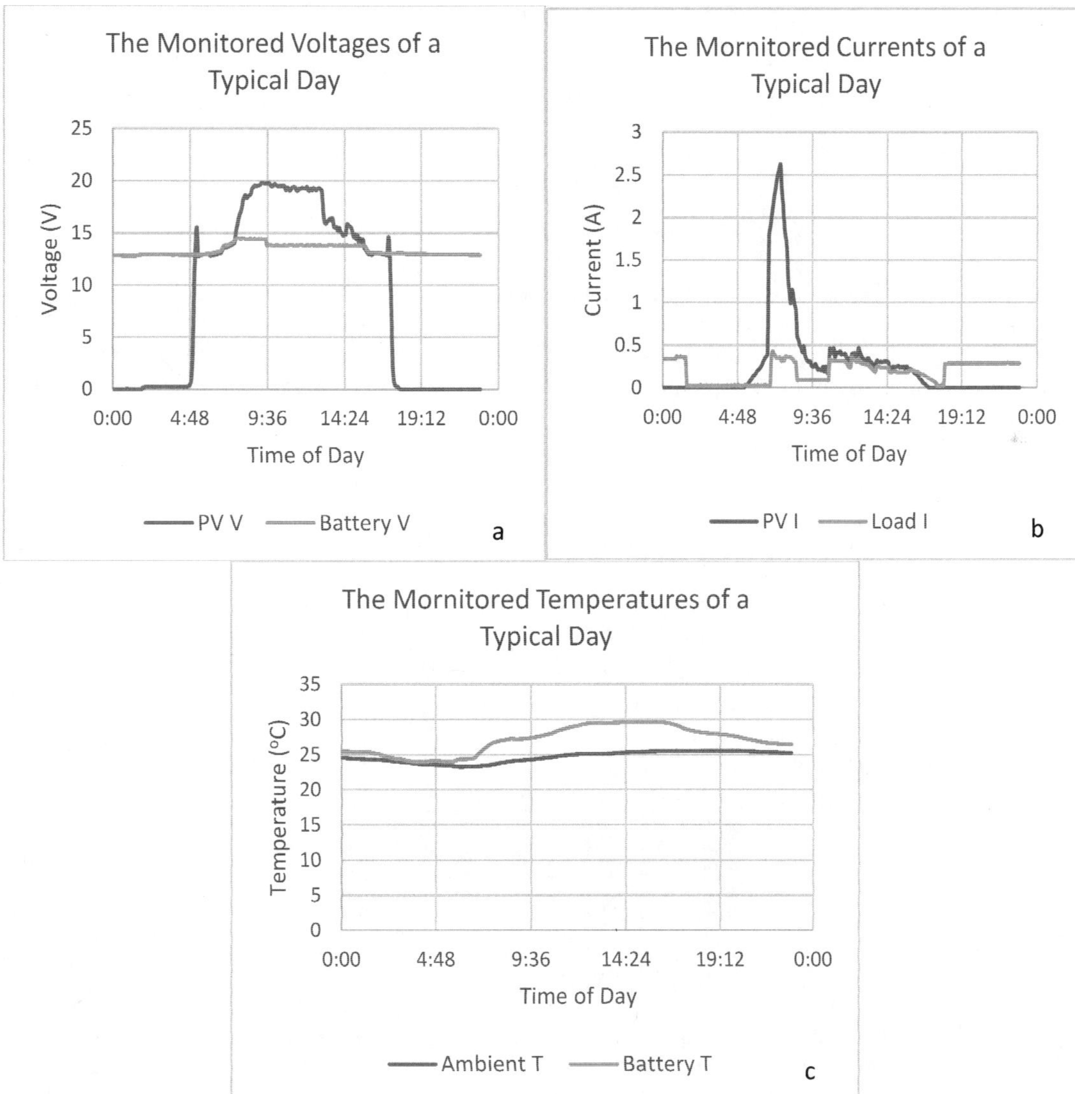

Figure 7. Monitored voltages (a), currents (b), and temperatures (c) for a typical day.

The monitored temperatures of the battery and ambient for a typical day in October are depicted in figure 7(c). It can be seen that the battery temperature is near the ambient temperature when the battery is not charging nor discharging. Its temperature is greatly increased during both charging and discharging periods.

To determine the usage data, usage was defined as a particular user port being used for more than 15 consecutive minutes at a time, and the current of each port was calculated accordingly. A least-squares linear regression analysis was used to determine the voltage to current conversion coefficients that were used for the determination of the load current of each of the five ports. Figure 8 shows histograms of the average per day use of each of the four monitored S4 ports and the unit as a whole.

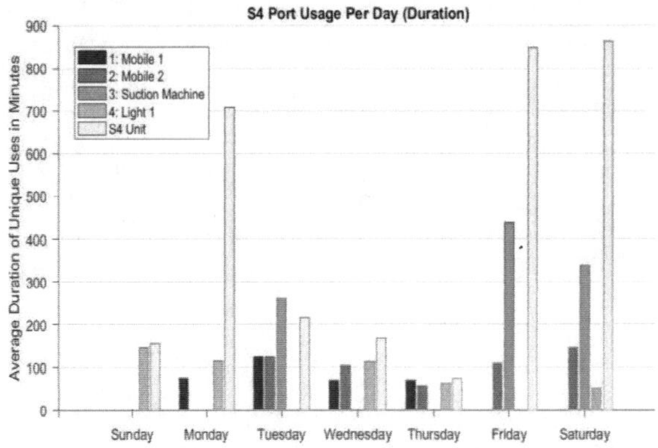

Figure 8. Average S4 port and S4 unit usage throughout a week.

Figure 8 shows that the S4 is used on Saturdays for longer periods of time than the rest of the week. In addition, out of the monitored ports, the port for charging the suction machine is used most. Overall the usage of the S4 seems to be low, which is expected for the first month of installation as the hospital staff learn to use the equipment.

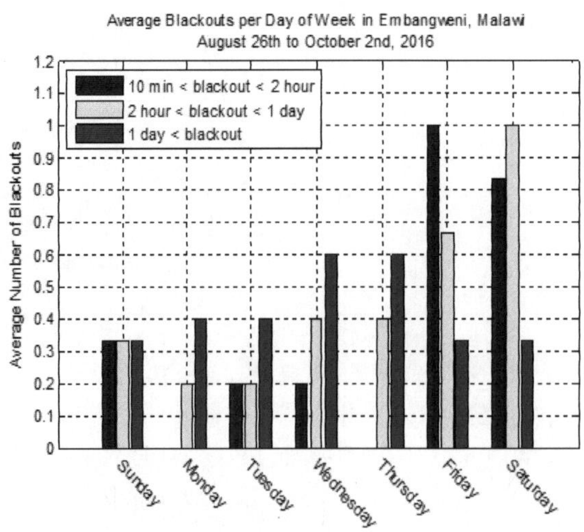

Figure 9. Average number of blackouts during the week lasting more than 10 minutes, 2 hours, and 1 day.

One and a half months of data was acquired from the wall power data logger. After analysing these data using the MATLAB program, a histogram was generated as seen in figure 9, which shows the average number of blackouts lasting between ten minutes and two hours, between two hours and a day, and more than a day. It can be seen from figure 8 that there were more frequent short blackouts on Fridays and Saturdays while Blackouts on Wednesday's and Thursday's last more than a day. The S4 is used for long periods of time on Fridays and Saturdays which could be due to the need to recharge devices that were used the previous days when there was no grid power available.

No concrete trends can be established with a month's worth of data. However, it is evident that the acquisition of data using the designed data loggers was successful, and the data loggers have been able to function under real world conditions. The data files are able to be sent via email with ease due to the widespread use of smartphones in Malawi. Additionally, the data are able to show the voltage and current changes of the panels and battery throughout the day, usage data of the devices, and power outages of the region. Thus, with additional data analysed, trends may be able to be observed.

6. Conclusions

The S4 data acquisition system and the wall power data acquisition system were effective in monitoring the status of PV systems. The two data logging systems provided a means of recording PV system parameters, S4 usage data, as well as power outage data that was low-cost and required little addition of auxiliary sensors. Atmega328 microcontrollers allow for a cost effective and versatile method for acquiring and recording data. The use of SD cards enables low cost data acquisition of remote systems where availability of computers and installation of servers is not feasible. With increasing usage of the S4 and larger data sets acquired from the data loggers, it may be possible to observe trends. Usage statistics will help improve future system development and possibly establish correlations between the usage of the S4 and regional blackouts in Malawi. Thus far, a baseline can be established for the performance of the system. Once future data sets are analysed, any decrease in overall performance can be observed. Utilizing Atmega328 microcontrollers and readily available breakout boards (e.g. real time clock board, microSD card board) can lower the cost of data acquisition systems and allow for increased customization.

References

[1] Woyte A, Richter M, Moser D, Mau S, Reich N and Jahn U 2013 Monitoring of photovoltaic systems: Good practices and systematic analysis *Proc. 28th European Photovoltaic Solar Energy Conference* (Villepinte) 3686-94

[2] Biswas L, Rahman H and Haque S 2012 A low cost characteristics analyzer of PV module *2nd International Conference on the Developments in Renewable Energy Technology (ICDRET)* (Dhaka) 1-5

[3] Mahjoubi A, Mechlouch R F and Brahim A B 2011 A low cost wireless data acquisition system for a remote photovoltaic (PV) water pumping system *Energies* **4** 68-89

[4] Jucá S C, Carvalho P C and Brito F T 2011 A low cost concept for data acquisition systems applied to decentralized renewable energy plants *Sensors* **11** 743-56

[5] Mukaro R and Carelse X 1999 A microcontroller-based data acquisition system for solar radiation and environmental monitoring *IEEE Trans Instrumen Measure* **48** 1232-8

[6] Akposionu K N and Nwokoye A O C 2012 Design and fabrication of a low-cost data logger for solar energy parameters *J Energy Technol Policy* **2** 12

[7] Fuentes M, Vivar M, Burgos J, Aguilera J and Vacas J 2014 Design of an accurate, low-cost autonomous data logger for PV system monitoring using Arduino™ that complies with IEC standards *Solar Energy Mater Solar Cells* **130** 529-43

[8] Koutroulis E and Kalaitzakis K 2003 Development of an integrated data-acquisition system for renewable energy sources systems monitoring *Renew Energy* **28** 139-52

[9] Purwadi A, Haroen Y, Ali F, Heryana N, Nurafiat D and Assegaf A 2011 Prototype development of a low cost data logger for PV based LED street lighting system *International Conference on Electrical Engineering and Informatics* (Bandung) 1244-8

[10] Kalaitzakis K, Koutroulis E and Vlachos V 2003 Development of a data acquisition system for

IOP Conf. Series: Earth and Environmental Science **93** (2017) 012048 doi:10.1088/1755-1315/93/1/012048

remote monitoring of renewable energy systems *Measurement* **34** 75-83

[11] Chiandone M, Cleva S, Pavan A and Sulligoi G 2009 Monitoring application of electrical and climate data for PV systems using Linux RTAI *International Conference on Clean Electrical Power* (Capri) 259-62

[12] Chowdhury S, Day P, Taylor G A, Chowdhury S P, Markvart T and Song Y H 2008 Supervisory data acquisition and performance analysis of a PV Array installation with data logger *Power and Energy Society General Meeting - Conversion and Delivery of Electrical Energy in the 21st Century* (Pittsburgh) 1-8

[13] Mahzan N N, Omar A M, Mohammad S Z and Mohd M Z 2013 Design of data logger with multiple SD cards *IEEE Conference on Clean Energy and Technology* (Langkawi) 175-80

IOP Conf. Series: Earth and Environmental Science **93** (2017) 012049 doi:10.1088/1755-1315/93/1/012049

Numerical investigation of the thermal and electrical performances for combined solar photovoltaic/thermal (PV/T) modules based on internally extruded fin flow channel

Y C Deng[1,2], Q P Li[1] and G J Wang[1]

[1]Beijing Institute of Residential Building Design & Research CO., LTD, Beijing, China

E-mail: dyc613@163.com

Abstract. A solar photovoltaic/thermal (PV/T) module based on internally extruded fin flow channel was investigated numerically in this paper. First of all, the structures of the thin plate heat exchanger and the PV/T module were presented. Then, a numerical model of the PV/T module considering solar irradiation, fluid flow and heat transfer was developed to analyze the performance of the module. Finally, the steady electrical and thermal efficiencies of the PV/T module at different inlet water temperatures and mass flow rates were achieved. These numerical results supply theory basis for practical application of the PV/T module.

1. Introduction

Energy crisis and environmental pollution have attracted broad attention. Therefore, adjustment the energy composition and development renewable energy have gradually become an important part of human sustainable development. In all kinds of renewable energy resources, solar energy has the advantages of enormous energy, longevity, cleanness, and cheapness. And in various application techniques of solar energy, the solar photovoltaic generation has been considered to be one of the best among the known electricity generation owing to the highest electricity energy levels.

The important technologies for the application of photovoltaic (PV) cell include the following two points: First, gain the high efficient and low-cost photovoltaic materials. Second, provide using conditions for the PV cell to reach optimal conversion efficiency. The former belongs to the PV materials technology, while the latter could be realized by high efficient heat transfer technology. Several researchers conducted material feature studies to improve the electrical performance of the PV cell. Ghanwat, *et al* [1,2] successfully obtained hierarchical TiO_2 nanostructures and synthesized SnS_2 thin film to improve the PV electrical efficiency. However, several researchers conducted PV heat dissipation and waste heat recovery, which is solar PV–thermal (PV/T) technology, to make the PV work at the optimal conditions. In the commercial applications, the electricity efficiency of a monocrystalline silicon solar PV cell is about 6～15%, whereas the remaining radiant energy of the sun is absorbed by the PV cell and converted into heat, which makes the PV electrical efficiency decrease with the increasing PV temperature. Moreover, the service life of a PV cell also decreases at elevated temperature. If the PV could had been cooled in time, the waste heat could had been collected and utilized, which might have improved the electrical efficiency, might have greatly increased the integrated use efficiency of the solar energy, and might have extended the lifespan of the PV cell.

Nowadays, the main PV/T technology is to make circulating water flow the copper or aluminium

IOP Conf. Series: Earth and Environmental Science **93** (2017) 012049 doi:10.1088/1755-1315/93/1/012049

tubes welded on the PV panels to co-generate heat and electricity. Chow [3] analyzed the performance of a single glazed flat plate PV/T collector by an explicit dynamic model. Touafek, *et al* [4] studied a galvanized steel tube-sheet PV/T module. Moreover, several scholars conducted studies for improving the performance of the tube-and-sheet PV/T system. Zondag, *et al* [5] analyzed nine different designs of the PV/T modules. The results showed that the channel-below-transparent-PV design had the best efficiency, but the PV-on-sheet-and-tube design was considered to be a better alternative owing to the higher annual efficiency and easier manufacture. Tripanagnostopoulos [6] and Abu Bakar, *et al* [7] studied the performance of the different forms of the PV/T collectors with water or air circulation. Liang, *et al* [8] tested the efficiency of the PV/T collector filled with graphite. The experimental results showed that the mean electrical efficiencies of the PV/T collector filled with graphite and the traditional PV module were 6.46% and 5.15%, respectively. Ibrahim, *et al* [9,10] analyzed the performance of the spiral flow tube sheet PV/T system. It produced a thermal efficiency of 54.6%, a PV electrical efficiency of 13.8%, and a PV/T total efficiency of 68.4%. To further increase the performance, numerous new types of PV/T modules have been developed. Chow, *et al* [11-13] proposed a flat-box metallic absorber to replace the tube-and-sheet design to get higher efficiency and durability. Pei, *et al* [14,15] studied the performance of the heat-pipe PV/T system under different operating and structural conditions by experiment and simulation. Moradgholi, *et al* [16] investigated the performance of the thermosiphon-type heat-pipe PV/T system when the heat pipes were filled with different working fluids. Deng, *et al* [17] presented a novel PV/T module based on the micro heat pipe array. The PV/T technology offers the benefits of high efficiency, high reliability, light weight, and frost-resistance during winter. Ziapour and Khalili [18] performed the performance study to verify the new design of the PV panel combined with the type of the wickless heat pipe solar water heater. In this enhanced design, each wickless heat pipe is assumed in the shape of the loop mini tube comprised of the flow boiling process in it.

The current study presents a PV/T technology based on the thin plate heat exchanger with internally extruded fin flow channel [19]. The simulations of the fluid flow and heat transfer were conducted on the PV/T module to achieve the electrical and thermal performance. The simulation results were analyzed to supply the basis for the practical application of the PV/T module.

2. PV/T module

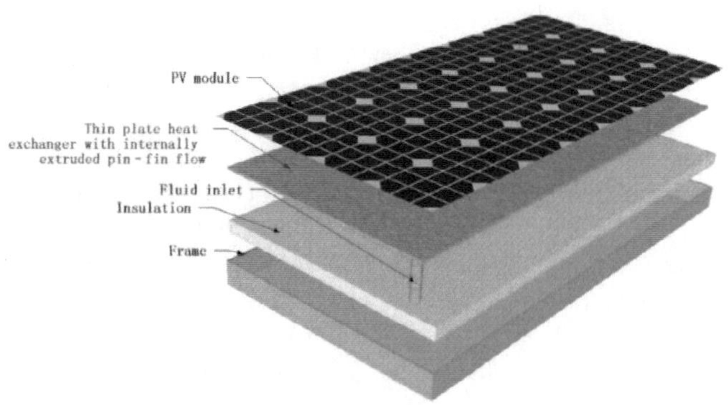

Figure 1. Structure schematic of the PV/T module.

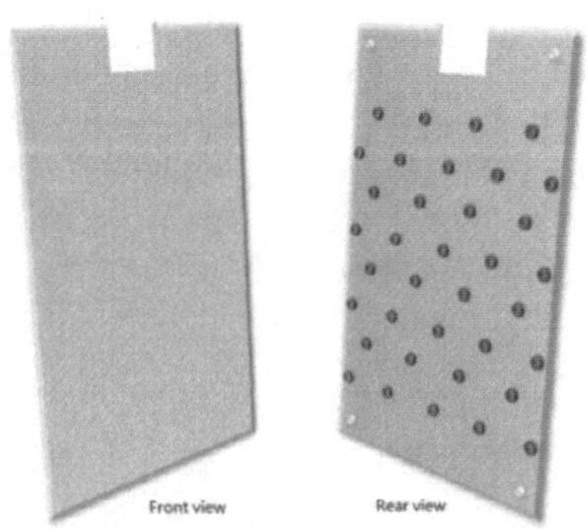

Figure 2. Structure schematic of the thin plate heat exchanger.

The basic configuration of the PV/T module with internally extruded fin flow channel is shown in figure 1. It combines PV module, thin plate heat exchanger, insulation layer, and frame. When the PV/T module is in operation, the thin plate heat exchanger transfers the heat absorbed by the PV module to the circulating water. The thin plate heat exchanger is shown in figure 2. It constructed by two parallel metal thin plates, one of which forms arrays of pin-fin corrugations through machinery mould extrusion, while another thin sheet remains smooth and is fixed with the PV panel by the metal retaining clip. Two thin plates join together by laser-welding technology, and forming a water flow channel. One fluid inlet at the bottom and one fluid outlet on the top of the thin plate heat exchanger respectively connects with the water pipes. Therefore, the thin plate heat exchanger has a better heat transfer property owing to the absorber with fins and the cross flow over the pin fins.

3. Numerical simulation of the PV/T module

The heat transfer mechanisms of the PV/T are shown in figure 3. When a light ray solar cell surface, some of the solar radiation energy converts into electricity, some is reflected by the PV surface, and some converts into heat. Part of the heat is transferred to the water inside the thin plate heat exchanger, and part of the heat is transferred to the surroundings through convection and radiation. The simulation of the PV/T with water flow was set up, which included the CFD modelling of the solar irradiation, the circulating water flow and the heat transfer inside the heat exchanger, the conduction of the insulation layer, and the heat exchange with the surroundings. Figure 4 shows the model geometry of the PV/T module, while the specifications of the different components of the PV/T studied are summarized in table 1.

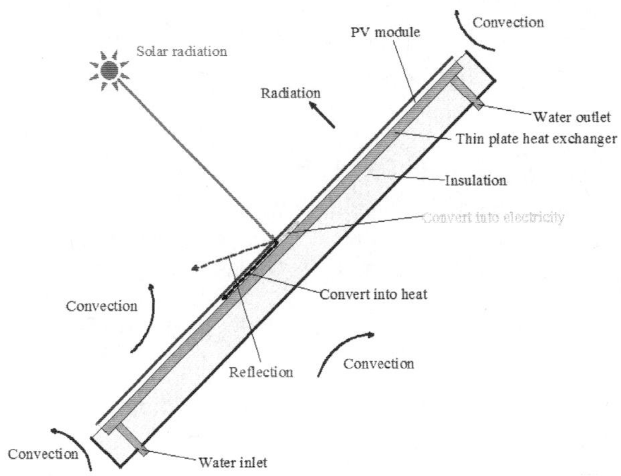

Figure 3. Schematic diagram of the heat transfer in PV/T module.

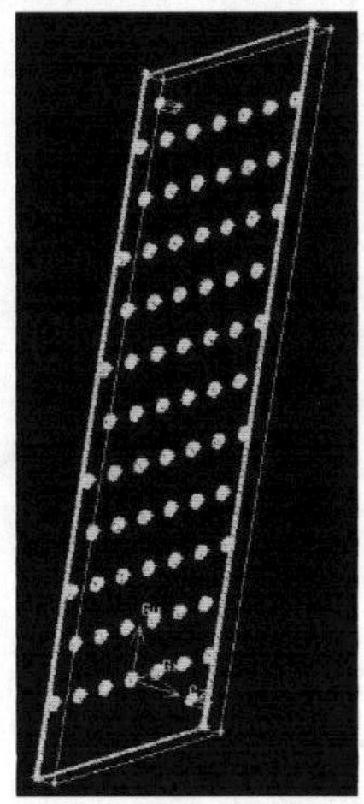

Figure 4. Model geometry.

Table 1. Basic characteristics and specifications of the PV/T module.

Components		Dimensions: 1580×808×50 mm
PV module	Glass cover	Material: low iron tempered glass Dimensions: 1580×808×3 mm
	Monocrystallinesilicon solar cell	Solar cell temperature coefficient:0.0045 K^{-1} Thickness: 1 mm Absorbability of solar energy: 0.95 Electrical efficiency in standard conditions: 15%
	TPT	Dimensions: 1580×808×3 mm Thickness: 3 mm
	EVA	Thickness: 0.5 mm
Thin plate heat exchanger		Type: With internally extruded pin–fin flow channel

	Material: Stainless steel Thickness: 5 mm Inlet/Outlet pipe diameter: 20 mm
Insulation layer	Material: polyurethane Thickness: 40 mm
Framing	Material: aluminium

3.1. Numerical model
The conservation of mass equations, momentum equations and energy equations are established, and the standard k-epsilon two-equation turbulent model is employed. The thermo physical properties of different components of the PV/T module are shown in table 2.

Table 2. List of thermo physical properties of different components of PV/T module.

Physical property parameters	Glass	EVA	PV	TPT	Stainless steel	Insulation	Aluminium
Density (kg/m³)	2500	900	1650	1475	7930	30	2719
Heat capacity (J/kg·K)	840	2300	700	1130	500	1380	871
Heat conductivity coefficient (W/m·K)	0.76	0.5	150	0.614	16.3	0.04	202.4

3.2. Boundary conditions
The equations are subjected to several boundary conditions as follows:

- The glass cover is convective thermal boundary condition, which allows considering the effect of the convective flow outside the PV/T and the heat radiation between glass cover and sun. This is done by calculating the convective heat transfer coefficient dependent on the outside velocity as follows:

$$h=5.7+3.8\upsilon \quad (1)$$

Where h is the convection heat transfer coefficient, $W/m^2 \cdot K$; υ is the ambient wind speed, m/s.

- For the power generation of the solar cell, the following empirical formula is used.

$$E=G\eta_N[1-\varphi_a(T_b-25)] \quad (2)$$

Where G is the solar irradiance, W/m^2.η_N is the electrical efficiency in standard conditions, namely the PV cell electrical efficiency under the condition of solar irradiance of 1000 W/m^2 and ambient temperature of 25°C, %. φ_a is the PV cell temperature coefficient, K^{-1}. T_b is the PV cell temperature, °C.

Except converting into electricity, all the solar irradiation that is absorbed presents in the PV cell as internal heat sources in the simulation process. The internal heat sources are determined by the way of iteration.

- The boundary condition for the water inlet is "mass flow rate" and the outlet is "outflow".
- For the side walls and bottom walls, the boundary conditions are considered to be convective boundary condition. These walls don't participate in solar radiation and heat radiation.
- All other surfaces are ''wall'', and the walls between the two zones are defined as "coupled".

3.3. Mesh formation
The grid is generated by CFD pre-processor. Through independent verification for the grid, a grid with 694406 cells is used for the current work.

3.4. Numerical method
A commercial computational fluid dynamics program is used to solve the coupled equations. The

solver used is the segregated solver, and the SIMPLE algorithm is used for calculating the pressure-velocity coupling. The second-order is selected as the discretization method for pressure, and the second order-upwind discretization method is used for the solving of momentum, energy, and turbulence equations.

3.5. Data processing

The thermal and electrical performances of the PV/T module are described by the thermal efficiency η_{th} and the electrical efficiency η_e. Based on the simulation results, the calculation formulas are as follows.

$$\eta_e = \frac{E}{A_c G} = \frac{G\eta_N[1-\varphi_a(T_b-25)]}{A_c G} \tag{3}$$

$$\eta_{th} = \frac{Q_u}{A_c G} \tag{4}$$

The performance of the PV/T module is described as the total efficiency η_o, which sums the thermal and electrical efficiencies. The calculation formula is as follow.

$$\eta_o = \eta_e + \eta_{th} \tag{5}$$

Where E is the generated power of the PV/T module (W); A_c is the aperture area (m^2); G is the solar irradiance (W/m^2); Q_u is the useful heat transferred to the circulating water.

4. Results and discussion

The steady state fluid flow and heat transfer in the PV/T module at different water temperatures (20, 30, 40°C) were simulated with the CFD models. The PV/T module is placed at a tilt angle of 45°. The ambient air temperature, wind speed, solar irradiance and mass flow rate are 20°C, 2 m/s, 800 W/m^2 and 0.02 kg/(m^2·s), respectively. The results of simulation on the heat transfer were obtained in detail, and the efficiencies could be calculated at different inlet cooling water temperatures. The simulation results are shown in figures 5 and 6.

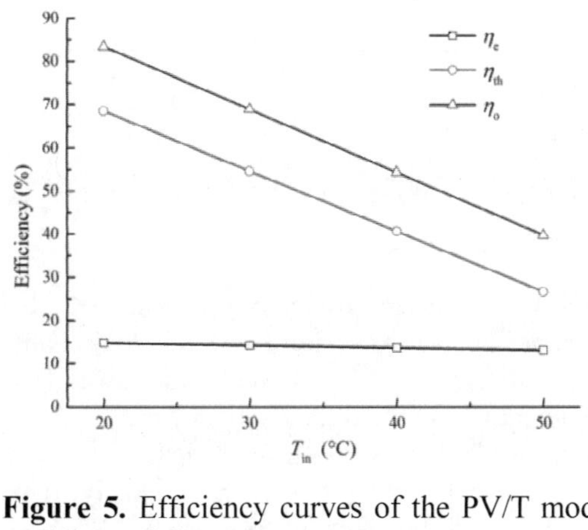

Figure 5. Efficiency curves of the PV/T module at different inlet water temperatures.

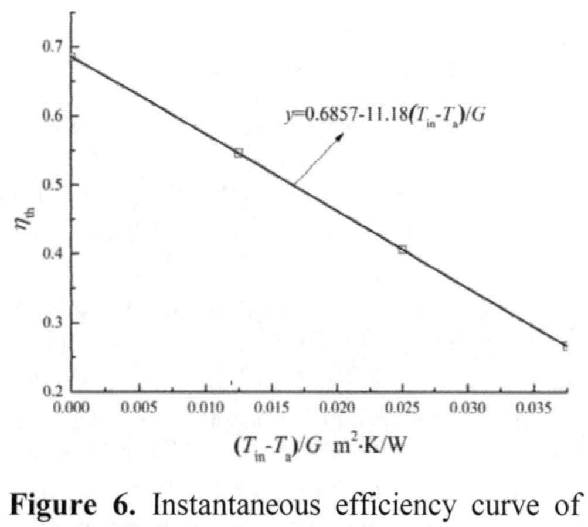

Figure 6. Instantaneous efficiency curve of the PV/T module.

Figure 5 presents the electrical efficiency, thermal efficiency and total efficiency curves of the PV/T module at different inlet water temperatures. As shown in the figure, the efficiencies decrease with the increasing inlet water temperatures. When the inlet water temperature is 20°C, the electrical efficiency, thermal efficiency, and total efficiency are 14.86%, 68.54%, and 83.40%, respectively. When the inlet water temperature is 50°C, the electrical efficiency, thermal efficiency and total efficiency are 13.12%, 26.63%, and 39.75%, respectively. The electrical efficiency and thermal efficiency fall by 11.71% and 61.15%, respectively. As the inlet water temperatures increase, the operating temperatures of the PV/T module increase, which makes the heat loss of the PV/T through the top surface increase and the useful heat reduces. Therefore, the thermal efficiency of the PV/T will reduce with the increasing inlet water temperatures. Similarly, the electrical efficiency will reduce with the increasing PV temperatures as shown formula (2).

By using a curve fitting technique, the numerical instantaneous thermal efficiency of the PV/T module is obtained in Figure 6. It is represented as a linear relation of the parameter $[(T_{in}-T_a)/G]$. The intercept of the efficiency line with the y-axis represents the maximum instantaneous thermal efficiency, while the slope represents the heat loss coefficient. The numerical maximum instantaneous efficiency is 68.57%, and the heat loss coefficient is 11.18 W/(m^2·K). The thermal efficiency decreases rapidly with the increasing parameters $[(T_{in}-T_a)/G]$. The heat loss coefficient of the PV/T module is larger owing to the more heat loss between the glass cover and the surroundings. To decrease the heat loss, some studies had added the air gap between the glass cover and the PV cell, which could improve the thermal efficiency but impede the cooling of PV cell.

The steady state fluid flow and heat transfer in PV/T at different water mass flow rates (0.02, 0.05, 0.10, 0.15, 0.20 kg/(m^2·s))were simulated with the CFD models. The PV/T is placed at a tilt angle of 45°. The direct solar irradiance has an intensity of 800 W/m^2 perpendicular to the module. The ambient air temperature, wind speed and inlet water temperature are 20℃, 2 m/s and 30°C, respectively. The simulation results are shown in figures 7 and 8.

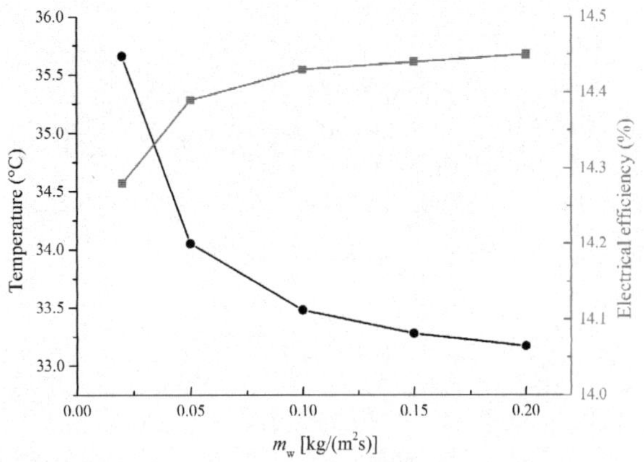

Figure 7. PV temperatures and PV/T electrical efficiencies at different water mass flow rates.

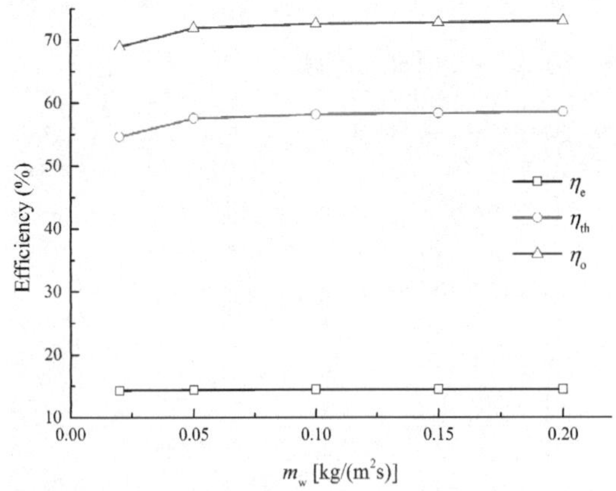

Figure 8. Efficiencies of PV/T module at different water mass flow rates.

The figures show that the PV temperature decreases and the electrical efficiency increases with the increasing water mass flow rates. The PV temperatures at the mass flow rate of 0.02 kg/(m^2·s) and 0.20 kg/(m^2·s) are 35.7°C and 33.2°C, respectively, with a difference of 7.5%. When the mass flow rate is 0.02 kg/(m^2·s), the electrical efficiency, thermal efficiency, and the total efficiency are 14.28%, 54.64%, and 68.92%, respectively. When the mass flow rate is 0.20 kg/(m^2·s), the electrical efficiency, thermal efficiency, and the total efficiency are 14.45%, 58.55%, and 73.00%, respectively. The efficiencies increase with the increasing mass flow rates. The electrical efficiency increases 1.2%, the thermal efficiency increases 7.2%, and the total efficiency increases 5.9%. The increases of efficiency are limited. Therefore, the appropriate mass flow rate should not be too large, lest improve the pressure loss. The optimum mass flow rate is about 0.10 kg/(m^2·s).

5. Conclusions

In order to increase the electrical and thermal efficiencies of the PV/T module, a solar photovoltaic

/thermal (PV/T) module with internally extruded fin flow channel was present in the study. Then, a numerical model of the PV/T module was developed to analyze the performance of the module. The numerical results show that the numerical maximum instantaneous efficiency is 68.57%, and the heat loss coefficient is 11.18 W/(m^2·K). The electrical efficiency, thermal efficiency and total efficiency of the PV/T module decrease with the increasing inlet water temperatures, and increase with increasing mass flow rates. The optimum mass flow rate is about 0.10 kg/(m^2·s).

Acknowledgements
The project was financially supported by Beijing Natural Science Foundation (3172041). The authors are grateful for the support of these sponsors.

References

[1] Khot K V, Ghanwat V B, Bagade C S, *et al* 2016 Synthesis of SnS_2 thin film via non vacuum arrested precipitation technique for solar cell application *Mater Lett* **180** 23-6
[2] Desai N D, Mali S S, Mane R M, *et al* 2016 Influence of deposition temperature on the optical, structural, morphological, compositional and photoelectrochemical properties of TiO_2 thin films *J Mater Sci: Mater Electron* **27** 11739-50
[3] Chow T T 2003 Performance analysis of photovoltaic-thermal collector by explicit dynamic model *Sol Energy* **75** 143-52
[4] Touafek K, Khelifa A and Adouane M 2014 Theoretical and experimental study of sheet and tubes hybrid PV/T collector *Energy Convers Manage* **80** 71-7
[5] Zondag H A, de Vries D W, van Helden W G J, *et al* 2003 The yield of different combined PV-thermal collector designs *Sol Energy* **74** 253-69
[6] Tripanagnostopoulos Y 2007 Aspects and improvements of hybrid photovoltaic/thermal solar energy systems *Sol Energy* **81** 1117-31
[7] Mohd Nazari A B, Mahmod O, MahadzirHj D, *et al* 2014 Design concept and mathematical model of a bi-fluid photovoltaic/thermal (PV/T) solar collector *Renew Energy* **67**(SI) 153-64
[8] Liang R B, Zhang J L, Ma L D, *et al* 2015 Performance evaluation of new type hybrid photovoltaic/thermal solar collector by experimental study *Appl Therm Eng* **75** (SI) 487-92
[9] Adnan I, Ahmad F, Kamaruzzaman S, *et al* 2014 Efficiencies and improvement potential of building integrated photovoltaic thermal (BIPV/T) system *Energy Convers Manage* **77** 527-34
[10] Ahmad F, Kamaruzzaman S, Mohammad H Y, *et al* 2014 Performance analysis of photovoltaic thermal (PV/T) water collectors *Energy Convers Manage* **78** 641-51
[11] Chow T T, He W and Ji J 2006 Hybrid photovoltaic-thermosyphon water heating system for residential application *Sol Energy* **80** 298-306
[12] He W, Chow T T, Ji J, *et al* 2006 Hybrid photovoltaic and thermal solar-collector designed for natural circulation of water *Appl Energy* **83** 199-210
[13] Chow T T, Pei G, Fong K F, *et al* 2009 Energy and energy analysis of photovoltaic–thermal collector with and without glass cover *Appl Energy* **86** 310-6
[14] Pei G, Fu H D, Zhang T, *et al* 2011 A numerical and experimental study on a heat pipe PV/T system *Sol Energy* **85** 911-21
[15] Pei G, Fu H D, Zhu H J, *et al* 2012 Performance study and parametric analysis of a novel heat pipe PV/T system *Energy* **37** 384-95
[16] Moradgholi M, Nowee S M and Abrishamchi I 2014 Application of heat pipe in an experimental investigation on a novel photovoltaic/thermal (PV/T) system *Sol Energy* **107** 82-8
[17] Deng Y C, Quan Z H, Zhao Y H, *et al* 2015 Experimental research on the performance of household-type photovoltaic–thermal system based on micro-heat-pipe array in Beijing *Energy Convers Manage* **106** 1039-47
[18] Ziapour B M and Khalili M B 2016 PVT type of the two-phase loop mini tube thermosyphon

solar water heater *Energy Convers Manage* **129** 54-61
[19] Zhang X X, Xu Y and Xu J H 2013 Solar photoelectric photo-thermal integrated assembly and power generation system (China Inventive Patent: CN201310582985.9)

IOP Conf. Series: Earth and Environmental Science **93** (2017) 012050 doi:10.1088/1755-1315/93/1/012050

Simulation of hybrid propulsion system using LSRG and single cylinder engine

C Han[1,2], K Ohyama[1,3] and W Q Wang[2,3]

[1]3-30-1 Wajirohigashi, Higahsi-ku, Fukuoka, Japan
[2]Nanjing University of Science and Technology, China

E-mail: ohyama@fit.ac.jp/camp2134@126.com

Abstract. Nowadays, more and more people are beginning to use hybrid vehicles (HVs). The drive system of HVs needs to produce the electric energy with the electric generator and gearbox powered by an engine. Therefore, the structure becomes complex and the cost is high. To solve this issue, this research proposes a new drive system design that combines the engine and a linear switched reluctance generator (LSRG). When the engine is operating, the LSRG can simultaneously assist the engine's mechanical output or can generate power to charge the battery. In this research, three research steps are executed. In the first step, the LSRG is designed according to the size of normal engine. Then, finite element analysis is used to get the data of flux linkage and calculate the inductance and translator force. Finally, Simulink models of control system are constructed to verify the performance of LSRG.

1. Introduction

Nowadays, global warming and air pollution have been caused in part by the rapid development of vehicles. In order to reduce the negative effects of using vehicles, more and more researchers are working on designing clean energy vehicles. Some researches focus on the natural gas consumption in vehicle. However, there are problems about the climate implications of increasing natural gas in vehicles and methane leakage from natural gas infrastructure [1]. Another strategy is the electric vehicles (EVs), compared with fuel energy resources, the EVs are environment-friendly, however, they are underpowered due to the limitation of batteries. Therefore, EVs are mainly used for small vehicles and short distance purposes [2]. The HVs have been the subject of intense scholarly debate [3].

However, the drive system of HVs needs to produce the electric energy with the use of an electric generator and gearbox powered by an engine. Therefore, the structure becomes complex and the cost is high. Normal HVs have the engine, generator, motor, and so on as shown in figure 1 [4].

According to the compositions, if a generator or motor can be a part of the engine, it will accomplish the purpose of simplifying the structure. That means the generator can work together with the engine.

Switched Reluctance Generators (SRGs) use the reluctance torque that is produced by the change of magnetic resistance in magnetic circuits. Figure 2 is the basic structure of the SRG.

In the SRG, coils are only installed in the stators, coils and permanent magnet are not used in the rotor. The stator and rotor are made from grain oriented silicon steel plates. Based on above, the SRG has the simple structure and low cost. Even in high-temperature environment, there is no issue with demagnetization [5]. The rotor of normal SRG rotates about a fixed axis as shown in figure 2.

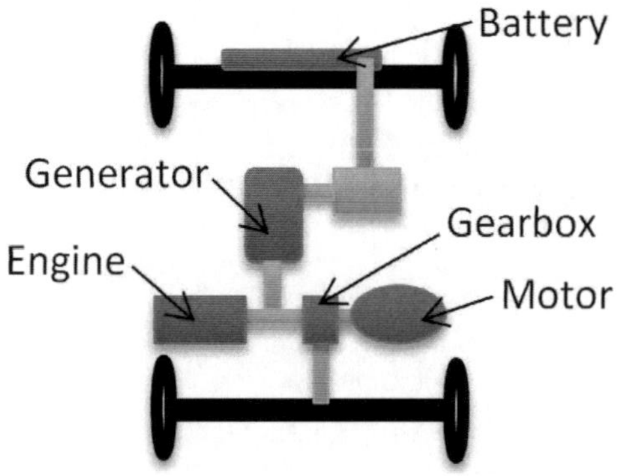

Figure 1. Hybrid vehicle compositions.

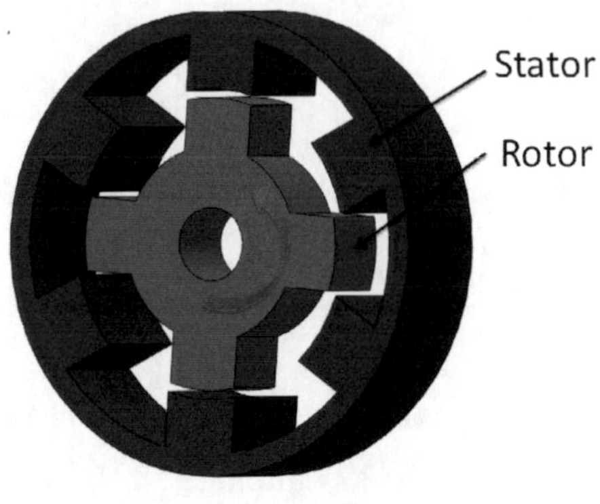

Figure 2. SRG structure.

A Linear Switched Reluctance Generator (LSRG) is deformed from the SRG as shown in figure 3. The processing is cutting and unrolling the basic SRG to obtain a LSRG. The rotor can be treated as a translator, and the translator moves in a linear [6]. The LSRG will be an effective candidate when high-speed and high-precision are necessary conditions, since the LSRG has a quicker response and more sensitivity compared with the normal SRG.

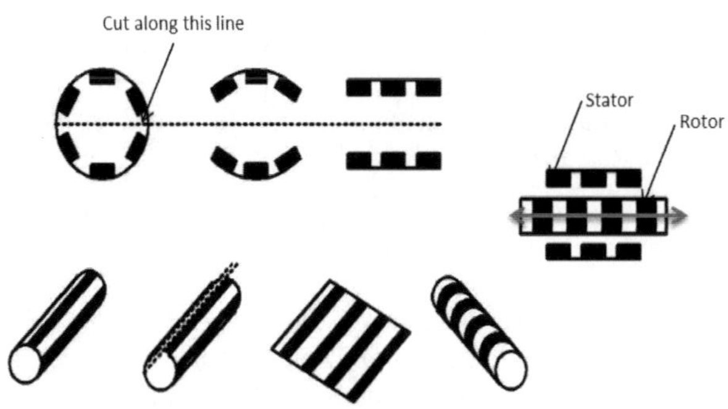

Figure 3. Cutting and unrolling processing of LSRG.

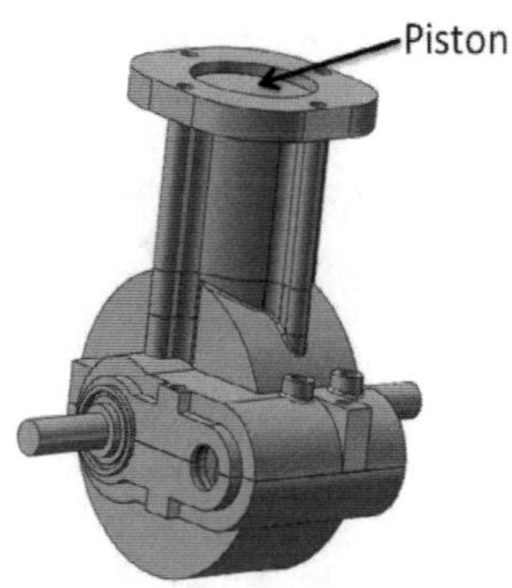

Figure 4. Single cylinder engine for combining with LSRG.

On the other hand, the LSRG is difficult to control, and maybe it has a high force ripple because of its nonlinear magnetic circuit. Also the LSRG obeys the same principle with that of SRGs. In other words, the magnetic flux must always pass through the minimum path with which the magnetic resistance becomes the minimum one.

A Single Cylinder Engine (SCE) represents the basic structure of complex engines. The SCE has a simple structure, and it is shown in figure 4. This SCE does not include an intake system since it is used only for experimental testing.

This paper presents the new structure to combine the LSRG with the SCE into a hybrid propulsion system. When the SCE is operated, the LSRG can simultaneously assist the SCE's mechanical output or can generate power charge a battery. Here, the LSRG is designed by using finite element method (FEM) magnetic field analysis software JMAG. Moreover, the controlling parts are tested by using MATLAB Simulink.

2. Design of new LSRG structure

According to the basic LSRG, the new structure is designed. One phase of LSRG is chosen as shown in figure 5. In order to combine with the SCE, the shape of stator is designed to adapt to the SCE. The rotor, which is the translator of LSRG, is linked with the piston of SCE through a non-magnetic shaft.

The new structure of LSRG is shown in figure 6. It is comprised of the translator and a 4-pole stator. The stator of LSRG is exactly the same as the stator of SRG. The translator is treated as the rotor of SRG [7]. In order to avoid to damage the coil of LSRG by the lubricating oil of SCE, the coil is fixed on the outside of SCE.

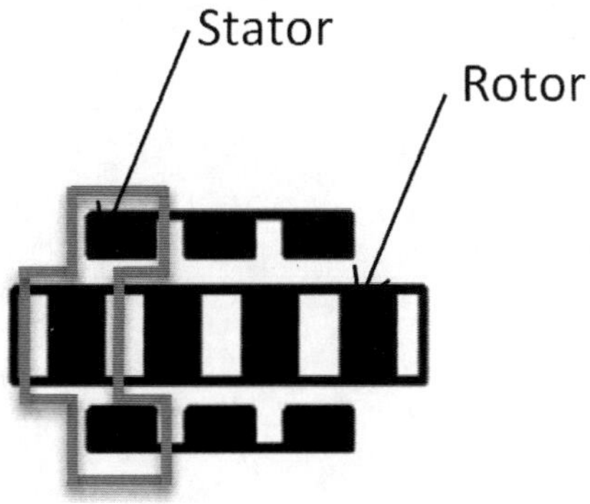

Figure 5. Basic LSRG.

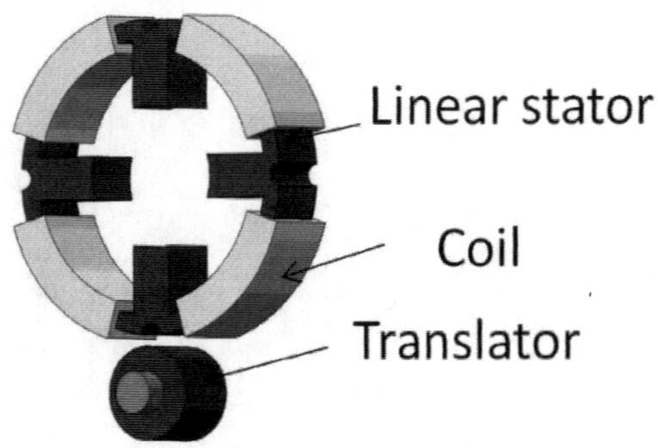

Figure 6. Propose structure for LSRG.

All parts of the LSRG coil are given an electric charge at the same time in a way that creates two pairs north and south magnetic poles.

Hence, the magnetic flux would have short paths in the translators. The translator is fixed on a non-magnetic shaft under the piston of SCE. When the piston moves, the translator is moved simultaneously. The cross section of LSRG is shown in figure 7.

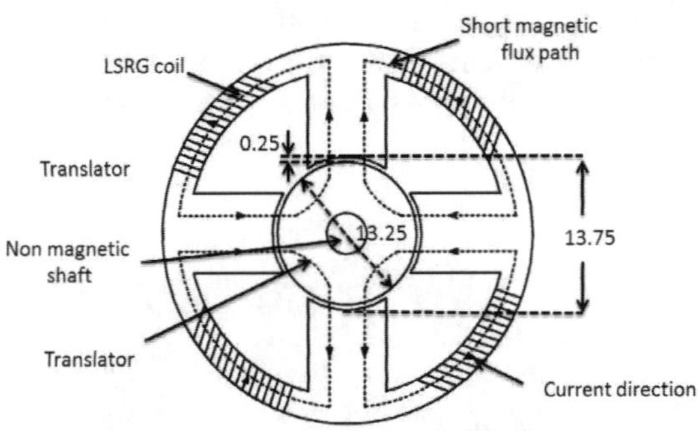

Figure 7. Front view of the LSRG.

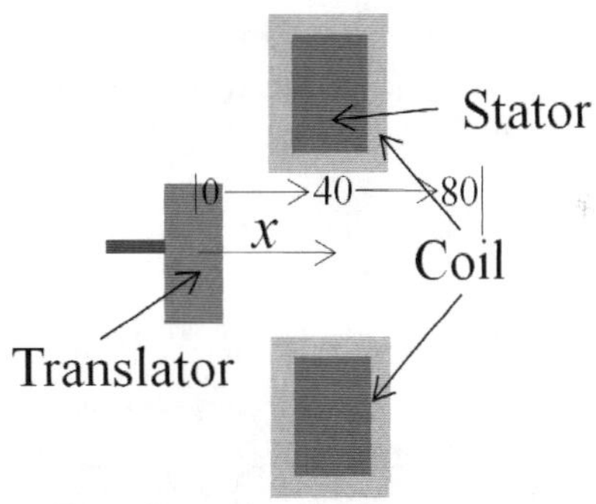

Figure 8. Side view of LSRG.

Table 1. The parameters of the LSRG.

Element	Style
Stator outer diameter	13.75 [mm]
Translator outer diameter	13.25 [mm]
Air gap	0.25 [mm]
Number of stator windings	440 [turn]

It is necessary to consider the motion of translator. Figure 8 shows the side view of LSRG. The x is the distance away from the initial position. This translator does reciprocating motion along with the piston.

Based above all, add the part of the LSRG into single cylinder engine shown in figure 9.

The parameters of LSRG are shown in table 1. The important sizes of stator outer diameter, translator outer diameter, and air gap are used to build a 3-D model in magnetic field analysis software JMAG. Also the details of the element listed in table 1 are shown in figure 7.

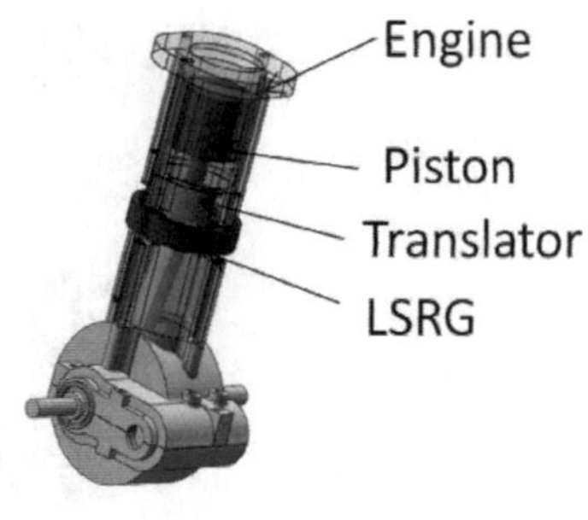

Figure 9. Single cylinder engine LSRG.

3. Finite element analysis

3.1. Magnetic simulation

The modified 3-D model is built by using FEM magnetic field software JMAG as shown in figure 10. The total distance of the translator's movement is 80 mm. When the piston is at the top of cylinder, the distance x is 0 mm. And when the piston is at the bottom of cylinder, the x is 80 mm. These positions are called unaligned position. When the x is 40 mm, it is called the aligned position.

The 3-D model is subdivided into small triangle elements. In order to improve the accuracy of FEM analysis, different mesh sizes are chosen for different areas. That means the corner and surfaces need finer meshes to satisfy higher accuracy requirements [8]. Figure 11 illustrates the meshed 3-D model.

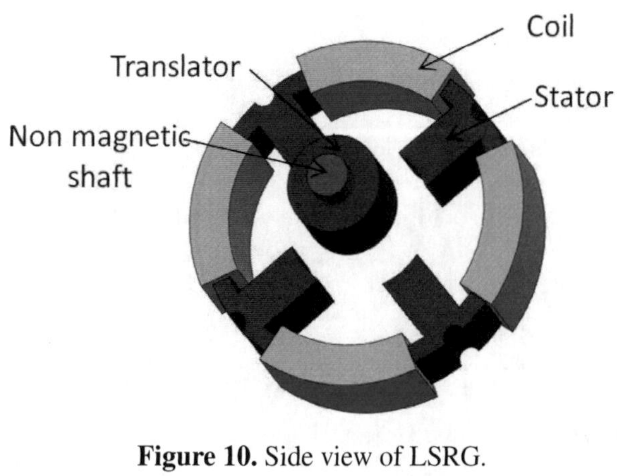

Figure 10. Side view of LSRG.

Figure 11. Finite-element mesh of LSRG.

The 3-D model is simulated with FEM magnetic field software JMAG. Figure 12 shows the finite element flux-plot when the current is 5 A. The same calculations are iterated for each position from the top to bottom. Also same calculations are executed to simulate the inductance for different currents and different positions. The current is changed from 0 A to 5 A, and the position is changed from 0 mm to 80 mm. Figure 13 illustrates the magnetic flux density through FEM magnetic field analysis. According to the calculated magnetic flux linkages, the inductance data are obtained as shown in figure 14.

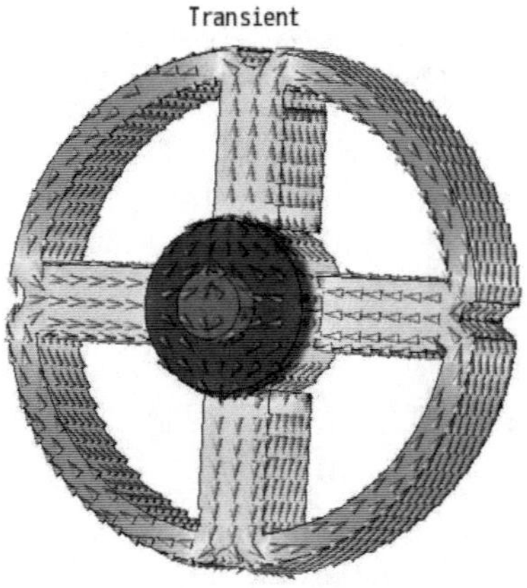

Figure 12. Simulated FE magnetic flux plot.

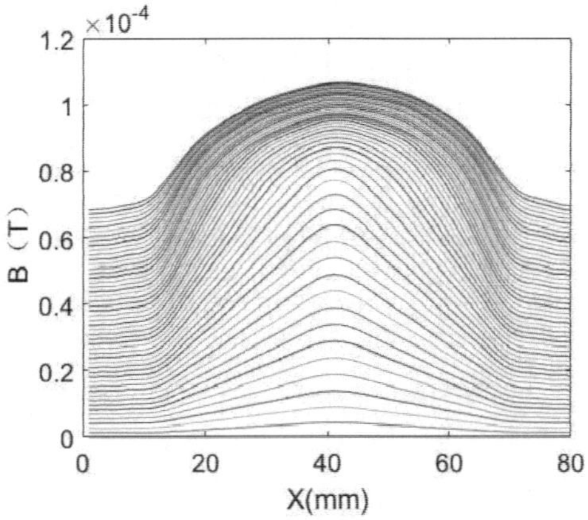

Figure 13. Magnetic flux linkage curve of LSRG.

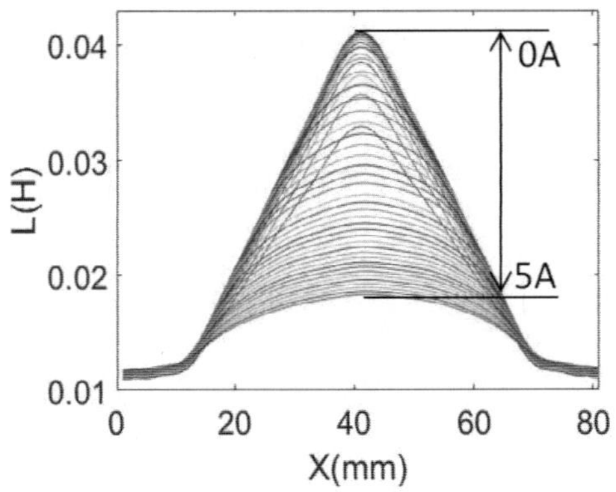

Figure 14. Inductance curve of LSRG.

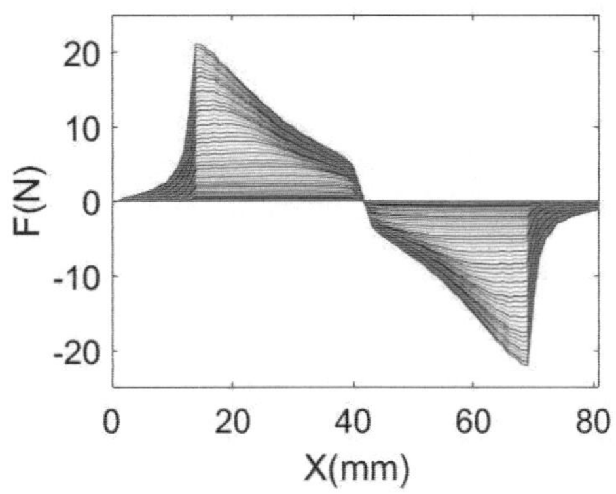

Figure 15. Calculated results of translation force.

3.2. *Calculation of static translation force*

At translator position x, the magnetic co-energy W' is calculated according to the following equations [9].

$$W'(x,i) = \int_0^i \Psi(x,i) \cdot di \tag{1}$$

Then, the translation force is obtained by:

$$F(x,i) = \frac{\delta W'(x,i)}{\delta x} \tag{2}$$

Where $\Psi(x,i)$ is the magnetic flux linkage for different current and position, i is the excitation current.

For small intervals of positions (Δx) and currents (Δi), a practical expression can be made by:

$$F(x,i) \approx \frac{\Delta W'}{\Delta x} = \frac{\Delta i}{\Delta x}\left[\sum_0^i \Psi(x+\Delta x,i) - \sum_0^i \Psi(x,i)\right] \tag{3}$$

The translation forces for different positions and currents are showed in figure 15.

4. MATLAB simulink model

The simulation model of LSRG consists of four parts, as shown in figure 16: the LSRG model, mechanical system model, controller model, and the converter model. The LSRG model is based on the electrical and mechanical equations of the LSRG. The mechanical system model is used to calculate the velocity and the position of translator. The converter model is treated as the power converter which has ideal switching. The controller model can generate drive signals to the converter.

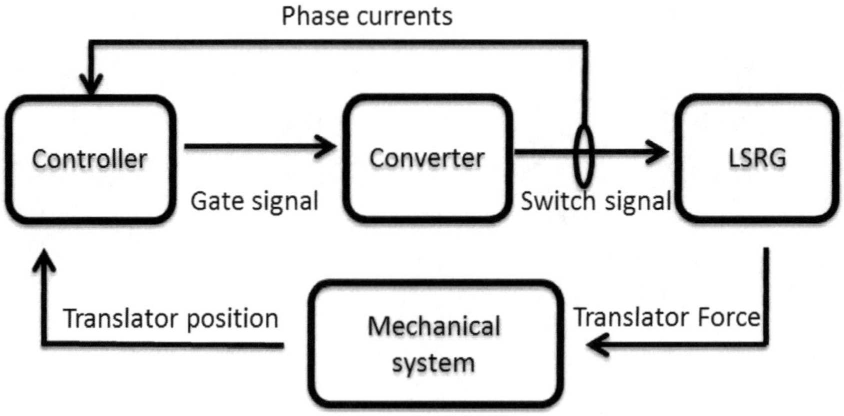

Figure 16. Simulation model of LSRG.

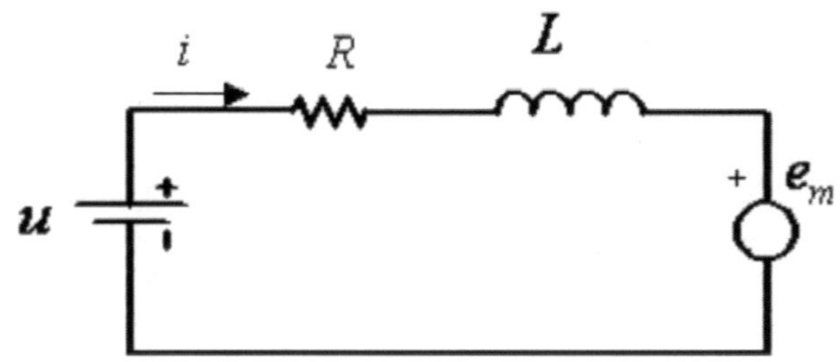

Figure 17. Equivalent electric circuit of LSRG.

4.1. LSRG model

Figure 17 illustrates the electrical equivalent circuit of LSRG. u is DC voltage, i is the phase current, R is the phase resistance, L is the inductance, e_m is the back electromotive force.

According to the equivalent electric circuit of LSRG, the voltage equation of the stator can be calculated by following equations [10]:

L can be calculated by:

$$L = \frac{\partial \Psi}{\partial i} \tag{4}$$

Where Ψ is the flux linkage, i is the phase current.

e_m is calculated by:

$$e_m = \frac{\partial \Psi}{\partial x}\frac{dx}{dt} \tag{5}$$

Where x is the position of the translator.

According to equations (4) and (5):

$$u = Ri + \frac{\partial \Psi}{\partial i}\frac{di}{dt} + \frac{\partial \Psi}{\partial x}\frac{d\theta}{dt} \tag{6}$$

The model is built based on the voltage equation. The data of $\partial\Psi/\partial\mathrm{i}$ and $\partial\Psi/\partial\mathrm{x}$ are obtained from the FEM magnetic field analysis and calculation with MATLAB.

4.2. Mechanical system model

In reality, the external load is added from the engine, which gives the force to the translator. In this model case, the external load is chosen as a constant value of 1500 N. The velocity can be calculated with equation (7).

$$V = \int \frac{F_L - F - F_f}{m} dt \tag{7}$$

Where F_L is the load force, F is the translator force of LSRG, F_f is the friction force, m is the mass of the piston and translator.

4.3. Controller model

The purpose of the controller model is to detect if the translator is between the on-position and off-position. When in the setting position, this model can deliver the signals to the converter for switching the IGBTs as shown in figure 18.

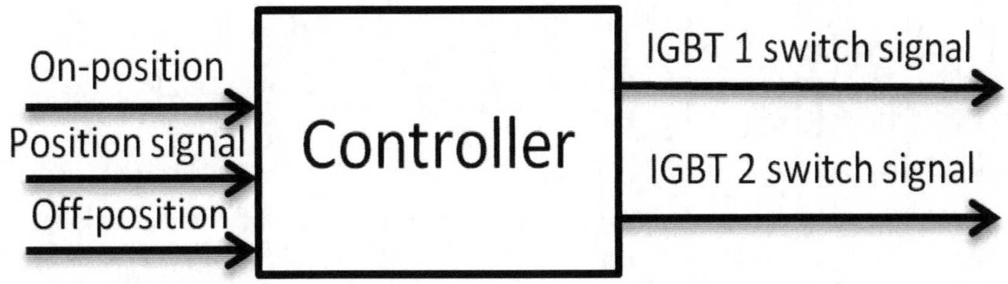

Figure 18. Controller model.

4.4. Converter model

The asymmetric half-bridge converter is used to excite the coils of LSRG according to the switching signal from the controller model. The excitation voltage is given to the LSRG with IGBTs [11]. The asymmetric half-bridge is shown in figure 19. This circuit has three operation modes as shown in figure 20. Figure 20(a) is the excitation state, figure 20(b) is regenerating state, and figure 20(c) is the soft switching state.

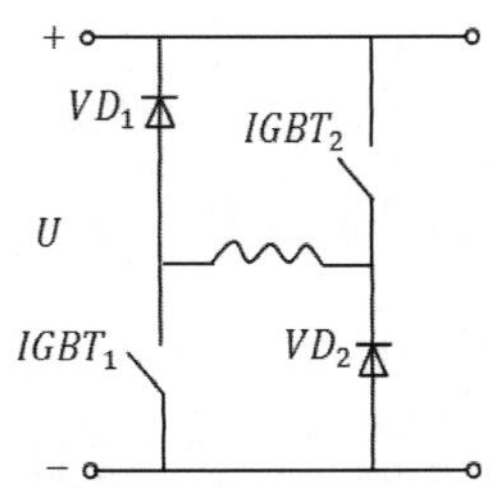

Figure 19. Asymmetric half-bridge circuit.

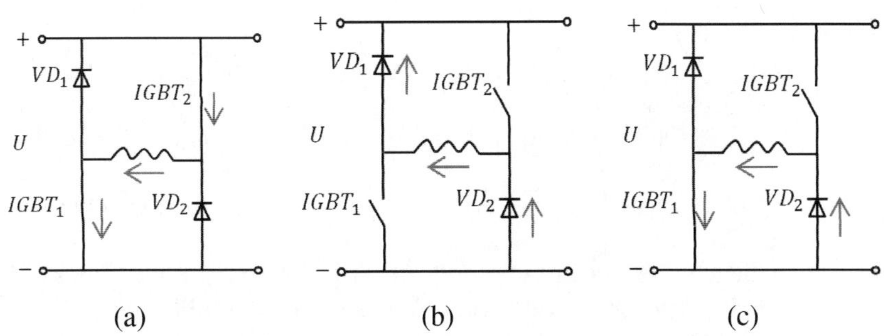

Figure 20. Asymmetric half-bridge circuit operation modes.

5. Simulation results of MATLAB simulink

The simulation results are shown in figures 21-24. The position signal, flux linkage, phase voltage signal, and phase current signal are shown, respectively. In the simulation results, the LSRG is excited between the on-position and off-position. When the voltage is positive, the LSRG is in the excitation state. When the voltage is negative, the LSRG is in the regenerating state [12]. The current results are consistent with known theories, when the LSRG generates electricity.

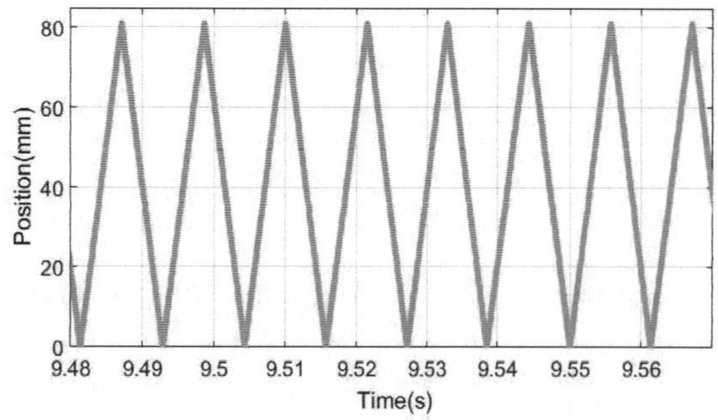

Figure 21. Position signal of LSRG.

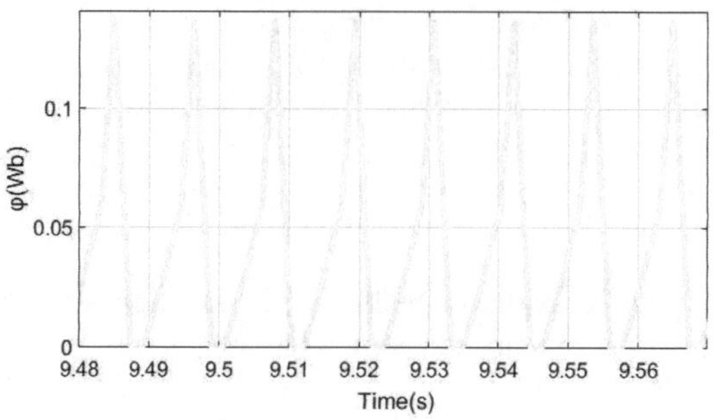

Figure 22. Flux linkage of LSRG.

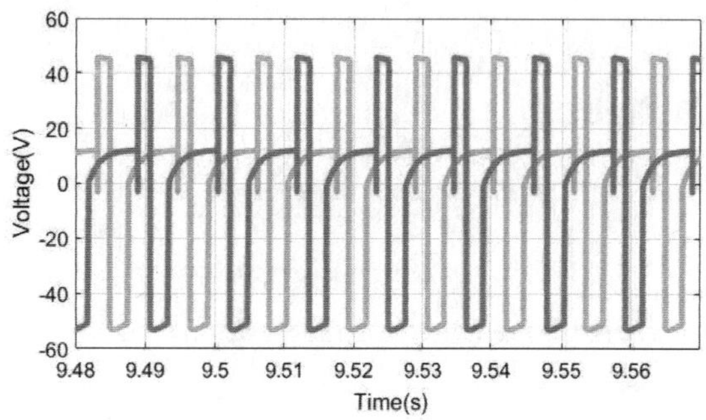

Figure 23. Phase voltage signal of LSRG.

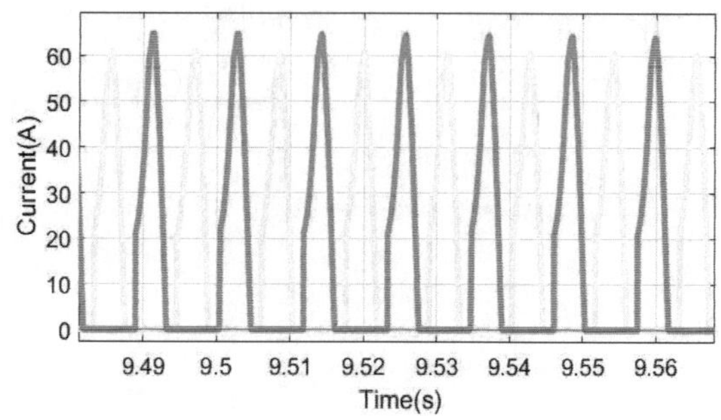

Figure 24. Phase current of LSRG.

6. Conclusion

In this paper, the new LSRG structure is proposed to be combined with the SCE. When the piston is moving in SCE, the LSRG can assist the SCE's mechanical output or can generate power to charge the battery. Through the FEM magnetic analysis, the data of flux linkage and inductance are obtained, and the translator forces are calculated with the MATLAB. Also the LSRG system is simulated with MATLAB Simulink to verify the basic operations of LSRG.

References

[1] Alvarez R A, *et al* 2012 Greater focus needed on methane leakage from natural gas infrastructure *Proceedings of the National Academy of Sciences* **109** 6435-40

[2] Chan C C 2007 The state of the art of electric, hybrid, and fuel cell vehicles *Proceedings of the IEEE* **95** 704-18

[3] Sciarretta A and Lino G 2007 Control of hybrid electric vehicles *IEEE Control Systems* **27** 60-70

[4] Schouten N J, Mutasim A S and Naim A K 2002 Fuzzy logic control for parallel hybrid vehicles *IEEE Transactions on Control Syst Technol* **10** 460-8

[5] Sakthivel P, Chandrasekar V and Arumugam R 2011 Design of a 250 W, low speed switched reluctance hub motor for two wheelers *IEEE 2011 1st International Conference on Electr Energy Syst (ICEES)* pp 176-81

[6] Boldea I and Syed A N 1997 Linear electric actuators and generators 1997 *IEEE International Electric Machines and Drives Conference Record* MA1/1.1_MA1/1.5.

[7] Nezamabadi, Mohammad M, Ebrahim A and Hossein T 2016 Design, dynamic electromagnetic analysis, FEM, and fabrication of a new switched-reluctance motor with hybrid motion *IEEE Transactions on Magn* **52** 1-8

[8] Nezamabadi, Mohammad M, Ebrahim A and Hossein T 2016 Design, dynamic electromagnetic analysis, FEM, and fabrication of a new switched-reluctance motor with hybrid motion *IEEE*

Transactions on Magn **52** 1-8
[9] Amoros J G, Baldui B M and Pere A G 2011 Simulation of linear switched reluctance motor drives *Proceedings of the 2011-14th European Conference on Power Electronics and Applications (EPE 2011)* pp 175-8
[10] Ray W F, *et al* 1986 High-performance switched reluctance brushless drives *IEEE Trans on Ind Appl* **4** 722-30
[11] Bae H-K, *et al* 2000 A linear switched reluctance motor: Converter and control *IEEE Trans on Ind Appl* **36** 1351-9
[12] Nakazawa Y, Ohyama K, Fujii K, *et al* 2010 Excitation interval control of switched reluctance motor considering derivative of inductance *Power Electron and Motion Control Conference IEEE* T5-84-9

Study on the electrical performance of underwater photovoltaic modules

L Li[1,3], Y T He[1,3], S H Wang[1] and C Zhang[1,2]

[1]School of Physics & Electronic Science, Chu Xiong Normal University, Chu Xiong, 675000, Yun nan Province, China
[2]Chuangyuan Fangda Electric Co., Ltd, Tang Shan, 063000, He Bei Province, China

E-mail: solarcell2008@163.com/526667060@qq.com

Abstract. Light intensity directly affects the output current of photovoltaic (PV) modules. The output characteristics of underwater photovoltaic modules are directly affected by light intensity. In this paper, utilizing the matrix simulation of light, and solar module analyzer (PROVA) test equipment, the electrical parameters of polycrystalline silicon PV modules are tested underwater. The relationship between the output characteristics of the PV module and water depth is analyzed. The results show that the output characteristics of PV modules is insensitive to increasing water depth. For the vertically irradiated photovoltaic module, the short-circuit current of the silicon photovoltaic module decreased by 2.39 A, the power decreased by 2.57 W, and the efficiency decreased by 5% when the water depth was increased by 1 cm. Similarly, the when the dip angle is 30 degrees, 45 degrees, 60 degrees, with an increase of water depth by 1cm, the PV module of short-circuit current, power and efficiency were decreased by 2.05 A, 2.1 W, 8%; 1.85 A, 2.07 W, 6.93%; 1.05 A, 0.74 W, 3.67%, respectively. Relative to the 0 degree angle, when the dip angle is 30 degrees, 45 degrees, 60 degrees angle, the percentage drop in PV module of short circuit current is14%, 23%, 56% and the power is 18.3%, 19.5%, 71.2% respectively.

1. Introduction

The development of solar energy photovoltaic power systems has made remarkable strides in recent years. At the same time, solar module development has changed rapidly. This development makes the test of solar module more standardized. From component IV feature testing to component mechanical damage testing, and from component thermal performance testing to temperature impact testing on component performance, the standards are very strict. But Photovoltaic modules often need to work outdoors for a long time, under a variety of climate conditions. In particular, the effect of rain on these devices is very critical. When a photovoltaic module is saturated in rain water, the incident light entering the water will refract and reflect, and at the same time, it will be absorbed and scattered during propagation. Compared with the photovoltaic module in the air, the power generation efficiency of the water photovoltaic module decreases with increasing water depth. Therefore, both for the photovoltaic modules test standard, and for photovoltaic engineering design, it is highly important to explore the influence of water depth on the output performance of PV modules.

Photovoltaic components in the water were patented in the United States in 2006 [1]. In 2008, Wang [2] investigated the performance of silicon solar cells operated in liquids. In 2014, Saurabh Mehrotra [3] studied the performance of a solar panel with water immersion cooling technique. In

IOP Conf. Series: Earth and Environmental Science **93** (2017) 012051 doi:10.1088/1755-1315/93/1/012051

2016, at the University of Shanghai for Science and Technology, Chen [4] studied the application characteristics of surface water cooling solar photovoltaic modules. Starting from the propagation law of light in water, this paper explores the relationship between water depth change and the power and output efficiency of photovoltaic modules. In water, the IV output characteristics of polysilicon photovoltaic modules under different dip angles were tested by the solar module analyzer (PROVA). The findings reveal that the law of water depth and solar cell output characteristics play a key role in evaluating the relationship between the output characteristics of underwater photovoltaic modules and water depth.

2. Theoretical model of underwater photovoltaic modules

The heat transfer model of an underwater PV module is shown in figure 1. The heat transfer energy of underwater photovoltaic module mainly includes the energy of incident solar radiation, the output power of the photovoltaic module, the heat transferred by the upper surface layer of the PV module, the heat stored in the lower surface layer of the PV module, and the PV module output power. According to the principle of conservation of energy, the unsteady energy equation of underwater PV component has the formula (1):

$$C_{pv}\frac{dT}{dt} = A(1-\rho)I_s\tau - Q_1 - Q_2 - P_{out} \qquad (1)$$

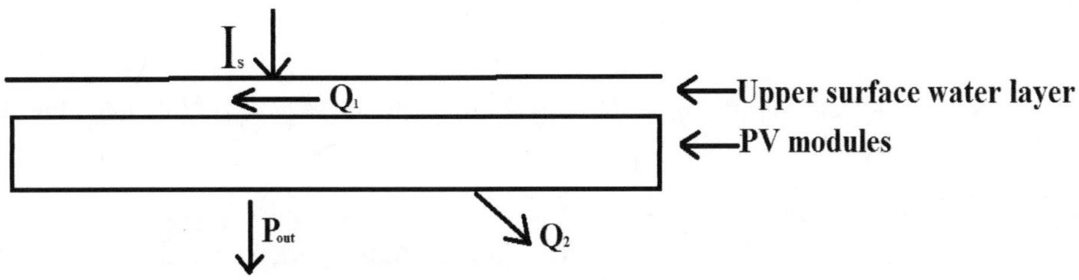

Figure 1. Heat transfer model of underwater photovoltaic module.

In formula (1), C_{pv} represents the total heat capacity of a single cell assembly, with C_{pv}=2918 J/K; T represents the operating temperature of the PV module, K; The T approximation is equal to the average temperature at the front and back of the PV module; A represents the area of PV modules, with A=0.53 m^2; ρrepresents water reflectance of solar radiation; with ρ=0.1; τ represents the transmissivity of solar radiation in the water, with τ=0.8; Is represents the intensity of solar radiation projected onto the surface of a module, W/m^2; Q1 represents the heat transferred by the surface water of a photovoltaic module; Q_2 represents the heat transmitted by the lower surface of the PV module, W; Pout represents the power output of photovoltaic modules, W.

2.1. Surface heat conduction of an underwater photovoltaic module

The upper surface heat transfer of an underwater photovoltaic module satisfies Fourier's law of heat conduction. One-dimensional flat wall heat conduction, i.e., Q_1, λ, and A, does not vary with temperature.

$$Q_1 = -\lambda A\frac{dT}{dx} \qquad (2)$$

In formula (2), λ is thermal conductivity, w/ (m·k); A is the PV module area, m^2 boundary condition of temperature is that heat generated by the PV module and the water temperature at the top of the PV modules.

2.2. The lower surface heat conduction of an underwater photovoltaic module

Unlike the upper surface heat conduction, the lower surface heat conduction of the PV module can be considered as a process of the gradual change of the heat source from temperature T_2 to the water temperature, thus:

$$Q_2 = \lambda A \frac{dT_2}{dx} \quad (3)$$

In formula (3), dx is the distance between the temperature of the backplane of the PV module and the average temperature of the water temperature; T_2 is the temperature of the PV module backplane. boundary condition of temperature is that heat generated by the PV module and Mean water temperature.

2.3. *Output power of PV modules*

The output power of the PV module is determined by the fill factor model constant [5]; it can be expressed by formula (4):

$$P_{out} = C_{ff} \frac{I_s ln(CI_s)}{T} \quad (4)$$

In the formula, C_{ff} is the fill factor model constant, C_{ff}, =1.22 K·m^2, K; C is constant, C=10^6, m^2/W.

2.4. *The conversion efficiency of a PV module*

The conversion efficiency of a PV module is defined by its formula (5):

$$\eta_e = \frac{P_m}{AI_s} \quad (5)$$

In the formula, η_e represents the PV module conversion efficiency, %; P_m represents the maximum output power of PV module, W.

2.5. *Absorption of light waves in water*

At the same time, according to the Law of Lambert [6], when light incidents a transparent medium, the light and medium undergo a molecular interaction so that the intensity will be weakened. The weakened light intensity is related to the absorption coefficient and the passing path. The available expression (6) describes this phenomenon:

$$I_s = I_0(1 - \mathrm{R})e^{-\alpha_a d} \quad (6)$$

Where I_0 is the incident light intensity, I_0=800 W/m^2; α_a is the absorption coefficient, α_a= 5.6197 m^{-1} [7]; d is the length of the light through the path.

3. Experiment device and test process

Figure 2. Experiment device.

Underwater photovoltaic module testing systems consists of a solar simulator, a photovoltaic module, a solar module analyzer, as well as other equipment. The test system is shown in figure 2. Among them are the solar simulator which consists of a matrix-type light emitting device, a host power control cabinet, a high voltage trigger controller, an optical radiation sensor, a temperature sensor, a solar analog transmitter monitor, and solar simulation management software. The technical parameters of the simulated light source are shown in table 1.

Table 1. The technical parameters of the simulated light source.

Model	TMC-PD8	Spectral rating	Class B
Light unevenness	≤5%	Irradiation adjustment range	600-1000 W/m^2
Light Instability	≤5%	Light source output mode	parallel light

According to the theoretical model of underwater photovoltaic modules, this experiment selects the polysilicon photovoltaic module tested by the Solar Energy Ltd. The component parameters are shown in table 2. In the process of experiment, the PV modules need to be intruded into the water-filled containers. The maximum distance between the upper surface of the water and the PV module is H, and the maximum value of the distance is 15 cm. Then, in turn, the capacity of the water in the container is reduced, and the water level in the container is reduced to 1 cm height until H equals 0 cm. The simulated light source used in the test is the TMC-PD15 artificial solar simulator produced by the Dongguan Green Light new energy technology company. The simulated light source is composed of 16 independent xenon light sources. The PV module is tested using the Taiwan solar module analyser (PROVA) to collect information about the work of PV modules.

Table 2. The parameters of polycrystalline silicon PV module of photovoltaic solar energy limited company.

Model	GHM80 W	operating current	4.55 A
Maximum power	80 W	open circuit voltage	21.6 V
Operating voltage	17.6 V	short circuit current	5.35 A

4. Experimental results analyses

4.1. The relationship between water depth and output current of the PV module

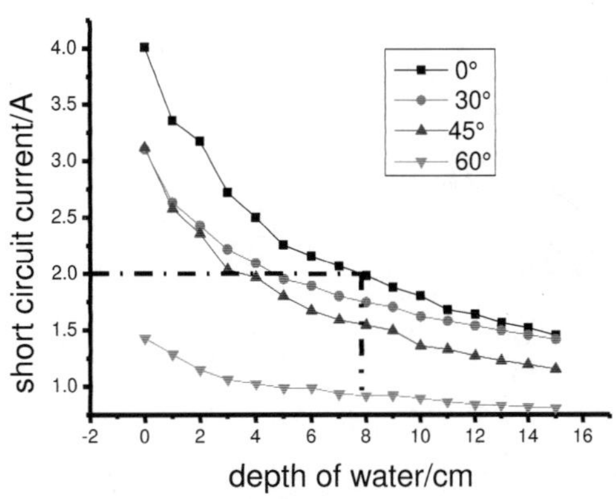

Figure 3. The relationship of depth and short current for a PV module.

In order to reveal the relationship between the water depth, the output current and the angle, it is necessary to draw water depth, current and angle diagrams. Figure 3 shows the relationship between water depth and the short circuit current of PV module. With increasing water depth, the short circuit current of the PV module decreases, and the deeper the water depth, the lesser the change of short-circuit current. On the other hand, with an increase of dip angle, the intensity of incident light decreases, and the decrease of short-circuit current is evidently weakened. For example: when the incident angle is 0 degrees, and water depth varies from 0cm to 15 cm, the average drop in unit water depth current is 2.39 A/cm. With the increase of incident angle, the change of current decreases. When the incident angle is 30 degrees, 45 degrees and 60 degrees, the current changes are 2.05 A/cm, 1.85 A/cm, 1.05 A/cm. On the whole, the current change of PV module exhibits an exponential distribution. When the incident angle is 0 degrees and the water depth is 8cm, the short circuit current of the PV module diminishes by half.

4.2. Influence of water depth on IV characteristics of photovoltaic modules

The change of short circuit current will directly lead to the change of output power of the PV module. At the same irradiation conditions, the water depth will affect the light intensity of the surface of the PV module. The deeper the water, the weaker the intensity of the surface of the PV module. Figure 4 shows the IV characteristics curve of polysilicon PV module with perpendicular irradiation at differing water depths. There is a direct relationship between the depth and the PV module short-circuit current. When the depth of the water is increased, the effect of current is gradually weakened. For perpendicular irradiation, when the water depth is 5 cm, the density of current is reduced by nearly half for the PV module. The greater the water depth, the less significant the current changes, and intuitively, the output power of PV modules decreases with the increase of water depth, among which the former 5 cm power decreases the fastest. Figure 5 shows the relationship between water depth and output power of a PV module with perpendicular irradiation. With an average water depth of 1 cm, there is a 4.78 W reduction in the output power of photovoltaic module, which is far greater than the average value 2.57 W. When the angle of incident light is 30 degrees, the output power of PV module is reduced by 2.1 W/cm. When the incident light angle is 45 degrees, the output power of the PV module changes to 2.07 W/cm, while when the incident light angle is 60 degrees, the output power of the PV module changes to 0.74 W/cm. As the light intensity directly affects the current change of the PV module, the output power of the PV module is mainly reduced by the decrease of incident light intensity, resulting in a smaller output current of PV module. For 30 and 45-degree angles, the incident light will produce a strong reflection, and the light entering the solar cell will be reflected and weaken, leading to a significant decrease in the current.

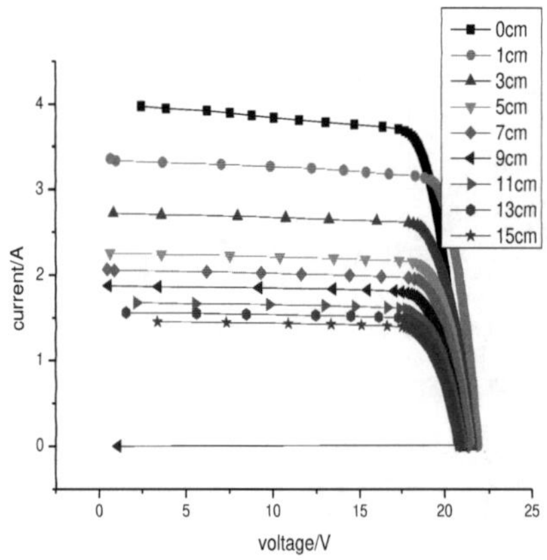

Figure 4. The IV characteristics curve of a polysilicon PV module for perpendicular irradiation at different water depths.

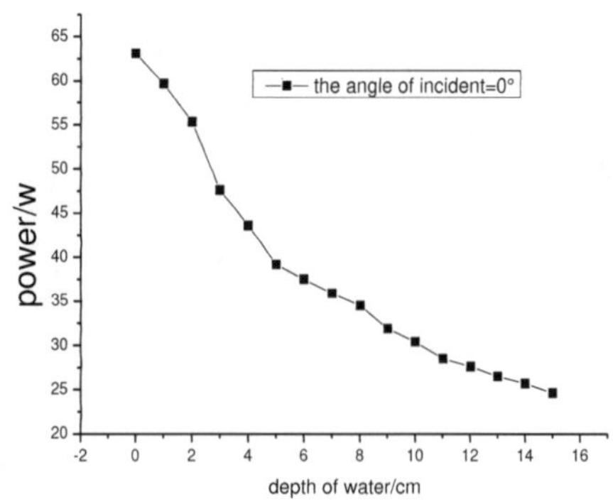

Figure 5. Relationship between water depth and output power of a PV module with perpendicular irradiation.

4.3. Relationship between water depth and output efficiency of PV module

When the PV module short-circuit current changes, the output efficiency and power will change accordingly. This rule is depicted in figure 6. This figure gives the relationship between the variation of water depth and the output efficiency of PV modules, and shows that there is an inherent correlation between the water depth and the conversion efficiency and output power.

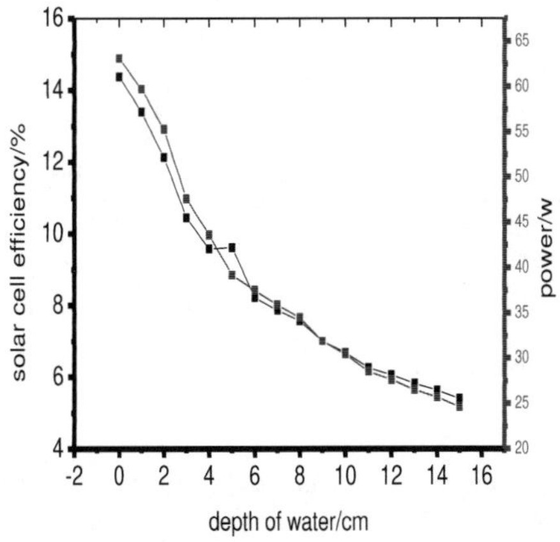

Figure 6. The relationship between the variation of water depth and the output efficiency of PV modules.

In fact, the photoelectric conversion efficiency of a solar cell refers to the ratio of the maximum output power of the PV module to the light incident upon the cell. This relationship can be expressed by formula (7) [8]:

$$\eta = \frac{I_m V_m}{AP_{in}} = \frac{P_m}{AP_{in}} \tag{7}$$

When the incident light intensity is constant, the photoelectric conversion efficiency of the solar cell is linearly related to the maximum power.

For incident angles of 0 degrees, 30 degrees, 45 degrees, and 60 degrees, when the water depth increases to 1 cm, the output efficiency and power change are reduced to 9.07%/cm, 40.8 W/cm; 8%/cm, 36.3 W/cm; 6.93%/cm, 31.7 W/cm; 3.67%/cm, 16.6 W/cm, respectively.

5. Simulation and measurement analysis

The short-circuit current of PV module is simulated and compared in this section. The relationship between water depth and short circuit current of photovoltaic module is also explored. When the water depth is 0 cm, the current test result of the PV module is 4.01 A, and the simulation current result is 3.79 A. The test results of PV modules match well with the simulation results. Figure 7 compares short-circuit current simulation and test results of PV modules.

According to equation (4), the output power of the PV module is calculated and can be compared with the measured results. Figure 8 is a comparison between simulated and measured output power of an underwater photovoltaic module. It can be seen from figure 8 that the output power of the PV module is well matched with the output curve of the PV module. The output power shows a downward trend with increasing water depth. At a measured water depth of 0 cm, PV module output power is 63.1 W, the analogy PV module output power is 61.2 W. At a measured water depth of 16 cm, PV module output power is 24.6 W, and the analogy PV module output power is 25.4 W.

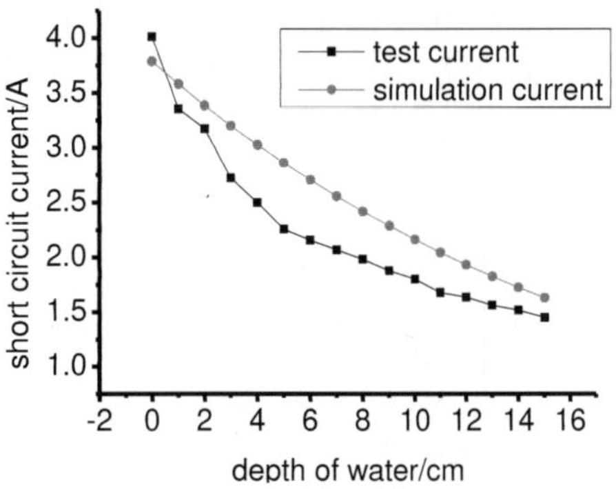

Figure 7. Comparison of simulated and measured short current of underwater photovoltaic module.

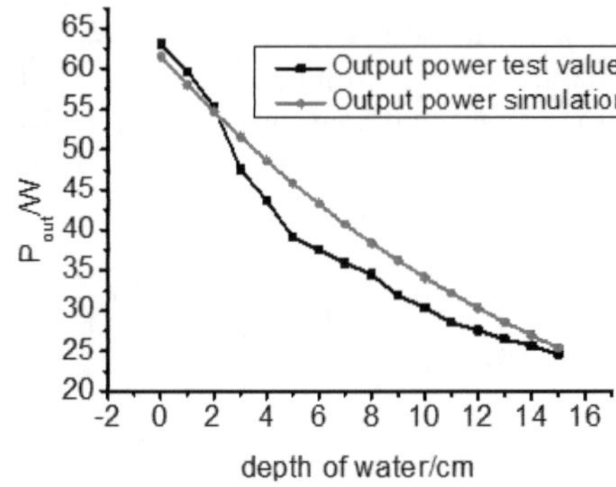

Figure 8. Comparison of simulated and measured output power of underwater photovoltaic module.

6. Conclusions

In order to analyze the output characteristics of polysilicon photovoltaic modules in water, this paper uses the Taiwan solar module analyzer (PROVA) to test the output characteristics at 0 to 15 cm depth of water under the simulated light source. The experimental results are analyzed and studied in detail, and the following conclusions are obtained:

- The water depth directly affects the output power of PV modules. When the incident light is direct, the output power of the PV module drops by half for a water depth of 5 cm. The PV output power of unit water depth is decreased by 2.57 W/cm.
- The output power of the PV module is affected by the short circuit current, and if the water depth of the PV module increases by 1cm each time, the short-circuit current will decrease by 2.39 A/cm and the output efficiency will decrease 9.07%/cm.
- Finally, in this study, the output power of underwater photovoltaic module was also simulated, and the simulation results were in good agreement with the test results.

Acknowledgments

This research was financially supported by research project of teaching reform in Chuxiong Normal University and Yunnan Applied Basic Research Projects.

References

[1] Mook J R W J 2006 Solar panels with liquid super concentrators exhibiting wide fields of view (patent US 2006/0185713 A1)
[2] Wang Y P, Fang Z L, Zhu L, *et al* 2009 The performance of silicon solar cells operated in liquids *Appl Energ* **86** 1037-42
[3] Mehrotra S, Rawat P, Debbarma M, *et al* 2014 performance of a solar panel with water immersion cooling technique International *J Sci, Environ Technol* **3** 1161-72
[4] Chen J B, Yu H Z and Yao J S 2016 Research on application characteristics of surface water cooling solar photovoltaic modules *Acta Energiae Solaris Sinica* **37** 1768-72
[5] Liu J M 2010 *The Utilization of Solar Energy: Principle, Technology, Engineering* (Beijing: PublishingHouse of Electronics Industry)
[6] Yao Q J 2002 *Optics Course* (Beijing: Higher Education Press)
[7] Deng R R, He Y Q, Qin Y, *et al* 2012 Pure water absorption coefficient measurement after eliminating the impact of suspended substance in spectrum from 400 nm to 900 nm *J Remot Sens* **16** 176-84
[8] Zhao F X and Wei Y Z 1985 *Solar Cells and Applications* (Beijing: National Defense Industry Press)

Design and implementation of a 38 kW dish-Stirling concentrated solar power system

J Yan[1], Y D Peng[1,3], Z R Cheng[1], F M Liu[1] and X H Tang[2]

[1]Hunan Provincial Key Laboratory of Health Maintenance for Mechanical Equipment, Hunan University of Science and Technology, Xiangtan 411201, Hunan Province, China

[2]Solar Business Department of Xiangtan Electric Manufacturing Group, Xiangtan 411101, Hunan Province, China

E-mail: ydpeng@hnust.edu.cn

Abstract. Dish-Stirling concentrated solar power system (DS-CSP) is an important pathway for converting solar energy into electricity at high efficiency. In this study, a rated power 38 kW DS-CSP system was developed (installed in Xiangtan Electric Manufacturing Group). The heat engine adopted the alpha-type four cylinders double-acting Stirling engine (Stirling Biopower Flexgen S260). The absorber flux distribution simulation was conducted using ray tracing method and then the 204 m^2 parabolic dish concentrator system (diameter is 17.70 m and focal length is 9.49 m) with single concentrator plus single pillar supporting has been designed and built. A water-cooled disc target and an absorber imitation device were adopted to test the tracking performance of the dish concentrator system, homogeneity of the focal spot and flux distribution of the absorber. Finally, the S260 Stirling engine was installed on the focal position of the dish concentrator and then the net output power date of the 38 kW DS-CSP system was tested. The absorber overheating problem on the DS-CSP system performance was discussed when the DS-CSP system was installed in different locations. The testing result shows that this system achieved the net output power of 38 kW and solar-to-electricity efficiency (SEE) of 25.3% with the direct normal irradiation (DNI) at 750 W/m^2. The net output power can further increase to 40.5 kW with the SEE of 26.6% when the DNI reaches up to the maximum of 761 W/m^2. The net output power of the 38 kW DS-CSP system has a linear function relationship with the DNI. The fitting function is Net power output=0.1003×DNI－36.129, where DNI is at the range of 460~761 W/m^2. This function could be used to predict the amount of the 38 kW DS-CSP system annual generation power.

1. Introduction

Solar thermal power generation utilizes the sun as a heat source which can drive a heat engine for power production by concentrating technologies [1]. At present, Concentrated Solar Power technologies majorly include the solar tower system, Dish-Stirling concentrated solar power system (DS-CSP), parabolic trough system and Fresnel system [2]. The DS-CSP system is characterized by high solar-electricity efficiency, scalable power generation and high modularity, which has been regarded as a promising CSP technology for highly efficient solar power generation [3,4].

DS-CSP system consists of a parabolic concentrator and a Stirling engine system [5,6]. The parabolic concentrator provides highly concentrated solar irradiation as the heat source for the Stirling engine to produce electricity with a high efficiency which in turn achieves a high solar-to-electricity efficiency [2,3,7-14]. The DS-CSP unit with world maximum output power was built by Schlaich-Bergermann und Partner Company (SBP). The diameter of the single dish parabolic concentrator is 17.0 m with a metal tensioned membrane structure. The maximum output power can reach 53 kW (direct normal irradiation (DNI) is 1000 W/m^2) with the 4-275 kinematic Stirling engine (rated power 50 kW) and the solar-to-electricity efficiency (SEE) is 23.0% [8,9]. Five types of DS-CSP system was reviewed by Mancini *et al* [3] at 2003 with system rated power between 9.5~25 kW and SEE value between 19.0%~29.4%. Among them, the Stirling Engine Systems (SES) company 25 kW DS-CSP system got the highest efficiency (29.4%). The SES company established a 150 kW power station in 2005 (rated power is 25 kW), and the SEE of the Serial # 3 can reach to 31.25% on 31 January 2008 and create a new world record [9]. In September 2009, SES company built the first demonstrated plant in Arizona, United States, which was composed of 60-SunCatcher™ dish units with each unit of 25 kW (1.5 MW in total). The plant started production in January 2010 with annual solar-to-electricity gross efficiency equaled to 26% [10]. In year of 2012, a 100 kW DS-CSP demonstration station was built in the Ordos city in Nei Mongolia of China with the unit power of 10 kW. Baharoon *et al* [2] reviewed the development of the CSP technology and presented the DS-CSP power station projects under building or planning (Infinia Corp [12], Capacity 1.5 MW, and start production year 2013). From the recently review[1-3,13,14], most existing DS-CSP systems are with a rated power below 25 kW, but only the SBP company has a system with rated power of 50 kW. Very few literatures have been found for the DS-CSP system rated power between 25 kW to 50 kW which is a power range worth of exploring. In this study, we will develop the rated power 38 kW large DS-CSP system, and future application in the large-scale solar power station or remote areas for distributed power supply.

The development of the DS-CSP system generally includes the design of the Stirling engine, cavity absorber, and the parabolic dish concentrator [13]. The Stirling engine development process includes the mechanical structure design [15,16] (e.g, the design of the piston motion mechanism, regenerator and cooler), thermodynamic calculation [17], fluid dynamics analysis [18], and the design of the generator and control system. This is a complex project. In this paper, a mature Stirling engine (Stirling Biopower Flexgen S260) was applied to the 38 kW DS-CSP system development. The S260 Stirling engine is α-type four cylinders double-acting Stirling engine and working fluid is hydrogen. The S260 Stirling engine was originally applied to biomass thermal power generation, and its heat absorber is enclosed cylindrical cavity structure. We need to redesign the absorber structure in order to apply the S260 Stirling engine to the solar thermal power generation.

In this study, a rated power 38 kW DS-CSP system was developed and tested. The heat engine adopted the alpha-type four cylinders double-acting Stirling engine (Stirling Biopower Flexgen S260). The 17.70 m parabolic dish concentrator has been designed and implementation. The paper structure and main contents are as follows. In section 2, the flux distribution of absorber simulation was conducted by using ray tracing model for analyzing. In section 3, the single dish plus single pillar supporting parabolic dish concentrator system was designed. In section 4, the installation of the concentrator mirror and optical performance test of the concentrator were described. In section 5, the operation data of the 38 kW DS-CSP system were analysed. Lastly, a summary and conclusions were drawn.

2. Absorber flux distribution

The parabolic dish concentrator parameters (radius and focal length) determination needs to consider the performance matching of the concentrator and the Stirling engine. This includes two aspects, one is that the concentrator should provide enough solar energy for the Stirling engine, which means energy matching of the concentrator and the Stirling engine. The other is that the absorber flux distribution needs to be reasonable.

2.1. Basic parameters of dish concentrator

The global coordinate system *O-xyz* is built at the top of point *O* on the parabolic surface, the *z* axis points to the focal point F. The surface of the reflect mirror can be defined as $x^2 + y^2 = 4fz$, where *f* is focal radius.

The concentrator structure composed of twelve truss beams in uniform distribution. The mirror is not installed between the two truss beams in the below part to avoid the mechanical interference between the concentrator and the supporting pillar (diameter 800 mm). The angle θ_m is 30°. The supporting truss was fixed on the concentrator centre position. The mirror can't be installed in this centre position (the corresponding radius R_2=0.90 m). While the section size of Stirling heat engine is 1200 mm×1400 mm (wide × high, Figure 1), the shadow caused by the Stirling engine is located in the circular area radius, which can not cause additional shadows for the reflection mirror. There is a certain gap between the concentrator mirror units. This gap is 20 mm along the radius direction and 25 mm along the circumferential direction which means reflection mirror gap rate k_m is 0.15, and the convenience of mirror installation and also reduces the wind load by allowing part of the wind passing through the gap (figure 4(a)). The effective concentrator aperture area A_{dish} is expressed as:

$$A_{dish} = \frac{2\pi - \theta_m}{2}(R_1^2 - R_2^2)(1-k_m) \tag{1}$$

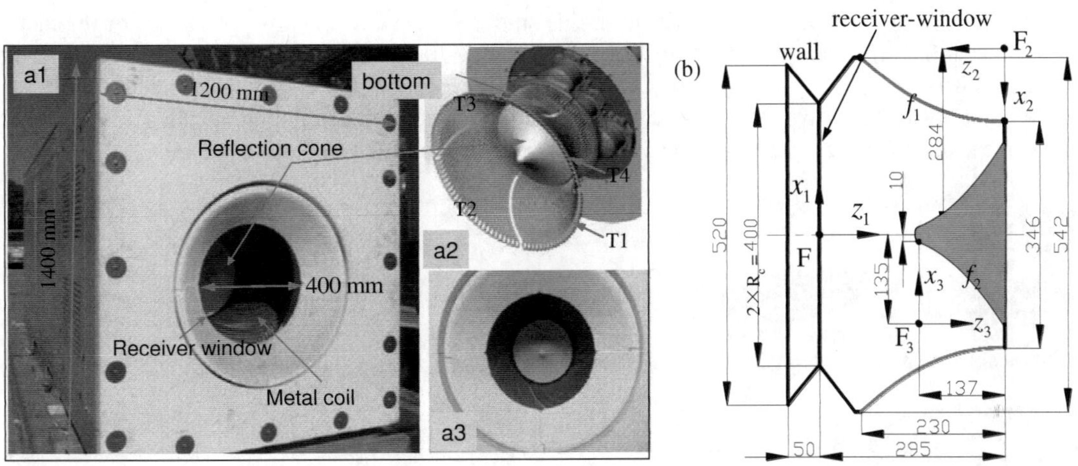

Figure 1. Structure and geometrical parameters of the heat absorber: a1) picture of the Stiring Biopower Flexgen S260, a2) CAD design model of the heat absorber, a3) picture of the heat absorber front view and b) geometrical parameters of the heat absorber.

According to the energy conservation equation (input energy equals output energy) of the DS-CSP, the rated power E_{Ra} is given as:

$$E_{Ra} = A_{dish} \cdot W_0 \cdot \eta_{DS} \tag{2}$$

Where W_0 is the direct normal irradiation (unit is W/m^2); η_{DS} is the designed value of the DS-CSP solar-to-electricity efficiency (SEE).

In this work, the 38 kW DS-CSP system will build in the Xiangtan electric group factory, China (east longitude 112.95°, north latitude 27.87°). The Xiangtan region solar irradiation ability has been evaluated and finally determined the DNI design value is 750 W/m^2. The operated DS-CSP system SEE value (19.0%~29.4%) [1-3] has been referenced and the η_{DS} will be directly set at 25% that is a typical efficiency value in the DS-CSP system engineering applications. So the concentrator radius R_1 is selected as 8.850 m to achieve a rated power of 38 kW for the DS-CSP system.

2.2. *Structure and parameters of the absorber*

In this paper, an engine with four cylinders double-acting Stirling engine and hydrogen as the working medium (Stirling Biopower Flexgen S260, α-type) is used. The absorber structure parameters are shown in Figure 1 and the absorber are composed of a metal coil and a secondary reflection cone.

The flux distribution of absorber is conducted using a simulation software OptisWorks 2012 based on Monte Carlo Ray Tracing Method [19,20]. Due to the metal coil arranged densely in the absorber and the tube diameter is small, hence it is difficult to directly calculate the flux distribution on the metal coil surface. In this research, the metal coil is simplified into a red surface f_1 and the secondary reflection cone is simplified into a gray surface f_2. In Figure 1(b), the coordinate system F-$x_1y_1z_1$ parallels to the coordinate system O-xyz is established on the focal point F. The coordinate system F_2-x_2z_2 and F_3-x_3z_3 are established respectively for calculating the generating lines of the metal coil and the secondary reflection cone and can be expressed as:

$$x_2 = -0.0020(z_2)^2 + 0.0337z_2 + 111.12 \tag{3}$$

$$x_3 = -0.0042(z_3)^2 - 0.3730z_3 + 125.41 \tag{4}$$

2.3. Absorber flux distribution

The solar half angle 4.65 mrad is considered in this study. We assume the absorption rate of the metal coil surface is 0.75 and reflectance of the cone surface is 0.70 and W_0 is 800 W/m^2. The concentrator reflection mirror using the thickness 4.0 mm silver mirror glass and reflectivity is 0.95. The totally ray tracking number is 20 Mrays during the calculating process according to the independence study. The results of the absorber flux distribution are shown in Figure 2 and Figure **3**.

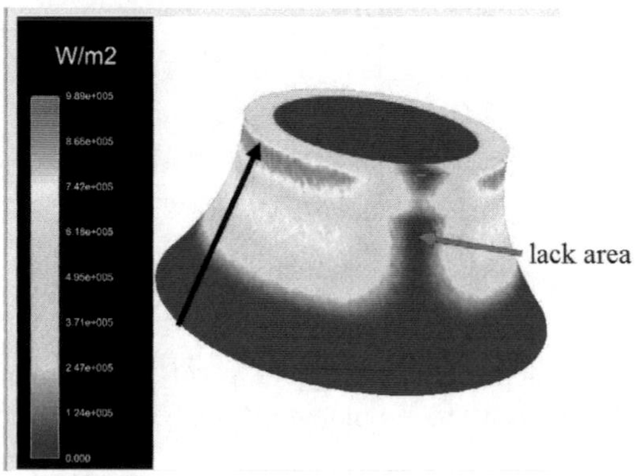

Figure 2. The flux distribution on the metal coil surface in heat absorber (R_1=8.85 m, f=9.49 m).

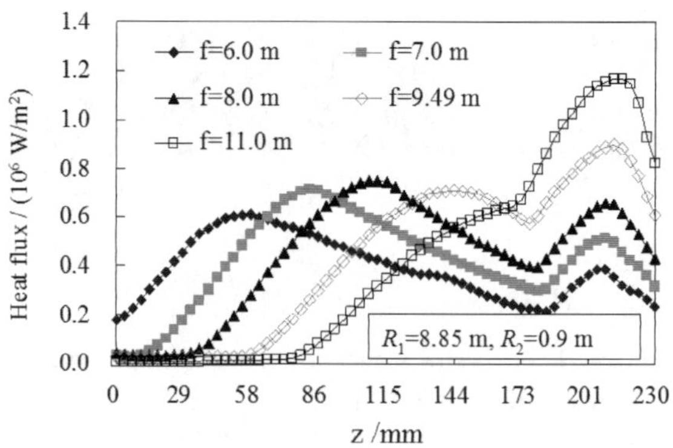

Figure 3. The flux distribution on the metal coil surface (the cross section where the arrow is).

As seen from Figure 2, the 'lack area' exists in the metal coil surface flux distribution, which is resulted from the existence of θ_m. The solar energy is concentrated on the bottom area of the metal coil (Figure 1(a)), which shows two peaks distribution (figure 3). Figure 3 is the flux distribution of the cross section where the arrow is shown in Figure 2. The concentrator radius and focal length significantly affect the flux distribution on the metal coil surface, but still has two peaks distribution characteristics. The focal length is larger and metal coil surface energy is more concentrated to the bottom area when the dish concentrator radius is fixed. This is easy to form the flow stagnation zone at the bottom area, which in turn reduces the convection heat loss. Considering the weight of the Stirling engine (~1600 kg), larger focal length f will lead to significant deformation of the support truss which will further affect the absorber receive the solar energy. Finally, we select the concentrator focal length f is 9.49 m (rim angle is 50°) when the dish concentrator rim angle is approaching to 45°.

3. DS-CSP system frame structure design

Dish concentrator frame structure was mainly used to support the reflection mirror and Stirling engine for realizing the concentrator tracking the sun position. This frame structure design includes the concentrator structure, supporting structure, Stirling engine supporting structure and two-axis tracking mechanism. In this paper, such factors as the manufacturing, construction, maintenance and environmental climate (such as wind and dust) are comprehensive considered by us. The 17.70 m single dish plus single pillar supporting parabolic dish concentrator system was designed.

3.1. Concentrator selection

In the design of the big dish concentrator structure, the SBP Company used single dish concentrator with tensioned membrane structure [9], which ensured a high quality mechanical structure while the concentrating performance under wind load is difficult to guarantee. This makes the structure difficult to be implemented in the windy climate area in China. The use of the multi-surface dish concentrator can increase the drafty performance [20],[22], but the gap among the mirror units reduced the active concentrating areas. In order to guarantee a certain concentrating area, space size of concentrator will be larger which is not suitable for large concentrators.

Therefore, we choose the single parabolic surface concentrator which normally consists of the rectangle or sector mirror unit (use glass reflective mirror). Using rectangle mirror units can generally form an approximate parabolic surface concentrator, such as the Serial #3 unit in SES Company [2]. However, the sector mirror units could ensure a accuracy parabolic surface concentrator [23],[24], which can effective reduce the number and the type of the concentrator mirror units. Such as the SunCatcher^{TM} unit improving designed by the SES company, the number of the mirror units are only 40 [2]. In this paper, our single concentrator will use this sector mirror unit (Figure 4(a)).

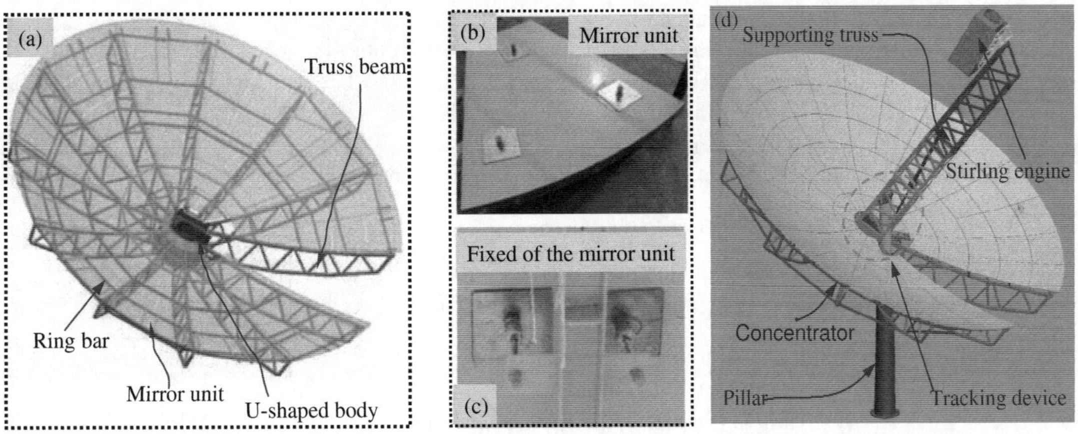

Figure 4. DS-CSP frame structure: a) CAD design model for the concentrator, b) mirror units structure, c) installation of the mirror unit, and d) CAD image of the DS-CSP system.

3.2. Supporting structure selection

Regarding the support structure of the dish concentrator system, the large dish concentrator usually uses the disk rail support due to the superior bearing characteristics in existing literatures [9][25][26]. For example, the Australian National University successively established a 400 m^2 and a 500 m^2 dish concentrator [25][26], and the SBP company implemented the 50 kW DS-CSP unit with such support structure [9]. The disk rail has advantages in stress characteristics. However, the disk rail support still has some disadvantages. For instance, the structure should be protected from the dust and sand and additional weights are needed to balance the dish concentrator system. While the single pillar support form shows advantages in easy construction and good sealing performance, which requires less maintenance and the balance of the system can be easily achieved by the structure parts mass optimize allocation. Due to the advantages of such structure, it is widely used in various CSP projects [2,3]. Hence, the 38 kW DS-CSP system in this study also utilize the single pillar support (Figure 4(d)).

3.3. Structure design of dish concentrator system

Modular design is considered in present study in order to have convenience in manufacturing, transportation, and installation. Concentrator is connected with twelve truss beams to the centre U-shaped body by bolt (Figure 4(a)). The adjacent truss beams are connected through the ring bar which further forms a space truss structure to support the mirror units. Mirror unit is an aluminium honeycomb panel matrix with polyurethane pasted on the upper surface. The polyurethane is bended into parabolic surface with the reflection mirror surface adhesive to the polyurethane surface (Figure 4(b)). This structure can improve the rigidity of the mirror unit structure. The mirror unit is fixed in the space truss structure by the ball joint bolt and the space position of the mirror unit is controlled by adjusting the length of the bolt (Figure 4(c)).

The single dish plus pillar supporting structure is designed for the DS-CSP system and is shown in figure 8(d). The front end of the supporting truss is fixed on the concentrator U-shaped body. The Stirling engine is installed at the end of the supporting truss (absorber is at the focal position of the concentrator), and the supporting truss is connected with the tracking device mounted on the top of the pillar. The tracking device uses the elevation angle plus azimuth angle two-axis tracking method (Figure 4(d)). The tracking device is a combination of the azimuth angle rotating mechanism and lead screw expand and contract mechanism. Both of these two mechanisms are equipped with worm-gear motor to lock the frame structure when the external load applied. The specialty of this design is that the lead screw is jointed at the lower part of the pillar (normally at the top of the pillar, Serial#3 unit for example [2]). The novel design could lead to the benefit that the lead screw will not cover the focal axis of the concentrator and it is convenient for the calibration of the concentrator focal axis when the Stirling engine is installed. Moreover, it can increase the lead screw expand and contract mechanism transmission angle which in turn reduce the push-pull force of the lead screw.

It is crucial to make sure that there is no mechanical interference between the lead screw and the support truss when the lead screw is expanding and contracting. The concentrated system gravity center should be coincidence with the supporting pillar axis which could be achieved by optimizing the allocation of the total mass weight of the concentrator and the supporting truss. This can be effectively obtained in the Pro/E software platform. The main parameters of the DS-CSP frame are shown in table 1.

Table 1. The main parameters and performance index for the 38 kW DS-CSP system.

Name	Numerical or model
System design value	Rated power is 38 kW, DNI design value is 750 W/m^2, solar-to-electricity efficiency design value is 25%.
Stirling system	Stirling Biopower Flexgen S260, α-type, four cylinders double-acting Stirling engine, working medium is hydrogen, mass is 1600 kg, receiver window diameter is 400 mm.
Dish concentrator	Aperture area is 204 m^2, diameter is 17.70 m, focal length is 9.49 m, rim angle is 50.0°
Number of mirror units	164
Total mass	25.2 tons (Do not include the foundation)
Tracking mode	Elevation-azimuth double-axis tracking device, open loop control system
Output function	Net power output=0.1003×DNI-36.129; DNI∈[460, 761] W/m^2
Peak power	40.5 kW (The effective total aperture area is 200 m^2, DNI is 761 W/m^2, and SEE is 26.9%)

4. Concentrator system installation and test

4.1. Installation of the mirror unit

The optical performance of the concentrator is directly decided by the precision of the mirror unit installation. A laser positioning device is used to guide the installation of mirror units (Figure 5(a)) and the schematic of the laser device are shown in Figure 5(b).

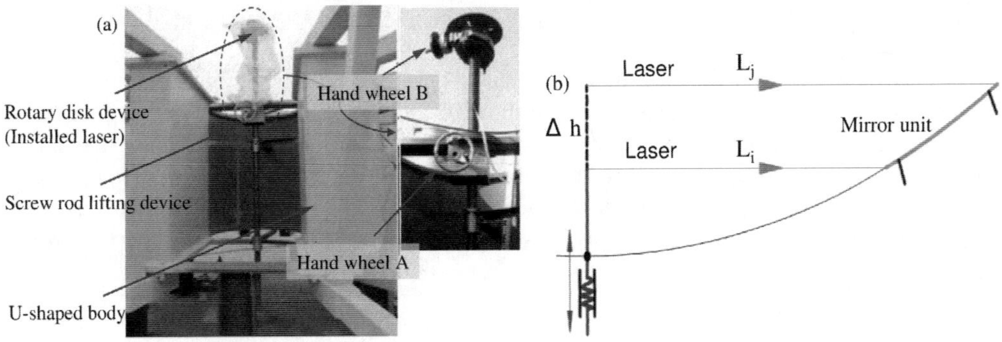

Figure 5. The installation and positioning of mirror units: a) laser positioning device and b) working principle.

Figure 6. The construction site: a) mirror units installation, b) connecting process of concentrator to the pillar, c) rear view and d) front view.

The detailed implementation steps are as listed follows: 1) pose concentrator truss structure open upward and mark the about-to-install parabolic mirror's focal axis (Z-axis). The make the screw axis coincidence with the z-axis and fix the laser device on the center of the U-shaped body, 2) rotating the hand wheel A to lift up or down the screw to mark parabolic reflection mirror surface top point (the origin of coordinate O) and then rotate the hand wheel B to point the original position (X-axis) at the parabolic reflection mirror surface (hand wheel A and hand wheel B also have the rules that can accurate fix the desired position), 3) adjust the hand wheel A and B to make laser beam align to the install mirror angular point to the ideal installation mirror unit position, and 4) tune the length of the mirror unit ball joint bolt to make sure that the angular point of the mirror unit is aimed at the laser beam. Repeat the step 3) and step 4) to determine the position of the four mirror unit angular points to complete the installation of the mirror units. Once all the mirror units are installed (Figure 6(a)), fixing the concentrator with the support truss and then combine them with the tracking device on the top of the pillar (Figure 6(b)). The pillar is mounted around the concentrator construction area (Figure 6(a)). The DS-CSP system is built in the factory of the Xiangtan Electric Manufacturing Group (Figure 9).

4.2. Concentrator performance test

4.2.1. Flux distribution on focal plane. In order to determine the concentrator of the focal spot distribution and tracking performance, the simple and effective method is to install a flat receiver in the concentrator focal plane [25] for receiving the solar energy reflected from the concentrator (to form a focal spot). The performance of the concentrator was evaluated by the focal spot brightness distribution and geometrical shape.

In this paper, a water-cooled circular target (diameter is 720.0 mm) is designed to calibrate the focal plane in terms of test the focal spot geometry and two-axis tracking performance (Figure 7). Under the sun tracking condition, a CCD camera (Sony A99) is used to take focal spot image from the water-cooled target plane (Figure 7(c)). Since the water-cooled target surface is lack of Lambert features, we can't precisely know the absolute focal spot flux distribution. However, it is sufficient for us to investigate the geometry and size of the focal spot. The results show that the focal spot is in approximate circular shape and the focal spot center is coincident with the center of the water-cooled disk target. This indicates that the sun tracking performance is well (the tracking control is open loop based on the astronomical algorithm). The diameter of the focal spot flare highlighted area is 272.0 mm (not including the halo area) which is less than the diameter of the absorber receiver window (R_c=400 mm). The absorber can receive most of the concentrated solar energy which can be proved in Figure 9.

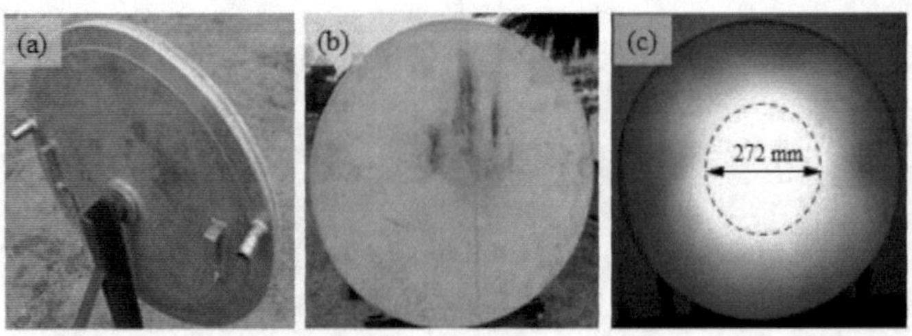

Figure 7. Pictures of the water-cooled target test: a) water-cooled target back view, b) water-cooled target front view, and c) focal spot (focal plane position).

4.2.2. Flux distribution on absorber imitation device. The absorber metal coil flux distribution homogeneity performance influenced the Stirling engine running performance, which including two aspects: (1) the absorber metal coil is composed of four integrated tubes (Figure 1 a2), each integrated tube connected with one Stirling engine piston chamber for providing energy to the piston movement. When four integration tubes receive the difference total heat, the four piston thrust will be different which can cause unstable phenomenon in the S260 Stirling engine operation process. Therefore, the requiring of the four integration tubes receiving the equal total energy. This is a macro requirement for the metal coil flux distribution homogeneity. (2) The metal coil working in high temperature and high pressure environment. But the metal coil hydrogen working medium heat exchanging ability is limited, the local metal coil area receiving too much energy will cause the metal coil burning or the thermal stress fracture accident caused by large temperature gradient [24]. Hence, the microcosmic requirement for the flux distribution homogeneity on the metal coil is to avoid local "hot spots".

However, water-cooled target experiment can determine the focal plane flux distribution and it can not reveal the absorber flux distribution on the metal coil surface. Therefore, in order to ensure the safe and smooth operation of the S260 Stirling engine, we put forward a new method for evaluating the absorber surface energy distribution. An absorber imitation device (metal material is Q235) is designed and implemented based on the shape of the S260 Stirling engine absorber cavity (Figure 1(b)). The absorber imitation device is installed on the focal position of the dish concentrator (figure 12(a)) and heating performance test (Figure 8(b) and Figure **8**(c)) was conducted. It can be seen that melting was detected on the surface of the absorber imitation device and local surface temperature could be larger than 1493°C (the metal Q235 melting point is 1493°C). The brightness distribution within the absorber imitation device has the similar behaviour with the results shown in Figure 2(a) and the energy mainly focus on the absorber bottom area. The absorber inner surface receiving the solar energy distribution is a circular, the energy distribution is homogeneous and without obvious local "bright spot", which reached to the absorber energy uniformity requirements. Besides, the brightness distribution in Figure 8(c) shows that the reflection cone area has the extremely lower irradiance due to the reflective cone surface reflected the incoming sunlight into metal coil surface and increase the energy of the metal coil surface which shows the functionality of the reflection cone.

It indicates that both the installation precision and two-axis tracking precision of the dish concentrator system meet the operating requirements and Stirling engine can be installed and then the DS-CSP system can make trial operations.

Figure 8. Test of the absorber imitation device.

5. Power generation operation and analysis

The Stirling engine can be started only when the absorber working temperature is high enough (400℃ for the hydrogen (working fluid of the engine) for the 38 kW DS-CSP system in present study) which is the precondition for the operation of the DS-CSP system.

As shown in Figure 9, the first field test of the 38 kW DS-CSP system was conducted on 19th December 2014 with the two inner loops of the mirror covered with shelters. Due to the discontinuity of solar irradiation resulted from the clouds on that day, the hydrogen temperature will decrease accordingly and the Stirling engine will stop when the hydrogen temperature is lower than 400℃. The Stirling engine will be started again then the hydrogen temperature in the metal coil is up to 400℃ once the cloud passing away. A series of the Stirling engine start-stop process prove that the 38 kW DS-CSP system can automatic fit cloud cover weather. The Stirling engine automatic start-stop process in controlled by control system and the automatic control is depended on the feedback of the metal coil hydrogen temperature. The absorber detailed picture during the working condition is given in Figure 9 which can be seen that most of the absorber aperture could capture most of the concentrated solar irradiation. The adjustment and stable test of the DS-CSP control system are accomplished in 8 months. The relationship between the DS-CSP generation power and the DNI is measured and presented in figure 10 under clear weather with the total concentrated area is 200 m^2 (real time the DNI measurement is achieved by using the GEONICA METEODATA-3000 system, this system was from Geonica company). The main parameters and performance index for the 38 kW DS-CSP system is shown in table 1.

Figure 9. Pictures for the DS-CSP system power operation for the first time (19 December 2014).

As seen from the power generation curve of the DS-CSP system in figure 10(a) (October 26th, 2015), the DS-CSP system continuous operation for 6.5 hours and the power generation is larger than 35 kW during the time from 11:00 to 15:00 (excluded the cloud cover time). When the DNI maximal value is 761 W/ m^2 the DS-CSP system power can reach up to 40.5 kW and the SEE is 26.6% (at 13:25). In the figure 10(b) (October 15th, 2015), during the time period of 11:00 to 13:10, the DNI basically stabilized around 750 W/m^2 and the DS-CSP power value is larger than 38 kW and SEE is 25.3%. The DS-CSP system testing operating SEE value is higher than the design value 25%. Results show that the concentrator design can effectively match the S260 Stirling engine power and the design of the absorber for such Stirling engine is reasonable, which also validates that the S260 Stirling can be well coupled to a solar dish concentrator for power generation. In addition, the concentrator has some optical losses due to the dust coverage, which indicates that the DS-CSP system could achieve even higher SEE. Note that when the elevation angle is small the DS-CSP system will be shadowed by the nearby buildings that the DS-CSP system can only operate up to 15:41 in figure 10(a).

According to the operating data, the relationship between the DS-CSP system net power output and the DNI value is obtained in figure 10(c). The data originated from four days' operation (8th, 15th, 17th, and 26th October 2015) and the unstable data are not included during the Stirling engine shut-up and shut-down. It can be seen that the DS-CSP system net power output has linearly proportional relationship with the DNI value, and this is the same to the DS-CSP system power output curve fitted by Mancini *et al* [3] (DS-CSP system power output curve comparison can be seen in Figure 11). The 38 kW DS-CSP system net power output fitting linear function is y=0.1003x-36.129, where y is the net power output of the DS-CSP system (unit is kW), x is the DNI value. The range of the x is between 460 W/m^2 to 761 W/m^2 (figure 10(c)). This net power output function can be used for 38 kW DS-CSP system output power under different DNI values. Only when we know the power station region solar irradiation data throughout the year can we evaluate the annual power generation according to the net power output function and then assess the economy of the power station construction and operation. This is a very important evaluation work during the early construction period of the business DS-CSP system.

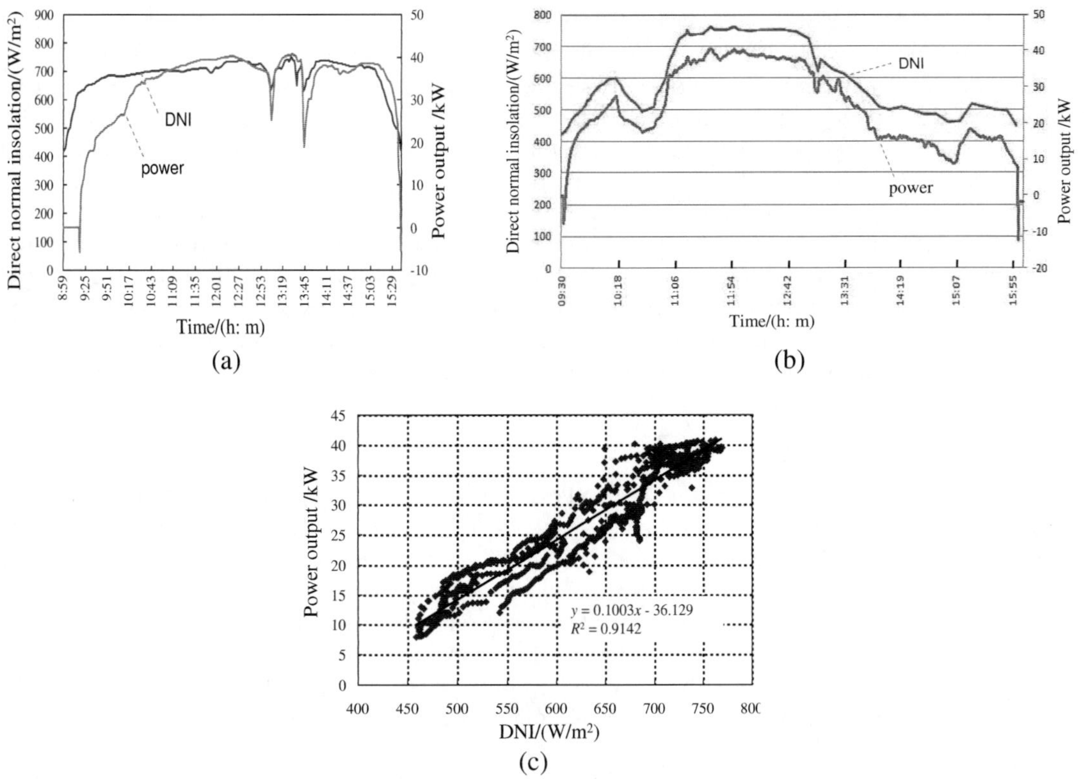

Figure 10. Operating data for the 38 kW DS-CSP system (concentrated area is 200 m2): a) power generation curve on October 26th, 2015, b) power generation curve on October 15th, 2015, and c) DS-CSP power generation curves for 8th, 15th, 17th, and 26th October 2015.

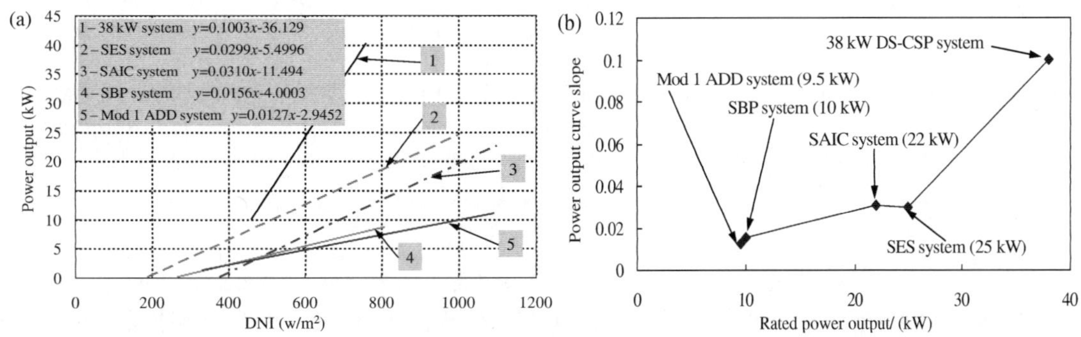

Figure 11. DS-CSP system power output curve comparison: a) power output curves for various systems, and b) power output curve slope comparison. The curves are all acquired form literature (Mancini *et al* [3]) and the test data for our 38 kW DS-CSP system was obtained from the date (8th, 15th, 17th, and 26th October 2015). Where x is the DNI value and y is the power output of the DS-CSP system, the range of x is between the line's both endpoints respectively.

In Figure 11(a), the 38 kW DS-CSP system net output power reached to 38 kW when the DNI value was 750 W/m^2 and the absorber metal coil would be overheated-damaged when the DNI value was over 750 W/m^2, while other four kinds of DS-CSP system could reach the rated output power only when the DNI was about 1000 W/m^2. The Stirling engine has not been full played when the DNI value was less than 1000W/m^2. Since the DNI value is changing with the time, the absorber metal coil overheated and the Stirling engine lack of heat supply problems would happen. For the metal coil overheated problem, we have figured out a solution: installing the glass cover on the cavity absorber (Figure 1 a1) receiver window position to form a closed space in the absorber. When the metal coil temperature is high, pump the air into the absorber to properly cool the metal coil (made the cold air into one side of the absorber cavity, made the hot air out of the other side). The outflow hot air could also be used to heat the water that can achieve the solar power joint supply. This case could fit the single family or farm offering electric power and heating. Besides, developed the absorber with energy storage function is the best way to solve this problem, because it will not only satisfy the commercial power station energy storage requirements of DS-CSP systems but also could solve the absorber energy overheated-damaged problems. For the DS-CSP system lack of solar energy supply problems, it could be solved by the method of gas and solar energy complementary power generation. The key is to design an absorber that not only could receive the solar energy but also could receive the heat from the gas combustion.

6. Conclusions

In this paper, the rated power 38 kW DS-CSP system was developed. The parabolic concentrator aperture area is 204 m^2 with 8.85 m in radius and 9.49 m in focal length and the heat engine adopted the Stirling Biopower Flexgen S260. This paper forms a general method and process of the structure design, mirror installation and optics testing, which could provide the reference for the DS-CSP system design and development. The outcome has been concluded in the following points:

The 38 kW DS-CSP system has a well recorded working performance which shows the net output power as high as 38.0 kW under the DNI of 750 W/m^2 and the SEE of 25.3% accordingly. The maximum recorded net output power is 40.5 kW and the corresponding SEE is 26.6% with measured DNI of 761 W/m^2.

The net power output of the 38 kW DS-CSP system is linearly proportional to the DNI value and the fitting function is Net power output=0.1003×DNI−36.129, DNI∈[460, 761] W/m^2. The linear function can predict the annual total generation power of the 38 kW DS-CSP system, which could guide the DS-CSP system station construction economic evaluation.

The absorber imitation device testing method was presented, which could assess the flux distribution on the absorber cavity surface. The test results show that the brightness distribution within the absorber imitation device has the similar behaviour with the absorber flux distribution simulation result and the energy mainly focused on the absorber bottom area. While the reflection cone area reports much lower luminance, the reflection cone could effectively reflect the incoming sunlight into metal coil surface and this could conducively improve the absorber thermal efficiency.

7. Future work

The space size and mass of the dish concentrator in DS-CSP system are large. The structure deformation is easy to occur under the effect of wind load and gravity load further to affect the optical performance of the dish concentrator [27]. In order to reduce the mass and manufacturing cost of dish concentrators, it is necessary to carry out lightweight design of the frame structure on the basis of ensuring optical performance. Next, in the large-scale DS-CSP power station, the proximity of the DS-CSP system will generate light shading and cloud cover against the sunlight, which will affect the normal operation of the DS-CSP system. Therefore, the layout optimization of the DS-CSP system in large-scale power stations is especially important. Besides, large dish concentrators are usually assembled with many mirror units forming the parabolic reflecting mirrors and each mirror unit must be accurately fixed in a present position within the truss structure. If facet alignment error exists after the mirror unit is installed, the shape accuracy of the parabolic concentrator surface can be affected [28]. Therefore, correctly aligning the mirror units in dish a concentrator while minimizing installation error is therefore of great importance to overall performance. The development of an effective mirror alignment method or alignment tool to guide mirror facet alignment during operation would be extremely valuable [29].

Acknowledgments
This work is supported by the National Natural Science Foundation of People's Republic of China (No. 51576061, 51641504), Hunan Province Natural Science Foundation of People's Republic of China (No. 2016JJ2052), and Hunan Province Postgraduate Innovation Project Foundation of People's Republic of China (No. CX2016B549, CX2017B628).

References

[1] Poullikkas A *et al* 2010 Parametric analysis for the installation of solar dish technologies in Mediterranean regions *Renew Sust Energ Rev* **14** 2772-83
[2] Baharoon D A *et al* 2015 Historical development of concentrating solar power technologies to generate clean electricity efficiently–A review *Renew Sust Energ Rev* **41** 996-1027
[3] Mancini T *et al* 2003 Dish-Stirling systems: An overview of development and status *J Sol Energy Engineering* **125** 135-51
[4] Bravo Y *et al* 2012 Environmental evaluation of dish-Stirling technology for power generation *Sol Energ* **86** 2811-25
[5] Sandia 2004 Stirling to build solar dish engine power plant, goal is to deploy solar dish farms with 20,000 units producing energy https://share.sandia.gov/news/resources/releases/2004/renew-energy-batt/Stirling.html
[6] Karabulut H *et al* 2009 An experimental study on the development of a β-type Stirling engine for low and moderate temperature heat sources *App Energ* **86** 68-73
[7] Stine W B 1993 International survey of parabolic dish/Stirling engine electrical power generation technology *ASME* (New York, USA) 421-7
[8] Stine W B and Diver R B 1994 A compendium of solar dish/Stirling technology (No. SAND93-7026) *Sandia national labs albuquerque NM*.
[9] Koshaim B 1986 Report: Fifty KW solar Membrane Concentrator *The SOLERAS Program, Saudi Arabian National Center for Science and Technology*
[10] News Releases 2008 Sandia, Stirling energy systems set new world record for solar-to-grid conversion efficiency https://share.sandia.gov/news/resources/releases/2008/solargrid.html
[11] Maricopa Solar Project 2013 http://www.nrel.gov/csp/solarpaces/project_detail.cfm/projectID=58
[12] Directory: InfiniaCorp 2016 http://peswiki.com/directory:infiniacorp
[13] Hafez A Z *et al* 2016 Solar parabolic dish Stirling engine system design, simulation, and thermal analysis *Energ Convers Manage***126** 60-75.
[14] Hafez A Z *et al* 2017 Design analysis factors and specifications of solar dish technologies for different systems and applications *Renew Sust Energ Rev* **67** 1019-36
[15] Cheng C H and Yang H S 2012 Optimization of geometrical parameters for Stirling engines based on theoretical analysis *App Energ* **92** 395-405
[16] Hooshang M *et al* 2015 Optimization of Stirling engine design parameters using neural networks *Renew Energ* **74** 855-66
[17] Ahmadi M H *et al* 2017 Thermal models for analysis of performance of Stirling engine: A review *Renew Sust Energ Rev* **68** 168-84
[18] Xiao G *et al* 2017 Design optimization with computational fluid dynamic analysis of β-type Stirling engine *App Therm Eng* **113** 87-102
[19] He Y L *et al* 2011 A MCRT and FVM coupled simulation method for energy conversion process in parabolic trough solar collector *Renew Energ* **36** 976-85
[20] Tan M H and Chong K K 2016 Influence of self-weight on electrical power conversion of dense-array concentrator photovoltaic system *Renew Energ* **87** 445-57
[21] Huang X *et al* 2014 Development of a multi-layer and multi-dish model for the multi-dish solar energy concentrator system *Sol Energy* **107** 617-27
[22] Giannuzzi A *et al* 2015 Enhancing the efficiency of solar concentrators by controlled optical aberrations: Method and photovoltaic application *App Energ* **145** 211-22

[23] Wang K *et al* 2015 Experimental study on a coiled tube solar receiver under variable solar radiation condition *Sol Energy* **122** 1080-90
[24] Xiao G *et al* 2016 A model-based approach for optical performance assessment and optimization of a solar dish *Renew Energ* **100** 103-13
[25] Lovegrove K *et al* 2011 A new 500 m 2 paraboloidal dish solar concentrator *Sol Energy* **85** 620-6
[26] Lovegrove K *et al* 2003 Paraboloidal dish solar concentrators for multi-megawatt power generation *ISES Solar World Conference* (Goteborg, Sweden)
[27] Yan J *et al* 2015 Research on thermal-structural-optical multi-field coupling modeling and concentrating performance predication of solar dish system *Journal of Mechanical Engineering (in Chinese)* **51** 138-51
[28] Yan J *et al* 2016 Correlative characteristics between focal spot of solar dish concentrator mirror unit and posture error *Acta Optica Sinica (in Chinese)* **36** 1122003
[29] Yan J *et al* 2017 Solar concentrator mirror unit supporting/adjusting structure and posture alignment *Acta Optica Sinica (in Chinese)* **37** 0522001

Post-evaluation of a ground source heat pump system for residential space heating in Shanghai China

Y Lei[1,2], H W Tan[1,2,3,5] and L Z Wang[4]

[1]School of Mechanical and Energy Engineering, Tongji University, Shanghai 201804, China
[2]Research Center of Green Building and New Energy, Tongji University, Shanghai, 200092, China
[3]UNEP-Tongji Institute of Environment for Sustainable Development 200092, China
[4]Shanghai Research Institute of Building Sciences, Shanghai 201108, China

E-mail: hw_tan@tongji.edu.cn

Abstract. Residents of Southern China are increasingly concerned about the space heating in winter. The chief aim of the present work is to find a cost-effective way for residential space heating in Shanghai, one of the biggest city in south China. Economic and energy efficiency of three residential space heating ways, including ground source heat pump (GSHP), air source heat pump (ASHP) and wall-hung gas boiler (WHGB), are assessed based on Long-term measured data. The results show that the heat consumption of the building is 120 $kWh/m^2/y$ during the heating season, and the seasonal energy efficiency ratio (SEER) of the GSHP, ASHP and WHGB systems are 3.27, 2.30, 0.88 respectively. Compared to ASHP and WHGB, energy savings of GSHP during the heating season are 6.2 $kgce/(m^2.y)$ and 2.2 $kgce/(m^2.y)$, and the payback period of GSHP are 13.3 and 7.6 years respectively. The sensitivity analysis of various factors that affect the payback period is carried out, and the results suggest that SEER is the most critical factor affecting the feasibility of ground source heat pump application, followed by building load factor and energy price factor. These findings of the research have led the author to the conclusion that ground source heat pump for residential space heating in Shanghai is a good alternative, which can achieve significant energy saving benefits, and a good system design and operation management are key factors that can shorten the payback period.

1. Introduction

After a period of heavy snow and freezing weather in Southern China at the beginning of 2013, the debate over whether the government should provide central heating in the south of China is becoming more intense. Up to now, the central heating systems cover about 70% of the urban building area in Northern China (Yellow River basin and its north) but less than 5% of the urban building area in Southern China (Yangtze River basin, Hunan, Guizhou, and Yunnan etc.) [1]. During the winter season, the indoor as functions of residential buildings without central heating systems are fluctuated as functions of the change of outdoor temperature in China's hot summer and cold winter region. And the average indoor temperature is only 10-17°C, which is far from ASHRAE comfort zone [2,3]. Therefore, space heating in winter is becoming an urgent need for residents in those areas.

For the purpose of finding a cost-effective way of residential space heating in China's hot summer and cold winter region, energy efficiency and economic feasibility of three residential space heating ways, including ground source heat pump (GSHP), air source heat pump (ASHP) and wall-hung gas boiler (WHGB), are assessed based on Long-term measured data in the present study.

2. Systems description

2.1. Building

The testing building is located in Shanghai (31.20°N, 121.21°E). It is a three-storey private residential house, with a total gross floor area of 280 m^2 and air-conditioned area of 180 m^2. Figure 1 shows the second-floor plan of the building. The average low temperatures in the area in January are about 4.3℃. There are approximately 1648 heating degree days (18℃ base). The peak value of the heat consumption was calculated as 20 kW, equivalent to 110 W/m^2. The occupancy time of the building is from 17:00 to 08:00 next day for the working days and 24 h for weekend.

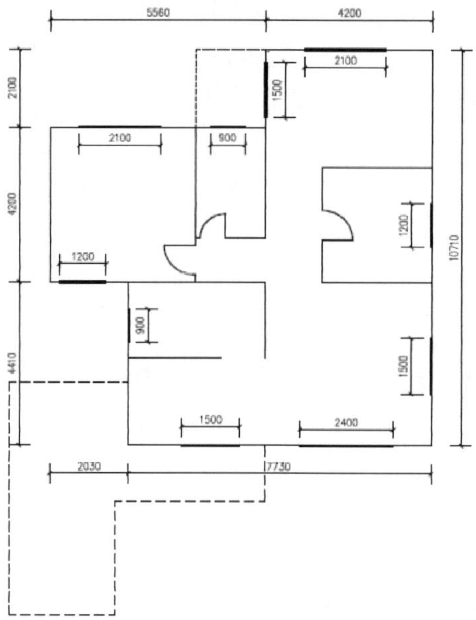

Figure 1. The second-floor plan of the building.

2.2. Space heating systems

2.2.1. GSHP heating system. The GSHP heating system installed in the testing building is mainly composed of three parts of ground heat exchanger (GHE), mechanical equipment (including heat pump and circulating pumps), and radiant floor heating terminal units. The detailed specifications and main technical parameters of the GSHP heating system were given in table 1.

Table 1. The parameters of the main components in GSHP heating system.

Ground heat exchanger (GHE)	
Type of GHE	Vertical single U-bend
Material of U-bends	HDPE high density polyethylene
Size of U-bends	Φ32 mm×3.2 mm
Depth of boreholes	5×90 m
Mechanical equipment	
Heat pump	Rated heating capacity 16 kW /inverter / DAKIN EEHS-6P

circulation pumps	Nmax=0.27 kW, Hmax=12.8 m, Qmax=4 m^3/h GROUNDFOS 25-125-180

2.2.2. Reference systems. Two conventional heating systems, which are the most commonly used for space heating in Shanghai, were selected as reference system in this study. One is a wall-hung gas boiler (WHGB) with a heating capacity of 20 kW, and the other is a heating system composed by eight split air source heat pumps (ASHP). The heating capacity of a single ASHP unit is about 3.5 kW.

2.3. Data acquisition system

A network data-acquisition system was designed to monitor the key operating parameters of GSHP system and ASHP system. The following data were regularly recorded every 5 minutes for GSHP system and ASHP system, respectively.

The parameters monitored for GSHP system as follows:

- Electrical power input to the compressor and circulating pumps by using watt-hour meters;
- Supply /return water temperature and water flow rates of the GHE by using an ultrasonic calorimeter with resistance thermometers (Pt100).
- Supply /return water temperature and water flow rates of the radiant floor heating system by using an ultrasonic calorimeter with resistance thermometers (Pt100).
- Ambient and indoor air temperatures by using resistance thermometers (Pt100).

The parameters monitored for ASHP system as follows:

- Electrical power input to the compressor and fans by using a watt-hour meter;
- Supply /return air temperatures of outdoor and indoor unit by using resistance thermometers (Pt100);
- Circulating air flow rates of outdoor and indoor unit by using hot-wire anemometers.

3. Analysis methods

3.1. Energy consumption and efficiency

3.1.1. Heat consumption. A calorimeter was used to monitor the thermal energy supplied to the building. Therefore, the heat consumption of the building in the $\Delta\tau$ interval starting at τ_0 time was integrated by the following formulas.

$$Q = \int_{\tau_0}^{\tau_0+\Delta\tau} Q(\tau)d\tau \tag{1}$$

$$Q(\tau) = m_{w,\tau}c_{p,w}\left(t_{sw,\tau} - t_{rw,\tau}\right) \tag{2}$$

Where $Q(\tau)$ is the instantaneous heat consumption, $m_{w,\tau}$ is the instantaneous mass flow rate of water, $c_{p,w}$ is the specific heat capacity of water, $t_{sw,\tau}$ and $t_{rw,\tau}$ represent the instantaneous temperature of supply water and return water, respectively. And the heat consumption per unit area over the heating season was calculated by using equation (3).

$$q_s = \frac{Q|_{\tau_{start}}^{\tau_{end}}}{A} \tag{3}$$

Where A is the total space heating area, $Q|_{\tau_{start}}^{\tau_{end}}$ is the total heat consumption over the heating season, τ_{start} and τ_{end} represent the space heating start and stop times, respectively.

3.1.2. GSHP system performance calculation. The energy consumption of the GSHP heating system consists of the electricity consumption of the heat pump units and water circulating pumps. The electricity consumption is monitored by electricity meters through integrating instantaneous power consumption ($N(\tau)$) by using equation (4).

$$N = \int_{\tau_0}^{\tau_0+\Delta\tau} N(\tau)d\tau \tag{4}$$

The coefficient of performance (COP) of the GSHP unit ($COP_{GSHP,\ unit}$) and the whole GSHP heating system ($COP_{GSHP,\ sys}$) were calculated by equations (5) and (6), respectively.

$$COP_{GSHP,unit} = \frac{Q}{N_{GSHP}} \tag{5}$$

$$COP_{GSHP,sys} = \frac{Q}{N_{GSHP}+N_{pump}} \tag{6}$$

Where N_{GSHP} is the electricity consumption of GSHP unit, and N_{pump} is the electricity consumption of water circulating pumps ,including circulating pumps of ground coupling loop and circulating pump of radiant floor heating loop.

The seasonal primary energy ratio of GSHP heating system ($SPER_{GSHP,sys}$) was calculated by

$$SPER_{GSHP,sys} = \varepsilon \cdot SEER_{GSHP,sys} \tag{7}$$

Where ε is the power generation efficiency, according to statistics [4], the power generation efficiency in the present study is set to 0.31. $SEER_{GSHP,sys}$ is the seasonal energy efficiency ratio of GSHP heating system, which was calculated by equation (8)

$$SEER_{GSHP,sys} = COP_{GSHP,sys}|_{\tau_{start}}^{\tau_{end}} \tag{8}$$

3.1.3. Reference system calculation. A split air source heat pump (ASHP) heating system was investigated experimentally. The following parameters were monitored for a long period with 5-minute intervals, including the power consumption of the system (N_{ASHP}), supply air temperature ($t_{sa,\tau}$), return air temperature ($t_{ra,\tau}$), circulating air flow ($m_{a,\tau}$), ambient temperature ($t_{a,\tau}$), etc. The instantaneous COP of the ASHP heating system was calculated using the following formula:

$$COP_{ASHP,sys}(t_{a,\tau}) = m_{a,\tau}c_{p,a}(t_{sa,\tau} - t_{ra,\tau})/N_{ASHP} \tag{9}$$

Considering the impact of ambient temperature on the COP of the ASHP heating system, the seasonal energy efficiency ratio of ASHP heating system ($SEER_{ASHP,sys}$) was estimated as weighted average of $COP_{ASHP,sys}$ on different ambient temperature conditions, using the following formula:

$$SEER_{ASHP,sys} = \sum \omega(t_{a,\tau}) \cdot COP_{ASHP,sys}(t_{a,\tau}) \tag{10}$$

Where $\omega(t_{a,\tau})$ is the probability of ambient temperatures which are between $t_{a,\tau}$ and $(t_{a,\tau} + 0.5)$°C during the whole heating season. The seasonal primary energy ratio of ASHP heating system ($SPER_{ASHP,sys}$) was calculated by

$$SPER_{ASHP,sys} = \varepsilon \cdot SEER_{ASHP,sys} \tag{11}$$

The energy efficiency ratio of wall-hung gas boiler (WHGB) heating system is equal to the primary energy ratio of WHGB heating system at about 0.85-0.90 [6], and it does not vary with changes of the ambient temperature. The seasonal energy efficiency ratio and the seasonal primary energy ratio of WHGB heating system ($SEER_{WHGB,sys}$ and $SPER_{WHGB,sys}$) were set to 0.88.

We assumed that both the building heat consumption over the heating season by using reference heating systems (ASHP or WHGB) and GSHP heating system are the same. So, the energy consumption (EC) and primary energy consumption (PEC) of three heating systems were calculated by equations (12) and (13) respectively.

$$EC = Q|_{\tau_{start}}^{\tau_{end}}/SEER \tag{12}$$

$$\mathrm{PEC} = Q|_{\tau_{start}}^{\tau_{end}}/(29.3 \cdot SPER) \tag{13}$$

IOP Conf. Series: Earth and Environmental Science **93** (2017) 012053 doi:10.1088/1755-1315/93/1/012053

Where value of 29.3 is the calorific value of coal equivalent.

3.2. Economic analysis

To determine whether GSHP would produce any economic benefits for home owners, an economic analysis was carried out to compare GSHP with ASHP and WHGB system.

3.2.1. Initial cost. According to actual situation of the project construction and estimated using data from local contractors, equipment suppliers and construction estimating software, the investment cost, including material, labor, and taxes, for the three heating systems compared are given in

3.2.2. Table *2* [6].

Table 2. Initial investment of three heating systems.

Systems	Item	Cost (Yuan)
GSHP	Ground heat exchanger system	19,000
	Equipment	25,000
	radiant floor heating system	28,000
	Total	**72,000**
ASHP	Equipment	38,000
	installation and debugging	14,000
	Total	**52,000**
WHGB	Equipment	20,000
	radiant floor heating system	28,000
	Total	**48,000**

3.2.3. Annual operating cost. The annual operating costs of a space heating system include the costs of fuel and maintenance of equipment. The annual operating costs (AC) were given as:

$$\mathrm{AC} = EC \cdot FP + MC \tag{14}$$

Where FP is fuel price, MC is equipment maintenance costs. Values of FP and MC are given in Table 3 and 4, respectively.

Table 3. Fuel prices.

Fuel type	Unit cost
Electricity	0.67 Yuan/kWh
Natural gas	2.9 Yuan/m^3

Table 4. System maintenance costs.

Systems	Maintenance costs
GSHP	100 Yuan/y
ASHP	400 Yuan/y
WHGB	200 Yuan/y

3.2.4. Economic comparison. To study the economic feasibility of a system, different methods could be used to evaluate the different figures of merit of the systems. Some examples are: Net present value method, the internal rate of return method, the annual cost and other methods [8]. In this study, the present worth (PW) method has been used to compare the cost effectiveness of the three heating systems.

The PW method refers to all the events in an economy, both incomes and expenditures, as one figure at one point in time. When comparing different alternatives for a given revenue or cost saving, the alternative with the least present worth of all expenditures or annual charges is to be preferred.

IOP Conf. Series: Earth and Environmental Science **93** (2017) 012053 doi:10.1088/1755-1315/93/1/012053

Therefore, comparing the present worth of alternative A with alternative B, the payback period (PP) of alternative A would be the point-in-time when the present worth of B over that of A. The present worth (PW) was given as:

$$\mathrm{PW} = IC + AC \cdot \frac{1}{i}[1 - (1 + i)^{-n}] \tag{15}$$

Where IC is the initial cost, AC is the annual operating cost, i is interest rate, n is number of interest periods.

3.2.5. Sensitivity analysis. The PW method can be used for assessing the economic feasibility of different heating systems, but the result is subject to many uncertainties. Such as the building heat consumption, the system energy efficiency is uncertain due to the impact of the human behavior and test instruments accuracy, and fuel price also has some uncertainty. Thus, analysis of these uncertainties impact on the economy of the heating system is also very important.

Sensitivity analysis is very useful when attempting to determine the impact the actual outcome of a particular variable will have if it differs from what was previously assumed. By creating a given set of scenarios, the analyst can determine how changes in selected variables will impact the target variable. In this respect, the sensitivity analysis of the key factors that affect the payback period is carried out. For sensitivity analysis, three key parameters, including building heat consumption (q_s), the seasonal energy efficiency ratio of GSHP heating system ($SEER_{GSHP,sys}$) and fuel price (FP), were to be increased and decreased by 20% of their original value.

4. Results and discussions

4.1. Variation of ambient temperature and heat consumption

The ambient temperature is a key factor for building heat consumption and performance of heating systems. In order to observe and find out the interaction of energy consumption and coefficients of heating systems performance with ambient temperatures, the ambient temperature was measured continuously during the heating season. The measured data show that the ambient temperature was changing in the range of -2 ~20℃. Figure 2 represents the distribution of probability of ambient temperatures over the heating season. As we can see from the figure, the outdoor temperature is mainly distributed in the range of 2~13°C, accounting for a proportion more than 90%. And the weighted average value of the ambient temperature during the whole heating season is 7.96℃.

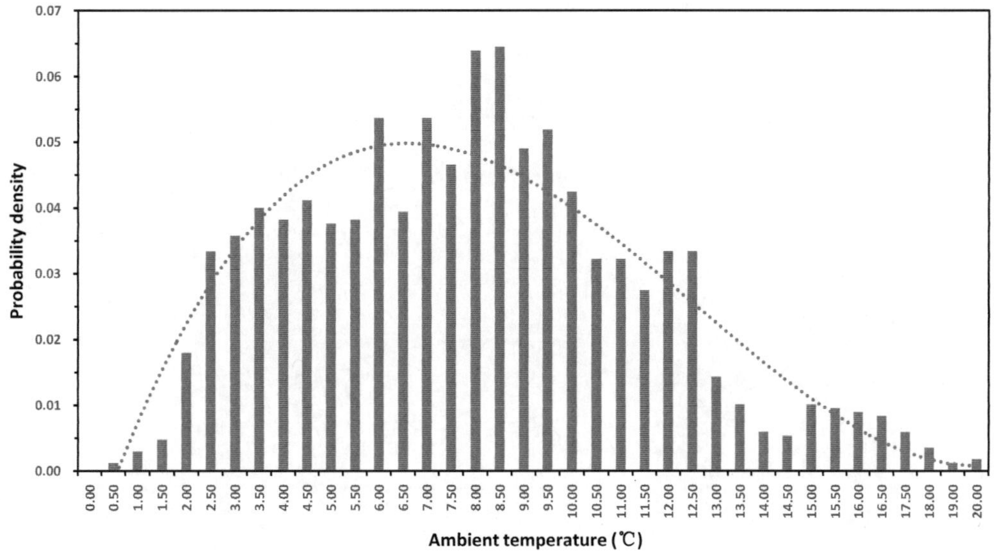

Figure 2. Probability density of ambient temperature during heating season.

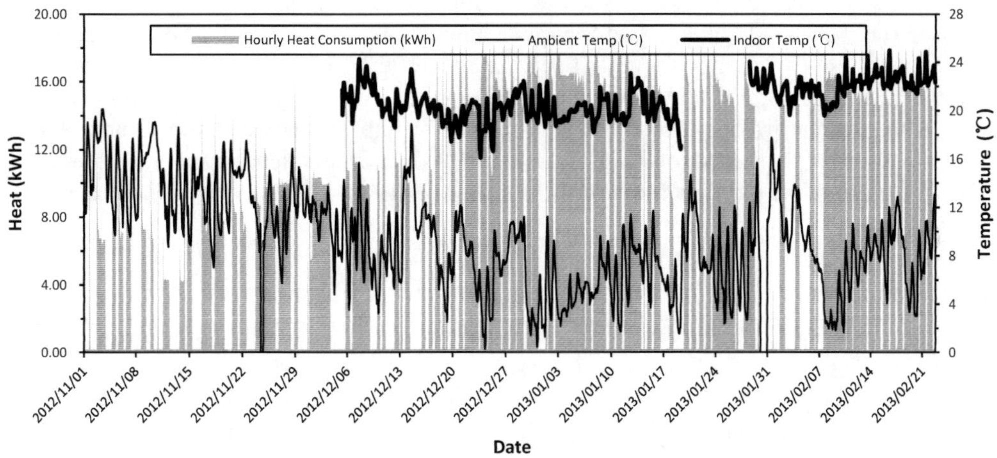

Figure 3. Variation of heat consumption with ambient and indoor temperature.

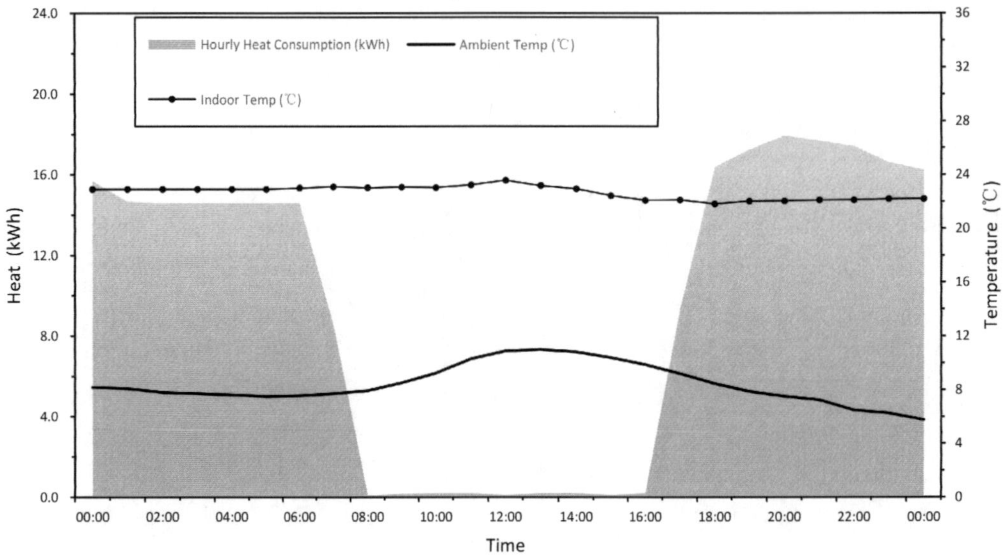

Figure 4. Variation of heat consumption in a typical day.

The variations of heat consumption, ambient temperature and indoor temperature during the heating season are shown in Figure 3. And Figure 4 shows the hourly variation of heat consumption in a typical day. As these two figures suggested that the indoor temperature is mainly maintained at 19-23℃, and the heat consumption and outdoor temperature has a very significant correlation. Figure 5 indicates that the average value of heat load is a linear function of ambient temperature. During the heating season, the maximum heat load is about 80 W/m^2 when the ambient temperature reached 0 ℃, and the heat consumption per unit area over the heating season calculated by equation (3) is 120 kWh/m^2.

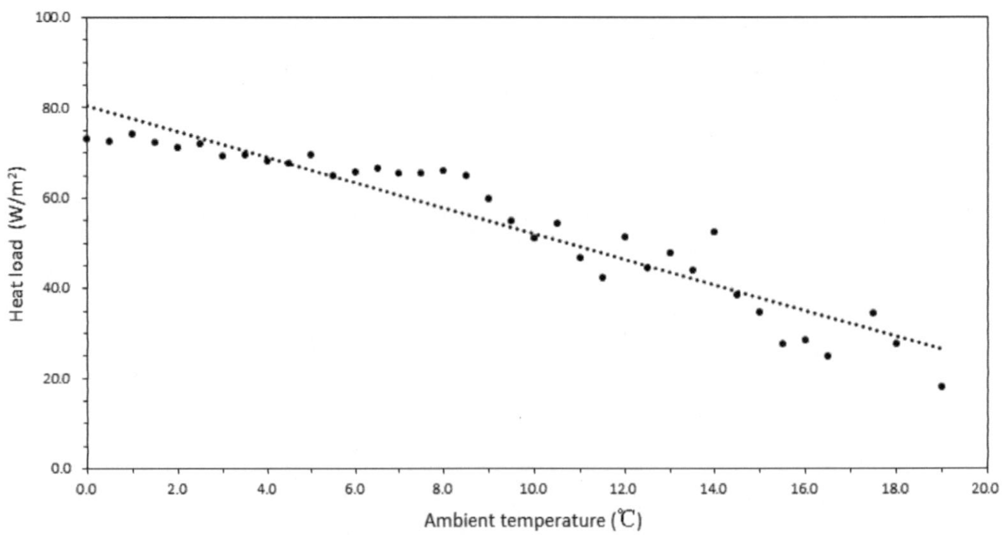

Figure 5. Average heat consumption versus ambient temperature.

4.2. Performance of GSHP heating system

The energy performance of a GSHP heating system can be influenced by three primary factors: (a) the heat pump machine, (b) the circulating pump and (c) the ground coupling. The heat pump is the largest single energy consumer in the system. Its performance is a function of the efficiency of the machine and the water temperature produced by the ground coupling [7].

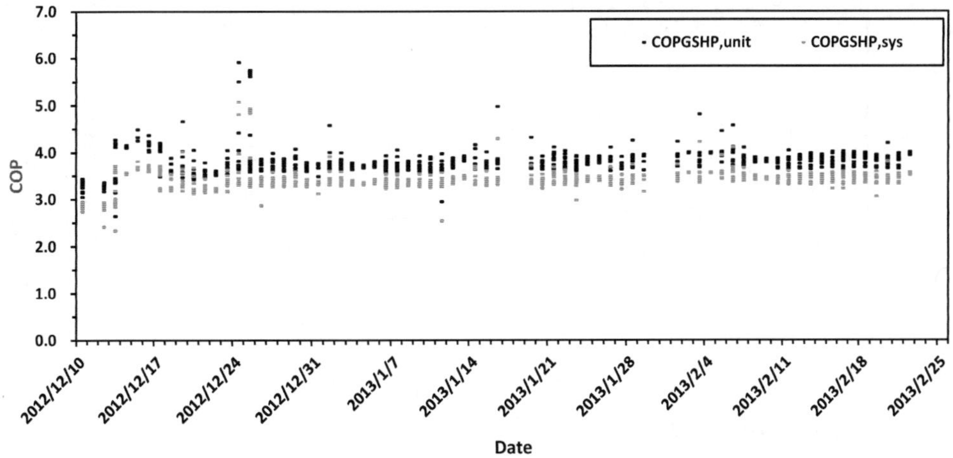

Figure 6. Unit and system COP values of GSHP during a heating period.

Based on the measured data, the COP of the heat pump unit ($COP_{GSHP,unit}$) and the whole system ($COP_{GSHP,sys}$) during the heating season were calculated by using equations (5) and (6). Figure 6 shows the hourly average values of $COP_{GSHP,unit}$ and $COP_{GSHP,sys}$ during a long-term heating period, from Dec. 2012 to Feb. 2013. It is clear from this figure that the values of $COP_{GSHP,unit}$ and $COP_{GSHP,sys}$ are relatively stable except some outlier points. The highest COP values were observed as 4.65 for the heat pump and 4.16 for the system at the beginning of the heating season, when the water temperature produced by the ground coupling reached a relatively high value at 17.2℃. About one month later since the starting time of heating, the water temperature produced by the ground coupling stabilized at 12℃ to 13.5℃ and the $COP_{GSHP,unit}$ and $COP_{GSHP,sys}$ varied between 3.51-4.25 and 3.2-3.6 respectively. The value of the seasonal energy efficiency ratio of GSHP heating system ($SEER_{GSHP,sys}$) and the seasonal primary energy ratio of GSHP heating system ($SPER_{GSHP,sys}$) which calculated by equations (8) and (7), is 3.27 and 1.01, respectively.

4.3. Performance of ASHP heating system

To determine the actual heating performance of an air source heat pump, a split ASHP was investigated experimentally under different weather conditions. The COP of the ASHP heating system was calculated using the equation (9), and Figure 7 shows the average values of $COP_{ASHP,sys}$ under different ambient temperatures at 0.5℃ interval. As the figure shows, the $COP_{ASHP,sys}$ with ambient temperature is a linear relationship. The $COP_{ASHP,sys}$ reached 3.57 when the ambient temperature get 12.5℃, but only 1.50 when the outdoor temperature dropped to 2 ℃.

According to this function between $COP_{ASHP,sys}$ and ambient temperature, and the above-mentioned distribution probability of ambient temperatures over the heating season, the seasonal energy efficiency ratio of ASHP heating system ($SEER_{ASHP,sys}$) and the seasonal primary energy ratio of ASHP heating system ($SPER_{ASHP,sys}$) calculated by equations (10) and (11) are 2.30 and 0.71, respectively.

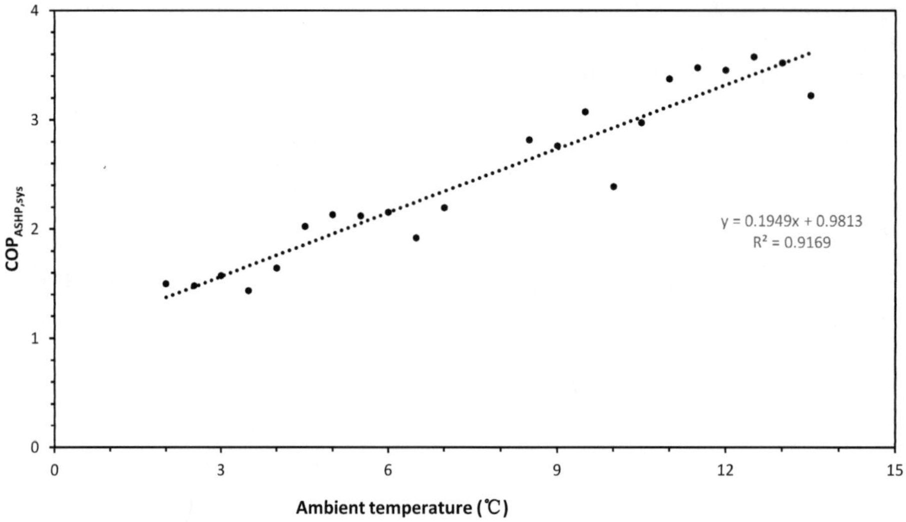

Figure 7. Variation of $COP_{ASHP,sys}$ versus the ambient temperature.

4.4. *Energy efficiency comparison*

Table 5. Energy efficiency comparison.

Item	GSHP	ASHP	WHGB
q_s (kWh/m^2)	120	120	120
SEER	3.27	2.30	0.88
SPER	1.01	0.71	0.88
EC (kWh(e)/m^2, Nm^3(g)/ m^2)	36.7(e)	52.2(e)	15.8(g)
PEC (kgce/m^2)	14.5	20.7	16.7

Based on the above-calculated results, the comparison of energy efficiency of three heating systems is given in table 5. As the table shows, the GSHP heating system has the highest seasonal primary energy ratio (SPER), followed by WHGB. Compared with ASHP and WHGB, the annual energy saved per unit heating area by GSHP is 6.2 kgce and 2.2 kgce, respectively. This has confirmed that GSHP is the most energy efficient way for space heating in Shanghai.

4.5. *Economic feasibility evaluation*

The present worth (PW) of three heating systems has been calculated by using equation (16) with a lifetime of 25 years and an interest rate of 5.6%. Figure 8 illustrates the variation of the present worth (PW) of three systems versus the operation life. As can be seen from the figure, at the twenty-five years, PW value of the GSHP system is the minimum and that of WHGB is the maximum. This result proves that the GSHP heating system is to be preferred among three heating systems. Compared with ASHP and WHGB heating system, the payback periods of GSHP system are 13.3 and 7.6 years.

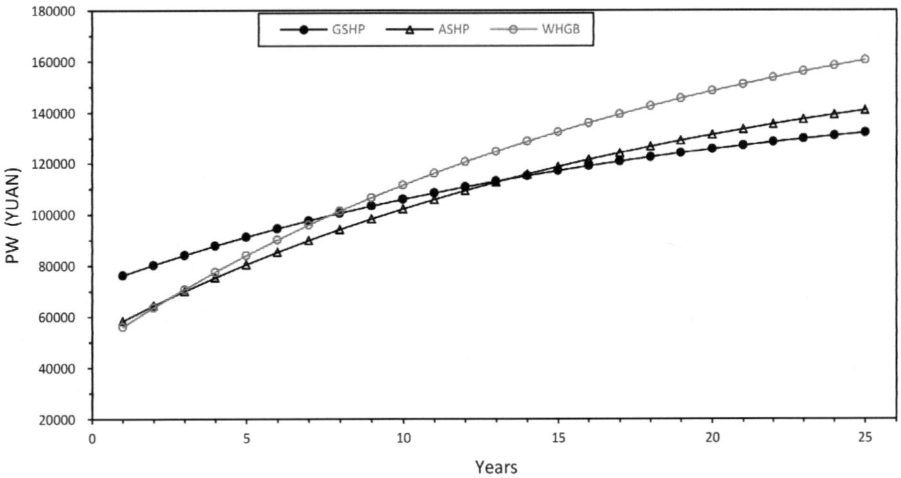

Figure 8. Variation of the present worth versus the operation life of the system.

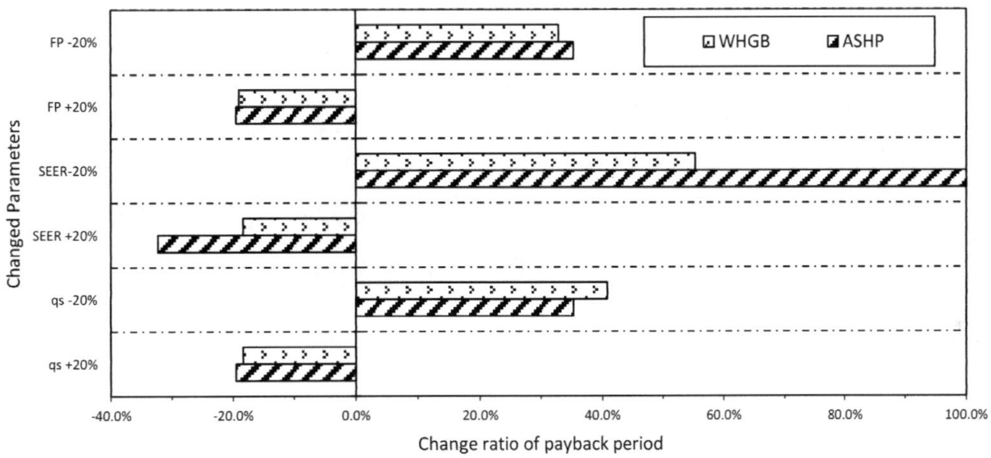

Figure 9. Sensitivity analysis of the payback period of GSHP compared with WHGB and ASHP.

To determine which parameters are the key drivers of the payback period of GSHP system, a sensitivity analysis is carried out. Three key parameters, including building heat consumption (q_s), the seasonal energy efficiency ratio of GSHP heating system ($SEER_{GSHP,sys}$) and fuel price (FP), were to be increased and decreased by 20% of their original value. The calculated results are shown in Figure 9. It is clear from this figure that the payback period of GSHP compared with WHGB and ASHP decreased with one of the parameters (q_s, $SEER_{GSHP,sys}$ and FP) increased. Among these three key parameters, the $SEER_{GSHP,sys}$ is the most important parameter affecting the payback period of GSHP both compared with WHGB and ASHP.

5. Conclusions

Energy efficiency and economic feasibility analysis of three heating systems for residential building in Shanghai was carried out in the present study. From the results of the experimental study, the following conclusions may be drawn:

- The average ambient temperature during the whole heating season is 7.96 ℃. The peak heat load of the residential building is about 80 W/m^2, and the total heat consumption is 120 kWh/m^2/y.

- The values of seasonal energy efficiency ratio (SEER) of the GSHP, ASHP and WHGB heating systems are 3.27, 2.30 and 0.88, respectively. And the corresponding values of seasonal primary energy ratio are 1.01, 0.71 and 0.88, respectively. Compared to ASHP and WHGB, energy savings of GSHP during the heating season are 6.2 kgce/(m^2.y) and 2.2 kgce/(m^2.y). Those results indicate that the GSHP heating system is the most efficient way for residential space heating in Shanghai.
- Comparing with ASHP and WHGB, the present worth (PW) value with 25 years lifetime of GSHP is the minimum, and the payback periods are13.3 and 7.6 years, respectively. This proves that the GSHP heating system is to be preferred among three heating systems.
- The sensitivity analysis results suggested that the seasonal energy efficiency ratio of GSHP heating system is the most important parameter affecting the payback period of GSHP heating system. Therefore, a good system design and operation management are the key factors that can shorten the payback period.

Acknowledgements
Financial support of this study by the research fund of the China- U.S. Clean Energy Research Center (CERC) under grant No. 2010DFA72740-05 is gratefully acknowledged.

Appendix A

Nomenclature			
A	The total heating space area [m^2]	SEER	Seasonal energy efficiency ratio [kWh/kWh]
AC	The annual operating costs [Yuan]	SPER	Seasonal primary energy ratio [kWh/kWh]
ASHP	Air source heat pump	t	Temperature [℃]
COP	The coefficient of performance[kWh/kWh]	WHGB	Wall-hung gas boiler
EC	Energy consumption [kWh]	i	interest rate [%]
FP	Fuel price [Yuan/kWh, Yuan/Nm^3]	τ	instantaneous time [h]
GSHP	Ground source heat pump	Δτ	time interval [h]
IC	The initial cost [Yuan]	Subscripts	
m	Mass flow rate [kg/h]	a	ambient
MC	Maintenance costs [Yuan]	pump	pump
n	Number of interest periods [year]	ra	return air
N	Electricity consumption [kWh]	rw	return water
ε	The power generation efficiency [%]	unit	unit
PEC	Primary energy consumption [kgce]	s	the heating season
PP	The payback period [year]	sa	supply air
PW	The present worth [Yuan]	sw	supply water
Q	Heat consumption [kWh]	sys	system
q	Heat consumption per unit area [kWh/m^2]	w	water

References

[1] CESSTU (Centre for Energy Saving Studies of Tsinghua University) 2009 *2009 Annual Report on China Building Energy Efficiency* (Beijing: China Architecture and Building Press)
[2] Yoshino H, *et al* 2006 Indoor thermal environment and energy saving for urban residential buildings in China *Energy Buildings* **2006** 1308-19
[3] ASHRAE Fundamentals Handbook 2001(SI) Chapter 9 Indoor Environmental Health 9.12

Hyperthermia Fundamentals 2001
[4] State Grid Energy Research Institute 2010 *Analysis Report on World Energy and Electricity Statistics* (Beijing: China Electric Power Press)
[5] China Academy of Building Research 2008 Evaluation Guidelines for Renewable Energy Demonstration Project
[6] Han W G, Jiang Y and Guo F 2005 Energy consumption analysis of several kinds of heating systems *HV and AC* **2005** 106-10
[7] Hepbasli A, Akdemir O and Hancioglu E 2003 Experimental study of a closed loop vertical ground source heat pump system *Energy Convers Manage* **44** 527-48
[8] Jasbir S A 2004 *Introduction to Optimum Design* 2nd ed (San Diego: Elsevier Academic)

IOP Conf. Series: Earth and Environmental Science **93** (2017) 012054 doi:10.1088/1755-1315/93/1/012054

A study of alternative refrigerants for the refrigeration and air conditioning sector in Mauritius

R K Dreepaul[1, 2]

[1]Faculty of Engineering and Sustainable Development, Universite de Mascareignes, Rue de la Concorde, Camp Levieux, Rose Hill, Mauritius

E-mail: rdreepaul@udm.ac.mu

Abstract. The most frequently used refrigerants in the refrigeration and air conditioning (RAC) sector in Mauritius are currently hydrochlorofluorocarbons (HCFC) and hydrofluorocarbons (HFC). However, because of their strong influence on global warming and the impact of HCFCs on the ozone layer, refrigerants such as ammonia (NH_3), carbon dioxide (CO_2) and Hydrocarbons (HC), having minimal impact on the environment, are being considered. So far, HCs have only been safely used in domestic refrigeration. Ammonia has been used mainly for industrial refrigeration whereas CO_2 is still under study. In this paper, a comparative study of the various feasible alternatives is presented in a survey that was undertaken with major stake holders in the field. The retrofitting possibility of existing equipment was assessed and safety issues associated with each refrigerant were analysed. The major setback of hydrocarbons as a widely accepted refrigerant is its flammability which was considered as a major safety hazard by the majority of respondents in the survey and the main advantages are the improved equipment coefficient of performance (COP) and better TEWI factor. This resulted in a 12 % drop in energy consumption. Despite the excellent thermodynamic properties of ammonia, its use has mainly been confined to industrial refrigeration due to its toxicity. In Mauritius, the performance of ammonia in air conditioning is being evaluated on a pilot basis. The major setback of carbon dioxide as a refrigerant is the high operating pressure which is considered a safety hazard. The high initial investment cost and the lack of qualified maintenance technician is also an issue. The use of CO_2 is mainly being considered in the commercial refrigeration sector.

1. Introduction

Mauritius has a mild tropical climate throughout the year and there is relatively little seasonal variation in temperature. January and February are the warmest months with an average day maximum temperature of 29.2 degrees Celsius while July and August are the coolest months with an average night temperatures dropping down to 16.4 degrees Celsius [1]. The average relative humidity which is around 60% in winter period reaches up to 95% in coastal regions during summer [1].

Demand for air conditioning is therefore high in Mauritius especially in summer. Refrigeration is also employed on large scale for processing, storage and transportation of food. The RAC sector in Mauritius will be described in more details in section 3.

Global warming and depletion of the ozone layer are two phenomena directly related to the refrigeration and air conditioning market. CFC's and HCFC's have pervaded the RAC sector since their advent in early 1930's. They gained popularity due to their high stability, very good thermodynamic characteristics and they were not flammable.

IOP Conf. Series: Earth and Environmental Science **93** (2017) 012054 doi:10.1088/1755-1315/93/1/012054

CFC's and HCFC's have been widely used in domestic refrigerators, commercial refrigerating equipment, industrial refrigeration, room air conditioning and water chillers. Nonetheless because of increasing environmental issues, their usage in the various sectors has been controlled.

The Montreal Protocol came into force on 1st January 1989 and its objective was to reduce the manufacture of Ozone depleting Substances (ODS) and hence curb down their consumption so as bring down their quantity in the air thereby protecting the environment.

With the Montreal Protocol a program was devised for phasing out the use of CFC's, HCFC's and other ODS by member nations [2, 3]. As per this agreement, small developing countries are grouped under article 5 nations having a consumption of 0.3 kg ODS per individual. As per the agreement, article 5 nations have an obligation to cease using CFCs by the year 2010, while HCFCs must be phased out by the year 2040 [4].

Mauritius has already put a ban on CFC's and use of HCFC's will be stopped in 2025 which is prior of the required schedule for article 5 countries. Import of new equipment that contains HCFC's has also been stopped since 2013.

A new generation of refrigerants, the HFCs, was developed in the 1990's to be used in place of CFC's and HCFC's. HFCs do not contain chlorine atoms; they are thus harmless for the ozone layer. This made them gain rapid popularity in the RAC sector and are widely being used to replace HCFC's. But in spite of having zero ODP, all HFC's have a very high GWP. HFC's can therefore not be considered as good substitutes because their influence on global warming is significantly high [5].

2. Background

Synthetic refrigerants containing halogen atoms are known to be emitted from RAC applications [6]. They were developed from organic gases such ethane and methane by replacing hydrogen atoms with chlorine or fluorine. The group consists of CFC's, HCFC's and HCF's. These substances have been shown to have direct impact on global warming and depletion of ozone layer [7].

2.1. Destruction of the ozone layer

Ozone which occurs naturally in the Earth's atmosphere is an unstable gas with an irritating smell. Ozone occurs 15 to 50 Kilometres above the Earth's surface, when oxygen molecules are broken down by the effect of UV rays. The concentration of ozone varies with altitude, with a peak in the stratosphere about 25-30 km from the Earth's surface. This is known as the ozone layer. The Ozone layer has the important role of filtering harmful UV rays reaching Earth. It therefore protects life on earth.

Molina and Rowland [8] through their research showed that refrigerants that contain chlorine are able to rise up in the atmosphere, where they catalyse destruction of ozone molecules.

2.2. Global warming

Global warming is a phenomenon whereby the mean temperature of the air, oceans and land masses is increasing. Statistics shows that the earth's atmospheric temperature is on the rise, and this has been attributed mostly to human activity, such as usage of RAC equipment running on HCFC's and HFC's with a high GWP.

Global warming occurs due to the fact that infrared emissions emitted by the earth are absorbed by the atmosphere thus raising the overall surface temperature of the globe [9].

The Kyoto agreement was proposed in December 1997 and enforced in 16th February, 2005. Its objective is to curb down release of greenhouse gases such as CO_2 and HFC's which includes R404a, R407c, R410a and R134a.

HCFC's and HFC's are powerful greenhouse gases (GHGs) which contains the fluorine atom. Their effective GWP is thousands of times greater than that of CO_2. Moreover, HFC refrigerants last longer in the atmosphere in comparison to chlorine containing refrigerants. HFCs are thus being controlled under the Kyoto protocol and under European Union regulation No. 517.20, 2014 [7]. In Mauritius the target is to bring a gradual reduction in HFC's in the forthcoming years.

2.3. Alternatives to HFC refrigerants

Having established that HFC's are hazardous refrigerants for the environment and they are hence facing ban under the EU regulations and under the Kyoto protocol, this section discusses about alternatives refrigerants that have been identified to replace HFC's.

Alternatives that would eventually take the place of halogenated refrigerants would be required to have zero ODP and low GWP. Low toxicity, low price, good chemical stability and good thermal properties would be a definite advantage. Naturally occurring substances like NH_3, HCs and CO_2 could be substitutes to HFCs and HCFCs. Table 1 gives the values of ODP and GWP of frequently used refrigerants.

Table 1. ODP and GWP of common refrigerants (Calm and Hourahan, 2001).

Refrigerant	Group	Atmospheric life	ODP	GWP
R11	CFC	130	1	4000
R12	CFC	130	1	8500
R22	HCFC	15	0.05	1500
R134a	HFC	16	0	1300
R404a	HFC	16	0	3260
R410a	HFC	16	0	1720
R717	NH3	-	0	0
R744	CO2	-	0	1
R290	HC	<1	0	8
R606a	HC	<1	0	8

2.3.1. Hydrocarbons. P.J. van de Weyde was the first to demonstrate the use of hydrocarbons as a refrigerant in 1866. Hydrocarbons such as Ethane, propane and isobutane were commercially available in the 1920s [11]. However, the substances available at that time were unrefined and contained a significant amount of contaminants. The pressure-temperature correlation was also not satisfactory due to poor technical ability to purify HCs.

At that time refrigeration equipment was not manufactured to the same precision and standards that we have nowadays. So these old systems were not very reliable and leakage of refrigerant was quite high, especially through the seals on the compressor driving shaft, thus giving rise to fire and health hazards. Hence, when synthetic refrigerants were developed in the 1930's, they were very easily accepted at the detriment of HCs due to their chemical stability and very low flamability.

But with the adverse environmental impacts of synthetic refrigerants, HCs made a successful comeback in the mid-1990s [7, 12]. HCs have been extensively used in domestic refrigeration and low capacity commercial refrigeration and air conditioning equipment mainly in Asia and Europe.

Table 2. ODP and GWP of common synthetic and HC Refrigerants [10].

Refrigerant	R12	R22	R134a	R600a	R290
Class	CFC	HCFC	HFC	HC	HC
Atmospheric life	130	15	16	<1	<1
ODP	1.0	0.07	0	0	0
GWP	8500	1700	1300	8	8

In comparison to HCFCs and HFCs, HCs provides zero ODP and very low GWP as shown in table 2. They also offer good efficiency, lower charge sizes, good compatibility with mineral oils, lower compressor discharge temperatures and more efficient heat transfer [13]. But they are highly flammable and this limits their usage in large systems.

The two most commonly used hydrocarbons are isobutane (R600a used in domestic refrigeration) and propane (R290 used in low charge commercial equipment). Hydrocarbons are generally non-toxic.

Hydrocarbons has the advantage of providing an increased efficiency of 14 -20 % in vapour compression systems and since its density is only 40 % that of HFCs the refrigerant charge used is much less [5].

Maclaine-Cross and Leonardi, 1997, stated that if a refrigerator is charged below 50 grams, and the risk of explosion are inexistent. The use of HCs in small commercial equipment and domestic refrigeration is therefore on the rise. HC usage in bigger capacity equipment is limited due to risks of fire and explosion [15].

2.3.2. Ammonia (R717). Ammonia has widely been used in refrigeration equipment for more than a century and has been employed in a large range of industrial activities [16]. It is one of the oldest refrigerants and it is very commonly used in industrial processing and storage of food.

The GWP and ODP of ammonia is zero. Although it is toxic, it has been safely used on a large scale throughout the world in applications such as food processing and air conditioning. Ammonia has also been employed in confined systems in various commercial applications and water chillers [17].The thermodynamic characteristics of ammonia make it a very efficient refrigerant in vapour compression systems.

Despite the fact that it is very poisonous at relatively low concentrations, fatal accidents resulted from ammonia are not common, because even at very low concentrations it can be easily be identified by its pungent smell. With the introduction of the screw compressors, which overcomes the problem of the high discharge temperature and the advent of plate thermal exchangers operating on volumes of refrigerant, production of simple systems operating on ammonia has been made achievable.

Owing to its environment friendly properties, ammonia is a candidate for much wider use if safety measures and training of technicians are well implemented. The RAC sector especially in Northern Europe is showing new interest in ammonia applications for its market [18].

However, due to its flammability and corrosiveness to skin, its usage in warmer regions which commonly use bigger chillers close to highly populated areas is limited [16]. In Mauritius, only a few minor accidents concerning ammonia have been reported.

2.3.3. Carbon dioxide (R-744). Carbon dioxide was formerly used in refrigeration equipment before the advent of CFCs. But due to the popularity of CFCs, its use was discontinued. However, use of Carbon dioxide is again being considered as refrigerant due to its environment friendly characteristics [19]. The GWP of CO_2 is 1, it does not affect the ozone layer, and it is non-toxic and is not flammable. As CO_2 works at a higher pressure than HFCs, this implies that it needs new operating configurations and parts. Properties such as high density, high latent heat, good conductivity, good cooling capacity per volume and lower value of viscosity make CO_2 an excellent refrigerant [20].

With comparison to other refrigerants, CO_2 has a lower critical point and a higher operating pressure and this poses a major drawback [21]. However, CO_2 can be operated in two different configurations: firstly in the supercritical mode and secondly on the subcritical mode.

Use of CO_2 as refrigerant has found its way in a number of applications including automotive air conditioning and heat pumps [16]. However, the high working pressure of CO_2 implies the redesign of components such as compressors and thermal exchangers and also when it is used as the lower-stage fluid in cascade systems employing common fluids such as HFC, NH3 or HCs in the higher stage. Cascade systems operating with CO_2 have been used since the last two decades and they have shown rapid development in recent years. Today it is estimated that more than 125–140 such installations are being used in the commercial sector in Europe whereas around 150 are used in agro food processing [16]. In Mauritius, CO_2 is being used in retail applications such as bottle coolers.

3. Survey of refrigerants and equipment used in the RAC sector in Mauritius

A survey was conducted among the major stake holders in the RAC sector to study the types of equipment and refrigerant currently being employed in Mauritius. The RAC sector in Mauritius involves the following uses:

IOP Conf. Series: Earth and Environmental Science **93** (2017) 012054 doi:10.1088/1755-1315/93/1/012054

- Domestic refrigeration
- Commercial refrigeration (which includes retail food, cold storage and transport refrigeration)
- Industrial process refrigeration which involves food processing
- Motor vehicle air conditioning (MVAC)
- Domestic and commercial air conditioning (including chillers)

The objective of the survey was to determine,

- Type of equipment currently being used in the RAC sector in Mauritius.
- Refrigerants and refrigerant amounts currently being used in Mauritius.
- Knowledge of the problems associated with the use of HCFCs and HFCs
- Knowledge of technicians about alternative refrigerants to HCFCs and HFCs and proper recovery and retrofitting techniques of refrigerants on the field.
- Emissions of refrigerant from the various sectors.

3.1. Methodology
A survey questionnaire was designed and distributed among the large end-users of refrigeration and air conditioning (RAC) in Mauritius as well as suppliers and contractors in the field. Since it was impossible to contact all companies involved in the field, a sample was made with the major companies/group representing each sector of RAC. In addition, interviews were also carried out with certain respondents to complement information given on the survey questionnaire.

A brief description of each of these sectors is given below in the light of data gathered during the survey.

3.2. Domestic refrigeration sector
In Mauritius, the number of households is increasing with the rising population. It is projected that the number of individual household will increase from 330,000 presently to 420,000 in the year 2020 [1]. 85 % of households have a refrigerator and this figure is likely to rise to 90% or more in 2020 [1].

3.3. Domestic air conditioning
In Mauritius, the domestic air conditioning market is growing rapidly since the late 1990s. The main reason behind this has been the decrease in price of air conditioning units, coming from the south East Asian countries and Egypt which have flooded the market. Air conditioning units for the domestic market are the 'split' unit type. The cooling capacities offered vary from 9,000 BTU to 32,000 BTU. The number of small companies offering domestic air conditioning units for sale, together with an attractive after sales service, has gone up significantly in the recent years. Domestic air conditioning systems for the domestic market are not presently being manufactured in Mauritius.

3.4. Commercial refrigeration sector
Small to large display cabinets are used in the retail food sector such as in supermarkets and retail shops etc, to display foodstuffs offered for sale. In Mauritius, the shopping trend has shifted from small foodstuff stores to super markets and hyper markets. In the last twenty years, we have seen a very rapid growth of groups of supermarkets and hypermarkets. All these together now share the major part of the retail food market in Mauritius. Display cabinets and refrigerated furniture in various size and capacity are used to display frozen and chilled food. Each store is also equipped with a number of cold rooms for their stocks of frozen and food.

3.5. Industrial refrigeration
Industrial refrigeration involves the processing of food such as chicken, fish, meat and ice cream, in mass under a controlled temperature environment. In Mauritius, food processing plants offering a variety of fresh and frozen food are quite widespread and have existed since a relatively long time. For example, the 'Food and Allied ltd' which specialises in the production of chicken and 'Maurilait Productions ltd' which produces ice cream and other dairy products have been in operation since the

late 1960s. Other stake holders in this sector are the 'Froids des Mascareignes ltd' involved in production of Seafood, the 'Margarine Industries ltd' and 'Happy World Foods ltd'. This sector also employs an important number of cold rooms for stocking their product prior to distribution in sales outlets. Refrigeration equipment in this sector is often custom designed to suit the needs of the company.

3.6. Commercial air conditioning

In Mauritius, the commercial air conditioning sector embraces primarily large buildings such as offices, hotels and shopping centres. Water chillers are the most widespread equipment for air conditioning large office buildings and hotels, although split units are also widely employed.

With the government policy to increase the number of tourists' arrivals in Mauritius, the number of hotels has known a very rapid growth in recent years. And since all of the luxurious hotels are built on the coast of the island where temperatures is often above 32°C, air conditioning equipment is extensively used. In the bigger hotels, it is primarily water chillers units that are being used. Smaller hotels opt for split unit air conditioning systems that costs cheaper to install and to maintain. The hotel sector has now become the primary user of commercial air conditioning equipment in Mauritius.

3.7. Motor vehicle air conditioning

The automobile sector in Mauritius has known a very important rise in the last twenty years. The number of motor vehicles (excluding motor cycles and auto cycles) registered in Mauritius were 300 124 in June 2016 out of which 244 401 were cars and dual purpose vehicles [1]. Air conditioning in cars has become popular since the mid-1990s. Air conditioning which was once offered as option in standard cars now forms an integral part of the vehicle. In fact, all new and re-conditioned cars now offered for sale is equipped with air conditioning. This sector was not considered in the survey.

4. Analysis of survey data

4.1. Type of equipment, refrigerants and typical charge size per sector

Table 3 below displays the various types of equipment used in the different RAC sector in Mauritius. Refrigerants currently employed together with the typical charge sizes are also given.

Table 3. Type of equipment, refrigerant and typical charge size.

Sector	Type of equipment used	Refrigerants currently being used	Typical refrigerant charge (kg)
Domestic Refrigeration	Domestic refrigerators	R-134a Hydrocarbons (limited)	0.05 – 0.25
Domestic Air Conditioning	Window and split type air conditioning unit	R-22, R-134a	0.5 – 1.5
Commercial Refrigeration	Display cabinets and refrigerated or chilled furniture. Cold Rooms:	R-134a, R-502, R-404A, R408C, R-22	0.2 – 5 15 - 25
Industrial Food Processing	Large chest freezers and cold rooms	R-502, R-22, Ammonia	5 - 250
Commercial air Conditioning	- Large split type air conditioning units - Water chiller systems	R-22, R-407A, R-134a, R 410a, R407c	1.2 – 2.0 125 – 270
Vehicle Air Conditioning	Vehicle air conditioning system	R-134a	0.8 – 1.0

IOP Conf. Series: Earth and Environmental Science **93** (2017) 012054 doi:10.1088/1755-1315/93/1/012054

- **Hotel Sector:**

The hotel sector in Mauritius comprises of 116 hotels ranging from 40 to 350 rooms [1] and most of them are located on the coastal regions where temperatures are highest. As such hotels present a high demand for air conditioning.

Hotels where the rooms are clustered in large buildings mainly use chiller units whereas hotels which have individual bungalows mainly use splits air conditioning systems. Smaller hotels with number of rooms ranging from 40 to 75 favour the use of split units as they are easier to maintain. Figure 1 compares the percentage of hotels using chillers, split units and both.

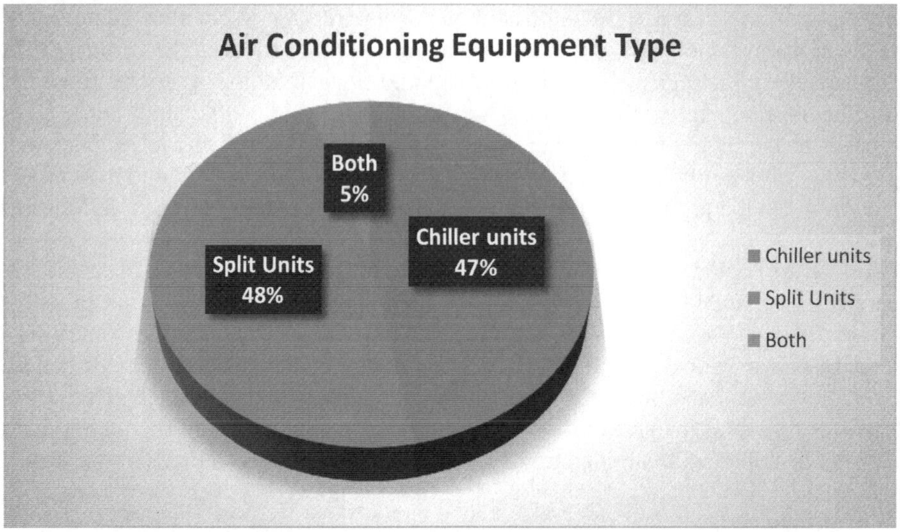

Figure 1. Type of air conditioning equipment in the hotel sector.

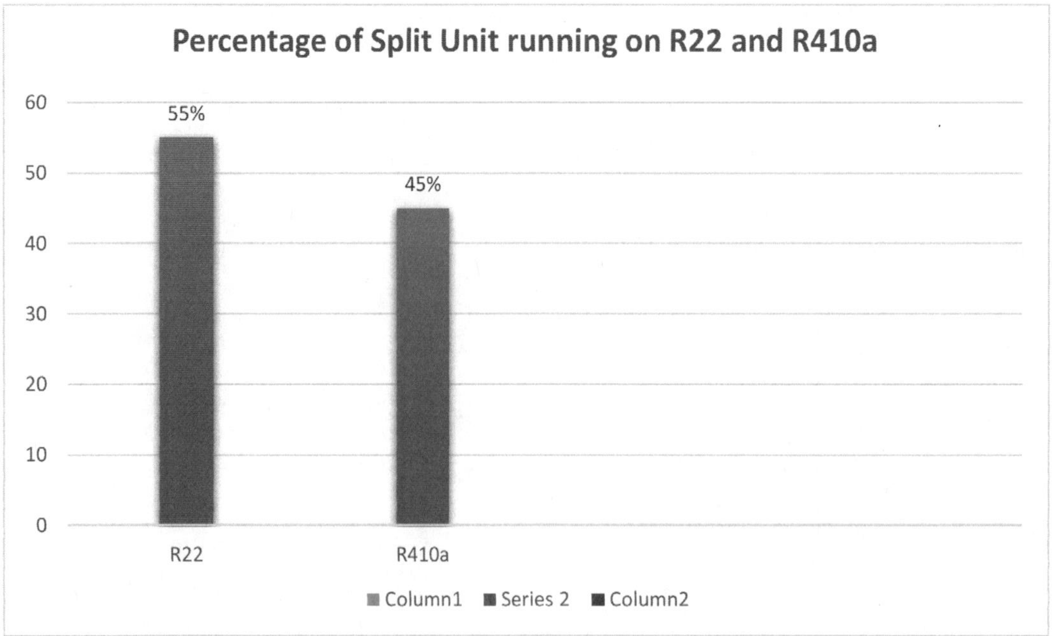

Figure 2. Percentage of split units by type of refrigerant in the hotel sector.

HCFCs, particularly R22, are still being used in the tourism industry, mainly for servicing of RAC equipment. HCFC consumption in the tourism sector accounts for 10 to 15% of the total refrigerants used in Mauritius [22].

An important number of the hotels surveyed have split air conditioners using R22. Figure 2 shows the refrigerant by percentage being used in split units in the hotel sector in Mauritius. It is obvious that with the policy of Mauritian government of shifting to the import of R410a split air conditioner, it may be foreseen that split air conditioner with HCFCs as refrigerant will disappear from the Hotel Sector in the coming 9 years. There is an estimated 3000 split unit room air conditioners in hotels in Mauritius with a stored amount of 3 tons of HCFC 22 [22].

This survey has shown that through renovations and in some cases reconstruction, hotel industry in Mauritius is gradually moving away from HCFCs. However, the import of HCFC's remains significant as shown in figure 3 below for maintenance of existing equipment.

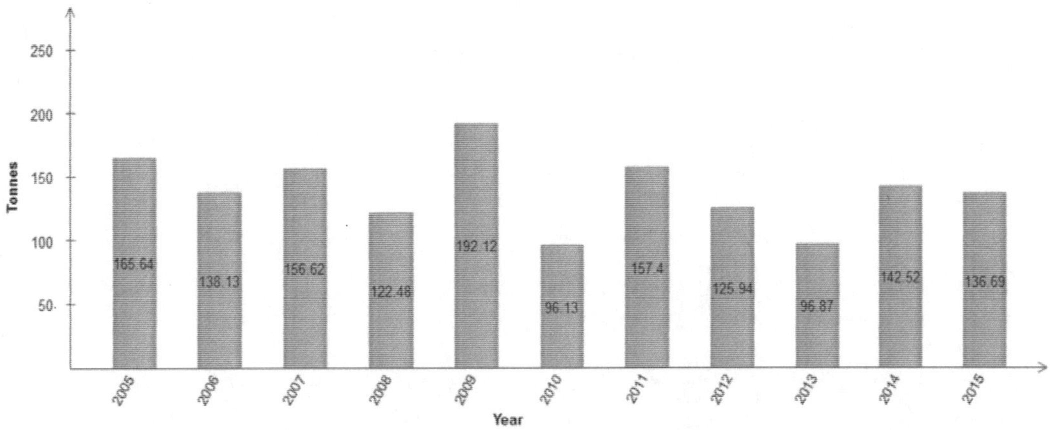

Figure 3. Consumption of HCFC's in the RAC sector in Mauritius (2005-2015) [1].

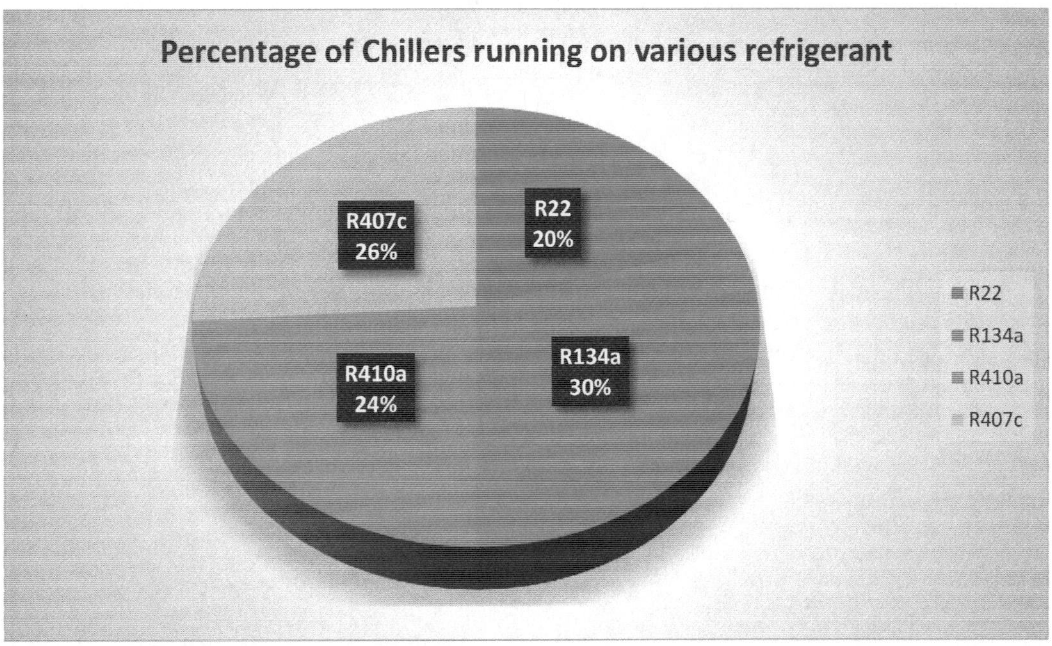

Figure 4. Percentage of chillers by type of refrigerant in the hotel sector.

This is because the commercial and industrial sector is still widely using R22 in existing equipment and they need to purchase this refrigerant for maintenance and repair work on their systems.

- **Chillers:**

Some chillers in the hotel sector still run on R22. The survey identified 15 with an average charge of 50 kg per chiller. Other chillers in this sector are on the HFC namely, R134a, R407c and R410a. Figure 4 shows the percentage of chillers running on each refrigerant.

It is estimated that 750 kg to 1 ton of HCFC 22 are being used in chillers in hotels in Mauritius.

- **Commercial Refrigeration:**

The commercial refrigeration sector consists of hypermarkets and supermarkets. Smaller shops were not considered in this survey. Equipment used are mainly display cabinets and freezers for displaying and selling frozen food and also bottle coolers. All supermarkets have cold rooms for keeping their stocks of perishable foodstuff.

Refrigeration equipment in the surveyed supermarkets in Mauritius operates on R22, R134a and R404a. However, old refrigerator systems working on R22 are gradually being replaced by systems working on HFC refrigerants. Figure 5 gives the percentage of equipment operating on R22, R404a and ammonia.

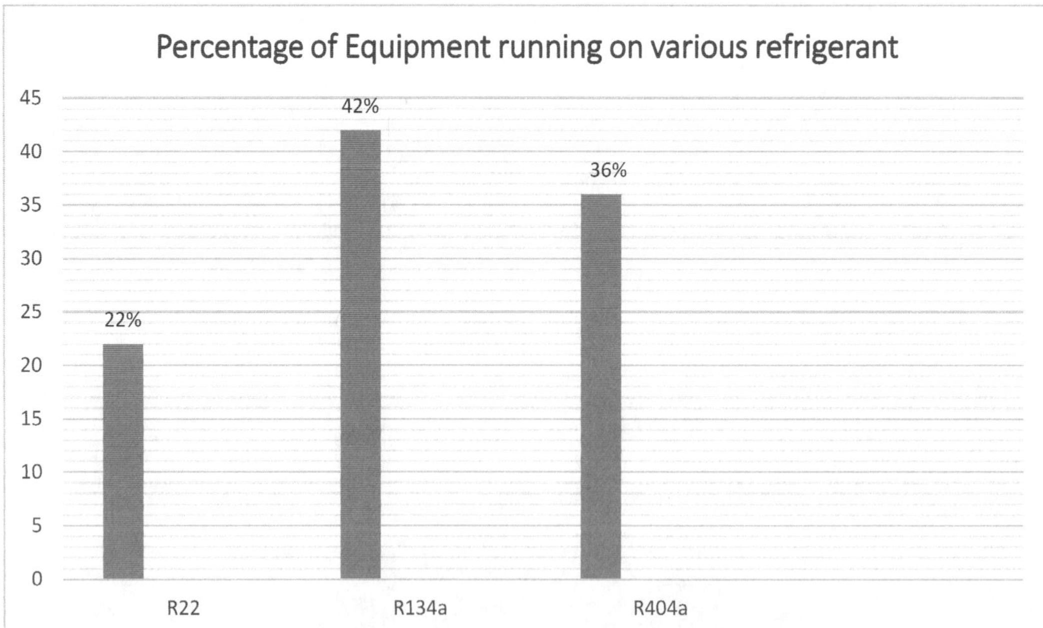

Figure 5. Percentage of equipment by type of refrigerant in the commercial refrigeration sector.

Table 4. Use of R22 in supermarkets in Mauritius.

S/N	Equipment using R22	No of units	Charge per unit (Kg)	Total Charge (kg)
1.	Freezers	34	20	680
2.	Refrigerators	75	5	375
	(Chill Cabinet Unit)	52	10	520
		225	25	5625
3.	Cold room	87	15	1305
		1	250	250
	Total			8755

Source: [22]

Table 4 illustrates an estimated use of HCFC refrigerant (R22) in the supermarkets in Mauritius [20].

- **Industrial Refrigeration:**

This sector employs large cold rooms and blast freezers to rapidly freeze food. The survey showed that three refrigerants namely R22, R404a and R717 are being used. Figure 6 shows the percentage of equipment with their respective operating refrigerants.

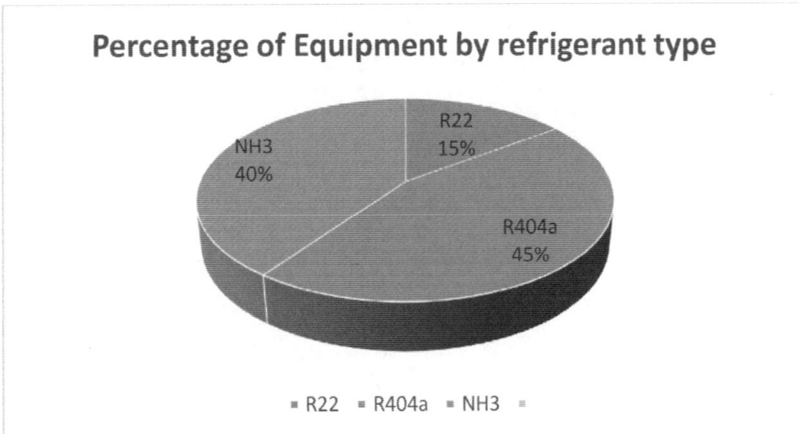

Figure 6. Percentage of equipment by refrigerant type in the industrial refrigeration sector.

4.2. Refrigerant recovery during servicing and disposal of equipment

Table 5 shows the status of refrigerant recovery during maintenance and disposal of equipment.

Table 5. Recovery of refrigerant.

Sector	Recovery of Refrigerant during maintenance
Domestic refrigeration and air conditioning	No
Commercial Refrigeration	Yes
Industrial refrigeration	Yes
Commercial Air conditioning	Yes

4.2.1. Domestic refrigeration/air conditioning sector. In the domestic RAC sector, servicing and repair is done by private mechanics that very often have little knowledge about the theory of refrigeration and air conditioning. They are often trained on the field itself. Out of ignorance, these mechanics most of the times allow the refrigerant to be evacuated in the atmosphere during repair of equipment.

Most of the repair mechanics are also not equipped with a refrigerant recovery unit. The relatively high price of recovery units may be the reason behind this.

At the end of their lifecycle, most domestic refrigerator and air conditioners are scrapped off as solid waste by households. The refrigerant is not recovered at disposal. The same applies to office air conditioners.

Whilst it is not worthwhile to recover the refrigerant for domestic refrigerators as they contain very little refrigerant (50-100 grams in the majority of cases), recovery of refrigerant from air conditioning units can be interesting and economical as the refrigerant can be re-used if recycled.

4.2.2. Commercial/industrial refrigeration and commercial air conditioning sectors. The equipment in these sectors requires mechanics that are more qualified. It is therefore easier to pass information about environmental aspects of refrigeration and air conditioning. Companies that carry out repairs and maintenance are much better equipped than private mechanics that work for their own. Recovery of refrigerant during repairs in these sectors is therefore much better done. The same thing applies to recovery of refrigerant in equipment completing their lifecycle.

However, three respondents have admitted that they often evacuate important quantities of refrigerant in the atmosphere due do shortage of recovery cylinders.

However, one major impediment to the recovery of refrigerant is that there is no recycling plant for refrigerants in Mauritius. So many companies, after having recovered refrigerants for some time, do not know what to do with them and how to dispose them.

4.3. Consumption of HCF's in Mauritius

Figure 7 shows the variation in the import of HFC's in Mauritius.

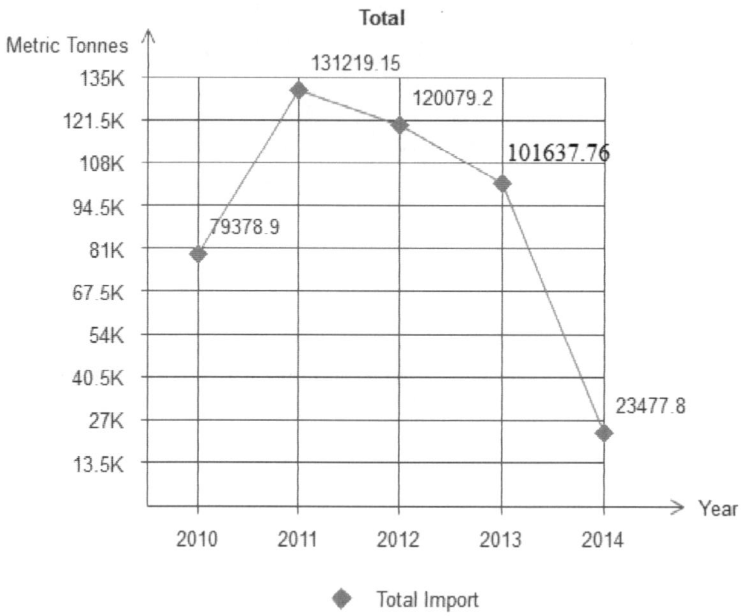

Figure 7. Import of HFC's for the RAC sector in Mauritius, [1].

It should be noted that all import of HFC's is used for charging new equipment and for top-up during maintenance of existing RAC equipment. The graph shows an overall decrease in the import of HFC's in Mauritius. This is an indication that there is less new equipment running on HFC's that is being installed from the period 2011 to 2014. The imported gases in 2014 onwards are thus being used solely for top-up purposes due to leakage in the refrigeration systems.

4.4. Knowledge of alternative refrigerants to HCFCs and HFCs

Most of the respondents were well aware of the problems associated with depletion of the ozone layer and that CFCs and HCFCs were largely responsible for that. However, for many of them ozone depletion and the global warming phenomenon were the same thing and they often mixed issues. But most of them were aware of new replacement refrigerants for CFCs and HCFCs.

Concerning proper retrofitting techniques for new refrigerants many of the technicians were not familiar with the procedures and precautions to be taken. Most of the respondents preferred that the retrofitting be carried out by the representative of the supplier of the equipment.

The proposed replacement refrigerants per sector are presented in table 6 below.

Table 6. Replacement refrigerants per sector.

Sector	Type of equipment used	Refrigerants currently Being used	Replacement refrigerants (post montreal protocol)	Remarks
Domestic Refrigeration	Domestic refrigerators	R-134a Hydrocarbons	R-134a, Hydrocarbons (R-600a)	Hydrocarbons are easily available
Domestic Air Conditioning	Window and split type air conditioning unit	R-22, R-134a	R-407C, R404A, Hydrocarbons	Hydrocarbons are easily available
Commercial Refrigeration	Display cabinets and refrigerated or chilled furniture. Cold Rooms	R-134a, R-502, R-404A, R-22	R-134a, R-404A, Hydrocarbons	Hydrocarbons are easily available
Industrial Food Processing	Large chest freezers and cold rooms	R-22, Ammonia	R-134a, R-404A, R-407A, R-410A, Ammonia, CO_2 (long term)	CO_2 require special high pressure piping and equipment
Commercial air Conditioning	- Large split type air conditioning units - Water chiller systems	R-22, R-407A, R-134a	R-134a, R-407A, Hydrocarbons	
Vehicle Air Conditioning	Vehicle air conditioning system	R-134a	R-134a, CO_2 (long term), Hydrocarbons	CO_2 require special equipment

5. Discussion and recommendations

5.1. Options for retrofitting existing equipment

Small systems operating on R22 have been successfully retrofitted to alternative refrigerants which do not affect the ozone layer, but it is not always possible to replicate the exact operating conditions of R22 and therefore conversions often involve replacing certain components. Therefore prior to any large scale retrofitting of equipment the following should be given due consideration:

- Cost of replacement with a new refrigerant and components which will improve system efficiency,
- The age of the plant,
- The possible hazards due to retrofitting

5.2. Alternative Refrigerants for the RAC sector in Mauritius

It is established that HFC's are not good solutions when considering environmental safety the challenge is to find better alternatives from an energy efficiency and environment point of view. In the survey presented in section 3, it is clear that in Mauritius the focus is now on natural refrigerants such as hydrocarbons and ammonia.

Ammonia which had been employed in large industrial refrigeration applications and food processing industry partly in replacement of the formerly widely used CFC's refrigerant in this sector is now being called upon to be used as a replacement for HCFC's and HFC's. Figures show that ammonia equipment is on the rise. Based on this, it can be safely forecasted that ammonia will become increasingly popular in the long run due to its excellent thermodynamic properties and higher COP with smaller heat exchanger footprints [18].

HC's have also been successfully used in small refrigerator display units in commercial applications but is not popular. Tests conducted recently on domestic AC units and automobile AC units when retrofitted from R22 and R134a respectively showed up to 18% energy efficiency and 12.9% improvement in the COP achieving design temperatures in a shorter time [11]. The only drawback of HC's is their high flammability.

CO_2 as refrigerant is not popular yet in Mauritius. The only few applications has been in soft drinks dispensing machines and bottle coolers. The Ministry of Environment recently donated a CO_2 / ammonia cascade system to the Universite des Mascareignes so that the University can promote CO_2 and ammonia as a safe and efficient refrigerant to the local market.

6. Conclusions

This study shows that in the aftermath of the Montreal protocol which was successfully implemented in Mauritius, HFC refrigerants such as R134a, R404a, R407C and R410 are the predominating refrigerant in the Mauritian RAC sector. HCFC R22 is also being used to a smaller extent in split AC units and large industrial and commercial applications. Survey results show that usage of R22 is slowly declining.

Results of the survey also show that stake holders on the market are well aware of the issue of global warming as related to HFC and HCFC refrigerants and are well acquainted with the fact that natural refrigerants are the alternative solutions. However, the hazards associated with these natural refrigerants such high flammability for HC's, toxicity for NH_3 and high working pressure for CO_2 are posing as barriers in their rapid implementation. Mauritius is also facing a lack of qualified technologists to work with these new technologies especially CO_2. High investment costs for CO_2 and ammonia are also a major setback.

High energy efficiency, good thermodynamic properties together with smaller charges, low price and good COP make HC's an excellent alternative for small and medium refrigeration and air conditioning applications.

NH_3 is well accepted in large industrial refrigeration applications whereas CO_2 is being used on a small scale in bottle coolers and is likely to pick momentum in the coming years in supermarket refrigeration as prices get lower and more technicians acquire adequate training.

References

[1] Central Statistics Office 2015 Mauritius, Digest of environmental statistics
[2] IPCC Climate Change 2007 the Physical Science Basis, The Working Group I
[3] United Nations Environmental Programme 1987 Montreal protocol on substances that deplete the ozone layer (Final Act, New York, United Nations)
[4] Powell R L 2002 CFC Phase out have we met the challenge *J Flu Chem* **114** 237-50
[5] IPCC Fourth Assessment Report 2012
[6] Mohanraj M, Jayaraj S and Muraleedharan C 2009 Environment friendly alternatives tohalogenated refrigerants-A review *Int J Greenhouse Gas Control* **3** 108-19
[7] Choudhari C S and Sapali S N 2017 Performance assessment of natural refrigerant R290 as a substitute to R22 in refrigeration systems *Energy Procedia* **109** 346-52
[8] Molina M J and Rowland F S 1974 Stratospheric sink for chlorofluoromethane: Chlorine atom catalyzed destruction of ozone *Nature* **249** 810-2
[9] McCulloch A and Lindley A A 2003 From mine to refrigeration: A life cycle inventory analysis of the production of HFC134a *Int J Refrigeration* **26** 865-72
[10] Calm J M and Hourahan G C 2001 Refrigerant data summary *Eng Sys* **18** 74-88
[11] Dreepaul R K 2016 An assessment of hydrocarbons as an alternative refrigerant: The Mauritian context *Int J Global Energ Issues* **39** 222-40
[12] Ki-Jung P and Dongsoo J 2006 Thermodynamic performance of HCFC22 alternative refrigerants for residential air-conditioning applications, www.sciencedirect.com, 17 October
[13] Zhang W, Yang Z, Li J, Ren C X, Li D, Wang J, Zhang X and Wu W 2013 Research on the

IOP Conf. Series: Earth and Environmental Science **93** (2017) 012054 doi:10.1088/1755-1315/93/1/012054

flammability hazards of an air conditioner using refrigerant R-290 *Int J Refrige* **36** 1483-94
[14] Maclaine-cross I L and Leonardi E 1997 Why hydrocarbons save energy *AIRAH Journal* **51** 33-7
[15] Kondo S, Takizawa K, Takahashi A, Tokuhashi K, Mizukado J and Sekiya A 2009 Flammability limits of olefinic and saturated fluoro-compounds *J Hazard Mater* **171** 613-8
[16] Bhatkar V W, Kripalini V M and Awari G K 2013 Alternative refrigerants in vapour compression refrigeration cycle for sustainable environment: A review of recent research *Int J Environ Sci Technol* **10** 871-80
[17] Fernando P, Palm B, Ameel T, Lundqvist P and Granryd E (2008b) A mini channel aluminium tube heat exchanger –Part III: Condenser performance with propane *Int J Refrig* **31** 696-708
[18] Pearson A 2008 Refrigeration with ammonia *Int J. of Refrigeration* **31** 545-51
[19] Sawalha S 2005 Using CO_2 in supermarket refrigeration *ASHRAE Journal* **47** 26-30
[20] Ge Y T and Cropper R T 2009 Simulation and performance evaluation of finned-tube CO_2 gas coolers for refrigeration systems *Appl Therm Eng* **29** 957-65
[21] Wu Y L, Shiue Y D, Chien K H and Wang C L 2011 A study on the performance of drinking fountain using CO_2 heat pump under different system pressures *Appl Mech Mater* **44-47** 2433-7
[22] National Ozone unit, 2016 Report, Mins of Environment, Mauritius

Design and optimization of zero-energy-consumption based solar energy residential building systems

D L Zheng[1,3], L J Yu[1,3] and H W Tan[2]

[1]School of Mechanical Engineering, Shanghai Jiao Tong University, Shanghai 200240, China
[2]Research Center of Green Building and New Energy,Tongji University, Shanghai 200092, China

E-mail: zhengdonglin@secsc.cn

Abstract. Energy consumption of residential buildings has grown fast in recent years, thus raising a challenge on zero energy residential building (ZERB) systems, which aim at substantially reducing energy consumption of residential buildings. Thus, how to facilitate ZERB has become a hot but difficult topic. In the paper, we put forward the overall design principle of ZERB based on analysis of the systems' energy demand. In particular, the architecture for both schematic design and passive technology is optimized and both energy simulation analysis and energy balancing analysis are implemented, followed by committing the selection of high-efficiency appliance and renewable energy sources for ZERB residential building. In addition, Chinese classical residential building has been investigated in the proposed case, in which several critical aspects such as building optimization, passive design, PV panel and HVAC system integrated with solar water heater, Phase change materials, natural ventilation, etc., have been taken into consideration.

1. Introduction

In recent years, energy consumption of residential buildings in China has shown a tendency of substantially increase. The main cause of the above-mentioned tendency lies in the following two aspects: Firstly, with the development of residents' living standards, customers' demands for energy consumption are continuously increasing. Secondly, the continuous and rapid growth of newly-built residential buildings in China has given rise to a substantial increase of total amount of residential buildings, thus significantly increasing the total energy consumption. According to the released *China Energy Statistical Yearbook in 2015* [1], China's total residential electricity consumption reached 698.92 billion kWh in 2013, corresponding to 12.4% increment as compared to that in 2012; Considering the ever increasing residential energy consumption, energy-saving techniques have attracted more and more attentions. Compared with the developed countries, the development of energy saving buildings in China is still in its infant stage. Not until 1986 did China promulgate and implement the first regulation referred to as '*Energy conservation design standard for new heating residential buildings (JGJ26—86)*', aiming at propelling the energy saving of buildings to attain an energy-saving rate of 30%. In 1995, the Ministry of Construction promulgated the second energy saving Standard '*Energy conservation design standard of new heating residential buildings (JG9 26—95)*', followed by proposing '*Design Standard for Energy Efficiency of Residential Buildings in Hot Summer and Cold Winter Zone (JGJ134-2001)*' and '*Design Standard for Energy Efficiency of residential Buildings in summer-hot and winter-warm zone (JGJ75-2003)*' in 2001 and 2003, respectively. In all the above-mentioned standards, the government made compulsory requirements on energy-saving design and raised the requirement on energy-saving rate to 50%. In 2010, such as '*Design Standard for energy Efficiency of residential Buildings in severe cold and cold zone (JGJ26-2010)*' and '*Design Standard for Energy Efficiency of Residential Buildings in Hot Summer and Cold Winter Zone (JGJ134-2010)*', which required an energy-saving rate up to 65%, were promulgated in some related areas. Despite all this, there still exists a gap between the standard in the developed countries as England, the U.S., and Germany and that in China.

In 2009, the United States' government released the report titled '*US federal zero energy consumption, high-performance green building administrative order*'. By 2020, residential buildings may be facilitated a zero net energy-oriented design, By 2030, new residential buildings is to achieve zero energy consumption England, on the other hand, has led the study on 'zero energy buildings'. In 2006, the England government put forward to establish an environment-friendly future in a conference concerning the sustainable development of residence, zero carbon emissions were achieved by all new buildings by 2019 [2]. Furthermore, European Council for an Energy Efficient Economy has proposed the target of making 'nearly Zero-Energy residential buildings (ZERBs)' [3]. Thus, people all over the world are working towards the aspect of 'zero-energy-consumption' buildings.

In recent years, many scholars have carried out the studies on zero energy buildings. For instance, the most of the existing ZEB definitions [4] and roadmap towards net zero and positive-energy buildings [5] have been presented. Furthermore, Shady Attia reveal potential challenges and opportunities for integrating optimization tools, and also assess the gaps and needs for integrating building performance optimization tools in net zero energy buildings (NZEB) design [6]. Meanwhile, Liping Wang employed simulation method to research zero energy house design in UK [7]. In addition, Ravi S. Srinivasan used a 'Renewable Energy Balance' (REB) in environmental building design as a tool to maximize renewable resource use [8] and Ming Jin develop micro grid optimal dispatch with demand response (MOD-DR), which fills in the gap by coordinating both the demand and supply sides [9]. Apart from it, an energy storage system was designed in residential buildings with photovoltaic [10] and PV systems and ground source-heat pumps were found to affect the specific energy demand of buildings [11]. Multi-Attribute Decision Making methods have been used for achieving ZERB [12] and building cluster emulator were also applied in Net-zero energy building clusters [13]. Many scholars have respectively conducted researches on zero-energy-consumption perspectives as concept, calculation, development, application and economic analysis. However, without probing into the systematic design approaches, we have developed the solar energy housing from experimental zero-energy buildings to that of the direction of integrated design of variant aspects.

In this paper, through the analysis on influential factors of ZERB, we probe and investigate the systematic thinking and approaches for the design of zero-energy-consumption solar energy residences. The structure of this paper is as follows: Section 2 describes design method for ZERB by focusing on passive technology and solar systems. In section3, a case is studied on residential building scheme, simulation analysis and energy balance. Section 4 discusses intelligent micro-networks and incremental costs of zero-energy-consumption buildings. Section 5 summarizes five design steps to ZERB.

2. Design method for zero-energy-consumption buildings

A zero-energy building may reduce the energy consumption of a building to the level as low as possible relying on the advanced materials and designs, together with technologies of energy conservation, thus supplementing the needed remaining energy by using renewable energy.

2.1. Energy flow analysis for residential buildings

To achieve a ZERB design, the most important step is to analyse the critical issues such as energy type, flow direction and the other aspects of the building. Figure 1 shows the energy flow diagram of residential energy supplying systems. While the system is divided into energy utilization units at different ends, energy flow can be further divided into four parts, i.e. inputting, converting, delivering and utilization. The inputting side of energy needs to consume the primary energy, while the generating side of energy is derived from the renewable energy (such as household solar-thermal utilization, photovoltaic power generation and biomass power generation).

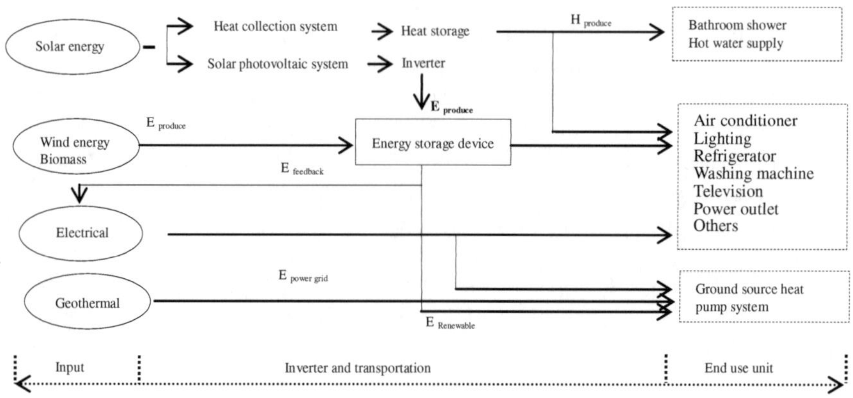

Figure 1. Residential energy flow diagram.

In the following, ZERB is used to denote the total energy generated through renewable energy in the building that is larger than the total consumption of various purchased energy of the building during the statistical period of one year, but sometimes you may also need energy exchange with the grid to obtain instantaneous balance.

2.2. Significant factors influencing energy consumption of residential buildings

Building factors mainly involve shape coefficient, direction of the building and natural ventilation, etc, as well as window-wall ratio and thermal performance parameters of building enclosure. And equipment factors mainly involve system efficiency of air-conditioners. Furthermore, other factors include matching problem between the air-conditioning system and cooling/heating loads of the building, the influence on air-conditioners of heat source inside electrical equipment, and the energy efficiency of household appliances.

As a complex energy system, a ZERB possess direct and indirect influential factors of energy consumption. To improve the entire energy efficiency of the building. The influential factors of energy consumption can be divided into a variety of factors, including outer factors, building factors, equipment factors, habits and other factors. Thus, by finding out the related factors of energy system, corresponding measures will be adopted to both reduce consumption and increase surplus energy.

2.3. Design paradigm of ZERB

To design ZERBs, we should adopt integrated methods. Figure 2 shows the design of ZERB relies on systematic ideas, firstly, it is suggested to consider the energy system while designing the buildings, secondly, making the design scheme match the energy system. Thirdly, the building scheme with passive technology optimization as well as advanced energy-saving technology can reduce energy consumption. Fourthly, you could use tools to simulate energy consumption. Fifth, utilization of resources such as solar energy is highly encouraged.

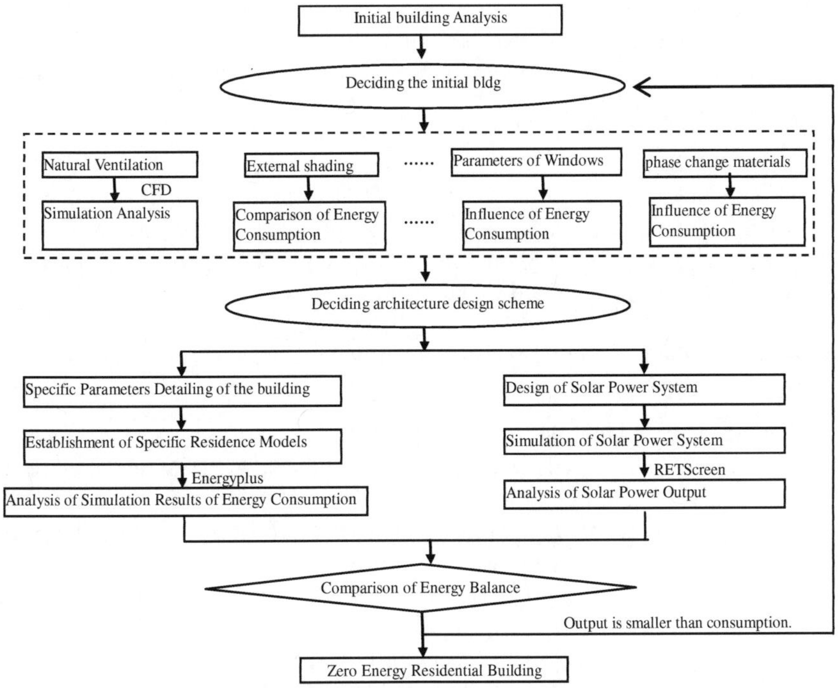

Figure 2. Design paradigm for ZERB.

2.3.1. Passive technology. When designing ZERB, passive technology may be implemented; the following aspects shall be taken into consideration.

- Passive heating. Through design of various aspects (such as building orientation, surrounding environment setting), the building is capable of utilizing solar energy to the greatest extent for providing heat in winter, whilst mitigating an extremely high temperature in summer days.
- Passive cooling. We may store cold capacity generated during the day, while open air-vents at night and release stored cold in structural elements into the air through ventilation at night, thus achieving passive refrigeration in the building.
- Sun-shading Technology. Outer sun-shading can be designed either by adopting fixed sun-shading structure or by employing movable sun-shading structure. Besides outer sun-shading, the light guiding and sun-shading system as well as changeable glass can also be utilized.
- Natural Ventilation. We properly decide design of the building such as direction of the building, direction, opening and size of windows.

2.3.2. Scheme optimization. In this part, we discuss effects on reducing energy consumption from aspects such as windows performance, outer sun-shading, natural ventilation and PCM utilization. Meanwhile, we may select relevant performance parameters and reduce indoor cooling loads and heating loads.

2.3.3. Equipment selection. The typical furniture such as cooking equipment, refrigerators, televisions, washing machines and air-conditioners, there also exist other devices such as household entertainment device, ovens, sterilizing cabinets and drying machines etc. Using low-power electrical equipment will have an active effect on saving energy. Efficient air-conditioners or heat pump sets are highly welcome. Currently, the standards of energy efficiency level for household appliances have been established.

2.3.4. Energy simulation. Simulation will be performed on energy usage of the building for the whole year. By adopting comparison demonstration with multi-schemes, we predict formation of energy consumption. Furthermore, simulation software to simulate energy consumption, together with simulation analysis on cooling, heating, ventilation, lighting and other units that consume power in the building.

2.3.5. Solar energy systems. Solar energy is divided into a variety of categories, including photovoltaic, photo-thermal, cooling and thermal utilization, cooling and heating etc, some technology summary you could see table 1. It is necessary to optimize the design of solar energy system on the basis of analysis results of energy balance by combining actual conditions. Furthermore, renewable energy sources can be utilized, with energy be supplied to the greatest extent.

Table 1. Technology summary of solar energy in residential buildings.

Category	Form	Technical Characteristics
Photovoltaic Power Generation	Different Forms	Amorphous silicon thin-film solar: Possible to integrate with external walls of the buildings; with about 8% conversion efficiency; Mono-crystalline silicon: With about 18% conversion efficiency of solar energy and high cost. Poly-crystalline silicon: With about 12% conversion efficiency of solar energy and low cost.
Photo-thermal Utilization	Solar Collector System	Collect solar radiation to heat water by utilizing solar collectors (including natural-circulation-type solar water heaters and forced-circulation-type solar water heating system).
Cooling and Thermal Utilization	Solar-wall Air Collector System	Waste heat generated from photovoltaic batteries can be collected to heat water and heat in the cavity of the solar wall can be taken away through water circulation. Meanwhile, waste heat of air evacuated from the cavity of the solar wall can also be used.

Note: PVs efficiency is from product performance information

2.3.6. Energy balance. On the basis of energy simulation, we analyse the system and the form of usable/renewable energy sources in the residential building. Meanwhile, we conduct calculation on energy balance to supplement needed remaining energy. Since solar energy represents the development direction of residential buildings in the future, we will focus on the utilization of solar energy in residential buildings in this paper.

3. Case study

We have built a solar energy residence with an area of no more than 74 m^2, which applies anti-woo roof structure elements of Chinese classical building. The curve roof can not only maximize PV panels' power generation efficiency, but also benefit both ventilation and cooling. Meanwhile, the proposed design integrates many passive and active energy-saving technologies. For instance, the external walls and interior surfaces are both made of eco-bamboo, while the walls are fitted with a new high-performance vacuum insulation board and a vacuum and hollow three-layer glass window with high insulation level. In addition, PCM are applied to the surfaces of interior walls to both ease the change of air temperature and increase living comfort. Last but not the least, heat recovery unit or thermal duct is equipped in the back of PV components set to recover heat and ensure high efficiency of power generation, together with solar hot water heaters and heat pump units that are integrated into multi-heat sources equipment systems.

3.1. Initial determination of residential building

We arrange a courtyard in the southwest of the site considering the fact that the special exhibition site is closed to the river and the limited footprint. First of all, the house is connected with the river. Furthermore, the living spaces indoor surround this courtyard. Secondly, connected with the prevailing wind direction, a courtyard in the southwest of the site is beneficial to improve the outdoor wind environment. In addition, the south-to-north direction of the bedroom's layout is helpful to get sufficient sunshine, while the living room faces the main landscape in the west.

3.2. Deepening residential building scheme

In the proposed design, we should consider the meteorological conditions as well as the surrounding buildings. In light of the fact that local temperature difference between day and night in the summer is large, and outdoor wind speed and relative humidity is suitable, natural ventilation technology, table 2 is the thermal performance parameter index of the envelope in the case.

Table 2. Thermal parameter components of building envelope.

construction	Walls	Floor	Windows	Doors	Ceiling
Heat transfer coefficient (w/m^2.K)	0.2	0.25	1.0	1.2	0.12

To further optimize the energy systems, the following four cases, i.e. windows performance, external sun-shading, natural ventilation and PCM utilization, are analysed through the simulation.

After considering the way of using three-layer Low-E glass filling argon gas to take place of common double-layer Low-E glass, energy consumption of cooling of sensible heat parts can be saved up to 450 kWh (and that of heating can be saved up to 143 kWh).

We use external sun-shading to prevent solar radiation. The energy consumption of cooling in the summer is capable of decreasing up to 176 kWh after adopting this measure.

PCM are used to both achieve heat storage/release and improve indoor thermal comfort. Heat storage and release of PCM will make cooling (heating) load in the summer (winter) decrease by about 20%, with energy consumption of cooling and heating decreased by 105 kWh and 131 kWh, respectively.

As ventilation of the building in the summer night had been considered in the energy consumption simulation before, the energy-saving is mainly reflected by the fact that the amount of air conditioner energy consumption can be saved up to 57 kWh.

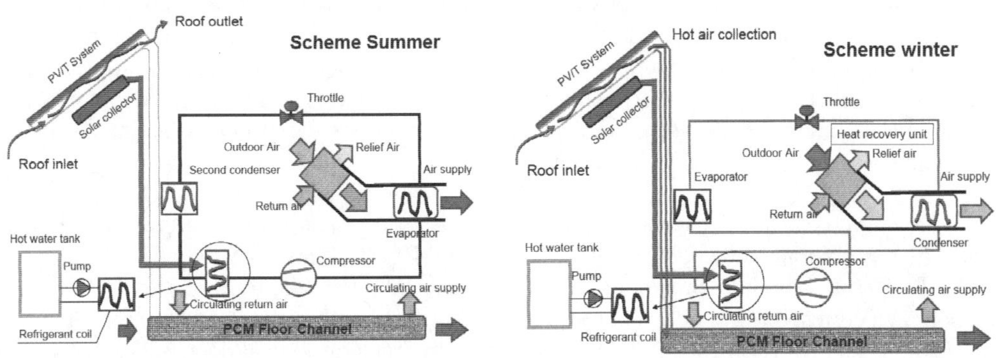

Figure 3. Schematic of HVAC system, PV/T system and hot water system for the proposed house.

We combine local meteorological conditions together with resources and select efficient and reasonable energy systems, figure 3 describes air-source heat pump with heat recovery equipment in the summer and recover may waste heat of the condenser to store in the living hot water tank that can be employed for heating in the winter, while outdoor fresh air is pre-cooled and pre-warmed by indoor back-wind through the heat exchangers in summer or winter for reducing fresh air load. Furthermore,

PCM are installed in the floor ventilation tubes so as to not only circulate for accumulation of cold outdoors in the summer night and then make it cycle to release cold quantity indoors in the day, but also circulate and reserve heat by leading hot air around the roof to the tubes in the winter days and make it cycle to release heat indoors. As for cooling load in the summer and heating load in winter, we select air conditioner with the cooling capacity of up to 3.5 kW.

3.2.1. Building envelope. Envelope insulation is critical to the building energy consumption, and building a high-performance envelope constitutes a cost-efficiency strategy. Thus, vacuum insulation panel (VIP) is used in the proposed house by significantly reducing the thickness of the walls and guaranteeing the best thermal effects, because VIPs are 7-10 times than traditional insulation materials in thermal effects. In order to keep envelope insulation at a good condition, insulation materials must be kept dry and thus can be functioning normally. In addition, the products minimize air infiltration and resist to water penetration, whilst allowing moisture to migrate freely and safely into the atmosphere.

3.2.2. HVAC System integrated with solar water heater. A heat pump and a hot water tank are integrated, in which the exhausted heat is used to as an assistant heat source or heat water if insufficient solar radiation is met. For the ventilation side, the needed outdoor air can be delivered based on the required reading data by the CO_2 monitor located in the return section, and thus can be treated after heat exchange with indoor exhaust air. HVAC system includes several critical functionalities, such as air distribution, outdoor unit, indoor unit, heat exchange unit, PCM channels, temperature/humidity sensor, and CO_2 sensor. The characteristic of PCM is elaborated on as follows: Latent heat is120kJ/kg, Melting temperature is around 21℃.

3.2.3. PV/Thermal system. PV panels are installed in a polyline curved roofs in the house. However, most of the solar energy is lost in a heat-emission manner, which otherwise weakens PV performance. Wasted heat produced by PV can be released in summer and collected to heat floor panel in winter, thus establishing a so called 'PV/T system'. Furthermore, the collected heat from PV/T system is used with PCM floor channels to both provide hot air to bedroom and maintain a relative high temperature in winter night.

3.2.4. Solar hot water system. The house uses solar thermal energy to meet almost all hot water demand. On rainy or cloudy days, solar collector cannot meet the demand of providing hot water, while the integrated HVAC heat pump will be operated automatically for heating water. No matter how good the weather is, solar collector with the heat pump can always supply sufficient hot water for the house. Furthermore, temperature stratification is formed in the water tank, thus substantially improving the heating efficiency of solar water system.

3.2.5. Photovoltaic system. PV panels are installed on two south-facing slope roofs as well as the south wall, and its conversion efficiency can be up to 13.4%. PV system consists of several components, including PV Array, Module Supporter, Junction Box, Inverter, and Power Panel. Note that PV installation area corresponds to 60 m^2 in roofs and 13 m^2 in a south wall. Thus, the overall capacity is about 10 kW. Using RETScreen software and according to meteorological data etc., we observed that PV systems can generate electricity up to 13362 kWh/a.

Table 3 is the simulation result. From table 3, July being the month with highest energy, while that in the winter days becomes lower. Furthermore, power-generating capacity in every month of the whole solar photovoltaic power generation system can be obtained, which of June is about 1500 kWh. For instance, in June, daily power-generating capacity under normal lighting conditions is about 52 kWh, while that in cloudy days is only 13 kWh. In addition, PV system can produce more energy than consumption, with extra energy flowing into the utility grid.

Table 3. Annual PV energy production (unit /kWh).

Month	Roof			South wall	Total energy (kWh)
	35 degree inclination slope	23 degree inclination slope	11 degree inclination slope	30 degree inclination slope	
January	213.72	222.32	188.07	50.57	674.68
February	271.08	292.57	259.46	65.08	888.19
March	336.01	378.92	354.01	82.1	1151.04
April	344.54	406.83	399.69	85.78	1236.84
May	363.73	442.62	450	91.71	1348.06
June	414.05	511.68	529.34	105.09	1560.16
July	428.79	526.6	540.87	108.57	1604.83
August	412.6	492.97	489.49	103.26	1498.32
September	360.87	413.98	394.05	88.79	1257.69
October	263	290.32	264.92	63.7	881.94
November	212.76	224.39	193.39	50.61	681.15
December	184.81	190.93	160.11	43.61	579.46
Total	3805.96	4394.13	4223.42	938.89	13362.4

3.3. Simulation parameters setting

Parameter settings of energy-consumption are shown in table 4, in which indoor temperature and humidity set points are in strict accordance with the requirements of the competition. It is shown that heat insulation performance of various structures is superior to the same type of products. Furthermore, air conditioning system selects air source heat pump systems with heat recovery device, whose heat recovery efficiency can be up to 70%. When the accumulation of heat of solar collector is insufficient for hot water requirement, the heat quantity recovered will be finally adopted as a supplement. Two aspects are considered when we buy household appliances: Firstly, electrical power consumption is kept as low as possible; secondly, the size of household appliances is kept as small as possible.

Table 4. Module input.

Location		Climate file		Type
Set point				
Air Temperature (℃) 23-25	Air Relative Humidity (%RH) 45-55	Condition space 54	Floor area (M^2) 54	space volume (M^3) 63
HVAC System				
Equipment	Cop	Fan Efficiency	HRV Efficiency	DHW
Air sourced heap pump	3.5	70%	70%	Solar sourced/heat pump recovery
Occupant				
medium level 4 people		Lighting w/m^2 3		

Additional, we choose four cases' (see table 5, in which we adopt different kinds of measures) to simulate, Four cases' simulation results are given by figure 4, figure 4 tells us when we adopt Case4 will be reduce energy consumption by 40% compared with the original design.

Table 5. Module input of four cases.

Items	Original design	Case 1	Case 2	Case 3	Case 4
Use low-E triple skin glazing	No	Yes	Yes	Yes	Yes
Add External sun-shading	No	No	Yes	Yes	Yes
Add PCM (floor, ceiling)	No	No	No	Yes	Yes
Add Natural Ventilation	No	No	No	No	Yes

Note: Conditioned space area is 54 m^2. There are 4 people in the Bamboo House. We keep the temperature to be 23-25℃and humidity to be about 55%RH.

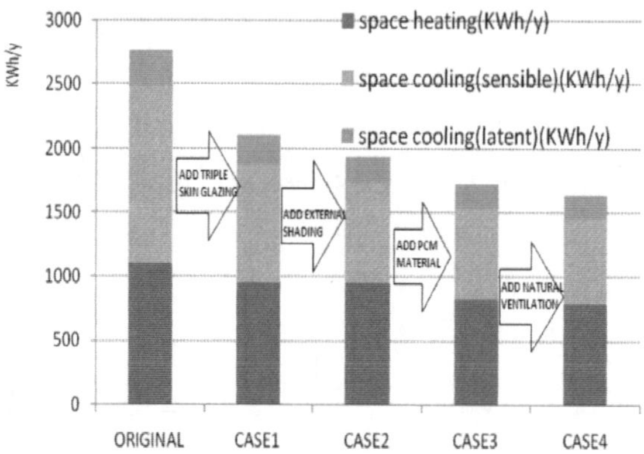

Figure 4. Simulation results after optimizing the building and energy system.

3.4. Energy Balance

Annual power-generating capacity of PV panels is 13362 kWh (see table 3), while annual electricity demand is only 5289 kWh. The power-generating capacity of PV system is 2.5 times as high as that of annual theoretical electricity consumption, and power-generating capacity of every month is higher than electricity demand of the corresponding month. Although, the total energy generated is larger than the total consumption of various purchased energy of the building during the statistical period of one year, here is one point to explain, when the solar energy can't meet with the demand instantaneously, it is necessary to obtain energy from the grid.

4. Discussion

This paper focuses on the design steps of ZERB systems. However, intelligent micro-networks and incremental costs shall be taken into account when we design.

The ZERB in the future is an autonomous system with capabilities of self-control, self-protection and self-management, which can operate not only parallel with the external power grid but also alone. It is regarded as an important part of the intelligent power grid, which has been proposed recently to encourage local balancing of energy. It's also helpful to relate the design of ZERB to the design of zero-energy community, which will encourage local balancing of energy [9]. Furthermore, It should also consider the latest development in IoT-enabled smart controls, Jin M develops Virtual Occupancy Sensing which you could use smart meters to indicate your presence in order to improve energy efficiency [14,15].

Nowadays, costs of renewable energy are still higher than conventional power, therefore, we need consider cost control objectives as much as possible. So when we design renewable energy, we should match necessary loads and make incremental cost of ZERB minimum. As for economic analysis of ZERB, Jarek Kurnitskia studied cost optimal and nearly zero energy performance calculations for residential buildings for nearly ZERB's [16].

5. Conclusion

This paper shows that ZERB can be achieved from three aspects: Firstly, it is capable of reducing energy consumption at the terminals. Secondly, it can improve energy efficiency. Thirdly, it increases the generated energy at the side of renewable energy. More important, this paper shows us that ZERB design should consider the following five aspects: (1) Optimization of building schematic design. (2) Optimization of passive building design. (3) Energy Simulation. (4) Utilization of renewable energy sources. (5) Energy balancing analysis.

Future work is to develop ZERB technology, which include intelligent micro-networks and costs of zero-energy-consumption.

References

[1] Statistics Bureau of People's Republic of China 2017 *China Energy Statistical Yearbook 2015* (China Statistics Press)
[2] Department for Communities and Local Government, Greener Homes for the Future 2008 https://www.gov.uk/government/news/baroness-hanham-sees-greener-homes-of-the-future
[3] European Council for an Energy Efficient Economy 2011 Nearly zero energy buildings: Achieving the EU 2020 target
[4] Marszal A J, Heiselberg P, Bourrelle J S, Musall E, Voss K, Sartori I, *et al* 2011 Zero energy building – a review of definitions and calculation methodologies *Energy and Buildings* **43** 971-9
[5] Kolokotsa D, Rovas D, Kosmatopoulos E and Kalaitzakis K 2011 A roadmap towards intelligent net zero- and positive-energy buildings *Solar Energy* **85** 3067-84
[6] Attia S, Hamdy M, O'Brien W and Carlucci S 2013 Assessing gaps and needs for integrating building performance optimization tools in net zero energy buildings design *Energy and Buildings* **60** 110-24
[7] Wang L P, Gwilliam J and Jones P 2009 Case study of zero energy house design in UK *Energy and Buildings* **41** 1215-22
[8] Srinivasan R S, Braham W W, Campbell D E and Curcija C D 2012 Re(De)fining net zero energy: Renewable emergy balance in environmental building design *Building and Environment* **47** 300-15
[9] Jin M, Feng W, Liu P, Marnay C and Spanos C 2017 MOD-DR: Microgrid optimal dispatch with demand response *Applied Energy* **187** 758–76
[10] Vieira F M, Moura P S and Almeida A T D 2017 Energy storage system for self-consumption of photovoltaic energy in residential zero energy buildings *Renewable Energy* **103** 308-20
[11] Thygesen R and Karlsson B 2017 An analysis on how proposed requirements for near zero energy buildings manages pv electricity in combination with two different types of heat pumps and its policy implications – a swedish example *Energy Policy* **101** 10-9
[12] Zavadskas E K, Antucheviciene J, Kalibatas D and Kalibatiene D 2017 Achieving nearly zero-energy buildings by applying multi-attribute assessment *Energy and Buildings* **143** 162-72
[13] Li X and Wen J 2017 Net-zero energy building clusters emulator for energy planning and operation evaluation *Computers Environment & Urban Systems* **62** 168-81
[14] Jin M, Jia R and Spanos C 2017 Virtual occupancy sensing: Using smart meters to indicate your presence *IEEE Trans Mobile Computing* **99** 1
[15] Jin M, Bekiaris-Liberis N, Weekly K and Spanos C J 2016 Am Bayen occupancy detection via environmental sensing *IEEE Trans Automation Science & Engineering* **99** 1-13
[16] Kurnitski J, Saari A, Kalamees T, Vuolle M, Niemela J and Tark T 2011 Cost optimal and nearly zero (nZEB) energy performance calculations for residential buildings with REHVA definition for nZEB national implementation *Energy and Buildings* **43** 3279-88

A method for evaluating photovoltaic potential in China based on GIS platform

L Z Wang[1,3], H W Tan[2], L Ji[1] and D Wang[1]

[1]Shanghai Institute of Building Science, Shanghai 201108, China
[2]Research Center of Green Building & New Energy, Tongji University, Shanghai 200092, China

E-mail: 2009wanglizhen@tongji.edu.cn

Abstract. Solar photovoltaic systems are commonly utilized in China. However, the associated research is still lack of its resource potential analysis in all regions in China. Based on the existed data about solar radiation and system conversion efficiency data, a new method for distributed photovoltaic potential assessment has been presented. The experiment of three kinds of solar photovoltaic system has been set up for the purpose of analyzing the relationship between conversion efficiency and environmental parameters. This paper fits the relationship between conversion efficiency and solar radiation intensity. This method takes into account the amount of solar radiation that is effectively generated and drives away the weak values. With the spatial analysis function of geographic information system (GIS) platform, frequency distribution of solar radiation intensity and PV potential in China can be derived. Furthermore, analytical results show that monocrystalline-silicon PV generation in the north-western and northern areas have reached a level of more than 200 kWh/(m^2.a), making those areas be suitable for the development of PV system. However, the potential for southwest areas reaches a level of only 130 kWh/(m^2.a). This paper can provide the baseline reference for solar energy development planning.

1. Introduction

Photovoltaic power generation plays an important role. Large-scale development and application of photovoltaic systems will gradually optimize the energy structure. Photovoltaic (PV) technology has been used for many years. In China, there are four types of solar energy resources. The key problem is the actual solar energy potential in actual engineering. Consequently, Jaroslav Hofierk and Yosoon Choi reported city-wide application of distributed PV needs to be involved at planning stage of community scale, instead of individual buildings [1,2]. The exploration of necessary technical support on regional energy planning is in urgent need.

Scholars use different methods to analyze solar energy resources. Salvador Izquierdo *et al* studied the potential of large-scale photovoltaic energy systems with available roof area distribution [3]. Rylatt M, Stuart Gadsden *et al* analyzed the solar energy resources of the photovoltaic energy system with the TRNSYS and 3D urban models [4], and they also carried out the study of solar potential with GIS [5,6]. Ramachandra T V, *et al* used the GIS system to map the potential map [7]. Sadik Kucuksari with GIS found the optimal size and locations for PV project [8]. Juan M used multi-objective decision and GIS to explore the solar energy potential of Karnataka state [9]. Ramachandra T V, *et al* focuses talukwise mapping of renewable energy (solar, wind and small hydroenergy) potential for Karnataka using GIS [10]. Bent Sθrensen, *et al* used satellite data based on the use of GIS systems to map solar energy resources map [11], and carried out the application of solar photovoltaic research. In China, researches focus on solar energy resources and solar cell module performance research. Zhou Y used the total solar radiation and sunshine hours on the spatial distribution of solar energy resources and resource potential assessment [12]. Liao Z L established a simplified mathematical model for nonlinear engineering of batteries [13].

The rest of the paper explains the proposed framework in details as follows: In section 2, the proposed method of calculating the potential of photovoltaic power generation efficiency based on solar radiation and real time. Experiment is set up to analyze the relationship between the conversion efficiency and the environmental parameters of monocrystalline silicon (MS), polysilicon silicon (PS) and thin film (TF) modules under the same environmental conditions. In section 3, a case is studied on PV potential the solar radiation intensity frequency and PV generation potential spatial distribution. Section 4 summarizes main conclusions.

2. Methodology

2.1. *Evaluation method of photovoltaic resource potential*

In China, solar energy utilization is based on the division of the four types of regions according to solar radiation quantity (RQ). I areas (resource-rich zone): annual radiation is in the 6700-8370 MJ/m^2. II areas (more resource-rich): annual radiation is in the 5400-6700 MJ/m^2. III areas (resources with the general): annual radiation is in the 4200-5400 MJ/m^2. IV areas: annual radiation in the 4200 MJ/m^2 below. This area is the least solar energy resources in China. The characteristic index is solar radiation intensity (RI) per unit area and the number of sunshine hours. In order to simplify the calculation, it is found that the annual output energy of the PV system usually used equation (1) in engineering applications [14].

$$E_{pv,out} = A_{pv,s}\eta_e RI \tag{1}$$

In fact, this simplified calculation is theory. The electricity generation efficiency of PV system is mainly influenced by RI and the temperature of the PV panel, and RI determines the generation current. The temperatures of the PV panel mainly influence the power generation voltage, which are determined by the outdoor air temperature and RI [15]. In fact, the conversion efficiency is the change value with different RI. This paper presents a real-time output power that the annual output energy of the PV system is more accurate, such as equation (2).

$$E_{pv,out} = A_{pv,s} \bullet e_{pv} = A_{pv,s} \int_{1}^{8760} RI(t)\,\eta(t)\,dt \tag{2}$$

Taking the radiation quantity as the characteristic index, the traditional solar energy map is divided into four parts, which can neither identify the diversity within the same resource partition, nor show the distinction when applied with specific technologies. In this paper, the resource of photovoltaic potential is considered conversion efficiency of solar energy resource potential,which is affected by the two factors of RI (t) and η(t).

2.2. *Experimental testing*

2.2.1. Summary of PV System. In order to scientifically estimate the actual regional adaptability of PV panels, an experimental platform was performed to find out the relationship between conversion efficiency and solar radiation intensity (RI) of three kinds of PV panels (Figure 1. The diagram of the experimental system).

This on-line monitoring platform focuses on PV panels of monocrystalline silicon (MS), polycrystalline silicon (PS) and thin film (TF) PV panels, which designed with basic capacity of about

1 kW (table 1). Monitored data collected by the data acquisition system (DAS) are radiation intensity, outdoor air temperature, outdoor wind speed, surface temperature of PV modules, power generation and so on (table 2).

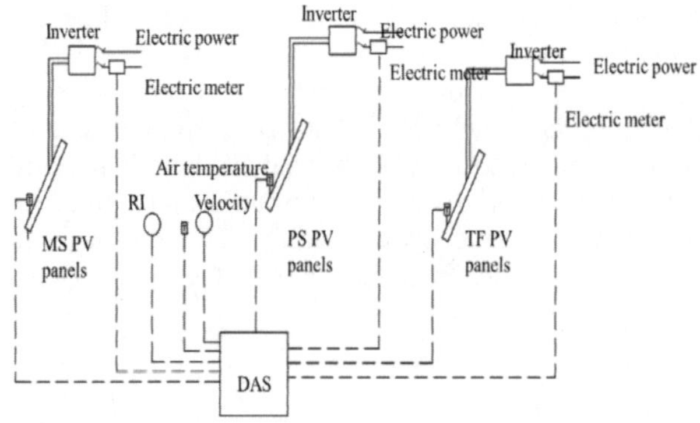

Figure 1. The diagram of the experimental system.

Table 1. Characteristic parameters of designed PV modules.

PV module type	Size of modules (m)	The number of panels	Installed capacity (W)
MS	1.580×0.808	6	1020
PS	1.655×0.992	6	1320
TF	1.330×1.100	10	1150

Table 2. Main test instrument.

Instruments	The performance parameters
Ultrasonic anemometer	wind speed: 0.01 m/s, measuring range 0 ~ 70m/s, ±0.3 m/s wind Direction: 1°, measurement range 0° ~ 360° accuracy ±2°
TBQ-2 Irradiation meter	the test range is 0 ~ 2000W/m^2, Signal output 0 ~ 20 mV
T-type thermocouple Temperature sensor	measurement range -40～350℃, ±0.5%accuracy
Electric meter	Active level 1, RS485, MODBUS—RTU

2.2.2. Results of the experiments. Based on the experimental results, we got the data such as RI and the outdoor air temperature and wind speed (Figure 2. Radiation intensity, outdoor air temperature and wind velocity in Shanghai).

The data of the power generation and radiation intensity are analyzed. The experimental error of operating conditions is eliminated. The conversion efficiency of PV system is calculated according to the equation (3).

$$\eta(t) = \frac{E_{pv}(t)}{RI(t) \times A_{pv,s}} \times 100\% \tag{3}$$

As addressed in Figure 3. Conversion efficiency and radiation intensity of three types of PV panels, the conversion efficiency tends to be steady when RI exceeds 200 W/m^2, based on which, the frequency analysis can be done to identify the stable PV solar resource partitions with the dividing line of 200 W/m^2. It should be mentioned that the unsteady scope of RI below 200 W/m^2 also makes contribution to the power consumption, so it is a conservative method in this paper to evaluate the PV-related solar energy. With the errors eliminated, these real-time data are gathered and dealt with according to equations (2) and (3) to get the real-time efficiency. On this basis, the relationship of conversion efficiency can be fitted as table 3.

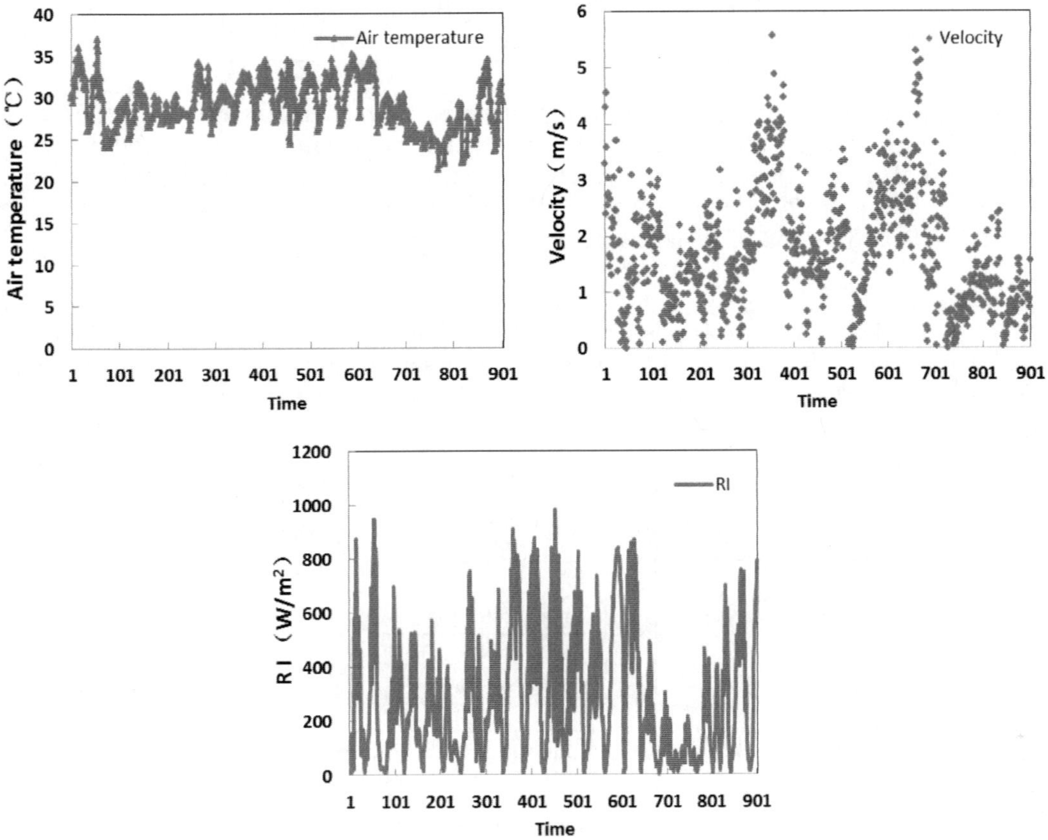

Figure 2. Radiation intensity, outdoor air temperature and wind velocity in Shanghai.

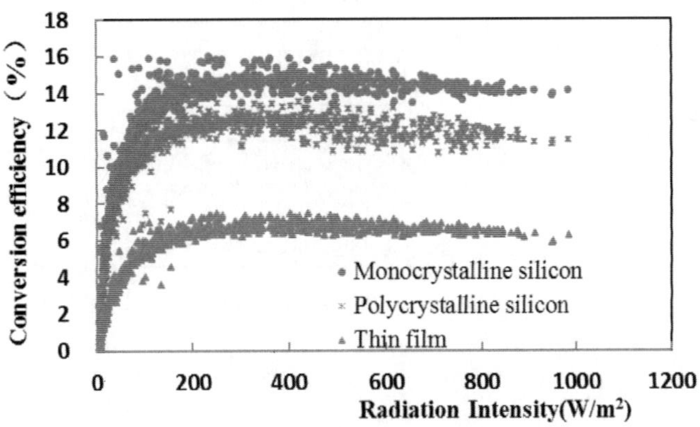

Figure 3. Conversion efficiency and radiation intensity of three types of PV panels.

Table 3. Relationship of conversion efficiency and RI for three photovoltaic modules.

PV module type	The limits of RI (W/m^2)	Relationship of conversion efficiency and RI
MS	10≤RI≤200	3.5781ln[RI(t)]-4.0401
	RI≥200	-0.0006 RI(t)+14.798
PS	10≤RI≤200	3.4155ln[RI(t)]-5.0545
	RI≥200	-0.0012 RI(t)+12.71
TF	10≤RI≤200	1.9076ln[RI(t)]-3.4561
	RI≥200	-0.0003 RI(t)+6.7909

Figure 3 shows under the same environment parameters, the highest efficiency of MS panel, PS and TF, PV panel in turn. There is a strong correlation between different types, and conversion efficiency and RI are the same principle. Under low intensity (less than 100 W/m^2), the conversion efficiency is growing rapidly. With the increase of RI, conversion efficiency growth is slow. When RI is about 200 W/m^2, conversion efficiency is gradually stabilized.

2.3. Geodatabase

PV system utilization evaluation database is based on the GIS map of the administrative map, and multi-attribute which affects the meteorological data, the resource distribution and energy price into the GIS database, the structure is shown in Figure 4. The flow chart of data processing in GIS database.

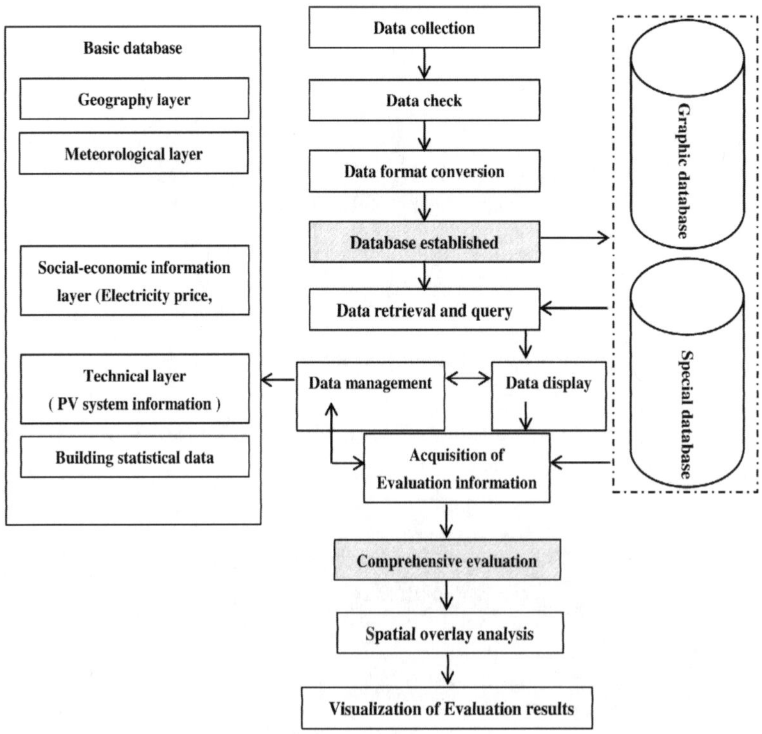

Figure 4. The flow chart of data processing in GIS database.

3. Case study

3.1. Frequency analysis of solar RI

The author uses the meteorological data of 270 weather stations in the special meteorological data set of *Building Thermal Environment Analysis in China*, which includes the annual radiation quantity, the annual level of radiation parameters of each site. The total number of radiation hours is [4300, 4550] hours. Using renewable energy on-line monitoring system and PV module experimental data, we study the characteristics of solar radiation and power generation system, etc.

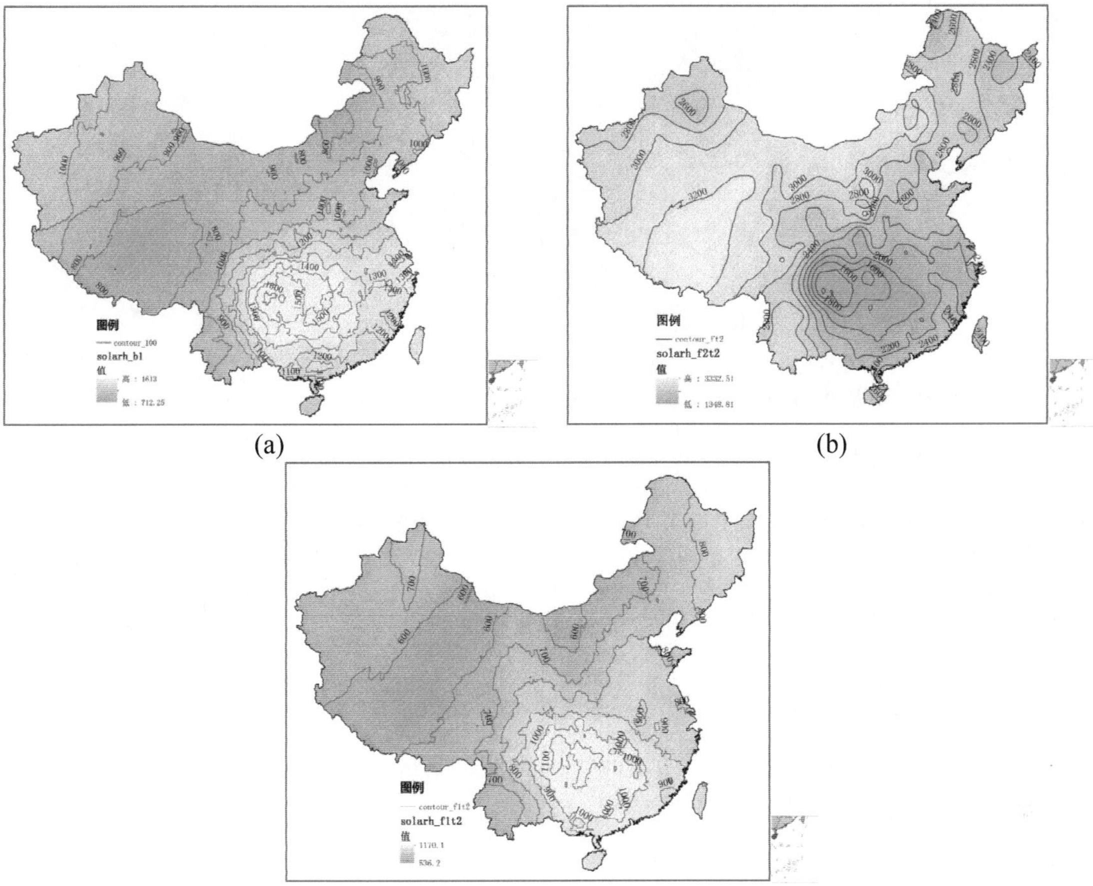

(a) (b)

(c)

Figure 5. Spatial distribution of PV power potential (a) Sunshine hours between 0-100 W/m^2, (b) Sunshine hours between 100-200 W/m^2 and (c) Sunshine hours above 200 W/m^2.

Solar resource evaluation modified by the PV technical parameters. The data yielded by the testing imply a steady zone of conversion efficiency, when the RI exceeds 200 W/m^2. Divided into two sections of 0-200 W/m^2 and above 200 W/m^2, frequency analysis of horizontal RI is carried out by writing a program with Microsoft SQL, taking advantage of hourly horizontal RI of 270 weather stations. The results are shown in Figure 5. Spatial distribution of PV power potential (a) Sunshine hours between 0-100 W/m2, (b) Sunshine hours between 100-200 W/m2 and (c) Sunshine hours above 200 W/m2(a)-5(c). The results of frequency analysis can be divided into 3 partitions with Kriging interpolation algorithm [16], which can identify the technical-related disparity of solar PV resources.

In figure 5, radiation frequency of 0-100 W/m^2 in most of the southwest area achieves 1300-1600 hours, which exceeds 1/3 of the total radiation hours. The radiation frequency above 200 W/m^2 in most of the northwest area achieves 2600 hours.

Taking MS module system as an example, the distribution of the photovoltaic power generation potential is shown in figure 6 by using the equation (2) and the GIS platform.

The annual power production varies from 127 to 267 kWh/m^2, covering the mainland of China. Figure 6. Annual power generation of MS panel per square meter of roof area shows the potential distribution of power generation. Northwest and north have a high PV utilization potential.

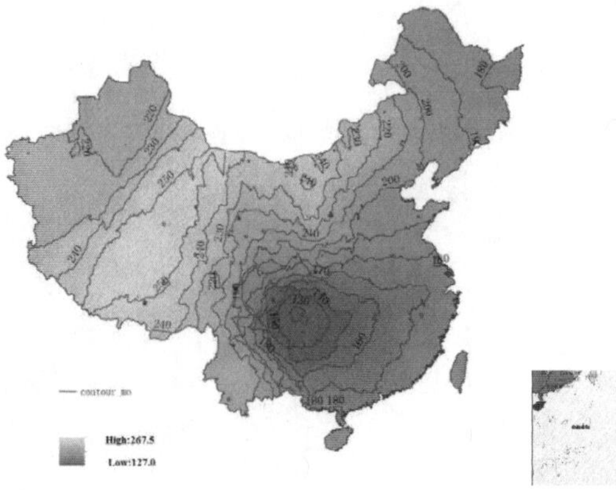

Figure 6. Annual power generation of MS panel per square meter of roof area.

3.2. Evaluation method of solar photovoltaic resource potential in Shanghai

This paper takes Shanghai City as an example, by using the method above, we can get the conclusion that there are 8760 hours in a whole year, 4418 hours of RI is zero, 28.9% of the rest of the time are in the inefficient zone, but only nearly 17.7% are high efficient area as shown in figure 7.

According to equations (2) and (4), annual capacity of MS, PS and TF PV panels installed on the building roof in Shanghai are 180.9, 151.1, 82.3 kWh/(m^2. a).

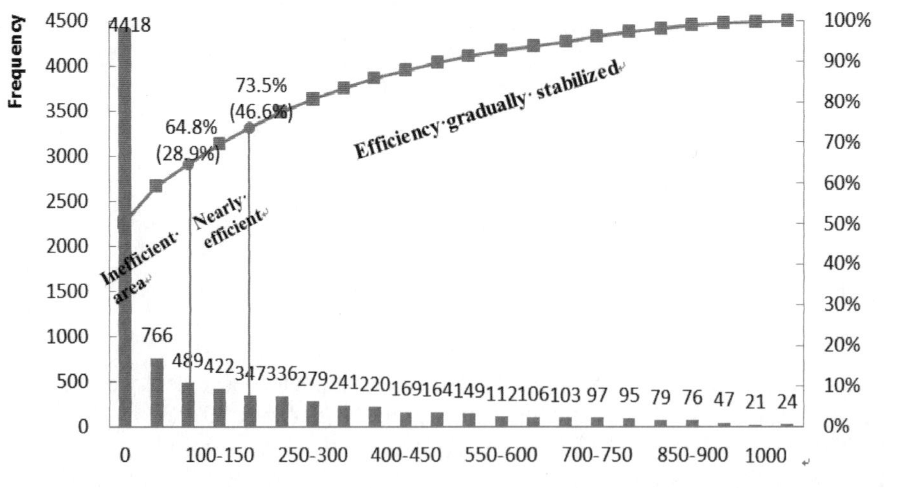

Figure 7. Frequency and the cumulative probability distribution of annual radiation intensity in Shanghai.

4. Conclusions

As demonstrated in the results section, this paper puts forward the regional potential analysis with the help of a GIS platform. In the future research, we will investigate the building roof in all studied cities. There are still further issues needed to be discussed as follows.

- The experimental platform of solar photovoltaic system was established, and the data of the test results were analyzed and analyzed to obtain the photovoltaic efficiency equation of MS, PS and TF cell module system.

- With the spatial analysis of GIS platform, RI analysis in all parts in China showed that the radiation intensity of more than 200 W/m^2 was more than 2600 hours in the northwest and most parts of the north and about 1600 hours in the southwest.
- With the monocrystalline silicon module as an example, the potential of photovoltaic generation in China is distributed at [127,267] kWh/m^2.a, which is up to 200 kWh/m^2.a. The northwest and north regions, suitable for the development of PV system.
- RI distribution in Shanghai has been studied, and 71.1% of sunshine hours are more than 100 W / m^2, which are near-efficient and efficient.
- With the implementation of the distributed solar PV subsidy policy, the adaptability of solar PV systems should be further analyzed throughout the country according to local radiation resource conditions, energy prices and government subsidy policies.

Acknowledgments
This research was supported by Shanghai Science and Technology Committee (Project number: 17DZ2292900). The authors also appreciate the anonymous reviewers for their detailed and constructive comments.

Appendix

Nomenclature		RQ	solar radiation quantity
Apv, s	the surface area of the solar cells of the system,m^2	$RI(t)$	The radiation intensity for t time, W/m^2
e_{pv}	annual electricity generation per unit i area	$\eta(t)$	the real-time conversion efficiency for t time, %
$E_{pv,out}$	the annual power generation for installed pv panel,W	η_e	the mean annual power conversion efficiency coefficient for PV system
MS	PV panels of monocrystalline silicon	η_{mono}	the real-time conversion efficiency for MS PV system, %
PS	polycrystalline silicon	η_{thin}	the real-time conversion efficiency for TF PV system ,%
RI	solar radiation intensity	η_{poly}	the real-time conversion efficiency for PS PV system, %
TF	thin film PV panels		

References

[1] Hofierka J and Kaňuk J 2009 Assessment of photovoltaic potential in urban areas using open-source solar radiation tools *Renew Energy* **34** 2206-14
[2] Choi Y, Rayl J, Tammineedi C, *et al* Brownson 2011 PV Analyst: Coupling ArcGIS with TRNSYS to assess distributed photovoltaic potential in urban areas *Solar Energy* **85** 2924-39
[3] Izquierdo S, Rodrigues M and Fueyo N 2008 A method for estimating the geographical distribution of the available roof surface area for large-scale photovoltaic energy-potential evaluations *Solar Energy* **82** 929-39
[4] Rylatt M, Gadsden S and Lomas K 2001 GIS-based decision support for solar planning in urban environments *Computer Environ Urban Syst* **25** 579-603
[5] Gadsden S, Rylatt M, Lomas K, *et al* 2003 Predicting the urban solar fraction: A methodology for energy advisers and planners based on GIS *Energy & Buildings* **35** 37-48
[6] Gadsden S, Rylatt M and Lomas K 2003 Putting solar energy on the urban map: A new GIS-based approach for dwellings *Solar Energy* **74(5)** 397-407
[7] Ramachandra T V and Shruthi B V 2007 Spatial mapping of renewable energy potential *Renew Sust Energy Reviews* **11** 1460-80
[8] Kucuksari S, Khaleghi A M, Hamidi M, *et al* 2004 An Integrated GIS, optimization and

simulation framework for optimal PV size and location in campus area environments *Appl Energy* **113** 1601-13

[9] Sánchez-Lozano J M, Teruel-Solano J, Soto-Elvira P L, *et al* 2013 Geographical information systems (GIS) and multi-criteria decision making (MCDM) methods for the evaluation of solar farms locations: Case study in south-eastern Spain *Renew and Sust Energy Reviews* **24** 544-56

[10] Ramachandra T V and Shruthi B V 2007 Spatial mapping of renewable energy potential *Renew. & Sust. Energy Reviews* **11(7)** 1460-80

[11] Bent Sθrensen 2001 GIS management of solar resource data *Solar Energy Mater & Solar Cells* **67(1-4)** 503-9

[12] Zhou Y, Wu W X, Hu Y, *et al* 2010 Spatial distribution characteristics and resource potential assessment of solar energy resources in Northwest China *J of Natural Resources* **10 (25)** 1738-49

[13] Liao Z L and Ruan X G 2009 Nonlinear engineering mathematical model of silicon solar cell with arbitrary light intensity and temperature *Acta Solarellica Sinica* **30(4)** 430-5

[14] Liu S M and Hong W 2009 Design and construction of solar photovoltaic power generation system *Biography* (Beijing: Science Press)

[15] Fanney A H, Dougherty B P and Davis M W 2001 Measured performance of building integrated photovoltaic panels *J of Solar Energy Engin Trans ASME* **123** 187-93

[16] Zhang X and Srinivasan R 2009 GIS-based spatial precipitation estimation: A comparison of geostatistical approaches *JAWRA J of the American Water Resources Association* **45(4)** 894-906

IOP Conf. Series: Earth and Environmental Science **93** (2017) 012057 doi:10.1088/1755-1315/93/1/012057

Experimental study of mass diffusion coefficients of hydrogen in dimethyl phosphate and n-heptane

Y Guo[1,2], L K Zhu[1], Y P Zhang[1], J Liu[1] and J S Guo[1]

[1]Turbine Institute Shandong Electric Research Institute, Jinan 250002, Shandong Province, China

E-mail: guoying20048@hotmail.com

Abstract. In this study, a laser holographic interferometer experimental system was developed for studying the gas-liquid mass diffusion coefficient. Then the experimental system's uncertainty was analyzed to be at most ±0.2%; therefore, this system was reliable. The mass diffusion coefficient of hydrogen in dimethyl phosphate and n-heptane was measured at atmospheric pressure in the temperature range of 273.15–338.15 K. Then, the experimental data were used to fit the correlations of the mass diffusion coefficient of hydrogen in dimethyl phosphate and n-heptane with temperature.

1. Introduction

The mass diffusion of hydrogen in liquid hydrocarbons plays an important role in many industrial applications. For example, hydrogen is used as a fuel for internal combustion engines [1]; during combustion, it diffuses into the lubricating oil in the engine. In chemical industries, various hydrogenation technologies such as hydrocracking [2] and hydrofining [3] are used to improve the base oil of lubricants to realize high viscosity, low volatility, and improved safety [4]. In the hydrogenation process, hydrogen diffuses into the base oil of the lubricants. Even in the thermal power industry, hydrogen used for cooling the generator rotor diffuses into the lubricating oil and sealing oil [5]. In some of these applications, the diffused mass transfer is advantageous, because hydrocracking improves the lubricant properties and diffusion can promote the lubricant upgrading process. However, diffusion can also be detrimental. For example, when hydrogen is used as a coolant, diffusion will increase hydrogen leakage. Therefore, it is important to study the diffusion coefficient of hydrogen in liquid hydrocarbons.

At present, experimental studies of hydrogen diffusion have mainly focused on hydrogen diffusion into solid materials [6-8]. Few studies have investigated hydrogen diffusion into liquids, especially liquid hydrocarbons. Some studies have used mathematical approaches to provide equations for predicting the diffusion coefficient of hydrogen in liquid hydrocarbons [9,10]. In this study, we have developed a laser holographic interferential experimental platform for studying the gas-liquid diffusion coefficient. Then, by using this experimental system, we have measured the mass diffusion coefficient of hydrogen in dimethyl phosphate and n-heptane, the main components of lubricating oil and seal oil, at atmospheric pressure in the temperature range of 273.15–338.15 K. Then, we used the experimental data to fit the correlations of the mass diffusion coefficient of hydrogen in dimethyl phosphate and n-heptane with temperature. These correlations can be used as a reference in future engineering applications.

2. Experimental principle and experimental system

2.1. Experimental principle

For one-dimensional diffusion of two solutions in the vertical direction, the relation between concentration with time and space is given by equation (1) [11], and the concentration distribution curves with time and space are shown in figure 1.

$$(\frac{\partial c}{\partial t}) = D_{12}(\frac{\partial^2 c}{\partial z^2}) \quad (1)$$

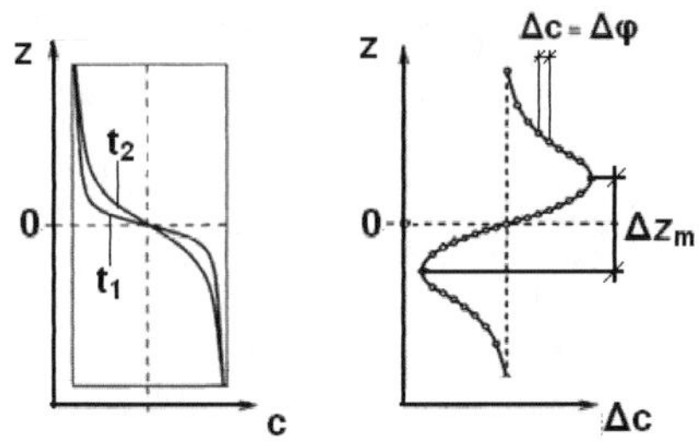

Figure 1. Concentration distribution with time and space.

In equation (1), D_{AB} (unit: m^2/s) is the mass diffusion coefficient; c(unit: kg/m^3), the concentration; z (unit: m), the distance in the vertical direction; and t (unit: s), the time.

Upon substituting the initial and boundary conditions into equation (1) and performing a series of derivations [12], the final formula for calculating the diffusion coefficient D_{AB} for one-dimensional diffusion can be obtained as given in equation (2).

$$D_{AB} = \Delta z_m^2 \frac{t_1 / t_2 - 1}{8 t_1 \ln(t_1 / t_2)} \quad (2)$$

In equation (2), t (unit: s) is the time at which the two holograms are obtained, and ΔZm (unit: m) (figure 1) is the vertical distance between the two extreme points of concentration that can be obtained using interference fringe processing.

2.2. Experimental system

The experimental system for digital holographic interferometry consists of an optical interference system, a diffusion cell system, and a constant temperature water system.

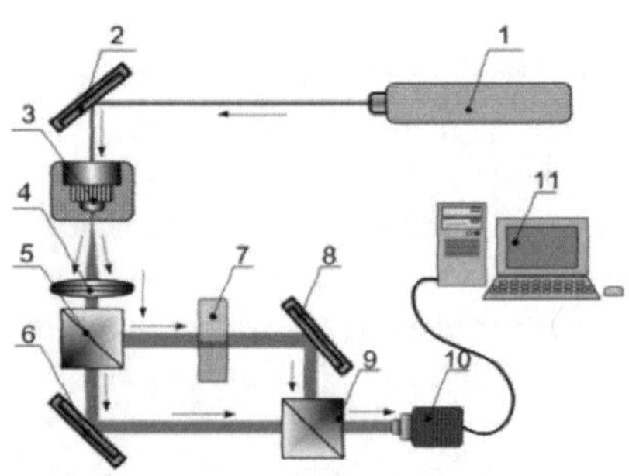

Figure 2. Optical interference system.
1, laser launcher; 2, reflector; 3, filter; 4, doublet lens; 5, light splitter prism; 6, reflector; 7, diffusion cell; 8, reflector; 9, light splitter prism; 10, image acquisition camera; 11, computer.

Figure 2 shows the laser holographic interferometry system used for measuring the diffusion coefficient. A 632.8 nm He-Ne laser is used in this experiment. The He-Ne laser beam is expanded into a parallel beam of light with ~50-mm diameter by using a spatial filter and a collimated lens. This parallel beam then passes through an aperture with 30-mm diameter such that its center uniform brightness is preserved. Then, the beam is split using a prism into the object light and reference light beams. The object light which contains the information of concentration is used to analyze diffusion, it interferes with the reference beam, and the resulting interference pattern is captured using a charge-coupled device CCD. In the experimental system, each component is placed on an anti-vibration platform to reduce the vibrations and improve the measurement accuracy.

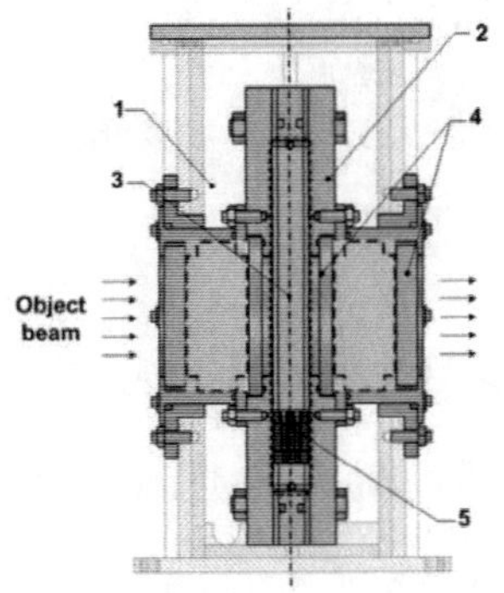

Figure 3. Diffusion subsystem.
1, water bath; 2, diffusion cell; 3, diffusion zone; 4, observation window; 5, honeycomb.

Figure 3 shows the diffusion system used for measuring the gas-liquid diffusion coefficient. This system consists of a thermostatic water bath, diffusion cell, diffusion zone, and observation window. The thermostatic water bath helps ensure that diffusion occurs under a constant temperature. The diffusion cell contains a diffusion zone within which the diffusion of two substances occurs. The observation window is the channel through which the object light is passed; it can also be used to monitor the injection process and to observe the diffusion process.

To test the gas-liquid diffusion coefficients at different temperatures, the diffusion resulting should be controlled accurately. Figure 4 shows the thermostatic water bath system used for controlling the diffusion temperature.

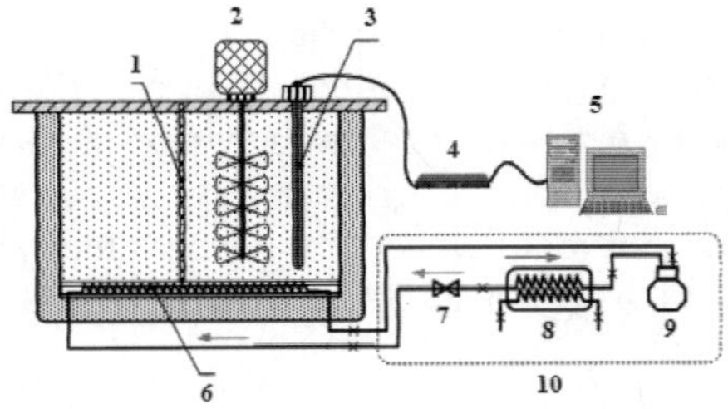

Figure 4. Thermostatic water bath subsystem.
1, rectifier grid; 2, stirrer ; 3, heating rod; 4, temperature controller; 5, computer; 6, evaporator; 7, throttle valve; 8, condenser; 9, compressor; 10, refrigeration system.

This system consists of a thermostatic environment box, a heating device, and a refrigeration device. The heating rod can be heated to a temperature of 80°C, and the evaporator of the refrigeration system at the bottom of the box can reduce the temperature to 0°C. The stirrer and rectifier grid are used to mix the water to achieve a uniform temperature. The temperature controller connects the heating rod and the computer to control the heating temperature and record the water temperature on the computer. The system can be used to set a temperature in the range of 0–80°C with 0.1°C accuracy.

The temperature, pressure, and other operational data were collected using an NI-SCXI data acquisition system controlled by a data acquisition program written using Labview. This system consists of an SCXI-1000 universal case, an SCXI-1102 signal conditioning module, and an SCXI-1581 analog output module. Figure 5 shows the signal acquisition process.

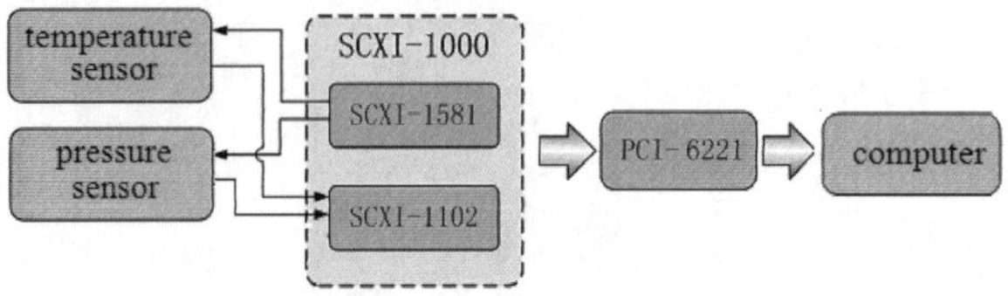

Figure 5. Experimental data acquisition system.

In order to evaluate the reliability of the experimental system, the mass diffusion coefficient of KCl in aqueous solution was measured, the relative deviations of the experimental results with literature values are within 1.30% [13]. The uncertainty of the physical quantities measured using this system are also determined. In this paper, the uncertainty of the mass diffusion coefficient measurement system depends on the temperature, time at which hologram is obtained, and the distance between two extreme points of concentration. According to the International Organization for Standardization [14], the combined experimental uncertainty of these three is given as

$$U = k \cdot u_c = k\sqrt{\sum (u_i)^2} \tag{3}$$

where u_i is the uncertainty of every variable in the experiment; u_c, is the synthetic uncertainty of all variables; and k is the confidence coefficient of the total uncertainty, set as 2 in this study, with the degree of confidence being 95%. Using equation (3), the uncertainty of temperature in this experiment can be calculated to be ±0.16 K, and the uncertainty of mass diffusion coefficient can be calculated as ±0.2% [13].

2.3. Interference fringe processing

Figures 6 and 7 show the interference fringe processing. The interference fringe recorded using the CCD was first denoised by the wavelet method, and then, a Fourier transform was used to obtain the frequency spectrum of the interference fringe, and then take out the first-order frequency form the spectrum with a rectangular window function and then the wrapped phasic difference can be obtained using the first-order frequency to reconstruct the object wave at two different moments. Then, it is unwrapped to obtain the continual phasic difference, from which the value of ΔZ_m can be determined to calculate the diffusion coefficient using equation (2). Details of the image processing principles and processes are given in reference [15].

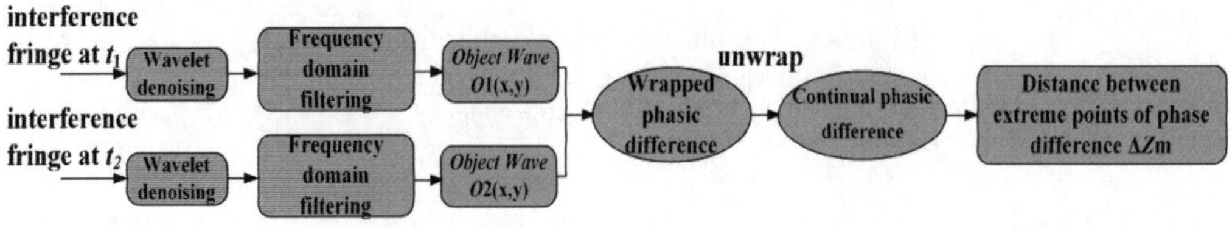

Figure 6. Flowchart of interference fringe processing.

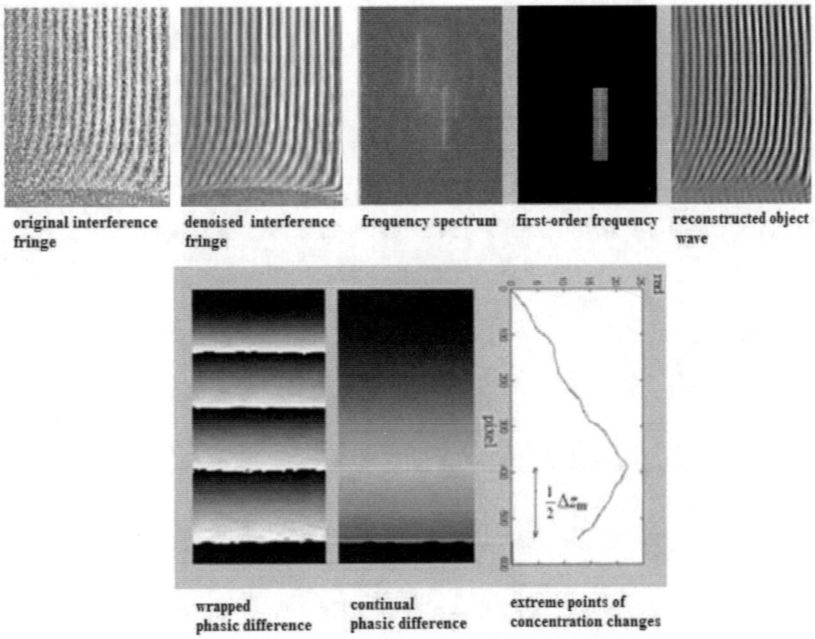

Figure 7. Interference fringe processing.

3. Experimental results and analysis

The mass diffusion coefficients of hydrogen in dimethyl phosphate and n-heptane were measured in the temperature range of 278.15–338.15 K and at atmospheric pressure of 0.1 MPa. Table 1 and figure 8 show the experimental results and uncertainty of all data.

Table 1. Experimental diffusion coefficients of hydrogen in n-Heptane and dimethyl phosphate.

T/K	**Hydrogen** (1) + **n-Heptane** (2)		**Hydrogen** (1) + **Dimethyl Phosphate** (2)	
	$10^5\ D_{AB}$/cm^2·s^{-1}	$10^5/u_{D_{AB}}$ cm^2·s^{-1}	$10^5\ D_{AB}$/cm^2·s^{-1}	$10^5 u_{D_{AB}}$/cm^2·s^{-1}
278.15	8.990	±0.0023	8.148	±0.0021
281.15	9.043	±0.0022	8.204	±0.0023
284.15	9.093	±0.0021	8.246	±0.0022
287.15	9.172	±0.0021	8.311	±0.0021
290.15	9.248	±0.0022	8.359	±0.0021
293.15	9.336	±0.0022	8.408	±0.0022
296.15	9.442	±0.0021	8.485	±0.0023
299.15	9.576	±0.0021	8.558	±0.0023
302.15	9.709	±0.0020	8.637	±0.0022
305.15	9.836	±0.0022	8.740	±0.0021
308.15	9.990	±0.0023	8.853	±0.0023
311.15	10.161	±0.0021	8.979	±0.0021
314.15	10.330	±0.0022	9.105	±0.0022
317.15	10.491	±0.0022	9.249	±0.0023
320.15	10.694	±0.0021	9.418	±0.0022
323.15	10.896	±0.0024	9.558	±0.0023
326.15	11.139	±0.0021	9.707	±0.0023
329.15	11.370	±0.0022	9.881	±0.0023
332.15	11.600	±0.0022	10.094	±0.0021

335.15	11.870	±0.0020	10.307	±0.0023
338.15	12.130	±0.0021	10.549	±0.0022

D_{AB} is the experimental result; $u_{D_{AB}}$ is the uncertainty of the diffusion coefficient.

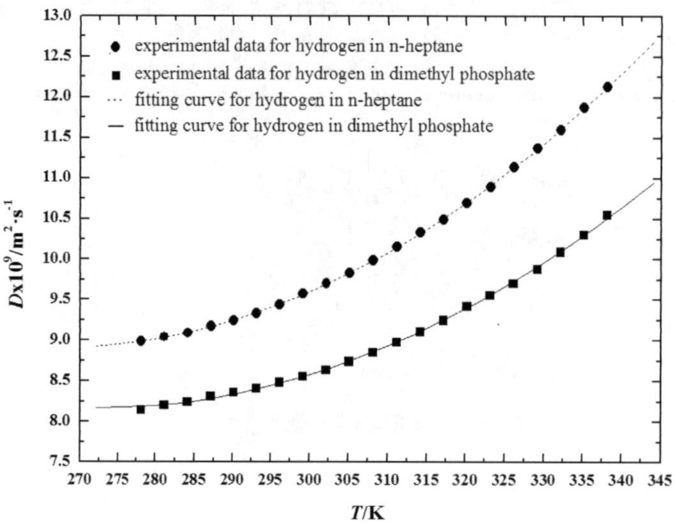

Figure 8. Experimental diffusion coefficients of hydrogen in n-heptane and dimethyl phosphate.

Figure 8 shows that as the temperature increases, the diffusion coefficients of hydrogen in the two liquid hydrocarbons increase. Under the same temperature condition, the mass diffusion coefficient of hydrogen in n-heptane is larger than that of hydrogen in dimethyl phosphate. This can be attributed to the inverse relation between diffusion coefficient and molecular size.

For engineering applications, a polynomial for calculating the mass diffusion coefficient of hydrogen in dimethyl phosphate and n-heptane was fitted as given by equation (4).

$$D_{12}/\mathrm{cm}^2\cdot\mathrm{s}^{-1} = \mathrm{C}_1 + \mathrm{C}_2\cdot(T/\mathrm{K}) + \mathrm{C}_3\cdot(T/\mathrm{K})^2 \quad (4)$$

Where T is the absolute temperature; C_1, C_2, and C_3 are the fitting parameters; and the fitted values are shown in table 2.

Table 2. Parameters of polynomial (4) fitted by experimental data.

	C_1	C_2	C_3	RMSD
hydrogen in n-heptane	5.30553×10^{-4}	-3.40312×10^{-6}	6.37003×10^{-9}	5.493×10^{-8}
hydrogen in dimethyl phosphate	4.67295×10^{-4}	-2.97017×10^{-6}	5.44822×10^{-9}	8.871×10^{-8}

The root mean square deviation (RMSD) between the results calculated using equation (4) and the experimental values listed in table 3 is given by equation (5) as

$$\mathrm{rmsd} = \left[\sum_{j=1}^{N}(x_j^{\exp} - x_j^{\mathrm{cal}})^2/(N-1)\right]^{1/2} \quad (5)$$

Where N is the number of experimental points, x_j^{cal} is the diffusion coefficient calculated according to equation (4), and $x_j^{\exp}$ is the diffusion coefficient measured experimentally.

4. Conclusion

In this study, a laser holographic interferometer experimental system was developed for studying the gas-liquid mass diffusion coefficient. Each subsystem of the experimental platform is introduced in detail, and then, the reliability and the accuracy of the system were confirmed by uncertainty analysis and experimental verification. By using this experimental system, the mass diffusion coefficient of hydrogen in dimethyl phosphate and n-heptane was measured at atmospheric pressure in the temperature range of 273.15–338.15 K. Then, we used the experimental data to establish correlations of the mass diffusion coefficient of hydrogen in dimethyl phosphate and n-heptane with temperature. These correlations can be used as a reference in future engineering applications.

Acknowledgments

We are thankful and acknowledge the support under the funding system for scientific research programme of Shandong Electric Research Institute.

References

[1] Michelotti A C and Silva J C D 2016 Design innovation in dynamic coupling of starting system for internal combustion engines *J Braz Soc Mech Sci Eng* **38** 177-88
[2] Peng C,Yang X J, Fang X C, Huang X L, Cheng Z M, Zeng R H and Guo R 2015 Development of light cycle oil (LCO) hydrocracking technology over a commercial W-Ni based catalyst *China Petrol Process Petrochem Technol* **2015** 30-6
[3] Gulyaeva L A, Shmel'Kova O I, Khavkin V A, *et al* 2016 Hydrofining of vegetable, synthetic, and petroleum blendstocks to produce diesel fuel *Petrol Chem* **56** 883-7
[4] Bitter J H and Stellwagen D R 2014 Structure-performance relations of molybdenum and tungsten carbide catalysts for deoxygenation *Green Chem* **17** 582-93
[5] Pimerzin A A, Tomina N N, Nikul'shin P A, Maksimov N M, Mozhaev A V, Ishutenko D I and Vishnevskaya E E 2015 Catalysts based on molybdenum and tungsten heteropoly compounds for the hydrotreatment of oil fractions *Catal Ind* **7** 30-7
[6] Moshref-Javadi M, Edris H, Shafyei A and Salimi-Jazi H 2016 Diffusion behavior of hydrogen through thermally sprayed coating of 316L stainless steel *Int J Hydrogen Energ* **42** 6409-19
[7] Li Q, Yi X Y, Lu Y, Li G S and Chen W L 2016 The law of the hydrogen ion diffusion coefficient in acid rock reaction *J Pet Sci Eng* **146** 694-701
[8] Yukawa H, Zhang G X, Watanabe N, Morinaga M, Nambu T and Matsumoto Y 2009 Hydrogen diffusion coefficient during hydrogen permeation through Nb-based hydrogen permeable membranes *Defect Diffus Forum* **817** 225-30
[9] Yang X and Oyeniyi W O 2016 Kinetic monte carlo simulation of hydrogen diffusion in tungsten *Fusion Eng Des* **114** 113-7
[10] Xiong X J 1997 An equatijon for the prediction of diffusivities of hydrogen in liquid hydrocarbons *Petrol Process Petrochem* **28** 57-61
[11] Chhaniwal V K, Anand A, Girhe S, Patil D, Subrahmanyam N and Narayanamurthy C S 2003 New optical techniques for diffusion studies in transparent liquid solutions *J Opt A: Pure Appl Opt (Special Edition)* **5** 29-37
[12] Guo Y, He M G, Zhong Q and Zhang Y 2008 Mass diffusion of coefficients of dimethyl carbonate in heptanes and in air at T (278.15 to 338.15 K) *J Chem Eng Data* **53** 2861-4
[13] He M G, Guo Y and Zhong Q 2009 Determination of mass diffusion coefficients of oxygenated fuel additives in air using digital real-time holographic interferometry *J Phys Conf Series* **147(1)** 1328-38
[14] Keith D McCroan 1995 *Guide to the Expression of Uncertainty in Measurement* Corrected and Reprinted (ISO: Genevese)
[15] He M G, Guo Y, Zhong Q and Zhang Y 2009 A new method of processing holographic interference fringe gathered in procedure of diffusion *Int J Thermophys* **30** 1823-37

Halbach array-based design and simulation of disc coreless permanen-magnet integrated starter generator

Y B Li[1], Z X Yang[1,2], W Chen[1] and Q Y He[1]

[1]Institute of Systems Science and Engineering, Henan Engineering Laboratory of Wind Power Systems, Henan University of Science and Technology, No. 48 Xiyuan Road, Luoyang 471003, China

E-mail: zxyang@haust.edu.cn

Abstract. The functional performance, such as magnetic flux leakage, power density and efficiency, is related to the structural characteristics and design technique for the disc permanent magnet synchronous generators (PMSGs). Halbach array theory-based magnetic circuit structure is developed, and Maxwell3D simulation analysis approach of PMSG is proposed in this paper for integrated starter generator (ISG). The magnetization direction of adjacent permanent magnet is organized in difference of 45 degrees for focusing air gap side, and improving the performance of the generator. The magnetic field distribution and functional performance in load and/or unload conditions are simulated by Maxwell3D module. The proposed approach is verified by simulation analysis, the air gap flux density is 0.66T, and the phase voltage curve has the characteristics of a preferable sinusoidal wave and the voltage amplitude 335V can meet the design requirements while the disc coreless PMSG is operating at rated speed. And the developed magnetic circuit structure can be used for engineering design of the disc coreless PMSG to the integrated starter generator.

1. Introduction

With the development of automotive integration, the integrated starter and generator (ISG) is deemed fit as sustainable alternatives in electricity generation and dynamic equipment [1,2]. ISG is capable of reducing the fuel consumption of hybrid electric vehicle (HEV) which has the function of automatic start stop, power compensation, and high efficiency power output. The disc coreless permanent magnet synchronous generator (PMSG) machine is widely studied in HEV because of their simple structures and low costs compared with general PMSG [3].

The Halbach array is a new type of permanent magnet arrangement, which is especially suitable for the rotor surface adhered the permanent magnet [4]. When the permanent magnet is arranged in Halbach array, the most notable feature is that the air gap flux increases, while the rotor yoke flux decreases, and the sinusoidal air gap magnetic field is obtained [5,6]. These are very beneficial for reducing the motor volume and increasing the power density. These make it susceptible to disc coreless PMSG [7].

There are two methods of electromagnetic analysis for disc coreless PMSG. They are named respectively numerical techniques and analytical techniques [8,9]. The magnetic equivalent circuits (MEC) calculation is a kind of numerical techniques. It is a method to divide the uneven magnetic field in space into the equivalent magnetic circuit of many segments. ANSYS Maxwell 3D is a type of analytical techniques that uses finite element analysis (FEA) to solve electromagnetic problems [10].

In order to improve the design efficiency and precision, the field circuit coupled finite element method is used. The electromagnetic design procedure of disc coreless PMSG involves three steps. Firstly, MEC as a calculation method, the initial size of the disc coreless PMSG is calculated. Secondly, the model built by SolidWorks and imported to the Maxwell. Thirdly, the disc coreless PMSG electromagnetic characteristics are verified by Maxwell 3D analysis [11]. To reduce the weight of the generator and improve the utilization ratio of permanent magnet, the structural design and simulation analysis is presented.

2. Electromagnetic field design and analysis

The disc PMSG also called axial flux permanent magnet machines (AFPM), which is different from cylindrical motor [12]. The magnetic circuit distribution of disc PMSG is very complex. The equivalent magnetic circuit calculation assumes that the magnetic flux in each section of the magnetic circuit is along the cross section and the length distribution. In fact, the existence of a nonuniform distribution of magnetic fields becomes equivalent to multiple magnetic circuits. This will reduce the amount of time, especially in initial structure design. The mechanical and electromagnetic design was accomplished with an example of the rated speed 3000 r/min, the rated power 6 Kw and the rated voltage 300 V disc coreless PMSG. The parameters are shown in table 1. By using the field circuit coupled finite element method, the disc coreless PMSG based on the Halbach array were analyzed.

Table 1. ISG AFPM parameter.

Parameters	Values
Rated Power	6 KW
Rated Voltage	300V
Rated Speed	3000 r/min
Frequency	400 Hz
Rated Power Factor	0.9
Operating Temperature	75°C

2.1. Electromagnetic field calculation

Because of the special structure, the AFPM appears the uneven distribution of the flux density and magnetic circuit at different radius length. In order to accurately calculate the magnetic field distribution, it is necessary to use MEC method to calculate the three-dimensional magnetic field. The main parameters, such as air gap magnetic density distribution and magnetic flux leakage coefficient, can be calculated by three-dimensional magnetic field calculation [13].

ANSYS Maxwell3D can visually show the distribution of electromagnetic fields. The theoretical principle of Maxwell is finite element analysis. The electromagnetic fields have been analysed by Maxwell formulas which obtain a guaranteed solution with appropriate boundary conditions in a finite region of space [14]. Maxwell's formulas are shown as formula (1)-(4).

$$\begin{cases} \nabla \times H = J + \dfrac{\partial D}{\partial t} & (1) \\ \nabla \times E = -\dfrac{\partial B}{\partial t} & (2) \\ \nabla \times D = \rho & (3) \\ \nabla \times B = 0 & (4) \end{cases}$$

Generally, the electromagnetic field of the PMSG varies with the different media. There are the following electromagnetic relations.

$$\begin{cases} D = \xi E & (5) \\ J = \sigma E & (6) \\ B = \mu H & (7) \end{cases}$$

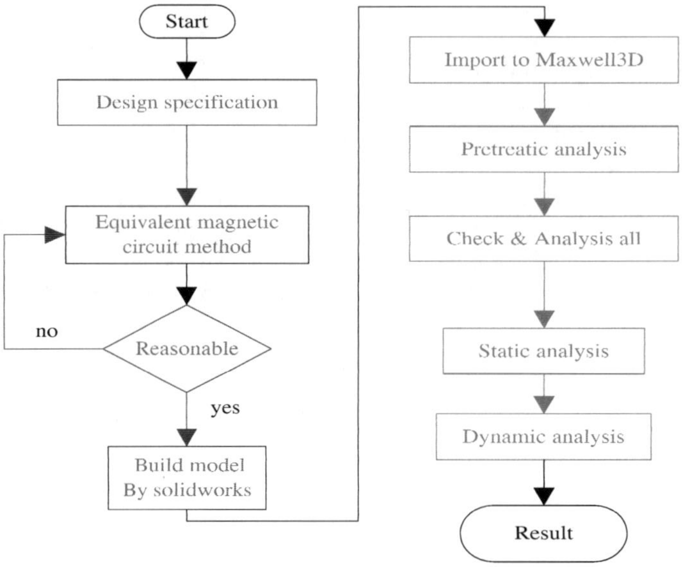

Figure 1. Framework for the design and electromagnetic analysis of PMSG.

As shown in the framework, the electromagnetic field calculation and analysis of dics coreless PMSGs is presented. It can be summed up in three parts. First, initial electromagnetic parameters and geometric dimensions are calculated by MEC calculation. Second, the model is established by SolidWorks and imported into the Maxwell 3D. Last, the operation situation of the generator is obtained by Maxwell 3D. The calculation of electromagnetic field requires the initial estimation of generator line load and magnetic load. Parametric initial estimated will be constantly validated during the calculations. Maxwell 3D is a mighty tool which can be verified the optimum value of the MEC calculation results.

2.2. Structure and electromagnetic analysis

As shown in figure 2(a), the disc coreless PMSG is composed by double-sided rotor and single stator. The Halbach PM rotor symmetrically distribute on both sides to form a closed magnetic circuit structure, thereby reducing the leakage flux of the motor, and improving the utilization of permanent magnets. The rotor consists of 64 PMs bonded the rotor yoke with the material of NdFeB35. The direction of the permanent magnet magnetization is 45 Halbach array as figure 2(b). The rotor array-based Halbach has the magnetic shielding property, which increases the air gap of magnetic field intensity and reduces the yoke of magnetic field intensity. Therefore, the yoke thickness is reduced. The stator PMSG is composed by 24 rhomboidal coils with sticked epoxy as figure 2(c).

It can be seen from the analysis of MEC that the main path of the magnetic circuit is composed of the circumferential path and the axial path. The axial magnetic circuit is composed by the magnetic flux flowing from one PM through the air gap, stator and the yoke to another PM.

The air gap magnetic field is divided into three parts of the air gap magnetic field. As shown in figure 3, the area of I is the yoke region. II describes the air gap region. III represents air gap and the winding region.

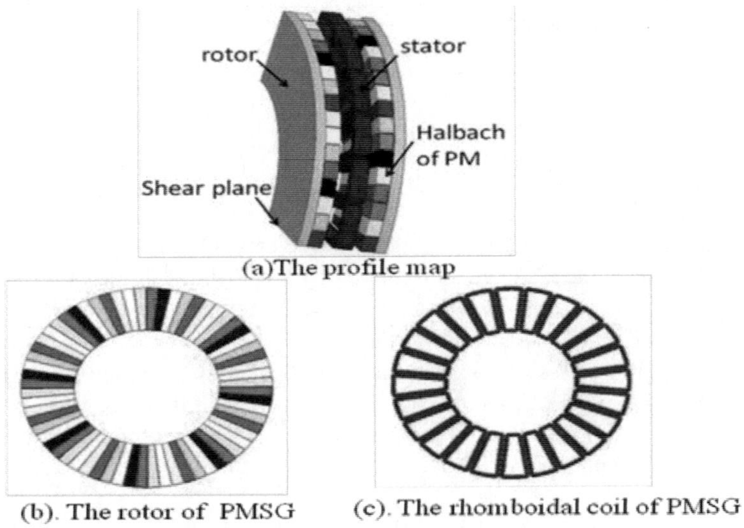

Figure 2. The structure of disc coreless permanent magnet synchronous generator.

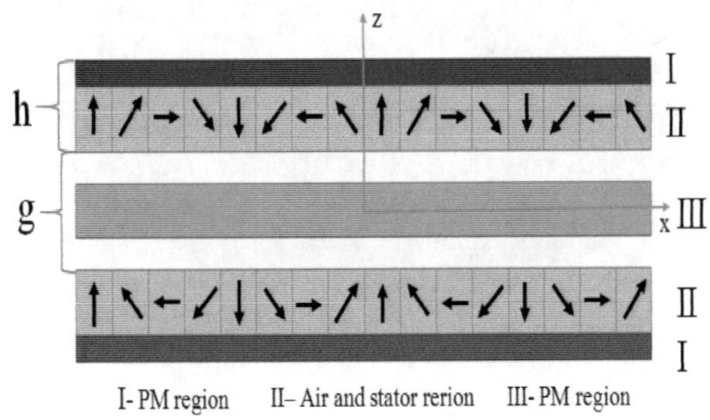

Figure 3. 2D model of AFPM.

The air gap magnetic field of the disc coreless PMSM is generated by the interaction between the permanent magnet region II and the region III, and the formula (8) represents the axial component of the air gap magnetic flux density [15].

$$B_\delta\left(x,z\right)=B_\delta\frac{2}{m_n g}\cosh\left(m_n z\right)e^{mn^x}\cos\left(m_n x\right) \tag{8}$$

Where B_δ describes the air gap magnetic flux density, h is the thickness of the permanent magnet, g is the sum of the equivalent air gap length of the motor which is equal to the length of the air gap and the thickness of the winding. The m_n is calculated by $m_n = (2n-1)\,\pi/\tau_p$

This machine is a kind of electromagnetic conversion device. It can mutually convert the mechanical energy into electric energy storaged as the second power source. The coil cuts the magnetic field lines to produce three-phase voltage. E represents the electromotive force generated by the disc motor.

$$E=\frac{1}{4}\Omega N a_i k_w B_\delta(D_0^2-D_i^2)/a \tag{9}$$

Where D_0 is outer diameter of permanent magnet, D_i is inner diameter of permanent magnet, Ω describes the mechanical angular velocity, and N describes the number of conductors per slot.

The output power is an important parameter for the performance of PMSG. It is affected by the change of the load, where $\cos\varphi$ is power factor.

$$P = \sqrt{3}UI\cos\varphi \tag{10}$$

2.3. Modeling design

The design of AFPM mainly includes the size of model structure and the choice of material. The model is composed of double-sided rotor and coreless stator. The stator winding consists of many single-layer rhomboidal coils. The windings are fixed together with a composite material of epoxy resin [16]. Figure 2(c) describes the stator model builded by the RMxprt in the rapid establishment of the Maxwell 3D.

The rotor has 8 pairs of PMS which are respectively pasted on the rotor yokes. Each pole permanent magnet is composed of 8 sections. The magnetization vector angle of adjacent PM is 45°. The rotor model is established by SolidWorks and then imported into Maxwell 3D. Through the translation and rotation of the stator matching, the geometry model is established. Figure 4 shows the geometry model of PMSG after matching.

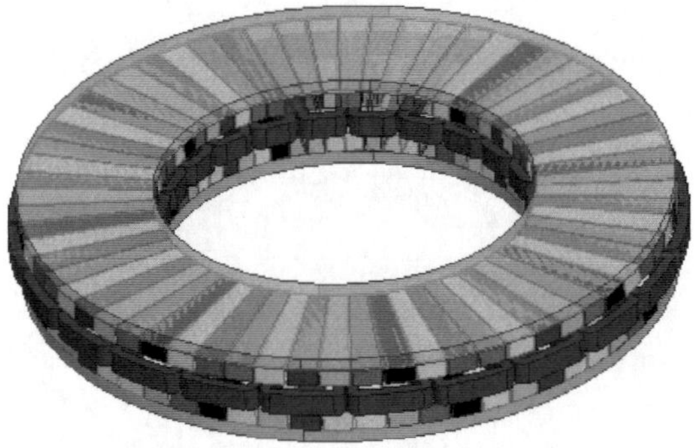

Figure 4. The model of AFPM.

According to table 2, the rotor and stator structures are determined, and all material characteristics have been applied to each component, such as permanent magnets, air-gap, rotors and stator. The double-sided rotor is free to move and the stator is stationary. The materials of rotor select the steel-1008. The permanent magnets used in the generator are NdFeB magnets with Br =1.28 T, Hc = 970 kA/m and (BH) max=310.4 kJ/m^3 and maximum operating temperature of 140℃ [17]. In order to increase the distribution of the magnetic flux density in the air gap, the two pairs of poles are distributed by N-S, which is beneficial to the magnetic properties of Halbach array. What's more, compared with 90°, 60° and 45°, the Halbach array permanent magnet is found to be stable at 45°. After selecting the material of the permanent magnet, the magnetization direction of the permanent magnet is specified. It is necessary to establish a local coordinate to define the complex Halbach array permanent magnet.

Table 2. ISG design parameters.

Part	Parameters	Values
Windings	Number of windings	24
	Number of phase	3
	Thickness of coil	13 mm
	Winding type	rhomboidal coil
	Turns per pole	20
	Wire diameter of windings	1.08 mm
Rotor	Outer diameter of PM	320 mm

Inner diameter of PM	185 mm
Number of pole-pairs	8
Thickness of PM	11 mm
Material of PM	NdFe35
Air gap	1 mm

3. Results and analysis

Ansoft Maxwell 3D analysis has two parts, one is static magnetic field analysis, and the other is transient field analysis. The static simulation verifies the rationality of magnetic circuit and the external characteristics are calculated by transient simulation.

As shown in figure 5 is the mesh of disk Coreless PMSG. Figure 6 is the number of the mesh of each part of the generator. It is clearly shown in Maxwell3D when we click the solution data. The maximum mesh size is 39 mm and the minimum is 0.001147 mm. The total number of meshes is 478145.

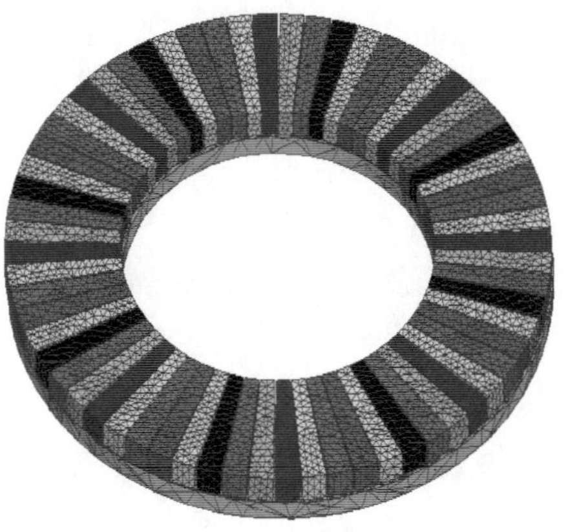

Figure 5. Mesh generation of rotor.

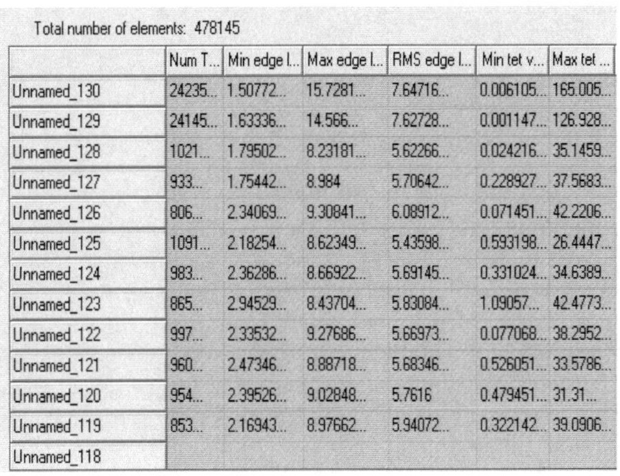

Total number of elements: 478145

	Num T...	Min edge l...	Max edge l...	RMS edge l...	Min tet v...	Max tet ...
Unnamed_130	24235...	1.50772...	15.7281...	7.64716...	0.006105...	165.005...
Unnamed_129	24145...	1.63336...	14.566...	7.62728...	0.001147...	126.928...
Unnamed_128	1021...	1.79502...	8.23181...	5.62266...	0.024216...	35.1459...
Unnamed_127	933...	1.75442...	8.984	5.70642...	0.228927...	37.5683...
Unnamed_126	806...	2.34069...	9.30841...	6.08912...	0.071451...	42.2206...
Unnamed_125	1091...	2.18254...	8.62349...	5.43598...	0.593198...	26.4447...
Unnamed_124	983...	2.36286...	8.66922...	5.69145...	0.331024...	34.6389...
Unnamed_123	865...	2.94529...	8.43704...	5.83084...	1.09057...	42.4773...
Unnamed_122	997...	2.33532...	9.27686...	5.66973...	0.077068...	38.2952...
Unnamed_121	960...	2.47346...	8.88718...	5.68346...	0.526051...	33.5786...
Unnamed_120	954...	2.39526...	9.02848...	5.7616	0.479451...	31.31...
Unnamed_119	853...	2.16943...	8.97662...	5.94072...	0.322142...	39.0906...
Unnamed_118						

Figure 6. The number of mesh.

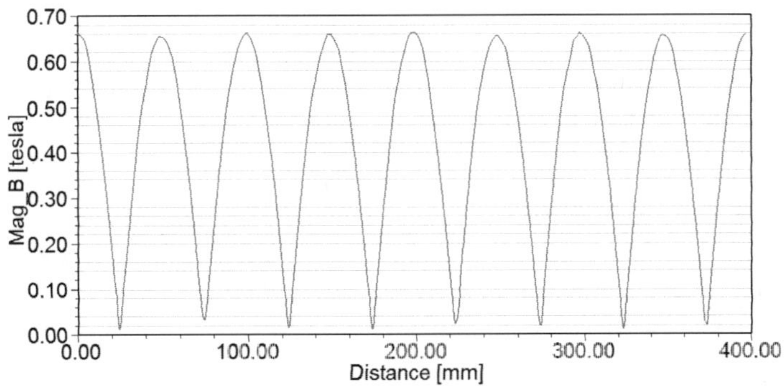

Figure 7. Air gap flux density.

Figure 7 appears the air gap flux density. As expected, the air gap flux density of disc coreless PMSG based on the Halbach is periodical. The peak value is about 0.6626 T. Compared with the value of experience, it is reasonable.

Figure 8 describes the vector distribution of magnetic field intensity. It can be clearly seen the magnetic field direction of each PM. In addition, the intensity of the arrow represents the magnetic field.

Figure 9 shows the magnetic flux density distribution. It can be seen that the magnetic field is enhanced in the air gap, which proves the Halbach array has the ability of gathering magnetic.

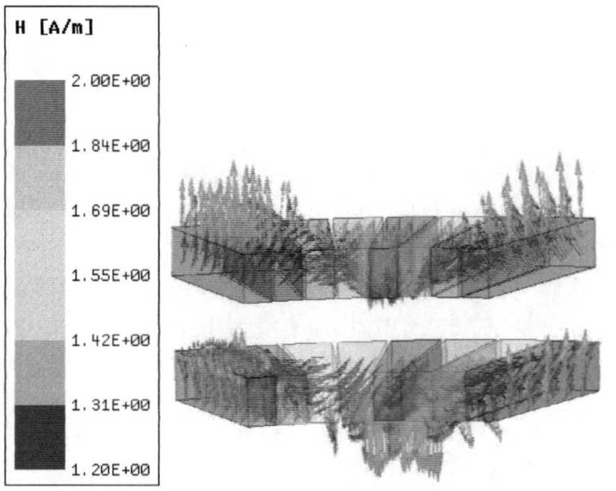

Figure 8. Vector distribution of magnetic field intensity.

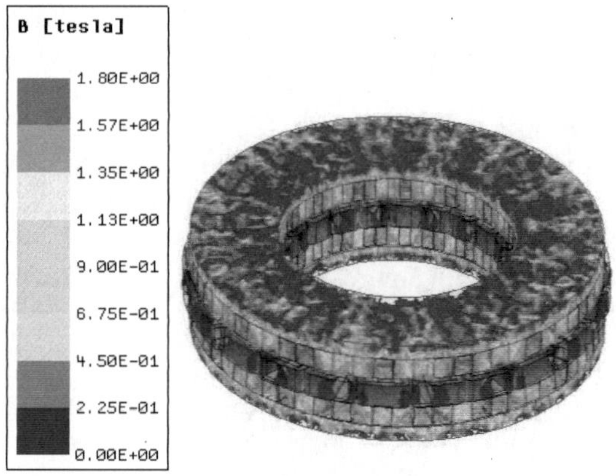

Figure 9. Magnetic flux density distribution of generator.

The Phase voltage presents very perfect sinusoidal characteristics in the figure 10. The amplitude of the voltage is about 335 V.

In this paper, a three-phase load circuit is simulated to analyze the characteristics of PMSG under load condition. The load current is shown in figure 11. The maximum amplitude of the current is 12.5 A.

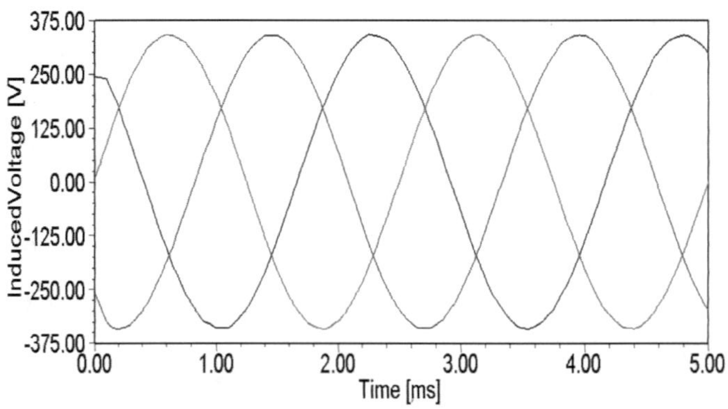

Figure 10. Phase voltage.

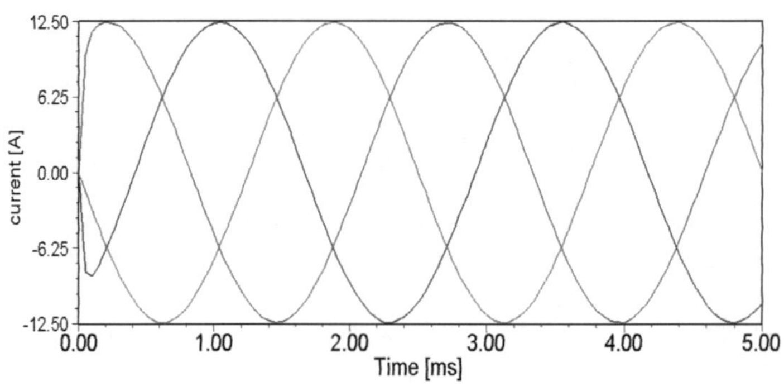

Figure 11. Phase current.

The output power can be obtained from the phase voltage and the current. The rated power factor of PMSG is not a constant value. It is generally less than 1. The generator power factor is 0.9. After calculation, the output power of the generator is 6.53 kW, more than rated power. It is obvious that the generator designed in this paper meets the design requirements. The waveform presents sinusoidal characteristic, which reduces the amplitude of harmonics and vibration and noise of PMSG.

4. Conclusion

The working principles and design features of Halbach permanent magnet disc coreless PMSG are discussed in this paper. Based on the coupled field circuit design of disc coreless PMSG is used in ISG Hybrid electric vehicle. Ansoft Maxwell3D simulated and analyzed the distribution of the magnetic field and the output characteristics. The results show that the disc coreless PMSG with Halbach array-based PM can meet the design requirements, while the PMSG is operating at rated speed, the output power, voltage and current can achieve practical engineering. The field circuit coupling method is used to reduce the design cycle and improve the work efficiency.

They are put forward in order to be further researched:

- The calculation parameters based MCE are not accurate, and the parameters in this paper are verified by simulation.
- The parameter analysis is in the case of ideal values, without considering the effects of loss, so the optimization results have certain limitations.
- The above performance analysis is ready for further prototyping

Acknowledgments

The financial support from the National Natural Science Foundation of China (NSFC) under the grant No. 70671035 and No. 71071078 is acknowledged gratefully.

References

[1] Lulhe A M and Date T N 2015 A technology review paper for drives used in electrical vehicle (EV) & hybrid electrical vehicles (HEV) *2015 International Conference on Control* **5** 632-6

[2] Liu C H, Chau K T and Jiang J Z 2010 A permanent-magnet hybrid brushless integrated starter-generator for hybrid electric vehicles *IEEE Transactions on Industrial Electronics* **12** 4055-64

[3] Yin A, Chen W, Zhao H, Lu R and Feng R 2011 Parameter optimization of ISG hybrid electric vehicle based on genetic algorithm *Automotive Engineering* **10** 834-7

[4] Wang X Y, Zhao F, Du J J and Qi L X 2007 Design of multi disc coreless permanent magnet synchronous motor based on Halbach array *Micromotors* **40** 433-8

[5] Gieras J F, Wang R J and Kamper M J 2008 *Axial Flux Permanent Magnet Brushless Machines* (New York: Kluwer Academic Publishers)

[6] Mahmoudi A, Kahourzade S and Rahim N A 2014 Design analysis and prototyping of a novel-structured solid-rotor-ringed line-start axial-flux permanent-magnet motor *IEEE Transactions on Industrial Electronics* **4** 1722-34

[7] Chen E K, Zhao L T, Yang S, Li Y J, Bao R N and Yan J H 2013 Study on simulation of disc permanent magnet synchronous motor *Micromotors* **46** 331-40

[8] Zhang Y M, Qiao D Z and Gao J X 2010 Research and application of Halbach array permanent magnet Analysis instrument **2** 5-10

[9] Li C, Zhang Y J and Jing L B 2013 Researches on an exact analytical method of Halbach-array permanent-magnet motors with semi-closed slots *Proceedings of the CSEE 2013* **33** 788-94

[10] Li Y H, Dou M F and Zhang C L 2013 Analytical calculation of air gap magnetic field for Halbach array PM motor *Micromotors* **3** 18-26

[11] Zhang Z, Muyeenb S M, Al-Durrab A, Nilssena R and Nysveen A 2014 Multiphysics 3D modelling of ironless permanent magnet generators *Energy Procedia* **53** 34-43

[12] Zhao P, Wang Z X, Yang S, Jia J, Zhang G L and Sun W 2014 The analysis of

three-dimensional electromagnetic field of coreless disc permanent magnet brushless DC motor *Shandong Electric Power* **2** 11-9

[13] Uygun D, Cetinceviz Y and Bal G 2016 Optimization study on a 0.6 kW PMSG for VAWTs and determination of open and short circuit *International Journal of Hydrogen Energy* **2016** 55-60

[14] Li G D, Yu H F, Wang X Y and Zhang B 2015 Magnet optimization of disc coreless permanent magnet synchronous motor based on Halbach *Micromotors* **48** 635-41

[15] Gieras J F, Wang R J and Kamoer M J 2005 *Axial Flux Permanent Magnet Brushless Machines* (Dordrecht: Kluwer Academic Publishers)

[16] Qiu H, Duan Q, Yao L, Dong Y, Yi R, Cui G Z and Li W L 2016 Analytical analysis of sleeve permeability for output performance of high speed permanent magnet generators driven by micro gas turbines *Applied Mathematical Modelling* **1** 1-12

[17] Li X M, Yang Z X, Li Y B, Chen W and Zhang L P 2016 Performance analysis of permanent magnet synchronous generators for wind energy conversion system *2016 International Conference on Advanced Mechatronic Systems* **2016** 26-34

Risk analysis of urban gas pipeline network based on improved bow-tie model

M J Hao[1,4], Q J You[2] and Z Yue[3]

[1]Capital University of Economics and Business, Beijing 100070, China
[2]Beijing Urban System Engineering Research Center, Beijing 100089, China
[3]Capital University of Economics and Business, Beijing 100070, China

E-mail: 18810598300@163.com

Abstract. Gas pipeline network is a major hazard source in urban areas. In the event of an accident, there could be grave consequences. In order to understand more clearly the causes and consequences of gas pipeline network accidents, and to develop prevention and mitigation measures, the author puts forward the application of improved bow-tie model to analyze risks of urban gas pipeline network. The improved bow-tie model analyzes accident causes from four aspects: human, materials, environment and management; it also analyzes the consequences from four aspects: casualty, property loss, environment and society. Then it quantifies the causes and consequences. Risk identification, risk analysis, risk assessment, risk control, and risk management will be clearly shown in the model figures. Then it can suggest prevention and mitigation measures accordingly to help reduce accident rate of gas pipeline network. The results show that the whole process of an accident can be visually investigated using the bow-tie model. It can also provide reasons for and predict consequences of an unfortunate event. It is of great significance in order to analyze leakage failure of gas pipeline network.

1. Introduction

With the widespread use of natural gas, more and more gas pipelines are laid in urban areas. In China, the construction of gas pipeline network reached its peak in recent years. Due to a variety of reasons, gas pipeline network accidents frequently occurred. These accidents have aroused great concern in society. Urban gas pipeline, oil pipeline, communication, drainage, heat [1] and other networks are laid crossing each other. Some of the pipeline networks are in serious aging but still overloaded. This can greatly increase the probability of an accident. The accident rate of gas pipeline follows the rule of "bathtub curve", which is manifested in the fact that accidents prone to happen in the early and late stages [2] of the life cycle of gas pipelines. Most of China's urban gas pipeline networks are in their early and late stages [3]. And gas pipeline network is a major hazard source in urban areas. In the event of an accident, there could be grave consequences. These accidents could bring great loss to lives, property, and society.

It is of great significance in order to conduct a necessary risk analysis and studies of urban gas pipeline network. The bow-tie model introduced in this paper represents risk identification, risk analysis, risk assessment, and risk control as in one chart. This is a highly visible, straightforward structured approach. It is easy to understand and master, and easy to use and operate, so it is ideal for conducting a risk analysis of urban gas pipeline network. Moreover, the bow-tie model conducts a risk analysis of urban gas pipeline by combining fault tree analysis (FTA), event tree analysis (ETA) [4], pipeline failure probability, and its consequences together. Reasons for pipeline failure are located by analyzing fault trees. Then risks can be prevented after they are identified and prevention and control measures are adopted. Possible consequences of an accident can be predicted by analyzing event trees. Then mitigation measures can be adopted to minimize losses to guarantee the safety of people's lives and property. The improved bow-tie model introduced in this paper analyzes accident causes from four aspects: human, materials, environment and management; it also analyzes the consequences from four aspects: casualty, property loss, environment and society. Then it quantifies the causes and consequences to conduct a risk analysis of urban gas pipeline network.

2. Introduction to the bow-tie model and its improvement

2.1. Development of bow-tie model

The bow-tie model appeared in the 1970s. Later David Gill developed it and named it Bow –tie Diagram [5]. It was proposed in university course notes as a systematic model and was constantly reviewed. At the end of the last century, the bow-tie model was successfully applied to analyze accidents by explosion [6]. At the beginning of this century, NASA used the bow-tie model in risk management, promoting its development in the field. The Bow-tie model is widely used in risk analysis and management of oil, natural gas and other industries because of its practical and highly visualized features.

2.2. Bow-tie model principle

The bow-tie model is a practical, easy-to-use risk analysis and evaluation method [7]. It combines FTA with ETA, accident prevention measures and control mitigation measures for the first time. It analyzes causes and consequences of an accident using quality features of FTA and ETA [8]. It identifies risk factors at all stages. Therefore it can inform accident development and results of risk analysis [9]. FTA takes an accident as top event and finds the direct and indirect reasons leading to the top event. Then it analyzes all the reasons qualitatively. It provides the basis for the analysis of accident causes. ETA begins with the top event that can lead to an accident, taking into account of the successes and failures of events in terms of time development. ETA can analyze a variety of accident models and their consequences that may occur in complex systems. In a bow-tie diagram, the accident is situated in the center with FTA to its left and ETA to its right. From right to left, FTA finds out accident causes layer by layer, identifies risk factors, and develops prevention control measures. Regarding the central accident as top event, ETA infers the consequences of an accident according to different conditions. As a result, mitigation measures are developed. Bow-tie model shows the whole process of an accident as well as the links between its parts [10]. A bow-tie model is shown in figure 1.

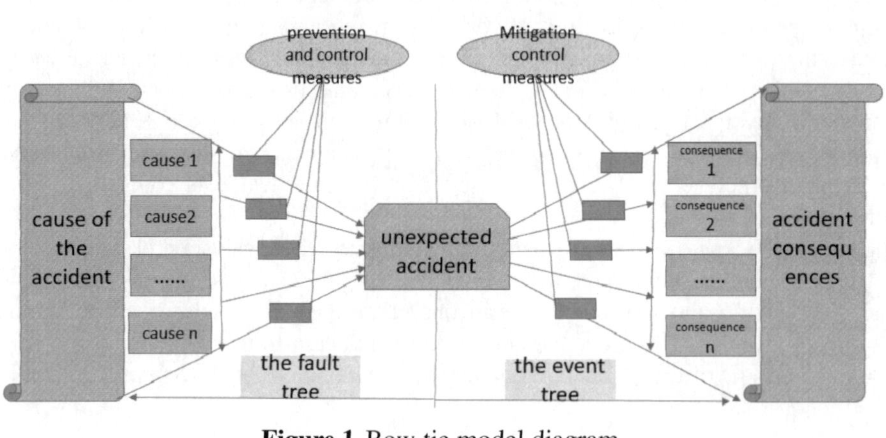

Figure 1. Bow-tie model diagram.

2.3. Improved bow-tie model

People usually choose the traditional one when applying the bow-tie model. However, it lists accident causes and consequences indiscriminately, making it difficult to identify then shoot problems. Therefore, the author proposes the improved bow-tie model. The improved bow-tie model analyzes accident causes from four aspects: human, materials, environment, and management. Then it develops prevention control measures accordingly to comprehensively identify higher risk factors with focuses on prevention and control. It also analyzes accident consequences from four aspects: casualties, property losses, environmental and social consequences. Then effective emergency rescue could be carried out to reduce the damage index using mitigation measures put forward by the model. The traditional bow-tie model only qualitatively analyzes events and their consequences. When conducting a risk analysis of urban gas pipeline network, the author made a statistical analysis on the causes and consequences and determined their quantities. Quantification can visibly show the causes more likely leading to gas pipeline network accidents and their more damaging consequences. When producing statistics on accident consequences, the type of an accident is determined by the aspect which has sustained the most damages. There are four aspects to consider, which are the previous listed casualty, property loss, environment and society. For example, a gas pipeline network failure is the most disastrous regarding casualties. When preparing statistics, the accident is categorized as one of accidents with casualties. The improved bow-tie model is shown in figure 2.

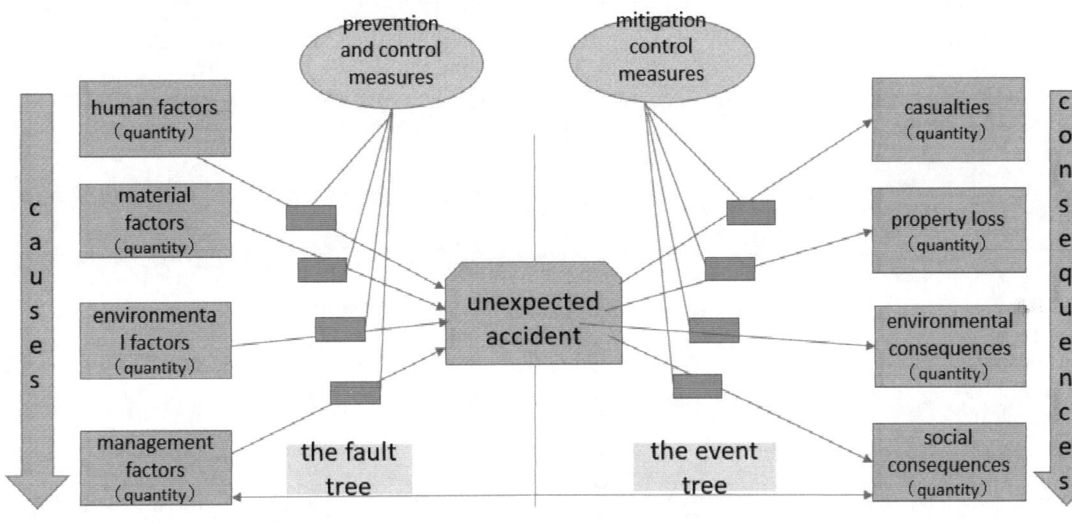

Figure 2. Improved bow-tie model diagram.

3. Analysis of urban gas pipeline failure

3.1. Fault tree analysis of gas pipeline failure

At present, in many cities and towns of China, a large amount of aged gas pipeline networks are still running. And a large number of new pipeline networks continue to be built. This has led to many gas pipeline network failures. They have brought grave losses to people's lives, property loss, environment, and society. Therefore, it is necessary for the author to conduct an analysis of urban gas pipeline network failure.

In order to conduct a fault tree analysis of urban gas pipeline network failure, the author has built a fault tree as shown in figure 3.

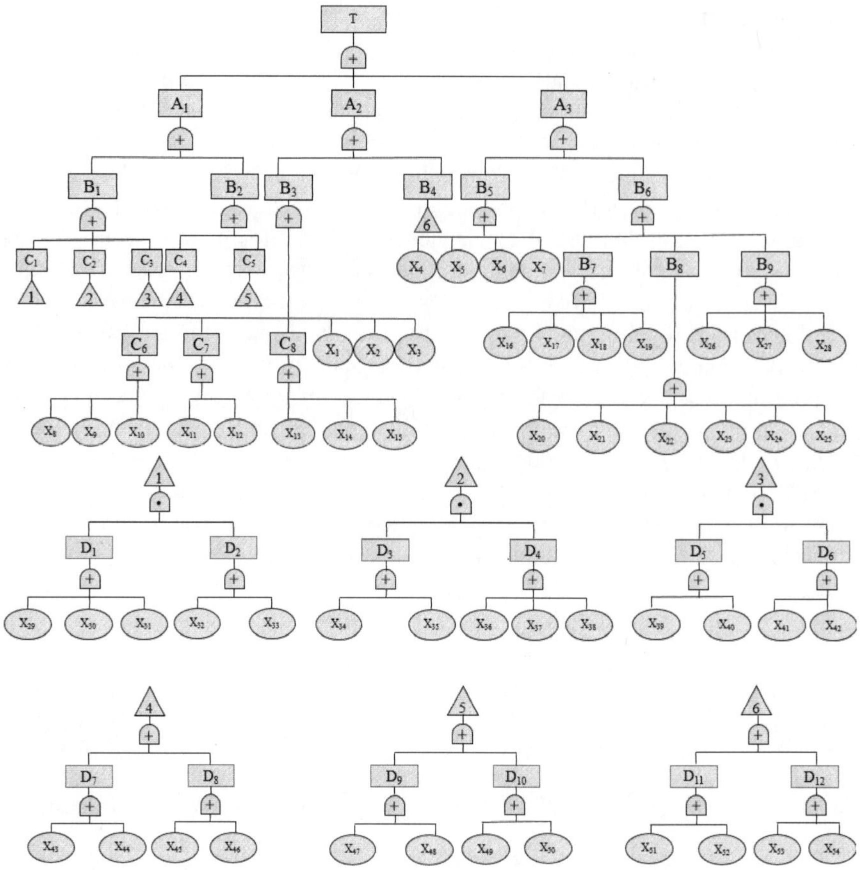

Figure 3. Leakage accident tree of gas pipe network.

Risk factors are identified by fault tree analysis of urban gas pipeline network failure. The events represented by specific symbols are shown in table 1.

Table 1. Urban gas pipeline network leakage accident tree symbol representative event.

serial number	basic event	serial number	basic event	serial number	basic event
T	pipeline leakage	X_1	natural disasters	X_{34}	acidic medium
A1	pipeline perforation	X_2	unintentional damage	X_{35}	water in pipe network
A2	pipeline rupture	X_3	deliberate destruction	X_{36}	poor quality of internal corrosion coating
A_3	related accessories leaked	X_4	internal overpressure	X_{37}	corrosion inhibitor failure
B_1	pipeline corrosion	X_5	film or pressure tube vibration damage	X_{38}	aging and damage of inner coating
B_2	pipeline defect	X_6	film aging	X_{39}	corrosive gases

					in the atmosphere
B_3	third party damage	X_7	gasket damage	X_{40}	high humidity and temperature
B_4	mechanical failure	X_8	pipeline network is not clear	X_{41}	coating repair, replacement is not timely
B_5	voltage regulator leakage	X_9	illegal construction	X_{42}	coating aging
B_6	valve leakage	X_{10}	construction error	X_{43}	uneven grain size
B_7	stem defect	X_{11}	not found in time	X_{44}	improper selection
B_8	screw defect	X_{12}	not timely treatment	X_{45}	uneven deformation
B_9	valve defect	X_{13}	uneven bottom	X_{46}	serious welding defects
C_1	buried corrosion	X_{14}	failure to take necessary protective measures	X_{47}	welding defect
C_2	internal corrosion	X_{15}	not according to actual traffic intensity design	X_{48}	welding material defect
C_3	atmospheric corrosion	X_{16}	stem deformation by external force	X_{49}	mandatory installation
C_4	pipeline quality defect	X_{17}	stem severely corroded	X_{50}	large fault between pipe segments
C_5	improper construction	X_{18}	stem wear	X_{51}	security system design is not reasonable
C_6	construction failure	X_{19}	filler filling is not standard	X_{52}	system design safety factor is small
C_7	illegal tying	X_{20}	nuts loosening	X_{53}	soil settlement
C_8	traffic damage	X_{21}	bolt pretightening force is not uniform	X_{54}	installation stress
D_1	soil corrosion environment	X_{22}	Insufficient bolt pretightening force		
D_2	external corrosion failure	X_{23}	incorrect gasket mounting		
D_3	internal corrosion environment	X_{24}	gasket aging		
D_4	internal	X_{25}	insufficient gasket		

	corrosion failure		
D_5	corrosive atmosphere	X_{26}	improper valve body
D_6	failure of protective measures	X_{27}	severe corrosion of valve body
D_7	material defect	X_{28}	valve manufacturing defects
D_8	poor rolling process	X_{29}	chemical corrosion environment
D_9	pipeline welding	X_{30}	electrochemical corrosion environment
D_{10}	pipeline installation	X_{31}	microbial corrosion environment
D_{11}	low pressure bearing capacity	X_{32}	cathodic protection failure
D_{12}	pipe network under high stress	X_{33}	coating failure

By analyzing the fault tree of urban gas pipeline network failure layer by layer, the basic event could be found. First, it needs to find key control points and key prevention stages of the accident. Then, it is possible to qualitatively analyze the probability of the accident. Finally, appropriate prevention and control measures could be developed based on the analysis results.

3.2. Event tree analysis of gas pipeline failure

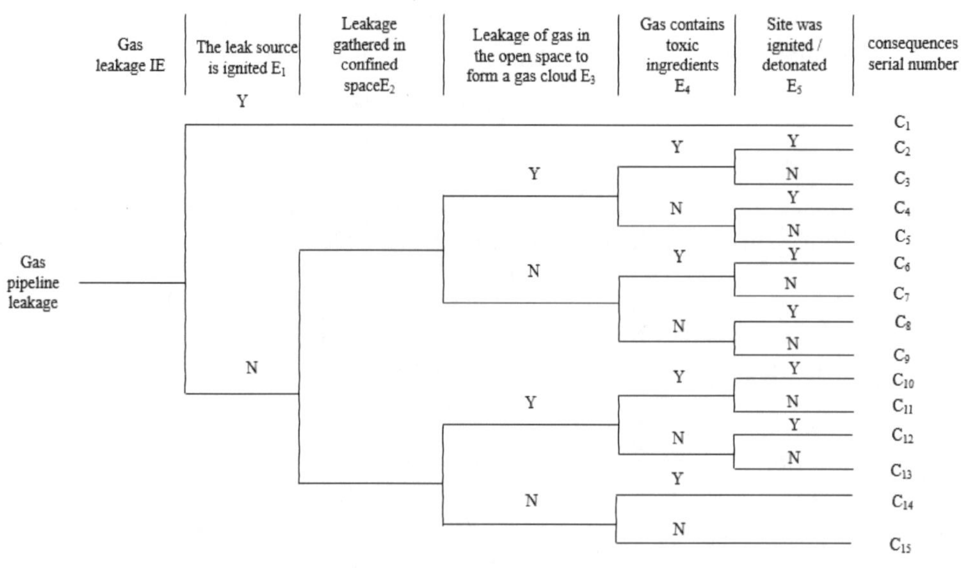

Figure 4. Gas pipeline leakage consequence event tree.

IOP Conf. Series: Earth and Environmental Science **93** (2017) 012059 doi:10.1088/1755-1315/93/1/012059

In different environment and under different conditions, the accident consequences will be different as well. By analyzing the event tree of urban gas pipeline network failure, accident consequences under different environmental conditions are found. Gas pipeline leakage consequence event tree is shown in Figure 4.

Events in the event tree of gas pipeline network failure are represented by symbols shown in table 2.

Table 2. Urban gas pipeline network leakage incident tree event.

event sequence	consequence event	event sequence	consequence event
C_1	jet fire, fire	C_9	Explosion hazard, suffocation
C_2	confined space explosion, steam cloud explosion, poisoning	C_{10}	steam clouds explode, fire
C_3	poisoning, suffocation, explosion hazard	C_{11}	poisoning, explosion hazard
C_4	confined space explosion, steam cloud explosion	C_{12}	steam clouds explode, fire
C_5	suffocation, explosion hazard	C_{13}	gas loss, explosion hazard
C_6	confined space explosion, fire, poisoning	C_{14}	poisoning, gas loss
C_7	Poisoning, explosion hazard, suffocation	C_{15}	gas loss, diffusion
C_8	confined space explosion, fire		

By analyzing the event tree, it is found that a pipeline network failure can cause grave consequences such as fire, vapor cloud explosion, suffocation, and poisoning in different environmental conditions. Especially in densely populated urban areas, these consequences could bring immeasurable losses to people and environment, creating uproar in society.

4. Application of improved bow-tie model in risk analysis of urban gas pipeline network failure

4.1. Building of improved bow-tie model of gas pipeline network failure

In order to better understand urban gas pipeline network failure, it needs to build an improved bow-tie model to comprehensively analyze it.

- Identify the top event

A wide variety of pipeline networks are laid out in densely populated urban areas. Aged pipeline network, occupied infrastructure, and others all could cause pipeline failure. As a result, the gas pipeline network failure is identified as the top event (or an unexpected event).

- Identify the initiating events

By analyzing the fault tree of gas pipeline failure, the event causes are identified. According to the improved bow-tie model, the initiating events are divided into four categories: human, materials, environment and management. This paper suggests analyzing the event causes from four aspects: 1) human-made mistakes when operating; 2) flaws in pipeline network; 3) disasters in environment; 4) lack of regular checkup by the management.

- Develop prevention and control measures

In view of each type of accidents that have been identified, appropriate prevention and control measures need to be developed to prevent accidents and reduce the accident rate.

- Identify the outcome risk

A pipeline network failure can cause grave consequences such as fire, explosion, suffocation, and poisoning. According to the improved bow-tie model, the accident consequences are to be analyzed and identified from four aspects: casualties, property losses, environmental and social consequences.

- Develop mitigation and control measures

According to the probable consequences of the accident, corresponding mitigation measures need to be developed in order to minimize losses caused by the accident and to mitigate damages done to people, property, environment, and society.

- In light of steps performed above and the statistics of 223 incidents of gas pipeline network failure of city X in the period of two years, the author has built an improved bow-tie diagram of gas pipeline network failure as shown in figure 5.

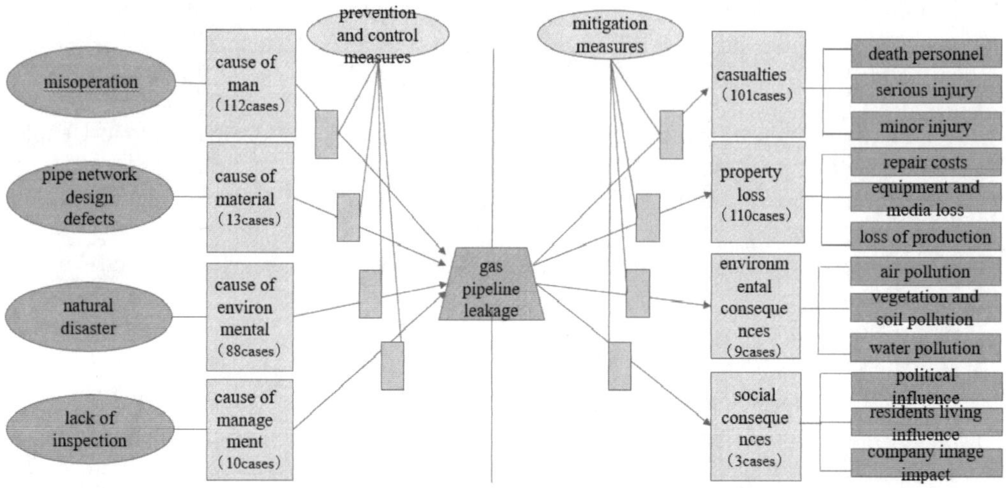

Figure 5. Improved bow-tie model for leakage accident of gas pipeline network.

4.2. Risk Analysis Based on Improved Bow-tie Diagram

- Human-made mistakes when operating

By conducting a risk analysis of urban gas pipeline network failure based on improved bow-tie model, it is found that if people make mistakes leading to accidents, it is usually because of lack of safety knowledge, awareness, skills, and incompetence on the part of personnel. They can manifest in forms of improper and even illegal operations and constructions. Among them, inappropriate operation is the most common. Under certain conditions, these risk factors made by human could lead to gas pipeline failure, fire, explosion, and other major disasters, resulting in grave consequences.

- Prevention measures

To avoid human-made mistakes, appropriate prevention and measures need to be developed. Regular safety training and education can make employees take safety issues more seriously by exposing them to more safety lessons. Regular assessment of employees can help make sure that tasks are manned by capable people with certificates.

- Flaws in pipeline network

Accident causes regarding flaws in pipeline network are generally design flaws, technical defects, welding and construction defects, pipe, equipment, device, too accessory defects, and corrosion failure. They manifest in the pipes made of inferior materials and techniques, pipeline network aging, design and construction flaws, poor network design, and cracks in shoddy welding work. Among them, design flaw in pipeline network is the most serious.

- Prevention measures

Strict quality control is recommended to ensure the quality of the pipe network. Implementing strict guidelines and using quality materials in welding work can improve its overall quality. Then serious welding defects can be avoided. Clear design of pipeline network can prevent accidents from happening.

- Disasters in environment

Environmental causes leading to urban gas pipeline network failure include interference by other pipeline networks, human-made environment such as theft of devices and equipment, external environment such as lack of protection measures, and natural environment such as floods and earthquakes. Among them, natural disasters are the most serious.

- Prevention measures

Strengthening the supervision and protection of the network can prevent the theft of equipment and devices. Adequate protection should be given to the pipeline network. There should be emergency preparedness and contingency plan in place in case of an emergency to reduce accident loss and control damages to the pipeline network.

- Lack of regular checkup by the management

Defects in the management of pipeline network also can cause accidents in pipeline network. Command defects in the organization, safety training and education deficiencies, fire management defects, and lack of regular inspection and checkups can all lead to accidents. Among them, inadequate inspection is the most serious.

- Prevention measures

Effective organization and command system are needed. Employees are required to attend trainings on safety management to recognize responsibilities towards management and protection. Regular and effective inspections of pipeline network should be carried out to reduce the accident rate.

In case of an urban gas pipeline network failure, it will bring grave consequences to people's lives and safety, society, and environment. It is very likely to cause deaths, serious and minor injuries, and environmental pollution, which will affect the normal life of residents and cause adverse social effects. In order to reduce the accident rate and loss, we should pay attention to promote public safety awareness and detect and address problems in time.

5. Conclusions

Urban gas pipeline network failure is a major safety concern. In the event of an accident, there could be grave damages and consequences. This paper uses the improved bow-tie model to conduct a risk analysis of urban gas pipeline network failure and draws the following conclusions:

- The bow-tie model perfectly combines quality features of both FTA and ETA. It clearly analyzes causes and consequences of an accident, then develops prevention and mitigation measures accordingly;
- The improved bow-tie model conducts a risk analysis of urban pipeline network failure by analyzing accident causes and consequences from four aspects respectively. The four aspects of accident causes are people, materials, environment, and, management; the four aspects of accident consequences are casualty, property loss, environment, and, society.
- The improved bow-tie model is ideal for risk analysis of urban gas pipeline network failure. It can analyze its cause risks and consequence risks. And it can develop prevention and mitigation measures accordingly to eliminate these causes and consequences. The whole process can be shown dynamically.
- Through the use of improved bow-tie model for urban gas pipeline network risk analysis, we can know that human causes are the most important factors in the causes of gas pipe network leakage, followed by environmental factors such as corrosion, we should focus on these two aspects of prevention and control to reduce the incidence of accidents better. Gas pipe network leakage, resulting in casualties, property loss is more serious, we should mainly from these two aspects to develop mitigation measures to reduce the impact of accidents.

The improved bow-tie model is easy-to-understand, clear, and practical, so that people can understand the causes and consequences of an accident simply by reading the figures it provides. It is of great significance to the analysis of gas pipeline network failure.

Acknowledgments
This project was supported by Beijing Natural Science Foundation (No.4172023) and Beijing financial

IOP Conf. Series: Earth and Environmental Science **93** (2017) 012059 doi:10.1088/1755-1315/93/1/012059

project (No. PXM2017_178215_000001). We also thank the anonymous reviewers for their insightful and valuable hints.

References

[1] Chen Y C 2016 Study on the risk evaluation model of urban pipeline Bowtie based model *China Safety Sci & Technol* **12** 148-52
[2] Wang T, Zhang P and Liu G 2003 Reliability analysis of urban natural gas pipeline network *Oil and Gas Stor and Transport* **22** 15-8
[3] Yu X C 2013 Quantitative risk assessment of urban gas pipeline leakage based on fuzzy Bow-tie model *Natural Gas Ind* **33** 134-9
[4] Li T T 2013 Research on the combination of three evaluation models based on FTA, ETA and Bow-tie and its application *Value Eng* **32** 35-7
[5] Jia P M 2014 Application of Bow-tie technology in city gas pipeline risk management *Ind Safety and Environ Protection* **40** 14-7, 39
[6] U S 2000 Department of energy dose reduction technique (HNF-63 16 -FP., 5)
[7] Lu Z 2015 The runway landing aircraft out of the risk analysis model based on Bow-tie *Ind Safety and Environ Protection* **41** 4-8
[8] Yang Y 2012 Application of Bow-Tie analysis technology in FPSO construction project *Project Management Technol* **2012** 100-3
[9] Zeng Q J 2014 The successful application of risk analysis tool Bow-Tie in the cross operation of west to east gas pipeline *Petrochemical Safety and Environ Protection Technol* **30** 10-3, 9, 5
[10] Liu J J 2016 Analysis of heavy landing event based on improved bow-tie model *Traffic Info and Security* **34** 57-62

3.0 MeV proton-irradiation induced non-radiative recombination center in the GaAs middle cell and the GaInP top cell of triple-junction solar cells

J L Wang[1], T C Yi[1], Y Zheng[1], R Wu[1] and R Wang[1,2]

[1]Key Laboratory of Beam Technology and Materials Modification of Ministry of Education, College of Nuclear Science and Technology, Beijing Normal University, Beijing 100875, China

E-mail: wangr@bnu.edu.cn

Abstract. 3.0 MeV proton-irradiation effects on the GaAs middle cell and the GaInP top cell of n^+-p GaInP/GaAs/Ge triple-junction (3J) solar cells have been analyzed using temperature-dependent photoluminescence (PL) technique. The E5 (E_c - 0.96 eV) electron trap in the GaAs middle cell, the H2 (E_v + 0.55 eV) hole trap in the GaInP top cell are identified as the proton irradiation-induced non-radiative recombination centers, respectively, causing the performance degradation of the triple-junction solar cells. The GaAs middle cell is less resistant to proton irradiation than the GaInP top cell.

1. Introduction

Recently, GaInP/GaAs/Ge triple-junction (3J) solar cells are the main power generator used in space satellites [1]. It consists of three subcells, i.e, GaInP top cell, GaAs middle cell and Ge bottom cell, having the potential to provide high efficiency. For space applications, exposure to proton irradiation degrades the performance of solar cells, probably resulting in space emission failure [2]. Therefore, the understanding of the mechanisms responsible for the irradiation degradation is important for the use of solar cells in space. The radiation response of the 3J solar cells is complicated due to three subcells with different proton irradiation effects. Exactly, The proton effects of the Ge bottom cell are not considered here as the Ge bottom cell does not contribute to the voltage output or affect the current degradation of the 3J solar cells [3] and effects of the GaInP top cell and the GaAs middle cell are investigated.

In fact, the degradation of solar cells induced by proton irradiation has been found to be directly related to the concentrations of the defects which play the role of non-radiative recombination centers for low irradiation fluences (typically below 10^{16} cm^{-2}) [4]. The proton-induced defects in solar cells have been analyzed by deep-level transient spectroscopy (DLTS) [5]. However, the defects acting as non-radiative recombination centers have not been identified among all the detected defects. Photoluminescence (PL) is a potential method used to analyze non-radiative recombination centers of each individual subcell in multi-junction solar cells independently of the others [6]. In particular, temperature-dependent photoluminescence could not only identify non-radiative recombination centers among all the proton irradiation induced defects, but also reveal their activation process in solar cells [7,8]. Therefore, in this work, to clarify the proton effects on GaInP/GaAs/Ge triple-junction solar cells, temperature-dependent photoluminescence was applied to recognizing the 3.0

MeV proton-irradiation induced non-radiative recombination centers in the GaAs middle cell and the GaInP top cell.

2. Experiments

The samples were n^+-p GaInP/GaAs/Ge triple-junction solar cells grown by the metal-organic chemical vapor deposition (MOCVD) technique. The detailed structure of the solar cells is shown in [9]. The solar cells were irradiated with 3.0 MeV proton beams with fluence of 3×10^{11} cm^{-2}, avoiding significant sample heating during irradiation.

In the PL analysis of the GaAs middle cell and the GaInP top cell, a 730 nm (1.70 eV) red diode laser (100 mW) and a 532 nm (2.33 eV) green solid state laser (75 mW) were used as the excitation sources, respectively. The laser beam diameter were both about 3.0 mm. Photoluminescence in the GaAs middle cell was transferred to a grating monochromator with 600 grooves/mm grating blazed at 750 nm and that in the top cell was transferred to another grating monochromator with 600 grooves/mm grating blazed at 500 nm. The luminescence of the middle cell was detected by Si photodetector and that of the top cell was detected by photomultiplier with a lock-in amplifier.

The temperature range of 10 - 300 K was controlled by ARS - 4 HW closed-cycle cryogenic refrigerator equipped with a digital thermometer controller (Lake Shore, 355 temperature controller). The temperature step is 20 K for the regions of 10 - 100 K and 200 - 300 K, and 10 K for the region of 100 - 200 K. The temperature stability is 0.1 K or more.

3. Result and discussion

Figure 1 shows the PL spectrum of the GaAs middle cell and the GaInP top cell of unirradiated and 3.0 MeV proton-irradiated GaInP/GaAs/Ge triple-junction solar cells at room temperature. The band gap of GaInP top cell is approximately 1.88 eV at room temperature, and that of GaAs middle cell is approximately 1.39 eV, decreasing about 30 meV compared with the band gap of GaAs (1.42 eV), probably attributed to Indium doping in triple-junction solar cells [10]. It can also be observed from figure 1 that the PL intensity of both the middle cell and the top cell shows a great degradation after irradiation, but the PL intensity of the middle cell degrades much more obviously than that of the top cell. This indicates that the GaAs middle cell is less radiation-tolerant than the GaInP top cell.

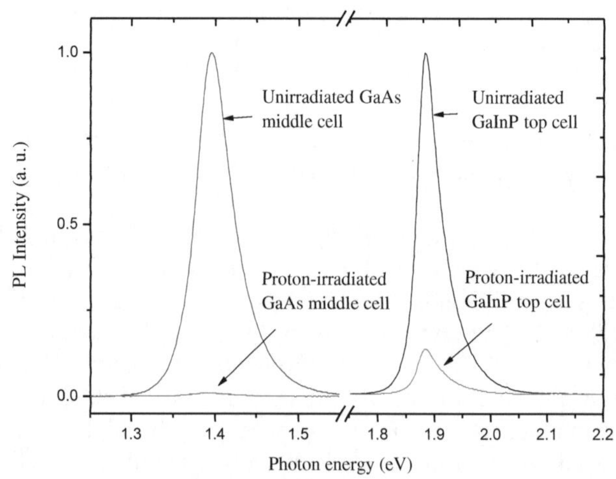

Figure 1. Room temperature PL spectrum of the GaAs middle cell and the GaInP top cell of GaInP/GaAs/Ge triple-junction solar cells.

Typical PL spectrum of the GaAs middle cell and the GaInP top cell of the triple-junction solar cells irradiated with 3.0 MeV protons with fluence of 3×10^{11} cm^{-2} at 10 K, 160 K, 210 K and 290 K are presented in figures 2 and 3. There are two peak emissions with peak energy of 2.00 eV and 1.96 eV at 10 K in the spectrum of the top cell. The peak emission at 2.00 eV is probably attributed to band-to-band transition. The spectrum of the middle cell also consists of two peak emissions, i.e., 1.49 eV (exciton emission) and 1.47 eV (donor-to-acceptor pair emission) at 10 K [11]. The photon energy of the band-edge-related PL peak decreases with increasing temperature for the two subcells, following Varnish equation [12].

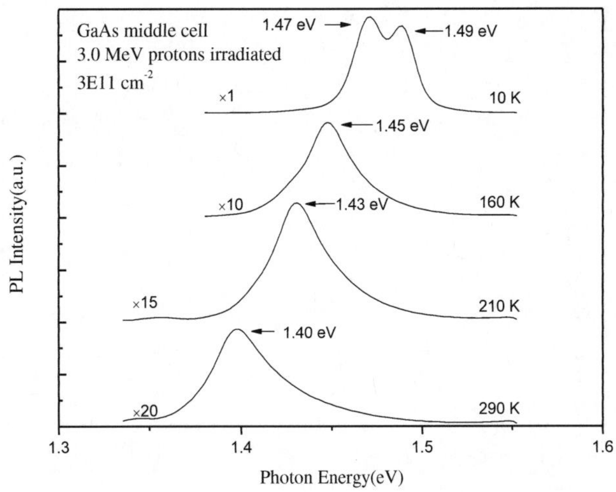

Figure 2. Typical PL spectrum of the GaAs middle cell of the triple-junction solar cells irradiated by a 3.0 MeV protons fluence of 3×10^{11} cm^{-2} at 10 K, 160 K, 210 K and 290 K.

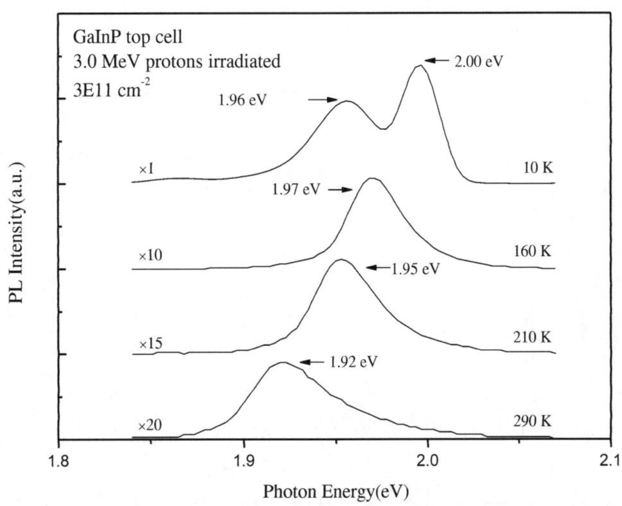

Figure 3. Typical PL spectrum of the GaInP top cell of the triple-junction solar cells after irradiated with 3.0 MeV protons with a fluence of 3×10^{11} cm^{-2} at 10 K, 160 K, 210 K and 290 K.

As the temperature increases, the PL intensity of the GaAs middle cell and the GaInP top cell decreases monotonously. The PL emission efficiency [13] $\eta(T)$ has the following dependence on temperature:

$$\eta(T) = \left[1 + \kappa\, exp\left(-\frac{E}{k_B T}\right)\right]^{-1} \tag{1}$$

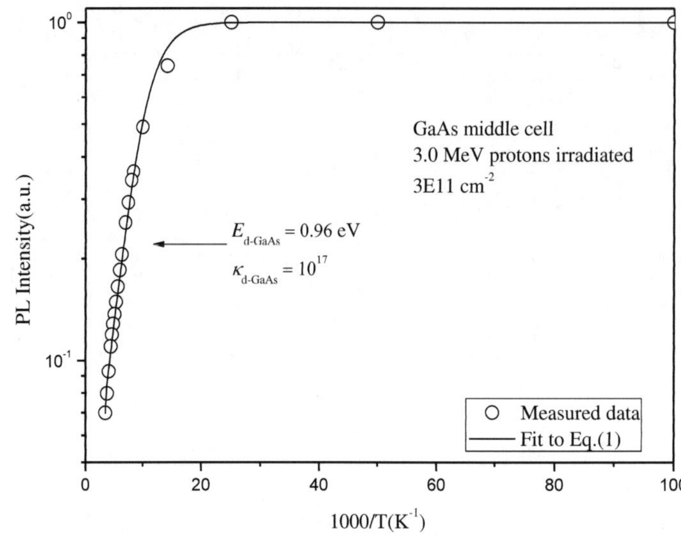

Figure 4. Arrhenius plot of temperature-dependent PL intensities of the GaAs middle cell of the triple-junction solar cells irradiated by a 3.0 MeV protons fluence of 3×10^{11} cm^{-2}.

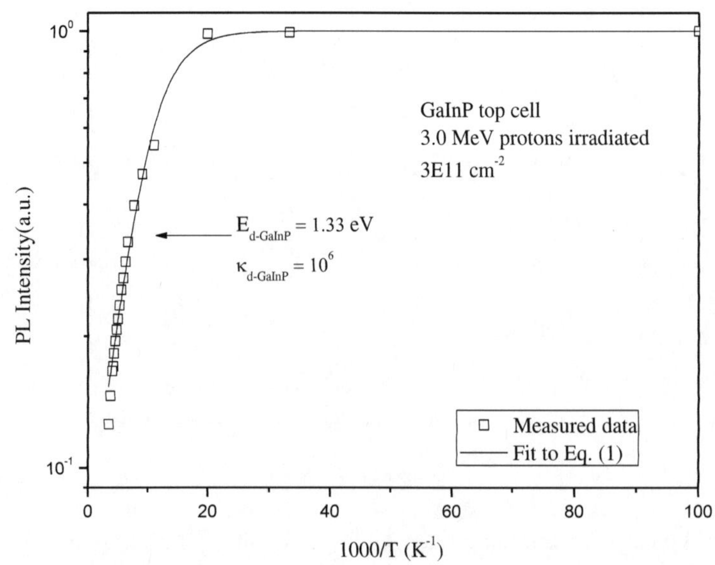

Figure 5. Arrhenius plot of temperature-dependent PL intensities of GaInP top cell of the triple-junction solar cells after irradiated with 3.0 MeV protons with a fluence of 3×10^{11} cm^{-2}.

where κ is the ratio of the non-radiative to radiative probability for non-radiative recombination mechanism at room temperature, E is the thermal activation energy, and k_B is the Boltzmann constant. The data in Arrhenius plots shown in figures 4 and 5 seem to exhibit an exponential region, both suggesting that a thermally activated non-radiative recombination mechanism is presented. The solid lines represent the fits of the measured data of the two subcells to equation (1) and the fitting parameters are $\kappa_{d-GaAs} = 10^{17}$, E_{d-GaAs}= 0.96 eV, $\kappa_{d-GaInP} = 10^6$, $E_{d-GaInP}$= 1.33 eV, respectively.

The PL quenching behaviors take place with the thermal activation of the defects in the p-type base layers [14,15]. For electron trap, the thermal activation energy (E_d) equals to its ionization energy (E_t), that is, $E_d = E_t$. For hole trap, the thermal activation energy (E_d) equals to the difference between band gap energy (E_g) and the ionization energy (E_t), i.e., $E_d = E_g - E_t$. Therefore, in the GaAs middle cell, we yield a defect level near 0.96 eV below the conduction band from the Arrhenius plot. It has been suggested that the defect could be consistent with E5 (E_c - 0.96 eV) electron trap, an effective nonradiation recombination center, probably related to an As vacancy-interstitial pair [16]. Furthermore, in the GaInP top cell, a defect level near 1.33 eV below the conduction band is yielded, corresponding to H2 (E_v + 0.55 eV) defect, located at 0.55 eV above valence band. This is consistent with the results of Nethaji Dharmarasu *et al* [5] that H2 hole trap (0.51 ± 0.09 eV) was identified as recombination centers in the 3.0 MeV proton-irradiated p-type GaInP, playing an important role in determining the minority carrier lifetime.

The large value of κ_{GaAs} indicates that the E5 electron trap in GaAs middle cell is a highly efficient non-radiative recombination center. Meanwhile, κ_{GaAs} is eleven orders of magnitude higher than κ_{GaInP}, showing that the GaAs middle cell is less resistant to proton irradiation than the GaInP top cell, which is thought to be due to higher migration energies of Ga- and As- related defects than that of In- and P- related defects [17].

4. Conclusion

Temperature-dependent PL measurements for studying non-radiative recombination center induced by 3.0 MeV proton irradiation in the GaAs middle cell and the GaInP top cell of GaInP/GaAs/Ge triple-junction cells at the temperature range of 10 - 300 K is demonstrated. E5 electron trap in the GaAs middle cell, located at 0.96 eV below the conduction band, and H2 hole trap in the GaInP top cell, located at 0.55 eV above valence band are identified as the non-radiative recombination centers, respectively. The E5 electron trap in the GaAs middle cell is a highly efficient non-radiative recombination center, and the GaAs middle cell shows less proton radiation resistance than the GaInP top cell.

Acknowledgments

This study was supported by the National Natural Science Foundation of China under Grant Nos. 10675023, 11075018, 11375028, and 11675020.

References

[1] Yamaguchi M, Takamoto T, Araki K, *et al* 2005 Multi-junction III–V solar cells: Current status and future potential *Sol Energy* **79(1)** 78-85

[2] Gao X, Feng Z Z, Cui X Y, *et al* 2014 Performance evaluation and prediction of single-junction and triple-junction GaAs solar cells induced by electron and proton irradiations *IEEE Trans Nucl Sci* **61(4)** 1838-42

[3] Imaizumi M and Ohshima T 2013 Radiation effects in solar cells *Proc. of SPIE* **8725** 872515

[4] Mbarki M, Sun G C and Bourgoin J C 2004 Prediction of solar cell degradation in space from the electron–proton equivalence *Semicond Sci Technol* **19(9)** 1081-5

[5] Dharmarasu N, Yamaguchi M, Bourgoin J C, *et al* 2002 Majority- and minority-carrier deep level traps in proton-irradiated n^+/p-InGaP space solar cells *Appl Phys Lett* **81(1)** 64-6

[6] Zazoui M, Mbarki M, Aldin A Z, *et al* 2003 Analysis of multijunction solar cell degradation in space and irradiation induced recombination centers *J Appl Phys* **93(9)** 5080-4

[7] Reshchikov M A 2014 Temperature dependence of defect-related photoluminescence in III-V and II-VI semiconductors *J Appl Phys* **115** 115205-120
[8] Wang J L, Yi T C, Zheng Y, *et al* 2017 Temperature-dependent photoluminescence analysis of 1.0 MeV electron irradiation-induced nonradiative recombination centers in n^{+}- p GaAs middle cell for GaInP/GaAs/Ge triple-junction solar cells *Chin Phys Lett* **34(7)** 076106
[9] Wang R, Liu Y and Sun X 2008 Effects of 0.28–2.80 MeV proton irradiation on GaInP/GaAs/Ge triple-junction solar cells for space use *Nucl Instr Meth Phys Res* **266(5)** 745-9
[10] Takamoto T, Agui T, Ikeda E, *et al* 2001 High-efficiency InGaP/$In_{0.01}Ga_{0.99}As$ tandem solar cells lattice-matched to Ge substrates *Sol Energy Mater Sol Cells* **66(1)** 511-6
[11] Sugimoto H, Tajima M, Hase S, *et al* 2005 Luminescence analysis of radiation effects in multijunction solar cells for space *IEEE* 830-3
[12] Varnish Y P 1967 Temperature dependence of the energy gap in semiconductors *Physica* **34(1)** 149-54
[13] Romero M J, Araújo D, GarcíA R, *et al* 1999 Spatial distribution of radiation-induced defects in p^{+}–n InGaP solar cells *Appl Phys Lett* **74(25)** 3812-4
[14] Klasens H A 1946 Transfer of energy between centres zinc sulphide phosphors *Nature* **158** 306 -7
[15] Schön M 1942 Zum Leeuchtmechanismus der Kristallphosphore *Z. Phys* **119** 463-71
[16] Bourgoin J C and Zazoui M 2002 Irradiation-induced degradation in solar cell: Characterization of recombination centres *Semicond Sci Technol* **17(5)** 453
[17] Dharmarasu N, Khan A, Yamaguchi M, *et al* 2002 Effects of proton irradiation on n^{+}p InGaP solar cells *J Appl Phys* **91(5)** 3306-11

A new type of coil structure called pan-shaped coil of wireless charging system based on magnetic resonance

Z K Yue[1,3], Z Z Liu[1], Y J Hou[2], H Zeng[1], L H Liang[1] and S Cui[1]

[1]School of Electrical Engineering, Shandong University, Jinan 250061, Shandong Province, China
[2]Energy Research Institute, Shandong Academy of Sciences, Jinan 250014, Shandong Province, China

E-mail: yuezengkai2016@163.com/ liuzhizhen@sdu.edu.cn

Abstract. The problem that misalignment between the transmitting coil and the receiving coil significantly impairs the transmission power and efficiency of the system has been attached more and more attention. In order to improve the uniformity of the magnetic field between the two coils to solve this problem, a new type of coil called pan-shaped coil is proposed. Three-dimension simulation models of the planar-core coil and the pan-shaped coil are established using Ansoft Maxwell software. The coupling coefficient between the transmitting coil and the receiving coil is obtained by simulating the magnetic field with the receiving coil misalignment or not. And the maximum percentage difference strength along the radial direction which is defined as the magnetic field uniformity factor is calculated. According to the simulation results of the two kinds of coil structures, it is found that the new type of coil structure can obviously improve the uniformity of the magnetic field, coupling coefficient and power transmission properties between the transmitting coil and the receiving coil.

1. Introduction

Electric vehicles (EVs) and plug-in hybrids (PHVs) are gaining popularity as they provide superior energy efficiency and have the potential to reduce carbon dioxide emissions. Compared with wired charging technology, wireless charging technology is easy to operate and has no risk of exposure to sparks, thus it is able to adapt to harsh environments. In recent years, researches and applications of wireless charging technology have been raising more and more attention [1].

The basic working principle of the wireless power transmission technology based on magnetic coupling resonant is that using the resonance generated in the circuit to achieve wireless transmission of electrical energy. The system generally uses a series-series resonant structure. The energy transmitter is a resonator consisting of a transmitting coil and a resonant capacitor in series [2]. And the energy receiver is a resonator consisting of a receiving coil and a resonant capacitor in series. As shown in figure 1, the energy coupling mechanism of the electric vehicle wireless charging system is mainly composed of a transmitting coil and a receiving coil and this structure is the type of planar-core coil structure which is commonly used [3]. The key to improving the output power and efficiency of the wireless charging device is to improve the coupling coefficient between the two coils. At present, methods used commonly to improve the coupling coefficient of these two coils are improving the shape of the coil and increase the core. In [4] and [5], the author respectively designs a double-D (DD) coil and a DD-quadrature coil to improve the coupling coefficient.

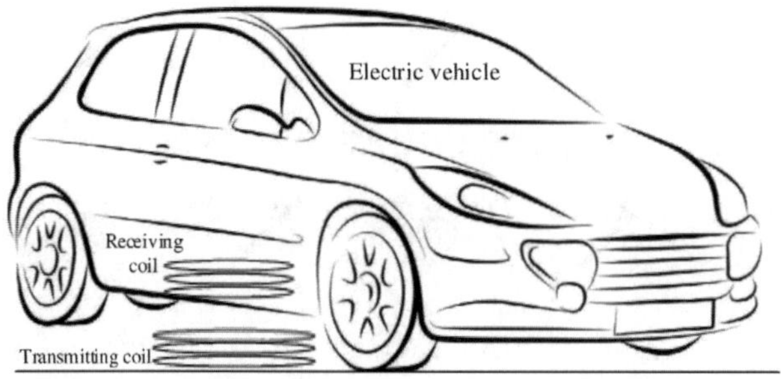

Figure 1. Wireless charging display for electric vehicle.

In addition, the wireless charging technology is facing a serious challenge that the misalignment between the transmitting coil and the receiving coil can significantly impair the power transfer efficiency of the system [6]. However, the positioning of the receiving coil does not need to be very accurate if the magnetic field generated by the transmitting coil is uniform so that the effect of the misalignment on the transmission efficiency can be avoided.

Many researches on producing a uniform magnetic field by designing different coil structures have been done in recent years. In [7], the authors propose a structure of a coil array composed of a number of small coils, which are tightly arranged in certain geometries. Diao Y, *et al* [8] proposed a rectangular coil structure and provided a function to calculate the turn distribution of the coil. In [9], a hybrid coil structure comprising a concentrating coil and a helical coil was proposed. However, the magnetic field strength of most structural coils is not constant and the uniformity of the magnetic field depends on the size of the receiving coil. So the solutions in these studies will be meaningless if the receiving coil is too small. In addition, a potential problem with these studies is that they are not designed for the size of the coils used for electric vehicles, which are either with a small coil size or with distance of only a few millimeters between the transmitter and receiver.

This paper presents a new type of pan-shaped coil structure, which is used to improve the uniformity of the magnetic field between the transmitting coil (at the parking point) and the receiving coil (in the car) and solve the problem that misalignment between the two coils can significantly impair the power transfer efficiency of the system. Firstly, the coupling coefficient and the magnetic field uniformity factor are introduced. Then, the planar-core coil and the new type of pan-shaped coil are modeled and magnetic field characteristics such as the coupling coefficient are simulated with Ansoft Maxwell software. Finally, magnetic field simulation results of the two kinds of coil structures are compared and analyzed. We evaluate the effect of the pan-shaped coil structure by comparing their magnetic field evenness factor and the coupling coefficient.

2. Analysis of system theory

2.1. Coupling coefficient

Two electromagnetic coupling coils are shown in figure 2. N_T and N_R represent the number of turns of the transmitting coil and the receiving coil. L_T and L_R are respectively the self-inductance of the transmitting coil and receiving coil. M is the mutual inductance between the coils. r_T and r_R are respectively the radii of the transmitting coil and receiving coil. k and D are the coupling coefficients and the distances between the two coils.

The degree of flux coupling between the two coils is expressed by the coupling coefficient. And the formula is shown in equation (1). It is necessary to increase the coupling coefficient between the transmitting coil and the receiving coil in order to improve the power transmission efficiency in the electric vehicle wireless charging system.

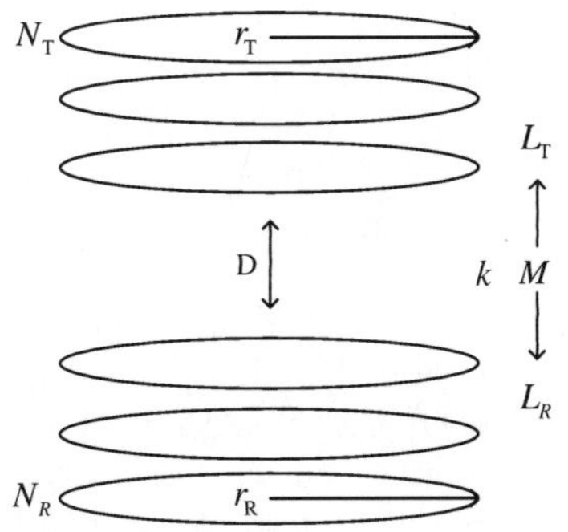

Figure 2. Configuration of electromagnetic coupled coils.

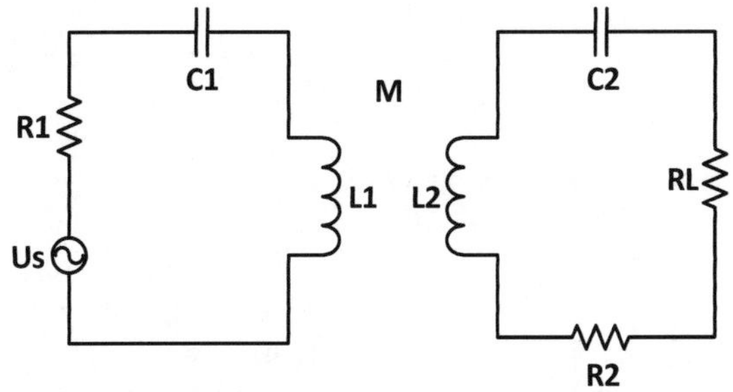

Figure 3. The series-series resonant compensation structure.

$$k = \frac{M}{\sqrt{L_T L_R}} \tag{1}$$

2.2. Uniform magnetic field

In this paper, a new type of coil structure is designed to improve the uniformity of the magnetic field between the transmitting coil and the receiving coil, which is aimed to solve the problem that misalignment between the transmitting coil and the receiving coil can significantly impair the power transfer efficiency of the system. The uniformity of the magnetic field generated by the transmitting coil can be measured as the maximum percentage difference strength along the radial direction of the receiving coil. This value is tentatively defined as the uniformity factor which is expressed as a parameter named λ_U according to equation (2) as follow.

$$\lambda_U = \frac{\max(H) - \min(H)}{\max(H)} \tag{2}$$

2.3. Analysis of series-series (PSSS) resonant compensation structure

The equivalent circuit diagram of the series-series (PSSS) resonant compensation structure commonly used in the wireless charging system is shown in figure 3. U_S is the AC voltage source. L_1 and L_2 are respectively the self-inductance of the transmitting coil and receiving coil. M is the mutual inductance between the coils. C_1 and C_2 represent the values of resonant compensation capacitors. R_1 and R_2 are respectively the resistance of the transmitting coil and receiving coil. R_L is the load of the wireless charging system.

According to the equivalent circuit diagram, the state equation of the system can be obtained as shown in equation (3). And the equations calculating transmission power and efficiency of the resonant compensation structure are respectively shown in equations (4) and (5).

$$\begin{bmatrix} U_S \\ 0 \end{bmatrix} = \begin{bmatrix} R_1 + j\omega L_1 - j\dfrac{1}{\omega C_1} & -j\omega M \\ -j\omega M & R_2 + j\omega L_2 - j\dfrac{1}{\omega C_2} + R_L \end{bmatrix} \begin{bmatrix} I_1 \\ I_2 \end{bmatrix} \tag{3}$$

$$P = \frac{U_S^2 k^2 L_1 L_2 \omega^2 R_L}{\left(R_1 R_2 + k^2 L_1 L_2 \omega^2\right)^2} \tag{4}$$

$$\eta = \frac{k^2 L_1 L_2 \omega^2}{k^2 L_1 L_2 \omega^2 + R_1 R_2} \tag{5}$$

3. Conventional planar-core coil

As a carrier for wireless power transmission, the core enhances the degree of magnetic field coupling between the inductive coils and plays an important role in improving the transmission power and efficiency in the high-power and long-range wireless power transmission. The thin plate-shaped core is widely used for its small size, light weight, easy to fix, and the large equivalent area and it has achieved some successes in the field of wireless energy transmission. At present, the resonant coil structure using the planar spiral structure or the three-dimensional spiral structure will affect the normal driving if the size of the coil in the electric vehicle wireless charging system is too large or the car chassis is too low caused by the vertical height of the coil. So this kind of inductance coil generally uses the type of planar structure. The model of the coil is shown in figure 4.

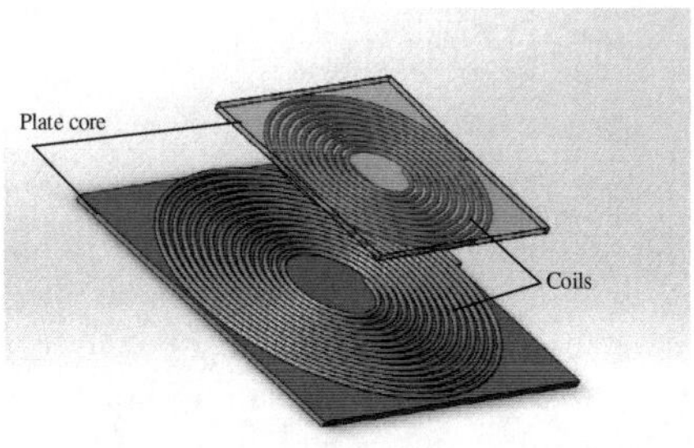

Figure 4. Model of conventional planar-core coil.

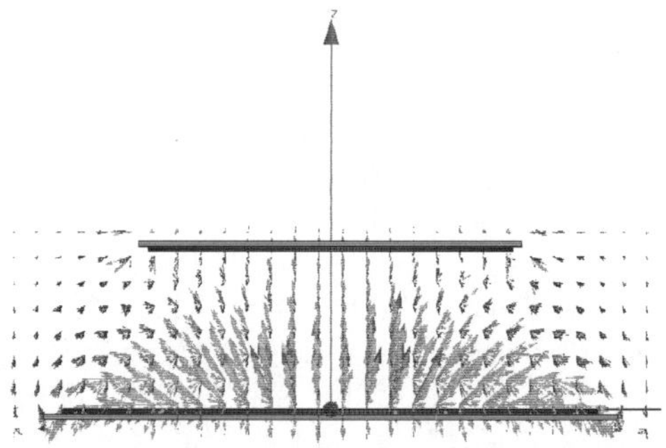

Figure 5. The magnetic field distribution between the two coils.

This type of coil structure is asymmetrical in which the size of the transmitting coil is larger than the size of the receiving coil so that the variation of the coupling coefficient can be reduced when the receiving coil shifts in the magnetic field coverage of the transmitting coil. In addition, the coupling coil is laid on the planar core, which can effectively increase the self-inductance value of the transmitting coil and the receiving coil and the mutual inductance between the two coils. Compared with the coil structure composed of two or more air-core coils, the uniformity of the spatial alternating magnetic field is improved and the magnetic field strength of the space is also enhanced so that the transmission distance of the system is increased and the energy transmission power and efficiency of the system are improved effectively.

We model this coil structure using the Ansoft Maxwell software for magnetic field analysis, in which the maximum radius of the transmission coil is 175 mm and that of the receiving coil is 125 mm. The size of the two cores are $360\text{mm}\times 360\text{mm}\times 6\text{mm}$ and $260\text{mm}\times 260\text{mm}\times 6\text{mm}$ respectively and the distance between the two coils is 120 mm. The distribution of the magnetic field between the coils and the distribution of the magnetic field in the radial direction of the receiving coil are shown in figures 5 and 6 respectively through performing a magnetic field simulation.

As shown in equation (6), the uniformity factor of the magnetic field between the two coils is 0.602 by calculation. In addition, the coupling coefficient value between the two coils in this coil structure is 0.15889 by calculation using the simulation software.

$$\lambda_{\text{U1}} = \frac{425.1 - 169.4}{425.1} = 0.602 \tag{6}$$

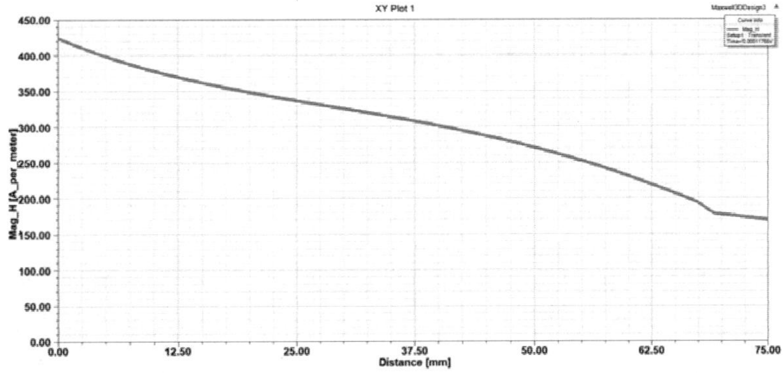

Figure 6. The distribution of the magnetic field in the radial direction of the receiving coil.

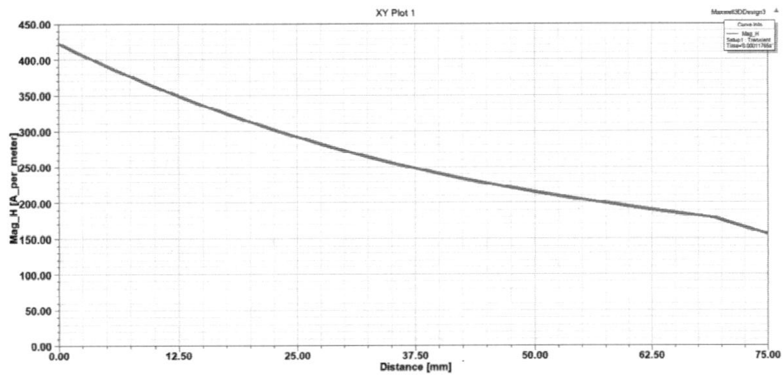

Figure 7. The distribution of the magnetic field in the radial direction of the receiving coil when misaligned by 5 mm.

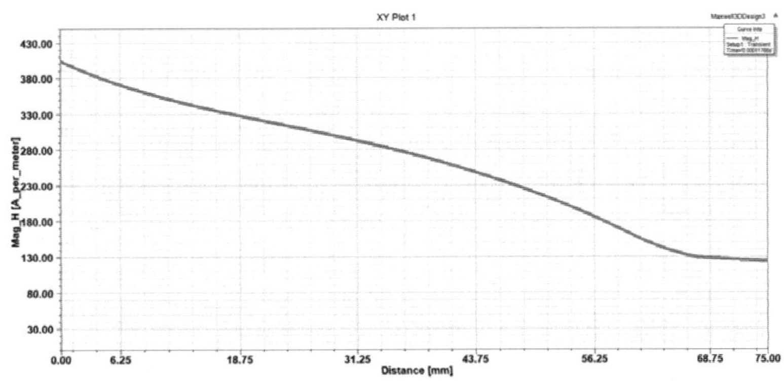

Figure 8. The distribution of the magnetic field in the radial direction of the receiving coil when misaligned by 10 mm.

In addition, we simulate the magnetic field and calculate the magnetic field uniformity and coupling coefficient when the receiving coils are horizontally misaligned by 5 mm and 10 mm in order to see the effect on the transmission efficiency of the system produced by the conventional planar-core coil. The distribution of the magnetic field between the two coils and the distribution of the magnetic field in the radial direction of the receiving coil are shown in figures 7 and 8 respectively. The uniformity factors of the magnetic field between the two coils are 0.629 and 0.688 respectively and the coupling coefficients are 0.15857 and 0.15770 respectively. It can be seen that the magnetic field uniformity and the coupling coefficient between the two coils become worse when the receiving coil is shifted.

This type of coil structure is then used in the wireless charging system based on magnetic resonance, of which the resonant compensation structure is series-series. And the effect is analyzed by using Ansoft Maxwell software. First of all, we build the external circuit according to the schematic diagram shown in figure 3. We get the inductance values of the transmitting coil and the receiving coil. And then we calculate the value of resonance compensation capacitor using the series resonant theory. The simulation chart of the external circuit is shown in figure 9. And we obtain the voltage and current waveforms of the transmitting coil and the receiving coil which are shown in figures 10 and 11 respectively.

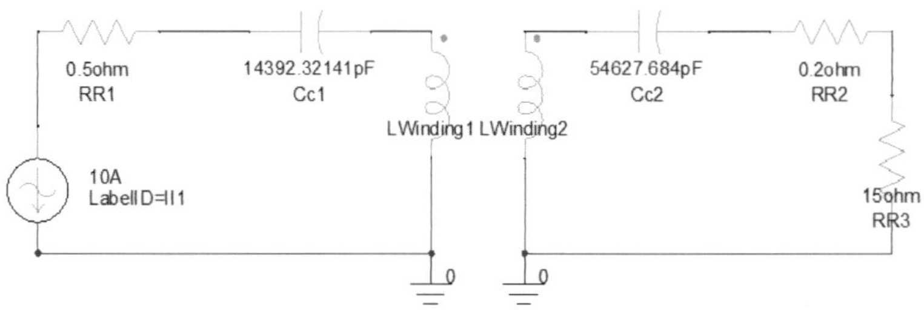

Figure 9. The simulation chart of the external circuit.

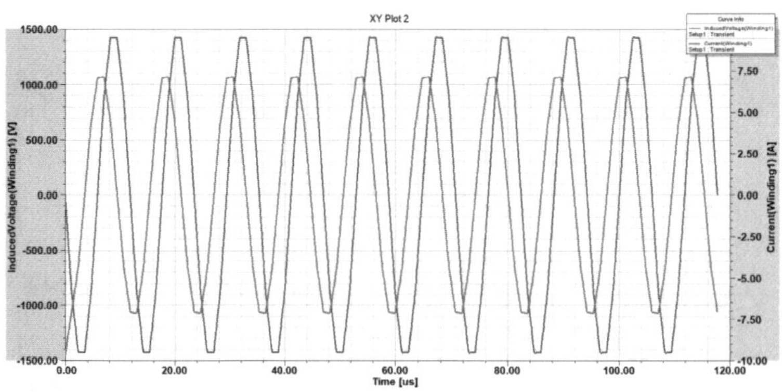

Figure 10. The voltage and current waveforms of the transmitting coil.

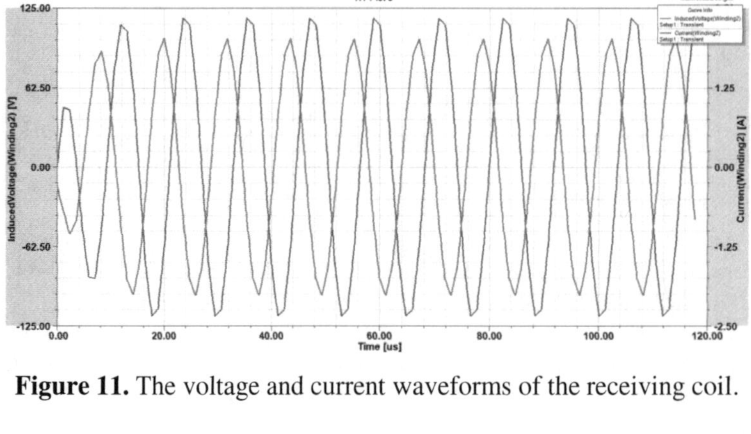

Figure 11. The voltage and current waveforms of the receiving coil.

4. New type of pan-shaped coil

The new type of pan-shaped coil consists of multi-turn circular coils aligned along the vertical axis. And it can improve the uniformity of the magnetic field between the transmitting coil and the receiving coil to solve the problem that misalignment between the two coils can significantly impair the transmission power and efficiency of the system. The illustration of the new type of pan-shaped coil is shown in figure 12. This type of coil can produce a relative smooth uniform magnetic field because ripples in the field strength curve along the radial direction can be eliminated by introducing height differences among adjacent turns.

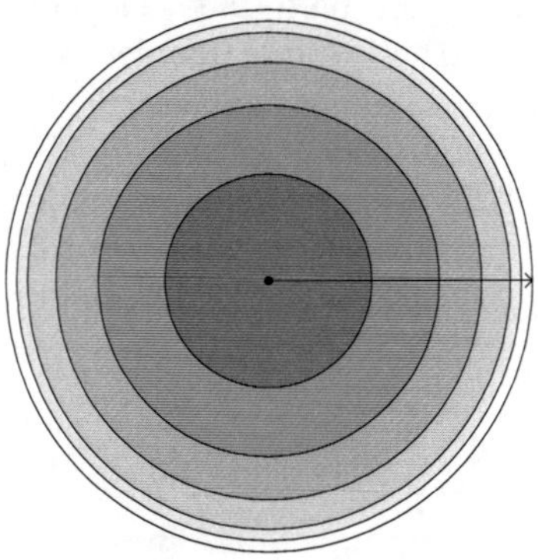

Figure 12. Illustration of the new type of pan-shaped coil.

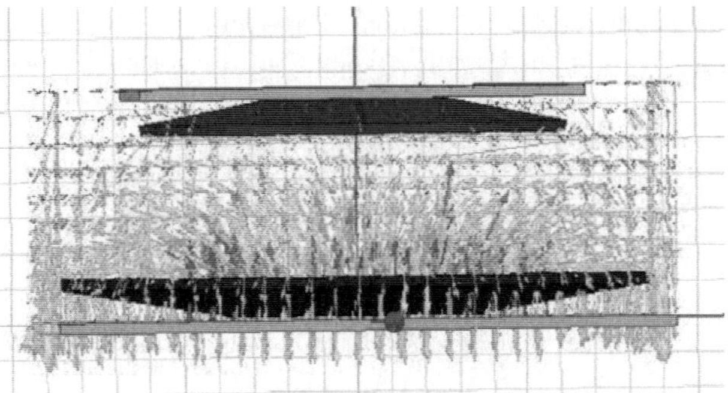

Figure 13. The magnetic field distribution between the two coils.

Similarly, we model this coil structure using the Ansoft Maxwell software for magnetic field analysis, in which the maximum radius of the transmission coil is 175 mm and that of the receiving coil is 125 mm, which are the same as the size of the conventional planar-core coil. The distance between the two coils is 120 mm. Especially, in order to reflect the advantages of the type of pan-shaped coil by contrasting with the conventional planar-core coil, we set the same excitation and use exactly the same cores as that of the conventional planar-core coil. The size of the two cores are $360\text{mm}\times 360\text{mm}\times 6\text{mm}$ and $260\text{mm}\times 260\text{mm}\times 6\text{mm}$ respectively and the distance between the two coils is 120 mm. We also make the inductance values of the transmitting coil and the receiving coil of the two types of coil structure equal by adjusting the number of turns. The distribution of the magnetic field between the coils and the distribution of the magnetic field in the radial direction of the receiving coil are shown in figures 13 and 14 respectively through performing a magnetic field simulation.

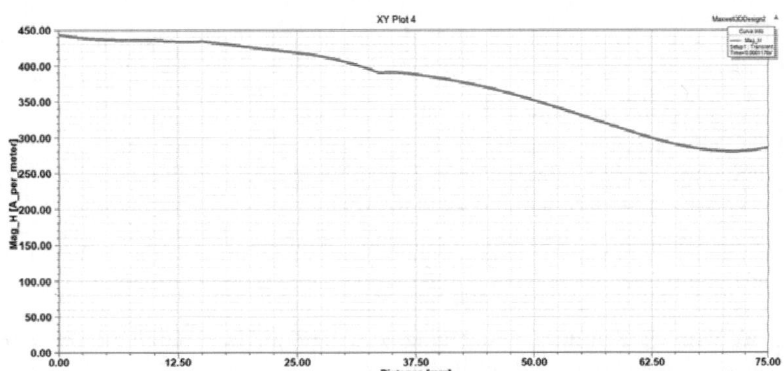

Figure 14. The distribution of the magnetic field in the radial direction of the receiving coil.

As shown in equation (7), the uniformity factor of the magnetic field between the two coils is 0. 0.365 by calculation. In addition, the coupling coefficient value between the two coils in this coil structure is 0.19242 by calculation using the simulation software.

$$\lambda_{U2} = \frac{442.5 - 281.0}{450.1} = 0.365 \quad (7)$$

It can be seen that the magnetic field uniformity of the new type of pan-shaped coil is improved obviously. And the coupling coefficient between the two coils is still improved though the amount of core is relatively smaller than the conventional planar-core coil.

We also simulate the magnetic field and calculate the magnetic field uniformity and coupling coefficient when the receiving coil are horizontally misaligned by 5 mm and 10 mm in order to see the effect on the transmission efficiency of the system produced by the new type of pan-shaped coil. The distribution of the magnetic field between the coils and the distribution of the magnetic field in the radial direction of the receiving coil are shown in figures 15 and 16 respectively. The uniformity factors of the magnetic field between the two coils are 0.382 and 0.404 respectively and the coupling coefficients are 0.19194 and 0.19077 respectively. It can be seen that the magnetic field uniformity and the coupling coefficient between the two coils are improved obviously compared with the conventional planar-core coil when the receiving coil is misaligned.

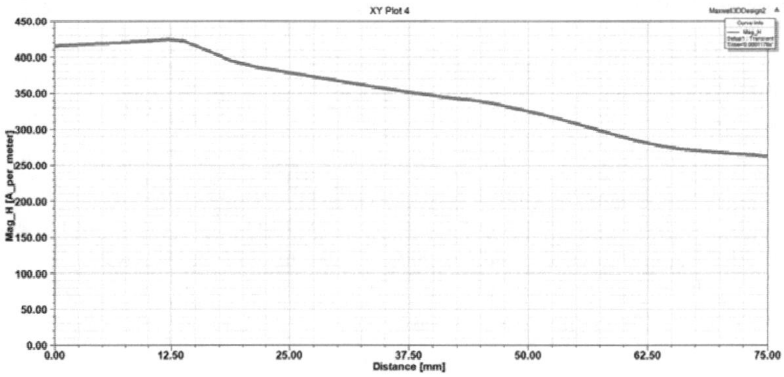

Figure 15. The distribution of the magnetic field in the radial direction of the receiving coil when misaligned by 5 mm.

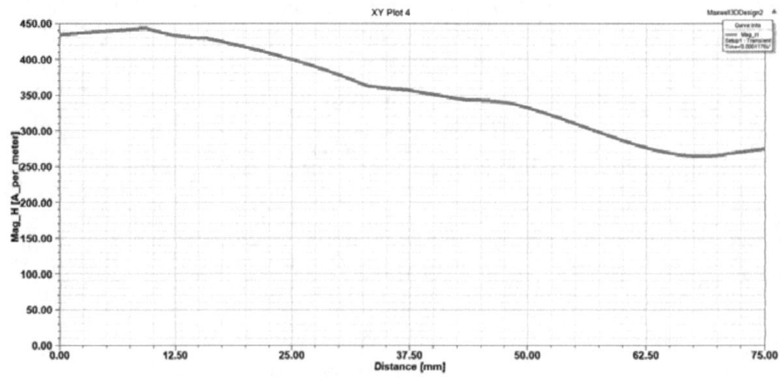

Figure 16. The distribution of the magnetic field in the radial direction of the receiving coil when misaligned by 10 mm.

This new type of coil structure is also used in the wireless charging system based on magnetic resonance, of which the resonant compensation structure is series-series. And the effect is analyzed by using Ansoft Maxwell software with the same external circuit according to the schematic diagram shown in figure 3. We obtain the voltage and current waveforms of the transmitting coil and the receiving coil which are shown in figures 17 and 18 respectively. Then we calculate the transmission power and efficiency of the resonant compensation structure according to equations (4) and (5). By comparison, we found that the power transmission properties of the new type of pan-shaped coil have been greatly improved than the conventional planar-core coil.

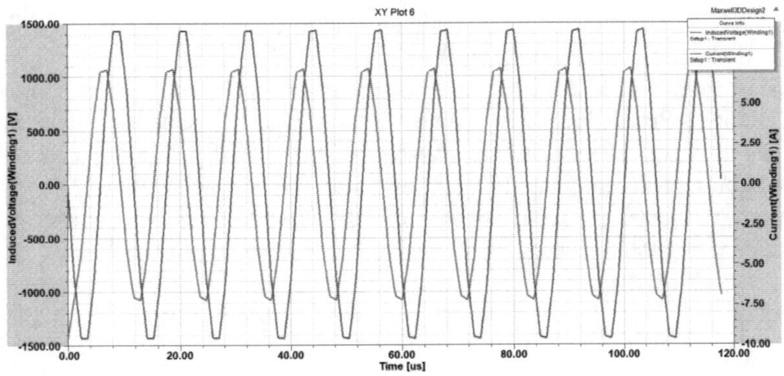

Figure 17. The voltage and current waveforms of the transmitting coil.

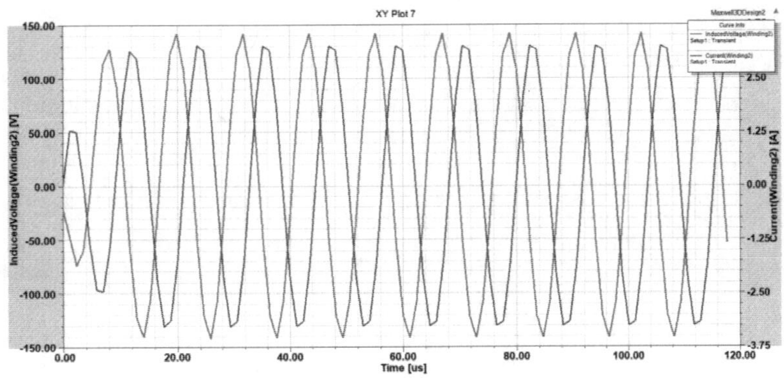

Figure 18. The voltage and current waveforms of the receiving coil.

5. Conclusions

This paper proposes a new type of coil structure called pan-shaped coil of wireless charging system based on magnetic resonance which can produce a smooth uniform magnetic field because ripples in the field strength curve along the radial direction can be eliminated by introducing height differences among adjacent turns. The pan-shaped coil can improve the uniformity of the magnetic field between the transmitting coil and the receiving coil and solve the problem that misalignment between the two coils can significantly impair the transmission power and efficiency of the system. In order to reflect the superiority of the pan-shaped coil, this paper also introduces the conventional planar-core coil and simulates the magnetic field of the coil with the receiving coil misaligned or not using the Ansoft Maxwell software. Compared with the simulation results of the two coils, it can be seen that this new type of coil structure can obviously improve the uniformity of the magnetic field, the coupling coefficient and the power transmission properties between the transmitting coil and the receiving coil.

References

[1] Zeng H, Liu Z Z, *et al* 2017 Optimization of magnetic core structure for wireless charging coupler IEEE Transactions on Magnetics **53** 61-2

[2] Qu X D 2016 Reaearch on wireless charing system of electric vehicle based on magnetic resonance 9-18

IOP Conf. Series: Earth and Environmental Science **93** (2017) 012061 doi:10.1088/1755-1315/93/1/012061

[3] Guo Y and Zhu C B 2014 Magnetic resonant wireless power transmission technology based on planar core Journal of Harbin Institute of Technology **46** 523-6

[4] Budhia M, Boys J T, Covic G A and Chang-Yu H 2013 Development of a single-sided flux magnetic coupler for electric vehicle IPT charging systems IEEE Trans. Ind. Electron **60** 1318-28

[5] Budhia M, Covic G A, *et al* 2011 Development and evaluation of single sided flux couplers for contactless electric vehicle charging IEEE ECCE **47** 10614-21

[6] Krishnan S, Bhuyan S, *et al* 2012 Frequency agile resonance-based wireless charging system for electric vehicles Proceeding of IEEE International Electric Vehicle Conference (IEVC) 1-4

[7] Xu Q, Wang H, *et al* 2013 A novel mat-based system for position-varying wireless power transfer to biomedical implants IEEE Transactions on Magnetics **49** 84774-9

[8] Diao Y, Shen Y and Gao Y 2011 Design of coil structure achieving uniform magnetic field distribution for wireless charging platform Proceeding of Power Electronics Systems and Applications (PESA) 1-5

[9] Liu X and Hui S Y 2008 Optimal design of a hybrid winding structure for planar contactless battery charging platform IEEE Transactions on Power Electronics **23** 1455-63

IOP Conf. Series: Earth and Environmental Science **93** (2017) 012062 doi:10.1088/1755-1315/93/1/012062

Operation strategy for grid-tied DC-coupling power converter interface integrating wind/solar/battery

H L Jou[1, 4], J C Wu[2], J H Lin[1], W N Su[3], T S Wu[3] and Y T Lin[3]

[1]Department of Electrical Engineering National Kaohsiung University of Applied Sciences Kaohsiung, Taiwan
[2]Department of Microelectronics Engineering National Kaohsiung Marine University Kaohsiung, Taiwan
[3]Institute of Nuclear Energy Research, AEC, EY, Taiwan

E-mail: hljou5519@gmail.com

Abstract. The operation strategy for a small-capacity grid-tied DC-coupling power converter interface (GDPCI) integrating wind energy, solar energy and battery energy storage is proposed. The GDPCI is composed of a wind generator, a solar module set a battery bank, a boost DC-DC power converter (DDPC), a bidirectional DDPC power converter, an AC-DC power converter (ADPC) and a five-level DC-AC inverter (DAI). A solar module set, a wind generator and a battery bank are coupled to the common DC bus through the boost DDPC, the ADPC and the bidirectional DDPC, respectively. For verifying the performance of the GDPCI under different operation modes, computer simulation is carried out by PSIM.

1. Introduction

The conventional fossil fuels have resulted in the problems of greenhouse emissions and climate change and thus damaged the earth's environment seriously [1,2]. Paris agreement on global reduction of greenhouse emissions to relieve the climate change has promoted to use the renewable energies worldwide [3]. The wind energy and solar energy are the most attractive renewable energy sources now. However, the wind energy and solar energy have the characteristics of instability, intermittence and hard to be predicted. Battery energy storage can be integrated to make the energies of wind generator and solar module set more stable and reliable [4-7].

Power converter interface is one of the key technologies for the applications of the wind energy and solar energy effectively. The power converter interface for integrating wind energy, solar energy and battery energy storage can be divided into the AC-coupling topology [8,9] and DC-coupling topology [8-12]. Figure 1 shows the diagram of AC-coupling topology, and the diagram of DC-coupling topology is shown in figure 2.

As seen in figure 1, it consists of a wind generator, a solar module set, a battery bank, an AC-DC power converter (ADPC), a DC-DC power converter (DDPC), a bidirectional DDPC and three DC-AC inverters (DAIs). The output AC voltage of wind generator is rectified by an ADPC, and a DAI for converting the DC power to AC power injecting into the grid or supplying to the load. The module set is connected to a DDPC and a DAI connecting in cascade for converting the solar power to AC power injecting into the grid or supplying to the load. A combination of a bidirectional DDPC and a DAI is

IOP Conf. Series: Earth and Environmental Science **93** (2017) 012062 doi:10.1088/1755-1315/93/1/012062

used for charging or discharging the battery bank. In this circuit, three individual DAIs are needed to couple the energies of wind energy, solar energy and battery energy storage at the AC grid. It is known as AC-coupling topology.

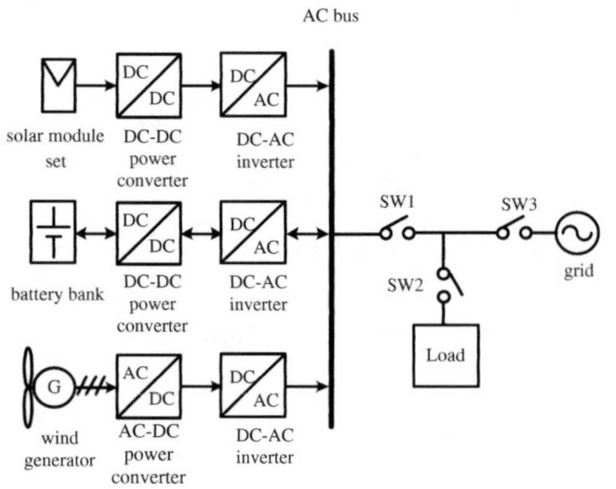

Figure 1. Diagram of AC-coupling topology for integrating the wind/solar/ battery.

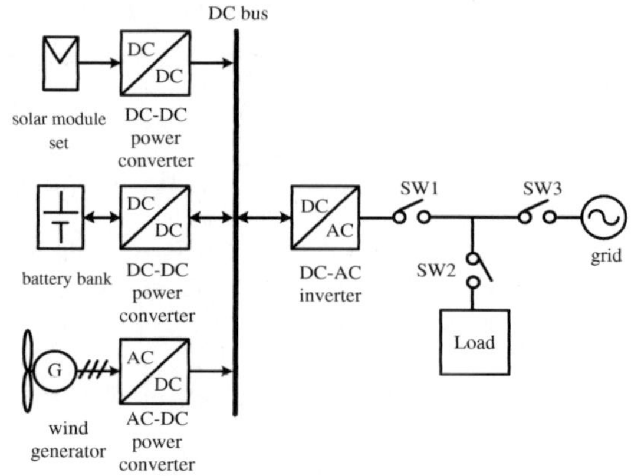

Figure 2. Diagram of DC-coupling topology for integrating the wind/ solar/ battery.

As seen in figure 2, it consists of a wind generator, a solar module set, an ADPC, a DDPC, a bidirectional DDPC and a DAI. The output of wind generator is converting to the DC power by an ADPC. A DDPC has the function of regulating the voltage and the output power of sola module set. The battery bank is connected to a bidirectional DDPC for charging or discharging the battery power. The outputs of ADPC, DDPC and bidirectional DDPC are connected to a common DC bus of DAI which is operated as an interface to the grid or the load. It is known as DC-coupling topology. In this circuit, only a DAI is shared by three energy sources.

As can be seen in figures 1 and 2, it can be found that six power converters are required in AC-coupling topology and only four power converters are used in the DC-coupling topology. Hence, it can save two DAIs in the DC-coupling topology although the capacity of DAI should be enlarged slightly. Moreover, the energy is only processed by two power conversion stages in the DC-coupling topology but is processed by four power conversion stages in the AC-coupling topology when the charging power of battery bank is from the output powers of wind generator and solar module set. Therefore, the DC-coupling topology has the advantages of simplified hardware, lower cost and higher energy efficiency.

In this paper, operation strategy for small-capacity grid-tied DC-coupling power converter interface (GDPCI) integrating the wind energy, solar energy and battery energy storage is presented. The output powers of wind generator, solar module set and battery bank are coupling to the common DC-bus of a simplified five-level inverter through an ADPC, a boost DDPC and a bidirectional DDPC, respectively. The performance of GDPCI under different operation modes is validated by the simulation of PSIM.

2. Circuit topology

Figure 3 is the power circuit for the proposed small-capacity GDPCI integrating the solar energy, wind energy and battery energy storage. It is composed of a permanent magnetic synchronous generator (PMSG), a solar module set, battery bank, an ADPC, a boost DDPC, a bidirectional DDPC and a bidirectional five-level DAI. The ADPC is connected to PMSG, and it is comprised of a three-phase rectifier and a boost DDPC due to the consideration of cost in the application of small-capacity wind power generator. The bidirectional DAI is a five-level inverter which can save two power electronic switches as compared with the conventional five-level T-type power converter.

3. Operation strategy

The operation of proposed small-capacity GDPCI can be divided into five operation modes based on the charge/discharge of battery bank. Modes 1, 2, 3 are the charging modes, and modes 4 and 5 are the discharging modes. Table 1 shows the conditions of different modes. The power flow diagrams for five operation modes are shown in figures 4-8.

Table 1. Conditions of different operation modes.

mode	conditions	charge/discharge
1	$P_{wind}+P_{solar}<P_{bat}$	charge
2	$P_{bat}+P_{load}>P_{wind}+P_{solar}>P_{bat}$	charge
3	$P_{bat}+P_{load}<P_{wind}+P_{solar}$	charge
4	$P_{load}>P_{wind}+P_{solar}+P_{bat}$	discharge
5	$P_{load}<P_{wind}+P_{solar}+P_{bat}$	discharge

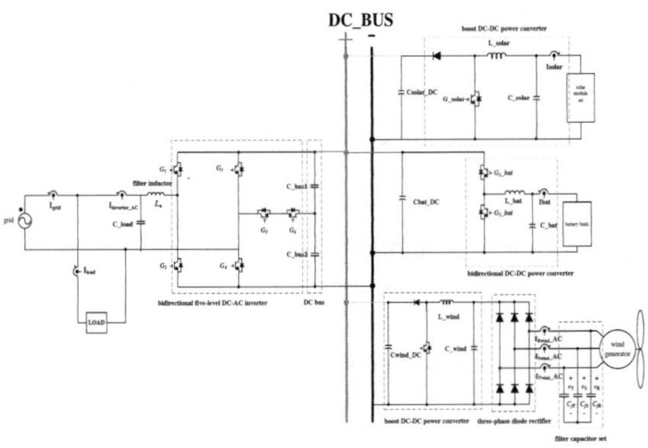

Figure 3. Power circuit for proposed small-capacity GDPCI.

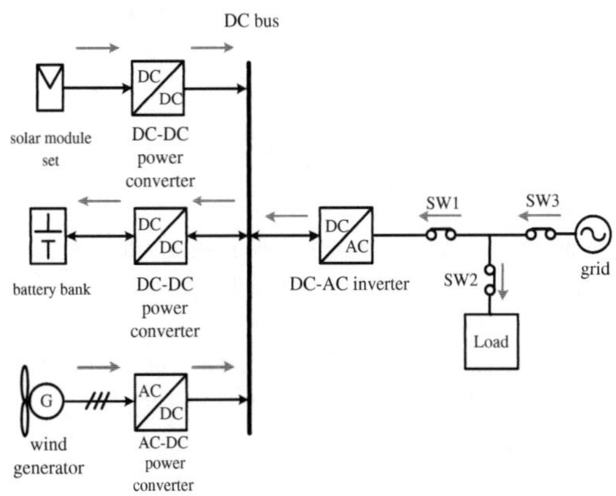

Figure 4. Power flow diagram for mode 1.

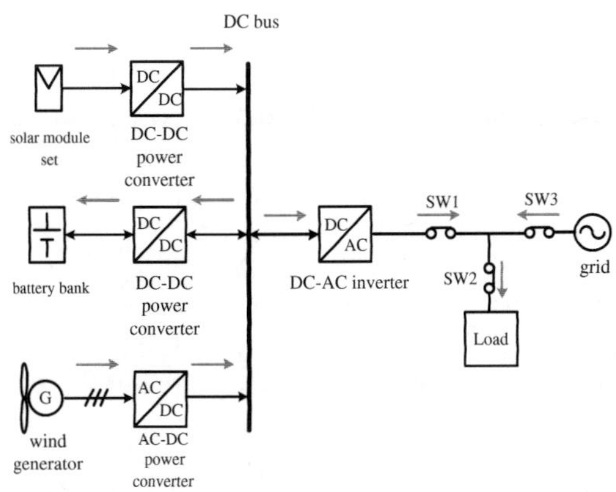

Figure 5. Power flow diagram for mode 2.

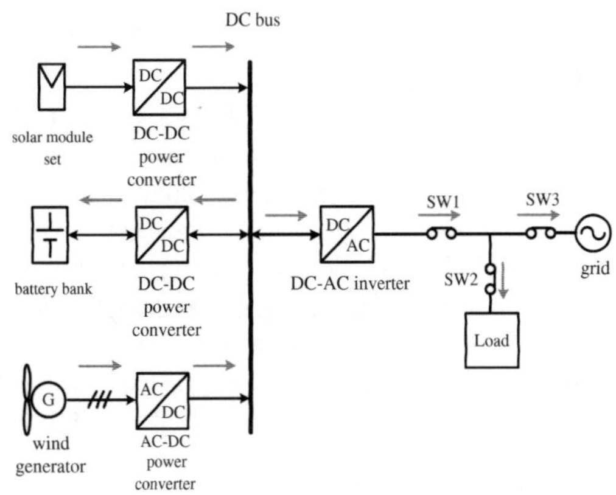

Figure 6. Power flow diagram for mode 3.

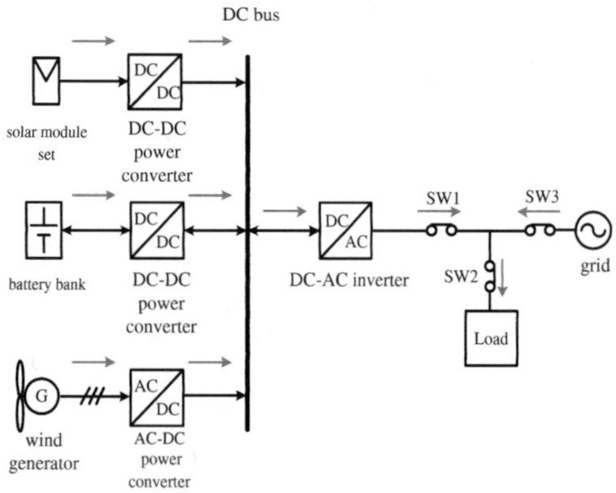

Figure 7. Power flow diagram for mode 4.

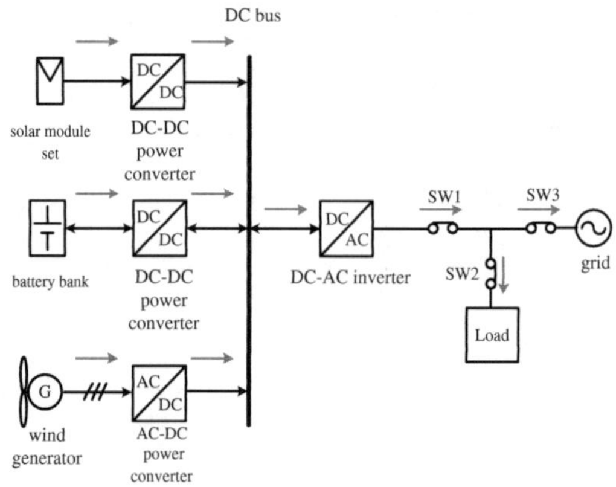

Figure 8. Power flow diagram for mode 5.

3.1. Charge

- mode 1:

The generated power of wind solar is smaller than the demanded power of battery bank. In this condition, the insufficient charging power is supplied by the grid. The power flow diagram of the GDPCI is shown in figure 4.

- mode 2:

The generated wind power r and solar power is larger than the demanded power of battery bank but smaller than the demanded power of battery bank and the load. In this condition, the generated wind generator and solar module set is charged to the battery bank, and the additional generated power is supplied to the load. The insufficient demanded load power is supplied by the grid. The power flow diagram of the GDPCI is shown in figure 5.

- mode 3:

The generated power of wind and solar is larger than the demanded power of battery bank and the load. In this condition, the generated wind power and solar power is charged to the battery bank and supplied to the load, and the additional power is fed to the grid. The power flow diagram of the GDPCI is shown in figure 6.

3.2. Discharge

- mode 4:

The demanded power of load is larger than the generated power of wind and solar power and the output power of the battery bank in this condition. The generated power of wind and solar as well as the output power of the battery bank is supplied to the load, and the grid supplies the insufficient demanded power of load. The power flow diagram of the GDPCI is shown in figure 7.

- mode 5:

The generated power of wind and solar power and the output power of the battery bank are larger than the demanded power of the load in this condition. The generated wind power, solar power and the output power of the battery bank are supplied to the load, and the additional power is fed to the grid. The power flow diagram of the GDPCI is shown in figure 8.

4. Simulation results

Computer simulation by PSIM is used to validate the function of the GDPCI under different operation modes. The battery bank is charged by the hybrid constant-current (CC)/constant-voltage (CV) strategy. The value of CC is 15 A, and the value of CV is 270 V.

Figure 9 shows the simulation result during the transient from mode 1 to mode 2. The demanded load power is 5 kW, the output power of wind generator is 1 kW and the output power of solar is 1.5 kW. It can be found that the battery bank is charged by 15 A before 3.47 s. The battery voltage reaches

270 V at 3.47 s, and the battery bank is changed from the CC to the CV. The generated power of wind and solar is charged to the battery bank when the generated power is small than the charging power of battery bank. The current charging to the battery bank is decreased gradually when the battery bank is charged by CV. The demanded power of the battery bank is then smaller than the generated power of wind and solar at about 3.52 s. And then, the additional generated power is supplied to the load and the grid supplies the insufficient load power. The operation of GDPCI is changed from mode 1 to mode 2.

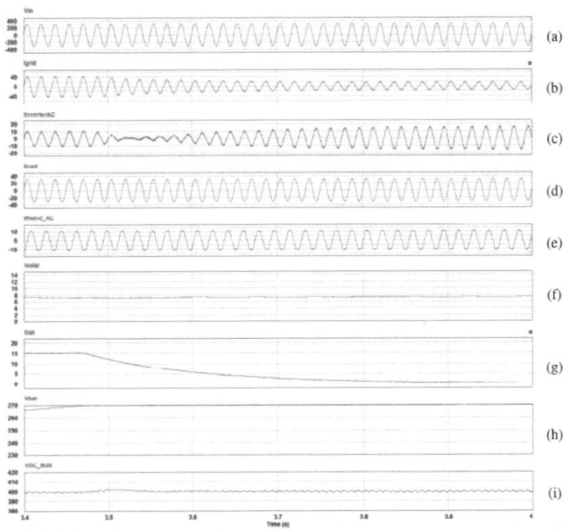

Figure 9. Simulation result during the transient from mode 1 to mode 2, (a)v_{in}, (b)i_{grid}, (c)$i_{inverter\_AC}$, (d) i_{load}, (e) i_{Rwind_AC}, (f) i_{solar}, (g) i_{bat}, (h) v_{bat}, (i)V_{dc_bus}.

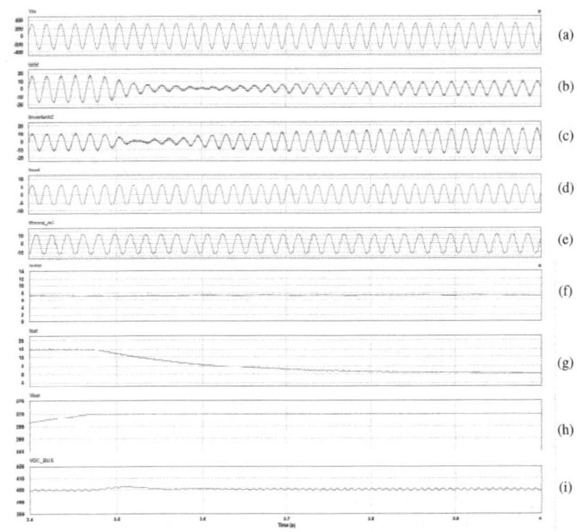

Figure 10. Simulation result simulation result during the transient from mode 1 to mode 3, (a) v_{in}, (b) i_{grid}, (c) $i_{inverter\_AC}$, (d) i_{load}, (e) i_{Rwind_AC}, (f) i_{solar}, (g) i_{bat}, (h) v_{bat}, (i) V_{dc_bus}.

Figure 10 shows the simulation result during the transient from mode 1 to mode 3. The demanded power of load is 1 kW, the output power of wind generator is 1 kW and the generated power of solar is 1.5 kW. It can be found that the battery bank is charged by CC 15 A before 3.47 s. The voltage of battery bank reaches 270 V at 3.47 s, and the battery bank is changed from CC to CV. The battery current is decreased gradually when the battery bank is charged by the CV. The demanded power charged to the battery bank is smaller than the generated power of wind and solar at about 3.52 s. Then, the additional generated power t is supplied to the load. The charging power of the battery bank is still decreased gradually, and the additional generated power after absorbing by the load is fed to the grid at about 3.6 s. The operation of GDPCI is changed from mode 1 to mode 3.

Figure 11 shows the simulation result for mode 4. The demanded load power is 5 kW. The output power of wind generator and solar module set are respectively 1 kW and 1.5 kW and the power discharging by the battery bank is 2 kW. The generated wind power, and solar power as well as the discharging power of battery bank are supplied to the load, and the grid supplies the insufficient load power. The power of battery bank is turned off while the voltage of battery bank reached 200 V for protecting over-discharging of battery bank.

Figure 12 shows the simulation result during the transient form mode 1 to mode 5. The demanded load in this condition is 2 kW. The output power of wind generator is 1 kW, and the output power of solar module set is 1.5 kW. The battery bank is charged by a constant current of 15 A before 2.6 s, and the GDPCI is thus operated in mode 1. The battery bank is changed from CC charging of 15 A to constant power (CP) discharging of 2 KW at 2.6 s. The operation of GDPCI is changed from mode 1 to mode 5.

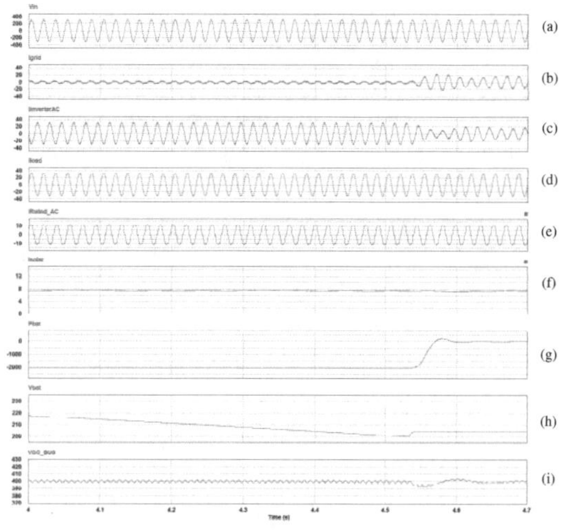

Figure 11. Simulation result for mode 4, (a) v_{in}, (b) i_{grid}, (c) $i_{inverter\_AC}$, (d) i_{load}, (e) i_{Rwind_AC}, (f) i_{solar}, (g) i_{bat}, (h) v_{bat}, (i) Vdc_bus.

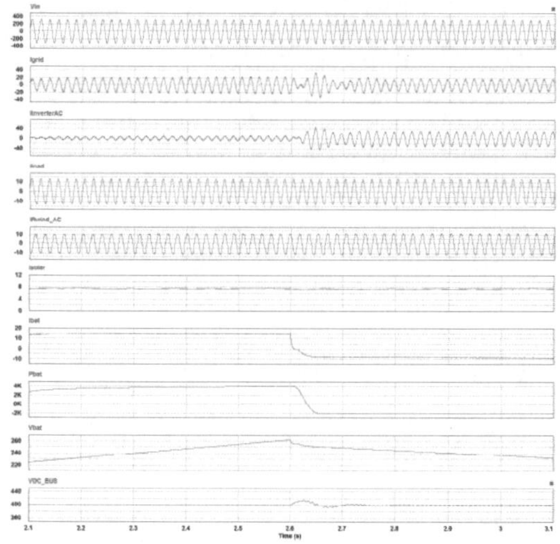

Figure 12. Simulation result during the transient form mode 1 to mode 5, (a) v_{in}, (b) i_{grid}, (c) $i_{inverter\_AC}$, (d) i_{load}, (e) i_{Rwind_AC}, (f) i_{solar}, (g) i_{bat}, (h) v_{bat}, (i) Vdc_bus.

5. Conclusions

The number of power converters using in the grid-tied AC-coupling GDPCI for integrating wind energy, solar energy and battery energy storage is six, and that using in the GDPCI is only four. Besides, the number of power conversion stages for charging the battery bank by the wind energy or solar energy is four in the AC-coupling topology, and that is only two in the DC-coupling topology. Therefore, the DC coupling topology is superior to the AC-coupling topology due to the advantages of simplified hardware, lower cost and higher energy efficiency.

The operation of GDPCI integrating small-capacity wind energy, solar energy and battery energy storage are divided into five operation modes based on the charge/discharge of battery bank. Three modes are for charging battery bank, and two modes are for discharging battery bank. Computer simulation verifies the operation strategy of the GDPCI under different operation modes.

Acknowledgments

The authors would like to express their acknowledgments for the financial support of Institute of Nuclear Energy Research Atomic Energy Council under the contract NL1060284. This paper is commissioned to study, but the paper does not represent the opinions of the commission unit.

References

[1] Pendergast D 2006 Kyoto and beyond: Development of sustainable policy *IEEE EIC Climate Change Technol* 1-3

[2] Arai J, Iba K, Funabashi T, Nakanishi Y, Koyanagi K and Yokoyama R 2008 Power electronics and its applications to renewable energy in Japan *IEEE Circuits and Syst Mag* **8** 52-66

[3] Dimitrow R S 2016 The Paris agreement on climate change: Behind closed doors *Global Environ Politics* **16** 1-11

[4] Manimekalai P, Harikumar R and Raghavan S 2013 An overview of batteries for photovoltaic (PV) systems *Int J of Computer Appl* **82** 28-32

[5] Díaz-González F, Sumper A, Gomis-Bellmunt O and Villafáfila-Robles R 2012 A review of energy storage technologies for wind power applications *Renew and Sust Energy Reviews* 2154-72

[6] Simões M G, Busarello D T C, Bubshait A S, Harirchi F, Pomilio J A and Blaabjerg F 2016

Interactive smart battery storage for a PV and wind hybrid energy management control based on conservative power theory *Int J of Control* **89** 850-70

[7] Masaud T M, Lee K and Sen P K 2010 An overview of energy storage technologies in electric power systems: What is the future? *North American Power Symposium (NAPS)* pp 1-6

[8] Nejabatkhah F and Li Y W 2014 Overview of power management strategies of hybrid AC/DC microgrid *IEEE Power Electron Society* **30** 7072-89

[9] Badwawi R A, Abusara M and Mallick T 2015 A review of hybrid solar PV and wind energy system *Smart Sci* **3** 127-38

[10] Zhang F G, Chen X Y, Yin X J and Wang Z S 2013 An improved capacity ratio design method based on complementary characteristics of wind and solar *Electr Mach and Syst (ICEMS)* 405-408

[11] Valenciaga F and Puleston P F 2005 Supervisor control for a stand-alone hybrid generation system using wind and photovoltaic energy *IEEE Trans on Energy Conver* **20** 398-405

[12] Zhang Y, Jia H J and Guo L 2012 Energy management strategy of islanded microgrid based on power flow control *IEEE PES Innovative Smart Grid Techn*ol (ISGT) 1-8

Study on frequency characteristics of wireless power transmission system based on magnetic coupling resonance

L H Liang[1,3], Z Z Liu[1], Y J Hou[2], H Zeng[1], Z K Yue[1] and S Cui[1]

[1]School of Electrical Engineering, Shandong University, Jinan 250014, Shandong Province, China
[2]Energy Research Institute, Shandong Academy of Sciences, Jinan 250061, Shandong Province, China

E-mail: L_LeoH1251@163.com/ liuzhizhen@sdu.edu.cn

Abstract. In order to study the frequency characteristics of the wireless energy transmission system based on the magnetic coupling resonance, a circuit model based on the magnetic coupling resonant wireless energy transmission system is established. The influence of the load on the frequency characteristics of the wireless power transmission system is analysed. The circuit coupling theory is used to derive the minimum load required to suppress frequency splitting. Simulation and experimental results verify that when the load size is lower than a certain value, the system will appear frequency splitting, increasing the load size can effectively suppress the frequency splitting phenomenon. The power regulation scheme of the wireless charging system based on magnetic coupling resonance is given. This study provides a theoretical basis for load selection and power regulation of wireless power transmission systems.

1. Introduction

Wireless power transmission technology (WPT), also known as non-contact power transmission technology is a power supply mode that transfers power from the power supply to the electrical equipment by means of a space medium. WPT avoids the connection between the electrical equipment and the grid and it achieves full electrical isolation. The transmission process is safe, reliable, and flexible, providing the possibility for man to get rid of the wire. WPT is a basic research and an applied research of multi-disciplinary cross-over of electromagnetic field, power electronics, high-frequency electron, electromagnetic induction and coupled-mode theory. It is a revolutionary progress in energy transmission [1]. WPT is important in military, aerospace, oilfield, mine, underwater operations, industrial robots, electric vehicles, wireless sensor networks, medical devices, household appliances and RFID identification. And WPT also has important scientific and research value and practical significance in the development of electromagnetic theory.

In 2007, Marin Soljacic of MIT proposed the principle of magnetic coupling resonant wireless power transmission technology. They used the theory to light up a 60 W bulb in the 2 m range successfully [2]. Since then, wireless power transmission technology has become a hot topic of domestic and foreign scholars.

At present, there are three main technologies for wireless power transmission: radio frequency or microwave wireless power transmission technology, electromagnetic induction wireless transmission technology and magnetic coupling resonant wireless power transmission technology (MCR-WPT).

Compared to the first two technologies, MCR-WPT has the following advantages: MCR-WPT has a larger transmission distance; Because of the use of strong coupling resonant technology, MCR-WPT can achieve higher reception power and efficiency; The system uses magnetic field coupling rather than electric field, and has no harm to the human body; MCR-WPT has good penetrability and is not affected by nonmetallic obstructions. Therefore, the magnetic coupling resonant wireless energy transmission system has become a new direction of development of wireless energy transmission technology, and more and more researchers carry out research work in this direction.

The key to the wireless energy transmission technology based on magnetic coupling resonant is that the system should be in a resonant state, to obtain a larger reception power and efficiency. Therefore, maintaining the stability of the resonant frequency is a key problem of the magnetic coupling resonant wireless power transmission technology. However, the resonant frequency of the magnetic coupling resonant wireless power transmission system is not only affected by the coil radius, coil shape and other factors, but also by the distance between the coil and the impact of the load and the resonant frequency of the system will split. [3-6] analyze the frequency splitting characteristics from different aspects. [7-9] propose the problem of efficiency reduction caused by frequency tracking to solve the problem of frequency splitting. [10] analyzes the maximum power transmission efficiency of a wireless power transmission system.

This paper considers the resonant frequency, load resistance and coil resistance and an equivalent model based on magnetic coupling resonant wireless power transmission is established. The effect of load resistance on the received power and energy transmission efficiency is analyzed, and the frequency characteristics of received power and efficiency are obtained. A magnetic coupling resonant wireless power transmission system is designed. The frequency characteristics of different load resistance values are simulated and verified.

2. System model and theoretical analysis

Magnetic coupling resonant wireless power transmission is based on the concept of near field coupling. The basic principle is that two objects with the same resonant frequency can achieve efficient energy exchange, while the energy exchange between no resonant objects is very weak. Figure 1 is a block diagram of a magnetic coupling resonant wireless power transmission system.

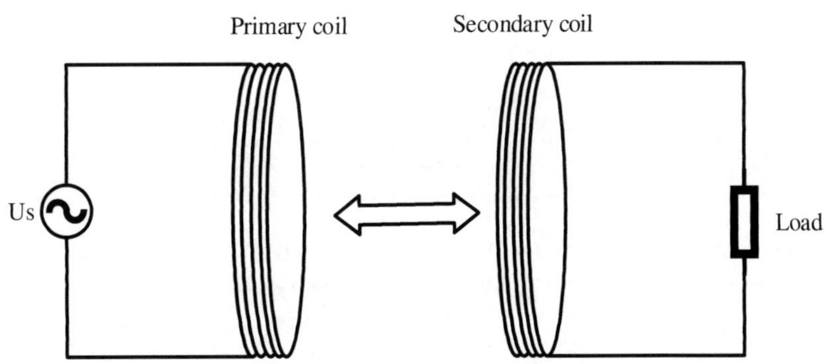

Figure 1. Block diagram of magnetic coupling resonant wireless power transmission system.

As can be seen from figure 1, U_S is the three-phase grid voltage which is rectified and inverted equivalent to become the primary power source for wireless power transmission. The wireless power transfer between the transmitting coil and the receiving coil through the resonance coupling of the magnetic field of the space. This is the core of the entire wireless power transmission system and the level and characteristics of energy transfer are primarily determined by this.

The equivalent circuit based on the magnetic coupling resonant wireless power transmission system is shown in figure 2. U_S is the equivalent voltage source of the primary side of the grid voltage obtained by rectification and inverter. R_1,R_2 are the primary and secondary side coil equivalent resistance, respectively. R_L is the load resistance. C_1, C_2 are capacitors and L_1, L_2 are the coil equivalent inductances.

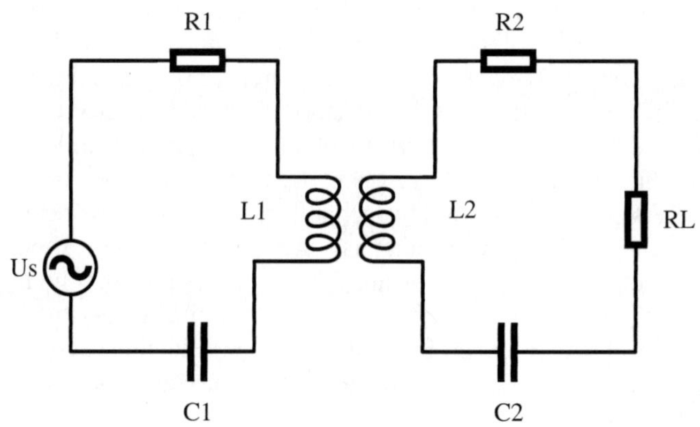

Figure 2. Equivalent circuit of wireless power transmission system.

At present, there are three main theories for quantitative analysis of wireless energy transmission, which are coupled-mode theory, circuit theory and band-pass filter theory. The paper uses the internal equivalent method of the circuit theory of the system for theoretical analysis.

When the coupling coefficient k between the two coils satisfies the formula (1), the system will appear frequency splitting phenomenon [12].

$$k \geq \frac{1}{\sqrt{Q_1 Q_2}}\left[\frac{1}{2}\left(\frac{Q_1}{Q_2}+\frac{Q_2}{Q_1}\right)\right]^{\frac{1}{2}} \tag{1}$$

Where Q_1 and Q_2 are the quality factors of the transmitting coil and the receiving coil, respectively.

$$\begin{cases} Q_1 = \frac{\omega L_1}{R_1} \\ Q_2 = \frac{\omega L_2}{R_L} \end{cases} \tag{2}$$

Where ω is the input frequency of the supply voltage.

Joint formulas (1) and (2) and formula (3) can be obtained by transformation.

$$R_L \leq \sqrt{\left(2\omega^2 k^2 - \frac{R_1^2}{L_1^2}\right)L_2^2} \tag{3}$$

When R_L satisfies equation (3), the system will appear frequency splitting phenomenon. So the minimum or critical load value of the system to prevent frequency splitting is:

$$R_{Lmin} = \sqrt{\left(2\omega^2 k^2 - \frac{R_1^2}{L_1^2}\right)L_2^2} \tag{4}$$

It can be seen from the formula (4), for the different magnetic coupling coefficient k, there is a minimum value R_{Lmin} of the load. When the load is greater than R_{Lmin}, the system will not appear frequency splitting phenomenon.

Let the current flow through the transmitting coil and the receiving coil be I_1 and I_2 respectively. The mutual inductance between the two coils is expressed as $M = k\sqrt{L_1 L_2}$. According to Kirchhoff's theorem,

$$\begin{cases} U_s = Z_1 I_1 + j\omega M I_2 \\ 0 = j\omega M I_1 + Z_2 I_2 \end{cases} \tag{5}$$

Where Z_1, Z_2 are the impedance of the transmitting coil and the receiving coil, respectively.

$$\begin{cases} Z_1 = R_1 + j\omega L_1 + \frac{1}{j\omega C_1} \\ Z_2 = R_L + R_2 + j\omega L_2 + \frac{1}{j\omega C_2} \end{cases} \tag{6}$$

When the system is in the resonant state, joint formulas (5) and (6) and we can get receipt power on the load as,

$$P_L = I_2^2 R_L = \frac{(\omega M U_s)^2}{R_1[(R_L+R_2)+(\omega M)^2]^2} R_L \tag{7}$$

Since the equivalent resistance of the coil is small, its influence is negligible. Calculate partial differential of equation (7).

$$\frac{\partial P_L}{\partial R_L} = 0 \tag{8}$$

When the received power reaches its maximum value, the best match load is

$$R_{Lmax} = \frac{k^2 L_2}{C_2 R_1} \tag{9}$$

It can be seen from equation (9) that the best matching load is not only related to the coil parameters, but also influenced by the magnetic coupling coefficient k.

3. Simulation and analysis

To verify the correctness of the theoretical analysis, this paper uses MATLAB to simulate. In this paper, we set up a simulation circuit model based on the two-coil structure of the circuit in SIMULINK. The parameters of the simulation circuit are set as follows. L_1=233.6 μH, L_2=140 μH, C_1=15 pF, C_2=25 pF, R_1=0.5 Ω, R_2=0.2 Ω, U_s=200 V.

For a magnetic coupling coefficient k, the wireless power transmission system will exhibit three states depending on the load: No frequency split phenomenon, which is called under-coupling; Just appear the phenomenon of frequency splitting, known as critical coupling; There is a frequency splitting phenomenon, known as over-coupling. Respectively, take k = 0.05, 0.1, 0.15, 0.2, 0.25, 0.3. The frequency characteristics of the received power and efficiency at different loads are obtained by simulation as figure 3.

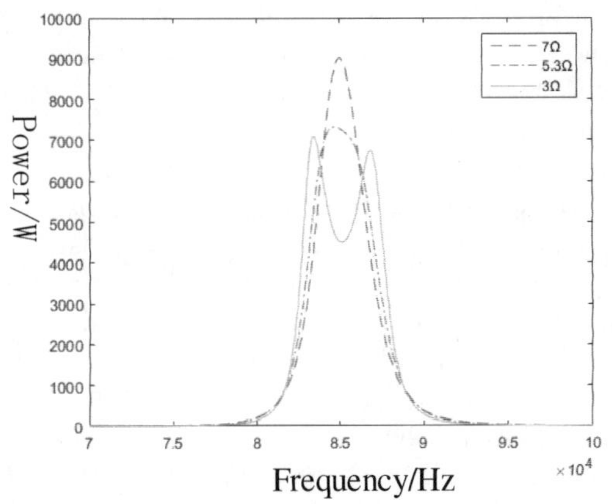

(a) k=0.05

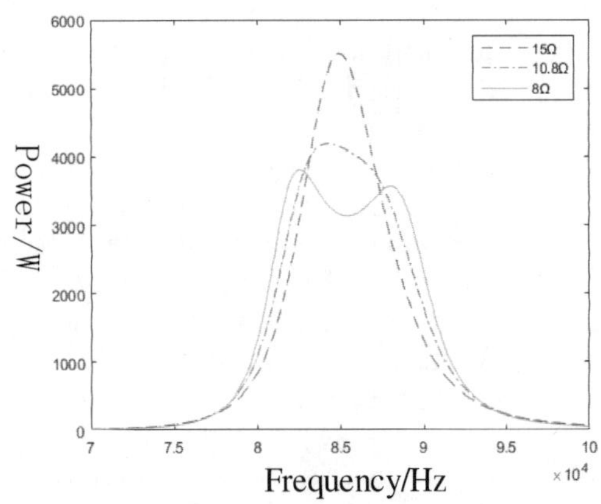

(b) k=0.1

IOP Conf. Series: Earth and Environmental Science **93** (2017) 012063 doi:10.1088/1755-1315/93/1/012063

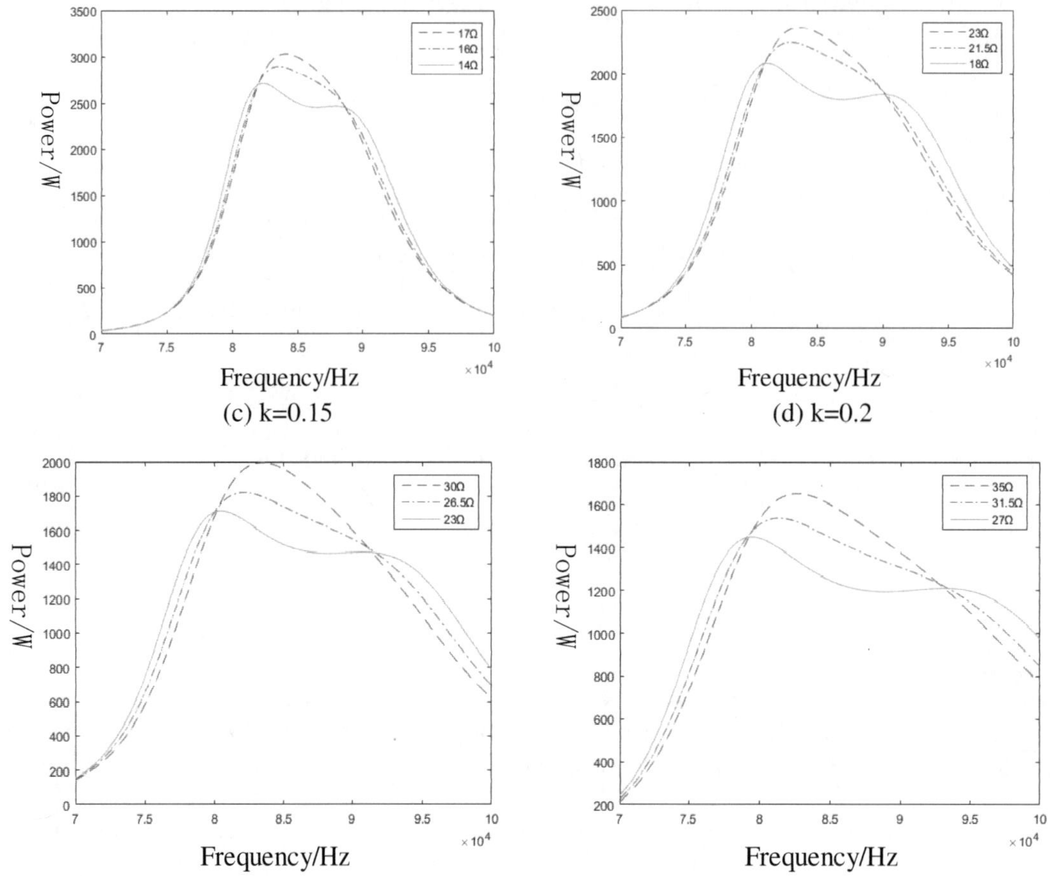

(c) k=0.15 (d) k=0.2

(e) k=0.25 (f) k=0.3

Figure 3. The frequency characteristic of received power.

From figure 3, when the load resistance is lower than the critical resistance, the frequency characteristics of the received power of the wireless power transmission exhibit two peaks. At this point, the system appears frequency splitting phenomenon. When the load resistance is greater and equal to the critical resistance, the received power of the system has only one peak with the frequency change. So, increasing the load resistance can achieve the purpose of suppressing frequency splitting.

Figure 4 shows the frequency characteristic curves of the system transmission efficiency corresponding to the three typical load resistance values when k is different. When the load resistance value changes from the undercounted state to the over-coupled state, the frequency characteristic of the efficiency gradually becomes smooth from the spikes. As the load resistance decreases, the transmission efficiency of the system increases. When the frequency of the system is split, the transmission efficiency of the system maintains a high value over a wide frequency range. However, when the load resistance is too low, the system transmission efficiency appears multiple peaks. This provides a reference for the power regulation of the wireless charging system.

Table 1 gives a comparison of the theoretical and simulated values of the minimum load resistance R_{Lmin} for suppressing frequency splitting. From table 1, the difference between the theoretical calculated value and the simulated value of the critical resistance R_{Lmin} is small and the error is kept within 2%. The simulation results verify the correctness of the theoretical analysis.

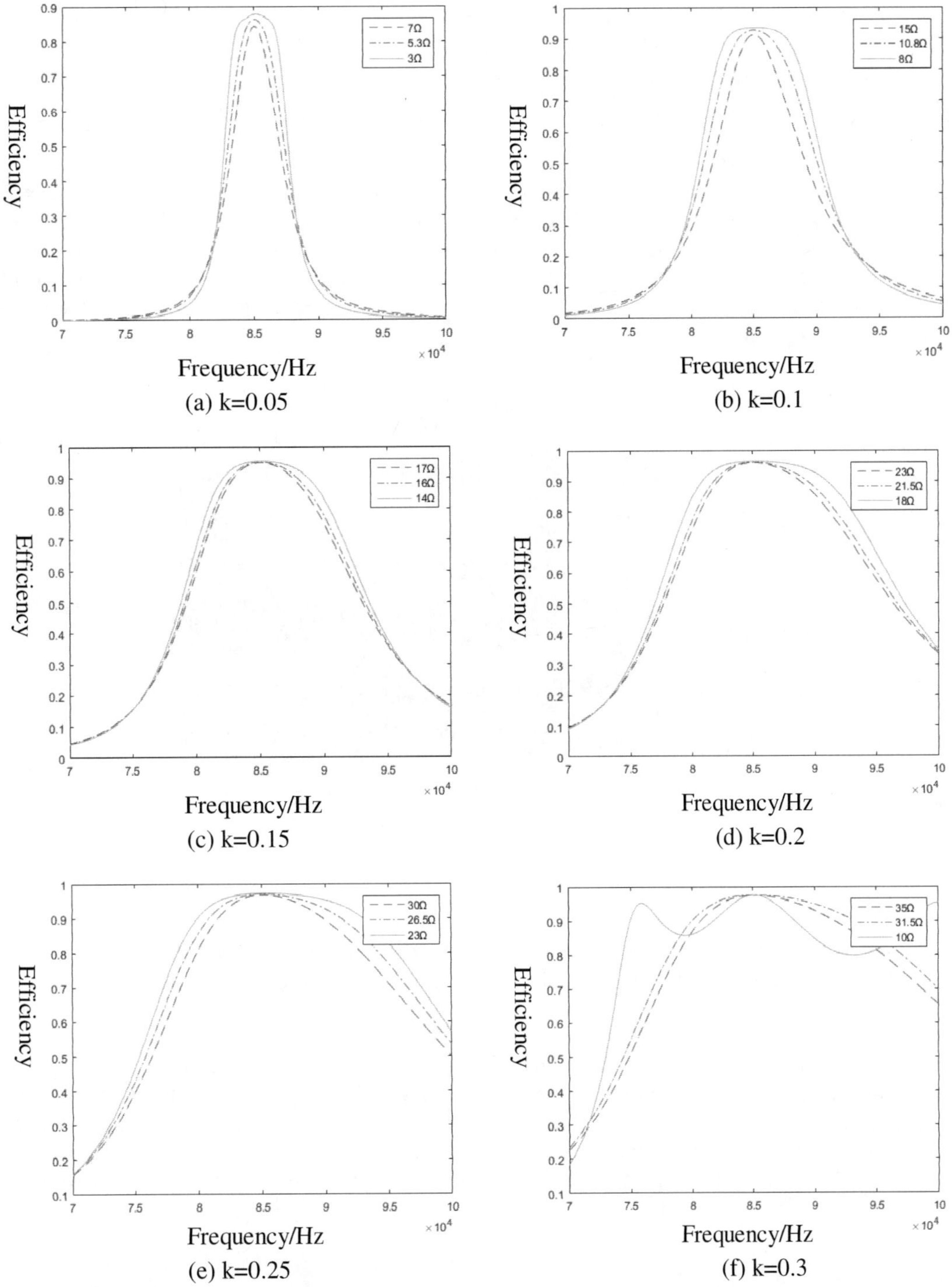

Figure 4. The relationship between transmission efficiency and system frequency.

Table 1. Comparison of theoretical value and simulation value of critical resistance R_{Lmin}.

k	Theoretical Value /Ω	Simulation Value /Ω	Error/%
0.05	5.27	5.3	0.57

0.1	10.57	10.8	2.10
0.15	15.85	16	0.95
0.2	21.14	21.5	1.70
0.25	26.42	26.5	0.30
0.3	31.71	31.5	0.66

4. Experimental verification

To verify the correctness of the theoretical analysis of received power and efficiency frequency characteristics, this paper builds a wireless charging system based on magnetic coupling resonance. The resonant frequency of the system is 85 kHz. The equivalent inductance of the transmitting and receiving coils are $L_1 = 233.6\,\mu H$, $L_2 = 140\,\mu H$. Capacitance are $C_1 = 15\,pF$, $C_2 = 25\,pF$. Coil equivalent resistance are $R_1 = 0.5\,\Omega$, $R_2 = 0.2\,\Omega$. The equivalent supply voltage is $U_s = 200\,V$. The experimental platform is shown in figure 5.

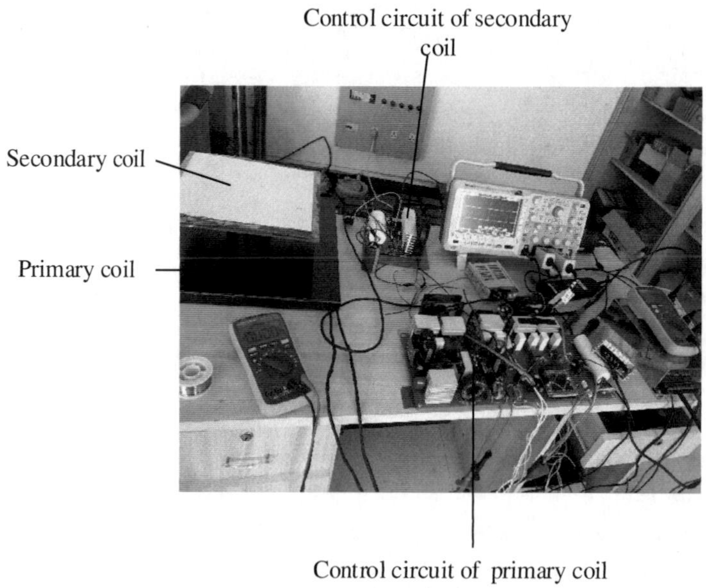

Figure 5. Experimental platform of wireless energy transmission system based on magnetic coupling resonance.

Both the transmitting coil and the receiving coil adopt a circular coil structure. During the experiment, the diameter of the coils is kept constant, and the coil of the transmitting system is placed in parallel with the coil of the receiving system. The distance between the transmitting coil and the receiving coil is maintained at 9 cm. The magnetic coupling coefficient of the system is k = 0.15. Figure 6 shows the frequency characteristics of the system received power and transmission efficiency when the load is 5 Ω and 10 Ω, respectively. The experimental results are in good agreement with the theoretical analysis, which verifies the correctness of the theoretical analysis.

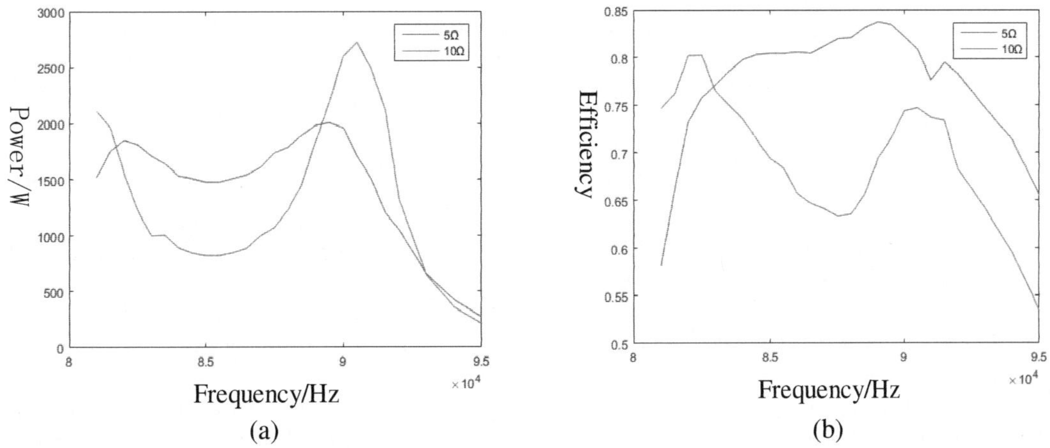

Figure 6. Experimental frequency characteristics curve of received Power and efficiency, (a) Frequency characteristics of power of two kinds of loads and (b) Frequency characteristics of efficiency of two kinds of loads.

5. Discussion on power regulation scheme based on wireless charging technology

From the previous analysis, we know that we can adjust the load to change the frequency characteristics of the system, so that the system avoids running in the frequency split state. However, for the wireless charging system, the load is a DC battery of which equivalent impedance is small and cannot be adjusted. Therefore, by adjusting the load resistance to improve the frequency characteristics of the method is no longer applicable. However, the analysis of frequency characteristics provides theoretical support for power regulation. There are two solutions to the wireless charging system to solve the power regulation.

- The frequency characteristic of the received power has a frequency splitting phenomenon, and the frequency characteristic does not appear frequency splitting. From the analysis above, the system at the resonant frequency can still achieve the maximum efficiency. And the efficiency can be maintained at a higher value in a wide frequency range near the resonant frequency. Therefore, we can adjust the system frequency near the resonant frequency, find the power to meet the system requirements. In this way, the system's demand for power and efficiency can be met.
- When the equivalent resistance of the battery is too small, the frequency characteristics of the system will also appear frequency splitting phenomenon. In this case, it is difficult to meet the requirements by adjusting the frequency of the system to adjust the power. So, we can set the frequency of the system at the resonant frequency, so that the system achieves maximum energy transfer efficiency. The output power is then adjusted by adjusting the input voltage of the system, thus enabling the system to meet the power and efficiency requirements.

6. Conclusion

In this paper, the theory and circuit model of wireless energy transmission system based on magnetic coupling resonant are established. The circuit model is analysed, and the frequency characteristics of the magnetic coupling resonant wireless transmission system are analysed. The equation of the minimum value of the load required to suppress the frequency splitting is deduced. The results show that for different coupling coefficients, increasing the load can suppress the occurrence of frequency splitting. The theory is simulated and experimentally verified. Simulation and experimental results verify the correctness of theoretical analysis. The power regulation based on resonant wireless charging system is discussed, and the power regulation scheme is proposed.

References

[1] Zeng H, Liu Z Z, Chen H X, et al 2016 Optimization of unipolar magnetic couplers for EV wireless power chargers *Lop Conference* **40** 1012010
[2] Kurs A, Karalis A, Moffatt R, et al 2007 Wireless power transfer via strongly coupled magnetic resonances *Science* **317** 583483
[3] Kim J W, Son H C, Kim K H *et al* 2011 Efficiency analysis of magnetic resonance wireless power transfer with intermediate resonant coil *IEEE Antenn. & Wirel. Propag. Lett.* **10** 3389-92
[4] Lee W S, Son W I, Oh K S, *et al* 2012 Contactless energy transfer systems using antiparallel resonant loops *IEEE T. on Ind. Electr.* **60** 1350-9
[5] Imura T and Hori Y 2011 Maximizing air gap and efficiency of magnetic resonant coupling for wireless power transfer using equivalent circuit and neumann formula *IEEE T. on Ind. Electr.* **58** 104746-52
[6] Tak Y, Park J and Nam S 2009 Mode-based analysis of resonant characteristics for near-field coupled small antennas *IEEE Antenn. & Wirel. Propag. Lett.* **8** 41238-41
[7] Kim N Y, Kim K Y and Kim C W 2012 Automated frequency tracking system for efficient mid-range magnetic resonance wireless power transfer *Microw. & Opt. Technol. Lett.* **54** 61423-6
[8] Park J, Tak Y, Kim Y, *et al* 2011 Investigation of adaptive matching methods for near-field wireless power transfer *IEEE T. on Antenn. & Propag.* **59** 51769-73
[9] Sample A P, Meyer D T and Smith J R 2011 Analysis, experimental results, and range adaptation of magnetically coupled resonators for wireless power transfer *IEEE T. on Ind. Electr.* **58** 2544-54
[10] Yang L, Zhang Y, Yang Q, *et al* 2016 Analysis and experimental validation on maximum power and efficiency in wireless power transfer system via coupled magnetic resonances *T. of China Electr. Soc.* **31** 218-24
[11] Zeng H, Liu Z Z, Hou Y J, *et al* 2017 Optimization of magnetic core structure for wireless charging coupler *IEEE T. on Magn.* **53** 68000804
[12] Xue H *et al* 2015 Research of load characteristics of wireless power transmission system based on magnetic coupling resonance *Ind. and Mine Automat.* **41** 366-9

Study on efficiency of different topologies of magnetic coupled resonant wireless charging system

S Cui[1,3], Z Z Liu[1], Y J Hou[2], H Zeng[1], Z K Yue[1] and L H Liang[1]

[1]School of Electrical Engineering, Shandong University, Jinan 250061, Shandong Province, China
[2]Energy Research Institute, Shandong Academy of Sciences, Jinan 250061, Shandong Province, China

E-mail: cuishuo218@126.com/liuzhizhen@sdu.edu.cn

Abstract. This paper analyses the relationship between the output power, the transmission efficiency and the frequency, load and coupling coefficient of the four kinds of magnetic coupled resonant wireless charging system topologies. Based on mutual inductance principle, four kinds of circuit models are established, and the expressions of output power and transmission efficiency of different structures are calculated. The difference between the two power characteristics and efficiency characteristics is compared by simulating the SS (series-series) and SP (series-parallel) type wireless charging systems. With the same parameters of circuit components, the SS structure is usually suitable for small load resistance. The SP structure can be applied to large load resistors, when the transmission efficiency of the system is required to keep high. If the operating frequency deviates from the system resonance frequency, the SS type system has higher transmission efficiency than the SP type system.

1. Introduction

As early as 1914, Tesla put forward the idea of using wireless mode for power transmission [1]. In 2007, the MIT research team, using two identical copper coils, successfully lightened a 60W bulb two meters outside, which provided ideas and new methods to study the wireless power transmission technology [2,3].

Magnetic coupled resonant wireless power transmission technology uses the principle of resonance. By setting the parameters of the device reasonably, the coil has the same resonant frequency as the whole system. Under the power supply of the resonant frequency, the system can achieve an "electric resonance" state. In this condition, energy can be transmitted efficiently between the transmitter and the receiver [4]. The magnetic coupled resonant wireless charging system consists of three parts: power supply, energy transmission device (coil resonator) and energy receiver. At present, the research about wireless charging technology focuses on topology and transmission coil innovation [5].

There are two basic structures of magnetic coupled resonant wireless charging system: two coil structure and multi coil structure [6]. The two coil structures of magnetic coupled resonant wireless charging system is divided into four different topologies [7]. On the basis of different topologies, different capacitors or inductors can be incorporated into the system circuit loop to form a new topology [8,9]. The four common topology structures are: transmitting coil series resonant, receiving coil series resonant (SS); transmitting coil series resonant, receiving coil parallel resonant (SP); transmitting coil parallel resonant, receiving coil series resonant (PS); transmitting coil parallel resonant, receiving coil parallel resonance (PP).

2. Analysis of system theory

2.1. The basic principle of a wireless charging system

The basic principle of a wireless charging system: The control part of the transmitting end completes the rectification and inversion of the electric energy supplied by the power section and converts it into a high-frequency alternating current. Then, the converted high-frequency alternating current is transmitted to the controller by electromagnetic conversion between the transmitting coil and the receiving coil. Finally, through the adjustment and rectification of the receiving end controller, the high-frequency alternating current is converted into direct current.

2.2. Topology model of wireless charging system

The analytical model of magnetic coupled resonant wireless power transmission can be established by equivalent circuit theory [10,11]. The circuit theory model of a wireless charging system can be divided into four kinds, as shown in figure 1.

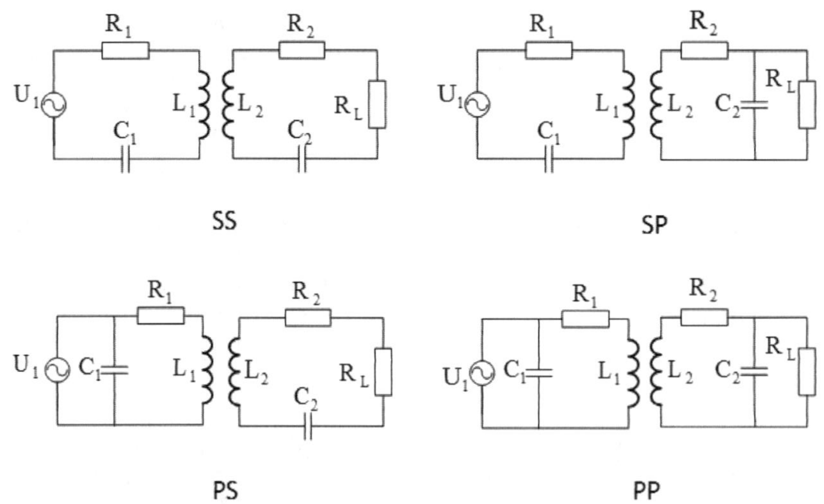

Figure 1. Four kinds of topology model of wireless charging system.

As shown in figure 1 above, U_1 is the voltage source, C_1 and C_2 are the compensation capacitors of the transmitter and receiver of the system. L_1 and L_2 are transmitter coils and receiver coils. R_1 and R_2 are the equivalent internal resistances of the transmitting coils and receiving coils, while R_L is the equivalent load at the receiving end. Because the internal resistance of the power supply is much smaller than the coil's internal resistance and the load resistance, it is negligible.

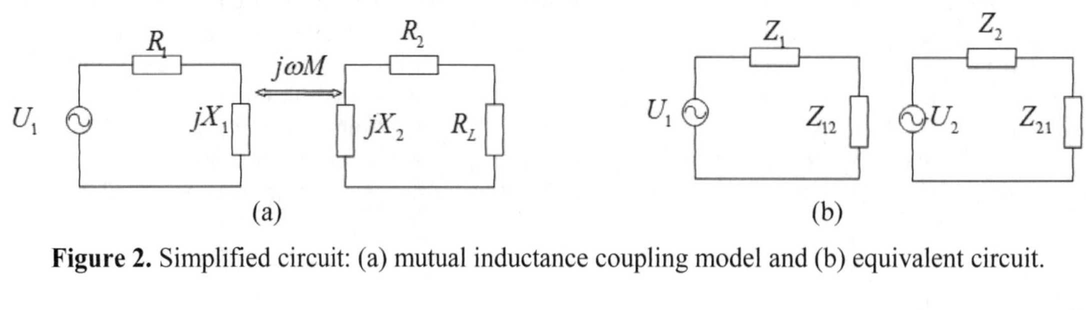

Figure 2. Simplified circuit: (a) mutual inductance coupling model and (b) equivalent circuit.

These four kinds of topology structure circuits can be simplified as the mutual inductance coupling model circuit and equivalent circuit according to the mutual inductance principle, as shown in figure 2.

As shown above, figure 2(a) is a simplified mutual inductance coupling model. jX_1 and jX_2 are the equivalent reactance of the loop at the transmitter and receiver respectively, and $j\omega M$ is the mutual inductance of both ends of the coil (ω is frequency and M is mutual inductance). Figure 2(b) is a simplified equivalent circuit. Among them, Z_1 and Z_2 are self-impedance of the transmitter and receiver coil circuit, Z_{12} is impedance reflected from receiver circuit to transmitter circuit, Z_{21} is impedance reflected from the transmitter circuit to receiver circuit. While U_2 is the equivalent voltage source reflected from transmitter circuit to receiver circuit.

According to the circuit theory, it can be concluded that:

$$\begin{cases} Z_1 = jX_1 + R_1 \\ Z_2 = jX_2 + R_2 + R_L \end{cases} \tag{1}$$

$$\begin{cases} Z_{12} = \omega^2 M^2 / Z_2 \\ Z_{21} = \omega^2 M^2 / Z_1 \end{cases} \tag{2}$$

$$U_2 = j\omega M U_1 / Z_1 \tag{3}$$

For type SS, the jX_1 and jX_2 can be computed as follows:

$$\begin{cases} jX_1 = j\omega L_1 + 1/j\omega C_1 \\ jX_2 = j\omega L_2 + 1/j\omega C_2 \end{cases} \tag{4}$$

Due to the characteristics of series resonance, $\omega^2 CL=1$, in the SS type structure, the resistance values of each impedance element can be computed by expression (5).

$$\begin{cases} Z_1 = R_1 \qquad Z_{12} = \omega^2 M^2 / R_2 + R_L \\ Z_2 = R_2 + R_L \; Z_{21} = \omega^2 M^2 / R_1 \end{cases} \tag{5}$$

The I_1 and I_2 are currents flowing through the transmitting coil loop and the receiving coil circuit, respectively. The expression is as follows:

$$\begin{cases} I_1 = \dfrac{U_1}{Z_1 + Z_{12}} \\ I_2 = \dfrac{U_2}{Z_2 + Z_{21}} \end{cases} \tag{6}$$

The transmission power and efficiency analysis of SS structure can be obtained:

$$\begin{cases} P_{in} = U_1 I_1 = \dfrac{U_1^2 (R_2 + R_L)}{R_1 (R_2 + R_R) + \omega^2 M^2} \\ P_{out} = I_2^2 R_L = \dfrac{\omega^2 M^2 U_1^2 R_L}{\left[R_1 (R_2 + R_L) + \omega^2 M^2\right]^2} \end{cases} \tag{7}$$

$$\eta = \frac{P_{out}}{P_{in}} = \frac{\omega^2 M^2 R_L}{(R_2 + R_L)\left[R_1(R_2 + R_L) + \omega^2 M^2\right]} \tag{8}$$

Similarly, the other three topologies can be computed using the above methods. There is something different: For the wireless charging system with type SS and PS, the current flowing through the load is consistent with the current flowing through the coil. But for the wireless charging system which is parallel to the receiving end, like type SP and PP, the current flowing through the load is different from the current flowing through the coil.

The following expression (9) uses $\mathbf{I}_L$ to represent the current flowing through the load, in the SP and PP type wireless charging systems. The expression of output power is expressed as equation (10).

$$I_L = \frac{I_2}{1/j\omega C_2 + R_L} \cdot \frac{1}{j\omega C_2} = \frac{I_2}{1 + j\omega C_2 R_L} \tag{9}$$

$$P_{out} = I_L^{\,2} R_L \tag{10}$$

The output power, input power and efficiency of the remaining three topologies can be calculated as follows:

$$\begin{cases} P_{in\text{-}SP} = \dfrac{U_1^2\left(R_2 + j\omega L_2 + R_L // 1/j\omega C_2\right)}{\left(R_1 + j\omega L_1 + 1/j\omega C_1\right)\left(R_2 + j\omega L_2 + R_L // 1/j\omega C_2\right) + \omega^2 M^2} \\ P_{out\text{-}SP} = \dfrac{\omega^2 M^2 U_1^2 R_L}{\left[\left(R_1 + j\omega L_1 + 1/j\omega C_1\right)\left(R_2 + j\omega L_2 + R_L // 1/j\omega C_2\right) + \omega^2 M^2\right]^2} \cdot \dfrac{1}{(1 + j\omega C_2 R_L)} \end{cases} \tag{11}$$

$$\eta_{SP} = \frac{\omega^2 M^2 R_L}{\left[\left(R_1 + j\omega L_1 + 1/j\omega C_1\right)\left(R_2 + j\omega L_2 + R_L // 1/j\omega C_2\right) + \omega^2 M^2\right]\left(R_2 + j\omega L_2 + R_L // 1/j\omega C_2\right)} \cdot \frac{1}{(1 + j\omega C_2 R_L)}$$

$$\begin{cases} P_{in\text{-}PS} = \dfrac{U_1^2 (R_2 + R_L)}{\left[(R_1 + j\omega L_1) // 1/j\omega C_1\right](R_2 + R_L) + \omega^2 M^2} \\ P_{out\text{-}PS} = \dfrac{\omega^2 M^2 U_1^2 R_L}{\left\{\left[(R_1 + j\omega L_1) // 1/j\omega C_1\right](R_2 + R_L) + \omega^2 M^2\right\}^2} \end{cases} \tag{12}$$

$$\eta_{PS} = \frac{\omega^2 M^2 R_L}{\left\{\left[(R_1 + j\omega L_1) // 1/j\omega C_1\right](R_2 + R_L) + \omega^2 M^2\right\}(R_2 + R_L)}$$

$$\begin{cases} P_{in\text{-}PP} = \dfrac{U_1^2\left(R_2 + j\omega L_2 + R_L // \frac{1}{j\omega C_2}\right)}{\left[(R_1 + j\omega L_1) // \frac{1}{j\omega C_1}\right]\left(R_2 + j\omega L_2 + R_L // \frac{1}{j\omega C_2}\right) + \omega^2 M^2} \\ P_{out\text{-}PP} = \dfrac{\omega^2 M^2 U_1^2 R_L}{\left\{\left[(R_1 + j\omega L_1) // \frac{1}{j\omega C_1}\right]\left(R_2 + j\omega L_2 + R_L // \frac{1}{j\omega C_2}\right) + \omega^2 M^2\right\}^2} \cdot \dfrac{1}{(1 + j\omega C_2 R_L)} \end{cases} \quad (13)$$

$$\eta_{PP} = \frac{\omega^2 M^2 R_L}{\left\{\left[(R_1 + j\omega L_1) // \frac{1}{j\omega C_1}\right]\left(R_2 + j\omega L_2 + R_L // \frac{1}{j\omega C_2}\right) + \omega^2 M^2\right\}\left(R_2 + j\omega L_2 + R_L // \frac{1}{j\omega C_2}\right)} \cdot \frac{1}{(1 + j\omega C_2 R_L)}$$

From the above expression, output power and transmission efficiency are variable functions about the frequency (*w*), mutual inductance (*M*), capacitance (*C*), inductance of coil (*L*), load resistance (R_1 and R_2) and loss resistance (R_L). Changing the parameters will inevitably affect the output power and efficiency. However, for a certain system, if the parameters of the circuit components have been determined, the output power, input power and efficiency of the system will be changed according to the frequency of power supply voltage (*w*) and the mutual inductance (*M*).

2.3. Comparison of four structures

Comparing the power and efficiency formulas of the four different structures, points can be summarized: The maximum efficiency expression for type SS and PS is the same in form. While in the system, with the load and the capacitor connected in parallel, like type SP and PP, the current flowing through the load is different from the current flowing through the coil. The relationship between the current flowing through the coil and the current flowing through the load must be considered when calculating output power and efficiency.

In the maximum efficiency expressions of type SS and PS, the difference is nothing but the parameter of Z_1. This feature is the same as the difference between type SP and PP.

It can be seen from the formula that the series structure is much simpler than the parallel structure. What's more, the type PP and PS systems require the use of current sources that are infrequently used. In conclusion, the series structure of the system has more advantages than the parallel structure in application.

3. Simulation

The following six graphs show the frequency characteristics of the SS and SP wireless charging systems after simulation by using MATLAB.

3.1. Simulation parameter setting

The internal resistances of the transmitter coil and the receiving coil are set to 0.5 ohms and 0.2 ohms respectively. The inductance of the coil at the transmitter is set to 248 μH , and the inductance of the coil at the receiving end is set to 149 μH . The resonance frequency of the system is set to 85 kHz, and the coupling coefficient is set to 0.182. The 60-ohm-resistance is selected as the load resistance. The voltage source is selected as a three-phase alternating current of 220 V.

3.2. Simulation of SS type topology

Firstly, the influence of the load, the coupling coefficient and the operating frequency on the resonant network of the wireless charging system with SS topology is analysed.

The following figure 3 shows the influence of the coupling coefficient on the transmission efficiency.

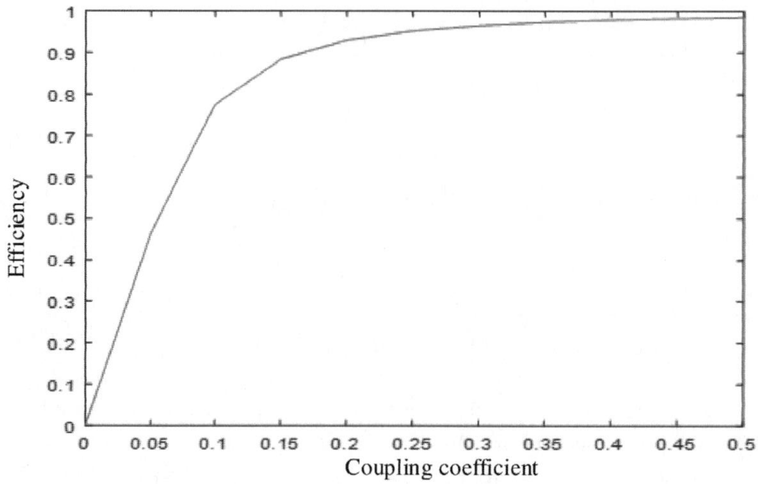

Figure 3. The relationship between the efficiency and coupling coefficient (SS).

As shown in the figure above, the transmission efficiency rises with the increase of the coupling coefficient.

The influence of load resistance on the transmission efficiency of the system is shown in figure 4.

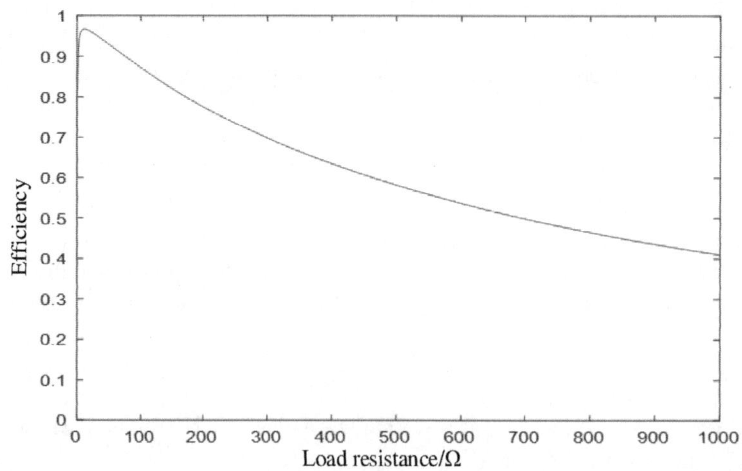

Figure 4. The influence of load on the transmission efficiency of the SS system.

The curve in the graph shows that the system transmission efficiency decreases as the load increases when the load resistance exceeds a certain value.

The simulation diagram of the relationship between the system transmission efficiency and the operating frequency is as follows. Turn the system operating frequency to a variable parameter that changes the range between 80 kHz-95 kHz as shown in figure 5.

The maximum efficiency point of the system deviates from the resonant frequency. But there is still a high efficiency at the resonant frequency.

3.3. Simulation of SP type topology

The following picture shows the frequency characteristics of the SP wireless charging system. Figure

6 shows the influence of the coupling coefficient on the transmission efficiency.

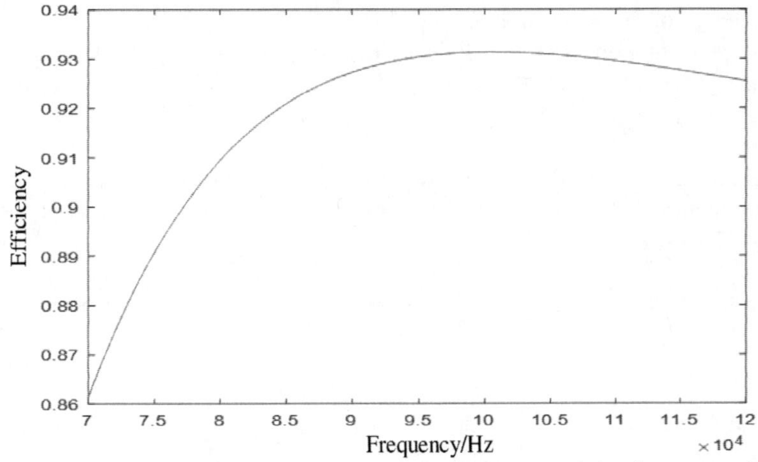

Figure 5. The relationship between the efficiency and the frequency (SS).

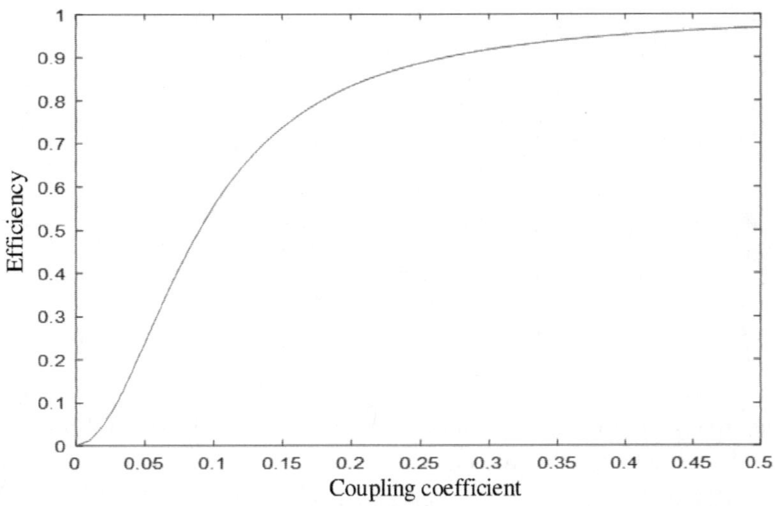

Figure 6. The relationship between efficiency and coupling coefficient (SP).

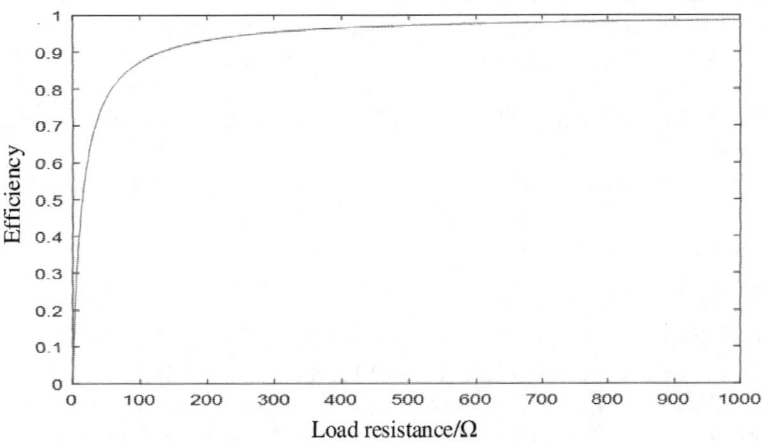

Figure 7. The influence of load on the transmission efficiency of the SP system.

From the above figure, we can see that the efficiency of coupling coefficients is very similar between the SP resonant network and the SS resonant network, and the transmission efficiency increases as the coupling coefficient increases.

As shown in Figure 7, the transmission efficiency of the system keeps increasing with the increase of the load resistance. When the resistance is over 100 ohms, the efficiency is slowly increased and is stabilized at around ninety percent. At this point, there are great differences between the SP type system and the SS system.

The last graph is concerned with the relationship between the efficiency of the SP structure system and the operating frequency. The other parameters remain the same, allowing the operating frequency to vary between 80 kHz-95 kHz.

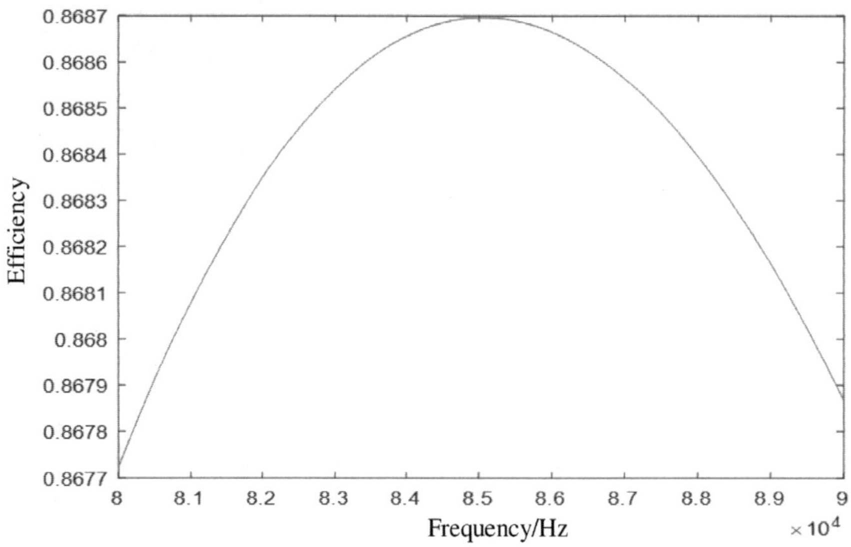

Figure 8. The relationship between the efficiency and the frequency (SP).

Figure 8 is a frequency efficiency characteristic diagram of an SP system. As shown in figure 8, the maximum efficiency point is close to the resonant frequency 85 kHz. But with the decrease of the operating frequency, the efficiency will drop rapidly.

4. Conclusion

For the four topologies of a wireless charging systems, series of systems are much simpler than parallel systems. The PP and PS type systems require the use of current sources which are infrequently used. Therefore, type SS and SP are more suitable for practical applications than the other two structures.

By comparing the previous SS and SP topologies, the simulation of a wireless charging system can be obtained: There is no great difference between the SS system and the SP type system. Compared with type SS, the SP topology is better suited for systems with a relatively large load resistance because it can maintain high efficiency under heavy load conditions. However, the SS type system is able to withstand large frequency fluctuations while maintaining high efficiency.

References

[1] Huang X L, Tan L L, *et al* 2013 Overview of research and application of wireless charging technology *Transactions of China Electrotechnical Society* **28** 101-11
[2] Zhao Z M, Zhang Y M, *et al* 2013 Recent advances in magnetically coupled resonant charging technology *Proceedings of The Chinese Society for Electrical Engineering* **33** 31-13
[3] Jang Y and Jovanović M M 2003 A contactless electrical energy transmission system

forportable-telephone battery chargers *IEEE Transactions on Industrial Electronics* **50** 3520-7

[4] Qu X D 2016 Reaearch on wireless charing system of electric vehicle based on magnetic resonance 1-18

[5] Zeng H, Liu Z Z, *et al* 2017 Optimization of magnetic core structure for wireless charging coupler *IEEE Transactions on Magnetics* **53** 61-2

[6] Sun Z F, Ding E J, *et al* 2015 Magnet coupled resonant automotive wireless power transmission system research and design 9-41

[7] Wang G D, Qiao Z P, *et al* 2015 Researchon topology of magnetically-coupled resonant wireless power transmission *Ind and Mine Autom* **41** 558-63

[8] Li J F, Liao C L, *et al* 2015 Decoupling method of maximum efficiency and transferring power for electric vehicle wireless charging system via LCCL circuit *Transactions of China Electrotechnical Society* **30(S)** 1199-203

[9] Liu C, Guo Y, *et al* 2015 Characteristics analysis and experimental verification of the double LCL resonant compensation network for electrical vehicles wireless power transfer *Transactions of China Electrotechnical Society* **30** 15127-35

[10] Liu Z F, Liu R, *et al* 2015 Study on wireless electric energy transmission with series type of magnetic coupling resonance mode *Modern Electron Technique* **38** 17127-32

[11] Huang X L, Ji Q J, *et al* 2013 Study on series-parallel model of wireless power transfer via magnetic resonance coupling *Transactions of China Electrotechnical Society* **28** 3171-6

Research on application model of blockchain technology in distributed electricity market

S Cheng[1,2], B Zeng[1], Y Z Huang[1]

[1] North China Electric Power University, Changping District, Beijing, China

E-mail:15810868303@163.com

Abstract. In the context of current energy Internet, the emergence of a large number of energy productive consumers will create a new business model. In the decentralized electricity market, the cost of traditional centralized solution construction, management and maintenance is too high, and it is difficult to support the collection, transmission, reception, storage and analysis of massive data. To provide a solution to this phenomenon, we apply the blockchain technology to this distributed electricity market to achieve peer to peer transactions in the power systems. The blockchain technology which is very popular nowadays will be used in power system to establish a credible direct transaction between devices. At first, this article analyzes the future direction of the development of power systems, studies the characteristics of decentralized power systems and summarizes the main issues in the development process. Then, we analyze the basic characteristics of blockchain and put forward a new transaction framework in consideration of problems existing in current energy market. The transaction framework is based on the blockchain technology in the distributed electricity market and includes the pricing method, the power transaction system architecture, various modules of the trading system and the details of the whole transaction system runtime. This framework provides a viable solution for increasingly complex energy transactions.

1. Introduction

1.1. Background

At present, the rapid development of large data and Internet technology is driving the whole social form and infrastructure changes. This sort of change not only contains different types of economic systems, but also includes the corresponding management model. In this flattened social system, the boundaries between consumers, businesses, governments and regulators are becoming increasingly blurred [1]. All nodes can understand the actual operation of the system through a flat system network structure and can also affect the operation of the system. What's more, with the development of energy Internet, the advancement of electricity market reform and the popularization of distributed power generation, energy storage, electric vehicle and demand side response, a large number of consumers will evolve into prosumers. The emergence of the prosumers also makes the current power system in the supply and demand boundaries gradually blurred. The current power system operation and management models are gradually unable to meet the various types of energy unit "plug and play" access mode [2].

1.2. Decentralized power system resources

Decentralized power generation (DG) is a form of power generation that is directly connected to the distribution network or users' meter installation. It includes photovoltaic cell power generation, wind power generation and fuel cell power generation. According to the different power generation technology, decentralized power generation can be divided into renewable energy decentralized power generation, modular decentralized power generation and combined heat and power generation. With the development of new battery energy storage, power generation cost with decentralized power generation device is getting lower and lower, and some have even been close to that with the traditional centralized power plant, therefore the application scope of decentralized power generation will be more and more widely. In addition, the emergence of electricity retail market will also be a strong promotion of decentralized energy production development. Decentralized energy production with its clean environmental protection and local power generation service users and many other advantages, will have an increasing market share. With the development of distributed renewable energy and distributed energy storage permeability, the boundaries of supply and demand in traditional energy and power systems are becoming increasingly blurred, and scattered energy production leads to the emergence of a large number of productive consumers.

Figure 1 describes the energy systems that include a variety of new energy sources. In the future of the electric energy system, there will be a great possibility of the following situation: in the sunny weather it can produce a lot of electricity, while in the cold windless weather there will be the situation lack of electricity. In some areas, decentralized power systems may also imply constraints on the local grid [3].

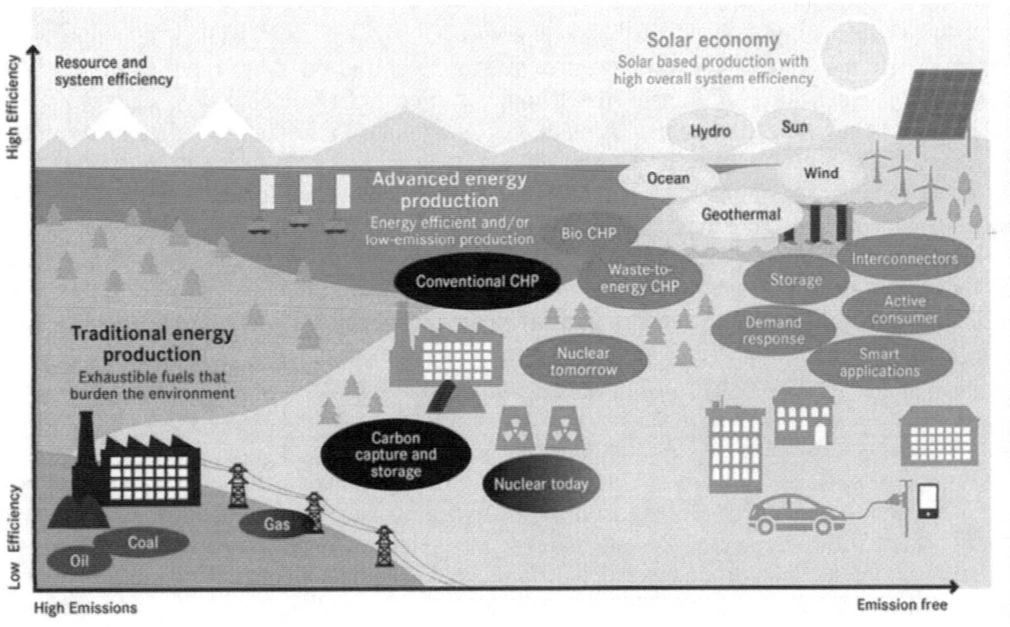

Figure 1. Future energy system framework.

With the infiltration of decentralized power systems and the increasing power data, how to build a new, reliable and effective energy trading model is very urgent. The key to solving such problems is to ensure that the balance of electricity production and consumption, which is essential to increase the flexibility of electricity production and consumption [4]. If a power consumer can autonomously allocate electricity, making the allocation effective, self-sustaining and free from external control, then the entire power system will become much more flexible. There are already similar application cases, such as the use of blockchain technology between adjacent households for power exchange.

1.3. Contributions

To summarize, in order to cope with the problems raised above, this paper constructs a decentralized trading platform based on the blockchain technology, which will realize the decentralized pricing and P2P (Peer-to-Peer) transactions in the distributed power market, and realize an effective trading model to deal with the increasing volume of power transaction information and the increasingly blurred development trend of supply and demand.

In our view, the contributions of this paper include:

- We analyze the future direction of the development of power systems, study the characteristics of decentralized power systems and summarize the main issues in the development process.
- Based on the analysis of possible problems in the decentralized power energy system, we propose the application of the blockchain technology to the future transaction mode in electric power transaction. In addition, by comparing the traditional power transaction model and the blockchain based transaction model, we sum up the advantages of the latter to adapt to modern power development.
- We propose a transaction framework in decentralized electricity market based on the blockchain technology. First, we set up a power trading model, and clearly demonstrated its trading process. Then, we develope a pricing mechanism under the framework of this transaction to promote the balance between supply and demand among users.

2. Blockchain technology

2.1. The overview of blockchain technology

As a distributed accounting system, the blockchain has the characteristics of high reliability and de-centralization, so that it can be well adapted to the decentralized system structure [5]. Shared economic and ecological environment based on the block chain technology can promote the development of decentralized systems. Through data encryption, timestamp, intelligent contract and other technical means, we can achieve the peer-to-peer transactions (P2P) in the absence of third-party trusted endorsement. The introduction of blockchain technology can reduce the barriers to entry trading systems, while ensuring the transparency and quick access to asset transactions.

Energy Internet technology is the focus of current academic attention. The third industrial revolution, represented by energy Internet technology, is booming with its core research areas expanding from the smart grid, the energy Internet to the global energy Internet [6,7]. Energy Internet is based on the existing power grid. Through advanced power electronics and information technology, a large number of new power network nodes which consist of distributed energy collection devices, distributed energy storage devices and a variety of load can be connected to achieve energy flow, two-way flow of energy information on the shared network.

This paper will start from the development trend of power transaction in the background of energy Internet, combine the existing problems and the problems that may arise in future development of electricity market with the relevant characteristics of blockchain, and propose a kind of point-to-point power trading framework based on the blockchain technology in the electricity market, including the power transaction system architecture, the various modules of the trading system, and the details of the entire transaction system runtime.

The concept of Bitcoin was first proposed in 2008, when it was used as a way of securely conducting electronic financial transactions without the need for centralized authorization. In the Bitcoin trade network, there is only one complete chain of transactions, the main chain, which is made up of a large number of blocks. Since each transaction is relatively fragmented, in order for better statistical transactions, the blocks are introduced. The bitcoin system creates a block every ten minutes or so, which contains all the transaction information for that period. Each block also contains the ID of the previous block, so that each block can find its previous node, which can be pushed from the latest block to the starting node, thus forming a complete transaction chain.

The blockchain is an emerging technology based on decentralized computing and data storage, protected by a combination of cryptographic signatures and distributed sharing mechanisms. Even in the presence of network attacks and communication interruption, participants on the blockchain network can achieve a general agreement on the state of the system [8]. This contrasts sharply with the traditional architecture. In the traditional architecture, the central coordinator defines the state of the system, but may receive the effect of attack or misconduct. The basic architecture of the blockchain is shown in figure 2. Figure 2 shows the block B^0 to B^{t+1} with the detailed information of B^t. Blocks are connected by encrypted hashes, protecting content from being changed and allowing the history to be audited. Participants for P2P networks areM_i^t (i=1,2,…,N). The message that the network broadcasts contains commands (control actions, account extraction, etc.) of the system status, and each node can use the verification function to verify the feasibility of each message.

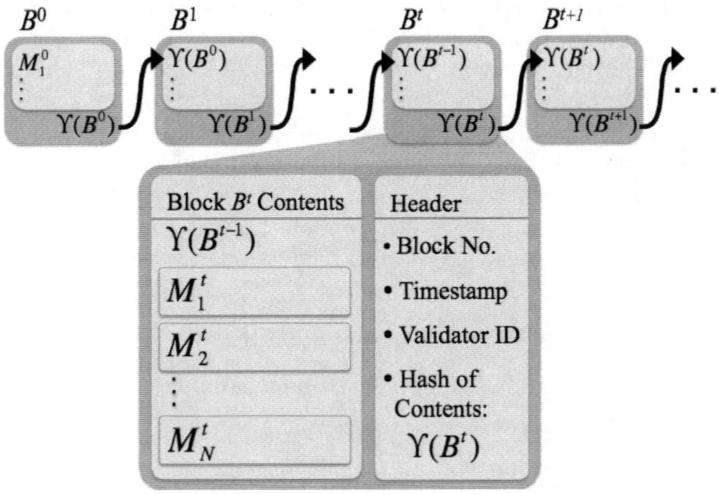

Figure 2. Symbolic representation of data in blockchain.

The blockchain header contains a timestamp, a concise encrypted hash with the contents of the previous block, and the result of a validation test that is difficult to forge under computational and economic constraints. The new blocks are broadcast to the network. Their validity will be checked, and the nodes will be consistent in the system's update status. When the state transition function can execute the computer code embedded in the transmission M^i, the effect of the blockchain can be further extended. These smart contracts can be checked and reviewed very transparently and can be guaranteed to be implemented on the web. The pillars of the block chain structure are the contents of the block loop connection, validating the new block with the P2P consensus and verifying the communication with the encrypted signature. These key processes can provide a representation of the system state without intervention of the trusted central authority. While this architecture introduces some computational expenses, it can guarantee the eternal invariance of transaction history, the transparency and verifiability of the transaction, and the ability to adapt the system to the coordination between parties that do not trust each other.

2.2. The basic characteristics of blockchain

As can be seen in the above content, the blockchain has some very obvious features. Table 1 compared the existing energy transactions and energy transactions based on blockchain.

Table 1. Comparison between existing electricity transactions and transactions based on blockchain.

Trading Platform	Existing electricity trading process	The transaction process based on the blockchain

Data Object	Non-sequential data. In accordance with the transaction, the central database generates specific transaction data.	Sequential data. Combine the data block in the form of chain and chronological order.
Transaction Structure	The trading entity completes the transaction through a trusted central trading body, such as an electricity trading centre.	There is no trusted central organization. And there is no need for mutual trust between the transactional entities. Transactions are traded through smart contracts and the blockchain trading system.
Information Extension	The information extension is not high. The transaction information is stored in the centralized transaction mechanism. Only the central organization can check the information. If the central organization is attacked, it may lose some or all of the transaction information.	The information is highly accessible. The blockchain can record all the transaction records from the start of the coin system. Therefore, each transaction is recorded and maintained by all the nodes. Each transaction object can view all the transaction information. Thus, the transaction process has a good traceability.
Data Security	Security is low. Once the central authority is attacked, the data may be lost or tampered with, which will result in serious consequences.	Security is relatively high. The data is stored in each node of the network. It is maintained by all nodes.

3. Power transaction framework based on blockchain technology

3.1. Transaction process

The transaction system based on the blockchain technology is decentralized, whose transaction information can be tracked to ensure the privacy of the transaction node information. This part will propose a power trading framework based on the blockchain technology in the distributed electricity market.

In the future, without central pricing agencies in the distribution network or regional energy Internet market, the pricing of power resources are real-time. Here is an example if we assume a transaction scenario and analyze the transaction chain based on the blockchain [9]. Assuming that there are power purchase node A and supply node B in the system, the basic flow of the transaction is as follows:

- Node A predicts that its distributed renewable energy will fluctuate in a short time [t0,t1], or suddenly the power load increases, requiring more power supply. It will send his own purchase demand information in the P2P network. The information release requires node A to process its own public key A-*Publickey* and private key A-*Privatekey* with the encryption function to generate the address for transaction authentication and information transmission, as follows:

$$T_{A-add} = Hash(A - Publickey); A - Publickey = Hash(A - Privatekey) \tag{1}$$

Among them, T_{A-add} stands for the address of node A, $Hash$ for the encryption process to generate node address, and T_{A-add} for the result that is calculated after SHA256 and Base58. Then the node A broadcasts the purchase request information to the network:

$$M_{in.A} = (E_1||[t_0, t_1]||A - Publickey||T_{A-add}) \tag{2}$$

Among them, E_1 stands for the purchase demand, $[t_0, t_1]$ for the period to purchase electricity resources. All the participating nodes in the transaction system are able to receive the transaction request information $M_{in.A}$ issued by the node A.

- User B accepts the transaction request issued by A, and on the basis of analysis of its own energy demand and energy supply, it is considered that the corresponding energy resource can be provided for user A in $[t_0, t_1]$. Node B will also generate the corresponding transaction authentication and information transmission address T_{B-add} to send the response message:

$$M_{A-add.B} = (E_2||[t_0, t_1]||P_0||LT_{ID}Sign.B(E_1, [t_0, t_1], LT)||B - Publickey||T_{B-add}) \quad (3)$$

Among them, E_2 is the power supplied by B. P_0 is the quotation of B for the electricity transaction. LT_{ID} is the record address of the last transaction information in the blockchain at Node B. $Sign.B(E_1, [t_0, t_1], LT)$ is the encryption signature B through $B - Publickey$ on the transaction response and the contents of last transaction. Because there are multiple power suppliers and multiple quotation, the information is encrypted with $A - Publickey$ and sent to user A.

- After receiving the information sent by Node B, user A decrypt the message digest sent by B with its own private key $A - Privatekey$ to get the entire content of $M_{A-add.B}$. At the same time, it gets the content information in the signature $Sign.B$ with $B - Publickey$. In addition, the LT_{ID} and the corresponding Merkel tree can quickly locate the last transaction, verify whether there is a public key for B in this transaction, and extract the transaction information LT to compare with the signature content to ensure that the transaction is real, which can also verify that $M_{A-add.B}$ is actually sent by node B.

The subjective behavior of the user will affect the effectiveness of the power resources, but the user can obtain the deviation between the amount of the agreement and the amount of the settlement in all transactions on the basis of the transaction information. And then according to the amount of deviation he can estimate the quality of the supplied power to determine whether he should amend his purchase amount. In addition, if there are multiple power resource supply nodes or individual supply nodes cannot meet the demand, then the EMS of node A needs to optimize his own purchase combination strategy according to the resource quotation and the available quantity of different nodes. In order to simplify the description of the information exchange and transaction process, we assume that the power supply of node B can meet the needs of node A, that is, $E_1 = E_2$.

- If A does not accept the quotation of B, it may be required that B provides a secondary offer, or terminate the transaction. In the process of price agreement, nodes A and B can still use the public key to encrypt the information and then send information to the target address.

If A accepts the quotation of B, A will send the contract script information to B. In the transaction, because after the transaction is completed, he overall output / energy characteristics of the user A and B will change, it is necessary to report the second level of EMS to adjust optimization boundaries of the user A and B. At the same time, in order to ensure that the IDR resources will not be "secondary payments", the script information must contain the signature of the regional energy Internet operator. The regional EMS should ensure that a user can only trade the power resource for one time in a transaction settlement cycle, that is, the output / energy characteristics reported to the second level EMS can only be modified once in one cycle to ensure the effectiveness of the transaction. Therefore, the transaction information structure of the contract transaction should be:

$$Script_1 = (Sign.A||Sign.B||Sign.DSO||(E_2, [t_1, t_2], P_0)) \quad (4)$$

After receiving $Script_1$, the regional EMS will modify the output / energy curve reported by the users A and B. Till the time t_0, the script will trigger the energy resource of node B, and then EMS will read the actual output / energy curve of B to record. To the end of the transaction t_1, A and B need to confirm again:

$$Script_2 = (Sign.A||Sign.B||Sign.DSO||(E_3)) \quad (5)$$

Among them, E_3 is the actual amount of electricity trading. Once $Script_2$ is signed by three parties, then A needs to pay in accordance with the previous agreement to the cost of B. After the completion of the transaction, the transaction information written in formula 1 will be broadcast to the whole network. After all nodes have verified the transaction information, they can be written to the current block.

3.2. Pricing method

Through the entire power trading platform, we use Nbitcoin to carry out the sale of electricity. Nbitcoin uses distributed databases throughout the entire P2P transaction network nodes to manage currency issuance, record currency transactions and account balance information. In addition, it uses cryptographic design verification to repeat consumption to ensure the security of all aspects of currency circulation. In this trading market, unlike the technology that relies on the forecast and the market, the system operator charges the users according to the actual power usage situation and awards the users according to the actual situation of the electric power, so as to realize the whole transaction system based on the disperse digital currency, that is, the Nbitcoin.

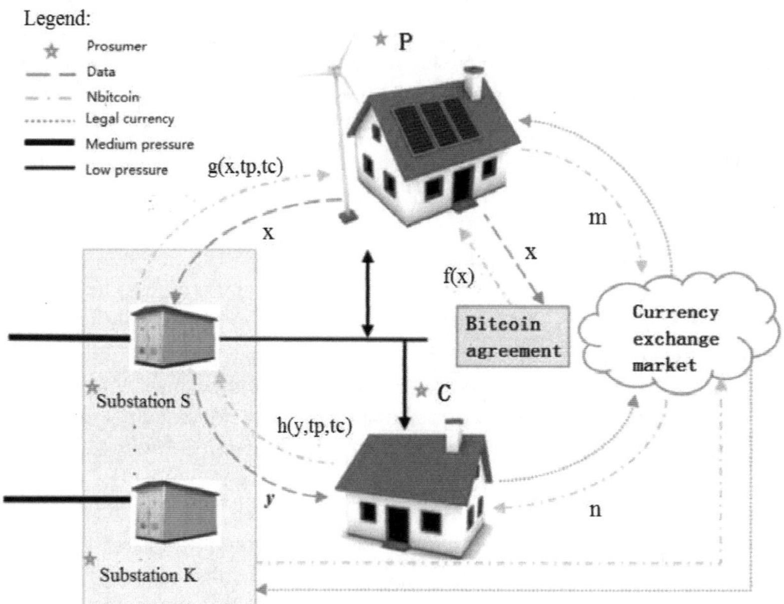

Figure 3. Example scenario.

Below we present a sample scenario, as shown in figure 3, which will help us visualize the whole process [10,11]. A user P generates a certain amount of electrical energy and feeds it to the grid while broadcasting the information to all nodes in the network. These nodes then update the P's disclosure using the monetary function $f(x)$ based on the rules of the distributed electricity market trading system recording. Thus, the function $f(x)$ is responsible for generating and then entering the decentralized market in which the currency algorithm is generated. In addition, the local substation S connected to P measures the total energy in the gap to produce t_p and total consumption t_c. The substation then transfers the $g(x, t_p, t_c)$ currency calculation results publicly to the users. $g(x, t_p, t_c)$ is the production price function defined by the EMS. Therefore, the user P that injects energy x can acquire a certain amount of electronic money. And the amount of money received depends only on the functions f and g, and is not associated with the monetary value on the market. At any given time period, P can join the exchange market to sell its electronic currency to convert the legal currency. Here we will elaborate on this power trading system using Nbitcoin for the realization of electricity transactions

In the process of power transactions, each user can first meet his own needs using the energy that he produces, and then the excess energy can be sent to the grid. The information on local electricity production and consumption is sent to the power substation of the energy Internet operator via a smart meter, which is then used to determine the rate of return on the energy produced by the user, in order to motivate the agent to balance supply and demand by means of reducing production and consumption peak. For example, in low demand or high output period, the cost of energy consumption is very low. Similarly, low supply and high demand will also promote the price rise. Therefore, every 15 minutes, we use the following methods to determine energy consumption and productivity. The price function paid to the electricity producer forms a bell curve, defined as follows:

$$g(x, t_p, t_c) = \frac{x \cdot q_{t_p=t_c}}{e^{\frac{(t_p-t_c)^2}{a}}} \tag{6}$$

Where $q_{t_p=t_c}$ is the maximum reward rate obtained by the power producer for the power of x when the total supply t_p is equal to the total demand t_c and is defined by EMS; a is zoom factor when $t_p \neq t_c$. When the total energy production is equal to the total consumption, the function reaches the peak, $g = x \cdot q_{t_p=t_c}$. In other words, when $t_p \gg t_c$ or $t_p \ll t_c$, the producer is almost unable to get any reward.

In the actual operation, according to the actual situation, if we assume that the maximum incentive rate of 5%, the scaling factor of 2000, the bell curve of price function with the total supply and aggregate demand will be shown in figure 4. If we change the maximum reward rate to 7%, and zoom factor remains unchanged, the price function will be shown in figure 5. If the maximum incentive rate remains unchanged, the zoom factor becomes 500, the price function drawn graph will be shown in figure 6. As can be seen in the figure, when the total power supply in the electricity market is far greater than the total demand for electricity, the price of the electricity unit paid to the electricity producers is almost zero. As the total demand for electricity in the market increases, the electricity price increases. When the total supply is equal to the aggregate demand, the price reaches the highest value, and if the total demand for electricity increases further, the electricity price will drop, which will cause the electricity producers to produce more electricity to maintain the balance between supply and demand to make their own energy production profit reach the maximum, which is the basic principles of electricity price function. Under this principle, through the role of price mechanism, the balance of power supply and demand is promoted and the stability of the electricity market is maintained. When the maximum incentive rate is changed, the electricity price will produce a positive correlation. When the scaling factor is changed, the steepness of the curve changes, and with the decrease of the scaling factor, the steepness of the price curve becomes higher. That means that the smaller the scaling factor, the greater the impact of electricity prices on the market.

$h(\cdot)$ is the price function that consumers need to pay when they buy electricity:

$$h(y, t_p, t_c) = \frac{y \cdot r_{t_c \gg t_p} \cdot t_c}{t_c + t_p} \tag{7}$$

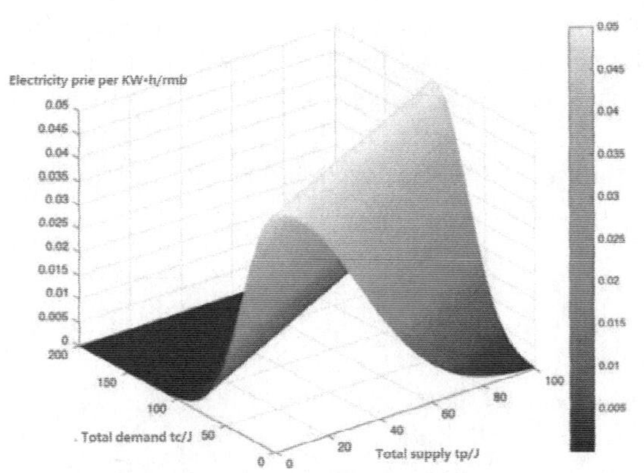

Figure 4. Price function bell curve.

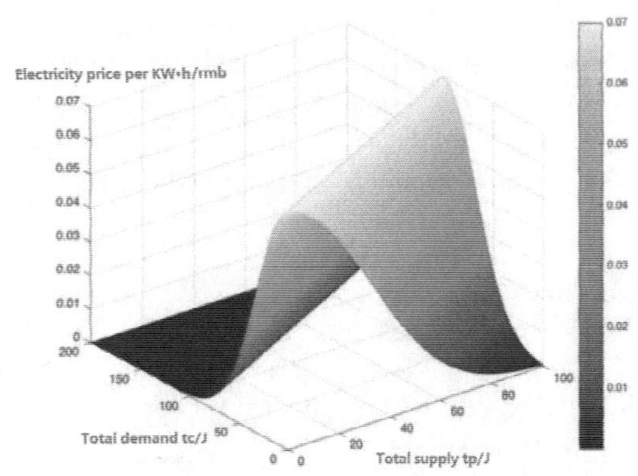

Figure 5. The bell curve after changing the

maximum reward rate.

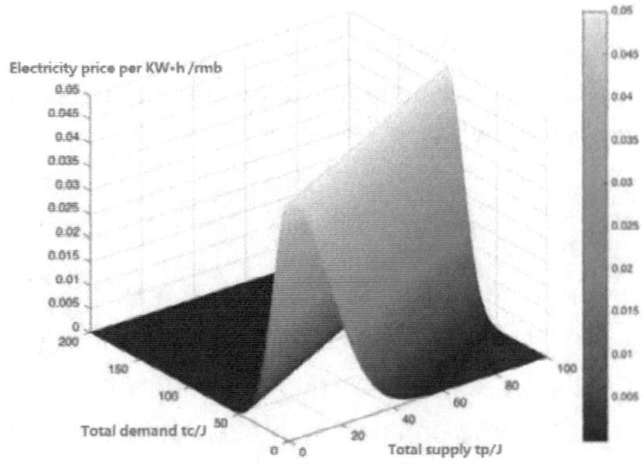

Figure 6. The bell curve after changing the scaling factor.

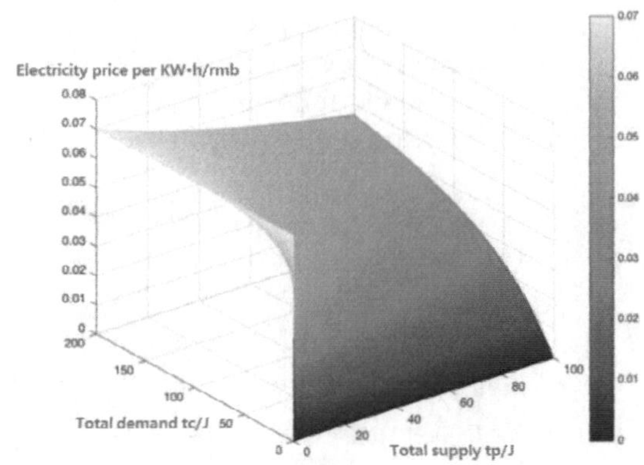

Figure 7. The bell curve of consumers' payment.

Where $r_{t_c \gg t_p}$ is the highest cost that the buyer needs to pay when the energy supply is low. When $t_c = t_p$, the buyer should pay $\frac{r_{t_c \gg t_p}}{2}$ per KW • h. When$t_c \ll t_p$, h approaches 0, thus when the production is too much, the spending of the power consumers is almost 0, which can promote energy consumption to transfer over-production period to balanced period. The price function curve that consumers need to pay when they buy electricity is shown in figure 7.

Similar to the Bitcoin trading system, the Nbitcoin currency in the electricity trading market is based on the blockchain technology that is not controlled by any central authority in its monetary value. Futhermore, the bitcoin in this trading system is generated by injecting electrical energy into the grid. The price function $f(x)$ depends only on the amount x of renewable energy injected into the grid. This number is broadcast to the network, for each node in the network to track each transaction and earnings situation. But the actual transaction user information is not open, that is, all income and transactions are anonymous. Nbitcoin is obtained by the function $f(x)$, defined as follows:

$$f(x) = b \cdot x \tag{8}$$

Where b is a constant indicating the ratio of the bitcoin that can be obtained through injecting a certain amount of electricity x by the electricity producer.

4. Conclusion

In the context of the energy Internet, the emergence of a large number of energy productive consumers will create a new business model, which means that energy P2P transactions will spring up, and decentralized power trading models will also be a big change. Based on the development trend of power transaction in energy Internet, we analyze the basic concepts and characteristics of distributed energy system and blockchain. Then, considering the problems existing in the current electricity market transaction and the problems that may arise in the future, we put forward the framework of the distributed electricity market based on the blockchain technology.

References

[1] Munsing E, Mather J and Moura S 2017 Blockchains for decentralized optimization of energy resources in microgrid networks (Working Papers)
[2] Nakamoto S 2009 Bitcoin: A peer-to-peer electronic cashsystem (Consulted, 2008)
[3] Mattila J, Seppala T, Naucler C, Stahl R, Tikkanen M, Badenlid A and Seppala J 2016 Industrial blockchain platforms: An exercise in use case development in the energy industry *Etla Working Papers* **43** 10-1

IOP Conf. Series: Earth and Environmental Science **93** (2017) 012065 doi:10.1088/1755-1315/93/1/012065

[4] Bahrami S and Sheikhi A 2016 From demand response in smart grid toward Integrated demand response in smart energy hub *IEEE Transactions on Smart Grid* **7** 650-8

[5] Swan M 2015 *Blockchain: Blueprint for a New Economy* 1st ed (O'Reilly Media)

[6] Zhang Y and Wen J 2016 The IoT electric business model: Using blockchain technology for the internet of things *Peer-to-Peer Networking and Applications* **2016** 1-12

[7] Tanaka K and Nagakubo K 2017 Blockchain-based electricity trading with Digitalgrid router *IEEE International Conference on Consumer Electronics* (Taiwan) pp 201-2

[8] Zhang Y and Wen J 2015 An LoT electric business model based on the protocol of bitcoin *International Conference on Intelligence in Next Generation Networks* pp 184-91

[9] Kosba A, Miller A, Shi E, Wen Z and Papamanthou C 2016 Hawk: The blockchain model of cryptography and privacy-preserving smart contracts *Security & Privacy* **2016** 839-58

[10] Mihaylov M, Jurado S, Moffaert K V, Avellana N and Nowe A 2014 NRG-X-Change: A novel mechanism for trading of renewable energy in smart grids *International Conference on Smart Grids & Green It Systems* **2014** 101-10

[11] Kilanc G P and Or I 2017 A decision support tool for the analysis of pricing, investment and regulatory processes in a decentralized electricity market *Energy Policy* 2008 **36** 3036-44

IOP Conf. Series: Earth and Environmental Science **93** (2017) 012066 doi:10.1088/1755-1315/93/1/012066

Multi-time scale energy management of wind farms based on comprehensive evaluation technology

Y P Xu[1,3], Y H Huang[1], Z J Liu[2], Y F Wang[1], Z Y Li[2] and L Guo[2]

[1]State Key Laboratory of Operation and Control of Renewable Energy and Storage Systems, China Electric Power Research Institute, Beijing, China
[2]State Grid Jilin Electric Power Company Limited, Changchun, Jilin, China

E-mail: xuyanping33@163.com

Abstract. A novel energy management of wind farms is proposed in this paper. Firstly, a novel comprehensive evaluation system is proposed to quantify economic properties of each wind farm to make the energy management more economical and reasonable. Then, a combination of multi time-scale schedule method is proposed to develop a novel energy management. The day-ahead schedule optimizes unit commitment of thermal power generators. The intraday schedule is established to optimize power generation plan for all thermal power generating units, hydroelectric generating sets and wind power plants. At last, the power generation plan can be timely revised in the process of on-line schedule. The paper concludes with simulations conducted on a real provincial integrated energy system in northeast China. Simulation results have validated the proposed model and corresponding solving algorithms.

1. Introduction

Power systems are viewed as typical integrated energy systems converting various primary energy to electricity [1,2]. In China, power systems with the characteristic of extremely complex structure are dispatched and controlled at province scale. In recent years, Power systems in many provinces of China tend to become more complicated due to rapid increase in wind power penetration [3-7]. While wind power is of superiority in reducing CO_2 emissions, it is inherently fluctuating and difficult to predict accurately. So large-scale integrations of wind power inevitably pose great challenges to the schedule of power systems, such as increases in operation costs, operation risk level, and difficulties in determining spinning reserve (SR) capacity. In recent years, wind curtailments in 'Three North' areas (Northeast, North China and Northwest) of China have become very serious [8,9].

Efficient energy management of wind farms can effectively reduce negative impacts of wind power integration on power systems. It is emphasized that wind power is a part of power systems, so energy management of wind farms is also a part of energy schedule of the whole power systems. In [10,11], unit commitment (UC) and dispatch with uncertain wind power generation have been studied, however, without considering features of heating units and heating load. As a result, conclusions obtained in these two literatures may mislead power grid planning and operation of real systems in China, due to the fact that heating units exist widely in 'Three North' areas of China and they always operate in winter. Lowery *et al* [12] and Ela *et al* [13] studied impacts of high penetration levels of intermittent generation on power systems at multiple timescales, but did not give a useful energy management strategy for wind farms. In those formulations proposed in [10-13], wind power is considered as a non-dispatched random power because production costs of wind power cannot be evaluated effectively.

Based on the virtual production cost of wind farm, a multi time-scale formulation (day-ahead 72-hour schedule, intraday 4-hour schedule, and on-line 15-min schedule) [14] is proposed to optimize energy management strategy of wind farm in this paper, in which, including different time scale features of wind/water resource, electrical/heat load characteristics, thermo-electric coupling characteristics of different thermal units, and transmission capacity. The day-ahead schedule is adopted to optimize UC decisions of thermal power generators for maximizing ability of accommodating wind power generated from wind farms with higher operational level preferentially. Remaining parts of this paper are organized as follows: Section 2 presents the novel comprehensive evaluation system of each wind farm. Section 3 describes the framework of multi-time scale energy management of wind farms. Numerical results based on a practical provincial integrated energy system in northeast China are provided in Section 4. Section 5 concludes the paper with discussions.

2. A novel comprehensive wind power evaluations system

The electricity production process of wind farms is totally different from that of thermal generators because they do not consume any fossil fuels. That is to say, short-term production costs can be ignored. In order to accommodate wind energy generated from wind farms with higher operational level preferentially, a novel comprehensive wind power evaluation system is proposed to quantify economic properties of each wind farm in both intraday and on-line schedule.

The related evaluation indexes accounting for main factors that influence energy management of integrated energy systems are listed as follows: 1) Forecast Accuracy Rate (FAR), 2) Forecast Qualified Rate (FQR), 3) Forecast Reported Rate (FRR), 4) Low Voltage Ride through Capability(LVRT), 5) Active Power Control Ability (APC), 6) Reactive Power Control Ability (RPC), 7) LPPD (Limited Power of the Previous Day), 8) Wind Power Forecasting System (WPFS), 9) Wind Power Integration Testing (WPIT), 10) Wind Farm Equipment Accident (WFEA).

The formulations for each index are presented in table 1, in which weighting coefficients were chosen based on the operation of each integrated energy system:

Table 1. Formulations for each index.

Serial number	Index name	Specific formulation	weighting coefficient
1	FAR	$r_1 = \left[1 - \sqrt{\frac{1}{T \times \mathrm{Wcap}_{i,w}} \sum_{t=1}^{T} (P_{i,w}^{t^*} - P_{i,w}^{t})^2}\right] \times 100\%$	WC_1
2	FQR	$r_2 = \frac{1}{T}\sum_{t=1}^{T} M^t \times 100\%$ $\begin{cases} M^t = 1 & [1 - \frac{1}{T}(P_{i,w}^{t^*} - P_{i,w}^{t})] \times 100\% \geq 80\% \\ M^t = 0 & [1 - \frac{1}{T}(P_{i,w}^{t^*} - P_{i,w}^{t})] \times 100\% < 80\% \end{cases}$	WC_2
3	FRR	$r_3 = \frac{d_s}{d_c} \times 100\%$	WC_3
4	LVRD	$r_4 = \begin{cases} 1 & \text{with LVRD ability} \\ 0 & \text{without LVRD ability} \end{cases}$	WC_4
5	APC	$r_5 = \begin{cases} 1 & \text{with active power control ability} \\ 0 & \text{without active power control ability} \end{cases}$	WC_5
6	RPC	$r_6 = \begin{cases} 1 & \text{with reactive power control ability} \\ 0 & \text{without reactive power control ability} \end{cases}$	WC_6
7	LPPD	$r_7 = \sum_{t=1}^{T_{preday}} (P_{i,w}^{t^*} - P_{i,w}^{t})$	WC_7
8	WPFS	$r_8 = \begin{cases} 1 & \text{with wind power forecasting system} \\ 0 & \text{without wind power forecasting system} \end{cases}$	WC_8
9	WPIT	$r_9 = \begin{cases} 1 & \text{with testing} \\ 0 & \text{without testing} \end{cases}$	WC_9
10	WFEA	$r_{10} = \begin{cases} 1 & \text{without accident} \\ 0 & \text{with accident} \end{cases}$	WC_{10}

t is the index of hour, T denotes total evaluation time, $\mathrm{Wcap}_{i,w}$ is the operation capacity of wind farms,

IOP Conf. Series: Earth and Environmental Science **93** (2017) 012066 doi:10.1088/1755-1315/93/1/012066

$P_{i,w}^{t}$ is the output power of wind farm w at bus i during period t, $P_{i,w}^{t*}$ is wind power forecasting output based on forecasting technology, d_s is number of days that data were transferred successfully, d_c is the total counted days, T_{preday} denotes total evaluation time .

The scoring model for each index is illustrated in figure 1 with detailed formulations shown in table 2.

Figure 1. The scoring model for each index. a and b are thresholds, and the normalized values are adopted for all scores. (a) Rose semi-ladder model and (b) Fall semi-ladder model.

Table 2. Scoring model formulations for each index adopted for all scores.

Serial number	Scoring model name	Specific formulation	Explanation
1	Rose semi - ladder model	$Score(x)=\begin{cases}0, 0\le x<a\\ \frac{x-a}{b-a}, a\le x<b\\ 1, b\le x\end{cases}$	It can be applied to indexes which are the bigger the better.
2	Fall semi - ladder model	$Score(x)=\begin{cases}1, 0\le x<a\\ 1-\frac{x-a}{b-a}, a\le x<b\\ 0, b\le x\end{cases}$	It can be applied to the indexes which are the smaller the better.
3	Integer scoring model	$Score(x)=x$	It can be applied to the indexes which are with or without this function.

For the wind farms undergoing technical reformation, their indices such as LVRD, APC and RPC ability are gradually improving, rose semi-ladder model could be applied to those indices during incipient stage to promote technical renovation of wind farms. After renovation, integer scoring model should be applied in scoring the involved wind farms.

The final score of each windfarm is shown in the formulation below,

$$S_{i,w}=\sum_{c=1}^{10}WC_c * Score_c \tag{1}$$

where w denotes the index of the wind farm, i is the index of the bus, $S_{i,w}$ is the comprehensive performance score of the wst wind farm in ist bus, WC_c is the weight coefficient of the cst index. The $Score_c$ is performance score of the cst index. The novel evaluation system is finally developed when synthesizing all wind farm final scores.

3. Methodology of multi-time scale energy management modelling

3.1. Objective function

The original objective of day-ahead schedule, intraday schedule optimization and on-line schedule optimization is the same, which is the maximize accommodation of wind power with higher operational level. The optimization objective is shown below in equation (2),

$$F = \max \sum_{t=1}^{T} \sum_{i=1}^{B} \sum_{w=1}^{N_w} S_{i,w} P_{i,w}^{t} \tag{2}$$

where, t is the index of hour, T is the set of optimization steps of day-ahead schedule, B is the total number of buses. N_w denotes the total number of wind farms, $P_{i,w}^{t}$ is the power output of the w st wind farm in the i st bus at the t st hour.

3.2. Constraints

- Thermal unit constraints

$$P_{i,g}^{\min} X_{i,g}^{t} \le P_{i,g}^{t} \le P_{i,g}^{\max} X_{i,g}^{t} \tag{3}$$

$$P_{i,g}^{t+1} - P_{i,g}^{t} \le \Delta P_{i,g}^{up} \tag{4}$$

$$P_{i,g}^{t} - P_{i,g}^{t-1} \le \Delta P_{i,g}^{down} \tag{5}$$

$$P_{i,BPg}^{t} = \alpha_{i,BPg} H_{i,BPg}^{t} \tag{6}$$

$$\alpha_{r,CGg} H_{i,CGg}^{t} \le P_{i,CGg}^{t} \le P_{i,CGg}^{\max} - \beta_{i,CGg} H_{i,CGg}^{t} \tag{7}$$

$$Y_{i,g}^{t} + \sum_{j=0}^{k} Z_{i,g}^{j+i} \le 1 \tag{8}$$

$$Z_{i,g}^{t} + \sum_{j=0}^{k} Y_{i,g}^{t+j} \le 1 \tag{9}$$

$$X_{i,g}^{t} - X_{i,g}^{t-1} - Y_{i,g}^{t} + Z_{i,g}^{t} = 0 \tag{10}$$

$$Y_{i,g}^{t} + Z_{i,g}^{t} \le 1 \tag{11}$$

$$Q_{i,g}^{\min} X_{i,g}^{t'} \le Q_{i,g}^{t} \le Q_{i,g}^{\max} X_{i,g}^{t'} \tag{12}$$

where g denotes the index of thermal unit, $P_{i,g}^{\max}$ and $P_{i,g}^{\min}$ are respectively upper and lower bound of output power of the thermal generator g at bus i, $Q_{i,g}^{\max}$ and $Q_{i,g}^{\min}$ are respectively upper and lower bounds of reactive power output for generator g, $Q_{i,g}^{t}$ is a variable indicating the reactive power output of unit g. $\Delta P_{i,g}^{up}$ and $\Delta P_{i,g}^{down}$ are respectively ramp-up rate limit and ramp-down limit of generator g at bus i, $P_{i,BPg}^{t}$ and $P_{i,CGg}^{t}$ are positive variables that indicate the output of back pressure generator and extraction turbo-generator during heating period, $\alpha_{i,BPg}$, $\alpha_{r,CGg}$ and $\beta_{i,CGg}$ are ratios between heat power output and electricity power output; $H_{i,BPg}^{t}$ and $H_{i,CGg}^{t}$ are the heat load provided by back pressure generator and extraction turbo-generator during period t, respectively, $X_{i,g}^{t}$ denotes on/off statues of unit g at bus i during period t, $Y_{i,g}^{t}$ is a binary decision variable that indicates whether unit g at bus i will be started up or not during period t (1/0 for start-up/not start-up), $Z_{i,g}^{t}$ is a binary decision variable that indicates whether unit g will be shut down or not during period t (1/0 for shut down/not shut down).

Those constraints on thermal units listed above include unit generation capacity constraints illustrated as equation (3), unit ramping-up constraints illustrated as equation (4), unit ramping-down constraints illustrated as equation (5), output of heating units during heating period constraints illustrated as equations (6) and(7), unit minimum-up time constraints illustrated as equation (8), unit minimum-down time constraints illustrated as equation (9), and the logical state of unit start-up, unit shut-down and unit operations constraints illustrated as equations (10) and (11), unit reactive power output as equation (12).

- Hydro unit constraints

$$P_{i,RO}^{t} = P_{i,RO}^{t^*} \tag{13}$$

$$P_{i,RS}^{\min} \le P_{i,RS}^{t} \le P_{i,RS}^{\max} \tag{14}$$

$$WQ_{i,RS}^{\min} \le \sum_{j=0}^{k} P_{i,RS}^{t+j} \le WQ_{i,RS}^{\max} \tag{15}$$

$$W^{t} - W_{\max} \le P_{PSG}^{t} \cdot \eta_g - P_{PSP}^{t} \cdot \eta_p \le W^{t} - W_{\min} \tag{16}$$

$$\sum_{t=1}^{Tday} (P_{PSG}^{t} \cdot \eta_g - P_{PSP}^{t} \cdot \eta_p) = W_0 - W_{end} \tag{17}$$

$$0 \le a_i^t + b_i^t \le 1 \tag{18}$$

$$b_i^t \times P_{PSG\min}^{t} \le P_{PSG}^{t} \le b_i^t \times P_{PSG\max}^{t} \tag{19}$$

$$\begin{cases} P_{PSP}^{t} = a_i^t \times N_{PS}^{t} \times P_{PSPunit}^{t} \\ 0 \le N_{PS}^{t} \le N_{PS\max}^{t} \end{cases} \tag{20}$$

where $P_{i,RO}^{t}$ is a positive variable that indicates the output power of the runoff generator at bus i during the period t, $P_{i,RO}^{t^*}$ is the theoretical output of the runoff generator, $P_{i,RS}^{t}$ is a positive value that indicates the output power of the reservoir generator, $P_{i,RS}^{\max}$ and $P_{i,RS}^{\min}$ are respectively output power upper and lower bound of the reservoir generator, $WQ_{i,RS}^{\max}$ and $WQ_{i,RS}^{\min}$ are maximum and minimum electric quantity of reservoir stations, $W_{\max}$ and $W_{\min}$ are respectively maximum and minimum water storage capacity of the pumped-storage station, W^{t} is the positive variable that denotes water storage capacity of the pumped-storage station during the period t, η_g and η_p are water/electricity conversion coefficients in generating stage and pumping stage, respectively. P_{PSG}^{t} and P_{PSP}^{t} are positive values that respectively indicate generated power and pumping power of pumped-storage generator, a_i^t is a binary decision variable that indicates weather a pumped-storage generator is in the state of pumping or not (1/0 for pumping power/not pumping power), b_i^t is a binary decision variable that indicates weather a pumped-storage generator is in the state of generating power or not during period t (1/0 for generating power/not generating power), $P_{PSG\max}^{t}$ and $P_{PSG\min}^{t}$ are respectively output power upper and lower bound of pumped-storage generator, N_{PS}^t is the integer variable indicating the number of operating pumped-storage generators when the station is in the state of pumping, $N_{PS\max}^t$ denotes the total number of pumped-storage generators, $P_{PSPunit}^t$ is maximum capacity of a single pumped-storage generator.

The hydro unit constraints listed above include the runoff generator generation capacity constraints illustrated as equation (13), the reservoir generator generation capacity constraints illustrated as equation (14), the electric quantity of reservoir stations constraints illustrated as equation (15), the pumped-storage capacity constraints illustrated as equations (16) and (17), the logical state between generating and pumping constraints illustrated as equation (18), the pumped-storage generator generation capacity constraints illustrated as equation (19), and the pumped-storage generator pumping capacity constraints illustrated as equation (20) (once the unit is in both pumping and start-up state, it should be operated at maximum capacity).

- System constraints

$$\sum_{w=1}^{N_W} P_{i,w}^t + \sum_{g=1}^{N_G} X_{i,g}^t P_{i,g}^t + P_{i,RO}^t + \sum_{rs=1}^{N_{RS}} P_{i,rs}^t + \sum_{PS=1}^{N_{PS}} (P_{i,PS}^t - Q_{i,PS}^t) + \sum_{j=1}^{B} L_{i,j}^t = P_{i,l}^t \tag{21}$$

$$H_{i,CGg}^t + H_{i,BPg}^t = H_{i,l}^t \tag{22}$$

$$\sum_{i=1}^{B}(\sum_{g=1}^{N_G} P_{i,g}^{\max} X_{i,g}^t + \sum_{w=1}^{N_W} R_{i,w}^t) \geq \sum_{r=1}^{R} P_{i,l}^t + R_P \tag{23}$$

$$\sum_{i=1}^{B}(\sum_{g=1}^{N_G} P_{i,g}^{\min} X_{i,g}^t - \sum_{w=1}^{N_W} R_{i,w}^t) \geq \sum_{i=1}^{B} P_{i,l}^t - R_N \tag{24}$$

$$L_{i,j}^t \leq \left|L_{i,j,\max}\right| \tag{25}$$

$$L_{i,j}^t = U_i^t U_j^t (G_{ij}\cos\delta_{ij}^t + B_{ij}\sin\delta_{ij}^t) - U_i^{t^2} G_{ij} \tag{26}$$

$$0 \leq P_{i,w}^t \leq P_{i,w}^{t^*} \tag{27}$$

$$P_{i,w}^t + P_{i,g}^t + P_{i,ro}^t + P_{i,rs}^t - P_{i,l}^t = U_i^t \sum_{j\in i} U_j^t \left(G_{ij}\cos\delta_{ij}^t + B_{ij}\sin\delta_{ij}^t\right) \tag{28}$$

$$Q_{i,w}^t + Q_{i,g}^t + Q_{i,ro}^t + Q_{i,rs}^t - Q_{i,l}^t = U_i^t \sum_{j\in i} U_j^t \left(G_{ij}\cos\delta_{ij}^t - B_{ij}\sin\delta_{ij}^t\right) \tag{29}$$

$$U_{i,\min} \leq U_i^t \leq U_{i,\max} \tag{30}$$

$$\delta_{ij}^t \leq \left|\delta_{ij,\max}^t\right| \tag{31}$$

where $P_{i,l}^t$ is the value of electrical load of bus i during period t, $H_{i,l}^t$ is the value of heat load of bus i during period t, $R_{i,w}^t$ is a wind power credible capacity during period t. R_P and R_N are respectively the system positive and negative reserve demand during period t, $L_{i,j}^t$ denotes the power flow on transmission line from bus i to j, $\left|L_{i,j,\max}\right|$ is a power flow limit on transmission line which links buses i and j. U_i^t is a positive value that denotes voltage at bus i at period t, $U_{i,\max}$ and $U_{i,\min}$ are respectively maximum and minimum voltage at bus i, G_{ij} and B_{ij} are real and imaginary parts of the admittance matrices, δ_{ij}^{t} is a variable that denotes the angular phase difference between buses *i* and *j*.

Hourly system constraints listed above include system electric power balance constraints illustrated as equation (21), system heat power balance constraints illustrated as equation (22), system spinning reserve constraints illustrated as equations (23) and (24), constraints on grid security illustrated as equations (25) and (26) and wind power output constraints illustrated as equation (27). Hourly system constraints in intraday schedule listed above include constraints on active/reactive power equilibrium of all buses illustrated as equations (28) and (29), constraints on bus voltage illustrated as equation (30), constraints on angular phase illustrated as equation (31).

- Heating load constraints

$$\sum_{BPg=1}^{N_{BPG}} X_{i,BPg}^{t} H_{i,BPg}^{t} + \sum_{CGg=1}^{N_{CGg}} X_{i,CGg}^{t} H_{i,CGg}^{t} = H_{i,l}^{t} \tag{32}$$

Where $X_{i,BPg}^{t}$ and $X_{i,CGg}^{t}$ denotes on/off statues of back pressure generator *BPg* and extraction turbo-generator *CGg* at bus *i* during period *t*, $H_{i,l}^{t}$ is the sum value of heating load of bus *i* during period *t*.

3.3. The model of multi-time scale energy management

The flowchart of the multi time-scale formulation ((day-ahead 72-hour schedule, intraday 4-hour schedule, and on-line 15-min schedule)) is illustrated in figure 2.

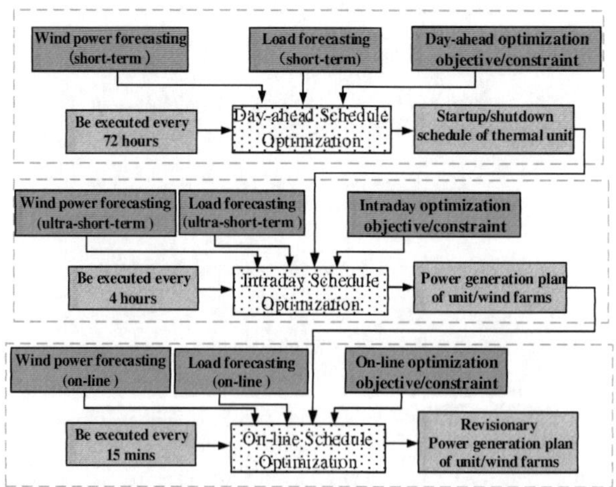

Figure 2. Flow diagram for multi-time scale energy management.

The objective function listed as equation (2) and constraints listed as equations (3)-(11), (13)-(25), (27) and (32) constitute the model of intraday schedule, which is a process of mixed-integer linear programming (MILP). Start-up status of thermal units during the whole optimization period are determined and optimization results of present period should be extracted and saved as initial values of next 72 hours. A computationally efficient mixed-integer linear formulation for the UC problem of thermal units is applied in the day-ahead schedule model [15].

Integer variables are not included in the intraday schedule model because UC decisions have been determined in day-ahead schedule model. In the intraday schedule, branch power flows and node voltages are calculated exactly based on the AC power flow formulation. The process of intraday schedule should be executed every 1 hour. The objective function listed as equation (2) and constraints listed as equations (3), (12)-(20) and (25)-(32) constitute the model of intraday schedule. The model is a QCP problem and can be solved by IPOPT solver in GAMS.

The wind power is inherent fluctuating and poses great challenges on real-time frequency regulation. Therefore, on-line schedule optimization is necessary, in which power generation plan can be timely revised to guarantee active-power balance. The formulation of on-line schedule optimization is basically the same with that of intraday schedule optimization. The only difference between on-line schedule optimization and intraday schedule optimization is that theoretical wind power provided by on-line prediction systems is more accurate. The on-line schedule optimization should be executed every 15 minutes.

4. Case studies

An actual provincial power system is selected as an example system in this section. The methodology proposed in this paper is utilized to optimize the energy management of wind farms based on the novel comprehensive evaluation system.

4.1. Data for simulations

The grid topology of the example system is illustrated in figure 3.

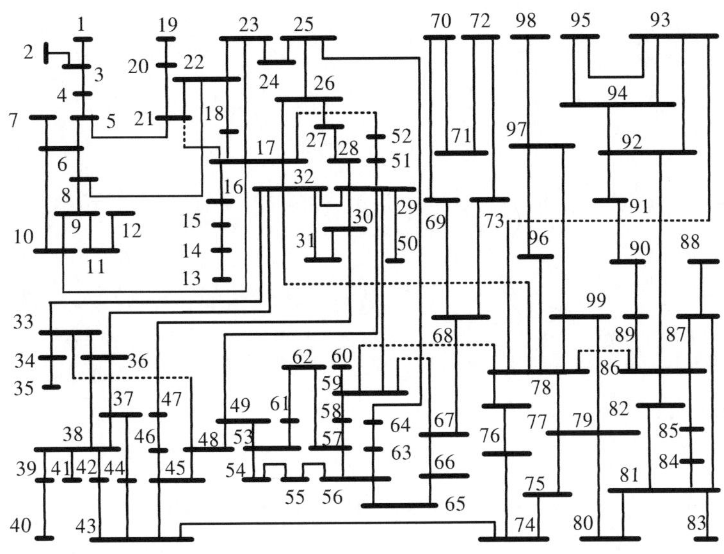

Figure 3. Topology of case system.

The case system has 99 buses. Briefed portfolio information for conventional generation and wind farms are given in table 3. The installed capacity of the maximum unit (660 MW) is selected as the amount of spinning reserve.

Table 3. Briefed generation portfolio information.

Generation unit type	Number of units	Bus locations	Total capacity /(MW)
Condensate gas generator	40	17,22,29,32,36,38,56,60,72 ,78,79,88,91,92,93,98,99	9,146
Back pressure generator	3	43,62	350
Extraction turbo-generator	25	6,29,30,44,48,49,50,55,62, 68,75,84,86	6,160
Reservoir generator	2	50,91	200

Pumped-storage generator	2	76	300
Wind farms	/	1,8,9,16,21,34,40,42,52	3,150

4.2. Illustrative examples

The day-ahead startup/shutdown schedules of thermal units optimized in day-ahead schedule are illustrated in table 4.

Table 4. Day-head start up/shutdown schedule.

Units	Periods																							
	1	2	3	4	5	6	7	8	9	10	11	12	13	14	15	16	17	18	19	20	21	22	23	24
T1-T4	1	1	1	1	1	1	1	1	1	1	1	1	1	1	1	1	1	1	1	1	1	1	1	1
T5	0	0	0	0	0	0	0	0	0	0	0	0	0	0	0	0	0	0	0	0	0	0	0	0
T6-T7	1	1	1	1	1	1	1	1	1	1	1	1	1	1	1	1	1	1	1	1	1	1	1	1
T8-T9	0	0	0	0	0	0	0	0	0	0	0	0	0	0	0	0	0	0	0	0	0	0	0	0
T10-T12	1	1	1	1	1	1	1	1	1	1	1	1	1	1	1	1	1	1	1	1	1	1	1	1
T13-T18	0	0	0	0	0	0	0	0	0	0	0	0	0	0	0	0	0	0	0	0	0	0	0	0
T19-T37	1	1	1	1	1	1	1	1	1	1	1	1	1	1	1	1	1	1	1	1	1	1	1	1
T38	1	1	1	1	1	1	1	1	1	1	1	1	1	1	1	1	1	1	1	1	1	0	0	0
T39-T48	1	1	1	1	1	1	1	1	1	1	1	1	1	1	1	1	1	1	1	1	1	1	1	1
T49-T50	0	0	0	0	0	0	0	0	0	0	0	0	0	0	0	0	0	0	0	0	0	0	0	0
T51-T52	1	1	1	1	1	1	1	1	1	1	1	1	1	1	1	1	1	1	1	1	1	1	1	1
T53-T55	0	0	0	0	0	0	0	0	0	0	0	0	0	0	0	0	0	0	0	0	0	0	0	0
T56-T59	1	1	1	1	1	1	1	1	1	1	1	1	1	1	1	1	1	1	1	1	1	1	1	1
T60-T63	1	1	1	1	1	1	1	1	1	1	1	1	1	1	1	1	1	1	1	1	1	1	1	1
T64	0	0	0	0	0	0	0	0	0	1	1	1	1	1	1	1	1	1	1	1	1	1	1	1
T65	0	0	0	0	0	0	0	0	0	0	0	0	0	0	0	0	0	0	0	0	0	0	0	0
T66-T67	1	1	1	1	1	1	1	1	1	1	1	1	1	1	1	1	1	1	1	1	1	1	1	1
T68	0	0	0	0	0	0	0	0	0	0	0	0	0	0	0	0	0	0	0	0	0	0	0	0

As shown in table 4, the maximum thermal power start-up capacity of the province is 9,782.5 MW. The thermal unit T64 is started at 10.00 a.m., while T38 is shut down at 22:00 in order to meet peak load regulation demand.

Based on day-ahead startup/shutdown schedule, the power generation plan for all thermal power generating units, hydroelectric generating sets and wind power plants are shown in figure 4.

It can be observed from figure 4 that all heating units keep their minimum outputs in the heating period to guarantees largest accommodation space for wind power. Most of the hydro power outputs coincide with peak load period. The pumped-storage generator is pumping water during valley load period in order to reduce wind curtailment.

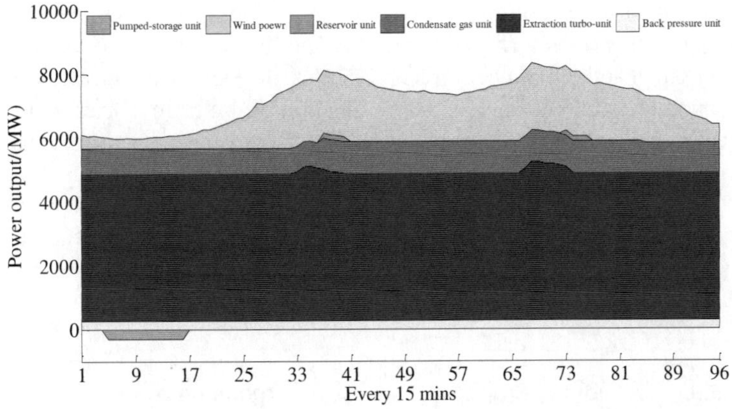

Figure 4. Day-ahead start up/shutdown schedule.

The score of each wind farm determined by the comprehensive evaluation system proposed in the paper is illustrated in table 5. Wind farm outputs with and without using the evaluation system are presented in figure 5, in which only W17 (with higher score) and W25 (with lower score) are selected among all wind farms to show for demonstration purpose.

Table 5. The score of each wind farm.

Wind Farms	Score	Wind Farms	Score	Bus	Score	Wind Farms	Score	Wind Farms	Score
W1	63.93	W7	62.24	W13	30.00	W19	65.76	W25	35.00
W2	60.63	W8	55.59	W14	70.58	W20	70.74	W26	59.70
W3	52.50	W9	68.10	W15	71.17	W21	73.92	W27	79.62
W4	50.04	W10	42.29	W16	74.93	W22	59.63	W28	78.38
W5	56.37	W11	42.21	W17	81.17	W23	68.54	W29	61.00
W6	56.37	W12	66.00	W18	60.53	W24	65.97	W30	69.77

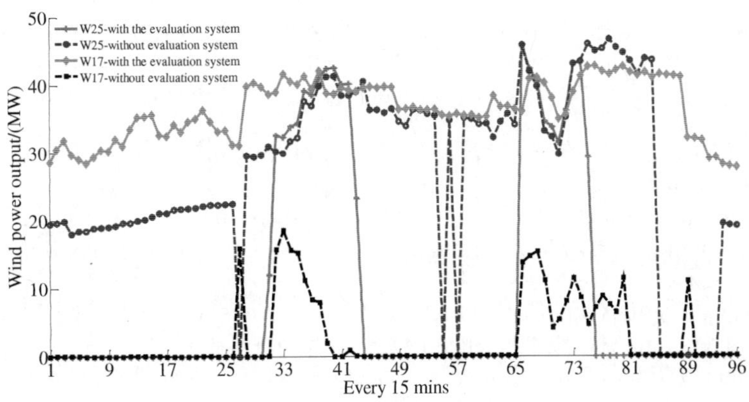

Figure 5. Wind power output with and without using the evaluation system.

Figure 5 shows that wind power generated from wind farms with higher technical levels like W17 would be accommodated preferentially. On the contrary, if the wind curtailment occurs, wind farms with lower technical levels like W25 would be curtailed preferentially.

5. Conclusions

A detailed model taking into account comprehensive factors for provincial integrated energy system has been presented, especially the various forms of generators and loads. A novel comprehensive evaluation system has been proposed to quantify economic properties of each wind farm to make the energy management more economical and reasonable. A combination of multi time-scale schedule method has been proposed to develop a novel energy management. The paper concludes the simulations conducted on a practical provincial power system in northeast China. Simulation results have validated the proposed model and corresponding solving algorithms. Results show that the proposed model is more suitable for real power systems than those of conventional methods. The presented optimization results not only provide guidance for the energy management of large-scale complex power systems featured with integrated energy supply, but also promote technical renovation of wind farms.

Acknowledgments
This work is supported by Technology Project of State Grid Corporation of China.

References

[1] Tascikaraoglu A, Erdinc O, Uzunoglu M and Karakas A 2014 An adaptive load dispatching and forecasting strategy for a virtual power plant including renewable energy conversion units *Appl Energy* **119** 445-53
[2] Liu Z-Y 2015 *Global Energy Internet* (Beijing: China Electric Power Press)
[3] GWEC 2015 Global wind report 2014 (Istanbul: GWEC) http://www.gwec.net/publications /global-wind-report-2/global-wind-report-2014-annual-market-update
[4] Meibom P, Barth R, Hasche B, Brand H, Weber C and O'Malley M 2011 Stochastic optimization model to study the operational impacts of high wind penetrations in Ireland *IEEE Trans on Power Syst* **26(3)** 1367-78
[5] Faias S, de Sousa J, Reis F S and Castro R 2012 Assessment and optimization of wind energy integration into the power systems: application to the Portuguese system *IEEE Trans on Sustain Energy* **3(4)** 627-35
[6] Yang X, Song Y-H, Wang G-H and Wang W-S 2010 A comprehensive review on the development of sustainable energy strategy and implementation in China *IEEE Trans on Sustain Energy* **1(2)** 57-65
[7] Azizipanah-Abarghooee R, Niknam T, Amin Bina M and Zare M 2015 Coordination of combined heat and power-thermal-wind photovoltaic units in economic load dispatch using chance constrained and jointly distributed random variables methods *Energy* **79** 50-67
[8] National Energy Administration 2014 Wind power industry motioning, 2015 http://www. nea.gov.cn/2015-02/12/c_133989991.htm
[9] Brand A J, Gibescu M, Kling W L, *et al* 2007 Impacts of wind power on thermal generation unit commitment and dispatch *IEEE Trans on Energy Convers* **22(1)** 44-51
[10] Wang Q-F, Guan Y-P and Wang J-H 2012 A chance-constrained two-stage stochastic program for unit commitment with uncertain wind power output *IEEE Trans on Power Syst* **27(1)** 206-15
[11] Wang X-N, Palazoglu A and El-Farra N H 2015 Operational optimization and demand response of hybrid renewable energy systems *Appl Energy* **143** 324-35
[12] Lowery C and O'Malley M 2012 Impact of wind forecast error statistics upon unit commitment *IEEE Trans on Sustain Energy* **3(4)** 760-8
[13] Ela E and O'Malley M 2012 Studying the variability and uncertainty impacts of variable generation at multiple timescales *IEEE Trans on Power Syst* **27(3)** 1324-33
[14] Zhang B-M, Wu W-C, Zheng T Y and Sun H-B 2011 Design of a multi time scale coordinated active power dispatching system for accommodating large scale wind power penetration *Automation of Electric Power System* **35(1)** 1-6 (in Chinese)
[15] Carrión M and Arroyo J M 2006 A computationally efficient mixed-integer linear formulation for the thermal unit commitment problem *IEEE Trans on Power Syst* **21(3)** 1371-8

Numerical studies on heat transfer and pressure drop characteristics of flat finned tube bundles with various fin materials

Y Peng[1], S J Zhang[1,3], F Shen[1], X B Wang[2], X R Yang[2] and L J Yang[2]

[1]State Power Investment Central Research Institute, South Zone of the Future Science & Technology Park, Beijing, China
[2]Key Laboratory of Condition Monitoring and Control for Power Plant Equipment of Ministry of Education, School of Energy Power and Mechanical Engineering, North China Electric Power University, No. 2 Beinong Road, Changping District, Beijing, China

E-mail: zhangshengjun@snptc.com.cn

Abstract. The air-cooled heat exchanger plays an important role in the field of industry like for example in thermal power plants. On the other hand, it can be used to remove core decay heat out of containment passively in case of a severe accident circumstance. Thus, research on the performance of fins in air-cooled heat exchangers can benefit the optimal design and operation of cooling systems in nuclear power plants. In this study, a CFD (Computational Fluid Dynamic) method is implemented to investigate the effects of inlet velocity, fin spacing and tube pitch on the flow and the heat transfer characteristics of flat fins constructed of various materials (316L stainless steel, copper-nickel alloy and aluminium). A three dimensional geometric model of flat finned tube bundles with fixed longitudinal tube pitch and transverse tube pitch is established. Results for the variation of the average convective heat transfer coefficient with respect to cooling air inlet velocity, fin spacing, tube pitch and fin material are obtained, as well as for the pressure drop of the cooling air passing through finned tube. It is shown that the increase of cooling air inlet velocity results in enhanced average convective heat transfer coefficient and decreasing pressure drop. Both fin spacing and tube pitch engender positive effects on pressure drop and have negative effects on heat transfer characteristics. Concerning the fin material, the heat transfer performance of copper-nickel alloy is superior to 316L stainless steel and inferior to aluminium.

1. Introduction

Air-cooled heat exchangers are widely used in various industries such as thermal power plants, air conditioning, refrigeration, chemical engineering, etc. The heat transfer coefficient is the dominant factor that could affect the heat transfer performance of air-cooled heat exchangers. In addition, the air-side thermal resistance of the heat exchangers is the main contributor for the total thermal resistance of the heat exchanger [1]. Generally, flat fins are arranged on the outer surface of the round tubes in order to increase the heat transfer area on the air side of heat exchangers, improve the thermal performance effectively and optimize the structure size and weight of the air-cooled heat exchanger.

There have been many studies on the heat transfer characteristics and pressure drop of the finned tubes. He *et al* [2] conducted three dimensional numerical studies on laminar flow and heat transfer of a plate fin-and-tube heat exchanger to reveal the effects of five factors: Reynolds (Re) number, fin pitch, tube row number, spanwise and longitudinal tube pitch. It was found that the Nusselt number increased with the increase of Re number an optimum fin pitch at which the Nusselt number became maximum existed, and the increase of the number of tube rows led to decrease of the average Nusselt number. Kang *et al* [3] carried out a systematic experimental investigation on heat transfer and pressure drop characteristics of four types of plate fin-and-tube heat exchanger surfaces: the plain plate, the slotted plain, the wavy fin with triangular cross section and the wavy fin with sinusoidal cross section. They found that the Nusselt number of the slotted plain fin surface and of the plain plate fin were the greatest and the lowest respectively, while the thermal performance of those two with the wavy fins was between them. Three-dimensional simulations were made to investigate heat transfer and fluid flow characteristics of a four row plain fin-and-tube heat exchanger considering in-lined and staggered arrangements by means of the CFD commercial software ANSYS CFX [4], which revealed the effects of Reynolds number, fin pitch and tube pitch on the overall heat transfer and friction factor for plate fin-and-tube heat exchangers.

On the other hand, air-cooled heat exchangers could also be utilized in nuclear power plants to remove the core decay of residual heat from inside of the containment under severe accident circumstances. Component standards in nuclear power field differ from the ones in traditional industries (such as conventional thermal power plants), especially in issues concerning materials. In terms of environmental conditions in nuclear power plants, like irradiation and corrosion, it is necessary to pay attention to material selection for the fins and tubes of air-cooled exchangers. With respect to these conditions, three kinds of materials, namely 316L stainless steel, copper-nickel alloy and aluminium, are selected as candidates. Copper for example plays the important role in the material choice due to its favorable corrosion resistance, ductility, electric and thermal conductivity. However, its expensive price is the main restrictive factor, so seeking alternative materials is the trend in current research. Li [5] carried out a numerical simulation method to study the heat transfer coefficient of different materials of finned tube bundles in dry conditions, on the basis of the same boundary condition. As a result, the required quantity and cost of different heat exchanger materials were obtained, as well as the theory basis for material replacement of finned tube bundles was provided. For the material selection of finned tube bundles, the price of aluminium is much less than the copper, so comparison of the performance of these two materials is essential, Chen [6] analyzed the variations of heat transfer coefficient and heat transfer rate and it had proved that it is feasible to replace the copper with aluminium. In addition, environmental conditions and especially the blowing wind is also a basic factor for the heat transfer characteristics of finned tube bundles. For such and air-cooled condenser of a power plant, Yang [7] established the physical and mathematical models for the fluid and heat flow characteristics and obtained various results for the flow resistance and average convection heat transfer coefficient with different windward velocities. Hu *et al* [8] presented a new and reliable model to predict the flow and heat transfer characteristics of an air-cooled condenser under the influence of natural wind based on POD (Proper Orthogonal Decomposition) methods in order to account for the three-dimensional velocity and temperature fields when the natural wind directions ranging from 0 to 90 degrees. The heat transfer characteristics of stainless steel at different wind velocities were analyzed by Shi [9]; the results showed that increasing inlet velocity appropriately can improve the heat transfer performance of the finned tube bundles. Kong *et al* [10] performed a three-dimensional numerical study on the variations of friction factor, heat flow rate and convection heat transfer coefficient of flat and slotted finned tube bundles under various wind velocities, which indicated that heat flow rate and convection heat transfer coefficient raised continuously, while the friction factor decreased with the lower wind speed.

The studies mentioned above are few and there is still the need for further studies on the finned tube materials in conjunction with different wind speed at home and abroad. In the present study, the heat transfer performance of fins and tubes, constructed of three different materials (copper-nickel alloy, 316L stainless steel, and aluminium) and for different wind velocities are investigated, in order to benefit for the design of air-cooled exchangers in nuclear power plants. To this end, several simulations in order to numerically predict the heat transfer characteristics and pressure drop of fin-and-tube heat exchangers for different materials and various inlet velocities are carried out.

2. Models

2.1. Physical model and boundary conditions

Figure 1 shows the schematic diagram of the flat fin-and-tube bundles, which are arranged in a staggered pattern. In doing so, the existence of fins and the tube arrangement can greatly the heat transfer surface area and enhance the flow disturbance, contributing to the enhancement of heat transfer performance of air-cooled heat exchangers.

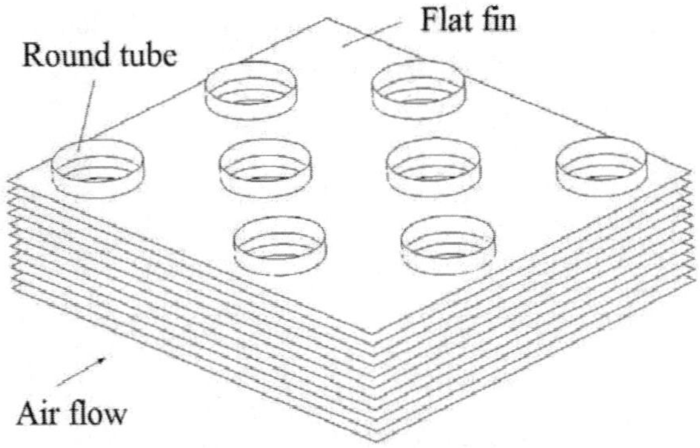

Figure 1. Schematic diagram of flat finned tube bundles [10].

The schematic diagram and geometric details of the flat finned tube bundles used in this study are described in figure 2, while. The corresponding dimensions are listed in table 1.

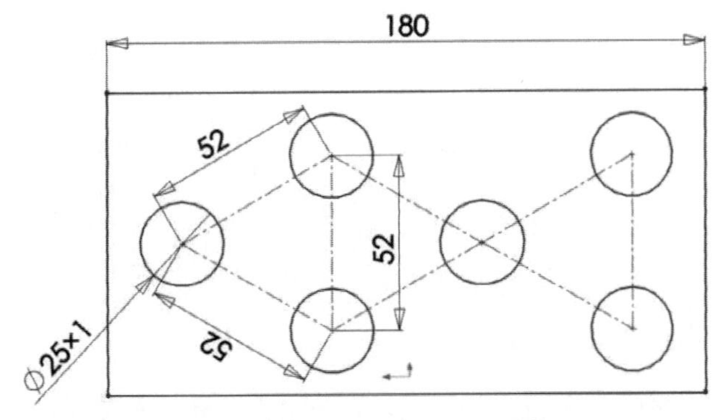

Figure 2. Schematic diagram of the tube arrangement.

Table 1. Dimensions of the flat finned tubes.

Parameter	Value
Base tube (outer diameter and thickness)/mm×mm	Φ25×1
Fin (width and thickness)/ mm×mm	180×0.12
Fin-pitch/mm	2.4, 2.8, 3.2, 3.6, 4.0
Tube row number	4
Tube pitch/mm	34, 40, 46, 52

Figure 3 shows the main computational domain for the flat fin-and-tube configuration and its boundaries, in which inlet, outlet, symmetry, wall and periodic boundary conditions are applied at the corresponding parts of the boundary. In order to ensure uniform air flow at the inlet and avoid backflow at the outlet that could pose negative effects on computational convergence, extended areas are added in the computational domain, as is depicted in figure 4. The inlet and outlet part extends 270 mm and 900 mm in the X-direction of the computational domain, respectively, so that uniform inlet velocity distribution could be applied to the inlet as well as the fully-developed outflow boundary to the outlet. A velocity inlet boundary condition was implemented at the inlet of the computational domain with inlet velocity ranging from 0.25 m/s to 2 m/s and constant air inlet temperature 323.15 K. No-slip conditions for momentum and constant temperature 363.15 K are set for the inner surfaces of the base tube while at its outer surfaces coupled thermal conditions are imposed. Hence, the temperature distribution both on the solid surface and the fluid side can be calculated simultaneously with the flow field by solving the conjugate problem. In addition, the Z-direction is periodic and the Y-direction is symmetry boundary condition.

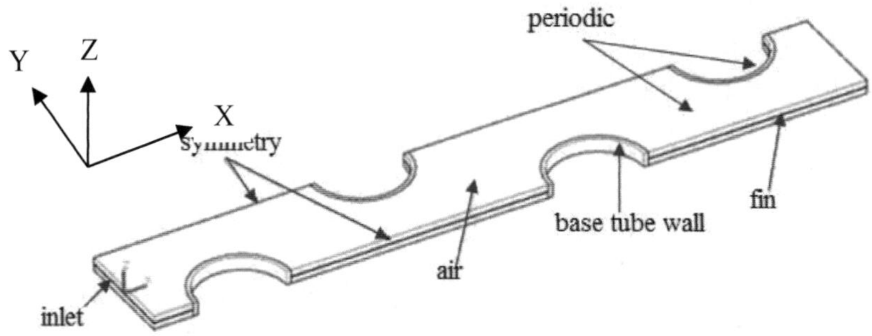

Figure 3. Boundary conditions for finned tube bundles.

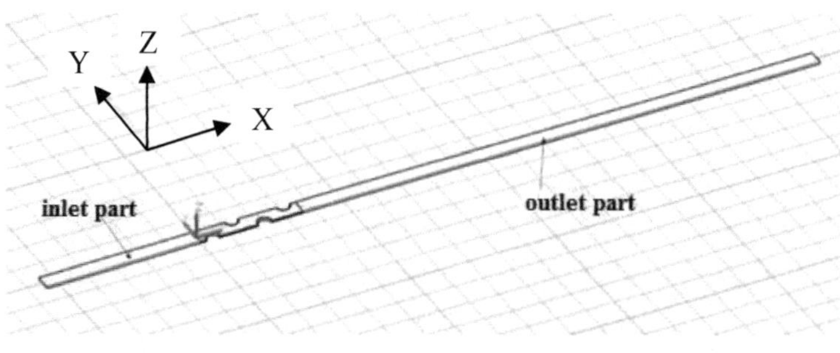

Figure 4. Extended domains for finned tube bundles.

The computational mesh was created by the commercial software GAMBIT. For the inlet and outlet part, hexahedral structured mesh was adopted since they are regular hexagons, as shown in figure 5. In the fin-and-tube region, the relative coarse hexahedral unstructured grids were generated for the flow channel, and the mesh near the walls was refined to capture flow and heat transfer details accurately around the tube. The schematic mesh around the tube-and-fin is presented in figure 6.

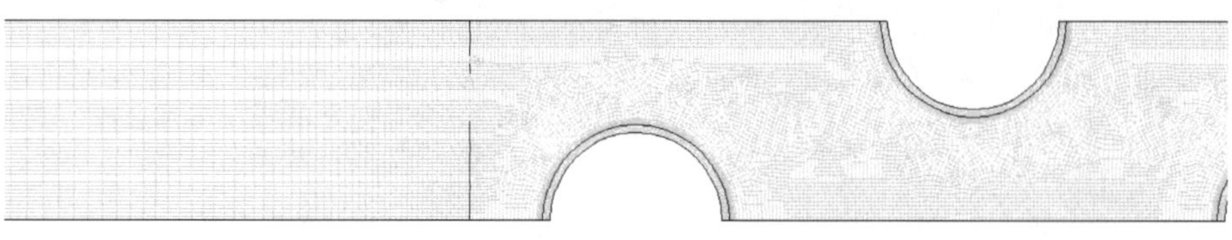

Figure 5. Schematic mesh for the input part.

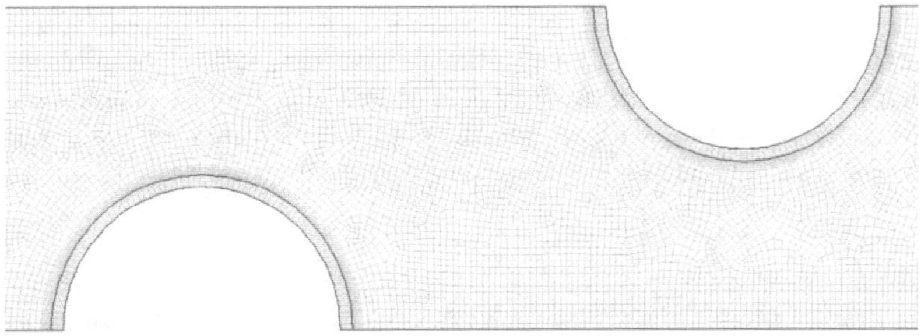

Figure 6. Schematic mesh around the fin-and-tube domains.

2.2. Governing equations

Numerical studies are conducted based on the assumptions that the fluid is incompressible, its physical properties are constant, and the flow is steady-state turbulent. The governing equations including continuity, momentum and energy equations for the computational domain can be expressed in tensor form as follows.

Continuity equation:

$$\frac{\partial \rho u_i}{\partial x_i} = 0 \tag{1}$$

Momentum equation:

$$\frac{\partial}{\partial x_i}(\rho u_i u_j) = -\frac{\partial p}{\partial x_i} + \frac{\partial \tau_{ij}}{\partial x_j} + \frac{\partial}{\partial (x_j)}\left(-\rho \overline{u_i' u_j'}\right) \tag{2}$$

where

$$\tau_{ij} = \mu\left(\frac{\partial u_i}{\partial x_j} + \frac{\partial u_j}{\partial x_i} - \frac{2}{3}\delta_{ij}\frac{\partial u_k}{\partial x_k}\right) \tag{3}$$

According to Boussinesq hypothesis, the turbulent shear stress can be written as the following form in which the Reynolds stress is analogous to the shear stress of laminar flow.

$$-\rho \overline{u_i' u_j'} = \mu_t\left(\frac{\partial u_i}{\partial x_j} + \frac{\partial u_j}{\partial x_i} - \frac{2}{3}\delta_{ij}\frac{\partial u_k}{\partial x_k}\right) - \frac{2}{3}\delta_{ij}\rho k \tag{4}$$

Energy equation:

$$\frac{\partial}{\partial x_i}(\rho u_i H) = \frac{\partial}{\partial x_i}\left(\lambda\frac{\partial T}{\partial x_i} + \frac{\mu_t}{\sigma_t}\frac{\partial H}{\partial x_i}\right) + u_i\frac{\partial P}{\partial x_i} + S_h \tag{5}$$

Turbulence kinetic energy equation:

$$\frac{\partial}{\partial x_i}(\rho k u_i) = \frac{\partial}{\partial x_j}\left(\left(\mu + \frac{\mu_t}{\sigma_k}\right)\frac{\partial k}{\partial x_j}\right) + G_k + G_b - \rho\varepsilon - Y_M + S_k \tag{6}$$

Turbulence dissipation rate equation:

$$\frac{\partial}{\partial x_i}(\rho \varepsilon u_i) = \frac{\partial}{\partial x_j}\left(\left(\mu + \frac{\mu_t}{\sigma_\varepsilon}\right)\frac{\partial \varepsilon}{\partial x_j}\right) + C_{1\varepsilon}\frac{\varepsilon}{k}(G_k + C_{3\varepsilon}G_b) - C_{2\varepsilon}\rho\frac{\varepsilon^2}{k} + S_\varepsilon \tag{7}$$

In these equations, G_k represents the generation of turbulence kinetic energy due to mean velocity gradients, $G_b$represents the generation of turbulence kinetic energy due to buoyancy, Y_M represents the contribution of the fluctuating dilatation in compressible turbulence to the overall dissipation rate, $C_{1\varepsilon}$, $C_{2\varepsilon}$, and $C_{3\varepsilon}$ are constants, σ_k and σ_ε are the turbulent Prandtl numbers for k and ε respectively and S_k and S_ε are the source terms.

IOP Conf. Series: Earth and Environmental Science **93** (2017) 012067 doi:10.1088/1755-1315/93/1/012067

2.3. Numerical method

The governing equations mentioned above are solved by the commercial CFD software Fluent 16 [11] with the finite volume method. The continuity, momentum and energy equations are used second-order upwind spatial discretization scheme, and the SIMPLE algorithm is applied to the pressure-velocity coupling iterative computation. The convergence criterion for residual values of governing equations was set to 10^{-6}.

2.4. Parameter definition

For the flow and heat transfer characteristics calculation, the following parameters defined below in order to improve the understanding on post processing of the results.

$$h = \frac{\emptyset}{A \cdot \Delta T_m} \quad (8)$$

$$\Delta T_m = \frac{T_o - T_{in}}{ln\left(\frac{T_w - T_{in}}{T_w - T_o}\right)} \quad (9)$$

$$\emptyset = c_p m_a (T_o - T_i) \quad (10)$$

$$\Delta P = P_{in} - P_{out} \quad (11)$$

$$KL = \frac{\Delta P}{\frac{1}{2}\rho u^2} \quad (12)$$

where T_o and T_{in} are the area-weighted average temperatures at the outlet and inlet surfaces respectively, T_w is the fluid-side wall temperature, m_a is the air mass flow rate, P_{in} and P_{out} represent the inlet and outlet pressures respectively and KL is the friction factor.

3. Results and discussion

3.1. Effects of inlet velocity and material

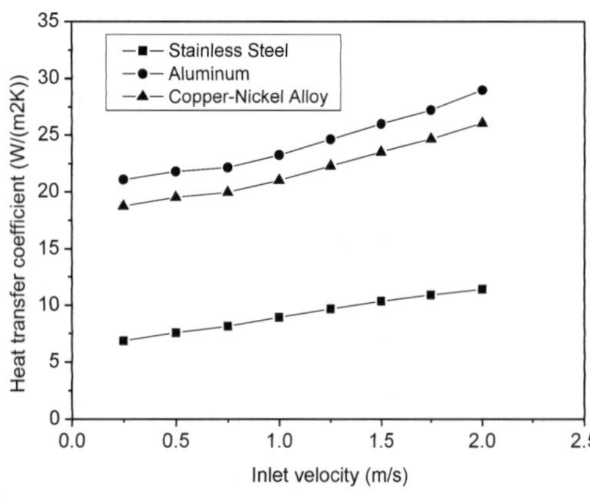

Figure 7. The effects of inlet velocity on heat transfer coefficient for different fin and tube materials.

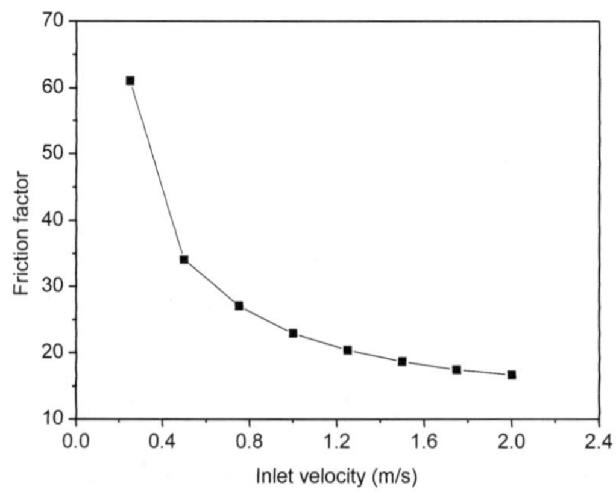

Figure 8. The effects of inlet velocity on friction factor.

The effects of the air velocity on heat transfer characteristics with constant fin spacing 3.2 mm and tube pitch 52 mm are investigated for the following values of inlet velocity: 0.25 m/s, 0.5 m/s, 0.75 m/s, 1 m/s, 1.25 m/s, 1.5 m/s, 1.75 m/s and 2 m/s. Figure 7 shows the numerical results obtained for stainless steel, aluminium and copper-nickel alloy fin-and-tubes with different inlet air velocity. The heat transfer coefficient increases with the increase of inlet velocity. The maximum values for stainless steel, aluminium and copper-nickel alloy in this speed condition are approximately 11.34, 28.87 and 24.57 W/(m^2·K) respectively, while the corresponding minimum values are 6.88, 21.08 and 18.74 W/(m^2·K). It can be seen clearly that the aluminium is the most effective material on heat transfer, and the copper-nickel alloy works better than the stainless steel. In particular, the heat transfer coefficient of aluminium is about 60.7% higher than that of stainless steel at the same inlet velocity.

The pressure drop characteristics are independent of the fin-and-tube material according to the results of friction factor calculations for three kinds of materials. Since the friction factor for different materials is almost identical, only the friction factor of stain steel is presented in figure 8. The friction factor decreases sharply under the value of 1 m/s inlet velocity, and then declines gradually when the inlet velocity is higher than 1 m/s. It is can be concluded that the inlet velocity has a great impact on pressure drop characteristics.

3.2. Effects of fin spacing and material

Figure 9 compares the heat transfer coefficient of flat finned tube bundles for the various different materials of construction for fin spacing ranging from 2.4 mm to 4.0 mm and inlet velocity 0.5 m/s. The heat transfer coefficient decreases with the increase of fin spacing. The maximum values of the heat transfer coefficient are 7.998, 27.470 and 23.853 W/(m^2·K) for stainless steel, aluminium and copper-nickel alloy, respectively. The results also show that the heat transfer coefficient of aluminium is 243% higher than that of stainless steel at the same inlet velocity and fin spacing.

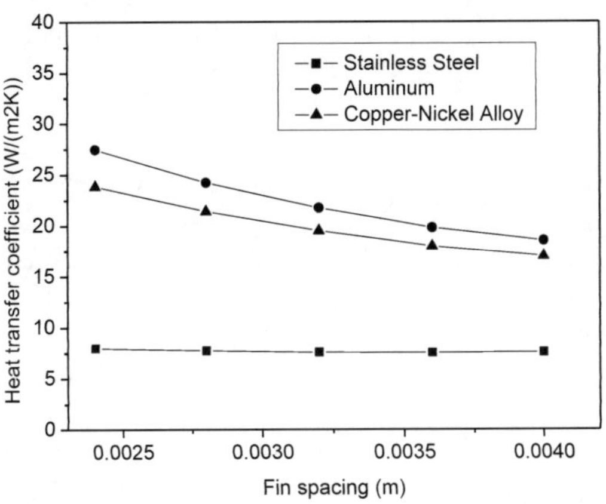

Figure 9. The effects of fin spacing on heat transfer coefficient (inlet velocity 0.5 m/s, tube pitch 52 mm).

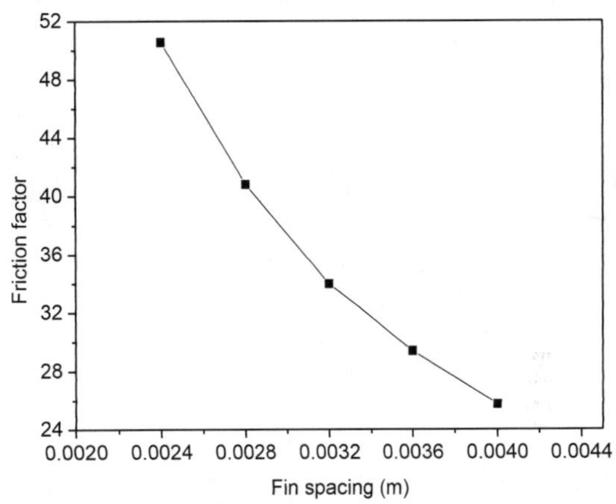

Figure 10. The effects of fin spacing on friction factor (inlet velocity 0.5 m/s, tube pitch 52 mm).

Since the simulated pressure drop coefficient corresponding to the three different materials are closely identical, only one friction factor curve is presented. The effects of fin spacing at the constant velocity on friction factor are depicted in figure 10. The friction factor decreases steeply when the fin spacing is increasing. It is shown that the increase of fin spacing contributes clearly to the reduction of the friction coefficient and the pressure drop (since the velocity is constant).

3.3. Effects of Tube pitch and material

The variation of the heat transfer coefficient for the three different materials as the tube pitch ranges from 34 mm to 52 mm with constant inlet velocity 1.0 m/s is depicted in figure 11. Along with the increase of tube pitch, the heat transfer coefficient decreases gradually. The thermal performance of copper-nickel is better than that of stainless steel, but inferior to that of aluminium. Figure 11 shows that the heat transfer coefficient of aluminium is 67% higher than that of stainless steel at the same inlet velocity and fin spacing.

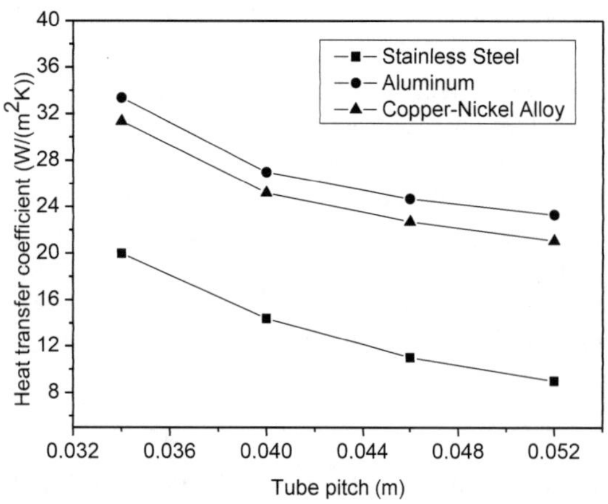

Figure 11. The effects of tube pitch on heat transfer coefficient (inlet velocity 1 m/s, fin spacing 3.2 mm).

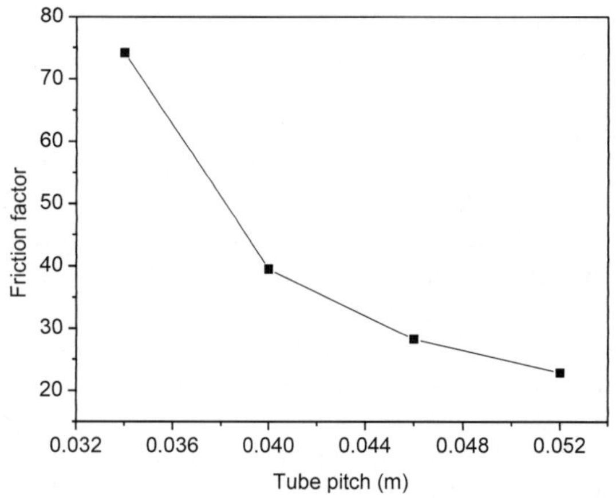

Figure 12. The effects of tube pitch on friction factor (inlet velocity 1 m/s, fin spacing 3.2 mm).

As before, the friction factor is given only for one type of material. Figure 12 shows the effects of tube pitch on friction factor. The friction factor decreases steeply when the tube pitch is ranging from 34 mm to 40 mm and drops gently afterwards. The result shows that the increase of tube pitch leads to reduction of the pressure drop.

4. Conclusions

The heat transfer and pressure drop characteristics for different fin-and-tube materials at various inlet velocities, fin spacing and tube pitches were numerically investigated. To this end, the flow and temperature fields were numerically simulated and the heat transfer coefficient and the friction factor were numerically predicted. The most effective material for heat transfer was proved to be aluminium, the heat transfer coefficient of which is much higher than that of stainless steel. Between the other two materials tested, copper-nickel alloy works better than the stainless steel in terms of the heat transfer coefficient. The heat transfer coefficient increases with the increase of inlet velocity, and decreases with the increase of fin spacing as well as the tube pitch for all the materials. The friction factors are almost identical for all the materials, i.e. pressure drop is independent of the fin-and-tube material. The friction factor has a clear decrease when the inlet velocity, fin spacing, and tube pitch increase. The heat transfer coefficient and pressure drop variation with various materials at different inlet velocity, fin spacing and tube pitch is expected to provide the useful information for reference in fin-and-tube material selection in engineering application.

Nomenclature

Nomenclature		Greek symbols	
A	area (m^2)	μ_t	turbulent viscosity (kg/(m·K))
x,y,z	coordinates (m)	μ	dynamic viscosity (Pa s)
u	velocity (m/s)	τ	shear stress (Pa)
P	pressure (Pa)	ρ	density (kg/m^3)
H	enthalpy (J/kg)	ε	Kinetic energy dissipation rate
S	source term	k	turbulent kinetic energy (m^2/s^2)
T	temperature (K)	λ	thermal conductivity (W/(m K))
h	heat transfer coefficient ($W/(m^2 \cdot K)$)	in	inlet
ΔT_m	logarithmic mean temperature difference (K)	O	outlet
ΔP	pressure drop (Pa)	σ	turbulent Prandtl number
c_p	specific heat at constant pressure (J/(kg·K))	q	heat flow rate (W)

m_a	air flow mass rate (kg/s)
KL	friction factor

Acknowledgments
This work is supported by the National Key Science and Technology Project (2015ZX06004004).

References
[1] Wang C-C, Chang Y-J, Hsieh Y-C and Lin Y-T 1996 Sensible heat and friction characteristics of plate fin-and-tube heat exchangers having plane fins *Int J Refrig* **19** 223-30
[2] He Y L, Tao W Q, Song F Q and Zhang W 2005 Three-dimensional numerical study of heat transfer characteristics of plain plate fin-and-tube heat exchangers from view point of field synergy principle *Int J Heat Fluid Fl* **26** 459-73
[3] Kang H J, Li W, Li H Z, Xin R C and Tao W Q 1994 Experimental study on heat transfer and pressure drop characteristics of four types of plate fin-and-tube heat exchanger surfaces *J Therm Sci* **3** 34-42
[4] Bhuiyan A A, Amin M R and Islam A K M S 2013 Three-dimensional performance analysis of plain fin tube heat exchangers in transitional regime *Appl Therm Eng* **50** 445-54
[5] Li J 2014 Numrical simulation study of finned tube heat exchanger performance with different materials (Guangzhou: Guangzhou University)
[6] Li C 2008 Analysis about aluminum replaced of copper in heat exchanger manufacture The 13th Academic Conference of Heat Pump and Systematic Energy-saving Technology (Gansu, China) p 4
[7] Yang L J, Jia S N, Bi Y D, Du X Z and Yang Y P 2012 Numerical study on flow and heat transfer characteristics of finned tube bundles for air-cooled heat exchangers of indirect dry cooling systems in power plants *Proceedings of the Chinese Society for Electrical Engineering* 50-8
[8] Hu H, Du X, Yang L and Yang Y 2014 POD based modeling on flow and heat transfer of air-cooled condenser influenced by natural wind *Int J Heat Mass Tran* **74** 431-40
[9] S. Xinwu, 2012 The analysis of flow resistance and heat transfer characteristics of the stainless steel fin tube heat exchanger (Guangzhou: South China University of Technology)
[10] Kong Y Q, Yang L J, Du X Z and Yang Y P 2016 Air-side flow and heat transfer characteristics of flat and slotted finned tube bundles with various tube pitches *Int J Heat Mass Tran* **99** 357-71
[11] Fluent A 2011 *Ansys Fluent Theory Guide* (ANSYS Inc., USA) **15317** 724-46

Sizing and economic analysis of stand alone photovoltaic system with hydrogen storage

N D Nordin[1,2] and H A Rahman[1,2,3]

[1]Centre of Electrical Energy Systems, Institute of Future Energy, Universiti Teknologi Malaysia, 81300, Johor, Malaysia
[2]Faculty of Electrical Engineering, Universiti Teknologi Malaysia, 81300, Johor, Malaysia

E-mail: hasimah@fke.utm.my

Abstract. This paper proposes a design steps in sizing of standalone photovoltaic system with hydrogen storage using intuitive method. The main advantage of this method is it uses a direct mathematical approach to find system's size based on daily load consumption and average irradiation data. The keys of system design are to satisfy a pre-determined load requirement and maintain hydrogen storage's state of charge during low solar irradiation period. To test the effectiveness of the proposed method, a case study is conducted using Kuala Lumpur's generated meteorological data and rural area's typical daily load profile of 2.215 kWh. In addition, an economic analysis is performed to appraise the proposed system feasibility. The finding shows that the levelized cost of energy for proposed system is RM 1.98 kWh. However, based on sizing results obtained using a published method with AGM battery as back-up supply, the system cost is lower and more economically viable. The feasibility of PV system with hydrogen storage can be improved if the efficiency of hydrogen storage technologies significantly increases in the future. Hence, a sensitivity analysis is performed to verify the effect of electrolyzer and fuel cell efficiencies towards levelized cost of energy. Efficiencies of electrolyzer and fuel cell available in current market are validated using laboratory's experimental data. This finding is needed to envisage the applicability of photovoltaic system with hydrogen storage as a future power supply source in Malaysia.

1. Introduction

World's electricity production mostly depends on coal, oil and natural gas as power supply. Due to limited fuels reserves, fuels' unstable prices and global warming risks, the interest in renewable energy (RE) sources increases significantly. The importance of RE is not only to reduce dependence on fossil fuels, but it is also environmently-friendly, free, and abundantly available [1].

Compared to other alternatives electricity generation available in Malaysia, photovoltaic (PV) system is the most promising RE. However, due to intermittence of solar irradiation, standalone photovoltaic (SAPV) system needs a backup energy storage system to ensure continuous power supply.

Among wide varieties of technology options for electrical energy storage system, hydrogen storage (HS) has been receiving a lot of attention due to clean quality, insightful environmental benefits, and its' regenerative hydrogen feature. Integrating PV with HS leads to construction of non-polluting generation system and reducing emissions. Introduction of HS in a small-scale power application could be one of the big breakthroughs to introduce hydrogen economy as an environmental friendly electrical storage system.

In SAPV system, hydrogen is produced from water by electrolysis powered by electricity generated from PV, and then stored in hydrogen tank. When load demand is higher than generated PV generation, the stored hydrogen and air are fed into a fuel cell (FC), then the reverse process of electrolysis takes place, and energy is supplied to consumers [2,3].

However, despite being high potential and endowed with high irradiation [4], SAPV with HS system is still being explored and has not being successfully applied in Malaysia due to high initial cost and requirement of large space area. Hence, further study on design and feasibility of SAPV with HS is necessary. Before any installation, it is crucial to ensure the recommended SAPV and HS system will not oversize/undersize, and designers have to investigate system viability in preliminary design. In order to efficiently and economically utilize solar energy, optimization in system sizing is necessary so that the proposed system can operate in optimum condition in terms of investment and reliability [5]. Designers also have to forecast electricity consumption and generation in order to ensure the proposed system can fulfill energy demand.

During the past ten years, more information is available on optimization of SAPV with HS system. A method to determine optimum design, control strategy, economic and performance analysis of SAPV with HS system at Aswan, Egypt was proposed [6]. The paper iteratively found optimum size for system's components that able to match four different load profiles (load profile of January, April, July, and October). By using autonomous simulation built from operation control strategies and energy balance mathematical modeling, loss of power supply probability (LPSP) was used to analyze system reliability. Detailed economic analysis was performed and presented in levelized cost of energy (LCOE). However, the sizing method is too complex and will need long calculation time.

In 2012, another authors [7] presented a direct mathematical approach to find optimal size of SAPV with HS and battery system based on load consumption and irradiation curve from selected site [6]. By using the calculated system size, the whole system was simulated using mathematical model in Matlab Simulink. Detailed energy flow was provided and energy flow to battery and HS depended on value of limit charge and discharge power. This work presented an extensive and insightful simulation results, but no economic analysis to validate system feasibility was performed.

In 2013 [2], the authors determined the size of SAPV with hydrogen and battery system using Simulink Design Optimization Method existed in Matlab. However, the paper mainly focused on modeling and control strategies for the system. The optimization was done only to satisfy load and maintain hydrogen energy reserve and battery's state of charge, with simple total cost comparison between SAPV/battery and SAPV/hydrogen storage system. System feasibility was not performed on the proposed system.

This paper presents sizing and economic analysis of SAPV system with HS using intuitive method. By using this method, sizing task becomes much simpler by using direct mathematical approach to find system size. A case study is conducted using Kuala Lumpur meteorological data and typical rural area daily load profile of 2.215 kWh. Finally, an economic analysis using LCOE is calculated to determine the system feasibility and to economically utilize the PV and HS system [5,8]. The results are compared to the calculated results using published method in 2014 [9], based on same initial cost rating, load and PV configurations. A sensitivity analysis is performed to investigate the effect of electrolyzer and fuel cell efficiencies towards levelized cost of energy in the future. Current PV panel and hydrogen storage efficiencies are validated using pre-processed measured data from an installed photovoltaic system in Universiti Teknologi Malaysia and laboratory's experimental data. The contribution of this paper is to introduce the integration of PV with HS, using prior information, assessment, and extensive sizing steps on each component with respect to Malaysia's local environment. This paper also envisage the applicability of photovoltaic system with hydrogen storage as a future power supply source in Malaysia

2. System configuration

The proposed configuration for the SAPV with HS system in Malaysia is shown in figure 1. During daytime, PV array converts solar irradiation into DC power. If PV array produces higher power than load demand, the surplus power will be transferred to electrolyzer. Electrolyzer uses the excess power to decomposition water into hydrogen and oxygen gasses. The hydrogen gas will be stored in hydrogen tank, and oxygen releases to the air. However, during insufficient PV generation, fuel cell draws hydrogen from tank and oxygen from the air, and converts both gasses into water to produce additional power for load. DC bus voltage is nominal voltages for PV array, HS, and DC appliance. Inverter converts DC power into AC power to match AC bus and AC appliances.

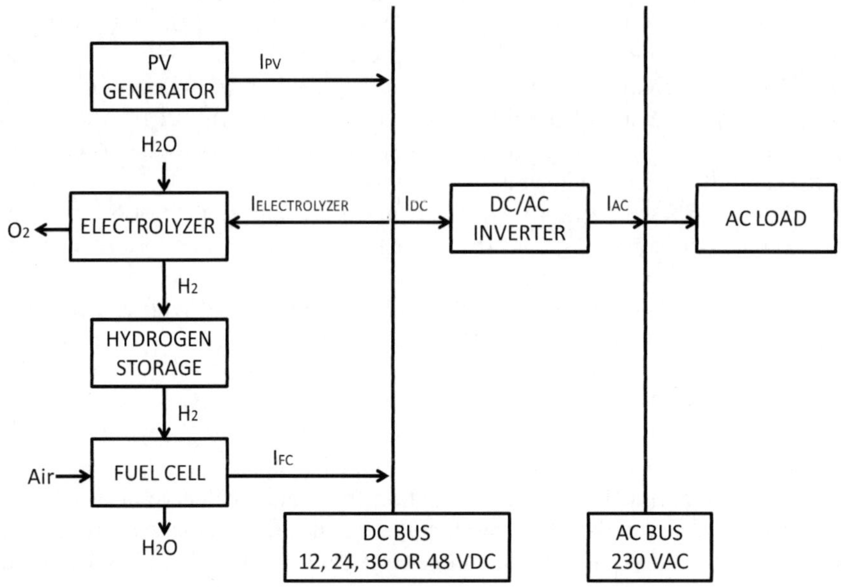

Figure 1. System configuration for SAPV with HSS.

2.1. Solar energy resources forecast

Accuracy in predicting PV generation plays a great significance in system design, where energy output from the proposed system is predicted by an inclusive study on site's meteorological condition [10]. In this paper, meteorological data was imported from Meteonorm 6.1. Table 1 shows the annual global irradiation for each states in Malaysia. In the other hand, the average monthly annual irradiation and PSH data for Kuala Lumpur is presented in table 2 [11]. It is shown that the state's annual global irradiation is 1655.59 kWh/m^2, maximum average annual irradiance over a year, H_{Smax} is 205.511 W/m^2 (March) and the lowest monthly daily's peak sun hour, PSH_{lowest} is 4.081 h/day (November).

Table 1. Annual global irradiation in Malaysia [11].

State	Global Irradiation, kWh/m²
Kuala Lumpur	1655.59
Labuan	1792.10
Putrajaya	1654.60
Terengganu	1771.49
Johor	1637.32
Kedah	1795.90
Kelantan	1782.71
Negeri Sembilan	1650.40
Pahang	1675.50
Perak	1760.00
Perlis	1788.50

Pulau Pinang	1794.90
Sabah	1733.23
Sarawak	1579.58

Table 2. Monthly average annual irradiation and daily's peak sun hour in Kuala Lumpur [11].

Month	Average Annual Irradiance, W/m²	PSH (h)
January	179.839	4.319
February	202.976	4.871
March	205.511	4.945
April	199.861	4.800
May	194.892	4.681
June	187.639	4.513
July	188.978	4.535
August	186.962	4.487
September	187.778	4.513
October	190.860	4.584
November	171.944	4.127
December	170.027	4.081

2.2. Load demand

Table 3 presents load profile for a house in rural area in Malaysia. It is assumed that the daily load remains the same throughout the year [12]. The AC load profile of 2.215 kWh consists of different types of home appliances. The hourly load profile is as illustrated in figure 2 with an assumption that the daily energy requirement remains the same throughout the year [9,12-15].

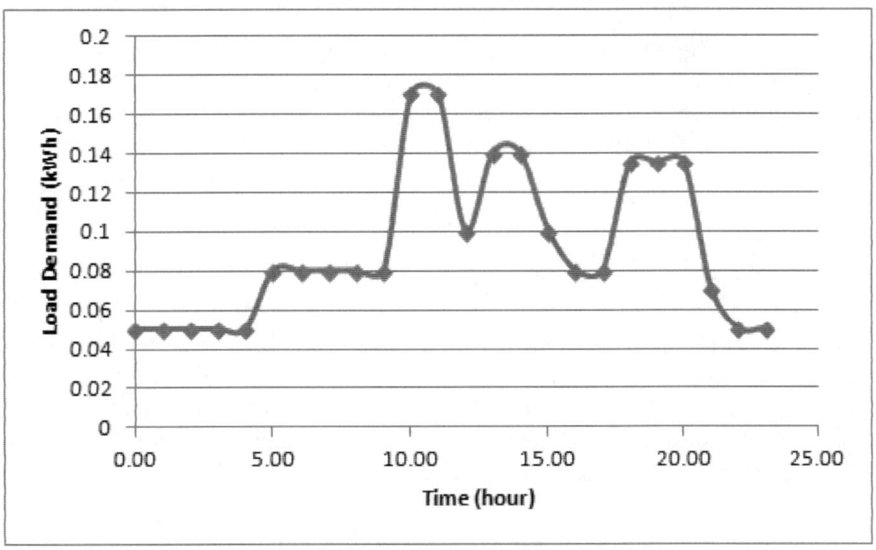

Figure 2. Hourly load profile [9, 12-15].

Table 3. Daily demands for a house of rural area [9,12-15].

Appliance	Voltage (V)	Power (W)	Daily Usage (h)	Energy (Wh)
Flourescent Lamp 1	230	20	10	200
Flourescent Lamp 2	230	40	4	160

TV	230	60	5	300
Refrigerator	230	50	24	1200
Radio Cassette	230	10	11	110
Ceiling Fan	230	60	2	120
Desk Fan	230	25	5	125
Total		265		2215

3. System design

This section consists of design process used to determine the size of SAPV components and each calculation involved in details. The analysis will be done in Amp-hour method. The advantages of Amp-hour analysis is taking account of the real world behavior of SAPV system components [16].

3.1. Estimate load demand

Design process started with estimation of total daily load requirement. First, a list of all lights and appliances, as well as loads' individual power, P_{load} and usage (in hours per day) is required. Modeling of load demand, E_{load} is calculated using equation (1). Extra amount of energy is needed to cover losses due to inefficiencies in system components. System charge requirement, Q_{load} (Ah) is calculated by equation (2), where AC losses, $_{AClosses}$ is 0.35 [9,16]. Equation (3) shows calculation of daily charge demand required, Q_{load_req}, where V_{DCbus} is DC bus voltage.

$$E_{load} = \sum P_{load} * t \tag{1}$$

$$Q_{load} = E_{load} * \left(1 + \eta_{AClosses}\right) \tag{2}$$

$$Q_{req} = \frac{E_{req}}{V_{DCbus}} \tag{3}$$

3.2. PV array selection and sizing

Before sizing the PV arrays, designers need to estimate solar energy available at the site. By selecting the month with the lowest PSH (h), PSH_{lowest}, is actually avoiding for the system to be undersize. The array sizing must match with total daily energy requirement, so by using the value obtained from equation (3), Q_{req} (Ah), system charging current from PV array, I_{charge} (A) is calculated using equation (4) [16].

$$I_{charge} = \frac{Q_{req}}{PSH_{lowest}} \tag{4}$$

The modules connection and arrangement is determined based on preselect PV model [17]. The number of parallel string, N_p is calculated using equation (5), where I_{pv_mp} is module's current maximum power under standard test condition. Next, equation (6) is used to calculate series-connected modules in each parallel string, $N_{mod/strg}$, where V_{mod_rated} is module's rated voltage. Lastly, total PV modules, N_{pv} can be calculated using equation (7) [16]:

$$N_p = roundup\left(\frac{I_{charge}}{I_{pv\_mp}}\right) \tag{5}$$

$$N_{mod/strg} = \frac{V_{mod\_rated}}{V_{DCbus}} \tag{6}$$

$$N_{pv} = N_p * N_{mod/strg} \tag{7}$$

3.3. Proton exchange membrane electrolyzer (PEME) sizing

The preferable/most common technologies used in water electrolyzers are Alkaline and Proton Exchange Membrane Electrolyzer (PEME). Commercially available PEME can produce from 400 to 7900 kg of hydrogen each year with pressure generation up to 13.8 bar. Meanwhile, alkaline electrolyzer produces 2400-71000 kg over a year. The advantage of PEME technology is it has high energy efficiency for each stack (75%-88%) with 5% as the minimum required output flow for a safe operation [18].

An electrolyzer decomposes water into hydrogen and oxygen by passing an electrical current (DC) between two electrode separated by an aqueous electrolyte with good ionic conductivity. The reactions take place at the anode and the cathode of electrolyzer as follows [19].

Anode: $$H_2O \rightarrow \frac{1}{2}O_2 + 2H^+ + 2e^- \tag{8}$$

Cathode: $$2H^+ + 2e^- \rightarrow H_2 \tag{9}$$

Storage charge capacity required, Q_{storeq} is calculated based on equation (10) [16,20], where N_c is battery's reserved days, and DOD is battery's depth of discharge. N_c usually set as 1- 4 days, and it is advised that DOD should not be over 60%. Energy storage consumption, E_{stored} is the amount of energy consumed to produce required hydrogen, and calculated using equation (11), where η_{ely} is electrolyzer efficiency, and η_{fc} is fuel cell efficiency (η_{ely} is 0.75 and η_{fc} is 0.5).

$$Q_{storeq} = \frac{Q_{req} * N_c}{DOD} \tag{10}$$

$$E_{stored} = \frac{Q_{stpreq} * V_{DCbus}}{\eta_{ely} * \eta_{fc}} \tag{11}$$

Meanwhile, the rated power for the electrolyzer, P_{Ely} is equal to the maximum excess in the PV generation power over the minimum load power; where A_{PV} is PV panel area, and η_{PV} is PV panel efficiency [7,18].

$$P_{Ely} = H_{S\max} * A_{PV} * \eta_{PV} \tag{12}$$

3.4. H_2 storage tank sizing

Hydrogen can be stored in liquid or gas condition. Liquid hydrogen is stored in cryogenic tanks and gaseous hydrogen stored in either medium or high-pressure cylinders or near atmospheric pressure in metal hydrides.

The hydrogen formed by the electrolyzer during each month is calculated by assuming the produced hydrogen is stored completely before it supplies power to the residential load. The amount of hydrogen produced annually is the total excess of PV power transformed into volume, which is expressed in m^3 [7].

Equation (13) is the hydrogen mass produced from specific electrical energy, m_{h2} where HHV is higher heating values of hydrogen (33 kWh/kg). The hydrogen's required volume capacity in Normal cubic meters unit (Nm^3), V_{h2} is calculated in the next equation, where ρ_{h2} is hydrogen density and its value is 0.08988 kg/m^3 [18].

$$m_{h_2} = \frac{E_{storeq}}{HHV} \tag{13}$$

$$V_{h_2} = \frac{m_{h_2}}{\rho_{h_2}} \tag{14}$$

3.5. Proton exchange membrane fuel cell (PEMFC) sizing

The reverse reaction occurred in the electrolyzer takes place in fuel cell. Hydrogen in the anode is ionized releasing electrons and protons. Electrons flow to the cathode through a circuit producing electric current. Protons diffuse through a polymer electrolyte membrane and react at cathode with oxygen and electrons to form water. The chemical reactions happened in PEMFC are as below [19].

Anode: $$H_2 \rightarrow 2H^+ + 2e^- \tag{15}$$

Cathode: $$\frac{1}{2}O_2 + 2H^+ + 2e^- \rightarrow H_2O \tag{16}$$

The following equation is fuel cell sizing by considering maximum AC load power in the household, where M_{fc} is margin coefficient of the fuel cell, P_{ACload} is total power from AC load demand and η_{inv} is inverter efficiency [7].

$$P_{fc} = M_{fc} * \frac{P_{ACload}}{\eta_{inv}} \tag{17}$$

3.6. Inverter selection and sizing

Inverter is expected to deliver maximum AC load in household. Hence, inverter power rating, P_{inv} is selected using equation (18), where 1.25 is set as oversized factor [21].

$$P_{inv} = P_{ACload} * 1.25 \tag{18}$$

4. Economic analysis

Economic analysis is performed using Malaysian Ringgit, (*MYR).* The viability of proposed system can be determined by using economic analysis. The designer will know whether the investment is feasible or not. In this paper, the life cycle cost (LCC) and levelized cost of energy (LCOE) is used to analyze system practicality. LCC is the sum of installation cost, operating and maintenance of an item for a period of time, and replacement cost in present value [22],

$$\begin{aligned} LCC = {} & C_{pv} + C_{inv} + C_{ely} + C_{h2} + C_{fc} + \\ & C_{install} + C_{cont} + C_{invrep} + C_{O\&M\_20\,years} - C_{salvage} \end{aligned} \tag{19}$$

where C_{pv} is PV initial array cost, C_{inv} is inverter initial cost, C_{ely} is electrolyzer initial cost, C_{h2} is hydrogen tank initial cost, C_{fc} is fuel cell initial cost, $C_{install}$ is support structure and installation cost, C_{cont} is system contingencies value (1% from initial cost), C_{invrep} is present values for inverter replacement, $C_{O\&M\_20years}$ is present worth of operation and maintenance cost for 20 year and $C_{salvage}$ is system's salvage value (20% from initial cost).

Initial price of PV panel and inverter is presented by equations (20) and (21). Initial price of electrolyzer, hydrogen storage and fuel cell is presented by equations (22)-(24). Initial price for support structure and installation cost is calculated using equation (25) and initial price's contingencies value is calculated using equation (26).

$$C_{pv} = \text{Price}_{pv} * N_{pv} \tag{20}$$

$$C_{inv} = \text{Price}_{inv} * N_{inv} \tag{21}$$

$$C_{ely} = \text{Price}_{ely} * P_{ely} \tag{22}$$

$$C_{h2} = \text{Price}_{h2} * V_{h_2} \tag{23}$$

$$C_{fc} = \text{Price}_{fc} * P_{fc} \tag{24}$$

$$C_{install} = \text{Price}_{install} * N_{pv} * P_{pv\_mp} \tag{25}$$

$$C_{cont} = \frac{1}{100} * C_{pv} + C_{inv} + C_{ely} + C_{h2} + C_{fc} + C_{install} \tag{26}$$

where N_{inv} is number of inverter and P_{pv_mp} module's maximum power output under standard test condition.

All components considered to have 20 years of lifetime, beside inverter (Inverter replacement is 10 years) [23,24]. C_{invrep} is calculated by equation (28), where N is component's lifetime and i is market rate. $C_{O\&M}$ is operation and maintenance cost for each year and it is predicted to be 1% from total initial cost.

$$C_{invrep} = C_{inv}\left(\frac{1}{1+i}\right)^N \tag{27}$$

$$C_{O\&M\_20years} = (C_{O\&M}) * \left[\frac{(1+i)^N - 1}{i(1+i)^N}\right] \tag{28}$$

Market interest rate, i is calculated using equation (29), where i' is real interest rates determined by local bank. i' is calculated as equation (30) [25]:

$$i = i' + \overline{f} - i'\,\overline{f} \tag{29}$$

$$i' = BLB - 2\% \tag{30}$$

where BLB is Base Lending Rates (Malaysia BLR is 6.60%), and $\overline{f}$ is inflation rate (Malaysia inflation rates is 2014 is 3.4%) [26,27].

LCOE (RM/kWh) is defined as the average cost per kWh of electrical energy produced by PV system [28]. It is calculated by dividing annualized life cycle cost, LCC_{1year} with total useful electrical energy generated, $E_{PV,}$ as calculated by the following equations [29]. In economic analysis calculation, market prices for the proposed system components are summarized in table 4 [8,18,24,30-33].

$$LCC_{1year} = \frac{LCC}{\left[\frac{(1+i)^N}{i(1+i)^N}\right]} \tag{31}$$

$$LCOE = \frac{LCC_{1year}}{E_{PV}} \tag{32}$$

$$E_{PV} = N_{PV} * P_{mp\_STC} * PSH_{year} * \eta_{losses} \tag{33}$$

Table 4. Components' pricing [9, 18, 24].

Components	Symbol	Unit	RM/Unit
Solar Panel	$Price_{pv}$	1	886.17
Inverter	$Price_{inv}$	1	560.71
PEME	$Price_{ely}$	RM/W	6.60
H2 Storage Tank	$Price_{h2}$	RM/Nm3	2280
PEMFC	$Price_{fc}$	RM/W	1.38
Support Structure & Installation cost	$Price_{install}$	RM/W	4.00

5. Results and analysis

This section shows the system sizing result and economic analysis for a project lifetime of 20 years.

IOP Conf. Series: Earth and Environmental Science **93** (2017) 012068 doi:10.1088/1755-1315/93/1/012068

5.1. System sizing results

Table 5 below shows technical data of each component. A combination of 1.12 kWh PV modules, 335.4 W PEME, 324.3 Nm^3 H_2 Tank, 285.0 W PEMFC, and 340 W inverter were chosen as proposed system.

Table 5. Proposed system sizing.

Daily System Energy Requirement		
Total Daily AC Energy Demand	2215.00	Wh
AC losses	775.25	Wh
Total Daily System Energy Requirement	2990.25	Wh
System Voltage	24.00	VDC
Daily System Charge Requirement	124.59	Ah
Sizing and Choosing the Modules		
Lowest Daily's Peak sun hours	4.08	h
System Design Charging Current	30.53	Amps
Details of Selected Modules		
Model	Kyocera KD140	
Type	Multicrystalline	
Peak Watts	140.00	Wp
Rated Voltage	12.00	VDC
Maximum Power Rated Current	7.91	Amps
Short Circuit Current	8.68	Amps
PV Efficiency	20.40	%
Number of Parallel String	4	
Number of Modules/ String	2	
Num of Modules	8	
PV Array Area	8.016	m^2
Hydrogen Storage Sizing and Selection		
Daily System Charge Requirement	124.59	Ah
Reserve Days	2	day
Maximum DOD	50	%
Required System Storage Capacity	498.38	Ah
Details of PEME Sizing		
Electrolyzer Efficiency, η_{elz}	75	%
Energy Storage of Electrolyzer	22653.4	Wh
Rated Power of Electrolyzer	335.41	W
Details of H_2 Tank Sizing		
Mass of Hydrogen	0.97	kg
Hydrogen Storage Capacity	10.75	Nm^3

Details of PEMFC Sizing		
Fuel Cell Efficiency, η_{fc}	50	%
Rated Power of Fuel Cell	284.95	W
Inverter Sizing and Selection		
Maximum Power AC Load	265	W
Inverter Rating	340	W
Details of Selected Inverter		
Model	Cotek	SK350-224
Rated Voltage	24/230	VDC/VAC
Rated Power	350	W
Inverter Efficiency	93	%

5.2. Economic analysis results

Table 6 below shows the economic analysis of the proposed system, as presented in previous section. Discount / interest rates value used in economic analysis is 4.60%, inflation rates value used is 3.40% and net discount rates value is 8.00%. Based on the analysis, initial cost for equipment, installation and structure is RM 39,559.79. The present value calculated for operation and maintenance is RM 3,884.06, and replacement cost is RM 259.72. The salvage value obtained for the whole system is RM 7,911.96. Hence, LCC for 20 years is RM 35,791.61, and annual LCC is RM 3,645.46. E_{PV} for the proposed system is 1828.305 kWh. Therefore, from the result, LCOE calculated is RM 1.98/kWh.

Table 6. Economic analysis.

Initial Cost		
	Components	RM
Energy Equipment	PV Module(s)	7088.48
	Electrolyzer	2213.70
	H_2 Storage Tank	24518.63
	PEMF	393.23
	Inverter	560.70
Balance of Equipment	Module Support Structure & System Installation	4480.00
Miscellaneous	Contingencies	392.55
Total Initial Cost		39647.29
O&M Cost		
	Components	RM
Annual O&M		396.48
Total Operation & Maintenance		3892.7
Replacement Cost		
Components		RM
Inverter	LCC $inverter_{n10}$	259.72
Inverter Replacement		259.72
Salvage		
Components		RM

Salvage Value	Salvage	7929.46
Result		RM
LCC Cost for 20 Years		35870.25
LCC Cost / Year		3653.47
Cost of Energy		1.98

5.3. Comparison of SAPV with AGM battery system sizing

The current results are compared with conventional sizing method and economic analysis of SAPV system with Absorbent Glass Mat (AGM) battery, which published in 2014 [9]. The finding shows that the LCC for the proposed system is RM 34,232. Meanwhile, LCC for a year is RM 3,206.82 and LCOE for the respective case is RM 1.74/kWh. Table 7 shows the comparison of current findings with previously published results.

Table 7. Economic analysis.

	SAPV with Hydrogen Storage	SAPV with AGM Battery [9]
LCC Cost for 25 Years (RM)	35791.61	34231.98
LCC Cost / Year (RM/year)	3645.46	3206.81
Cost of Energy (RM/kWh)	1.98	1.74

Based on the table, the LCC, annual LCC and LCOE for system with HSS has higher expenditure compared to the published findings. AGM battery needs to be replaced every 5 years. Meanwhile, PEME, H_2 Tank and PEMFC do not have to be replaced because the lifetime is longer, which is 20 years [24]. However, the initial cost needed for HSS is very high because fuel cell efficiency is quite low (50%). Due to that, bigger HSS capacity is needed to have back-up energy similar with AGM battery during insufficient PV generation.

6. Sensitivity analysis

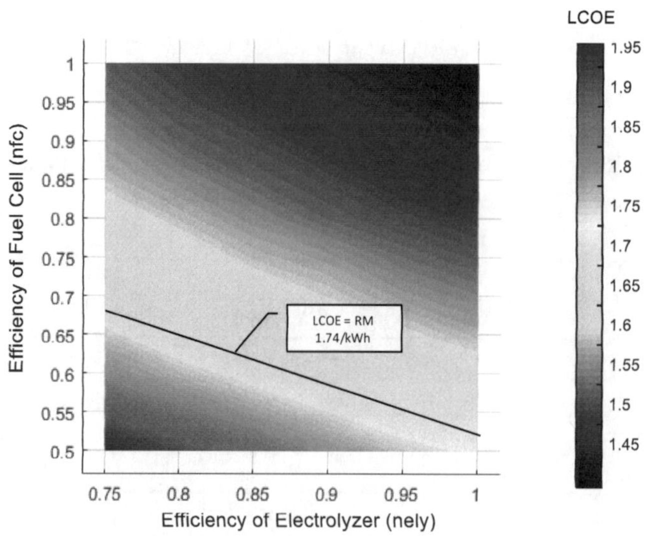

Figure 3. Effect of electrolyzer efficiency and fuel cell efficiency on levelized cost of energy.

The sensitivity of system's LCOE will be affected by η_{elz} and η_{fc}. In the future, enhancing η_{elz} and η_{fc} is crucial to ensure SAPV with HS can compete with conventional SAPV with AGM Battery system. A benchmark value of η_{elz} and η_{fc} to acquire LCOE equal or lower than conventional system will be obtained from this analysis. A range of 0.75 - 1.00 is considered for η_{elz} and a range of 0.50 - 1.00 is for η_{fc}. The effect of η_{elz} and η_{fc} on LCOE is shown in figure 3 above.

Based on figure 3, for a specific LCOE value (eg: RM1.74 kWh), the relation between η_{elz} and η_{fc} is approximately linear. Hence, the result can be simplified by using relationship between product of η_{elz} and η_{fc} ($\eta_{elz}*\eta_{fc}$) and LCOE, as represented in figure 4. From figure 4, it shows that in order to obtain an LCOE lower than SAPV with AGM battery, $\eta_{elz}*\eta_{fc}$ has to be higher than 0.52.

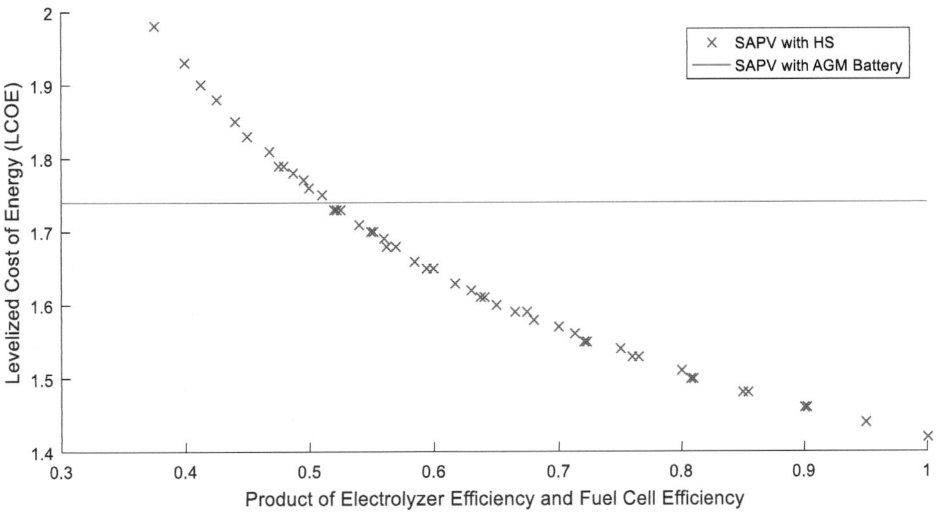

Figure 4. Effect of $\eta_{elz}*\eta_{fc}$ on levelized cost of energy.

7. Validation of electrolyzer and fuel cell efficiencies

Validation of efficiency for PEME and PEMFC available in current market is performed and analyzed in this section. Electrical efficiency of electrolyzer, η_{ely} is verified by experimental setup and the result is shown in the figure below. By using constant input voltage and current on electrolyzer (1.81VDC and 1.11ADC), time required to produce each 5 cm^3 of hydrogen is measured, as illustrated in figure 5. η_{elz} is calculated as equation (34), where the η_{ely} actual value obtained was 75.8%, which is within range claimed by previous reference and validated the value use in this analysis (0.75) [11].

$$\eta_{ely} = \frac{E_{hydrogen}}{E_{electrical}} = \frac{V_{H_2} * LHV}{V * I * t} = \frac{0.00003cm^3 * 9.9 * 10^6 \ J/m^3}{1.81V * 1.11A * 195s} = 0.758 = 75.8\% \quad (34)$$

PEMFC efficiency was also calculated based on results obtained from experiment. To obtain constant output voltage and current from fuel cell (2.87VDC and 0.337ADC), time required to consume each 5 cm^3 of hydrogen is measured, as illustrated in figure 6. η_{fc} is calculated as equation (35) below. Unfortunately, η_{fc} actual value obtained was 43.8%, which is lower than efficiency value used in this sizing method (0.5). Hence, it is advised to make additional review in the future to search for better fuel cell with higher efficiency value in order to obtain more reliable system.

$$\eta_{fc} = \frac{E_{electrical}}{E_{hydrogen}} = \frac{V * I * t}{V_{H_2} * LHV} = \frac{2.87V * 0.337A * 112s}{0.000025cm^3 * 9.9 * 10^6 \ J/m^3} = 0.438 = 43.8\% \quad (35)$$

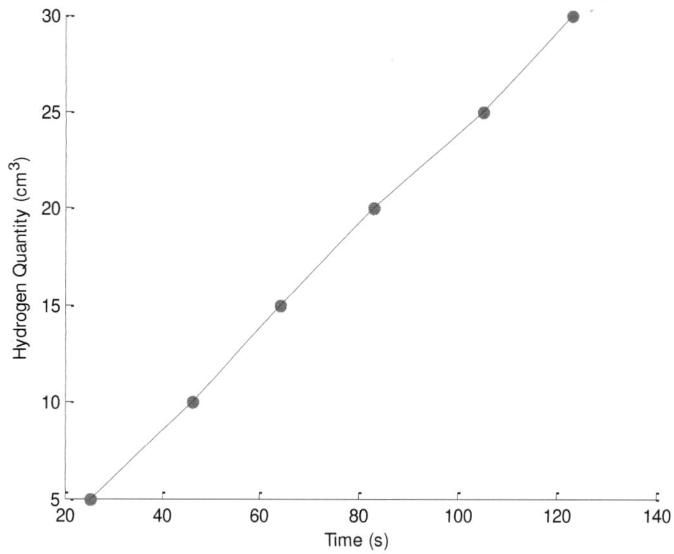

Figure 5. Volume of hydrogen's produced from PEME vs time diagram.

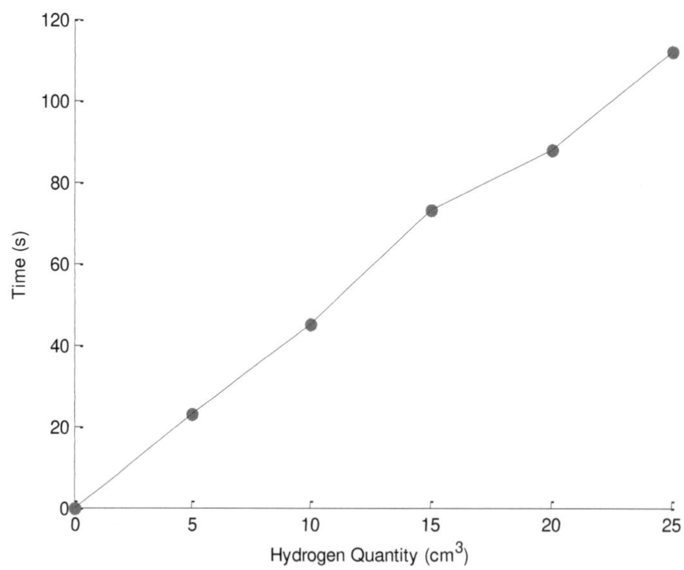

Figure 6. Hydrogen consumption of PEMFC vs time vs power produced.

8. Conclusion

Sizing and economic analysis of SAPV with HSS has been presented. A small scale household demand and daily's average solar irradiation was used to design off-grid PV system. Economic assessment is employed to calculate the life cycle cost and levelized cost of energy analysis. It shows the LCC for proposed system is RM 35,791.61. Meanwhile, annual LCC is RM 3,645.46 and LCOE value obtained is RM 1.98 kWh.

However, compared to economic analysis of SAPV with AGM batteries from the previously published results [9], the expenditure for SAPV with AGM battery is lower, where LCC for the system is RM 34,232, annual LCC is RM 3,206.82 and the LCOE for the respective cases is RM 1.76/kWh. Based on current findings, even though hydrogen energy storage has high energy density per mass, the system has low efficiency as a storage system and still expensive. Due to the low efficiency storage, designers need to allocate hydrogen storage capacity higher than AGM battery. However, as an emerging commercialize product, the manufacturers and researchers can focus on how to increase storage system efficiency for the future usage. The HSS round-up trip must improve dramatically before they can offer the same overall energy efficiency as batteries.

Therefore, a sensitivity analysis is performed to find a benchmark value of electrolyzer and fuel cell efficiencies to acquire acceptable levelized cost of energy. From the analysis, in order to obtain an LCOE lower than SAPV with AGM battery, $\eta_{elz} * \eta_{fc}$ has to be higher than 0.52. Besides, the electrolyser efficiency and fuel cell in current market are validated using laboratory's experimental data. This finding is needed to envisage the applicability of photovoltaic system with hydrogen storage as a future power supply source in Malaysia.

Acknowledgments
The authors wish to acknowledge the supports for this research paper received from Malaysian Ministry of Higher Education (MOHE) and Universiti Teknologi Malaysia leading to this article under the Tier 1 grant (vote no: Q.J130000.2523.14H02).

References

[1] Mellit A and Kalogirou S A 2008 Artificial intelligence techniques for photovoltaic applications: A review *Progress in Energy and Combustion Science* **34** 574-632

[2] Castañeda M, Cano A, Jurado F, Sánchez H and Fernández L M 2013 Sizing optimization, dynamic modeling and energy management strategies of a stand-alone PV/hydrogen/battery-based hybrid system *International Journal of Hydrogen Energy* **38** 3830-45

[3] Pellow M A, Emmott C J, Barnhart C J and Benson S M 2015 Hydrogen or batteries for grid storage? A net energy analysis *Energy & Environmental Science* **8** 1938-52

[4] Mustapa S I, Peng L Y and Hashim A H 2010 Issues and challenges of renewable energy development: A Malaysian experience *Energy and Sustainable Development: Issues and Strategies (ESD), 2010 Proceedings of the international conference on* 1-6

[5] Zhou W, Lou C, Li Z, Lu L and Yang H 2010 Current status of research on optimum sizing of stand-alone hybrid solar–wind power generation systems *Applied Energy* **87** 380-9

[6] Elbaset A A 2011 Design, modeling and control strategy of PV/FC hybrid power system *J. Electrical Systems* **7** 270-86

[7] Jallouli R and Krichen L 2012 Sizing, techno-economic and generation management analysis of a stand alone photovoltaic power unit including storage devices *Energy* **40** 196-209

[8] Barsoum N and Vacent P 2007 Balancing cost, operation and performance in integrated hydrogen hybrid energy system *First Asia International Conference on Modelling & Simulation* 14-8

[9] Nordin N D and Rahman H A 2014 Design and economic analysis in stand alone photovoltaic system *2014 IEEE Conference on Energy Conversion (CENCON)* 152-7

[10] King D L, Boyson W E and Kratochvil J A 2002 Analysis of factors influencing the annual energy production of photovoltaic systems *Photovoltaic Specialists Conference, 2002. Conference Record of the Twenty-Ninth IEEE* 1356-61

[11] 23rd February *PVsyst: Software for Photovoltaic System*. Available: http://www.pvsyst.com/en/

[12] Shen W 2009 Optimally sizing of solar array and battery in a standalone photovoltaic system in Malaysia *Renewable Energy* **34** 348-52

[13] Nordin N D and Rahman H A 2016 A novel optimization method for designing stand alone photovoltaic system *Renewable Energy* **89** 706-15

[14] Nordin N D and Rahman H A 2015 An optimization method for designing stand alone photovoltaic system using iterative method *2015 IEEE International Conference on Smart Energy Grid Engineering (SEGE)* 1-6

[15] Nordin N and Rahman H A Assessment of conventional and improved SAPV sizing methods using statistical indicator

[16] M. Hankins, *Stand-alone Solar Electric Systems: The Earthscan Expert Handbook for Planning, Design and Installation*: Earthscan, 2010.
[17] (20 May 2014). *PV Modules in a String*. Available: https://nationalvetcontent.edu.au/alfresco/d/d/workspace/SpacesStore/4e1e0e97-bc55-4c44-a692-4d31f4d0d934/13_02/content_sections/learn_about/08_solar_page_010.htm
[18] H. Tebibel and S. Labed, "Design and sizing of stand-alone photovoltaic hydrogen system for HCNG production," *International Journal of Hydrogen Energy,* vol. 39, pp. 3625-3636, 2014.
[19] Ipsakis D, Voutetakis S, Seferlis P, Stergiopoulos F and Elmasides C 2009 Power management strategies for a stand-alone power system using renewable energy sources and hydrogen storage *International Journal of Hydrogen Energy* **34** 7081-95
[20] Nafeh A 2009 Design and economic analysis of a stand-alone PV system to electrify a remote area household in Egypt *Open Renewable Energy Journal* **2** 33-7
[21] Sulaiman S I, Rahman T K A, Musirin I and Shaari S 2011 Sizing grid-connected photovoltaic system using genetic algorithm, *2011 IEEE Symposium on Industrial Electronics and Applications (ISIEA)* 505-9
[22] Applasamy V 2011 Cost evaluation of a stand-alone residential photovoltaic power system in Malaysia *2011 IEEE Symposium on Business, Engineering and Industrial Applications (ISBEIA)* 214-8
[23] Sullivan W G, Wicks E M and Luxhoj J T 2009 *Engineering economy*: Pearson Prentice Hall
[24] Zoulias E and Lymberopoulos N 2007 Techno-economic analysis of the integration of hydrogen energy technologies in renewable energy-based stand-alone power systems *Renewable Energy* **32** 680-96
[25] Park C S, Kumar P and Kumar N 2004 *Fundamentals of engineering economics*: Pearson/Prentice Hall
[26] 1 December 2014 *Bank lending rates* Available: http://www.blr.my/
[27] 2017 August *Malaysia inflation rates* Available: http://www.tradingeconomics.com /malaysia/inflation-cpi
[28] Kamel S and Dahl C 2005 The economics of hybrid power systems for sustainable desert agriculture in Egypt *Energy* **30** 1271-81
[29] Näsvall D 2013 Development of a model for physical and economical optimization of distributed PV systems
[30] 20 May 2014 *Sealed-AGM batteries* Available: http://www.wholesalesolar.com /products.folder/battery-folder/Surretterolls.html#SealedAGMBatteries
[31] 20 May 2014 *140 Watt black frame solar panel* Available: http://www.ecodirect.com/Kyocera-KD140GX-LFBS-140-Watt-PV-Panel-p/kyocera-kd140gx-lfbs.htm
[32] 20 May 2014 *Blue sky energy* Available: http://www.blueskyenergyinc.com/pricing/
[33] 20 May 2014 *350 Watt 24 volt inverter / pure sine wave* Available: http://www.ecodirect.com/Cotek-SK350-124-350W-24V-Pure-Sine-Inverter-p/cotek-sk350-124.htm

Research on fault characteristics about switching component failures for distribution electronic power transformers

Z X Sang[1, 3], J Q Huang[2], J Yan[1], Z Du[1], Q S Xu[2], H Lei[2], S X Zhou[1] and S C Wang[2]

[1]State Grid HBEPC Economic & Technology Research Institute, Wuhan 430077, China
[2]State Grid Laboratory for Hydro-thermal Power Resources Optimal Allocation & Simulation Technology, Wuhan 430077, China

E-mail: sang@hust.edu.cn

Abstract. The protection is an essential part for power device, especially for those in power grid, as the failure may cost great losses to the society. A study on the voltage and current abnormality in the power electronic devices in Distribution Electronic Power Transformer (D-EPT) during the failures on switching components is presented, as well as the operational principles for 10 kV rectifier, 10 kV/400 V DC-DC converter and 400 V inverter in D-EPT. Derived from the discussion on the effects of voltage and current distortion, the fault characteristics as well as a fault diagnosis method for D-EPT are introduced.

1. Introduction

With the development of power electronic device and technology, the electronic power transformer (EPT), or solid state transformer (SST) and power electronic transformer (PET) has attracted more attention recently, although its concept has been mentioned in 1970s for the first time [1]. Working as a power electronic device, the EPT becomes a flexible energy source [2].

The reliability of EPT is highly related to its power electronic devices as their failures may cause the problem to the EPT and then the power grid; hence the fault characteristics and fault location scheme should focus on the power electronic converters, especially power switches in EPT.

Failures of switches are mainly classified to short and open circuit faults [3]. The short circuit usually happens very fast which requires a triggering circuit protection [4]. The protection of open circuit faults for power converters are studied in [5,6].

The original cause of open circuit fault is always a short circuit fault-induced rupture [5], gate circuit failure [6] and loose wire [7]. When open circuit fault happens, the power device might continue working steadily, however in an abnormal state which could possibly increase the voltage stress or current stress on other switches.

The previous achievements on fault protection of power switch are mainly based on driver circuit [8,9], and configuration design of EPT [10,11], applications in power grid [12], energy management systems which take EPT as a component in power system [13]. A new topological circuit structure with multi-windings medium frequency transformer is introduced to improve the reliability of EPT in [14], but the study mainly focuses on protection schemes that could achieve redundancy in cascaded H-bridge (CHB) converter in EPT, rather than the nature of fault characteristics. In [15], the analysis of a MMC based EPT is introduced.

In this paper, the characteristics of D-EPT after the switching component failures are proposed to locate the failure spot and protect the D-EPT system. Section 2 presents the basic circuit schematics and control method of D-EPT. The fault characteristics of D-EPT switching components are presented in Section 3, and then verified by simulation in Section 4. The failure switch location scheme is discussed in Section 5, and conclusion is presented in the end.

2. Basic topology and control methods

2.1. Basic topology and application

The topology in figure 1(a) shows basic sections that include AC-DC converters, DC-DC converters and DC-AC converters in Distribution Electronic Power Transformer.

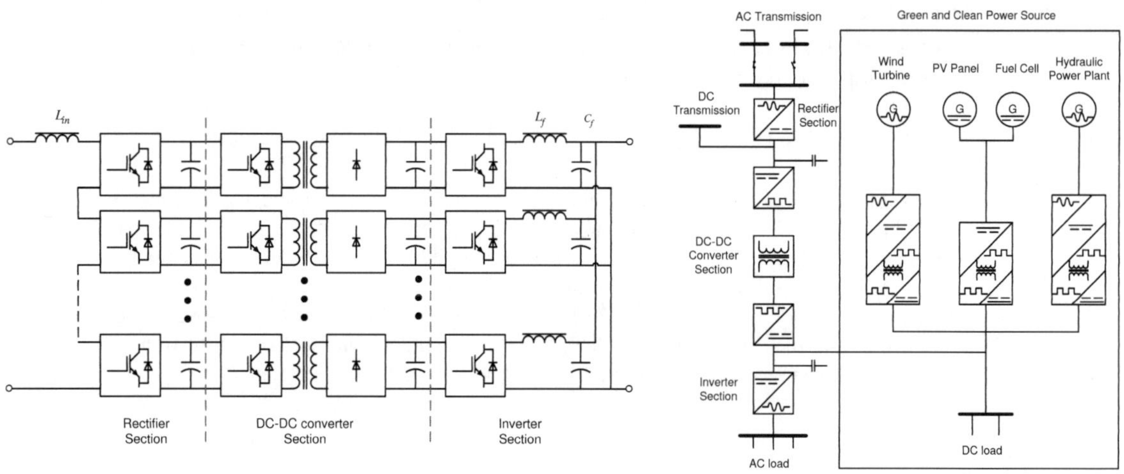

(a) (b)

Figure 1. The design of D-EPT in (a) Topology; (b) Substation.

Cascaded by a series of H-bridge converters, the DC-AC converters that form the rectifier section of D-EPT transfers input voltage to DC voltage of DC-DC converter section.

The DC-DC converter section consists of a DC-AC converter, an Isolated High Frequency Transformer (IHFT) and an AC-DC converter. The DC output voltage of rectifier section will be transferred into high frequency AC voltage by the H-bridge converter, and then transmitted from the primary side of IHFT to its secondary side, at last rectified to DC voltage for inverter section by diode rectifier.

The DC-AC converters that form the inverter section are parallelled, providing AC output for AC load.

The EPT shown in figure 1(b) can be set in a substation which links the transmission lines and the renewable sources. DC and AC loads as well as DC and AC sources can be perfectly compatible in the EPT based substation as EPT has been designed to access both DC and AC system.

2.2. Control methods of converters in different sections

The fault characteristics are highly related to the configuration and control scheme for a power electronic device, hence the control methods of rectifier section, DC-DC converter section and inverter section are of great importance for the analysis of fault characteristics of D-EPT.

The rectifier section of D-EPT which connects transmission system and HVDC system in figure 1(b) works as a PFC, making the I_{in} signal in figure 2 sinusoidal and V_{dc} signal including Rectifier Section DC-Link (RSDCL) voltage (10 kV) as well as Inverter Section DC-Link (ISDCL) voltage

(400 V) controllable.

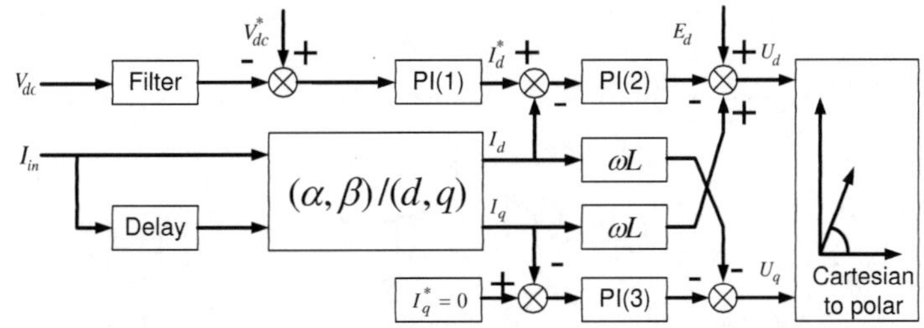

Figure 2. The rectifier section control strategy of D-EPT.

The rectifier section model is presented below:

$$\begin{cases} L\dfrac{dI_d}{dt} = E_d - U_d + \omega L I_q \\ L\dfrac{dI_q}{dt} = -U_q + \omega L I_d \end{cases} \tag{1}$$

where L is the inductance and ω is the angular velocity.

The H-bridge converter of DC-DC Converter Section will first transform RSDCL voltage to AC voltage, and then transmit it to diode rectifier through the IHFT, and rectified to DC voltage in ISDCL at last; hence the model of the DC-DC converter section is given by:

$$V_{RSDC} = k \times V_{ISDC} \tag{2}$$

where k indicates winding ratio of IHFT; V_{RSDC} is RSDCL voltage; V_{ISDC} is ISDCL voltage .

The H-bridge inverters in inverter section are parallelled, and controlled by the same triggering signal, which is presented as:

$$\begin{cases} C_f \times \dfrac{dv_o}{dt} = i_L - i_o \\ L_f \times \dfrac{di_L}{dt} = v_i - v_o \end{cases} \tag{3}$$

where C_f is the capacitance of the output filter; L_f is the inductance of the output filter; v_i is the input AC voltage and i_L is the input AC current of the output filter; v_o is the output AC voltage and i_o is the output AC current of the output filter.

Based on equation (3), the single phase control strategy of the inverter section is shown in figure 3.

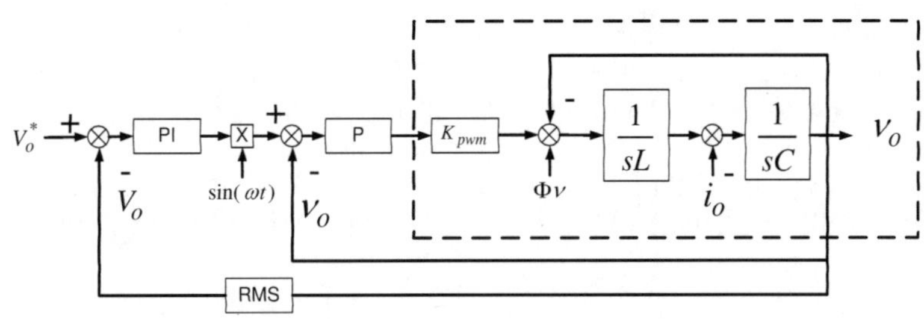

Figure 3. The inverter section control strategy of D-EPT.

3. Switching component failure classification

The short circuit and open circuit could happen on any switching components of D-EPT, including diode, IGBT and its body diode, which have been marked on figure 4.

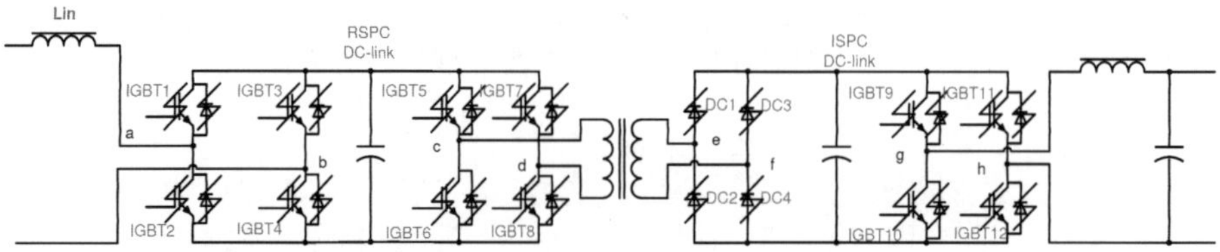

Figure 4. Switching component failures in D-EPT.

The IGBT module contains an IGBT and a body diode, which normally do not fail together. As a result, the fault could happen on IGBT only, or on the whole IGBT module including both IGBT and body diode. Based on the situation illustrated in figure 4, 14 possible faults for IGBT and diode can be summarized in figure 5, which are listed as A1, A2, A3, B1, C1, C2, C3, D1, E1, F1, G1, G2, G3, and H1. As the fault effects are similar for A3 and B1, C3 and D1, G1 and G2, G3 and H1 fault, hence B1, D1, G1 and H1 faults are analyzed instead of total 8 fault cases.

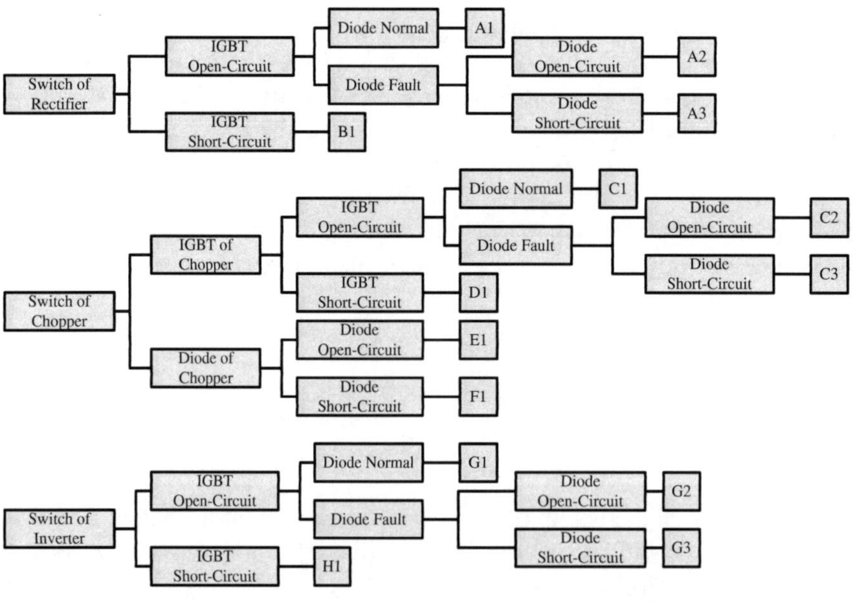

Figure 5. 14 possible faults of switches in D-EPT.

3.1. Single IGBT3 failure (A1)

Due to the unipolar PWM control scheme in figure 2, all IGBTs in the H-bridge will cause same effect to the D-EPT. Before the A1 fault, the current goes through IGBT1 and IGBT3 in the positive half-cycle of input voltage. When IGBT3 is open circuit with its body diode still working properly, the current goes through RSDCL, IGBT1 body diode and IGBT4 body diode. Due to the abnormal operation status, the RSDCL voltage of rectifier with fault A1 is different from the other normal rectifiers.

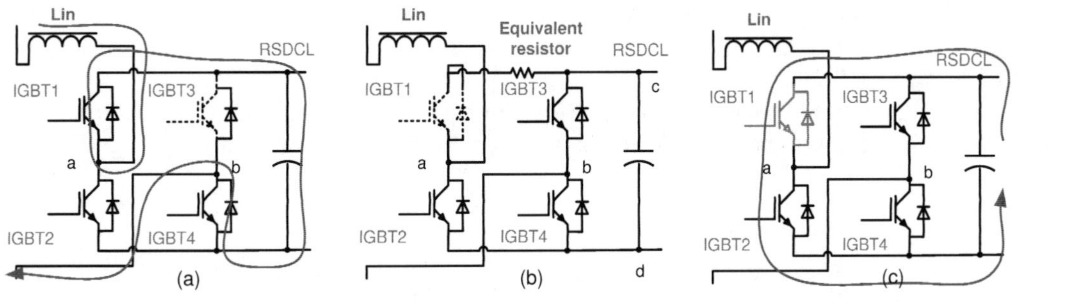

Figure 6. Equivalent circuit of (a) Fault A1 (b) Fault A2 (c) Fault B1.

3.2. *IGBT1 module failure (A2)*

It can be seen from figure 6(b), IGBT2 equals to an equivalent resistor in A2 fault, and V_{IGBT2} before and after the open circuit fault can be calculated by KVL as:

$$V_{IGBT2\_normal} = V_{da} = V_s + L_{in}\frac{di}{dt} = V_{RSDCL} \quad (4)$$

$$V_{IGBT2\_fault} = V_{da} = V_s + L_{in}\frac{di}{dt} = V_{RSDCL} + V_{eq} \quad (5)$$

where V_s is the input AC voltage, $L_{in}(di/dt)$ is the inductor voltage, V_{eq} is voltage on open-circuited IGBT1 and V_{RSDCL} is the RSDCL voltage.

The input inductor voltage $L_{in}(di/dt)$ will increase largely because of the equivalent resistance of IGBT1. As a result, RSDCL voltage will be reduced.

3.3. *Single IGBT1 failure (A3/B1)*

It can be seen from figure 6(c), the B1 fault will cause RSDCL short circuit, hence RSDCL voltage and OSDCL voltage will decrease, accompany with damage on IGBT1 and IGBT2 during the fault.

3.4. *Single IGBT5 failure (C1)*

It can be seen from figure 7(a) that the IHFT not saturated as the IHFT-IGBT5-RSDCL-IGBT8 path can still clear the flux through body diodes of IGBT 5 and IGBT 8.

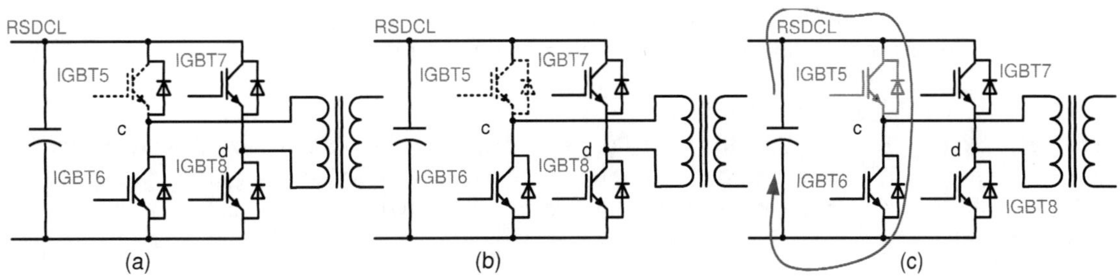

Figure 7. Equivalent circuit of DC-DC converter section for (a) Fault C1 (b) Fault C2 (c) Fault D1.

3.5. *IGBT5 module failure (C2)*

It can be seen from figure 7(b) that the IHFT-IGBT5-RSDCL-IGBT8 path to clear the flux is blocked due to C2 fault, and the IHFT core will saturate at last.

3.6. *Single IGBT5 failure (C3/D1)*

The D1 fault will cause the voltage drop on RSDCL voltage and ISDCL voltage, as well as the short circuit damage on all components wherever the short circuit current flows.

3.7. Diodel failure (E1)
As shown in figure 8(a), the current through RD1 will be zero in E1 fault. Hence the current used to flow in e point-ISDCL-f point loop will flow in f point-ISDCL-e point loop in the next half duty cycle.

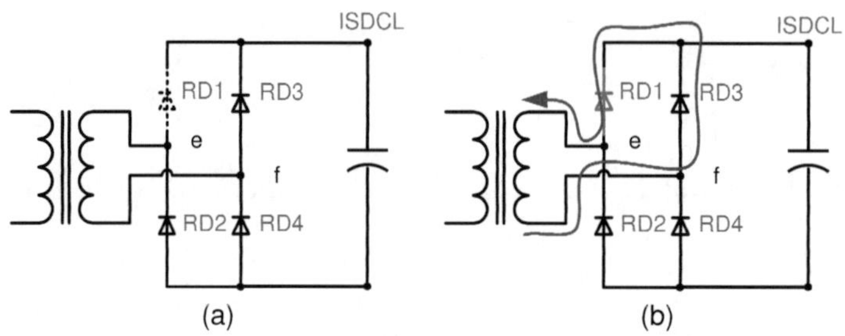

Figure 8. Equivalent circuit of DC-DC converter section for (a) Fault E1 (b) Fault F1.

3.8. Diodel failure (F1)
The IHFT will be over current in F1 fault, and the short circuit current is shown in figure 8(b). The IGBTs and rectifier diodes will be over current and the RSDCL voltage and ISDCL voltage will both decrease.

3.9. Single IGBT9 failure (G1/G2)
It can be seen from figure 9(a) that the open circuit on IGBT9 will stop the connection between ISDCL and the load, resulting in distorted current for the faulty inverter.

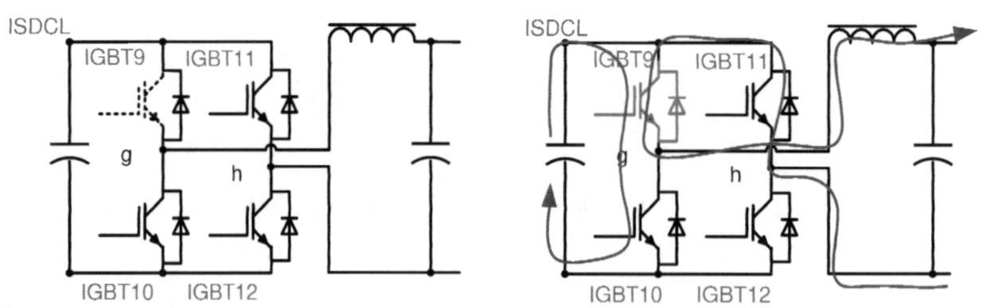

Figure 9. Equivalent circuit of inverter section for (a) Fault G1 (b) Fault H1.

3.10. Single IGBT failure (G3/H1)
The short circuit fault of IGBT9 in figure 9(b) will lead to 2 short circuit loops, including ISDCL-IGBT9-g point-IGBT10 loop and output filter-g point-IGBT9-IGBT11-h point loop. As a result, the ISDCL voltage and output voltage will both decrease, and IGBTs of the faulty inverter will be damaged.

4. Simulation verification on failure switches in D-EPT
The proposed D-EPT in figure 1(a) transforms the input phase voltage 5770 V (L-L voltage is 10 kV) to 220 V AC voltage (L-L voltage is 380 V). The rectifier section consists of 4 CHB, whose DC-link voltages are 9 kV in total and 2250 V for each. The DC-DC converter section transmits the RSDCL voltage (2250 V) to ISDCL voltage (375 V) by its H-bridge converters, IHFT and diode rectifier. The rectifier section at last provides an output with 220 V AC voltage. The system parameters are designed as inductor 30 mH, RSDCL capacitor 3 mF, ISDCL capacitor 22.4 mF and load resistor 2 ohm. The failure switching components are in accordance with the fault analysis above, and all 14 faults that are

marked in figure 5 are triggered at 0.8 s, before which the D-EPT works properly.

The distortion on voltage and current from all switching components faults are listed in table 1. The Failure Record (FR) about FR1, FR2, FR3 and FR4 is defined as under voltage, over voltage, over current, and currents unbalance.

Table 1. Effects of switching components faults in D-EPT.

Rectifier Section	**A1**	**A2**	**A3/B1**	**DC-DC Section**	**C1**	**C2**	**C3/D1**	**E1**	**F1**	**Inverter Section**	**G1/G2**	**G3/H1**
P F	<1	<1	<1	P F	1	1	<1	1	<1	P F	1	<1
I_{in}			FR3	I_{in}			FR3		FR3	I_{in}		FR3
V_{RS1}	FR1	FR1	FR1	V_{RS1}			FR1		FR1	V_{RS1}		FR1
V_{RS2}	FR1	FR1	FR2	V_{RS2}			FR2		FR2	V_{RS2}		FR1
V_{RS3}	FR1	FR1	FR2	V_{RS3}			FR2		FR2	V_{RS3}		FR1
V_{RS4}	FR1	FR1	FR2	V_{RS4}			FR2		FR2	V_{RS4}		FR1
V_{IS1}	FR1	FR1	FR1	V_{IS1}			FR1		FR1	V_{IS1}		FR1
V_{IS2}	FR1	FR1	FR2	V_{IS2}			FR2		FR2	V_{IS2}		FR1
V_{IS3}	FR1	FR1	FR2	V_{IS3}			FR2		FR2	V_{IS3}		FR1
V_{IS4}	FR1	FR1	FR2	V_{IS4}			FR2		FR2	V_{IS4}		FR1
IGBT1		FR2	FR3	IGBT5			FR3		FR3	IGBT9		FR3
IGBT2		FR2	FR3	IGBT6			FR3			IGBT10		FR3
IGBT3				IGBT7	FR3	FR3		FR3	FR3	IGBT11		FR3
IGBT4				IGBT8	FR3	FR3		FR3		IGBT12		FR3
				IHFT1		FR3			FR3	Io_1	FR4	FR4
				IHFT2						Io_2	FR4	FR4
				IHFT3						Io_3	FR4	FR4
				IHFT4						Io_4	FR4	FR4

4.1. *IGBT open circuit in rectifier section (A1)*

The saber simulation in figure 10 shows that the RSDCL voltage (V_RSDCL) and ISDCL (V_ISDCL) are increasing, and input current I_in is distorted.

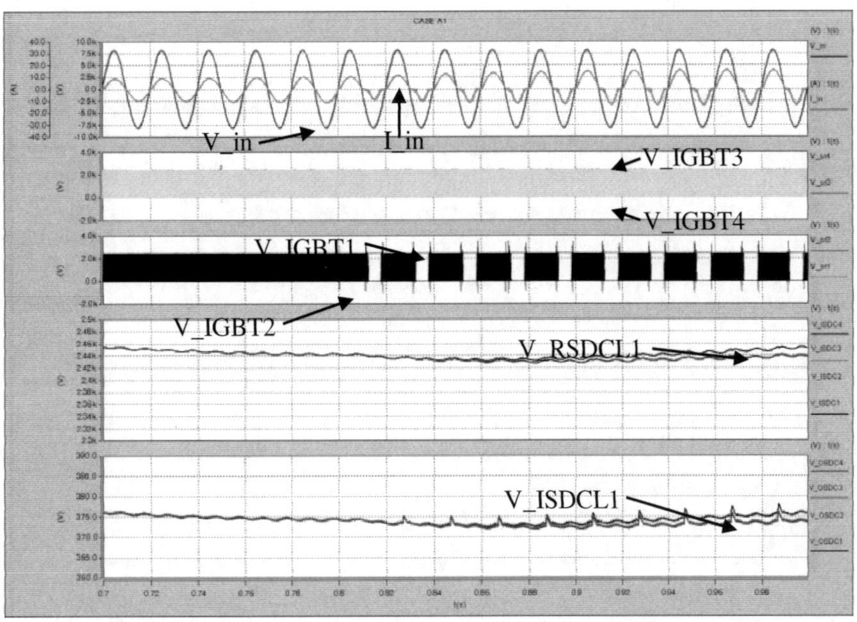

Figure 10. Simulation on rectifier IGBT open circuit fault A1.

4.2. IGBT and body diode open circuit in rectifier section (A2)

The simulation in figure 11 shows that the voltage across IGBT1 and IGBT2 (V_IGBT1 and V_IGBT2) are around 2 kV before fault A2, and increase to 50 kV after the fault, which are much higher than the rated voltage of the IGBT. It also can be seen in figure 11 that the input current I_in is distorted, and the RSDCL voltage as well as ISDCL voltage decreases largely when the fault happens.

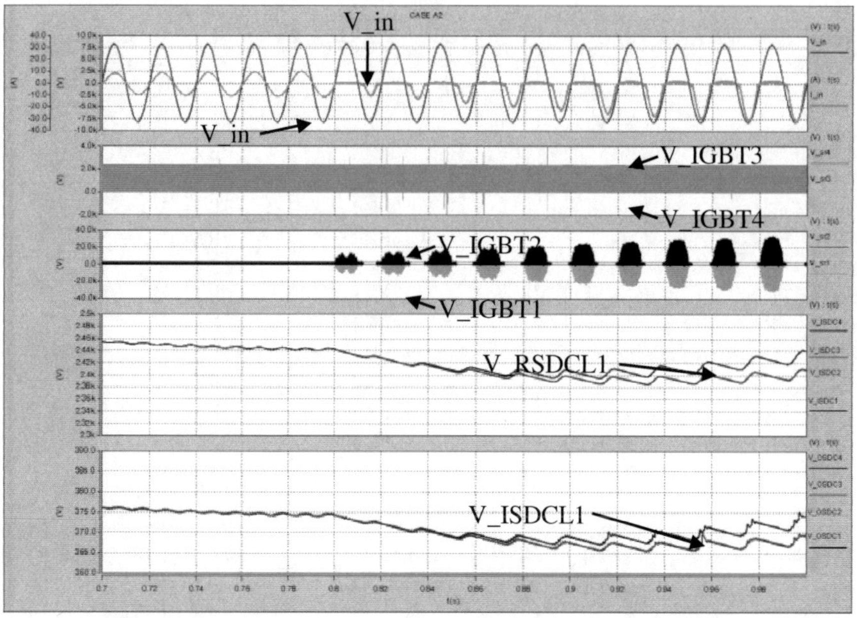

Figure 11. Simulation on rectifier IGBT open circuit fault A2.

4.3. IGBT short circuit in rectifier section (A3/B1)

It can be seen from the simulation in figure 12 that the current of IGBT1 is over 1MA, input current I_in is over 75 A and the RSDCL voltage (V_RSDCL) decreases to zero, as well as the ISDCL voltage (V_ISDCL).

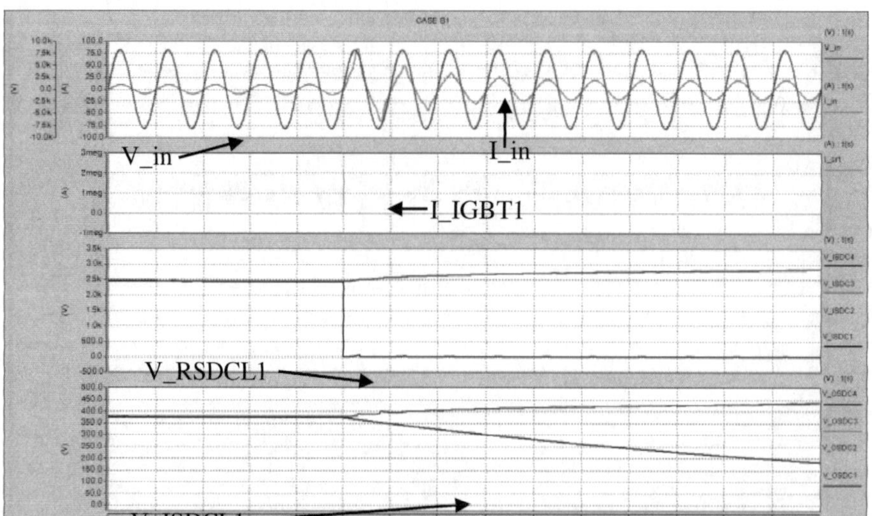

Figure 12. Simulation on rectifier IGBT short circuit fault A3/B1.

4.4. IGBT open circuit in DC-DC converter section (C1)

It can be observed from simulation in figure 13 that only negative current remains for IHFT current after the fault. As the absolute value of the IHFT current is doubled, it might do harm to the IGBTs.

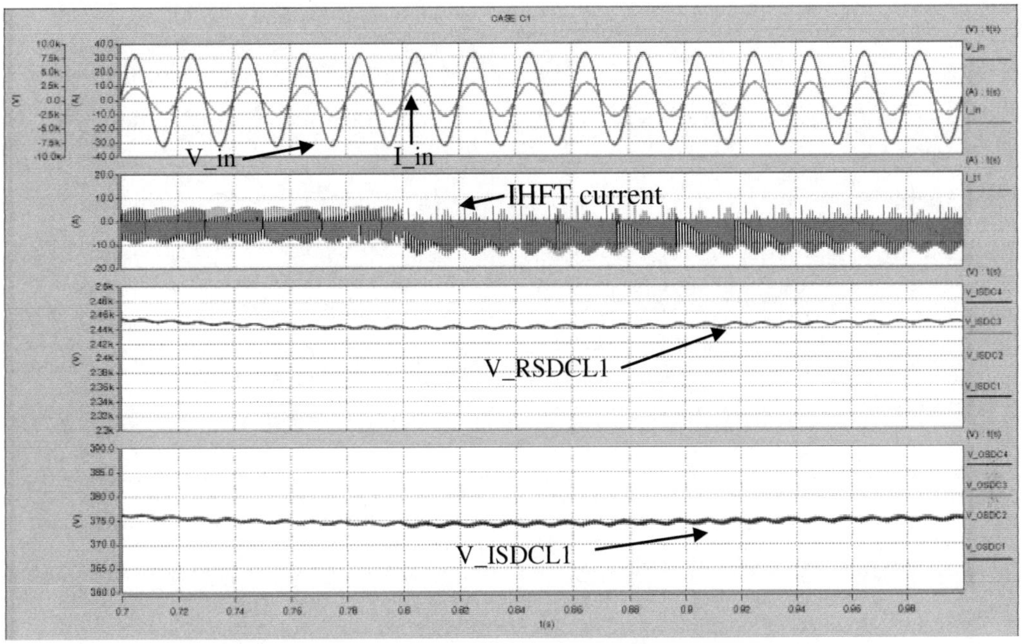

Figure 13. Simulation on DC-DC converter IGBT open circuit fault C1.

4.5. IGBT and body diode open circuit in DC-DC converter section (C2)

As shown in figure 14, the beginning of C2 fault is similar to C1 fault, however, the IHFT core will then saturate because C2 fault does not have a path to reset the magnetizing current as C1 fault does.

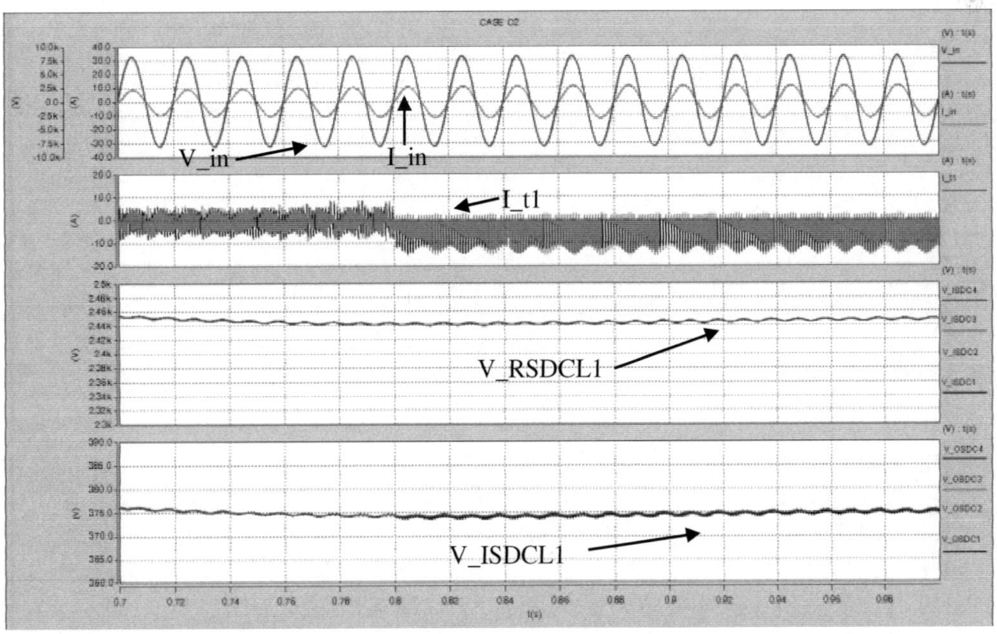

Figure 14. Simulation on DC-DC converter IGBT open circuit fault C2.

4.6. IGBT short circuit in DC-DC converter section (C3/D1)

The short circuit fault on IGBT5 causes a 2000 kA over current on IGBT5 and IGBT6, as well as a under voltage on RSDCL and ISDCL, which is shown in figure 15. As the short circuit fault has discharge the energy in RSDCL capacitor, the IHFT current I_t1 decreases largely.

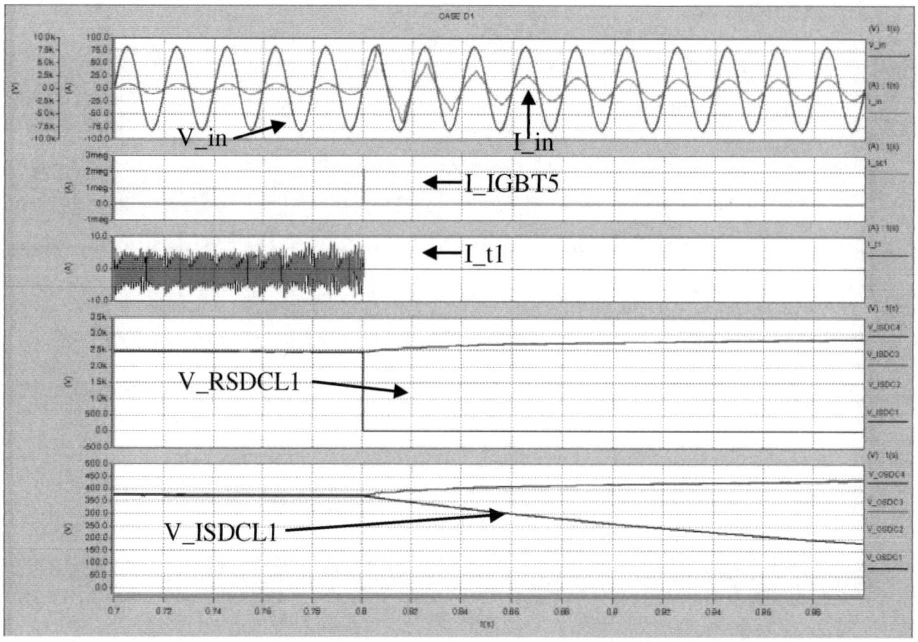

Figure 15. Simulation on DC-DC converter IGBT short circuit fault D1.

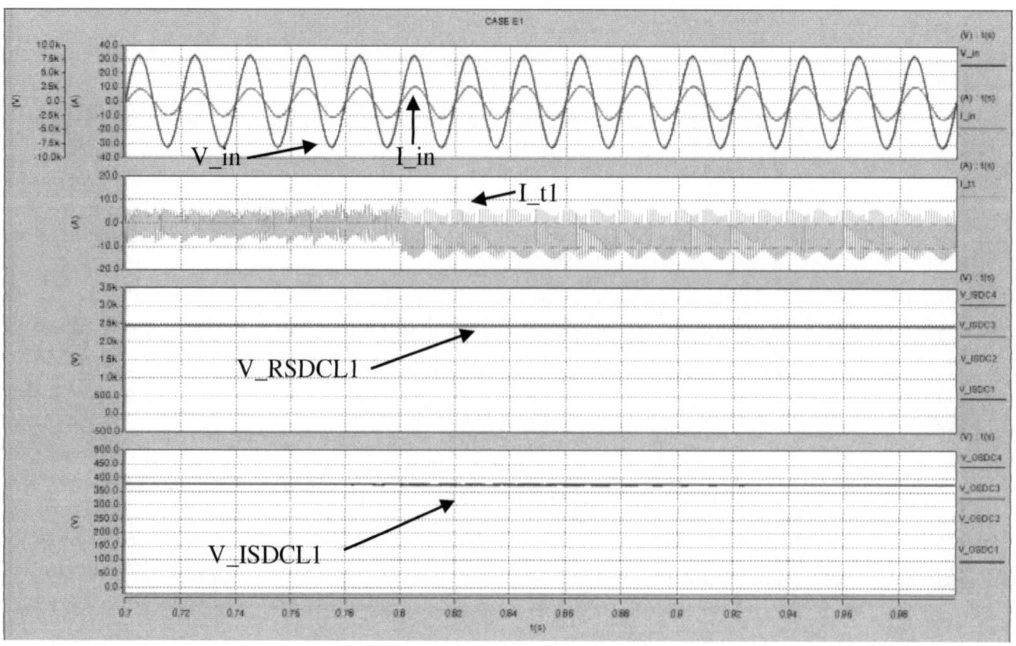

Figure 16. Simulation on DC-DC converter rectifier diode open circuit fault E1.

4.7. Rectifier diode open circuit in DC-DC converter section (E1)

The IHFT current I_t1 only flows in the negative half cycle due to the open circuit fault on Diode1, which is shown in figure 16. As the IHFT current I_t1 is almost doubled due to E1 fault, the Diode2 and Diode3 may get over current.

4.8. Rectifier diode short circuit in DC-DC converter section (F1)

As shown in figure 17, the fault F1 causes serious over current on rectifier diode (I_dc1) as well as transformer (I_t1), and under voltage on RSDCL as well as ISDCL.

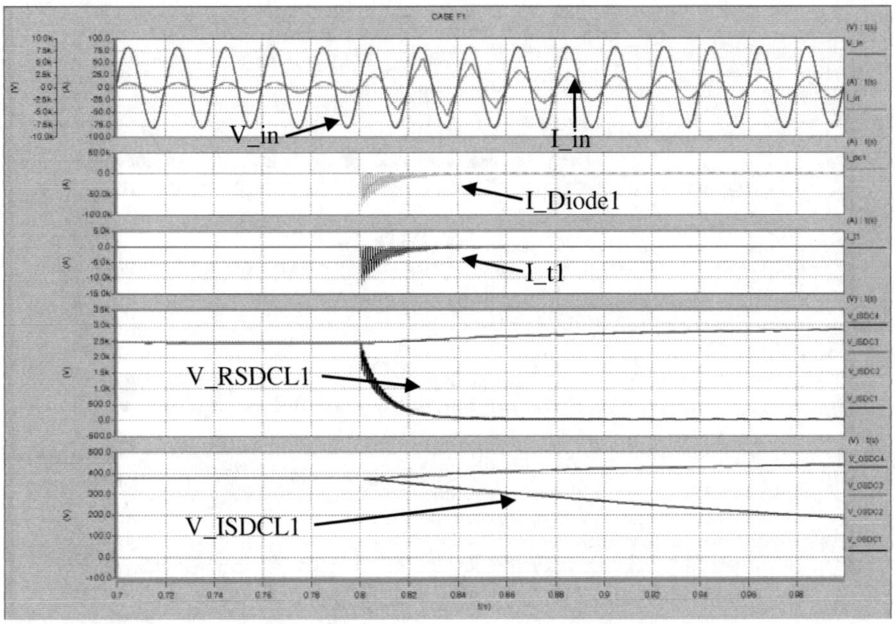

Figure 17. Simulation on DC-DC converter rectifier diode short circuit fault F1.

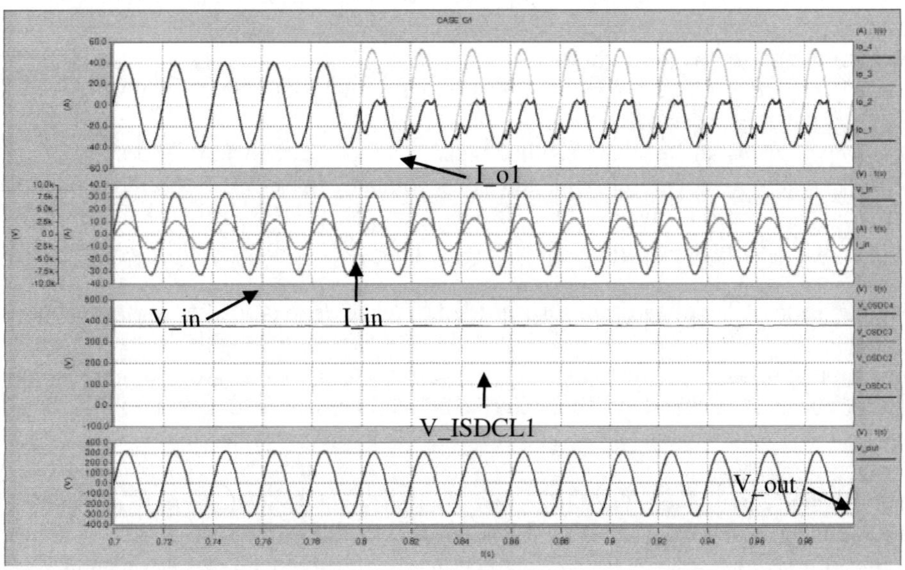

Figure 18. Simulation on inverter IGBT open circuit faults G1/G2.

4.9. IGBT open circuit in inverter section (G1/G2)

As shown in figure 18, the fault G1 or G2 forces output current of faulty H-bridge (Io_1) only flow in the negative cycle, which creates a dc component in other normal H-bridges of inverter.

4.10. IGBT short circuit in inverter section (G3/H1)

As shown in figure 19, the fault H1 causes over current on IGBT9, whose current is over 100 kA. Four RSDCL voltages and ISDCL voltages all decrease largely, as the faulty IGBT9 connects all four H-bridges of inverter, hence the output voltage of inverter is distorted.

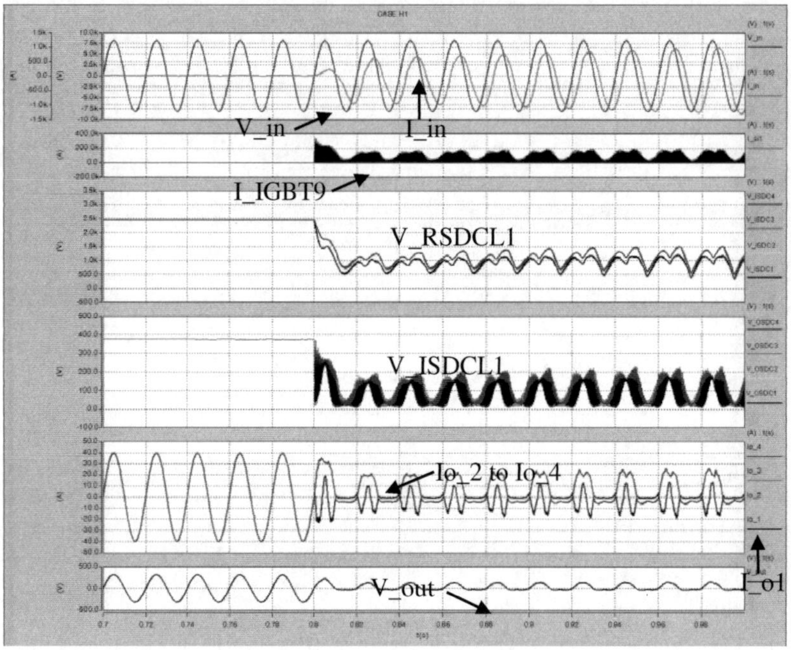

Figure 19. Simulation on inverter IGBT short circuit fault G3/H1.

5. Fault detection and protection scheme

The fault location scheme in figure 20 which is based on summary in table 1 shows that all faults can be located by judging the key statuses.

The most important parameter that can distinguish the open circuit fault and short circuit fault is input current (I_{in}), whose over current status always accompany by IGBT short circuit fault. Combined with abnormal statuses such as over voltage or over current on other parts, the failure IGBT will be located immediately. A PFC algorithm on the I_{in} in figure 20 is also needed to distinguish faults D1, H1, and F1.

The monitor on ISDCL voltages are also needed, as many faults lead to voltage abnormality on DC-link voltages.

The fault simulation above shows that the short circuit faults usually cause serious damage in a very short time, hence an immediate short circuit protection is needed to detect and stop the short circuit fault on IGBT. As a result, the hardware protection circuit embedded in IGBT gate driver is needed.

To detect the short circuit fault, the IGBT driver measures the Collector-Emitter-Voltage (V_{ce}) after the response time according to the IGBT turn-on characteristics, which are shown in figure 21. The driver will turn off the corresponding IGBT if the dynamic threshold V_{cethx} is lower than V_{ce}. Then, the driver circuit will transfer a fault status to the DSP of D-EPT though I/O pins which connect driver circuit and DSP, hence following measures will then be taken according to the fault message on location and classification.

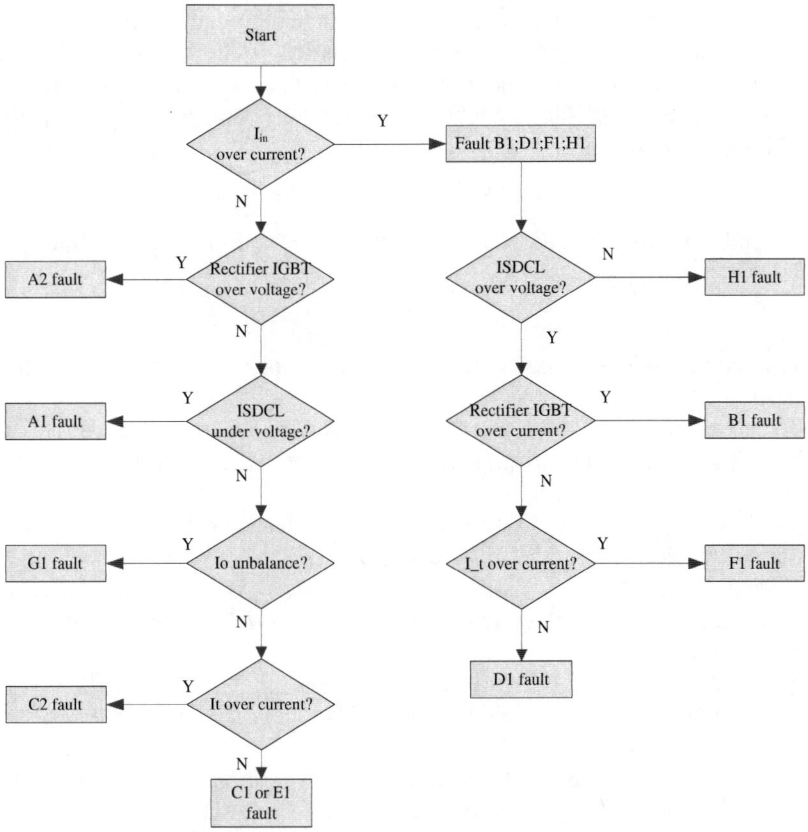

Figure 20. Flowchart for fault location scheme.

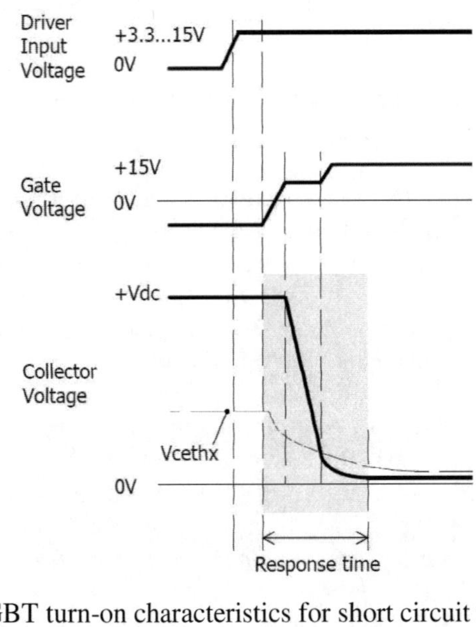

Figure 21. IGBT turn-on characteristics for short circuit fault detection.

6. Conclusions

The proposed detection and protection scheme based on the fault characteristics analysis for power converters in D-EPT is presented. The possible faults on switching components including IGBTs and diodes are discussed and then verified by Saber simulation, with results listed in table 1. The protection scheme detects the statuses on key parameters such as ISDCL voltage, IHFT current, and output AC voltage to locate the failure switches, and then take remedies to prevent the consequent damages to D-EPT and the power grid.

Acknowledgments
This work was supported by the State Grid Hubei Electric Power Company Science and Technology Project (521538160010).

References

[1] McMurray Power Converter Circuits Having a High-Frequency Link. U.S. Patent 3,517,300, 23 June 1970
[2] Fuentes C D and Rojas C A 2017 Experimental validation of a single DC bus cascaded H-bridge multilevel inverter for multistring photovoltaic systems *IEEE Trans. on Power Electron.* **64** 930-4
[3] Gandelli A, Leva S and Morando A P 2000 Topological considerations on the symmetrical components transformation *IEEE Trans. Circuits Syst. I, Fundam. Theory Appl.* **47** 1202-11
[4] Zhao T, Zeng J, Bhattacharya S, Baran M E and Huang A Q 2009 An average model of solid state transformer for dynamic system simulation *Proc. IEEE Power Energy Soc. Gen. Meet.* (Calgary AB Canada) pp 1-8
[5] Shukla A and Ghosh A 2010 Flying-capacitor-based chopper CIRCUIT for DC capacitor voltage balancing in diode-clamped multilevel inverter *IEEE Trans on Ind Electron* **57** 2249-61
[6] Han G H, Kim J H and Go B S 1999 The effective protection circuit in inverter microwave oven *Proc. IEEE Int. Conf. Electron.Circuits syst.* **3** 1423-6
[7] Yu Q and Parisella J 2007 Frequency diagnostic universal fault protection for current fed parallel resonant electronic ballast *IEEE Trans. Power Electron.* **22** 881-8
[8] Sinha G, Hochgraf C, Lasseter R H, Divan D M and Lipo T A 1995 Fault protection in a multilevel inverter implementation of a static condenser *Proc. IEEE Ind. Appl. Soc.* **3** 2557-64
[9] Zhang J M, Wang Z H and Shao S 2017 A three-phase modular multilevel DC–DC converter for power electronic transformer applications *IEEE Journal of Emerging and Selected Topics in Power Electronics* **5** 1
[10] Amini M and Moallem M 2016 A fault-diagnosis and fault-tolerant control scheme for flying capacitor multilevel inverters systems *IEEE Trans. on Power Electron* **64** 1818-26
[11] Sang Z X, Mao C X and Wang D 2014 Staircase control of hybrid cascaded multi-level inverter *Electric Power Components and Systems* **42** 23-34
[12] Huang A Q, Crow M L, Heydt G T, Zheng J P and Dale S J 2011 The future renewable electric energy delivery and management (FREEDM) system *The Energy Internet Proc. IEEE* **99** 133-48
[13] Fan H and Li H 2011 A distributed control of input-series-output-parallel bidirectional DC-DC converter modules applied for 20 kVA solid state transformer *Proc. 26th Annu. IEEE Appl. Power Electron. Conf. Expo.* (Fort Worth TX USA) pp 939-45
[14] Nie S X, Mao C X and Wang D 2016 Fault tolerant design for electronic power transformer *Proceedings of Power and Energy Engineering Conference (APPEEC)* (Xi'an, China) pp 692-6
[15] Zhou T D and Xu Y H 2017 Fault characteristic analysis and simulation of power electronic transformer based on MMC in distribution network *Proceedings of 2017 IEEE International Conference on Energy Internet (ICEI)* (Beijing, China) pp 332-7

IOP Conf. Series: Earth and Environmental Science **93** (2017) 012070 doi:10.1088/1755-1315/93/1/012070

Conservation voltage regulation (CVR) applied to energy savings by voltage-adjusting equipment through AMI

B-R Lan[1], C-A Chang[2,3,4], P-Y Huang[3], C-H Kuo[3], Z-J Ye[3], B-C Shen[3] and B-K Chen[3]

[1]Industrial Technology Research Institute, Hsinchu, Taiwan
[2]Department of Electrical Engineering, Tatung University, Taipei, Taiwan
[3]Institute of Nuclear Energy Research, Taoyuan City, Taiwan

E-mail: d10302001@ms.ttu.edu.tw

Abstract. Conservation voltage reduction (CVR) includes peak demand reduction, energy conservation, carbon emission reduction, and electricity bill reduction. This paper analyzes the energy-reduction of Siwei Feeders with applying CVR, which are situated in Penghu region and equipped with smart meters. Furthermore, the applicable voltage reduction range for the feeders will be explored. This study will also investigate how the CVR effect and energy conservation are improved with the voltage control devices integrated. The results of this study can serve as a reference for the Taiwan Power Company to promote and implement voltage reduction and energy conservation techniques. This study is expected to enhance the energy-reduction performance of the Penghu Low Carbon Island Project.

1. Introduction

Conservation voltage reduction (CVR) is energy-saving measure that, in accordance with the voltage regulated standards at the users' end (e.g. ANSI C84.1 for the United States of America; article 36 of the Electricity Act of Taiwan) and by lowering the operating voltage of distribution feeder, allows reduction of the system's demand for electricity [1,2]. This approach is known to have the following advantages: to hinder the peak load; to reduce electrical energy loss; to defer investment on new equipment, and to lower operating cost of electrical power generation, and eventually to save energy and reduce carbon emission.

Power companies can reduce the electricity demand through the application of CVR. However, voltage drop occurs during the power transmission process, as shown in figure 1. To avert overly low of voltage at the feeder's end, the power company must boost the voltage at the feeder's end as shown in figure 2. If the operator may lower the voltage at the end of the feeder, along with voltage-adjusting equipment located at the middle and downstream of the feeder, then it will allow the voltage at the users' end to fall within a reasonable range, and this will expand the scope for managing the voltage.

Some studies also point out that the application of CVR does appear effective in terms of energy savings. Table 1 compiles results of tests run by power companies from different geographical zones, on applications of CVR on residential and commercial electrical loads: This study have observed from these test results that every 1% of reduction in voltage, it will also reduce the demand of power consumption by 0.8 to 1.38% [3]. This study can therefore determine that CVR does enhance energy-saving effectiveness; however, the aforementioned studies and documents did not discuss the impact of CVR on either loss from transmission line or transformer.

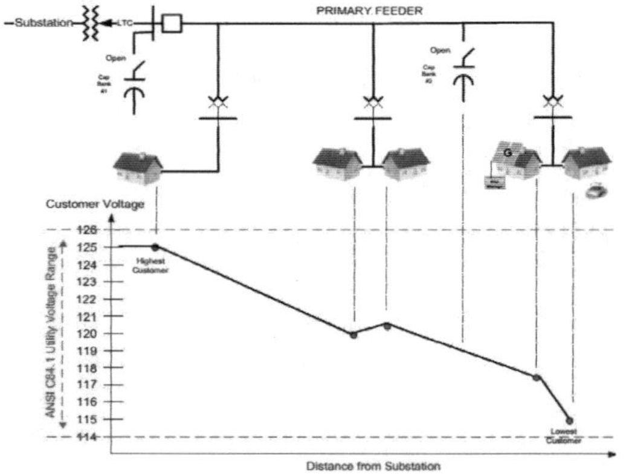

Figure 1. Feeder voltage drop without voltage-adjusting equipment.

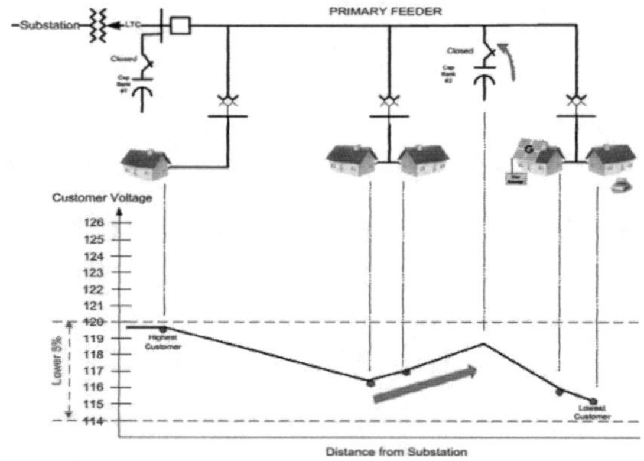

Figure 2. Feeder voltage drop with voltage-adjusting equipment.

Table 1. Test Results of CVR in the US.

Utility	Percent Demand Reduction for 1% Voltage Reduction	
	Residential	Commercial
American Electric Power 1	0.80 %	0.78 %
American Electric Power 2	0.90 %	0.86 %
Consumers Power Company	0.83 %	1.38 %
San Diego Gas & Electric Co.	1.14 %	0.08 %

In 2013, Bureau of Energy, Ministry of Economic Affairs of Taiwan began to promote and implement the "Penghu Smart Grid Demo Site" project, with the Siwei Feeder as one of the automated feeders of the project. In line with of the development of "Low-Carbon Island Project", the authority aims to install smart meters and build Smart Homes within the zones covered by the Siwei Feeder; the infrastructure will work together with Distribution Management System (DMS), Meter Data Management System (MDMS), and Demand Response Management System (DRMS), etc. [4]. The monitoring, management and control functions over grid catered by the smart grid do not only provide real-time data on power consumption, support demand response, accelerate detection and repair of failures [5], but also help power companies to carry out CVR in a more effective manner in order to achieve a better energy savings.

This study has chosen the Siwei Feeder for a simulation of CVR in order to analyze the losses of transmission lines and distribution transformers at different levels of voltage reduction, then further studies the changes on energy-saving performance after adding up voltage-adjusting equipment. Through the application of CVR, this study will enhance the outcome in terms of energy savings and carbon emission reduction of the Low-Carbon Island Project and the installation of smart grid, and eventually promote the power company's performance.

2. CVR in combination with smart grid and voltage-adjusting equipment

Before implementation of CVR, the power company needs to understand the voltage allocation on the feeders in order to assure that the voltage at the users' end complies to the voltage regulated standards. Generally, there are two ways to find out the voltage of the electrical power distribution system: the first is to measure the voltage at the feeder's end with a meter, and send the data back to the power company with communication systems. The second way is to measure the voltage and current output at the substation, then in collocation with the circuit model to come up with the estimation of voltage at every end of the feeder. However, neither of the above methods allows power companies to gain a precise and accurate observation of the voltage at the users' end of the feeder, thus in order to prevent adverse effects on electrical devices' lifecycle and efficiency brought by overly low voltage. Thus, power companies are often reluctant to carry out measures of significant voltage reduction, resulting in limited energy-saving performance of CVR [6].

With the Advanced Metering Infrastructure (AMI) of smart grid and the Distribution Automation System (shown in figure 3), one may measure and collect voltages at users' ends along the feeder with the AMI system, then transfer the data collected back to the internal data base of the power company. When the voltage at user's end disrespects the voltage-regulated standards, the user's meter will send an alarm message. Then the Distribution Management System (DMS) will analyze the collected data for decision-making purpose, and send the commands to the voltage-adjusting equipment of the power distribution system, allowing the company to monitor and control simultaneously the voltage at the users' end, and increase the voltage when it is too low [4]. In this case, the power company, in compliance with the voltage-regulated standards at the users' end, reduces the system voltage as much as possible to maximize the energy-saving performance. The integrated volt/var control for CVR should rely on the distribution system state estimation to provide the on-line model for effective VCR control and monitoring. And the utility should consider the feeder load variations would lead to uncertain effects on CVR controls.

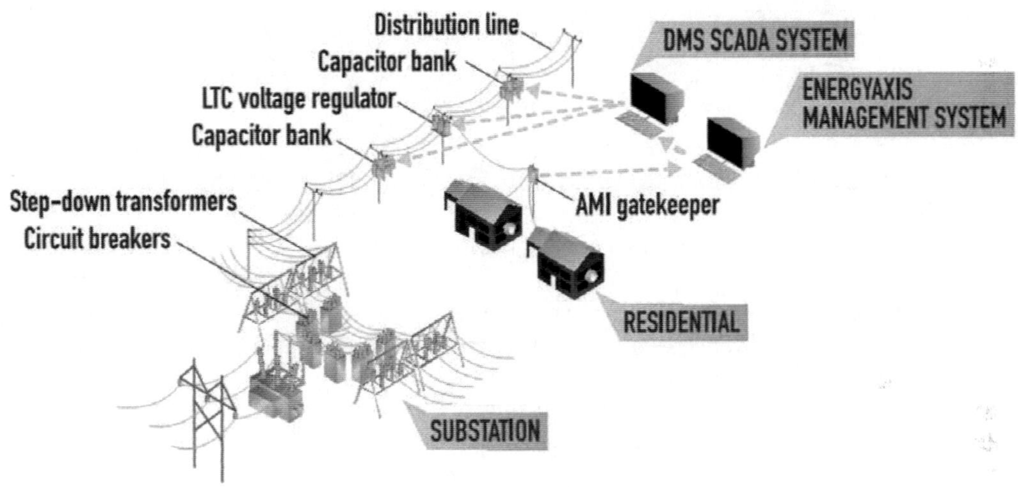

Figure 3. CVR with smart grid and voltage-adjusting equipment.

3. Presentation of simulation system

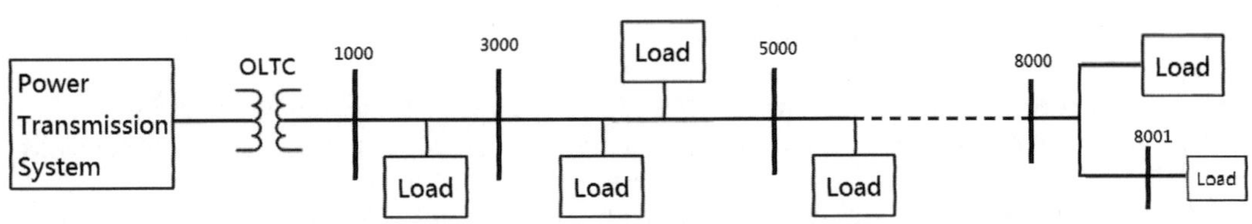

Figure 4. Siwei feeder of Penghu.

This study has chosen the Siwei Feeder as sample system, with the estimated peak and off-peak loads on 2018 as base for further analysis. Figure 4 shows the single-line diagram of Siwei Feeder, alimented by Penghu Magong S/S, it is an underground electrical power distribution system of 11.4 kV. In order to study the impact of CVR on transformer loss, this study adopted the data and typical characteristics collected from the handbook, "Underground Distribution System of Taiwan Power Company" to calculate the series reactance via equation (1) [7]. This study then came up with impedance parameters of transformers with different capacities (table 2), and entered the data into the transformer model (figure 5).

Equation of voltage regulation rate is expressed as follows:

$$VR\%(pf) \approx Rs\%\cos\theta + Xs\%\sin\theta \quad (1)$$

Table 2. Taipower-impedance data of typical transformers used in underground distribution system.

Transformer Capacity (kVA)	Resistance in Series (%)	Reactance in Series (%)	Core Loss (W)	Exciting Current (%)
25	1.03	2.29	65	0.75
50	0.93	2.52	110	0.68
100	0.78	3.03	170	0.64
167	0.75	3.17	153	0.60

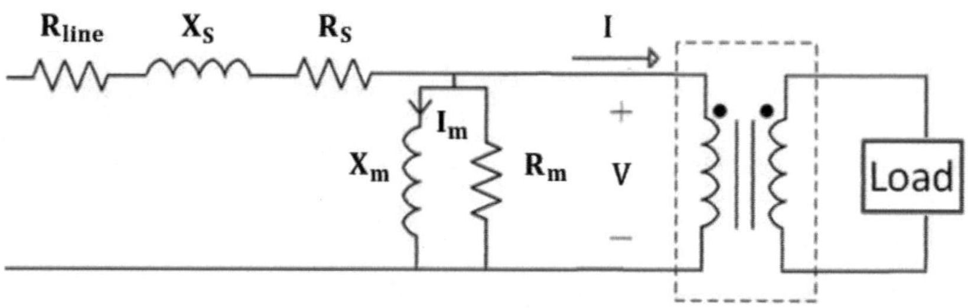

Figure 5. Transformer model.

In terms of the power load model, given the fact that there are mostly residential and commercial users in Penghu, this study has consulted the ZIP model for typical residential and commercial loads from the analysis report of power load models by General Electric (USA). This study selected load model data of ZIP which can present performances of Penghu's load, and the load characteristic varies with the voltage are shown as equations (2) and (3) [8,9].

$$P = P_0\left[0.1934\left(\frac{V_i}{V_0}\right)^2 + 0.6897\frac{V_i}{V_0} + 0.1169\right] \quad (2)$$

$$Q = Q_0\left[3.3383\left(\frac{V_i}{V_0}\right)^2 - 3.1939\frac{V_i}{V_0} + 0.8556\right] \quad (3)$$

where P, Q: Demand for active and reactive power at operation voltage, V_i and P_0, Q_0: Demand for active and reactive power at nominal voltage, V_0.

4. Case study-energy saving performance and permissible range for voltage reduction

First, this study established a simulation system of the Siwei Feeder to analyze the power demands of each component of the feeder under different operating voltages, including the load power, the line loss, the core and copper loss of transformers, with data of minimum load voltage recorded and analysis results shown in table 3.

The results from table 3 indicate that, when the measure of CVR is taken at the Siwei Feeder, the peak load is higher than off-peak and makes the lowest voltage of peak load pattern less than off-peak pattern's; thus the power demand decreases along with the lowered voltage (equation (2)) while system loss also decreases. Hence, the measure of CVR may reduce power load and system loss, and enhance the energy savings.

Table 3. Siwei Feeder-power demands, load power, losses, and minimum load voltage at different operating voltage.

Load	**Annual peak**					**Annual off-peak**				
Voltage at Feeder's Sending End (p.u.)	**0.96**	**0.97**	**0.98**	**0.99**	**1**	**0.96**	**0.97**	**0.98**	**0.99**	**1**
Total Demand of Feeder (kW)	3771	3813	3855	3897	3939	1113	1125	1138	1150	1163
Load Power (kW)	3721	3762	3804	3845	3887	1095	1107	1119	1132	1144
System Loss (kW)	**50.35**	**50.78**	**51.22**	**51.67**	**52.13**	**17.86**	**18.17**	**18.51**	**18.85**	**19.18**
Line Loss (kW)	21.69	21.77	21.85	21.94	22.02	1.97	1.98	1.99	2.01	2.02
Copper Loss (kW)	14.34	14.39	14.44	14.5	14.56	1.32	1.32	1.33	1.34	1.35
Core Loss (kW)	14.32	14.62	14.93	15.23	15.55	14.57	14.87	15.19	15.5	15.81
Minimum Voltage of feeder (p.u.)	**0.936**	**0.946**	**0.956**	**0.966**	**0.975**	**0.952**	**0.962**	**0.972**	**0.982**	**0.992**

The application of CVR must assure that the voltage at users' end complies to "article 36 of Taiwan's Electricity Act: the voltage regulation factor of electrical power supply shall not exceed 5%." Examining the off-peak data in table 3, if the operating voltage is reduced by 4%, the minimum voltage of the feeder will be higher than 0.95 p.u. However, if the operating voltage is reduced by 3% during peak hours, the minimum load voltage of the feeder will become less than 0.95 p.u., which does not respect the voltage-regulated standards. Without the intervention of voltage-adjusting equipment installed at the middle and down stream of the feeder, the maximum permissible voltage reduction range at the receiving end of the Siwei Feeder will be 2%.

4.1. Impact of CVR on distribution system losses

Distribution losses include line loss, transformer core loss and copper loss. The line loss and transformer copper loss are dependent on the load current, while transformer core loss is dependent on the operating voltage. Table 3 indicates that since the off-peak load current, hence, the line loss and copper loss on the Siwei Feeder are less than during the peak hours; however, given that the voltage of the feeder is higher during this period, the core loss will be higher than the peak hours. If the data of the losses are further analyzed, core loss is the most recognizable among the losses generated by CVR, which as shown in figures 6 and 7, is visibly better than line and copper loss. Therefore, an effective reduction in transformer core loss will enhance a better energy-saving performance.

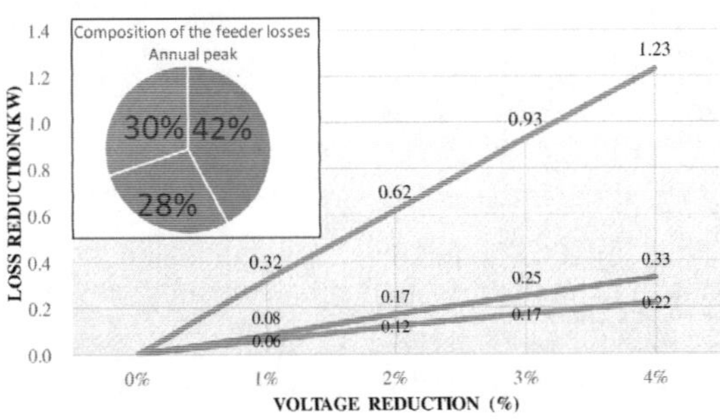

Figure 6. Distribution system losses from CVR – annual peak.

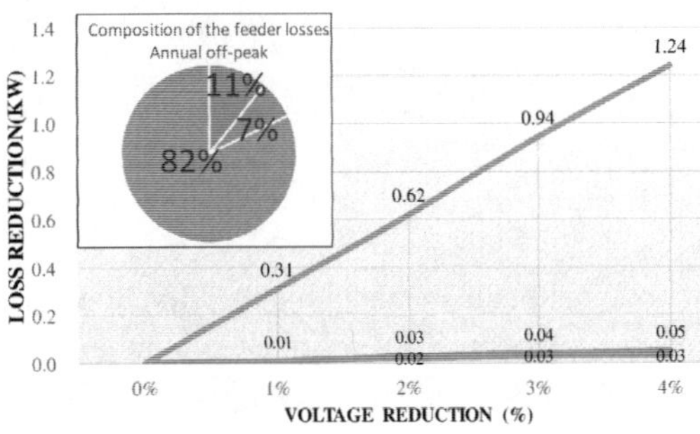

Figure 7. Distribution system losses from CVR – annual off-peak.

4.2. CVR combined with voltage-adjusting equipment

This study brings out measures in order to enhance the energy savings by reducing the operating voltage at the substation, then increase the voltage with control devices (e.g. capacitors or voltage regulator) installed at the middle and lower reaches of the feeder. This allows the voltage at users' end to remain within normal operating range and to increase operating flexibility of CVR. The following sections explain the position of installation of capacitors and voltage regulators and the result analysis.

4.2.1. Capacitor installation. This study consulted the guidelines on capacitor installation described by "Underground Distribution System of Taiwan Power Company" [10] and capacitor standards defined by IEEE (Std. 18) [11], and considered the demand and distribution of system reactive power, then This study decided to install capacitor on Bus #5000 at the Siwei Feeder. Given the fact that TaiPower switch the capacitors off on off-peak, hence this study will not discuss the voltage-regulation performance of capacitors on annual off-peak load. The results of using capacitors to improve the overly low voltage at the Siwei Feeder are shown in table 4.

Table 4. Results of using voltage-adjusting equipment to improve overly low voltage.

Load	Voltage-Adjusting Equipment	Siwei Feeder Operating Voltage 0.97 p.u.
Annual peak	No Device	0.9458 p.u.
	Capacitor	0.9469 p.u.
	Voltage Regulator	0.9504 p.u.
Annual off-peak	No Device	0.9620 p.u.
	Voltage Regulator	0.9620 p.u.

4.2.2. Voltage regulator. This study considered the specification provided by American voltage regulator manufacturers [12] as well as the distribution of overly low voltage throughout the Siwei Feeder, and this study decided to install voltage regulators on the Siwei feeder (as shown in figure 8), with voltage regulators of different phases installed as shown in table 5.

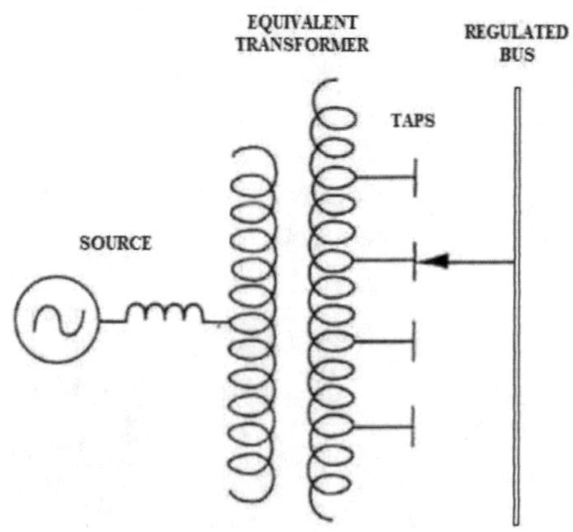

Figure 8. Voltage regulator model.

Table 5. Installation location and installed capacity of voltage regulators on the feeder.

Location	Rated Capacity (KVA)		
	Phase A	Phase B	Phase C
Bus #5000	57	114	57

Table 6 reveals that when the CVR of the Siwei Feeder is done by 3%, it has the terminal voltage without voltage regulators. During off-peak periods, the system does not manifest overly low voltage and the regulator does not carry out any adjustment. However, during peak periods, the measure of

CVR causes the minimum voltage of the feeder to become too low and violates the voltage regulated standards. With the installation of voltage regulators, the equipment detects when the voltage is too low and adjusts the system's minimum voltage to above 0.95 p.u. Therefore, the increment of CVR at the Siwei Feeder could go up to 3% with the voltage regulators installed.

Table 6. Comparative results of different operation at the Siwei Feeder.

Annual peak	Voltage at the Sending End	Load (kW)	Minimum Load Voltage (p.u.)	Demand Reduced
	1.00 p.u.	3939.3	0.9754	—
	0.98 p.u.	3855.0	0.9557	84.3 kW (2.14%)
	0.97 p.u. with voltage regulator	3822.7	0.9504	116.6 kW (2.96%)
Annual off-peak	1.00 p.u.	1163.2	0.9920	—
	0.98 p.u.	1137.9	0.9721	25.3 kW (2.18%)
	0.97 p.u.	1125.3	0.9620	37.9 kW (3.26%)

4.3. Comparison of results in energy savings of different operation

Table 6 concludes load demands, minimum voltage under different operation, and the energy savings during peak and off-peak loads of the Siwei Feeder. Without any voltage-adjusting equipment, the permissible range of voltage reduction at the Siwei Feeder is 2%, and it can hinder the peak demand by 2.14%, and decrease peak load capacity correspondingly by 84.3 kW. If voltage regulators at the Siwei Feeder are installed, the permissible voltage reduction range may increase from 2% to 3%, raising the flexibility of voltage reduction measure to enhance better energy savings. The energy-saving performance will increase from 2.14% to 2.96% during peak periods, with peak load reduced from 84.3 kW to 116.6 kW.

This study took the data collected during peak and off-peak loads to evaluate the operating benefits generated by the reduced power consumption derived from CVR. For power companies, the operating deficits reduced from CVR represents the "difference between the fuel cost and the electricity selling price". Hence this study uses the USD: TWD exchange rate in 2017 and the difference between the cost of power generation in Penghu (Cplant, 0.197 USD/kWh) and the selling price of TaiPower (Pselling, 0.098 USD/kWh), as the operating deficit per power consumption unit. The following is the method of the analysis.

Based on the daily power consumption of peak and off-peak days, and proportion of energy savings as shown in "Demand Reduced" of table 6, this study estimated the power consumption that could be reduced on peak and off-peak days. Then, this study used equation (4) to estimate the annual reduced deficit. The results of energy-saving evaluations carried through the aforementioned methods in table 7. If voltage regulators are installed at the Siwei Feeder to reduce the operating voltage by 3%, the energy savings will become more since power company can reduce the annual power consumption of 702,427 kWh, and the corresponding operating deficit is 69,479 USD.

$$\text{Annual reduced deficit} = 8760 \times 0.5 \times \sum_{i=peak,off-peak}(p_i \times \eta_i) \times (P_{selling} - C_{plant}) \quad (4)$$

Table 7. Operating benefits and power consumption reduced by CVR at the Siwei feeder.

Voltage Reduction	Load Pattern	Daily Power Consumption (p_i)	Energy Saving Proportion (η_i)	Annually Reduced Consumption	Annually Reduced Deficit
2%	Peak	77,904 kWh	2.1%	482,502 kWh	47,725 USD
	Off-Peak	45,812 kWh	2.2%		
3% with Voltage Regulators	Peak	77,904 kWh	3.0%	702,427 kWh	69,479 USD
	Off-Peak	45,812 kWh	3.3%		

5. Conclusion

CVR is an energy-saving measure through reducing the operating voltage of the feeder, in compliance with the voltage regulated standards, to reduce the system demand. This study adopted sample system with smart meters installed along the Siwei Feeder in Penghu to analyze the impact of CVR. The analysis result indicates that under CVR, there will also be less copper, core, and line losses. Moreover, along with the increment in voltage reduction, the core loss reduction becomes more visible. This study suggested that the utility should use high-efficiency transformers, which have lower transformer losses. Lowering the energy consumption would result in reduced operating costs and less harmful emissions for utility operation. Installation of voltage-adjusting equipment at appropriate locations of the feeder will increase larger range of voltage reduction and enhance better energy savings, and eventually reduce more operating deficits.

Via CVR, lower the load demand is carries out and the power company also can reduce power supply which implies to lower Penghu's operating deficits generated from the relatively high costs of fuel cost. This study reveals that the application of CVR can effectively save energy, and reduce overall system losses so that the power company will not have to make extra investments to upgrade its distribution system to bring up power transmission efficiency, and eventually increase operating performance. In the future, the application of smart grid combined with voltage-adjusting equipment will help power companies to increase the range of voltage reduction.

References

[1] Schneider K P, Tuffner F K, Fuller J C and Singh R 2010 Evaluation of conservation voltage reduction (CVR) on a national level Pacific Northwest National Laboratory

[2] Electricity Delivery & Energy Reliability 2012 Application of automated controls for voltage and reactive power management initial results Department of Energy

[3] Taylor C W 1994 *Power System Voltage Stability* (New York: McGraw Hill)

[4] Sierra Energy Group 2010 AMI and DA convergence: Enabling energy savings through voltage conservation

[5] McLean M 2012 Three DA initiatives that leverage AMI investment Silver Spring Networks

[6] Ellens W, Berry A and West S 2012 A quantification of the energy savings by conservation voltage reduction 2012 IEEE International Conference on Power System Technology (POWERCON) (Auckland) pp 1-6

[7] Underground Distribution System of Taiwan Power Company lesson 4, Taiwan Power Company

[8] Price W W, *et al* 1988 Load modeling for power flow and transient stability computer studies IEEE Trans. on Power Systems **3** 180-7

[9] Zhi Z D 1994 Transient stability studies of the Jen-wuco-generation facility (Tatung University)

[10] Underground Distribution System of Taiwan Power Company lesson 4, Taiwan Power Company, September 2009

[11] IEEE Standard for Shunt Power Capacitors, IEEE Standard 18, 2012

[12] Single-phase pad-mounted step voltage regulator, Cooper Power Systems

Sensor Transmission Power Schedule for Smart Grids

C Gao[1], Y H Huang[2], J Li[3, 4], X D Liu[2]

[1]School of Business, North China Electric Power University, Changping District, Beijing 100192, China.
[2]China Electric Power Research Institute, Changping District, Beijing 102206, China.
[3]School of Automation, Beijing Institute of Technology, Beijing 100081, China.

E-mail: sunnydannie@gmail.com

Abstract. Smart grid has attracted much attention by the requirement of new generation renewable energy. Nowadays, the real-time state estimation, with the help of phasor measurement unit, plays an important role to keep smart grid stable and efficient. However, the limitation of the communication channel is not considered by related work. Considering the familiar limited on-board batteries wireless sensor in smart grid, transmission power schedule is designed in this paper, which minimizes energy consumption with proper EKF filtering performance requirement constrain. Based on the event-triggered estimation theory, the filtering algorithm is also provided to utilize the information contained in the power schedule. Finally, its feasibility and performance is demonstrated using the standard IEEE 39-bus system withphasor measurement units (PMUs).

1. Introduction

Aiming at sustaining long-term energy supply, renewable energy has attracted widespread attention. Renewable energy such as wind and solar needs a new-generation power grid infrastructure called smart grid to accommodate to the power network. Smart grid utilizes automate control, intelligent and information technology to realize a reliable and sustainable energy management system (EMS) [1–3].

Among the EMS, accurate estimation for the dynamic status of smart grid plays the key role, which is hard to be realized by conventional supervisory control and data acquisition (SCADA) systems. To satisfy the accurate and dynamic estimation needs, the real-time phasor measurement unit (PMU) is invented utilizing the GPS time-stamped technology to provide high frequency accurate synchronous phasor data. Based on PMU, Wide-area measurement systems (WAMS) is established to accomplish the synchrophasor real-time state estimation (RTSE) [4, 5]. The structure of WAMS is show in Fig. 1[6].

Amount of research on RTSE has been developed based on the observation data of PMU [7– 12]. Due to the nonlinearity of power system, nonlinear filter is wildly applied in RTSE. Extended Kalman filter (EKF), which linearizes the system functions to realize the iteration of covariance in nonlinear filtering process, was firstly introduced to estimate the dynamic states of smart grid [7]. However, the higher order terms of Taylor series translates into linearization error during the EKF process. Thus, when the system's nonlinearity is strong, EKF will lead low filtering accuracy and even filtering divergency. The unscented Kalman filter (UKF) was introduced by Singh and Pal to solve the strong nonlinear power system's RTSE problem, which utilized the sigma points to approximate the distribution rather the system function [8]. Compared with EKF, UKF performs better when power system is strong nonlinear and the noise the Gaussian. However, UKF ineluctably suffers from the numerical stability, which makes UKF an unreliable solution for power system's RTSE. Particle filter (PF), which utilizes Monte Carlo Methodology to approximate the condition distribution, is accurate and reliable enough for nonlinear system's estimation [10]. Nevertheless, the computing power needed by PF is unpractical, thus PF can only be treated as the decade-to-be solution. Thus, EKF is wildly studied in dynamic states estimation of power systems.

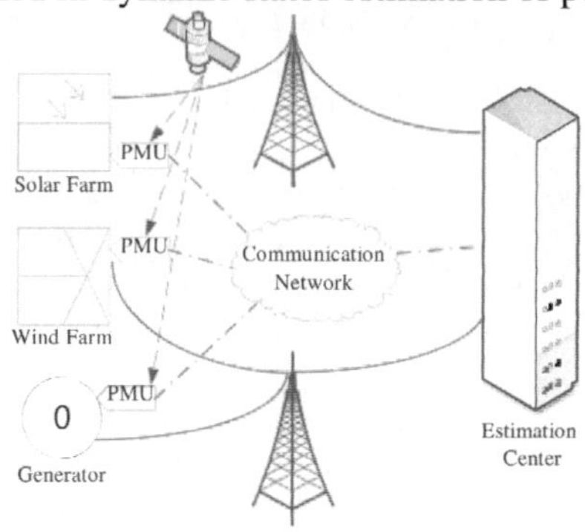

Figure 1. System structure of RTSE for WAMS based on PMU.

A usual assumption in RTSE research is that the communication channel in Fig. 1 is ideal. However, with widespread development of wireless sensors, the communication channel between smart grid and estimation center is susceptible to environmental influence. Amid all the influence, transmission failure, which means that the observation transmitted by the smart grid is not received by the estimation center, is the one that has the most serious impact on the estimation. The probability of transmission failure typically has a positive correlation with transmission power. Because the transmission power is usually provided by limited on-board batteries in smart grid, an sensor transmission power schedule should be designed to balance the transmission failure probability and the transmission power consumption [13–15].

This paper focuses on the sensor transmission power schedule problem in RTSE of power system. Considering the nonlinearity of power system, the transmission power schedule is analyzed based on EKF. The rest of this paper is organized as follows. Section 2 introduces the dynamic states estimation and the transmission power schedule problem of power system. Section 3 designs the transmission power schedule that minimizes energy consumption and guarantee proper estimation performance by determining transmission power according to the innovation of observation. Also, the filtering algorithm is provide, which utilizes the information in transmission power schedule to promote estimation performance. Section 4 utilizes IEEE 39-bus system to demonstrate the feasibility of the proposed method. The following standard notations are adopted throughout this paper. The norm of vector $\|x\|_\infty$ stands for the maximum norm $\|x\|_\infty = \max(|x1|, \cdots, |xn|)$. The expectation of random number x is described as E(x), and the condition expectation of random number x given y is described as E ($x|y$). The probability of an event is described as Pr $(\cdot)$.

2. System description and problem statement

As most smart grid is nonlinear system, a general nonlinear model is set up as,

$$\begin{aligned} x_{k+1} &= f(x_k) + \omega_k \\ y_k &= h(x_k) + v_k \end{aligned} \tag{1}$$

where $x_k \in \mathrm{R}^n$ and $y_k \in \mathrm{R}^p$ are the system state and measurement output, respectively. The functions of

$f(x)$ and $g(x)$ are continuously diff erentiable at $\hat{x}$. The process noise $\omega_k \in R^n$ and measurement noise $v_k \in R^n$ are white sequences with zero means. Their covariance matrices $E[\omega_k \omega^T] = Q_k \delta_{kj} > 0$ and $E[v_k v^T] = R_k \delta_{kj} > 0$. The initial distribution of x_0 is Gaussian with zero mean and covariance matrix P_0. Moreover, x_0, ω_k and v_k are assumed to be independent with each other.

The measurement y_k is transmitted over a wireless fading channel by the wireless emission infrastructure in the smart grid, which is shown in Figure 1. A binary variable γ_k is denoted to reflect whether y_k is successfully received by the estimation center as,

$$\gamma_k = \begin{cases} 1, & \text{y}_\text{k} \text{ is received} \\ 0, & \text{y}_\text{k} \text{ is not received} \end{cases} \tag{2}$$

At the time instance k, both the receiving states γ_k and the measurement y_k make up the observation information for estimation as

$$Y_k = \{\gamma_k, y_k\}. \tag{3}$$

It should be noticed that y_k is meaningless when $\gamma_k = 0$, because estimation only get a pure noise at time instance k. Moreover, the trigger contained information set is defined as

$$F_k \,\square\, \{Y_1, Y_2, \ldots, Y_k\} \tag{4}$$

The status of γ_k depends on both the communication channel and the communication power. According to the law of large number, the noise in communication channel can be described by an Additive White Gaussian Noise (AWGN). At the same time, the observation signal is modulated by Quadrature Amplitude Modulation (QAM) [16]. Therefore, by the communication theory, the symbol error rate (SER) can be approximated as,

$$SER = \exp(-\alpha \frac{M_k}{N_0 W}) \tag{5}$$

here N_0 is the AWGN noise power spectral density, W is the channel bandwidth, α is the constant determined by the number of byte in the observation signal, and M_k is the transmission power of QAM. Amount these parameter, N_0, W and α are all positive constant and determined off-line in practice, while M_k is the only parameter need to be scheduled. According to (5) and [16], the probability of transmission failure is that,

$$P\{\gamma_k = 0\} \,\square\, \lambda_k = (1-\theta)^{M_k}, \tag{6}$$

where θ is the constant parameter satisfying

$$\theta = 1 - \exp(\frac{-\alpha}{N_0 W}). \tag{7}$$

It should be noticed that $\theta \in (0, 1)$, thus, the transmission failure probability satisfies $\lambda \in (0, 1)$.The task of sensor transmission power schedule is calculating the minimum transmission power M_k that guarantee proper estimation performance.

3. EKF based transmission power schedule and filtering algorithm

This section establish the transmission power schedule based on EKF, also the filtering algorithm under the designed schedule is also proposed to promote the estimation accuracy inspired by the event-trigger filter theory.

3.1. EKF theory

In this part, the two-step EKF is introduced, and execute the filtering algorithm when the observation y_k is received. For convenience, the one step state prediction and the prior respective error covariance matrices are denoted as $\hat{x}_{k|k-1} = E(x_k|F_{k-1})$ and $P_{k|k-1} = Cov(x_k|F_{k-1})$, respectively. At the same time, the state estimation and the posterior respective error covariance matrices are denoted as $\hat{x}_{k|k-1} = E(x_k|F_k)$ and $P_k = Cov(x_k|F_k)$ respectively.

Then, if the previous time estimation result $\hat{x}_{k|k-1}$ and P_{k-1} are already know, the iteration of EKF includes the prediction and measurement update steps. The prediction step is described as

$$\begin{aligned}\hat{x}_{k|k-1} &= f(\hat{x}_{k-1}) \\ \hat{P}_{k|k-1} &= A_k \hat{P}_{k-1} A_k^T + Q\end{aligned} \tag{8}$$

where A_k is got by Taylor expansion of the continuous function $f(x)$ as,

$$A_k = \frac{\partial f}{\partial x}\Big|_{x=\hat{x}_{k-1}}. \tag{9}$$

Moreover, if the observation y_k is received by the estimation center, the measurement update step satisfies that,

$$\begin{aligned}\hat{x}_k &= \hat{x}_{k|k-1} + K_k(y_k - h(\hat{x}_{k|k-1})) \\ \hat{P}_k &= \hat{P}_{k|k-1} - K_k C_k \hat{P}_{k|k-1}\end{aligned} \tag{10}$$

where C_k is the first order Taylor polynomial of the continuous function $h(x)$ as,

$$C_k = \frac{\partial h}{\partial x}\Big|_{x=\hat{x}_{k|k-1}}, \tag{11}$$

and K_k satisfies that,

$$K_k = \hat{P}_{k|k-1} C_k^T (C_k \hat{P}_{k|k-1} C_k + R)^{-1} \tag{12}$$

3.2. Transmission power schedule

It can be noticed from (10) that diff erent value of y_k has diff erent influence on the measurement update step of **EKF** process. In filtering theory, $(y_k - \hat{y}_{k|k-1})$ is known as the “innovation” of y_k, which evaluate the importance of the observation y_k.

To obtain an appropriate estimation accuracy, the transmission power should determined by the innovation of y_k. When the innovation is big, the information contained in y_k is important for filter, thus paying plenty of transmission power is worth. On the contrary, y_k with small innovation is less important for estimation, which dose not need too much transmission power. Before utilizing the innovation of y_k, the weighted average can be carried out by

$$\varepsilon_k = (y_k - \hat{y}_{k|k-1})^T Z_k (y_k - \hat{y}_{k|k-1}), \tag{13}$$

where Z_k is the weighted matrix to be designed and $\hat{y}_{k|k-1}$ is the prior condition mean of y_k as,

$$\hat{y}_{k|k-1} = E(y_k | F_{k-1}) = h(f(\hat{x}_{k-1})) \tag{14}$$

The transmission power schedule in this paper is designed as,

$$M_k = \frac{\varepsilon_k N_0 W}{2\alpha}. \tag{15}$$

Thus the transmission failure probability under the condition that observation equals y_k in can be deduced as,

$$\begin{aligned} \mathrm{P}(\gamma_k = 0 \mid y_k) &= (\exp(\frac{-\alpha}{N_0 W}))^{\frac{\varepsilon_k N_0 W}{2\alpha}} \\ &= \exp\left(\frac{-\alpha}{N_0 W} \cdot \frac{(y_k - \hat{y}_{k|k-1})^T Z_k (y_k - \hat{y}_{k|k-1})/2) N_0 W}{2\alpha} \right) \\ &= \exp\left(-\frac{(y_k - \hat{y}_{k|k-1})^T Z_k (y_k - \hat{y}_{k|k-1})}{2} \right) \end{aligned} \tag{16}$$

Because y_k is the random variable, and its prior condition distribution can be approximated by the Gaussian distribution that

$$p(y_k \mid \mathrm{F}_{k-1}) \sim \mathrm{N}(\hat{y}_{k|k-1}, P_{y_k}), \tag{17}$$

where, P_{y_k} is the prior condition covariance of y_k calculated by

$$P_{y_k} = C_k (A_k \hat{P}_{k-1} A_k^T + Q) C_k^T + R \tag{18}$$

Thus the expectation of the transmission failure probability is calculated by

$$\begin{aligned} \mathrm{P}(\gamma_k = 0 \mid \mathrm{F}_{k-1}) &= \int_{\square^p} \frac{1}{(2\pi)^{p/2} \det(P_{y_k})^{1/2}} \exp\left(\frac{-(y_k - \hat{y}_{k|k-1})^T P_{y_k}^{-1} (y_k - \hat{y}_{k|k-1})}{2} \right) \mathrm{P}(\gamma_k = 0 \mid y_k) dy_k \\ &= \int_{\square^p} \frac{1}{(2\pi)^{p/2} \det(P_{y_k})^{1/2}} \exp\left(\frac{-(y_k - \hat{y}_{k|k-1})^T (P_{y_k}^{-1} + Z_k)(y_k - \hat{y}_{k|k-1})}{2} \right) dy_k \\ &= \frac{1}{\det(I_{p\times p} + P_{y_k} Z_k)^{1/2}} \end{aligned} \tag{19}$$

According to related work with EKF, the performance of EKF depends on the expectation of transmission failure probability. To get proper performance of EKF, it must be satisfied that $\mathrm{P}(\gamma_k = 0|F_k{-}1) \leq \lambda_c$, where λ_c is the gate depended on different system [17]. Due to (19), if Z_k is selected as

$$Z_k = \rho P_{y_k}^{-1}, \tag{20}$$

the expectation of transmission failure probability can be calculated as,

$$\mathrm{P}(\gamma_k = 0 \mid \mathrm{F}_{k-1}) = 1/(1+\rho)^{p/2}. \tag{21}$$

If λ_c is determined and the aim of transmission power schedule is to minimize energy consumption as much as possible, ρ can be calculated as,

$$\rho = \lambda_c^{-2/p} - 1. \tag{22}$$

3.3. Filtering algorithm in estimation center

Section 3.2 established the transmission power schedule in smart grid, this section will provide the corresponding filtering algorithm in estimation center. From Section 3.2 it can be noticed that the observation y_k can either successful or failed to be received by the estimation center. When $\gamma_k = 1$, filtering algorithm can be realized by EKF in Section 3.1. Otherwise, when $\gamma_k = 0$, filtering algorithm should be delicately design by takeing advantage of event-trigger theory to make full use of information containing in the transmission power schedule.

The iteration under $\gamma_k = 0$ is same with EKF as,

$$\begin{aligned} \hat{x}_{k|k-1} &= f(\hat{x}_{k-1}) \\ \hat{P}_{k|k-1} &= A_k \hat{P}_{k-1} A_k^T + Q \end{aligned} \tag{23}$$

where A_k is

$$A_k = \frac{\partial f}{\partial x}\Big|_{x=\hat{x}_{k-1}} . \tag{24}$$

However, the measurement update step is quite diff erent from EKF, because the observation y_k is not available for estimation center. Even though, the information contained in the transmission power schedule can be utilized by the filtering algorithm. Noticing that (16) is same with the trigger strategy in [18], the stochastic event-triggered filter in [18] can be utilized to realize the filtering algorithm when $\gamma_k = 0$, which is shown as,

$$\begin{aligned} \hat{x}_k &= \hat{x}_{k|k-1} \\ \hat{P}_k &= \hat{P}_{k|k-1} - \tilde{K}_k C_k \hat{P}_{k|k-1} \end{aligned} \tag{25}$$

where C_k is that,

$$C_k = \frac{\partial h}{\partial x}\Big|_{x=\hat{x}_{k|k-1}} , \tag{26}$$

And $\tilde{K}$ satisfies that,

$$\tilde{K} = \hat{P}_{k|k-1} C_k^T (C_k \hat{P}_{k|k-1} C_k + R + Z^{-1})^{-1} \tag{27}$$

Comparing the posterior condition covariance in (25) utilizing the transmission power schedule information with the prior condition covariance. It can be found that the posterior condition covariance is smaller than the prior condition covariance due to the existence of the term $\tilde{K}_k C_k \hat{P}_{k|k-1}$, which actually reflects the influence of the information contained in transmission powerschedule.

Take advantage of the binary variable γ_k, the filtering algorithm regardless of the receiving of the observation y_k can be united into

$$\begin{aligned} \hat{x}_k &= \hat{x}_{k|k-1} + \gamma_k \hat{K}_k (y_k - h(\hat{x}_{k|k-1})) \\ \hat{P}_k &= \hat{P}_{k|k-1} - \hat{K}_k C_k \hat{P}_{k|k-1} \end{aligned} \tag{28}$$

where $\hat{K}_k$ satisfies

$$\hat{K} = \hat{P}_{k|k-1} C_k^T (C_k \hat{P}_{k|k-1} C_k + R + (1-\gamma_k) Z^{-1})^{-1} \tag{29}$$

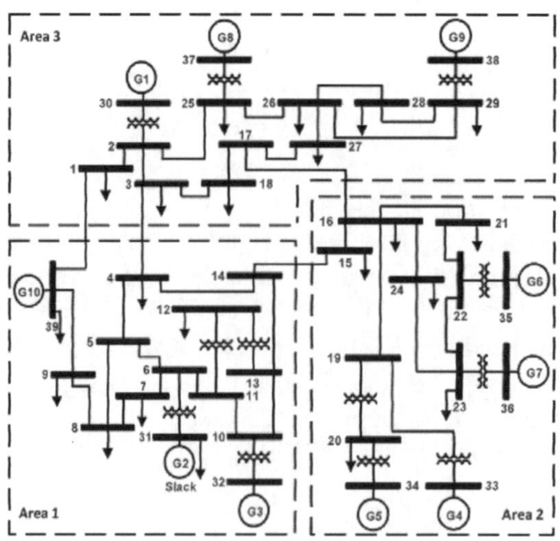

Figure 2. IEEE 39-bus 10-generator (New England) system.

4. Case study and simulation

IEEE 39-bus system, which is also know as the 10-generator New England system, is wildly applied to evaluate the estimation performance of smart grid [8, 10]. Fig. 2 describes the constitution of IEEE 39-bus system, where each grid is a generator. The parameters of generators quotes from [10]. The simulation executes 15 seconds after a fault breaks out at the line connecting the bus 14 to bus 15. By selecting the sampling interval as 0.02*s*, 750 discrete time interval estimation is performed.

In Fig. 1, each generater consists a smart grid, and its model can be described as,

$$\begin{aligned}
E_{d'}(k+1) &= \frac{\Delta t}{T'_{qo}}\left(-E'_d(k)+(X_q-X'_q)i_q(k)\right)+E'_d(k),\\
E_{q'}(k+1) &= \frac{\Delta t}{T'_{do}}\left(-E'_q(k)-(X_d-X'_d)i_d(k)+E_{fd}(k)\right),+E'_q(k)\\
\delta(k+1) &= \Delta t\left(\omega(k)-\omega_s\right)\omega_b+\delta(k),\\
\omega(k+1) &= \frac{\Delta t}{2H}\omega_s\left(T_m-E'_d(k)i_d(k)-E'_q(k)i_q(k)-(X'_q-X'_d)i_d(k)i_q(k)-D(\omega(k)-\omega_s)\right)+\omega(k).
\end{aligned} \tag{30}$$

This simulation pays attention on the smart of the generator 2 in Fig. 1, which is consisted by the generator and IEEE DC1A AVR. The generater can be described by (30), and the AVR can be modeled as,

$$\begin{aligned}
E_{fd}(k+1) &= \frac{\Delta t}{T_E}\left(-\left(K_E+A_x e^{B_x E_{fd}(k)}\right)E_{fd}(k)+V_R(k)\right)+E_{fd}(k),\\
R_f(k+1) &= \frac{\Delta t}{T_F}\left(-R_f(k)+\frac{K_F}{T_F}E_{fd}(k)\right)+R_f(k),\\
V_R(k+1) &= \frac{\Delta t}{T_A}\left(-\frac{K_A K_F}{T_F}E_{fd}(k)-V_R(k)+K_A R_f(k)+K_A\left(v_{ref}-v(k)\right)\right)+V_R(k).
\end{aligned} \tag{31}$$

PMU measures the magnitude and phase for both voltage of the smart grid and current flowing through generator. For convenience, the measurement data of PMU is denoted as $v(k)$, $\theta(k)$, $i_d(k)$, $i_q(k)$. By Kirchhoff Law, the observation function can be established as

$$\begin{aligned}
i_d(k) &= \frac{1}{X'_d}\left(E'_q-v(k)\cos(\delta(k)-\theta(k))\right),\\
i_q(k) &= \frac{1}{X'_q}\left(-E'_d+v(k)\sin(\delta(k)-\theta(k))\right).
\end{aligned} \tag{32}$$

The system noise is set as Gaussian noise with the covariance matrix $diag[10^{-18}, 10^{-18}, 10^{-9}, 10^{-9}, 10^{-9}, 10^{-9}, 10^{-9}]$and the observation noise is set as Gaussian noise with covariance matrix $diag[10^{-10}, 10^{-8}, 10^{-10}, 10^{-8}]$. The estimator is placed at the estimation center, and the observations is transmitted by the pow- er schedule in Section 3.2.

For convenient, this section denotes the filtering algorithm in Section 3.3 is denoted as ET-EKF. The gate for the expectation of transmission failure probability is selected as $\lambda_c = 90\%$, which means that only 10% of the measurement by PMU is received by the estimation center.

The simulation result is shown in Fig. 3. It can be seen that ET-EKF keeps stable in the statistics sense and provides accuracy filtering results, even only one-tenth observations are received by the estimation center.

5. Conclusion

Considering the RTSE problem for WAMS under practical communication channel, the event-trigger based transmission power schedule is designed. Moreover, the corresponding filter is designed based on EKF, which utilizes the information contained in the power schedule to promote the filtering accuracy. The simulation verifies that the power schedule and filter designed in this paper can reduce the transmission power and achieve high accuracy estimation result at the same time.

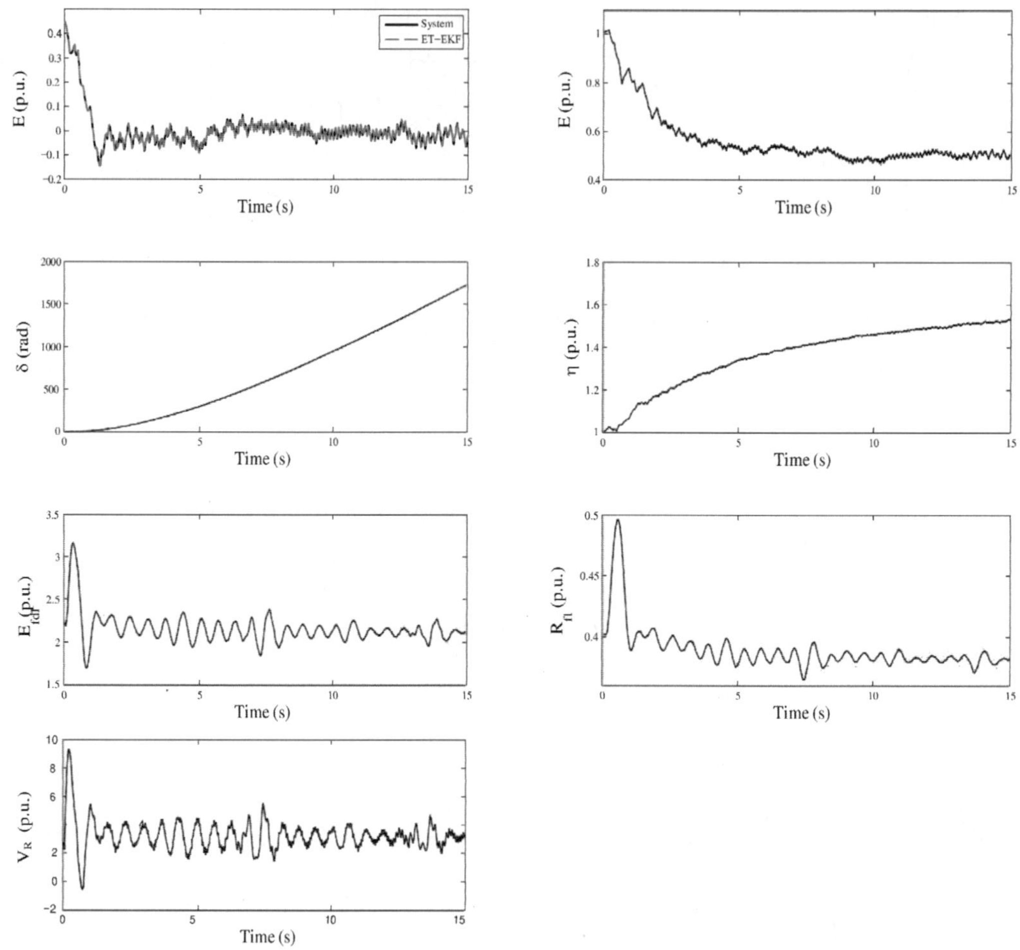

Figure 3. States and estimation results of generator when $\lambda = 10\%$ and plot of PMU transmissions.

References

[1] G ungor V C, Sahin D, Kocak T, Ergut S, Buccella C, Cecati C and Hancke G P 2011 *IEEE Trans. Ind.Inform.* **7** 529–539
[2] Ma R, Chen H H, Huang Y R and Meng W 2013 *IEEE Trans. Smart Grid* **4** 36–46
[3] Erol-Kantarci M and Mouftah H T 2015 *IEEE Commun. Surveys Tuts.* **17** 179–197
[4] Liu J, Tang J, Ponci F, Monti A, Muscas C and Pegoraro P A 2012 *IEEE Trans. Smart Grid* **3** 915–924
[5] Mohammadi M B, Hooshmand R A and Fesharaki F H 2016 *IEEE Trans. Smart Grid* **7** 84–93
[6] Ge Y, Flueck A J, Kim D K, Ahn J B, Lee J D and Kwon D Y 2015 *IEEE Trans. Smart Grid* **6** 2088–2097
[7] Ghahremani E and Kamwa I 2011 *IEEE Trans. Power Syst.* **26** 2556–2566 ISSN 0885-8950
[8] Singh A K and Pal B C 2014 *IEEE Trans. Power Syst.* **29** 794–804 ISSN 0885-8950
[9] Qi J, Sun K, Wang J and Liu H 2016 *IEEE Trans. Smart Grid* ISSN 1949-3053

doi:10.1109/TSG.2016.2580584

[10] Emami K, Fernando T, Iu H H C, Trinh H and Wong K P 2015 *IEEE Trans. Power Syst.* **30** 2665–2675 ISSN 0885-8950

[11] Leite J B and Mantovani J R S 2016 *IEEE Trans. Smart Grid* **7** 366–375

[12] Chen Y C, Wang J, Domnguez-Garca A D and Sauer P W 2016 *IEEE Trans. Smart Grid* **7** 2507–2515 ISSN 1949-3053

[13] Caire G, Taricco G and Biglieri E 1999 *IEEE Transactions on Information Theory* **45** 1468–1489 ISSN 0018-9448

[14] Quevedo D E, Ahl´en A, Leong A S and Dey S 2012 *Automatica* **48** 1306–1316

[15] Li Y, Quevedo D E, Lau V and Shi L 2013 Online sensor transmission power schedule for remote state estimation *52nd IEEE Conference on Decision and Control* pp 4000–4005 ISSN 0191-2216

[16] Marques A G, Wang X and Giannakis G B 2008 *IEEE Trans. Signal Process.* **56** 4446–4457 ISSN 1053-587X

[17] Liu X, Li L, Zhen L, Fernando T and Iu H H C 2016 *IEEE Trans. Circuits Syst. II, Exp. Briefs* ISSN 1549-7747 doi:10.1109/TCSII.2016.2578956

[18] Han D, Mo Y, Wu J, Weerakkody S, Sinopoli B and Shi L 2015 *IEEE Trans. Autom. Control* **60** 2661–2675

Phase separation of bio-oil produced by co-pyrolysis of corn cobs and polypropylene

D Supramono[1], Julianto[1], Haqqyana[1], H Setiadi[1, 2] and M Nasikin[1]

[1]Department of Chemical Engineering, Faculty of Engineering, Universitas Indonesia, Depok, 16424, Indonesia

E-mail: setiadi@che.ui.ac.id

Abstract. In co-pyrolysis of biomass-plastics, bio-oil produced contains both oxygenated and non-oxygenated compounds. High oxygen composition is responsible for instability and low heating value of bio-oil and high acid content for corrosiveness. Aims of the present work are to evaluate possibilities of achieving phase separation between oxygenated and non-oxygenated compounds in bio-oil using a proposed stirred tank reactor and to achieve synergistic effects on bio-oil yield and non-oxygenated compound layer yield. Separation of bio-oil into two layers, i.e. that containing oxygenated compounds (polar phase) and non-oxygenated compounds (non-polar phase) is important to obtain pure non-polar phase ready for the next processing of hydrogenation and used directly as bio-fuel. There has been no research work on co-pyrolysis of biomass-plastic considering possibility of phase separation of bio-oil. The present work is proposing a stirred tank reactor for co-pyrolysis with nitrogen injection, which is capable of tailoring co-pyrolysis conditions leading to low viscosity and viscosity asymmetry, which induce phase separation between polar phase and non-polar phase. The proposed reactor is capable of generating synergistic effect on bio-oil and non-polar yields as the composition of PP in feed is more than 25% weight in which non-polar layers contain only alkanes, alkenes, cycloalkanes and cycloalkenes.

1. Introduction

Co-pyrolysis of corn-cobs and polypropylene plastic is intended to obtain more bio-oil yield and more non-oxygenated compound contents in bio-oil by free radicals interaction of both charges and by hydrogen transfer in which plastic charge is hydrogen donor which transfers hydrogen to biomass charge as hydrogen acceptor [1]. Oxygenated compounds in bio-oil is undesirable for its application as a fuel because they leads to storage instability and low heating value for direct combustion. High carboxylic acids exert high corrosiveness [2]. The present work aims to assess possibilities of attaining phase separation of bio-oil into layers oxygenated compounds (polar phase) and non-oxygenated compound (non-polar phase) as products of a proposed stirred tank reactor and of achieving synergistic effects on yields of bio-oil and non-oxygenated compound layer separated from the bio-oil. This separation is advantageous in obtaining pure non-polar phase for proceeding to hydrogenation or used directly for combustion [3]. There has been no research work on co-pyrolysis of biomass-plastic leading to the phase separation of bio-oil [4]. In co-pyrolysis processing, most of types of reactors used are fixed bed reactors and auger reactors [1]. The selection of reactors for the production of bio-oil should consider the pyrolysis characteristics of feedstock. In the case of feedstock of plastics and biomass, the liquid yield of the plastic and biomass pyrolysis may depend on temperature and heating rate of the pyrolysis. According to Lopez *et al* [5] in their experiment of steam gasification of waste plastics, in which pyrolysis of the plastics was part of gasification, the liquid yield reduced as the temperature of gasification was increased. They worked at 800°C and 900°C. In terms of the effect of heating rate, Hoseini and Dastanian [6] found that liquid yield of the plastic pyrolysis was improved by pyrolysis at lower heating rate. Their work took heating rates spanning from 6 to 14°C/minute using low density polyethylene (LDPE), polypropylene (PP) and polyethylene terephthalate (PET) with final temperature of 500°C. At the lowest heating rate, the liquid yield of LDPE and PP was approximately 80%, while that of PET was 40%. The result of Hoseini and Dastarian' work suggests that the pyrolysis be carried out at quite low heating rate [6]. For the biomass pyrolysis, in general, Bridgwater [7] noted that higher heating rate of biomass pyrolysis would result in higher yield of liquid product.

Co-pyrolysis of biomass-plastics may produce non-oxygenated compounds in large proportions [8]. Therefore, it is worth investigating how to separate non-oxygenated (non-polar) compounds from oxygenated (polar) compounds in the least cost before the non-polar compounds are further processed to produce bio-fuel. Some researchers in co-pyrolysis suggested that polar and non-polar compounds are difficult to be separated as a result of interactions of radicals derived from co-pyrolysis reactions of their raw materials [4,9]. This difficulty raises efforts to upgrade bio-oil by extracting low molecular weight (MW) oxygenated compounds from bio-oil by adding ethyl acetate [10] or diethyl ether [11] or by water [12]. This separation leaves some large MW oxygenated compounds and hydrocarbons in the bio-oil thus improving stability and heating value of the bio-oil. However, the undertaking could not obtain the quality approaching that of fossil fuels because the bio-oil still contains oxygenated compounds of high MW. Therefore, the bio-oil obtained is restricted to specific uses. An alternative methodology of producing bio-oil by co-pyrolysis is needed so that bio-oil will have quality similar to that of fossil fuels, which contain only hydrocarbons, and will be flexibly produced in terms of its average and range of MW. The undertaking uses phase separation may be an option.

2. Phase separation

In the realm of fluid dynamics, phase separation has been investigated in relation to viscosity and viscosity asymmetry of components of liquids. Lamorgese and Mauri [13] in their simulation work on macroscopically quiescent liquid mixture found that phase separation is easier as the mixture has low average viscosity when the mixture is quenched below a critical point to minimise its mixing free energy. In the co-pyrolysis context, it is suggested that bio-oil have low MW. In terms of viscosity asymmetry, Lamorgese and Mauri [14] found that for the quenching mixture which has a tendency to minimise mixing free energy, at fixed fluidity coefficient of the dispersed phase, increasing viscosity ratio of the dispersed phase to the continuous phase from the unity ratio leads to faster phase separation. On the other hand, at fixed fluidity coefficient of the continuous phase, increasing viscosity ratio leads to slower phase separation. The fluidity coefficient is defined as the ratio of capillary to viscous forces, where the capillary force combines convection and diffusion. This coefficient is proportional to Marangoni number (Reynolds number times Prandtl number) [15]. These two extreme conditions, i.e. the fixed fluidity of the dispersed phase and that of the continuous phase do not occur when vapour phase of co-pyrolysis product is immediately quenched in the condensers. In real co-pyrolysis work, in-between condition occurs because both the oxygenated and non-oxygenated compounds simultaneously make convection by flowing down from the inlet of the condensers as the vapour to the outlet of the condenser as the liquid to enter a bio-oil container. The non-oxygenated compound phase was the dispersed phase, while the oxygenated compound phase was continuous phase in the current work. Whatever conditions, the viscosity asymmetry may induce phase separation. In order to attain phase separation between polar compounds and non-polar compounds, a reactor design with better heat transfer is mandatory which enables to produce bio-oil with relatively low MW range or that with different viscosities between oxygenated and non-oxygenated compounds.

The reduction of viscosity and the enhancement of viscosity ratio between polar and non-polar compounds in bio-oil may be tailored by adjusting the flow of nitrogen gas to the co-pyrolysis feedstock and stirring the pyrolysis feedstock. Heat flux to the reactor feedstock may be improved by the presence of nitrogen bubbles in the plastic melt, the shear stress at the wall of reactors, and the shear stress in the plastic melt, the latter two of which can be induced by stirring. Zhang *et al* [16] by their simulation based on experimental database of turbulence intensities in the cores and near the walls of vertical ducts from various researchers concluded that the introduction of gas bubbles to large ducts can induce high local turbulence in the cores, and to small ducts lower local turbulence intensities in the cores, but low local turbulence near the wall. In the case of turbulence on the vertical duct wall, simulations by Tryggvason and Lu [17] and by Mattson and Mahesh [18] found that as the bubbles went up the duct, initially bubbles approached the wall, but then went away from the wall. This indicates that turbulence near the wall occurrs at upstream position only.

Simulation by Abbassiei [19] suggested that when bubbles rise in low viscosity liquid, the bubble shape and hydrodynamics are dominated by inertial and surface tension forces and the influence by viscous force is neglected. In this case, bubbles are not in spherical shapes and they generate turbulence in the liquid resulting in wobbling/zig-zag motions in the liquid. However, in the viscous liquid environment, the bubble shape and motion are dominated by viscous force in which bubbles retain their spherical shape and no turbulence is generated around the bubbles. By increasing the velocity of bubbles in the plastic melt and reducing the melt viscosity, the average size of the bubbles will increase [20]. Both high velocity and large average size bubbles are favourable to generate turbulence in the plastic melt, which corresponds to the increasing drag and lift forces [21].

In terms of the effect of bubbles on heat transfer, according to Tokunaga *et al* [22], heat transfer coefficient by the bubbles is much lower (two order of magnitudes lower) than that by liquid. Therefore, high void fraction of nitrogen bubbles in the plastic melt is undesirable. Experiment database acquired by Zhang *et al* [16] from various researchers suggested that bubble void fraction of max 15% is sufficient to exert high turbulence intensities in the liquid for large diameter ducts. However, the nitrogen bubbles are not expected to reduce plastic melt viscosity, because it requires gas weight fraction of 0.005 or approximately 5 times volume of plastic melt to allow the melt viscosity drop of 20%. Therefore, heat transfer in the plastic melt can rely on improved turbulence intensity in the presence of nitrogen bubbles, but not on reduced bulk viscosity. With sufficient turbulence, it is expected that Biot number, which measures the ratio of the convective heat transfer in the plastic melt to the conductive heat transfer in the biomass particles, becomes high to allow the biomass particle temperature to be approached by plastic melt temperature [23].

Heat transfer enhancement in a co-pyrolysis reactor may adopt that in plastic extruders because they involve in plastic heating. At early process of co-pyrolysis, there is melting conduction, which is a slow process because of very low thermal conductivity of plastics [24]. To accelerate this process, small plastic pellets and good mixing have been suggested [24]. Good mixing can be obtained by rotating the screw of the extruder an optimum rpm (rotation per minute). Rubbing plastic pellets on a heating surface of the extruder as a result of screw rotation induces pellet melting. Large heat conduction near the hot surface and heat convection in the plastic bulk can be enhanced by generating velocity gradient across the radius of the extruder as a result of extruder screw rotation [23].

3. Heat transfer in co-pyrolysis

Two types of reactors generally considered in co-pyrolysis are fixed beds and auger reactors. Fixed bed reactors are mostly designed in the form of tubular reactor [10,4,25] to produce bio-oil in semi-batch mode. No papers have analysed the heat transfer aspect of the fixed bed reactors. Data of different fixed bed reactors suggest that the largest diameter of the reactor ever made is 40 mm and the highest ratio of the height to the diameter is 46. Such a low reactor diameter was designed to overcome low thermal conductivity of the plastic melt in which the heat transfer is governed by heat conduction from the reactor wall to the pyrolysis material. The values of thermal conductivity of PP are between 0.125-0.210 W/(m.K) depending on molecular weight of PP, the higher the molecular weight of PP plastic, the lower is the thermal conductivity of the plastic [26]. Those values of PE are between 0.25-0.33 W/(m.K) [26]. All these values are less than that of water, i.e. about 0.591 W/(m.K). Nitrogen gas in the fixed bed reactor is merely used as sweep gas to bring the pyrolysis vapour out of the reactors and not considered to improve the heat transfer. The reactor may have had a large temperature gradient in radial direction, if the diameter of the tubular fixed bed reactor had been larger. This heat conduction constraint impedes efforts of scaling up the reactor into commercial scale. It seems that such a heat transfer mode tends to produce bio-oil with some high MW compounds or that with high content of aromatics [10,4,25]. The presence of high MW compounds implies that the heat transfer in the reactors were unable to attain high extent of pyrolysis. The presence of aromatics indicates the combination of plastic free radicals did not sufficiently occur as a result of poor mixing [27]. If the design was maintained for commercial purposes, it required multiple tubular reactors, which are not desirable in terms of investment and maintenance.

Auger reactors are designed to produce bio-oil from co-pyrolysis continuously. To facilitate the continuity of the pyrolysis, at the front end of the reactor there is hopper to feed the feedstock of the pyrolysis [28]. Inside the reactor, there is a screw to move the feedstock to the rear end of the reactor while it is being heated during its movement to produce different phases of products, i.e. condensable gas, non-condensable gas, char and wax. In most designs, heat is transferred radially from the wall of the reactors into the materials of pyrolysis. Gas nitrogen as a sweep gas flows through the rear end of the reactor to bring out gases through a series of condensers where the condensable gas is converted into bio-oil. The designs of reactors suggests that the heat transfer to the plastic melt is governed by laminar heat convection in radial direction while plastic melt and biomass particles loaded in the plastic melt flow in axial direction. It seems that heat flux in the auger reactors is not sufficiently high to heat up pyrolysis materials. The bio-oil obtained by previous researchers still contained mixed oxygenated and non-oxygenated compounds with some high MW compounds or that with high content of aromatics [4,28,29].

A reactor design alternative to the fixed bed and auger reactors is worth proposing to overcome the heat transfer constraints in these types of reactors. Considering elucidation of heat transfer above, a stirred tank reactor equipped with nitrogen flow to induce local turbulence may be a better option for co-pyrolysis of biomass and plastics. In this case, the heat transfer is dominated by turbulent heat convection, which is different from the previous reactors, which used heat conduction and laminar heat conduction. Temperature and heating rate of the feedstock, which is contributed by the heat flux, determines the extent of the reactor feedstock cracking [30].

4. Experimental

Corn cobs obtained from an agricultural waste were ground and sieved to a particle size of 8 – 50 mesh. The particles were then dried in an oven so that the moisture content reached $\leq$ 10%. Meanwhile, the PP granules used in the experiment were of 3 mm average diameter. Unlike the corn cobs, the PP granules were not pre-treatment prior to pyrolysis.

Table 1. Proximate and ultimate analysis of the feedstock (air-dried basis).

Analysis Parameter	Corn Cobs (%weight)	Polypropylene (%weight)
Proximate analysis		
Air-dried moisture	11.02	0.11
Ash	5.07	0.00
Volatile matter	69.31	99.89
Fixed carbon	14.60	0.00
Ultimate analysis		
C	41.07	87.18
H	6.54	12.70
N	0.76	0.03
Total sulphur	0.09	0.07
O*	46.47	0.02
H/(C+H+O+N) atomic ratio	0.07	0.13

* by difference

Ultimate analysis and proximate analysis of feedstock are given in table 1. Data of proximate analysis show that corn cobs are composed of mainly volatile matter, which allows high yield of bio-oil [31]. Data of ultimate analysis show that atomic ratio of H in corn cobs is nearly a half of that in PP. In co-pyrolysis of corn cob particles and PP granules, the corn cob particles are deficient in hydrogen, whereas PP granules contain more hydrogen as seen in their atomic ratios of hydrogen. The hydrogen excess provided by PP exerts two effects against bio-oil produced by biomass alone, the first, it improves hydrogen atmosphere in pyrolysis reactions resulting more bio-oil yield [32] and the second, it partly removes oxygen radicals derived from biomass pyrolysis thus improving non-oxygenated compound composition in bio-oil [33].

To perform co-pyrolysis in the stirred tank, feedstock charges of corn cob particles and PP granules were varied at 0%, 12.5%, 25%, 37.5%, 50%, 62.5%, 75%, 87.5%, and 100% weight of PP. A schematic of the co-pyrolysis experimental rig is shown in figure 1. In all runs of co-pyrolysis, 250 grams of feedstock were charged to the stirred tank reactor at heating rate of 5°C/minute from ambient temperature to 500°C, with a holding time of 10 minutes after the peak temperature to allow further pyrolysis of the remaining feedstock in the reactor. The flow of nitrogen gas was maintained at 500 ml/minute, which had been proven obtaining maximum oil yield using plastic feedstock. A faster flow of nitrogen is likely to lead to a shorter residence time in the vapor phase for interaction between free radicals of biomass and plastic to attain a synergetic effect. Conversely, if the flow rate of nitrogen is too low, it could trigger a secondary pyrolysis reaction in the reactor due to an excessively long residence time leading to production more char [7,34]. The rotation rate of the two-paddle stirrer was 100 rpm. The resulting pyrolysis vapor was condensed in a series of two water-cooled bulb condensers and the bio-oil produced was collected in a vacuum flask. Cooling water at 15°C was used in the condensers. At the end of each experiment run, the liquid product was weighed, separated into 2 samples for that bearing phase separation and analysed using GCMS analyser.

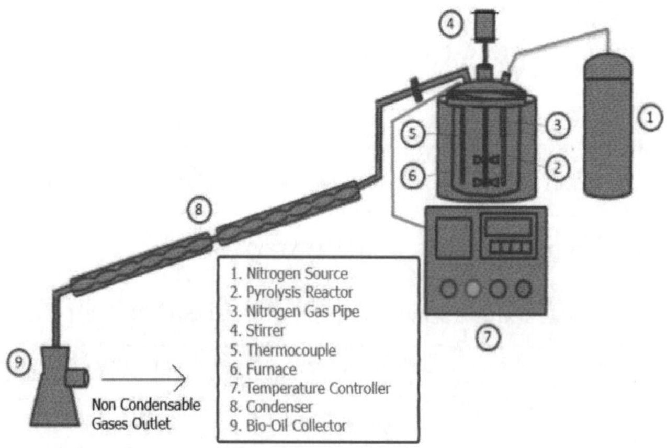

Figure 1. Experimental rig of co-pyrolysis.

5. Results and discussion

5.1. Synergistic effect on bio-oil yield

Experiment of co-pyrolysis corn cob particles and PP plastic have resulted bio-oil samples with low yields from co-pyrolysis using feed containing 0 and 12.5% PP and yields increasing at increased PP composition in the feedstock from 12.5% to 87.5%. The bio-oil yield slightly reduces as the pyrolysis uses pure PP. The effect of PP composition on the bio-oil yield is described by figure 2. Comparing the bio-oil yields obtained experimentally from co-pyrolysis and those calculated based on experimental yields using pure feedstock and proportional to the compositions of PP in the feeds will give insight about synergistic effect of the co-pyrolysis on the bio-oil yield. Figure 3 shows that the synergistic effect is favoured as the PP composition is more than 25%. Setting heating rate as low as 5°C/min also assists to retain plastic melt in the reactor in close contact with biomass particle which is favourable to attain high oil yield [35]. Therefore, in co-pyrolysis involving high portion of PP in the feedstock, the yield of bio-oil increases and that of char reduces as the PP composition in feed increases The regime where co-pyrolysis using feeds containing PP 25% and less is subsequently called as regime 1, and that using feeds containing PP 37.5% and more as regime II.

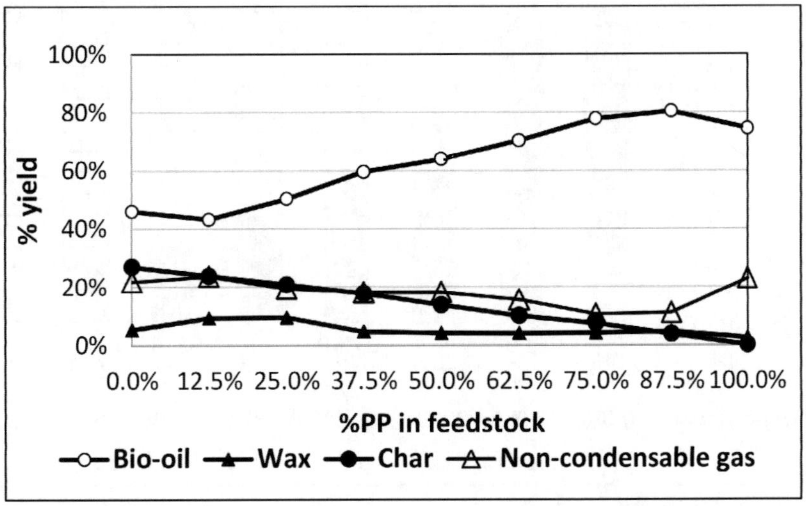

Figure 2. Yields of co-pyrolysis products as function of PP compositions in feeds.

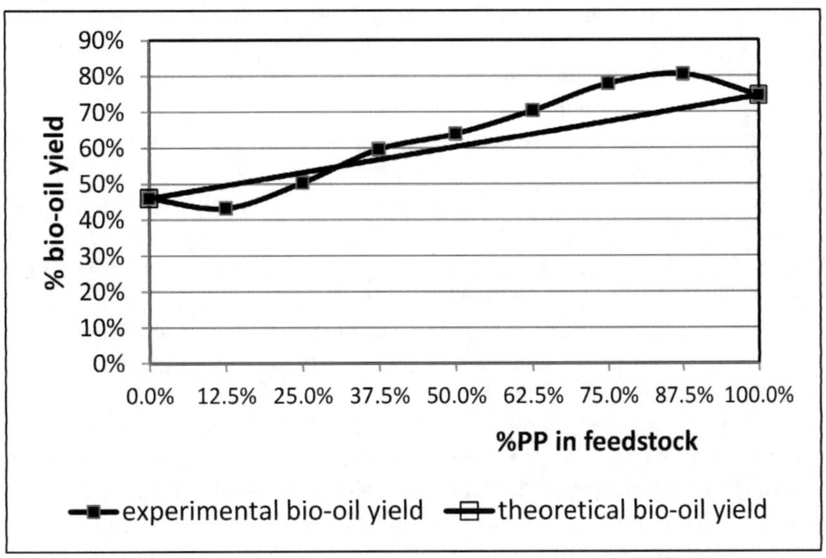

Figure 3. Synergistic effect of co-pyrolysis on bio-oil yield.

5.2. Synergistic effect on non-oxygenated phase yield

In terms of hydrogen balance, plastics as hydrogen donor transfer hydrogen atoms to biomass to produce bio-oil with more non-oxygenated compounds [33]. Squir *et al* [36] and Matzakov [37] found that as the free radicals of a hydrogen acceptor exceed the capacity of the hydrogen acceptor to catch hydrogen atoms, then the free radicals combine or attack available large molecules to bear cross-linking reactions which produce secondary char. Figure 4 shows the result of the hydrogen transfer as well as the phase separation of the bio-oil in the present work where the upper layer of the bio-oil container composes non-oxygenated compounds, while the lower layer oxygenated compounds. Comparison of PP compositions in the feeds and corresponding non-polar % volume also in regime II exhibits the existence of synergistic effect on the non-polar bio-oil where % volume of non-polar layer is always higher than %PP in the feedstock.

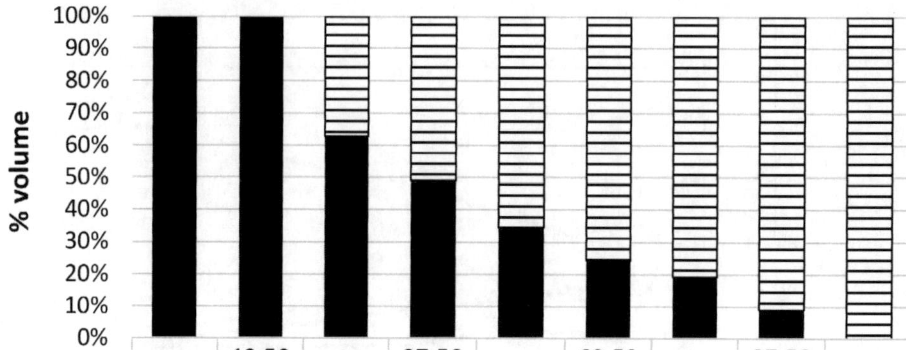

	0%	12.50 %	25%	37.50 %	50%	62.50 %	75%	87.50 %	100%
%volume of non-polar	0.0%	0.0%	37.2%	50.9%	65.5%	75.7%	80.8%	91.0%	100.0%
%volume of polar	100.0%	100.0%	62.8%	49.1%	34.5%	24.3%	19.2%	9.0%	0.0%

Figure 4. Synergistic effect of co-pyrolysis on non-oxygenated compound phase.

Measurement of viscosity of polar layers from co-pyrolysis in the present work using various feed compositions demonstrated that they have very similar viscosities with average of 1.1 cStokes and % standard deviation of 2.7%. Those values for non-polar layers are 2.03 cStokes and 6.4%. Comparison of average viscosities of polar and non-polar layers shows that there is viscosity asymmetry between the polar layers and the non-polar layers where the viscosities of the polar layers are smaller than those of the non-polar layers. The magnitudes of these average viscosities are considered low because they are within the range of gasoline viscosity of approximately 0.7 cStokes and diesel oil viscosity of approximately 2.7 cStokes. Both low viscosities of polar and non-polar layers and viscosity asymmetry of these layers in the present work induced the phase separation between polar and non-polar layers [13,14]. This is very advantageous because physical separation between these layers shortly after the formation of bio-oil is no longer needed. In contrast, most of the previous co-pyrolysis works have never reached such phase separation thus requiring complicated bio-oil phase separation because of introduction of new solvent [10-12].

The viscosity difference may be inferred by figures 5 and 6. Figures 5 and 6 show % total area against retention time interval limit extracted from GCMS analysis of the polar layer and the non-polar layers of bio-oil samples, respectively. The value 1 is the upper limit of the retention time interval from 0 to 1 minute, 2 the upper limit of the interval from 1 to 2 minutes, and so on. Comparison between these figures demonstrates that the maximum retention times of the non-polar layers are about twice those of the polar layers. This indicates roughly that the longest carbon chains of compounds in the non-polar layers are longer than those in the polar layers. Figure 5 indicates that qualitatively the polar layers compose mostly shorter carbon-chains of oxygenated compounds described by the left skewness of the curves, whereas figure 6 indicates the total area roughly descends gradually from the highest at the lowest retention time to the highest retention time. The presentation of curves in figures 5 and 6 suggests that the carbon-chain distributions in the polar layers of bio-oil samples taken from co-pyrolysis of various PP compositions in the feeds are very similar, but fluctuated in the non-polar layers. These levels of similarities are also implied by the values of %standard deviation of various viscosities, which were 2.7% for viscosities of the polar layers and 6.4% for those of the non-polar layers.

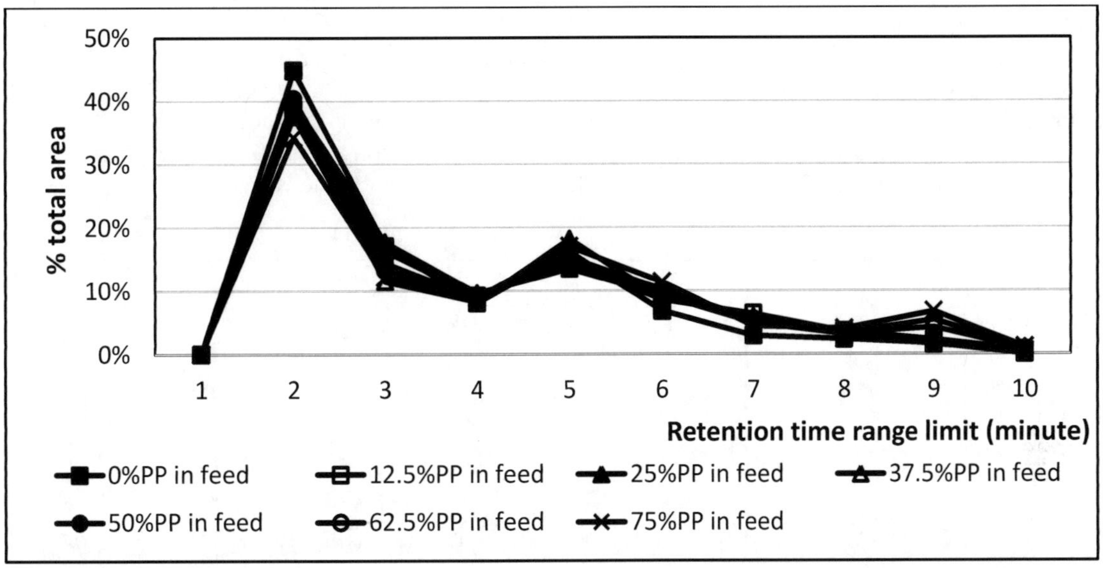

Figure 5. Total area percentages of oxygenated compounds acquired from GCMS analysis.

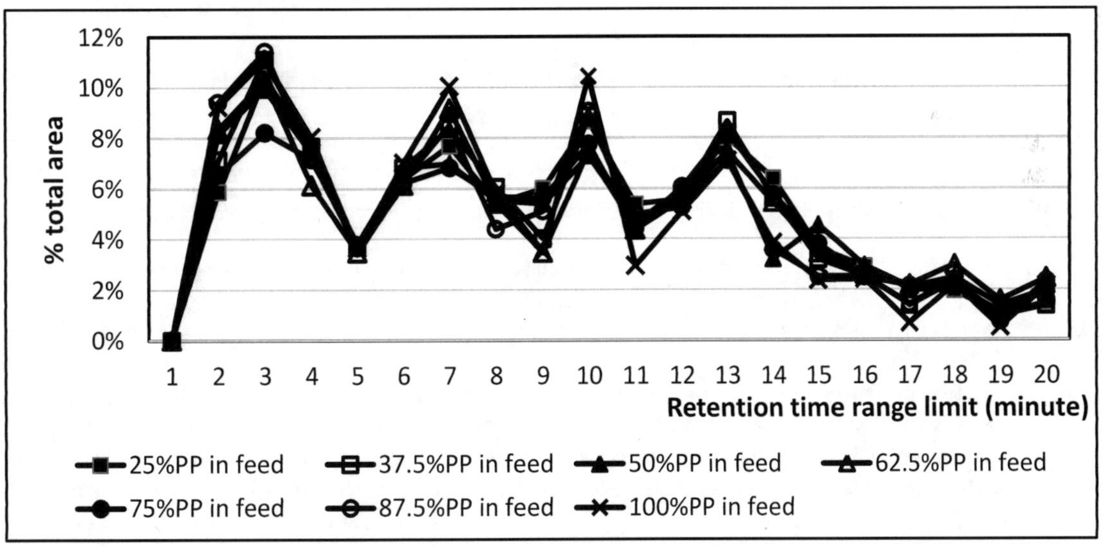

Figure 6. Total area percentages of non-oxygenated compounds acquired from GCMS analysis.

5.3. Bio-oil composition

Figures 7 and 8 exhibit the compound compositions based on functional groups for oxygenated compounds (in the polar layers of bio-oil samples) and types of carbon chains for non-oxygenated compounds (in the non-polar layers of bio-oil samples). Figure 7 shows that main compounds in the polar layers are carboxylic acids, furans, ketones, guaiacols, and phenols. The compositions of these compounds vary at different compositions of PP in feeds, but in general the rank of compositions from the highest is carboxylic acids, furans and interchangeably between ketones, guaiacols and phenols.

Carboxylic acids are obtained from pyrolysis of cellulose, hemicellulose and lignin, while furans from pyrolysis of cellulose and hemicellulose. Both carboxylic acids and furans are formed at mild condition (approximately 300°C) [38]. Guaiacols and phenols are obtained from pyrolysis of lignin at more severe condition approaching 500°C [38]. Therefore, most of compounds were produced at mild conditions. Approximate analysis data in table 1 show that nearly 70% of biomass contains volatile matters, which are mostly from pyrolysis of cellulose and hemicellulose [39] and most of them seem to be cracked at mild condition. Ketones and aldehydes are mainly produced from pyrolysis of levoglucosan (originated from cellulose as parent structure) [40]. These compounds are readily hydrogenised towards more stable groups [41] leading to their small compositions in the bio-oil. Abella *et al* [40] found that activation energy towards the formation of aldehydes is high and their presence in the bio-oil therefore depends on how heat flux to biomass particles occurred.

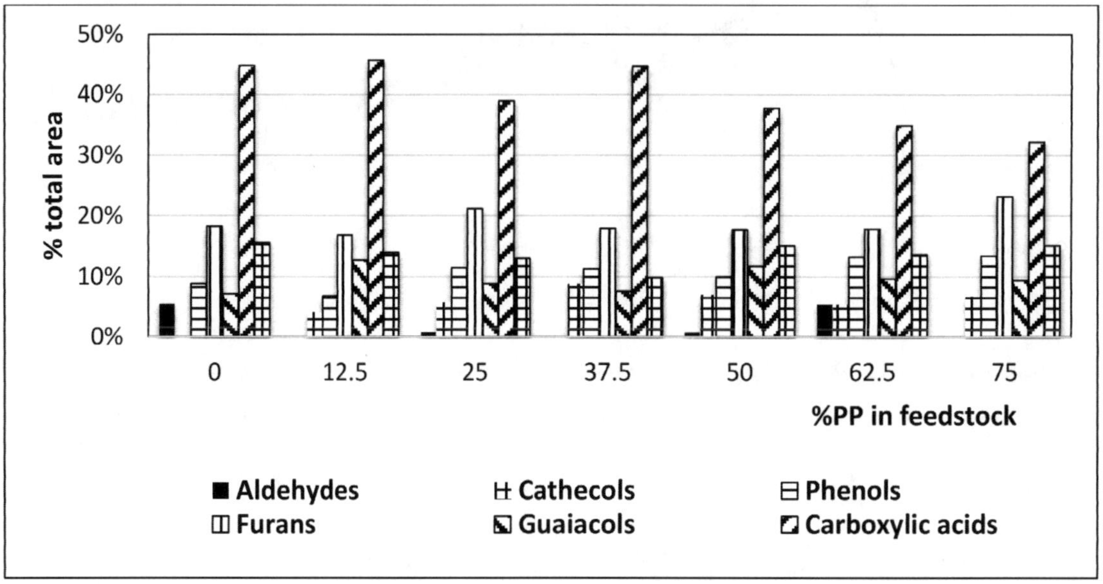

Figure 7. Compositions of oxygenated compounds in bio-oil acquired from GCMS analysis.

Figure 8 exhibits the compositions of non-oxygenated compounds in the non-polar layers of bio-oil. GCMS analysis noted that there are four main types of hydrocarbons in the non-polar layers, i.e. alkanes, alkenes, cycloalkanes and cycloalkenes, which were produced from PP plastic pyrolysis. In case of thermal cracking, the cyclisation leading to the formation of cycloalkanes and cycloalkenes is promoted by free radical mechanism [42]. No aromatics have been produced from plastic degradation in the co-pyrolysis. This indicates that there was no methyl branching leading to aromatisation or methyl groups combine each other to form oligomers as non-condensable gas, which otherwise are contributed by the presence of Pt-containing catalyst at 525°C. [27]. Crosslink of the plastic free radicals may have also not occurred in regime II as described by the diminishing wax yield as the PP composition was increased [43]. For regime II, the compositions of hydrocarbons taken from co-pyrolysis at various PP compositions in the feds are similar or nearly identical. It seems that part of plastic free radicals give away their hydrogen to biomass free radicals to add up non-oxygenated compounds. This phenomenon has been predicted to occur by Bhattacharya *et al* [28] and Chen *et al* [44].

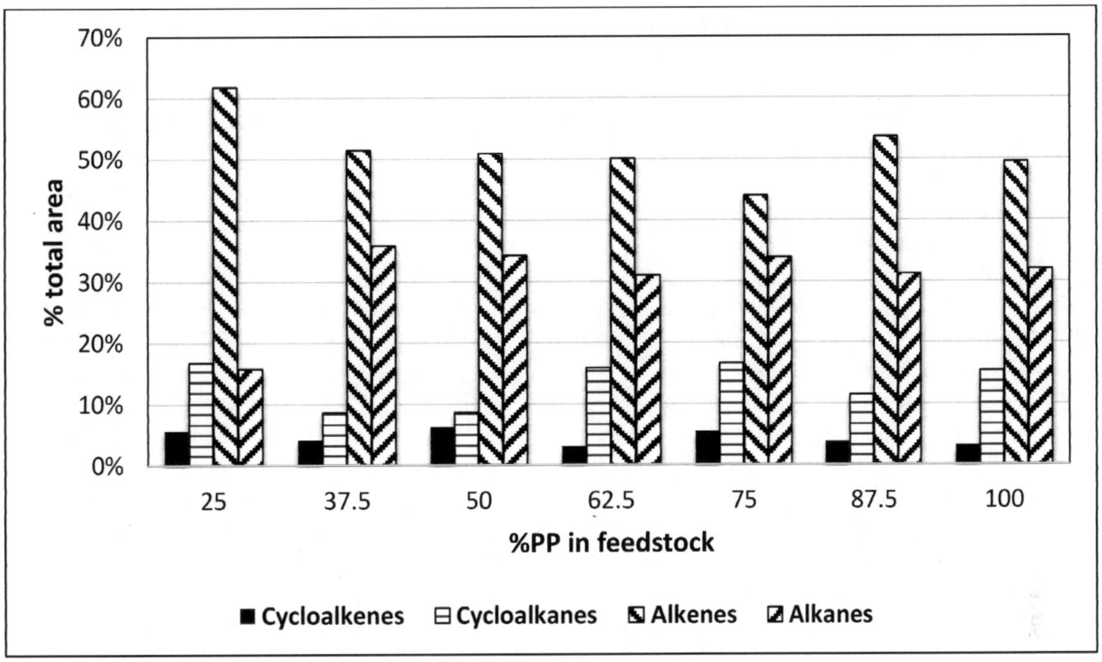

Figure 8. Compositions of non-oxygenated compounds in bio-oil acquired from GCMS analysis.

6. Conclusions

The present work of co-pyrolysis of corn cob particles and PP granules in the proposed stirred-tank reactor results in some conclusions as follows:

- The reactor is capable of producing bio-oil with low viscosity and viscosity asymmetry leading to phase separation between oxygenated and non-oxygenated compounds
- Synergistic effect has been achieved on yields of bio-oil and non-polar layer of the bio-oil for co-pyrolysis using feeds containing more than 25% weight of PP.
- Using settings of stirrer rotation of 100 rpm and flow of nitrogen of 500 mL/min, non-polar layer contains only alkanes, alkenes, cycloalkanes and cycloalkenes. This layer may be used as a biofuel due to the absence of oxygenated compounds.

Acknowledgments
Authors wish to express their gratitude to PITTA research grant 2017 from Universitas Indonesia for financing the present work.

References

[1] Abnisa F and Wan Daud W M A 2014 A review on co-pyrolysis of biomass: An optional technique to obtain a high-grade pyrolysis oil *Energy Conver and Management* **87** 71-85
[2] Khor K H, Lim K O and Zainal Z A 2009 Characterization of bio-oil: A by-product from slow pyrolysis of oil palm empty fruit bunches *American J of Appl Sci* **6** 9 1647-52
[3] Xiu S and Shahbazi A 2012 Bio-oil production and upgrading research: A review *Renew and Sust Energy Reviews* **16** 4406-14
[4] Martinez J D, Veses A, Mastral A M, Murillo R, Navarro M V and Puy N 2014 Copyrolysis of biomass with waste tyres: Upgrading of liquid bio-fuel *Fuel Process Technol* **119** 263-71
[5] Lopez G, Erkiaga A, Amutio M, Alvarez J, Barbarias I, Bilbao J and Olazar M 2013 Steam gasification of waste plastics in a conical spouted bed reactor *The 14th International Conference on Fluidization – From Fundamentals to Products, eds, ECI Symposium Series* () pp

[6] Hoseini S M F and Dastanian M 2013 Predicting pyrolysis products of PE, PP, and PET using NRTL activity coefficient model *Hindawi Publishing Corporation Journal of Chemistry* 487676
[7] Bridgwater A V 2012 Review of fast pyrolysis of biomass and product upgrading *Biomass Bioenergy* **38** 68-94
[8] Supramono D, Jonathan, Haqqyana, Setiadi and Nasikin M 2016 Improving bio-oil quality through co-pyrolysis of corn cobs and polypropylene in a stirred tank reactor *Int J of Technol* **8** 1381-91
[9] Onal E, Uzun B B and Putun A E 2014 Bio-oil production via co-pyrolysis of almond shell as biomass and high density polyethylene *Energy Conver and Management* **78** 704-10
[10] Brebu M, Ucar S, Vasile C and Yanik J 2010 Co-pyrolysis of pine cone with synthetic polymers *Fuel* **89** 1911-8
[11] Uçar S and Karagöz S 2009 The slow pyrolysis of pomegranate seeds: The effect of temperature on the product yields and bio-oil properties *Journal of Analytical and Appl Pyrolysis* **84** 151-6
[12] Park L K, Ren S, Yiacoumi S, Ye X P, Borole A P and Tsouris C 2016 Separation of switch grass bio-oil by water/organic solvent addition and pH adjustment *Energy Fuels* **30** 2164-73
[13] Lamorgese A G and Mauri R 2006 Mixing of macroscopically quiescent liquid mixtures *Physics of Fluids* **18** 044107
[14] Lamorgese A G and Mauri R 2017 Effect of viscosity ratio on structure evolution during mixing/demixing of regular binary mixtures *Chem Engin Transactions* **57** 1225-30
[15] Lamorgese A G and Mauri R 2011 Liquid mixture convection during phase separation in a temperature gradient *Physics of Fluids* **23** 034102
[16] Zhang H, Yokomine T and Kunugi T 2015 Turbulence modulation of the upward turbulent bubbly flow in vertical ducts *Nuclear Engin Technol* **47** 513-22
[17] Tryggvason G and Lu J 2015 Direct numerical simulations of bubbly flows *Mech Engin Reviews* **2** 1-14
[18] Mattson M and Mahesh K 2011 Simulation of bubble migration in a turbulent boundary layer *Physics of Fluids* **23** 045107
[19] Abbassi W, Besbes S, El Hajem M, Ben Aissia H, Champagne J Y and Jay J 2017 Influence of operating conditions and liquid phase viscosity with volume of fluid method on bubble formation process *European J of Mech B/Fluids* **65** 284-98
[20] Komarov S V and Sano M 1998 Bubble liquid behavior and heat transfer bath *ISIJ International* **38** 1045-52
[21] Schwarz M P 1996 Simulation of gas injection into liquid melts 1996 *App. Math. Modelling* **20** 41-51
[22] Tokunaga H, Iguchi M and Tatemichi H 1995 Heat transfer between bubbles and molten wood's metal *ISIJ International* **35** 21-5
[23] Khaghanikavkani E and Farid M M 2011 Thermal pyrolysis of polyethylene: Kinetic study *Energy Sci and Technol* **2** 1-10
[24] Chung C I 2007 *Extrusion of Polymers, Theory & Practice* 2nd ed. (Cincinnati: Hanser Publishers)
[25] Abnisa F 2015 Study on pyrolysis of oil palm solid wastes and co-pyrolysis of palm shell with plastic and waste tyre (Universiti Malaya)
[26] Ramsey III J C, Fricke A L and Caskey J A 1973 Thermal conductivity of polymer melts *J of Appl Polymer Sci* **17** 1597-605
[27] Uemichi Y, Makino Y and Kanazuka T J 1989 *Anal. Appl. Pyrol.* **16** 229-38
[28] Bhattacharya P, Steele P H, Hassan E B M, Mitchell B, Ingram L and Pittman Jr C U 2009 Wood/plastic copyrolysis in an auger reactor: Chemical and physical analysis of the products *Fuel* **88** 1251-60
[29] Rotliwala Y C, Shah J A, Dhingra N, Parekh D B and Parikh P A 2013 Plastics/biomass

copyrolysis in an auger reactor: Physico-chemical analysis of the products *7th International Symposium on Feedstock Recycling of Polymeric Mater* (New Delhi, India) pp 23-6

[30] Tchapda A H and Pisupati S V 2014 A Review of thermal co-conversion of coal and biomass/waste, *Energies* **7** 1098-148

[31] Chouhan A P S and Sarma A K 2013 Critical analysis of process parameters for bio-oil production via pyrolysis of biomass: A review *Recent Patents on Engin* **7** 1-17

[32] Sharypov V I, Beregovtsova N G, Kuznetsov B N, Baryshnikov S V, Cebolla V L, Weber J V, Collura S, Finqueneisel G and Zimny T J 2006 Co-pyrolysis of wood biomass and synthetic polymers mixtures: Part IV: Catalytic pyrolysis of pine wood and polyolefinic polymers mixtures in hydrogen atmosphere *J. Anal. Appl. Pyrol.* **76** 265-70

[33] Sharypov V I, Beregovtsova N G, Kuznetsov B N, Membrado L, Cebolla V L, Marin N and Weber J V 2003 Co-pyrolysis of wood biomass and synthetic polymers mixtures. Part III: Characterisation of heavy products *J. Anal. Appl. Pyrol* **67** 325-40

[34] Kato Y, Enomoto R, Akazawa M and Kojima Y 2016 Characterization of Japanese cedar bio-oil produced using a bench-scale auger pyrolyzer *Springer Plus* **5** 177

[35] Fei J, Zhang J, Wang F and Wang J 2012 Synergistic effects on co-pyrolysis of lignite and high-sulfur swelling coal *Journal of Analytical and Appl Pyrol* **95** 61-7

[36] Squir K R, Solomon P R, Carangelo R M and DiTaranto M B 1986 Tar evolution from coal and model polymers: 2. The effects of aromatic ring sizes and donatable hydrogens *Fuel* **65** 833-43

[37] Matzakos A N 1992 Fundamental mechanism of coal pyrolysis and char combustion (Rice University)

[38] de Wild P 2011 Biomass pyrolysis for chemicals (Rijksuniversiteit Groningen)

[39] Nunn T R, Howard J B, Longwell J P and Peters W A 1985 Product compositions and kinetics in the rapid pyrolysis of sweet gum hardwood *Ind. Eng. Chem. Proc. Des. Dev.* **24** 836-44

[40] Abella L, Nanbu S and Fukuda K 2007 A Theoretical study on levoglucosan pyrolysis reactions yielding aldehydes and a ketone in biomass *Memoirs of the Faculty of Engineering* Kyushu University **67** 2

[41] Laurent E, Pierret C, Grange P and Delmon B 1991 Control of the deoxygenation of pyrolytic oils by hydrotreatment *Proc. of 6th Conference on Biomass for Energy Ind and Environ* (Athens, Greece) pp 665-71

[42] Miskolczi N and Nagy R 2012 Hydrocarbons obtained by waste plastic pyrolysis: Comparative analysis of decomposition described by different kinetic models *Fuel Processing Technol* **104** 96-104

[43] Ma C, Yu J, Wang B, Song Z, Xiang J, Hu S and Sun L 2016 Influence of zeolites and mesoporous catalysts on catalytic pyrolysis of brominated acrylonitrile–butadiene–styrene (Br-ABS) *Energy Fuels* **30** 4635-43

[44] Chen W, Shi S, Zhang J, Chen M and Zhou X 2016 Co-pyrolysis of waste newspaper with high-density polyethylene: Synergistic effect and oil characterization *Energy Conver and Management* **112** 41-8

Optimized design and control of an off grid solar PV/hydrogen fuel cell power system for green buildings

C Ghenai[1,3] and M Bettayeb[2]

[1]Department of Sustainable and Renewable Energy Engineering, University of Sharjah, PO Box 27272, Sharjah, United Arab Emirates
[2]Department of Electrical and Computer Engineering, University of Sharjah, PO Box 27272, Sharjah, United Arab Emirates

E-mail: cghenai@sharjah.ac.ae

Abstract. Modelling, simulation, optimization and control strategies are used in this study to design a stand-alone solar PV/Fuel Cell/Battery/Generator hybrid power system to serve the electrical load of a commercial building. The main objective is to design an off grid energy system to meet the desired electric load of the commercial building with high renewable fraction, low emissions and low cost of energy. The goal is to manage the energy consumption of the building, reduce the associate cost and to switch from grid-tied fossil fuel power system to an off grid renewable and cleaner power system. Energy audit was performed in this study to determine the energy consumption of the building. Hourly simulations, modelling and optimization were performed to determine the performance and cost of the hybrid power configurations using different control strategies. The results show that the hybrid off grid solar PV/Fuel Cell/Generator/Battery/Inverter power system offers the best performance for the tested system architectures. From the total energy generated from the off grid hybrid power system, 73% is produced from the solar PV, 24% from the fuel cell and 3% from the backup Diesel generator. The produced power is used to meet all the AC load of the building without power shortage (<0.1%). The hybrid power system produces 18.2% excess power that can be used to serve the thermal load of the building. The proposed hybrid power system is sustainable, economically viable and environmentally friendly: High renewable fraction (66.1%), low levelized cost of energy (92 $/MWh), and low carbon dioxide emissions (24 kg CO_2/MWh) are achieved.

1. Introduction

The rapid growth of the world energy demand has raised concerns over the heavy environmental impacts (climate change and global warming due to greenhouse gas emissions from fossil fuel combustion), depletion of the energy resources (fossil fuels – coal, oil and natural gas) and future energy supply shortage. Clean and renewable energy systems and the development of alternative fuels are needed to meet the future energy demand for buildings, transportation, and industrial sectors [1]. The energy consumption for both residential and commercial buildings has increased steadily for the last years and has reached an average of 40% - 50% of the total energy consumptions. In some countries, the energy consumption for buildings has exceeded the energy consumptions for other sectors such as transportation and industrial applications. In 2016, and based on the U.S. Energy Information Administration (EIA) estimates, about 40% of the total U.S. energy consumptions was consumed by residential and commercial sectors. In the United Arab Emirates in particular and Gulf

Cooperation Countries located in the desert region, the energy consumption for building can reach 80% (mainly energy consumption from air conditioning system) of the total energy consumption [2]. This upward trend in energy demand in building will continue in the future due to the population and economic growth, the increase in the demand of the building services and comfort level and new building construction boom. The building sector has a key role to play when it comes to energy efficiency and the development and integration of renewable energy systems.

One way to increase the share of renewable energy in the energy mix is the development of micro grid power systems. Micro grid power systems use different types of renewable energy resources (wind, solar, biomass, hydro); energy generators powered with fossil, alternative or renewable fuels; energy storage devices; loads for residential, commercial and industrial applications; electric vehicles charging stations; and power conditioning units such as inverters and rectifiers. The micro grid energy system can be connected to the utility grid or can operate separately off grid as stand-alone power system. The locally controlled micro grid power system provides more independence from the grid, a backup to the utility grid and security of the energy supply in case of emergency caused by major storms and natural disasters, offers more reliability of the energy supply and reduces the transmission losses. Several studies can be found in the literature on the use of micro grid power systems for different applications [3-7]. Simulation, modeling and optimization using HOMER software (Hybrid Optmization Model for Electric Renewable) and Simulink [8-13] have been used to identify the optimal off-grid options.

The focus of this study is the use of micro-grid energy system to power commercial building. Fossil fuel was the primary fuel used to generate electricity for buildings. New energy systems using renewable energy resources (solar, wind, biomass, and biofuels) can be used as alternative to the conventional fossil fuel based power systems. A building powered by renewable energy systems is more sustainable solution to the high energy consumption for building (use of renewable resources – no depletion of natural resources; clean energy systems – reduction of GHGs emissions; reduction of the operational costs – use of natural resources). Hazem *et al* [14] used optimal design of stand-alone hybrid photovoltaics and fuel cell power system without battery storage to supply the electric load demand of the city of Brest in France. The study focused on the economic performance and is mainly based on the loss of the power supply probability concept. The Homer based simulation and optimization study showed that the hybrid power systems based on solar PV and fuel cells are a viable alternative to diesel generators. The fuel cell generator can efficiently complement a fluctuating renewable resource such as solar or wind to satisfy the energy demand of the city. Simulation of the operation of the PV-wind-fuel cell system was performed by Attala et al. [15]. The system architecture was based on 40 KW PV, 2 wind turbine, 6 KW Fuel cell, 10 KW Converter, and 10 kW electrolyzer. HOMER software was used to size and optimize the hybrid system. The results showed that the combination of Solar PV/Fuel cell and wind turbine is viable solution to meet the energy demand.

The principal objective of this study is to model, simulate, and optimize the design of an off grid hybrid power system based on solar PV and fuel cell in order to meet the desired electric load of the commercial building. The goal is to design a stand-alone renewable power system with high renewable fraction (increase the penetration of renewable resources for the energy mix), low levelized cost of energy and with low environmental impacts (reduce the carbon dioxide foot print).

2. Stand-alone hybrid power system modeling

The off grid hybrid power system, shown in table 1 consists of a solar PV array, Fuel Cell, Electrolyzer for hydrogen production, tank for the storage of compressed hydrogen, backup generator, energy storage system or battery bank and converter (DC/AC inverter). The proposed hybrid system is used to meet the building AC load. The building AC load is based on the building energy consumption from the air conditioning system, lighting and other equipment (see figure 1). The proposed hybrid off grid power system is modeled as follows:

Table 1. Hybrid power system components and specifications.

System Component	Description
Solar PV	**Type**: Canadian Solar Max Power CS6U-330 with maximum power capacity $P_{PV})_{max}$ = 1200 kW **Module**: polycrystalline, Nominal Maximum Power = 330 W, Operating Voltage V_{mp}= 37.5 V, Operating Current I_{mp} = 8.80 A, Open Circuit Voltage V_{OC} = 45.9 V, Open circuit Current I_{SC} = 9.31 A, Efficiency = 16.97%, Operating Temperature 45°C, and Derating factor f_{PV} = 80% **Cost per 1 kW**: Capital = \$1200/kW, Replacement = \$1200/kW, O&M = \$3/year **Life time** = 25 years
Fuel Cell	**Type**: PEM Fuel Cell (DC) with Maximum Power Capacity $P_{FC})_{max}$ = 200 kW Fuel: Hydrogen, Electrical Efficiency η_{FC} = 70% **Cost per 1** kW: Capital = \$400, Replacement = \$400, O&M = \$0.01/hour **Life time** (hours)= 50,000
Electrolyzer	**Type**: Generic electrolyzer (DC), Efficiency η_{EZ} = 90% with Maximum Power Input Capacity $P_{EZ})max$ = 200 kW **Cost per 1 kW:** Capital = \$100, Replacement = \$100, O&M = \$8/year **Life time** = 15 years
Hydrogen Tank	**Type**: Hydrogen tank with a Maximum Capacity of 200 kg. **Cost per 1 kg:** Capital = \$0.5, Replacement = \$100, O&M = \$10/year , Hydrogen fuel cost = \$1/kg **Life time** = 25 years
Backup Generator	**Type**: Genset, Electrical AC bus with Maximum Power Capacity $P_{BG})_{max}$ = 100 kW, Fuel = Diesel, HHV = 43.2 MJ/kg, density = 820 kg/m^3 , Minimum Load Ratio = 25%. Cost per 1 kW: Capital = \$300, Replacement = \$300, and O&M = 0.01 \$/hr. Diesel Fuel cost = \$0.6/liter Life time (hours) = 15,000
Battery	**Type**: Lead Acid Battery 1 kWh, Voltage = 12 V, 4 batteries in series: 12 x 4 = 48 V with a Maximum of 1200 strings in parallel. Cost per 1 kWh: Capital = \$200, Replacement = \$200 per battery, O&M = \$5/year. Lifetime = 12.5 years
Converter: Inverter/Rectifier	Leonics S219CPH with Maximum Power Capacity of $P_{InvIn})max$ = 1200 kW

System Component	Description
	Voltage = 48 VDC, Efficiency = 96% Cost per 1 kW: Capital = \$40, Replacement = 40\$, O&M = 10\$/year Lifetime = 25 years

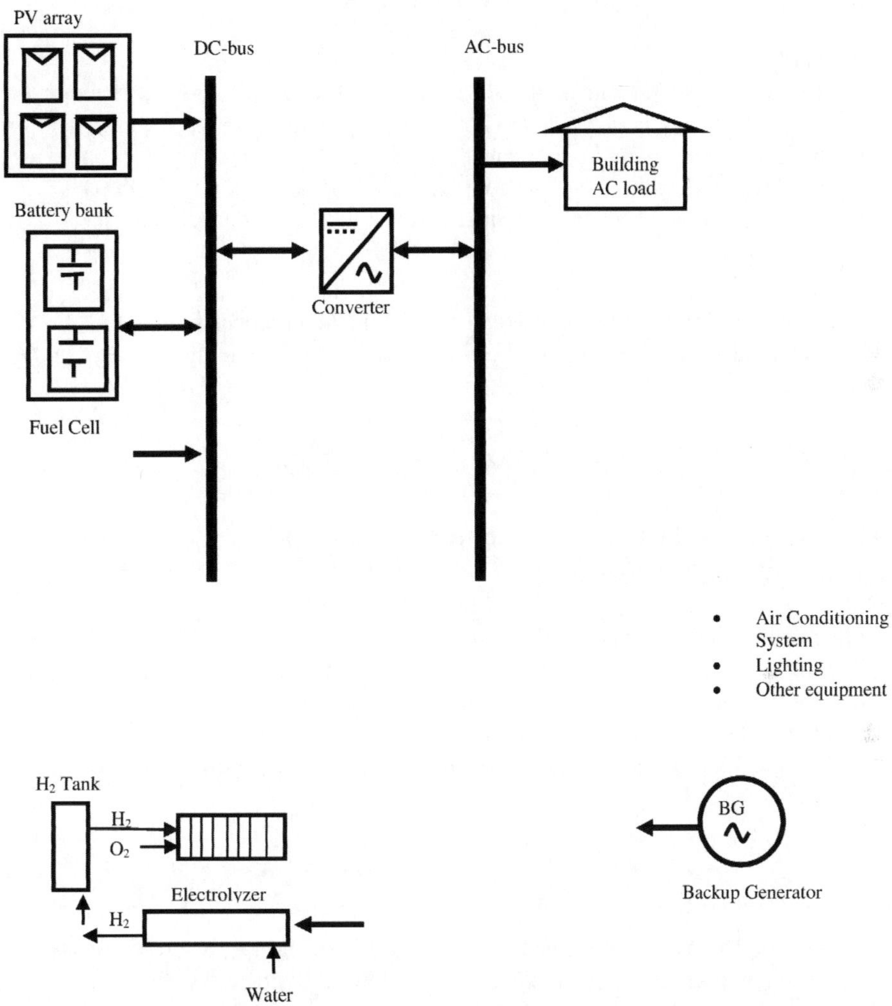

Figure 1. Hybrid power system.

2.1. Solar PV array

The power output (P_{PV}) from the PV array is expressed as

$$P_{PV} = P_{PV,STC} f_{PV} \left(\frac{G}{G_{STC}} \right) \left[1 + \alpha_P (T_c - T_{c,STC}) \right] \tag{1}$$

where, $P_{PV,STC}$ is the power output under standard test conditions (STC), f_{PV} is the PV derating factor (accounting for power reduction due to soiling of the panels, wiring losses, shading, and aging), G is the current irradiance, G_{STC} is the irradiance under STC, α_P is the temperature coefficient of the power, T_c is the PV module temperature, and $T_{c,STC}$ is the PV module temperature under STC. The standard test conditions (STC) are: G_{STC}=1000W/m^2, $T_{c,STC}$=25°C, and no wind.

The efficiency of the PV module (conversion of the sunlight into electricity at its maximum power point under standard test conditions) is given by:

$$\eta_{m,STC} = \frac{P_{m,STC}}{A_{PV}\ G_{STC}} \tag{2}$$

where A_{PV} is the surface area of the PV module [m^2].

2.2. Fuel cell

The fuel cell converts chemical fuel to electricity through a chemical reaction in which the fuel is oxidized and electricity is generated. The fuel cell is modeled as a power generator with hydrogen as the main fuel stored in hydrogen tank. Part of the consumed chemical fuel power which is not converted into electric power (P_{FC}) is converted to thermal stack power (P_{FCT}).
The electric stack power P_{FC} is the product of stack voltage and stack load:

$$P_{FC} = U_{Stack}\ I = U_{SC}\ N\ I \tag{3}$$

where U_{SC} is the average single cell voltage and N is the number of cells.

The electrical efficiency of the fuel cell (conversion of the chemical energy in the hydrogen fuel into electricity) is given by:

$$\eta_{FC} = \frac{P_{FC}}{\dot{m}_{H_2}\ HHV_{H_2}} \tag{4}$$

where $\dot{m}_{H_2}$ (kg/s) and HHV_{H2} are respectively the mass flow rate and the higher heating value (HHV = 120 MJ/kg) of the hydrogen fuel.

2.3. Electrolyser for hydrogen production

An electrolyzer consumes electricity to generate hydrogen via the electrolysis of water. The electrolysis is used for the decomposition of water into oxygen and hydrogen gas by passing an electric current through the water. The hydrogen produced is compressed and stored in the hydrogen storage tank to be used in the fuel cell for power generation as shown in figure 1. The power consumption of the electrolyzer is given by:

$$P_{EZ} = \frac{\dot{m}_{H_2} HHV_{H_2}}{\eta_{EZ}} \tag{5}$$

here, P_{EZ} is the DC power consumed in the electrolyzer, and η_{EZ} is the efficiency of the electrolyzer. A hydrogen tank is needed to store the hydrogen produced by the electrolyzer for later use in a hydrogen-fueled generator or fuel cell. The model assumes that during the process of adding hydrogen to the tank, no electricity is consumed and the tank experiences no leakage.

2.4. Battery bank

The charge or the discharge power into or out of the battery is regulated by the output power from the hybrid generation system (Solar PV and Fuel Cell) and the load demand at a given time. At any time, the battery state of charge (SOC) is expressed as follows

$$\mathrm{SOC}(t) = \mathrm{SOC}(0) + \eta_c \sum_{k=0}^{t} P_{CB}(k) + \eta_d \sum_{k=0}^{t} P_{DB}(k) \tag{6}$$

Where, SOC(0) is the initial SOC of the battery, P_{CB} is the charged power, P_{DB} is the discharged power, η c is the charging efficiency, η d is the discharging efficiency. It is noted that a battery string or bank comprises a number of cells/batteries connected in series to produce a battery or battery string with the required usable voltage/potential e.g. 6 V, 12 V, 24 V, 48 V, 110 V.

2.5. Generator

The generator is the back-up power supply to be used in case the load demand cannot be satisfied by the PV array, Fuel Cell and the battery bank. The generator power output is function of the fuel consumption and is modeled by means of a linear equation:

$$F_{BG} = \alpha_0 \ P_{BGR} + \alpha_1 P_{BG} \tag{7}$$

where F_{BG} is the generator fuel (Diesel, Biodiesel, Biogas, Natural gas, etc…) consumption (l/hr), α_0 is the generator fuel intercept coefficient (L/hr/kW$_{rated}$), P_{BGR} is the rated capacity of the generator (kW), α_1 is the generator fuel curve slope (L/hr/kW$_{Output}$) and P_{BG} is the power output (kW) from the generator. The electrical efficiency of the generator is the ratio of the electrical energy coming out divided by the chemical energy of the fuel going in (product of the fuel mass flow rate and the higher heating value of the fuel HHV).

2.6. Converter

The converter is connected between the DC-bus and the AC-bus as shown in figure 1. It has a bidirectional operation depending on the power flow, where it works as an inverter to transfer the power from the PV array, battery bank (discharge) and Fuel Cell to the AC load, and as a rectifier in case of charging the battery bank from the backup generator. The converter is modeled based on its rated capacity (P_{Inv}) and efficiency (N_{Inv}), which are assumed to be constant throughout its operating range. Furthermore, its rated DC voltage specifies the number of PV modules and batteries in each string of the PV array and the battery bank, respectively.

$$P_{InvOut} = P_{InvIn} \eta_{Inv} \tag{8}$$

2.7. Load

The power AC load is specified by the power consumption of the building (Air Conditioning System, Lighting, Equipment, etc…). The PV array, fuel cell, battery bank, and the backup generator are used to supply the electrical demand of the building.

2.8. Optimization and control strategies

An optimization analysis is used to find the best possible hybrid power system configuration based on the desired constraints at the lowest total net present cost. To find the optimal possible configuration, a search space was created for each component (PV, Fuel cell, Electrolyzer, Battery, backup Generator, Converter and hydrogen tank) using the size of each component (lower and upper capacity limits) and the dispatch strategy. Many different system configurations were simulated to find the feasible solutions to satisfy the desired constraints. The calculations were performed for each simulation time step (60 minutes), which corresponds to 8760 simulations per year.

Two dispatch or control strategies were selected for this study: the load following (LF) and cycle charging (CC). In the load following strategy, the generator will operate only to meet the primary demand load. The renewable power sources are used to charge the storage bank. In the cycle charging, the generator will operate at maximum capacity to meet the demand load and the surplus of electrical production is used to charge the battery and run the electrolyzer.

In the combined design optimization and control strategies of the hybrid power system, the fraction of the solar energy and other renewable resources from the total generated energy is specified by the renewable fraction (f_{ren}) coefficient calculated by

$$f_{ren} = 1 - \frac{E_{nonren}}{E_{cons}} \tag{9}$$

where, E_{nonren} is the nonrenewable energy production, and E_{cons} is the total energy consumed by the load.

3. Load profiles, renewable resources and hybrid power system component selection and sizing

3.1. Daily and seasonal profiles of the building

The University of Sharjah administration building is used as a case study for the proposed stand-alone hybrid power system. This Building is 18 years old and has a total area of approximately 14,092 m^2. The building consists of four floors (ground floor plus three levels). The AC load is specified by the energy (kWh) or power (kW) consumption of the building as shown in figures 2 and 3. An energy audit has been performed at selected buildings at the University. The results of the energy audit showed that the energy consumption of the administration building at the University of Sharjah is 80% from the air conditioning system, 10% lighting and 10% for other equipment. The University of Sharjah is located in Sharjah City in the United Arab Emirates. This is a desert region where the air conditioning system is running most of the time during the year. The energy consumption is mainly related to the cooling of the building. The average energy consumption per day is about 6,540 kWh based on the results of the energy audit. In addition to the seasonal profile shown in figure 2, the daily average profile is needed for the hourly simulation of the performance of the hybrid system. The daily average power consumption for four days (January 15, April 15, July 15, and November 15) is shown in figure 3.

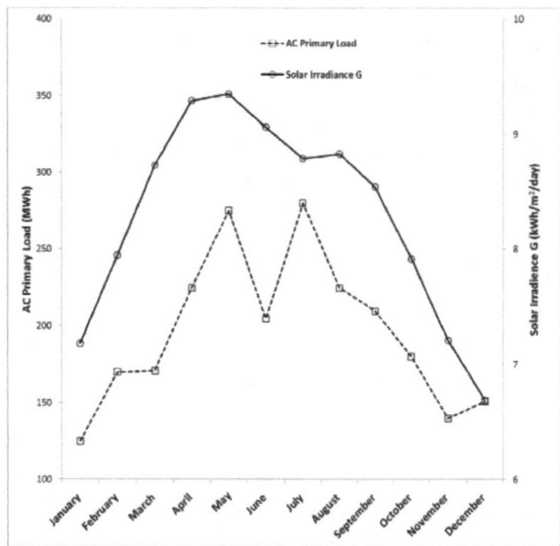

Figure 2. Monthly AC primary load profiles (Year 2015) and Solar Irradiance G.

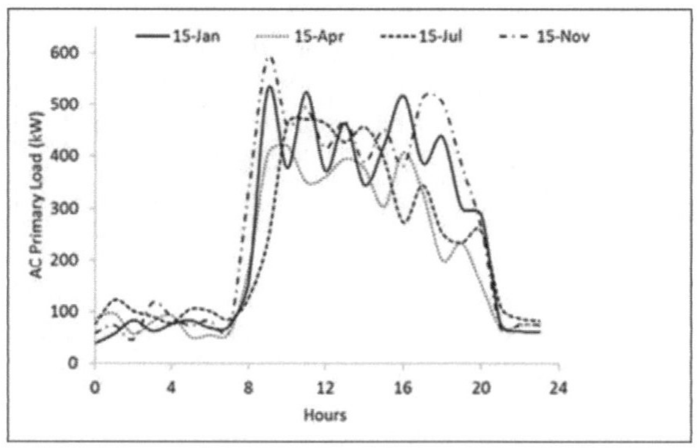

Figure 3. Daily AC primary load profiles.

3.2. Renewable resources

The monthly average solar global horizontal irradiance in Sharjah is illustrated in figure 2. The annual average solar radiation in Sharjah is 8.30 kWh/m^2/day. The solar radiation is available throughout the year, therefore considerable amount of solar power output can be generated using solar PV to serve the load of the building.

3.3. Hybrid power system components and sizing

The hybrid off grid power system is composed of solar flat PV system, fuel cell, electrolyzer, hydrogen fuel tank, backup generator, battery bank, converter (DC/AC inverter), and controller. A description of the selected parts for the hybrid power system is summarized in table 1. The effects of temperature on the performance of the PV system were also included in this study. The performance of the PV system decreases with (1) an increase of the ambient temperature (high temperature in Sharjah – Desert region) and (2) the accumulation of dust on the solar panels. The accumulation of dust on the solar panel and its effect on the power produced was included in the derating factor f_{PV}.

4. Results

The simulation results include the production (Solar PV array, Hydrogen Fuel Cell and Backup Diesel Generator) and the consumption of electricity by the system (primary loads of the building and Electrolyzer for hydrogen production), the annual pollutants emitted by the power system, and cost summary (capital, fuel, operation and maintenance O&M, replacement of the parts of the system and salvage). The daily and monthly average electrical production, and the net present cost over the life (25 years) of the hybrid power system are presented in this paper. The hybrid power system was designed to meet the energy demand of the commercial building that consumes 6,540 kWh/day. Thousands of hourly simulation and optimization runs were performed to calculate the energy to and from each component. The optimized results based on the lowest cost of energy (LCOE) of the hybrid power system were obtained using the optimization search space. The simulation and optimization methodology is based on the microgrid power system model selected in this study. Based on the selected design configuration (Solar PV, fuel cell, and back generator), search space (maximum power capacity off each component) and the daily power load requirement for the building, the simulation is performed for each hour of the year to investigate the technical feasibility and life cycle cost. In the optimization process, simulation of different configurations (single component such as solar PV only or combination of two or more components such as solar PV/fuel cell or solar PV/fuel cell/generator) is carried out to find the most appropriate system configuration that satisfies the technical constraints and the lowest life cycle cost. Based on this simulation and optimization methodology, the best three configurations (System 1, System 2 and System 3) exhibiting the lowest cost of energy are described in what follows: first configuration (System 1) uses two renewable systems (Solar PV and Fuel Cell) and one backup generator using Diesel or fossil fuel. The second configuration (System 2) uses only renewable and clean energy systems (Solar PV and Fuel Cell) to generate electricity. The third one (System 3) uses Solar PV in combination with Diesel generator (fossil fuel power system). Table 2 summarizes the energy production, consumption, the excess energy, the energy lost in the system (battery and inverter), and the unmet load for the three systems. It is noted that based on the results of the simulations and the technical constraint (lowest cost of energy), all the three systems use the cycle charging control strategies for the power generation (the generator will run at maximum capacity to meet the AC primary load and the excess power is used for the power input of the electrolyzer and to charge the battery). The results show that System 1 offers the lowest solar PV (1,025 kW), storage (267 strings in parallel) and inverter (864 kW) system capacities compared to the systems 2 and 3.

The total electrical production from the hybrid power system (system 1) is 3,055.95 MWh/year with 2,247.88 MWh/year from the PV system (73.6%), 723.22 MWh/year from the Fuel Cell (23.7%) and 84.85 MWh/year (2.7%) from the backup Diesel generator. The hybrid power system (system 1) meets all the AC primary load of 2,384.75 MWh/year for the building but also produces some excess power (557.06 MWh/year) that can be used to meet the thermal load of the building (the thermal load of the building was not included in the present study). The mean power output from the PV array is 257 kW with 4,345 hours per year of operation. This represents a mean output of 6,159 kWh/d and a capacity factor of 25%. A mean electrical power output of 118 kW was produced from the fuel cell with hydrogen fuel consumption of 31,821 kg/year, 6142 hours per year of operation, and a mean electrical efficiency of 68%. The backup Diesel generator produced a mean electrical power of 88 kW with 28,505 kg/year of Diesel fuel consumption, 967 hours per year of operation and 30% thermal efficiency. For the energy storage, the nominal capacity is 1,068 kWh (string size of 4 and 267 strings in parallel). The energy in for the battery bank is 69,640 kWh/year and the energy out is 55,713 kWh/year. This represents 13,928 kWh/year of loss through the battery bank. The inverter capacity was 864 kW but the mean power output from the inverter was 263 kW. This represents a capacity factor of 30% for the 8,760 hours per year of operation. The energy in for the inverter is 2,385,728 kWh/year and the energy out is 2,299,899 kWh/year. The inverter power loss is 95,829 kWh/year. The optimized hybrid Solar PV/Fuel Cell/Generator/Battery/Inverter power system produces 3,055.95 MWh/year. From the total energy produced from the hybrid system, 78.1 % is used to meet the AC primary load of the building, 18.2 % is the excess power (can be used to meet the thermal load of the building), 3.6% is the system losses in both the battery bank during charging and discharging process and the inverter (conversion form DC to AC power), and 0.08% is the unmet load. The renewable fraction of the hybrid power system (system 1) is 66.1%. Figure 4 shows a comparison between the three systems of the total energy production from the solar PV, Fuel cell, and Diesel generator; the energy consumption (AC primary load of the building), the losses in the battery and inverter, the unmet load and the renewable fractions. For the daily performance of the system, the results show that during the day most the energy is produced from the PV system where the solar irradiance is high and during the night most of the energy is produced from the fuel cell.

Table 2. Comparison of the energy production and consumption, excess energy, systems losses and unmet load for the three systems.

Production Consumption Losses	System 1 *Solar PV* *Fuel cell* *Backup Generator*	System 2 *Solar PV* *Fuel cell*	System 3 *Solar PV* *Backup Generator*
Total production	3,055.95 MWh/year (100%)	3,154.16 MWh/year (100%)	3,207.008 MWh/year (100%)
Solar PV	Maximum Power 1.025 MW	Maximum Power 1.077 MW	Maximum Power 1.2 MW
	Production 2,247.88 MWh/year (73.60%)	Production 2,359.786 MWh/year (74.81%)	Production 2,630.391 MWh/year (82%)
Fuel Cell	Maximum Power 200 kW	Maximum Power 200 kW	No
	Production 723.22 MWh/year (23.70%)	Production = 794.381MWh/year (25.19%)	
Backup Generator	Maximum Power = 100 kW	No	Maximum Power = 100 kW
	Production = 84.85 MWh/year (2.7%)		Production = 576.617 MWh/year (18%)
AC primary Load	2,389.1 MWh/year (78.10%)	2,389.1 MWh/year (75.70%)	2,389.1MWh/year (74.50%)
Excess Power	557.06 MWh/year (18.22%)	633.2 MWh/year (20.07%)	675 MWh/year (21.04%)
System losses - Battery and Inverter	109.9 MWh/year (3.59%)	131.2 MWh/year (4.16%)	140.78 MWh/year (4.38%)
Unmet Load	2.35 MWh/year (0.08%)	2.306 MWh/year (0.07%)	2.378 MWh/year (0.07%)

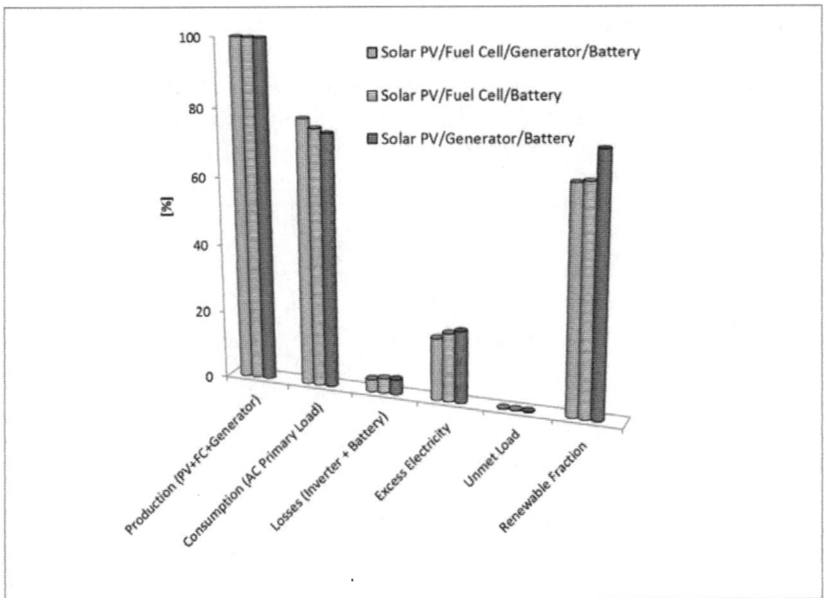

Figure 4. Energy production, consumption, losses, unmet load and renewable fraction of the off grid hybrid power systems.

A comparison study was performed concerning the performance and the cost of the hybrid power systems for the best three Systems (see table 3). It is shown clearly from these results that system 1 with Solar PV/Fuel Cell/Generator/Battery offers the best option with respect to the cost of energy (92 $/MWh). The system offers a high renewable fraction (66.1%) and is the second rated with respect to the carbon dioxide emissions (24 kg CO_2/MWh). The system will meet all the energy demand for the building with 18.2 % excess electricity that can be used for the thermal load of the building. A summary of the capital, replacement, fuel, and operation and maintenance costs for the PV/Fuel Cell/Diesel Generator/Battery/Inverter hybrid power system was also determined. The initial total capital of the hybrid power system (system 1) is $1.54 Millions ($1.23 Millions for the PV system) with a payback period of 6.83 years. The cost of energy (COE) for the proposed system is 92 $/MWh.

Table 3. Hybrid power system performance comparison.

Systems	**Cost of Energy ($/MWh)**	**Renewable Fraction (%)**	**Excess Electricity (%)**	**CO_2 Emissions (kg/MWh)**
Base line system (grid – fossil fuel power system)	120	0	0	632
1: Solar PV/Fuel Cell/Generator/Battery/Inverter	92	66.1	18.2	24
2: Solar PV/Fuel Cell/Battery/Inverter	99	66.7	20.1	0
3: Solar PV/Generator/Battery/Inverter	169	75.8	21	157.6

5. Conclusions

Simulation, modelling and optimization were performed in this study to design a hybrid off grid power system to meet the electric loads of a commercial Building (Administration Building at the University of Sharjah). The micro grid hybrid power system includes solar PV, fuel cell, electrolyzer, backup generator, battery bank and inverter. The building AC primary loads, solar resources in Sharjah, the technology options, component cost, and constraints were determined. An integrated optimization design and control strategy was used to test the performance and the cost of the proposed hybrid micro grid system. The results show that the solar PV/Fuel Cell/Generator/Battery/Inverter off grid power system meets the yearly electrical demand of the selected building. The proposed hybrid system offers the best option for power generation compared to the other system architectures. The hybrid power system based on solar PV and hydrogen fuel cell offers the best penetration of renewable resources (66.1% renewable fraction), low levelized cost of energy (92 \$/MWh), and low carbon dioxide (CO_2) emissions (24 kg CO_2/MWh).

References

[1] Ghenai C and Janajreh I 2013 Comparison of resource intensities and operational parameters of renewable, fossil fuel, and nuclear power systems *Int. J. of Thermal & Environmental Engineering* **5** 95-104
[2] Alkhateeb E, Abu Hijleh B, Rengasamy E and Muhammed S 2016 Building refurbishment strategies and their impact on saving energy in the united arab emirates *Proceedings of SBE16* (Dubai, 17-19 January, Dubai-UAE)
[3] Heydari A and Askarzadeh A 2016 Optimization of a biomass-based photovoltaic power plant for an off-grid application subject to loss of power supply probability concept *Applied Energy* **165** 601-11
[4] Sen R and Bhattacharyya S C 2014 Off-grid electricity generation with renewable energy technologies in India: An application of HOMER *Renewable Energy* **62** 388-98
[5] Lan H, Wena S, Hong Y-Y, Yu D C and Zhang L 2015 Optimal sizing of hybrid PV/diesel/battery in ship power system *Applied Energy* **158** 26-34
[6] Gan L K, Shek J K H and Mueller M A 2015 Hybrid wind–photovoltaic–diesel–battery system sizing tool development using empirical approach, life-cycle cost and performance analysis: A case study in Scotland *Energy Conversion and Management* **106** 479-94
[7] Merabet A, Ahmed K, Ibrahim H, Beguenane R and Ghias A 2017 Energy management and control system for laboratory scale microgrid based wind-PV-battery *IEEE Trans. Sustain. Energy* **8** 145-54
[8] Ghenai C and Janajreh I 2016 Design of solar-biomass hybrid microgrid system in Sharjah *Energy Procedia* **103** 357-62
[9] Lambert T, Gilman P and Lilienthal P 2006 Micropower system modeling with HOMER Chap. 15 *Integration of Alternative Sources of Energy* F A Farret and M G Simoes eds. (John Wiley & Sons)
[10] Yilmaz S, Ozcalikb H R, Aksua M and Karapınara C 2015 Dynamic simulation of a PV-diesel-battery hybrid plant for off grid electricity supply *Energy Procedia* **75** 381-7
[11] Ghenai C, Salameh T, Merabet A and Hamid A K 2017 Modeling and optimization of hybrid solar-diesel-battery power system *7th International Conference on Modeling, Simulation, and Applied Optimization (ICMSAO)*
[12] Hussain M N, Janajreh I and Ghenai C 2016 Multiple source sustainable hybrid micro-grid for urban communities: A case study in UAE *Energy Procedia, Elsevier* **103** 419-24
[13] Wies R W, Johnson R A, Agrawal A N and Chubb T J 2005 Simulink model for economic analysis and environmental impacts of a PV with diesel-battery system for remote villages *IEEE Trans. Power Syst.* **20** 692-700
[14] Mohamed O H, Amirat Y, Benbouzid M and Elbast A 2014 Optimal design of PV/fuel cell hybrid power system for the city of brest in France *IEEE ICGE* (Sfax, Tunisia) pp 119-23
[15] Attala A, Taha A T M and Ryad A 2014 PV, wind and fuel cell hybrid optimization generation for AL-Arish site *International Journal of Recent Development in Engineering and Technology* **2**

Research on Automatic Positioning System of Ultrasonic Testing of Wind Turbine Blade Flaws

Q X Liu[1,3], Z H Wang[2], S G Long[2], M Cai[2], M Cai[1], X Wang[1, 2], X Y Chen[1], J L Bu[1]

[1] Zhuzhou Times New Material Technology CO., LTD, Zhuzhou, Hunan, 412007, China
[2] Faculty of Material Science and Engineering, Xiangtan University, Xiangtan, Hunan 411105, China

E-mail: liuqx2@csrzic.com

Abstract. Ultrasonic testing technology has been used essentially in non-destructive testing of wind turbine blades. However, it is fact that the ultrasonic flaw detection method has inefficiently employed in recent years. This is because the testing result will illustrate a small deviation due to the artificial, environmental and technical factors. Therefore, it is an urgent technical demand for engineers to test the various flaws efficiently and quickly. An automatic positioning system has been designed in this paper to record the moving coordinates and the target distance in real time. Simultaneously, it could launch and acquire the sonic wave automatically. The ADNS-3080 optoelectronic chip is manufactured by Agilent Technologies Inc, which is also utilized in the system. With the combination of the chip, the power conversion module and the USB transmission module, the collected data can be transmitted from the upper monitor to the hardware that could process and control the data through software programming. An experiment has been designed to prove the reliability of automotive positioning system. The result has been validated by comparing the result collected form LABVIEW and actual plots on Perspex plane, it concludes that the system possesses high accuracy and magnificent meanings in practical engineering.

1. Introduction

Wind energy is one of the growing renewable energy resources. It is going to have a remarkable share in the energy market. According to the collected statistical data, it is estimated that wind power could give about 12% of global electricity supply by 2020 and achieve above 20% by 2030 [1] [2] [3].

Wind turbine system consists of numerous mechanical and electrical components including the wind turbine blade, rotor, gearbox, shaft, generator, bearings, tower, pitch and yaw system [2] [4]. Among these components, wind turbine blade is regarded as one of the most critical components [5] [6].The common defects usually can be founded by Non-Destructive Testing (NDT) during the blade manufacturing process, which can cause a series of problems during normal operation [7] [8].

NDT is the procedure of testing, inspecting, estimating structure and components without damaging the integrality of the structure. In this case, the structure could still be used whenever NDT has been applied. Contrastively, other inspection methods are destroyable and restricted to be used in need, while they are not widely applied for the materials, elements and structures.

Detractive testing methods are usually to decide the physical characteristics of material like tensile strength, ductility and fatigue toughness. However, the material discontinuities and property differences are efficiently used by NDT. At present, NDT are widely utilized in manufacturing and in-service detecting in order to make sure the serviceability of the system, maintain the quality level and avoid a sudden failure.

There are a number of NDT techniques, such as visual [9] [10] [11], sonics, ultrasonics [12] [13] [14] [15] [16] [17], optic [18] [19], electromagnetics [20] [21] and optical thermography [22] [23]. In particular, the normal working principles applied in ultrasonic testing is same as the principles used in Naval SONAR and fish finders. To be specifically, Ultra-high frequency sound is introduced into the component for inspecting, a part of sound could reflect back to the processing unit when the sound hits the component with different material and different acoustic impendence [12] [14].

Technically, sound goes through the structure by using the ultrasonic probe which converts electrical impulses from the ultrasonic testing machine into sound waves, hence converts sound back into electric impulses. After that, they could be illustrated on the LCD screen for visual demonstration. Therefore, the distance could be measured by operator from probe to the reflector. In general, the experienced operator can decide the discontinuity which caused the reflector as the ultrasound could not go though air and liquid [12] [14].

On the other hand, it is clear that wind turbine blade is one of the main components. Regarding to the large outline and complex procedure, the blade is easy to generate the flaws or wrinkles during the placement, which could result in significant influences on the performance of blade [15] [16]. Especially, the strength might reduce about 50%, the fiber might deviate 15° and flaws might appear when the glass fibers are slightly loose. After that, the blade fracture could turn up when the wrinkles grow up on the main composite layer, which will result in a destructive damage [15] [16]. In this case, the manufacturing processes of wind turbine blade are the manual operations, which are the main reason to generate a series of unpredictable factors that essentially affect the quality of wind turbine blade. It is fact that the flaws on wind turbine blade are commonly caused by the negligence of workers. As a result, it is necessary to use automatic positioning measurement system of Ultrasonic NDT to find out the flaws, which enable to give an indication without destroying the integrity of wind turbine blade. Hence, the lifetime of wind turbine blade could be guaranteed at a high level [17].

In particular, the automatic positioning measurement system of Ultrasonic NDT is a convenient way to find out the flaws of structure, which is recently used to inspect the detection of defects of wind turbine blades. In the process of automatic positioning measurement system, the first task is to determine the detection location. In this case, the method is to manually draw coordinates on the surface of the test object, record the specific locations and set up an equi-spaced grid of monitoring points. This approach is a time and energy consuming as it is hard to operate when the surface of the wind turbine is sufficiently large.

In this paper, an automatic positioning system is developed to track the two-dim coordinates of the trajectory. An experiment setup will be designed to validate the reliability of automatic positioning system. This system has great performance than the traditional manual operation approach, especially in automation and efficiency of the measurement procedure.

2. System design

The automatic positioning measurement system is shown in Figure 1, which includes the Light Emitting Diode (LED), lens assembly, optical sensors, microprocessor and RF modules. The LED will reflect the surface of the testing object, irradiate through the lens assembly and finally reach the optical sensor. The optical sensor uses the Agilent ADNS-3080 optical chip to capture the continuous microscopic images of the surface of the test object. Then, it compares and determines the movement direction and location of the system, which finally sends the data back to the microprocessor. The microprocessor is a MCS-51 microcontroller chip. It receives and transmits the real time data via the RF module to the upper computer that uses the corresponding LABVIEW software to take subsequent detecting procedures according to the displacement situations. The automatic positioning system also includes a power conversion module which could convert the 5V DC power supply to a stable 3.3V and power the optical sensor chips.

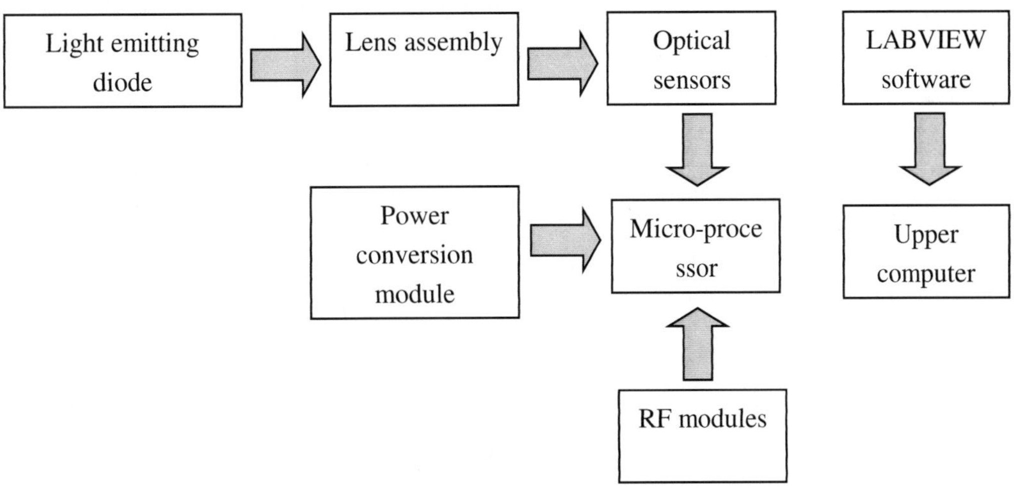

Figure 1. The schematic diagram of the automatic positioning measurement system

3. Implementation

In operation, the operator is asked to hold the collection device, also move in front of the surface of the blade under test in automatic positioning system. The system will automatically measure the two-dim coordinates of the movement and send them to the upper computer. The upper computer will analyze these coordinates and obtain the sampling distance to decide where the sampling distance is qualified. If so, the position will be the sampling position and the procedure of automatic positioning is completed.

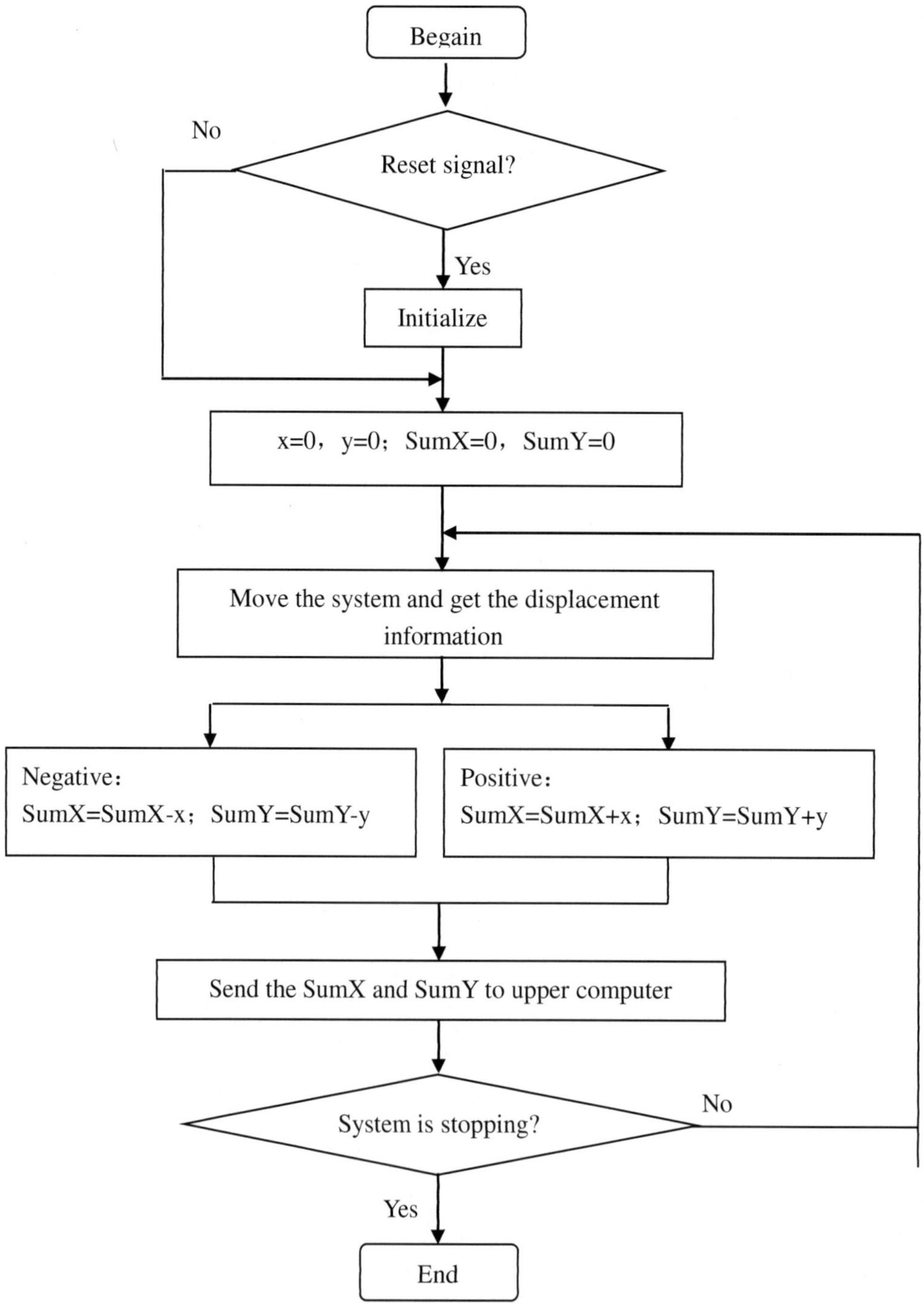

Figure 2. The flow chart of automatic positioning measurement system

The detailed implementation of this system is shown in Figure 2. The system will immediately decide whether to receive the reset signal from the control terminal since the system starts to work. If yes, the internal parameter of the optoelectronic chip will be reinitialized. In this case, the horizontal direction of the chip will be defined as the X-axis, the vertical direction will be determined as the Y-axis. The initial position of the photosensitive eye of the chip will be set as the origin position (0, 0), the input displacement change will be set as x = 0 and y = 0. More importantly, the overall displacement value will be defined as Sum X = 0 and Sum Y = 0.

When the system moves, the optical sensors will obtain the microscopic surface image captured by the lens and lighting systems through the image acquisition system. Using the photographic eye, thousands of frames will be captured per second and stored according to the chronological order of the image being photographed.

The digital signal processor will process the information in the register in order and implement image processing. Meanwhile, it will determine the positive and negative directions as well as vary x, y values. If it is negative (positive) direction, the displacement changes would be subtracted (added). After that, the sum X and sum Y will be sent to the distance measuring device through the RF module to determine whether to stop or not. If the system is stopped, the control program is completed. Once the pin NCS of optical sensor chip is pulled down, the four-wire serial interface will be activated and transmit the data to the microprocessor. The microprocessor will process the changes and obtain the coordinates from origin to the stopping location, while transmit these data to the upper computer through the RF module.

The data are transformed into a DLL file and called by LABVIEW, they will be finally displayed on the upper computer program panel. Once sampling distance is qualified, it stops moving and then performs collection mission. The following sections will present the experiment procedures and results obtained.

4. Experimental analysis

In the experiment, a 30cm×30cm Perspex plane was designed and used to be scanned by using Ultrasonic NDT. The plane was meshed and 5 plots were marked for ensuring the positioning device movements. That is, (300,0), (100,100), (150,150), (250,100), (0,300). In this case, Ultrasonic NDT includes the automatic positioning device, ultrasonic device and TH-606 analyzer, as Figure 3 shows. Firstly, Automatic positioning device and ultrasonic device were combined together and were connected to TH-606 analyzer. After that, TH-606 analyzer was turned on, while the positioning device was slowly moved on the plane and operated by single person. The moving path was in Z-shape in order to cover the whole plane.

IOP Conf. Series: Earth and Environmental Science **93** (2017) 012074 doi:10.1088/1755-1315/93/1/012074

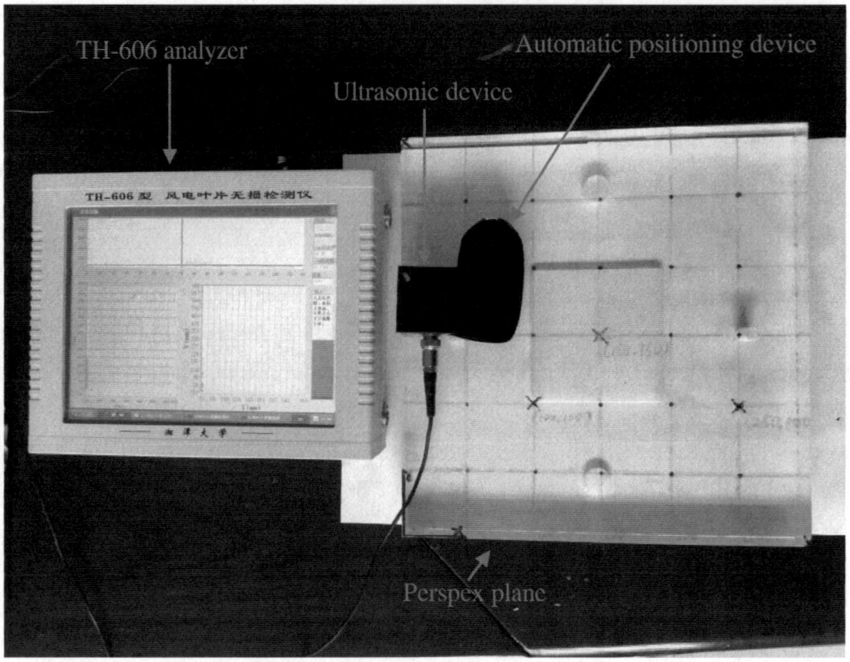

Figure 3. The experiment setup

To be specifically, TH-606 is the equipment that analyses the data from the ultrasonic probe and automatic positioning device, while it gives clear images on the screen to show the probe location and the situation of flaw detection. In this case, the result collected from the automatic positioning device will be essential.

5. Result

After the experiment, the result of automatic positioning device from LABVIEW could be observed in Figure 4. Comparing with the five plots in Figure 5, the automatic positioning device moves successfully and each plot has been scanned precisely, which also indicates that this device could provide very concise scanning movements and analyze efficiently in LABVIEW.

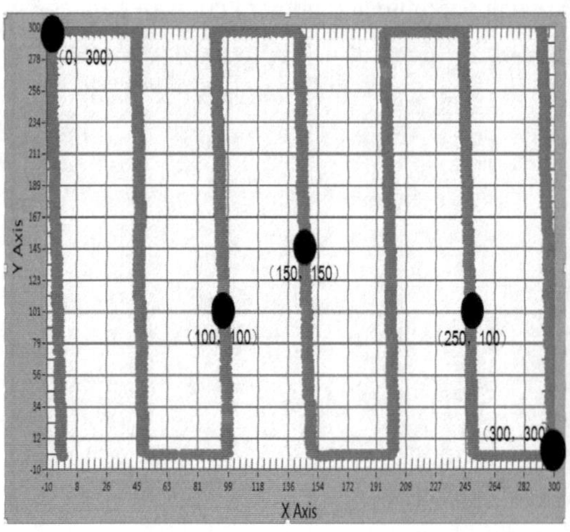

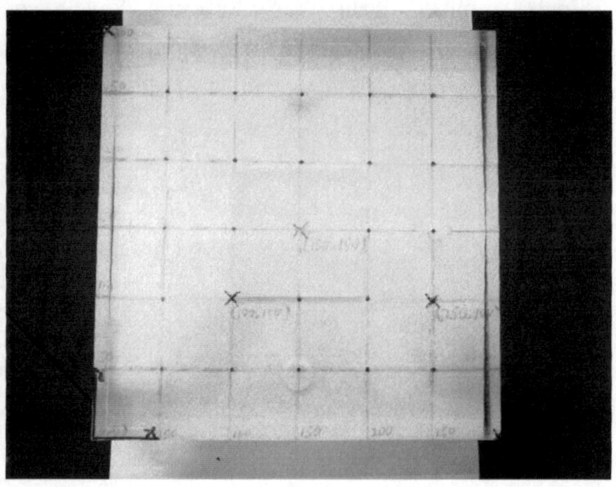

Figure 4. The result of automatic positioning device from LABVIEW

Figure 5. The five plots on Perspex plane

In other words, since targeting path has been selected and the positioning device has been moved, the system will display the real-time movement information on the upper computer panel (TH-606 analyzer) and generate a roadmap with a high accuracy movement.

6. Conclusion

The paper has clearly explained that the importance of using the automatic positioning measurement system of Ultrasonic NDT. That is, the quality of wind turbine blade could be improved and the flaws could be successfully found out by using this technique. Also, it has outlined the key components of automatic positioning measurement system. In this case, the system consists of LED, lens assembly, optical sensors, microprocessor and RF modules. In particular, the system could measure the two-dim coordinates automatically. Meanwhile, the system can immediately decide whether to receive the reset signal from the control terminal since the system starts to work. In addition, the experiment was set to prove the reliability and accuracy of automatic positing system. The result shows that automatic positioning system moves successfully, while 5 plots have been scanned precisely and accurately. More importantly, it implies that the automatic positioning measurement system could capture the real time position of the movements and store the target position in the flaw detection of wind turbine blades. This measurement system can be simply operated by a single person. Indeed, it has the number of advantages such as highly efficiency and easy to use. Also, it can be used in many fields besides wind turbine blade.

7. Reference

[1] Soua S., Van L P, Perera A, Gan T H and Bridge B 2013 Determination of the combined vibrational and acoustic emission signature of a wind turbine gearbox and generator shaft in service as a pre-requisite for effective condition monitoring *Renew. Energy* **51** 175-81

[2] Liu W Y, Tang B P, Han J G, Lu X N, Hu N N and He Z Z 2015 The structure healthy condition monitoring and fault diagnosis methods in wind turbines: a review Renew. *Sustain. Energy Rev.* **44** 466-72

[3] Chehouri A, Younes R, Ilinca A and Perron J 2015 Review of performance optimization techniques applied to wind turbines *Appl. Energy* **142** 361-88

[4] Yang W, Tavner P J, Crabtree C J and Wilkinson M 2010 Cost-effective condition monitoring for wind turbines *IEEE Trans Ind. Electron* **57** 263-71

[5] Yang B, Sun D 2013 Testing, inspecting and monitoring technologies for wind turbine blades: a survey. *Renew. Sustain. Energy Rev.* **22** 515-26

[6] Zhou H F, Dou H Y, Qin L Z, Chen Y, Ni Y Q and Ko J M 2014 A review of full-scale structural testing of wind turbine blades. *Renew. Sustain. Energy Rev.* **33** 177-87

[7] Drewry M, Georgiou G 2007 A review of NDT techniques for wind turbines *Insight* **49** 137-41

[8] Li D, Ho S C M, Song G, Ren L and Li H 2015A review of damage detection methods for wind turbine blades *Smart Mater Struct* **24** 033001

[9] Marsh G 2011 Meeting the challenge of wind turbine blade repair. *Reinf Plast* **55** 6-32

[10] Cripps D 2011 The future of blade repair *Reinf Plast* **55** 28–32

[11] Wallace J, Dawson M O&M 2009 strategies: wind turbine blades *Renew. Energy Focus* **10** 36-41

[12] Juengert A and Grosse C U 2009 Inspection techniques for wind turbine blades using ultrasound and sound waves. Nantes, France: Non-Destructive Testing in Civil Engineering

[13] Jasiūnienė E, Raišutis R, Šliteris R, Voleišis A and Jakas M 2008 Ultrasonic NDT of wind turbine blades using contact pulse-echo immersion testing with moving water container *Ultrasound* **63** 28–32

[14] Chakrapani S K, Dayal V, Krafka R and Eldal A 2011 Ultrasonic testing of adhesive bonds of thick composites with applications to wind turbine blades *AIP Conf Proc*, **1430** 1284-90

[15] Ye G, Neal B, Boot A, Kappatos V, Selcuk C and Gan T H 2014 Development of an ultrasonic NDT system for automated in-situ inspection of wind turbine blades *Proceedings of the 7th European workshop on structural health monitoring* Nantes, France

[16] Gan T H, Soua S, Dimlaye V and Burnham K 2012 Real-time monitoring system for defects detection in wind turbine structures and rotating components *Proceedings of the 18th world conference on nondestructive testing* 16-20

[17] Raišutis R, Jasiūnienė E and Žukauskas E 2008 Ultrasonic NDT of wind turbine blades using guided waves. *Ultragarsas* **63** 7-11

[18] Zhang H and Jackman J 2013 A feasibility study of wind turbine blade surface crack detection using an optical inspection method *Proceedings of the 2013 International conference on renewable energy research and applications(ICRERA)* Madrid: IEEE 847-852

[19] Zhang H and Jackman J 2014 Feasibility of Automatic detection of surface cracks in wind turbine blades *Wind Eng*. **38** 575-86

[20] Mook G, Lange R and Koeser O 2001 Non-destructive characterisation of carbon-fibre-reinforced plastics by means of eddy-currents *Compos. Sci. Technol*. **61** 865-73

[21] Jeong H, Hsu D K and K Liaw P 1998 Anisotropic conductivities of multiphase particulate metal–matrix composites *Compos. Sci. Technol.* **58** 65-76

[22] Melnyk S, Tuluzov I and Melnyk A 2014 Method of remote dynamic thermographic testing of wind turbine blades *Proceedings of the 12th International conference on quantitative infrared thermography*. Bordeaux

[23] Hahn F, Kensche C W, Paynter R J H, Dutton A G, Kildegaard C and Kosgaard J 2002 Design, fatigue test and NDE of a sectional wind turbine rotor blade *J. Thermoplast. Compos. Mater.* **15** 267-77

IOP Conf. Series: Earth and Environmental Science **93** (2017) 012075 doi:10.1088/1755-1315/93/1/012075

An experimental study on the application of polyalcohol solid-solid phase change materials in solar drying with cross-corrugated solar air collectors

W F Gao[1,3], W X Lin[1,2], T Liu[1] and M Li[1]

[1]Solar Energy Research Institute, Yunnan Normal University, Kunming 650500, P.R. China

[2]School of Engineering and Physical Sciences, James Cook University, Townsville, QLD 4811, Australia

E-mail: wenfenggao@163.com

Abstract. In this paper, two identical solar driers with the same cross-corrugated solar air collectors and drying chamber were developed, one with phase-change materials (PCMs) and the other without PCMs. These two solar drying systems were tested in typical sunny and cloudy days in Kunming and their thermal performances were analyzed. The experimental results show that the temperature changing is smoother in the collector with the PCMs, which is beneficial for the drying as the useful drying time was prolonged. The same trend was also found in the chamber with the PCMs. The PCMs in solar drying system was found to play a role in temperature regulating. There were several cycles of heat charging-discharging in a cloudy testing day while the temperatures on collectors and in chambers with the polyalcohol PCMs is higher than each phase-change temperature. Nevertheless, there was only one cycle of heat charging-discharging in a sunny testing day. The collector with PCMs has higher daily useful heat gain than the collector without PCMs.

1. Introduction

Solar drying technology has long been recognized as a very promising renewable energy technology, especially for rural regions and developing countries, as evidenced by its utilization in a wide range of applications such as the drying of crops, timber, fish, paddy, fruit, vegetable, etc. With the urgent request for a substantial reduction of global greenhouse gas emission to avoid a catastrophic climate change, the significance to develop and expand this promising renewable energy technology is becoming more evident and many new and improved solar drying facilities and technologies have been developed.

A solar air collector is a key component in a solar drying facility and its thermal performance dictates the efficiency and effectiveness of the drying facility. It is therefore essential for a solar drying facility to be equipped with a high thermal performance solar air collector. Nevertheless, due to the poor thermal conductivity and small heat capacity of air, the convective heat transfer rate inside the air flow channel formed by the absorbing plate and the bottom plate in a solar air collector, where the air is heated, is not high, resulting in a relatively poor thermal performance of the collector. A great deal of effort has been made to increase this rate [1]. Among many effective ways to augment the convective heat transfer rate in channel flows, those by increasing the heat transfer surface area and turbulence inside the channel with fins or corrugated absorbing surfaces and/or bottom surfaces have been the most popular [2] and many studies have been carried out on them [3,4]. A cross-corrugated solar air collector with a wavelike absorbing plate and a wavelike bottom plate which are positioned crosswise has attracted many interests attributed to its high thermal performance. For example, Piao *et al* [5] investigated experimentally the natural, forced and mixed convective heat transfer in a cross-corrugated channel solar air collector; Noorshahi *et al* [6] conducted a numerical study on the natural convection in a corrugated enclosure with mixed boundary conditions; Karim and Hawlader [7] evaluated, both analytically and experimentally, the thermal performance of a v-groove solar air collector under the meteorological conditions of Singapore; and more recently, we have made a comprehensive study on the thermal performance of v-groove and cross-corrugated solar air collectors under a wide range of configurations and operating conditions [8,9].

For many solar drying products, such as timber, vegetables, crops, fruits, etc., a lower temperature than that achieved by a high thermal performance solar air collector is usually required. Furthermore, to be cost effective, it is usually desirable for the solar drying facility to overcome the intermittence and diurnal nature of solar radiation. To meet this two-fold aim, this paper proposes an effective way to equip a high thermal performance solar air collector with heat storage materials that have suitable working temperatures for these low-temperature solar drying applications.

Among different types of heat storage, phase-change heat storage has the advantages of larger energy storage density, smaller volume, higher efficiency, fixed charging and discharging temperatures, etc. Many studies have been carried out using paraffin wax as Phase Change Material (PCM) in solar dryer [8-12]. Solid-solid phase-change heat storage is the process in which solid materials absorb a great deal of heat at a fixed temperature (charging temperature) and transfer from a low-symmetry crystal structure to a high-symmetry one, in which the materials store a great deal of heat which can then be discharged reciprocally. Compared to solid-liquid phase-change materials (PCMs), solid-solid PCMs have the advantages of smaller volume change during the phase-change process, no leakage, smaller erosion to the device, longer lifespan, etc., which make them particularly suitable for solar air collectors.

One of the most promising solid-solid PCMs are polyalcohols, polyethylenes, layered perovskites, etc. Among polyalcohols, neopentyl glycol (NPG, $C_5H_{12}O_2$), pentaerythritol (PE, $C_5H_{12}O_4$), trihydroxy methyl-aminomethane (TAM, $C_4H_{11}O_3N$), pentaglycerol (PG, $C_5H_{12}O_3$), etc., are the prevailing subjects of investigations. Benson [1], Son and Morehouse [13] made extensive experimental investigations into the heat storage performance of solid-solid polyalcohols including PE, NPG and trimethylol ethane and found that the mixtures of several different polyalcohols have the advantage of having adjusted transition temperatures which makes them attractive for low and medium-temperature utilizations such as in solar drying applications considered here. More recently, we have conducted a comprehensive experiment study on the heat storage performances of polyalcohols NPG, TAM, PE, and AMPD and their mixtures as solid-solid phase-change materials for solar energy applications [14-16].

In this paper, we carried out an experimental study on combining the advantage of the high thermal performance of the cross-corrugated solar air collector with the suitable working temperature of polyalcohols to construct solar drying facilities for low-temperature solar drying applications.

2. Experimental setup and method

2.1. Solar air collectors and the drying chamber

The experimental study was carried out in the solar drier sketched in figure 1. Air was pumped into the cross-corrugated solar air collector, where it was heated by solar radiation, before it passed through the drying chamber to carry away the moisture contained in the product to be dried in the chamber.

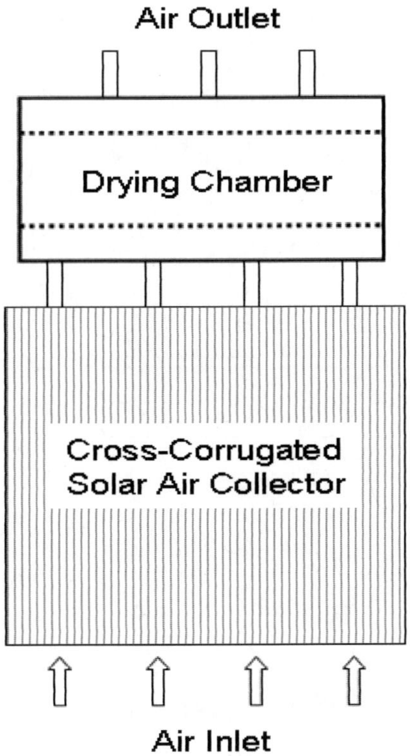

Figure 1. Schematic of the solar drying facility with a cross-corrugated solar air collector.

The cross-corrugated solar air collector was 1 m wide and 2 m long and consisted of a wavelike absorbing plate and a wavelike bottom plate, which were positioned crosswise to form the cross-corrugated air flow channel, as sketched in figure 2. The collector had a single flat glass cover, and the wavelike bottom plate was attached by a back insulation underneath. The air was heated by the absorbed solar radiation on the absorbing plate. Our previous studies [8, 9, 16] demonstrated that this type of cross-corrugated solar air collectors have significantly improved thermal performance compared to flat-plate solar air collector (with about 16% increase in efficiency).

The drying chamber, 0.8 m wide and 1 m long inside, was positioned higher than the collector and supported by a metal frame. On the top of the chamber, 3 chimneys were installed for the exiting of the air containing the removed moisture. Three mesh trays made of perforated stainless steel panels, on which the materials to be dried laid, were placed with equal spaces in the vertical direction inside the chamber.

Two identical such solar driers were constructed and installed in Kunming, China. In one drier, there were no PCMs used in the collector and the drying chamber. In the other drier, however, polyalcohol PCMs were used both in the collector and in the chamber. The locations of the PCMs in the collector, which were in direct contact with the absorbing plate, were inside the cross-corrugated air channel as sketched in figure 3, whereas the PCMs were placed on the bottom floor in the drying chamber.

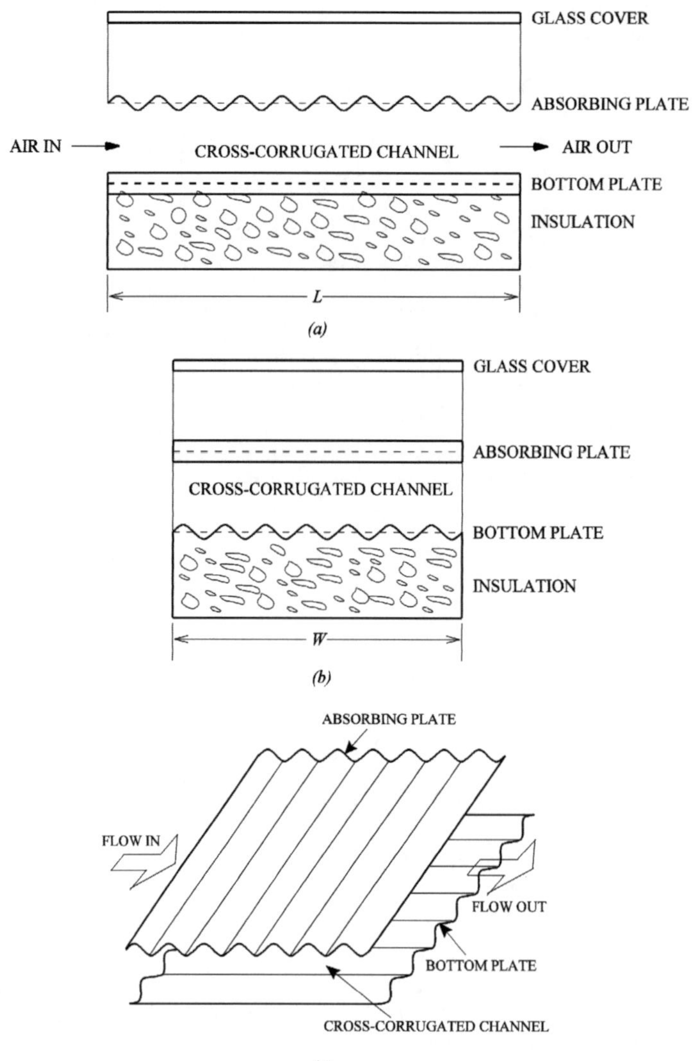

Figure 2. Schematic of the cross-corrugated solar air collector which has the cross-corrugated absorbing plate and bottom plate. The wavelike shape of the absorbing plate is along the flow direction and that of the bottom plate is perpendicular to the flow direction. (a) View on the cross section perpendicular to the flow direction; (b) View on the cross section along the flow direction; and (c) Isometric view of the cross-corrugated absorbing plate and bottom plate.

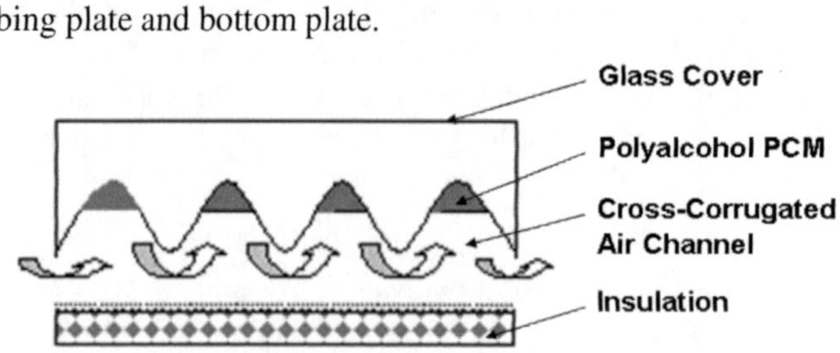

Figure 3. Schematic of the cross-corrugated solar air collector with polyalcohol phase-change materials.

2.2. PCM preparation
In order to choose appropriate PCMs for experiments, the data of phase-change temperature and enthalpy for the binary mixtures of polyalcohols NPG/TAM, NPG/PE and NPG/PG which approach the working temperature of solar drying are listed in table 1. It is found that the phase-change temperature and enthalpy decrease with the decrease of the weight ratio of NPG in the binary mixtures. It is also found that binary mixtures with higher weight ratios of NPG are more suitable for lower temperature (30-50℃) solar thermal applications, and those with higher weight ratios of TAM or PG are more suitable for higher temperature (70-150℃) solar thermal applications.

Table 1. Phase-change temperatures and enthalpies of polyalcohol mixtures.

Polyalcohol mixtures	Weight percentage x (%)	Phase-change temperature (°C)	Phase-change enthalpy (J/g)
TAMxNPG1-x	0	44.09	116.54
	20	37.86	95.23
	35	36.48	82.50
	50	38.45	38.46
	65	36.45	12.88
	80	39.54	6.34
	100	133.83	270.31
PExNPG1-x	0	44.0	116.5
	16	34.0	99.03
	20	33.5	57.7
	28	33.5	26.0
	30	33.5	13.4
	40	33.0	13.0
	60	35.0	10.7
	80	32.0	N/A
	90	35.0	N/A
PGxNPG1-x	0	44.09	116.54
	30	31.14	73.55
	50	38.09	55.28
	70	40.32	27.79
	89	36.62	6.18
	100	81.76	172.58

The PG was selected as PCMs for the solar collector because it has higher working temperature and NPG was selected as PCMs for the drying chamber as it has lower working temperature. To improve the poor heat conductive performance of PCMs in practical applications, the aluminum powder was well mixed with the prepared polyalcohols by a weight ratio of 1:1.

2.3. Experimental methodology
To measure the thermal performance of the solar driers, a series of sensors were placed in the two driers. In each drier, three PT-100 temperature sensors were placed at the lower, middle and upper section on the absorbing plate of the collector along the air channel direction to measure the temperatures in the channel; three PT-100 temperature sensors were also placed inside the drying chamber with equal spaces to measure the temperatures in the chamber. The average values from the respective three sensors were taken to be the average temperatures of the collector and the chamber respectively. Two PT-100 temperature sensors were placed at the outlets of the collector and the chamber in each drier. One PT-100 temperature sensor was used to measure the ambient air

temperature. A CM11 pyranometer manufactured by Kipp & Zonen with the measuring error less than 1% was used to measure the solar radiation fallen on the collector's surface. The temperatures and solar radiation from these sensors and the pyranometer were collected and recorded every 5 minutes by a data logger.

A fine leaf plant named Solidago Decuurrens, which is usually dried as dried flowers, was selected and placed onto each tray of the two chambers with the same amount before the testing day. There were 15 kg of Solidago Decuurrens in each chamber. The tests were carried out under several different weather conditions. In every testing day, the experiment started at 9:00 am and ended at 6:00 pm. The data under two typical weather conditions, that is, in a sunny day and in a cloudy day, were selected for analysis in the subsequent section.

3. Experimental results and discussions

The time series of the measured average collector temperature, the outlet temperature, and the ambient temperature in a typical sunny testing day are presented in figure 4 for both the solar driers with and without the polyalcohol PCMs. It is seen that the average temperatures of both collectors rose promptly after the testing started, and attained their maxima at about 1:00 pm at which time the incident solar radiation also reached its maximum, as shown in figure 5, where the time series of solar radiation incident on the collector surface in a typical sunny testing day and in a typical cloudy testing day are presented, respectively. The PCMs in the collector started charging heat while the temperature of the collector was more than 82℃, which was higher than the phase-change temperature of the PG. Consequently, the rate of the temperature increase on the collector with the PCMs became smaller than that on the collector without PCMs. The temperatures of both collectors declined promptly with the fall of the incident radiation after it attained its maximum at 1:00 pm; but after 3:00 pm, the rate of the temperature drop on the collector with the PCMs became slower than that on the collector without PCMs, in particular during the period of 4:00-5:00 pm of the testing period, as the PCMs were discharging heat due to the collector temperature being lower than the phase-change temperature of the PG. This trend can also be seen from the time series of the outlet temperature of the collector as shown in figure 4.

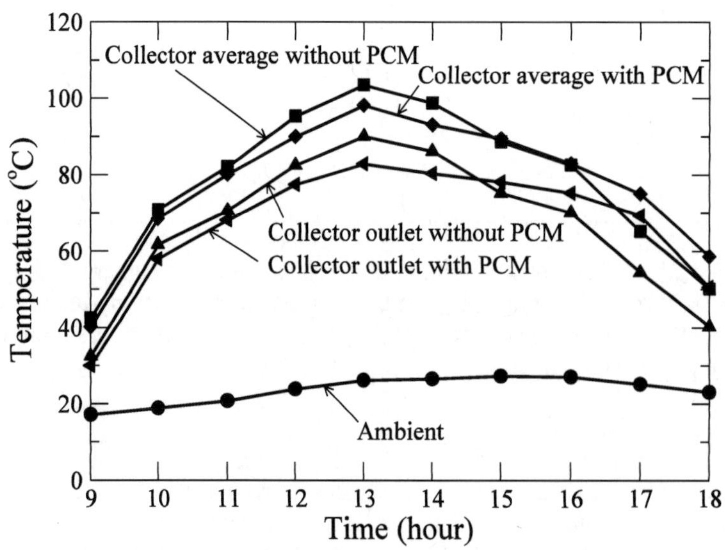

Figure 4. Measured collector average and outlet temperatures, both with and without the polyalcohol PCMs, and the ambient temperature in a sunny testing day.

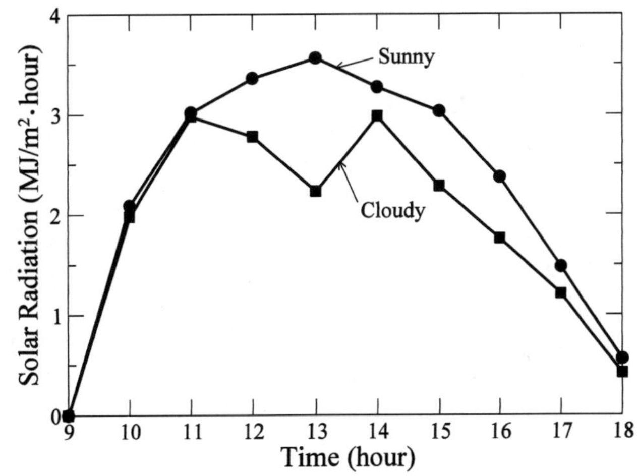

Figure 5. Time series of solar insolation incident on the collector surface both in a sunny testing day and in a cloudy testing day.

Figure 6 presents the measured average temperatures and outlet temperatures in the drying chambers with and without the polyalcohol PCMs in a typical sunny testing day. It is seen that the PCMs started charging heat while the temperature in the chamber was more than 45℃, which is higher than the phase-change temperature of the NPG, leading to slower rate of the temperature rise in the chamber with the PCMs than that in the chamber without PCMs. The temperature in the chamber with the PCMs gradually decreased after 13:00 pm in the sunny testing day as the PCMs started discharging heat after that instant when the temperature in the chamber with the PCMs began to fall below 40℃. The actual drying time was prolonged one hour in the chamber with PCMs than that in the chamber without PCMs. Although the temperature drop in the chamber with the PCMs had the similar trend as that in the collector with the PCMs, however, the rate of the drop was smaller, as clearly shown in figures 4 and 6. The outlet temperatures of the chambers were lower than those inside the chambers. Obviously, this is a result that the drying plants had absorbed a significant portion of heat.

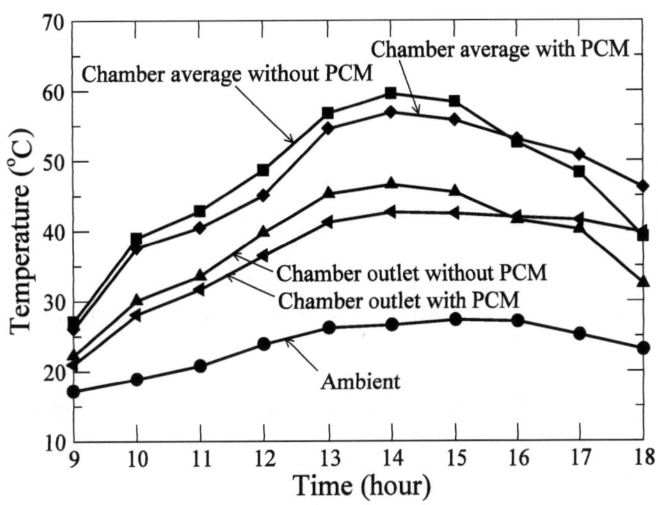

Figure 6. Measured average and outlet temperatures in the drying chamber, both with and without the polyalcohol PCMs, and the ambient temperature in a sunny testing day.

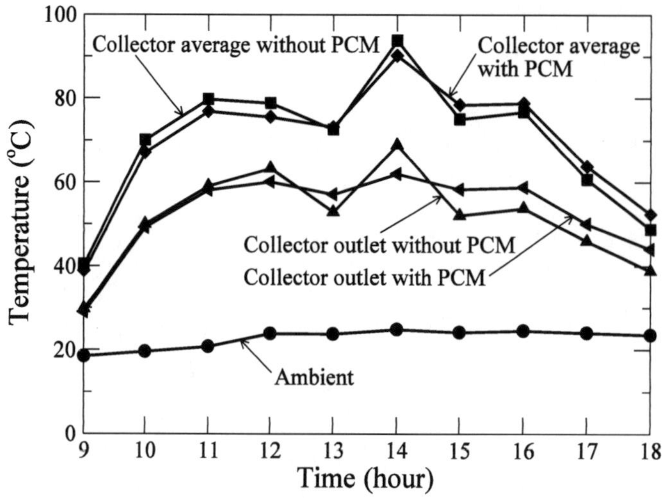

Figure 7. Measured collector average and outlet temperatures, both with and without the polyalcohol PCMs, and the ambient temperature in a cloudy testing day.

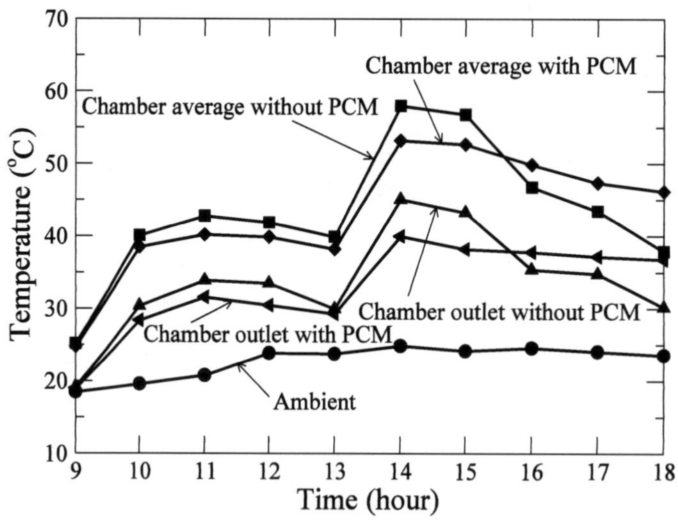

Figure 8. Measured average and outlet temperatures in the drying chamber, both with and without the polyalcohol PCMs, and the ambient temperature in a cloudy testing day.

The time series of the measured average collector temperature, the outlet temperature, and the ambient temperature in a typical cloudy testing day are presented in figure 7 for both the solar driers with and without the polyalcohol PCMs, whereas the measured average temperatures and outlet temperatures in the drying chambers with and without the polyalcohol PCMs in a typical cloudy testing day are shown in figure 8. It is clear that there were several charge-discharge processes in the typical cloudy day due to the intermittence of the incident solar radiation, in contrast to the single charge-discharge process in the typical sunny day. When the working temperatures in collectors with and without PCMs are compared, it can also be concluded that the collector with PCMs has lower working temperature, but it is high enough for the drying applications. This means that the collector with PCMs has smaller heat loss and higher thermal efficiency than the collector without PCMs. Meanwhile, the temperature changing is much smoother in the drier with PCMs, which benefits the drying and the useful drying time was prolonged. Furthermore, it is clear that the PCMs in solar drying system play an important role in the temperature regulating of the drying applications.

4. Conclusions

The polyalcohol PCMs has the advantages of larger energy storage density, smaller volume, higher efficiency, fixed charging and discharging temperatures, among others. It can be used in solar thermal applications which require appropriate phase-change temperatures and enthalpies. In this paper, the cross-corrugated solar air collector with the PG PCMs and the chamber with the NPG PCMs in solar drying systems were developed and the characters of drying systems were investigated and compared to the identical solar drying system without PCMs under sunny and cloudy testing conditions.

The experimental results show that there were several charging-discharging processes in the solar drier with PCMs in a typical cloudy testing day while there was only one charging-discharging process in the solar drier with PCMs in a typical sunny day. The cross-corrugated solar air collector with PCMs has lower heat loss and higher heat efficiency than the collector without PCMs. The temperature changing is more smoother which benefit for drying and the useful drying time has been prolonged. It has similar trend in the chamber with PCMs as the collector with PCMs. The PCMs in solar drying system play a role in temperature regulating.

Acknowledgments

The authors would like to acknowledge the support from the National Natural Science Foundation of China (51469035, 11662021) and the Australian Research Council (ARC, DP160102134).

References

[1] Benson D K, Burrows R W and Webb J D 1986 Solid-state phase-transitions in pentaerythritol and related polyhydric alcohols *Solar Energy Materials* **13** 133-52
[2] Goldstein L and Sparrow E M 1977 Heat/mass transfer characteristics for flow in a corrugated wall channel *ASME J Heat Transf* **99** 187-95
[3] Meyer B A, Mitchell J W and El-Wakil M M 1982 Convective heat transfer in Vee-trough linear concentrators *Solar Energy* **28** 33-40
[4] Stasiek J A 1998 Experimental studies of heat transfer and fluid flow across corrugated-undulated heat exchanger surfaces *Int. J. Heat Mass Transfer* **41** 899-914
[5] Piao Y, Hauptmann E G and Iqbal M 1994 Forced convective heat transfer in cross-corrugated solar air heaters *ASME J. Solar Energy Eng* **116** 212-4
[6] Noorshahi S, Hall III C A and Glakpe E K 1996 Natural convection in a corrugated enclosure with mixed boundary conditions *ASME J. Solar Energy Eng* **118** 50-7
[7] Karim M A and Hawlader M N A 2006 Performance evaluation of a v-groove solar air collector for drying applications *Applied Thermal Eng* **26** 121-30
[8] Gao W F, Lin W X and Lu E R 2000 Numerical study on natural convection inside the channel between the flat-plate cover and sine-wave absorber of a cross-corrugated solar air heater *Energy Conversion & Management* **41** 145-51
[9] Gao W F, Lin W X, Liu T and Xia C F 2007 Analytical and experimental studies on the thermal performance of cross-corrugated and flat-plate solar air heaters *Applied Energy* **84** 425-41
[10] Khadraoui A E, Bouadila S, Kooli S, Farhat A and Guizani A 2017 Nocturnal usage of solar air collector with PCM *Journal of Cleaner Production* **148** 37-48
[11] Shalaby S M and Bek M A 2014 Experimental investigation of a novel indirect solar dryer implementing PCM as energy storage medium *Energy Conversion and Management* **83** 1-8
[12] Jain D and Tewari P 2015 Performance of indirect through pass natural convective solar crop dryer with phase change thermal energy storage *Renewable Energy* **80** 244-50
[13] Son C H and Morehouse J H 1991 An experimental investigation of solid-state phase-change materials for solar thermal storage *ASME Journal of Solar Energy Engineering* **113** 244-9
[14] Wang X W, Lu E R, Lin W X and Wang C Z 2000 Micro-mechanism of heat storage in a binary

IOP Conf. Series: Earth and Environmental Science **93** (2017) 012075 doi:10.1088/1755-1315/93/1/012075

systems of two kinds of poly-alcohols as a solid-solid phase change material *Energy Conversion & Management* **41** 135-44

[15] Wang X W, Lu E R, Lin W X, Liu T, Shi Z S, Tang R S and Wang C Z 2000 Heat storage performance of the binary systems NPG/PE and NPG/TAM as a solid-solid phase change material *Energy Conversion & Management* **41** 129-34

[16] Gao W F, Lin W X, Liu T and Xia C F 2007 An experimental study on the heat storage performances of polyalcohols NPG, TAM, PE, and AMPD and their mixtures as solid-solid phase-change materials for solar energy applications *Int. J. Green Energy* **4** 301-11

Analysis and evaluation of the applicability of green energy technology

Z J Xu[1,2] and Y K Song[1]

[1]College of Management, Capital Normal University, Beijing, 100089, China

E-mail: xzjccc@sina.com/sylviasong@163.com

Abstract. With the seriousness of environmental issues and the shortage of resources, the applicability of green energy technology has been paid more and more attention by scholars in different fields. However, the current researches are often single in perspective and simple in method. According to the Theory of Applicable Technology, this paper analyzes and defines the green energy technology and its applicability from the all-around perspectives of symbiosis of economy, society, environment and science & technology etc., and correspondingly constructs the evaluation index system. The paper further applies the Fuzzy Comprehensive Evaluation to the evaluation of its applicability, discusses in depth the evaluation models and methods, and explains in detail with an example. The author holds that the applicability of green energy technology involves many aspects of economy, society, environment and science & technology and can be evaluated comprehensively by an index system composed of a number of independent indexes. The evaluation is multi-object, multi-factor, multi-level and fuzzy comprehensive, which is undoubtedly correct, effective and feasible by the Fuzzy Comprehensive Evaluation. It is of vital theoretical and practical significance to understand and evaluate comprehensively the applicability of green energy technology for the rational development and utilization of green energy technology and for the better promotion of sustainable development of human and nature.

1. Introduction

With the shortage of resources and the environmental crisis, the development and utilization of green energy technology have become the focus of current society and the analysis and evaluation of its applicability are increasingly highly valued by experts and scholars in different fields and disciplines. Since its applicability involves many aspects of economy, society, environment and science & technology, experts and scholars have conducted some relevant researches, but their perspectives are often single and methods simple [1,2].

In fact, the applicability is the essential requirement of green energy technology. It is very important to analyze and evaluate the applicability of green energy technology correctly and comprehensively. It is of vital theoretical and practical significance to explore the problems for the rational development and utilization of green energy technology, for the better realization of the industrialization, networking and intelligence of green energy technology and for the promotion of sustainable development of human and nature.

For this purpose, this paper analyzes and defines the green energy technology and its applicability guided by the Theory of Applicable Technology [3] from the all-around perspectives of symbiosis of economy, society, environment and science & technology etc. Correspondingly, the evaluation index system is constructed with certain structures by a number of independent indexes which combine the qualitative with quantitative and reflect respectively the requirements of applicability on the basis of the investigation of literature at home & abroad and consulting experts [4-8]. This paper further applies the extensively used and effective Fuzzy Comprehensive Evaluation (FCE) [9,10] to the analysis and evaluation of the applicability of green energy technology, discusses in depth its evaluation models and methods, and explains in detail with an example.

2. Definitions of the green energy technology and its applicability

According to the Theory of Applicable Technology, we hold that the green energy technology is the applicable energy technology and refers to the energy technology without or with very little pollution to the ecological environment. It is essentially a friendly, ecological and symbiotic technology which is compatible and coordinated with the natural environment tolerance and the renewable natural resources [11]. In general, there are two types of the green energy technology. One is a new energy technology to develop clean and pollution-free technology, such as solar energy, wind energy, tidal energy etc.; the other is the energy technology that can fully utilize waste, convert harm into good and improve the environment, such as various biomass energy.

From the all-around perspectives of symbiosis of economy, society, environment and science & technology, the applicability of green energy technology means that it is applicable and symbiotic with economy, society, environment and science & technology etc., it can meet their various requirements and can reap the better, diversified and comprehensive benefits. We hold that the applicability of green energy technology can be specifically analyzed and evaluated from the following aspects, and their requirements of applicability can be described, characterized and evaluated separately by a number of independent indexes [7,8]. Limited to space, the specific meaning of each index is not explained in the following.

2.1. Economic rationality

The green energy technology should be the energy technology of lower capital costs and high economic efficiency, and it should be economical, reasonable and symbiotic with the economic system. The applicability of economic rationality are described and evaluated by 5 indexes such as energy unit price, profit, energy resources quantity, market share rate and industrialized degree.

2.2. Social adaptability

The green energy technology should be the energy technology of lower social costs and better social benefits, and it should be human, social and symbiotic with the human society. The applicability of social adaptability are described and evaluated by 5 indexes such as supply security level, use security level, working conditions improvement level, increased employment rate and life quality improvement level.

2.3. Environmental coordination

The green energy technology should be the energy technology of low ecological costs and good ecological efficiency, and it should be friendly, ecological and symbiotic with the natural environment. Generally speaking, the applicability of ecological coordination are described and evaluated by 4 indexes such as pollutant decrement degree, waste recycling degree, major ecological destruction complaints and energy geography distribution.

2.4. Technical effectiveness

The green energy technology should be the energy technology of more efficient and beneficial, and it should be smart, interconnected and symbiotic with the technological system. The applicability requirements of technical effectiveness are described and evaluated by 4 indexes such as energy conversion rate, durable years, technical difficulty and technical maturity.

Therefore, we can construct the comprehensive evaluation index system of applicability of green energy technology. It is composed of 4 factors and 18 indexes which are independent, as shown in table 1.

Table 1. Evaluation index system of the applicability of green energy technology.

Applicability of green energy technology			
Economic rationality (C_1)	Social adaptability (C_2)	Environmental coordination (C_3)	Technical effectiveness (C_4)
Energy unit price (c_{11}) Profit (c_{12}) Energy resources quantity (c_{13}) Market share rate (c_{14}) Industrialized degree (c_{15})	Supply security level (c_{21}) Use security level (c_{22}) Working conditions improvement level (c_{23}) Increased employment rate (c_{24}) Life quality improvement level (c_{25})	Pollutant decrement degree (c_{31}) Waste recycling degree (c_{32}) Major ecological destruction complaints (c_{33}) Energy geography distribution (c_{34})	Energy conversion rate (c_{41}) Durable years (c_{42}) Technical difficulty (c_{43}) Technical maturity (c_{44})

3. Comprehensive evaluation models of the applicability of green energy technology

Reasonable selection and determination of the evaluation model is an important prerequisite for scientific analysis and evaluation of the applicability of green energy technology. Due to the diversity and comprehensiveness of applicability of green energy technology, we think it is more effective, feasible and appropriate to analyze and evaluate the applicability of green energy technology by selecting and applying FCE.

The basic principle of FCE is to determine the rating criteria and weights of the evaluation factor and index, to describe the fuzzy boundary of each index by membership degree, to construct the fuzzy evaluation matrix by the transformation principle of fuzzy set, and finally to determine the evaluation level of the object through the multi-layer compound operation.

The evaluation model of the applicability of green energy technology can be obtained by FCE, see equation (1). Generally, it can be referred as the comprehensive evaluation model of generalized fuzzy operation, abbreviated as $M(\overset{\bullet}{*}, \overset{+}{*})$[10]:

$$B = W \circ R = (b_1, b_2, \ldots, b_m) \tag{1}$$

In which,

$$b_j = (w_1 \overset{\bullet}{*} r_{1j}) \overset{+}{*} (w_2 \overset{\bullet}{*} r_{2j}) \overset{+}{*} \ldots \overset{+}{*} (w_n \overset{\bullet}{*} r_{nj}) \quad (j=1,2,\ldots,m)$$

B is the comprehensive evaluation value obtained by the generalized fuzzy operatio $\overset{\bullet}{*}$ n "and" and operation $\overset{+}{*}$ "or" with weight $W = \{w_j\}$ $(j=1,2,\ldots,m)$ and transformation matrix $R = \{r_{ij}\}$ $(i=1,2,\ldots,n, j=1,2,\ldots,m)$.

W is the weight of index, which refers to the measurement of the role and importance of evaluation index in the evaluation index system. It represents the ability and contribution degree of index and embodies the value orientation of the evaluator. The methods of obtaining weight include expert consultation method, rating method, experience the look-up table method, statistical test method, grey correlation method, comprehensive evaluation of inverse problem calculation method, Analytic Hierarchy Process(AHP) and other variants [8]. We choose AHP in the following example.

r is the single factor evaluation value, which refers to the rating ratio or membership degree of the index belonging to each evaluation grade. R is the fuzzy evaluation transformation matrix obtained by the single factor evaluation. There are two ways to obtain r: one is Experiential Look-up Tables Method; the other is Statistical Method of Peer Review [8]. We choose the latter in the following example.

In theory, there are infinitely many kinds of fuzzy operators, but only five specific models have been proposed up to now, see table 2.

Table 2. Formula and operator models of fuzzy compound operation.

Type M / Operation b_j / Operator	$M^{(\wedge,\vee)}$	$M^{(\bullet,\vee)}$	$M(^{\bullet,\oplus})$	$M(^{\wedge,\oplus})$	$M^{(\bullet^k,\wedge)}$
	$b_j = \max[\min(a_1, r_{1j}), \min(a_2, r_{2j}), \cdots \min(a_n, r_{nj})]$	$b_j = \max[a_1 r_{1j}, a_2 r_{2j}, \cdots a_n r_{nj}]$	$b_j = [1, \sum_{i=1}^{n} a_i r_{ij}]$	$b_j = \min[1, \sum_{i=1}^{n} \min(a_i r_{ij})]$	$b_j = \min[r_{1j}^1, r_{2j}^2, \cdots r_{nj}^n]$
$\overset{\bullet}{*}$	$\overset{\bullet}{*} \to \wedge$ $\alpha \wedge \beta = \min(\alpha, \beta)$	$\overset{\bullet}{*} \to \bullet$ $\alpha \bullet \beta = \alpha \times \beta$	$\overset{\bullet}{*} \to \bullet$ $\alpha \bullet \beta = \alpha \times \beta$	$\overset{\bullet}{*} \to \wedge$ $\alpha \wedge \beta = \min(\alpha, \beta)$	$\overset{\bullet}{*} \to \bullet^k$ $\alpha \bullet^k \beta = \beta^\alpha$
$\overset{+}{*}$	$\overset{+}{*} \to \vee$ $\alpha \vee \beta = \max(\alpha, \beta)$	$\overset{+}{*} \to \vee$ $\alpha \vee \beta = \max(\alpha, \beta)$	$\overset{+}{*} \to \oplus$ $\alpha \oplus \beta = \min(1, \alpha + \beta)$	$\overset{+}{*} \to \oplus$ $\alpha \oplus \beta = \min(1, \alpha + \beta)$	$\overset{+}{*} \to \wedge$ $\alpha \wedge \beta = \min(\alpha, \beta)$

The different operators of fuzzy compound operation in table 2 have different characteristics, which play different roles and reflect different value orientation of evaluators, see table 3.

Table 3. Characteristics of operators in fuzzy compound operation.

Operators / Characteristics	$M^{(\wedge,\vee)}$	$M^{(\bullet,\vee)}$	$M(^{\bullet,\oplus})$	$M(^{\wedge,\oplus})$	$M^{(\bullet^k,\wedge)}$
Types	Main factor determinant	Main factor prominent	Weighted average	Main factor prominent	Weighted average
Weight role	Non prominent	Prominent	Prominent	Non prominent	Prominent
R information	Incomplete	Incomplete	Complete	More complete	Complete
Comprehensive degree	Low	Low	High	High	Low

It is of great significance for fuzzy comprehensive evaluation to select the operators of fuzzy compound operation. The choice of compound operator depends on the scope of the analysis. We consider five specific operators are used respectively in the following different situations [8]: a. M $^{(\wedge,\vee)}$ is the operator of main factor determinant type, which is suitable for evaluating only the key factors, that is to say, the optimum of the key factors is considered as the best case for the comprehensive evaluation; b. $M(^{\bullet,\oplus})$ is the operator of weighted average type, which is suitable for evaluating things according to the comprehensive requirements of each factor; c. M $^{(\bullet,\vee)}$ and $M(^{\wedge,\oplus})$ are the operators of main factor prominent type, which are suitable for both to consider the overall but also focus on the evaluation of the situation; d. M$^{(\bullet^k,\wedge)}$ is the operator of weighted average type, which is suitable for the weight of all the factors based on the weight of the index, with the smallest weight even as the best requirements to evaluate the situation of things. We choose M$^{(\wedge,\vee)}$, M$^{(\bullet,\vee)}$ and M($^{\wedge,\oplus}$) in the following example.

4. Comprehensive evaluation methods of the applicability of green energy technology

In the following, we illustrate the basic process and method of analyzing and evaluating the applicability of green energy technology by FCE with an example [12].

4.1. Determine the tender object set E= {E$_k$} (k=1,2,…,q)

A green industrial region in a city conducts a public tender for the 3 tender companies of the green energy technology E_1, E_2, E_3 (q=3) and evaluates respectively their applicability of green energy technology.

4.2. Obtain the index set C= {c$_{ij}$} (i=1,2,…,n)

Through the research of related literature at home and abroad and consulting experts, we construct the evaluation index system of the applicability of green energy technology, see table 1.

4.3. Obtain the comment set G= {g$_l$} (l=1,2,…,s)

According to the identified evaluation criteria, we determine the rating criteria of applicability of green energy technology, and obtain the corresponding comment set G= {g_l} = {very good, good, pass, poor, very poor} (s=5).

4.4. Determine the weight set W= {w$_j$} (j=1,2,…,m)

The distribution of weight is the key to the whole comprehensive evaluation. Here we choose AHP and invite 7 experts of energy technology to determine the weight of the various indexes at all levels, see table 4.

Table 4. Summarization of the evaluation data of applicability of green energy technology.

Evaluation indexes and weight				Tender company's applicability, evaluation grades and rating ratio														
				Company E$_1$'s applicability					Company E$_2$'s applicability					Company E$_3$'s applicability				
Factor level(C_i)	Weight (W_i)	Index level(c_{ij})	Weight (W_{ij})	G_1	G_2	G_3	G_4	G_5	G_1	G_2	G_3	G_4	G_5	G_1	G_2	G_3	G_4	G_5
C_1	0.3	c_{11}	0.20	1/7	4/7	2/7	0	0	1/7	3/7	2/7	1/7	0	1/7	2/7	2/7	1/7	1/7
		c_{12}	0.25	2/7	3/7	1/7	1/7	0	0	2/7	3/7	1/7	1/7	1/7	4/7	1/7	1/7	0
		c_{13}	0.25	2/7	3/7	2/7	0	0	1/7	2/7	3/7	1/7	0	2/7	4/7	1/7	0	0
		c_{14}	0.15	2/7	4/7	1/7	0	0	0	3/7	3/7	1/7	0	2/7	3/7	2/7	0	0
		c_{15}	0.15	1/7	3/7	2/7	1/7	0	1/7	2/7	2/7	1/7	1/7	1/7	4/7	2/7	0	0
C_2	0.25	c_{21}	0.25	2/7	4/7	1/7	0	0	1/7	3/7	2/7	1/7	0	1/7	3/7	2/7	1/7	0
		c_{22}	0.25	1/7	4/7	2/7	0	0	1/7	2/7	2/7	1/7	1/7	2/7	3/7	1/7	1/7	0
		c_{23}	0.20	2/7	3/7	1/7	1/7	0	1/7	3/7	2/7	1/7	0	2/7	4/7	1/7	0	0
		c_{24}	0.15	1/7	3/7	3/7	0	0	0	1/7	3/7	2/7	1/7	2/7	3/7	2/7	0	0
		c_{25}	0.15	2/7	2/7	3/7	0	0	0	1/7	3/7	3/7	0	1/7	3/7	1/7	1/7	1/7
C_3	0.25	c_{31}	0.30	2/7	3/7	2/7	0	0	1/7	2/7	2/7	1/7	1/7	2/7	4/7	1/7	0	0
		c_{32}	0.25	1/7	4/7	2/7	0	0	1/7	1/7	2/7	2/7	1/7	1/7	3/7	2/7	1/7	0

IOP Conf. Series: Earth and Environmental Science **93** (2017) 012076 doi:10.1088/1755-1315/93/1/012076

		c_{33}	0.25	3/7	3/7	1/7	0	0	1/7	3/7	2/7	1/7	0	2/7	3/7	2/7	0	0
		c_{34}	0.20	2/7	3/7	1/7	1/7	0	0	2/7	3/7	1/7	1/7	1/7	4/7	1/7	1/7	0
C_4	0.2	c_{41}	0.30	4/7	2/7	1/7	0	0	1/7	3/7	1/7	1/7	0	1/7	3/7	2/7	1/7	0
		c_{42}	0.25	3/7	3/7	1/7	0	0	0	3/7	2/7	1/7	1/7	1/7	2/7	2/7	1/7	1/7
		c_{43}	0.20	2/7	3/7	1/7	1/7	0	1/7	2/7	3/7	1/7	0	2/7	4/7	1/7	0	0
		c_{44}	0.25	2/7	3/7	2/7	0	0	1/7	2/7	2/7	1/7	1/7	1/7	3/7	2/7	1/7	0

4.5. Obtain the evaluation matrix R= {r_{ij}} (i=1,2,...,n, j=1,2,...,m)

We can obtain the evaluation transformation matrix R = {r_{ij}} through the single factor evaluation. Single factor evaluation value r_{ij} is the rating ratio or membership degree of index c_i belonging to the evaluation grade g_j.

The data for the weight, the evaluation grade and the rating ratio are summarized in table 4.

4.6. Obtain the final comprehensive evaluation results

By evaluating comprehensively on each level, we can obtain the final comprehensive evaluation results. We first choose the operation $M^{(\wedge,\vee)}$ (take $\wedge$(min) and $\vee$(max) instead of $\overset{\bullet}{*}$ and $\overset{+}{*}$ in equation (1) to calculate. For the sake of space saving, the calculation process is omitted, and only the results are given here.

Evaluation matrix R_{1E1} of Company E_1's applicability of green energy technology in single factor C_{1E1} is:

$$R_{1E1}=\begin{bmatrix}1/7 & 4/7 & 2/7 & 0 & 0\\ 2/7 & 3/7 & 1/7 & 1/7 & 0\\ 2/7 & 3/7 & 2/7 & 0 & 0\\ 2/7 & 4/7 & 1/7 & 0 & 0\\ 1/7 & 3/7 & 2/7 & 1/7 & 0\end{bmatrix}=\begin{bmatrix}0.14 & 0.57 & 0.29 & 0 & 0\\ 0.29 & 0.43 & 0.14 & 0.14 & 0\\ 0.29 & 0.43 & 0.29 & 0 & 0\\ 0.29 & 0.57 & 0.14 & 0 & 0\\ 0.14 & 0.43 & 0.29 & 0.14 & 0\end{bmatrix}$$

Comprehensive evaluation value B_{1E1} of Company E_1's applicability of green energy technology in single factor C_{1E1} is:

$$B_{1E1}={}_1oR_{1E1}=[0.2\ 0.25\ 0.25\ 0.15\ 0.15]o\begin{bmatrix}0.14 & 0.57 & 0.29 & 0 & 0\\ 0.29 & 0.43 & 0.14 & 0.14 & 0\\ 0.29 & 0.43 & 0.29 & 0 & 0\\ 0.29 & 0.57 & 0.14 & 0 & 0\\ 0.14 & 0.43 & 0.29 & 0.14 & 0\end{bmatrix}=[0.25\ 0.25\ 0.25\ 0.14\ 0]$$

Similarly, comprehensive evaluation values B_{2E1}, B_{3E1}, B_{4E1} of Company E_1's applicability of green energy technology in single factor C_{2E1}, C_{3E1}, C_{4E1} are:

$B_{2E1}=W_2\,o\,R_{2E1}=[0.25\ \ 0.25\ \ 0.25\ \ 0.14\ \ 0.14]$

$B_{3E1}=W_3\,o\,R_{3E1}=[0.29\ \ 0.30\ \ 0.29\ \ 0.14\ \ 0\,]$

$B_{4E1}=W_4\,o\,R_{4E1}=[0.3\ \ 0.29\ \ 0.25\ \ 0.14\ \ 0]$

Thus we obtain the evaluation matrix R_{E1} of Company E_1's applicability of green energy technology in all the single factors:

$$R_{E1} = \begin{bmatrix} B_{1E1} \\ B_{2E1} \\ B_{3E1} \\ B_{4E1} \end{bmatrix} = \begin{bmatrix} 0.25 & 0.25 & 0.25 & 0.14 & 0 \\ 0.25 & 0.25 & 0.25 & 0.14 & 0.14 \\ 0.29 & 0.30 & 0.29 & 0.14 & 0 \\ 0.3 & 0.29 & 0.25 & 0.14 & 0 \end{bmatrix}$$

Finally, we reach the comprehensive evaluation result B_{E1} of Company E_1's applicability of green energy technology in all the single factors, which is in the overall evaluation system:

$$B_{E1}=W\text{o}R_{E1}=[0.3\ 0.25\ 0.25\ 0.2]\text{o}\begin{bmatrix} 0.25 & 0.25 & 0.25 & 0.14 & 0 \\ 0.25 & 0.25 & 0.25 & 0.14 & 0.14 \\ 0.29 & 0.30 & 0.29 & 0.14 & 0 \\ 0.3 & 0.29 & 0.25 & 0.14 & 0 \end{bmatrix}=[0.25\ 0.25\ 0.25\ 0.14\ \ 0.14].$$

Similarly, we reach the comprehensive evaluation results B_{E2} and B_{E3} of Company E_2 and E_3's applicability of green energy technology in all the single factors, which is in the overall evaluation system.

$B_{E2} = W \text{ o } R_{E2} = [0.20\ \ 0.20\ \ 0.25\ \ 0.15\ \ 0.14]$

$B_{E3} = W \text{ o } R_{E3} = [0.20\ \ 0.25\ \ 0.25\ \ 0.14\ \ 0.14]$

For the sake of clarity, the above final comprehensive evaluation results B_{E1}, B_{E2} and B_{E3} can be converted into comprehensive evaluation values Z_{E1}, Z_{E2} and Z_{E3} by the evaluation level $G = \{G_1, G_2, G_3, G_4, G_5\} = [2, 1, 0, -1, -2]$.

$$Z_{E1}=B_{E1}\cdot G=[0.25\ 0.25\ 0.25\ 0.14\ \ 0.14]\cdot\begin{bmatrix} 2 \\ 1 \\ 0 \\ -1 \\ -2 \end{bmatrix} = 0.33$$

$Z_{E2} = B_{E2}\cdot G = 0.22$

$Z_{E3} = B_{E3}\cdot G = 0.27$

If the operations $M(\bullet,\vee)$ and $M(\wedge,\oplus)$ are chosen, the corresponding comprehensive evaluation values of the 3 companies' applicability of green energy technology are shown in table 5.

Table 5. Multi-operator values of comprehensive evaluation.

Operators Values of comprehensive evaluation Z Tender companies E	$M(\wedge,\vee)$	$M(\bullet,\oplus)$	$M(\wedge,\oplus)$
E_1's applicability of green energy technology	0.33	0.42	0.78
E_2's applicability of green energy technology	0.22	0.29	0.43
E_3's applicability of green energy technology	0.27	0.36	0.67

4.7. Reach the conclusion of evaluation by sorting and selecting

According to the final evaluation results, we can sort, select and reach the conclusion of evaluation of the 3 tender companies' applicability of green energy technology.

From the final comprehensive evaluation value, we can clearly see that in the applicability of green energy technology, Company E_1's technology is better than Company E_3's technology, Company E_3's technology is better than Company E_2's technology. Therefore, Company E_1's applicability of green energy technology is the best and will be the winner.

5. Conclusion

The applicability of green energy technology is complex, systematic and dynamic. It needs to be analyzed and evaluated from various aspects of economy, society, environment and science & technology, which is a problem of multi-object, multi-factor, multi-level and fuzzy comprehensive evaluation. Thus it is feasible and effective to analyze and evaluate the applicability of green energy technology by FCE which plays an irreplaceable important role. Compared with the traditional evaluation method, the advantage of FCE is that it can quantify some indexes with unclear and difficult quantitative, combine qualitative analysis with quantitative analysis, subjective value judgment and objective data, and then it is more suitable for solving the problems of comprehensive evaluation. Of course, the defect of FCE is mainly that the amount of calculation of FCE is relatively large and complex in the case of a large number of indexes. In short, we believe that with the scientific methods such as FCE, the problems of the applicability of green energy technology will be increasingly explored in depth to provide the important theoretical guidance for the rational development and utilization of green energy technology and the better promotion of its industrialization, networking and intelligence.

Acknowledgments

This paper is funded by the Project of Soft Science (2017) from Technology Talent Development Service Center, Ministry of Science and Technology.

References

[1] Miao J M 2013 Review of world clean energy development *Journal of Shanxi Agricultural University (Social Science Edition)* **7** 694-5
[2] Zheng X P 2015 Re-exploration on the evaluation system of green energy technology applicability *Southern Building* **3** 119-23
[3] Peng P 2003 Revelation to the theory of applicable technology *Journal of Wuhan University of Technology (Social Science Edition)* **6** 7-12
[4] Pezzey J C V 1992 Sustainable development concepts: An economic analysis *World Bank Environment Paper* **2** 46-52
[5] Qiu B X 2012 *Taking the Ideal and Reality into Account Practice Demonstration Built on Indicator System of China's Low-carbon Eco-city* (Beijing: China Architecture & Building Press)
[6] Gu S T 2016 Evaluation on the utilization system of rural clean energy (Wuhan: Huazhong University of Science and Technology)
[7] Li J Z *et al* 2010 Application of fuzzy comprehensive evaluation in the evaluation of green energy *Journal of Hydropower and Energy Science* **6** 171-4
[8] Xu Z J 2011 Fuzzy comprehensive evaluation on ability of symbiotic innovation in sustainable technology *Advanced Materials Research* **3** 1364-9
[9] Wang P Z 1980 Introduction to fuzzy mathematics *Practice and Understanding Mathematics* **2** 9-14
[10] Liu Y Y *et al* 1983 Mathematical models of comprehensive evaluation *Fuzzy Mathematics* **1** 7-13
[11] Xu Z J and Song Y K 2013 Analysis of sustainable technology innovation benefits based on symbiosis theory *Development of Industry and Economy* **10** 898-902
[12] Cai Y N *et al* 2014 Concept and practice of eco-city planning-A case study of Haizhu ecological city in Guangzhou *Southern Architecture* **6** 88-94

A distributed monitoring system for photovoltaic arrays based on a two-level wireless sensor network

F P Su[1,2], Z C Chen[1,2,3], H F Zhou[1,2,3], L J Wu[1,2], P J Lin[1,2], S Y Cheng[1,2] and Y F Li[1,2]

[1]Institute of Micro/Nano Devices and Solar Cells, College of Physics and Information Engineering, Fuzhou University, Fuzhou 350116, China
[2]Jiangsu Collaborative Innovation Centre of Photovoltaic Science and Engineering, 213164 Changzhou, China

E-mail: zhicong.chen@fzu.edu.cn/ zhhafa@163.com

Abstract. In this paper, a distributed on-line monitoring system based on a two-level wireless sensor network (WSN) is proposed for real time status monitoring of photovoltaic (PV) arrays to support the fine management and maintenance of PV power plants. The system includes the sensing nodes installed on PV modules (PVM), sensing and routing nodes installed on combiner boxes of PV sub-arrays (PVA), a sink node and a data management centre (DMC) running on a host computer. The first level WSN is implemented by the low-cost wireless transceiver nRF24L01, and it is used to achieve single hop communication between the PVM nodes and their corresponding PVA nodes. The second level WSN is realized by the CC2530 based ZigBee network for multi-hop communication among PVA nodes and the sink node. The PVM nodes are used to monitor the PVM working voltage and backplane temperature, and they send the acquired data to their PVA node via the nRF24L01 based first level WSN. The PVA nodes are used to monitor the array voltage, PV string current and environment irradiance, and they send the acquired and received data to the DMC via the ZigBee based second level WSN. The DMC is designed using the MATLAB GUIDE and MySQL database. Laboratory experiment results show that the system can effectively acquire, display, store and manage the operating and environment parameters of PVA in real time.

1. Introduction

The energy challenge and environmental problem is becoming increasingly serious all over the world. Therefore, the renewable solar energy is gradually developed and used by more and more countries [1-5]. According to the latest announcement of the National Energy Board of China, by the end of 2016, China's new PV installed capacity was 34.54 million kilowatts, the cumulative installed capacity was up to 77.42 million kilowatts and the amount of both new and cumulative installed capacity are at the first place in the world. PV arrays of PV power plants are usually installed in unattended outdoor fields such as desert, coast and on the roof [6], which are supposed to run for a long time. However, due to harsh outdoor environment, PV arrays are prone to suffer various faults [7-9]. In order to support automatic fault detection and facilitate the maintenance, it is significant to deploy distributed real-time monitoring systems for the PV arrays. The available monitoring systems for PV arrays can be classified into wired monitoring and wireless monitoring. Wired monitoring system mainly utilizes RS485 and CAN bus serial interface for local communication, and then achieves remote communication via the Internet [10, 1]. Due to the complicated wiring, deployment of wired monitoring systems is laboured intensive and expensive. Therefore, many wireless monitoring systems have been proposed and developed [5,6,12,13].

Wireless sensors network (WSN) has been widely used in various fields due to its high flexibility, excellent maintainability and scalability [14-19]. It also has been applied for monitoring PV power plants in real-time [2-6,18,19]. Xu and Wang applied the WSN to collect surrounding environment information and status data of PV power plant [6,19]. Papageorgas *et al* fuse computation and communication technologies for real-time monitoring and characterization of PV panel operational status based on wired networking combined with ZigBee [3]. Another case is that a WSN for smart monitoring of PV systems at module level is proposed, and the efficiency losses and related causes are estimated based on the acquired data [4]. In addition, Hua *et al* utilize the ZigBee to deliver the monitoring data of PV modules from different power stations to monitoring centre located in cloud [2]. A web-based and ZigBee-based monitoring system for grid-connected PV converter is presented by Shariff *et al*, in which a web-application is also developed to make the monitored data accessible via Internet [5]. An on-line distributed monitoring system based on ZigBee WSN was designed to monitor the electrical parameters and environmental parameters of PV modules in [20].

In order to achieve fine monitoring, all operating parameters of PV modules and PV strings in a PV array are expected to be monitored, including operating voltage, current, temperature and irradiance etc. However, PV arrays usually consist of a lot of PV modules, strings and sub-arrays, which will require a lot of wireless sensors leading to a high cost. Therefore, the cost of wireless sensors is an important issue. In this paper, we propose a new distributed monitoring system based on a low cost two-level WSN. The first level WSN refers to the communication between the wireless sensors on PV modules and the corresponding wireless sensors on the combiner box, which is implemented by the low cost nRF24L01 transceiver. The second level WSN refers to the communication among the wireless sensors on combiner boxes and the sink node, which is realized by the CC2530 based flexible ZigBee network. It is worth noticing that the work in this paper is based on [20] and we make further improvements. The rest of paper is organized as follows. Section 2 presents the system architecture. Section 3 introduces the design of sensing nodes installed in PV modules. The design of sensing and routing nodes deployed on combiner boxes of PV sub-arrays is described in Section 4. Section 5 demonstrates the realization of data management centre. Section 6 illustrates the system prototype and experimental results. Finally, Section 7 outlines the conclusion.

2. System architecture

2.1. System design

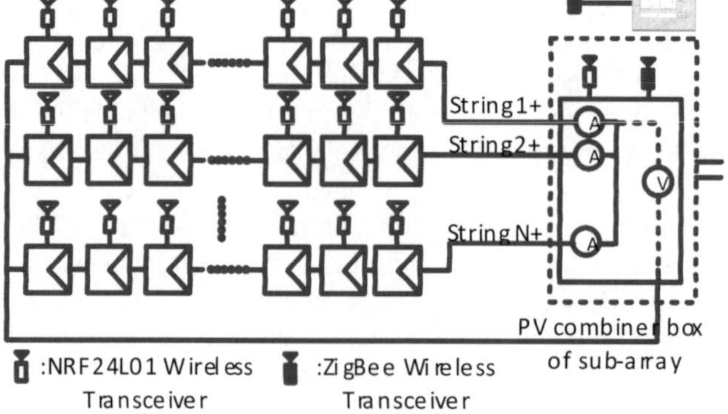

Figure 1. Block diagram of the WSN based distributed monitoring system for PVA.

As shown in figure 1, the distributed monitoring system for PV array mainly consists of a large number of PV module (PVM) sensing nodes installed in the junction box of each PVM, PV sub-array (PVA) sensing and routing nodes installed in the combiner box of each PV sub-array, a sink node connected to data management centre (DMC), and a DMC running on a local host computer.

The PVM nodes are designed to monitor the operating voltage and backplane temperature of each PVM. The monitored data are sent to the corresponding PVA nodes via the first level nRF24L01 based WSN. The PVA nodes are used to monitor the current and voltage of each PV string, environment irradiance and temperature around the combiner box of PV sub-array. In addition, the PVA nodes merge their monitored data with the one received from their PVM nodes, and then send them together to the sink node via a ZigBee based second level WSN. Because PV modules are usually connected in series to form PV strings, the current of PV modules should be equal to the current of their PV strings. Moreover, if PVM nodes acquire the operating PV module current directly, the current sensors need to be connected to the PV module, which will affect the wiring of PV modules and increase the cost of wireless sensors. Therefore, the operating current of PVM is not monitored by the PVM nodes to reduce the cost and complexity. The sink node passes the received data to the DMC via the serial port for data storage, display, and processing. Figure 2 further illustrates the distributed monitoring system involving two PV sub-arrays.

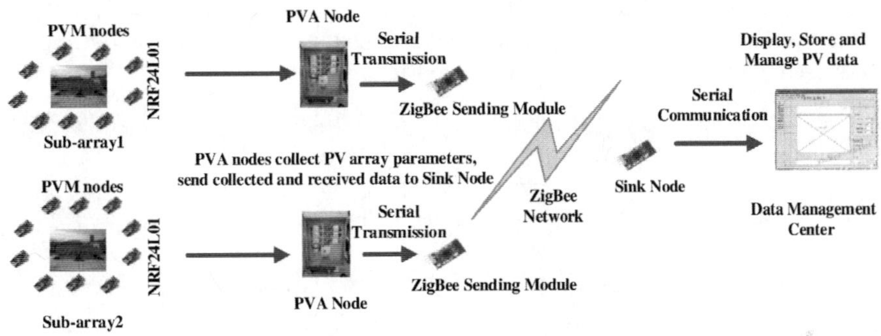

Figure 2. Illustration of the distributed monitoring systems involving two photovoltaic sub-array.

2.2. Wireless sensor network topology

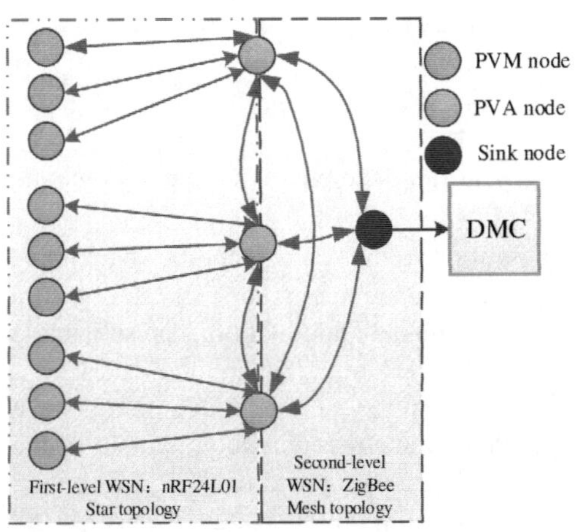

Figure 3. The wireless sensor networks topology of proposed system.

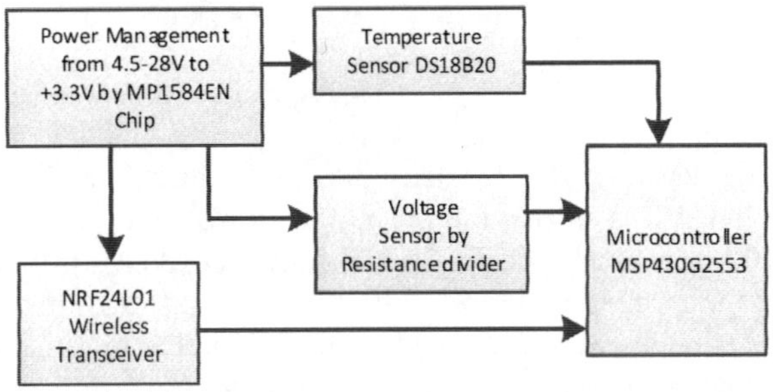

Figure 4. The block diagram of PVM circuit.

The proposed system adopts two-level WSN for data transmission. Due to low cost and low power consumption, the first-level single hop star NRF24L01 wireless network is applied to transmit and receive the data in PV module via the Serial Peripheral Interface (SPI) communication. The second-level multi-hop mesh ZigBee wireless network is employed to deliver the PVM data, PV string data and environment parameters to DMC for further display and storage. As shown in figure 3, The PVM nodes are integrated with NRF24L01 WSN to send the PVM parameters, both NRF24L01 and ZigBee WSN are embedded in PVA nodes to receive data from PVM nodes and transmit data to sink node, respectively. The sink node delivers parameters to the DMC via serial communication.

3. Design of PVM node

The block diagram of PVM circuit is shown in figure 4. The hardware of PVM nodes consists of power management module, temperature and voltage sensors, microcontroller and wireless transmission module.

3.1. Power management module

The PVM node is powered by a battery together with the PVM. During the daytime, the PVM can generate electricity and thus it is used to power the PVM node and charge the battery as well, while the battery can maintain the basic operation of the PVM node during the night time. Since the open-circuit voltage of the PVM under Standard Test Condition (STC) is about 21.5 V which is much higher than the operating voltage of PVM node, the efficient stepping-down DC-DC voltage converter MP1584EN is used to generate the 3.3 V power supply.

3.2. Temperature and voltage sensors

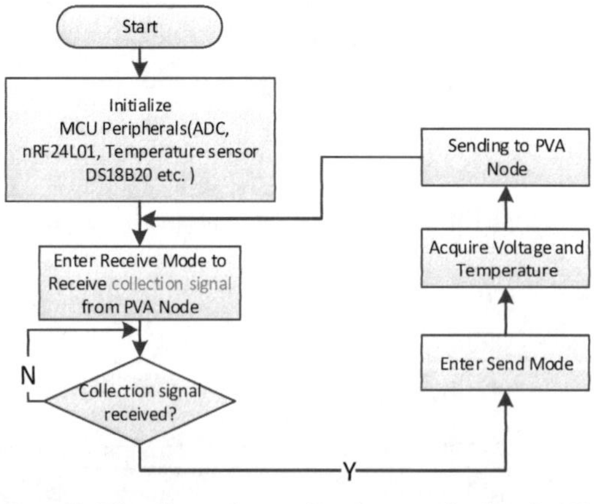

Figure 5. The flow chart of software design in PVM.

The PVM nodes are designed to monitor the PVM operating voltage and module backplane temperature in real time. The low-power microcontroller MSP430G2553 which is integrated with a 10-bit resolution analog-digital converter (ADC) is used to achieve the data acquisition. The schematic diagram of the adopted voltage sensor is shown in figure 5. Since the operating voltage of PVM is usually about twenty volts but still much higher than the input range of the ADC, therefore a potentiometer implemented by two precise resistors is used to linearly scale down the voltage to fit the input range of the ADC inside the microcontroller. In this circuit, the ratio is set to be 1:10. Moreover, a capacitor is used to filter the noise, and two diodes are used to limit the output voltage between -0.7 V and 4 V. PVM usually run under the sunlight for a long time, the PVM panel temperature will be very high. Therefore, the DS18B20 temperature sensor is used. The flow chart of software design in PVM is shown in figure 5. Firstly, the MCU peripherals such as ADC, NRF24L01, and temperature sensor are initialized. Secondly, the NRF24L01 is set as receiving mode to receive the collection signal from PVA node. Once the signal is received completely, the NRF24L01 is set as sending mode, and MCU acquire the voltage and temperature. Finally, the MCU send acquired data to PVA node via NRF24L01 network.

3.3. NRF24L01 WSN
In order to store the parameters collected by the PVM in the DMC, the system employs the low cost, low power consumption NRF24L01 wireless transceiver to deliver data, and sends the PVM's parameter to their corresponding PVA node for the further communication. The sending and receiving flow chart of NRF24L01 wireless transceiver in proposed system is shown in figure 6. In order to ensure the synchronization of PVM acquisition nodes in time, the broadcast mechanism of NRF24L01 in this network is utilized to ensure that the PVA node receives synchronized data. Therefore, it is obvious that the sending part is set as receiving mode to receive the collect signal firstly, and the receiving part is set as sending mode to send the collect signal via the broadcast mechanism of NRF24L01.

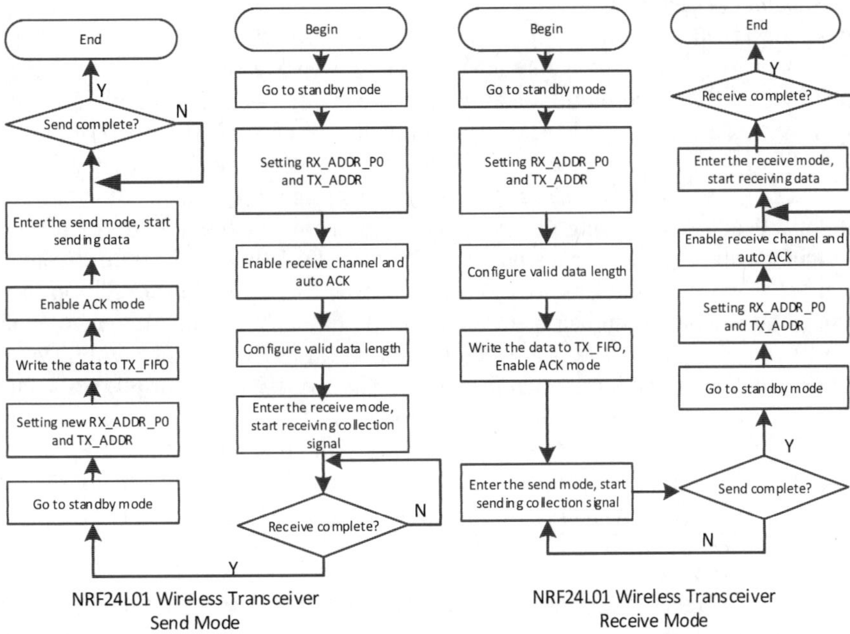

Figure 6. The sending and receiving flow chart of NRF24L01 wireless transceiver.

4. Design of PVA node
Figure 7 shows the block diagram of PV string circuit. The hardware of PVA nodes consists of power management module, voltage sensors, current sensors, temperature sensors and irradiance sensors, microcontroller and two wireless transmission module.

4.1. Power management module
The PVA node is powered by electricity, the electricity voltage can be employed for sub-array voltage collection and gradually reduced to the required working voltage via the different power stepping-down DC-DC voltage converter. The LM7805 chip is used to generate the 5 V power supply for current sensors and irradiance sensors. And the LM1084IS-3.3 chip is applied to generate the 3.3 V power supply for microcontroller, NRF24L01 wireless transceiver and ZigBee wireless transmission module. During the conversion, several capacitors are used to filter the noise.

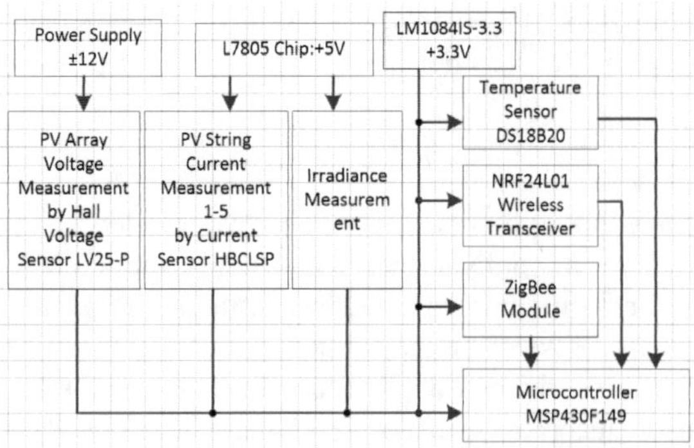

Figure 7. The block diagram of PV string circuit.

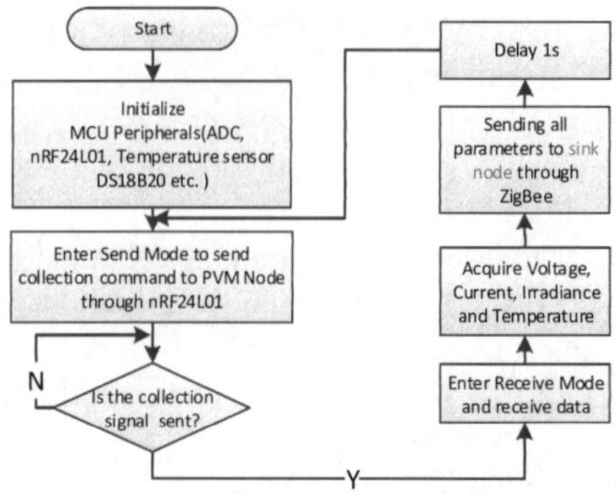

Figure 8. The flow chart of software design in PVA.

4.2. PV string parameters acquisition

The PVA nodes are designed to real-time monitor the PV string operating voltage, operating current, irradiance, environment temperature in combiner box. The low-power microcontroller MSP430F149 which is integrated with a 12-bit resolution analog-digital converter (ADC) is applied to achieve the data acquisition. PVA nodes adopt the isolated Hall voltage sensor LV25-P and Hall current sensor HBC06LSP for PV string operating voltage and operating current acquisition. Because the PVA consists of three strings in parallel, the system only designs a string voltage acquisition and five strings current acquisition, the extra two strings current acquisition, one of them is used to measure the irradiance, the other is applied for expansion in the future. DS18B20 temperature sensor is also applied to collecting the working temperature of combiner box. Due to the linear relationship between short-circuit current and irradiance, and temperature has little influence on the irradiance which can be neglected, so the PVA nodes obtain the irradiance by measuring a PVM's short-circuit current. The flow chart of software design in PVA is shown in figure 8. Firstly, the MCU peripherals such as ADC, NRF24L01, and temperature sensor are initialized. Secondly, the NRF24L01 is set as sending mode to send the collection signal to PVM node. Once the signal is sent completely, the NRF24L01 is set as receiving mode, and MCU acquire the voltage, current, irradiance and temperature. Finally, the MCU merges received data with acquired data and sends to sink node through ZigBee network.

5. Design of the DMC software

The DMC running on a host computer is designed by Matlab GUIDE, and the DMC includes the serial communication, parameters displaying, curve plot and data storage. The data uploaded from PVA nodes is received via CC2530 based ZigBee module, and is stored in MySQL database. Meanwhile, the PV data such as temperature, irradiance, voltage and current etc. is displayed and plotted in the interface. The flow chart of the DMC is shown in figure 9, and figure 10 shows the interface of DMC.

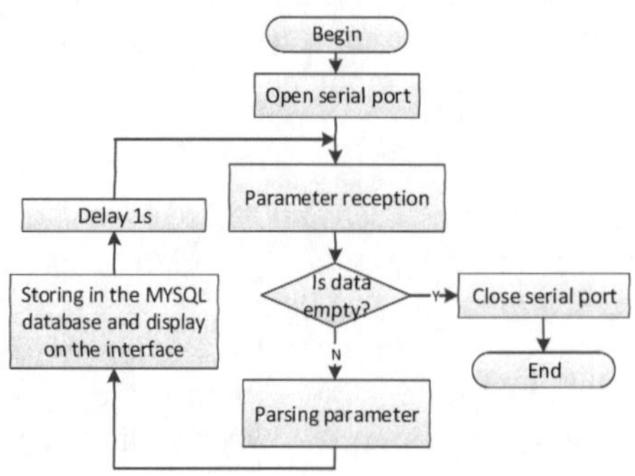

Figure 9. The flow chart of DMC.

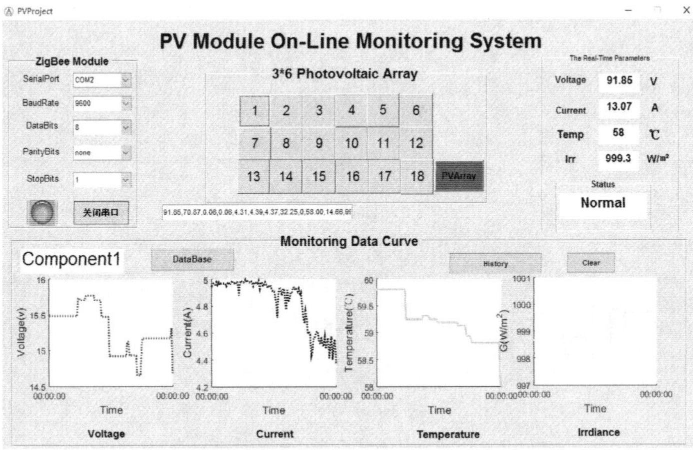

Figure 10. The interface of DMC.

6. System prototype and experimental results

In order to verify that the designed system can run normally, the laboratory PV array on the 6th floor is selected as the laboratory object, which consists of three PV strings in parallel and each string consists of six PVM in series. The open-circuit voltage of the single PV module under STC is about 21.5 V and the short-circuit current is about 6.03 A under STC.

The test procedure is as follows. Firstly, the PVM nodes are installed in the junction box of the PVM and also powered by the PVM. The PVA node is deployed in the PV combiner box, the node collects real-time data per second. Secondly, the data collected by PVM nodes and PVA node are transmitted to the sink node via two-level WSN. Finally, the sink node sends the received data to DMC via the serial port to store and display. The physical pictures of the PVM and PVA node are shown in figures 11 and 12, respectively. The real-time parameters of PVM and PV strings are illustrated in figures 13 and 14, respectively. The part of real-time PV data stored in database is shown in figure 15. During the parameter acquisition, the selected PV array collects and transmits the parameters in different irradiance.

Figure 11. The physical picture of the PVM node.

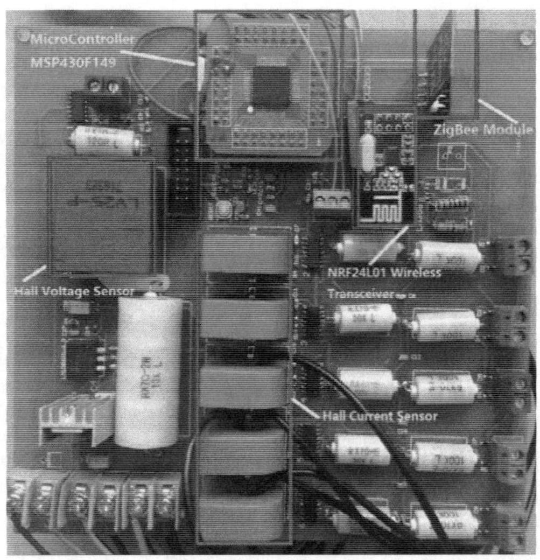

Figure 12. The physical picture of the PVA node.

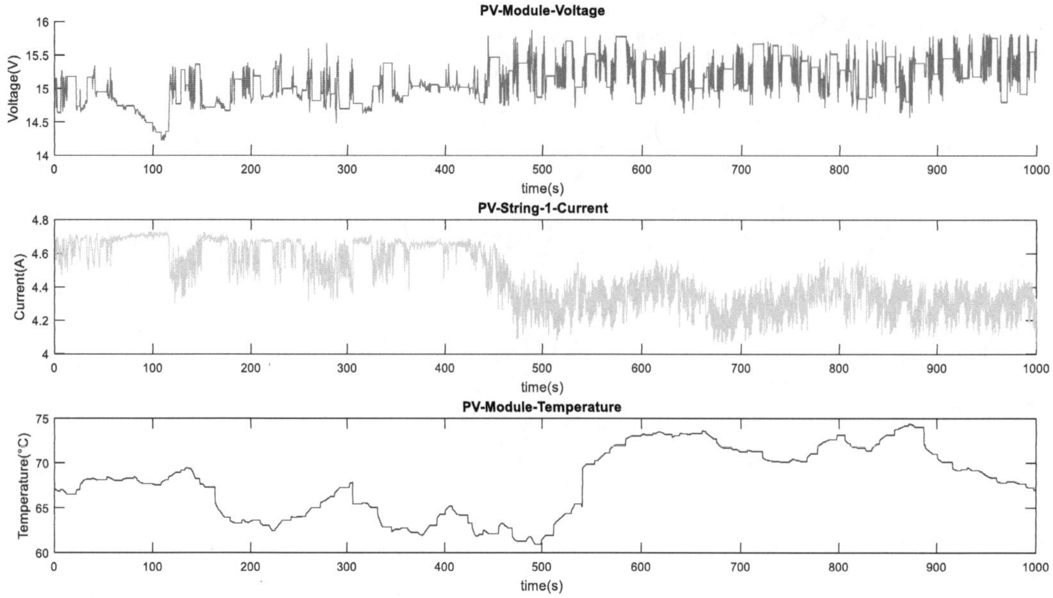

Figure 13. The real-time parameters of PVM.

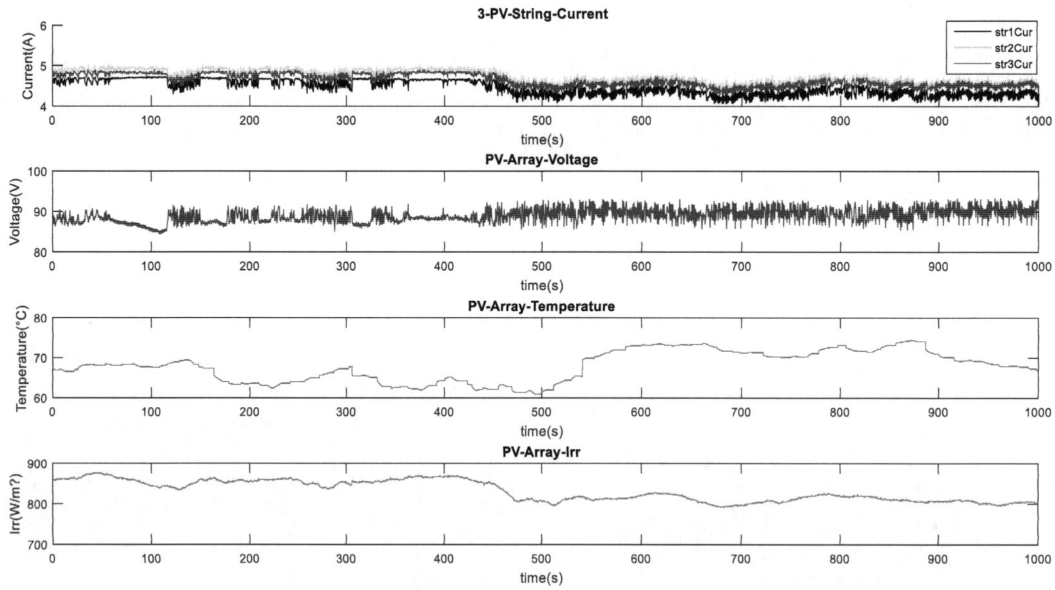

Figure 14. The real-time parameters of PVA.

Compared the collected data by the proposed system with the data collected by multimeter, thermometer and illuminometer, they are mostly similar to each other. Several tests show that the data is normal and reliable in the distance between the fifth floor and the sixth floor in the laboratory. When the communication distance of the node is increased (such as > 100 m), a small amount of packet loss occurs. The distance between the PVM in the PVA and the different types of nodes will not exceed the packet loss distance generally, so it doesn't affect the normal operation of the proposed monitoring system.

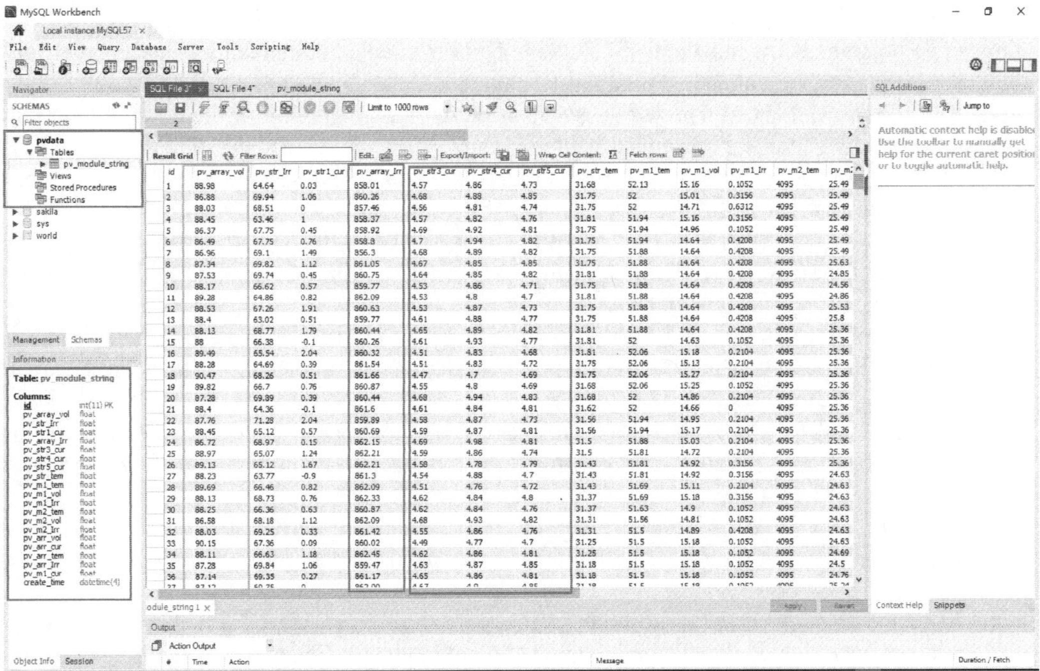

Figure 15. The real-time PV data stored in database.

7. Conclusions

A distributed monitoring system for PVA based on NRF24L01 and ZigBee two-level WSN is proposed in this paper. The proposed system consists of the sensing nodes installed on PVM, sensing and routing nodes installed on combiner boxes, data management centre running on a host computer. Firstly, PVM nodes collect the parameters of PVM and transmit them to PVA node via the first level single hop nRF24L01 WSN. Secondly, PVA nodes receive the data from PVM nodes and collect the parameters of PV strings, then merge all data to the sink node via multi-hop ZigBee WSN. Finally, the sink node sends the data to the DMC via serial communication. The laboratory experiment results demonstrate that the proposed system can effectively collect, display, store and manage the working status and environment parameters of PV array real-time.

Acknowledgments

The authors would like to acknowledge the supports by the National Natural Science Foundation of China (Grant Nos. 61574038, 61601127, and 51508105), the Fujian Provincial Department of Science and Technology of China (Grant Nos. 2015H0021, 2015J05124 and 2016H6012), the Science Foundation of Fujian Education Department (Grant No. JAT160073), and the Fujian Provincial Economic and Information Technology Commission of China (Grant No. 830020 and 83016006).

References

[1] Chouder A and Silvestre S 2010 Automatic supervision and fault detection of PV systems based on power losses analysis *Energy Convers Manage* **51**(10) 1929-37

[2] Hu T, Zheng M H, Tan J J, Zhu L and Miao W 2015 Intelligent photovoltaic monitoring based on solar irradiance big data and wireless sensor networks *Ad Hoc Netw* **35** 127-36

[3] Papageorgas P, Piromalis D, Antonakoglou K, Vokas1 G, Tseles2 D and Arvanitis K G 2013 Smart solar panels: In-situ monitoring of photovoltaic panels based on wired and wireless sensor networks *Energy Procedia* **36** 535-45

[4] Ando B, Baglio S, Pistorio A, Tina M G and Ventura C 2013 Sentinella: A WSN for a smart monitoring of PV systems at module level *Measurements and Networking Proceedings*

(M&N), IEEE International Workshop on IEEE 36-40
[5] Shariff F, Rahim N A and Hew W P 2015 Zigbee-based data acquisition system for online monitoring of grid-connected photovoltaic system *Expert Syst Appl* **42**(3) 1730-42
[6] Xu X L and Wang H 2011 The wireless sensor network construction of the photovoltaic power system ZigBee *Procedia Engineering* **15** 2511-5
[7] Drews *et al* 2007 Monitoring and remote failure detection of grid-connected PV systems based on satellite observations *Sol Energy* **81**(4) 548-64
[8] Chen Z C, Wu L J, Cheng S Y, Lin P J, Wu Y and Lin W C 2017 Intelligent fault diagnosis of photovoltaic arrays based on optimized kernel extreme learning machine and IV characteristics *Appl Energ* **204** 912-31
[9] Chen Z C, Wu L J, Lin P J, Wu Y and Cheng S Y 2016 Parameters identification of photovoltaic models using hybrid adaptive Nelder-Mead simplex algorithm based on eagle strategy *Appl Energ* **182** 47-57
[10] Zdanowicz T and Roguszczak H 2000 Design and application of autonomous data acquisition systems used as calibration and qualification tool in photovoltaics *Optoelectron Rev* **8**(4) 356-67
[11] Li K F and Luo W P 2013 Research on the remote monitoring system of photovoltaic power station based on CAN bus and GPRS with mechanical properties *Adv Mater Res Trans Tech Publications* **644** 60-3
[12] Xu X L and Qiao D E 2011 Remote monitoring and control of photovoltaic system using wireless sensor network *2011 International Conference on Electric Information and Control Engineering (ICEICE) IEEE* 633-8
[13] Batista N C, Melício R, atias J C O M and Catalão J P S 2013 Photovoltaic and wind energy systems monitoring and building home energy management using ZigBee devices within a smart grid *Energy* **49** 306-15
[14] Wang Z J and Wang Q H 2010 The application of ZigBee technology in transmission lines cladding ice alert system *Electr PowElec* **29**(13) 82-4
[15] Wang W B and Song Y 2010 Design and research of smart switch control system based on zigbee (in Chinese) *Electric Automatization* (in Chinese) **32**(3) 57-9
[16] Tan H Q and Wang Q 2010 Design on solar radiation based on ZigBee technology (in Chinese) *Instrumentation Technique and Sensor* (in Chinese) **6** 36-8
[17] Huang C C 2015 Study on cloud-dust based intelligent maximum performance analysis system for power generation with solar energy *Sci Iran* **22**(6) 2170s
[18] Ibrahim R, Chung T D, Hassan S M, Bingi K and Salahuddin B K S 2017 Solar energy harvester for industrial wireless sensor nodes *IEEE International Symposium on Robotics and Intelligent Sensors* 111-8
[19] Wang H and Xu X L 2013 Wireless sensor network in the photovoltaic power generation monitoring system *Indo J Electr Eng Co* **11** 4774-9
[20] Lin H W, Chen Z C, Wu L J, Lin P J and Cheng S Y 2015 On-line monitoring and fault diagnosis of PV array based on BP neural network optimized by genetic algorithm *International Workshop on Multi-disciplinary Trends in Artificial Intelligence* (Springer, Cham) pp 102-12

IOP Conf. Series: Earth and Environmental Science **93** (2017) 012078 doi:10.1088/1755-1315/93/1/012078

On-line monitoring system of PV array based on internet of things technology

Y F Li[1,2], P J Lin[1,2,3], H F Zhou[1,2,3], Z C Chen[1,2], L J Wu[1,2], S Y Cheng[1,2] and F P Su[1,2]

[1]Institute of Micro/Nano Devices and Solar Cells, College of Physics and Information Engineering, Fuzhou University, Fuzhou 350116, China
[2]Jiangsu Collaborative Innovation Centre of Photovoltaic Science and Engineering, Changzhou 213164, China

E-mail: linpeijie@fzu.edu.cn/zhhafa@163.com

Abstract. The Internet of Things (IoT) Technology is used to inspect photovoltaic (PV) array which can greatly improve the monitoring, performance and maintenance of the PV array. In order to efficiently realize the remote monitoring of PV operating environment, an on-line monitoring system of PV array based on IoT is designed in this paper. The system includes data acquisition, data gateway and PV monitoring centre (PVMC) website. Firstly, the DSP-TMS320F28335 is applied to collect indicators of PV array using sensors, then the data are transmitted to data gateway through ZigBee network. Secondly, the data gateway receives the data from data acquisition part, obtains geographic information via GPS module, and captures the scenes around PV array via USB camera, then uploads them to PVMC website. Finally, the PVMC website based on *Laravel* framework receives all data from data gateway and displays them with abundant charts. Moreover, a fault diagnosis approach for PV array based on Extreme Learning Machine (ELM) is applied in PVMC. Once fault occurs, a user alert can be sent via E-mail. The designed system enables users to browse the operating conditions of PV array on PVMC website, including electrical, environmental parameters and video. Experimental results show that the presented monitoring system can efficiently real-time monitor the PV array, and the fault diagnosis approach reaches a high accuracy of 97.5%.

1. Introduction

Clean energy, especially solar energy, has become more and more popular and important due to their substantial benefits in many aspects [1]. Solar energy is regarded as an important renewable energy in the world, a large number of grid-connect photovoltaic (PV) power plants are widely distributed worldwide. However, the PV power plants are generally built in outdoor harsh environment, such as desert and roof. The PV power plants are prone to failure under such environmental conditions [2-4]. Accordingly, it is essential to monitor PV system for ensuring efficient power delivery.

The traditional monitoring systems mainly include manual investigation and remote wired monitoring. There exists some drawbacks such as time-consuming and wiring complexity [5], which is difficult to meet the monitoring requirement in the large-scale PV array. As the term Internet of things (IoT) is expanding, there is a huge increase in applications such as smart home, intelligent city, wisdom campus and so on [6]. The IoT reduces the effort of human by introducing machine to machine interaction, which is applied to effectively facilitate the integration of material production, service management [7,8]. It also has been applied for designing the remote monitoring system of PV array [9-12]. Soham A *et al* employs IoT technology to remotely monitor PV power plants, GSM/GPRS is used for communication between data logger and server [9]. Nalamwar H S *et al* propose a smart automated monitoring and controlling system for the PV panel via the Internet through an embedded GSM module [10]. Putta S *et al* present a smart power monitoring and control system through IoT using cloud data storage [11]. A solar remote monitoring system in the architecture of IoT is developed, moreover the system is installed to identify and eliminate faults of the plant [12].

All these works demonstrate that a monitoring and control system based on Iot is practical, flexible and expansible to supervise the system operation. Moreover, for monitoring PV power plants in different dimensions, it is necessary to obtain the surrounding environment of PV power plants. Thus, video surveillance is essential, yet which is not involved in the aforementioned systems. Therefore, a novel on-line monitoring system for PV arrays based on IoT is designed and implemented. The proposed system consists of data acquisition, data gateway and PV monitoring center (PVMC) website. The system uses DSP as microcontroller to collect indicators. Raspberry Pi3 has advantages of small size, fully customization and excellent scalability [13], thus it is applied to the data gateway to capture the video around PV array, obtain geographic information, store and upload data to PVMC. Moreover, *Laravel* is a free, open source PHP web application framework, which is suitable for designing web applications [14]. Hence, the PVMC website is developed by using *Laravel* PHP framework, JavaScript and MySQL database to support multiple platforms. In addition, the Extreme Learning Machine (ELM) is applied to PVMC to diagnose faults due to few adjustment parameters needed, extremely fast learning speed and good generalization performance [15-17]. Once fault occurs, users alerts can be sent through E-mail, and then users can browse the video content very efficiently to check out actual situation of PV array.

The rest of paper is organized as follows. Section 2 presents the system architecture. Section 3 introduces how PV array data is acquired. The data gateway is described in Section 4. Section 5 demonstrates the realization of PVMC website in detail. Section 6 illustrates the experimental results. Finally, Section 7 outlines the conclusion.

2. System architecture

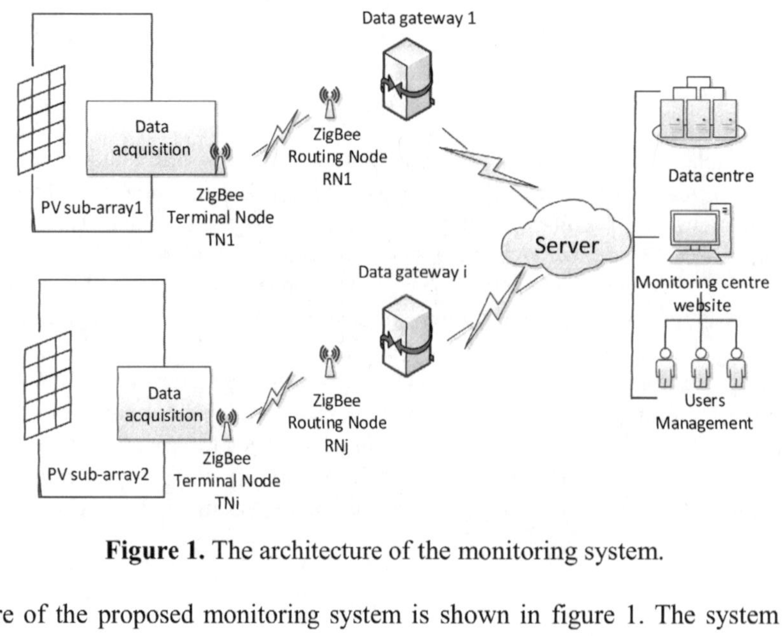

Figure 1. The architecture of the monitoring system.

The architecture of the proposed monitoring system is shown in figure 1. The system includes data acquisition part and data gateway, and PVMC website for the entire PV array. The system acquires the temperature, irradiance, voltage and current of PV sub-array$_i$, and transmits to the data gateway via the ZigBee WSN. Then, the gateway stores the data, and uploads them to the PVMC website. In addition, the monitoring video for PV sub-array$_i$ is sent to the PVMC website via the RTMP. PVMC website achieves data storage, user management, video surveillance and fault diagnosis. There are rich charts displayed on the PVMC website, which make it conveniently for users keep abreast of the situation of indicators of the PV array, such as daily power generation and the total historical power generation.

3. Data acquisition

As a well-known embedded processor, DSP-TMS320F28335 has plentiful hardware resources, such as 12-bits resolution analog-to-digital converters (ADC), serial communication interface (SCI) and abundant I/O ports [18], which make it ideal to be the microprocessor in the proposed monitoring system. The block diagram of data acquisition is shown in figure 2. The Hall sensor is used to collect voltage and current due to its high sensitivity, excellent anti-interference and low power consumption. Temperature sensor DS18B20 is applied to sensing the temperature in PV array via single bus protocol. ZigBee module is utilized to transmit the PV data to the data gateway. Irradiance sensor FZD-V1-2000 with fast response and high sensitivity is carried out to measure the irradiance. In order to ensure data transmission reliability, Md5 cryptographic algorithm with a 16 bit hash function is applied to the proposed system. It generates the public key for each node in the sensor network for authentication.

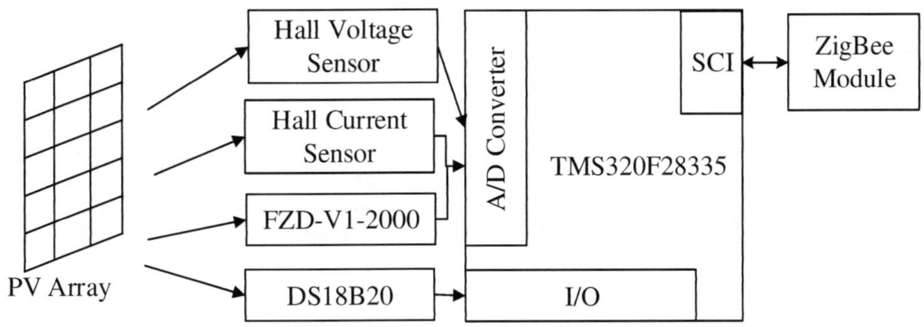

Figure 2. The block diagram of data acquisition.

4. The design of data gateway

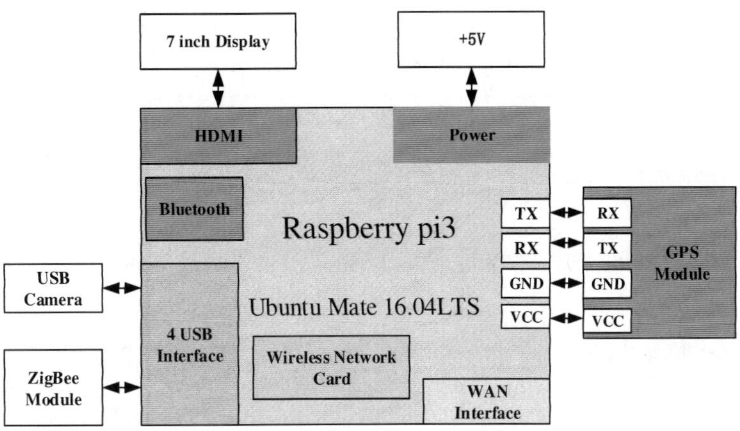

Figure 3. The block diagram of data gateway.

The block diagram of data gateway is shown in figure 3. It includes Raspberry Pi 3, GPS module,

ZigBee serial transmission module and USB camera. The Raspberry Pi 3 has advantages such as cost-effective, excellent interconnection, small size, powerful and low power consumption. In addition, wireless network card and Bluetooth module are embedded into Raspberry Pi 3. Hence, the data gateway applies the Raspberry Pi 3 installed with Ubuntu Mate16.04 LTS operating system as the processor. The data gateway receives the PV data via the ZigBee network, and employs the GPS module to obtain the geographical location of the PV array. The camera is applied to capture live video in the PV array. Then, the gateway connects to the Internet via wireless network card and uploads the PV data, GPS data and video to the PVMC website.

5. PV array monitoring centre website

To make it convenient for users to view monitoring information via web browsers, the B/S architecture is selected to design the PVMC. A monitoring centre website is designed on the base of the PHP *Laravel* framework that is running in the LNMP (Linux + Nginx + MySQL + PHP) installed on a server. Meanwhile, the Postfix mail server is installed to provide E-mail service for informing users the fault warning. The framework of the monitoring centre website is shown in the figure 4, the PVMC website consists of client layer, application layer and data layer. In the client layer, users can visit website via the web browser. The *Laravel* 5.4 is applied as the framework in the application layer, and e-mail, maps and video surveillance services is integrated in this layer to achieve data visualization, user management, and message management. The data layer selects MySQL as the database management platform to store PV data. The gateway communicates with the PVMC by representational state transfer (REST) API design patterns in website design. Moreover, the gateway integrates software development kit (SDK) coded by Python. As shown in figure 4, the database contains data access layer (DAL) and data service level (DSL).

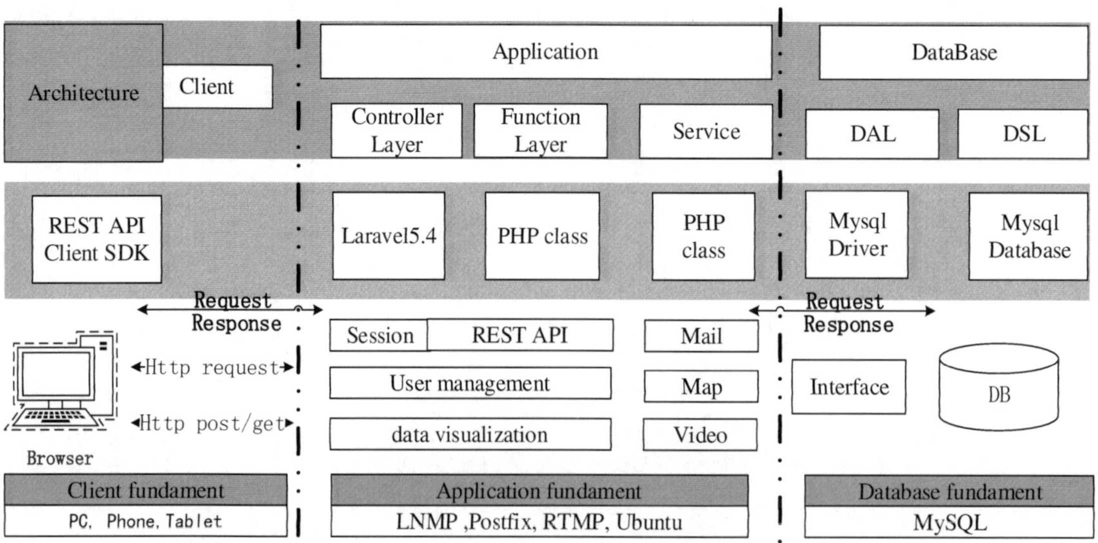

Figure 4. The framework of the on-line PV monitoring centre website.

5.1. Laravel framework

Laravel is an accessible and powerful web application framework which provides powerful tools that are needed for large-scale, robust applications [19]. Therefore, the *Laravel* 5.4 framework is selected to design the PVMC website. The *Laravel* 5.4 framework structure is shown in the table 1 as below.

Table 1. Laravel5.4 framework structure.

Directory name	Contents
app	Including the Controller and Model classes, most of the business is carried out in the directory
bootstrap	Framework entrance directory
Config	Framework configuration file directory
database	Define the database-related table structure
Public	The entry file directory in the website
Resource	Resource file directory that contains web page view files
routes	Define the website routing directory
storages	File storage directory
tests	Unit test directory
vendor	The directory contains the *Laravel* source code and the third party dependency package

5.2. Database design

The MySQL database is employed to store data in website, the table structure of database is shown in figure 5. There are 4 tables which represent *PV Array* table, *PV station* table, *Raspberry Pi* table, and *Users* table, respectively. PV Station table applys the *station_id* as primary key, and employs *user_id*, *pvarray_id*, *device_uuid* as foreign keys. Furthermore, the table stores GPS geographic information. PV array table employs the *pvarray_id* as the primary key, and it stores the indicators about PV array such as *Current*, *Voltage*, *Power*, *Temp* and *Irr*. *Raspberry Pi* table utilizes the *device_uuid* as the primary key, which stores the parameters about Raspberry Pi 3 such as *ip_address*, *cpu_temp*, *disk_space_usage* etc. Users table adopts the *user_id* as the primary key, and it stores the information of users, including *api_key, avatar, phoneNumber, user_name, is_active, email* and *roles*. The *api_key* and *api_token* are certification marks for uploading data, the *is_active* represents the active state of user accounts.

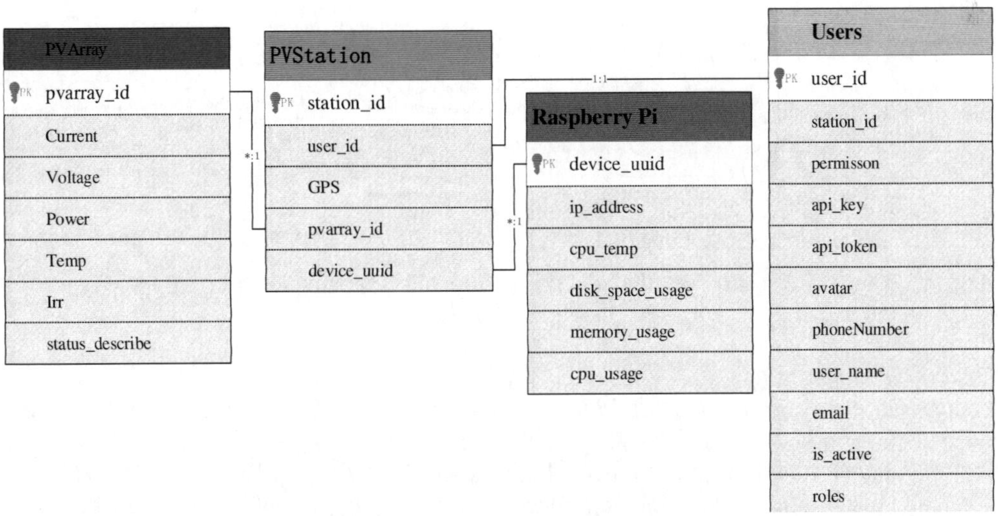

Figure 5. The table structure of the database.

5.3. REST API design patterns for PVMC

Representational State Transfer (REST) architectural style is commonly applied to the networking

protocol design of APIs for modern web service [20]. Therefore, the monitoring system is designed for the communication between the data gateway and PVMC based on REST API designed patterns. The flow chart of the client is shown in figure 6. Developers firstly create an account, then the obtained *api_key* and *api_token* are used as the certification mark for data uploading. Once ZigBee module receives data, the data will be updated to the PVMC. Figure 7 shows the flow chart of the server, the server is implemented based on Laravel5.4 framework, the steps is as follows:
Step1: create the pvdata table, Controller and Model by following commands

```
php artisan migrate:make create_pvdata_table
php artisan make:controller PostDataController
php artisan make:model PVData
```

Step2: As laravel5.4 has completed the API of REST architectural style, the important codes is as below:

```
Routes: group (['prefix' => 'api/v1'], function (){
        Route::resource ('postData','PostDataController');
});
```

Step3: Setting different state of the JSON format response information for debugging according to different client's request status.

```
return \Response::json([
'status'=>'success',
'status_code'=> 200,
'data' =>$lessons->toArray();
]);
```

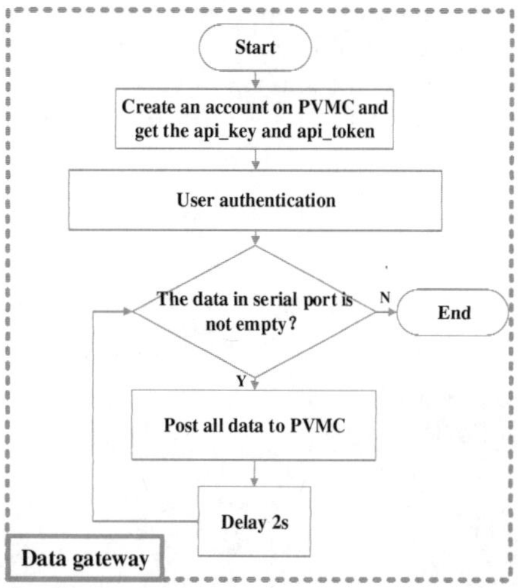

Figure 6. The flow chart of the data gateway.

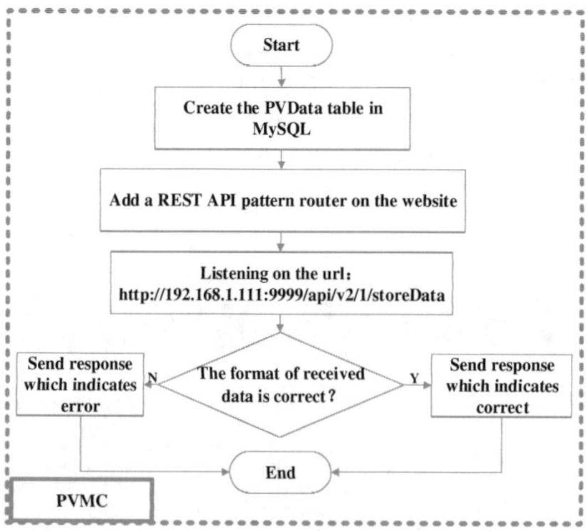

Figure 7. The flow chart of the PVMC.

5.4. Video surveillance

Video surveillance plays a significant role in PV monitoring system. RTMP is a TCP-based protocol for data transmission between a flash player and a server. The congestion control of the RTMP is much better than real-time transport protocol (RTP) based on UDP [21]. Hence, the RTMP is applied to transmit audio and video stream, the *libffmpeg* and *libv4l* library are installed in Raspberry Pi 3 for RTMP uploading. And *libffmpeg* includes encoding and decoding library of audio and video such as H.264 and *aac* library. Therefore, video and audio stream can be encoded by using FFmpeg library [22]. The block diagram of video surveillance is shown in figure 8. Video is captured via the camera embedded in Raspberry Pi 3 to collect the surrounding environment of the PV array. The original video stream is encoded by H.264 library. Meanwhile, audio stream is obtained by microphone and is encoded by *aac* library. The encoded audio stream and video stream are merged to the audio and video package. The package is pushed to the Nginx server and decoded by H.264 and *aac* library, then the original video and audio stream can be played via the player *Video.js*.

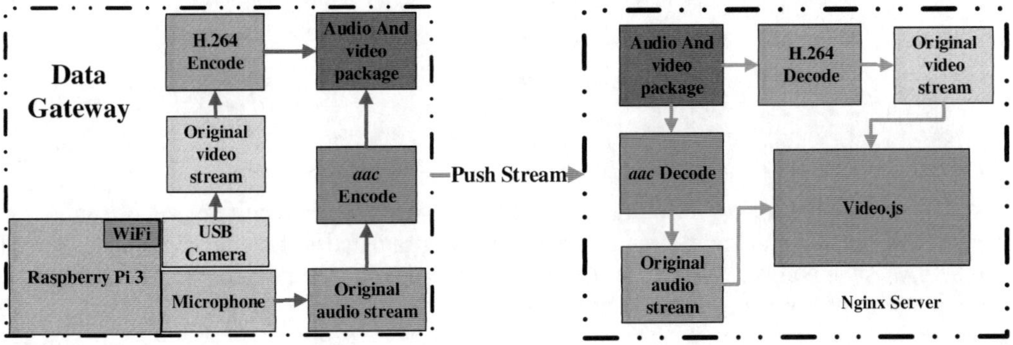

Figure 8. The block diagram of data gateway.

5.5. PV fault diagnosis approach

The training process of ELM is fast and simple, and the parameter selection is easy [23]. Therefore, ELM is applied to train the fault diagnosis network model of PV array. The block diagram of ELM fault diagnosis network model of PV array is shown in figure 9. It includes input layer, a single-hidden layers and output layer.

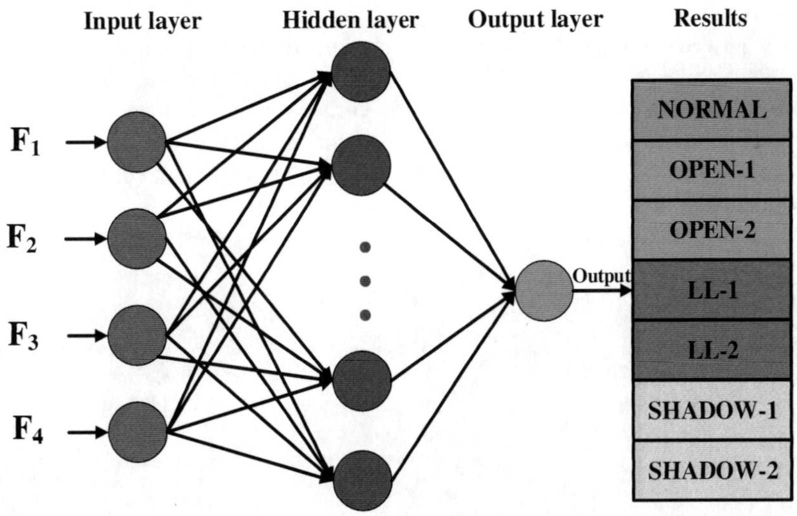

Figure 9. The block diagram of ELM fault diagnosis network model of PV array.

To create the fault diagnosis model, the labelled dataset should be prepared. Here, seven conditions of PV array are studied, including NORMAL (fault-free) condition and six typical faults. The six faults denotes as OPEN-1, OPEN-2, LL-1, LL-2, SHADOW-1 and SHADOW-2. The OPEN-1 and OPEN-2 mean open-circuit occurs in one PV string and two PV strings, respectively; LL-1 and LL-2 denote short-circuit in one PV module and two PV modules, respectively; SHADOW-1 and SHADOW-2 are a PV module and two PV modules are partial shaded, respectively. The corresponding label the claimed seven conditions are 1 to 7, respectively. Firstly, four PV data features are selected as the input attributes for the input layer as follows:

$$F_1 = \frac{V_{\text{MPP}}}{V_{\text{OC-STC}} \times N_P} \tag{1}$$

$$F_2 = \frac{I_{\text{MPP}}}{I_{\text{SC-STC}} \times N_S} \tag{2}$$

$$F_3 = \frac{G_i}{1000} \tag{3}$$

$$F_4 = \frac{T_i}{100} \tag{4}$$

where V_{MPP} is the maximum power point (MPP) voltage, I_{MPP} is the MPP current, $V_{\text{OC-STC}}$ and $I_{\text{SC-STC}}$ is the open-circuit voltage and short-circuit current of PV array under standard test condition (STC), respectively; N_{P} is the number of string in parallel of PV array, N_{S} is the number of module in serial of each string. G_i is the irradiance of PV array, T_i is the temperature of PV array. Hence, the four attributes are F_1, F_2, F_3 and F_4.

Secondly, the number of hidden layer nodes (which is defined as N) greatly influence the performance of ELM. But the classification accuracy rate tends to be stable, when N is enough to describe the nonlinear relationship between input and output [22]. Hence, the N is set by the ELM performance curve.

Finally, the train data set is put into the ELM to create the PV array fault diagnosis model. Then, the PV array condition can be predicted by putting the online data that acquired by the monitoring system into the model.

6. Experimental results

A laboratory PV power plant and data acquisition equipment are shown in figure 10. The PV array of the PV power plant consists of three PV strings in parallel and each string consists of six PV modules in series. Irradiance sensor is installed on the same corner of the PV array to ensure the accuracy of irradiance acquisition. The temperature sensor is tied on the back of the PV module to measure the temperature. Data acquisition module collects PV array voltage and current. The data are transmitted by ZigBee module DL-20.

Figure 10. The overview of PV array.

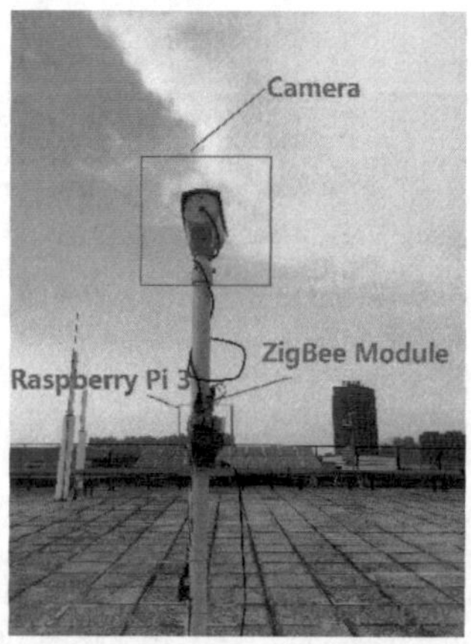

Figure 11. The overview of data gateway.

The overview of the gateway is shown in figure 11. Due to the waterproof and high night vision, the camera W03A is applied by the gateway to collect the video around PV array. The ubox-GPS module is selected to locate the position of PV array. And ZigBee module is employed to receive data from the data acquisition part. Raspberry Pi 3 with built-in wireless card is used to communicate with the PVMC.

The homepage of the PVMC is shown in figure 12, which displays the real-time indicators in PV array, and the Baidu map is embedded to obtain geographic information of PV array. In addition, the hardware resources usage of Raspberry in gateway is also shown as below, such as disk storage usage, memory usages and CPU usage.

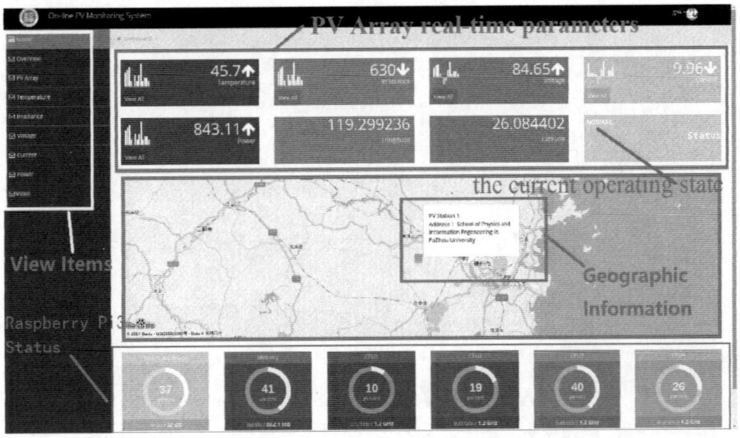

Figure 12. The home page of PVMC.

The figure 13 shows the line chart of indicators of PV array, including temperature, irradiance, PV array voltage and current. And figure 14 shows the parameters by the form of table view. Moreover, the data can be exported in Excel format.

The video captured is shown in figure 15. According to tests, there is about 2s delay due to the limitations of RTMP protocol, while it does not affect system performance.

To verify the performance of PV fault diagnosis approach, we collect 2500 samples and divide them into 5 groups randomly, in which 4 groups are used as train dataset and the rest one is applied as the test dataset. The influence of hidden layer node numbers on classification accuracy is shown in figure 16. It is obviously that the classification accuracy tends to stable after $N > 200$, therefore we choose $N = 210$ as the node number to train the ELM model. The radar map of accuracy after 10 independent runs is shown in figure 17, we can easily find that the total average accuracy is up to 97 %. As can be seen in the figure 12, the current operating state is normal.

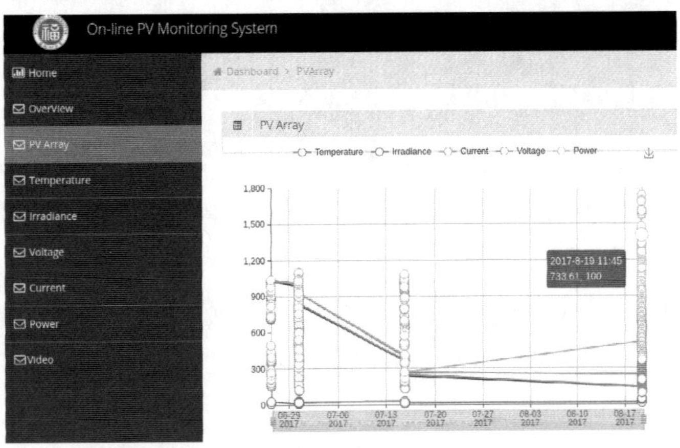

Figure 13. The line chart of the PV data.

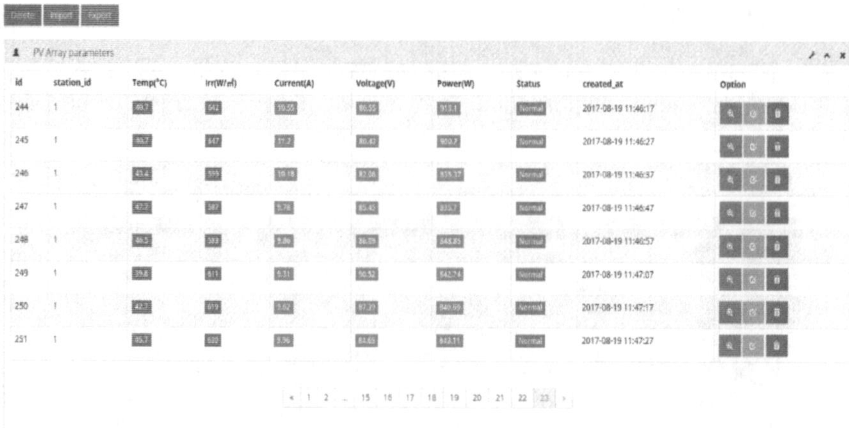

Figure 14. The table view of the PV data.

Figure 15. The video surveillance in PVMC.

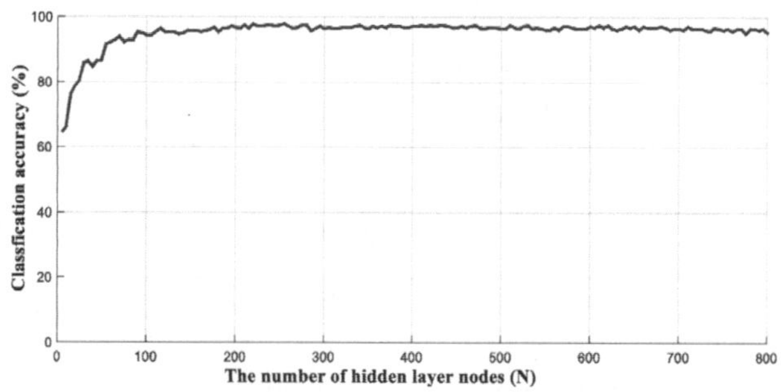

Figure 16. The influence of hidden layer node numbers on classification accuracy.

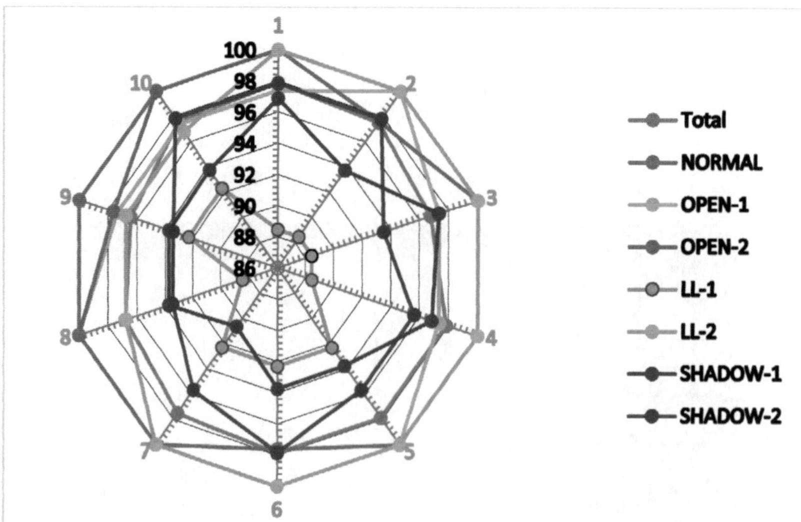

Figure 17. The radar map of accuracy after 10 runs.

7. Conclusions

An on-line Internet of Things (IoT) based remote monitoring system for PV arrays is proposed, and a fault diagnosis approach based on ELM is studied. The proposed system consists of data acquisition, data gateway and PVMC website. Data acquisition part successfully collect data and transmit them to data gateway, the data gateway also capture video and acquire position information then upload them to PVMC website for supervision. In addition, a fault diagnosis approach based on ELM is implemented on PVMC website. The experimental results demonstrate that the proposed monitoring system provide users with web-based interface, in which users can browse the real-time PV indicators, geographic information, in-site monitoring video and operating condition. In addition, PVMC can send warning information to users via E-mail once PV array fault occurs. The proposed system can be a promising solution for intelligent management of PV power plants.

References

[1] Hu T, Zheng M, Tan J, Zhu L and Miao W 2015 Intelligent photovoltaic monitoring based on solar irradiance big data and wireless sensor networks *Ad Hoc Netw* **35** 127-36

[2] Lin P J, Lin Y H, Chen Z C, Wu L J, Chen L C and Cheng S Y 2017 A density peak-based clustering approach for fault diagnosis of photovoltaic arrays *Int J Photoenergy* 4903613

[2] Lin P J, Cheng S Y, Yeh W C, Chen Z C and Wu L J 2017 Parameters extraction of solar cell models using a modified simplified swarm optimization algorithm *Sol Energy* **144** 594-603

[3] Chen Z C, Wu L J, Lin P J, Wu Y and Cheng S Y 2016 Parameters identification of photovoltaic models using hybrid adaptive Nelder-Mead simplex algorithm based on eagle strategy *Appl Energ* **182** 47-57

[4] Madeti S R and Singh S N 2017 Monitoring system for photovoltaic plants: A review *Renew and Sustain Energ Rev* **67** 1180-207

[5] Raza A, Ikram A A, Amin A and Ikram A J 2016 A review of low cost and power efficient development boards for IoT applications *Future Technologies Conference IEEE* 786-90

[6] Kelly S D T, Suryadevara N K and Mukhopadhyay S C 2013 Towards the Implementation of IoT for Environmental Condition Monitoring in Homes *IEEE Sens J* **13** 3846-53

[7] Lazarescu M T 2013 Design of a WSN platform for long-term environmental monitoring for IoT applications *IEEE Jon Em Sel Top C* **3(1)** 45-54

[8] Adhya S, Saha D, Das A, Jana J and Saha H 2016 An IoT based smart solar photovoltaic remote monitoring and control unit *International Conference on Control, Instrumentation Energy & Communication IEEE*

[9] Nalamwar H S H S, Ivanov M A and Baidali S A 2017 Automated intelligent monitoring and the controlling software system for solar panels *Journal of Physics: Conference Series* **803** 012107

[10] Sindhuja P and Balamurugan M S 2015 Smart power monitoring and control system through internet of things using cloud data storage *J Mater Sci Technol* **8** 1-7

[11] Xu X L and Wang H 2012 Construction of solar PV power generation remote monitoring system in the architecture of internet of things *Advanced Materials Research* **347** 178-82

[12] Maksimovic M, Vujovic V, Davidovic N, Milosevic V and Perisic B 2014 Raspberry Pi as Internet of things hardware: Performances and constraints *IcETRAN Conference Design Issues* 1-6

[13] Olanrewaju R F, Islam T and Ali N 2015 An empirical study of the evolution of PHP MVC framework *Advanced Computer and Communication Engineering Technology* (Springer, Cham) pp 399-410

[14] Chen Z C, Wu L J, Cheng S Y, Lin P J, Wu Y and Lin W C 2017 Intelligent fault diagnosis of photovoltaic arrays based on optimized kernel extreme learning machine and IV characteristics *Appl Energ* **204** 912-31

[15] Huang G B, Zhu Q Y and Siew C K 2006 Extreme learning machine: Theory and applications *Neurocomputing* **70** 489-501

[16] Huang G B, Wang D H and Lan Y 2011 Extreme learning machines: A survey *Int J Mach Learn Cyb* **2** 107-22

[17] Liu J, Hua Z, Chen W and Wu X 2015 Multi-channel current control system for three degree of

freedoms permanent magnet spherical actuator *Information and Automation, International Conference on IEEE* 2598-603

[18] Hen R Y 2014 Design and implementation of web based on Laravel framework *International Conference on Computer Science and Electronic Technology* 301-4

[19] Zhou W, Min L and Wu C 2014 REST API design patterns for SDN northbound API *2014 28th International Conference on Advanced Information Networking and Applications Workshops (WAINA) IEEE* 358-65

[20] Zhao P, Li J W, Xi J X and Gou X R 2012 A mobile real-time video system using RTMP *2012 Fourth International Conference on Computational Intelligence and Communication Networks (CICN) IEEE* 61-4

[21] Chang Q, Yang Z H and Song Y X 2011 A scalable mobile live video streaming system based on RTMP and HTTP transmissions *International Conference, Conmputer Science and Information Engineering (CSIE)* 113-8

[22] Liu C X, Fang J J, Liu Y X and Lu Y J 2017 The field terrain recognition based on extreme learning machine using wavelet features *2017 IEEE International Conference on Mechatronics and Automation (ICMA) IEEE* 1947-51

A grid-connected single-phase photovoltaic micro inverter

X Y Wen[1], P J Lin[1,2,3], Z C Chen[1,2], L J Wu[1,2] and S Y Cheng[1,2,3]

[1]Institute of Micro/Nano Devices and Solar Cells, College of Physics and Information Engineering, Fuzhou University, Fuzhou 350116, China
[2]Jiangsu Collaborative Innovation Center of Photovoltaic Science and Engineering, 213164 Changzhou, China

E-mail: linpeijie@fzu.edu.cn/ sycheng@fzu.edu.cn

Abstract. In this paper, the topology of a single-phase grid-connected photovoltaic (PV) micro-inverter is proposed. The PV micro-inverter consists of DC-DC stage with high voltage gain boost and DC-AC conversion stage. In the first stage, we apply the active clamp circuit and two voltage multipliers to achieve soft switching technology and high voltage gain. In addition, the flower pollination algorithm (FPA) is employed for the maximum power point tracking (MPPT) in the PV module in this stage. The second stage cascades a *H*-bridge inverter and LCL filter. To feed high quality sinusoidal power into the grid, the software phase lock, outer voltage loop and inner current loop control method are adopted as the control strategy. The performance of the proposed topology is tested by Matlab/Simulink. A PV module with maximum power 300W and maximum power point voltage 40V is applied as the input source. The simulation results indicate that the proposed topology and the control strategy are feasible.

1. Introduction

Recently, distributed photovoltaic (PV) power plants have become increasingly popular all over the world. Compared to the traditional centralized inverter or string inverter, photovoltaic micro-inverter has the advantage of achieving maximum power point tracking (MPPT) in single PV module [1-3]. Due to the characteristics of independence, single or several micro inverters fault would not affect the normal operation of entire PV power plant. In addition, micro inverter can locate the fault point quickly and accurately, so the PV micro inverter can improve the reliability of PV plants and the output efficiency of PV modules greatly. As the core of photovoltaic power generation system of power conversion, PV inverter's reliability and energy conversion efficiency have attracted great attention. For this, there is an effective way for improving the reliability and efficiency of the photovoltaic inverter to enhance topology and grid control strategy.

According to the number of transform stage, the topology of micro inverter can be divided into single-stage and multi-stage inverters [4,5], among which the two-stage converter is mainly applied. Generally, the first stage DC-DC realizes the MPPT and boost, and the second stage DC-AC realizes the inverter and grid-connected. According to whether the power transformer is supplied in convertor, micro inverters can be subdivided into isolated or non-isolated [6,7]. The former has the advantages of high voltage gain and simple circuit, but the volume is large and the efficiency is low. Since there has no transformer, the non-isolated inverter has the advantages of high efficiency, small size, light weight and low cost. However high voltage gain should be solved urgently because of the low output voltage of PV modules. Consequently, we propose a micro inverter, whose first stage DC boost contains non-isolation, coupled inductor and active clamp module in the first stage as well as adopts *H*-bridge topology in second stage. Moreover, the flower pollination algorithm (FPA) is applied as the global MPPT technology. Finally, a simulation based on Matlab/Simulink is carried out to test the feasibility of the proposed topology and control strategy.

2. Analysis of system structure and working principle

2.1. System structure

The proposed inverter consists of DC-DC boost, DC-AC conversion, voltage and current acquisition, data processing and MPPT controller. The system block diagram is shown in figure 1. The flower pollination algorithm (FPA) is used to track maximum power point. The *H*-bridge topology is adopted in the second stage. The amplitude of the inverter's output current is controlled by the inner current loop and the outer voltage loop. The output filter is LCL filter.

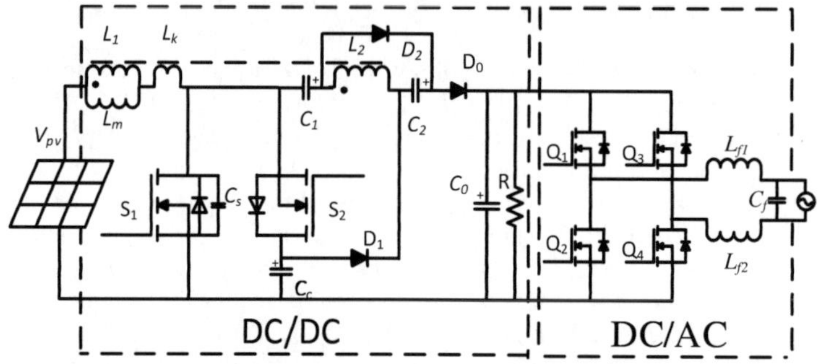

Figure 1. The block diagram of the proposed topology.

2.2. Operation principle of front stage DC-DC converter

The proposed DC-DC converter applies the non-isolated boost circuit with coupled inductor and active clamp module, which can obtain a higher voltage gain than basic boost circuit at the same duty ratio [8]. The coupled inductor is equivalents to ideal transformer (L_1, L_2), magnetizing inductance L_m and leakage inductance L_k. The active clamp module consists of the clamp switch S_2 and the clamp capacitor C_c, as shown in figure 1. The input voltage range of DC/DC converter is designed as 30-65 V, and the output DC bus voltage is about 400 V.

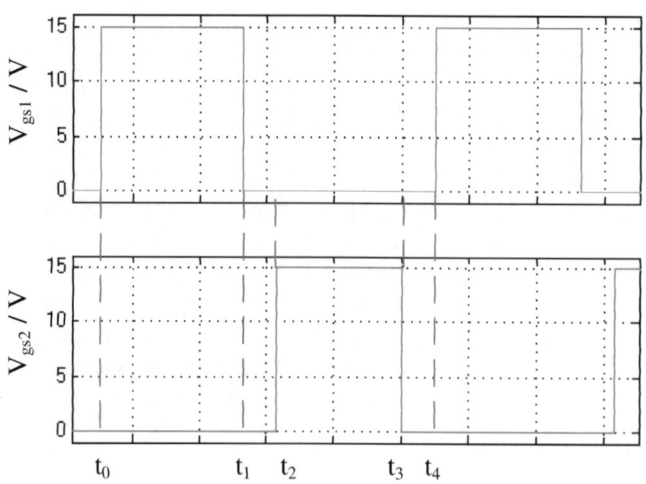

Figure 2. Switching sequence of S_1 and S_2.

The proposed topology operates under the continuous conduction mode (CCM). The switching sequence of DC/DC converter is shown in figure 2, which includes four operating periods, i.e, t_0-t_1, t_1-t_2, t_2-t_3 *and* t_3-t_4, and the corresponding operating mode is shown in figure 3. The steady-state analysis of the four mode is described as follows.

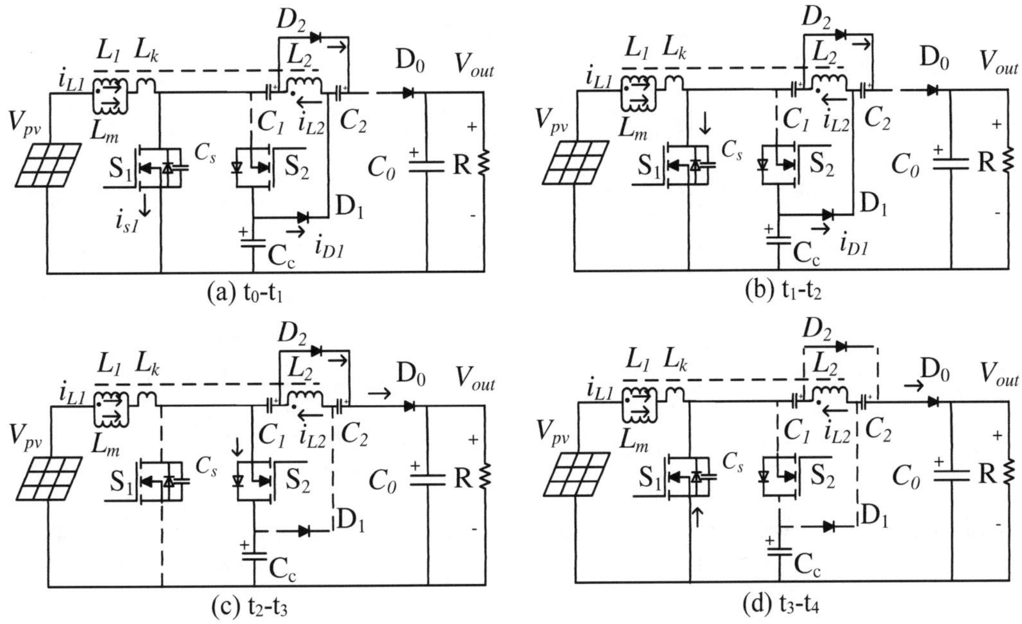

Figure 3. The four operating mode of DC/DC converter.

Mode 1(t_0-t_1*)*: once PV module output voltage exceeds 30 V, the power switching S_1 and S_2 begin to be controlled by a set of complementary PWM signals with different duty cycles. When the main switch S_1 is turned on, and the clamp switch S_2 is turned off, the primary current i_{L1} and secondary current i_{L2} of the coupling inductance start to increase linearly. The energy stored on the clamp capacitor C_c through D_1, L_2 to C_1; L_2's energy through D_2 to C_2; and D_0 is turned off. At the same time, the DC bus capacitor C_0 provides energy to the second stage. According to the capacitance volt-seconds balance theorem, the average charge current value of the capacitor equals to the average discharge current value as shown in equation (1).

$$\bar{I}c_{1\_discharge} = \bar{I}c_{2\_discharge} = \bar{I}_{D_0} = \bar{I}_0 \tag{1}$$

The secondary peak current can be expressed as:

$$I_{L_2\_on\_peak} = \frac{2\left(I_{C_{1\_charge}} + I_{C_{2\_charge}}\right)}{D} = \frac{4\bar{I}_0}{D} \tag{2}$$

The voltage of leakage inductance L_k can be expressed as:

$$V_{L_{k\_on}} = L_K N \frac{I_{L_2\_on\_peak}}{DT} = \frac{4N\bar{I}_0 L_k f_s}{D^2} \tag{3}$$

The voltage of the capacitor C_1 and C_2 can be expressed as:

$$V_{c1} = V_{Cc} + N(V_{in} - V_{L_{K\_on}}) \tag{4}$$

$$V_{c2} = N(V_{in} - V_{L_K\_on}) \tag{5}$$

Mode 2(t_2-t_3): when the main switch S_1 is turned off and the clamp switch S_2 is turned on, then the primary current of coupling inductance begins to charge C_c via S_2. Meanwhile, the secondary current of coupling inductance i_{L2} begin to decline. Then D_1 is cut off, and the energy stored on C_1 and C_2 begins to charge C_0 through D_2 and D_0 for providing to the second stage.

The secondary peak current can be expressed as:

$$I_{L_2\_off\_peak} = \frac{2\bar{I}_{D_0}}{1-D} = \frac{2\bar{I}_0}{1-D} \tag{6}$$

The voltage of leakage inductance L_k can be expressed as:

$$V_{L_k\_off} = L_K N \frac{I_{L_2\_off\_peak}}{(1-D)T} = \frac{2N\bar{I}_0 L_k f_s}{(1-D)^2} \tag{7}$$

Mode 3(t_1-t_2/t_3-t_4): when the main switch S_1 and S_2 are turned off together, there are two cases. Case 1(t_1-t_2): L_k and parasitic capacitance C_s in S_1 induce resonance. Due to the small value of C_s, V_{cs} can quickly rises to V_{Cc}, leading to the anti-parallel diode of switch S_2 turn on, which can clamp V_{cs} at V_{Cc}. Since the anti-parallel diode of S_2 has been turned on, the S_2 can achieve zero-voltage-switching (ZVS) when there is a drive signal at t_2, which greatly reduces the switching loss; Case 2(t_3-t_4): At t_3 the switch S_2 turned off, the C_s in S_1 begins to supply power to the load. Because C_s is very small, the voltage of S_1 drops to zero rapidly. Then the anti-parallel diode of S_1 is turned on, which makes the S1 working at zero voltage switching possible [9]. At t_4, the DC/DC converter returns to the working state of mode 1.

According to the Thévenin's theorem, the output voltage is:

$$V_{out} = V_{Cc} + V_{c1} + N(V_{Cc} - V_{pv} - V_{L_k\_off}) + V_{c2} \tag{8}$$

where V_{Cc} equals V_{pv} / (1-D) according to the inductance volt-seconds balance theorem. The voltage gain can be calculated by substituting the V_{Cc}, equations (6) and (7) into equation (8) as follows,

$$\frac{V_{out}}{V_{pv}} = \frac{(2+2N-ND)}{1-D} \frac{1}{1+\frac{8N^2L_kf_s}{RD^2}+\frac{2N^2L_kf_s}{R(1-D)^2}} \tag{9}$$

where N is the ratio between the secondary side and the primary side of coupling inductance, D is the duty cycle of the main switch S_1, R is the equivalent resistance of the second stage, L_k is the inductance value of the leakage inductance, f_s is the switching frequency. Obviously, the voltage gain is related to N, D, R, L_k, f_s.

2.3. Maximum power point tracking control algorithm

According to the output P-V characteristic curve of the photovoltaic module, the maximum power point varies with the change of irradiance and module temperature. There are a variety of MPPT algorithms, such as open circuit voltage method, perturb and observe method, conductance increment method and global intelligent optimization algorithm [10]. It is difficult for the traditional algorithms to judge the global maximum power point in the situation of multi-peak under partial shadow. Therefore, the evolution algorithm FPA is adopted [11].

FPA algorithm is a global group optimization algorithm, which follows the natural pollen propagation behaviour. The algorithm includes two processes of global pollination and local pollination [12]. In the global pollination stage, pollinators carry pollen for Levy flight, which can reach far places, and ensure that the most polluted and pollinated flowers of the environment. The global pollination is defined as follows:

$$x_i^{t+1} = x_i^t + C_p \cdot L\left(x_i^t - g^*\right) \tag{10}$$

where x_i^t represents the ith solution at t^{th} generation, $g*$ is the best flower, C_p is the step coefficient of *Levy* flight, L is *Levy* flight distance. The typical *Levy* flight in a two-dimension plane is depicted in figure 4. After searching an area with many smaller steps, it will shift to another area by long distance jumps. Comparing to other evolution algorithms, these small steps, occasionally large-step and long jumps may increase the accuracy and speed of pollination search significantly in some cases, especially for multimodal, nonlinear problems.

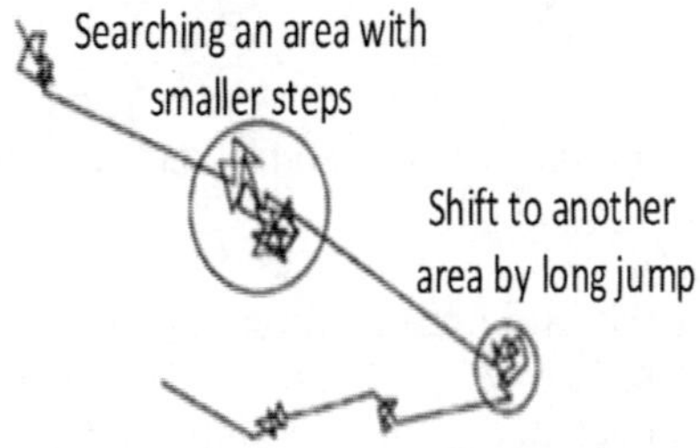

Figure 4. A Levy flight in 2-dimensional plane.

Local pollination is abiotic media self-pollination criteria, which can be expressed as follows,

$$x_i^{t+1} = x_i^t + \varepsilon\left(x_j^t - x_k^t\right) \tag{11}$$

where x_j^t and x_k^t are pollens from different flowers of the tth generation, respectively; Mathematically, if x_j^t and x_k^t come from the same species or selected from the same population, this will become a local random walk if we draw ε from a uniform distribution in [0,1].

Most flower pollination activities can occur at both local and global scale. In practice, adjacent flower patches or flowers in neighbourhood are more likely to be pollinated by local flower pollens than those far away. For this, we can use a switch probability p_o to switch between common global pollination and intensive local pollination. From simulations, we have found that $p_o = 0.8$ works better for most applications. When the random number r uniform distributed between [0,1] is greater than p_o for the global pollination, otherwise for the local pollination.

2.4. System control strategy

Figure 5 shows the micro-inverter control system structure. V_{pv} and I_{pv} are the PV array input voltage and input current, respectively. Sampling the voltage of V_{pv} and the current of I_{pv} is used for MPPT operation. By the logic circuit to generate different duty cycle D_{new} PWM signal to control the first stage power switch S_1, to trick maximum power point. V_g is the grid voltage sampling value. through the software phase-locked loop (SPLL) to get the current reference phase sinΘ; DC bus voltage V_{dc} and bus voltage reference value V_{dref} error value are calculated by the outer ring PI after the current reference amplitude i *, i * and sinΘ to get the reference current i_{ref}; i_g for the inverter output current sampling value. The error of i_{ref} and i_g is calculated by PI calculation and logic circuit. The SPWM signal is used to control the switching of the power at the second stage, and then the current signal with the same frequency and phase as grid voltage is generated [13,14].

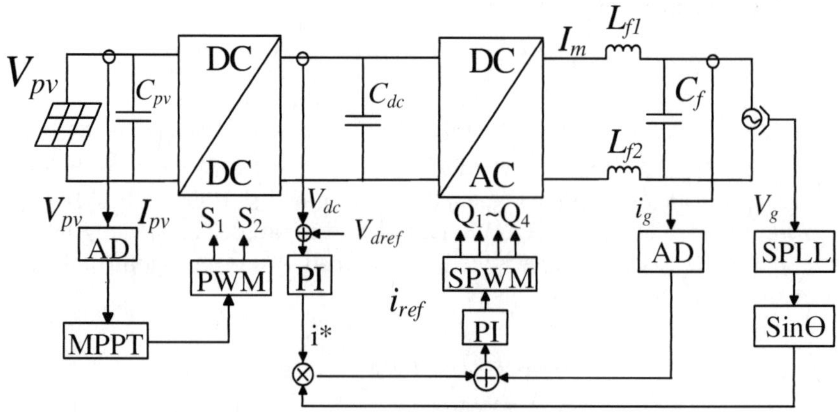

Figure 5. The structure of the control system.

3. Simulation analysis

In order to verify the feasibility of the proposed topology and control strategy, a 3×2 simulated PV array is established based on the SM55 in Matlab/Simulink. The mathematical model of PV array is based on the literature [15]. The parameters of power circuit and the PV array are set in tables 1 and 2, respectively. The ratio between primary and secondary of coupling inductance is 1:2. The switching frequency is 100 kHz and the dead time is 100 ns. The parameters of FPA is set as: $p_o = 0.8$, $n = 4$, and termination condition: *Maxgen*=30 or ΔD <0.01. The *H*-bridge adopts unipolar modulation.

Table 1. Parameters of power circuit.

Parameters	value	Parameters	value
Grid voltage V_g/V	220	Input filter capacitor C_{pv}/uF	980
Grid frequency f_g/Hz	50	DC bus capacitor C_{dc}/uF	1320
Switching frequency f_s/kHz	100	Filter inductance L_{f1}/mH	2.5
Coupling inductance L_m/uH	375	Filter inductance L_{f2}/mH	2.5
leakage inductance L_k/ uH	10	Filter capacitor C_f/uF	4.7
Voltage gain capacitor C_1/C_2/uF	2.2/4.4	The clamp capacitor C_c/uF	1

Table 2. Parameters of the PV module and PV array.

Parameters	PV module	PV array
Open voltage V_{oc}/V	21.7	65.1
Operating voltage V_{mp}/V	17.4	52.2
Short-circuit current I_{sc}/A	3.40	6.80
Operating current I_{mp}/A	3.16	6.32
Peak power P_{mp}/W	55	330

Note: the parameters in the table are measured under STC with irradiance 1000W/m^2 and temperature 25°C, AM 1.5.

Under the input voltage of 40 V and the duty ratio of 60%, the output voltage of the proposed DC-DC converter reaches 325 V, while that of the traditional boost [16] is 100 V. The DC-DC voltage gain of the proposed and traditional boost is shown in figures 6 and 7, respectively. The voltage gain of the proposed topology is 8.125, while that of the traditional boost is only 2.5, which verifies the proposed DC-DC converter with high voltage gain performance.

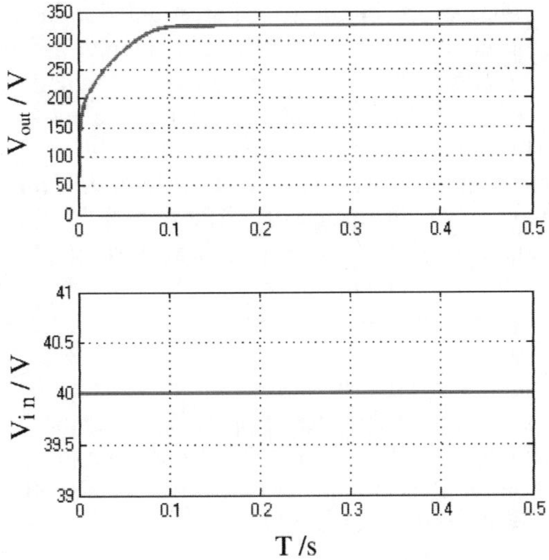

Figure 6. The proposed DC-DC voltage gain.

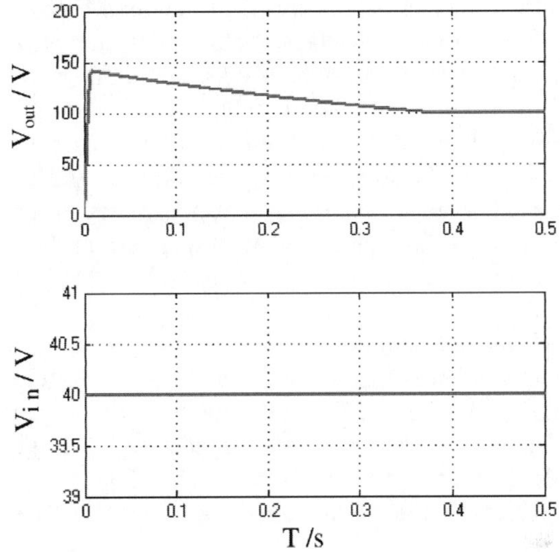

Figure 7. The traditional boost DC-DC voltage gain.

To verify the performance of the MPPT, the PV array is simulated under STC with partial shading. The MPPT control process is shown in figure 8. The proposed approach is tracking for the MPP from t=0 s to 0.6 s. The FPA algorithm generates a different number of flower pollens during which corresponds to different duty cycles during this period. Therefore, we can see the voltage and power of PV array output gradually increased, accompanying with the slightly drop of PV array output current. The maximum power point is tracked at about 1.2 s and the output power is 280 W. DC bus voltage in the 1.2 s is stable at 400 V. At 0.6 s the system is successful in connecting the power grid, also the output current and grid voltage achieve the same frequency in phase, seen in figure 9.

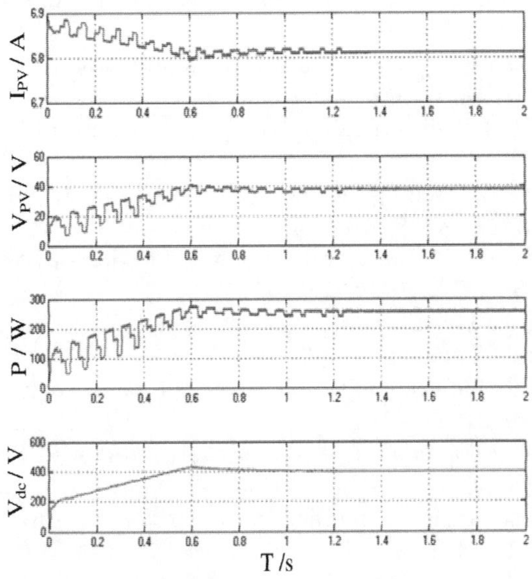

Figure 8. MPPT tracking process.

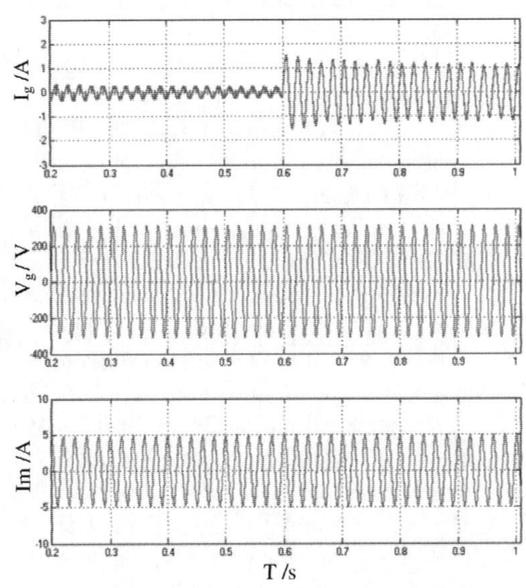

Figure 9. Grid connection process.

4. Conclusions

In this paper, we propose a two-stage PV grid-connected micro-inverter topology and control strategy. In the first stage, the working principle of high voltage gain DC-DC circuit based on coupled inductor, active clamp and voltage multiplier is analysed, and the condition of high efficiency is realized by suppressing the reverse recovery of diode and applying the technology of zero-voltage switching. Furthermore, the global flower pollination algorithm is adopted to improve the efficiency of photovoltaic array, and the simulate results verify the feasibility of the above scheme. In the second stage, the *H*-bridge topology is adopted, and the software phase lock, outer voltage loop and inner current loop are taken as control strategy. Finally, it has realized the global MPPT, high-boosting voltage gain and produced an output current that has a shape similar to a sinusoidal waveform in phase with the grid voltage.

Acknowledgments

This work was supported by the National Natural Science Foundation of China (Grant No. 61574038 and Grant No. 31300473), the Science Foundation of Fujian Education Department (Grant No. JAT160073), the Science Foundation of Fujian Science & Technology Department (Grant No. 2015H0021 and Grant No. 2015J05124), and the Scientific Research Foundation for the Returned Overseas Chinese Scholars, State Education Ministry (Grant No. LXKQ201504).

References

[1] Sher H A and Addoweesh K E 2012 Micro-inverters-Promising solutions in solar photovoltaics Energy for Sust Dev **16** 389-400
[2] Tan K S F, Rahim N A, Hew W P, *et al* 2014 Comparison and analysis of single-phase transformerless grid-connected PV inverters IEEE Trans Pow Electr **29** 5358-69
[3] Jiang S, Cao D, Li Y, *et al* 2012 Grid-connected boost-half-bridge photovoltaic microinverter system using repetitive current control and maximum power point tracking IEEE Trans Pow Electr **27** 4711-22
[4] Islam M, Mekhilef S and Hasan M 2015 Single phase transformerless inverter topologies for grid-tied photovoltaic system: A review Renew Sust Energ Rev **45** 69-86
[5] Hassaine L, OLias E, Quintero J, *et al* 2014 Overview of power inverter topologies and control structures for grid connected photovoltaic systems Renew Sust Energ Rev **30** 796-807
[6] Hasan R, Mekhilef S, Seyedmahmoudian M, *et al* 2017 Grid-connected isolated PV microinverters: A review Renew Sust Energ Rev **67** 1065-80
[7] Ahmad Z and Singh S N 2017 Comparative analysis of single phase transformerless inverter topologies for grid connected PV system Sol Energy **149** 245-71
[8] Cha W J, Kwon J M and Kwon B H 2016 Highly efficient step-up dc–dc converter for photovoltaic micro-inverter Sol Energy **135** 14-21
[9] Li W and He X 2011 Review of nonisolated high-step-up DC/DC converters in photovoltaic grid-connected applications IEEE Trans Ind Electron **58** 1239-50
[10] Kandemir E, Cetin N S and Borekci S 2017 A comprehensive overview of maximum power extraction methods for PV systems Renew Sust Energ Rev **78** 93-112
[11] Yang X S 2012 Flower pollination algorithm for global optimization International Conference on Unconventional Computation and Natural Computation **7445** 240-9
[12] Alam D F, Yousri D A and Eteiba M B 2015 Flower pollination algorithm based solar PV parameter estimation Energ Convers Manage **101** 410-22
[13] Petreuş D, Daraban S, Ciocan I, *et al* 2013 Low cost single stage micro-inverter with MPPT for grid connected applications Sol Energy **92** 241-55
[14] Oncu S and Karafil A 2017 Pulse density modulation controlled converter for PV systems International Journal of Hydrogen Energy **42** 17823-30
[15] Shell S M 55 Photovoltaic solar module http://www.aeet-service.com/pdf/shell/Shell-Solar_SM55.pdf

IOP Conf. Series: Earth and Environmental Science **93** (2017) 012079 doi:10.1088/1755-1315/93/1/012079

[16] Hasaneen B M and Mohammed A A E 2008 Design and simulation of DC/DC boost converter Power System Conference 12th International Middle-East IEEE pp 335-40

Research on temperature control and influence of the vacuum tubes with inserted tubes solar heater

L X Xiao[1,2], Y T He[1] and J Q Hua[1]

[1]Department of Physics and Electronic Science, Chuxiong Normal University, Chuxiong 675000, Yunnan Province, China

E-mail: hyt_127@163.com/ xlx@cxtc.edu.cn

Abstract. A novel snake-shape vacuum tube with inserted tubes solar collector is designed in this paper, the heat transfer characteristics of the collector are analyzed according to its structural characteristics, and the influence of different working temperature on thermal characteristics of the collector is studied. The solar water heater prototype consisting of 14 vacuum tubes with inserted tubes is prepared, and the hot water storage control subsystem is designed by hysteresis comparison algorithm. The heat characteristic of the prototype was experimentally studied under hot water output temperature of 40-45℃, 50-55℃ and 60-65℃. The daily thermal efficiency was 64%, 50% and 46%, respectively. The experimental results are basically consistent with the theoretical analysis.

1. Introduction

Solar collector, including solar water heater, air heater, etc., can be categorized into two types of flat type and vacuum tube according to adopted structure of the collector. Wherein, vacuum tube solar collector can be divided into two types of all-glass vacuum tube and glass - metal vacuum tube [1]. All-glass vacuum tubes are widely used in solar water heaters and air collectors due to their technical maturity and low cost [2]. In vacuum tube solar water heater system, the vacuum tube is directly connected to the thermal insulation water tank, and the water is heated in the vacuum tube before heat transfer to the water in the tank via convection and heat conduction. As hot water in the vacuum tube has small thermal convection strength, dirt will appear at the bottom of the vacuum tube after long-term use, causing secondary pollution. To this end, the application of glass-metal vacuum tubes collector in solar water heater has been studied. However, it has not been widely used because of the limitation of process and cost [3-5]. In addition, on the basis of all glass vacuum tubes, the vacuum tube with inserted tubes air collector has been studied and used. For example, Chang W *et al* proposed an efficient all-glass vacuum tube with inserted tubes air collector [6]. The results show that the collector has average efficiency of 49.5% and 63.7% in cloudy and sunny days, respectively, and the instantaneous efficiency of the collector increases with the increase of inlet mass flow. Yuan Y L *et al* designed a new type of vacuum tube with inserted tubes air collector, tested and analyzed the instantaneous heat collection characteristics under different seasons. The study found that stability of the air collector under stable working condition is about 60% in different seasons [7]. Yan S Y *et al* studied structural characteristic and non-steady-state efficiency of glass vacuum tube with inserted tubes solar collector, and obtained ideal non-steady-state efficiency equation [8]. Zeng D Q *et al* studied the air collection performance and flow resistance of tubular solar collector array, and the results show that the heat collection temperature is 87.8°C and 109.3°C in winter and summer respectively, with the average heat collection efficiency at 47.4% [9]. Above research shows that the vacuum tube with inserted tubes designed on the basis of all glass vacuum tube has the characteristics of easy processing and low cost. Therefore, in this paper, a vacuum tube with inserted tubes solar water collector is designed, and the influence of heat transfer characteristics in different hot water storage control temperature the system is studied.

2. Structure and heat transfer mechanism of vacuum tube solar with inserted tubes collector

The vacuum tube with inserted tubes collector is the core device for solar thermal conversion and transfer. It is composed of interpolating snake-shape copper tube, bracket and thermal insulation cover on the basis of all-glass vacuum tube. The structural sectional view is shown in figure 1.

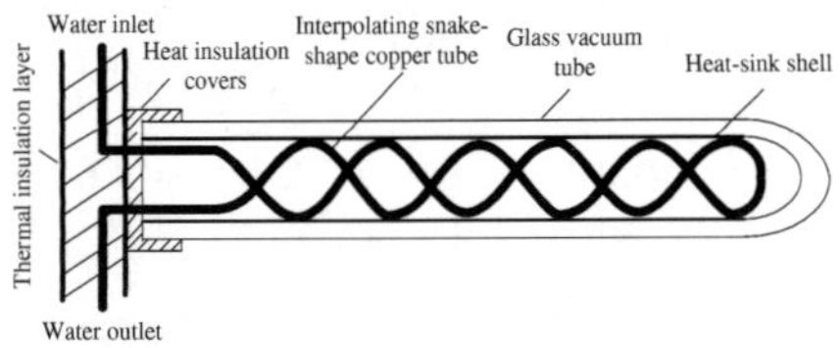

Figure 1. Structure chart of vacuum tube with tubes collector.

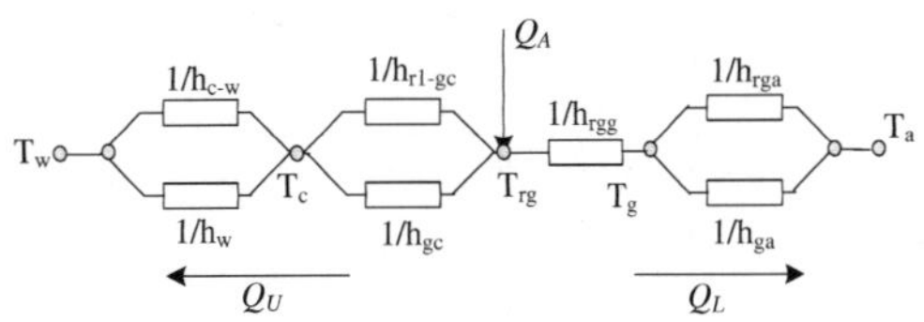

Figure 2. Thermal network diagram of inserted vacuum tube with inserted tubes collector.

In the vacuum tube with inserted tubes collector, the tube heat-sink shell of the glass vacuum tube absorbs the solar radiation energy and transfers it into heat energy. Then, the majority of the heat energy heats the fluid (water) in the copper tube via heat convection and radiation, while a small part passes through the thermal insulation cover on vacuum tube port and outer wall of the vacuum tube, radiating heat to the environment via radiation and convection. In accordance with the law of energy balance, when the solar collector is in the quasi-steady state, useful thermal energy equals to solar radiation absorbed by the collector tube minus heat dissipation of vacuum tube to the environment, as shown in formula (1). The thermal network diagram is shown in figure 2. In figure 2, T_a is the ambient temperature, T_g is the outer glass temperature of the vacuum tube, T_{rg} is the glass temperature of inner tube, T_c is the copper tube temperature, and T_w is the water temperature in the copper tube.

$$Q_u = Q_A - Q_L \tag{1}$$

In the above formula, Q_A is the solar energy absorbed by the vacuum tube per unit time, Q_L is the heat loss of the vacuum tube per unit time, and Q_u is useful thermal energy.
According to structural characteristics of the vacuum tube with inserted tubes collector and heat transfer-related theory [10], the useful thermal energy (unit in W) of a single vacuum tube with inserted tubes collector can be expressed as formula (2)

$$Q_u = D_{ro}LE(\tau\alpha)_e - \frac{A_{rg}(T_{rg}-T_a)\sigma(T_{rg}+T_g)(T_{rg}^2+T_g^2)\left(h_{rga}+\varepsilon_g\sigma(T_{rg}+T_g)(T_{rg}^2+T_g^2)\right)}{\left(\frac{1}{\varepsilon_{rg}}+\frac{A_{rg}}{A_g}(\frac{1}{\varepsilon_g}-1)\right)\left(h_{rga}+\varepsilon_g\sigma(T_{rg}+T_g)(T_{rg}^2+T_g^2)\right)+\left(\sigma(T_{rg}+T_g)(T_{rg}^2+T_g^2)\right)} \tag{2}$$

In the above formula, D_{ro} is the inner tube diameter of the vacuum tube, L is the effective length of the vacuum tube, E is the solar irradiance, $(\tau\alpha)_e$ is the effective product of the transmittance ratio of the outer tube of the vacuum tube and the absorption ratio of the inner tube absorption layer, ε_{rg} is inner tube radiation coefficient of the vacuum tube, A_{rg} is the inner tube area of the vacuum tube, A_g is the surface area of the outer tube of the vacuum tube, ε_g is the surface radiation coefficient of the outer tube of the vacuum tube, h_{rga} is the heat transfer coefficient between the inner and outer tubes, σ is the blackbody radiation coefficient, 5.67×10^{-8} (W/(m^2·K^4)).

In addition, according to the relationship between water absorption heat (unit w) and temperature variation, the relationship between the system hot water temperature and useful thermal energy can be expressed as formula (3).

$$t_l = \frac{\int_0^T Q_u dt}{mC} + t_i \tag{3}$$

According to formulas (2) and (3), the temperature characteristics of the output hot water of the vacuum tube with inserted tubes solar system under different working conditions are analyzed based on the design parameters shown in table 1. The analysis is shown in figure 3.

Table 1. Structural parameters table of vacuum tube with inserted tubes.

parameter name	Parameter value	parameter name	Parameter value
Inner tube radiation coefficient of the vacuum tube ε_{rg}	0.09	inner tube temperature of the vacuum tube T_{rg}	65°C
Outer tube radiation coefficient of the vacuum tube ε_g	0.88	Outer tube temperature of the vacuum tube T_g	35°C
Inner tube area of the vacuum tube A_{rg}	0.27 m^2	Outer tube area of the vacuum tube A_g	0.32 m^2
Effective product of the transmittance ratio of the outer tube of the vacuum tube and the absorption ratio of the inner tube absorption layer $(\tau a)_e$	0.82	ambient temperature T_a	25°C
		blackbody radiation coefficient σ	5.67×10^{-8} W/ /(m$^2\cdot$K^4)
Initial water temperature t_i	25°C	Water mass m	1.2 kg
solar irradiance E	800 W/m^2	Effective sunlight area of the vacuum tube $D_{ro}L$	0.097m^2
Number of vacuum tube in series N	14	Distance between tubes	60 mm

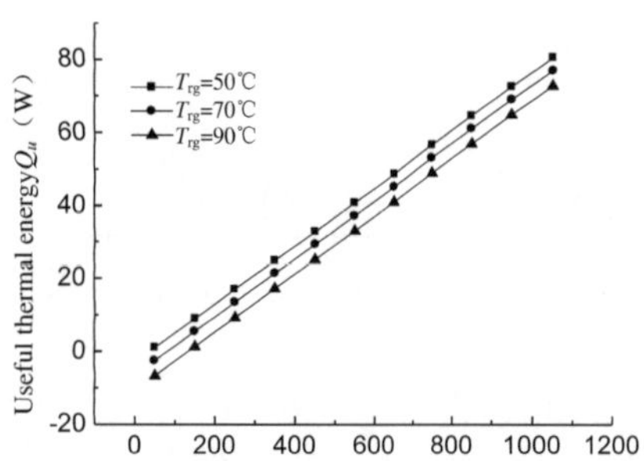

Figure 3.(a) Useful thermal energy

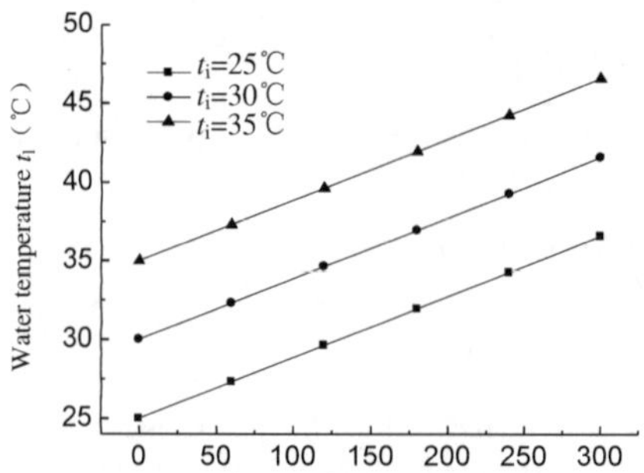

Figure 3.(b) Water temperature

Figure 3. Thermal characteristics of the vacuum tube with inserted tubes collector system (N = 14).

As can be seen from figure 3, for solar water heater using the vacuum tube with inserted tubes, the vacuum tube operating temperature directly affects thermal performance of the system. Taking into account the relationship between hot water temperature of the interpolating tube and operation of the vacuum tube, the analysis result is that the system hot water storage temperature provides theoretical support for influence of system thermal characteristics. Therefore, in the design of vacuum tube with inserted tubes heating system, the optimization design of hot water storage control mode is strengthened, and the influence of hot water control system and its hot water storage temperature on thermal characteristics of the system is studied for different hot water storage temperature.

3. Design and control principle of hot water storage controller

The vacuum tube with inserted tubes solar water heater control system is designed based on single chip AT89C2051. Using hysteresis control algorithm (shown in figure 4(a)), the solenoid valve controls the collector output hot water temperature in a specified set range (such as: 45°C-60°C), as shown in figure 4 (b). The control system consists of hardware and software.

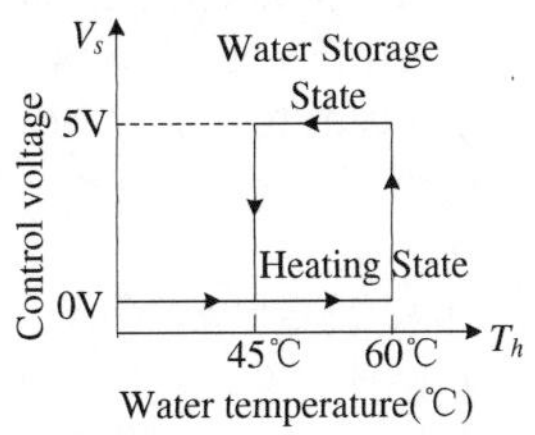

Figure 4. (a) Hysteresis control principle

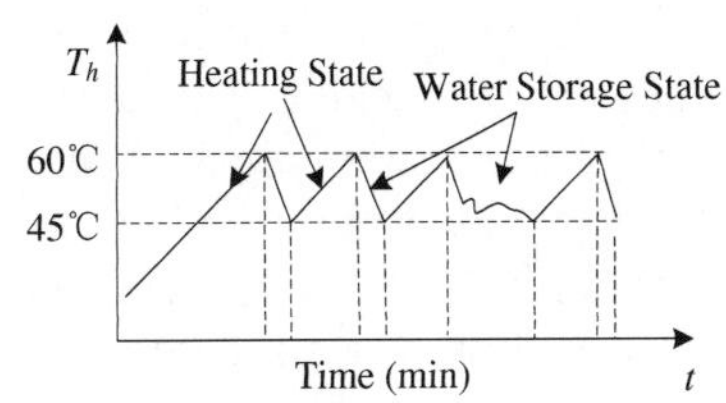

Figure 4. (b) Schematic diagram of system outlet temperature variation

Figure 4. Schematic diagram of hot water storage control process.

3.1. Hardware system design

The hardware system includes: temperature sensor (PT100), ADC0809, single chip AT89C2051 and its peripheral drive circuit. The control system hardware composition is shown in figure 5.

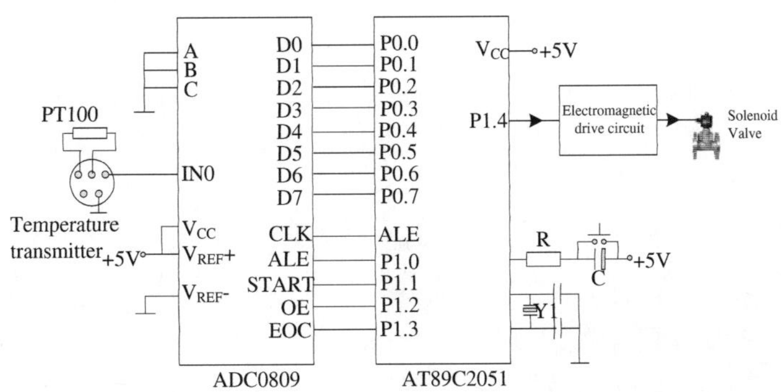

Figure 5. Constitutional diagram of hot water storage control system.

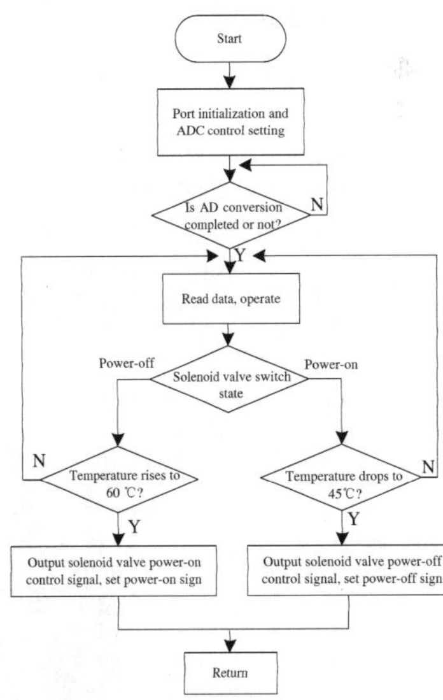

Figure 6. Main program flow chart.

In figure 5, the temperature sensor PT100 is arranged inside the output end of the interpolating copper tube for measuring water temperature. The temperature transmitter is used to convert the

PT100 resistance variation with the water temperature to voltage change information (1-5 V); ADC0809 is used to convert hot water temperature variation information output by temperature transmitter into 8-digit signal; single chip AT89C2051 serves as the core of the control system, which collects hot water temperature information, calculates and outputs solenoid valve power- on/off control signal under the control of the software system.

3.2. Software system design

On the basis of hardware design, the software flow chart is designed as shown in figure 6 for system control function. It consists of system initialization, AD conversion control and hot water storage control. Wherein, there are three states in hot water storage control process including: temperature control initial state, temperature rise holding state and storage control state.

Temperature control initial state: when the system begins operation or water temperature in the copper tube does not reach the minimum control temperature (45°C), the hot water storage controller operates in the temperature control initial state, solenoid valve power-off state flag bit is "1", solenoid valve is in power-off state, the vacuum tube heats the water, and the water temperature gradually rises.

Temperature rise holding state: When the hot water temperature in the outlet pipe reaches the minimum temperature (45°C) and power-off state flag bit of the solenoid valve in the hot water storage controller is "1", the hot water storage controller operates in temperature rise holding state, the solenoid valve is in power-off state, water temperature in the copper tube gradually rises. When the hot water temperature reaches the maximum control temperature (60°C), the temperature rise holding state ends, the hot water storage controller enters storage control state. Meanwhile, the solenoid valve power-off state flag bit is "0" and the solenoid valve power-on state flag bit is "1".

Storage control state: when the hot water temperature reaches the maximum control temperature (60°C), the hot water storage controller operates in the storage control state, the controller outputs solenoid valve power-on (high level) drive signal, hot water is stored in the incubator. Meanwhile, water temperature gradually decreases. When the water temperature lowers to 45°C, the hot water storage controller outputs solenoid valve power-off (high level) drive signal, hot water storage ends, the solenoid valve power-on state flag bit is "0", solenoid valve power-off state flag bit is "1", and the hot water storage controller operates in temperature rise state. In the case of solar illumination, the hot water storage controller circulates between temperature rise holding and storage control states, as shown in figure 4(b).

4. Experiment and analysis

A solar water heater prototype with collector array consisting of 14 vacuum tube with inserted tubes is designed. The distance between the two vacuum tubes is 60 mm in the collector array, and the prototype photo is shown in figure 7.

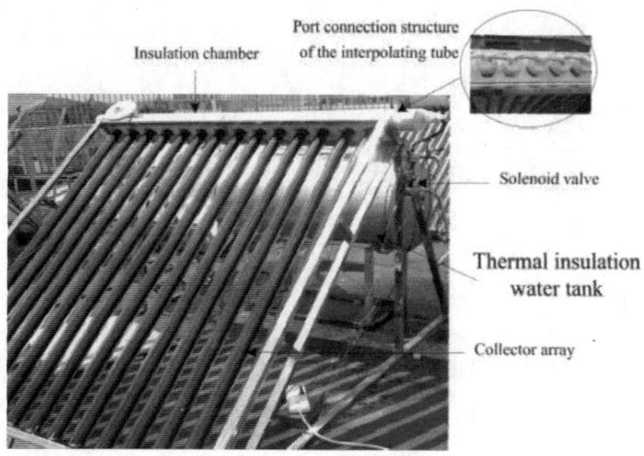

Figure 7. Photo of vacuum tube with inserted tubes solar water heater.

IOP Conf. Series: Earth and Environmental Science **93** (2017) 012080 doi:10.1088/1755-1315/93/1/012080

Set the temperature controller on April 14, 15, 19, 2017. The prototype outlet temperature was respectively controlled within the ranges of 40-45°C, 50-55°C and 60-65°C, and thermal characteristics of vacuum tube with inserted tubes solar heating system were tested experimentally. The test system consists of Altay multi-channel data acquisition system, TB-2 solar radiation general table, temperature transmitter and temperature sensor PT100, etc. The outlet flow rate is 17 ml/s during water storage. The test contents include: inlet water temperature of the vacuum tube array, internal temperature of the vacuum tube, outlet hot water temperature of the collector array, ambient temperature and solar irradiance, as shown in figures 8-10.

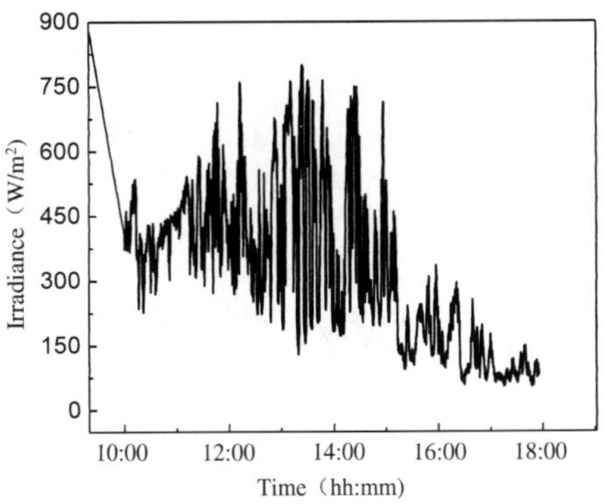

Figure 8. (a) Solar irradiance

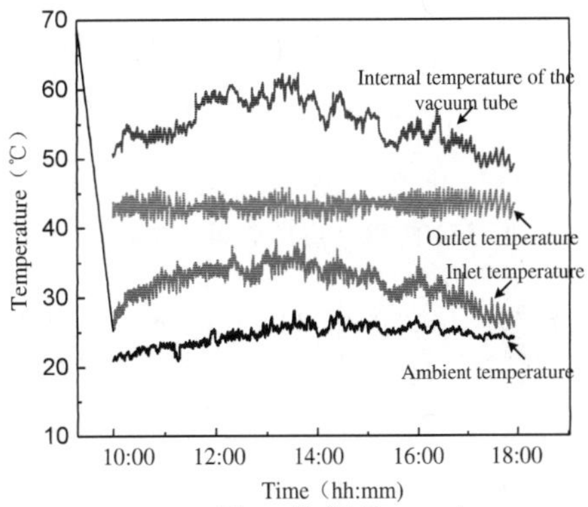

Figure 8. (b) Temperature

Figure 8. Thermal characteristics of the vacuum tube with inserted tubes solar collector system (outlet temperature 40-45℃, 2017.4.14).

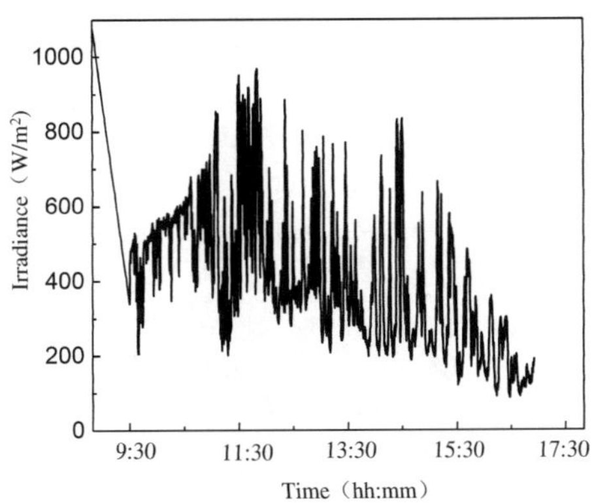

Figure 9 (a) Solar irradiance

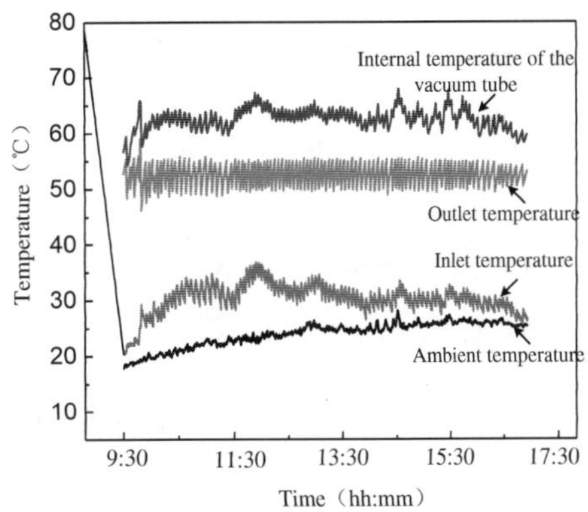

Figure 9. (b) Temperature

Figure 9. Thermal characteristics of the vacuum tube with inserted tubes solar collector system (outlet temperature 50-55℃, 2017.4.15).

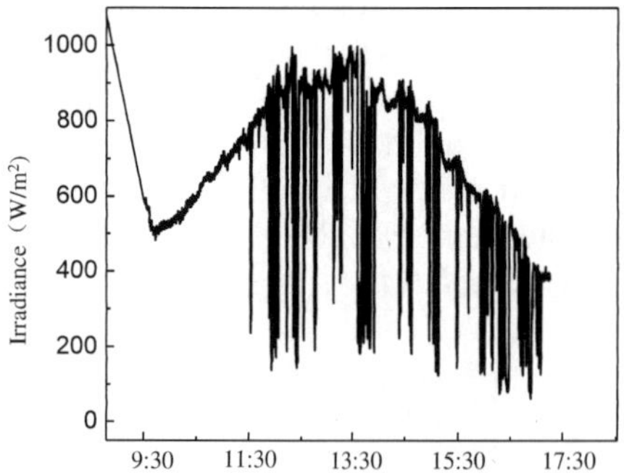

Figure 10 (a) Solar irradiance

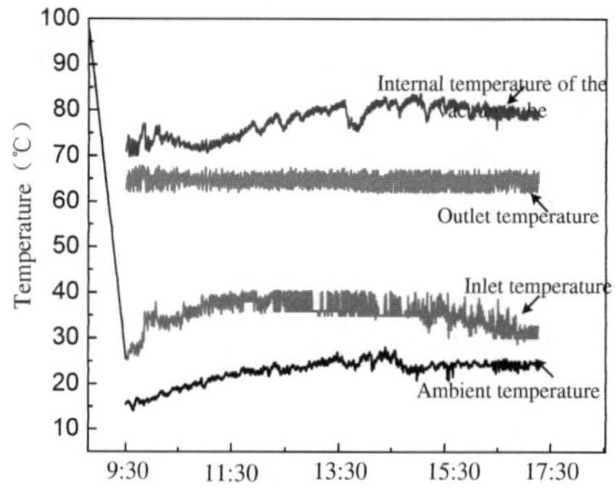

Figure 10. (b) Temperature

Figure 10. Thermal characteristics of the vacuum tube with inserted tubes solar collector system (outlet temperature 60-65℃, 2017.4.19).

During the test, the solar radiation amount (unit, J) of the collector array is obtained by averaging the solar irradiance. The expression is as shown in formula (4):

$$Q_u = \int_{t_0}^{t_l} A \cdot E dt \tag{4}$$

Where, A is the total area of the collector array, and E is the solar irradiance (unit, W / m^2).

At the same time, the available energy generated by system is calculated by the output hot water and expressed in formula (5):

$$Q_u = C \cdot m \cdot \Delta T \tag{5}$$

In the above formula, C is the specific heat capacity of water, m is the mass of output hot water, and ΔT is the difference between the outlet and inlet temperature.

According to the test results, the thermal characteristics of the prototype under three temperature conditions were statistically analyzed. The results are shown in table 2.

Table 2. Statistics table of prototype thermal characteristics under different outlet temperature.

Outlet temperature	Measurement time(t)	Difference outlet and inlet temperature(℃)	Output (kg)	Irradiation (J)	Thermal output (J)	Daily efficiency
40℃-45℃	10:00-18:00	11.93	224.00	17449641	11223744	64%
50℃-55℃	9:30-17:30	23.68	105.55	21147439	10497580	50%
60℃-65℃	9:30-17:30	28.78	130.00	34625989	15773940	46%

As can be seen from the statistical results, prototype thermal efficiency is related to the outlet hot water temperature. The system thermal efficiency reduces as the outlet temperature rises. The main reason is increase of the working temperature and heat loss after outlet temperature rise. In addition, the system hot water output during the test concerns weather conditions such as system operating temperature, solar irradiance. Therefore, based on users' different requirements for output hot water temperature, the system operation can be optimized by adjusting the hot water storage temperature.

5. Conclusions

The designed vacuum tube with inserted tubes solar collector, being ordinary all glass vacuum tube - based, has the characteristics of simple structure, low cost. According to the working principle of vacuum tube with inserted tubes solar water heater, the hot water storage control subsystem (hardware and software) with hysteresis comparison algorithm is designed. For the solar heating system prototype composed of 14 vacuum tubes with inserted tubes, the daily thermal efficiency of the prototype is 64%, 50% and 46% respectively under the output hot water temperature of 40-45°C, 50-55°C and 60-65°C. The experimental results show that the prototype thermal efficiency is greatly affected by system outlet hot water temperature. The system thermal efficiency lowers as the outlet temperature rises, which is basically consistent with the theoretical analysis. The results will provide experimental support for reasonable optimization of working temperature of vacuum tube with inserted tubes heater. In the meantime, statistics result of the hot water amount output by the prototype during the test shows good practicability of vacuum tube with inserted tubes solar heating system which has simplified installation requirements and simpler usage due to the separation of thermal insulation water tank and the collector.

Acknowledgments
The authors thank National Natural Science Foundation of China (No. 51566001) and Education department Major Project Foundation of Yunnan province (No. ZD2014014) for their financial supply.

References

[1] Luo Y J and He Z I, *et al* 2014 Solar Energy Utilization Technology (Chemical Industry Press, Beijin,China)
[2] Kumar S, Kumar A and Yadav A 2014 Experimental analysis of thermal performance of evacuated tube and flat-plate solar air collectors at different air flow rates *Int.J.Sustain.Eng.* **8(4)** 1-14
[3] Dabra V, Yadav L and Yadav A 2013 The effect of tilt angle on the performance of evacuated tube solar air collectorexperimental analysis *Int.J.Eng.Sci.Technol.* **5(4)** 100-10
[4] Wang P Y, Li S F and Liu Z H 2015 Collecting performance of an evacuated tubular solar high-temperature air heater with concentric tube heat exchanger *Energy Convers. Manage.* **106** 1166-73
[5] Wei B C, Yin Zh Q, Wang J and Huang K Z 2014 Stress analysis of glass-metal sealing structure of high temperature solar vacuum collector *Acta Energiae Solaris Sinica* **35(5)** 761-7
[6] Chang W, Li M, Wang Y F, Luo X and Hong Y R 2015 The experimental research on thermal performance of the evacuated tubular solar air collector with inserted tubes system *J.Anhui.Agric.Sic.* **43(12)** 269-73
[7] Yuan Y L, Li Y, Dai Y J and Wang R Z 2010 Experimental study on vacuum tube with inserted tubes solar air collector *Acta Energiae Solaris Sinica* **31(11)** 1429-33
[8] Yan S Y, Tian R, Hou S and Zhang L 2008 Analysis on non-steady-state efficiency of interpolating heat pipe collector with glass vacuum tube *J.Eng.Thermophys* **29(2)** 323-6
[9] Zeng D Q, Li H and Dai Y J 2014 Study on the heat collection performance and flow resistance of interpolating solar air collector array *Acta Energiae Solaris Sinica* **35(3)** 546-52
[10] Zhang H F 2004 *Thermal Utilization Theory and Computer Simulation of Solar Energy* (Northwestern Polytechnic University Press,USA)

New solar cell and clean unit system platform (CUSP) for earth and environmental science

A Ishibashi[1, 5], T Matsuoka[2], R Enomoto[3] and M Yasutake[4]

[1]Nanostructure Physics Laboratory, Research Institute for Electronic Science, Hokkaido University, Sapporo, Hokkaido 001-0020, Japan
[2]Physics of Electronic Materials, Institute for Materials Research, Tohoku University, Sendai, Miyagi 980-8577, Japan
[3]Institute for Cosmic Ray Research, the University of Tokyo, 5-1-5 Kashiwa, Chiba 277-8582, Japan
[4]Department of General Medicine & Health Science, Graduate School of Medicine, Nippon Medical School, Bunkyo-ku, Tokyo 113-8602, Japan

E-mail: i-akira@es.hokudai.ac.jp

Abstract. We have investigated InGaN-based multi-striped orthogonal photon-photocarrier propagation solar cell (MOP^{3}SC) in which sunlight propagates in a direction being orthogonal to that of photocarriers generated by the sunlight. Thanks to the orthogonality, in MOP^{3}SC, absorption of the sunlight and collection of the photocarriers can be simultaneously and independently optimized with no trade-off. Furthermore, by exploiting the degree of freedom along the photon propagation and using multi-semiconductor stripes in which the incoming photons first encounter the widest gap semiconductor, and the narrowest at last, we can convert the whole solar spectrum into electricity resulting in the high conversion efficiency. For processing MOP^{3}SC, we have developed Clean Unit System Platform (CUSP), which turns out to be able to serve as clean versatile environment having low power-consumption and high cost-performance. CUSP is suitable not only for processing devices, but also for cross-disciplinary fields, including medical/hygienic applications.

1. Introduction

Functional new devices and systems in materials/biological science, information technology, renewable energy, and environmental engineering have been of increasing importance. In research and development of those highly functional devices and systems, it seems convenient for us to investigate the systems in atom-bit-energy/environment (ABE^2) four-dimensional space [1]. We have been studying quantum-cross devices in atom-bit (AB)-plane and multi-striped orthogonal photon-photocarrier- propagation solar cell (MOP^{3}SC) in bit-energy (BE)-plane. We are at full notice that the high efficiency solar cell serves as a key device of electronics, i.e., a ubiquitous energy source for communication and the data, obtained exploiting highly clean environment realized by clean unit system platform (CUSP) [2], being available through information technology (IT) network would mean much for all of us in keeping ourselves in good health condition especially in a country like Japan where elder people is increasing in number as a generation. This would be the case for other large-population countries, in coming decades. Although many kinds of solar cells [3-5] have been studied, and high conversion efficiency solar cells have attracted much interest, it is not easy so far for conventional solar cells, including tandem solar cells, to convert the whole optical spectrum of the Sun into electricity. In this paper, we report the current status of our waveguide-coupled MOP^3SC whose solar cell parts, being made based on InGaN [6], are to be processed in CUSP and that of the potentially important CUSP application as a platform for sleep analysis.

In conventional solar cells, the photon propagation direction is orthogonal to a pn junction, and the diffusion direction of photo-carriers is parallel to that of the photon propagation. Thus, in general, especially for indirect bandgap materials, we need a thick layer to fully absorb the solar light, while we have to make the layer thin enough to collect photo-generated carriers as much as possible, because the carrier lifetime is finite. Thus, the conventional solar cells are with serious trade-off in determining the semiconductor layer thickness between the light absorption and the photocarrier collection. Here importance of edge injection of photons into the planer pn junction has been noticed [7]. For this purpose, since the edge is, by definition, very small compared to the planer size of the solar-cell, we need some apparatus to turn the spatially propagating three-dimensional (3D) photons into two-dimensional (2D) photons confined in a thin region that matches the size of the edge. For the 3D to 2D transformation, diffraction scheme plus planer waveguide seemed effective [8], but in reality it is not, because of symmetry issues. The structure used to cause the diffraction has spatial (left-right) symmetry, thus a half of 3D photons, after diffraction, become left-going 2D photons in the planer waveguide, while the rest half the right-going photons [8]. Because of time-reversal symmetry, the left-going photons, are now able to be considered as the time-reversed version of right-going photons that has turned into 2D from 3D by diffraction as time goes by, and then they turn into 3D from 2D by the same diffraction-casing structure, i.e., diffracted back to 3D space form the planer waveguide. Full 2D confinement is very difficult in the symmetric diffraction scheme.

2. A new high efficiency solar cell

Situations are dramatically changed in the structure shown in figure 1, i.e., the case of MOP^3SC with an asymmetric redirection waveguide (ARW) [9]. With one example of the ARW, as shown in the bottom left inset of figure 1, the incoming photons first meet periodic parabola mirror, the photons make focus in the asymmetric waveguide, and, being fully reflected, propagate to the right-hand side only. The guided 2D-photons eventually get into the edge of the wide-gap semiconductor, then medium gaped ones, and finally the narrow gap material, in parallel to the pn junction planes as shown in the bottom left inset of figure 1. The direction of the photon propagation is orthogonal to that of the photocarriers that moves perpendicular to the pn junction plane. Thus, we call this system multi-striped orthogonal photon-photocarrier-propagation solar cell (MOP^3SC), and it is connected to the ARW, in this example, consisting of the periodically placed parabolic mirrors and the 2D waveguide as shown in figure 1. In this solar cell, photons being absorbed in the lateral direction which is perpendicular to the direction of the carrier drift/diffusion, the aforementioned serious trade-off can be lifted, thanks to the orthogonality of photon-photocarrier propagation directions. We can enjoy the freedom to make the stripe wide enough to absorb laterally all the photons keeping the semiconductor layer (the p/n electrode distance) thin (short) enough to allow most of photocarriers to reach out to the contact metals. As stated above, by placing those multiple semiconductor stripes, neighboring to each other, with different band-gaps in such an order that the incoming guided photons first encounter the widest gap semiconductor, then medium-gap semiconductors, and the narrowest at last, we can virtually achieve full conversion of sunlight spectrum into electricity. Note that our system being a tandem structure connected, not in serial, but in parallel, usage of many bandgaps is easier as long as materials with those bandgaps available. In this sense, for the MOP^3SC of ours, InGaN [6,10] is a promising candidate among many inorganic semiconductors. Ideally, we could obtain the energy conversion efficiency as high as ~60 % and 75% when using, respectively, four and ten stripes of semiconductors for the concentrated sunlight [7].

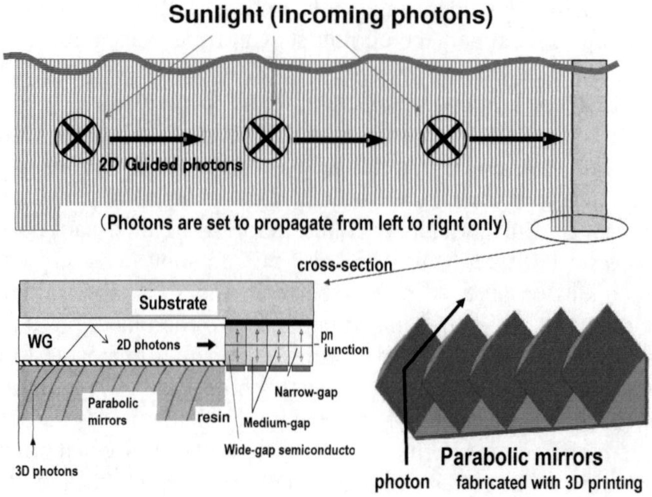

Figure 1. Waveguide (WG)-coupled multi-striped orthogonal photon-photocarrier- propagation solar cell (top view), cross-section (bottom left), and periodic parabolic mirrors (bottom right). WG size is tens of centimeters, and total width of the multi-striped semiconductor is tens to hundreds of microns.

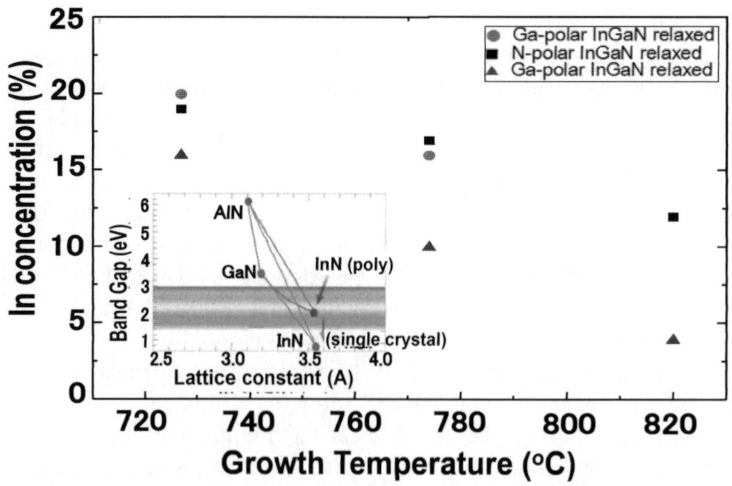

Figure 2. Growth temperature dependence of Indium composition in InGaN layers.

The usage of such a number of semiconducting materials would be unrealistic in conventional serially connected tandem structures, that number of pn junctions and accompanying tunnel junctions are very difficult, but is realistic in the InGaN-based MOP^{3}SC, because such many number of stripes, i.e., various In concentrations in the lateral direction, could be provided in a single epitaxial process, for example, by changing the substrate temperature in that lateral direction in growing InGaN. Figure 2 shows the growth temperature dependence of in composition in InGaN layers [10]. Thus, by deliberately giving temperature difference laterally in the substrate, we would be able to obtain graded In composition in a single InGaN growth. Then we are to put electrode in the direction perpendicular to the temperature gradient, i.e., the direction along which the gradient of Indium composition is zero. The number of the electrodes depends on how large the composition difference is, and also on practical lithographical dimensions. The number can be much larger than that of component cells in

IOP Conf. Series: Earth and Environmental Science **93** (2017) 012081 doi:10.1088/1755-1315/93/1/012081

the conventional tandem solar cell. InGaN-based MOP[3]SC with different band-gaps covering the whole solar spectrum could successfully be made in a planer manner, which would be of interest for integrating with ARW. As shown in figure 1, the multi-striped solar cell structure is to be placed at the edges of ARW. The waveguide-coupled MOP[3]SC serves as a concentration photovoltaic system typically operating under a few hundreds to a thousand suns depending on the ARW size. Note that when the ARW's front layer that the incoming 3D sunlight encounters first is a sheet with an integrated parabola cross-section [9] we can regard all the incoming photons virtually impinge the periodic parabola mirrors at a right angle.

3. CUSP, a platform for processing MOP[3]SC and for obtaining big data

For processing the MOP[3]SC, clean unit system platform (CUSP) [2] which was, originally, developed for uniting bottom-up and top-down systems [11] are well suited, since CUSP, having fan-filter-unit (FFU) in full feedback configuration, is fully capable of providing us with highly clean space, no matter how dusty its ambient is. Dust- and microbe-free environment as good as US FED 209D class 1 ~100 is obtained with tent-type CUSP (T-CUSP). Sleeping in T-CUSP has made Kinetosomnogram (KSG) possible [12]. Based on the analysis of the peaks in the particle-count as well as its autocorrelation, KSG can evaluate body-movements of people in sleep. We have made it possible to extract useful information on the sleep quality from the time-dependence of the particle count and its autocorrelation.

Speaking of usage of net or tent-like structure, Japan has a long tradition of "*Kaya*" a net to prevent mosquitos from getting close while people is sleeping especially for summer. We have expanded the concept of *Kaya* by replacing the target, i.e., mosquitos by particles/dusts, and developed the tent type CUSP (T-CUSP) as shown in figure 3(a). People can sleep inside T-CUSP as usual with futon or mattress.

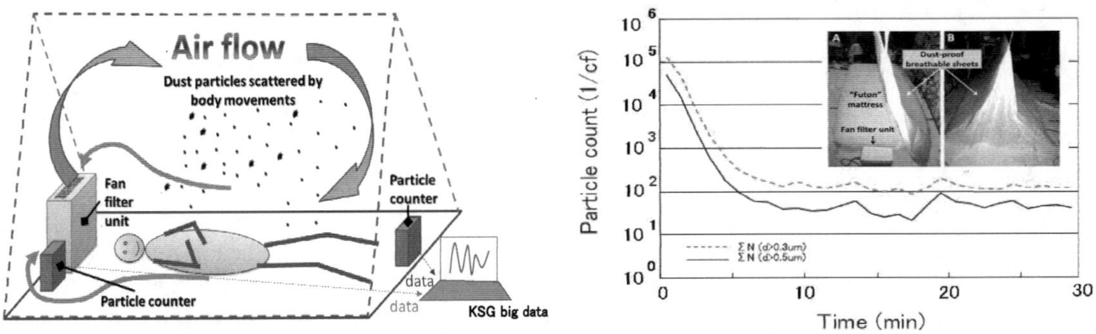

(a) (b)

Figure 3. Cleanliness of T-CUSP (a) and Cleanliness of T-CUSP (b).

The particle (dust) density $n(t)$ in T-CUSP shown in figure 3(a) is governed by the equation,

$$V\frac{dn(t)}{dt} = S\sigma - n(t)F + n(t)F(1-\gamma) \\ = S\sigma - \gamma F n(t), \tag{1}$$

where V and S, σ, and N_0 are, respectively, the volume of the closed space defined by the tent and futon placed on the floor, area of the inner surface of the closed space, rate of particles coming from the unit area (of the inner surface of the closed space) per unit time, and the particle density of the ambient space where the tent is placed. F is the flow rate of the fan filter unit set in the closed space, and γ is the filtrating efficiency of the FFU.

Equation (1) can be solved exactly quite easily [11] but what we are interested in is the steady-state particle count, *n,* being obtained just by putting LHS of equation (1) i.e., $dn/dt = 0$, and it is given by

$$n = \frac{S\sigma}{\gamma F} \tag{2}$$

This is the dust density obtained as time goes by in CUSP. The T-CUSP size is compact, i.e., just enough to cover futon/mattress, which makes the volume of the closed space or the T-CUSP small and, as seen from equation (1), the conversion time scales with V/F, and low particle state is established quickly when V is small. The flow rate F of the FFU used in T-CUSP is typically about 1 cubic meters per minute and T-CUSP is 2 m long, 1m wide, and 1m high, giving $V \sim 2$ m x 1 m x 1 m /2 ~ 1 m^3. The cleanliness provided is shown in figure 3(b). The solid line shows the number per cubic feet of particles whose size is 0.5 um or larger. The cleanliness reaches class 100 in 5 minutes. Note that super cleanrooms have been used almost exclusively for semiconductor device processing, and it would be ridiculous to even try to sleep in super cleanroom with ordinary bedclothes. With T-CUSP, however, it has become quite affordable to sleep in such clean-space. Users can enjoy non-invasive, non-contact natural sleeping in very clean air, being free from nuisance of using a mask, under lower risk of suffering from PM2.5 [13,14] and/or diseases mediated by airborne microbes.

As seen in figure 3(b), in T-CUSP the air quality is improved from 50000-150000/cubic feet (cf) to 1-100/cf in 5 minutes for sum of the particles whose diameter is 0.5 μm or larger. A bout of body rolling in T-CUSP causes a surge of the air-borne particles with a peak of 3000-10000/cf in a minute (top figure in figure 4), and raising an arm or a leg does the same with a peak of 1000-2000/cf. These data suggest that the T-CUSP is suitable for assessing bio-kinetic signals in relation to body movements during sleep. Each surge of particle counts in the KSG appears to have a corresponding arousal response (stage W) in the hypnogram obtained by PSG as shown in figure 4. Moreover, there is a significant peak of power spectral density at 80-100 minutes suggesting of REM periods.

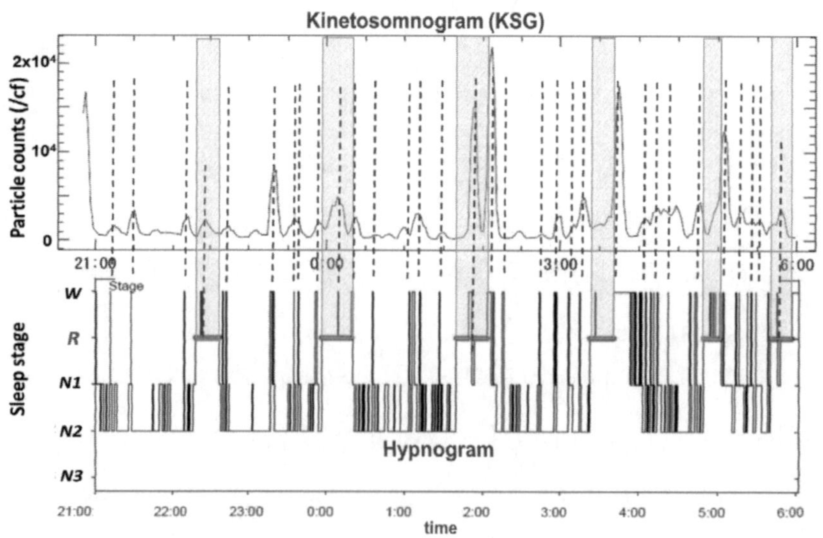

Figure 4. Comparison between kinetosomnogram (KSG) and hypnogram obtained by PSG.

Now let us denote those data that are human-related and of consciousness, respectively, as (↑ , ↑). Point of sales (POS) is an example in this category. In this definition, internet of things (IOT) data, being of things and therefore not human-related and of no consciousness, are characterized by (↓ , ↓). Then the aforementioned KSG produces the data having the property of (↑ , ↓), since the data are of human and of no consciousness when they are asleep. Usually it is hard to connect data characterized by (↑ , ↑) and those characterized by (↓ , ↓). When we are with data of (↑ , ↓), however, situations changes. By coupling KSG data with existing data we can generating the data base with the features of (↑ , ↑) (↑ , ↓) (↓ , ↓), a new data set with which we can handle the data on the Earth in brain-related, body-related, and IOT-related, highly sophisticated manner with no flips in the direction of the arrows across the data boundary (parentheses). Further, for detecting γ rays, a very sensitive detector called γ I has been developed [15]. Thus, when fan-filter-unit with a thick sheath is installed, CUSP can help people, together with γ I not to inhale dusts and particles containing ^{137}Cs or other radioactive elements by providing them with the clean air to live in. CUSP in its full line-up will be the cleanroom for all of us.

4. Conclusion

In ABE^2 space we have discussed a high efficiency solar cell and the processing platform of it, CUSP, and application of CUSP. We have investigated InGaN-based multi-striped orthogonal photon-photocarrier propagation solar cell (MOP^{3}SC) in which the photons propagate in the direction orthogonal to that of the photocarriers. Because of the orthogonality, in the newly proposed solar cell, the absorption of light and the photo-carrier collection can be independently and simultaneously optimized without any trade-off. Furthermore, by exploiting the degree of freedom along the photon propagation and using multi-semiconductor stripes in which the incoming photons first encounter the widest gap semiconductor, and the narrowest at last, we can convert the whole solar spectrum into electricity resulting in the high conversion efficiency. When this multi-In striped structure is coupled to a waveguide, the further advancement can be expected. Not only the waveguide-coupled MOP^{3}SC can optimize the absorption of light and the photocarrier collection independently and convert virtually the whole spectrum of sunlight into electricity, but also the waveguide-coupled MOP^{3}SC is expected to serve as a highly efficient concentration photovoltaic system thanks to the low temperature rise (due to the minimal thermal dissipation) and the diffusive light convertibility when used, for example, with the integrated-paraboloid-sheet on the top of the waveguide. The waveguide-coupled MOP^{3}SC is also of potential interest as a high reliability system, because the high energy photons that can damage the bonding of the materials, being converted into electricity already at upstream in the InGaN layer with less Indium content, never go into the medium or narrow gap semiconductors. Thus, the waveguide-coupled MOP^{3}SC would serve as an ultimate high efficiency all-in-one system.

For processing MOP^{3}SC, we have developed Clean Unit System Platform (CUSP), which turns out to be able to serve as clean versatile environment having low power-consumption and high cost-performance, and is suitable not only for processing devices, but also for cross-disciplinary fields, including medical/hygienic applications. CUSP can provide us with dust- and microbe-free environment as good as US FED 209E class 10 to 100 in foldable tent-type CUSP (T-CUSP). Sleeping in T-CUSP has opened a window for *Kinetosomnogram*, which, based on the analysis of the peaks in the particle-counts and the autocorrelation of them, would contribute to evaluation of sleep quality and/or sleep disorders. The T-CUSP can reduce the risk of diseases mediated by air-borne microbes as well as the risk of suffering from particulate matters (PM) 2.5 or other foreign materials. We envisage that CUSP would work as a proactive contributor of well-being. Breathing in this pristine air quality of CUSP, while sleeping for example, people would benefit from positive and direct effects to better health in the era of big data. The CUSP would be the clean space for all of us.

Acknowledgments

This work is supported, in part, by Special Education & Research Expenses from Post-Silicon Materials and Devices Research Alliance, JST Seeds Innovation Program, Post-Silicon Materials and Devices Research Alliance, Nano-Macro Materials, Devices and System Research Alliance, Dynamic Alliance for Open Innovation Bridging Human, Environment and Materials, 2010-2012 Grant-in-Aid for Scientific Research (B) [22350077], 2013-2015 Grant-in-Aid for Scientific Research (B) [25288112], and 2016-2018 Grant-in-Aid for Scientific Research (B) [16H04221] from the Japan Society for the PROMOTION of Science (JSPS). This work is also supported by Gant-in-Aid for

Challenging Researches [15K15280] and [16K12698] from Japan Society for the Promotion of Science (JSPS).

References

[1] Ishibashi A, Enomoto R, Ishikawa T and Yasutake M 2017 Systems development in atom-bit-energy/environment (ABE2) space for a new solar-cell, medical and safety applications based on clean unit system platform (CUSP) *Proc. Collaborative Conference on Materials Research* 191-5
[2] Rahaman Md D, Kaiju H, Kawaguchi N and Ishibashi A 2008 Ultra-high cleanliness of ISO class minus 1 measured in triply connected clean-unit system platform *Jpn J. Appl. Phys.* **47** 5712-6
[3] Green M A, Emery K, Hishikawa Y, Warta W and Dunlop E D 2012 Solar cell efficiency tables (version 40) *Prog. Photovolt. Res. Appl.* **20** 606-14
[4] Suemori K, Miyata T, Hiramoto M and Yokoyama M 2004 Vertical junction type organic photovoltaic cells *Jpn. J. Appl. Phys.* **43** L1094-6
[5] Liu M, Jonson M B and Snaith H J 2013 Efficient planar heterojunction perovskite solar cells by vapour deposition *Nature* **501** 395-8
[6] Matsuoka T 1992 Current status of GaN and compounds as wide-gap semiconductor *J. Crystal Growth* **124** 433-8
[7] Ishibashi A, Kawaguchi N, Kondo K, Kaiju H and White S 2009 Spiral-heterostructure-based new high-efficiency solar cells *Proc. Int. Symp. Environment. Consci. Design and Inv. Manufact. (EcoDesign 2009)* 55-8
[8] Ishibashi A, Kobayashi H, Taniguchi T, Kondo K and Kasai T 2016 Optical simulation for multi-striped orthogonal photon-photocarrier-propagation solar cell with redirection waveguide *3D Res.* 7 33 doi:10.1007/s13319-016-0109-4
[9] Ishibashi A, Kobayashi H, Sawamura N, Kondo K and Kasai T 2017 Symmetric and asymmetric wave-guides for multi-striped orthogonal photon-photocarrier-propagation solar cells *IEEE-ICASI 2017, Meen, Prior & Lam (Eds)* 1477-9
[10] Choi J 2014 *PhD thesis* Tohoku University
[11] Ishibashi A 2016 An approach for uniting bottom-up and top-down systems and its applications *Int. J. Eng. Res. Sci.* **2** 103-14
[12] Ishibashi A and Yasutake M 2016 Clean unit system platform (CUSP) for medical/hygienic applications *Int. J. Eng. Res. Sci.* **2** 92-7
[13] Sakamoto K 2011 Measurement of ambient fine particles (PM2.5) *J. Jpn. Soc. Atmos. Environ.* **46** 61-9
[14] Takebayashi T, Asakura K and Yamada M 2011 Posure to PM2.5 and effects on human health: Implications for health risk assessment in Japan *J. Jpn. Soc. Atmos. Environ.* **46** 70-6
[15] Enomoto R 2013 http://openit.kek.jp/workshop/2013/d-sys/doc/Enomoto.pdf